Fundamentals of Jet Propulsion with Power Generation Applications

Fully updated and revised, the second edition of this introductory text on air-breathing jet propulsion focuses on the basic operating principles of jet engines and gas turbines. A state-of-the-art review of turboramjet engines, hypersonic applications, geared turbofans, and adaptive cycle engines, accompanies an examination of emissions and pollutants, greatly expanding the importance of power generation gas turbines in industrial applications, and ensuring that students will be introduced to the most current trends in the subject. With completely rewritten chapters on the operating characteristics of components and ideal and nonideal cycle analysis, additional SI units in numerous examples, new and expanded end-of-chapter problems, and updated accompanying software, this remains the ideal text for advanced undergraduate and beginning graduate students in aerospace and mechanical engineering.

Ronald D. Flack, Jr. is a Professor Emeritus, former Chair of Mechanical and Aerospace Engineering, and former Director of the Rotating Machinery and Controls (ROMAC) Industrial Research Program at the University of Virginia. He is currently a consulting professional engineer for BRG Machinery and is an ASME Fellow and member of the ASME IGTI Turbomachinery Committee. Professor Flack began his career as an analytical compressor design engineer at Pratt & Whitney Aircraft, and has taught numerous academic and industry courses on gas turbines, propulsion, turbomachinery, fluid mechanics, gas dynamics, and thermodynamics.

Cambridge Aerospace Series

Editors:
Wei Shyy and Vigor Yang

1. J. M. Rolfe and K. J. Staples (eds.): *Flight Simulation*
2. P. Berlin: *The Geostationary Applications Satellite*
3. M. J. T. Smith: *Aircraft Noise*
4. N. X. Vinh: *Flight Mechanics of High-Performance Aircraft*
5. W. A. Mair and D. L. Birdsall: *Aircraft Performance*
6. M. J. Abzug and E. E. Larrabee: *Airplane Stability and Control*
7. M. J. Sidi: *Spacecraft Dynamics and Control*
8. J. D. Anderson: *A History of Aerodynamics*
9. A. M. Cruise, J. A. Bowles, C. V. Goodall, and T. J. Patrick: *Principles of Space Instrument Design*
10. G. A. Khoury (ed.): *Airship Technology*, Second Edition
11. J. P. Fielding: *Introduction to Aircraft Design*
12. J. G. Leishman: *Principles of Helicopter Aerodynamics*, Second Edition
13. J. Katz and A. Plotkin: *Low-Speed Aerodynamics*, Second Edition
14. M. J. Abzug and E. E. Larrabee: *Airplane Stability and Control: A History of the Technologies that Made Aviation Possible*, Second Edition
15. D. H. Hodges and G. A. Pierce: *Introduction to Structural Dynamics and Aeroelasticity*, Second Edition
16. W. Fehse: *Automatic Rendezvous and Docking of Spacecraft*
17. R. D. Flack: *Fundamentals of Jet Propulsion with Applications*
18. E. A. Baskharone: *Principles of Turbomachinery in Air-Breathing Engines*
19. D. D. Knight: *Numerical Methods for High-Speed Flows*
20. C. A. Wagner, T. Hüttl, and P. Sagaut (eds.): *Large-Eddy Simulation for Acoustics*
21. D. D. Joseph, T. Funada, and J. Wang: *Potential Flows of Viscous and Viscoelastic Fluids*
22. W. Shyy, Y. Lian, H. Liu, J. Tang, and D. Viieru: *Aerodynamics of Low Reynolds Number Flyers*
23. J. H. Saleh: *Analyses for Durability and System Design Lifetime*
24. B. K. Donaldson: *Analysis of Aircraft Structures*, Second Edition
25. C. Segal: *The Scramjet Engine: Processes and Characteristics*
26. J. F. Doyle: *Guided Explorations of the Mechanics of Solids and Structures*
27. A. K. Kundu: *Aircraft Design*
28. M. I. Friswell, J. E. T. Penny, S. D. Garvey, and A. W. Lees: *Dynamics of Rotating Machines*
29. B. A. Conway (ed.): *Spacecraft Trajectory Optimization*
30. R. J. Adrian and J. Westerweel: *Particle Image Velocimetry*
31. G. A. Flandro, H. M. McMahon, and R. L. Roach: *Basic Aerodynamics*
32. H. Babinsky and J. K. Harvey: *Shock Wave–Boundary-Layer Interactions*
33. C. K. W. Tam: *Computational Aeroacoustics: A Wave Number Approach*
34. A. Filippone: *Advanced Aircraft Flight Performance*
35. I. Chopra and J. Sirohi: *Smart Structures Theory*
36. W. Johnson: *Rotorcraft Aeromechanics vol. 3*
37. W. Shyy, H. Aono, C. K. Kang, and H. Liu: *An Introduction to Flapping Wing Aerodynamics*

38. T. C. Lieuwen and V. Yang: *Gas Turbine Emissions*
39. P. Kabamba and A. Girard: *Fundamentals of Aerospace Navigation and Guidance*
40. R. M. Cummings, W. H. Mason, S. A. Morton, and D. R. McDaniel: *Applied Computational Aerodynamics*
41. P. G. Tucker: *Advanced Computational Fluid and Aerodynamics*
42. Iain D. Boyd and Thomas E. Schwartzentruber: *Nonequilibrium Gas Dynamics and Molecular Simulation*
43. Joseph J. S. Shang and Sergey T. Surzhikov: *Plasma Dynamics for Aerospace Engineering*
44. Bijay K. Sultanian: *Gas Turbines: Internal Flow Systems Modeling*
45. J. A. Kéchichian: *Applied Nonsingular Astrodynamics: Optimal Low-Thrust Orbit Transfer*
46. J. Wang and L. Feng: *Flow Control Techniques and Applications*
47. D. D. Knight: *Energy Disposition for High-Speed Flow Control*
48. Eric Silk: *Introduction to Spacecraft Thermal Design*
49. Antony Jameson: *Computational Aerodynamics*
50. Peretz Friedmann, George Lesieutre and Daning Huang: *Structural Dynamics: Theory and Applications to Aerospace and Mechanical Engineering*
51. Ronald D. Flack: *Fundamentals of Jet Propulsion with Power Generation Applications*

"The previous edition – one of the best on air-breathing propulsion – is now even better, with content reorganized to be even clearer, while retaining its good balance between underlying thermodynamic and fluid dynamic fundamentals and their application in analyzing components and systems. The result is a very strong introductory text on this subject."

Werner J.A. Dahm, *Arizona State University and Former Chief Scientist of the U.S. Air Force*

"The most detailed book on gas turbines I know. A tremendous treasure of knowledge and ideally suited to students and professionals who really want to understand the function of a jet engine or gas turbine."

Hanns-Jürgen Lichtfuss, *Technical University of Munich*

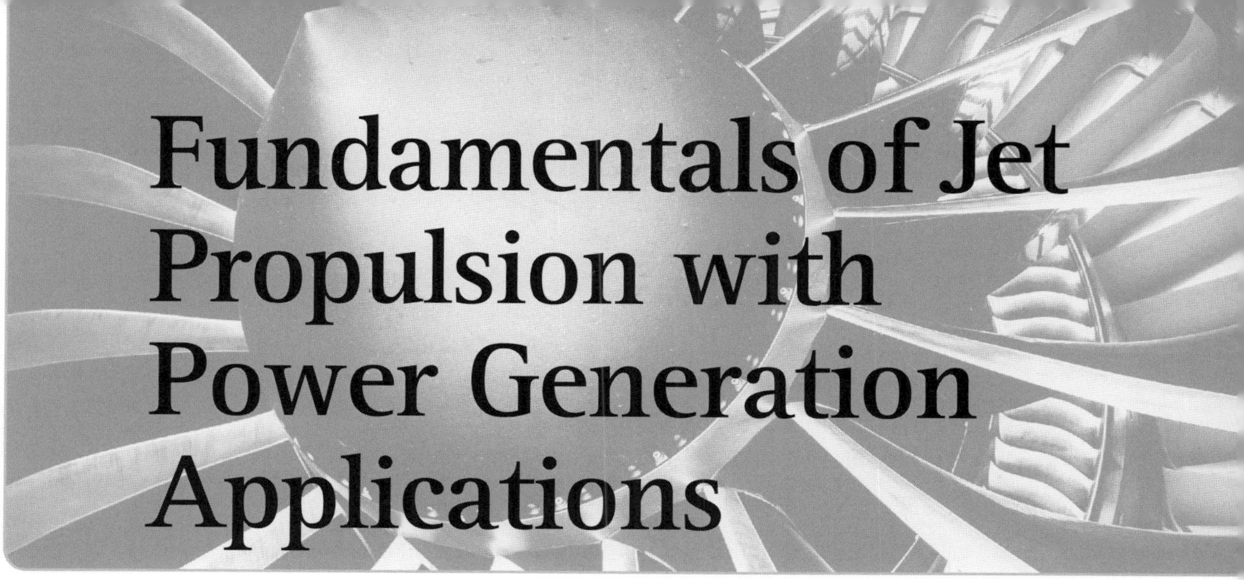

Fundamentals of Jet Propulsion with Power Generation Applications

RONALD D. FLACK
University of Virginia, BRG Machinery

CAMBRIDGE
UNIVERSITY PRESS

CAMBRIDGE
UNIVERSITY PRESS

Shaftesbury Road, Cambridge CB2 8EA, United Kingdom

One Liberty Plaza, 20th Floor, New York, NY 10006, USA

477 Williamstown Road, Port Melbourne, VIC 3207, Australia

314–321, 3rd Floor, Plot 3, Splendor Forum, Jasola District Centre, New Delhi – 110025, India

103 Penang Road, #05–06/07, Visioncrest Commercial, Singapore 238467

Cambridge University Press is part of Cambridge University Press & Assessment, a department of the University of Cambridge.

We share the University's mission to contribute to society through the pursuit of education, learning and research at the highest international levels of excellence.

www.cambridge.org
Information on this title: www.cambridge.org/highereducation/isbn/9781316517369

DOI: 10.1017/9781009043076

First published 2005
Second edition published 2024

Printed in the United Kingdom by TJ Books Limited, Padstow, Cornwall

A catalogue record for this publication is available from the British Library.

A Cataloging-in-Publication data record for this book is available from the Library of Congress.

ISBN 978-1-316-51736-9 Hardback

Additional resources for this publication at www.cambridge.org/flack2e.

Dedicated to Nancy

My best friend, my inspiration, my confidant, my love and soul mate since June 1966

CONTENTS

Appendices

PREFACE

My goal with this ongoing project is to repay the gas turbine industry for the rewarding profession it has provided for me over the course of more than five decades. Education remains a real passion for me and this textbook is one way I can archive and share experiences with a variety of "students." In addition to university courses, I also have taught numerous industry and government/military training classes, which are always a real joy as well. I wrote the first edition of this text thinking back to what I would have liked as an undergraduate student 50 years ago. Thus, this work has been tailored to be a very student-friendly text, whether it be a university, industrial, or other "student."

The first edition of this book was intended to serve primarily as an introductory text in air-breathing jet propulsion. It was directed at upper-level undergraduate students in mechanical and aerospace engineering. A basic understanding of fluid mechanics, gas dynamics, and thermodynamics was presumed; however, thermodynamics was reviewed, and an appendix on gas dynamics was included for reference. For the second edition the applications of "aeroderivatives," ground- or marine-based gas turbines such as those used for marine propulsion, ground transportation, or power generation, have been amplified. Although turbomachinery is not the primary target of the text, it is the book's secondary focus, and thus the fundamentals of, and some advanced topics in, compressors and turbines are also covered.

This text covers the basic operating principles of jet engines and power generation gas turbines. Both fundamental principles, mathematics and hardware, are addressed. Numerous examples based on modern units are included and are an integral part of the text so that students can grasp the methods and acquire an appreciation of different representative physical parameters. For this reason, development of "plug-and-chug" equations or "formulas" is deemphasized, and the solutions of all examples are logically and methodically presented. The examples are not intended to be side issues or optional reading. A student is expected to understand the individual steps of analyzing an entire gas turbine or an individual component. By using examples and homework problems a student is also expected to develop an appreciation of trend analysis; that is, if one component is changed by a known amount, how will the overall engine performance change? Both British and SI units are used in the examples. A strong and unique feature of the book is a capstone chapter (Chapter 12) that integrates the previous 11 chapters into a section on component matching. From this integrated analysis, gas turbine performance can be predicted for both on- and off-design conditions.

Subjects are treated with equal emphasis, and the parts of the book are interdependent in such a way that each step builds on the previous one. The presentation is organized into three basic areas as follows.

(1) Cycle analysis (Chapters 1, 2, and 3) – In these chapters, different engines and power units are defined and given a historical perspective. The fundamental thermodynamic

and gas dynamic behavior of the various components are covered. Ideal and nonideal analyses are performed on each type of gas turbine considered as a whole. Fundamental applicable thermodynamic principles are reviewed in detail. The performance of each individual component is assumed to be known at this point in the text. Trend studies and quantitative analysis methodologies are presented. The effects of nonideal characteristics are demonstrated by comparing performance results with those that would occur if the characteristics were ideal.

(2) Component analysis (Chapters 4–11) – In these chapters the components are studied and analyzed individually using thermodynamic, fluid mechanics, and gas dynamic analyses. Axial flow compressors, centrifugal compressors, axial flow turbines, combustors/afterburners/reheaters, diffusers/inlets, nozzles, ducts, exhausts, mixers, intercoolers, and regenerators are covered. Individual component performance can be predicted and analyzed, including on- and off-design performance and "maps," thus expanding on the fundamentals covered in cycle analyses. The effects on component performance of different geometries for the various components are covered. Some advanced topics are included in these sections.

(3) System analysis and matching (Chapter 12) – This chapter serves as a capstone chapter and integrates the component analyses and characteristic "maps" into generalized cycle analyses for both propulsion and power generation units. Individual-component performance and overall gas turbine performance are predicted and analyzed simultaneously. Both on- and off-design analyses are included, and prediction of gas turbine parameters such as the operating line and compressor surge margin is possible.

Every chapter begins with an introduction providing an historical overview and outlining the objectives of the chapter. At the end of every chapter, a summary reviews the important points and specifies which analyses a student should be able to perform. In addition, appendixes are included that review or introduce compressible flow fundamentals, some advanced compressor details that are summarized in the main body, and general concepts of iteration methods – all of which are a common thread throughout the text.

The text is well suited to independent study by students or practicing engineers. Several topics are beyond what a one-semester undergraduate course in gas turbines can include. For this reason, the book should also be a valuable reference text.

A suite of user-friendly .exe computer programs is available to instructors through the Cambridge website at www.cambridge.org/flack2e. The programs complement the text, but it can stand alone without the programs. I have used the programs in a variety of ways. I have found the programs (especially the cycle analysis, turbomachinery, and matching programs) to be most useful for design problems, and this approach reduces the need for repetitious calculations. This text has a brief review of the programs in Appendix M. A full set of instruction manuals is available with the software. The programs are as follows.

Atmosphere – Calculations or tables for standard atmosphere.
Ideal Gas – Very simple calculations for an ideal gas to eliminate repetitive computations. This is new.

Simple 1D – Compressible one-dimensional (1-D) calculations or tables for Fanno line, Rayleigh line, isentropic, normal shock flow, or constant static-temperature flows. This replaces multipage tables found in many texts. This can be used to generate printable tables.

General 1D – Computations for combined Fanno line, Rayleigh line, drag object, mixing flow, and area change.

Shock – Calculations for normal, planar oblique, or conical oblique shocks.

Nozzle – Calculations for shockless nozzle flow.

SSEffy – Single-stage efficiency for a compressor or turbine based on the total efficiency of the unit. The limit is the small stage or polytropic efficiency. This is new.

JetEngineCycle – Cycle analysis of ideal and real ramjets/scramjets, turbojets, turbofans, and turboprops. Student and instructor versions available.

PowerGTCycle – Cycle analysis of power generation gas turbines with regenerators, intercooling, inlet cooling, and reheat. Student and instructor versions available.

Turbomachinery – Mean-line turbomachinery calculations for axial and radial compressors and axial and radial turbines. Student and instructor versions available.

SLA – Educational three-dimensional (3-D) streamline analysis of axial flow compressors or turbines with radial equilibrium with several specifiable boundary-condition types.

SSAxCompPerf – Fundamental prediction of compressor-stage efficiency using cascade lift and drag characteristics and incidence flow conditions.

Kerosene – Adiabatic flame temperature of n-decane for different fuel-to-air mix ratios.

JetEngineMatch – Given diffuser, compressor, burner, turbine, and nozzle maps are matched to find overall turbojet engine performance and airframe drag maps are used to match engines with an aircraft.

PowerGTMatch – Given inlet, compressor, burner, turbine, regenerator, and exhaust maps are matched to find overall power generation gas turbine performance.

A solutions manual (PDF) to the more than 330 end-of-chapter problems is also available to instructors. Please contact Cambridge University Press at www.cambridge.org/flack2e.

The first edition was written from 1988 to 2004 and this second edition was written from 2017 to 2023. The bulk of the writing of the first edition was done at the University of Virginia, although a portion of the book was written at Universität Karlsruhe while I was on sabbatical (twice). Some chapters were used in my jet propulsion class starting in 1989, and I began to use full draft versions of the text starting in 1992. In the course of this extended use, students suggested many important changes, which I incorporated; more than 400 students were very important to the development of the first edition. I also used portions of draft versions of the text in a graduate-level turbomachinery course, and graduate students have also made very useful suggestions. I am indebted to the numerous students who contributed in this way. This edition was started in 2017 when it became apparent that updating, reorganizing, and adding more power generation applications would make this text more appealing and versatile to a wider range of "students."

A number of changes and upgrades have been initiated in the second edition.

(1) More is included on power generation. These changes take place mostly in Chapters 1, 2, 3, 8, 11, and 12. Regeneration, intercooling, inlet cooling, and reheat are all covered now. Software has been modified accordingly, especially for the cycle analysis and matching. More end-of-chapter problems are included for power generation.
(2) Chapters 2 and 3 from the first edition are now totally reorganized. A brief introduction to compressible flow/gas dynamics has been added to Chapter 2 along with the basic dynamics of the components. The review of gas dynamics is optional depending on the depth of previous coverage at different universities. This review was suggested by a few reviewers. Nonideal and ideal components are now covered simultaneously, including h–s and p–v diagrams.
(3) Chapter 3 presents many integrated ideal and nonideal cycle analyses. Likewise, nonideal and ideal cycle analyses are also covered simultaneously. As a result, comparing nonideal and ideal effects is much more streamlined, logical, and self-contained. Some redundancy has been eliminated from the first edition. Also, the long closed-form ideal cycle analysis derivations have been moved to an appendix. Final equations remain in the text but the derivations are in an appendix for the reader to follow at will.
(4) Many more SI examples and end-of-chapter problems are included. Students and instructors can chose which units can be used for many of the end-of-chapter. problems. Most of the software has been modified to include SI units.
(5) More examples and improved examples are in the text. These improvements are largely in Chapter 3.
(6) A brief introduction to advancing engines has been included, including hypersonic, turboramjet, pulse detonation, and adaptive cycle engines. Furthermore, the current geared turbofan is included.
(7) Turbomachinery fundamentals have been moved from an appendix to the main body (Chapter 4). Most preliminary undergraduate courses do not cover this.
(8) Compressor surge control has been expanded.
(9) Turbine blade coatings have been added to the turbine chapter.
(10) Brief introductions to chemical equilibrium/dissociation and emissions have been added to the combustion chapter (Chapter 9). More discussion of flame speed is also included.
(11) More discussion is included for inlets and exhausts/stacks for power generation gas turbines in Chapters 8 and 11.
(12) A brief identification of mixed-flow centrifugal compressor impellers and crossover design has been added to the centrifugal compressor chapter (Chapter 6).
(13) A comparison of axial flow and centrifugal compressors has been included.
(14) Details of derivations of advanced topics in the axial flow compressor chapter have been moved to appendices. These include radial equilibrium, streamline analysis, and performance based on cascade lift and drag data. Examples and discussion remain in the main body. These are considered to be advanced topics.
(15) Chapter objectives have been clearly identified for all chapters.

(16) Quick chapter questions are in an instructors' packet for instructors to administer before and after each chapter is covered. These are not to be graded but simply to give both students and an instructor an idea of what students know before proceeding and also to identify if the goals have been met afterwards for both students and instructors.
(17) Detailed titles have been added to all examples to make the content more obvious.
(18) A brief software user's guide is in the text. A complete user's guide is included with the software suite.
(19) Light color backgrounds are used for all examples to identify examples better.
(20) Some software has been modified to eliminate large printed tables (standard atmosphere, isentropic flow, etc.) usually found in other texts and also in the first edition of this text. Print files are available. Furthermore, tables are only valid for select values of specific heat ratios. The software can accommodate any value.
(21) Cycle analysis software has been improved by allowing the properties to be evaluated at temperatures other than the total temperature.

A table is presented here to give an instructor or independent reader a suggested schedule of lectures or learning schedule. This, of course, depends strictly on the background of courses covered before this course. It is intended only as a suggestion.

Lecture hours	Topic	Chapter
2	Introduction, Brayton cycle, engine classification	1
Optional	Compressible flow review	2
3	Ideal and nonideal components	2
6	Ideal and nonideal cycles	3
2	Turbomachinery fundamentals	4
6	Axial flow compressors	5
4	Centrifugal compressors	6
4	Axial flow turbines	7
3	Diffusers	8
3	Combustors, afterburners	9
3	Nozzles	10
2	Ducts/mixers/exhausts	11
4	Component matching	12
1	Review	
2	Exams	
45		

ACKNOWLEDGMENTS

I want to first acknowledge the advice of two sets of external and nameless reviewers for this edition along with suggestions from users and nonusers of the first edition. I also thank editors Helen Shannon, Steven Elliot, Lucy Edwards, copyeditor Beverley Lawrence, and others at CUP for their suggestions, support, and advice.

This ongoing project has been most fulfilling, and it has been a culminating point for me. Through the writing and the resulting input from both university and industrial students, I have become a better and more patient teacher. Acknowledgments and thanks are in order, starting with Mac Mellor and Sigmar Wittig, back in 1968, and then Doyle Thompson, in 1971, who all triggered my interest in gas dynamics and gas turbines with projects at Purdue. I also acknowledge and thank my compressor/fan group colleagues at Pratt & Whitney (West Palm Beach), who further ignited my interest in turbomachinery and propulsion while our team developed the compressor and fan for the early F100 engine. I also thank my colleagues at BRG Machinery, especially Hunter Cloud and Jim Byrne, whose association has allowed me to further my teaching beyond the university level. Certainly, thanks are due to my colleagues at both the University of Virginia and Universität Karlsruhe for their collegiality.

My dear and decades-long friends Sandy Maslen, Eric Maslen, Adéle Shaffer, Butch Shaffer, and the late Dave Brady are acknowledged but enough cannot be said. My late sister and living brother-in-law, Gwen and Jim Leighty, always let us feel like we were a part of their family and have not been only endearing relatives but also the best of friends. All of these friendships have brought great enjoyment to Nancy and me. They are the definition of everlasting friendships.

My late Uncle Pete (Harry K. Herr) was instrumental in me becoming an engineer. Since I was very young, he provoked my interest in engineering. His engineering professionalism and integrity greatly influenced my career. My remarkable late aunt, Judy Herr, was a role model as a family loving mother and wife, and unknowingly but positively impacted my life for family values.

My family has been a timeless inspiration to me. My son, Todd, is now a great adult who is also a best friend and is always enjoyable to be with. My daughter, Missy, has brought joy in years past. My granddaughters Mya and Maddie show me how endearing ambitious young adults can be. And then there are Dieter, Zell, Alice, Buddy, and Cody – one could not want more unconditional loving 24/7 (often office) companions.

Nancy is my loving soul mate and best friend since 1966. In the fall of 2022 I was again awakened onto how much of Nancy was imbedded into my soul. Both editions of this book and my professional success would never have come to fruition without her positive influence. She changed my life in 1966, and then always helped me to realize the true value of life and to keep the proper perspectives.

Sadly, Nancy Flack, Ron's soul mate, suddenly passed on May 5, 2023, during the final proofing stages of this textbook.

Part I
Cycle Analysis

General Electric GE90-115B (courtesy of GE Aerospace), available at https://www.ge.com/news/press-releases/ge-delivers-its-1000th-ge90-engine-emirates-airlines (Nov. 9, 2009).

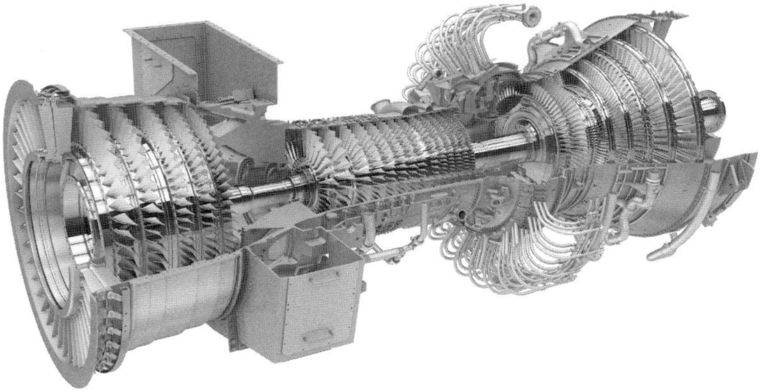

GE LM6000 aeroderivative gas turbine cutaway (courtesy of General Electric Gas Power).

1 Introduction

1.1 Introduction

Propulsion devices have existed for many centuries, and natural devices have developed through evolution. Most modern engines and gas turbines have one common denominator: compressors and turbines or "turbomachines." Several of the early turbomachines and propulsive devices will be described in this brief introduction before modern engines are considered. Included are some familiar names not usually associated with turbomachines or propulsion. Many of the early devices were developed by trial and error and represent early attempts at design engineering, and yet some were quite sophisticated for their time. Billington (1996), ASME (1997), Engeda (1998), St. Peter (1999), Wilson and Korakianitis (2014), and others all present very interesting introductions to some of this history supplemented by photographs.

In this chapter some of the history of gas turbines is presented as well as some of the performance characteristics of various types of current gas turbines. Both jet engines and power generation gas turbines are included. Before proceeding to the following chapters, the basic thermodynamic fundamentals for gas turbines are covered. The end result of this chapter is to provide the reader with a set of initial working equations to calculate thrust or net power of a gas turbine.

1.1.1 Objectives

After completing this chapter, the reader should be able to do the following.

- Understand the history of gas turbines
- Understand the fundamental thermodynamic processes of gas turbines
- Understand the types of gas turbines
- Understand the applications of propulsion gas turbines
- Understand the applications of power generation gas turbines
- Understand the types of performance characteristics of different gas turbines
- Understand the typical performance values of different gas turbines
- Understand the definition of a gas generator
- Understand performance enhancers for different gas turbines
- Get a feel for the future of gas turbines
- Apply the basic equations that can be used to calculate the performance of jet engines
- Apply the basic equations that can be used to calculate the performance of power generation gas turbines

1.1.2 History of Propulsion/Power Devices and Turbomachines

One of the earliest turbomachines was the aeolipile of Heron (often called "Hero" of Alexandria), as shown in Figure 1.1. This device was conceived around 100 BCE. It operated with a plenum chamber filled with water, which was heated to a boiling condition. The steam was fed through tubes to a sphere mounted on a hollow shaft. Two exhaust nozzles located on opposite sides of the sphere and pointing in opposite directions were used to direct the steam with high velocity and rotate the sphere with torque (from the moment of momentum) around an axis – a reaction machine. By attaching ropes to the axial shaft, Heron used the developed power to perform tasks such as opening temple doors.

Figure 1.1 Hero's Aeolipile, 100 BCE.

Figure 1.2 Dutch windmill (photo by R. Flack).

In about AD 1232, Wan Hu developed and tested the Chinese rocket sled, which was driven by an early version of the solid propulsion rocket. Fuel was burned in a closed container, and the resulting hot gases were exhausted through a nozzle, which produced high exit velocities and thus the thrust. Tragically, this device resulted in one of the earliest reported deaths from propulsion devices, for Hu was killed during its testing.

Leonardo da Vinci also contributed to the field of turbomachines with his chimney jack in 1500. This device was a turbine within the chimney that used the free convection of hot rising gases to drive a set of vanes rotationally. The rotation was redirected, using a set of gears, to turn game in the chimney above the fire. Thus, the game was evenly cooked. At the same time, da Vinci also contributed to turbomachinery development with his conception of a helicopter producing lift with a large "screw."

From the conceptions of Robert Hooke and others, windmills (Figure 1.2) – actually large wind turbines – were extensively used in the Netherlands for both water pumping and milling from the 1600s to the 1800s. These huge wind turbines (more than 50 m in diameter) made use of the flat terrain and strong and steady winds and turned at low rotational speeds. Through a series of wooden "bevel" gears and couplings, the torsional power was turned and directed to ground level to provide usable power. Some of the early pumping applications of "windmills" in the Netherlands used an inverse of a water wheel – that is, the "buckets" on the wheel scooped water up at a low level and dropped it over a dike to a higher level, thus, recovering land below sea level from flooding.

Giovanni de Branca developed a gas turbine in 1629 that was an early version of an impulse turbine. Branca used a boiling, pressurized vessel of steam and a nozzle to drive a set of radial blades on a shaft with the high-velocity steam. The rotation was then redirected with a set of bevel gears for a mechanical drive.

In 1687, Sir Isaac Newton contributed the steam wagon, which may be viewed as an early automobile. He used a tank of boiling water constantly heated by a fire onboard the wagon and a small nozzle to direct the steam to develop thrust. By adjusting the fire intensity, the valve on the nozzle, and the nozzle direction, he was able to regulate the exhaust velocity and thus the thrust level as well as thrust direction. Although the concept was viable, the required power exceeded that available for reasonable vehicle speeds. Thus, the idea was abandoned.

Denis Papin developed the first scientific conceptions of the principles of a pump impeller in a volute in 1689, although remains of early wooden centrifugal pumps from as early as the fifth century AD have been found. In 1754, Leonhard Euler, a well-known figure in mathematics and fluid mechanics, further developed the science of pumps and today has the ideal pump performance named after him – "Euler head." Much later, in 1818, the first centrifugal pumps were produced commercially in the United States.

Garonne developed a water-driven mill in 1730. This mill was an early venture with a water (or hydraulic) turbine. Water at a high hydrostatic head from a dammed river was used to direct water onto a conoid (an impeller) with a set of conical vanes and turn them. The rotating shaft drove a grinding mill above the turbine for grain preparation. The same concept was applied in 1882 in Wisconsin, where a radial inflow hydraulic turbine was used to generate electricity.

Gifford was the first to use a controlled propulsion device successfully to drive an "aircraft." In 1851, he used a steam engine to power a propeller-driven dirigible. The total load required to generate power was obviously quite large because of the engine size, combustion fuel, and water used for boiling, making the idea impractical.

In 1883, Carl de Laval developed the so-called Hero-type reaction turbine shown in Figure 1.3 utilized for early water turbines. Water flowed through hollow spokes, formed high-velocity jets normal to, and at the end of, the spokes, and was used to turn a shaft. This is the basic type of rotating sprinkler head used to convert potential energy from a static body of water to a rotating shaft with torque.

As another example, in 1897 de Laval developed the impulse steam turbine (Figure 1.4). This utilized jets of steam and turning vanes or blades mounted on a rotating shaft. The high-speed steam impinged on the blades and was turned, thus imparting momentum to the blades and therefore rotating the shaft and providing torque.

Over the next quarter century, rapid developments took place. Gas and steam turbines came into wide use for ships and power generation. For example, in 1891 the first steam turbine was developed by Charles Parsons. This device was a predecessor to the modern gas turbine. It had two separate components: the steam generator–combustor and the turbine. The generator–combustor developed high-pressure steam,

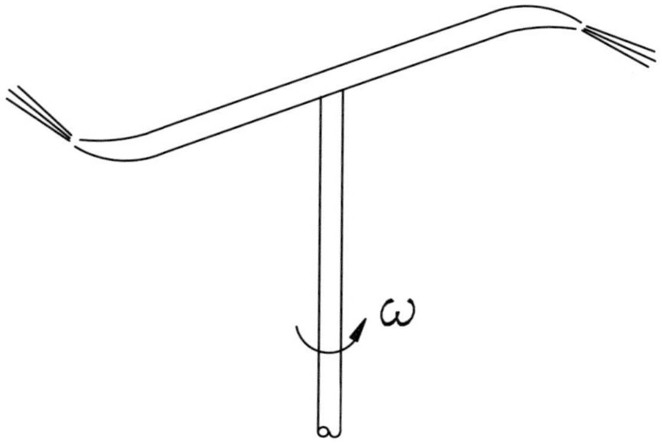

Figure 1.3 DeLaval "Hero" reaction turbine, 1883.

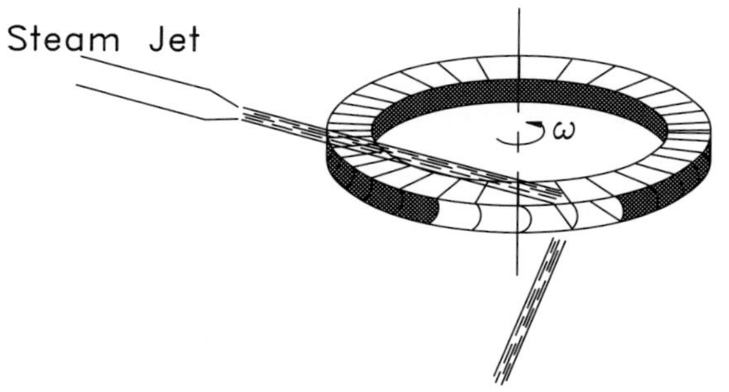

Figure 1.4 DeLaval impulse turbine, 1897.

Figure 1.5 Elling gas turbine rotor, 1924 (photo by Øyvind Hundseid).

which was directed as a high-velocity jet into the steam turbine. In the early 1800s, ship propellers or "screws," which are themselves a variety of turbomachines, were invented by Richard Trevithick and others. Parsons' steam turbine, rated at 2100 hp (1570 kW), was used to power such a propeller directly on the 100-ft (30.5-m)-long ocean vessel *Turbinia* in 1897 and drove it at 34 kt, which was a true feat if one considers that most seaworthy vehicles were slow-moving sail craft.

In 1912, a large (64-stage) steam-turbine facility was installed in Chicago and ran at 750 rpm to deliver 25 MW of electrical power. In the 1920s, several General Electric (GE) 40-MW units were put in service. These ran at 1800 rpm and had 19 stages. Although many refinements and advancements have been made to steam-turbine technology since this installation, the same basic design is still in use in power plants throughout the world.

In 1884, Norwegian engineer Ǽgidius Elling patented what was to become the first modern gas turbine with the final application intended to include aircraft propulsion. In 1903 he was able to run the first power generating gas turbine producing 11 hp. He continued development of his design including compressor cooling, regeneration, reheat, multiple shafts, and variable stators, and in 1932 built and successfully ran a gas turbine that produced 75 hp. A photograph of the rotor shaft from 1924 is shown in Figure 1.5, which was taken at the 2018 ASME Turbo Expo, Oslo, meeting. He had patents on two gas turbines in 1884 and 1906. A full documentation of this lost history was presented by Bakken, Jordal, Svyerud, and Veer (2004) and the 2018 ASME venue.

In the 1930s, simultaneous and strictly independent research and development were performed in Great Britain and Germany on gas turbines. In 1930, Sir Frank Whittle (Great Britain) patented the modern propulsion gas turbine (Figure 1.6). The engine rotated at almost 18,000 rpm and developed a thrust of 1000 lb_f (4450 N). It had a centrifugal flow compressor and a reverse-flow combustion chamber; that is, the

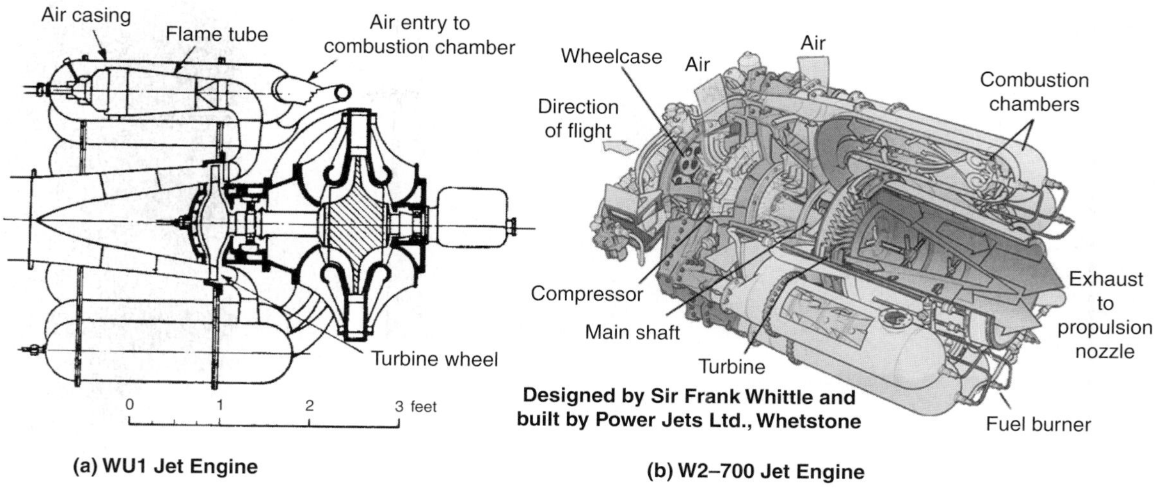

Figure 1.6 Whittle's jet engines ((a) from Fig. 105 of Lloyd (1945), (b) from Midland Air Museum website).

flow in the burner was opposite in direction to the net flow of air in the engine – a concept still used for small engines to conserve space. This gas turbine was first installed on an aircraft in 1941 after several years of development. Meher-Homji (1997a) reviews this early effort in great detail. Dunham (2000) reviews the efforts of A. R. Howell, also of Great Britain, which complemented the work of Whittle and development of the axial flow compressor.

In 1939, the first flight using a gas turbine took place in Germany. Hans von Ohain patented the engine for this aircraft in 1936 (Figure 1.7), which developed 1100 lb_f (4890 N) of thrust. This engine had a combination of axial flow and centrifugal compressor stages. In general, this gas turbine and further developmental engines were superior to the British counterparts in efficiency and durability. A few years later the German Junkers Jumo 004 (Figure 1.8), designed by Anselm Franz, was the first engine to be mass produced. Meher-Homji (1996, 1997b, and 1999) also presents an interesting review of these early developments. Other historical perspectives on turbomachines and propulsion are offered by Heppenheimer (1993), St. Peter (1999), and Wilson and Korakianitis (2014). Today Elling, Whittle, and von Ohain are all credited with the invention of the jet engine.

Also during the 1930s, the first high-speed turbopumps for rocket propulsion were developed in Germany for the V2. Hot exhaust gases from a combustor were expanded by turbines and drove the high-speed oxygen and hydrogen cryopumps, which in turn pumped or compressed the cryofluids, readying them for the combustor. The maiden voyage of the V2 occurred in 1940, and its introduction allowed for previously unattainable long-range delivery of warheads. This type of propulsion inspired modern rocket technology and is still the basic operating principle for modern rocketry.

In 1942, General Electric (GE) developed what is considered the first US jet engine, the GE I-A. This was a small engine that generated 1300 lb_f of thrust and can be seen to be a similar copy of the early Whittle engine. The GE I-A was developed into the larger GE J-31 (or I-16), which ran at 16,500 rpm, weighed 850 lb (385 kg), and developed 1650 lb_f (7340 N) of thrust (Figure 1.9). After this engine became commercially available, several standing manufacturers of other power generation equipment rapidly began developing such jet engines: Pratt & Whitney, Allison, Honeywell, Garrett, Avco, Solar, Volvo, Westinghouse, and Rolls-Royce, among others. Another very early engine in the

Figure 1.7 Replica of Ohain's Heinkel Turbostrahltriebwerk He-S3-B (1939) jet engine (courtesy of Deutsches Museum).

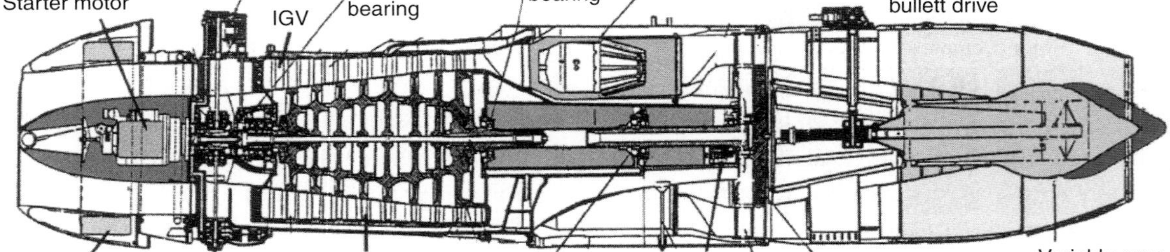

Figure 1.8 Junkers Jumo 004 (courtesy of Cyrus Meher-Homji).

United States was the Westinghouse J30 (Figure 1.10), which had a six-stage axial compressor, an annular burner, and a single-stage turbine and developed 1560 lb_f (6940 N) of thrust.

Today the largest engines are built by Pratt & Whitney (PW 4000 series), General Electric (GE 90, GE9X), and Rolls-Royce (Trent), and all of these manufacturers produce engines that develop thrusts in excess of

Figure 1.9 General Electric J31 (courtesy of the National Museum of the United States Air Force), on display at Wright Patterson Air Force Base.

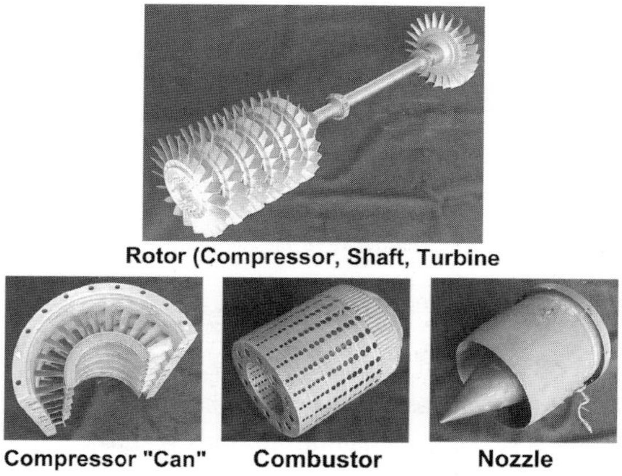

Figure 1.10 Disassembled Westinghouse J30. (photos by R. Flack).

110,000 lb_f (445,000 N). One version of the GE9X derivative (Figure 1.11) is under development and has been tested at over 130,000 lb_f (578,000 N). The Pratt and Whitney 4000-112 turbofan is one in such a series of engines (Figure 1.12). Since the 1950s, gas turbines, which are derivatives of jet engines, have found their way into automobiles (Parnelli Jones almost won the 1967 Indianapolis 500 with an Andy Granatelli turbine), naval and commercial ships and boats, and many electric-power generation units. Some Union Pacific trains in the 1950s used an oil-burning gas turbine (4500-hp, 3360-kW) and some other trains in the USA (Amtrack) and France (TGV) used gas turbines in the 1970s but electrical drives became more economical. Before proceeding, it is important to note that other very respectable books and references, provide complimentary material and sometimes different perspectives of analysis and development for both gas turbines and components thereof and have in fact provided inspiration for this book. These include Hesse and Mumford (1964), Treager (1979), Whittle (1981), Oates (1985, 1997), Pratt & Whitney (1988), Hill and Peterson (1992), Kerrebrock (1992), Rolls-Royce (1996), Greitzer (2009), Cumpsty (2010), Farokhi (2014), Cumpsty and Heyes (2015), Mattingly and Boyer (2016), Saravanamuttoo et al. (2017), and Kurzke and Halliwell (2018).

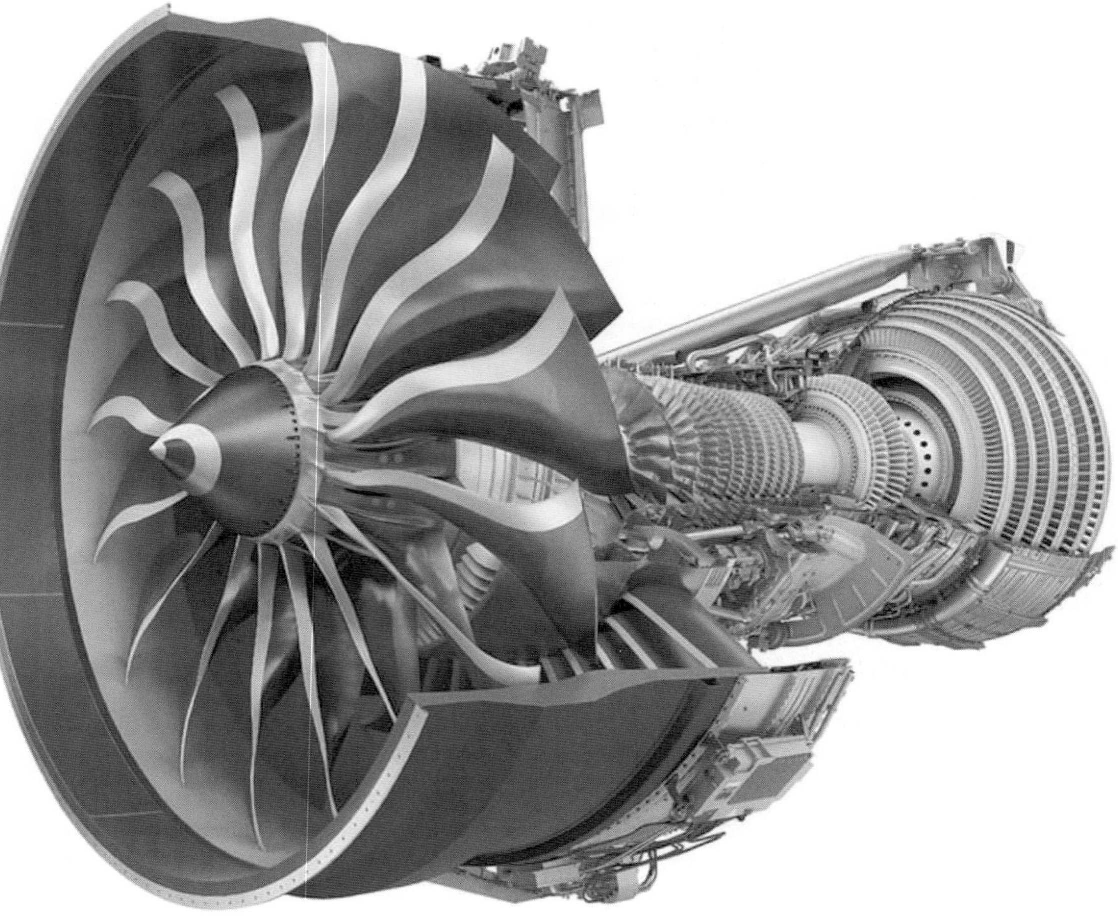

Figure 1.11 General Electric GE9x Turbofan (courtesy of GE Aerospace), available at https://blog.geaerospace.com/product/tooling-up-for-ge9x/ (May 8, 2018).

1.2 Brayton Cycle

A Brayton cycle is the basis for the operation of a gas turbine and can be used to approximate the cycle of all such units. A jet engine operates with an open cycle, which means fresh gas is drawn into the compressor and the products are exhausted from the turbine and not reused. Most power generation gas turbines also operate with the open cycle. A typical ideal system is illustrated in Figure 1.13. Shown are the compressor, combustor, and turbine. It is important to remember that the compressor and turbine are on the same shaft, and thus power is extracted from the fluid by the turbine and used to drive the compressor. Also shown in Figure 1.13 are the h–s and p–v diagrams for the cycle. The compression process, in which work is performed on the fluid mechanically and the pressure and enthalpy ideally increase isentropically (at constant entropy), is from 2 to 3. The combustion process, in which a fuel burns with the air, increases the enthalpy significantly from 3 to 4, and the process (3 to 4) is ideally isobaric. The expansion process, in which the pressure and enthalpy ideally decrease isentropically and energy is mechanically extracted from the fluid, is from 4 to 5.

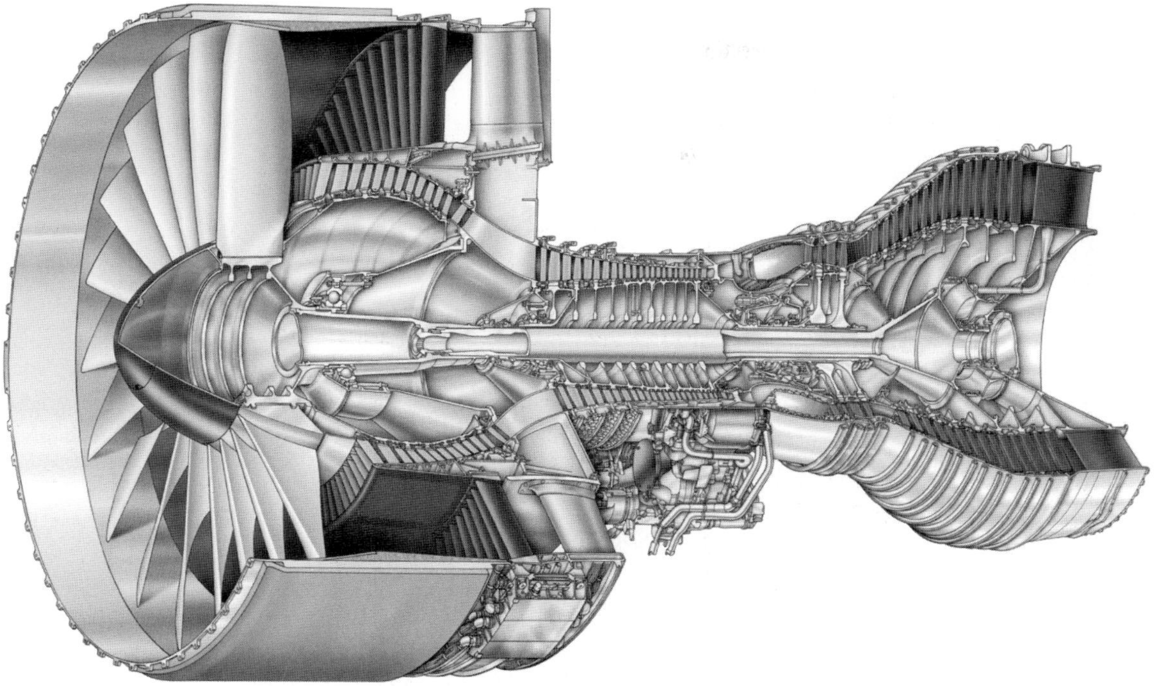

Figure 1.12 Cutaway of Pratt & Whitney 4000-112 (© 2022 Raytheon Technologies Corporation, Pratt & Whitney division. Reproduced with permission).

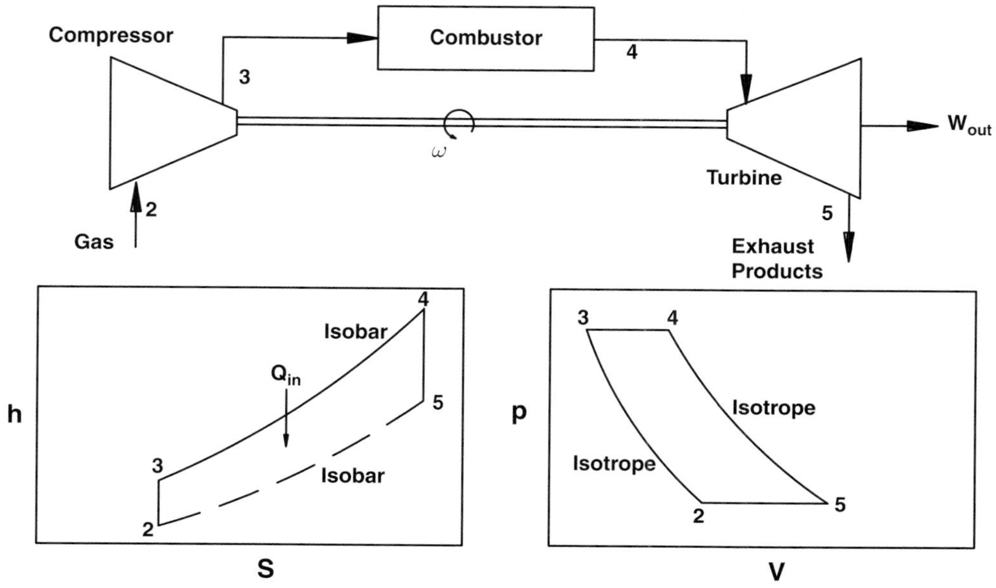

Figure 1.13 Open Brayton cycle.

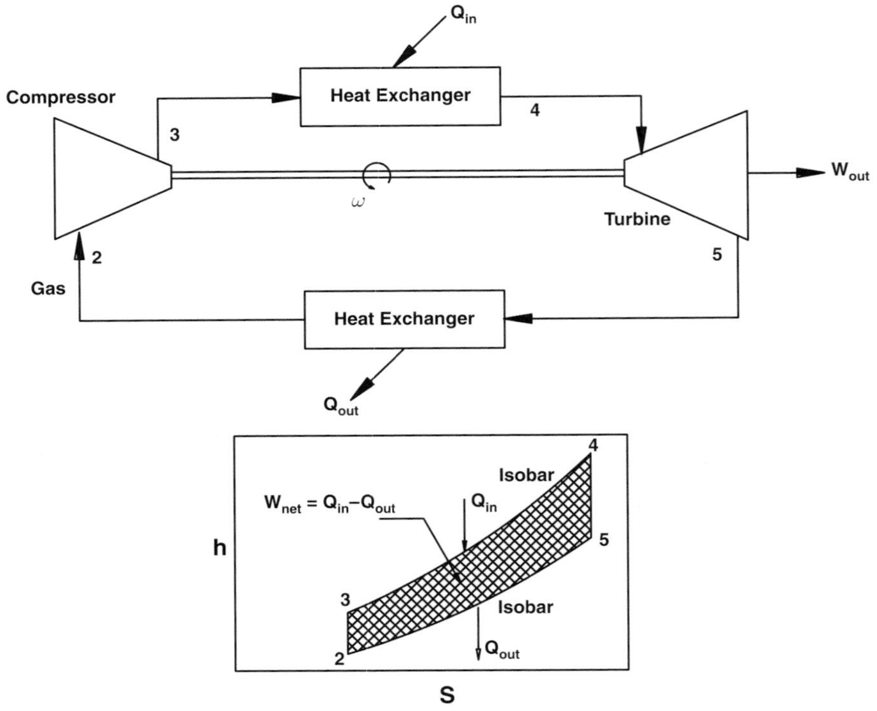

Figure 1.14 Closed Brayton cycle.

A few ground or power generation applications use closed Brayton cycles. In this process, the same gas is continually used in a recirculating process. Such a cycle is shown in Figure 1.14. The *h*–*s* diagram is also shown in Figure 1.14. For a closed cycle, the gas is heated with a heat exchanger from 3 to 4 (not a combustion process), power is derived by the turbine from 4 to 5, and hot gas products from the turbine are cooled with a heat exchanger and fed back into the compressor for process 5 to 2. The net power is shown with crosshatching on the *h*–*s* diagram, Figure 1.14. Typically, the heat exchanger is very heavy, which makes the closed cycle inappropriate for jet engines and results in an exhaust temperature that is too cool for propulsion purposes. For the closed cycle, the compression and expansion processes are ideally isentropic and the heat exchanger processes are ideally isobaric.

An especially important parameter used to evaluate the overall performance of a cycle is the thermodynamic efficiency. For a closed cycle this is given by

$$\eta_{th} = \dot{W}_{net}/\dot{Q}_{in}, \tag{1.2.1}$$

where $\dot{W}_{net}$ is the net usable power from the system, as shown with crosshatching in Figure 1.14, and $\dot{Q}_{in}$ is the heat that is input to the heat exchanger in process 3 to 4. The external heat input is the total area under the curve from station 3 to 4. Clearly the crosshatched area is significantly smaller than the total area under the curve thus the efficiency is less than unity. For the ideal case

$$\eta_{th} = 1 - \frac{h_5' - h_2}{h_4 - h_3'}, \tag{1.2.2}$$

where h is the static enthalpy. For an ideal gas, this becomes

$$\eta_{th} = 1 - \frac{T_5' - T_2}{T_4 - T_3'}, \tag{1.2.3}$$

where T is the static temperature and the primes (′) serve as reminders that the processes are ideal.

In Chapter 3 and Appendix H.7 this will be shown to be equal to

$$\eta_{th} = 1 - \left[\frac{v_3}{v_2}\right]^{\gamma-1} = 1 - \left[\frac{p_2}{p_3}\right]^{\frac{\gamma-1}{\gamma}} = 1 - \left[\frac{1}{\pi_c}\right]^{\frac{\gamma-1}{\gamma}}$$
$$= 1 - \left[\frac{T_2}{T_3}\right] = 1 - \frac{1}{\tau_c}, \tag{1.2.4}$$

where v is the specific volume, p is the pressure, γ is the ratio of specific heats, π_c is the pressure ratio, and τ_c is the temperature ratio for the compressor. It is important to note that, as the pressure or temperature ratio increases across the compressor, the thermodynamic efficiency increases. This will be discussed and demonstrated at many points throughout this book. However, note that, as the compressor's pressure ratio is increased, the size of the compressor becomes larger, adding to both the cost and weight. Thus, a tradeoff study is usually required for a given design and application. One should also remember from basic thermodynamics (Wark and Richards, 1999) that the Carnot and Ericsson cycles are the most ideal cycles possible and have higher efficiencies. But these do not have practical applications.

Now that we have presented the engineering definition of a Brayton cycle and reviewed the very basic thermodynamics, one question may challenge the reader: How can the general principle of a jet engine or power generation unit be explained in layman's terms? One explanation is as follows. A high nozzle pressure is required to develop high momentum or thrust for a jet engine and another turbine is required for extracting power for a power generation unit. The compressor initially works on the working fluid and increases the pressure of the fluid. Next, energy is added as heat in the combustor. Third, some energy is extracted mechanically by the turbine (thus reducing the temperature of the fluid) to drive the compressor, which also reduces the pressure of the fluid. However, more energy is added by the combustor than is extracted by the turbine. Consequently, the pressure and temperature exiting the turbine remain higher than the inlet pressure and temperature to the compressor. As a result, enough internal energy remains in the fluid to be converted to a high-velocity fluid (with high momentum) or thrust for a jet engine and enough energy remains to be converted to power for a power generation unit.

1.3 Classification of Jet Engines

Historically, jet engines and power units have evolved from simple conceptions to very complex designs. Yet, several different types of engines continue to be used. For a given application, one particular type may be significantly better than another. For example, one might expect a short-range "puddle jumper" to have different needs than a supersonic fighter. In this section, different types of engines will be defined and described. In Chapters 2 and 3, the basic principles (including thermodynamics and gas dynamics) of the components and of these engines are covered. Diagrams (such as h–s and p–v) will be included. From these principles, one can predict and compare the different operating characteristics of the various gas turbines and determine what parameters make some gas turbines more appropriate for some applications than others. In the following sections, example details of some current and previously used units are presented; some outdated gas turbines are discussed for historic purposes. Many of the engines have different submodels or "builds" for different applications, and for these cases typical characteristics are presented. Information was collected from engine data in *Aviation Week & Space Technology*, *Flight International*, Mattingly and Boyer (2016), Treager (1979), engine/power units and airframe manufacturers' materials, brochures, web pages, and military publications. Although every effort has been made by to be accurate, cross-checking information as much as possible, accuracy cannot be guaranteed because some gas turbines and manufacturers are forever changing.

1.3.1 Ramjet

The "simplest" jet engine is the ramjet, which is shown in Figure 1.15. This engine is used only in very-high-speed applications and is not capable of self-propelled takeoff.

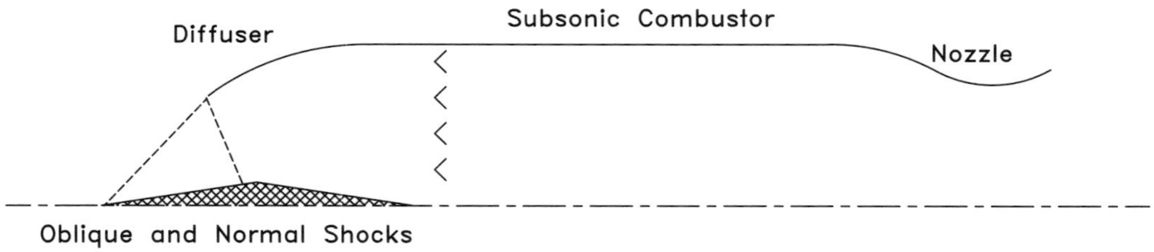

Figure 1.15 Schematic of a ramjet.

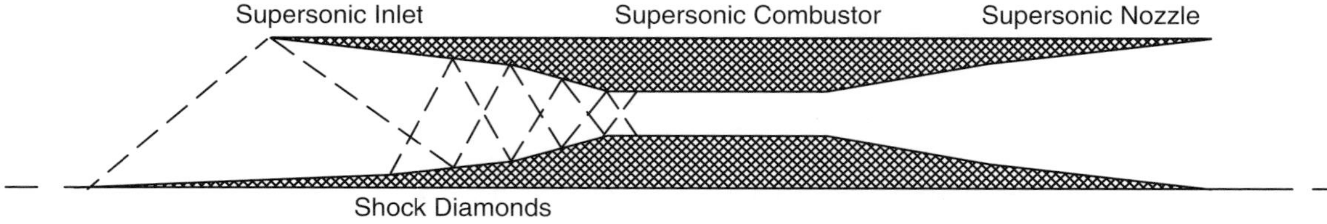

Figure1.16 Schematic of a scramjet.

The ramjet is simple because it has no moving parts. Basically, the engine moves relative to the air with a high velocity u_a. This engine typically operates at low supersonic speeds. Air enters the diffuser, where the air pressure is significantly increased owing to the high air speed. As a result, the inlet must be designed to decelerate the flow but with minimal losses. Oblique shock waves are present at the inlet for supersonic flow and the flow enters the combustor at subsonic speeds. As will be shown in Chapters 2 and 8, oblique shock waves incur smaller losses than normal shocks. In the combustor air mixes with the fuel and burns, thus increasing the temperature. Finally, the hot and expanded gases are accelerated and leave the engine through the nozzle, producing thrust. Because the engine must move relative to the ambient air to develop a pressure rise in front of the burner, this engine cannot operate statically and must be launched by some other means. The ramjet is basically for military rocket applications. It was first developed during the 1940s through 1960s. Marquardt (and later named Bomarc) was among the largest producers, with the RJ43-MA-7 producing 12,000 lb_f thrust and a thrust to weight ratio of 40. A thorough review of the history of ramjets is presented by Fry (2004).

1.3.2 Scramjet

The ramjet is also the basis for a scramjet (supersonic combustion ramjet), Figure 1.16. For this configuration a series of shock waves or "diamonds" are present in the inlet section. Again, oblique shock waves are used to slow the flow but minimize the losses as compared with normal shocks. Although no rotating parts are included in a scramjet, the supersonic combustion is far from simple, and significant research has been dedicated to this process. Similarly, the diffuser is not a simple device. The overall performance depends heavily on both of these components. Because the flow is supersonic in the combustion and other areas, the total or stagnation temperature is much higher than the static temperature, as will be discussed later. The thermal limits on the materials are basically dependent on the local static temperatures. This means that owing to the supersonic conditions the internal total temperatures are much higher by design than those of ramjets and other turbojets. As will be discussed in the next two chapters, the basic cycle analysis is essentially the same for ramjets and scramjets. However, that *does not* mean the designs are essentially the same.

1.3.3 Turbojet

The turbojet represents the next level of engine complexity. Such an engine is shown in Figure 1.17. The turbojet also has a diffuser, which decreases the fluid velocity and increases the incoming air pressure. Air enters the compressor next, where the fluid is worked on and the density is increased, and this process is accompanied by an elevation in pressure and a moderate increase in temperature. The air then enters the combustor, where the injected fuel burns with the air. The temperature and specific volume of the gas increase significantly. The turbine is used to extract some of the energy from the air, and this energy is used to drive the compressor; the turbine and compressor are on the same shaft. Finally, the hot and expanded air, which is still at moderate pressure, flows through the nozzle and is accelerated to a high velocity to produce thrust. It is important to note that the compressor/combustor/turbine combination is defined as the "gas generator" throughout this text and other references.

The configuration shown in Figure 1.17 is called a single-shaft or single-spool engine, which represents an older design. An example of this type is the GE J85-21 (Figure 1.18). This engine weighs 684 lb (310 kg), develops 3500 lb_f (15,600 N) of thrust at maximum power (without the afterburner; see

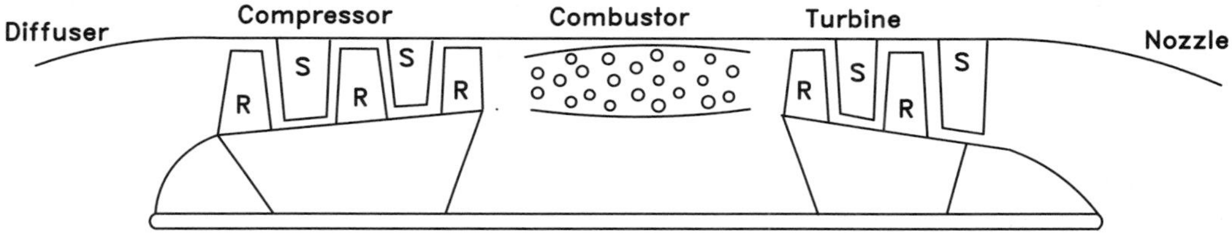

Figure 1.17 Schematic of a single-spool turbojet.

Figure 1.18 General Electric J85-21 Single Spool Turbojet (courtesy of GE Aerospace), from GE publication, "J85 Augmented Turbojet," distributed at ASME IGTI meeting circa 1980s; used in the first edition of this text.

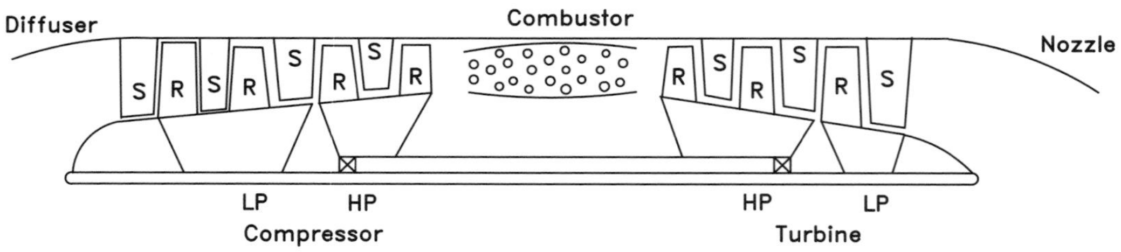

Figure 1.19 Schematic of a two-spool turbojet.

Figure 1.20 Pratt & Whitney J52 twin-spool turbojet (© 2022 Raytheon Technologies Corporation, Pratt & Whitney division. Reproduced with permission).

Section 1.3.4), ingests 52 lb_m/s (24 kg/s) of air, and uses 1.24 lb_m of fuel per hour to generate each lb_f of thrust (lb_m-fuel/lb_f-thrust-h will later be defined as the thrust-specific fuel consumption [*TSFC*] or 0.126 kg/N-h). One of the smallest turbojets is the Williams WR24-7 (used on a drone). This engine weighs 44 lb (20 kg), develops 160 lb_f (710 N) of thrust, ingests only 3 lb_m/s (1.4 kg/s) of air, and uses 1.2 lb_m-fuel/lb_f-thrust-h (0.12 kg/N-h).

A second turbojet configuration is shown in Figure 1.19. This is a two-spool or twin-spool engine. This type of engine is more complex than the single-spool engine but has better operating characteristics at off-design conditions. This configuration has, in fact, two separate consecutive compressors and two separate consecutive turbines on two different shafts, one of which is hollow. One turbine is used to drive one compressor, and the second turbine is used to drive the other compressor. The first compressor that works on the air is called the low-pressure compressor and is driven by the last turbine – the low-pressure turbine – which extracts energy from the air. The second compressor that works on the air is called the high-pressure compressor, and it is driven by the high-pressure turbine. Typically, the high-pressure shaft or spool rotates considerably faster than the low-pressure spool. An example of this type of engine is the Pratt & Whitney J52 (Figure 1.20). The J52-P408 version of this engine develops 11,200 lb_f (49,800 N) of thrust, weighs 2320 lb (1050 kg), and uses 140 lb_m/s (63 kg/s) of air and 2.1 lb_m-fuel/lb_f-thrust-h (0.21 kg-fuel/N-thrust-h).

Current turbojet developments include interesting mini-applications to model aircraft (van den Hout and Koullen, 1997). These small engines develop 9 lb_f (40 N) of thrust, consume 0.33 lb_m/s (0.15 kg/s) of air, and weigh 4 lb (1.8 kg).

1.3.4 Turbojet with Afterburner

A turbojet with an afterburner is an engine with an additional combustion chamber after the turbine. Such an engine is shown in Figure 1.21. The additional combustor adds the capability for nearly instantaneous additional thrust and acceleration. The process is sometimes called thrust augmentation. This is very important in military fighter craft applications. As a result, almost all fighter aircraft include afterburners in their engines. Basically, the additional burning increases the temperature and volume of the engine air and allows for more velocity in the exhaust nozzle, thus producing more thrust. However, using an afterburner is very inefficient thermodynamically. Therefore, afterburners are only used for short-term needs (e.g., for aircraft carrier takeoff or when a fighter is being tailed by an enemy aircraft). Afterburners are not usually found on transport aircraft with the exception of the Rolls-Royce/Snecma Olympus 593 for the now historical Concorde, which was one of the largest of this category (thrust: 31,350 lb_f [139,500 N] with the afterburner off and 38,050 lb_f [169,200 N] with the afterburner on; weight: 6600 lb [3000 kg]; total airflow: 415 lb_m/s [189 kg/s] fuel usage: 0.70 lb_m-fuel/lb_f-thrust-h [0.070 kg/N-h] with the afterburner off and 1.18 lb_m-fuel/lb_f-thrust-h [0.120 kg/N-h] with the afterburner on). Afterburners can be found on all basic types of turbojets (single-spool, twin-spool, etc.). Another example of a twin-shaft afterburning turbojet is the Pratt & Whitney J57. The schematic in Figure 1.22 clearly shows the two shafts and also identifies internal temperatures and pressures. Another example of this type of engine is the Pratt & Whitney J58 (Figure 1.23). This engine

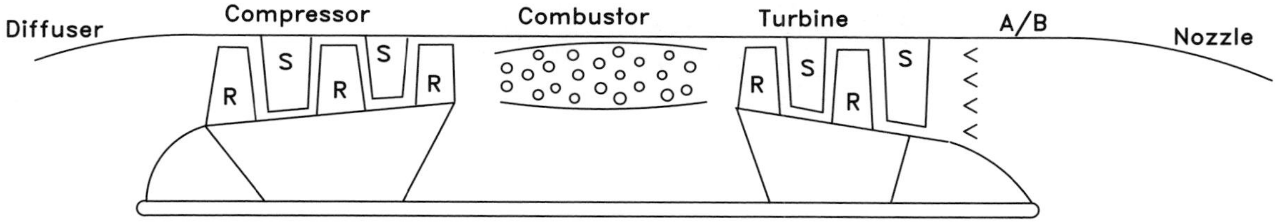

Figure 1.21 Schematic of a turbojet with afterburner (A/B).

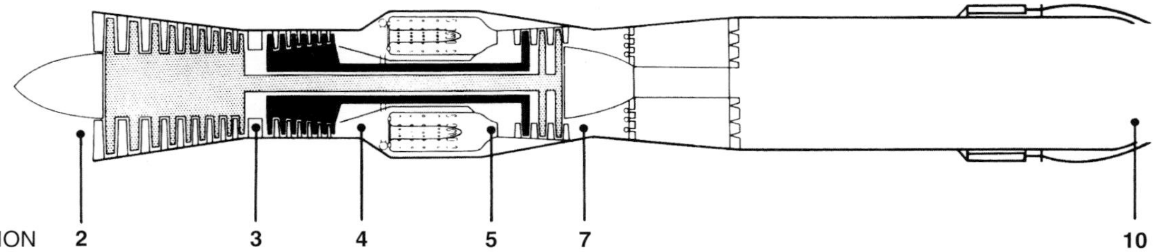

STATION	2	3	4	5	7		10
P_t (psia)	14.7	54.0	167.0	158.0	36.0	MAXIMUM (A/B)	31.9
T_t (°F)	59°	330°	660°	1570°	1013°		2540°
P_t (psia)	14.7	54.0	167.0	158.0	36.0	MILITARY	33.0
T_t (°F)	59°	330°	660°	1570°	1013°		1013°

Figure 1.22 Pratt & Whitney J57 afterburning turbojet showing internal conditions (© 2022 Raytheon Technologies Corporation, Pratt & Whitney division. Reproduced with permission).

develops up to 34,000 lb_f (151,000 N) with the afterburner on (sometimes called "wet" thrust, whereas "dry" thrust is with the afterburner off) and is used on the SR-71 Blackbird, a Mach 3 fighter. The J58 engine was actually a core engine for a more complex geometry due to the need for Mach 3 operation. This entire powerplant operated somewhat like a turboramjet which will be covered in more detail in Section 1.3.11. The previously discussed GE single-shaft J85-21 (Figure 1.18) also has afterburning capabilities. It develops 5000 lb_f (22,200 N) of thrust at maximum power (with the afterburner combusting) and uses 2.1 lb_m-fuel/lb_f-thrust-h (0.21 kg/N-h).

1.3.5 Turbofan

The turbofan is one level of complexity above a turbojet engine. It is a heavier powerplant but has better fuel economy than a turbojet. Two fundamental types of turbofans are used. Each will be described separately.

Turbofan with Fan Exhausted

A turbofan engine with fan exhausted is shown in Figure 1.24. Turbofans are always multispool engines. In the front of the engine, the air is first diffused. The air enters the fan, which compresses the air and increases the pressure somewhat. The air is then split at the "splitter," and a portion of it enters the low-pressure

Figure 1.23 Pratt & Whitney J58 afterburning turbojet (© 2022 Raytheon Technologies Corporation, Pratt & Whitney division. Reproduced with permission).

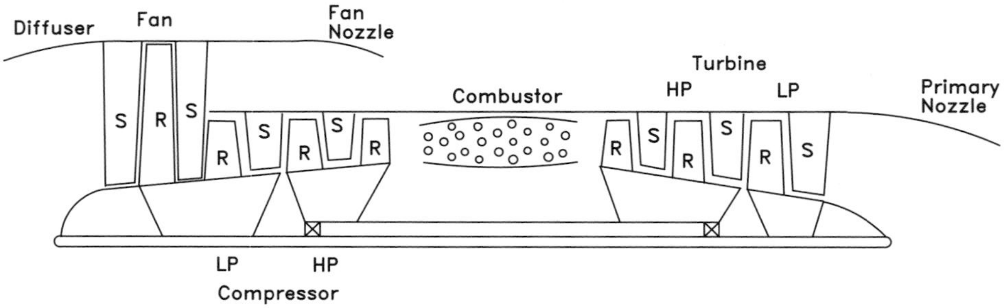

Figure 1.24 Schematic of a turbofan with fan exhausted.

Figure 1.25 General Electric CF6-80C2 Turbofan (courtesy of GEAerospace), from GE publication, "CF6-80C2 High Bypass Turbofan Engine," publication no. AE3258 (7/90), distributed at ASME IGTI meeting circa 1990.

compressor and continues down the "core" of the engine. Eventually this "core" air exhausts through the primary exhaust nozzle and produces thrust. The second stream of air is called the "bypass" air. In this engine type, the "bypass" air is accelerated in the fan nozzle, producing a second or additional thrust. The fan and usually the first few stages of the low-pressure compressor are driven by the low-pressure turbine. The high-pressure compressor extracts its energy from the high-pressure turbine. In some cases, three shafts are used. As will be discussed in the following chapters, modern commercial engines operate with "high bypass ratios" (mass flow rate of the bypassed air to mass flow rate through the core). Namely, the major portion of total air goes through the fan and fan nozzle. Such engines use fuel very efficiently. They tend to be heavier than low-bypass engines or turbojets with similar thrusts, however.

The GE CF6-80C2 (Figure 1.25) is an example of a twin-shaft fanjet and is used on the Boeing 767. The low- and high-pressure shafts rotate at about 3800 and 10,300 rpm, respectively. For the CF6-80E1A2 version (used on the Airbus 330), the developed thrust is 65,800 lb_f (293,000 N), and the engine uses about 0.33 lb_m-fuel/lb_f-thrust-h (0.034 kg/ N-h). The engine has a weight of 11,162 lb (5060 kg). The airflow rate for

Figure 1.26 Rolls-Royce RB211-535 exhausted turbofan (© Rolls-Royce plc 2022. All rights reserved).

the engine is about 1950 lb_m/s (885 kg/s). The GE 90-115B has a bypass ratio of about 9 and develops 115,300 lb (513,000 N) of thrust. It has a fan/compressor pressure ratio of 42. The two-stage low-pressure turbine drives the fan and four-stage low-pressure compressor. The six-stage high-pressure turbine drives the nine-stage high-pressure compressor. This engine powers the Boing 777. Another example of this type of engine is the Rolls-Royce RB211 series (Figure 1.26). This three-spool engine develops up to about 60,000 lb_f (267,000 N), but the thrust varies according to the particular model. The Rolls-Royce Trent 800 also powers the Boing 777 and is a larger engine that was developed based on the RB211. It develops 93,400 lb of thrust (415,000 N) and has a bypass ratio of 9.6 and a fan/compressor pressure ratio of 40.7. It also is a three-spool engine in which the five-stage low-pressure turbine (LPT) drives the large fan, the one-stage intermediate-pressure turbine (IPT) drives the eight-stage intermediate-stage compressor (IPC), and the one-stage high-pressure turbine (HPT) drives the six-stage high-pressure compressor (HPC). The three shafts typically rotate at about 3000, 7500, and 10,000 rpm. The three spools operate at different speeds, and a three-spool engine has better off-design operating characteristics than a twin-spool engine.

However, such an engine is more complex. Next, a schematic of the Pratt & Whitney PW4000 is shown with representative internal temperatures and pressures in Figure 1.27a, and the materials for the PW4084 are illustrated in Figure 1.27b. The PW4000-112 (Figure 1.12) also powers the Boing 777 with 99,000 lb (440,000 N) of thrust. Furthermore, one of the smallest engines in this category is the Williams/Rolls-Royce FJ44-1C, which weighs 45 lb (20 kg), develops 1600 lb_f (7100 N) of thrust, ingests 58 lb_m/s (26 kg/s) of air, and uses 0.46 lb_m-fuel/lb_f-thrust-h (0.047 kg/N-h). On the other end of the scale is the huge GE 90-115B, which generates commissioned thrusts up to 115,000 lb_f (512,000 N) (recent tests on the derivative engine GE9X have pushed this over 130,000 lb_f [578,000 N]), has a maximum diameter of 136 in. (3.45 m), and weighs 18,260 lb (8280 kg). As will be discussed in the following chapters, commercial engines such as these operate with "high bypass ratios" (mass flow rate of the bypassed air to mass flow rate through the core); the GE 90-115B has a bypass ratio of about 9. Again, the major portion of total air goes through the fan and fan nozzle and operates very efficiently although heavier than low-bypass engines or turbojets.

The GE GE9X is a very good example of some of the new developments (Figure 1.11). It powers the Boeing 777X with 110,000 lb_f thrust. In this design the bypass ratio was increased up to 10. The overall pressure ratio is 60. The efficiency of the compressor was increased by 2 percent due to advanced aerodynamic design capabilities! The first-stage fan blades are representative of many new large fans, in that the blade shapes have previously unconventional shapes, longer chords, and more twists than previous designs. Carbon-fiber composites are used for these fan blades, with steel leading edges and glass-fiber trailing edges. As a result, the fan is lighter and can rotate faster to better match the LP turbine. Some of the compressor stages are single-piece bladed disks (blisks). Three-dimensional "printing" reduces such complex manufacturing costs, a technique also being developed by other manufacturers. Similar advancements have been made with the combustor and turbines. Ceramic-matrix composites are used for parts of the combustor, allowing for higher temperatures. Also, parts of the high-pressure turbine are made from ceramic-matrix composites, reducing the needed cooling. Low-pressure turbine blades are manufactured from titanium aluminide, which is stronger and lighter than conventional nickel compounds, again allowing this component to rotate at higher speeds.

The Pratt & Whitney 7200 is an Engine Alliance (50 percent P & W and 50 percent GE) engine derived from the PW4000 and GE90 engines. It powers the Airbus A380 with 70,000 lb_f thrust. It has a bypass ratio of 8.8 and a total pressure ratio of 36. It has hollow titanium fan blades.

The Rolls-Royce Trent XWB is a three-spool engine (Figure 1.28) that powers the Airbus A350. It has a bypass ratio of 9.6 and develops thrusts up to 97,000 lb. It has a pressure ratio of 50. The fan blades are hollow titanium. Some of the compressor stages are single-piece bladed disks (blisks). Turbine blades are coated with a low-conductivity ceramic.

Overall, in the past 80 years gas turbines have grown remarkably. Jet engine thrusts have increased from 1000 lb_f to well over 100,000 lb_f. Thrust-to-weight ratios have increased from 1.6 to 8. Compressor pressure ratios have increased from 3 to 60. Bypass ratios have increased from 0 (simple turbojets) to 10. Turbine inlet temperatures have increased from 1500 °F to over 3000 °F. Blade technologies have developed into internally cooled turbine blades and hollow blades. New blade coatings have been advanced and new material types and processes have been developed, allowing higher rotational speeds. Afterburners have been added for military applications. Jet engine gas turbines have matured into complementary aeroderivative industrial power generation units. All of the above allowed for more powerful, reliable, and efficient gas turbines.

Turbofan with Fan Mixed

A turbofan with fan mixed is shown in Figure 1.29. It is similar to the turbofan with fan exhausted, but the bypass air is not directly exhausted. The secondary air is bypassed around the low- and high-pressure compressors, combustor, and the low- and

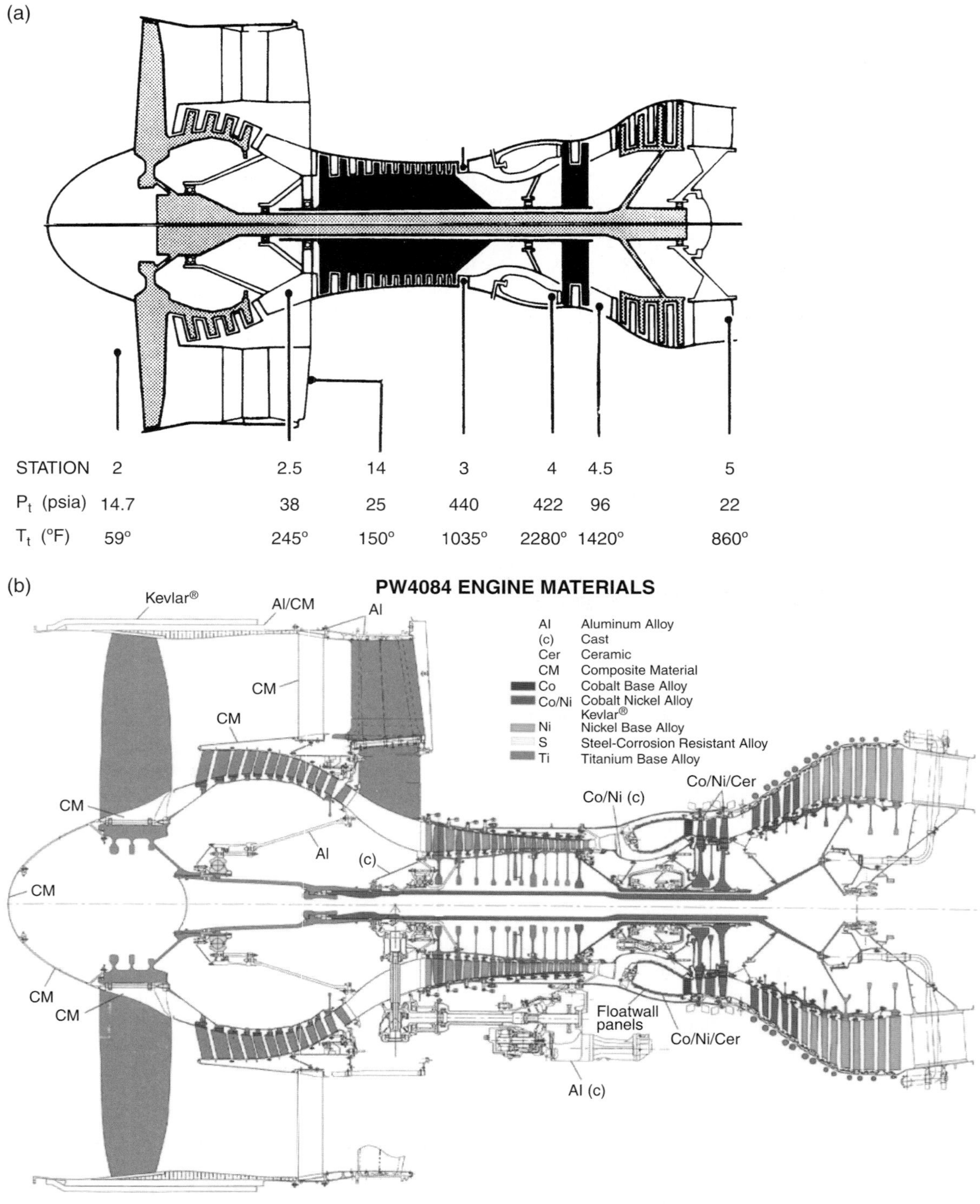

Figure 1.27 (a) Pratt & Whitney PW4000 exhausted turbofan showing internal conditions (courtesy of Pratt & Whitney). (b) Pratt & Whitney PW4084 exhausted turbofan showing materials (© 2022 Raytheon Technologies Corporation, Pratt & Whitney division. Reproduced with permission).

Figure 1.28 Rolls-Royce Trent XWB exhausted turbofan cutaway (© Rolls-Royce plc 2022. All rights reserved).

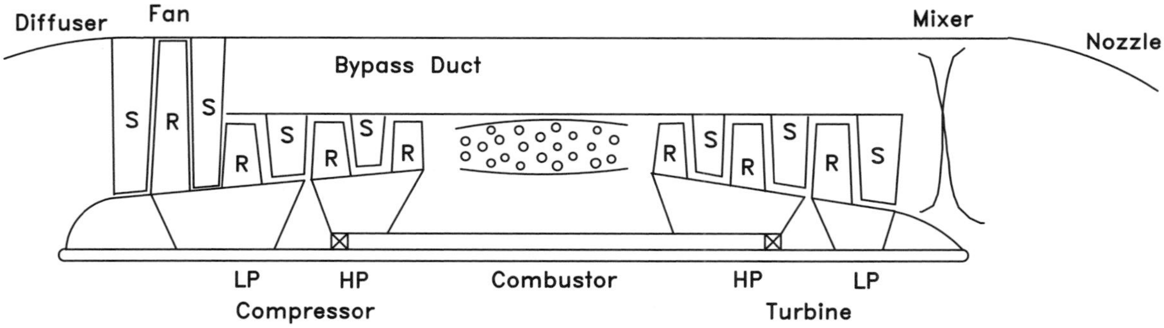

Figure 1.29 Schematic of a turbofan with fan mixed.

high-pressure turbines through a duct. The secondary air is then mixed with the turbine exhaust in a "mixer." The mixed air is then accelerated through the nozzle to produce the thrust. The Pratt & Whitney JT8D (Figure 1.30) is an example of this type. The JT8D-209 develops 18,500 lb_f (82,300 N) of thrust, weighs 4410 lb (2000 kg), and uses 0.51 lb_m-fuel/lb_f-thrust-h (0.052 kg/N-h). The fan and primary air flow rates are about 300 and 170 lb_m/s (136 and 77 kg/s), respectively. Another example of this configuration is the

Figure 1.30 Pratt & Whitney JT8D mixed turbofan (© 2022 Raytheon Technologies Corporation, Pratt & Whitney division. Reproduced with permission).

GE F118-GE-100, which produces 19,000 lb_f (84,500 N) of thrust and is used on the B2 stealth bomber.

In both of the fundamental types of turbofans, the secondary air is used as a source of low-pressure, low-temperature air. This air is bled off the fan and used as a driver for controllers, as cooling air for the turbine, and for other applications. These engine types produce thrust with better fuel economy than does the turbojet. As a result, almost all modern commercial transport and military aircraft use one of the two types of turbofans. Typically, commercial aircraft will use high bypass ratios (much more air flow in the fan than in the core) with an exhausted fan, and military aircraft will use low bypass ratios (approximately equivalent air flow rates) with a mixed fan.

Finally, a "hybrid" form of the two fundamental types of turbofan is sometimes used. Namely, the "cool" air flow is often bled from the compressor and used for cooling the downstream core. For this type of turbofan, a major portion of the air that enters the fan is exhausted through the fan exhaust, and the remainder of the air is eventually exhausted through the primary exhaust.

1.3.6 Turbofan with Afterburner

Afterburners can be added to turbofans just as they are to turbojets. Once again, the two fundamental types of turbofans are covered separately.

Turbofan with Fan Exhausted

A turbofan engine with fan exhausted is shown in Figure 1.31. As can be seen, one afterburner is in the core of the engine after the turbine, and a second (not shown) is rarely be used on the secondary or bypassed air between the fan and the secondary nozzle. The primary afterburner is strictly for the core flow. However, because most turbofans with an exhausted fan are used on commercial craft, such an afterburning geometry has found few applications.

Turbofan with Fan Mixed

Figure 1.32 depicts a turbofan engine with fan mixed. The afterburner is located downstream of the mixer for this configuration. This design is more common than the type shown in Figure 1.31 and is used for most of the military fighters. An example of this engine type is the Pratt & Whitney F100-PW-229 (Figure 1.33), which is used on the F-15 and F-16. This engine ingests 254 lb_m/s (115 kg/s) of air and produces 17,800 and 29,000 lb_f of thrust with the afterburner off and on, respectively (79,200 and 129,000 N). The bypass ratio is 0.40, and the engine weighs 3650 lb (1660 kg). The fuel usage is 0.74 and 2.05 lb_m/lb_f-h with the

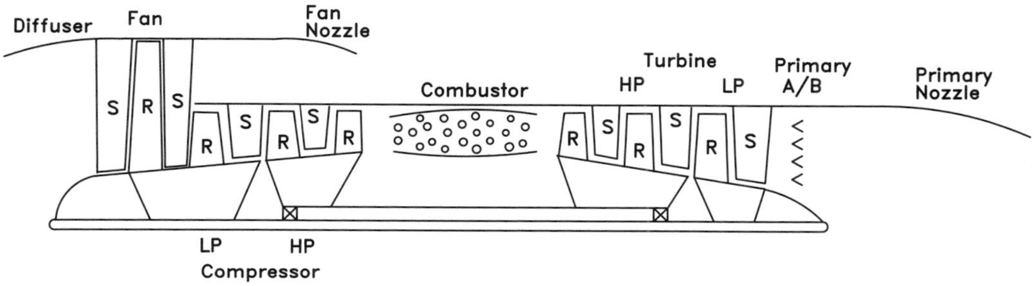

Figure 1.31 Schematic of a turbofan with fan exhausted and afterburner.

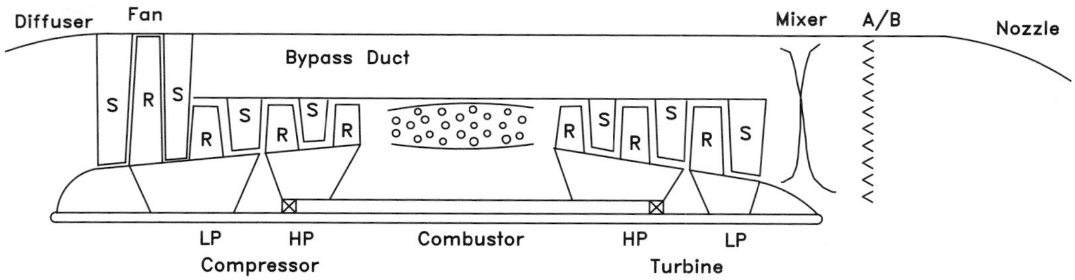

Figure 1.32 Schematic of a turbofan with fan mixed and afterburner.

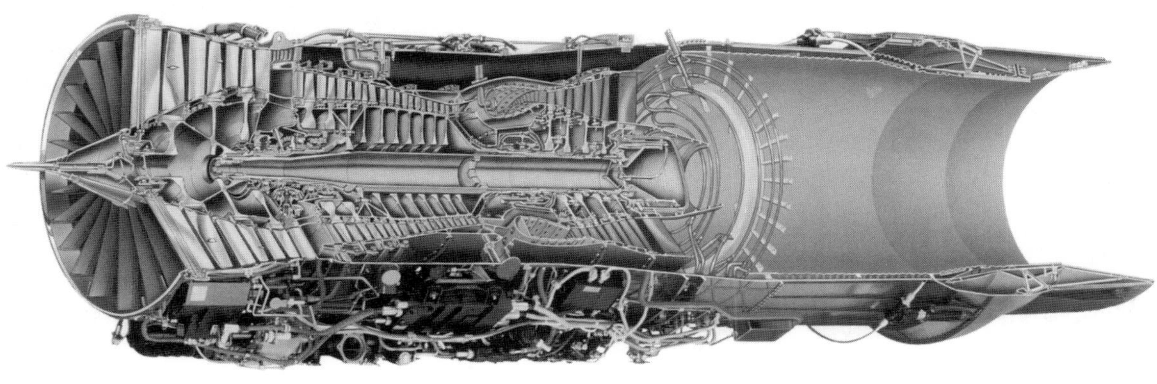

Figure 1.33 Pratt & Whitney F100 afterburning mixed turbofan (© 2022 Raytheon Technologies Corporation, Pratt & Whitney division. Reproduced with permission).

afterburner off and on, respectively (0.0755 and 0.225 kg/N-h).

Of course, these specifications vary from model to model in the F100 series. Another example of this configuration is the GE F110-GE-129 (an engine that is also used on the F-15 and F-16), which produces 17,000 and 32,000 lb_f of thrust with the afterburner off and on, respectively (75,600 and 142,000 N). The bypass ratio is 0.76, and the engine weighs 3980 lb (1805 kg). The fuel usage is 1.9 lb_m/lb_f-h with the afterburner on (0.19 kg/N-h). A smaller engine in this category is the F404-GE-400 (used on the Mach 1.8 F/A-18 Hornet), which produces 10,600 and 16,000 lb_f (47,100 and 71,200 N) of thrust with the afterburner off and on, respectively. The bypass ratio is 0.32, and the engine weighs 2195 lb (995 kg). The fuel usage is 1.85 lb_m/lb_f-h with the afterburner on (0.189 kg/N-h). An even smaller engine of this type is the Rolls-Royce Adour MK815, which produces 5500 and 8400 lb_f (24,500 and 37,400 N) of thrust with the afterburner off and on, respectively. The bypass ratio is 0.75, and the engine weighs 1630 lb (739 kg). Another engine of this variety is the Pratt & Whitney F135 (see Figure 1.34) for the joint strike fighter (JSF). This engine, which was developed for the Lockheed Martin F-35 Lightening II, combines vertical or short takeoff and landing with supersonic flight and has the highest thrust-to-weight ratio of any engine.

Once again, as with the turbojet, turbofan afterburners are used for rapid increases in thrust. They are thermodynamically inefficient and are thus not used in modern commercial aircraft. Fan-mixed engines with afterburners are very often used for military fighter applications. They operate with lower bypass ratios than commercial engines and thus are lighter in weight, but a large amount of air enters the afterburner, allowing for very large increases in thrust. More comparisons of engine types will be summarized in Section 1.3.12.

1.3.7 Geared Turbofan

High-bypass turbofan engines with a geared turbofan are currently being produced and tested by almost all of the manufacturers. In Figure 1.35 a cutaway from the Pratt & Whitney PW1000 series is shown. By using a planetary gearbox (Figure 1.35) between the low-pressure compressor and fan, the fan can rotate at a slower speed. As a result, much larger diameters and bypass ratios ("ultra high bypass") can be used (bypass ratios up to 12.5 for this series). Also, because the rotational speeds are controlled, the fan can operate more efficiency and with less noise. The PW1000G

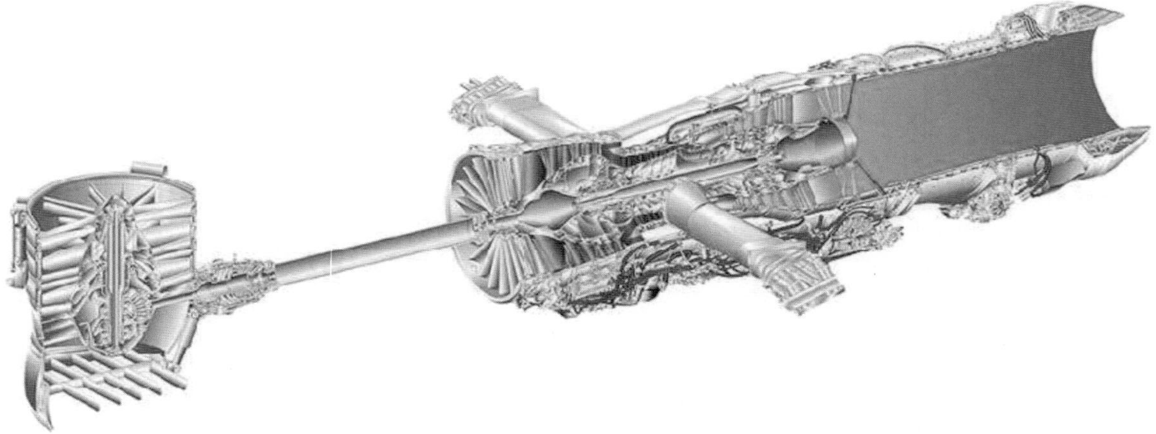

Figure 1.34 Pratt & Whitney F135 JSF engine (© 2022 Raytheon Technologies Corporation, Pratt & Whitney division. Reproduced with permission).

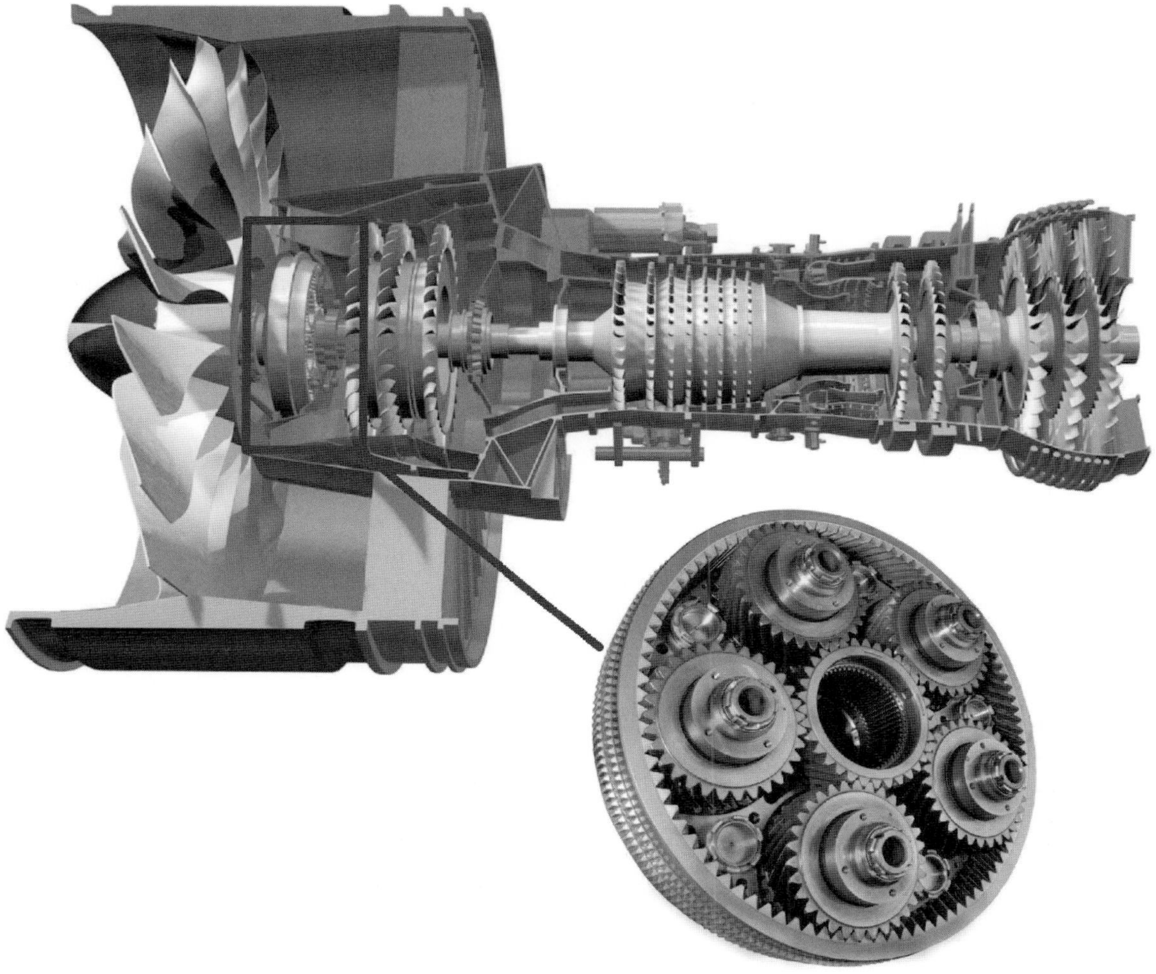

Figure 1.35 Pratt & Whitney geared turbofan cutaway and gearbox (© 2022 Raytheon Technologies Corporation, Pratt & Whitney division. Reproduced with permission).

develops thrusts up to 35,000 lb_f and is currently used on the Airbus 320, Airbus 220, and other aircraft. GE is currently developing an engine for a similar market and Rolls-Royce is developing a three-spool ultra high bypass engine for a larger aircraft market.

1.3.8 Turboprop

Another engine type is the turboprop, which is shown in Figure 1.36. For this type, the core of the engine is similar to a turbojet – namely, a diffuser, compressor, and turbine are used. The core airflow is accelerated through the exhaust nozzle, which produces one component of thrust. A second, and usually larger, component of thrust is obtained from the propeller. The power for the propeller is extracted from the turboshaft in the core – that is a part of the turbine work drives the propeller. A gearbox reduces the speed so that the propeller spins at a lower speed than the compressor. An example of a turboprop is the Rolls-Royce Tyne shown in Figure 1.37. Model 512 of this series produces 5500 total horsepower (4100 kW) and ingests 47 lb_m/s

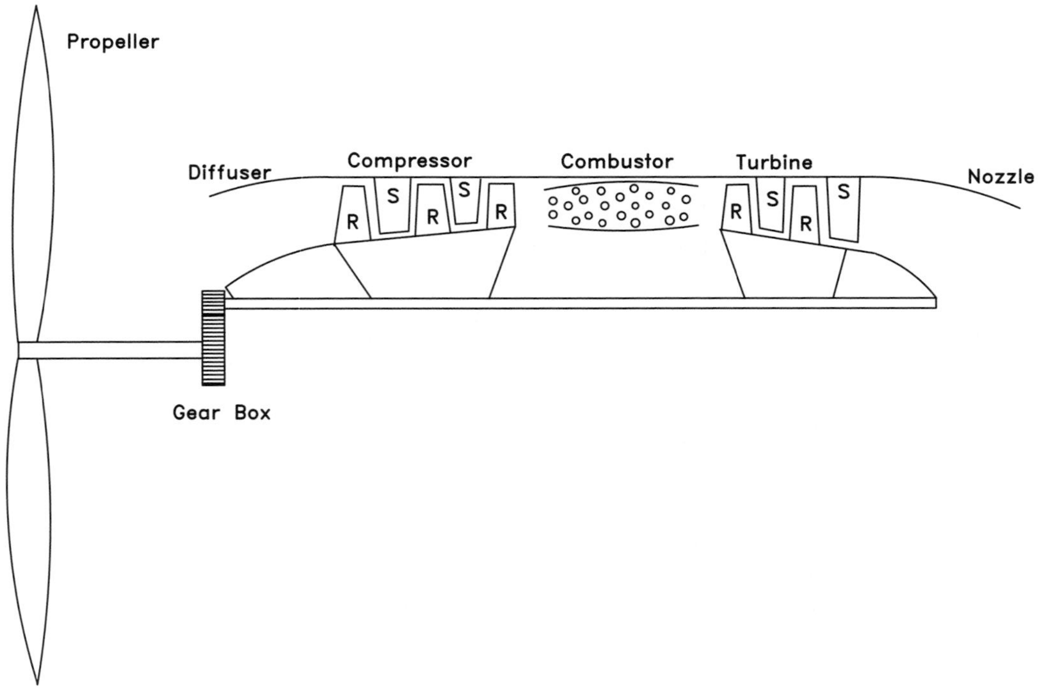

Figure 1.36 Schematic of a turboprop.

Figure 1.37 Rolls-Royce Tyne turboprop (© Rolls-Royce plc 2022. All rights reserved).

(21 kg/s) of air. Another example is the Honeywell TPE331 shown in a cutaway view in Figure 1.38 revealing the complex gear system. Model 12B of this series develops up to 1100 hp (820 kW), weighs 400 lb (180 kg), and uses 0.52 lb_m of fuel per hour to generate 1 hp (later defined as specific fuel consumption [*SFC*], which is also equal to 0.32 kg-fuel/kW-h). A third, smaller unit is the Rolls-Royce Allison 250-B17F,

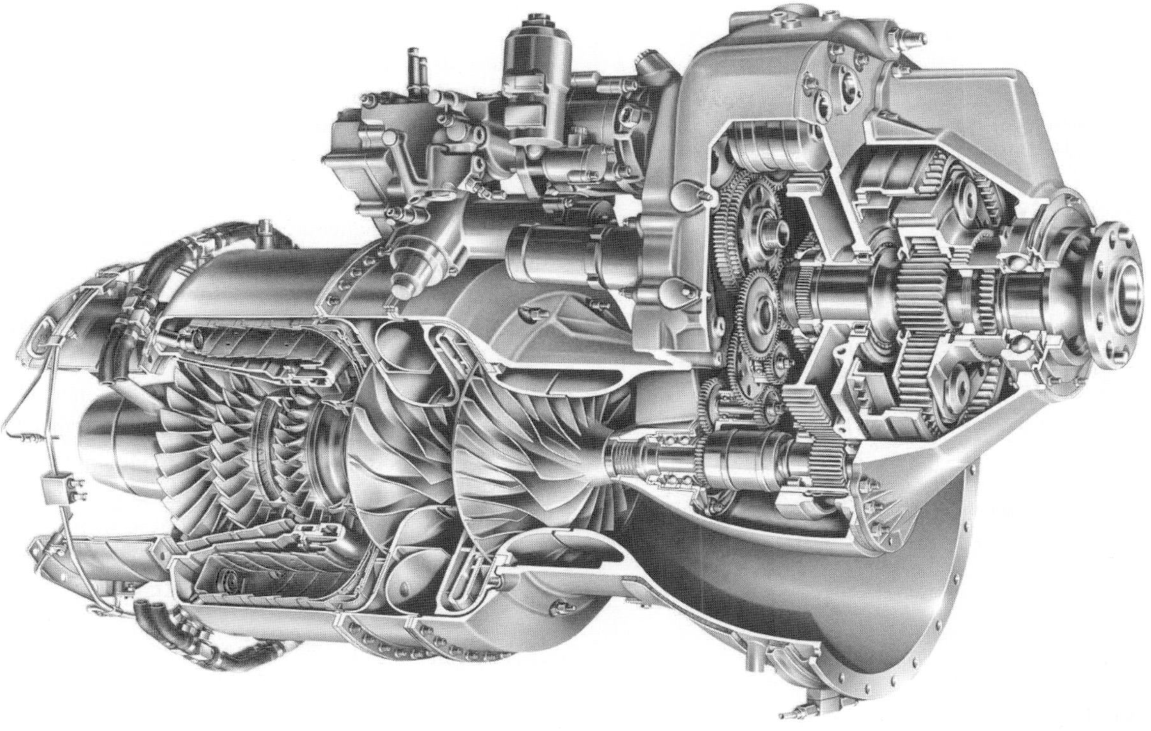

Figure 1.38 Honeywell TPE331 turboprop (courtesy of Honeywell).

which weighs 205 lb (93 kg), ingests 3.6 lb_m/s (1.6 kg/s) of air, produces 450 hp (336 kW), and uses 0.613 lb_m/hp-h (0.37 kg/kW-h) of fuel.

Because of the size and low speed of the propeller, turboprops cannot be used at high air speeds. However, at low speeds they are very efficient thermodynamically. As a result, turboprops are used on many light aircraft designed for short journeys.

1.3.9 Unducted Fan

The fundamental configuration of the turboprop engine is also used for the unducted fan (UDF), external fan engine, or prop-fan, which received considerable attention in the 1980s and early 1990s. This is an alternative to a turboprop, which has long propeller blades and is thus limited to low rotational speeds and has the drawback of a heavy gearbox. The UDF was developed during this period by NASA in concept with GE, Pratt & Whitney, Rolls-Royce, and others (Figures 1.39 and 1.40). This engine never went into production in the United States but was produced in the former Soviet Union. This engine type is a crossbreed between the turboprop, which demonstrates excellent low-aircraft-speed efficiency, and the conventional exhausted turbofan. The blades of a UDF are shorter than propeller blades and thus can be directly attached to the shaft without the need for a gearbox. More complexly designed blades are used as well, and they are highly "swept" (curved and twisted). Such designs can effectively operate at higher Mach numbers than turboprops. As noted, this engine never went into production in the United States (largely owing to decreased fuel prices in the early 1990s), and much of the advanced technology that was developed went into other advanced engine designs. For example, GE directly used such technologies in the GE 90 engine development.

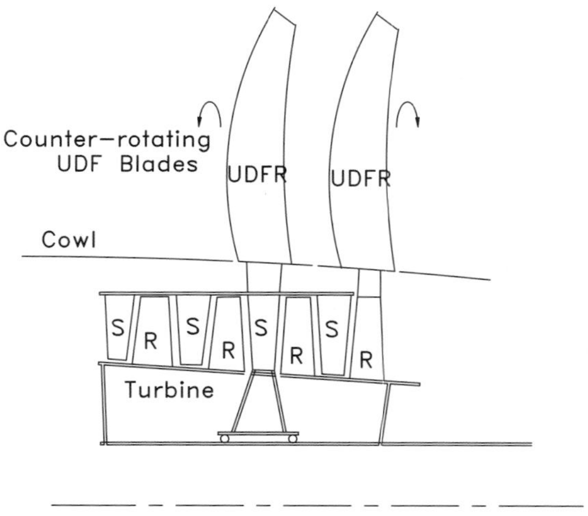

Figure 1.39 Schematic of an unducted fan.

1.3.10 Turboshaft

The last aircraft engine type is the turboshaft. It is basically the same as the turboprop except that thrust is not derived from the exhaust. The gas from the core exhausts is at a low velocity, and consequently additional thrust is not obtained. These engines are used largely for helicopter applications, although it is noteworthy that turboshaft engines are used to drive tanks and other ground vehicles with a transmission attached to the shaft in lieu of the rotating blades. An example of this engine is the Rolls-Royce/Turbomeca RTM322 shown in Figure 1.41. Depending on particular submodel, this engine develops up to 2800 hp (2090 kW) and weighs up to 540 lb (245 kg). The Pratt & Whitney Canada PW206A develops 621 hp (463 kW), and the output shaft rotates at 6240 rpm.

Figure 1.40 General Electric Unducted Fan "UDF" Engine (courtesy of GE Aviation), from GE publication, "Unducted Fan Engine (UDFTM)," publication no. AEBG-2966 (4/86), distributed at ASME IGTI meeting circa 1986; used in the first edition of this text.

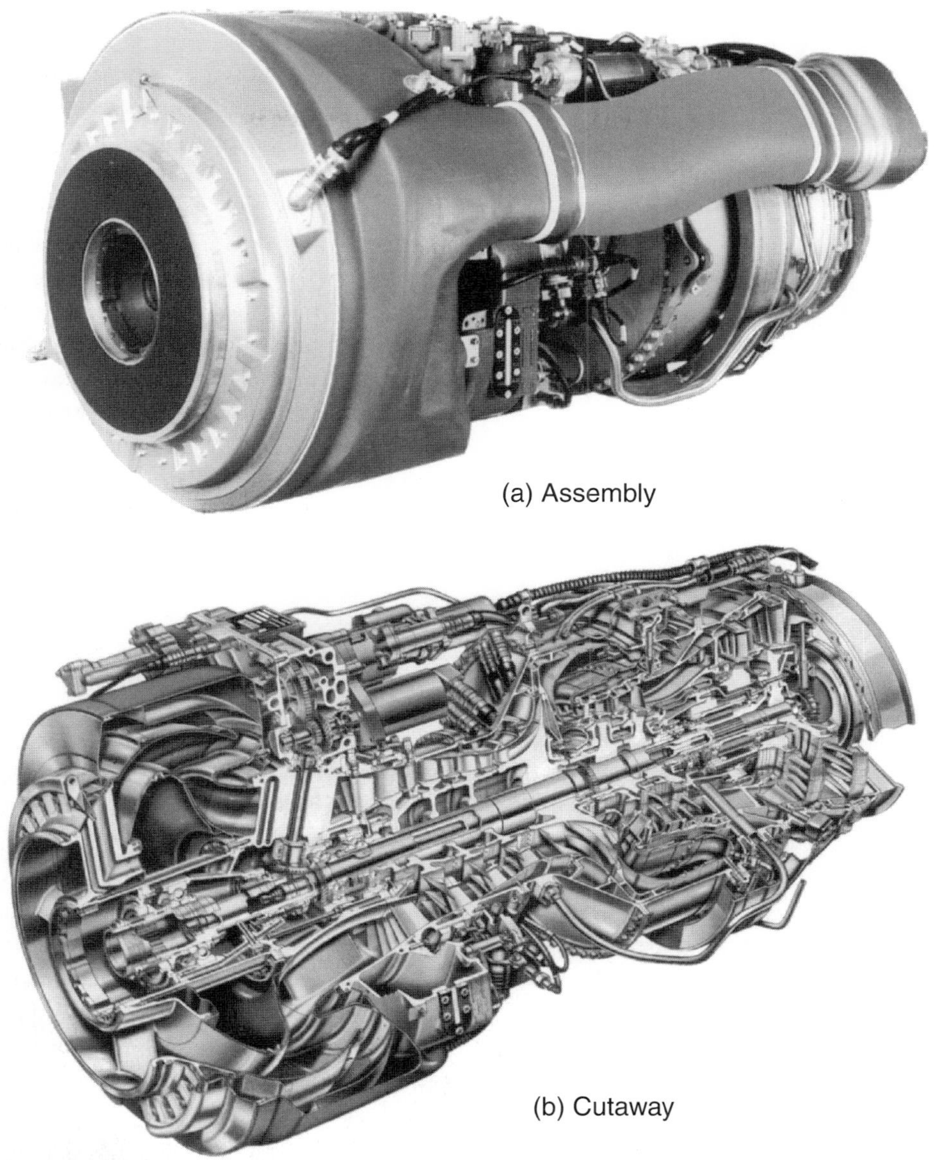

Figure 1.41 Rolls-Royce/Turbomeca RTM322 turboshaft (© Rolls-Royce plc 2022. All rights reserved).

Lastly, the GE T700-401 weighs up to 458 lb (208 kg), develops up to 1800 hp (1340 kW), and uses 0.46 lb_m/hp-h (0.28 kg/kW-h) of fuel. Many turboshafts have been developed concurrently with turboprops, and thus different designs of the same basic engine are used for fixed-wing and rotary aircraft.

1.3.11 Advancing Engine Designs

Adaptive Cycle Engines

A new concept engine is the GE ADVENT adaptive cycle engine. This engine is designed for military combat and bomber applications and is targeted for the Lockheed

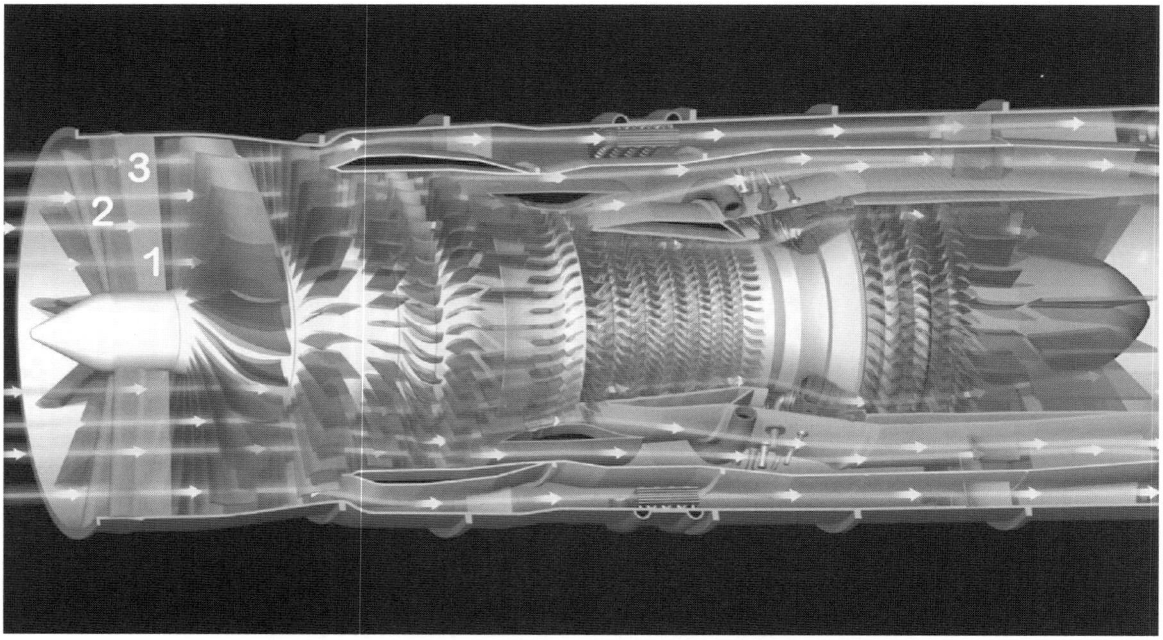

Figure 1.42 General Electric ADVENT Adaptive Cycle Engine Demonstrator (courtesy of GE Aviation), available at https://www.ge.com/news/reports/the-superjet (Mar. 4, 2014); numbers added.

Martin F35 Lightening II. It is currently in the second prototype mode and has a design thrust of 20,000 lb_f. Designs are also being developed for a 45,000 lb_f class of thrusts. The concept of this engine is fundamentally a ducted turbofan engine with a variable bypass ratio. This is accomplished with three fluid streams, as shown in Figure 1.42. For cruise conditions more air is directed into streams 2 and 3. This has the effect of a larger bypass ratio, which results in better fuel economy. This also results in a longer range, one of the important goals. For combat or other conditions when sudden increased thrust is needed more air is directed to streams 1 and 2, which effectively reduces the bypass ratio. This can be used with an afterburner where more air is heated in the augmenter allowing for much more thrust. The overall design also allows the cooler stream 3 to better cool the engine, especially under the hot conditions.

Turboramjet

The turboramjet is not a new concept but is gaining more recent attention with potential hypersonic applications. As the name implies it is a combination of a ramjet and turbojet. In Figure 1.43 the basics are shown. This engine is fundamentally operated in two modes. First in Figure 1.43a all of the airflow is directed into a turbojet gas generator. This would be for lower speeds including takeoff. The burner shown in Figure 1.43a can be used as an afterburner in this mode. In the second mode (Figure 1.43b) a flap is closed to the gas generator and all of the air is ducted around the gas generator. For this mode the burner becomes the primary combustor for the ramjet. This configuration would be used for supersonic/hypersonic flight conditions. Along with the scramjet, the turboramjet has the potential to launch aviation back into the supersonic arena again and then the hypersonic arena.

The Pratt & Whiney J58 and the SR-71 Blackbird were discussed in Section 1.3.4. For this engine and aircraft combination the engine is sometimes termed as a turboramjet, which is somewhat of a misnomer. For this application and design the gas generator is always in an active role but by varying degrees of air

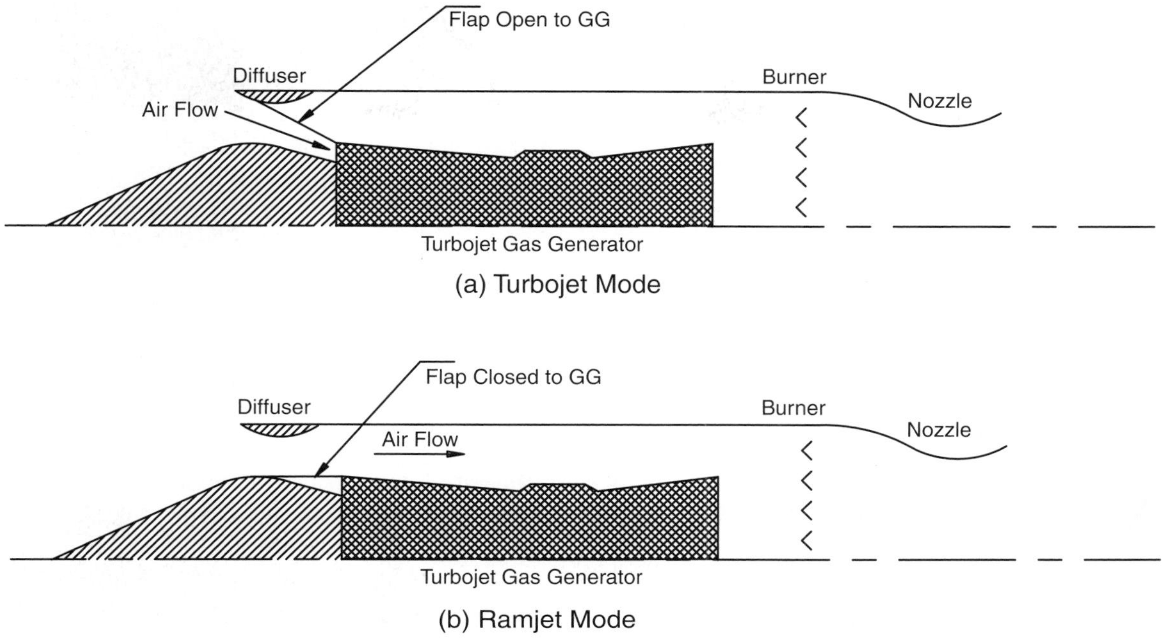

Figure 1.43 Turboramjet.

throughflow. As a result, the operation behaves with some of the characteristics of a turboramjet but is not by definition such a design.

Pulse Detonation Engines

Recent investigations of improving the gas performance from the traditional Brayton cycle have risen in popularity. The Humphrey cycle has attained particular attention. This thermodynamic cycle is the same cycle as the Atkinson cycle. The latter is typically associated with internal combustion engines while the former is more associated with gas turbine possibilities. The h–s and p–v diagrams are shown in Figure 1.44. Compared with the Brayton cycle, the constant pressure process 2 to 3 has been replaced by a constant volume process. Compared with the Brayton efficiency (Eq. (1.2.4)) the Humphrey efficiency is

$$\eta_{th} = \frac{1 - \frac{\gamma}{\tau_c}\left[(\tau_b)^{\frac{1}{\gamma}} - 1\right]}{\tau_b - 1}, \tag{1.3.1}$$

where γ is again the ratio of specific heats, τ_c is the temperature ratio for the compression process, and τ_b is the temperature ratio for the combustion process (Andriani and Ingenito, 2019). This Humphrey efficiency is higher than that for the Brayton cycle, particularly for lower combustion temperature ratios. For example, in Chapter 3 (Example 3.14) for the compression temperature ratio of 2.116, combustion temperature ratio of 2.388, and $\gamma = 1.35$, the Brayton efficiency is 52.7 percent. However, the Humphrey efficiency is 58.4 percent.

To achieve the constant-volume process, the pulse detonation process has received considerable attention. For this process a controlled spark pulse is used to ignite the fuel in a long combustion chamber. Fuel is injected only when the spark is active. This is a high-frequency occurrence. As a result, the fuel burns quickly and traverses the combustion tube as a supersonic shock at very high speeds and, as a result, the pressure increases dramatically due to the shock but the volume remains approximately constant

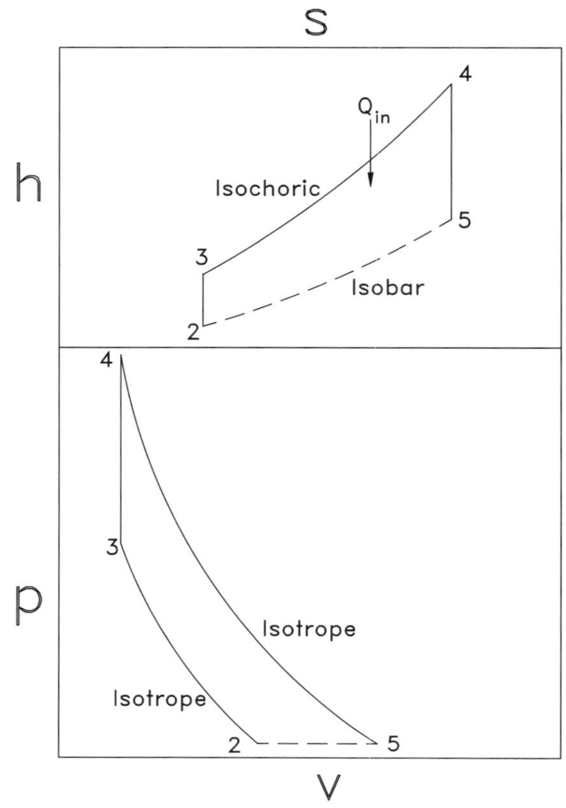

Figure 1.44 Humphrey/Atkinson cycle.

because the process is so fast. This process has several advantages. Not only is the efficiency improved, but the need for a rotating compressor as in conventional engines is eliminated.

The first test flight using this technology was by the US Air Force in 2008 at low speeds. Development has continued as reviewed by Pandey and Debnath (2016). Because of the characteristics of this engine, it has possibilities for hypersonic flight. Clearly more research is needed at this time to advance the state of the art.

Supersonic/Hypersonic Developments

Since the early 2000s, more attention has been paid to supersonic and hypersonic flight, especially since the last flight of the Concorde in 2003. NASA successfully tested the X-43A scramjet at speeds from Mach 6.8 to 9.6 in 2004. They flew more successful tests on the X-51 from 2010 to 2013. Aerion Corporation in conjunction with NASA-Langley teamed to develop a supersonic aircraft with test flights scheduled in 2024. Simultaneously, GE was developing the Affinity for this aircraft with prototypes revealed in 2020. Unfortunately, finances caused the Aerion work to be terminated and, as a result, GE ended developments as well in 2021. Currently, the only supersonic development work is by BOOM supersonic corporation for the aircraft with Rolls-Royce developing the engines. Reportedly, the engines are nonafterburning two-spool turbofans. Rollouts are predicted in 2025 with the first flights in 2026.

Other developments in hypersonic applications include nonairbreathing engines, in which the fuel is carried by the aircraft or missile. Since this text addresses only airbreathing gas turbines this topic is identified but not addressed or described in detail.

In 2021, the US Department of Defense and NASA initiated a long-term commitment to develop hypersonic missiles and aircraft. The goal is to fly in the upper atmosphere at hypersonic speeds. The initiative includes both reusable aircraft as well as the power plants. Some of the ensuing research projects are in the initial stages and others have been ongoing for the past few years. The NASA and DOD support should lead to a quantum jump in progress in the next few years.

1.3.12 Comparison of Jet Engine Types

Overall and in a brief summary, in the past 80 years gas turbines have grown remarkably. As discussed earlier, jet engine thrusts have increased from 1000 lb_f to well over 100,000 lb_f with thrust-to-weight ratios increasing from 1.6 to 8. Compressor pressure ratios have increased from 3 to 60 and bypass ratios have increased from 0 to 10. Turbine inlet temperatures have increased from 1500 °F to over 3000 °F and, as a result, blade technologies have developed into internally cooled turbine blades and hollow blades with new blade coatings and new material types allowing for higher rotational speeds. Afterburners were added decades ago for military applications. Geared turbofans have been more recently introduced. Furthermore, jet engine gas turbines have matured into complementary aeroderivative industrial power generation units. All of the above allowed for more efficient gas turbines.

Typical weights, powers, thrusts, and other parameters have been presented in the preceding sections for a spectrum of applications. Applications are presented so that one can appreciate the different uses. As noted, many of the engines have different submodels or "builds" for different applications. The reader is encouraged to routinely consult *Aviation Week and Space Technology* and *Flight International,* which both have approximately annual reviews of engine specifications as well as the Pratt & Whitney, Rolls-Royce, General Electric, Snecma, and other websites detailed in the reference list. One particularly convenient website is www.jet-engine.net, which categorizes specifications for turbofans, turbojets, and turboshafts for both commercial and military applications for engines up to 2005. An interesting observation on engine development and manufacture is that large corporations that were in competition two decades ago now are partners in some large engine development and manufacturing projects. For example, CFM is a venture between General Electric and Snecma; MTR is a joint effort with MTU (Motoren und Turbinen Union, Division of Daimler-Chrysler), Turbomeca, and Rolls-Royce; IAE is International Aero Engines (Pratt & Whitney, Rolls-Royce, Japanese Aero, and MTU); and the Engine Alliance is a cooperative effort with Pratt & Whitney and General Electric. Furthermore, companies have cooperated on individual engines: MTU has worked individually with General Electric, Pratt & Whitney Aircraft, and Rolls-Royce; IHI (Isakawajima-Harima Heavy Industries) has worked with Pratt & Whitney Aircraft and Rolls-Royce; BMW and Rolls-Royce have teamed, and so on. Furthermore, many companies have merged or been bought out, and thus the list of manufacturers changes annually. Companies have adopted the healthy strategy that engineering efforts should not be developed independently and that effective design ideas and research can be effectively transferred between corporations. The best example is probably the Engine Alliance, for which advanced technologies from the very large GE 90 and PW 4000 turbofan engines are merged and each member is engaged in building different parts of engines.

In Table 1.1, a general comparison of the different engine types is made. The table is self-explanatory, and one can make comparisons based on the different criteria. For example, it can be seen that a turboprop has the best fuel economy at low speeds and the turbojet has the best thrust-to-weight ratio. The information in this table is demonstrated with different applications and examples in Chapters 2 and 3 on cycle analysis. Certainly, this table must be considered as general trends. Actual applications of different engines affect the measurable trends. For example, if the same engine is integrated to the aircraft under the wings for one application and integrated on the fuselage for another application, the ground clearance will be different for the two applications. Engine attachment as well as inlet and design thereof are in general the responsibility of the aircraft manufacturer, of course working with the engine manufacturer. Other such differences for other measures and comparisons can occur.

As noted in the above discussions the very large high-bypass engines with exhausted fans are used for commercial aircraft or military transport applications because weight issues are not as important as for military combat applications. The high-bypass engines have much better fuel economy than lower-bypass-ratio engines or turbojets. On the other hand, the low-bypass-ratio engines with mixed fans are used for military combat applications. These have a higher thrust-to-weight ratio. They also lend themselves to the inclusion of an afterburner because all of the ingested air can be used for sudden thrust increases in the combusting exhaust.

1.4 Power Generation Gas Turbines

1.4.1 Simple Brayton Cycle

The next category of gas turbines to be discussed is not for aircraft applications but for power generation. A power generation unit of typical design is shown in

Table 1.1 **Comparison of engine types.**

	Ramjet	Scramjet	Turbojet [a]	Exhausted turbofan [a]	Mixed turbofan [a]	Turboprop
				(High bypass)	(Low bypass)	
Speed	Very high	Hypersonic	High	Moderate	Moderate/high	Low
Thrust/weight	Moderate	Moderate	High	Moderate/low	Moderate	Low
Fuel use/thrust	High	High	Moderate	Low	Moderate/low	Low
Thrust/airflow	High	High	High	Moderate	Moderate/high	Moderate
Ground clearance	NA	NA	Good	Moderate	Good	Low

[a] Afterburners increase speed, increase thrust/weight, increase fuel use/thrust.
NA = not applicable.

Figure 1.45. Three shaft arrangements are shown. In Figure 1.45a, the gas generator (compressor–burner–turbine) has two shafts and the load, power, or "free" turbine is a separate turbine. The turbine consists of the high- and low-pressure turbines, which drive the two compressors, and a "free" turbine, which is used for power generation.

In Figure 1.45b, the gas generator is a single shaft and a separate "free" turbine is used for power generation. Lastly, in Figure 1.45c, the entire unit is on a single shaft and the "leftover" portion of the power is used for power generation. As one can see, the unit is very much like a turbojet engine, although subtle differences exist in the design. An inlet bell replaces the diffuser, and the nozzle is replaced by an exhaust system. A power-extracting device is attached to one of the primary shafts or an auxiliary shaft. This device may be an electrical generator (which must run at constant rotational speed), ship propeller, offshore equipment, pumping equipment, or another mechanical power-extracting device. Until the late 1900s, aircraft engines dominated the gas turbine industry; however, since those years sales of power generation gas turbines complemented those for aircraft engines. One example of this type of unit is the Rolls-Royce Industrial RB211, for which the two-shaft gas generator is shown in Figure 1.46. As its designation implies, this electrical power generation unit is in fact a "derivative" of the aircraft engine (Figure 1.26). One can see many similarities between this gas turbine and the aircraft engine. This type of unit tends to be slightly heavier because weight is not a concern. The gas generator consists of a seven-stage intermediate pressure compressor driven by a single-stage intermediate pressure turbine and a six-stage HPC driven by a single-stage HPT. The free turbine is usually a Rolls-Royce RT62 (two stages) or an RT 61 (three stages). Depending on the "build" this unit can supply up to 44 MW (60,000 hp) at 39 percent efficiency. In Figure 1.46b, the RB211 is shown in a pump "train" in which the power is used to drive a multistage pump. The exhaust system can also be seen in this figure. The Rolls-Royce Trent is their largest power unit and delivers up to 69,800 hp (52,000 kW) with over 350 lb_m/s (160 kg/s) of air ingested. This gas turbine runs at about 42 percent thermodynamic efficiency. The Cooper-Rolls Coberra 6462 uses the RB211 gas generator, and the two shafts run at 9245 and 6555 rpm. The output power is 34,000 HP (25,350 kW) with an efficiency of 36.4 percent. The GE LM2500 is another example of this type of aeroderivative gas turbine based on the

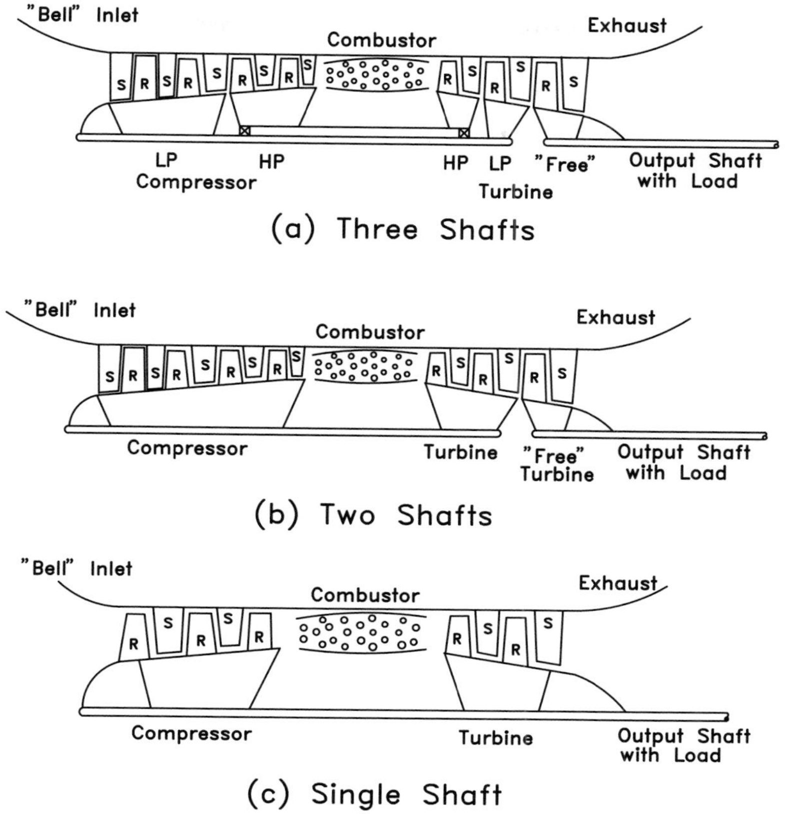

Figure 1.45 Schematics of a power generation gas turbine

GE CF6 (Figure 1.25) and is shown in Figure 1.47. Such gas turbines in this line have one shaft and generate up to 33,600 hp (25,000 kW) on the output shaft, ingest about 150 lb_m/s of air (68 kg/s), have a weight of 10,300 lb (4670 kg), and use 0.373 lb_m-fuel/hp-h (0.227 kg/kW-h) with a thermodynamic efficiency of 37 percent. A cutaway of the gas generator is shown on Figure 1.47b and a schematic with the exhaust is shown on Figure 1.47c. On Figure 1.47a, the linkages for compressor variable stators are shown. Variable stators are used to improve off-design performance of a compressor, as will be discussed in detail in Chapter 5. A larger unit, the LM 6000 (about 55,000 hp or 41,000 kW), is also a derivative of the GE CF6 aircraft engine. The LM 9000 is the newest and largest contribution from GE. It was derived from the GE90-115B Turbofan. It produces up to 75 MW (100,000 hp) at 42 percent efficiency. The Mitsubishi PW Power Systems (formerly Pratt & Whiney) gas turbine FT8 (Figure 1.48) is a derivative of the PW JT8D turbofan (Figure 1.30). It generates 26 MW (35,000 hp) at 39 percent efficiency. The complex exhaust system is noticed on this cutaway. The very large FT4000-140 is a derivative of the PW4000 (Figure 1.11) and produces 144 MW (190,000 hp) at 41.6 percent efficiency. In Figure 1.49, the Rolls-Royce Marine Tyne is shown. Once again, as indicated by the name, this seaworthy unit is a modification of an aircraft turboprop (Figure 1.37). One of the smallest is the Capstone C30, which delivers 39 hp (29 kW) with an air consumption of 0.7 lb_m/s (0.32 kg/s) at 25 percent efficiency.

As implied earlier in this section, many modern power

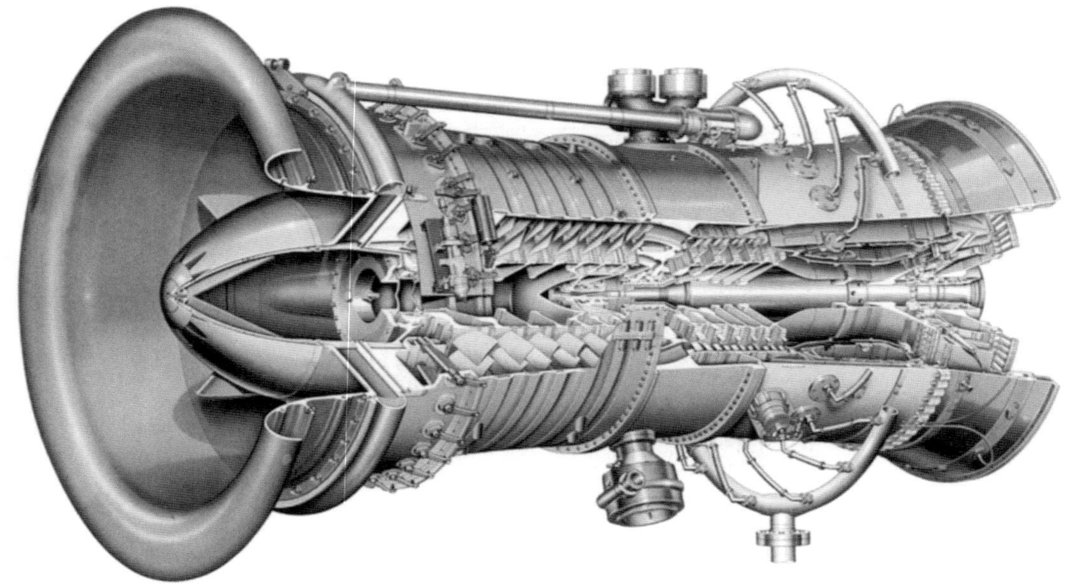

(a) Gas Generator Cutaway

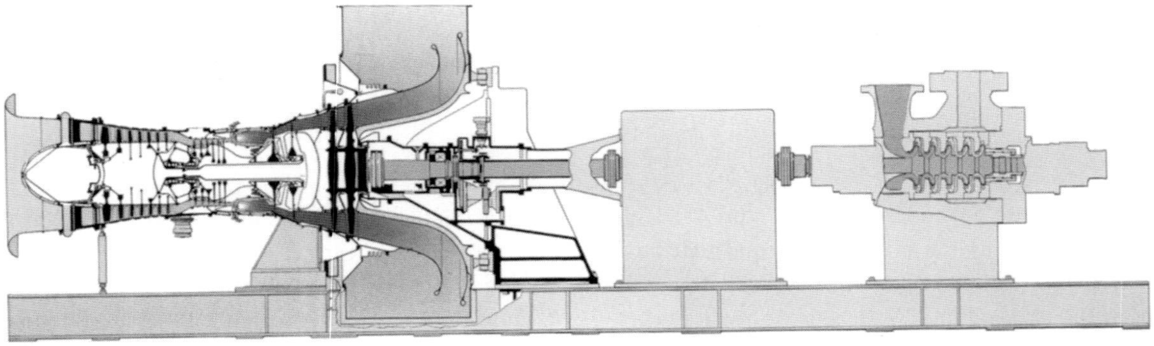

(b) Gas Turbine Driving a Pump

Figure 1.46 Rolls-Royce industrial RB211 (© Rolls-Royce plc 2022. All rights reserved).

generation units are "aeroderivatives" of the aircraft engines; namely, the design of the core of the engine is essentially the same as that of a jet engine. Thus, much of the development engineering costs entailed in the design of a new power generation unit are reduced because the original development was already completed for a propulsion unit (and sometimes paid for by a government). The cost of such a power generation unit can therefore be significantly reduced. Furthermore, because some parts are the same for both power generation and propulsion gas turbines, manufacturing needs and costs are reduced.

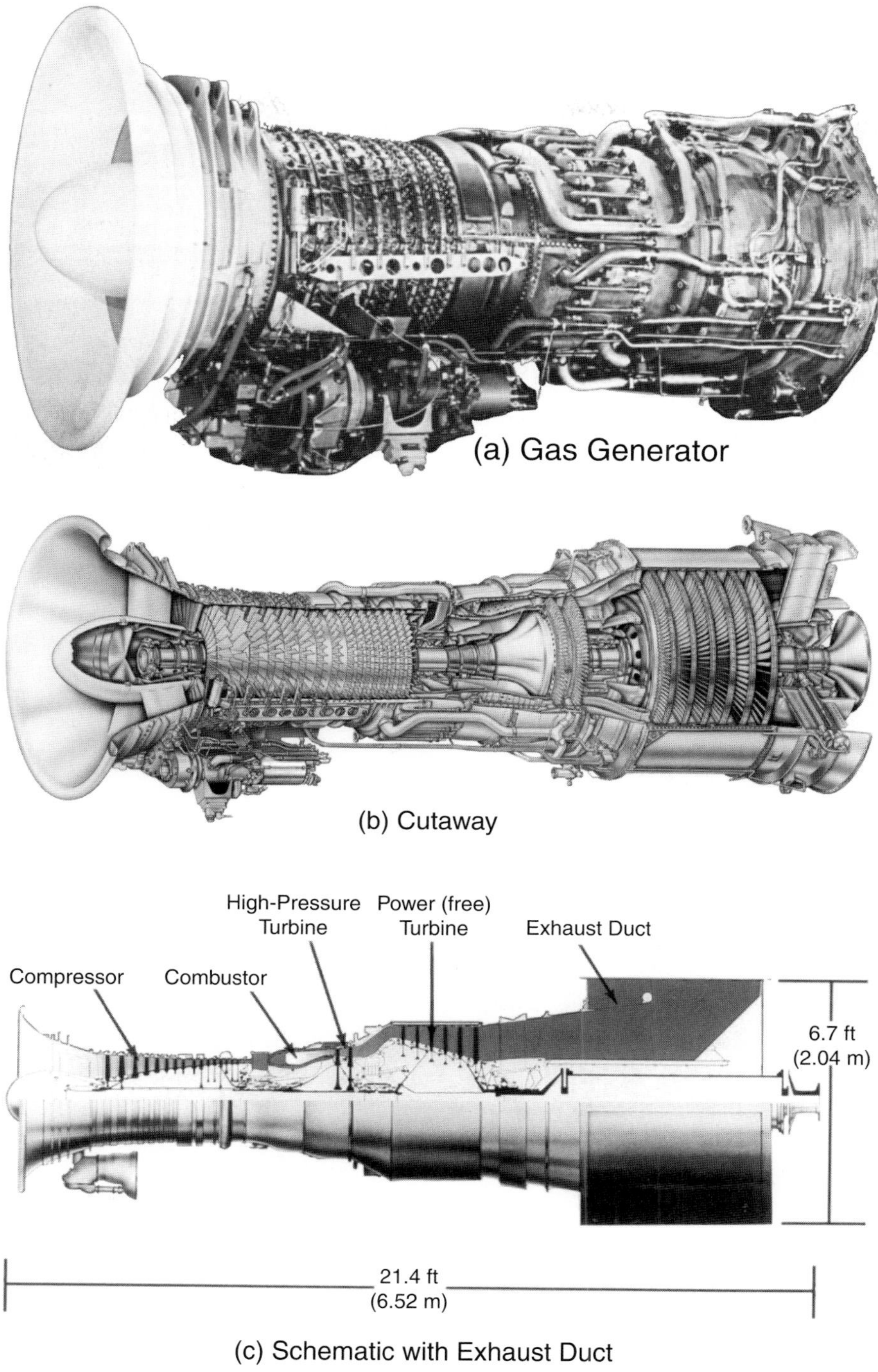

Figure 1.47 General Electric LM2500 gas turbine (courtesy of General Electric Gas Power).

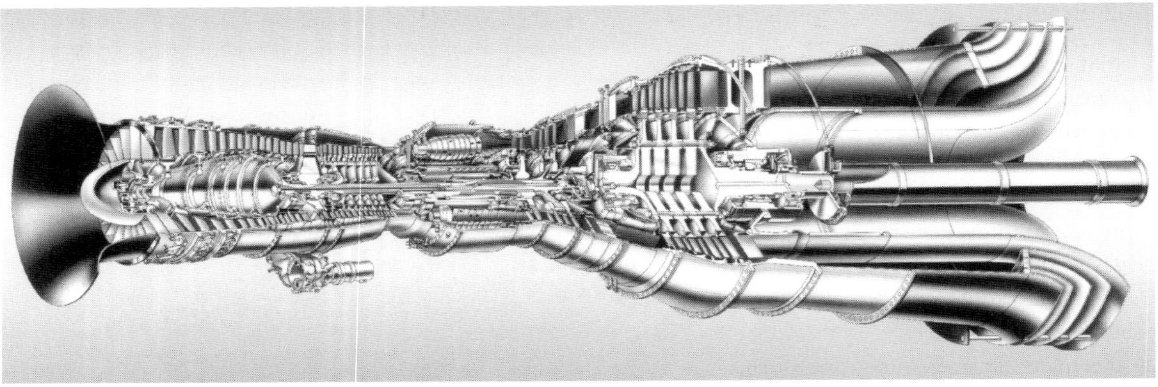

Figure 1.48 Mitsubishi/PW Power Systems FT8 (© 2022 Raytheon Technologies Corporation, Pratt & Whitney division. Reproduced with permission).

Figure 1.49 Rolls-Royce Marine Tyne gas turbine (© Rolls-Royce plc 2022. All rights reserved).

New applications of power generation units include the development of miniaturized and microminiaturized gas turbines to replace small household-size electrical generators and batteries. Some units under development are as small as 1 in. (25 mm) and produce a mere 50 W (0.067 hp) (Ashley, 1997).

1.4.2 Brayton Cycle with Regeneration

A variation of the Brayton cycle is a cycle with the addition of "regeneration." A diagram for this hardware is shown in Figure 1.50 along with the *h–s* diagram. This process utilizes the hot gas that is "thrown away" at the turbine exit with a simple open Brayton cycle. Using the warm air at station 5, a heat exchanger is used to warm the gas exiting from the compressor (station 3) before it reaches the inlet of the combustor. Again, the net power is shown with crosshatching. And the total heat input is the area under the curve from 3x to 4. This is less than that for the nominal simple Brayton cycle. Thus, for the same nominal exit temperature from the burner for the simple open Brayton cycle, less fuel is needed, $\Delta h_{3x-4} < \Delta h_{3-4}$, thus improving the thermodynamic efficiency, but the net power remains the same. Or, for the same amount of fuel, this cycle allows the exit temperature from the combustor to be higher than the burner exit temperature of the simple Brayton cycle, thus increasing the net power output. However, because of the additional heat exchanger, the weight is increased significantly. As a result, this type of gas turbine is not used for aircraft, although it has many important ground applications, including electric power generation. Importantly, one should note that if the temperature exiting the turbine, T_5, is less than that exiting the compressor, T_3, regeneration should NOT be used because increasing the inlet temperature to the burner will not be possible. In the cycle analysis in Chapter 3, checking these two values will be imperative.

1.4.3 Brayton Cycle with Intercooling

Another variation of the Brayton cycle that is often used for power generation is the addition of "intercooling." Such a cycle is shown in Figure 1.51. For this cycle the working gas is first compressed by a "low-pressure" compressor (process 2–2.5a), which increases the pressure to an intermediate value. Next, the gas is cooled (process 2.5a–2.5b), reducing the temperature of the gas. The gas is then further compressed (process 2.5b–3) until it reaches the nominal pressure. By examining the *h–s* diagram in Figure 1.51 the value of Δh for the simple cycle (no intercooling) is shown: $\Delta h_{\text{no IC}} = h_3 - h_2$. Also shown is the sum of Δh with intercooling for processes 2–2.5a and 2.5b–3:

$$\Delta h = \Delta h_{2-2.5a} + \Delta h_{2.5b-3} = (h_{2.5a} - h_2) + (h_3 - h_{2.5b}).$$

As can be seen in the detail figure for the same increase in pressure, the compression process with intercooling requires significantly less power:

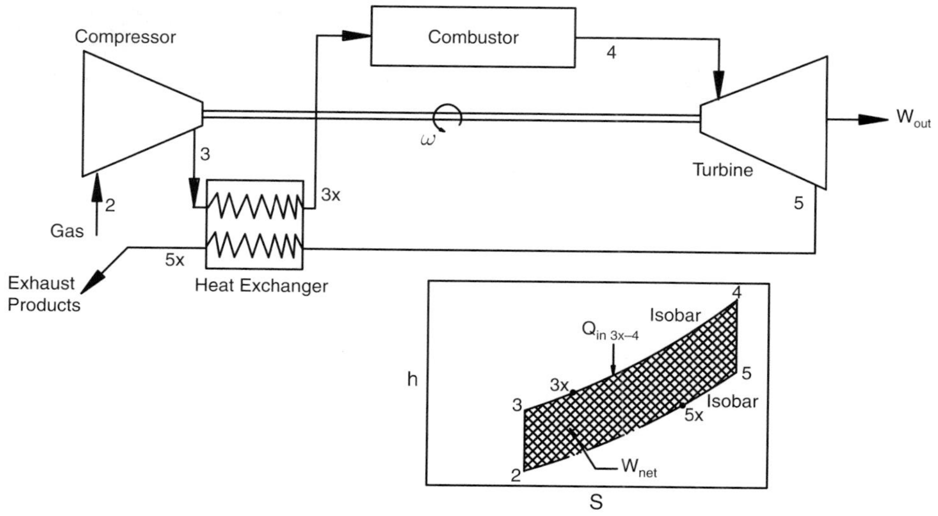

Figure 1.50 Brayton cycle with regeneration.

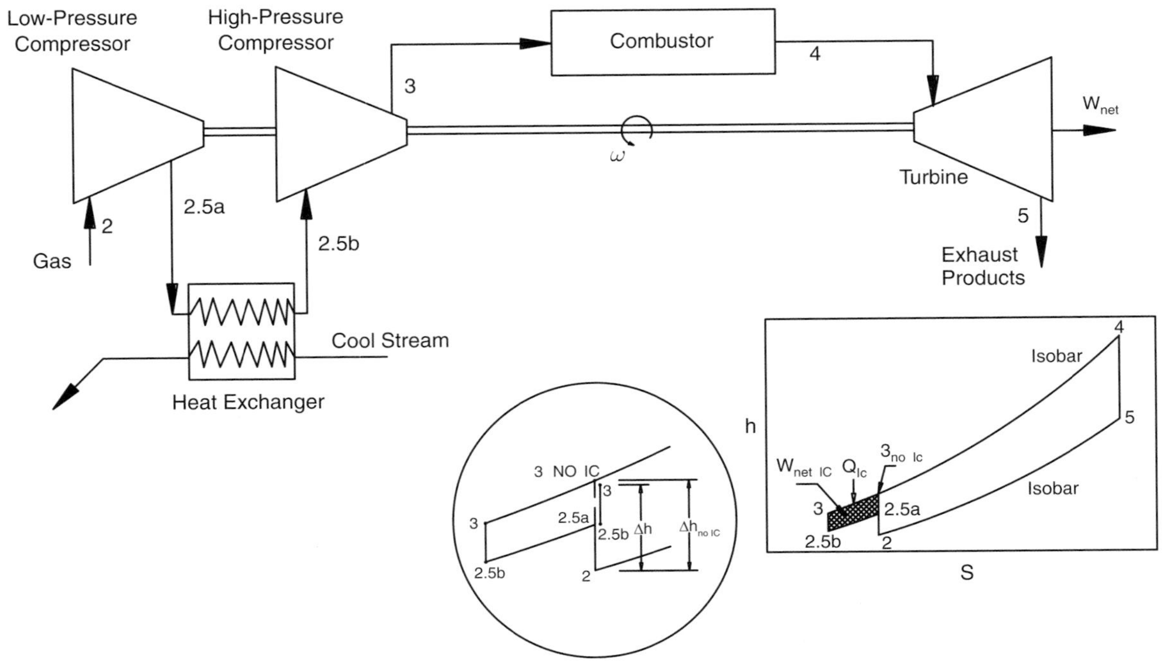

Figure 1.51 Brayton cycle with intercooling.

$\Delta h < \Delta h_{\text{no IC}}$. This will be further demonstrated with cycle analyses examples (Chapter 3) and additionally as will be shown with compressor design fundamentals (Chapter 5). Consequently, intercooling reduces the compressor power required for the cycle ($\Delta h < \Delta h_{\text{no IC}}$), thus increasing the net power but also increasing the required heat input to reach the same burner exit temperature. Next, we examine the ratio of the additional heat input, $\dot{Q}_{\text{IC}}$, which is the total area under the curve from 3 to $3_{\text{no IC}}$, to the net additional power generated by using intercooling, $\dot{W}_{\text{net IC}}$ (crosshatched area). However, this ratio is greater than the nominal or nonintercooling ratio of heat input to net power (area under the curve $3_{\text{no IC}}$–4 to area bounded by 2–$3_{\text{no IC}}$–4–5, or Figure 1.14). Also, if the two compressors have equal pressure ratios and thus temperature ratios, $\sqrt{\tau_c}$, the efficiency of only the added output power and added input energy is $\eta_{\text{th IC}} \equiv \dot{W}_{\text{net IC}}/\dot{Q}_{\text{IC}} = 1 - 1/\sqrt{\tau_c}$, which is clearly less than the overall efficiency for the nonintercooling case, $\eta_{\text{th}} = 1 - 1/\tau_c$. Thus, although the net output power increases, for this case the thermal efficiency will decrease somewhat. Also see Problem 3.101. As a result, intercooling and regeneration (and possibly reheat) are almost always used together to improve thermal efficiency. A second advantage for intercooling is that since the temperature is reduced for the intermediate stages, the exit temperature or last stage temperature is also reduced, thus reducing any thermal structure issues. This is particularly amplified for high-pressure compressors. As for the regenerator, because of the heat exchanger, the weight is increased significantly and is not used for aircraft applications.

1.4.4 Brayton Cycle with Inlet Cooling

For this cycle the working gas is first cooled by some means at the inlet. Often for low-humidity environments for power generation, water or a water–alcohol mix is sprayed at the inlet and the gas temperature is reduced by evaporative cooling. For propulsion applications, this is sometimes used for high-temperature/low-humidity locations for takeoff. This has a similar effect as intercooling. Namely, with a cooler inlet temperature less power is required to attain the same compressor exit pressure. The remainder of the cycle is the same as a simple Brayton cycle. The gas is then compressed by the compressor to

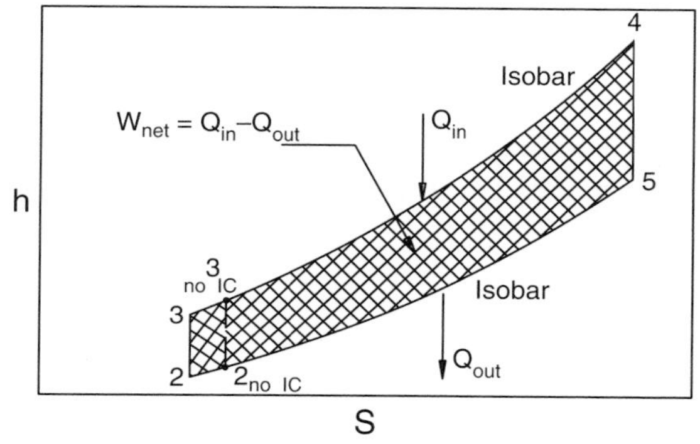

Figure 1.52 An h–s diagram for Brayton cycle with inlet cooling

some nominal pressure. The gas is heated to a nominal temperature in the combustor and enters the turbine. An h–s diagram is shown in Figure 1.52. The additional heat input needed to reach the same burner exit pressure (4) is the total area under the curve from 3 to $3_{no\ IC}$. The additional net power is the crosshatched area, 2–3–$3_{no\ IC}$–$2_{no\ IC}$. For this case, the additional process results in the total operation on the same two isobars. As a result, the proportion of input heat and net power is the same as for the nominal ratio. Thus, for power generation units, the net effect of inlet cooling is to increase the net power output and maintain the same overall thermal efficiency. For propulsion applications, the thrust is increased.

1.4.5 Brayton Cycle with Reheat

A fourth variation of the Brayton cycle that is sometimes used for power generation is the addition of "reheat." Such a cycle is shown in Figure 1.53a for the case of a second separate combustor. The h–s diagram is also shown in Figure 1.53a. For this cycle the working gas is first

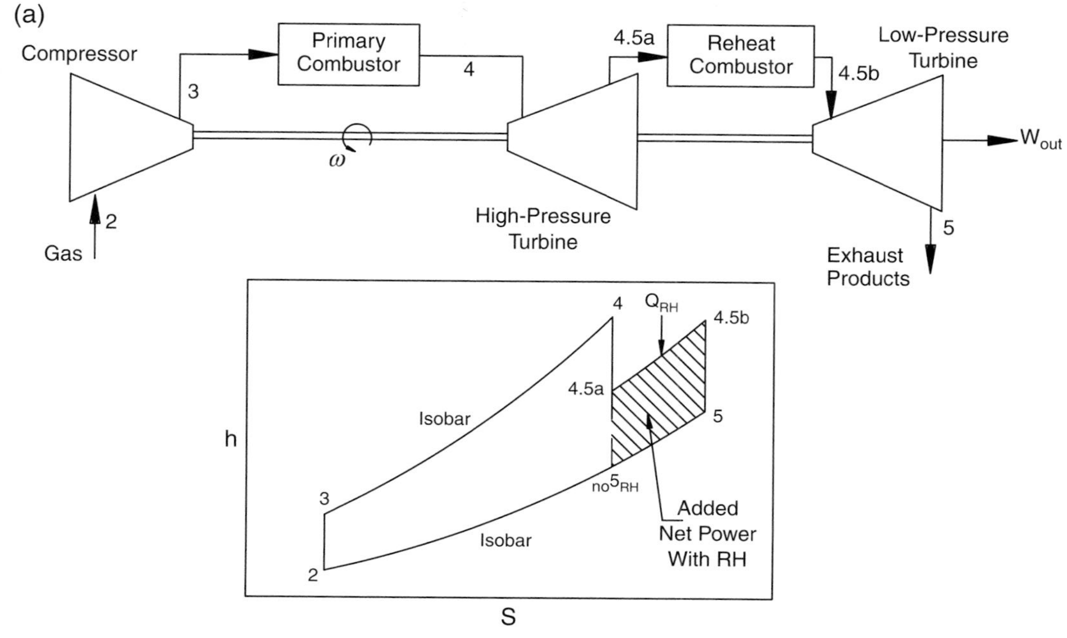

Figure 1.53 (a) Brayton cycle with reheat–separate burner. (b) Brayton cycle with reheat–heat exchanger.

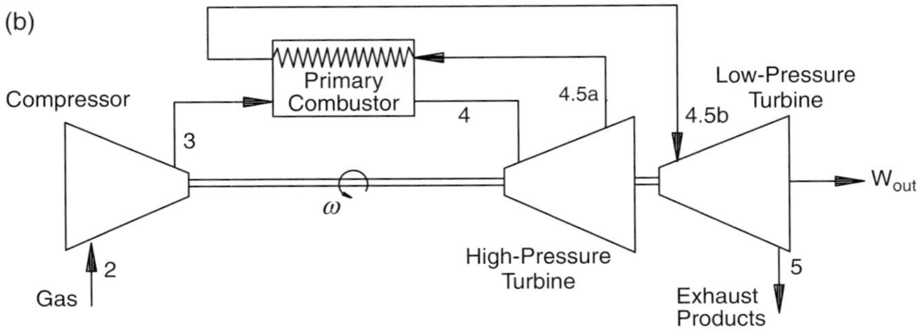

Figure 1.53 (cont.)

compressed by the compressor, which as before increases the pressure. Next, the gas is heated to a nominal temperature in the combustor and enters the high-pressure turbine, where energy is extracted and the temperature decreases. The gas then enters either the second combustor or in some cases a heat exchanger (Figure 1.53b). This again increases the temperature to some nominal value. The gas then enters the low-pressure turbine, where more energy is extracted. As a result, the reheat process significantly increases the net power output as shown on Figure 1.53a (crosshatched area). By examining the additional heat input (total area under the curve from 4.5a to 4.5b, Q_{RH}) the ratio of this quantity to the net power is higher than the nominal simple Brayton cycle ratio. As a result, the reheat cycle decreases the net thermal efficiency. Reheat is often used with regeneration and/or intercooling. However, because of the second combustor, the weight is increased significantly. As a result, this type of gas turbine is not used for aircraft. The "sequential combustion" Ansaldo Energia GT26 is shown in Figure 1.54, which in practice is a reheat unit. This gas turbine has 22 compressor stages with a pressure ratio of over 32. The turbine has fivestages and the reheat combustor is between stages 1 and 2. It produces 370 MW (496,000 hp) at 41 percent efficiency with an ingested flow rate of 741 kg/s.

1.4.6 Summary of Power Generation Performance Enhancers

The diagram for combining all four previous performance enhancing improvements of the Brayton cycle is shown in Figure 1.55 along with the combined *h–s* diagram. For this cycle the working gas is first cooled by some means at the inlet. It is then compressed by a low-pressure compressor to some intermediate pressure. This compressor is driven by a low-pressure turbine. Next, the gas is cooled by the intercooler and enters the high-pressure compressor, where the pressure rises to a nominal level. The gas is then "warmed" by the regenerator before entering the combustor. The gas is further heated to a nominal temperature in the combustor and enters the high-pressure turbine. Energy is extracted to drive the high-pressure compressor and the temperature decreases. The gas then enters the reheat combustor. This again increases the temperature to some nominal value. The gas then enters the low-pressure turbine, where more energy is extracted for the low-pressure compressor. The gas finally enters the "free" or power turbine, where power is used to drive an electric generator, auxiliary pumps or compressors, or some other auxiliary needs. Finally, still warm gas is then used in the regenerator before exiting the exhaust. Combining all enhancers in general both increases

Table 1.2 **Summary of power generation performance enhancers.**

	Net power	Efficiency
Simple Brayton	~ reference ~	~ reference ~
with regeneration	~ same ~	increase ↑
with intercooler	increase ↑	decrease ↓
with inlet cooling	increase ↑	~same~
with reheat	increase ↑	decrease ↓
with all four	increase ↑	increase ↑

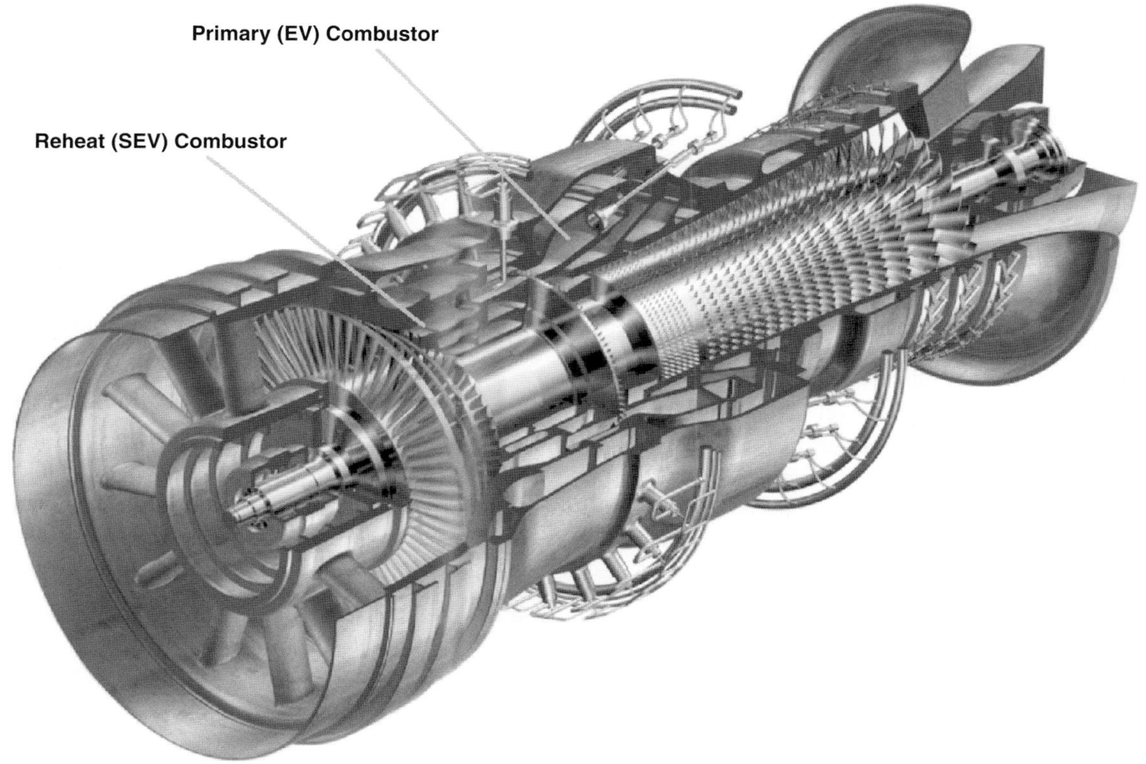

Figure 1.54 Cutaway of Ansaldo Energia GT 26 power generation gas turbine with reheat. (This picture is copyrighted by Ansaldo Energia s.p.a., who owns any related right. Any reproduction is forbidden without the written permission of the owner.)

the net output and also improves the overall efficiency, but also increases the net weight and maintenance requirements. Results are summarized in Table 1.2.

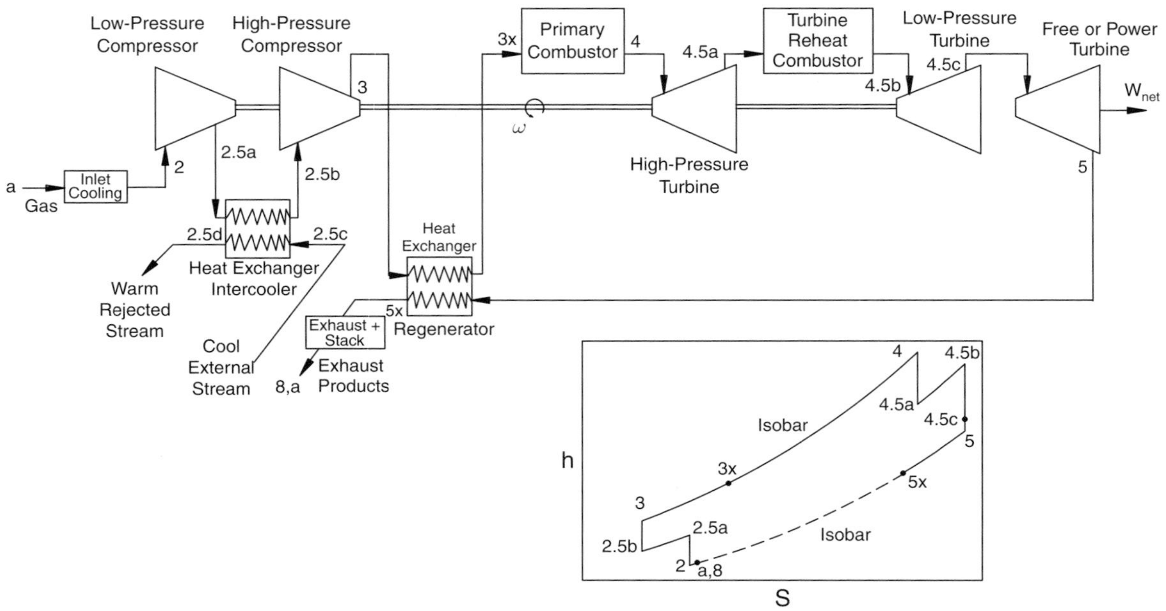

Figure 1.55 Power generation gas turbine with four performance enhancers.

1.4.7 Brayton Cycle with a Steam Topping Cycle

Another cycle that improves the performance of a simple Brayton cycle is the more complicated steam topping cycle. This cycle is also used with some stationary ground application gas turbines. The diagram for this system is shown in Figure 1.56. Like the regeneration cycle, this cycle is used because it takes advantage of the very hot gases that are dumped or thrown away as in a normal open Brayton cycle. The exhaust hot gases are used to heat water in a second or "topping" cycle. Thus, a certain amount of the energy is recovered and is used to drive a second turbine. This in fact produces two connected cycles. The first is an open Brayton cycle, and the second is a steam or Rankine cycle, which is a cycle of two phases. The h–s diagram for the two dependent cycles is also shown in Figure 1.56. The Brayton cycle ideally operates with two isentropic and two isobaric processes. The Rankine cycle ideally operates with two isentropic processes, one of which is nearly isovolumetric, and two isobaric processes. For the process d to a, saturated steam undergoes a change of phase and is condensed to a saturated liquid isothermally and isobarically. Two separate power outputs result.

1.5 Determination of the Net Output of a Gas Turbine

For a jet engine the most important characteristic is the thrust-producing capability. In this section, the basic equations that will allow the reader to calculate engine thrust are developed. These equations are developed based on the simple control volume approach in which the incoming and exiting properties are known and are used in Chapters 2, 3, and 12. Details of the internal flow are included in Chapters 4–11. The thrust equations are developed for three different engine types: the turbojet, the turbofan with an exhausted fan, and a turbporop. The thrust for all other engine types can be found from one of these cases but proving this will be left as an exercise for the reader.

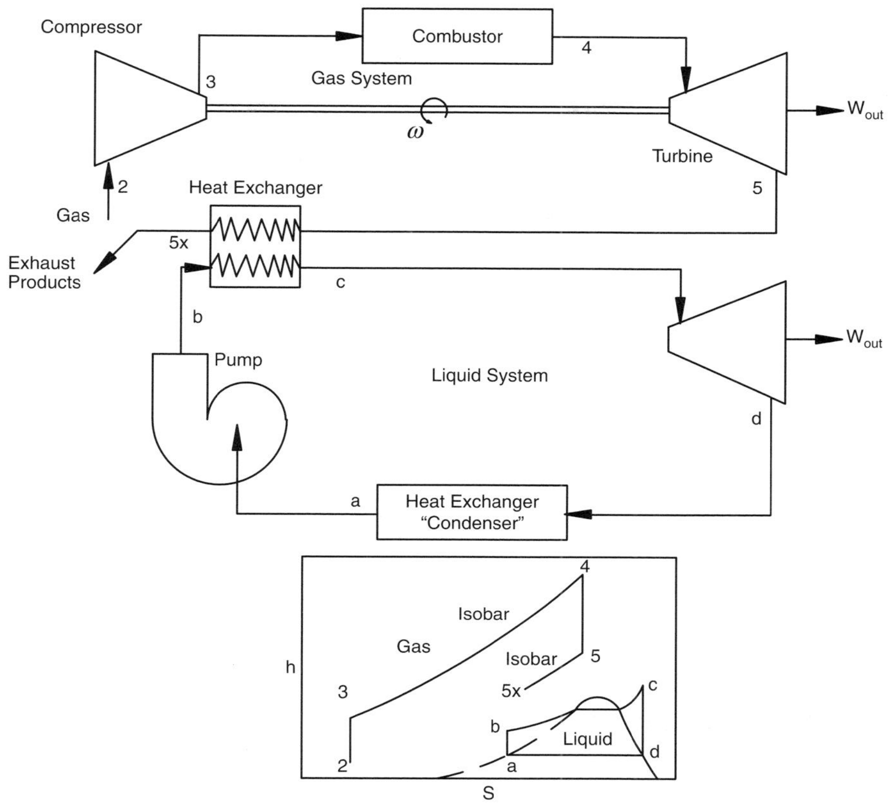

Figure 1.56 Steam topping cycle.

For a power generation gas turbine, the most important characteristic is the net power output. At this point of the text, it is a straightforward balance of the energy input by the combustor(s) and used by the compressor(s) and extracted by the turbine(s). Again, these energy equations are developed based on inlet and exiting properties and are used in Chapters 2, 3, and 12. Details of the internal flow are again included in Chapters 4–11.

1.5.1 Turbojet Thrust

Figure 1.57 shows a turbojet. At this point in the analysis, only the inlet and exit conditions of the engine are considered. The calculations of the different internal component processes are not undertaken, although they are obviously extremely important to the overall design and performance. A control volume is drawn around the engine as shown. The engine is moving in air to the left with velocity u_a. The large-diameter cylindrical control volume (CV1), which is fixed to the engine and thus is moving, cuts through the engine strut. Fuel is also injected into the engine through the control volume. One surface is at the exit plane of the engine, and the front plane is far from the inlet.

Writing the general linear momentum equation in rectangular coordinates yields

$$\sum F_s + \sum F_b = \frac{\partial}{\partial t} \int_{CV} V \rho d\mathcal{V} + \int_{CS} V \rho V \cdot dA. \quad (1.5.1)$$

However, if the flow is considered to be steady, the unsteady term $(\partial/\partial t)$ disappears. Next, one should consider applying the equation in the thrust (x) direction. The surface forces, F_s, are due to pressures on the

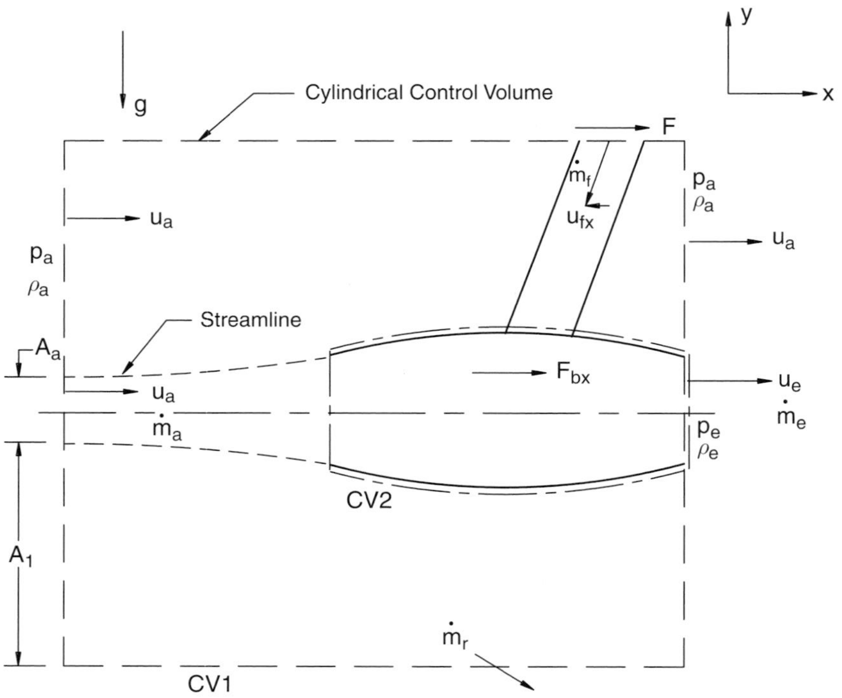

Figure 1.57 Control volume definition for a turbojet.

control volume and to the cutting of the strut, F, which is the force holding the engine in place. Thus, the force, F, represents the developed thrust. Because the front surface is far from the engine, the pressure is uniform and at atmospheric conditions, p_a, at this surface. At the front surface, the air flows at velocity u_a across an area $A_a + A_1$. Also, at the engine exit, the pressure in the engine is uniformly p_e, and the pressure outside of the engine is the atmospheric pressure, p_a. The body forces are F_b. For the momentum term, the velocity of the engine exhaust is uniformly u_e. The area of the engine exhaust is A_e. At the rear of the control volume, air exits at uniform velocity u_a across area A_2. The fuel is injected at a mass flow rate of m_f with velocity $-u_{fx}$ in the x-direction, but it does not usually contribute significantly to the flux portion of the momentum equation. A mass flow term, $\dot{m}_r$, can exit from the sides of the control volume owing to displacement of air by the engine, which must be evaluated and included in the analysis. Thus, evaluating each of the terms and rewriting the momentum equation yields

$$F + p_a A_a + p_a A_1 - p_e A_e - p_a A_2 + F_{bx} = -u_a \rho_a A_a u_a + u_e \rho_e A_e u_e - u_a \rho_a A_1 u_a + u_a \rho_a A_2 u_a + \dot{m}_r u_a - \dot{m}_f u_{fx}. \tag{1.5.2}$$

Next, one should realize that no body forces exist in the x-direction, $F_{bx} = 0$. Also, uniform flow is assumed across areas A_e and A_a. The mass fluxes across these inlet and exit surfaces, which are the gas flows into and out of the engine, respectively, are

$$\dot{m}_a = \rho_a A_a u_a \tag{1.5.3}$$

and

$$\dot{m}_e = \rho_e A_e u_e. \tag{1.5.4}$$

Also, by inspecting the control volume one can see the areas are related by

$$A_1 + A_a = A_2 + A_e. \tag{1.5.5}$$

Using the previous three equations with Eq. (1.5.2) yields

$$F + p_a A_2 + p_a A_e - p_e A_e - p_a A_2 = \dot{m}_e u_e - \dot{m}_a u_a + \dot{m}_r u_a - u_a^2 \rho_a (A_1 - A_2) - \dot{m}_f u_{fx}. \quad (1.5.6)$$

Because the flow is steady, the continuity equation yields (using a second control volume, CV2, for the engine only)

$$\dot{m}_e = \dot{m}_a + \dot{m}_f \quad (1.5.7)$$

and CV1 yields

$$\dot{m}_e + \rho_a u_a A_2 + \dot{m}_r = \dot{m}_a + \rho_a u_a A_1 + \dot{m}_f; \quad (1.5.8)$$

thus, using the preceding two equations, we obtain

$$\dot{m}_r = \rho_a u_a (A_1 - A_2). \quad (1.5.9)$$

Or, using Eqs. (1.5.6) and (1.5.9) and combining terms, we obtain

$$F = (\dot{m}_e u_e - \dot{m}_a u_a) + A_e(p_e - p_a) - \dot{m}_f u_{fx}. \quad (1.5.10)$$

Using this with Eq. (1.5.7) yields

$$F = \dot{m}_f u_a + \dot{m}_e (u_e - u_a) + A_e(p_e - p_a) - \dot{m}_f u_{fx}. \quad (1.5.11)$$

Often, the x-component of fuel velocity is negligible, resulting in

$$F = \dot{m}_f u_a + \dot{m}_e (u_e - u_a) + A_e(p_e - p_a). \quad (1.5.12)$$

This is the general equation for a turbojet engine. One further simplifying assumption can often be made. Usually, the fuel flow is much smaller than the airflow ($\dot{m}_f \ll \dot{m}_a$). As a result, with this assumption,

$$\dot{m}_e = \dot{m}_a = \dot{m}. \quad (1.5.13)$$

Thus, Eq. (1.5.12) becomes

$$F = \dot{m}(u_e - u_a) + A_e(p_e - p_a). \quad (1.5.14)$$

It is important to note that the resulting thrust has two contributions. One is due to the mass flux and flow acceleration. The second results from pressure differences between the engine exit and atmosphere. In Chapter 10, an engine will be shown to operate most efficiently (i.e., maximum thrust) when the pressures are equalized for the ideal case (no losses) and nearly equalized for the nonideal case. This analysis also applies to mixed-flow turbofans and ramjets because both have single entrances and single exhausts.

Example 1.1 Turbojet

A turbojet operates at sea level and moves at 800 ft/s (243.8 m/s). It ingests 250 lb_m/s (113.4 kg/s) of air and has negligible fuel flow. The diameter of the exit is 30 in. (0.762 m). The exit pressure is 22 psia (151.7 kPa), and the exit velocity is 1300 ft/s (396.2 m/s). Find the developed thrust.

Solution

From the problem statement,

$p_a = 14.7$ psia (101.3 kPa)
$p_e = 22$ psia (151.7 kPa)
$\dot{m} = 250\ \text{lb}_m/\text{s}$ (113.4 kg/s),
and because $32.17\ \text{lb}_m = 1$ sl, $\dot{m} = 7.8$ sl/s,
$A_e = \pi D_e^2/4 = 706\ \text{in.}^2\ (0.4554\ \text{m}^2)$
$u_a = 800$ ft/s (243.8 m/s)
$u_e = 1300$ ft/s (396.2 m/s).

Thus, using Eq. (1.5.14), we obtain

$$F = \dot{m}(u_e - u_a) + A_e\,(p_e - p_a)$$

$F = 7.8$ sl/s $(1300 - 800)$ ft/s $+ 706$ in.^{2} $(22 - 14.7)$ lb$_f$/in.^{2};

therefore, because lb$_f$ = sl-ft/s^2,

$F = 3900 + 5153 = 9053$ lb$_f$ (40,270 N). <ANS

As one can see, 3900 lb$_f$ (17,350 N) is due to mass flux, and 5153 lb$_f$ (22,920 N) is due to pressure differences.

1.5.2 Turbofan with a Mixed Fan Thrust

One should note that, although derived for a turbojet, the equations in Section 1.5.1 exactly apply for a fully mixed turbofan as well, because only one inlet and one exit exist. Deriving the equations is left as an exercise for the reader.

1.5.3 Turbofan with a Fan Exhaust Thrust

In Figure 1.58, a turbofan with a fan exhaust is shown. The analysis applies for either a turbofan with all of the fan air exhausted or for the hybrid turbofan (a part of the bypassed air separately exhausted). A cylindrical control volume is once again drawn that moves with the engine (CV1). The front surface is far away from the inlet, and the control volume has been drawn so that the rear control surface is at the core exhaust as well as the fan exit. The control volume once again cuts through the strut.

The general momentum equation is the same as for that of the turbojet (Eq. (1.5.1)). For the turbofan, applying this equation is slightly more complicated, however. The surface forces are again due to pressures acting on the surfaces, and the force F is due to cutting the strut. The pressure on the front surface of the control volume is uniform because it is far from the engine and at

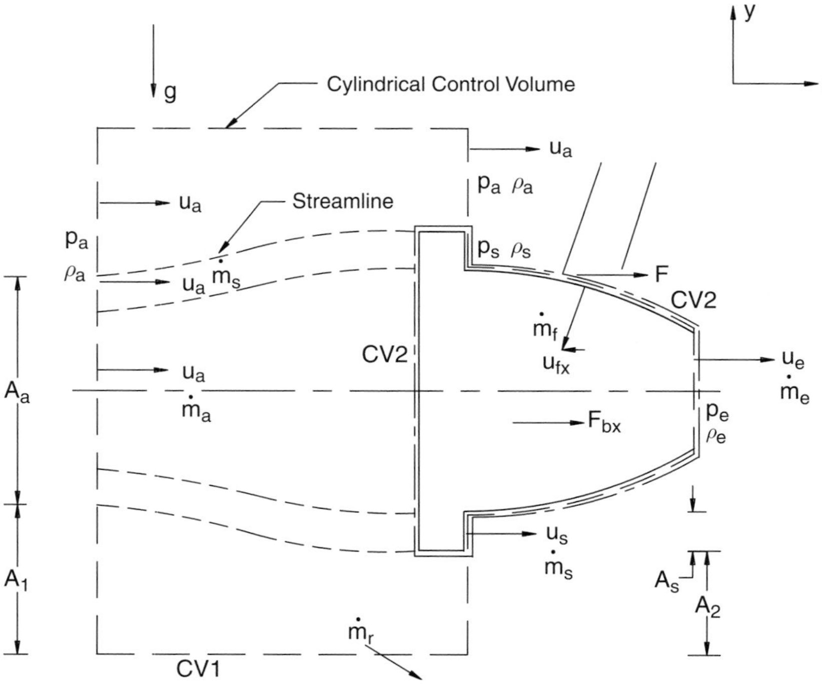

Figure 1.58 Control volume definition for a turbofan with fan exhausted.

atmospheric conditions p_a. At the front surface, air moves at velocity u_a, uniformly across area $A_a + A_1$. Part of this air enters the core, and part enters the fan. In the most general case (hybrid turbofan), the portion of air that enters the fan can be divided so that a portion exits through the fan exhaust and another part of it is mixed into the core and exits through the primary nozzle. Two exit areas must be considered. At the core exit, the pressure is p_e and the gas velocity is u_e. At the fan exhaust, the pressure is p_s and the air velocity is u_s. The areas of the core and fan exits are A_e and A_s, respectively. The pressure on other portions of the rear surface of the control volume is p_a. At the back of the control volume, air exits at velocity u_a across area A_2. Fuel is once again injected at a flow rate of $\dot{m}_f$. Evaluating each of the terms in Eq. (1.5.1) for steady flow yields

$$F + p_a A_1 + p_a A_a - p_a A_2 - p_s A_s - p_e A_e - p_a A_3 + F_{bx} = -u_a \rho_a A_a u_a + u_e \rho_e A_e u_e + u_s \rho_s A_s u_s - u_a \rho_a A_1 u_a + u_a \rho_a A_2 u_a + \dot{m}_r u_a - \dot{m}_f u_{fx}. \tag{1.5.15}$$

Once again, the body force in the x-direction, F_{bx}, is zero. By examination of the control volume, one can see that

$$A_1 + A_a = A_2 + A_3 + A_s + A_e. \tag{1.5.16}$$

Also, three different mass flow rates can be found as follows:

$$\dot{m}_a = \rho_a A_a u_a \tag{1.5.17}$$

$$\dot{m}_e = \rho_e A_e u_e \tag{1.5.18}$$

$$\dot{m}_s = \rho_s A_s u_s, \tag{1.5.19}$$

where $\dot{m}_a$ is the total flow entering the engine, $\dot{m}_e$ is the primary exhaust air flow, and $\dot{m}_s$ is the fan (or secondary) exhaust airflow. Thus, Eq. (1.5.15) becomes

$$F = A_e(p_e - p_a) + A_s(p_s - p_a) - \dot{m}_a u_a + \dot{m}_e u_e + \dot{m}_s u_s + \dot{m}_r u_a - u_a^2 \rho_a (A_1 - A_2) - \dot{m}_f u_{fx}. \tag{1.5.20}$$

The continuity equation can be used to show that using a second control volume, CV2, only around the engine

$$\dot{m}_e + \dot{m}_s = \dot{m}_a + \dot{m}_f, \tag{1.5.21}$$

and CV1 yields

$$\dot{m}_e + \dot{m}_s + \rho_a u_a A_2 + \dot{m}_r = \dot{m}_a + \dot{m}_f + \rho_a u_a A_1, \tag{1.5.22}$$

thus,

$$\dot{m}_r = \rho_a u_a (A_1 - A_2). \tag{1.5.23}$$

Using this with Eq. (1.5.20) yields

$$F = A_e(p_e - p_a) + A_s(p_s - p_a) + \dot{m}_e(u_e - u_a) + \dot{m}_s(u_s - u_a) + \dot{m}_f u_a - \dot{m}_f u_{fx}. \tag{1.5.24}$$

If the x-component of the fuel velocity is small,

$$F = A_e(p_e - p_a) + A_s(p_s - p_a) + \dot{m}_e(u_e - u_a) + \dot{m}_s(u_s - u_a) + \dot{m}_f u_a. \tag{1.5.25}$$

This is the general equation of the thrust of a turbofan engine with the fan exhausted. If the fuel flow is very small ($\dot{m}_f \ll \dot{m}_a$),

$$\dot{m}_a = \dot{m}_e + \dot{m}_s. \tag{1.5.26}$$

Equation (1.5.24) for this assumption reduces to

$$F = A_e(p_e - p_a) + A_s(p_s - p_a) + \dot{m}_e(u_e - u_a) + \dot{m}_s(u_s - u_a). \tag{1.5.27}$$

As can be seen, four terms arise from this equation. The first is thrust generated by pressure differences at the primary exhaust. The second is due to pressures at the fan exhaust. The third and fourth are due to mass fluxes in the primary exhaust and fan exhaust, respectively. A limiting case that could be considered is the fan becoming very small with a reduced flow rate, which is a turbojet. By examining Eq. (1.5.27), one can see that this does indeed reduce to Eq. (1.5.14), for $A_s = 0$ and $\dot{m}_s = 0$.

1.5.4 Turboprop Thrust

As indicated in Section 1.5.3, a turboprop is essentially a turbojet with a propeller. As a result, the derivation of the thrust equation is very similar to that for the turbojet. That is, the thrust is given by

$$F = \dot{m}_f u_a + \dot{m}_e(u_e - u_a) + A_e(p_e - p_a) + F_p, \tag{1.5.28}$$

where F_p is the additional thrust due to the propeller. Once again, the fuel flow is often much smaller than the airflow ($\dot{m}_f \ll \dot{m}_a$). As a result for this assumption,

Example 1.2 | Exhausted Turbofan

A turbofan (approximately the same size as a commercial turbofan engine) operates at sea level and moves at 269.7 m/s (885 ft/s). It ingests 121.1 kg/s (267 lb_m/s) of air into the core and five times this amount into the fan (the bypass ratio), which all exhausts through the fan exhaust. The fuel flow is negligible. The exit areas of the fan and core are 1.580 and 1.704 m^2 (2450 and 2642 $in.^2$), respectively. The exit pressures from the fan and core are 154.4 and 144.8 kPa (22.4 and 21 psia), respectively. The exhaust velocities from the fan and core are 328.6 and 362.7 m/s (1078 and 1190 ft/s), respectively. Find the thrust.

Solution

From the problem statement,

$p_a = 101.3$ kPa (14.7 psia) $u_a = 269.7$ m/s (885 ft/s)
$p_s = 154.4$ kPa (22.4 psia) $u_s = 328.6$ m/s (1078 ft/s)
$p_e = 144.8$ kPa (21 psia) $u_e = 362.7$ m/s (1190 ft/s)
$\dot{m}_e = 121.1$ kg/s (267 $lb_m/s = 8.3$ sl/s)
$\dot{m}_s = 5 \times 121.1$ kg/s $= 605.4$ kg/s (41.7 sl/s)
$A_s = 1.580\ m^2$ (2450 $in.^2$) $A_e = 1.704\ m^2$ (2642 $in.^2$).

Therefore, using Eq. (1.5.27), we obtain

$$F = A_e(p_e - p_a) + A_s(p_s - p_a) + \dot{m}_e(u_e - u_a) + \dot{m}_s(u_s - u_a)$$
$$F = 1.704\ m^2\ (144.8 - 101.3)\ kPa + 1.580\ m^2\ (154.4 - 101.3)\ kPa + 121.1\ kg/s\ (362.7 - 269.7)\ m/s + 605.4\ kg/s\ (328.6 - 269.7)\ m/s$$

and so, because N = kg-m/s^2 and N = Pa-m^2 and 1000 Pa = kPa,

$$F = 74{,}120_{p\text{-}core} + 83{,}900_{p\text{-}fan} + 11{,}260_{m\text{-}core} + 35{,}660_{m\text{-}fan} = 85{,}380_{core} + 119{,}560_{fan}$$

$F = 204{,}940$ N (46,075 lb_m). <ANS

By examining the four different terms, one can see that 85,380 N (19,200 lb_f) of thrust is generated by the core, and 119,560 N (26,880 lb_f) comes from the fan.

$$F = \dot{m}(u_e - u_a) + A_e(p_e - p_a) + F_p. \tag{1.5.29}$$

1.5.5 Power Generation Gas Turbine

For a gas turbine used for power generation the primary equation is a simple energy balance on the shaft. The net power or the power that can be used for driving a pump system, generating electricity among many other applications. For the gas generator with multiple compressors and turbines this yields:

$$\mathcal{P}_{net} = \sum \mathcal{P}_{turbine(s)} - \sum \mathcal{P}_{compressor(s)}. \tag{1.5.30}$$

For a simple gas turbine with a single compressor and single turbine this becomes:

$$\mathcal{P}_{net} = (\dot{m}_a + \dot{m}_f)(h_4 - h_5) - \dot{m}_a(h_3 - h_2); \tag{1.5.31}$$

if the fuel flow is very small ($\dot{m}_f \ll \dot{m}_a$):

$$\mathcal{P}_{net} = \dot{m}_a\ (h_4 - h_5) - \dot{m}_a\ (h_3 - h_2). \tag{1.5.32}$$

Example 1.3 | Power Generation Gas Turbine

An ideal power generation gas turbine approximating of a modern moderate size unit ingests $\dot{m} = 66.67\,\frac{\text{kg}}{\text{s}}\left(147\,\frac{\text{lb}_\text{m}}{\text{s}}\right)$ of air. The compressor has an inlet enthalpy $h_2 = 319\,\frac{\text{kJ}}{\text{kg}}$ and an exit enthalpy $h_3 = 675\,\frac{\text{kJ}}{\text{kg}}$. The turbine has an inlet enthalpy $h_4 = 1612\,\frac{\text{kJ}}{\text{kg}}$ and an exit enthalpy $h_5 = 762\,\frac{\text{kJ}}{\text{kg}}$. The fuel has a heating value of $42{,}800\,\frac{\text{kJ}}{\text{kg}}\left(18{,}400\,\frac{\text{BTU}}{\text{lb}_\text{m}}\right)$ and the fuel flow rate is $\dot{m}_\text{f} = 1.460\,\frac{\text{kg}}{\text{s}}\left(3.217\,\frac{\text{lb}_\text{m}}{\text{s}}\right)$. Find the net output power and net thermal efficiency, assuming the fuel flow is negligible.

Solution: Net Power and Thermal Efficiency

Now the power balance on the shaft is $\mathcal{P}_\text{net} = \dot{m}(h_4 - h_5) - \dot{m}(h_3 - h_2)$. The first term is the total power derived from the turbine and the second term is the power driving the compressor:

$$\mathcal{P}_\text{net} = \left[66.67\,\frac{\text{kg}}{\text{s}} \times (1612 - 762)\,\frac{\text{kJ}}{\text{kg}} - 66.67\,\frac{\text{kg}}{\text{s}} \times (675 - 319)\,\frac{\text{kJ}}{\text{kg}}\right] \times \frac{\text{kW}}{\text{kJ/s}}.$$

Here, $\mathcal{P}_\text{net}$ is the net power from the gas turbine, which can be used for electric generation or other applications:

$\mathcal{P}_\text{net} = 32{,}930\,\text{kW} = 32.93\,\text{MW}\;(44{,}160\,\text{hp}).$ <ANS

Next, the overall thermal efficiency is

$$\eta_\text{th} = \frac{\mathcal{P}_\text{net}}{\dot{Q}_\text{in}},$$

where the heat in is

$$\dot{Q}_\text{in} = \dot{m}_\text{f}\Delta H = 1.460\,\frac{\text{kg}}{\text{s}} \times 42800\,\frac{\text{kJ}}{\text{kg}} \times \frac{\text{kW}}{\text{kJ/s}} = 62{,}490\,\text{kW} = 62.49\,\text{MW}\;(83{,}800\,\text{hp})$$

or $\eta_\text{th} = \dfrac{32{,}930\text{k W}}{62{,}490\text{k W}} = 0.527$ or 52.7 percent. <ANS

1.6 Performance Measures

1.6.1 Propulsion Measures

One of the most important parameters quantifying aircraft engine performance is obviously thrust, F. If an engine does not generate enough thrust to overcome airframe drag, a given aircraft cannot fly. However, a quantity that is just as important as thrust is the thrust-specific fuel consumption (*TSFC*), which is defined as

$$TSFC = \frac{\dot{m}_{\text{f}_\text{t}}}{F}, \tag{1.6.1}$$

where $\dot{m}_{\text{f}_\text{t}}$ is the total mass flow rate of fuel (primary burner and afterburner). The *TSFC* is a measure of how much fuel is used for given thrust and allows for efficiency comparisons between different engines and engine types. It is somewhat comparable to the inverse of the miles-per-gallon (mpg) rating of an automobile. A well-suited engine will have a small value for *TSFC*. For a nonafterburning engine, this becomes

$$TSFC = \frac{\dot{m}_f}{F}, \tag{1.6.2}$$

where $\dot{m}_f$ is the mass flow rate of the primary burner fuel. The *TSFC* can conveniently be nondimensionalized using the ambient sound speed:

$$\overline{TSFC} = TSFC \times a_a. \tag{1.6.3}$$

A second quantity sometimes but less often used to evaluate fuel economy (but is more often used for rocket engines) is the specific impulse I. This is defined as

$$I = \frac{F}{g \times \dot{m}_{f_t}}, \tag{1.6.4}$$

where g is the gravitational constant. One can relate the specific impulse to *TSFC* by

$$I = \frac{1}{g \times TSFC}. \tag{1.6.5}$$

For the remainder of this book, *TSFC* (or the nondimensional value) will be used to evaluate fuel consumption.

Another dimensionless parameter used to characterize an engine's performance is the dimensionless thrust. This quantity is defined as

$$\overline{F} = \frac{F}{\dot{m}_t a_a}, \tag{1.6.6}$$

where $\dot{m}_t$ is the total air mass flow rate into the engine and a_a is again the local speed of sound. This dimensionless thrust represents the inverse to the physical size of an engine, including the diameter and weight. That is, for a given dimensional thrust, the dimensionless thrust decreases as the required mass flow (or physical size) increases and vice versa. In a sense, this nondimensional parameter represents the trends of the thrust-to-weight ratio.

Another performance characteristic for a jet engine that is sometimes but not often used is the propulsive efficiency. This is the ratio of thrust power to the rate of production-of-propellant kinetic energy. In equation form, this is defined as

$$\eta_k = \frac{F u_a}{\sum \dot{m}_{exit} u_{exit}^2 - \dot{m}_a u_a^2}. \tag{1.6.7}$$

For a turbojet or a mixed turbofan only, one exit stream is present; however, for an unmixed turbofan, two exit streams are exhausted. Note that this efficiency is not an overall thermal efficiency because the enthalpy change is not included. This equation has limited practicality. For example, for a simple turbojet, if the fuel flow is negligible and the exit pressure and ambient pressure are the same, the thrust is $F = \dot{m}(u_e - u_a)$. The efficiency parameter then becomes $\frac{u_a}{u_a + u_{exit}}$. One can maximize this (to unity) by setting u_{exit} equal to zero, which results in a negative thrust! Furthermore, if one sets $u_{exit} = u_a$, the efficiency is 50 percent, but the thrust is zero. Thus, this parameter is identified but will not be used in this book.

To evaluate the overall performance of an aircraft engine, the thrust, *TSFC*, nondimensional thrust, and sometimes the propulsive efficiency must all be considered. The four parameters cannot be treated independently. For different aircraft applications, various objectives will have to be satisfied. For example, for fighter aircraft the fuel consumption under duress is less important than for a commercial carrier, but the engine weight is extremely important. An optimization of the different parameters for a given application can be accomplished only with detailed cycle analyses, as will be shown in Chapter 3.

1.6.2 Power Generation Measures

For comparison, the most common evaluators for power generation gas turbines are the net power ($\mathcal{P}_{net}$), thermodynamic efficiency (η_{th}), heat rate ($\mathcal{HR}$), and specific fuel consumption (*SFC*) and nondimensional $\overline{SFC}$. The latter four are given as

$$\eta_{th} = \frac{\mathcal{P}_{net}}{\dot{Q}_{in}} \tag{1.6.8}$$

$$\mathcal{HR} = \frac{\dot{Q}_{in}}{\mathcal{P}_{net}} = \frac{\dot{m}_f \Delta H}{\mathcal{P}_{net}} \tag{1.6.9}$$

$$SFC = \frac{\dot{m}_f}{\mathcal{P}_{net}} \tag{1.6.10}$$

$$\overline{SFC} = a_a^2 \times SFC. \tag{1.6.11}$$

One should note that the efficiency is dimensionless. The heat rate is often used and is the inverse of the efficiency, but units are always included – it is not reported as dimensionless. The specific fuel consumption (*SFC*) is a measure of the fuel rate compared to the power output and the dimensionless value is used on occasion. To evaluate the overall performance of a power generation unit, one must consider the net power, *SFC*, and thermodynamic efficiency and or heat rate. The parameters also cannot be treated independently although they are directly related. For different applications, various objectives will have to be satisfied. As for an aircraft, an optimization of the different parameters for a given application can be accomplished only through detailed cycle analyses. It should be noted that the *SFC* is often used for turboprops as well.

1.7 Summary

In this chapter, a brief historical review of turbomachine and propulsion development was first presented and then the different types of aircraft jet engines were introduced. Early developments were initiated in Norway in the late 1800s and early 1900s including patents. Later developments culminated in the 1930s and 1940s simultaneously in Great Britain and Germany with the patents and operating prototypes of the basic modern jet engine designs. Interestingly, all three gas turbines were independently developed, and the designs were all strikingly different. The fundamental thermodynamic concepts of the Brayton cycle which is the basis for a gas turbine were reviewed. The different types of aircraft jet engines and gas turbines were introduced, including ramjets, scramjets, turbojets, turbofans (two types), turboprops and unducted fans, and turboshafts. As a part of these engines, the definition of the gas generator was introduced. Afterburners for thrust augmentation are used on turbojets and turbofans. Typical sizes and thrust or power levels were discussed for both modern and outdated engines to give the reader an appreciation of the variety of applications and historical trends. Cross-sectional views and photographs of engines and gas turbines were also presented. General comparisons were made to indicate which engines are best under different operating conditions. Limited discussions were included for currently being developed engines with future developments identified. Many modern power generation gas turbines were shown to be “aeroderivatives” of aircraft engines; that is, the core of the units is essentially the same. Power generation gas turbines were discussed and five performance-improving enhancers were discussed. Equations were also derived for turbojet and turbofan thrust if all of the inlet and exit conditions are known. Similarly, equations were also presented to allow evaluation of power generation gas turbines. Other parameters that allow overall evaluation of a propulsion gas turbine performance were defined: the thrust, the thrust-specific fuel consumption, *TSFC*, nondimensional thrust, and propulsive efficiency. Also, parameters that allow overall evaluation of a power generation gas turbine performance were defined: the net power, specific fuel consumption, *SFC*, and thermodynamic efficiency.

In the following chapters, thermodynamic and gas dynamic details of the individual components and internal flows are covered in increasing detail so that the operating conditions can be determined. First, in Chapters 2 and 3 a cycle analysis is presented in which the components perform both ideally and nonideally. The overall operating characteristics (thrust, *TSFC*, etc.) of jet engines are found and likewise operating characteristics of power generation units are found (net power, efficiency, *SFC*, etc.). For the ideal analyses, closed-form equations are found relating the overall gas turbine outputs for given components. Also, losses are included (but specified a priori) so that the overall nonideal performance of an engine can be determined. From the cycle analyses, trends are evident that can be used to compare the performances of different jet engine and power generation types. Similarly, component parameters are varied so that trends in overall performance can be maximized for a given engine type for a given application. Second, in Chapters 5–11, detailed component analyses are performed with which component losses can be predicted for the different component geometries and operating conditions (each component is covered in a separate chapter). Lastly, as a capstone in Chapter 12, component matching is included so that the dynamic interaction of the different nonideal components with variable losses can be found and the overall gas turbine performance can be predicted for both on-design and off-design conditions.

List of Symbols

A	Area
a	Sound speed
F	Force
g	Gravitational constant
h	Specific enthalpy
$\mathcal{HR}$	Heat rate
ΔH	Heating value
I	Specific impulse
$\dot{m}$	Mass flow rate
p	Pressure
$\mathcal{P}$	Power
$\dot{Q}$	Heat transfer rate
s	Entropy
SFC	Specific fuel consumption
T	Temperature
$TSFC$	Thrust-specific fuel consumption
u	Velocity
v	Specific volume
$\dot{W}$	Power
γ	Specific heat ratio
η	Efficiency
π	Pressure ratio
ρ	Density
τ	Temperature ratio

Subscripts

a	Freestream
b	Body (force)
c	Compressor
e	Primary exit
exit	Exit
f	Fuel
in	In-to component
k	Propulsive
out	Out-of component
net	Net out-of cycle
p	Propeller
r	Rejected
s	Secondary (bypassed)
s	Surface (force)
t	Total (summation)
th	Thermal
x	x-direction
1,2,3	Areas of control volume
1 5	Stations in cycle

Superscripts

$'$	Ideal

Problems

1.1 A turbojet operates at sea level and moves at 728 ft/s. It ingests 175 lb_m/s of air and has negligible fuel flow. The diameter of the exit is 29 in. The exit pressure is 6.75 psia, and the exit velocity is 2437 ft/s. Find the developed thrust.

1.2 A turbojet operates at 6096 m (20,000 ft) and moves at 222 m/s (728 ft/s). It ingests 79.4 kg/s (175 lb_m/s) of air and has negligible fuel flow. The diameter of the exit is 0.737 m (29 in.). The exit pressure is 46.5 kPa (6.75 psia), and the exit velocity is 743 m/s (2437 ft/s). Find the developed thrust.

1.3 A turbofan operates at sea level and moves at 272 m/s (893 ft/s). It ingests 68 kg/s (150 lb_m/s) of air into the core and 2.2 times this amount into the fan, which all exits through the fan exhaust. The fuel flow is negligible. The exit areas of the fan and core are 0.217 and 0.350 m^2 (336 and 542 in.^{2}), respectively. The exit pressures from the fan and core are 101 and 109 kPa (14.7 and 15.8 psia), respectively. The exhaust velocities from the fan and core are 697 and 355 m/s (2288 and 1164 ft/s), respectively. Find the thrust. What proportions arise from the fan and core?

1.4 A turbofan operates at sea level and moves at 272 m/s (893 ft/s). It ingests 68 kg/s (150 lb_m/s) of air into the core and 2.2 times this amount into the fan, which all is mixed into the core flow. The fuel flow is negligible. The exit area is 0.549 m^2 (851 in.^{2}). The exit pressure is 127 kPa (18.4 psia), and the exhaust velocity is 437 m/s (1435 ft/s). Find the thrust.

1.5 A turboprop operates at sea level and moves at 239 m/s (783 ft/s). It ingests 13.6 kg/s (30 lb_m/s) of air and has negligible fuel flow. The area of the exit is 0.187 m^2 (290 in.^{2}). The exit pressure is 101 kPa (14.7 psia), and the exit velocity is 190 m/s (625 ft/s). The thrust from the propeller is 11,200 N (2517 lb_f). Find the total developed thrust.

1.6 Show that a turbofan with all of the bypassed air mixed into the core flow results in the same equation for thrust as for a turbojet.

1.7 A turbofan operates at 30,000 ft and moves at Mach 0.75. It ingests 190 lb_m/s of air into the core and 1.2 times this amount into the fan, which all exits through the fan exhaust. The fuel flow is 4.66 lb_m/s. The exit areas of the fan and core are 936 and 901 in.^{2}, respectively. The exit pressures from the fan and core are 4.36 and 9.42 psia, respectively. The exhaust velocities from the fan and core are 1350 and 1880 ft/s, respectively. Find the thrust. What proportions arise from the fan and core? Do this problem for two conditions: first consider the fuel flow to be negligible; then, include the effect of finite fuel flow.

1.8 A turbojet operates at 18,000 ft and moves at Mach 0.80. The area of the exit is 361 in.^{2}. The exit pressure is 22.40 psia, and the exit velocity is 1938 ft/s. The developed thrust is 11,666 lb_f and the fuel flow rate is 4.04 lb_m/s. Find the air mass flow rate. Do this problem for two conditions: First consider the fuel flow to be negligible; then, include the effect of finite fuel flow.

1.9 A turbofan operates at 24,000 ft and moves at Mach 0.70. It has a bypass ratio of 3.00, and all of this air exhausts through the fan nozzle. The fuel flow is 3.77 lb_m/s. The exit areas of the fan and core are 1259 and 1302 in.^{2}, respectively. The exit pressures from the fan and core are 9.74 and 5.69 psia, respectively. The exhaust velocities from the fan and core are 1146 and 1489 ft/s, respectively. The thrust is 15,536 lb_f. Find the air mass flow rate in the core. Do this problem for two conditions: first consider the fuel flow to be negligible; then, include the effect of finite fuel flow.

1.10 A turboprop operates at 9000 ft and moves at 487 ft/s. The exit pressure is 10.5 psia, the exit velocity is 1137 ft/s, and the exit density is 0.000453 sl/ft^3. The fuel flow rate is 0.95 lb_m/s. The thrust from the propeller is 4243 lb_f, and the total developed thrust is 4883 lb_f. (a) What is the ingested air mass flow rate?

(b) If the exit nozzle is round, what is the nozzle exit diameter?

1.11 A turbojet operates at 20,000 ft and moves at Mach 0.75. The nozzle exit pressure is 19.70 psia, the exit velocity is 1965 ft/s, and the exit density is 0.000979 sl/ft^3. The developed thrust is 8995 lb_f, and the fuel flow rate is 3.16 lb_m/s. (a) Find the ingested air mass flow rate. (b) If the exit nozzle is round, what is the nozzle exit diameter?

1.12 A ramjet operates at 15200 m (50,000 ft) and moves at Mach 4. It ingests 40.8 kg/s (90 lb_m/s) of air and has a fuel flow of 1.70 kg/s (3.74 lb_m/s). The diameter of the exit is 0.805 m (31.7 in). The exit pressure is 11 kPa (1.60 psia), and the exit velocity is 1745 m/s (5726 ft/s). Find the developed thrust and *TSFC*.

1.13 A turbofan operates at 25,000 ft and moves at 815 ft/s. It ingests 1.2 times the amount of air into the fan than into the core, which all exits through the fan exhaust. The fuel-flow-to-core-airflow ratio is 0.0255. The exit densities of the fan and core are 0.00154 and 0.000578 sl/ft^3, respectively. The exit pressures from the fan and core are 10.07 and 10.26 psia, respectively. The developed thrust is 10,580 lb_f, and the exhaust velocities from the fan and core are 1147 and 1852 ft/s, respectively. (a) Find the ingested air mass flow rate for the core and *TSFC*. (b) What are the exit areas of the fan and core nozzles?

1.14 A turbofan operates at sea-level takeoff (SLTO). It has a bypass ratio of 8.00, and all of the secondary air exhausts through the fan nozzle. The fuel flow is 9.00 lb_m/s. The exit areas of the fan and core are 6610 and 1500 in.^{2}, respectively. The exit pressures from the fan and core are 14.69 and 14.69 psia, respectively. The exhaust velocities from the fan and core are 885 and 1255 ft/s, respectively. The core mass flow rate is 375 lb_m/s. Find the thrust and *TSFC*.

1.15 A power generation gas turbine ingests 54.4 kg/s (120 lb_m/s) of air. The compressor has an inlet enthalpy 288 kJ/kg and an exit enthalpy 457 kJ/kg. The turbine has an inlet enthalpy 1401 kJ/kg and an exit enthalpy 884 kJ/kg. The fuel has a heating value of 41,900 kJ/kg (18,000 BTU/lb_m) and the fuel flow rate is 1.274 kg/s (2.810 lb_m/s). Find the net output power, *SFC*, and net thermal efficiency.

1.16 A turbofan operates at 9144 m (30,000 ft) and moves at Mach 0.75. It has a bypass ratio of 4.00 and all of this air exhausts through the fan nozzle. The fuel flow is 2.360 kg/s (5.204 lb_m/s). The exit areas of the fan and core are 1.349 and 0.4858 m^2 (2092 and 753.4 in.^{2}). respectively. The exit pressures from the fan and core are 58.13 and 75.20 kPa (8.43 and 10.90 psia) respectively. The exhaust velocities from the fan and core are 324.2 and 587.1 mps (1064 and 1926 fps), respectively. The thrust is 123,600 N (27,800 lb_f). Find the air mass flow rate in the core.

1.17 A turboprop operates at 3048 m (10,000 ft) and moves at 145.1 mps (476 fps). The exit pressure is 119.5 kPa (17.34 psia), the exit velocity is 571.2 mps (1874 fps) and the exit density is 0.4946 kg/m^3 (0.000960 sl/ft^3). The fuel flow rate is 0.3423 kg/s (0.755 lb_m/s). The thrust from the propeller is 22,300 N (5011 lb_f) and the total developed thrust is 30,750 N (6913 lb_f). (a) What is the ingested air mass flow rate? (b) If the exit nozzle is round, what is the nozzle exit diameter?

1.18 A power generation gas turbine with regeneration and reheat ingests 54.4 kg/s (120 lb_m/s) of air. The compressor has an inlet enthalpy 288 kJ/kg and an exit enthalpy 457 kJ/kg. The primary turbine has an inlet enthalpy 1401 kJ/kg and an exit enthalpy 1235 kJ/kg. The free turbine has an inlet enthalpy 1401 kJ/kg and an exit enthalpy 1003 kJ/kg. The fuel has a heating value of 41,900 kJ/kg (18,000 BTU/lb_m), the primary fuel flow rate is 0.720 kg/s (1.588 lb_m/s), and the reheat fuel flow rate is 0.227 kg/s (0.500 lb_m/s). Find the net output power, *SFC*, and net thermal efficiency. Compare with the gas turbine without regeneration and reheat (Problem 1.15).

2 Components and Their Operating Characteristics: Ideal and Nonideal

2.1 Introduction

Chapter 1 identified the basic jet engine and power generation types and defined the important operating performance parameters of gas turbines. That chapter also briefly reviewed the fundamental thermodynamics of cycles and established that gas turbines consist of several important components. Our goal is to cover all basic jet engine and power generation configurations. Before proceeding to analyze complete gas turbines, the individual components must be analyzed. Both ideal and nonideal components are considered in this chapter. Ideal implies that there are no losses in the components. Nonideal includes such losses. In the following chapter components are assembled in a "cycle analysis" to make it possible to predict the overall gas turbine performance. Simple single- or two-term expressions are used to model the losses in each component in this chapter for simplicity. Each component is covered separately and in detail. Before proceeding with component analyzes, a brief review is provided for the thermodynamics of compressible fluid flow. More details of this review are covered in Appendix C. This review is provided for readers who have not had a comprehensive previous course in this subject and is considered to be optional.

2.1.1 Objectives

After completing this chapter, the reader should be able to do the following.

- Have reviewed thermodynamics and compressible flow (as needed)
- Understand the basic geometries of gas turbine components
- Understand the implications of ideal components
- Understand the thermodynamic principles of components
- Understand the thermodynamic differences between ideal and nonideal components
- Understand the definitions and application of the one- or two-term nonideal parameters of components
- Understand the idea and application of polytropic efficiency

2.2 Brief Summary of Thermodynamics and Compressible Flow (Optional)

One objective is to review the fundamental ideal and nonideal thermodynamic processes (see, for example, Keenan (1970) or Wark and Richards (1999)) and the gas dynamic processes (see, for example, Shapiro (1953), Zucrow and Hoffman (1976), and Anderson (1982)) and to explain the physical processes for each of the components. A brief summary of one-dimensional (1-D) compressible flow is first presented in this chapter if the reader needs to be refreshed. If the reader feels the need to better understand the fundamentals of compressible fluid flow, the reader is encouraged to further review in more detail the ideal gas equations and stagnation properties and other compressible flow processes in Appendix C, for they will be used extensively in the analysis in this and other chapters. Also note that an ideal gas assumption has been used in this and most other chapters.

The basic concepts of compressible flow are applied in all of the chapters herein. In general analyses of fundamental compressible steady state 1-D flow are used to predict total pressure losses and other characteristics in the different gas turbine components. Thus, to review very briefly some of the important characteristics, a summary of the thermodynamics and gas dynamics is presented here. Appendix C is referenced to include more and working details. Included in the processes are stagnation process, ideal gas properties, isentropic flow with area changes, frictional adiabatic with constant area (Fanno line) flow, flow with heat transfer and constant area (Rayleigh), normal and oblique shocks, flow with drag objects, flow with a drag object and heat transfer, mixing flows, and generalized 1-D flow. The latter allows the combination of two or more different fundamental phenomena. The flow is assumed to be 1-D, and the gas is assumed to be ideal. In Appendix C process diagrams and h–s diagrams are also presented.

2.2.1 Ideal Gas Equation and Stagnation Properties (from Appendix C.2)

First, one can relate the different pressures, temperatures, and densities by the ideal gas equation:

$$pv = p/\rho = \mathcal{R}T, \tag{2.2.1}$$

where p is the pressure, v is the specific volume, ρ is the density, T is the temperature, and $\mathcal{R}$ is the ideal gas constant. For air with a molecular weight of 28.97 $\text{lb}_\text{m}/\text{lb}_\text{mole}$ one finds that the ideal gas constant is 53.35 ft-$\text{lb}_\text{f}/(\text{lb}_\text{m}$-°R) (or 0.068559 BTU/$\text{lb}_\text{m}$-°R) (or 0.2871 kJ/kg-K). Furthermore, the specific heat ratio is

$$\gamma = \frac{c_\text{p}}{c_\text{v}}, \tag{2.2.2}$$

where c_v is the specific heat at constant volume and c_p is the specific heat at constant pressure.

The specific heats and ideal gas constant are related by:

$$c_\text{p} = \frac{\gamma\mathcal{R}}{\gamma - 1} \tag{2.2.3}$$

$$\mathcal{R} = c_\text{p} - c_\text{v} \tag{2.2.4}$$

$$\gamma = \frac{c_\text{p}}{c_\text{p} - \mathcal{R}}. \tag{2.2.5}$$

Next, consider a gas flowing with velocity u and with static pressure, enthalpy, and temperature p, h, and T, respectively. At some point the flow velocity is reduced to zero ($u = 0$). Applying the energy equation under steady state and adiabatic conditions yields:

$$h_\text{t} = h + \frac{1}{2}u^2. \tag{2.2.6}$$

In this equation h_t is called the total or stagnation enthalpy. If one considers an ideal gas:

$$dh = c_\text{p}dT, \tag{2.2.7}$$

so that

$$c_\text{p}T_\text{t} = c_\text{p}T + \frac{1}{2}u^2. \tag{2.2.8}$$

In this equation T_t is called the total or stagnation temperature. It is important to note that isentropic flow was not assumed – only adiabatic flow was assumed. Thus, as one reduces the flow velocity, the temperature rises. Next, the Mach number is defined by

$$M = \frac{u}{a}, \tag{2.2.9}$$

where a is the local speed of sound and for an ideal gas is given by:

$$a = \sqrt{\gamma\mathcal{R}T}. \tag{2.2.10}$$

Thus, by using this with Eq. (2.2.8) one finds

$$T_\text{t} = T\left[1 + \frac{\gamma - 1}{2}M^2\right]. \tag{2.2.11}$$

Again, one should remember that this is for an adiabatic process. Next, one can recall from thermodynamics that for an isentropic process (usually reversible (frictionless) and adiabatic):

$$\frac{p_\text{t}}{p} = \left[\frac{T_\text{t}}{T}\right]^{\frac{\gamma}{\gamma-1}}, \tag{2.2.12}$$

so that one can define the total or stagnation pressure, p_t:

$$p_t = p\left[1 + \frac{\gamma - 1}{2} M^2\right]^{\frac{\gamma}{\gamma-1}}. \tag{2.2.13}$$

This is for an isentropic process, which is more restrictive than the definition for the total temperature, which is only an adiabatic process.

Another important property is the entropy, s. A change in entropy is often used as a measure of a process irreversibility. Changes in entropy can be related to other property changes by the so-called *Tds* equations. These are:

$$Tds = du + p\,dv \tag{2.2.14}$$

and

$$Tds = dh - v\,dp \tag{2.2.15}$$

for the ideal gas

$$ds = c_v \frac{dT}{T} - \mathcal{R}\frac{dT}{v}. \tag{2.2.16}$$

and

$$ds = c_p \frac{dT}{T} - \mathcal{R}\frac{dT}{p}. \tag{2.2.17}$$

Thus, for the case of an ideal gas with constant specific heats the change in entropy for a process 1–2 is

$$\Delta s = c_v \ln\left(\frac{T_2}{T_1}\right) + \mathcal{R}\ln\left(\frac{\rho_1}{\rho_2}\right) \tag{2.2.18}$$

and

$$\Delta s = c_p \ln\left(\frac{T_2}{T_1}\right) - \mathcal{R}\ln\left(\frac{p_2}{p_1}\right). \tag{2.2.19}$$

2.2.2 Variable Specific Heats (from Appendix C.3)

For the ideal cycle analysis cases herein, the specific heats are considered constant throughout. However, for the nonideal cases one assumption to be relaxed is the constant specific heat assumption. Specific heats are approximated to be constant across each component; however, they are different for each component through the engine. The specific heats for a particular component are found based on the average temperature of that component. In applying the equations, the reader should evaluate the specific heats for each component independently.

An ideal gas approximation is still used in that $dh = c_p dT$. If the ideal gas assumption were to be eliminated, it would be necessary to use tabular data for the enthalpy as a function of both temperature and pressure as is done in the JANAF tables from, for example, Keenan, Chao, and Kaye (1983) or Chase (1998). Also, note that the gas is assumed to be pure air. With a combustor, this is obviously not the case. To perform a more accurate evaluation of the properties, one would need to find the enthalpy of the combustion products (which change as air goes through the engine) and then the mixed enthalpy of the air and other products. Once again, JANAF tables would be required for the combustion products. This is covered in detail in Chapter 9.

In Appendix C values of c_p and γ are tabulated for air between 360 °R and 5400 °R and between 1 and 100 atm. The values of c_p and γ are both very weak functions of pressure for temperatures below 3600 °R (2000 K). However, since c_p and γ both depend on temperature, the gas is considered to be ideal, not perfect. From this table one can find a least squares curve fit for c_p at one atmosphere, which is suitable for inclusion in computer analyses:

$$c_p = 0.2269807 e^{0.000097247\,T}, \tag{2.2.20}$$

where c_p is in BTU/lb$_m$-°R and T is in °R.

One should note that this is still an ideal gas approximation because one assumes $dh = c_p dT$. If one truly wanted to eliminate the ideal gas assumption, one would be required to use tabular data for the enthalpy as a function of both temperature and pressure, which is beyond the scope of this text.

Thus, for the nonideal case in this and other chapters, c_p and γ will be assumed to be constant across each component and evaluated at the average temperature of the component; this yields the best approximation for evaluating the change in properties between the end states. For the analyses herein the change in enthalpy is found by:

$$\Delta h = \bar{c}_p \Delta T, \tag{2.2.21}$$

where the value of $\bar{c}_p$ is evaluated at the average temperature of a component and the temperature difference is used to calculate the change in enthalpy.

2.2.3 Standard Atmosphere (from Appendix B)

Lastly, in this and following chapters, standard conditions will be used occasionally. Appendix B contains information on a standard atmosphere at different altitudes. Charts and figures are presented there. Also Atmosphere.exe can be used to find standard conditions at different altitudes. For reference, standard conditions at sea level are $p_{stp} = 14.69$ psia or 101.33 kPa and $T_{stp} = 518.7$ °R or 288.2 K.

2.2.4 Isentropic Flow with Area Change (from Appendix C.4)

The flow cross-sectional area can change axially in a number of components and ideally the flow is without friction or heat transfer, see Figure C.2. The nozzle, diffuser, inlet, and exhaust are the most obvious components in which the area changes. As the area changes, so do the Mach number, velocity, and other properties. Uniform flow enters a component with characteristics and properties at 1 (e.g., p_1, T_1, V_1, M_1, etc.) and leaves uniformly with all properties at 2 (e.g., p_2, T_2, V_2, M_2, etc.). For the isentropic process, the continuity and energy equations are used.

The flow is analyzed with a control volume approach in Appendix C.4. For an ideal gas the energy equation, continuity equation, and isentropic relationships of the properties are used. In Appendix C.4, 11 equations are summarized for the flow conditions. Sixteen variables are present, so five must be specified. For example, if M_1, T_1, p_1, A_1, and A_2 are specified, one can find M_2. For subsonic internal flow the Mach number decreases as the area increases (the usual case for a diffuser or exhaust). For subsonic internal flow the Mach number increases as the area decreases (the usual case for an inlet). For supersonic flow, however, the Mach number increases due to an increase in area, for example in the diverging section of a supersonic nozzle.

To aid in the solution of these equations, abbreviated tables are presented in Appendix D for two values of γ. In these tables T/T_t, p/p_t, ρ/ρ_t, and A/A^* are presented versus the Mach number. The superscript * indicates reference conditions at the sonic condition. Also, the software Isentropic.exe can be used to generate larger tables or specific individual results.

Example 2.1 | Isentropic Flow

Flow enters a diffuser at Mach 0.70. The inlet area is 0.5 m^2 and the exit area is 0.7 m^2. What is the exit Mach number for isentropic flow if $\gamma = 1.40$?

Solution

For $M_1 = 0.70$ the ratio $\left.\dfrac{A}{A^*}\right]_1$ is 1.0944 from tables (or Isentropic.exe).

Therefore, one can find the area at the sonic condition, which is only a reference condition (which might not occur in the actual conditions):

$A_1{}^* = 0.50/1.0944 = 0.4569\ \text{m}^2$.

But since the flow is isentropic, $A_1{}^* = A_2{}^*$.

Therefore, $\left.\dfrac{A}{A^*}\right]_2 = \dfrac{0.70}{0.4569} = 1.532$.

So now interpolating tables (or using Isentropic.exe) for this ratio yields $M_2 = 0.419$.

2.2.5 Fanno Line Flow (from Appendix C.5)

Boundary layers and other viscous flows are the primary means by which total pressure losses occur within inlets, exhausts, diffusers, and ducts. Such flow is also important in burners and afterburners, due to viscous flow through small orifices or holes in a burner wall. This flow will be analyzed based on Fanno line flow, which is constant-area flow with friction but no heat addition or loss (adiabatic); see Figure C.3. Fanno line flow is not isentropic. The Fanno line process is a constant-area process, which does not necessarily exactly represent a diffuser, duct. or burner. However, the basic principles apply so the mechanics can be used to reasonably predict and understand the total pressure losses. For the Fanno line process the continuity and energy equations are used. Again, details can be found in Appendix C.5 along with an h–s diagram, Figure C.4.

The eleventh equation is included, which is obtained from the momentum equation and which provides an empirical mechanism to include the effects of friction:

$$\int 4f\rho\frac{1}{2}V^2 d\left[\frac{x}{D}\right] = p_2 - p_1 + \rho_1 V_1 (V_2 - V_1), \quad (2.2.22)$$

where the left-hand side is integrated along the channel, and where f is the Fanning friction factor, and D is the effective diameter of the flow path. The friction factor is a function of the Reynolds number and wall roughness.

The equations from Appendix C.5 generate a solution. For subsonic internal flow the Mach number always increases due to friction as the gas flows through a channel. For supersonic flow, however, the Mach number always decreases due to friction. If the duct is long enough the flow will choke ($M = 1$) at length, L^*. Thus, a total of 11 equations were derived and 17 variables are present. Thus, six variables must be specified for a solution. For example, if M_1, T_1, L, D, f, and p_{t1} are specified, one can find p_{t2}.

To aid in the solution of these equations, abbreviated tables of T/T^*, $p_t/p_t{}^*$, $4fL^*/D$, and other variables versus Mach number are presented Appendix E (or software Fanno.exe). The quantity $4fL^*/D$ is an integral effect derived from the first term in Eq. (2.2.22). Once again, the superscript * indicates conditions at the sonic condition.

Example 2.2 | Fanno Line Flow

Flow with an incoming Mach number of 0.40 and an inlet total pressure of 28 psia enters the bypass duct of a turbofan. The duct is 80 in long with a hydraulic diameter of 4 in. The duct has a Fanning friction factor of 0.004. What is the total pressure ratio (p_{t2}/p_{t1}) if the specific heat ratio is 1.40?

Solution

From the tables for a Mach number of 0.40, one can find the nondimensional length-to-diameter and pressure parameters from the table (or Fanno.exe):

$$\left.\frac{4fL^*}{D}\right]_1 = 2.3085$$

and

$$\frac{p_{t1}}{p_t^*} = 1.5901,$$

thus one can find the total pressure at the sonic condition, which is only a reference condition (which might not occur in the actual conditions):

$p_t^* = 28/1.5901 = 17.609\,\text{psia}.$

Next, from the duct geometry one can find the parameter:

$$\left.\frac{4fL}{D}\right]_{1-2} = \frac{4 \times 0.004 \times 80}{4} = 0.3200,$$

thus one can find the corresponding length-to-diameter parameter at the exit:

$$\left.\frac{4fL^*}{D}\right]_2 = \left.\frac{4fL^*}{D}\right]_1 - \left.\frac{4fL}{D}\right]_{1-2} = 2.3085 - 0.3200 = 1.9885.$$

So interpolating the tables or using equations (or using Fanno.exe), the exit Mach number can be found:

$M_2 = 0.4191,$

as well as the ratio:

$$\frac{p_{t2}}{p_t^*} = 1.5316.$$

Thus, due to friction the Mach number increased about 5 percent.

The exit total pressure can be found:

$$p_{t2} = \frac{p_{t2}}{p_t^*} \times p_t^* = 1.5316 \times 17.609 = 26.97\,\text{psia},$$

and the pressure ratio for the duct is

$$\frac{p_{t2}}{p_{t1}} = \frac{26.97}{28} = 0.963.$$

2.2.6 Rayleigh Line Flow (from Appendix C.6)

In this subsection changes in properties and flow conditions due to heat transfer to the flow is discussed. This applies for example in burners where the temperature increases due to nonisentropic processes. This will be analyzed based on Rayleigh line flow, which is constant-area flow with heat transfer but no friction, see Figure C.5. The process is a constant area process. For the Rayleigh line the continuity and momentum equations are used and described in detail in Appendix C.6. The flow is analyzed with a control volume approach for an ideal gas. Thus, 10 independent equations are listed. Fourteen variables are in these equations. Therefore, if one specifies four variables, one can solve for the remaining 10. Typically, M_1, p_{t1}, T_{t1}, and T_{t2} are known. One should note that the energy equation was not utilized. These equations generate the h–s diagram in Figure C.6. For subsonic flow, as heat is added to the flow (the total temperature T_t increases) the Mach number increases until the flow chokes ($M = 1$). For supersonic flow, however, as heat is added, the Mach number decreases. One interesting phenomenon occurs as heat is added for the subsonic case. The static temperature, T, increases until the Mach number reaches $1/\sqrt{\gamma}$, at which point the temperature decreases. As noted above, the energy equation was not utilized. However, for the case of a burner, it can be used (and usually is) to relate the change in total temperature to fuel flow rate.

To aid in the solution of these equations, abbreviated tables of T/T^*, p_t/p_t^*, T_t/T_t^*, and other variables versus Mach number are presented in Appendix F (or software Rayleigh.exe). Once again, the superscript * indicates conditions at the sonic condition.

Example 2.3 Rayleigh Line Flow

Flow with an incoming Mach number of 0.25 and an inlet total pressure of 1800 kPa is heated from a total temperature of 700 K to 1400 K. What is the total pressure ratio if the specific heat ratio is 1.30?

Solution

From the Rayleigh line tables (or Rayleigh.exe) for a Mach number of 0.25 the temperature and pressure ratios can be found:

$$\left.\frac{T_t}{T_t^*}\right]_1 = 0.2482$$

$$\left.\frac{p_t}{p_t^*}\right]_1 = 1.2088.$$

Therefore, the total pressure and temperature at the sonic condition, which is not present but simply used as a reference, are:

$$p_t^* = \frac{1800}{1.2088} = 1489\text{ kPa}$$

$$T_t^* = \frac{700}{0.2482} = 2820\text{ K}.$$

Thus, the exit temperature ratio is

$$\left.\frac{T_t}{T_t^*}\right]_2 = \frac{1400}{2820} = 0.4965.$$

From the tables (or Rayleigh.exe) for this total temperature ratio:

$M_2 = 0.3887.$

And for this Mach number

$$\left.\frac{p_t}{p_t^*}\right]_2 = 1.1561;$$

thus, the total pressure at the exit is

$p_{t2} = 1.1561 \times 1489 = 1721\,\text{kPa}.$

Finally, the total pressure ratio due to heat addition (Rayleigh line flow) is

$$\frac{p_{t2}}{p_{t1}} = \frac{1721}{1800} = 0.956.$$

2.2.7 Normal Shocks (from Appendix C.7)

Shocks are the primary means by which total pressure losses occur outside of an inlet or in a nozzle. Normal shocks are a one-dimensional flow and lead to large total pressure losses. This flow is analyzed in Appendix C.7 in detail based on flow with no friction and no heat addition or loss, and a diagram is presented in Figure C.7. The normal shock process is also a constant-area process. For the normal shock process the continuity, linear momentum, and energy equations are used. Thus, the process is, in fact, a superposition of Fanno line and Rayleigh line processes as shown on the h–s diagram, Figure C.8. The incoming Mach number is always greater than one and the exiting Mach number is always less than one for a normal shock.

By assembling the equations, 11 equations and 14 variables are found. Thus if 3 variables are specified the remaining 11 can be found. Typically for a normal shock in engine applications, the incoming Mach number, total pressure, and total temperature are known. Thus, the exiting Mach number and total conditions can be found. Furthermore, the equations are combined to form six useful closed-form expressions in Appendix C.7.

To aid in the solution of these equations, abbreviated tables of M_2, p_{t2}/p_{t1}, p_2/p_1, and other variables are presented in Appendix G as functions of M_1. Also, the software NormalShock.exe can be used to generate larger tables or specific individual results.

Example 2.4 | Normal Shock

A "standing" normal shock is outside of diffuser. The Mach number before the shock is 1.50. The total pressure and total temperature are 50 psia and 500 °R, respectively. Find the exiting Mach number, exiting conditions, and total pressure ratio for the shock if $\gamma = 1.40$.

Solution

For $M_1 = 1.50$, using the tables (or NormalShock.exe) the exiting Mach number is

$M_2 = 0.7011.$

Also, the total pressure ratio is 0.9298 so that the exiting total pressure is

$p_{t2} = 0.9298 \times 50 = 46.5\,\text{psia}.$

Furthermore, since the process is adiabatic,

$T_{t2} = T_{t1} = 500\,°\text{R}.$

2.2.8 Oblique Planar Shocks (from Appendix C.8)

Oblique shocks incur lower losses in total pressure than do normal shocks. In Appendix C.8 planar two-dimensional (2-D) shocks are analyzed in detail. Although the flow is truly 2-D, it has many characteristics of 1-D flow. Flow before the shock is uniformly parallel to the axis and behind the planar shock flow is uniformly parallel to a wedge, see Figure C.9. Flow before the shock is again supersonic. It is important to note that the exiting Mach number can be subsonic or supersonic for the case of oblique shocks. A wedge with an angle of δ turns the flow and forms the planar shock. A control volume is again used. The continuity equation and momentum equations are used for both flow vectors normal to and parallel to the shock. The energy and ideal gas relationships are also used. Two resulting equations are named the Rankine–Hugoniot equations and relate pressure and density ratios across shocks. A resulting parameter is σ, the shock angle relative to the incoming flow. It is important to note that if the turning angle becomes too large or if the Mach number becomes too small the shock "detaches" and forms a normal shock.

In Appendix C.8, five equations result, and there are seven independent variables: σ, δ, M_1, M_2, p_2/p_1, p_{t2}/p_{t1}, and ρ_2/ρ_1. Therefore, if two are specified, the remaining five can be determined. For example, for a given incoming Mach number (M_1) and a given turning angle (δ), one can find the shock angle, exit Mach number, total pressure ratio, and all of the other variables. These equations can be easily programmed in a commercial math solver. To aid in the theoretical solution a set of figures are presented in Chapter 8 for $\gamma = 1.40$. Also, the software Shock.exe can be used to analyze planar oblique and other shocks.

It is also important to note that the above discussion was for oblique planar shocks. Conical oblique shocks are often common for supersonic inlets and although the same principles apply the application is more complex. Oblique shocks are covered in more detail in Chapter 8.

Example 2.5 | Oblique Shock

Flow with a Mach number of 3.00 enters an inlet with a wedge angle of 10°. Find the exiting Mach number, shock angle, and total pressure ratio for $\gamma = 1.40$.

Solution

By using Figure 8.6 (or Shock.exe) the figures and software indicate that two solutions exist: weak and strong shocks. Choosing which will depend on the downstream conditions, which are undefined at this point but will be discussed in Chapter 8. So both conditions will be considered.

Weak:

$M_2 = 2.505$

$\sigma = 27.4°$

$p_{t2}/p_{t1} = 0.963.$

Strong:

$M_2 = 0.489$

$\sigma = 86.4°$

$p_{t2}/p_{t1} = 0.330.$

The maximum turning angle for this Mach number (3.00) is 34.1°. The two solutions are quite different. The Mach numbers are subsonic and supersonic respectively. And the total pressure ratio is also much different.

2.2.9 Flow with a Drag Object (from Appendix C.9)

When an obstruction is in the flow the total pressure is also reduced because of the friction from the added turbulence and separation. Many primary burners and most afterburners have such obstructions in the form of flameholders to slow the fluid velocity. Mixers include such objects to enhance the folding and mixing of the two streams, for example for a mixed-flow turbofan, to ideally produce a uniform exit flow. A diagram is presented in Figure C.10. Thus, a method of including such losses is developed in Appendix C9.

For this process the continuity, momentum, and energy equations are used with a control volume approach. The process is a constant area with adiabatic flow, and with uniform properties at the entrance and exit. An ideal gas is assumed. An important parameter is the drag coefficient,

$$C_d \equiv \frac{F_d}{\frac{1}{2}\rho_1 V_1^2 A_d}, \tag{2.2.23}$$

where A_d is the frontal area of the object.

The method has 16 variables and 11 equations including:

$$p_1 + \rho_1 V_1^2 - \frac{1}{2} C_d \rho_1 V_1^2 \frac{A_d}{A} = \rho_1 V_1 \left[\frac{\mathcal{R} T_2}{\sqrt{2c_p(T_{t1} - T_2)}} + \sqrt{2c_p(T_{t1} - T_2)} \right]. \tag{2.2.24}$$

Thus, if five variables are specified the problem can be solved. Typically, A_d/A, C_d, M_1, p_{t1}, and T_{t1} will be known for a flameholder. If the inlet conditions are known, Eq. (2.2.24) can be used to solve for T_2 iteratively, after which all other exit conditions can be found. General characteristics of the process are that for subsonic flow the Mach number will increase and the total pressure will decrease.

2.2.10 Flow with a Drag Object with Heat Transfer (from Appendix C.10)

When an obstruction is in the flow, the total pressure is reduced due to the friction from the added turbulence and separation. When heat is added to the flow, the total pressure is also reduced because of the irreversible process. Thus, a method of combining such losses is developed in Appendix C.10; see Figure C.11. For this process the continuity and momentum are used with a control volume approach. The process is a constant area, with uniform properties at the entrance and exit. An ideal gas is assumed. An important parameter is again the drag coefficient. The method has 16 variables and 10 equations, including:

$$p_1 + \rho_1 V_1^2 - \frac{1}{2} C_d \rho_1 V_1^2 \frac{A_d}{A} = \rho_1 V_1 \left[\frac{\mathcal{R} T_2}{\sqrt{2c_p(T_{t2} - T_2)}} + \sqrt{2c_p(T_{t2} - T_2)} \right]. \tag{2.2.25}$$

Thus, if six variables are specified the problem can be solved. Typically, A_d/A, C_d, M_1, p_{t1}, T_{t1}, and T_{t2} will be known. If the inlet conditions are known, Eq. (2.2.25) can be used to solve for T_2 iteratively, after which all other exit conditions can be found. General characteristics of the process are that for subsonic flow the Mach number will increase and the total pressure will decrease.

2.2.11 Mixing Process (from Appendix C.11)

Another process that results in total pressure losses is the highly irreversible mixing process. For this occurrence, two streams at different temperatures, total temperatures, Mach numbers, flow rates, velocities, total pressures, etc., are irreversibly mixed to form one uniform stream. It is important to recognize that the two exit static pressures at the exit must match for flow stability. If the two Mach numbers are the same or if the two Mach numbers are very low, one approximation often made is that the two total pressures are the same. Two streams enter with properties 1 and 2 and one uniform stream leaves with all properties at 3 (for example p_3, T_3, V_3, M_3, etc.); see Figure C.12. One must be able to find the exiting conditions based on the two inlet conditions. For this process the continuity, momentum, and energy equations are once again used with the control volume approach.

Thus, in Appendix C.11 equations with 27 unknowns are present. However, 19 independent equations are available, including:

$$(p_1 + \rho_1 V_1^2)A_1 + (p_2 + \rho_2 V_2^2)A_2 = \dot{m}_3 \left[\frac{\mathcal{R}T_3}{\sqrt{2c_p(T_{t3} - T_3)}} + \sqrt{2c_p(T_{t3} - T_3)} \right]. \tag{2.2.26}$$

Thus, if one specifies eight variables, one can find the remaining 19. For example, for a mixer $\dot{m}_1$, $\dot{m}_2$, T_{t1}, T_{t2}, p_{t1}, p_{t2}, M_1, and M_2 will typically be known. Thus, if one examines Eq. (2.2.26), one can see that the left-hand side is usually known. Also, one can easily find T_{t3} and $\dot{m}_3$, respectively, so that the only remaining unknown is T_3, which can be found iteratively. General characteristics of the process are that, for subsonic flow, the Mach number will increase and the total pressure will decrease due to the irreversible process.

2.2.12 Mixing Process and a Mixing Enhancer (Drag Object) (from Appendix C.12)

Another process that results in total pressure losses is the mixing enhancer coupled with the mixing process. Again, two streams (for example, one from the bypass duct and one from the turbine exit) at different temperatures, total temperatures, Mach numbers, flow rates, velocities, total pressures, and so on are irreversibly mixed to form one uniform stream behind the enhancer. A diagram of this is shown in Figure C.13, which includes the control volume. Two streams enter with properties 1 and 2, and one uniform stream leaves with all properties at 3 (for example p_3, T_3, V_3, M_3, etc.). One must be able to find the exiting conditions based on the two inlet conditions. For this process the continuity, momentum, and energy equations are once again used.

Thus, in Appendix C.12 equations 27 unknowns are present. However, 19 independent equations are available, including:

$$\dot{m}_1\left(\frac{\mathcal{R}T_1}{V_1} + V_1\right) + \dot{m}_2\left(\frac{\mathcal{R}T_2}{V_2} + V_2\right) - \frac{1}{2}C_d A_R(\dot{m}_1 V_1 + \dot{m}_2 V_2) = \dot{m}_3\left(\frac{\mathcal{R}T_3}{\sqrt{2c_p(T_{t3} - T_3)}} + \sqrt{2c_p(T_{t3} - T_3)}\right). \tag{2.2.27}$$

For a mixer, $\dot{m}_1$, $\dot{m}_2$, T_{t1}, T_{t2}, p_{t1}, p_{t2}, M_1, and M_2 will typically be known. Also, T_{t3} and $\dot{m}_3$ can easily be found. Therefore, one can see that the left-hand side is usually known. Thus, the only remaining unknown in Eq. (2.2.27) is T_3, which can be found iteratively. After T_3 is found, all other parameters can be found from Eq. (2.2.27), for example M_3. General characteristics of the process are that, for subsonic flow, the Mach number will increase and the total pressure will decrease due to the irreversible process.

2.2.13 Generalized One-Dimensional Compressible Flow (from Appendix C.13)

In the previous subsections, area changes, friction, heat transfer, and drag were treated separately. However, for example, in an engine component several effects can occur simultaneously. Thus, in this section an analysis will be reviewed whereby several effects are included. The more advanced technique that will be used is the so-called generalized one-dimensional (1-D) flow method covered in most gas-dynamics texts. It is a differential equation analysis, whereby influence coefficients are derived. These influence coefficients relate changes of one variable on another. In Appendix C.13 the general method is covered in more detail.

In Figure C.14 the generalized geometry is shown. Included in the differential analysis are area changes, changes in total temperature by heat transfer, friction, and drag objects. To derive the influence coefficients a differential control volume is used and differential forms of the continuity, x-momentum, energy, and second law of thermodynamics are conventionally applied to the control volume. Resulting differential equations and a matrix of the so-called "influence coefficients" are tabulated in Appendix C.13. Incremental changes in Mach number, velocity, sound speed, temperature, density, pressure, total pressure, and entropy are related to area change, total temperature change, friction, and drag. Herein, results of the method will be used to determine and integrate the governing differential equations for flows with three specific combined effects. The method is versatile and can be used to simultaneously include area changes, heat transfer, friction, plus a differential drag object.

2.2.14 Flow with Combined Area Changes, Heat Transfer, and Friction (from Appendix C.16)

An analysis will be summarized whereby area changes, heat transfer, and friction are all included. The governing differential equations for flows with combined area changes, heat transfer and friction are as follows. The two most important equations are those for the Mach number and total pressure and these can be found from Appendix C.16. Namely:

$$\frac{dM^2}{M^2} = -2\left[\frac{1+\frac{\gamma-1}{2}M^2}{1-M^2}\right]\frac{dA}{A} + (1+\gamma M^2)\left[\frac{1+\frac{\gamma-1}{2}M^2}{1-M^2}\right]\frac{dT_t}{T_t} + 4f\gamma M^2\left[\frac{1+\frac{\gamma-1}{2}M^2}{1-M^2}\right]\frac{dx}{D} \tag{2.2.28}$$

$$\frac{dp_t}{p_t} = -\frac{\gamma M^2}{2}\frac{dT_t}{T_t} - \frac{1}{2}4f\gamma M^2\frac{dx}{D}. \tag{2.2.29}$$

In general, these can be numerically and simultaneously integrated between two end states to obtain the drop in total pressure due to both friction and heat transfer (an analytical integration is not possible). Namely, between any two states 1 and 2:

$$M_2^2 - M_1^2 = -2\int_{A_1}^{A_2} M^2\left[\frac{1+\frac{\gamma-1}{2}M^2}{1-M^2}\right]\frac{dA}{A} + \int_{T_{t1}}^{T_{t2}}(1+\gamma M^2)M^2\left[\frac{1+\frac{\gamma-1}{2}M^2}{1-M^2}\right]\frac{dT_t}{T_t} + 4f\gamma\int_0^L M^4\left[\frac{1+\frac{\gamma-1}{2}M^2}{1-M^2}\right]\frac{dx}{D} \tag{2.2.30}$$

$$p_{t2} - p_{t1} = -\frac{1}{2}\gamma\int_{T_{t1}}^{T_{t2}} p_t M^2\frac{dT_t}{T_t} - 2f\gamma\int_0^L p_t M^2\frac{dx}{D}. \tag{2.2.31}$$

As noted, these can be numerically integrated by small but finite steps. Such a numerical integration is included in the software General1D.exe.

2.2.15 Summary

In the previous subsections, important fundamental gas dynamics were reviewed for ideal gases, isentropic flow, Fanno line flow, Rayleigh line flow, shocks, and more general combined processes. Understanding these fundamentals is imperative before proceeding. Having reviewed a few of the important concepts of gas dynamics and using these fundamentals, we now proceed to discuss the operating characteristics of ideal and nonideal components followed by the integration of components to form the cycle analyses of the different engines.

2.3 Components: Ideal and Nonideal

This section covers each component separately as modules with the idea that a group of components will be analytically assembled to analyze a particular engine in the next chapter. In Figure 2.1, a sectional view of the Pratt & Whitney F100 is shown, illustrating most of the components covered in this section. Similarly, in Figure 2.2 a sectional view of the Pratt & Whitney FT8 power generation unit is shown, illustrating most of the components covered in this section. Enlargements of the components from these particular gas turbines are shown throughout this section. A few other gas turbine components are also shown and discussed. The basic physical processes of the components are discussed, and enthalpy–entropy (h–s) and pressure–volume (p–v) diagrams are used to reinforce the thermodynamic fundamentals.

A general propulsion engine is presented in Figure 2.3. No engine with this geometry exists in the current market because of the many features it incorporates. As will be seen, however, all engines are

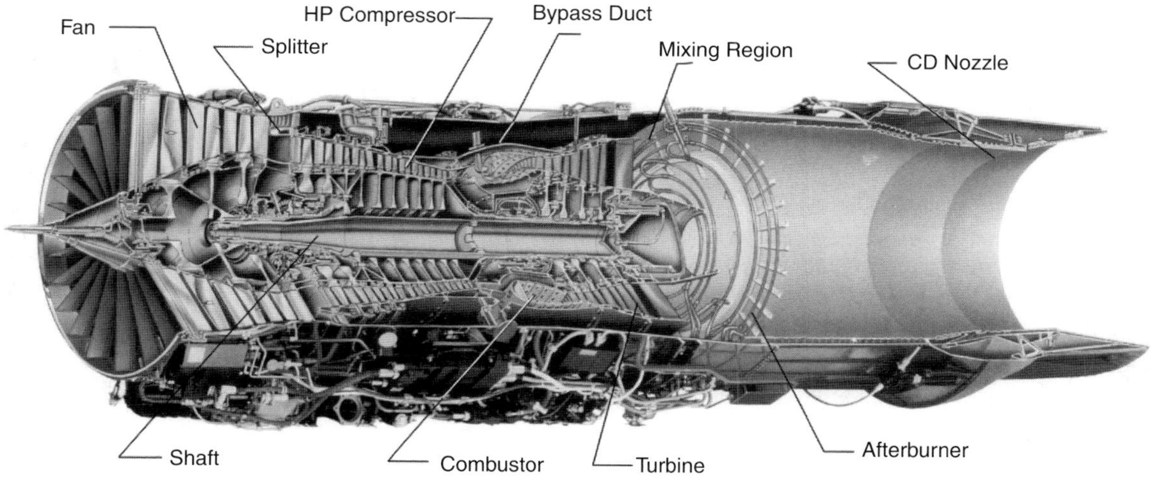

Figure 2.1 Sectional view of Pratt & Whitney F100 (© 2022 Raytheon Technologies Corporation, Pratt & Whitney division. Reproduced with permission).

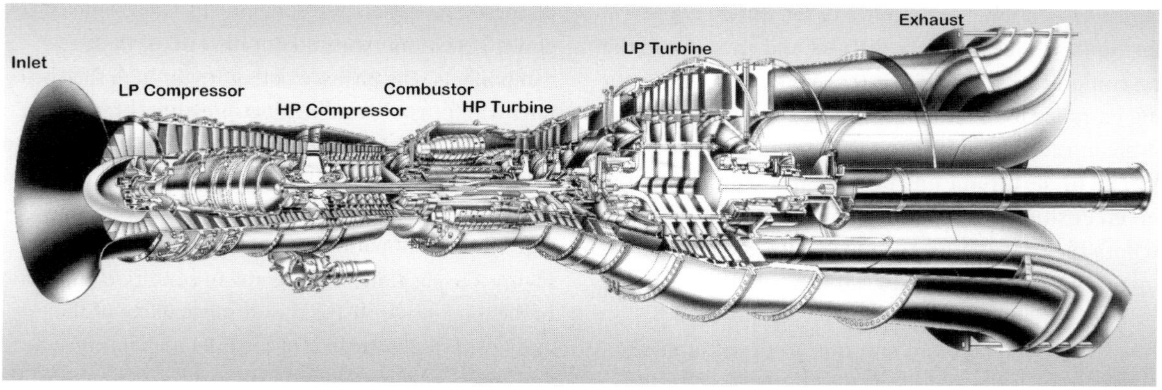

Figure 2.2 Sectional view of Pratt & Whitney FT8. (© 2022 Raytheon Technologies Corporation, Pratt & Whitney division. Reproduced with permission).

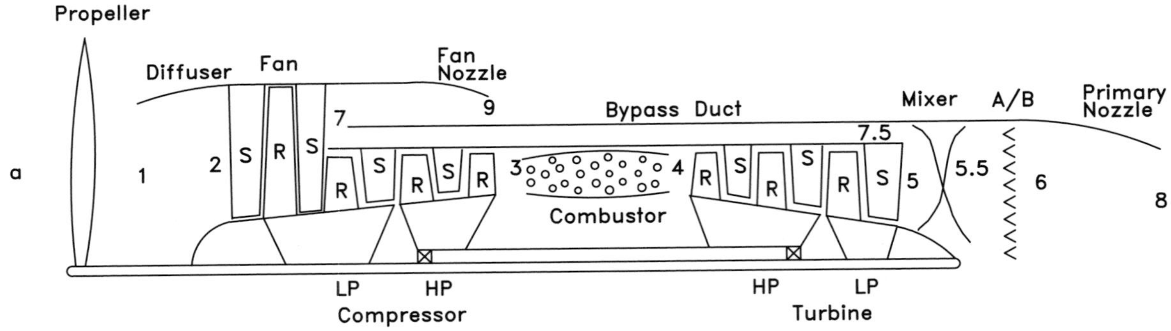

Figure 2.3 General jet engine description.

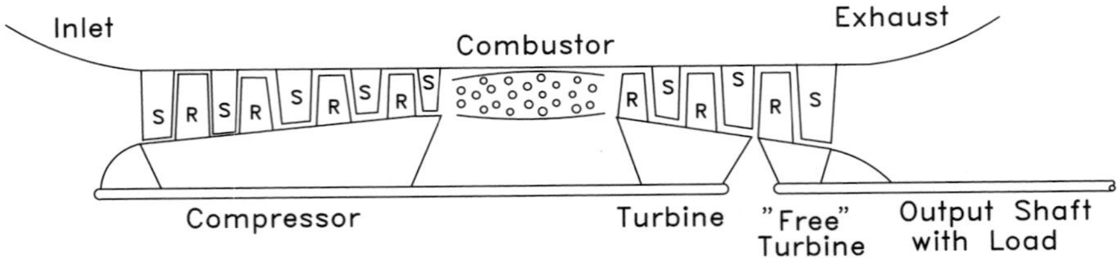

Figure 2.4 Description of a power generation gas turbine.

simplifications of this general engine. In the most general case, three individual flow paths exist. The primary path goes through the propeller, diffuser, fan and compressor, burner, turbine, mixer, afterburner, and primary nozzle. The second path includes the propeller, diffuser, fan, bypass duct, mixer, afterburner, and primary nozzle. In the third path, air flows through the propeller, diffuser, fan, and fan nozzle. In Figure 2.3, position a is the ambient condition, station 1 is the inlet to the diffuser, and position 2 is the exit of the diffuser and entrance to the fan compressor. Position 3 is the exit of the compressor and entrance to the combustor, and station 4 is the exit of the burner and entrance to the turbine. Station 5 is the exit of the turbine and primary flow entrance to the mixer. Position 7 is the exit of the fan and the entrance to both the fan nozzle and bypass duct. Station 9 is the fan nozzle exit. Position 7.5 is the bypass duct exit and the bypass flow entrance to the mixer. Position 5.5 is the exit of the mixer and entrance to the afterburner, and station 6 is the exit of the afterburner and entrance to the primary nozzle. Lastly, station 8 is the primary nozzle exit. For consistency between engine types, the nomenclature established here will be used for all engine types, each of which will be viewed as a simplification of this general engine.

In Figure 2.4 a simple schematic of a two-shaft power generation gas turbine is shown. The output shaft is called the free or power shaft. In Figure 2.5 a more-general power generation gas turbine that has performance-enhancing components is presented. Most power generation units do not have all of these components. The primary path goes through the "bell" inlet with possible cooling, low-pressure compressor, intercooler, high-pressure compressor, regenerator (cool path), primary burner, high-pressure turbine, reheat combustor, low-pressure turbine, free turbine, regenerator (hot path), and exhaust and stack. In Figure 2.5, position a is the ambient condition and inflow to the inlet, position 1 is the entrance plane of the inlet and position 2 is the exit of the inlet and entrance to the low-pressure compressor. Position 2.5a is the exit of the

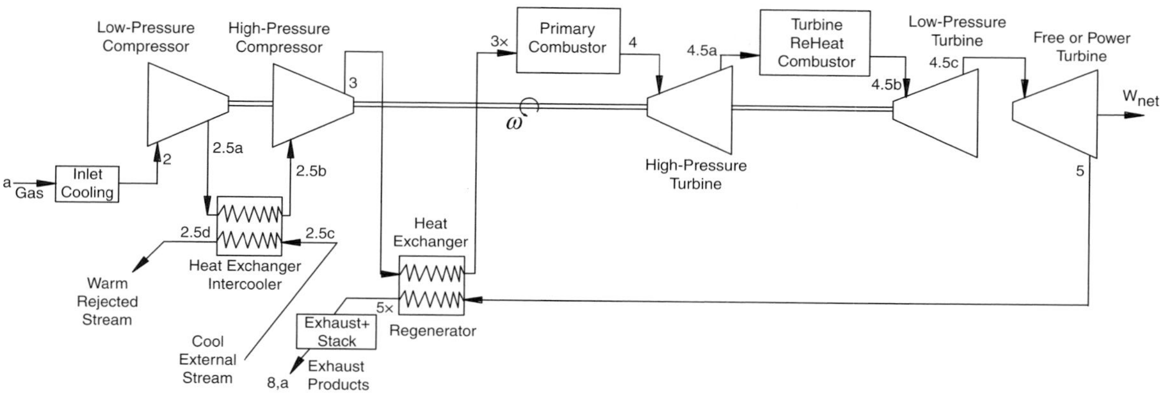

Figure 2.5 Diagram of an enhanced power generation gas turbine.

low-pressure compressor and entrance to the intercooler, and station 2.5b is the exit of the intercooler and entrance to the high-pressure compressor. Station 3 is the exit of the high-pressure compressor and entrance to the cool side of the regenerator. Station 3x is the exit of the regenerator and entrance to the primary combustor. Station 4 is the exit of the combustor and entrance to the high-pressure turbine. Station 4.5a is the exit of the high-pressure turbine and entrance to the reheat combustor. Station 4.5b is the exit of the reheat burner and entrance to the low-pressure turbine. Station 4.5c is the exit of the low-pressure turbine and entrance to the free turbine. Station 5 is the exit of the free turbine and entrance to the hot side of the regenerator. Station 5x is the exit of the regenerator and entrance to the exhaust and stack. Lastly, position 8 is the exit of the stack. For consistency between power generation types, the nomenclature established here will be used for types, each of which will be viewed as a simplification of this general unit.

The purpose of this section is to define the basic thermodynamic operating characteristics of each of the components in a general gas turbine shown in Figures 2.3, 2.4, and 2.5 and to introduce a few basic definitions. Next, the components will be assembled to compose the cycle analyses for different gas turbine configurations and yield the overall performance of a complete gas turbine. As will be seen, the ideal cycle analysis results in closed-form equations for the overall gas turbine characteristics. Also in this chapter, both ideal and nonideal components in which losses occur are used to evaluate the overall performance of an engine with various trend studies.

The adjective "ideal" implies that no losses occur in any of the components. It also suggests that a component is designed to operate at given conditions. An adiabatic process has no heat transfer, whereas an isentropic process is at constant entropy. In all cases in this book, an isentropic process is accomplished through an adiabatic and reversible process. Thus, the ideal components are assumed to have the following characteristics.

For the ideal case not only is the gas ideal, but the specific heats and the specific heat ratio are not functions of temperature and constant through a gas turbine. However, for the nonideal case these quantities are different for different components due to the temperature differences of the components. For the ideal case all of the components are assumed to operate without any losses. For the nonideal case these assumptions are relaxed, and the physical conditions that lead to losses are explained. Efficiency levels and losses are included for the different components so that more realistic predictions can be made for overall engine performance. Also, note that simple single- or two-term expressions are used to model the losses in each component in this chapter for simplicity. Each component is covered separately and in detail in this chapter. At this stage, these loss terms are specified a

External to inlet	Isentropic and adiabatic flow
Diffuser	Isentropic and adiabatic flow
Bell inlet	Isentropic and adiabatic flow
Compressor	Isentropic and adiabatic flow
Fan	Isentropic and adiabatic flow
Propeller	All propeller power generates thrust
Combustor	Constant static and total pressure
	Very low velocities
Turbine	Isentropic and adiabatic flow
Bypass duct	Isentropic and adiabatic flow
Bypass mixer	Isentropic and adiabatic flow
Afterburner	Constant static and total pressure
	Very low velocities
Nozzle	Isentropic and adiabatic flow
	Exit pressure matches atmospheric pressure
Exhaust stack	Isentropic and adiabatic flow
	Exit pressure matches atmospheric pressure
Overall	Thermally and calorically perfect gas
	Steady state
	Constant c_p, c_v, (and γ) throughout engine
	Negligible fuel flow
	No power loss by shaft

priori even though, in a real engine, the different component losses are dependent on the engine operating point and are thus dependent on each other. This advanced topic is the subject of Chapter 12, which addresses component matching. Even though most components operate with individually with relatively high efficiencies (upwards from 90 percent), when all the components are coupled the overall engine performance can be reduced drastically. However, in general, the performance trends are similar for both ideal and nonideal cases.

As stated in Section 2.2, simple one- and two-term parameters will be introduced for each of the engine components so that losses and nonideal effects can be included in the cycle analyses. Methods of determining or modeling these losses will not be covered; the losses and efficiencies are assumed to be known. More complex and refined analyses of the component losses are presented in each of the component chapters (Chapters 5–11). The determination of these parameters is the partial focus of these chapters. Herein, each component will be covered separately with the idea that a group of components will be assembled numerically to analyze a particular gas turbine.

2.3.1 Diffuser (Propulsion Gas Turbines)

Subsonic Diffuser

The first component to be considered is the subsonic diffuser or engine inlet. The purpose of the diffuser is to slow the fluid and to increase the pressure. Figure 2.6 shows the cutaway view of a diffuser of a large Rolls-Royce geared turbofan. The flow pattern in an ideal diffuser is depicted in Figure 2.7. As can be seen, the flow area increases and thus the flow velocity decreases from station 1 to 2. Recall that, for ideal cycle analyses, adiabatic and isentropic flows are assumed both outside (stations a to 1) and inside (stations 1 to 2) the diffuser. Ideally, the flow in the diffuser is uniform and the streamlines are smooth. Unfortunately, ideal flow is not the case and losses can occur both before the air reaches the diffuser and also once it enters the diffuser. Given the unfavorable pressure gradient, the boundary layer on the internal walls tends to grow quickly and can separate, as shown in Figure 2.7. This

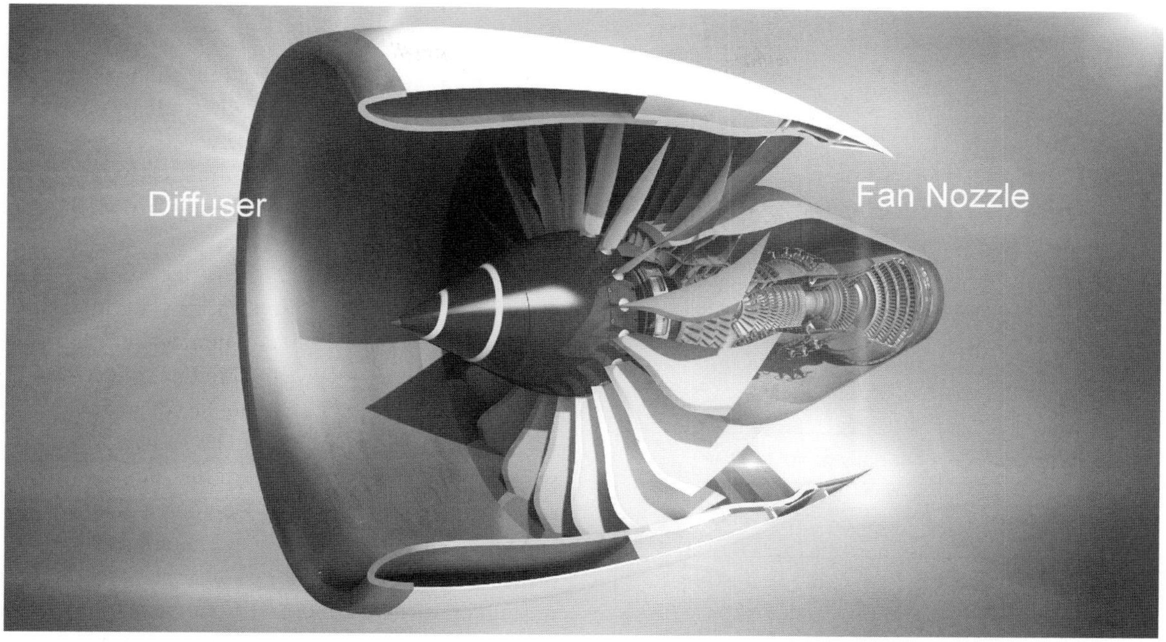

Figure 2.6 Inlet/diffuser and fan nozzle for Rolls-Royce geared turbofan ().

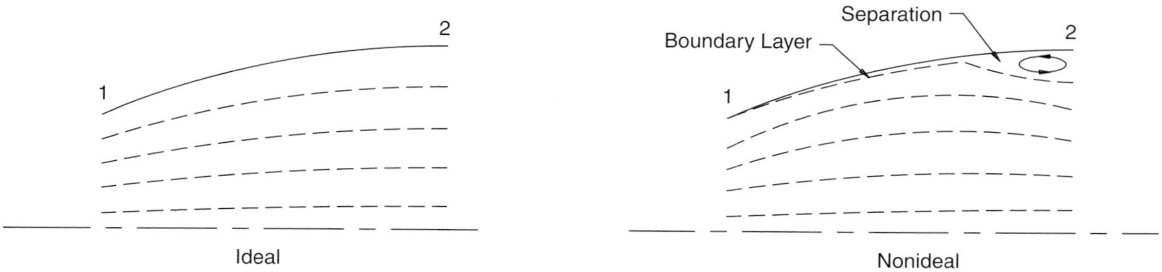

Figure 2.7 Streamlines and flow patterns in a diffuser.

can have three effects. First, owing to the wall shear, boundary layer turbulence, and internal freestream turbulence, the flow is nonisentropic because of viscous effects; thus, the total pressure is reduced. Second, if the flow separates, the flow area is decreased. Thus, the mainstream flow velocity remains high, and the static pressure does not increase as much as in the ideal case. Third, as the result of any separation that occurs, the flow can be highly nonuniform as it enters the compressor. As a result, compressor performance is compromised.

External streamlines for two flow conditions are shown in Figure 2.8. First, the streamline for high aircraft speed, or low ingested air flow rate, or both, is shown. Far away from the inlet, the encompassed area of the airflow is smaller than the diffuser inlet area. Thus, as it approaches the inlet, the flow decelerates and the pressure increases. Next, the streamline for low aircraft speed, high airflow rate, or both, is shown. This condition would occur for an aircraft on the runway or an engine on the test stand. Far away from the inlet, the encompassed area of the airflow is now larger than

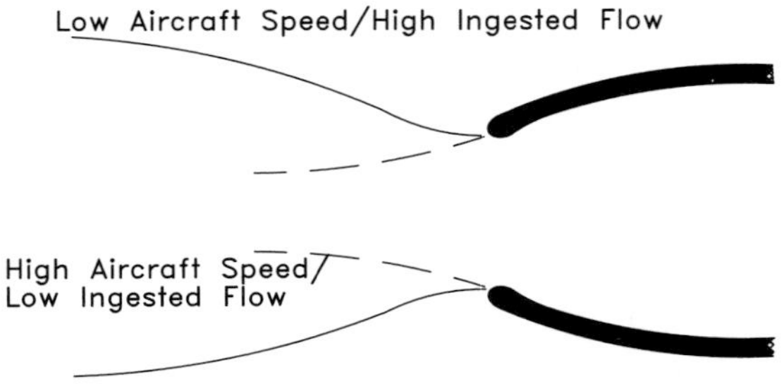

Figure 2.8 External flow pattern for two inlet conditions.

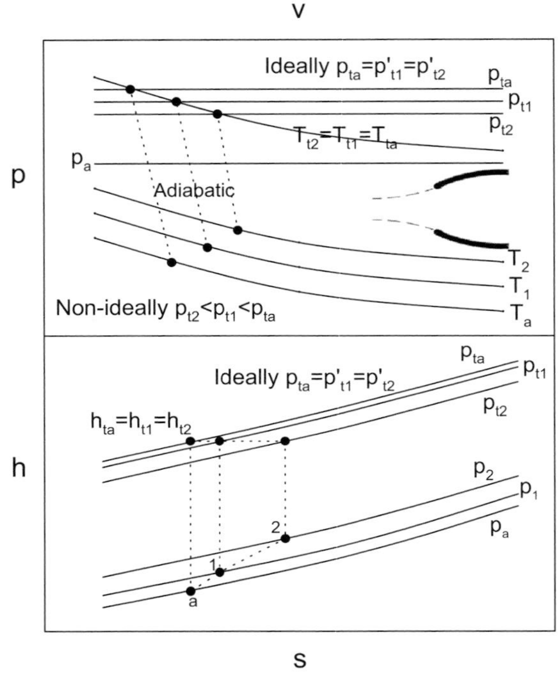

Figure 2.9 The h–s and p–v diagrams for a high-speed diffuser (flow decelerates between a and 1).

the diffuser inlet area. Thus, as the flow approaches the inlet, the flow accelerates and the pressure decreases. To allow for streamline curvature at the intersection of the inlet and for crossflows, the inlet is well rounded at the leading edge so that air enters with minimal separation for both cases.

The overall operation of the diffuser is best shown on h–s and p–v diagrams (Figures 2.9 and 2.10) for

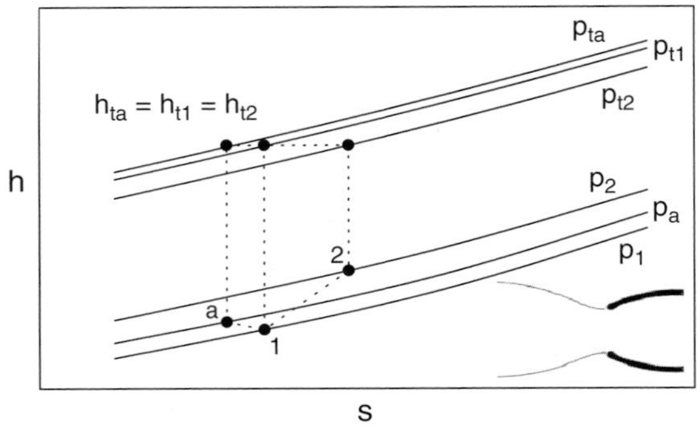

Figure 2.10 The h–s diagram for a low-speed diffuser (flow accelerates between a and 1).

two basic different operating conditions shown in Figure 2.8. For example, in Figure 2.9 the flow is shown to decelerate from a to 1, as the static pressure increases from a to 1 and also the enthalpy or temperature also increases. Stagnation conditions are also shown. All stagnation processes are by definition isentropic. Also, the flow velocity continues to decrease from 1 to 2 due to the increasing area. Thus, the pressure and temperature continue to increase. Both the ideal and nonideal processes are shown in this figure. The freestream air is at pressure p_a and can be decelerated isentropically to total enthalpy, h_t, and total pressure, p_{ta}. Ideally, this results in a total exit pressure $p'_{t2} = p_{ta}$. However, losses in the diffuser, the entropy increases from station a to station 2. Because the process is very nearly adiabatic (this assumption will not be relaxed), the total enthalpy (and total temperature) does not change from a to 2; thus, $h_{t2}, = h_{ta}$ and $T_{t2} = T_{ta}$. As a result of losses, however, the total pressure at station 2 is p_{t2}, which is less than $p'_{t2,}$ as shown in Figure 2.9. Similar conclusions are drawn when Figure 2.10 is considered for flow accelerating from a to 1.

Next, inasmuch as all processes are adiabatic (and not necessarily isentropic), for the flow outside and inside of the diffuser, the stagnation enthalpy is constant:

$$h_{t2} = h_{t1} = h_{ta}, \tag{2.3.1}$$

which for an ideal gas reduces to

$$T_{t2} = T_{t1} = T_{ta}. \tag{2.3.2}$$

Furthermore, outside the diffuser, for the definition of stagnation temperature (adiabatic process),

$$\frac{T_{ta}}{T_a} = \left[1 + \frac{\gamma - 1}{2} M_a^2\right], \tag{2.3.3}$$

where M_a is the freestream or aircraft speed Mach number. And for the definition of stagnation pressure (isentropic process),

$$\frac{T_{ta}}{T_a} = \left[\frac{p_{ta}}{p_a}\right]^{\frac{\gamma-1}{\gamma}}, \tag{2.3.4}$$

$$\frac{p_{ta}}{p_a} = \left[1 + \frac{\gamma - 1}{2} M_a^2\right]^{\frac{\gamma}{\gamma-1}}. \tag{2.3.5}$$

For an ideal diffuser the flow is also isentropic, and thus the stagnation pressure is constant:

$$p_{t2} = p_{t1} = p_{ta}. \tag{2.3.6}$$

Therefore, for an ideal diffuser,

$$\frac{T_{t2}}{T_a} = \left[1 + \frac{\gamma - 1}{2} M_a^2\right], \tag{2.3.7}$$

and

$$\frac{p_{t2}}{p_a} = \left[1 + \frac{\gamma - 1}{2} M_a^2\right]^{\frac{\gamma}{\gamma-1}}. \tag{2.3.8}$$

Thus, by knowing the freestream Mach number and freestream static properties, the stagnation (or total) temperature and pressure at the exit of and diffuser can be found for an ideal diffuser.

However, as discussed above, due to losses, the total pressure decreases both outside and inside the diffuser. The concept of total pressure recovery is introduced to quantify this loss. This is defined as

$$\pi_d = \frac{p_{t2}}{p_{ta}}. \tag{2.3.9}$$

Note that this recovery factor represents the loss from the freestream to the diffuser exit and can be expanded to

$$\pi_d = \left(\frac{p_{t2}}{p_{t1}}\right) \times \left(\frac{p_{t1}}{p_{ta}}\right). \tag{2.3.10}$$

The first term, or loss, is often defined to identify internal diffuser losses. This is the internal diffuser pressure recovery factor. Losses can also occur before the flow enters the diffuser, which is the second term, and which represents the loss from the far freestream to the diffuser inlet. For subsonic and shockless operation Eq. (2.3.10) will be redefined as

$$\pi_d = \pi_{d\text{-int}} \times \pi_{d\text{-ext}}. \tag{2.3.11}$$

For supersonic operation, additional losses can also occur due to shocks and will be discussed in the following section. Determining these losses will be deferred until Chapter 8. For subsonic flow the quantity $\pi_{d\text{-ext}}$ is approximately unity. As the freestream Mach number increases past unity, this loss can increase quickly if design precautions are not taken as discussed below. Consequently, by specifying either π_d or both $\pi_{d\text{-int}}$ and $\pi_{d\text{-ext}}$, the losses associated with the diffuser can be included in the analysis. For the remainder of this text π_d will be primarily used.

Supersonic Diffuser

The next component to be briefly discussed is the supersonic diffuser or engine inlet. For the cycle analysis the same modeling can be used as for the subsonic diffuser. Clearly the actual designs for the two diffuser types are much different. For ideal cycle analyses, adiabatic and isentropic flows are assumed both outside and inside the diffuser. Ideally, the flow in the diffuser is uniform and the streamlines are smooth. Ideal flow is not the case and losses can occur both before the air reaches the diffuser and also once it enters the diffuser. Furthermore, both oblique and normal shocks occur in supersonic diffusers as shown on Figures 1.15 and 1.16. These are generated by either 2-D wedges or conical spikes to slow the flow before entering the compressor or combustor. Such a cone is shown in Figure 8.8b for the Lockheed SR-71 Blackbird. Such shocks always incur significant losses in total pressure. Using a series of oblique shocks is better than a normal shock as total pressure losses are less as will be shown in Chapter 8. Also, owing to the wall shear, boundary layer turbulence, and internal freestream turbulence, the internal flow is nonisentropic because of viscous effects. Thus, the total pressure is reduced. Ideally, this results in a total exit pressure $p'_{t2} = p_{ta}$. However, because of losses in the diffuser, the entropy increases in the diffuser analysis.

Because the process is very nearly adiabatic (this assumption will not be relaxed for the supersonic diffuser), the total enthalpy (and total temperature) does not change. Next, inasmuch as all processes are adiabatic (and not necessarily isentropic), for the flow outside and inside of the diffuser the stagnation enthalpy is constant (as for the subsonic diffuser):

$$h_{t2} = h_{t1} = h_{ta}, \tag{2.3.1'}$$

As discussed previously, the total pressure decreases both outside and inside the diffuser due to losses. The concept of total pressure recovery is again used to quantify this loss. This is defined as

$$\pi_d = \frac{p_{t2}}{p_{ta}}, \tag{2.3.12}$$

which reduces to

$$\pi_d = \left(\frac{p_{t2}}{p_{t1}}\right) \times \left(\frac{p_{t1}}{p_{ta}}\right) = \pi_{d\text{-int}} \times \pi_{d\text{-ext}} = \pi_{d\text{-int}} \times \eta_r, \tag{2.3.13}$$

where η_r conventionally represents the loss from the far freestream to the diffuser inlet, and where $\pi_{d\text{-int}}$ is the internal diffuser recovery factor. The former is due to supersonic flow and ensuing shocks outside of, or in the plane of, the diffuser inlet. The latter is due to shocks within the diffuser and also other nonisentropic flows as discussed for subsonic operations. Consequently, by specifying either π_d or both $\pi_{d\text{-int}}$ and η_r, the losses associated with the diffuser can be included in the analysis. Because of shocks, both $\pi_{d\text{-int}}$ and especially η_r can be significantly less than unity for the supersonic cases. As noted above, for the remainder of this text π_d will be primarily used. It is important to note that although the modeling of the two diffusers are the same, in general the total pressure ratio, π_d, and external pressure or shock recovery ratio, η_r, are significantly less for supersonic diffusers than for subsonic diffusers due to the shocks. More analysis will be included in Chapter 8 for both supersonic and subsonic conditions.

2.3.2 "Bell" Inlet (Power Generation Gas Turbines)

The next component to be considered is the "bell" inlet. The purpose of the inlet is to draw air into the gas turbine. Figure 2.11 shows the inlet of the Pratt & Whiney FT8. The bell inlet for the Coberra 6000 is presented in Figure 2.47. As can be seen, the flow area decreases and thus the flow velocity increases. For ideal cycle analyses, adiabatic and isentropic flows are assumed both inside and outside the inlet. Ideally, the flow in the inlet is uniform and the streamlines are smooth. Unfortunately, as for a diffuser, ideal flow is not the case and losses can occur both before the air reaches the inlet and also once it enters the inlet. There is a boundary layer on the internal walls and turbulence in the atmosphere as shown on Figure 2.12. Owing to the wall shear, boundary layer turbulence, and freestream turbulence, the flow is nonisentropic because of viscous effects; thus, the total pressure is reduced. Also, the flow may not be uniform. As a result, compressor performance is compromised.

Figure 2.11 Bell inlet for Pratt & Whitney FT8 (from Figure 2.2).

The overall operation of the inlet is best shown on h–s and p–v diagrams (Figure 2.13). The flow is shown to accelerate from a to 1, as the static pressure decreases from a to 1 and the enthalpy or temperature also decreases. Stagnation conditions are also shown. All stagnation processes are by definition isentropic. Also, the flow velocity continues to increase from 1 to 2 due

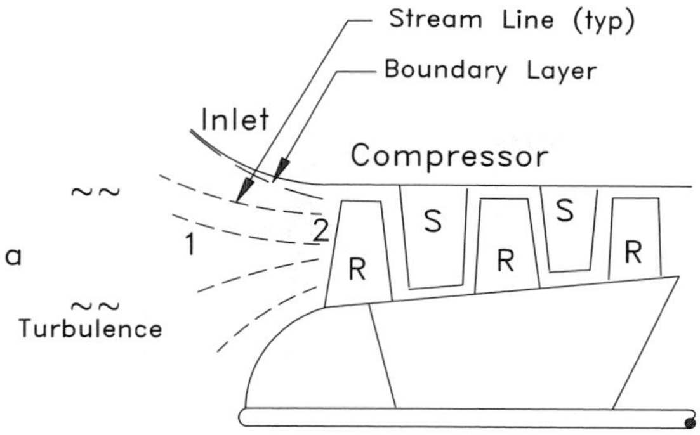

Figure 2.12 Bell inlet flow pattern.

to the decreasing area. Thus, the pressure and temperature continue to decrease. Both the ideal and nonideal processes are shown in this figure. The freestream air is at pressure p_a and can be decelerated isentropically to total enthalpy, h_{ta}, and total pressure, p_{ta}. However, for the inlet station a the static pressure, p_a, and total pressure, p_{ta}, are the same since the freestereeam velocity is zero. Ideally, this results in a total exit pressure $p'_{t2} = p_{ta}$. However, due to losses in the inlet, the entropy increases from station a to station 2. Because the process is very nearly adiabatic (this assumption will not be relaxed), the total enthalpy (and total temperature) does not change from a to 2; thus, $h_{t2} = h_{ta}$ and $T_{t2} = T_{ta}$. As a result of losses, however, the total pressure at station 2 is p_{t2}, which is less than $p'_{t2}(p_{ta})$ as shown in Figure 2.13.

Next, since all processes are adiabatic (and not necessarily isentropic), for the flow outside and inside of the inlet the stagnation enthalpy is constant:

$$h_{t2} = h_{t1} = h_{ta}, \tag{2.3.14}$$

which for an ideal gas reduces to

$$T_{t2} = T_{t1} = T_{ta}. \tag{2.3.15}$$

Furthermore, outside of the inlet, for the definition of stagnation temperature (adiabatic process),

$$\frac{T_{ta}}{T_a} = \left[1 + \frac{\gamma - 1}{2} M_a^2\right], \tag{2.3.16}$$

where M_a is the freestream Mach number. However, as just discussed the velocity is zero and as a result $M_a = 0$. So $T_{ta} = T_a$ and $h_{ta} = h_a$ as shown on the h–s diagram in Figure 2.13.

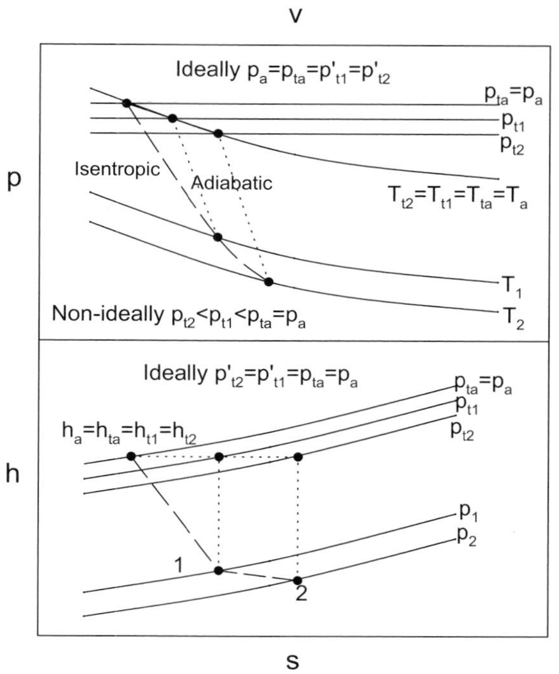

Figure 2.13 The h–s and p–v diagrams for a "bell" inlet.

For the definition of stagnation pressure (isentropic process),

$$\frac{T_{ta}}{T_a} = \left[\frac{p_{ta}}{p_a}\right]^{\frac{\gamma-1}{\gamma}}, \tag{2.3.17}$$

$$\frac{p_{ta}}{p_a} = \left[1 + \frac{\gamma - 1}{2} M_a^2\right]^{\frac{\gamma}{\gamma-1}}, \tag{2.3.18}$$

so that for $M_a = 0$, $p_{ta} = p_a$, also shown in Figure 2.13. For an ideal bell the flow is also isentropic, and thus the stagnation pressure is constant:

$$p_{t2} = p_{t1} = p_{ta}. \tag{2.3.19}$$

However, as discussed above, owing to losses, the total pressure decreases both outside and inside the inlet. The concept of total pressure recovery is introduced to quantify this loss. This is defined similar to that for the engine diffuser as

$$\pi_i = \frac{p_{t2}}{p_{ta}}. \tag{2.3.20}$$

Note that this recovery factor represents the loss from the freestream to the bell exit and can be expanded to

$$\pi_i = \left(\frac{p_{t2}}{p_{t1}}\right) \times \left(\frac{p_{t1}}{p_{ta}}\right). \tag{2.3.21}$$

The first term, or loss, is often defined to identify internal bell inlet losses. This is the inlet pressure recovery factor and is defined similar to that of the jet engine diffuser as

$$\pi_{i\text{-int}} = \frac{p_{t2}}{p_{t1}} \tag{2.3.22}$$

This expression represents losses only within the bell.

Losses can also occur before the flow enters the inlet. Determining these losses will be deferred until Chapter 8. However, this loss is the second term in Eq. (2.3.10) and is defined by

$$\pi_{i\text{-ext}} = \frac{p_{t1}}{p_{ta}}, \tag{2.3.23}$$

which represents the loss from the far freestream to the inlet plane. This quantity is usually unity or very close to unity. Thus, from Eqs. (2.3.21), (2.3.22), and (2.3.23) one finds that

$$\pi_i = \pi_{i\text{-int}} \times \pi_{i\text{-ext}}. \tag{2.3.24}$$

As a result, by specifying either π_i or both $\pi_{i\text{-int}}$ and $\pi_{i\text{-ext}}$ the losses associated with the bell inlet can be included in the analysis. For the remainder of this text only π_i will be used.

Also, on occasion inlet cooling is used to lower the total temperature at the inlet for power generation units and propulsion applications. This is often done in hot climates with low humidity where liquid water is sprayed into the inlet. The evaporation causes the total temperature to drop because heat is absorbed by the water to vaporize or change the phase of the liquid. This has the same fundamental effect as intercooling, as discussed in Chapter 1. Namely, by decreasing the inlet temperature less energy is needed by the compressor to develop the same exit pressure from the compressor. As will be seen later in this chapter, the pressure ratio of the compressor is a particularly important factor in the overall gas turbine performance.

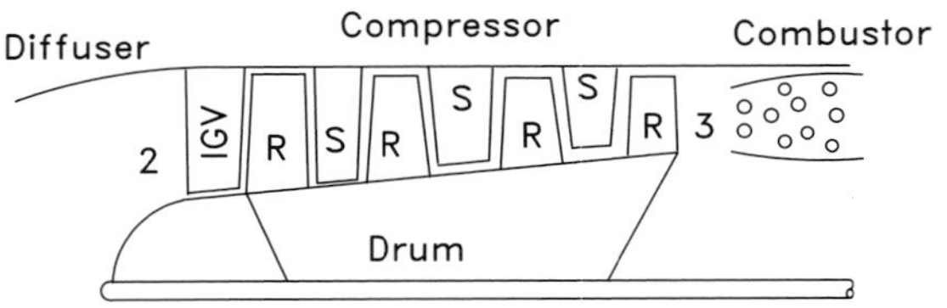

Figure 2.14 Schematic of compressor section.

2.3.3 Compressor

The next component to be considered is the compressor. A schematic is presented in Figure 2.14, and the Pratt & Whitney F100 compressor is shown in Figure 2.15. In these figures, sets of stationary airfoils are shown (called stator vanes). Directly downstream of these vanes are rotating or moving airfoils (called rotor blades), which are attached to the drum and shaft. The vanes are used to turn the flow in a compressor locally. The rotating blades turn the flow and add energy to the fluid, as will be discussed in Chapter 5. As is the case for any airfoil, air ideally flows across a vane or blade without incidence at the leading edge and is turned smoothly by the vane or blade. A compressor typically consists of 5–25 such sets (or rows) of rotor blades and stator vanes or stages. One should remember that, in general, as the compressor total pressure ratio increases, the performance of a gas turbine increases. However, as the compressor pressure ratio is increased, the number of compressor stages

Figure 2.15 F100 ten-stage compressor (from Figure 2.1).

increases, which adds to the cost and weight of the compressor. Furthermore, the burner needs a high inlet pressure (high compressor exit pressure) for a stable combustion.

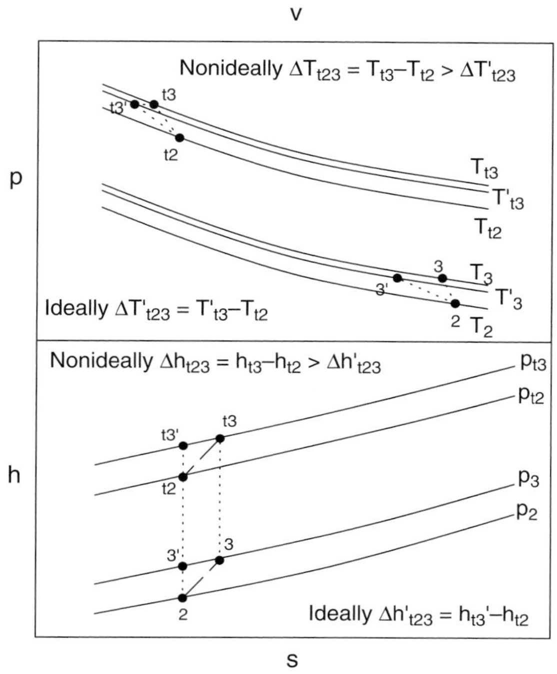

Figure 2.16 The h–s and p–v diagrams for a compressor.

The h–s and p–v diagrams for a compressor are presented in Figure 2.16. Ideally process 2 to 3 is isentropic, and both the enthalpy (and temperature) and pressure increase. Also shown are the total (or stagnation) pressures and total enthalpies (and temperatures) at conditions t2 and t3. Because energy is added to the flow, the total pressures and enthalpies also increase from stations 2 to 3. By definition, the stagnation processes are isentropic. Thus, as can be seen, ideally conditions 2, 3, t2, and t3 are all on the same isentropic process line. Next, one can define two quantities for the compressor, the total pressure and temperature ratios across the compressor (stations 2 to 3), as follows:

$$\pi_c = \frac{p_{t3}}{p_{t2}} \tag{2.3.25}$$

$$\tau_c = \frac{T_{t3}}{T_{t2}}. \tag{2.3.26}$$

As will be demonstrated later in this chapter, the total pressure ratio is a very important design parameter. Furthermore, if the flow is isentropic (ideal compressor), the total pressure and total temperature ratios can be ideally related as follows:

$$\pi_c = [\tau_c]^{\frac{\gamma}{\gamma-1}}. \tag{2.3.27}$$

Actually, isentropic flow does not occur in the compressor owing to boundary layers on the blades and case, turbulence, and other frictional losses, as shown in Figure 2.17. In fact, because the pressure increases as the

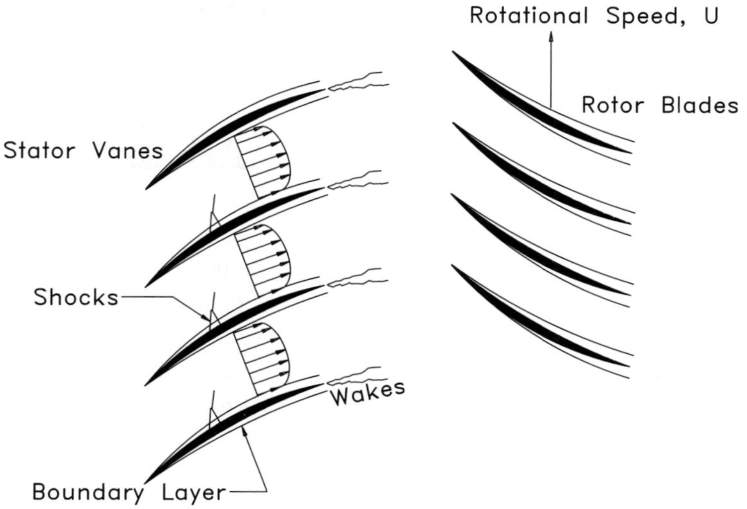

Figure 2.17 Top view of the rotor blades and stator vanes in a compressor stage.

air passes through the compressor, each row of blades acts like a diffuser (as will be discussed in Chapter 5). That is, an unfavorable pressure gradient is developed, which can lead to separation, viscous wakes, and reduced performance. Also, since the flow can be nonuniform both radially and circumferentially, it is not entirely incidence free to the leading edges of the rotor blades and stator vanes, which can also lead to separation on the blades, thus reducing compressor performance. Incidence increases, and so do the losses, if compressor operation deviates from the design point. The flow is also not steady within the compressor because of the wakes generated by every blade. For example, as a rotor blade passes behind every stator blade, the inlet velocity to the rotor blade fluctuates because of the wakes (called a jet-wake effect), which again can lead to flow incidence. Next, because the flow velocities can be high, regions of transonic or supersonic flow can occur. These can lead to partial choking in some of the passages or shocks. Shocks are highly nonisentropic. Also, because the blades have finite lengths, tip flows (and vortices) are generated. These secondary flows once again result in increased friction, which dissipates the energy and thus reduces performance. Lastly, as noted, any nonuniformities from the inlet will reduce compressor performance because of flow incidence to the blades. The details of these losses will be covered in Chapter 5.

A single parameter is introduced for the entire compressor to characterize all of these losses. This is the efficiency (η_c), sometimes referred to as the adiabatic efficiency, and is defined for the compressor as the ratio of the ideal power to actual power delivered to the fluid for a given total pressure rise. This is best illustrated by the h–s diagram (Figure 2.16). Shown on this diagram are the ideal and real processes. Both processes are operating between the same total pressures (p_{t2} and p_{t3}), which is an important consideration. The ideal case operates between total enthalpies h_{t2} and h'_{t3}, whereas the real case operates between total enthalpies h_{t2} and h_{t3}. As one can see, as a result of the losses and increase in entropy, the increase in total enthalpy for the non-ideal case is more than for the ideal case. That is, more power is required for the nonideal case than for the ideal case. Thus, h_{t3} is greater than h'_{t3}. It is very important to note that the efficiency is defined as the ratio of ideal power to the actual power for the same increase in total pressure. Thus, the efficiency is

$$\eta_c = \frac{\text{ideal power}}{\text{actual power}} = \frac{h'_{t3} - h_{t2}}{h_{t3} - h_{t2}}. \tag{2.3.28}$$

If the specific heats are constant across the compressor, this reduces to

$$\eta_c = \frac{T'_{t3} - T_{t2}}{T_{t3} - T_{t2}}. \tag{2.3.29}$$

Typically, modern compressors operate at design with peak efficiencies of 85 percent to 93 percent, although the efficiency is a function of the pressure ratio and other parameters. A compressor with a design efficiency of 80 percent would be considered poor. After great strides in the 1950s and 1960s to improve the peak compressor efficiencies, this number has only gradually improved over the past 15 years. Also, one should realize that the efficiency is a strong function of the operating conditions. At low or high flow rates, or at high or low rotational speeds, or both, the efficiency can be reduced significantly from the design value. This is discussed in detail in Chapter 5.

Combining Eqs. (2.3.29) and (2.3.27) yields

$$\begin{aligned}\pi_c \equiv \frac{p_{t3}}{p_{t2}} &= \left\{1 + \eta_c\left[\frac{T_{t3}}{T_{t2}} - 1\right]\right\}^{\frac{\gamma}{\gamma-1}} \\ &= \{1 + \eta_c[\tau_c - 1]\}^{\frac{\gamma}{\gamma-1}} = \{\tau'_c\}^{\frac{\gamma}{\gamma-1}},\end{aligned} \tag{2.3.30}$$

where $\tau'_c = \dfrac{T'_{t3}}{T_{t2}}$. Equation (2.3.30) thus defines the total pressure ratio of the compressor in terms of the total temperature ratio (which is related to the input work) and the efficiency of the compressor. If the efficiency is specified, this equation will allow the analysis to proceed. For the limiting case of an ideal compressor, η_c is unity and this equation reduces to Eq. (2.3.27). Note that, as for all of the components, γ will be determined based on the average temperature in the particular component of interest – in this case the compressor. Thus, the value of γ for the compressor will in general be different than that for the diffuser or any other component. Therefore, if the inlet conditions are known for a compressor and the total pressure ratio is specified, the exit conditions can be determined.

Polytropic Efficiency

A second and important method for quantifying the nonideal performance of a compressor is used. This method will not be used for the cycle analyses in this book. However, the reader should be aware of the method because several other texts, some manufacturers, and some regulatory commissions use it extensively. This method, which determines what is described as the polytropic efficiency, basically entails separating the dependence of the efficiency from the pressure ratio so that different compressors can be better compared. The advantage of using this approach is that the polytropic efficiency is approximately the same for all well-designed modern engines at the design point and is determined by the design of the single stages; this measurement is thus independent of the number of stages. One can therefore directly compare the polytropic efficiencies for compressors of different sizes and pressure ratios. The polytropic efficiency is often used in regulatory specifications. For example, the American Petroleum Institute (API) uses polytropic efficiency as the basis for a regulation to oversee the petroleum processing and other industries using compressor "trains" in the United States.

First, consider a single stage as shown in Figure 2.18. The inlet to the stage is station x and the exit is station y. For this single stage, the total pressure and total temperature ratios can be related by

$$\frac{p_{ty}}{p_{tx}} = \left\{1 + \eta_{ss}\left[\frac{T_{ty}}{T_{tx}} - 1\right]\right\}^{\frac{\gamma}{\gamma-1}}, \quad (2.3.31)$$

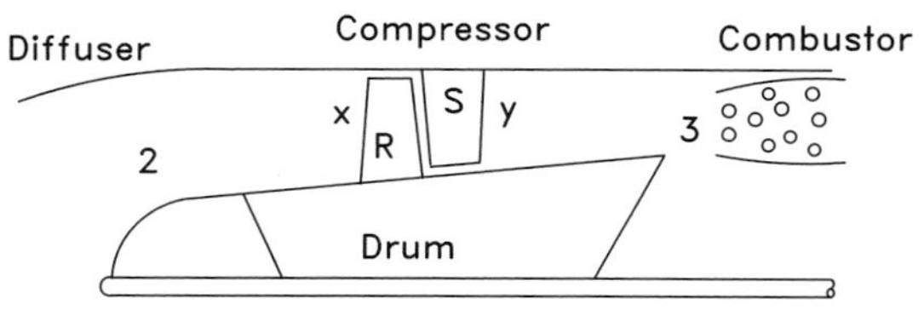

Figure 2.18 Single stage of a compressor.

where η_{ss} is the efficiency for the single stage. If one assumes that the entire compressor is composed of n identical stages and that all stages have the same single-stage efficiency and total pressure ratios, the following equation results for the total pressure ratio of the entire compressor:

$$\frac{p_{t3}}{p_{t2}} = \left[\frac{p_{ty}}{p_{tx}}\right]^n. \quad (2.3.32)$$

The total temperature rise for the compressor is described similarly as follows:

$$\frac{T_{t3}}{T_{t2}} = \left[\frac{T_{ty}}{T_{tx}}\right]^n. \quad (2.3.33)$$

Thus, using Eqs. (2.3.31)–(2.3.33) and solving for η_{ss}, one obtains

$$\eta_{ss} = \frac{\left[\frac{p_{t3}}{p_{t2}}\right]^{\frac{\gamma-1}{n\gamma}} - 1}{\left[\frac{T_{t3}}{T_{t2}}\right]^{\frac{1}{n}} - 1}. \quad (2.3.34)$$

For example, if a compressor has an overall efficiency of 0.88 and a total pressure ratio (p_{t3}/p_{t2}) of 25 with $\gamma = 1.40$ – and thus a total temperature ratio of 2.7142 – it is possible to determine the effect of the number of stages from Table 2.1. As can be seen, the most pronounced changes occur for values of n between 1 and 10. Lastly, one can now hypothetically allow the number of stages

Table 2.1 **Variation of single-stage efficiency with number of stages for $\eta_c = 0.88$, $\pi_c = 25$, and $\gamma = 1.40$.**

n	η_{ss}
1	0.88
2	0.9017
4	0.9117
6	0.9149
10	0.9174
20	0.9192
100	0.9207
∞	0.9211

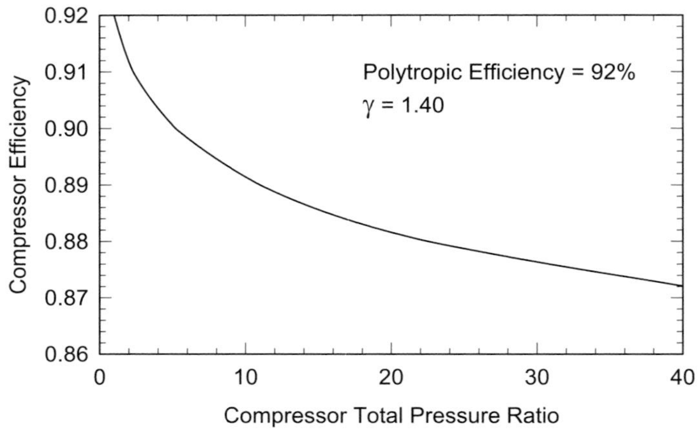

Figure 2.19 Compressor efficiency variation with pressure ratio.

to become infinite. This results in the definition of the so-called compressor polytropic or small-stage efficiency:

$$\eta_{pc} = \frac{\gamma - 1}{\gamma} \frac{\ln\left[\frac{p_{t3}}{p_{t2}}\right]}{\ln\left[\frac{T_{t3}}{T_{t2}}\right]}. \tag{2.3.35}$$

This equation can also be written as

$$\frac{p_{t3}}{p_{t2}} = \left[\frac{T_{t3}}{T_{t3}}\right]^{\eta_{pc}\frac{\gamma}{\gamma-1}}. \tag{2.3.36}$$

Using Eqs. (2.3.30) and (2.3.36), one can relate the overall efficiency and polytropic efficiency as follows:

$$\eta_c = \frac{[\pi_c]^{\frac{\gamma-1}{\gamma}} - 1}{[\pi_c]^{\frac{\gamma-1}{\gamma\eta_{pc}}} - 1}. \tag{2.3.37}$$

As noted, polytropic efficiency is a second and equally valid method for including compressor losses. To include the losses for a compressor, one must either specify the compressor efficiency (η_c) or the small-stage compressor efficiency (η_{pc}). If η_c is known it is possible to find η_{pc}, and vice versa. The value of η_c is always less than that of η_{pc}. As stated earlier in this section, all modern engines demonstrate approximately the same η_{pc} at the peak design condition for a well-designed compressor, and this is about 90 percent to 92 percent. The compressor efficiency is thus a function of the total pressure ratio of the compressor.

For example, the compressor efficiency is presented in Figure 2.19 for the total pressure ratio range of 1 to 40 for a polytropic efficiency of 92 percent and $\gamma = 1.40$. As can be seen, the compressor efficiency decreases with increasing pressure ratio, that is, the number of stages increases. The polytropic efficiency is an alternative method and is confused with the first approach by some because of the definition and the ensuing derivation. If one remembers that η_c and η_{pc} are dependent and that only one of these parameters needs to be specified, however, the method is straightforward. For the remainder of this book, the first approach (η_c) will be used.

2.3.4 Fan (Propulsion Gas Turbines)

For a turbofan, a portion of the gas passes through the core of the engine (compressor, combustor, and turbine, see Figure 2.3), and a portion of the gas is "bypassed," which passes through the fan and then through the bypass duct and/or fan nozzle. Such a fan section is shown in Figures 2.20 and 2.21. The h–s and p–v diagrams for a fan are identical in nature to those for a compressor except for station numbers. The only differences are in the relative increases in pressure and temperature for a compressor and fan; they are much higher for a compressor. Ideally, process 2 to 7 is isentropic, and both the temperature and pressure increase. Also, the stagnation pressures and total

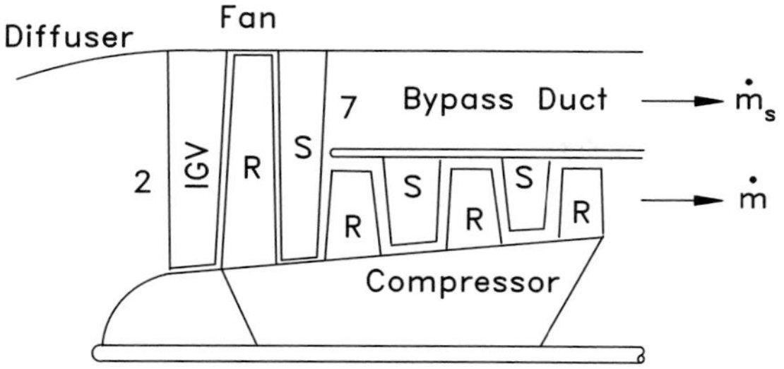

Figure 2.20 Schematic of a fan.

Figure 2.21 Three-stage F100 fan (from Figure 2.1).

temperatures are conditions t2 and t7, and these also increase from stations 2 to 7. Because the stagnation processes are isentropic, conditions 2, 7, t2, and t7 all are on the same isentropic process line.

The core gas flow will be referred to as $\dot{m}$, and the bypassed flow, or secondary flow, will be termed $\dot{m}_s$. Thus, the bypass ratio is the ratio of airflow through only the fan to that of the core and is defined by

$$\alpha = \frac{\dot{m}_s}{\dot{m}}. \tag{2.3.38}$$

This ratio is a very important design parameter. For modern aircraft, typical bypass ratios range from 0.5 to 10. The next quantities to be defined are the fan total pressure and total temperature ratios (stations 2 to 7). These are given by

$$\pi_f = \frac{p_{t7}}{p_{t2}} \tag{2.3.39}$$

and

$$\tau_f = \frac{T_{t7}}{T_{t2}}, \tag{2.3.40}$$

which for the ideal case can be related by the isentropic relationship

$$\pi_f = [\tau_f]^{\frac{\gamma}{\gamma-1}}. \tag{2.3.41}$$

It is very important to note that the total pressure and temperature ratios for the compressor (π_c and τ_c) include the property rises in both the compressor and fan. That is, the fan is also acting as the first stage(s) of the low-pressure compressor.

The physical losses and thermodynamic fundamentals of a fan are essentially the same as those for a compressor with a low total pressure ratio and exceptionally long blade lengths. As a result, the methodology for incorporating the losses into the fan analysis is fundamentally the same as that for the compressor. The h–s diagram is very similar except for the station numbers. With reference to Figure 2.20, the stations before and after the fan are 2 and 7, respectively. Thus, one obtains

$$\pi_f \equiv \frac{p_{t7}}{p_{t2}} = \left\{1 + \eta_f\left[\frac{T_{t7}}{T_{t2}} - 1\right]\right\}^{\frac{\gamma}{\gamma-1}}$$
$$= \{1 + \eta_f[\tau_f - 1]\}^{\frac{\gamma}{\gamma-1}} = \{\tau'_f\}^{\frac{\gamma}{\gamma-1}}. \tag{2.3.42}$$

This equation is used to characterize the fan, where η_f is the fan efficiency. The value of γ is determined by the average of the temperatures in the fan and will in general be different than the value for the compressor or any other component. As was true for the compressor, a polytropic efficiency for the fan can be defined. Although the fundamental thermodynamics for a fan are similar to those for a compressor owing to the large blade lengths of the fan, the blade design is considerably different, as is discussed in Chapter 5.

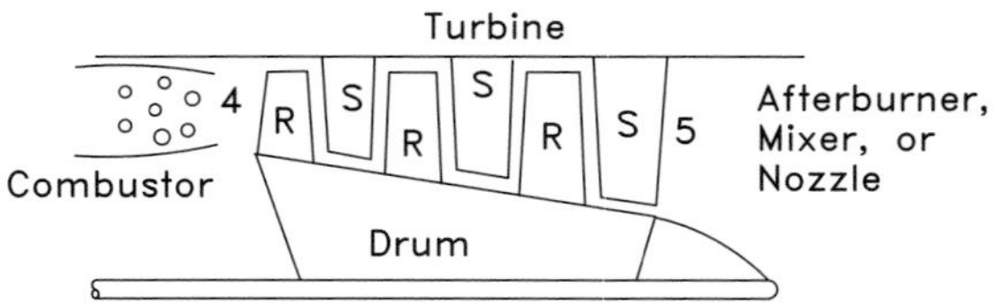

Figure 2.22 Schematic of a turbine.

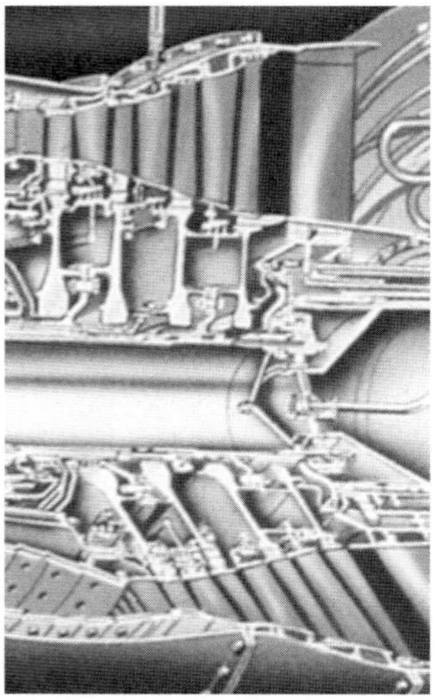

Figure 2.23 F100 four-stage turbine (from Figure 2.1).

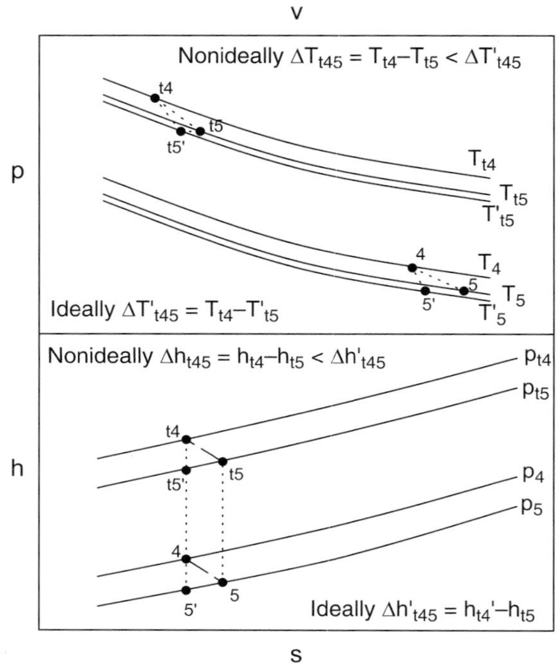

Figure 2.24 The *h*–*s* and *p*–*v* diagrams for a turbine.

2.3.5 Turbine

The next component to be considered is the turbine. A schematic is shown in Figure 2.22, and the F100 turbine is shown in Figure 2.23. Like a compressor, a turbine consists of a series airfoils: stator vanes (stationary) and rotating rotor blades. For this component, power is derived from the fluid with the rotor blades. The number of turbine stages is considerably less than for a compressor. The h–s and p–v diagrams for a turbine are shown in Figure 2.24. As is true for an ideal compressor, processes 4 to 5 are isentropic. For the turbine, however, the pressure and temperature drop as do the total pressure and temperature because energy is removed from the fluid. Once again, the static and total conditions at stations 4 and 5 all occur on the same isentropic process line. Two quantities can be defined for the turbine that are similar to those of a compressor: the total pressure and temperature ratios across the turbine (stations 4 to 5):

$$\pi_t = \frac{p_{t5}}{p_{t4}}. \tag{2.3.43}$$

Once again, because the flow is ideally isentropic in an ideal turbine one can ideally relate the total pressure and total temperature ratios:

$$\pi_t = [\tau_t]^{\frac{\gamma}{\gamma-1}}. \tag{2.3.44}$$

Ideally, a turbine (Figures 2.23 and 2.24) operates isentropically. As is true for a compressor, several losses contribute to the overall loss in a turbine. Among these losses are viscous effects due to the boundary layers on the vanes and blades, freestream turbulence, secondary flows, and incidence angles to the vanes and blades (because of inlet conditions, jet-wakes, and 3-D flows). Because the pressure drops through a turbine, it is more prone to partial choking (and ensuing shocks) than a compressor. Furthermore, because a turbine is directly downstream of the very hot combustor, cool air is often bled through and around the vanes and blades to avoid failures (called blade cooling; see Chapter 7). This injected cool air lowers the enthalpy of the fluid. Both the ideal and nonideal processes are shown in Figure 2.24. As is the case for the compressor, both processes are shown to operate between the same two total pressures. The ideal case operates between total enthalpies h_{t4} and h'_{t5}, whereas the nonideal case operates between h_{t4} and h_{t5}. Because of the losses and increase in entropy, the drop in enthalpy for the nonideal case is less than for the ideal case ($h'_{t5} < h_{t5}$). That is, less power is derived from the fluid for the nonideal case as compared with the ideal case. For the turbine, the efficiency is defined as the ratio of actual power to ideal power for the same total pressure ratio. This results in

$$\eta_t = \frac{\text{actual power}}{\text{ideal power}} = \frac{h_{t4} - h_{t5}}{h_{t4} - h'_{t5}}. \tag{2.3.45}$$

If the specific heats are constant across the turbine, the equation reduces to

$$\eta_t = \frac{T_{t4} - T_{t5}}{T_{t4} - T'_{t5}}. \tag{2.3.46}$$

Typically, because the pressure decreases in turbines, they operate with slightly higher efficiencies than do compressors. Typical peak values for turbine efficiencies at the design condition are 85 percent to 95 percent, but these depend on the pressure ratio and operating conditions. Finally, from Eqs. (2.3.43), (2.3.44), and (2.3.46), one finds

$$\pi_t \equiv \frac{p_{t5}}{p_{t4}} = \left\{1 - \frac{\left[1 - \frac{T_{t5}}{T_{t4}}\right]}{\eta_t}\right\}^{\frac{\gamma}{\gamma-1}} = \left\{1 - \frac{[1-\tau_t]}{\eta_t}\right\}^{\frac{\gamma}{\gamma-1}}$$

$$= \{\tau'_t\}^{\frac{\gamma}{\gamma-1}}, \tag{2.3.47}$$

where $\tau'_t = \frac{T'_{t5}}{T_{t4}}$. Equation (2.3.47) thus relates the total pressure and total temperature ratios for the nonideal case, which can be used for the total engine analysis with losses. Once again, the specific heat ratio is determined based on the average turbine temperature. The value of γ for the turbine is usually considerably lower than for a compressor owing to the higher operating temperature.

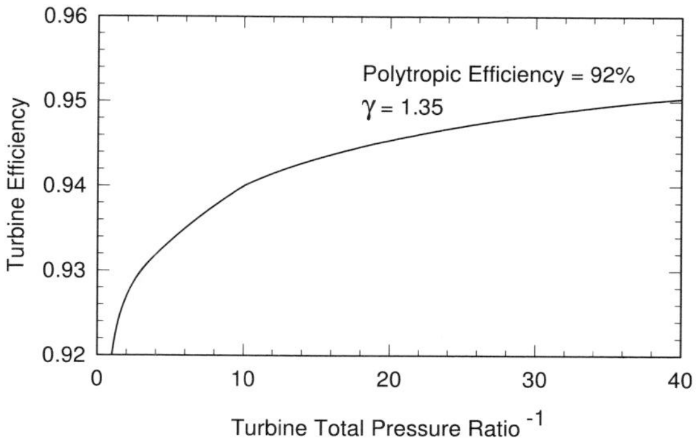

Figure 2.25 Turbine efficiency variation with pressure ratio.

Polytropic Efficiency

A turbine polytropic efficiency, η_{pt}, can be defined for the turbine in the same way as for the compressor basically to separate the efficiency from its dependence on the pressure ratio. Thus, different turbines can effectively be compared. This definition yields

$$\eta_t = \frac{1 - [\pi_t]^{\eta_{pt}\frac{\gamma-1}{\gamma}}}{1 - [\pi_t]^{\frac{\gamma-1}{\gamma}}}. \tag{2.3.48}$$

Again, for a given total pressure ratio, if the polytropic efficiency is known, the overall efficiency can be found, and vice versa. For this case, the polytropic efficiency is less than the overall turbine efficiency. As is the case for compressors, all well-designed modern engines demonstrate approximately the same η_{pt} at the design condition, and this is again about 90 percent. The turbine efficiency is thus a function of the total pressure ratio of the turbine. For example, the turbine efficiency is presented in Figure 2.25 for the pressure ratio range of 1 to 40^{-1} for a polytropic efficiency of 92 percent and $\gamma = 1.35$. As can be seen, turbine efficiency increases with increasing pressure ratio, whereas the efficiency of a compressor decreases with increasing pressure ratio.

2.3.6 Propeller (Propulsion Gas Turbines)

Propellers are used on turboprop engines (Figures 2.26 and 2.27) and behave somewhat like very large bypass-ratio (~20 to 100) fans. The basic operating principle of

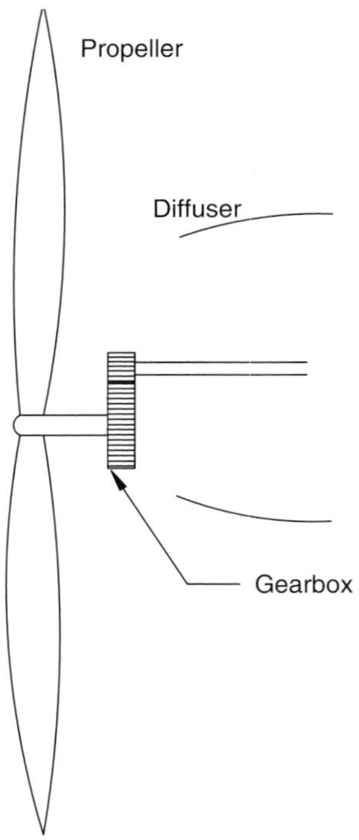

Figure 2.26 Schematic of propeller.

a propeller entails taking the axial flow of air into the propeller and accelerating that flow to a somewhat higher velocity by adding power derived from the turbine. Because the airflow through the propeller is large, considerable thrust is generated. Power is added to the flow by the rotating propeller blades, which are basically airfoils designed to efficiently accelerate the flow without any leading edge incidence (Theodorsen, 1948). Propellers are usually designed so that the blades can be turned or twisted at the hub to adjust the leading edge angle of the blades. Thus, a propeller can be used at various speeds and altitudes with minimal leading edge incidence by rotating the blades through various angles at different conditions. Blade settings at particular conditions are called "schedules." Blades can also be rotated to produce a negative thrust or reverse thrust for runway braking on landing. Propellers have large aspect ratios; that is, the tip diameter is much larger than the hub diameter. As a result, large stresses can be developed in a propeller because of the centrifugal forces. Propellers could not withstand the stresses if they were attached directly to the compressor shaft. Thus, gearboxes are used to reduce the rotational speed of the propeller – usually by a factor of 10 to 20. Such a gearbox is clearly seen in Figure 1.38.

For a turboprop engine, the propeller generates a major portion of the thrust, although some thrust is also derived from the jet. Also, the propeller requires a significant amount of power to generate this thrust. For the ideal analysis of the propeller, power delivered to the propeller, $\mathcal{P}_p$, will be related to the propeller thrust, F_p, by

$$\mathcal{P}_p = F_p u_a. \tag{2.3.49}$$

Namely, all of the propeller power generates thrust. A convenient dimensionless quantity is the propeller work coefficient (C_{w_p}), which was defined by Kerrebrock (1992) and is used in this text by

$$C_{w_p} = \frac{\mathcal{P}_p}{\dot{m} c_p T_a}. \tag{2.3.50}$$

Thus, if the propeller work coefficient, core mass flow rate, and aircraft speed are known, the thrust can be found.

Two other characteristics are often used to correlate propeller data. The first is the thrust coefficient, which is defined by

$$C_T = \frac{F_p}{\frac{1}{2}\rho_a u_a^2 A_p}, \tag{2.3.51}$$

where A_p is the circular frontal area swept by the rotating propeller. This is sometimes defined as

$$C_T = \frac{F_p}{\rho_a N_p^2 D_p^4}, \tag{2.3.52}$$

where N_p is the propeller rotational speed and D_p is the diameter of the propeller. The second coefficient is the power coefficient defined by

$$C_P = \frac{\mathcal{P}_p}{\frac{1}{2}\rho_a u_a^3 A_p} \tag{2.3.53a}$$

or sometimes by

$$C_P = \frac{\mathcal{P}_p}{\rho_a N_p^3 D_p^5}. \tag{2.3.53b}$$

Figure 2.27 Rolls-Royce Dart with a propeller on a test stand (© Rolls-Royce plc 2022. All rights reserved).

As discussed in Chapter 1, an alternative to a turboprop with long propeller blades is the unducted fan (UDF) or propfan, which was developed in the 1980s and early 1990s by NASA and General Electric, Pratt & Whitney, Rolls-Royce, and others (Figures 1.39 and 1.40), although this engine never went into production in the United States. Such an engine is a crossbreed between the turboprop, which demonstrates excellent low-aircraft-speed efficiency, and the conventional exhausted turbofan. The blades are shorter than a propeller and thus can be directly attached to the shaft without the need for a gearbox. More blades, which are highly swept, are used as well. Such designs can also effectively operate at higher flight Mach numbers than conventional turboprops.

In the ideal analysis, the assumption was made that all of the propeller power is used to generate thrust. In the nonideal case, this is not true. For example, because of the unbounded nature of the propeller and long blade lengths some of the flow exits the propeller with a radial component of velocity, whereas most of the flow enters the propeller with no radial velocity. Thus, some power is used to accelerate the flow from the axial direction to the radial direction without deriving any axial thrust. Second, because a diffuser is not used to direct the flow axially into the propeller, wind fluctuations and crossflows will result in incidence angles to the propeller blades, causing local separation and thus detracting from the propeller performance. Third, also because of the unbounded nature of the flow, more mixing and turbulence are generated; again power is wasted. Fourth, some of the power is used to generate noise, although this is much smaller than the other losses.

Thus, one can define the propeller efficiency as

$$\eta_p = \frac{\mathcal{P}_p}{\mathcal{P}'_p}, \tag{2.3.54}$$

where $\mathcal{P}_p$ is the power used for thrust and $\mathcal{P}'_p$ is the total power delivered to the propeller. Thus, the thrust from the propeller is given by

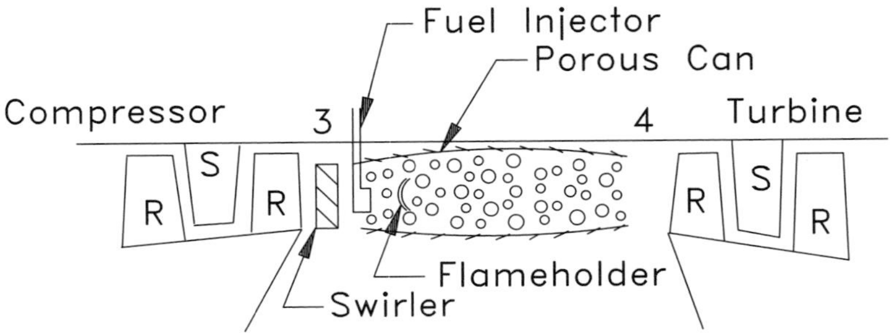

Figure 2.28 Schematic of a combustor.

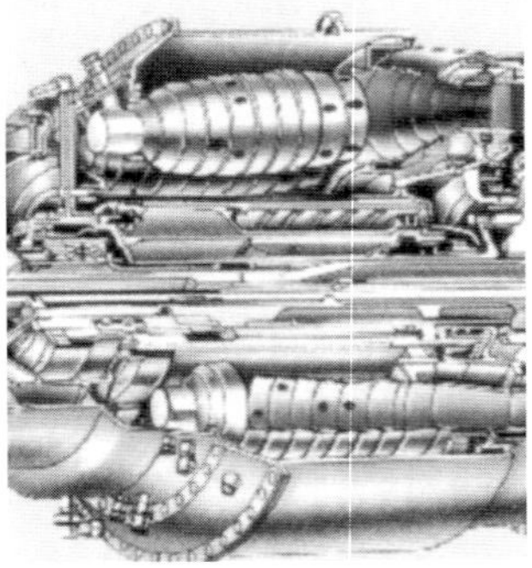

Figure 2.29 FT8 primary combustor (from Figure 2.2).

$$F_p = \eta_p C_{wp} c_p T_a \dot{m} / u_a, \quad (2.3.55)$$

where C_{w_p} is the propeller work coefficient as practically defined in Eq. (2.3.50). Other coefficients for the propeller are the thrust and power coefficients. Efficiencies vary widely owing to operating conditions, but maximum efficiencies are typically 70 percent to 90 percent. Propeller efficiencies are of course dependent on the particular designs and settings – especially the variable incidence. Efficiencies for propellers with fixed blades are compromised much faster when they are operated off of the design point (rotational speed, air speed, or both) than are propellers with variable blades as the result of increased leading edge incidence. More generally, propeller efficiencies are a function of (i) the propeller power coefficient, (ii) the "advance ratio" (freestream velocity to blade velocity) of the propeller at an operating speed, and (iii) the freestream Mach number. Some correlations are presented by Haines, MacDougall, and Monaghan (1946) and others. Such dependence is determined for particular designs and twist settings (or "schedules") from wind tunnel tests. A simple and practical model for propeller efficiency is presented by Mattingly, Heiser, and Daley (1987), who in general show in practice that, for a variable pitch propeller, the efficiency is at a maximum over a range of Mach numbers but deteriorates quickly as the freestream velocity goes to zero and also as the freestream Mach number approaches unity. When a nonideal analysis for an engine with a propeller is conducted, Eq. (2.3.55) becomes the working equation.

2.3.7 Combustor

A schematic of a primary combustor is shown in Figure 2.28, and the FT8 burner is shown in Figure 2.29. The design is complex and empirical. Fuel is injected into the air stream and gradually mixed and burned with the oxidant. The combustor or burner is often the only method by which energy is added to the gas turbine (the only other location is the afterburner for propulsion or reheat for power generation). Thus, an energy balance is necessary for the combustor (stations 3 to 4). The h–s and p–v diagrams for a combustor are shown in Figure 2.30. The process 3 to 4 is ideally isobaric (constant pressure) for an ideal process; the process is certainly not isentropic because the entropy increases owing to the irreversible combustion process. Also, for low gas velocities (as assumed), process 3 to 4 is ideally at constant total pressure (isotobaric) and is also not isentropic. Obviously, the static and total temperatures increase significantly in the combustor. As a result, the static and total specific volumes increase for process 3 to

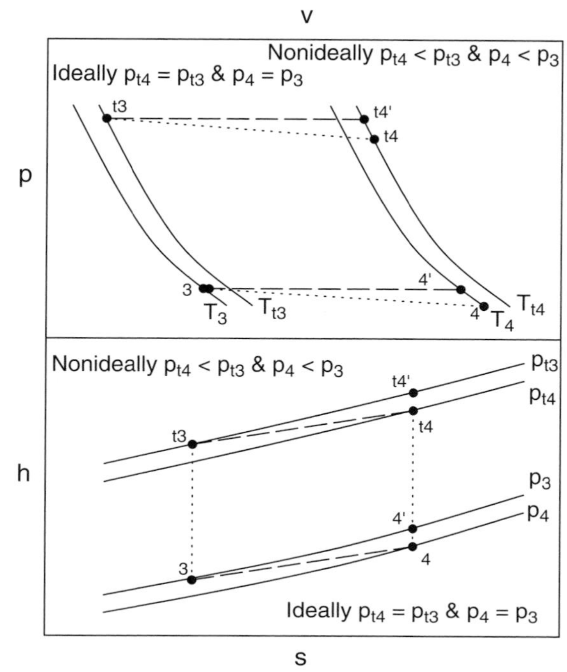

Figure 2.30 The h–s and p–v diagrams for a combustor.

4, and the fluid density decreases significantly across the combustor.

However, the combustion is not perfect and not all of the fuel is burned. As a result, the combustor exit temperature will not be as high as in the ideal case and some of the fuel will be wasted. That is the total enthalpy, h_{t4}, as shown in Figure 2.30, is less than the ideal value, h'_{t4}, and the static enthalpy h_4 is less than the ideal value, h'_{t4}. Second, because of the mixing and combustion processes in the combustor, which are both highly nonreversible and nonisentropic, the total pressure drops and the exit total pressure, p_{t4}, is less than the inlet value, p_{t3}, as shown in Figure 2.30. The two effects are somewhat independent in that two terms must be used to identify the combustor losses. However, as is shown in Chapter 9, in reality the two effects are somewhat inversely dependent. That is, if one wants to decrease the amount of wasted fuel, more mixing is required, which further reduces the total pressure, and vice versa.

Next, one needs first to examine the steady-state energy equation for the combustor:

$$\dot{Q} - \mathcal{P} = \Delta\dot{h}, \tag{2.3.56}$$

where $\dot{Q}$ is the heat energy rate added to the flow by combustion, $\mathcal{P}$ is the power derived from the flow, and $\Delta\dot{h}$ is the rate change in total enthalpy of the flow. The quantity $\Delta\dot{h}$ contains two contributions. The first is for the air passing through the burner. The second is for the fuel injected into the combustor. The entering air and exiting gas from the burner are both assumed to have specific heat c_p, which will be evaluated based on pure air. This assumption is valid for low fuel flow rates, as will be shown in Chapter 9. The fuel has specific heat c_{pf}. All gases leaving the burner are at T_{t4}. The air entering is at T_{t3}, whereas the fuel enters at T_{tf}. Applying this equation across positions 3 and 4 and realizing no work is produced by the burner, we find that for the ideal case

$$\dot{m}_f \Delta H = \dot{m}(h_{t4} - h_{t3}) + \dot{m}_f(h_{t4} - h_{tf}). \tag{2.3.57}$$

Or, using $dh = c_p\, dT$,

$$\dot{m}_f \Delta H = \dot{m}(c_p T_{t4} - c_p T_{t3}) + \dot{m}_f(c_p T_{t4} - c_{pf} T_{tf}). \tag{2.3.58}$$

In these equations, ΔH is the heating value of the fuel, and $\dot{m}_f$ is the flow rate of the fuel injected into the burner.

Imperfect Combustion

Ideally, the gas goes to complete combustion. In actuality, however, four nonidealities cause losses in the total temperature. First, the combustion is not complete because of poor or ineffective mixing or the fact that the mean velocity in the primary combustor is too high, and the mixture thus leaves the chamber before it has completely burned. Second, if cold fuel is injected that must therefore be warmed within the combustion chamber before burning, this again can result in fuel exiting the combustion region before it has burned. Third, because the temperatures are very high, some heat transfer can take place in the combustor and the total temperature can drop. Fourth, the fuel composition is not ideal. It is not strictly combustible because of additives and the presence of impurities. A simple, single parameter is introduced at this point to account for these four effects. This is called the primary burner efficiency η_b and is defined by, and used with, the simple heating value analysis:

$$\eta_b \dot{m}_f \Delta H = \dot{m}(h_{t4} - h_{t3}) + \dot{m}_f(h_{t4} - h_{tf}). \qquad (2.3.59)$$

Typical values for the primary burner efficiency range from 90 percent to 95 percent.

As is often the case, the heating value of the fuel is given for its inlet temperature. Thus, this equation reduces to

$$\eta_b \dot{m}_f \Delta H = \dot{m}(h_{t4} - h_{t3}) + \dot{m}_f h_{t4}. \qquad (2.3.60)$$

Rewriting this equation, and defining the engine parameter τ_b

$$\tau_b = \frac{T_{t4}}{T_{t3}} \qquad (2.3.61)$$

and, by assuming that the incoming and exiting gases are air, by using the value of c_p at the average temperature, and by assuming the value of c_p for the fuel is approximately that of air, one finds

$$\eta_b \frac{\dot{m}_f}{\dot{m}} \Delta H = c_p T_{t3}[\tau_b - 1] + \frac{\dot{m}_f}{\dot{m}} c_p T_{t3} \tau_b. \qquad (2.3.62)$$

Next, the fuel ratio for a nonafterburning propulsion engine and power generation gas turbine without reheat is defined as

$$f \equiv \frac{\dot{m}_f}{\dot{m}}. \qquad (2.3.63)$$

So solving for $\dot{m}_f$:

$$\dot{m}_f = \frac{\dot{m} c_p T_{t3}[\tau_b - 1]}{\eta_b \Delta H - c_p T_{t4}}. \qquad (2.3.64)$$

For many gas turbines the fuel ratio, f, is very small (on the order of 0.02). Thus, for the ideal analysis the second term of Eq. (2.3.62) is assumed to be negligible. Also, for the ideal case the efficiency is unity. Therefore, Eq. (2.3.64) becomes

$$\dot{m}_f = \frac{\dot{m} c_p T_{t3}[\tau_b - 1]}{\Delta H}. \qquad (2.3.65)$$

Note that the combustion analysis is greatly simplified at this point so that it can easily be incorporated into a cycle analysis. In Chapter 9, a more refined and much more accurate analysis referred to as the adiabatic flame temperature is covered. This method requires a chemical analysis (and tracking of all the reactants and products) as well as thermodynamics and will be covered in detail. For an initial engine cycle analysis, the heating-value method is sufficient.

Pressure Losses

The final working equation for the burner is given by

$$\pi_b \equiv \frac{p_{t4}}{p_{t3}}, \qquad (2.3.66)$$

which for the ideal case is unity because the total pressure is constant across the combustor:

$$p_{t4} = p_{t3}. \qquad (2.3.67)$$

However, the combustion process results in a loss in total pressure and nonideally p_{t4} is less than p_{t3}. This occurs for four reasons. First, a combustion process is taking place, which is in itself a highly irreversible (and nonisentropic) process. This reduces the total pressure. Second, the primary combustor has swirlers, flame holders (obstructions), and turbulence generators to encourage mixing between the fuel and air and induce regions of low velocity. Thus, more losses occur owing to increases in viscous effects. Third, within the primary burner itself, small holes (orifices) are used to distribute the air evenly in the combustion zone. Because air is forced through these small orifices, total pressure is lost. Fourth, the combustion process, which, because of the rapid density decrease is in itself a turbulence generator, induces more losses. Typical values for this pressure ratio are 0.92 to 0.98. As implied above, as the mixing is increased, generally π_b decreases and η_b increases. More discussion on pressure losses will be covered in Chapter 9.

2.3.8 Reheat Combustor (Power Generation Gas Turbines)

As discussed in Chapter 1, a reheat burner is sometimes used to improve the performance of a power generation gas turbine. The design is usually different but the performance is very similar to that of a primary combustor. For that reason, only the important equations will be presented. A diagram of the reheat combustor is shown in Figure 2.31 and a cutaway of the Ansaldo

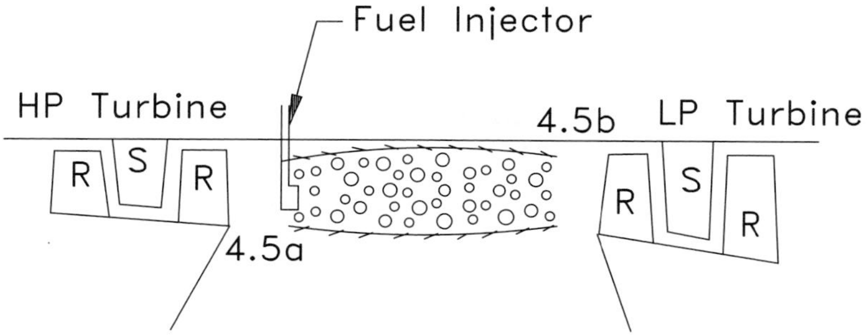

Figure 2.31 Diagram of reheat combustor.

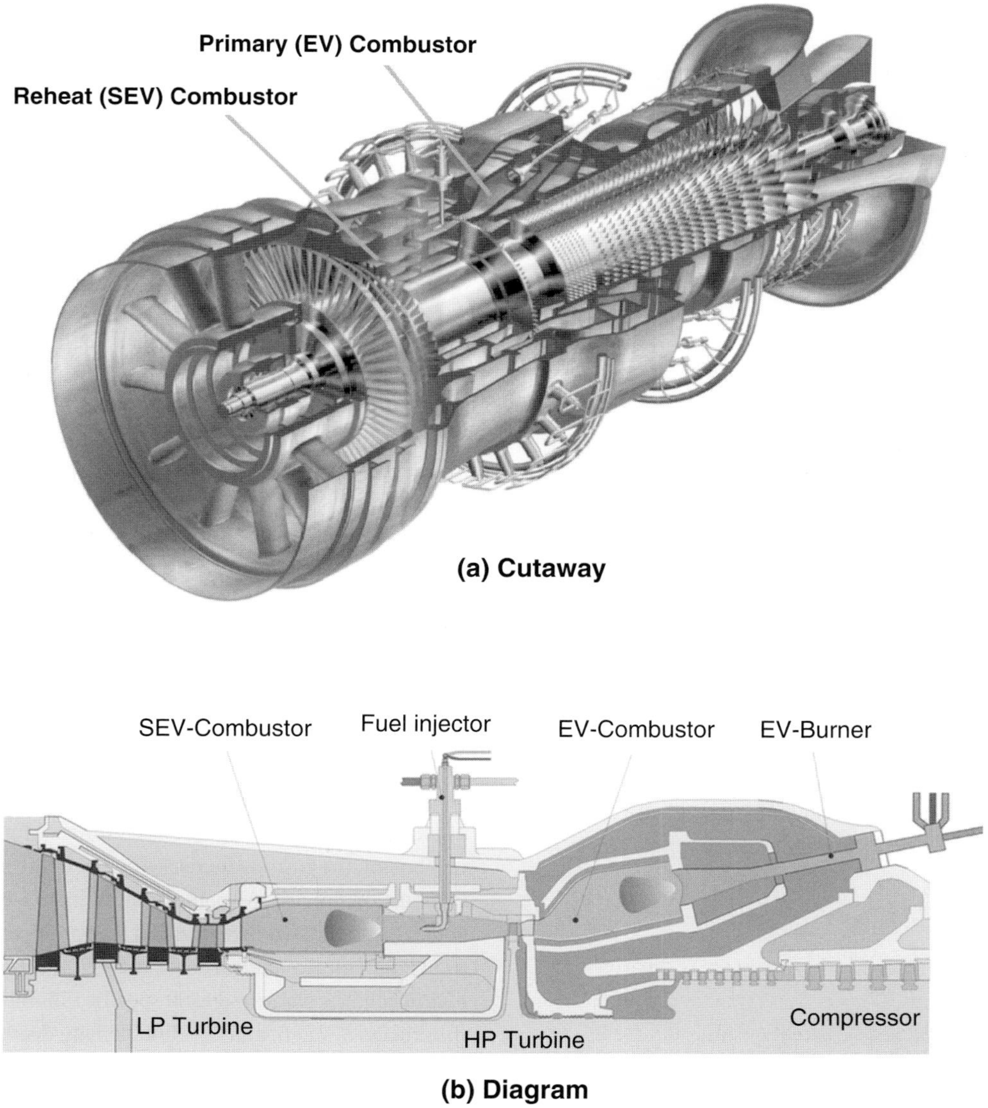

Figure 2.32 Reheat and primary combustors for Ansaldo Energia GT 26. (This picture is copyrighted by Ansaldo Energia s.p.a., who owns any related right. Any reproduction is forbidden without the written permission of the owner.)

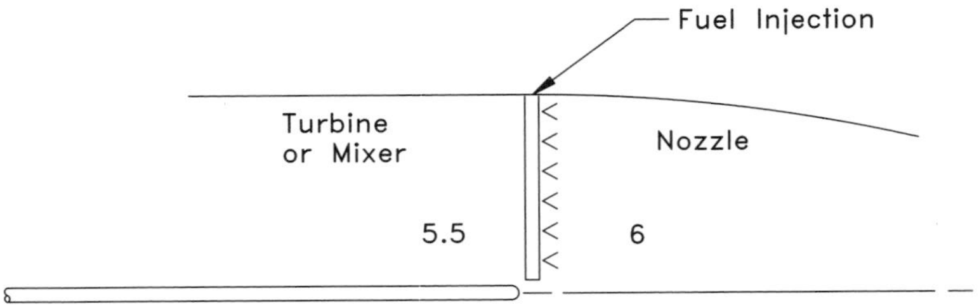

Figure 2.33 Schematic of an afterburner.

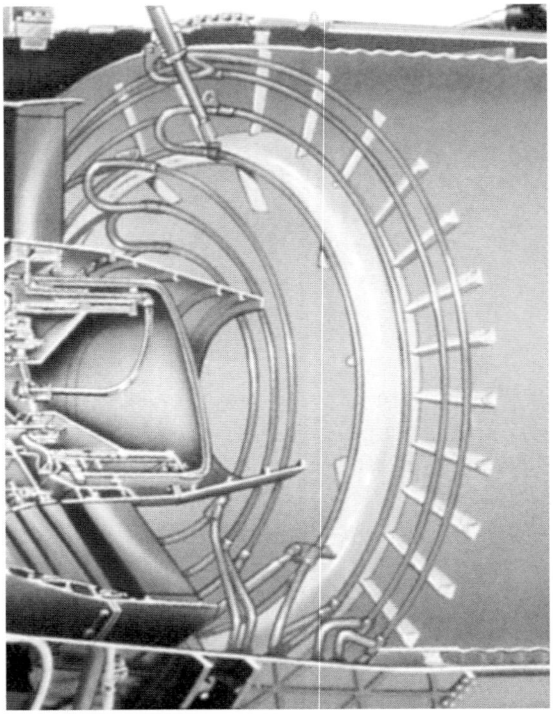

Figure 2.34 F100 afterburner (from Figure 2.1).

Energia GT 26 is shown in Figure 2.32. The primary and reheat burners are both shown in Figure 2.32.

As for the primary burner, for the reheat combustion:

$$\dot{m}_{\mathrm{RHf}} = \frac{\dot{m} c_{\mathrm{pRH}} T_{\mathrm{t4.5a}} [\tau_{\mathrm{RHb}} - 1]}{\eta_{\mathrm{RHb}} \Delta H - c_{\mathrm{pRH}} T_{\mathrm{t4.5b}}}, \tag{2.3.68}$$

where

$$\tau_{\mathrm{RHb}} = \frac{T_{\mathrm{t4.5b}}}{T_{\mathrm{t4.5a}}} \tag{2.3.69}$$

and

$$f_{\mathrm{RH}} \equiv \frac{\dot{m}_{\mathrm{RHf}}}{\dot{m}}. \tag{2.3.70}$$

As for the primary combustor due to f_{RH} being small, ideally Eq. (2.3.68) reduces to

$$\dot{m}_{\mathrm{RHf}} = \frac{\dot{m} c_{\mathrm{pRH}} T_{\mathrm{t4.5a}} [\tau_{\mathrm{RHb}} - 1]}{\Delta H}. \tag{2.3.71}$$

The same fuel is almost always used to both the primary combustor and reheat burner so that ΔH is the same for both. The final working equation for the reheat burner is given by the pressure ratio

$$\pi_{\mathrm{RHb}} \equiv \frac{p_{\mathrm{t4.5b}}}{p_{\mathrm{t4.5a}}}. \tag{2.3.72}$$

As noted in Chapter 1, in some cases the reheat process is a heat exchanger located in the primary combustor. However, in most cases it is a separate combustor and, for the sake of simplicity, it will be considered separate for the remainder of this text.

2.3.9 Afterburner (Propulsion Gas Turbines)

Figure 2.33 presents a schematic of an afterburner, and the F100 afterburner is shown in Figure 2.34. The design is much simpler than that for a primary combustor. Basically, fuel is injected and burned upstream of the nozzle. The process 5.5 to 6 is isobaric for an ideal afterburner and not isentropic. As is the case for a primary combustor, the entropy increases owing to the irreversible combustion process. Also, the process 5.5 to 6 is ideally at constant stagnation pressure. The static temperature and total temperatures increase markedly in the afterburner. As a result, the static and total specific volumes increase for process 5.5 to 6, and the fluid density decreases significantly across the afterburner. The h–s and p–v diagrams have the same trends as the primary combustor

(Figure 2.30). As is the case for the primary burner, an energy balance is needed for the afterburner so that the fuel rate can be related to the change in gas temperature. Following a procedure similar to that used for the combustor, one finds for stations 5.5 to 6 that for an ideal afterburner

$$\dot{m}_{f_{ab}} \Delta H = \dot{m}_{5.5}(c_p T_{t6} - c_p T_{t5.5}) + \dot{m}_{f_{ab}}(c_p T_{t6} - c_{pf_{ab}} T_{tf_{ab}}). \quad (2.3.73)$$

In this equation, $\dot{m}_{f_{ab}}$ is the flow rate of the fuel injected into the afterburner, and $\dot{m}_{5.5}$ is the mass flow rate of the gas entering the afterburner. Note that $\dot{m}_{5.5}$ will simply be $\dot{m}$ for an ideal turbojet, exhausted turbofan, or power gas turbine and will be ($\dot{m}+\alpha\dot{m}$) for an ideal mixed turbofan.

$$\frac{\dot{m}_{f_{ab}}}{\dot{m}_{5.5}} \Delta H = c_p T_{t5.5}[\tau_{ab} - 1] + \frac{\dot{m}_{f_{ab}}}{\dot{m}_{5.5}} c_p T_{t5.5}\left[\tau_{ab} - \frac{c_{pf_{ab}}}{c_p}\right], \quad (2.3.74)$$

where

$$\tau_{ab} = \frac{T_{t6}}{T_{t5.5}}. \quad (2.3.75)$$

One can next examine the quantity $\frac{\dot{m}_{f_{ab}}}{\dot{m}_{5.5}}$ and determine that for many engines this quantity is very small. Thus, for the ideal analysis, the second term in Eq. (2.3.74) will be assumed negligible:

$$\frac{\dot{m}_{f_{ab}}}{\dot{m}_{5.5}} \Delta H = c_p T_{t5.5}[\tau_{ab} - 1]. \quad (2.3.76)$$

For the nonideal afterburner two additional losses must be included. An afterburner is essentially a simple burner without most of the complex hardware of a primary burner. Fuel is injected into the flow upstream of the nozzle – usually with a series of spokes and circumferential rings to generate immediate increased thrust. The same two types of losses for a burner are also present for an afterburner. The physical processes and fundamental thermodynamic behavior are the same, although the actual designs are much different. For example, the afterburner combustion efficiency is defined by

$$\eta_a \dot{m}_{f_{ab}} \Delta H = \dot{m}_{5.5}(c_p T_{t6} - c_p T_{t5.5}) + \dot{m}_{f_{ab}}(c_p T_{t6} - c_{pf_{ab}} T_{tf_{ab}}). \quad (2.3.77)$$

Like the primary burner, this reduces to

$$\eta_a \dot{m}_{f_{ab}} \Delta H = \dot{m}_{5.5}(c_p T_{t6} - c_p T_{t5.5}) + \dot{m}_{f_{ab}} c_p T_{t6}, \quad (2.3.78)$$

where $\dot{m}_{5.5}$ is the total mass flow rate into the afterburner, which for a nonideal turbojet will be $\dot{m}(1+f)$ and for a mixed turbofan will be $\dot{m}(1+f)+\alpha\dot{m}$. The total fuel flow rate for an engine can also be realized as the sum of the primary and afterburner fuel flow rates, $\dot{m}_{ft} = \dot{m}_f + \dot{m}_{fab}$.

Also, the afterburner fuel ratio is defined by

$$f_{ab} \equiv \frac{\dot{m}_{f_{ab}}}{\dot{m}}, \quad (2.3.79)$$

so that

$$f_t = f + f_{ab}. \quad (2.3.80)$$

The final working equation for the afterburner is given by

$$\pi_{ab} = \frac{p_{t6}}{p_{t5.5}}, \quad (2.3.81)$$

which for the ideal case is unity because the total pressure is constant across the afterburner:

$$p_{t6} = p_{t5.5}. \quad (2.3.82)$$

However, the combustion process results in a loss in total pressure and nonideally p_{t6} is less than $p_{t5.5}$. As for a primary burner this occurs for four reasons. First, a combustion process is taking place, which is in itself a highly irreversible (and nonisentropic) process. This reduces the total pressure. Second, within the afterburner itself, spokes with spray bars are used to deliver fuel to the large burn zone. Again, these obstructions cause turbulence and pressure losses. Third, afterburner flame holders (obstructions) are used to encourage mixing between the fuel and air and induce regions of low velocity. Thus, more losses occur owing to increases in viscous effects. Fourth, the combustion process, which, because of the rapid density decrease is in itself a turbulence generator, induces more losses. As implied above, as the mixing is increased, generally π_b decreases and η_b increases. More discussion on pressure losses will be covered in Chapter 9.

However, for most of the engine operating time, an afterburner is not used. Thus, the added large pressure loss associated with a primary burner is not desired because this would also detract from the nominal operation of the engine. Thus, the design of an afterburner is simpler than that of a primary burner without most of the flow blockages used to induce good mixing. As a result, the efficiency is not as high as that of a primary burner. On the other hand, the pressure loss is not as large. Typical efficiencies are 80 percent to 90 percent, whereas typical total pressure ratios are 0.95 to 0.99 when the afterburner is ignited.

As noted earlier, afterburners are present on almost all military fighter craft but receive limited use. For normal engine operation (no afterburner fuel flow or combustion), a total pressure loss (Eq. (3.2.82) but not Eq. (3.2.78)) must still be included because the additional length of the engine and the presence of the fuel nozzles and struts in the flow path generate more viscous losses. However, when afterburner combustion is not occurring, this pressure ratio is usually higher than stated in the previous paragraph and is typically 0.98 to 1.00.

2.3.10 Shaft

In all gas turbines, except for a ramjet, power is derived from a turbine and delivered to drive all of the compressive devices (namely, a compressor and possibly a fan or propeller). In Figures 2.35 and 2.36, a turbojet and turbofan shaft are shown. Ideally, all of the derived power will be delivered to the compressive devices. Also, ideally, the mass flow rate remains constant through the engine core (i.e., the fuel flow rate is negligible), and thus the mass flow rate through the turbine is the same as in the compressor. Furthermore, the specific heat is constant through the gas turbine. Therefore, with the compressor, fan, propeller, and turbine power taken into consideration, an energy balance on the shaft for the most general ideal propulsion case (Figure 2.3) yields

$$\underbrace{\dot{m}c_p(T_{t3}-T_{t2})}_{\text{compressor}}+\underbrace{\dot{m}_s c_p(T_{t7}-T_{t2})}_{\text{fan}}+\underbrace{C_{w_p}\dot{m}c_pT_a}_{\text{propeller}}=\underbrace{\dot{m}c_p(T_{t4}-T_{t5})}_{\text{turbine}}, \tag{2.3.83}$$

where $\dot{m}_s$ is the bypassed air, which, when used with Eq. (2.3.38) yields

$$\dot{m}c_p(T_{t3}-T_{t2})+\alpha\dot{m}c_p(T_{t7}-T_{t2})+C_{w_p}\dot{m}c_pT_a = \dot{m}c_p(T_{t4}-T_{t5}). \tag{2.3.84}$$

Above, all of the derived power was assumed to be delivered to other components. A shaft, however, does rotate – often at high rotational speeds. Shafts are mounted in bearings and often a damping device

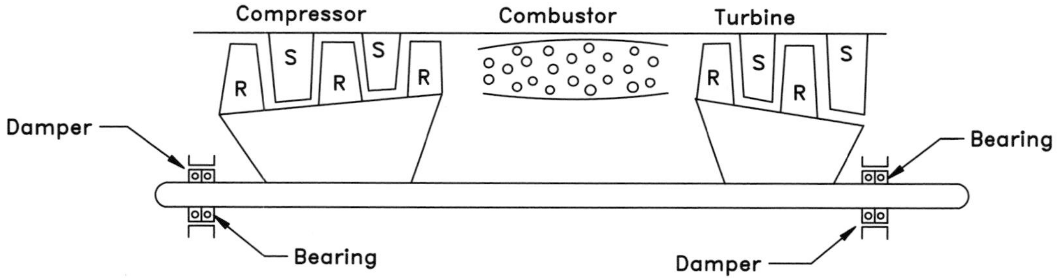

Figure 2.35 Schematic of the turbojet engine shaft.

Figure 2.36 F100 shaft (from Figure 2.1).

called a squeeze film damper to control shaft vibrations passively. Rolling element bearings are incorporated for aircraft applications, and hydrodynamic bearings are usually used in power gas turbines. Low-viscosity synthetic lubricants provide lubrication for the rolling element bearings, and mineral light turbine oils are used for the hydrodynamic bearings. As a result, these bearings and dampers parasitically dissipate some of the power due to viscous lubricant and surface friction. Also, in a turboprop engine, gears and associated bearings are usually used to reduce the speed of the propeller as discussed in Chapter 1. Planetary gears are also used on the modern high bypass engines as discussed in Chapter 1. Gears also demonstrate parasitic losses because of surface friction. In summary, not all of the power derived from the turbine is delivered to the compressive devices.

These losses can be accommodated in the analysis through the following definition of mechanical efficiency:

$$\eta_m = \frac{\mathcal{P}_c}{\mathcal{P}_t}, \tag{2.3.85}$$

where $\mathcal{P}_t$ is the power derived from all of the turbines and $\mathcal{P}_c$ is the total power delivered to all of the compressive devices (compressor, fan, propeller). Furthermore, the mass flow rate into the turbine is slightly higher than that for the compressor as a result of fuel added in the combustor with the fuel to air ratio of f. The general power equation is thus

$$\underbrace{\dot{m}c_{p_c}(T_{t3} - T_{t2})}_{\text{compressor}} + \underbrace{\alpha\dot{m}c_{p_f}(T_{t7} - T_{t2})}_{\text{fan}} + \underbrace{C_{w_p}\dot{m}c_{p_a}T_a}_{\text{propeller}} = \underbrace{\eta_m\dot{m}(1+f)c_{p_t}(T_{t4} - T_{t5})}_{\text{turbine}}. \tag{2.3.86}$$

This mechanical efficiency is usually high (between 98 percent and 100 percent). Thus, Eq. (2.3.86) can be used to include shaft losses in the nonideal analysis.

Similarly, for a power generation unit (Figure 2.5):

$$\underbrace{\dot{m}c_{pLPC}(T_{t2.5} - T_{t2})}_{\text{LPC}} + \underbrace{\dot{m}c_{pHPC}(T_{t3} - T_{t2.5})}_{\text{HPC}} = \underbrace{\eta_m\dot{m}(1+f)c_{pHPT}(T_{t4} - T_{t4.5a})}_{\text{HPT}} + \underbrace{\eta_m\dot{m}(1+f+f_{RH})c_{pLPT}(T_{t4.5b} - T_{t4.5c})}_{\text{LPT}} + \underbrace{\eta_m\dot{m}(1+f+f_{RH})c_{pPGT}(T_{t4.5c} - T_{t5})}_{\text{PGT(Free)}}. \tag{2.3.87}$$

2.3.11 Primary Nozzle (Propulsion Gas Turbines)

The purpose of the primary nozzle (stations 6 to 8) is to convert the moderate-temperature and moderate-pressure gas to a high-velocity gas that will produce thrust. Figure 2.37 presents a schematic of a primary nozzle and Figure 2.38 displays the variable F100 converging–diverging (CD) nozzle. In Figure 2.39, the h–s and p–v diagrams are depicted for a nozzle. Basically, the flow is accelerated from 6 to 8. As a result, the static temperature and pressure decrease through the nozzle. The specific volume markedly increases (density decreases) through the nozzle, which gives rise to the increase in fluid velocity. For the ideal case the acceleration process is isentropic and adiabatic as is the stagnation process. Thus, once again the conditions 6 and 8 and the stagnation conditions all occur on the same isentropic process line, and the total pressure and temperature are ideally constant through the nozzle. Also,

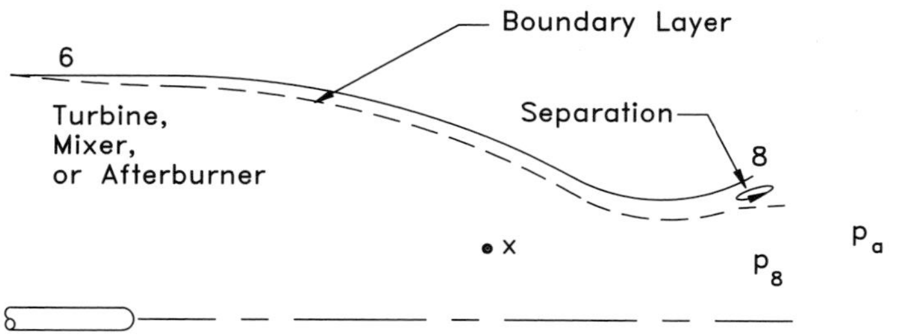

Figure 2.37 Schematic of a primary nozzle section.

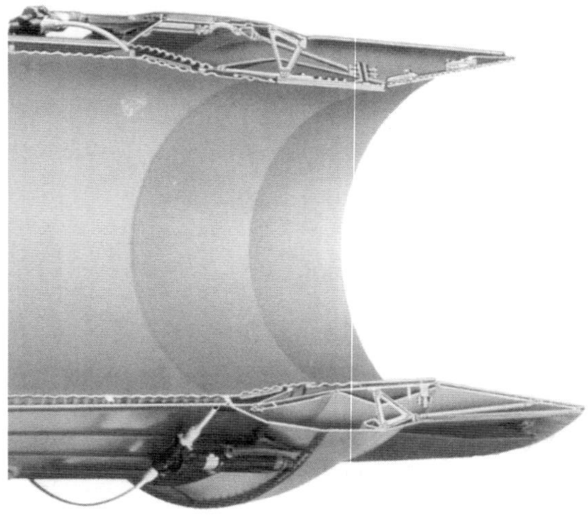

Figure 2.38 F100 variable CD nozzle (from Figure 2.1).

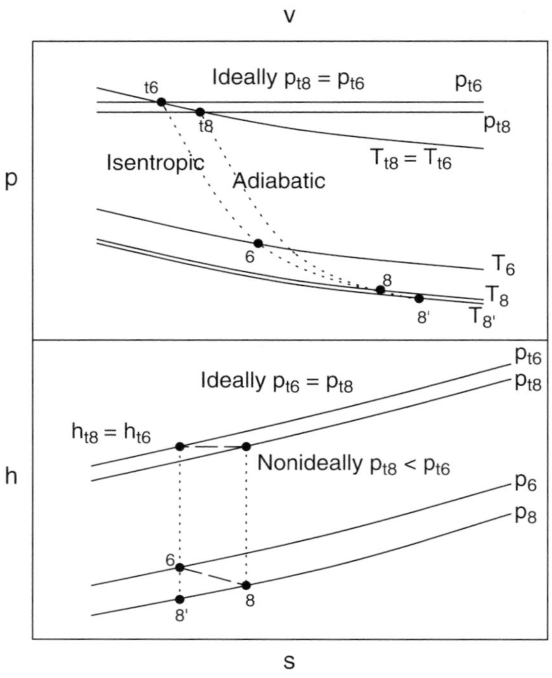

Figure 2.39 The h–s and p–v diagrams for a nozzle.

as will be discussed, ideally the nozzle exit pressure, p_8, is equal to the ambient pressure, p_a. This can only occur for the proper nozzle geometry and area ratios, however. Thus, for the ideal analysis, two assumptions are made for the nozzle flow. First, the flow is assumed to be frictionless and adiabatic. Second, the exit pressure is assumed to match the ambient pressure. For the nonideal analysis, these two assumptions are relaxed. At this point in the text only two nozzle types will be considered: fixed converging and variable converging–diverging (CD) nozzles. For the variable CD nozzles the nozzle areas will be chosen to match the nozzle exit pressure to the ambient pressure. Fixed CD nozzles can have imbedded shocks and require more complex analyses, and this will be covered in Chapter 10.

Exit Pressure

During actual or nonideal operation, the exit pressure may not match the ambient pressure. The nozzle areas may not be in the correct proportion, and thus the resulting exit and ambient pressures will be unequal. As is demonstrated in Chapter 10, for an efficient nozzle the maximum thrust is derived when these two pressures are the same.

If the exit pressure is greater than the ambient pressure, the flow is considered to be underexpanded. A positive thrust is thus derived from the pressure terms in the thrust equation. However, the exit velocity will not be as high as it would be the properly expanded case, and thus the momentum component of thrust will be lower than ideal. As a result, the two effects complement each other, but total thrust will be lower than ideal. On the other hand, if the exit pressure is lower than the ambient pressure, the flow is considered to be overexpanded, and negative thrust is derived from the pressure terms in the thrust equation. Now, the exit velocity will be higher than it would be in the properly expanded case, and thus the momentum component of thrust will be higher than ideal. Again, the two effects complement each other, but total thrust will again be lower than ideal. For the nonideal analysis, all three cases will be considered.

Pressure Losses

For the ideal case the flow was assumed to be isentropic. For the nonideal case, frictional losses will be included, although the flow is still considered to be adiabatic. In Figure 2.39, h–s and p–v diagrams are shown. The ideal case operates between end states 6 and 8' whereas the nonideal case operates between states 6 and 8. For the definition of nozzle efficiency presented in this section, both cases operate between the inlet total pressure p_{t6} and the exit static pressure p_8. For the ideal case, the entropy remains constant for the nozzle process. However, for the nonideal case the entropy increases owing to friction in the boundary layer, freestream turbulence, secondary flows, shocks in the nozzle, and possible flow separation. Because the flow is adiabatic, the total enthalpy remains constant in the nozzle; that is, h_{t6} is equal to h_{t8}. However, the entropy increase in the nozzle results in a drop in total pressure from inlet to exit (p_{t6} to p_{t8}), as shown in Figure 2.39. As a result, the nonideal static exit enthalpy, h_8, is greater than the ideal enthalpy, h'_8, which leads to a reduced exit velocity and in turn to a reduced thrust.

A nozzle efficiency is introduced to include these losses and is defined (only for adiabatic flow) by

$$\eta_n = \frac{h_{t6} - h_8}{h_{t6} - h'_8}, \tag{2.3.88}$$

which for the constant specific heat case results in

$$\eta_n = \frac{T_{t6} - T_8}{T_{t6} - T'_8}. \tag{2.3.89}$$

Now, by applying the energy equation (Eq. (2.2.6)) between states t6 and 8, one finds

$$u_8 = \sqrt{2(h_{t6} - h_8)}, \tag{2.3.90}$$

or, for the constant specific heat case

$$u_8 = \sqrt{2c_p(T_{t6} - T_8)}, \tag{2.3.91}$$

which is also

$$u_8 = \sqrt{2c_p\eta_n(T_{t6} - T'_8)}. \tag{2.3.92}$$

This is obviously less than the ideal value ($\eta_n = 1$) and leads to reduced thrust. Now, since the flow is still considered to be adiabatic,

$$\frac{T_8}{T_{t6}} = \frac{T_8}{T_{t8}} = \frac{1}{1 + \frac{\gamma - 1}{2}M_8^2}, \tag{2.3.93}$$

and for the ideal case

$$\frac{p_8}{p_{t6}} = \left[\frac{T'_8}{T_{t6}}\right]^{\frac{\gamma}{\gamma-1}}. \tag{2.3.94}$$

Thus, Eqs. (3.2.89), (3.2.93), and (3.2.94) yield

$$\frac{p_8}{p_{t6}} = \left[\frac{\frac{1}{1 + \frac{\gamma - 1}{2}M_8^2} - 1 + \eta_n}{\eta_n}\right]^{\frac{\gamma}{\gamma-1}}. \tag{2.3.95}$$

Thus, if the exit pressure is known, the exit Mach number can be found (on the assumption that the nozzle efficiency and upstream total pressure are known).

Now, for the ideal case $p_8 = p_a$ and $\eta_n = 1$ so that Eq. (2.3.95) becomes

$$p_{t6} = p_{t8} = p_a\left[1 + \frac{\gamma - 1}{2}M_8^2\right]^{\frac{\gamma}{\gamma-1}}. \tag{2.3.96}$$

Therefore, if the inlet total pressure and ambient pressure are known, the exit Mach number can be found. Using the exit Mach number, one can find the exit temperature from Eq. (2.3.93) if the inlet total temperature is known. Finally, the exit velocity can be found from the speed of sound (Eq. (2.2.10)) and Mach number:

$$u_8 = M_8\sqrt{\gamma\mathcal{R}T_8}. \tag{2.3.97}$$

Also, using the energy equation (Eq. (2.2.6)) across the nozzle for adiabatic flow yields

$$u_8 = \sqrt{2c_p(T_{t6} - T_8)}. \tag{2.3.98}$$

Equations (2.3.97) and (2.3.98) will yield identical results. Remember, the last three equations are for the ideal case with $\eta_n = 1$ but the exit pressure will not always match the ambient pressure.

Another nonideal effect that needs considered is nozzle choking. Before proceeding, one should realize that all of the preceding equations can be applied to an arbitrary location in the nozzle such as axial position x in Figure 2.37. For example, Eq. (2.3.95) can be used when p_8 is replaced with p_x and M_8 is replaced with M_x. Therefore, Eq. (2.3.95) can be applied at the condition where the Mach number is unity. One should also realize that this sonic condition may or may not exist in a nozzle but is a reference condition. For example, an unchoked converging nozzle will not have a sonic condition. Yet, one can still apply Eq. (2.3.95) with a Mach number of unity to determine what the conditions would be if the nozzle were choked and use this as the reference condition. Thus, by letting $M_x = 1.00$, one obtains

$$\frac{p*}{p_{t6}} = \left[1 + \frac{1-\gamma}{\eta_n(1+\gamma)}\right]^{\frac{\gamma}{\gamma-1}}, \tag{2.3.99}$$

where the superscript * indicates the sonic condition. This equation can thus be used to check a converging nozzle for choking. For example, if p_a is less than p^*, the nozzle is choked, $p_8 = p^*$, and $M_8 = 1$. In actuality for a choked converging nozzle with an efficiency less than unity, the nozzle exit Mach number is slightly less than one due to nonidealities and 2-D and 3-D flows as will be discussed in Chapter 10. However, to simplify the analyses in this book, the sonic condition will be used for the choked case because the difference has minimal effect on the engine thrust and *TSFC* (typically much less than 0.1 percent and conservative). However, if p_a is greater than p^*, the nozzle is not choked, M_8 is less than unity, and $p_8 = p_a$. Furthermore, when considering the temperature at the exit of a choked nozzle, from Eq. (2.3.93) one finds the following for a Mach number of unity:

$$\frac{T^*}{T_{t6}} = \frac{2}{1+\gamma}. \tag{2.3.100}$$

A relationship may also be desired between the area and the Mach number. Because mass flow is conserved between the sonic condition (if it exists in reality or not) and the exit condition,

$$\dot{m}_8 = \dot{m}^*. \tag{2.3.101}$$

Thus,

$$\rho^*\, u^*\, A^* = \rho_8\, u_8\, A_8,$$

or solving for A_8/A^* and using the ideal gas equation (Eq. (2.2.1)) yields

$$\frac{A_8}{A^*} = \frac{T_8 p^* u^*}{T^* p_8 u_8}. \tag{2.3.102}$$

By realizing that $u^* = a^*$ and that $u_8 = M_8\, a_8$, one obtains

$$\frac{A_8}{A^*} = \frac{T_8 p^* a^*}{T^* p_8 M_8 a_8}. \tag{2.3.103}$$

Using Eq. (2.2.10) results in

$$\frac{A_8}{A^*} = \frac{T_8 p^* \sqrt{T^*}}{T^* p_8 M_8 \sqrt{T_8}}. \tag{2.3.104}$$

Combining terms and multiplying and dividing by similar terms yields

$$\frac{A_8}{A^*} = \sqrt{\left[\frac{\frac{T_8}{T_{t6}}}{\frac{T^*}{T_{t6}}}\right] \times \left[\frac{\frac{p^*}{p_{t6}}}{\frac{p_8}{p_{t6}}}\right]}\, \frac{1}{M_8}. \tag{2.3.105}$$

Finally, one can apply Eq. (2.3.105) with (2.3.93), (2.3.95), (2.3.99), and (2.3.100) to yield

$$\frac{A_8}{A^*} = \frac{1}{M_8}\sqrt{\frac{\gamma+1}{2+(\gamma-1)M_8^2}}\left[\eta_n \frac{1+\frac{1-\gamma}{\eta_n(1+\gamma)}}{\frac{1}{1+\frac{\gamma-1}{2}M_8^2} - 1 + \eta_n}\right]^{\frac{\gamma}{\gamma-1}}. \tag{2.3.106}$$

This section has presented the equations that will both include losses and allow the nozzle to operate at off-design conditions. Typical nozzle efficiencies are 90 percent to 97 percent. Interestingly, for the non-ideal case ($\eta_n < 1$) one finds $\left(\frac{A_{min}}{A^*} < 1\right)$, as covered in Chapter 10. The application of all of the above equations requires careful thought on the part of the reader. One single, final equation cannot be derived because the operating conditions govern the form of the equations. For example, the application of the equations will be different for a choked converging nozzle than for an unchoked converging nozzle. For another example, the application of the equations will be different for a choked converging nozzle than for a supersonic converging–diverging (CD) nozzle. Some of the differences will be discussed in Examples in Section 3.3.2.

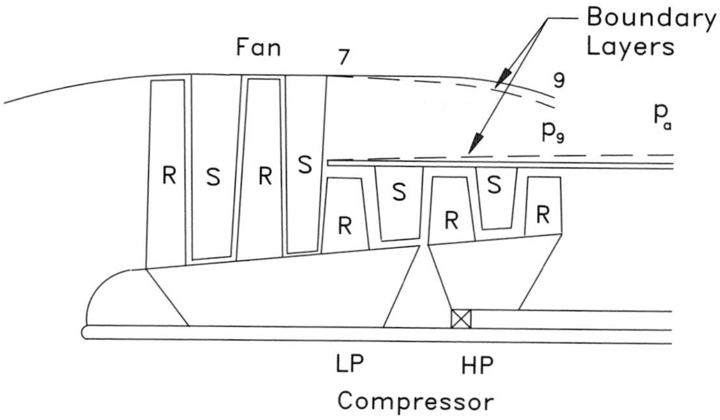

Figure 2.40 Schematic of a fan nozzle.

2.3.12 Fan Nozzle (Propulsion Gas Turbines)

Figure 2.40 is a schematic of a fan nozzle and a sectional view of the Rolls-Royce geared fan is shown in Figure 2.6. The fluid mechanics of this secondary nozzle, including physical losses, thermodynamics, and definitions, are fundamentally the same as for the primary nozzle. Similarly, h–s and p–v diagrams can be used. That is, the flow is accelerated from 7 to 9. The static temperature and pressure decrease through the fan nozzle. The specific volume significantly increases through the nozzle, so to increase the fluid velocity. Because the operation is similar to the primary nozzle, the details of the equation development will not be covered. The fan nozzle is almost always a fixed converging nozzle. The equations of interest are listed below.

The fan nozzle operates between stations 7 and 9 (Figure 2.40) with an efficiency of η_{fn}:

$$\eta_{fn} = \frac{h_{t7} - h_9}{h_{t7} - h'_9} \tag{2.3.107}$$

$$u_9 = \sqrt{2c_p(T_{t7} - T_9)} \tag{2.3.108}$$

$$u_9 = \sqrt{2c_p\eta_{fn}(T_{t7} - T'_9)} \tag{2.3.109}$$

$$\frac{T_9}{T_{t7}} = \frac{1}{1 + \frac{\gamma - 1}{2}M_9^2} \tag{2.3.110}$$

$$\frac{p_9}{p_{t7}} = \left[\frac{T'_9}{T_{t7}}\right]^{\frac{\gamma}{\gamma-1}} \tag{2.3.111}$$

$$\frac{p_9}{p_{t7}} = \left[\frac{\dfrac{1}{1 + \frac{\gamma - 1}{2}M_9^2} - 1 + \eta_{fn}}{\eta_{fn}}\right]^{\frac{\gamma}{\gamma-1}}. \tag{2.3.112}$$

So for the ideal case

$$p_9 = p_a, \tag{2.3.113}$$

and for the ideal case Eq. (2.3.112) reduces to

$$p_{t7} = p_a\left[1 + \frac{\gamma - 1}{2}M_9^2\right]^{\frac{\gamma}{\gamma-1}}. \tag{2.3.114}$$

Next, similar to the primary nozzle:

$$\frac{p^*}{p_{t7}} = \left[1 + \frac{1 - \gamma}{\eta_{fn}(1 + \gamma)}\right]^{\frac{\gamma}{\gamma-1}} \tag{2.3.115}$$

$$\frac{A_9}{A^*} = \frac{1}{M_9}\sqrt{\frac{\gamma + 1}{2 + (\gamma - 1)M_9^2}}\left[\eta_{fn}\frac{1 + \dfrac{1 - \gamma}{\eta_{fn}(1 + \gamma)}}{\dfrac{1}{1 + \frac{\gamma - 1}{2}M_9^2} - 1 + \eta_{fn}}\right]^{\frac{\gamma}{\gamma-1}}. \tag{2.3.116}$$

2.3.13 Bypass Duct (Propulsion Gas Turbines)

When a fan is present, and when at least a part of the flow is mixed into the turbine exit flow, a bypass duct (Figures 2.41 and 2.42) is used to direct the bypassed air. For the ideal cases, all of the bypassed air is considered to be either all exhausted or all mixed back

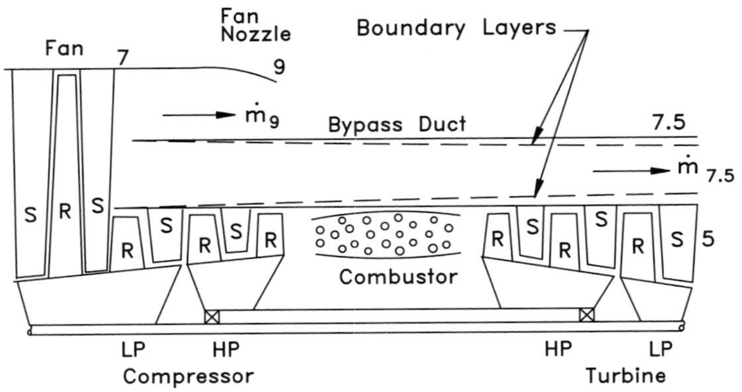

Figure 2.41 Schematic of a bypass duct.

Figure 2.42 F100 bypass duct (from Figure 2.1).

into the primary flow. This latter bypass process is from stations 7 to 7.5.

In practice, a portion of the air is sometimes bled and mixed. A new parameter, termed the duct split ratio, is defined as the ratio of secondary air that is mixed to the total secondary air. This is given by

$$\sigma = \frac{\dot{m}_{7.5}}{\dot{m}_{7.5} + \dot{m}_9}. \tag{2.3.117}$$

The quantities $\dot{m}_{75}$ and $\dot{m}_9$ are the mass fluxes at stations 7.5 and 9, respectively, as shown in Figure 2.41. The ratio σ is to unity for the fully mixed case and to zero for the totally exhausted case.

Also, ideally, the exit conditions of the duct are assumed to be identical to the inlet conditions. In actuality, however, this flow has frictional losses, as in any duct or pipe, including wall shear and freestream shear and turbulence. The flow is thus nonisentropic and has a drop in total pressure but is still considered to be adiabatic; thus, $h_{t7} = h_{t7.5}$. Since the process is adiabatic, the stagnation temperature is constant through the duct:

$$T_{t7.5} = T_{t7}. \tag{2.3.118}$$

The pressure loss is included in the cycle analysis simply as the ratio of the total pressures as follows:

$$\pi_u = \frac{p_{t7.5}}{p_{t7}}. \tag{2.3.119}$$

This ratio is typically high and near unity because bypass ducts are relatively smooth and short. Nonetheless, it should be included for completeness. For the ideal case this parameter is unity. More discussion on this and methods of approximating this loss will be covered in Chapter 11.

Next, a boundary condition is needed to match the bypassed air when it is mixed back into the primary flow. Once again, for flow stability, the physical process will naturally match the static pressures; that is,

$$p_{7.5} = p_5. \tag{2.3.120}$$

Applying this condition for the analysis at the mixer would require knowledge of the flow velocities or cross-sectional areas because the analyses of all of the other components use total pressures. However, if the Mach numbers are approximately the same for the two streams ($M_5 \cong M_{7.5}$), or if both Mach numbers are well below about 0.3, where compressibility effects can be ignored, an alternative condition can be used as follows:

$$p_{t7.5} \cong p_{t5}. \tag{2.3.121}$$

This is a simpler condition and will be used mostly herein because it facilitates the cycle analysis. However, if the two Mach numbers are approximately known or can be estimated, the true condition, $p_{7.5} = p_5$, can be used and the total pressures can be found from Eq. (2.2.13). This is an option in JetEngineCycle.exe. However, for ideal cases, Eq. (2.3.121) will always be used.

2.3.14 Bypass Mixer (Propulsion Gas Turbines)

As discussed in Chapter 1, for the mixed turbofan, the mixer combines the ducted and bypassed air (stream 7.5) into the turbine exit air (stream 5). A schematic is presented in Figure 2.43, and the F100 mixing region is shown in Figure 2.44. The mixer can be a series of blades to "fold" the cool bypassed air into the warmer gas from the turbine to induce a uniform flow. Ideally, the process is isentropic; thus, the stagnation pressure remains constant across the mixer. The average properties before the mixer are labeled as 5.2 and those after the mixer are stream 5.5. This mixing process of the two streams is highly irreversible and nonisentropic. Furthermore, mix enhancers are often used to further improve the mixing efficiency. The introduction of such enhancers increases viscous losses due to increased surface areas, blockage, and increasing turbulence and thus increases the nonisentropic behavior of the process. Good mixing is desired to ensure uniform properties. However, with improved mixing and resulting high turbulence levels, the entropy increases and the total pressure drops. Once again, a total pressure ratio is included to account for the frictional and mixing losses:

$$\pi_m = \frac{p_{t5.5}}{p_{t5.2}}. \tag{2.3.122}$$

Typically, this ratio is between 0.97 to 0.99 and again ideally this ratio is unity.

Also, one needs to find the total temperature of the exiting stream of gas from the mixer (into the afterburner or nozzle) because the two incoming streams are at different total temperatures. The energy equation is first applied to find

$$\dot{m}_5 h_{t5} + \dot{m}_{7.5} h_{t7.5} = \dot{m}_{5.5} h_{t5.5} \tag{2.3.123}$$

or using Eqs. (2.2.7), (2.3.38), (2.3.63), and (2.3.117) results in

$$\dot{m}(1+f)c_{pmc}T_{t5} + \dot{m}\sigma\alpha c_{pmu}T_{t7.5} = \dot{m}[(1+f)c_{pmc} + \sigma\alpha c_{pmu}]T_{t5.5} \tag{2.3.124}$$

or, finally,

$$T_{t5.5} = \frac{(1+f)c_{pmc}T_{t5} + \sigma\alpha c_{pmu}T_{t7.5}}{(1+f)c_{pmc} + \sigma\alpha c_{pmu}}, \tag{2.3.125}$$

where c_{pmu} is evaluated at the average temperature from stations 7.5 to 5.5 and c_{pmc} is evaluated at the average temperature from stations 5 to 5.5.

Lastly, for an ideal fully mixed turbofan (f is negligible, c_p is constant, and $\sigma = 1$), Eq. (2.3.125) reduces to

$$T_{t5.5} = \frac{\alpha T_{t7.5} + T_{t5}}{\alpha + 1}. \tag{2.3.126}$$

Thus, once the properties of the incoming streams are known, the exit conditions can be found. Models to predict the pressure loss will be covered in more detail in Chapter 11.

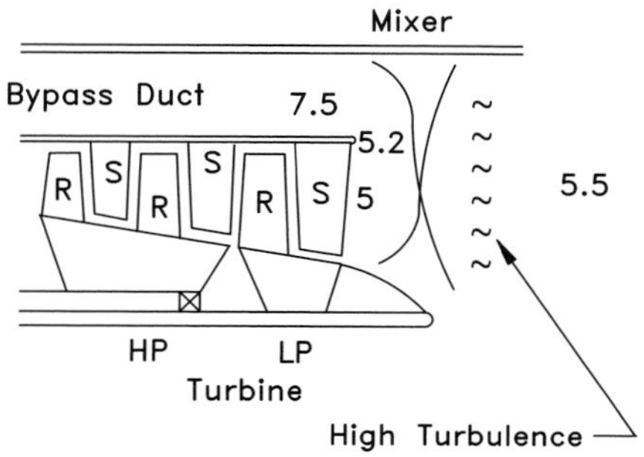

Figure 2.43 Mixer schematic.

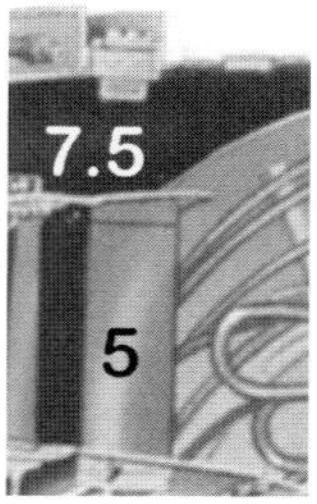

Figure. 2.44 F100 mixing region (from Figure 2.1).

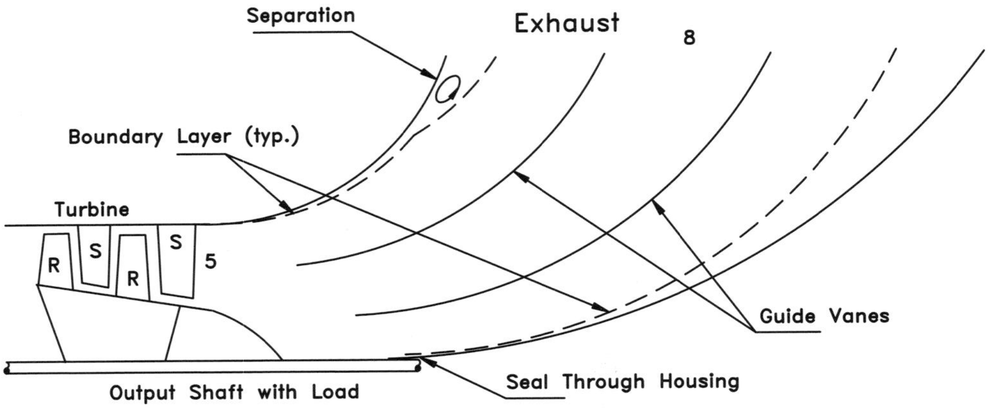

Figure 2.45 Schematic of a power generation gas turbine exhaust.

Figure 2.46 FT8 exhaust (from Figure 2.2).

2.3.15 Exhaust and Stack (Power Generation Gas Turbines)

One focus of this book is jet propulsion. However, power generation gas turbines are closely related. In fact, as discussed in Chapter 1, many power gas turbines for electric generation, marine or other applications are aeroderivatives. That is, much of the development work for such gas turbines was carried out as a part of jet engine design research. One of the largest differences between a power gas turbine and a jet engine is that a power generation gas turbine has an exhaust and stack with very low velocity and does not generate thrust compared with a nozzle with high velocities. Thus, because a nozzle and exhaust operate with different functions, they are treated differently and separately.

A simple schematic of an exhaust is shown in Figure 2.45. Figures 2.46 and 2.47 present the exhausts for the Pratt & Whitney FT8 and the Coberra 6000 gas turbine. The bell inlet can also be seen for the Coberra 6000. The reader should identify two major differences between this geometry and that of a jet engine. First, the exhaust section primarily takes axial flow out of the turbine and turns it mainly in a radial direction. The exhaust has a series of large vanes that smoothly direct the flow to the remainder of the exhaust system. Second, the bell inlet accelerates the flow axially into the compressor. Note also the large, two-stage auxiliary or free turbine used to generate mechanical power on this unit. An exhaust system is often quite long and may extend, for example, through ducting to a stack or tower. The velocity at the exit of the exhaust system is very low, and consequently the static and total pressures are nearly identical.

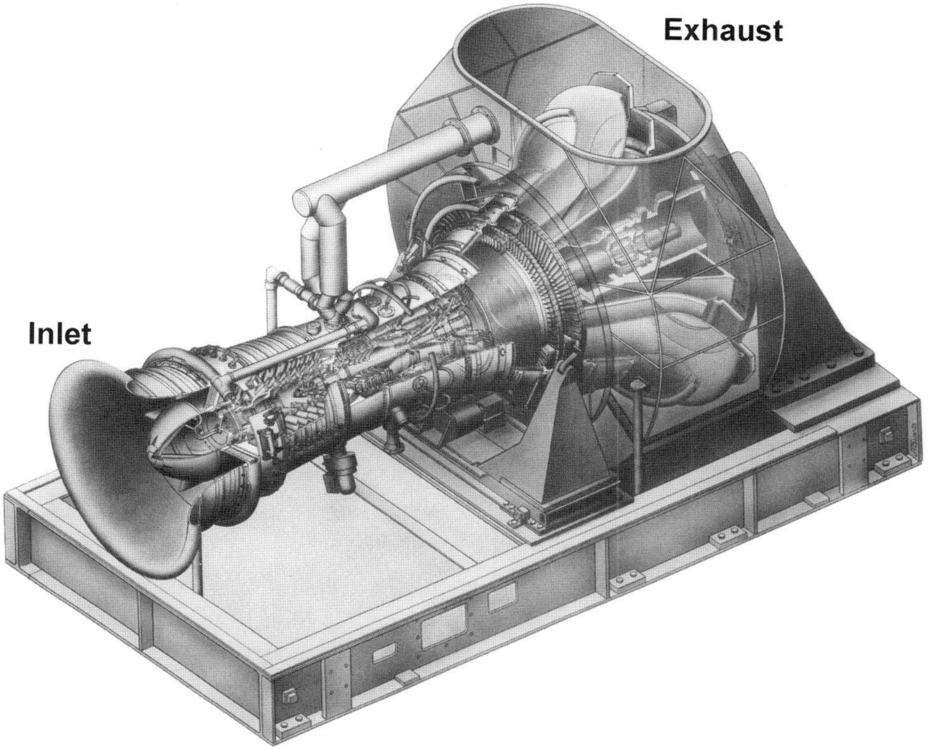

Figure 2.47 Coberra 6000 series gas turbine (© Rolls-Royce plc 2022. All rights reserved).

Exit Pressure

For the following discussions, the entrance of the exhaust is labeled as station 5. However, referring to Figure 2.5 the entrance can be station 5 or station 5x (with a regeneration unit) depending on the gas turbine design and configuration. Figure 2.48 displays the h–s and p–v diagrams for an exhaust for the case without regeneration. An ideal exhaust is assumed to be adiabatic and isentropic. The static pressure and temperature increase through the exhaust since the velocity is decreasing. Because flow is stable at the exit, the exit static pressure matches atmospheric pressure. And for the ideal cases, because the velocity is low at the exit, the exit total pressure matches atmospheric pressure:

$$p_{t8} = p_8 = p_a. \tag{2.3.127}$$

Thus, one can write for the ideal case:

$$p_{t8} = p_8 = p_{t5} = p_a \tag{2.3.128a}$$

$$T_{t8} = T_8 = T_{t5}. \tag{2.3.128b}$$

Pressure Losses

However, for the nonideal case total pressure is lost between the turbine and exit because of friction and internal dissipation in the fluid caused by wall friction, turbulence, and separation regions. As a result of this pressure drop, a turbine is not able to expand down to atmospheric pressure and thus is not able to derive as much power from the fluid as for the ideal case. The design of an exhaust should include smooth walls and gentle curves to reduce such losses. An exhaust is assumed to be adiabatic and, as can be seen in Figure 2.48, a total pressure loss between station 5 and 8 is realized owing to the entropy increase. Again, because flow is stable at the exit, the exit static pressure

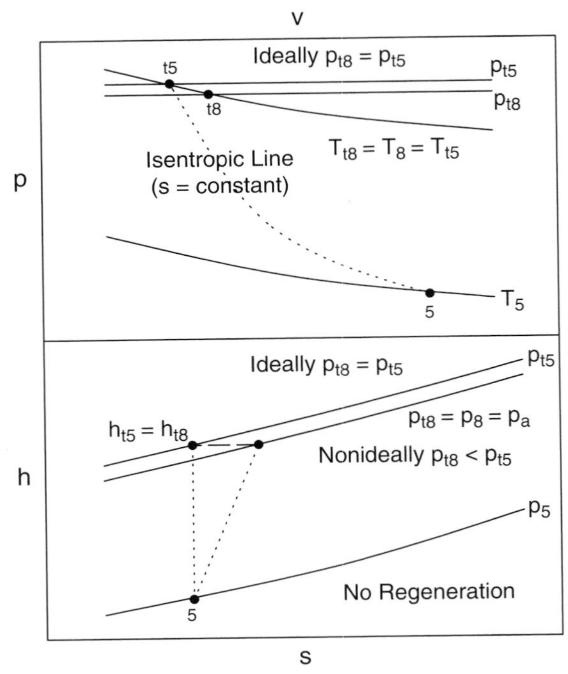

Figure 2.48 The h–s and p–v diagrams for power turbine exhaust.

matches atmospheric pressure. If the velocity is low at the exit (most cases), the exit total pressure again matches atmospheric pressure. Thus, for regeneration one can define an exhaust pressure ratio and write a working equation as follows:

$$\pi_e \equiv \frac{p_{t8}}{p_{t5x}} = \frac{p_8}{p_{t5x}} = \frac{p_a}{p_{t5x}}. \tag{2.3.129a}$$

Or for nonregeneration this reduces to

$$\pi_e \equiv \frac{p_{t8}}{p_{t5}} = \frac{p_8}{p_{t5}} = \frac{p_a}{p_{t5}}. \tag{2.3.129b}$$

This pressure ratio is usually high – on the order of 0.98 to 0.99. Models to predict this loss will be covered in more detail in Chapter 11.

2.3.16 Regenerator (Power Generation Gas Turbines)

As discussed in Chapter 1 a regenerator is sometimes used to improve the performance of a power generation gas turbine. The regenerator is essentially a heat exchanger

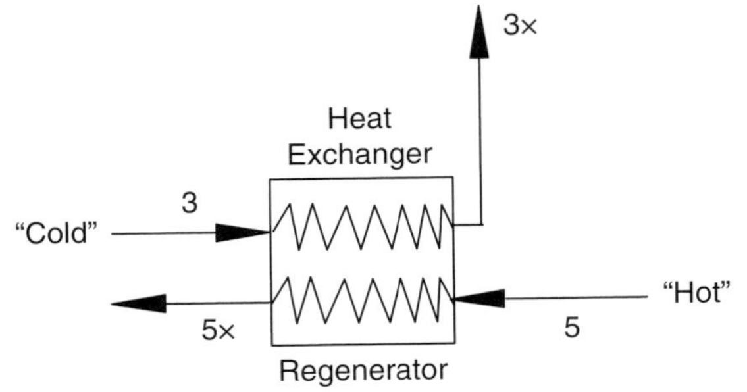

Figure 2.49 Diagram of a regenerator.

and thus a review of the basic operation is in order. The diagram is shown in Figure 2.49. Two streams enter a heat exchanger (HeX): one "hot" and another "cold."

Effectiveness

Owing to limited sizes and weights, heat exchangers are not perfect. Namely, the two exit stream temperatures will never match the two inlet stream temperatures. The effectiveness of a heat exchanger is defined as

$$\varepsilon = \frac{C_h(T_{hi} - T_{ho})}{C_{min}(T_{hi} - T_{ci})} = \frac{C_c(T_{co} - T_{ci})}{C_{min}(T_{hi} - T_{ci})}, \tag{2.3.130}$$

where C is defined as $C \equiv \dot{m} c_p$ and ε is the effectiveness.

The value of C_{min} is the minimum value of C_h or C_c. For the case of the regenerator:

$$\varepsilon_{RG} = \frac{C_c(T_{co} - T_{ci})}{C_{min}(T_{hi} - T_{ci})} = \frac{C_c(T_{t3x} - T_{t3})}{C_{min}(T_{t5} - T_{t3})} = \frac{C_{3-3x}(T_{t3x} - T_{t3})}{C_{min}(T_{t5} - T_{t3})} \tag{2.3.131}$$

or

$$T_{t3x} = T_{t3} + \varepsilon_{RG} \frac{C_{min}}{C_{3-3x}} (T_{t5} - T_{t3}), \tag{2.3.132}$$

again where C_{min} is the lesser of C_{3-3x} or C_{5-5x}. But since both streams are air with about the same value of c_p and the mass flow rate is approximately the same, a reasonable assumption is that:

$$C_{3-3x} \cong C_{5-5x} \cong C_{min}. \tag{2.3.133}$$

Furthermore, since usually $\dot{m}_3 < \dot{m}_5$ and $c_{p3} < c_{p5}$, C_{min} is equal to C_{3-3x}. As a result, for the regenerator, the total temperature entering the primary burner is:

$$T_{t3x} = T_{t3} + \varepsilon_{RG}(T_{t5} - T_{t3}). \quad (2.3.134)$$

Also doing an energy balance on the regenerator:

$$C_{5-5x}(T_{t5} - T_{t5x}) = C_{3-3x}(T_{t3x} - T_{t3}), \quad (2.3.135)$$

so solving for the exit temperature of the hot stream, T_{t5x}:

$$T_{t5x} = T_{t5} - \frac{C_{3-3x}}{C_{5-5x}}(T_{t3x} - T_{t3}) \cong T_{t5} - (T_{t3x} - T_{t3}). \quad (2.3.136)$$

One should note that T_{t5x} has minimal bearing on the overall performance of the gas turbine but does mildly affect the matching with the exhaust and/or stack.

Pressure Losses

Another important characteristic of a heat exchanger is that the total pressure will decrease as the fluid passes through the HeX. This is due to the internal "scrubbing" of the multipassages and thus highly viscous nonisentropic behavior of the flow. As a result:

$$p_{t3x} = \pi_{3-3x}\, p_{t3} \quad (2.3.137)$$

and

$$p_{t5x} = \pi_{5-5x}\, p_{t5}, \quad (2.3.138)$$

where π is, as before, the total pressure ratio of the particular stream and is less than unity. In any heat exchanger increasing the effectiveness tends to decrease the total pressure ratio.

The gas enters the primary burner with the pressure of p_{t3x}, so this is very important. However, as with T_{t5x}, the value of p_{t5x} has minimal importance to the overall performance, but does somewhat affect the calculations of flow in the exhaust and stack.

2.3.17 Intercooler (Power Generation Gas Turbines)

As discussed in Chapter 1 an intercooler is sometimes used to improve the performance of a power generation gas turbine. The intercooler is essentially a heat exchanger and thus a review of the basic operation is in order. The diagram is shown in Figure 2.50. As for the regenerator, two streams enter a heat exchanger (HeX), one "hot" and another "cold."

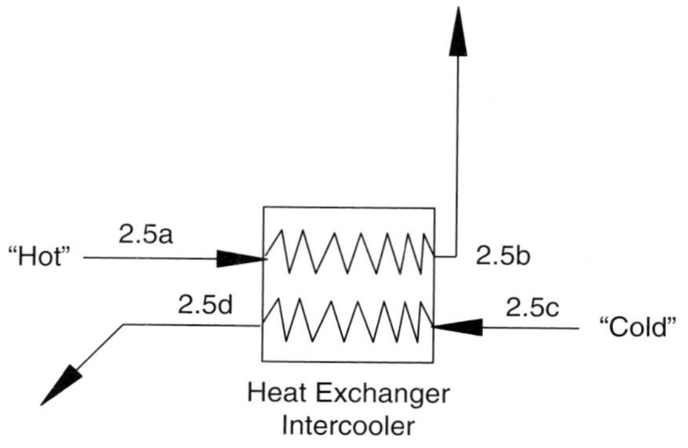

Figure 2.50 Diagram of an intercooler.

Effectiveness

The effectiveness is again defined as

$$\varepsilon = \frac{C_h(T_{hi} - T_{ho})}{C_{min}(T_{hi} - T_{ci})} = \frac{C_c(T_{co} - T_{ci})}{C_{min}(T_{hi} - T_{ci})} \quad (2.3.139)$$

and C is again defined as $C \equiv \dot{m}\, c_p$

The value C_{min} is the minimum value of C_h or C_c. For the case of the intercooler:

$$\varepsilon_{IC} = \frac{C_h(T_{hi} - T_{ho})}{C_{min}(T_{hi} - T_{ci})} = \frac{C_h(T_{25a} - T_{25b})}{C_{min}(T_{25a} - T_{25c})} = \frac{C_{25a-25b}(T_{25a} - T_{25b})}{C_{min}(T_{25a} - T_{25c})} \quad (2.3.140)$$

or

$$T_{25b} = T_{25a} - \varepsilon_{IC}\frac{C_{min}}{C_{25a-25b}}(T_{25a} - T_{25c}), \quad (2.3.141)$$

again where C_{min} is the lesser of $C_{25a-25b}$ or $C_{25c-25d}$.

For the intercooler the value of c_p for the coolant (for example, water) is usually much higher than that of the gas. Also, the coolant usually has a much higher density than that of the gas. As a result, the product of the specific heat and mass flow rate ($C \equiv \dot{m}c_p$) is usually higher than that of the gas so that

$$C_{min} = C_h = C_{25a-25b}. \quad (2.3.142)$$

As a result, for the intercooler, the total temperature entering the high-pressure compressor is:

$$T_{25b} = T_{25a} + \varepsilon_{IC}(T_{25c} - T_{25a}). \quad (2.3.143)$$

Now doing an energy balance on the intercooler:

$$C_{25a-25b}(T_{t25b} - T_{t25a}) = C_{25c-25d}(T_{t25c} - T_{t25d}), \quad (2.3.144)$$

so solving for the exit temperature of the cold stream, T_{t25d}:

$$T_{t25d} = T_{t25c} - \frac{C_{25a-25b}}{C_{25c-25d}}(T_{t25b} - T_{t25a})$$

$$\approx T_{t25c} - (T_{t25b} - T_{t25a}). \quad (2.3.145)$$

One should note that T_{t25d} has no bearing on the overall performance of the gas turbine.

Pressure Losses

As for the regenerator, another important characteristic of a heat exchanger is that the total pressure will decrease as the fluid passes through the HeX, due to the highly viscous nonisentropic behavior of the flow. As a result:

$$p_{t25b} = \pi_{25a-25b}\, p_{t25a} \quad (2.3.146)$$

and

$$p_{t25d} = \pi_{25c-25d}\, p_{t25c}, \quad (2.3.147)$$

where π is again the total pressure ratio of the particular stream.

The gas enters the high-pressure compressor with the pressure of p_{t25b}, so this is quite important. However, as with T_{t25d}, the value of p_{t25d} is not important to the overall performance.

2.3.18 Inlet Cooling

As discussed in Chapter 1, for this cycle the working gas is first cooled by some means at the inlet. Often for low humidity environments for power generation, water or a water–alcohol mix is sprayed at the inlet and the gas temperature is reduced by evaporative cooling. For propulsion applications, this is sometimes used in the diffuser for takeoff in high-temperature/low-humidity locations. This has three effects. First, it lowers the total temperature and, as a result, increases the total pressure at the inlet as shown by a generalized 1-D analysis presented in Appendix C.13. But second, because droplets are sprayed into the inlet, it has the same effect as adding a drag and mixing loss to the total pressure also as shown by a generalized 1-D analysis presented Appendix C.13 and in other compressible flow texts; see, for example, Shapiro (1953). And third, because the process is evaporative cooling, the process is irreversible, thus again decreasing the total pressure. To model an accurate prediction a full analysis is required. However, for the sake of this text, and since all three are very small, they will be assumed to offset each other. This has a similar effect as intercooling. With a cooler inlet temperature, less power is required to attain the same compressor exit pressure. The net effect of inlet cooling is to increase the net power output and maintain the same overall thermal efficiency for a power-generation application and to increase the thrust for propulsion applications. This enhancer is best defined as simply a temperature drop, $\Delta T_t = (T_{ta} - T_{t2})$, at the inlet or diffuser.

2.3.19 Summary of Nonideal Effects and Simple Parameter Models in Components

In the previous sections in this chapter, 17 components were covered and nonideal effects were discussed for each. For each component, one- or two-term parameters were introduced to incorporate the different types of losses. Other parameters were introduced to include operating characteristics. To conclude this section, these parameters are summarized in Table 2.2.

Table 2.2 **Component parameters.**

Component	Effect	Symbol
Diffuser [a]	Total pressure recovery	π_d
Bell inlet [b]	Total pressure ratio	π_i
Compressor	Efficiency	η_c
Fan [a]	Efficiency	η_f
	Bypass ratio	α
Turbine	Efficiency	η_t
Propeller [a]	Efficiency	η_p
Shaft	Efficiency	η_m
Combustor	Efficiency	η_b
	Total pressure ratio	π_b
Reheat Combustor [b]	Efficiency	η_{RHb}
	Total pressure ratio	π_{RHb}
Afterburner [a]	Efficiency	η_{ab}
	Total pressure ratio	π_{ab}
Primary nozzle [a]	Efficiency	η_n
Fan nozzle [a]	Efficiency	η_{fn}
Bypass duct [a]	Total pressure ratio	π_u
	Split ratio	σ
Bypass mixer [a]	Total pressure ratio	π_m
Exhausts and stack [b]	Pressure ratio	π_e
Intercooler [b]	Effectiveness	ε_{IC}
	Pressure ratio	$\pi_{25a-25b}$
	Pressure ratio	$\pi_{25c-25d}$
Regenerator [b]	Effectiveness	ε_{RG}
	Pressure ratio	π_{3-3x}
	Pressure ratio	π_{5-5x}
Inlet cooling	Temperature drop	$\Delta T_t = T_{ta} - T_{t2}$

[a] Propulsion only.
[b] Power generation only.

2.4 Summary

In this chapter the ideal and nonideal gas turbine cycle analyses were initiated to facilitate calculation of the ideal overall gas turbine performance in Chapter 3. First, a brief review of the thermodynamics of compressible fluid flow was presented. This included the definition of stagnation or total properties, isentropic flow, Fanno line flow, Raleigh line flow, normal and obliques shocks. More complex thermodynamics discussions for combinations of flow with area changes, friction, drag objects, and heat addition/removal were presented. The reader is encouraged to read this section (Section 2.2) and Appendix C, if unfamilar with the different processes.

Next, components (inlet, diffuser, compressor, fan, propeller, combustor, turbine, bypass duct, bypass mixer, afterburner, nozzle, power gas turbine exhaust, regeneration, intercooling, and reheat) were discussed, and the ideal and nonideal characteristics of the components were defined. The thermodynamic concepts were reviewed in detail, and the enthalpy–entropy (h–s) and pressure–volume (p–v) diagrams and operating equations were presented for each of the components for nonideal cases but ideal characteristics were also discussed. The physical phenomena that contribute to the component losses were discussed. One- and two-term specifiable losses and efficiencies were included in the operating equations and thermodynamic processes for all of the components. For the nonideal components and for the compressor and turbine, two different efficiencies were defined and discussed (conventional and polytropic). For the duct and mixer flow, stability was addressed. Simple models of losses for the components were chosen so that cycle analyses could be easily accomplished but would still enable the reader to make realistic predictions of engine performance. In addition, nonideal gas effects can be included with the addition of variable specific heats for the different components. In the next chapter, components are assembled to determine the ideal and nonideal cycle performance, for example thrust and power output, of different gas turbine types. Typical trend studies will be presented for different jet engine and power generation types.

List of Symbols

A	Area
A_d	Drag frontal area
A_R	Relative blockage or relative frontal area
a	Speed of sound
c_p	Specific heat at constant pressure
c_v	Specific heat at constant volume
C	Heat exchanger heat capacity rate
C_{w_p}, C_T, C_P	Coefficients for propeller
C_d	Drag coefficient
D	Diameter
F_d	Drag force
f	Fuel ratio
f	Fanning friction coefficient
h	Specific enthalpy
ΔH	Heating value
L	Length
$\dot{m}$	Mass flow rate
M	Mach number
n	Number of stages
p	Pressure
$\mathcal{P}$	Power
$\dot{Q}$	Heat transfer rate
$\mathcal{R}$	Ideal gas constant

s	Entropy
T	Temperature
u	Specific internal energy
u	Velocity
v	Specific volume
V	Velocity
x	Direction
α	Bypass ratio
γ	Specific heat ratio
ε	Heat exchanger effectiveness
η	Efficiency
π	Total pressure ratio
ρ	Density
σ	Split ratio
τ	Total temperature ratio

Subscripts

a	Freestream
a	Due to air only
ab	Afterburner
b	Primary burner
c	Compressor
c	Cold stream
d	Diffuser
e	Exhaust
ext	Outside of diffuser/inlet
f	Fan
f	Fuel
fn	Fan nozzle
h	Hot stream
HPC	High-pressure compressor
i	Inlet
int	Diffuser/inlet internal
IC	Intercooler
LPC	Low-pressure compressor
m	Mechanical (Shaft)
m	Mixer
min	Minimum heat capacity rate
mc	Mixer – core side
mu	Mixer – duct side
n	Nozzle
p	Propeller
p	Polytropic
p	Power
pc	Polytropic – compressor
pt	Polytropic – turbine
r	Shock recovery
RG	Regenerator
RH	Reheat
s	Secondary (bypassed)
ss	Single stage
t	Total (stagnation)
t	Total (summation)
t	Turbine
u	Duct
th	Thermal
w	Work
x	Inlet to single stage
y	Exit of single stage
1 ... 9, a,b, ...	Positions in components

Superscripts

′	Ideal
*	Choked
–	Average

Problems

2.1 Flow enters a nozzle at Mach 0.40. The inlet area is 0.7 m^2 and the exit area is 0.5 m^2. What is the exit Mach number for isentropic flow if $\gamma = 1.40$?

2.2 Flow with an incoming Mach number of 0.45 and an inlet total pressure of 25 psia enters the bypass duct of a turbofan. The duct is 60 in. long with a hydraulic diameter of 3.5 in. The duct has a Fanning friction factor of 0.004. What is the total pressure ratio (p_{t2}/p_{t1}) if the specific heat ratio is 1.38?

2.3 Flow with an incoming Mach number of 0.30 and an inlet total pressure of 1600 kPa is heated from a total temperature of 800 K to 1700 K. What is the total pressure ratio if the specific heat ratio is 1.32?

2.4 A "standing" normal shock is outside of a diffuser. The Mach number before the shock is 1.350. The total pressure and total temperature are 40 psia and 450 °R, respectively. Find the exiting Mach number, exiting conditions, and total pressure ratio for the shock if $\gamma = 1.38$.

2.5 Flow with a Mach number of 2.5 enters an inlet with a wedge angle of 13°. Find the exiting Mach number, shock angle, and total pressure ratio for $\gamma = 1.37$.

2.6 A compressor has an efficiency of 88 percent with a pressure ratio of 22 and specific heat ratio of 1.38. What is the polytropic efficiency?

2.7 A compressor has a polytropic efficiency of 93 percent with a pressure ratio of 24 and a specific heat ratio of 1.38. What is the compressor efficiency?

2.8 A turbine has an efficiency of 86 percent with a pressure ratio of 0.0455 and specific heat ratio of 1.34. What is the polytropic efficiency?

2.9 A turbine has a polytropic efficiency of 83 percent with a pressure ratio of 0.0417 and a specific heat ratio of 1.34. What is the turbine efficiency?

2.10 A compressor has an efficiency of 88 percent with a pressure ratio of 22 and specific heat ratio of 1.38. The compressor is comprised of low-pressure and high-pressure sections. Each have the same pressure ratio and efficiency. What is the pressure ratio and efficiency of each section?

2.11 A turbine has an efficiency of 86 percent with a pressure ratio of 0.0455 and specific heat ratio of 1.34. The turbine is comprised of high-pressure and low-pressure sections. Each have the same pressure ratio and efficiency. What is the pressure ratio and efficiency of each section?

2.12 Draw the h–s and p–v diagrams for an ideal gas generator.

2.13 Draw the h–s and p–v diagrams for a nonideal gas generator.

2.14 A regenerator in a power generation gas turbine operates with an effectiveness of 0.70. Air exits a high-pressure compressor at 481 K. Air exits the free turbine at 863 K. Find the inlet temperature into the primary burner. Assume the specific heat ratio, $\gamma = 1.35$.

2.15 An intercooler in a power generation gas turbine operates with an effectiveness of 0.70. Air exits the low-pressure compressor at 441 K at 60 kg/s. Water is used for the coolant and the specific heat is 4.2 kJ/kg-K and water flows at 26 kg/s. Find the inlet temperature for the high-pressure compressor. Assume the specific heat ratio for air, $\gamma = 1.35$.

2.16 A low-pressure compressor in a power generation gas turbine operates with a pressure ratio of 4.5 and efficiency of 90 percent. Air enters the compressor at 60 kg/s and at 288.2 K. Find the exit temperature and required power. Assume the specific heat ratio for air, $\gamma = 1.35$.

2.17 A low-pressure turbine in a power generation gas turbine operates with a pressure ratio of 0.641 and efficiency of 92 percent. Air enters the turbine at 61.6 kg/s and at 1500 K. Find the exit temperature and derived power. Assume the specific heat ratio for air, $\gamma = 1.35$.

2.18 A primary burner in a power generation gas turbine operates with a pressure ratio of 0.97 and efficiency of 96 percent. Air enters the burner at 60 kg/s and at 748.0 K. The fuel flow rate is 1.29 kg/s. The heating value of the fuel is 42,000 kJ/kg. Find the exit temperature. Assume the specific heat ratio for air, $\gamma = 1.35$.

3 Cycle Analysis: Ideal and Nonideal

3.1 Introduction

Chapter 1 identified the basic engine types and defined the important operating performance parameters of gas turbines. That chapter also briefly reviewed the fundamental thermodynamics of cycles and established that gas turbines consist of several important components. This chapter covers all basic jet engine and power generation configurations. Both ideal and nonideal components were considered in Chapter 2. Ideal implies that there are no losses in the components. Nonideal includes such losses. Components are now assembled in a “cycle analysis” to make it possible to predict the overall engine performance. As will be seen, the ideal cycle analysis results in closed-form equations for the engine characteristics and final equations are summarized in this chapter. Developing these equations (Appendix H) serves three purposes. First, doing so allows the reader to see how the equations are assembled to model an engine without having to deal with numerical details. Second, and more importantly, developing the equations allows the reader to observe how detail parameters affect the overall performance of an engine without parametrically varying numbers. Third, in some cases, analytically optimizing a performance characteristic is possible. On the other hand, nonideal cycle analysis does not result in closed-form equations owing to the increased complexity of the component equations of the engine. Nonideal efficiency levels and losses are included for the different components so that more realistic predictions can be made for overall engine performance. Also, note that the simple single- or two-term expressions used to model the losses in each component in Chapter 2 are used. Even though most components operate with individually with relatively high efficiencies (upwards from 90 percent), when all the components are coupled the overall engine performance can be reduced dramatically. However, in general, parametric studies and the performance trends are similar for both ideal and nonideal cases. This chapter presents quantitative examples to demonstrate the analysis and to give the reader a physical understanding of characteristics. Trend studies are also discussed to show the dependence of the overall characteristics on individual component parameters. At this stage the loss terms are specified a priori even though, in a real engine, the different component losses are dependent on the engine operating point and are thus dependent on each other. This advanced topic is the subject of Chapter 12, which addresses component matching. More complex and refined analyses of the component losses are presented in each of the component chapters (Chapters 4–11).

3.1.1 Objectives

After completing this chapter, the reader should be able to do the following.

- Integrate the thermodynamics of components into a cycle analysis of an entire gas turbine
- Understand the implications of the closed-form ideal cycle analysis equations
- Understand the effects of ideal vs. nonideal components on overall performance of a gas turbine
- Understand the differences on overall performance between different types of gas turbines

For selected equations in this chapter see equation indicators in Appendix H for full display. Appendix H is available as a downloadable resource on the Resources tab at www.cambridge.org/flack2e

- Understand how the different component parameters affect the overall performance of a gas turbine using parametric studies
- Understand how the different performance enhancers affect the overall performance of a gas turbine
- Understand how a cycle analysis can be used to perform a "first stage" design for a gas turbine with simple one or two term parameters for the components

3.2 Ideal Cycle Analysis Equations: Summary of Final Equations

In Section 3.1, ideal and nonideal components were described. In this section performance equations of different engines with ideal components will be summarized. Depending on the gas turbine type of concern, different combinations of the different components will be used. For the different jet engine types analytical closed-form equations for thrust and *TSFC* are derived in detail in Appendix H based on the ideal component equations developed in Section 2.3 and summarized here. A similar procedure is covered for power generation units. Deriving these equations allows the reader to see how the equations are assembled to model an engine without having to deal with numerical details. Second, and more importantly, developing the equations allows the reader to observe how detail parameters affect the overall performance of an engine without parametrically varying numbers. Third, in some cases, analytically optimizing a performance characteristic is possible. On the other hand, nonideal cycle analysis does not result in closed-form equations owing to the increased complexity of the engine but trend studies will be presented for both ideal and nonideal cases in the next section. One should remember the important assumptions for the ideal case include the following.

(a) Constant specific heat ratio.
(b) Negligible fuel flow, $f \equiv \dot{m}_f/\dot{m}$, is small.
(c) Exit pressure matches ambient conditions.
(d) Isentropic flow in all noncombusting components.
(e) Constant total pressure flow in all combusting components.
(f) Perfect combustion in all combusting components.

Before proceeding, the reader is cautioned that at first glance some of the resulting equations are long and rather intimidating. They are all derived in detail in Appendix H. However, they serve an important contribution. Namely, as will be shown, these equations often provide insight as to how a gas turbine may be improved or optimized. For nonideal cycle analysis, different optimal values will be found. However, with some of the insight provided by these equations, those parameters that lead to optimizations are identified for nonideal gas turbines.

3.2.1 Ramjet/Scramjet

In Figure 3.1 a general ramjet is considered and in Figure 1.16 a scramjet is shown. As indicated earlier, these are the simplest of jet engines as they do not have any rotating shafts. Also, for the application of cycle analysis, both the ramjet and scramjet can be treated in the same manner. This is not to say, however, that the designs are similar because they are markedly different as discussed in Chapter 1. It also should be noted that

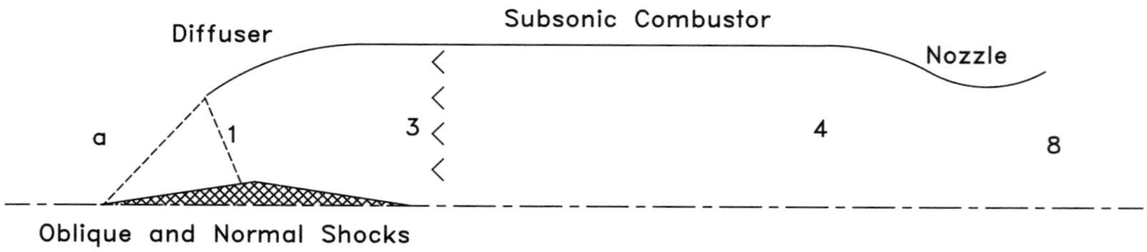

Figure 3.1 Ramjet.

the ideal analysis of ramjets and scramjets is subject to more scrutiny than for the other engine configurations. That is because, even with the best of conditions, shocks occur before or in the diffusers and burners resulting in total pressure losses (negating assumption (d) above. Regardless to provide an ideal benchmark, we will proceed with the above assumptions. In Figure 3.1 position a is the ambient condition, position 1 is the inlet to the diffuser, position 3 is the exit of the diffuser and inlet to the combustor, position 4 is the exit of the burner and inlet to the nozzle, lastly position 8 is the exit of the nozzle. The nondimensional thrust is given by

$$\frac{F}{\dot{m}a_a} = M_a\left(\sqrt{\tau_b} - 1\right), \qquad (3.2.1a) \; / \; (H.2.18)$$

where

$$\tau_b = \frac{T_{t4}}{T_{t3}}. \qquad (3.2.1b)$$

By examining this simple equation, one can see that the dimensional thrust is directly proportional to the mass flow rate. It is also directly proportional to the Mach number if all of the other parameters are constant. One limiting case is for a Mach number of zero; this equation indicates that the thrust is zero for this case. Moreover, as the burner total temperature ratio increases, so does the thrust. One should remember however, that T_{t3} is related to the freestream Mach number $\left(T_{t3} = T_{ta} = T_a\left[1 + \frac{\gamma - 1}{2}M_a^2\right]\right)$, so that independently varying M_a and τ_b is difficult. For the nondimensional *TSFC*, where a_a is the freestream sound speed:

$$TSFC\; a_a = \frac{c_p T_{ta}(\tau_b - 1)}{M_a \Delta H\left(\sqrt{\tau_b} - 1\right)}. \qquad (3.2.2) \; / \; (H.2.20)$$

By examining this equation, one can see that the *TSFC* is directly proportional to the total ambient temperature, but again one must remember that T_{ta} is related to the Mach number. Also, as the burner total temperature ratio increases, so does the *TSFC*. For a typical ramjet, the values of *TSFC* range from 1.5 to 3 $lb_m/h/lb_f$ (0.15 to 0.3 kg/kW/h).

One can differentiate Eq. (3.2.2) with respect to M_a and easily show that for constant altitude, heating value, and τ_b, the minimum or optimum value of *TSFC* occurs for:

$$M_{a_{opt}} = \sqrt{\frac{2}{\gamma - 1}}. \qquad (3.2.3) \; / \; (H.2.21)$$

One can now consider a special case, namely the *stoichiometric* condition. For this condition exactly the right proportion of fuel and air will be used so that *all* of the fuel and air are used in the combustion process and neither are left in the combustion products. This will be discussed in more detail in Chapter 9. For this case,

$$TSFC_{st}\; a_a = \frac{f_{st}}{M_a\left(\sqrt{\frac{f_{st}\Delta H\left(\frac{T_a}{T_{ta}}\right)}{c_p T_a} + 1} - 1\right)}. \qquad (3.2.4) \; / \; (H.2.24)$$

One should remember that the quantity T_{ta}/T_a is solely a function of the freestream Mach number. Thus, for a given fuel at stoichiometric conditions the value of *TSFC* is only a function of the Mach number and T_a (or a_a for a given γ). Typical values of the fuel ratio for stoichiometric conditions range from about 0.06 to 0.07. More will be discussed in Chapter 9.

3.2.2 Turbojet

In Figure 3.2 a general turbojet is shown. This is a much more complex machine than the ramjet because of the rotating shaft. Eight stations are identified. The freestream is position a. The inlet is position 1. Position 2 is the exit of the diffuser and compressor inlet. Station 3 is the compressor exit and burner inlet. The burner exit and turbine inlet is denoted as station 4. Station 5 is the turbine exit and afterburner inlet. The afterburner exit and nozzle inlet is position 6. Finally, station 8 is the nozzle exhaust. The nonafterburning case is presented in Figure 3.3.

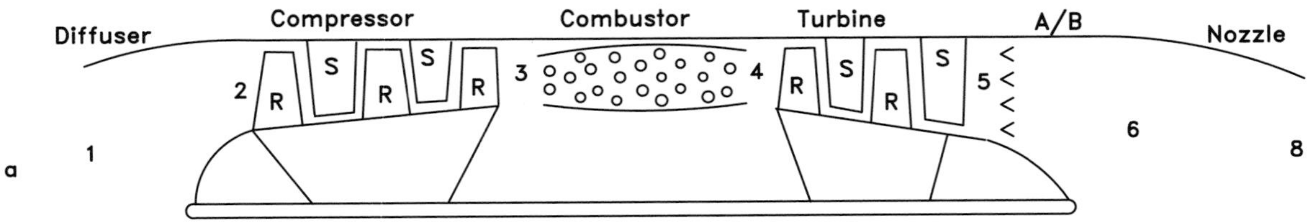

Figure 3.2 General turbojet.

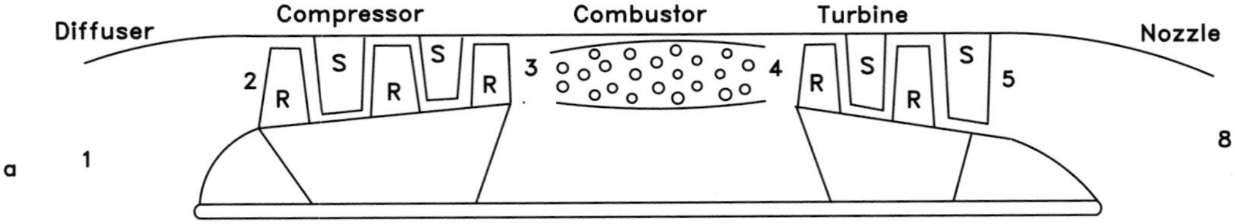

Figure 3.3 Nonafterburning turbojet.

Nonafterburning

For the ideal dimensionless thrust:

$$\frac{F}{\dot{m}a_a} = M_a\left[\sqrt{\left[\frac{\frac{T_{ta}}{T_a}}{\frac{T_{ta}}{T_a}-1}\right]\left[\frac{T_{t4}}{T_a}\frac{T_a}{T_{ta}}\frac{1}{\tau_c}-1\right][\tau_c-1]+\left[\frac{\frac{T_{t4}}{T_a}}{\frac{T_{ta}}{T_a}\tau_c}\right]}-1\right].$$

(3.2.5) / (H.3.16)

By examining this equation, one can see that, as for a ramjet, the dimensional thrust is directly proportional to the mass flow rate. It is also directly proportional to the Mach number if all of the other parameters are constant. However, one should remember that T_{ta}/T_a is strictly a function of Mach number and γ. Also, as the burner total temperature increases, so does the thrust. However, as the ambient static temperature decreases, the thrust increases. Also, in Eq. (3.2.5) the compressor total temperature ratio (which is directly related to the total pressure ratio, $\pi_c = [\tau_c]^{\frac{\gamma}{\gamma-1}}$) is seen to strongly influence the thrust performance. By differentiating this equation with respect to τ_c and setting to zero, one finds that the nondimensional thrust is maximized for a value of:

$$\tau_{c_{opt}} = \sqrt{(T_{t4}/T_a)/(T_{ta}/T_a)}. \qquad (3.2.6)\ /\ (H.3.17)$$

This can then be used to find the optimum compressor pressure ratio from $\pi_{c_{opt}} = [\tau_{c_{opt}}]^{\frac{\gamma}{\gamma-1}}$.
Finally, for the dimensionless *TSFC*:

$$TSFC\ a_a = \frac{\left[\frac{T_{t4}}{T_a}-\frac{T_{ta}}{T_a}\tau_c\right]\left[\frac{c_p\ T_a}{\Delta H\ M_a}\right]}{\left[\sqrt{\left[\frac{\frac{T_{ta}}{T_a}}{\frac{T_{ta}}{T_a}-1}\right]\left[\frac{T_{t4}}{T_a}\frac{T_a}{T_{ta}}\frac{1}{\tau_c}-1\right][\tau_c-1]+\left[\frac{T_{t4}}{T_a}\frac{T_a}{T_{ta}}\frac{1}{\tau_c}\right]}-1\right]}.$$

(3.2.7) / (H.3.20)

As the ambient static temperature decreases, the *TSFC* decreases. Careful examination also shows that the *TSFC* decreases as τ_c (or π_c) increases. For a typical nonafterburning turbojet, the values of *TSFC* range from 0.75 to 1.00 $lb_m/h/lb_f$ (0.07 to 0.1 kg/kW/h).

Afterburning

The nondimensional thrust with an afterburner is

$$\frac{F}{\dot{m}a_a} = M_a\left[\sqrt{\left[\frac{\frac{T_{t8}}{T_a}}{\frac{T_{ta}}{T_a}-1}\right]\left[1-\frac{\frac{T_{t4}}{T_a}\frac{T_a}{T_{ta}}}{\tau_c\left(\frac{T_{t4}}{T_a}-\frac{T_{ta}}{T_a}(\tau_c-1)\right)}\right]}-1\right].$$

(3.2.8) / (H.3.30)

This can be compared with the nonafterburning case (Eq. (3.2.5)). Again, the dimensional thrust is directly proportional to the Mach number and mass flow rate if all of the other parameters are constant. As the burner and/or afterburner total temperature increases, so does the thrust. Again, the compressor total temperature ratio is seen to strongly influence the thrust performance. By differentiating this equation relative to τ_c for a given value of T_{t8}, and setting to zero, one finds that the nondimensional thrust is maximized for a value of:

$$\tau_{c_{opt}} = [(T_{t4}/T_a) + (T_{ta}/T_a)]/[2(T_{ta}/T_a)], \qquad (3.2.9)\ /\ (H.3.31)$$

which can be related to the total pressure ratio. The optimum value for the afterburning turbojet is considerably higher than for the nonafterburning case.

For the nondimensional *TSFC*:

$$TSFC\ a_a = \frac{\left[\frac{T_{t8}}{T_a} - \frac{T_{ta}}{T_a}\tau_c\right]\left[\frac{c_p\,T_a}{\Delta H\,M_a}\right]}{\left[\sqrt{\left[\frac{\frac{T_{t8}}{T_a}}{\frac{T_{ta}}{T_a}-1}\right]\left[1 - \frac{\frac{T_{t4}}{T_a}\frac{T_a}{T_{ta}}}{\tau_c\left(\frac{T_{t4}}{T_a} - \frac{T_{ta}}{T_a}(\tau_c - 1)\right)}\right]} - 1\right]}. \qquad (3.2.10)\ /\ (H.3.32)$$

3.2.3 Turbofan

The general diagram for a turbojet or turbofan is shown in Figure 2.3. As indicated earlier, this engine has up to three basic flow paths. Two simplifications of the general turbofan are covered for the ideal analysis: the exhausted fan and the ducted/mixed fan. Both nonafterburning and afterburning cases are considered.

Fan Exhausted

The first type of a turbofan is shown in Figure 3.4. For this configuration two flow paths exist. For one path the air enters the diffuser and flows through the core and exhausts through the primary nozzle. For the second path air enters the diffuser flows through the fan and exits through the fan nozzle. Two separate subcases are considered here: without and with afterburner. In Figure 3.4 an afterburner is present in the core between stations 5 and 6. In Figure 3.5 the nonafterburning case is shown.

Nonafterburning The dimensionless thrust is:

$$\frac{F}{\dot{m}a_a} = M_a\left[\sqrt{\frac{\left[\frac{T_{t4}}{T_a}\frac{T_a}{T_{ta}}\frac{1}{\tau_c}\right]\left[\frac{T_{ta}}{T_a}\tau_c\left[1 - \left[\frac{T_{ta}}{T_a}\times\frac{T_a}{T_{t4}}\right][(\tau_c - 1) + \alpha\,(\tau_f - 1)]\right] - 1\right]}{\frac{T_{ta}}{T_a} - 1}} - 1\right] + \alpha M_a\left[\sqrt{\frac{\left[\frac{T_{ta}}{T_a}\tau_f - 1\right]}{\frac{T_{ta}}{T_a} - 1}} - 1\right]. \qquad (3.2.11)\ /\ (H.4.14)$$

Thus, the thrust is a strong function of compressor total pressure ratio, fan total pressure ratio, exit total temperature from the burner and bypass ratio. The nondimensional *TSFC* is

$$TSFC\ a_a = \frac{\left[\frac{T_{t4}}{T_a} - \frac{T_{ta}}{T_a}\tau_c\right]\left[\frac{c_p\,T_a}{\Delta H}\right]}{M_a\left[\sqrt{\frac{\left[\frac{T_{t4}}{T_a}\frac{T_a}{T_{ta}}\frac{1}{\tau_c}\right]\left[\frac{T_{ta}}{T_a}\tau_c\left[1 - \left[\frac{T_{ta}}{T_a}\times\frac{T_a}{T_{t4}}\right][(\tau_c - 1) + \alpha\,(\tau_f - 1)]\right] - 1\right]}{\frac{T_{ta}}{T_a} - 1}} - 1\right] + \alpha M_a\left[\sqrt{\frac{\left[\frac{T_{ta}}{T_a}\tau_f - 1\right]}{\frac{T_{ta}}{T_a} - 1}} - 1\right]}. \qquad (3.2.12)\ /\ (H.4.15)$$

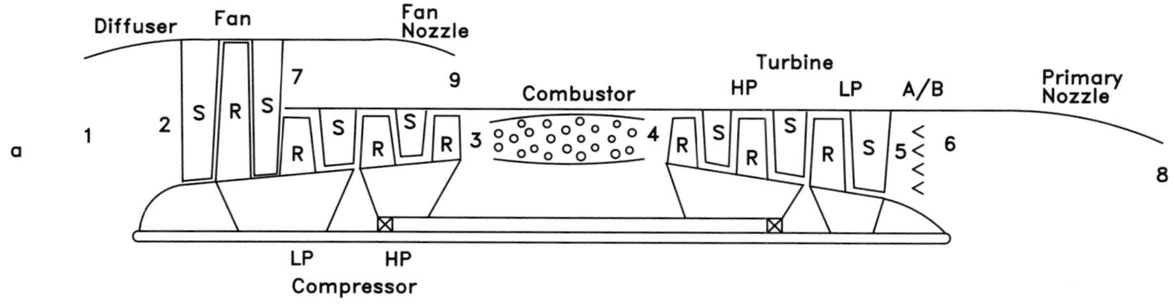

Figure 3.4 General turbofan with exhausted fan.

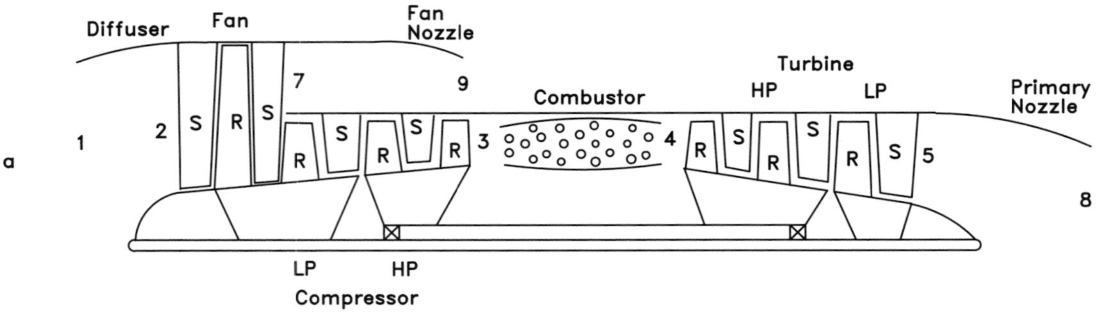

Figure 3.5 Nonafterburning turbofan with exhausted fan.

For a typical nonafterburning exhausted turbofan, the values of *TSFC* range from 0.35 to 0.75 $lb_m/h/lb_f$ (0.035 to 0.08 kg/kW/h).

If one chooses to maximize the thrust, one can differentiate F with respect to the fan total temperature ratio (again related to the fan total pressure ratio), τ_f, in Eq. (3.2.11), set equal to zero, and show that the value of τ_f is:

$$\tau_{f_{opt}} = \frac{1 + \frac{T_{t4}}{T_a} + \frac{T_{ta}}{T_a}[1 + \alpha - \tau_c] - \frac{\frac{T_{t4}}{T_a}}{\frac{T_{ta}}{T_a}\tau_c}}{\frac{T_{ta}}{T_a}[1+\alpha]}, \qquad (3.2.13) / (H.4.16)$$

which can be used to find the optimum total pressure ratio from $\pi_{f_{opt}} = [\tau_{f_{opt}}]^{\frac{\gamma}{\gamma-1}}$. This condition interestingly results in the two nozzle exit velocities being the same:

$$u_8 = u_9. \qquad (3.2.14) / (H.4.17)$$

Similarly, one can maximize the thrust by differentiating F with respect to the bypass ratio, α, set equal to zero and solve for α, which yields

$$\alpha_{opt} = \frac{1}{(\tau_f - 1)}\left\{\frac{\frac{T_{t4}}{T_a}}{\frac{T_{ta}}{T_a}}\left(1 - \frac{1}{\tau_c \frac{T_{ta}}{T_a}} - \frac{1}{4\frac{T_{t4}}{T_a}}\left[\sqrt{\frac{T_{ta}}{T_a}\tau_f - 1} + \sqrt{\frac{T_{ta}}{T_a} - 1}\right]^2\right) - (\tau_c - 1)\right\}. \qquad (3.2.15) / (H.4.18)$$

Through a series of parametric studies comparing engine types in the next subsection, a better parameter to nondimensionalize the thrust is the total flow rate ($\dot{m}_t = \dot{m}(1+\alpha)$).

Afterburning Figure 3.4 depicts this geometry for the afterburning case. One finds the dimensionless thrust:

$$\frac{F}{\dot{m}a_a} = M_a\left[\sqrt{\left[\frac{\frac{T_{t6}}{T_a}}{\frac{T_{ta}}{T_a} - 1}\right]\left[1 - \frac{1}{\left(\frac{T_{ta}}{T_a}\tau_c\left(1 - \left[\frac{T_{ta}}{T_a} \times \frac{T_a}{T_{t4}}\right][(\tau_c - 1) + \alpha(\tau_f - 1)]\right)\right)}\right]} - 1\right] + \alpha M_a\left[\sqrt{\left[\frac{\frac{T_{ta}}{T_a}\tau_f - 1}{\frac{T_{ta}}{T_a} - 1}\right]} - 1\right]. \qquad (3.2.16) / (H.4.19)$$

The thrust is a strong function of burner exit temperature, afterburner exit temperature, compressor pressure ratio, fan pressure ratio, and bypass ratio. Of course, the Mach number and mass flow rate are strong influences as well as for all other engines. Finally, for the nondimensional *TSFC*:

$$TSFC\ a_a = \frac{\left[\frac{T_{t6}}{T_a} \times \frac{T_a}{T_{ta}} + \alpha\tau_f - (1+\alpha)\right]\left[\frac{c_p T_a}{\Delta H}\right]\left[\frac{T_{ta}}{T_a}\right]}{M_a\left[\sqrt{\left[\frac{\frac{T_{t6}}{T_a}}{\frac{T_{ta}}{T_a} - 1}\right]\left[1 - \frac{1}{\left(\frac{T_{ta}}{T_a}\tau_c\left(1 - \left[\frac{T_{ta}}{T_a} \times \frac{T_a}{T_{t4}}\right][(\tau_c - 1) + \alpha(\tau_f - 1)]\right)\right)}\right]} - 1\right] + \alpha M_a\left[\sqrt{\left[\frac{\frac{T_{ta}}{T_a}\tau_f - 1}{\frac{T_{ta}}{T_a} - 1}\right]} - 1\right]}. \qquad (3.2.17) / (H.4.23)$$

Fan Mixed

The second type of fan is shown in Figure 3.6. For this configuration the fan exit bypassed air passes through a bypass duct and is mixed with the turbine exhaust before station 5.5. Two separate subcases are considered here: without and with afterburner.

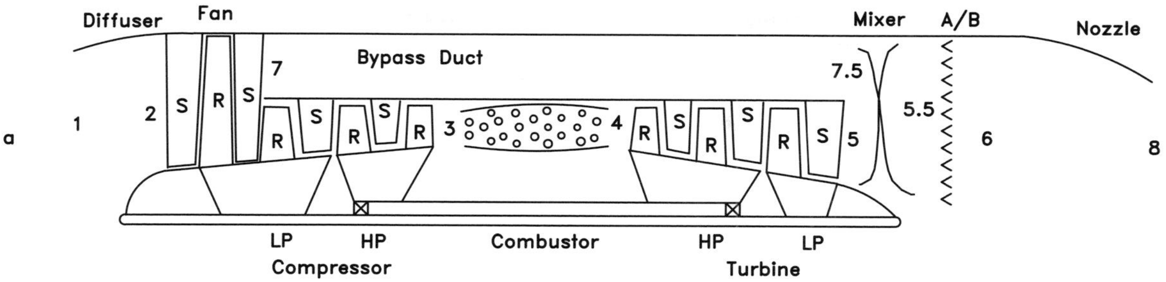

Figure 3.6 General turbofan with mixed fan.

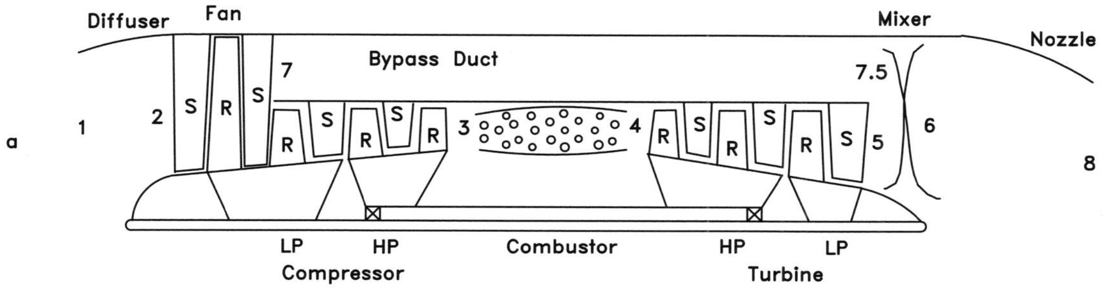

Figure. 3.7 Nonafterburning mixed turbofan.

Nonafterburning For the nonafterburning case stations 5.5 and 6 are one in the same and will be termed station 6 as shown in Figure 3.7. As for the turbojet, since only one exhaust gas stream or nozzle is present, one obtains the following for the dimensionless thrust:

$$\frac{F}{\dot{m}\,a_a} = M_a(1+\alpha)\left[\sqrt{\left[\frac{\alpha+\left[\frac{T_{t4}}{T_a}\right]\left[\frac{T_a}{T_{ta}}\right]\frac{1}{\tau_c}}{1+\alpha}\right]\left[\frac{\left[\frac{T_{ta}}{T_a}\right]\left[\frac{\tau_c+\left[\frac{T_{ta}}{T_a}\right]\left[\frac{T_a}{T_{t4}}\right]\tau_c(1+\alpha-\tau_c)}{1+\left[\frac{T_{ta}}{T_a}\right]\left[\frac{T_a}{T_{t4}}\right]\tau_c\alpha}\right]-1}{\left[\frac{T_{ta}}{T_a}\right]-1}\right]}-1\right],$$

(3.2.18) / (H.5.20)

where

$$\tau_f = \frac{\tau_c + \left[\frac{T_{ta}}{T_a}\right]\left[\frac{T_a}{T_{t4}}\right]\tau_c(1+\alpha-\tau_c)}{1+\left[\frac{T_{ta}}{T_a}\right]\left[\frac{T_a}{T_{t4}}\right]\tau_c\alpha}.$$

(3.2.19) / (H.5.19)

One should note that the fan total pressure ratio is *not* independent as it was for the exhausted case.

The thrust is a strong function of burner exit temperature, compressor pressure ratio, and bypass ratio. As for all other engines, the Mach number and mass flow rate are of course strong influences as well. Finding the nondimensionalized *TSFC* for the ducted fan is similar to that for the exhausted fan. For the *TSFC*:

$$TSFC\ a_a = \frac{\left[\frac{T_{t4}}{T_a}-\frac{T_{ta}}{T_a}\tau_c\right]\left[\frac{c_p T_a}{\Delta H}\right]}{M_a(1+\alpha)\left[\sqrt{\left[\frac{\alpha+\left[\frac{T_{t4}}{T_a}\right]\left[\frac{T_a}{T_{ta}}\right]\frac{1}{\tau_c}}{1+\alpha}\right]\left[\frac{\left[\frac{T_{ta}}{T_a}\right]\left[\frac{\tau_c+\left[\frac{T_{ta}}{T_a}\right]\left[\frac{T_a}{T_{t4}}\right]\tau_c(1+\alpha-\tau_c)}{1+\left[\frac{T_{ta}}{T_a}\right]\left[\frac{T_a}{T_{t4}}\right]\tau_c\alpha}\right]-1}{\left[\frac{T_{ta}}{T_a}\right]-1}\right]}-1\right]}.$$

(3.2.20) / (H.5.21)

For a typical nonafterburning mixed turbofan, the values of *TSFC* range from 0.35 to 0.75 $lb_m/h/lb_f$ (0.035 to 0.08 kg/kW/h).

Afterburning For the thrust for the afterburning mixed fan:

$$\frac{F}{\dot{m}a_a} = M_a(1+\alpha)\left[\sqrt{\frac{\left\{\left[\frac{T_{ta}}{T_a}\right]\left[\frac{\tau_c + \left[\frac{T_{ta}}{T_a}\right]\left[\frac{T_a}{T_{t4}}\right]\tau_c(1+\alpha-\tau_c)}{1+\left[\frac{T_{ta}}{T_a}\right]\left[\frac{T_a}{T_{t4}}\right]\tau_c\alpha}\right]-1\right\}\left[\frac{T_{t8}}{T_a}\right]}{\left[\frac{T_{ta}}{T_a}-1\right]\left[\frac{\tau_c + \left[\frac{T_{ta}}{T_a}\right]\left[\frac{T_a}{T_{t4}}\right]\tau_c(1+\alpha-\tau_c)}{1+\left[\frac{T_{ta}}{T_a}\right]\left[\frac{T_a}{T_{t4}}\right]\tau_c\alpha}\right]\left[\frac{T_{ta}}{T_a}\right]}}-1\right]. \tag{3.2.21 / H.5.35}$$

The thrust is a strong function of burner exit temperature, afterburner exit temperature, compressor pressure ratio, and bypass ratio. Also for the *TSFC*:

$$TSFC\ a_a = \frac{\left[\left(\frac{T_{t8}}{T_a}\right)-\left(\frac{T_{ta}}{T_a}\right)\right]\left[\frac{c_p\ T_a}{\Delta H\ M_a}\right]}{\left[\sqrt{\frac{\left\{\left[\frac{T_{ta}}{T_a}\right]\left[\frac{\tau_c + \left[\frac{T_{ta}}{T_a}\right]\left[\frac{T_a}{T_{t4}}\right]\tau_c(1+\alpha-\tau_c)}{1+\left[\frac{T_{ta}}{T_a}\right]\left[\frac{T_a}{T_{t4}}\right]\tau_c\alpha}\right]-1\right\}\left[\frac{T_{t8}}{T_a}\right]}{\left[\frac{T_{ta}}{T_a}-1\right]\left[\frac{\tau_c + \left[\frac{T_{ta}}{T_a}\right]\left[\frac{T_a}{T_{t4}}\right]\tau_c(1+\alpha-\tau_c)}{1+\left[\frac{T_{ta}}{T_a}\right]\left[\frac{T_a}{T_{t4}}\right]\tau_c\alpha}\right]\left[\frac{T_{ta}}{T_a}\right]}}-1\right]}. \tag{3.2.22 / H.5.37}$$

3.2.4 Turboprop

As indicated earlier, a turboprop is aimed at a different group of aircraft than the previously covered engines. This geometry is shown in Figure 3.8. Because of the less defined nature of the propeller flow (mass flow, pressure rise, and flow direction), a somewhat different approach to the analysis is used.

For the thrust:

$$\frac{F_t}{\dot{m}a_a} = \frac{C_{W_e}}{(\gamma-1)M_a}, \tag{3.2.23 / H.6.20}$$

where the work coefficient is

$$C_{W_e} \equiv \frac{\mathcal{P}_t}{\dot{m}h_a} = \left[\frac{T_{t4}}{T_a}\right]\left[1-\frac{\tau_5}{\left[\frac{T_{ta}}{T_a}\right]\tau_c}\right] - \left[\frac{T_{ta}}{T_a}\right](\tau_c-1)$$
$$+(\gamma-1)M_a^2\left[\sqrt{\frac{\left[\frac{T_{t4}}{T_a}\right]\left[\frac{T_a}{T_{ta}}\right]\left[\frac{\tau_5-1}{\tau_c}\right]}{\left[\frac{T_{ta}}{T_a}\right]-1}}-1\right], \tag{3.2.24 / H.6.17}$$

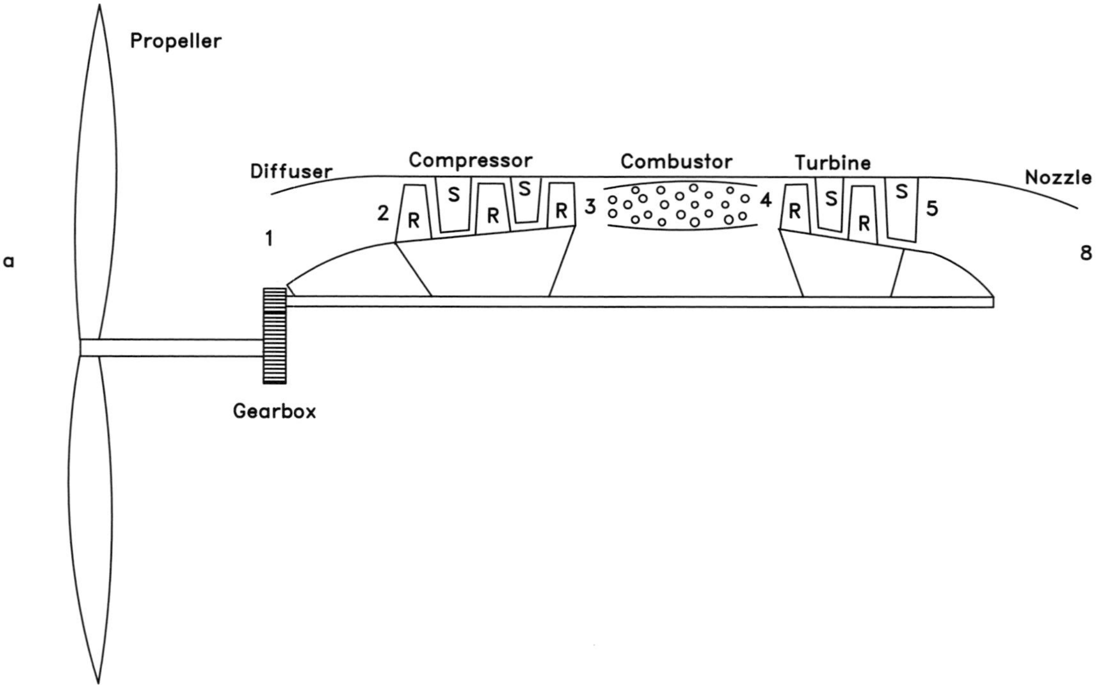

Figure 3.8 General turboprop.

and where

$$\tau_5 \equiv \left[\frac{T_{ta}}{T_a}\right]\tau_c\tau_t = [\pi_5]^{\frac{\gamma-1}{\gamma}} = \left[\frac{p_{t5}}{p_a}\right]^{\frac{\gamma-1}{\gamma}} = \frac{T_{t5}}{T_a} \times \frac{1}{\tau_b}.$$

(3.2.25) / (H.6.7, H.6.8, and H.6.9)

The nondimensional *TSFC* is

$$TSFC\ a_a = M_a(\gamma-1)\left[\frac{c_p T_a}{\Delta H}\right]\left[\frac{\left[\frac{T_{t4}}{T_a}\right] - \left[\frac{T_{ta}}{T_a}\right]\tau_c}{C_{W_e}}\right].$$

(3.2.26) / (H.6.26)

To find τ_5 more engine information is needed: for example the operating condition of the nozzle or more turbine information. However, if C_{W_p} is known by propeller testing and identification with the engine, where $C_{W_p} = \frac{\mathcal{P}_p}{\dot{m} c_p T_a}$, τ_5 can be found from

$$\tau_5 = \left(\frac{T_{ta}}{T_a}\tau_c\right)\left\{1 - \frac{T_a}{T_{t4}}\left[C_{W_p} + \frac{T_{ta}}{T_a}(\tau_c - 1)\right]\right\}.$$

(3.2.27) / (H.6.13)

For a typical turboprop, the values of *TSFC* range from 0.3 to 0.4 $lb_m/h/lb_f$ (0.03 to 0.04 kg/kW/h).

The thrust may also be maximized by using Eq. (3.2.24) since the thrust is proportional to the work coefficient. Namely, if one differentiates C_{W_e} with respect to τ_5, holding all other parameters fixed, and sets the result to zero, one obtains:

$$\tau_{5_{opt}} = 1 + \tau_c\left[\frac{T_{ta}}{T_a}\right]\left[\frac{T_a}{T_{t4}}\right]\left[\left[\frac{T_{ta}}{T_a}\right] - 1\right].$$

(3.2.28) / (H.6.27)

Although the analytical operations are on τ_5, in actuality the turbine operation is being optimized to deliver the best balanced performance between the propeller and exhaust. From this equation and Eq. (3.2.25) one can find the optimum value of π_5 which ideally results in $u_8 = u_a$, which implies that optimally the thrust from the turbojet is zero and that all of the thrust is provided from the propeller.

3.2.5 Power Generation Gas Turbine

The next category of gas turbines to be analyzed is for power generation. A simple power generation unit of general design is shown in Figure 3.9. Since the net power output is of primary concern, results of the fundamental equations are somewhat different.

For the nondimensional power output:

$$\frac{\mathcal{P}_{net}}{\dot{m} c_p T_a} = \left(\frac{T_{t4}}{T_a}\right)\left(1 - \frac{1}{\tau_c}\right) - (\tau_c - 1).$$

(3.2.29) / (H.7.8)

Thus, this is a function of only two variables. The first is the compressor total temperature ratio (which can be found directly from the total pressure ratio) and the second is the ratio of the combustor exit total temperature to the ambient temperature.

By differentiating the equation with respect to the total temperature ratio of the compressor, τ_c (which is

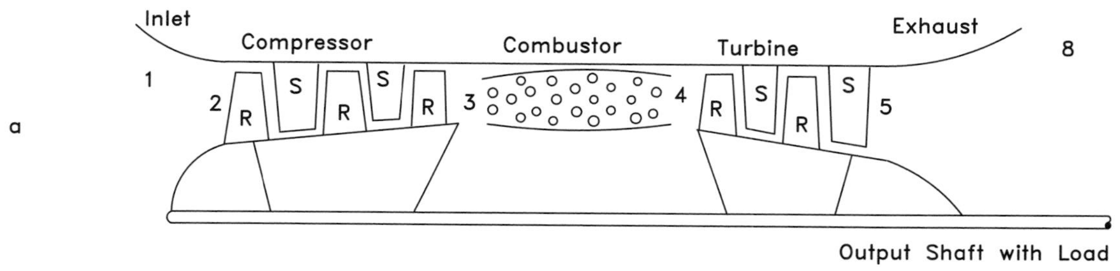

Figure 3.9 Simple power generation gas turbine.

directly related to the total pressure ratio), one can optimize the output power. This results in

$$\tau_{c_{opt}} = \sqrt{\frac{T_{t4}}{T_a}}. \qquad (3.2.30) \ / \ (H.7.9)$$

Also, for the nondimensional *SFC*:

$$SFC\ a_a^2 = (\gamma - 1)\left(\frac{c_p T_a}{\Delta H}\right)\left(\frac{\tau_c}{\tau_c - 1}\right). \qquad (3.2.31) \ / \ (H.7.16)$$

Next, the overall efficiency is:

$$\eta_{th} = \frac{\tau_c - 1}{\tau_c}. \qquad (3.2.32) \ / \ (H.7.19)$$

This reduced nondimensional variable is a function of only one variable: the compressor total temperature ratio (or total pressure ratio). Examination of this equation importantly shows that as the total temperature ratio (or pressure ratio) increases, so does the efficiency.

Lastly, heat rate is

$$\mathcal{HR} = \frac{\tau_c}{\tau_c - 1}. \qquad (3.2.33) \ / \ (H.7.21)$$

This is dimensionless but the reader should recognize that the heat rate is generally reported as kJ/(kW-hr) or BTU/(hp-h), so multiplication conversions are needed: 3600 and 2544, respectively.

Typical values of *SFC* for modern gas turbines are $0.35/\text{lb}_m\text{hp-h}(0.2/\text{kg/kW-h})$. Typical efficiencies are 40 percent to 45 percent.

3.2.6 Summary of Ideal Cycle Analysis Equations

We have now seen the final ideal equations derived in detail in Appendix H. In the previous sections some notes were made on the trends and optimization possibilities. In Section 3.3, numerous calculations and trend studies are presented to further amplify the findings. Also, along with ideal trends, comparisons will be made to nonideal characteristics.

3.3 Cycle Analysis: Studies and Trends

3.3.1 General Approach

Section 2.3 described ideal and nonideal components. In the previous section, closed-form ideal components were integrated for the propulsion thrust and *TSFC* and power generation *SFC* and efficiency for different gas turbine types. In Section 2.3 all of the different losses were covered for the nonideal components. For some components, single-term expressions were introduced that allowed nonideal effects to be included. For a few of the components, two-term expressions were required. However, because of the nonideal effects, combining equations would be a cumbersome, and in most cases, an impossible, task. Furthermore, more complex gas turbines can be assembled for the nonideal cases. Thus, each gas turbine type must be considered separately to solve nonideal problems. To determine the thrust and *TSFC*, *SFC*, and efficiency, one must step through the gas turbine component-by-component until all of the different parameters are determined. In some cases, iterative solutions are required in the step-by-step procedure because the nonideal effects do not allow the parameters to be directly determined.

As was noted in Section 2.3, the specific heat values for a given nonideal component should be evaluated at the average temperature of that component. This means that both the inlet and exit temperatures must be known. However, in practice this is not the case a priori. Usually only one temperature is known when the component is analyzed. Only after the analysis is done is the average temperature known. Thus, iterations are required for nearly all components and the use of a computer becomes almost imperative. That is, a value of c_p is first guessed for a given component, and the analysis is completed for the component. Next, the average temperature is found, which is used to find c_p from a table or from Eq. (2.2.20). The process is repeated until the assumed and calculated values of c_p agree within a small tolerance. As one would expect, such iterations require the aid of a personal computer for all of the examples herein, although the particular steps are not shown. Also, the values of specific heat and specific heat ratio should be evaluated at the average static temperature for each component. However, knowing the static values requires knowledge of the component Mach number

$$\left(T = \frac{T_t}{\left(1 + \frac{\gamma - 1}{2} M^2\right)} \right).$$

Therefore, for most of the examples herein the total temperature is used to find property values. However, a case study is presented later to demonstrate that the approximation is very reasonable for determining the overall performance. Regardless, in the cycle analysis software provided, an approximate Mach number can be used to evaluate the properties at an approximate static temperature. Two general approaches to iterative solutions are presented in Appendix I. Also, in Appendix I, some examples of iterative solutions are presented that correspond to some of the following engine examples.

In this section numerous ideal and nonideal examples will be worked in which each component will be analyzed and treated independently with step-by-step numerical computations. Also, numerous trend studies will be presented to demonstrate the effect the different component parameters have on the overall performance. In some cases optimal values can be found to maximize performance. Such optimizations were demonstrated also by the ideal equations in the previous section. However, the nonideal and ideal values will be shown to be different. Also, comparisons of ideal vs. nonideal will be made. Solutions for different gas turbine types and conditions can also be obtained using the cycle analysis software JetEngineCycle.exe and PowerGTCycle.exe.

Lastly, it is important to note that in many of the example problems and trend studies, component parameters such as the compressor pressure ratio and efficiencies are held constant while others are varied. However, in actuality all of the parameters vary as a function of flight or operating conditions. Namely, all of the components interact and determine how the parameters change. This is called component matching. For example, as the fuel ratio for a jet engine is changed, the compressor pressure ratio and efficiency changes as do the mass flow rate, flight Mach number, and other component parameters. This advanced topic is covered in Chapter 12, which is a capstone for this text.

3.3.2 Example Problems, Trend Studies, and Optimizations: Jet Propulsion

In this section, several examples are presented for jet propulsion. They do not cover the entire range of possible combinations of the different components. These examples do, however, cover all of the important component concepts that may be encountered in real analyses. All of the nonideal examples will be compared with ideal analyses for the same gas turbines. Trend studies are presented to indicate that the basic trends are similar for both ideal components and nonideal components. Also, complementary, or extended examples will be used for some examples to emphasize certain effects. The examples and trend studies presented are an integral portion of this chapter and should not be treated as supplementary or extraneous material. The example problems tend to be somewhat long; however, they follow a logical step-by-step or component-to-component process, so each component is analyzed in detail separately.

Ramjet/Scramjet

A general ramjet is considered in Figure 3.1. As indicated earlier, this is the simplest of jet engines because it does not have any rotating shafts. As noted earlier, for the application of cycle analysis, both the ramjet and scramjet can be treated in the same manner. However, the designs are markedly different as discussed in Chapter 1. Also, losses are considerably higher in the scramjet due to the increased in the number of shocks and much stronger shocks. In Figure 3.1, position a is the ambient condition, position 1 is the inlet to the diffuser, position 3 is the exit of the diffuser and inlet to the combustor, position 4 is the exit of the burner and inlet to the nozzle, and lastly position 8 is the exit of the nozzle. Figure 3.10 presents the h–s diagram for a ramjet system. Process a to 1 to 3 is a compression process in which the pressure is increased by the "ram" effect. Ideally this process is isentropic. Process 3 to 4 is a combustion process. Ideally this process is isobaric. Finally, process 4 to 8 is an expansion process that accelerates the flow velocity and decreases the pressure. Ideally this process is isentropic. Both states a and 8 are shown on separate pressure lines. When the pressure at state 8 is lower than atmospheric pressure the nozzle is overexpanded. If the pressure at state 8 is higher than atmospheric pressure the nozzle is underexpanded (shown in Figure 3.10). Ideally, p_a and p_8 are equal but do not constitute a process as in a closed cycle.

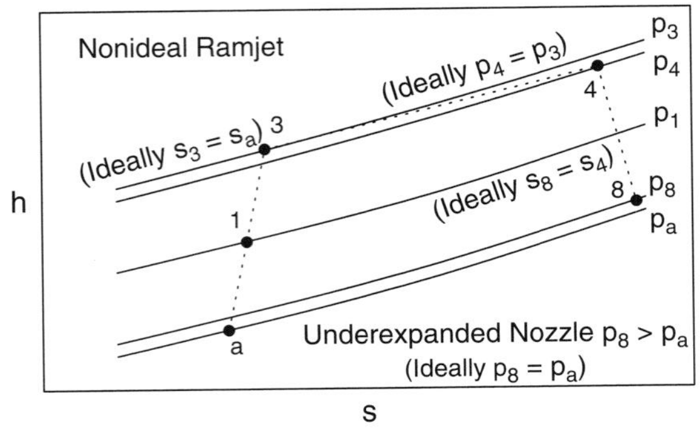

Figure 3.10 The h–s diagram for a ramjet.

The objective now is to determine the thrust for given component characteristics. Examples will be presented to demonstrate the steps.

Example 3.1 Ideal Ramjet

An ideal ramjet is traveling at Mach 3 at an altitude of 4572 m (15,000 ft). The external static temperature is 258.4 K (465.2 °R) and the external static pressure is 57.1 kPa (8.29 psia) as determined from Appendix B or Atmosphere.exe. The heating value of the fuel is 46,520 kJ/kg (20,000 BTU/lb$_m$). Air flows through the engine at 45.35 kg/s (100 lb$_m$/s). The burner exit total temperature is 1944 K (3500 °R). Find the thrust, fuel ratio, and *TSFC*. The specific heat ratio can be assumed to be 1.35.

Solution

For the solution one could simply use Eqs. (3.2.1a, b) and (3.2.2). However, these equations do not allow an engineer to obtain a "feel" of any of the internal numbers. Furthermore, for nonideal components (next example) a component-by-component analysis is necessary. Thus, for the solution all of the intermediate steps through all of the components will be performed in detail.

Diffuser

First, since the Mach number is given, $M_a = 3$, and the ambient temperature is $T_a = 258.4$ K (465.2 °R), one can find the speed of sound, a_a:

$$a_a = \sqrt{\gamma \mathcal{R} T_a} = \sqrt{1.35 \times 287.1 \frac{\text{J}}{\text{kg-K}} \times 258.4\ \text{K} \times \frac{\text{N-m}}{\text{J}} \times \frac{\text{kg-m}}{\text{s}^2\text{-N}}} = 316.4\ \text{mps}\ (1038\ \text{fps}).$$

Thus, the ramjet velocity is

$u_a = M_a a_a = 3 \times 316.4$ mps $= 949.3$ mps (3115 fps).

Next, for the stagnation temperature, T_{ta}:

$$\frac{T_{ta}}{T_a} = \left[1 + \frac{\gamma - 1}{2} M_a^2\right],$$

and so

$$T_{ta} = 258.4\ \text{K}\left[1 + \frac{1.35 - 1}{2} 3^2\right] = 665.6\ \text{K}\ (1198\ °\text{R}).$$

Thus, because the diffuser is ideally adiabatic the total temperature at the diffuser exit is:

$T_{t3} = T_{ta} = 665.6$ K(1198 °R).

Also, for the stagnation pressure p_{ta}:

$$\frac{p_{ta}}{p_a} = \left[1 + \frac{\gamma - 1}{2} M_a^2\right]^{\frac{\gamma}{\gamma - 1}}$$

so $p_{ta} = 57.1\ \text{kPa}\left[1 + \frac{1.35 - 1}{2} 3^2\right]^{\frac{1.35}{1.35-1}} = 2198\ \text{kPa}\ (318.8\ \text{psia}).$

Thus, because the diffuser, combustor, and nozzle are ideal, the stagnation pressure is constant through the entire engine:

$p_{t8} = p_{t4} = p_{t3} = 2198$ kPa (318.8 psia).

Burner

It is given that the burner exit total temperature is $T_{t4} = 1944$ K (3500 °R) and the heating value of the fuel is 46,520 kJ/kg. The specific heat at constant pressure is 1.107 kJ/(kg-K) for $\gamma = 1.35$.

Thus, the ideal fuel flow is given by the first law of thermodynamics (Eq. (2.3.58)):

$$\dot{m}_f = \frac{\dot{m} c_p [T_{t4} - T_{t3}]}{\Delta H} = \frac{45.35 \frac{\text{kg}}{\text{s}} \times 1.107 \frac{\text{kJ}}{\text{kg-K}} \times [1944 - 665.6]\ \text{K}}{46520 \frac{\text{kJ}}{\text{kg}}}$$

$$= 1.380\ \text{kg/s}\ (3.043\ \text{lb}_m/\text{s})$$

or

$$f = \frac{\dot{m}_f}{\dot{m}} = \frac{1.380}{45.35} = 0.03043,$$

which is small and consistent with the ideal assumption of a negligible fuel flow rate.

Nozzle

For an ideal analysis, the total temperature is constant from the burner through the nozzle (adiabatic):

$T_{t8} = 1944$ K (3500 °R).

Next, at the nozzle exit since for the ideal case the exit pressure matches the ambient pressure:

$p_8 = p_a = 57.1$ kPa (8.29 psia)

and from the definition of stagnation pressure,

$$p_{t8} = p_8 \left[1 + \frac{\gamma - 1}{2} M_8^2\right]^{\frac{\gamma}{\gamma - 1}}.$$

Thus, solving for the nozzle exit Mach number, M_8, yields

$$M_8 = \sqrt{\left[\frac{2}{\gamma - 1}\right]\left[\left[\frac{p_{t8}}{p_8}\right]^{\frac{\gamma-1}{\gamma}} - 1\right]} = \sqrt{\left[\frac{2}{1.4 - 1}\right]\left[\left[\frac{2198}{57.1}\right]^{\frac{1.35-1}{1.35}} - 1\right]} = 3.000,$$

which is identical to the inlet Mach number (as analytically shown in Appendix H).

Therefore, the nozzle exit temperature is

$$T_8 = \frac{T_{t8}}{\left[1 + \frac{\gamma - 1}{2} M_8^2\right]} = \frac{1944}{\left[1 + \frac{1.35 - 1}{2} 3.000^2\right]} = 755.0 \text{ K } (1359\ °\text{R}).$$

And the nozzle exit gas velocity is

$$u_8 = M_8 a_8 = M_8 \sqrt{\gamma \mathcal{R} T_8} = 3.000 \sqrt{1.35 \times 287.1 \frac{\text{J}}{\text{kg-K}} \times 755.0 \text{ K} \times \frac{\text{N-m}}{\text{J}} \times \frac{\text{kg-m}}{\text{s}^2\text{-N}}}$$

$$= 1622 \text{ mps } (5323 \text{ fps}).$$

Total Thrust and TSFC

Finally, the ideal thrust does not have a pressure thrust component and so

$$F = \dot{m}(u_e - u_a) = 45.35 \frac{\text{kg}}{\text{s}} (1622 - 949.3) \frac{\text{m}}{\text{s}} \frac{\text{N-s}^2}{\text{kg-m}}$$

$$= 30,530 \text{ N } (6866 \text{ lb}_f).$$ <ANS

So that the *TSFC* is

$$TSFC = \frac{\dot{m}_f}{F} = \frac{1.380 \frac{\text{kg}}{\text{s}} \times 3600 \frac{\text{s}}{\text{h}}}{30,530 \text{ N}}$$

$$= 0.1627 \frac{\text{kg}}{\text{h-N}} \left(1.595 \frac{\text{lb}_\text{m}}{\text{h-lb}_\text{f}} \right).$$ <ANS

Once again, simply using Eqs. (3.2.1a, b) and (3.2.2) could have eliminated all of these intermediate steps. However, the details of the internal operating conditions would be missing. For example, the supersonic exit Mach number would not have been seen nor would have been any of the intermediate pressures and temperatures. Furthermore, as will be seen below, for nonideal analyses performing these steps will be necessary because the equations will not reduce to closed-form equations as they do for an ideal case.

Example 3.2 Nonideal Ramjet

A ramjet is traveling at Mach 3 at an altitude of 4572 m (15,000 ft) and the external static temperature is 258.4 K (465.2 °R) and the external static pressure is 57.1 kPa (8.29 psia). The heating value of the fuel is 46,520 kJ/kg (20,000 BTU/lb_m). Air flows through the engine at 45.35 kg/s (100 lb_m/s). The burner exit total temperature is 1944 K (3500 °R). The burner has an efficiency of 90 percent and a total pressure ratio of 0.95. A converging nozzle is used and the efficiency is 95 percent. The total pressure recovery for the diffuser is 0.95. The specific heat and specific heat ratio should be determined at the average total temperature for each component. Find the thrust, fuel ratio, and *TSFC*. This is the nonideal performance calculation of Example 3.1.

Solution

Because this is nonideal, all of the intermediate steps through all of the components *must* be performed in detail.

Diffuser

First, since the Mach number is given, $M_\text{a} = 3$, and the ambient temperature is $T_\text{a} = 258.4$ K (465.2 °R), one can find the speed of sound, a_a:

$$a_\text{a} = \sqrt{\gamma \mathcal{R} T_\text{a}} = \sqrt{1.406 \times 287.1 \frac{\text{J}}{\text{kg-K}} \times 258.4 \text{ K} \times \frac{\text{N-m}}{\text{J}} \times \frac{\text{kg-m}}{\text{s}^2\text{ -N}}} = 322.9 \text{ mps } (1059 \text{ fps}),$$

where γ is evaluated at the freestream temperature. Thus, the ramjet velocity is

$$u_\text{a} = M_\text{a} a_\text{a} = 3 \times 322.9 \text{ mps} = 968.8 \text{ mps } (3179 \text{ fps}).$$

Next, for the stagnation temperature, $T_{\text{t}3}$:

$$\frac{T_{t3}}{T_a} = \left[1 + \frac{\gamma - 1}{2} M_a^2\right],$$

and so $\gamma_d = 1.366$ (iteratively found for the average total temperature for the diffuser) and $M_a = 3$. More of this iteration method and temperature implications will be covered in detail in Example 3.4 for the nonideal turbojet.

$$T_{t3} = 258.4 \text{ K}\left[1 + \frac{1.3660 - 1}{2} 3^2\right] = 684.2 \text{ K } (1232\ ^\circ\text{R}).$$

Also, for the stagnation pressure p_{ta}:

$$\frac{p_{ta}}{p_a} = \left[\left[1 + \frac{\gamma - 1}{2} M_a^2\right]\right]^{\frac{\gamma_d}{\gamma_d - 1}} = [2.647]^{\frac{1.3660}{1.3660 - 1}} = 37.83.$$

Next, the pressure recovery factor for the diffuser is 0.95, so that:

$$\frac{p_{t3}}{p_a} = \frac{p_{t3}}{p_{ta}}\frac{p_{ta}}{p_a} = \pi_d \frac{p_{ta}}{p_a} = 0.95 \times 37.83 = 35.94$$

and thus $p_{t3} = 35.94 \times 57.24$ kPa $= 2057$ kPa (298.3 psia).

Burner

Next, the total pressure ratio for the burner is 0.95, so the burner exit total pressure is

$$p_{t4} = \pi_b p_{t3} = 0.95 \times 2057 \text{ kPa} = 1954 \text{ kPa } (283.4 \text{ psia}).$$

It is given that the burner exit total temperature is $T_{t4} = 1944$ K (3500 °R) and the heating value of the fuel is 46,520 kJ/kg. The specific heat at constant pressure is 1.196 kJ/(kg-K) at the average total temperature of the burner.

Thus, the fuel flow is given by the first law of thermodynamics (Eq. (2.3.60)):

$$\Delta H\ \eta_b \dot{m}_f = \dot{m} c_{pb}(T_{t4} - T_{t3}) + \dot{m}_f c_{pb} T_{t4}$$

$$\text{So } \dot{m}_f = 43.35\frac{\text{kg}}{\text{s}} \times 1.196\frac{\text{kJ}}{\text{kg-K}} \times \frac{(1944 - 684.2) \text{ K}}{0.90 \times 46,520\frac{\text{kJ}}{\text{kg}} - 1.196\frac{\text{kJ}}{\text{kg-K}} \times 1944 \text{ K}}$$

$$= 1.728 \text{ kg/s } (3.810 \text{ lb}_m/\text{s})$$

$$\text{or } f = \frac{\dot{m}_f}{\dot{m}} = \frac{1.728}{45.35} = 0.03811,$$

which is small but larger than the ideal case due to burner inefficiency.

Nozzle

The total temperature is constant from the burner through the nozzle (adiabatic):

$$T_{t8} = 1944 \text{ K } (3500\ ^\circ\text{R}).$$

Since the nozzle is a fixed converging nozzle, one must first check to see if it is choked.

For choking, the fan nozzle exit pressure is $p_8^* = p_{t4}\left[1 + \dfrac{1-\gamma_n}{\eta_n(1+\gamma_n)}\right]^{\frac{\gamma_n}{\gamma_n - 1}}$ (Eq. (2.3.99));

thus for $p_{t4} = 1954$ kPa, $\eta_n = 0.95$ (given), and $\gamma_n = 1.2738$ (found iteratively as for the diffuser):

$$p_8^* = 1954 \text{ kPa} \times \left[1 + \frac{1 - 1.2738}{0.95 \times (1 + 1.2738)}\right]^{\frac{1.2738}{1.2738-1}} = 1002 \text{ kPa } (145.2 \text{ psia}),$$

but since $p_a = 57.24$ kPa, which is well below the exit pressure for choking, the nozzle is choked and the exit Mach number is identically unity.

Therefore, $p_8 = p_8{}^* = 1002$ kPa

and $T_8 = \dfrac{2T_{t8}}{1+\gamma_n}$ for $M_8 = 1.00$;

thus for $T_{t8} = 1944$ K, $T_8 = \dfrac{2 \times 1944 \text{ K}}{1 + 1.2738} = 1710$ K (3078 °R).

Now $u_8 = \sqrt{2c_{pn}\,(T_{t8} - T_8)}$

so that

$$u_8 = \sqrt{2 \times 1.335 \frac{\text{kJ}}{\text{kg-K}} \times (1944 \text{ K} - 1710 \text{ K}) \times 1000 \frac{\text{J}}{\text{kJ}} \times \frac{\text{N-m}}{\text{J}} \frac{\text{kg-m}}{\text{N-s}^2}}$$

$= 790.6$ mps (2594 fps).

Alternatively, one could have found u_8 by first finding the speed of sound at the exit $a_8 = \sqrt{\gamma_8 \mathcal{R} T_8}$, so

$$a_8 = \sqrt{1.274 \times 287.1 \frac{\text{J}}{\text{kg-K}} \times 1710 \text{K} \times \frac{\text{N-m}}{\text{J}} \times \frac{\text{kg-m}}{\text{N-s}^2}} = 790.6 \text{ mps};$$

but $M_8 = 1 = u_8/a_8$, so $u_8 = 790.6$ mps – the same as we have found previously.

Next, the exit area is given by $A_8 = \dfrac{\dot{m}_8}{\rho_8\, u_8}$,

where $\dot{m}_8$ is the total mass flow rate at 8 and is given by

$$\dot{m}_8 = \dot{m}_a + \dot{m}_f = 45.35 + 1.728 = 47.08 \frac{\text{kg}}{\text{s}} \left(103.8 \frac{\text{lb}_\text{m}}{\text{s}}\right).$$

Since both T_8 and p_8 are known, ρ_8 is found from the ideal gas equation

$$\rho_8 = \frac{p_8}{\mathcal{R} T_8} = \frac{1002 \text{ kPa} \times 1000 \dfrac{\text{Pa}}{\text{kPa}}}{287.1 \dfrac{\text{J}}{\text{kg-K}} \times 1710\text{K}} \times \frac{\text{N}}{\text{Pa-m}^2} \times \frac{\text{J}}{\text{N-m}} = 2.041 \text{ kg/m}^3 \ (0.003960 \text{ sl/ft}^3).$$

Thus, the nozzle exit area is $A_8 = \dfrac{47.08 \dfrac{\text{kg}}{\text{s}}}{2.041 \dfrac{\text{kg}}{\text{m}^3} \times 790.6 \dfrac{\text{m}}{\text{s}}} = 0.02918 \text{ m}^2 \ (45.22 \text{ in.}^2)$.

Total Thrust and TSFC

Finally, the thrust is

$$F = \dot{m}_8 u_8 - \dot{m}_a u_a + A_8(p_8 - p_a)$$

$$= 47.08\frac{\text{kg}}{\text{s}} \times 790.6\frac{\text{m}}{\text{s}}\frac{\text{N-s}^2}{\text{kg-m}} - 45.35\frac{\text{kg}}{\text{s}} \times 968.8\frac{\text{m}}{\text{s}}\frac{\text{N-s}^2}{\text{kg-m}}$$

$$+ 0.02918\ \text{m}^2 \times (1002 - 57.23)\text{kPa}\frac{\text{N-m}^2}{\text{Pa}} \times \frac{1000\ \text{Pa}}{\text{kPa}}$$

$$= -6717\ \text{N (momentum)} + 27,560\ \text{N (pressure)}$$

$$= 20,840\ \text{N}\ (4685\text{lb}_\text{f}). \qquad \text{<ANS}$$

All of the positive thrust is from the pressure difference.

So the *TSFC* is

$$TSFC = \frac{\dot{m}_{f_t}}{F} = \frac{1.728\frac{\text{kg}}{\text{s}} \times 3600\frac{\text{s}}{\text{h}}}{20,840\ \text{N}}$$

$$= 0.2986\frac{\text{kg}}{\text{h-N}}\left(2.927\frac{\text{lb}_\text{m}}{\text{h-lb}_\text{f}}\right). \qquad \text{< ANS}$$

Comparing with the ideal case, the thrust has decreased by 68 percent and the *TSFC* has increased by 84 percent. The losses have significantly affected the overall performance.

Extended Example 3.2.1 | Nonideal Ramjet: Adding a Variable CD Nozzle

In Example 3.2, the thrust and *TSFC* deteriorated tremendously from the ideal case. One has to question why so much. In this example the converging nozzle is replaced by a variable converging–diverging (CD) nozzle. Thus, p_8 is equal to p_a. All of the component calculations are the same up to the nozzle and will not be repeated

Solution

Nozzle

The inlet total temperature for the nozzle is again $T_{t4} = 1944$ K and the inlet total pressure is $p_{t4} = 1954$ kPa.

The nozzle is a variable CD nozzle so the exit pressure matches the ambient pressure and one does not need to check for choking as before.

Thus, $p_8 = 57.24$ kPa; and ideally

$$T_8{}' = T_{t4}\left[\frac{p_8}{p_{t4}}\right]^{\frac{\gamma_n-1}{\gamma_n}} = 1944\ \text{K}\left[\frac{57.24\ \text{kPa}}{1954\ \text{kPa}}\right]^{\frac{1.2738-1}{1.2738}} \quad \text{for } \gamma_n = 1.2738 \text{ (again, iterative)}$$
$$= 910.2\ \text{K}.$$

The nozzle efficiency is again 90 percent; thus the nozzle exit temperature is

$$T_8 = T_{t4} - h_n(T_{t4} - T_8{}') = 1944\ \text{K} - 0.90(1944\ \text{K} - 910.2\ \text{K}) = 1014\ \text{K}.$$

Now $u_8 = \sqrt{2\ c_{pn}\ (T_{t8} - T_8)}$

so that

$$u_8 = \sqrt{2 \times 1.335\frac{\text{kJ}}{\text{kg-K}} \times (1944\ \text{K} - 910.2) \times 1000\frac{\text{J}}{\text{kJ}} \times \frac{\text{N-m}}{\text{J}}\frac{\text{kg-m}}{\text{N-s}^2}}$$
$$= 1576\ \text{mps}.$$

Thrust and TSFC

Finally, the thrust is

$$F = \dot{m}_8 u_8 - \dot{m}_a u_a + A_8(p_8 - p_a)$$

$$F = 47.08\frac{\text{kg}}{\text{s}} \times 1576\frac{\text{m}}{\text{s}}\frac{\text{N-s}^2}{\text{kg-m}} - 45.35\frac{\text{kg}}{\text{s}} \times 968.8\frac{\text{m}}{\text{s}}\frac{\text{N-s}^2}{\text{kg-m}}$$
$$+ 0.1518\ \text{m}^2 \times (57.23 - 57.23)\ \text{kPa}\frac{\text{N-m}^2}{\text{Pa}} \times \frac{1000\ \text{Pa}}{\text{kPa}}$$
$$= 30{,}280\ \text{N (momentum)} + 0\ \text{N (pressure)}$$
$$= 30{,}280\ \text{N}.$$

<ANS

Now all of the thrust is from the momentum.

So the *TSFC* is

$$TSFC = \frac{\dot{m}_{f_t}}{F} = \frac{1.728\frac{\text{kg}}{\text{s}} \times 3600\frac{\text{s}}{\text{h}}}{30{,}280\ \text{N}}$$
$$= 0.2055\frac{\text{kg}}{\text{h-N}}.$$

<ANS

So now the thrust has decreased only 1 percent from the ideal case and the *TSFC* has increased by only 26 percent. Clearly for such an application a variable nozzle is needed.

It is now possible to perform the first of many trend studies. Before proceeding, however, note an important operating characteristic of all engines. One of the limiting quantities in modern gas turbines is the thermal limit of the materials. Thus, T_{t4} is essentially fixed by this consideration because this is the hottest region in the engine. For a ramjet this constitutes the nozzle, and for engines with rotating components the turbine inlet temperature is the limit. As a result, T_{t4}/T_a for a given altitude is fixed. Of primary importance for any engine are the developed thrust and *TSFC*.

In Figure 3.11, the dimensionless thrust, $(F/\dot{m}_a a_a)$, and dimensionless *TSFC*, $(TSFC\ a_a)$, for a ramjet are shown as functions of Mach number

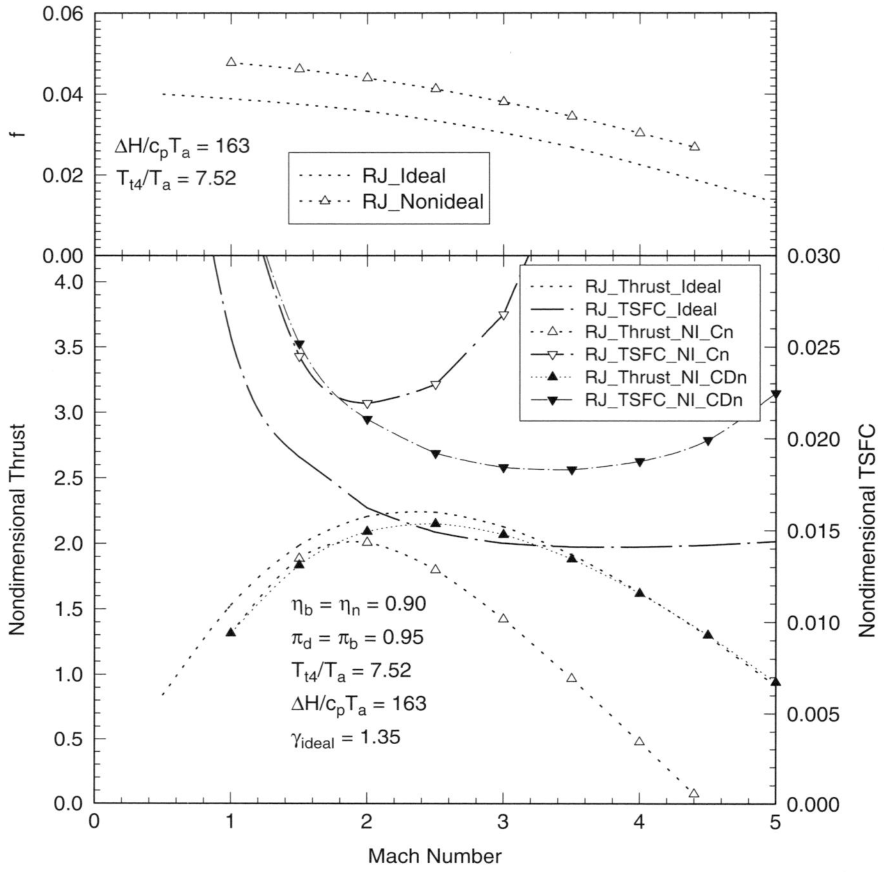

Figure 3.11 Effect of Mach number on dimensionless *TSFC*, dimensionless thrust, and fuel ratio of a ramjet.

for a fixed value of T_{t4} / T_a (i.e., $T_{t4} = 3500$ °R and $T_a = 465$ °R at 15,000 ft altitude. The fuel ratio is also shown. Both ideal and nonideal cases are shown. For the nonideal cases both converging and CD nozzles are included. For this trend study and all others to follow, the ideal specific heat ratio is 1.35. For this study the quantity $\Delta H/(c_p T_a)$ is 163 for $\Delta H = 20{,}000$ BTU/lb$_m$ and $T_a = 465\ 465$ °R. As can be seen, the thrust is maximized for one particular value of Mach number for each condition. For the ideal, non-ideal converging nozzle and nonideal CD nozzle the optimum values are 2.35, 1.90, and 2.45 respectively. On the other hand, the TSFC is minimized at a different and higher Mach number: 3.75, 2.05, and 3.35 respectively. This indicates that, for a given fuel and material operating limits, two optimum flight conditions exist as determined by which quantity is to be optimized. Note that, at low Mach numbers, the value of TSFC rapidly increases and the thrust rapidly decreases, which indicates that this engine should not be used at low velocities because the total pressure into it is too small owing to the lack of "ram" effect. As the Mach number becomes large, the exhaust velocity approaches the freestream velocity if T_{t4} is held constant; consequently, the resulting thrust decreases. These are very important findings. Similar trends are seen for both ideal and nonideal cycle analyses. As with the detailed example problems, the nonideal performance for the nonideal converging geometry is much poorer than the ideal case. However, the nonideal performance for the nonideal CD geometry is closer to the ideal case, especially at the higher Mach numbers.

A special case, namely the stoichiometric condition, is now considered. For this condition, exactly the right proportion of fuel and air will be used so that all of the fuel and air are consumed in the combustion process, and neither is left in the combustion products. Typically, the stoichiometric fuel ratio is 0.06 to 0.07 for jet fuels, which results in very high temperatures – too high for engines with turbines. However, ramjets can tolerate much higher temperatures because of the lack of rotating parts. For a given fuel at stoichiometric conditions, the value of *TSFC* was earlier shown to be a function only of the Mach number and T_a (or a_a for a given γ) for an ideal ramjet.

One can perform another trend study. In Figure 3.12, the dimensionless *TSFC* and thrust are plotted against the Mach number under stoichiometric conditions for ideal and nonideal conditions. Namely the fuel ratio, f, is fixed. The burner exit temperature is also plotted. As can be seen, a minimum *TSFC* is attained at one particular Mach number, which is dependent on ideal versus nonideal. For this condition, the maximum thrust is developed at the same Mach number. By comparing Figures 3.11 and 3.12, one can observe several characteristics. For example, under stoichiometric

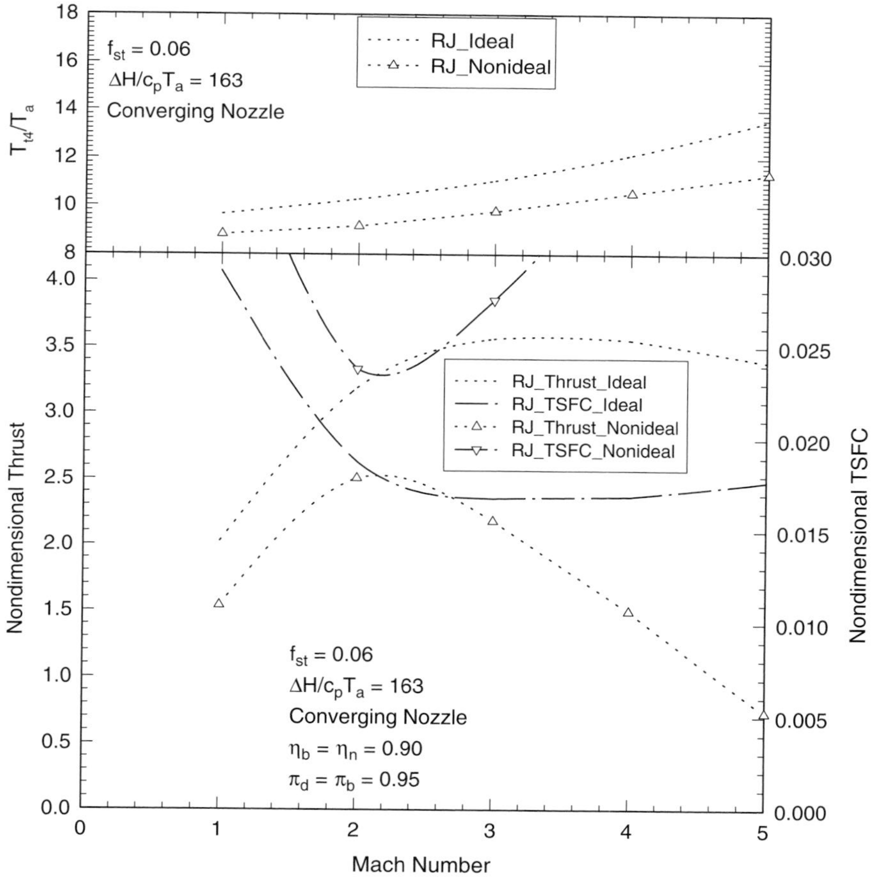

Figure 3.12 Effect of Mach number on dimensionless *TSFC* and dimensionless thrust of a ramjet operating stoichiometrically.

conditions, the optimum *TSFC* is higher than for fuel ratios that are below the stoichiometric condition. Therefore, this condition is less fuel efficient. However, the maximum thrust is higher than the nonstoichiometric value. An engine can thus develop more thrust when operated stoichiometrically. The optimum Mach number is different than either of the values found for the nonstoichiometric case. It is also important to recognize that as the Mach number increases, the burner exit temperature increases significantly for the stoichiometric condition.

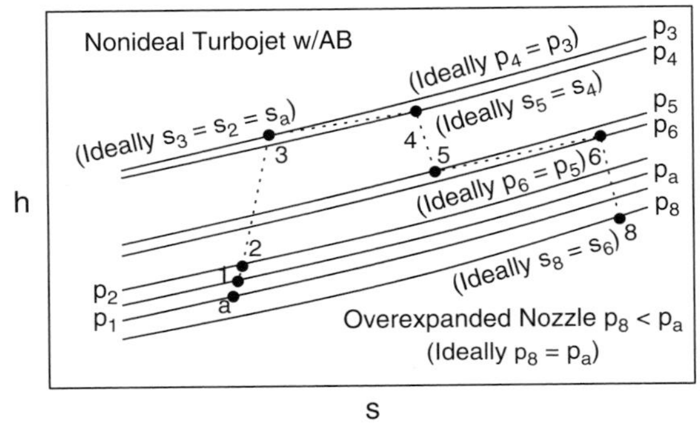

Figure 3.13 The h–s diagram for an afterburning turbojet.

Turbojet

A general turbojet with an afterburner is shown in Figure 3.2. The nonafterburning geometry is shown in Figure 3.3. This is a much more complex machine than the ramjet because of the rotating shaft. Eight stations are identified. The freestream is position a. The inlet is position 1. Position 2 is the exit of the diffuser and compressor inlet. Station 3 is the compressor exit and burner inlet. The burner exit and turbine inlet are denoted as station 4. Station 5 is the turbine exit and afterburner inlet. The afterburner exit and nozzle inlet are at position 6. Finally, station 8 is the nozzle exhaust. For a cycle analysis, single or multispool engines will yield exactly the same results because only the overall effects of the components are of importance. The h–s diagram for the general ideal turbojet with an afterburner is presented in Figure 3.13. Process a to 1 to 2 is a compression process during the diffusion. Ideally this process is isentropic. Process 2 to 3 is a compression process while power is required by the compressor. Ideally this is also an isentropic process. Process 3 to 4 is a combustion process. Ideally this process is isobaric. The expansion through the turbine is process 4 to 5, during which power is removed from the flow. Ideally, like the compression this is an isentropic process. Ideally, the afterburner process (5 to 6) is also isobaric. And finally, process 6 to 8 is an expansion in the nozzle where the flow velocity is accelerated. Ideally this is also an isentropic process. When the pressure at state 8 is lower than atmospheric pressure the nozzle is overexpanded (shown on Figure 3.13). If the pressure at state 8 is higher than atmospheric pressure the nozzle is underexpanded. Ideally the pressures at states 8 and a are the same. For the following examples both non-afterburning and afterburning configurations will be considered.

Example 3.3 Ideal Nonafterburning Turbojet

An ideal turbojet flies at sea level at a Mach number of 0.75. It ingests 165 lb_m/s (74.83 kg/s) of air and the compressor operates with a total pressure ratio of 15. The fuel has a heating value of 18,000 BTU/lb_m (41,870 kJ/kg) and the burner exit total temperature is 3000 °R (1667 K). Find the developed thrust and the *TSFC*. One can assume that the specific heat ratio is 1.35.

Solution

For the solution one could simply use Eqs. (3.2.5) and (3.2.7). Once again, all of the intermediate steps will be performed so that some of the internal parameter values can be appreciated.

Diffuser

First since the turbojet flies at sea level it can be determined from Appendix B or Atmosphere. exe that $p_a = 14.69$ psia (101.3 kPa) and $T_a = 518.7$ °R (288.2 K).

Thus, the speed of sound is

$$a_a = \sqrt{\gamma \mathcal{R} T_a} = \sqrt{1.35 \times 53.35 \frac{\text{ft -lb}_\text{f}}{\text{lb}_\text{m}\text{-°R}} \times 32.17 \frac{\text{lb}_\text{m}}{\text{sl}} 518.7\ \text{°R} \times \frac{\text{sl-ft}}{\text{s}^2\text{-lb}_\text{f}}} = 1096\ \text{fps}\ (334.0\ \text{mps})$$

and the jet velocity is $u_a = M_a a_a = 822.2$ fps (250.6 mps).

Now $\frac{T_{ta}}{T_a} = \left[1 + \frac{\gamma - 1}{2} M_a^2\right]$.

Thus, the inlet total temperature is

$$T_{ta} = 518.7\ \text{°R}\left[1 + \frac{1.35 - 1}{2} 0.75^2\right] = 569.7\ \text{°R}\ (316.5\ \text{K}).$$

Also for an ideal analysis, which is adiabatic, $T_{t2} = T_{ta}$.

So, the diffuser exit total temperature is $T_{t2} = 569.7$ °R (316.5 K).

Similarly, $\frac{p_{ta}}{p_a} = \left[1 + \frac{\gamma - 1}{2} M_a^2\right]^{\frac{\gamma}{\gamma - 1}}$.

Therefore, the inlet total pressure is

$$p_{ta} = 14.69\ \text{psia}\left[1 + \frac{1.35 - 1}{2} 0.75^2\right]^{\frac{1.35}{1.35 - 1}} = 21.11\ \text{psia}\ (145.6\ \text{kPa})$$

and for an ideal analysis, which is isentropic, $p_{t2} = p_{ta}$.

Thus, the diffuser exit total pressure is $p_{t2} = 21.11$ psia (145.6 kPa).

Also, for an ideal gas

$$\rho_a = \frac{p_a}{\mathcal{R} T_a} = \frac{14.69 \frac{\text{lb}_\text{f}}{\text{in.}^2} \times 144 \frac{\text{in.}^2}{\text{ft}^2}}{53.35 \frac{\text{ft-lb}_\text{f}}{\text{lb}_\text{m}\text{-°R}} \times 518.7\ \text{°R} \times 32.17 \frac{\text{lb}_\text{m}}{\text{sl}}} = 0.002376\ \text{sl/ft}^3\ (1.225\text{kg/m}^3).$$

Thus, if (not necessarily the case) the gas velocity at the diffuser inlet is the same as the jet velocity, the diffuser inlet frontal area is

$$A_{in} = \frac{\dot{m}}{\rho_a u_a} = \frac{165 \frac{\text{lb}_\text{m}}{\text{s}} \times 144 \frac{\text{in.}^2}{\text{ft}^2}}{0.002376 \frac{\text{sl}}{\text{ft}^3} \times 822.2 \frac{\text{ft}}{\text{s}} \times 32.17 \frac{\text{lb}_\text{m}}{\text{sl}}} = 378.1\ \text{in.}^2 (0.2440\ \text{m}^2),$$

which yields an inlet diameter of 21.88 in (0.5558 m). For this calculation the gas velocity at the diffuser inlet was assumed to be the same as the jet velocity. This may not, however, be the case and can be higher or lower than the freestream velocity, as determined from the engine operating point. More on this topic will be discussed in Chapter 8.

Compressor

The compressor total pressure ratio is given as 15. Thus, the compressor exit total pressure is: $p_{t3} = \pi_c p_{t2} = 15 \times 21.11$ psia $= 316.6$ psia (2182 kPa).

Also, because the compressor is isentropic, for an ideal case

$$\tau_c = \pi_c^{\frac{\gamma-1}{\gamma}} = 15^{\frac{0.35}{1.35}} = 2.018.$$

Thus, the total temperature at the exit of the compressor is

$$T_{t3} = T_{t2}\tau_c = 569.7\ ^\circ\mathrm{R} \times 2.018 = 1150\ ^\circ\mathrm{R}\ (638.9\ \mathrm{K}).$$

Burner

Also, because the burner is ideal, the total pressure is constant across the burner: $p_{t4} = p_{t3}$. Consequently, the exit total pressure is: $p_{t4} = 316.6$ psia (2182 kPa).

The exit total temperature is given as $T_{t4} = 3000$ °R (1667 K), and the heating value of the fuel is 18,000 BTU/lb$_m$. The specific heat at constant pressure is 0.2644 BTU/(lb$_m$-°R) for $\gamma = 1.35$.

The fuel flow is given by the combustor energy equation:

$$\dot{m}_f = \frac{\dot{m}c_p[T_{t4} - T_{t3}]}{\Delta H} = \frac{165\dfrac{\mathrm{lb_m}}{\mathrm{s}} \times 0.2644\dfrac{\mathrm{BTU}}{\mathrm{lb_m}\text{-}^\circ\mathrm{R}} \times [3000 - 1150]\ ^\circ\mathrm{R}}{18,000\dfrac{\mathrm{BTU}}{\mathrm{lb_m}}}$$

$$= 4.485\ \mathrm{lb_m/s}\ (2.034\ \mathrm{kg/s})$$

or $f = \dfrac{\dot{m}_f}{\dot{m}} = \dfrac{3.847}{165} = 0.02718$, which again is small and consistent with the ideal cycle assumption.

Turbine

Now, from the shaft energy balance for the ideal case,

$$\dot{m}c_p(T_{t4} - T_{t5}) = \dot{m}c_p(T_{t3} - T_{t2}).$$

Thus, solving for the turbine exit total temperature:

$$T_{t5} = T_{t4} - (T_{t3} - T_{t2}) = 3000 - (1150 - 569.7) = 2420\ ^\circ\mathrm{R}\ (1344\ \mathrm{K}),$$

which for an ideal analysis is also T_{t8}, since the nozzle is adiabatic.

For the ideal (isentropic) turbine the exit total pressure is

$$p_{t5} = p_{t4}\pi_t = p_{t4}[\tau_t]^{\frac{\gamma}{\gamma-1}} = 320.0\left[\frac{2420}{3000}\right]^{\frac{1.35}{0.35}} = 316.6 \times 0.4367 = 138.3\ \mathrm{psia}\ (953.6\ \mathrm{kPa}).$$

Nozzle

For an ideal (isentropic) nozzle the total pressure is constant: $p_{t8} = p_{t5} = 138.3$ psia (953.6 kPa). Thus at the exit since for the ideal case the exit pressure matches the ambient pressure: $p_8 = p_a = 14.69\ \text{psia}\ (101.3\ \text{kPa})$.

Consequently, the nozzle exit Mach number can be found from

$$p_{t8} = p_8\left[1+\frac{\gamma-1}{2}M_8^2\right]^{\frac{\gamma}{\gamma-1}}:$$

$$M_8 = \sqrt{\left[\frac{2}{\gamma-1}\right]\left[\left[\frac{p_{t8}}{p_8}\right]^{\frac{\gamma-1}{\gamma}}-1\right]} = \sqrt{\left[\frac{2}{1.35-1}\right]\left[\left[\frac{138.3}{14.69}\right]^{\frac{0.35}{1.35}}-1\right]} = 2.122.$$

Since this is supersonic, the nozzle will be converging–diverging.

Also, the exit nozzle temperature can be found from:

$$T_8 = \frac{T_{t8}}{1+\dfrac{\gamma-1}{2}M_8^2} = \frac{2326}{1+\dfrac{1.35-1}{2}2.122^2} = 1353\ ^\circ\text{R}\ (751.7\ \text{K}),$$

and the exit velocity is

$$u_8 = M_8a_8 = M_8\sqrt{\gamma\mathcal{R}T_8} = 2.122\sqrt{1.35\times 53.35\frac{\text{ft-lb}_\text{f}}{\text{lb}_\text{m}\text{-}^\circ\text{R}}\times 32.17\frac{\text{lb}_\text{m}}{\text{sl}}\times 1353\ ^\circ\text{R}\times\frac{\text{sl-ft}}{\text{s}^2\text{-lb}_\text{f}}}$$

$$= 2.112\times 1770 = 3758\ \text{fps}\ (1145\ \text{mps}).$$

From the ideal gas law:

$$\rho_8 = \frac{p_8}{\mathcal{R}T_8} = \frac{14.69\dfrac{\text{lb}_\text{f}}{\text{in.}^2}\times 144\dfrac{\text{in.}^2}{\text{ft}^2}}{53.35\dfrac{\text{ft-lb}_\text{f}}{\text{lb}_\text{m}\text{-}^\circ\text{R}}\times 1353\ ^\circ\text{R}\times 32.17\dfrac{\text{lb}_\text{m}}{\text{sl}}} = 0.0009111\ \text{sl/ft}^3\ (0.4696\ \text{kg/m}^3),$$

thus the nozzle exit area is

$$A_8 = \frac{\dot m}{\rho_8 u_8} = \frac{165\dfrac{\text{lb}_\text{m}}{\text{s}}\times 144\dfrac{\text{in.}^2}{\text{ft}^2}}{0.0009111\dfrac{\text{sl}}{\text{ft}^3}\times 3758\dfrac{\text{ft}}{\text{s}}\times 32.17\dfrac{\text{lb}_\text{m}}{\text{sl}}} = 215.7\ \text{in.}^2(0.1392\ \text{m}^2),$$

which is smaller than the inlet and yields a diameter of 16.52 in. (0.4200 m).

Thrust and TSFC

Finally, because the nozzle exit pressure is the same as the ambient pressure, the ideal thrust is:

$$F = \dot{m}(u_e - u_a) = 165\frac{\text{lb}_\text{m}}{\text{s}}(3758 - 822.2)\frac{\text{ft}}{\text{s}}\frac{\text{sl}}{32.17\ \text{lb}_\text{m}}\frac{\text{lb}_\text{f}\text{s}^2}{\text{sl -ft}}$$

$$= 15{,}060\ \text{lb}_\text{f}\ (66{,}990\ \text{N}). \qquad \text{<ANS}$$

Actually, the answer is $F = 15{,}057.3\ \text{lb}_\text{f}$ but the above is in keeping with the number of significant digits used in the above calculations. This thrust is referred to in Example I.1. The nondimensional thrust is

$$\frac{F}{a_a\dot{m}} = \frac{15{,}060\ \text{lb}_\text{f} \times 32.17\frac{\text{lb}_\text{m}}{\text{sl}} \times \frac{\text{sl -ft}}{\text{s}^2\ \text{lb}_\text{f}}}{1096\frac{\text{ft}}{\text{s}} \times 165\frac{\text{lb}_\text{m}}{\text{s}}} = 2.678.$$

Thus, the *TSFC* is

$$TSFC = \frac{\dot{m}_f}{F} = \frac{4.485\frac{\text{lb}_\text{m}}{\text{s}} \times 3600\frac{\text{s}}{\text{h}}}{15{,}060\ \text{lb}_\text{f}} = 1.072\ \text{lb}_\text{m}/\text{h}/\text{lb}_\text{f}\ (0.1093\ \text{kg/h/N}) \qquad \text{<ANS}$$

and $TSFC\, a_a = 1.072\frac{\text{lb}_\text{m}}{\text{h}/\text{lb}_\text{f}} \times 1096\frac{\text{ft}}{\text{s}} \times 32.17\frac{\text{lb}_\text{m}}{\text{sl}} \times 3600\frac{\text{s}}{\text{h}} = 0.01015.$

As before, using the closed-form equations could have eliminated all of these steps. However, the details would be again missing. Furthermore, this same example with details will be repeated including nonideal effects.

Extended Example 3.3.1 | Ideal Nonafterburning Turbojet: Using Optimal Compressor Ratio

As indicated in the ideal equations in Section 3.2.2, the thrust can be maximized by the right choice of compressor pressure ratio. Using Eq. (3.2.6), one finds $\tau_{c_{opt}} = 2.190$ which in turn yields $\pi_{c_{opt}} = 20.55$. For this ratio, the thrust is 15,130 lb_f, which is marginally better than above.

Example 3.4 | Nonideal Nonafterburning Turbojet

A turbojet flies at sea level at a Mach number of 0.75. It ingests 165 lb_m/s (74.83 kg/s) of air. The compressor operates with a pressure ratio of 15 and an efficiency of 88 percent. The fuel has a heating value of 18,000 BTU/lb_m (41,870 kJ/kg) and the burner total temperature is 3000 °R (1667 K). The burner has an efficiency of 91 percent and a total pressure ratio of 0.95, while the turbine has an efficiency of 85 percent. A converging nozzle is used and the nozzle efficiency is 96 percent. The total pressure recovery for the diffuser is 0.92 and the shaft efficiency is 99.5 percent. Find the developed thrust and *TSFC*. One should note that this is the same engine as in Example 3.3 with the exception of the nonideal effects.

Solution

Diffuser

First the standard conditions at sea level are $T_a = 518.7$ °R and $p_a = 14.69$ psia (from Appendix B). Next, the speed of sound is

$$a_a = \sqrt{\gamma_a \mathcal{R} T_a} = \sqrt{1.4029 \times 53.35 \frac{\text{ft-lb}_f}{\text{lb}_m\text{-°R}} \times 32.17 \frac{\text{lb}_m}{\text{sl}} \times 518.7\ \text{°R} \times \frac{\text{sl-ft}}{\text{lb}_f\text{-s}^2}}$$

$$= 1118 \text{ fps } (340.6 \text{ mps}),$$

so the jet velocity is $u_a = M_a\, a_a = 0.75 \times 1118$ fps $= 838.2$ fps (255.5 mps).

Next, $\frac{T_{ta}}{T_a} = 1 + \frac{\gamma_d - 1}{2} M_a^2 = 1 + \frac{1.3997 - 1}{2} \times 0.75^2 = 1.1124$ for $\gamma_d = 1.3997$ and $M_a = 0.75$

so that since the diffuser is adiabatic the total temperature at the diffuser exit is

$$T_{t2} = T_{ta} = 1.1124 \times 518.7\ \text{°R} = 577.0\ \text{°R}\ (320.6\ \text{K}).$$

Also, $\frac{p_{ta}}{p_a} = \left[\frac{T_{ta}}{T_a}\right]^{\frac{\gamma_d}{\gamma_d - 1}} = [1.1124]^{\frac{1.3997}{1.3997-1}} = 1.4522.$

Next, the pressure recovery factor for the diffuser is 0.92 so

$$\frac{p_{t2}}{p_a} = \frac{p_{t2}}{p_{ta}} \frac{p_{ta}}{p_a} = \pi_d \frac{p_{ta}}{p_a} = 0.92 \times 1.4522 = 1.3360.$$

Therefore, the diffuser exit total pressure is

$$p_{t2} = 1.3360 \times 14.69 \text{ psia} = 19.63 \text{ psia } (135.3 \text{ kPa}).$$

Compressor

Next, the total pressure ratio of the compressor is given as $\pi_c = 15$, so that the total pressure exiting the compressor is $p_{t3} = \pi_c\, p_{t2} = 15 \times 19.63$ psia $= 294.4$ psia (2030 kPa).

Ideally, $\tau_c{}' = [\pi_c]^{\frac{\gamma_c - 1}{\gamma_c}} = [15]^{\frac{1.3805-1}{1.3805}} = 2.1095$ for $\gamma_c = 1.3805$.

Therefore, ideally, $T_{t3}{}' = \tau_c{}' T_{t2} = 2.1095 \times 577.0$ °R $= 1217$ °R (676.2 K), but the compressor efficiency is given, and

$$\eta_c = 0.88 = \frac{T_{t3}{}' - T_{t2}}{T_{t3} - T_{t2}} = \frac{1217\ \text{°R} - 577.0\ \text{°R}}{T_{t3} - 577.0\ \text{°R}}$$

so solving for T_{t3} one finds the compressor total exit temperature is $T_{t3} = 1305$ °R (724.7 K).

Note: This specific heat ratio for the compressor was evaluated at 941 °R or 523 K, which is the resulting *average* total temperature of the compressor. As for all components, the average total temperature was not known a priori but was found by iteration. Namely, the compressor exit temperature depends on the specific heat ratio. But this ratio depends on the exit temperature.

Thus, one can find both values by a progressive iteration. Details of the “successive substitution” iteration are presented in Appendix I, Example I.3. Fortunately, the iteration process is quick to converge (four iterations for this case) and easily programmed. Actually two or three iterations could have been used with minimal affects on the net results. Also, one should note that although precise values of the specific heat ratio are desired, it should not become the dominant feature of examples or chapter problems.

One should also note that the static temperature would have best been used to evaluate the average temperature and resulting specific heat. However, for the lack of information about the static temperatures, or Mach numbers, velocities or areas at the different sections that would have allowed a calculation of static values, the total temperatures were used. This issue will be further addressed at the end of Extended Example 3.4.6.

Primary Combustor

Next, the total pressure ratio of the burner is given as 0.95, so the burner exit total pressure is $p_{t4} = \pi_b\, p_{t3} = 0.95 \times 294.4$ psia $= 279.7$ psia (1928 kPa) and the energy analysis is $\Delta H \eta_b \dot{m}_f = \dot{m} c_{pb}(T_{t4} - T_{t3}) + \dot{m}_f c_{pb} T_{t4}$, but the combustion efficiency is 91 percent and ΔH is 18,000 BTU/lb$_m$, so that solving for the fuel mass flow rate yields

$$\dot{m}_f = 165\frac{\text{lb}_m}{\text{s}} \times 0.2798\frac{\text{BTU}}{\text{lb}_m\text{-}^\circ\text{R}} \frac{3000\ ^\circ\text{R} - 1305\ ^\circ\text{R}}{0.91 \times 18{,}000\dfrac{\text{BTU}}{\text{lb}_m} - 0.2798\dfrac{\text{BTU}}{\text{lb}_m\text{-}^\circ\text{R}} \times 3000\ ^\circ\text{R}}$$
$$= 5.037\ \text{lb}_m/\text{s}\ (2.284\ \text{kg/s}),$$

where the burner specific heat is evaluated at the average burner temperature, which for this case is not an iteration (T_{t3} and T_{t4} are known). Since no afterburner is present, this is also the total fuel flow rate ($\dot{m}_{ft}$) one should also note that $\dot{m}_f/\dot{m} = 0.03053$, which is small. One can also compare the fuel flow rate to that from the ideal case (3.847 lb$_m$/s (1.744 kg/s)), which is 31 percent higher due to the combustor and compressor inefficiencies.

Turbine

Now the mechanical efficiency is given as 0.995, so the shaft power balance

$$\dot{m}\, c_{pc}(T_{t3} - T_{t2}) = \eta_m\, \dot{m}_t\, c_{pt}\, (T_{t4} - T_{t5})$$

becomes

$$165\frac{\text{lb}_m}{\text{s}} \times 0.2487\frac{\text{BTU}}{\text{lb}_m\text{-}^\circ\text{R}}(1305\ ^\circ\text{R} - 577.0\ ^\circ\text{R})$$
$$= 0.995 \times \left(165\frac{\text{lb}_m}{\text{s}} + 5.037\frac{\text{lb}_m}{\text{s}}\right) \times 0.2951\frac{\text{BTU}}{\text{lb}_m\text{-}^\circ\text{R}}\ (3000\ ^\circ\text{R} - T_{t5});$$

thus, solving for the turbine exit total temperature:
$T_{t5} = 2402\ ^\circ$R (1334 K).

Again, the specific heat of the turbine is iteratively found for the average turbine temperature. But the turbine efficiency is defined as $\eta_t = \dfrac{T_{t4} - T_{t5}}{T_{t4} - T_{t5}{}'}$ and given as 85 percent, so

$0.85 = \dfrac{3000\ °\text{R} - 2402\ °\text{R}}{3000\ °\text{R} - T_{t5}{}'}$; thus, solving for the ideal exit total temperature: $T_{t5}{}' = 2297\ °\text{R}$ (1276 K) and ideally $\tau_t{}' = \dfrac{T_{t5}{}'}{T_{t4}} = \dfrac{2297}{3000} = 0.7655.$

Now from the definition of turbine efficiency and the total pressure ratio $\tau_t{}' = [\pi_t]^{\frac{\gamma_t - 1}{\gamma_t}}$ so that the turbine total pressure ratio is $\pi_t = [\tau_t{}']^{\frac{\gamma_t}{\gamma_t - 1}} = (0.7655)^{\frac{1.3025}{1.3025-1}}$, where $\pi_t = 0.3165$ for $\gamma_t = 1.3025$ (evaluated at average turbine temperature).

Finally, the turbine exit total pressure is $p_{t5} = \pi_t\, p_{t4} = 0.3165 \times 279.7$ psia $= 88.57$ psia (610.7 kPa), and since no mixer is present $T_{t5.5} = T_{t5} = 2402\ °\text{R}$ (1334 K) and $p_{t5.5} = p_{t5} =$ 88.57 psia (610.7 kPa).

Primary Nozzle

Since no afterburner is present the inlet total temperature for the nozzle is $T_{t6} = T_{t5.5} = 2402\ °\text{R}$ (1334 K), and the inlet total pressure is $p_{t6} = p_{t5.5} = 88.57$ psia (610.7 kPa).

Since the engine has a fixed converging nozzle, one must first check to see if the nozzle is choked. For choking the nozzle exit pressure is

$$p_8^* = p_{t6}\left[1 + \frac{1 - \gamma_n}{\eta_n(1 + \gamma_n)}\right]^{\frac{\gamma_n}{\gamma_n - 1}}.$$

Thus, for $p_{t6} = 88.57$ psia, a nozzle efficiency of 96 percent and $\gamma_n = 1.3143$ (iteratively found for the average nozzle temperature)

$$p_8^* = 88.57\ \text{psia} \times \left[1 + \frac{1 - 1.3143}{0.96 \times (1 + 1.3143)}\right]^{\frac{1.3143}{1.3143-1}} = 46.81\ \text{psia}\ (322.8\ \text{kPa}).$$

But since $p_a = 14.69$ psia (101.3 kPa), which is well below the choking condition, p_8^*, the converging nozzle is choked and the exit Mach number is identically unity. Therefore, the exit pressure is $p_8 = p_8{}^* = 46.81$ psia (322.8 kPa) and $T_8 = \dfrac{2T_{t6}}{1 + \gamma_n}$ for $M_8 = 1$.

Thus, for $T_{t6} = 2402\ °\text{R}$ (1334 K), $T_8 = \dfrac{2 \times 2402\ °\text{R}}{1 + 1.3142} = 2076\ °\text{R}$ (1153 K).

Now the nozzle exit velocity is given by $u_8 = \sqrt{2\ c_{pn}\ (T_{t6} - T_8)}$, so that the exit velocity is

$$u_8 = \sqrt{2 \times 0.2867 \frac{\text{BTU}}{\text{lb}_\text{m}\text{-}°\text{R}} \times (2402\ °\text{R} - 2076\ °\text{R}) \times 32.17 \frac{\text{lb}_\text{m}}{\text{sl}} \times 778.16 \frac{\text{ft-lb}_\text{f}}{\text{BTU}} \times \frac{\text{sl-ft}}{\text{lb}_\text{f}\text{-s}^2}}$$

$$= 2164\ \text{fps}\ (659.5\ \text{mps}).$$

Next, the speed of sound at the exit is $a_8 = \sqrt{\gamma_8 \mathcal{R} T_8}$, so

$$a_8 = \sqrt{1.3142 \times 53.35 \frac{\text{ft-lb}_\text{f}}{\text{lb}_\text{m}\text{-}°\text{R}} \times 32.17 \frac{\text{lb}_\text{m}}{\text{sl}} \times 2076\ °\text{R} \times \frac{\text{sl-ft}}{\text{lb}_\text{f}\text{-s}^2}} = 2164\ \text{fps}\ (659.5\ \text{mps}).$$

Thus, as a check on the calculations, $M_8 = u_8/a_8 = 1$, as it should. Next, from the continuity equation the exit area is given by $A_8 = \dfrac{\dot{m}_8}{\rho_8\, u_8}$, where $\dot{m}_8$ is the total mass flow rate at 8 and is given by $\dot{m}_8 = \dot{m}_{a8} + \dot{m}_{ft}$ and $\dot{m}_{a8}$ is the air mass flow rate at 8 and is equal to $\dot{m}$, the ingested flow; so $\dot{m}_{a8} = 165$ lb$_m$/s (74.83 kg/s) such that the total mass flow rate at the exit is $\dot{m}_8 = 165 + 5.037 = 170.0$ lb$_m$/s (77.10 kg/s).

Since both T_8 and p_8 are known, ρ_8 is found from the ideal gas equation

$$\rho_8 = \frac{p_8}{\mathcal{R}T_8} = \frac{46.81\dfrac{\text{lb}_\text{f}}{\text{in.}^2} \times 144\dfrac{\text{in.}^2}{\text{ft}^2}}{53.35\dfrac{\text{ft-lb}_\text{f}}{\text{lb}_\text{m}\text{-}^\circ\text{R}} \times 2076\ ^\circ\text{R} \times 32.17\dfrac{\text{lb}_\text{m}}{\text{sl}}} = 0.001892\ \text{sl/ft}^3\ (0.9753\ \text{kg/m}^3).$$

Thus, the nozzle exit area is

$$A_8 = \frac{170.0\dfrac{\text{lb}_\text{m}}{\text{s}} \times 144\dfrac{\text{in.}^2}{\text{ft}^2}}{0.001892\dfrac{\text{sl}}{\text{ft}^3} \times 2164\dfrac{\text{ft}}{\text{s}} \times 32.17\dfrac{\text{lb}_\text{m}}{\text{sl}}} = 185.9\ \text{in.}^2 (0.1200\ \text{m}^2).$$

Total Thrust and TSFC

The thrust is given by

$$F = \dot{m}_8 u_8 - \dot{m}_{a8} u_a + A_8(p_8 - p_a)$$

so that

$$F = \frac{170.0\dfrac{\text{lb}_\text{m}}{\text{s}} \times 2164\dfrac{\text{ft}}{\text{s}}}{32.17\dfrac{\text{lb}_\text{m}}{\text{sl}}} - \frac{165\dfrac{\text{lb}_\text{m}}{\text{s}} \times 838.2\dfrac{\text{ft}}{\text{s}}}{32.17\dfrac{\text{lb}_\text{m}}{\text{sl}}} + 185.9\,\text{in.}^2 \times \left(46.81\dfrac{\text{lb}_\text{f}}{\text{in.}^2} - 14.69\dfrac{\text{lb}_\text{f}}{\text{in.}^2}\right).$$

Therefore $F = 7139$ lb$_\text{f}$ (momentum) + 5970 lb$_\text{f}$ (pressure), so that $F = 13{,}110$ lb$_\text{f}$ (58,310 N) and 54 percent of the thrust is from momentum.

Finally, $TSFC = \dot{m}_{ft}/F$, so that

$$TSFC = \frac{5.038\dfrac{\text{lb}_\text{m}}{\text{s}}}{13110\ \text{lb}_\text{f}} \times 3600\frac{\text{s}}{\text{h}}$$

$$= 1.383\frac{\text{lb}_\text{m}}{\text{h-lb}_\text{f}}\left(0.1411\frac{\text{kg}}{\text{h-N}}\right).$$

By comparing these results to those of the ideal case (Example 3.3), one finds that the thrust is 13 percent lower, but the *TSFC* is 29 percent higher due to the losses. In Figure 3.14 a scaled T–Δs diagram for this example calculated from the thermodynamics T–ds equation applied to each component $\left\{\Delta s = c_\text{p} \ln\left[\dfrac{T_\text{b}}{T_\text{a}}\right] - \mathcal{R}\ln\left[\dfrac{p_\text{b}}{p_\text{a}}\right]\right\}$ is shown.

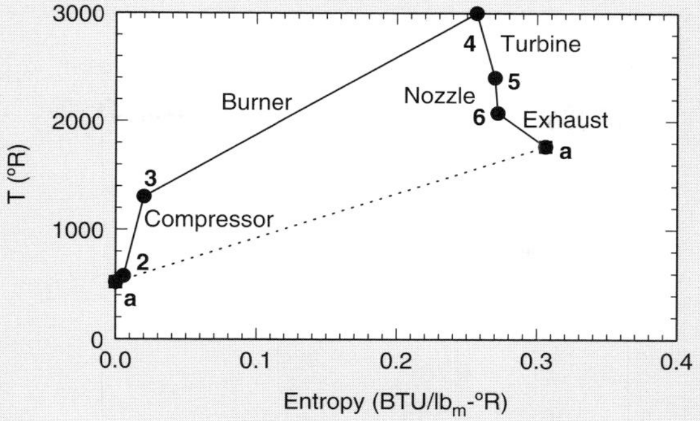

Figure 3.14 T–Δs diagram for the turbojet in Example 3.4.

Extended Example 3.4.1 Nonideal Nonafterburning Turbojet: Effects of Compressor Pressure Ratio

To complement the above example, a study with the parametric variation of the compressor pressure ratio was undertaken. Details of the calculations are exactly as above except for the parameter values. The results of the study are presented in Figure 3.15 in which the net thrust and *TSFC* are plotted as functions of the compressor pressure ratio. As found for the ideal case in Example 3.3.1, an optimum pressure ratio exists at which the thrust is maximized; for this nonideal case the optimum value of π_c is approximately 11.50 and the maximum thrust is 13,170 lb_f, marginally better than for $\pi_c = 15$. For a practical range of compressor pressure ratios (up to 40), the *TSFC* always decreases with increasing pressure ratio. One should also note that for a compressor ratio of 20 the *TSFC* is much lower than for the value of π_c that yielded the maximum thrust (1.323 vs. 1.446 $lb_m/h\text{-}lb_f$). Thus, if this engine is in the design phase, the designer must balance the need for the high thrust versus a low *TSFC*. The balance is further complicated by the fact that as the compressor pressure ratio is increased the number of compressor stages increases, which increases the cost and weight of the compressor.

It should be noted that for this nonideal case, although the performance is compromised compared with the ideal case and parameters are somewhat different, general trends are similar.

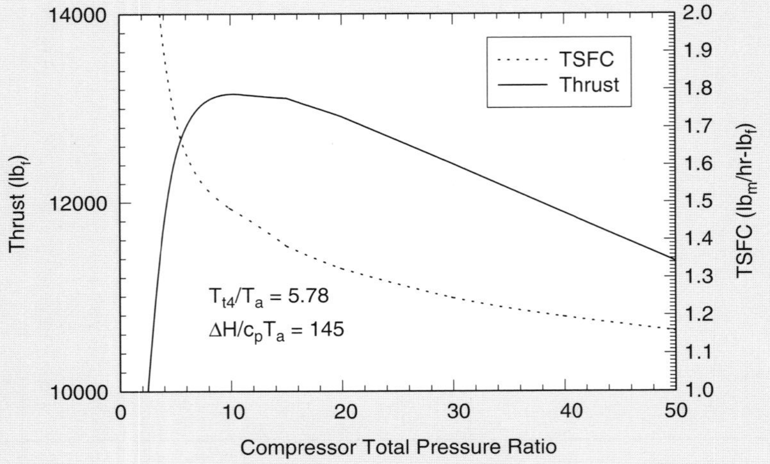

Figure 3.15 Effect of compressor pressure ratio on performance for the turbojet in Extended Example 3.4.1.

Extended Example 3.4.2 | Nonideal Nonafterburning Turbojet: Effects of Variable Converging–Diverging Nozzle

Although it was not asked for, one can replace the converging nozzle with a variable converging diverging nozzle (to match the exit pressure to the ambient pressure) and repeat the above calculations to see how much improvement results. All of the computations are the same as above up to the nozzle.

Solution

Primary Nozzle

Since no afterburner is present, the inlet total temperature for the nozzle is $T_{t6} = T_{t5.5} = 2402$ °R (1334 K) and the inlet total pressure is $p_{t6} = p_{t5.5} = 88.57$ psia (610.7 kPa).

The nozzle is a variable CD nozzle so the exit pressure matches the ambient pressure and one does not need to check for choking as previously.

Thus, $p_8 = 14.69$ psia (101.3 kPa).

Ideally $T_8{}' = T_{t6}\left[\dfrac{p_8}{p_{t6}}\right]^{\frac{\gamma_n-1}{\gamma_n}} = 1865\ °\text{R}\left[\dfrac{14.69\ \text{psia}}{88.57\ \text{psia}}\right]^{\frac{1.3142-1}{1.3142}}$ for $\gamma_n = 1.3142$ (again – iterative)

$= 1563$ °R (868.3 K).

The nozzle efficiency is again 96 percent, thus the nozzle exit temperature is
$T_8 = T_{t6} - \eta_n(T_{t6} - T_8{}') = 2402\ °\text{R} - 0.96(2402\ °\text{R} - 1563\ °\text{R}) = 1597\ °\text{R}\ (887.2\ \text{K})$.

Now the exit gas speed is $u_8 = \sqrt{2\ c_{pn}\ (T_{t6} - T_8)}$, so that

$$u_8 = \sqrt{2 \times 0.2867 \frac{\text{BTU}}{\text{lb}_\text{m}\text{-}^\circ\text{R}} \times (2402\ ^\circ\text{R} - 1563\ ^\circ\text{R}) \times 32.17 \frac{\text{lb}_\text{m}}{\text{sl}} \times 778.16 \frac{\text{ft-lb}_\text{f}}{\text{BTU}} \times \frac{\text{sl-ft}}{\text{lb}_\text{f}\text{-s}^2}}$$

$= 3400$ fps (1036 mps), which is much higher than the choked case above (2164 fps). Next the speed of sound at the exit is $a_8 = \sqrt{\gamma_8 \mathcal{R} T_8}$, so

$$a_8 = \sqrt{1.3142 \times 53.35 \frac{\text{ft-lb}_\text{f}}{\text{lb}_\text{m}\text{-}^\circ\text{R}} \times 32.17 \frac{\text{lb}_\text{m}}{\text{sl}} \times 1597\ ^\circ\text{R} \times \frac{\text{sl-ft}}{\text{lb}_\text{f}\text{-s}^2}} = 1898 \text{ fps } (578.5 \text{ mps}),$$

and thus $M_8 = u_8/a_8 = 3400$ fps/1898 fps $= 1.791$, which is considerably higher than above.

Next, the exit area is given by

$$A_8 = \frac{\dot{m}_8}{\rho_8\ u_8},$$

where $\dot{m}_8$ is the total mass flow rate at 8 and is given by $\dot{m}_8 = \dot{m}_{a8} + \dot{m}_{ft}$ and $\dot{m}_{a8}$ is the air mass flow rate at 8 and is equal to $\dot{m}$, so $\dot{m}_{a8} = 165$ lb_m/s (74.83 kg/s), such that $\dot{m}_8 = 165 + 5.037 =$ 170.0 lb_m/s (77.10 kg/s), as before.

Since both T_8 and p_8 are known, ρ_8 is again found from the ideal gas equation: $\rho_8 = 0.0007721$ sl/ft^3 (0.3980 kg/m^3),

thus the exit area is

$$A_8 = \frac{170.0 \frac{\text{lb}_\text{m}}{\text{s}} \times 144 \frac{\text{in.}^2}{\text{ft}^2}}{0.0007721 \frac{\text{sl}}{\text{ft}^3} \times 3400 \frac{\text{ft}}{\text{s}} \times 32.17 \frac{\text{lb}_\text{m}}{\text{sl}}} = 289.9 \text{ in.}^2\ (0.1870 \text{ m}^2).$$

Total Thrust and TSFC

Finally, the thrust is given by

$$F = \dot{m}_8 u_8 - \dot{m}_{a8} u_a + A_8(p_8 - p_a),$$

so that

$$F = \frac{170.0 \frac{\text{lb}_\text{m}}{\text{s}} \times 3400 \frac{\text{ft}}{\text{s}}}{32.17 \frac{\text{lb}_\text{m}}{\text{sl}}} - \frac{165 \frac{\text{lb}_\text{m}}{\text{s}} \times 838.2 \frac{\text{ft}}{\text{s}}}{32.17 \frac{\text{lb}_\text{m}}{\text{sl}}} + 289.9 \text{ in.}^2 \times \left(14.69 \frac{\text{lb}_\text{f}}{\text{in.}^2} - 14.69 \frac{\text{lb}_\text{f}}{\text{in.}^2}\right).$$

Thus, $F = 13{,}670$ (momentum) + 0 (pressure), or $F = 13{,}670$ lb_f (60810 N) and all of the pressure thrust has now been converted to momentum thrust.

Finally, $TSFC = \dot{m}_{ft}/F$, so that

$$TSFC = \frac{5.037 \frac{\text{lb}_\text{m}}{\text{s}}}{13672 \text{ lb}_\text{f}} \times 3600 \frac{\text{s}}{\text{h}} = 1.326 \frac{\text{lb}_\text{m}}{\text{h-lb}_\text{f}} \left(0.1353 \frac{\text{kg}}{\text{h-N}}\right).$$

As one can see for this operating point, the results are only marginally better. The thrust is 4.3 percent higher than for the fixed converging nozzle, and the *TSFC* is 4.1 percent lower. Thus, for this particular design one would probably not add the complexity and cost of a variable CD nozzle for the small improvement. However, depending on the range of operating conditions or "flight envelope," adding a variable CD nozzle could be advantageous. For example, if the engine was to be routinely used at different altitudes, nozzle inlet conditions, and thrust levels and speeds, a variable CD nozzle would probably be worth the extra engineering and cost.

Extended Example 3.4.3 **Ideal Nonafterburning Turbojet: Effect of Variable Specific Heats**

Next, to isolate only the effects of variable specific heats and specific heat ratios for the different components, one can repeat the computations for the case of all ideal components, but with variable specific heats. These calculations are shown below with all efficiencies and total pressure ratios of unity. Results are also compared with the results of Example 3.3 ($\gamma = 1.35$) to show the effects of the variable specific heats and to Extended Example 3.4.2 to isolate and to show the effects of efficiencies.

Diffuser

First, $a_a = \sqrt{\gamma_a \mathcal{R} T_a} = \sqrt{1.4029 \times 53.35 \dfrac{\text{ft-lb}_\text{f}}{\text{lb}_\text{m}\text{-°R}} \times 32.17 \dfrac{\text{lb}_\text{m}}{\text{sl}} \times 518.7\ \text{°R} \times \dfrac{\text{sl-ft}}{\text{lb}_\text{f}\text{-s}^2}}$

$= 1118$ fps (340.6 mps),

so that the air speed is $u_a = M_a\, a_a = 0.75 \times 1118$ fps $= 838.2$ fps (255.5 mps),

Next, $\dfrac{T_{\text{ta}}}{T_\text{a}} = 1 + \dfrac{\gamma_\text{d} - 1}{2} M_\text{a}^2 = 1 + \dfrac{1.3997 - 1}{2} \times 0.75^2 = 1.1124$ for $\gamma_\text{d} = 1.3997$ and $M_\text{a} = 0.75$,

so that $T_{\text{t}2} = T_{\text{ta}} = 1.1124 \times 518.7$ °R $= 577.0$ °R (320.6 K).

Also, $\dfrac{p_{\text{ta}}}{p_\text{a}} = \left[\dfrac{T_{\text{ta}}}{T_\text{a}}\right]^{\frac{\gamma_\text{d}}{\gamma_\text{d}-1}} = [1.1124]^{\frac{1.3997}{1.3997-1}} = 1.4522.$

Next, $\dfrac{p_{\text{t}2}}{p_\text{a}} = \pi_\text{d} \dfrac{p_{\text{ta}}}{p_\text{a}} = 1 \times 1.4522 = 1.4522,$

so that $p_{\text{t}2} = 1.4522 \times 14.69$ psia $= 21.33$ psia (147.1 kPa).

Thus, the total pressure into the compressor is 8.7 percent higher than for the nonideal case.

Compressor

Next, the total pressure ratio of the compressor is again given as $\pi_\text{c} = 15$, so that $p_{\text{t}3} = \pi_\text{c}$ and $p_{\text{t}2} = 15 \times 21.33$ psia $= 320.0$ psia (2206 kPa).

Ideally, $\tau_\text{c}' = [\pi_\text{c}]^{\frac{\gamma_\text{c}-1}{\gamma_\text{c}}} = [15]^{\frac{1.3827-1}{1.3827}} = 2.116$ for $\gamma_\text{c} = 1.3827$ (iteration).

Therefore, $T_{t3}' = \tau_c' T_{t2} = 2.116 \times 577.0\ °\text{R} = 1221\ °\text{R}$ (678.3 K), but the compressor efficiency is unity for this case and $\eta_c = 1.00 = \dfrac{T_{t3}' - T_{t2}}{T_{t3} - T_{t2}}$.

So, solving for T_{t3} one finds $T_{t3} = 1221°\text{R}$ (678.3 K).

This compares with 1305 °R for the nonideal case and indicates that considerably more power than ideal is being used to drive the compressor for the nonideal case.

Primary Combustor

For this case the total pressure ratio of the burner is unity, so $p_{t4} = \pi_b\, p_{t3} = 1.00 \times 320.0$ psia $=$ 320.0 psia (2206 kPa), which is 14.4 percent higher than for the nonideal case. This indicates that more pressure air is available for the turbine to expand (and thus derive more power) than for the nonideal case.

The energy analysis is $\Delta H \eta_b \dot{m}_f = \dot{m} c_{p_b} (T_{t4} - T_{t3}) + \dot{m}_f c_{p_b} T_{t4}$, but for this case $\eta_b = 100$ percent and $\Delta H = 18{,}000\ \text{BTU/lb}_\text{m}$ are given, so that solving for the fuel mass flow rate yields

$$\dot{m}_f = 165 \frac{\text{lb}_\text{m}}{\text{s}} \times 0.2787 \frac{\text{BTU}}{\text{lb}_\text{m}\text{-°R}} \frac{3000\ °\text{R} - 1221\ °\text{R}}{1.00 \times 17{,}800 \dfrac{\text{BTU}}{\text{lb}_\text{m}} - 0.2787 \dfrac{\text{BTU}}{\text{lb}_\text{m}\text{-°R}} \times 3000\ °\text{R}}$$

$$= 4.767 \text{lb}_\text{m}/\text{s}\ (2.162\ \text{kg/s}).$$

Again, here the burner specific heat is evaluated at the average burner temperature, which for this case is not an iteration. Since no afterburner is present, this is also the total fuel flow rate ($\dot{m}_{ft}$). The fuel flow is 5.6 percent lower than for the nonideal case due to the incomplete combustion for the nonideal case.

One should also note that $\dot{m}_f / \dot{m} = 0.02889$, which again is small.

Turbine

Now the mechanical efficiency is given as unity so the shaft power balance

$$\dot{m} c_{p_c} (T_{t3} - T_{t2}) = \eta_m \dot{m}_t c_{p_t} (T_{t4} - T_{t5})$$

becomes

$$165 \frac{\text{lb}_\text{m}}{\text{s}} \times 0.2477 \frac{\text{BTU}}{\text{lb}_\text{m}\text{-°R}} (1221\ °\text{R} - 577.0\ °\text{R})$$

$$= 1000 \times \left(165 \frac{\text{lb}_\text{m}}{\text{s}} + 4.767 \frac{\text{lb}_\text{m}}{\text{s}}\right) \times 0.2962 \frac{\text{BTU}}{\text{lb}_\text{m}\text{-°R}} (3000\ °\text{R} - T_{t5});$$

thus, solving for the turbine exit total temperature, $T_{t5} = 2477\ °\text{R}$ (1376 K). Note that the fuel flow rate is included for $\dot{m}_t$ for the shaft power balance, which for the ideal case (Example 3.3) was neglected.

Again, the specific heat of the turbine is iteratively found for the average turbine temperature. But the turbine efficiency is defined as $\eta_t = \dfrac{T_{t4} - T_{t5}}{T_{t4} - T_{t5}'}$ and is ideally unity for this case, so

$T_{t5}' = T_{t5} = 2477$ °R (1376 K) and $\tau_t' = \dfrac{T_{t5}'}{T_{t4}} = \dfrac{2477}{3000} = 0.8256.$

Ideally $\tau_t' = [\pi_t]^{\frac{\gamma_t - 1}{\gamma_t}}$, so that the turbine total pressure ratio is $\pi_t = 0.4368$ for $\gamma_t = 1.3011$ (evaluated at average turbine temperature).

Finally, the turbine exit total pressure is $p_{t5} = \pi_t\, p_{t4} = 0.4368 \times 320.0 = 139.8$ psia (963.9 kPa) and, since no mixer is present, $T_{t5.5} = T_{t5} = 2477$ °R (1376 K), and $p_{t5.5} = p_{t5} = 139.8$ psia (963.9 kPa), which is 58 percent higher than for the nonideal case. This indicates that much more high-pressure air is available to the nozzle to expand to a high-velocity exit gas, thus increasing thrust.

Primary Nozzle

Since no afterburner is present the inlet total temperature for the nozzle is
$T_{t6} = T_{t5.5} = 2477$ °R (1376 K)
and the inlet total pressure is $p_{t6} = p_{t5.5} = 139.8$ psia (963.9 kPa).
The nozzle is a variable CD nozzle so the exit pressure matches the ambient pressure and one does not need to check for choking.

Thus, $p_8 = 14.69$ psia (101.3 kPa), and ideally

$$T_8' = T_{t6}\left[\frac{p_8}{p_{t6}}\right]^{\frac{\gamma_n - 1}{\gamma_n}} = 2477\left[\frac{14.69\text{ psia}}{139.8\text{ psia}}\right]^{\frac{1.3113-1}{1.3113}}$$

$$= 1451\ °\text{R}\ (806.1\ \text{K}).$$

The nozzle efficiency is 100 percent, thus the nozzle exit temperature is $T_8 = T_8' = 1451$ °R (806.1 K). Now $u_8 = \sqrt{2\, c_{p_n}\,(T_{t6} - T_8)}$, so that

$$u_8 = \sqrt{2 \times 0.2888\frac{\text{BTU}}{\text{lb}_\text{m}\text{-}^\circ\text{R}} \times (2476\ °\text{R} - 1451\ °\text{R}) \times 32.17\frac{\text{lb}_\text{m}}{\text{sl}} \times 778.16\frac{\text{ft-lb}_\text{f}}{\text{BTU}} \times \frac{\text{sl-ft}}{\text{lb}_\text{f}\text{-s}^2}}$$

$$= 3851\ \text{fps}\ (1174\ \text{mps}),$$

which is close (2.5 percent) to the ideal case in Example 3.3 with $\gamma = 1.35$ ($u_8 = 3758$ fps), but which is 13 percent higher than the nonideal CD nozzle case (Extended Example 3.4.2, in which $u_8 = 3400$ fps), and 78 percent higher than for the nonideal converging nozzle. This difference in exit velocity directly increases the momentum thrust.

Next, the speed of sound at the exit is $a_8 = \sqrt{\gamma_8\, \mathcal{R}\, T_8}$, so

$$a_8 = \sqrt{1.3113 \times 53.35\frac{\text{ft-lb}_\text{f}}{\text{lb}_\text{m}\text{-}^\circ\text{R}} \times 32.17\frac{\text{lb}_\text{m}}{\text{sl}} \times 1451\ °\text{R} \times \frac{\text{sl-ft}}{\text{lb}_\text{f}\text{-s}^2}} = 1807\ \text{fps}\ (550.7\ \text{mps}).$$

Thus $M_8 = u_8/a_8 = 3851$ fps/1807 fps $= 2.131$, which is considerably higher than the nonideal CD nozzle previously, and obviously much higher than the choked converging nozzle case.

Next, the exit area is given by $A_8 = \dfrac{\dot{m}_8}{\rho_8 u_8}$, where $\dot{m}_8$ is the total mass flow rate at 8 and is given by $\dot{m}_8 = \dot{m}_{a8} + \dot{m}_{ft}$ and $\dot{m}_{a8}$ is the air mass flow rate at 8 and is equal to $\dot{m}$, so $\dot{m}_{a8} =$ 165 lb_m/s (74.83 kg/s), such that $\dot{m}_8 = 165 + 4.767 = 169.8\ lb_m/s\ (77.01\ kg/s)$.

Since both T_8 and p_8 are known, ρ_8 is again found from the ideal gas equation: $\rho_8 = 0.000850\ sl/ft^3\ (0.4381\ kg/m^3)$.

$$\text{Thus, } A_8 = \frac{169.8\dfrac{lb_m}{s} \times 144\dfrac{in.^2}{ft^2}}{0.000850\dfrac{sl}{ft^3} \times 3851\dfrac{ft}{s} \times 32.17\dfrac{lb_m}{sl}} = 232.1\ in.^2 (0.1500\ m^2).$$

Total Thrust and TSFC

Finally, the thrust is given by
$F = \dot{m}_8 u_8 - \dot{m}_{a8} u_a + A_8(p_8 - p_a)$,
so that

$$F = \frac{169.8\dfrac{lb_m}{s} \times 3851\dfrac{ft}{s}}{32.17\dfrac{lb_m}{sl}} - \frac{165\dfrac{lb_m}{s} \times 838.2\dfrac{ft}{s}}{32.17\dfrac{lb_m}{sl}} + 232.1\ in.^2 \times \left(14.69\frac{lb_f}{in.^2} - 14.69\frac{lb_f}{in.^2}\right).$$

Thus, $F = 16{,}030$ (momentum) + 0 (pressure), or $F = 16{,}030\ lb_f$ (71,280 N) and all of the thrust is momentum thrust.

Finally, $TSFC = \dot{m}_{ft}/F$, so that

$$TSFC = \frac{4.767\dfrac{lb_m}{s}}{16026\ lb_f} \times 3600\frac{s}{h} = 1.071\frac{lb_m}{h\text{-}lb_f}\left(0.1092\frac{kg}{h\text{-}N}\right).$$

As one can see, the thrust is 17.2 percent higher than for the nonideal components (Extended Example 3.4.2). The *TSFC* is 19.3 percent lower for two reasons; the fuel flow rate is lower (by 5.4 percent) and the thrust is larger. Furthermore, by comparing these results to the ideal case with a constant specific heat throughout the engine (Example 3.3), one finds that the thrust is interestingly 6.4 percent higher (better) and the *TSFC* is nearly the same. The primary cause here is the effect of variable specific heats and in Example 3.3 the value of γ was constant ($\gamma = 1.35$) throughout the engine. Also, a small contribution is that for the ideal case (Example 3.3) the fuel mass flow was neglected in the thrust calculations. The effects of variable specific heats and specific heat ratios for the different components are very important and should be included for most applications, especially large turbofans where the bypassed air is, in general, much cooler.

Extended Example 3.4.4 Nonideal Nonafterburning Turbojet: Effect of Constant Specific Heat

Next, to further identify the effects of variable specific heats and specific heat ratios for the different components, one can repeat the computations for the case of nonideal components as in Extended Example 3.4.2 (CD nozzle), but with a constant specific heat ratio of 1.35 throughout the engine. The details of these calculations are similar to above and will not be repeated. For this case the net thrust is 13,500 lb_f and the *TSFC* is 1.322 lb_f/lb_m-h. Comparing results to those in Extended Example 3.4.2 (13,670 lb_f), it can be seen that there are only a 1.2 percent difference in thrust and 0.4 percent difference in *TSFC*, indicating that one can resonantly use a constant specific heat ratio of 1.35 for this example.

Extended Example 3.4.5 Ideal Nonafterburning Turbojet: Effect of Including Fuel Flow

Next, we further identify the subtle effects of the ideal assumption of negligible fuel flow in Example 3.3 by repeating the computations for the case of Extended Example 3.4.3 with ideal components, namely all efficiencies and pressure loss coefficients are unity, and with a constant specific heat ratio of 1.35 throughout the engine. Again, the details of these calculations are similar to those done previously, and will not be repeated. For this case the net thrust is 15,750 lb_f and the *TSFC* is 1.0722 lb_f/lb_m-h. Comparing results with those in Example 3.3 (15,060 lb_f), it can be seen that there is a 4.6 percent increase in thrust and 0.1 percent decrease in *TSFC*. These subtle differences are due to the fact that in the ideal case of Example 3.3 the mass for of the fuel was neglected in the thrust and shaft power balance calculations. In this current case the added fuel flow component is included in the thrust equation thus increasing the thrust. Furthermore, in the calculation of the fuel flow rate in the combustor for the ideal case, the fuel component is neglected in the energy balance and, thus, is slightly lower than for the current case (4.485 vs. 4.692 lb_m/s).

Extended Example 3.4.6 Nonideal Nonafterburning Turbojet: Effect of Local Mach Number on Evaluation of Variable Specific Heats

Lastly, so far the average total temperature has been used to evaluate specific heats and specific heat ratios for the different components in the examples. Actually, a better method of evaluating these properties is by using the average static temperatures. As covered previously, the static and total temperatures are related by

$$T_t = T\left[1 + \frac{\gamma - 1}{2}M^2\right] \quad \text{or}$$

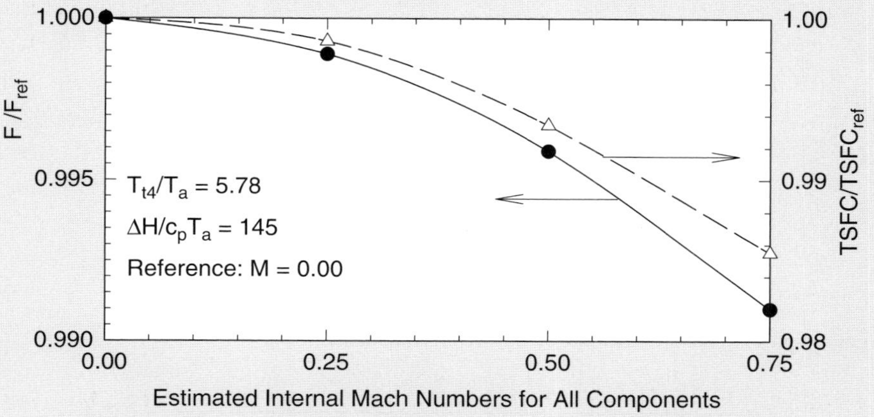

Figure 3.16 Effect of estimated component Mach numbers on calculated performance.

$$T = \frac{T_t}{\left[1 + \frac{\gamma - 1}{2}M^2\right]}.$$

Thus, the calculations from Example 3.4 were again repeated using the above equations to find the static temperature, T, for various Mach numbers and then evaluate c_p and γ. Results are presented in Figure 3.16. In this figure both normalized thrust and TSFC are presented for various estimated local Mach numbers for each component. The values have been normalized by the results from Example 3.4, namely using the total temperature for the evaluation of c_p and γ. Mach numbers up to 0.75 have been used, which is an upper limit for most components. As can be seen, results change minimally using static temperatures over this range – less than 1 percent and 2 percent for the thrust and *TSFC*, respectively. Regardless, the software JetEngineCycle.exe and PowerGTCycle.exe include the ability to include estimates of internal Mach numbers.

Nonafterburning Next, more general various parametric studies can be made for a turbojet. For example, one can study the effect of compressor pressure ratio on the overall performance. In Figure 3.17, a general study of this type is made for both ideal and nonideal cases. For the nonideal case all of the thermodynamic efficiencies are 0.90, all of the pressure loss ratios are 0.95 and the shaft efficiency is 1.00, which are simple but slightly different from the examples. It can seen that, for given operating conditions (M_a, T_{t4}/T_a, and $\dot{m}$), one particular value of π_c will maximize or optimize the thrust. Ideally this value is 20.55 and nonideally it is approximately 15.5. The nonideal thrust is about 15 percent lower than ideal, depending on the value of π_c.

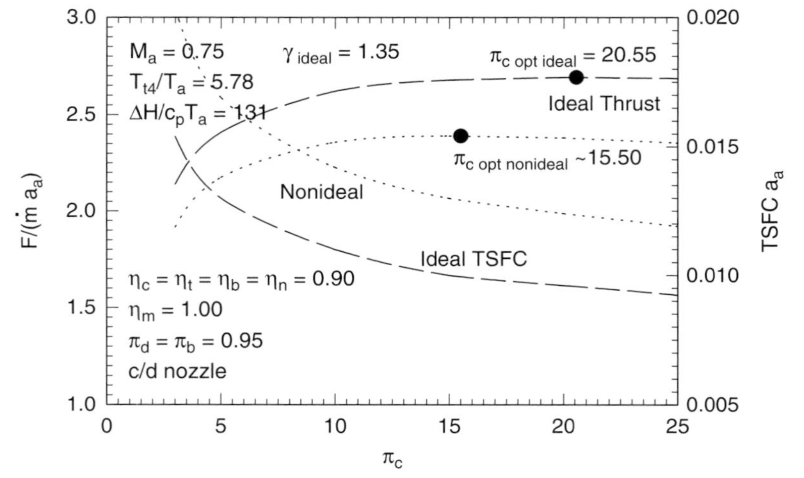

Figure 3.17 Effect of compressor total pressure ratio on the performance of a turbojet.

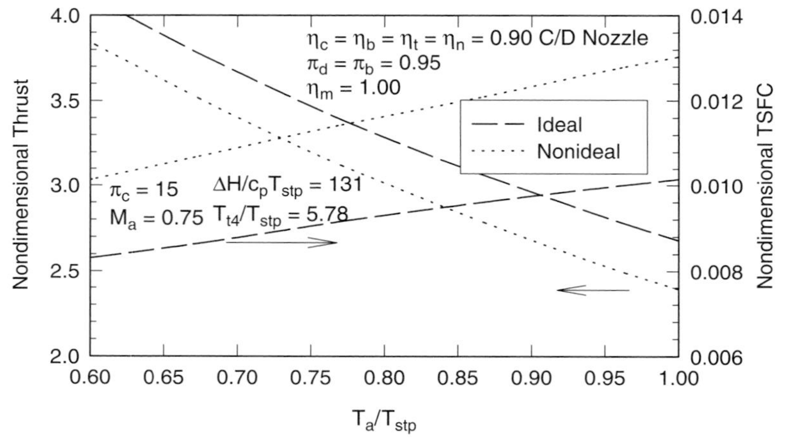

Figure 3.18 Effect of inlet static temperature on the performance of a turbojet.

The curves are relatively flat near the optimum compressor pressure ratio, indicating the there is considerable forgiveness if the optimum value is not used. The general trend of *TSFC* is also shown in this figure. The nonideal *TSFC* is about 15 percent higher than for the ideal case. As can be seen, this value always decreases with the compressor pressure ratio. For example, this general information could and would be used in a design process, because it indicates that, for a given engine size, the thrust can be optimized but the *TSFC* may be improvable. Again, realizing that if a somewhat higher value than optimum of π_c is used, the TSFC can be reasonably reduced without much loss in thrust. A designer would have to examine the specific information on the engine application before deciding on what value to use.

A second investigation is presented in Figure 3.18. The nondimensionalized thrust and *TSFC* are shown as functions of the nondimensionalized inlet temperature for both ideal and nonideal conditions. As can be seen, as the inlet temperature decreases (at constant inlet pressure and Mach number), the thrust increases and the

TSFC decreases. This indicates that the engine operates more effectively at low inlet temperatures and that the thrust of a given engine will be different as determined by the operating location. For example, the thrust at sea level will be higher at a pole than at the equator. Also, some engines are designed with an inlet water-spray injection system to take advantage of this phenomenon. A similar improvement in power generation gas turbines will be discussed later. As water is sprayed in the inlet, the air temperature decreases owing to the vaporization of the water, and the *TSFC* decreases. This behavior can be explained by referring to Eqs. (2.3.30) and (2.3.86). That is, as a given amount of energy is extracted from the fluid by the turbine and transferred to the compressor (a resulting compressor total temperature difference), the total temperature ratio will increase as the inlet temperature decreases for the same temperature difference. As the total temperature ratio increases, so does the total pressure ratio. As discussed in Chapter 1 (Section 1.4.4), for a power generation gas turbine as the total pressure ratio increases, the overall thermal efficiency increases. For the case of a jet engine, this translates into a lower *TSFC*.

Altitude also has a related effect on thrust. For example, as the altitude increases, the pressure and temperature decrease, but pressure decreases faster than temperature. As a result, the density decreases as the altitude increases, which decreases the intake air flow. The thrust directly decreases as the mass flow decreases as evidenced by Eq. (3.2.5). However, the thrust increases as the temperature decreases as discussed in the preceding paragraph. However, this effect is not as great as the effect of the density change and, as a result, thrust decreases with altitude increase. Furthermore, the temperature decreases until 36,000 ft (11,000 m) in a standard atmosphere (as shown in Appendix B) after which it remains constant. However, the pressure and density continue to decrease as the altitude increases. Thus, after 36,000 ft, the thrust decreases with altitude faster than at lower altitudes. As a result, many aircraft operate at around 36,000 ft.

A second study on the effects of temperature is presented in Figure 3.19. For this investigation, the turbine inlet total temperature was varied for a simple ideal turbojet. The Mach number was constant, and

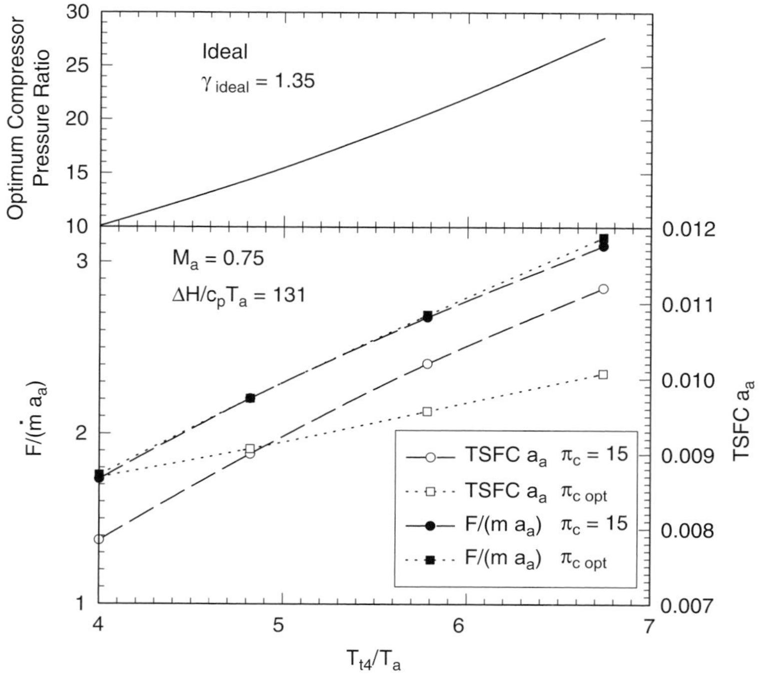

Figure 3.19 Effect of turbine inlet total temperature on the performance of a turbojet.

two different variations of compressor pressure ratio were considered. First, π_c was held constant at 15. As can be seen, the nondimensional thrust and *TSFC* monotonically increase. This implies that the thrust-to-weight ratio improves significantly, but unfortunately the *TSFC* also increases significantly. The second consideration is to optimize the pressure ratio of the compressor by Eq. (3.2.6). As can be seen for this case, once again, the nondimensional thrust increases dramatically, but now the *TSFC* still increases – albeit much less than with the constant π_c case. This implies that, in the design stage, simply increasing the turbine inlet temperature is not optimal and that the compressor must be matched to the turbine inlet temperature. This has further implications when all of the components are matched and many of the parameters are varying simultaneously. This will be further discussed in Chapter 12.

Another study is the effect of flight Mach number on performance. Such a general study is shown in Figure 3.20 for both ideal and nonideal cases for a turbojet and ramjet with CD nozzles. In this figure,

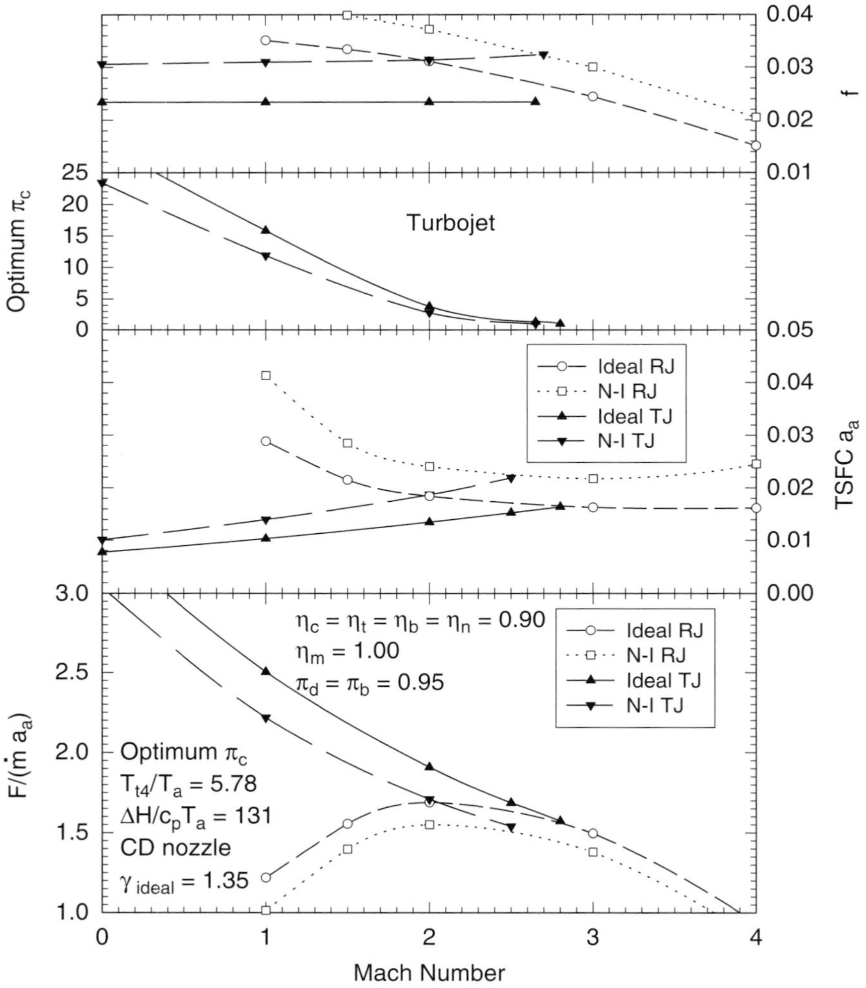

Figure 3.20 Effect of Mach number on the performance of a turbojet with optimum compressor pressure ratio.

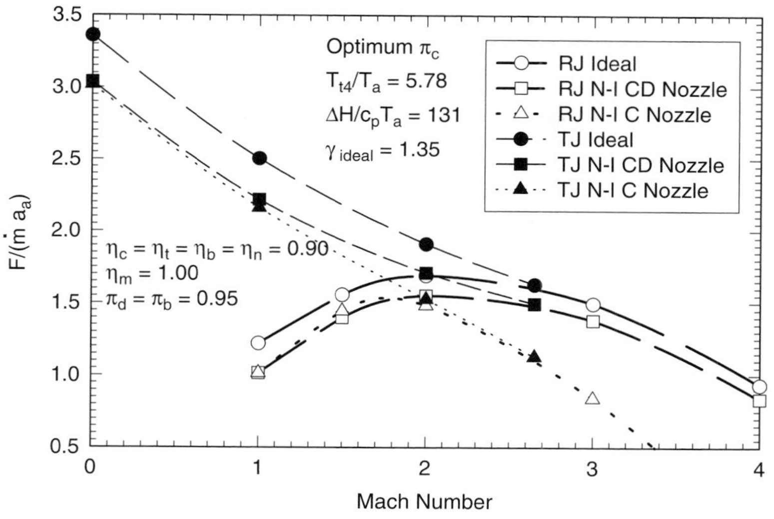

Figure 3.21 Effect of Mach number on the performance of a turbojet for two nozzle types.

the dimensionless thrust and *TSFC* are shown as functions of the Mach number for given values of T_{t4}/T_a and $\dot{m}$ and for the optimum compressor pressure ratio (which varies with M_a, as is also shown in Figure 3.20). As can be seen, the thrust always decreases with the Mach number and *TSFC* increases. Also, at a particular value of M_a, the pressure ratio in the compressor becomes unity (for this case $M_a \cong 2.84$, ideally; $M_a \cong 2.65$, nonideally). This indicates that the compressor is not energizing the fluid and the analysis is not valid for larger Mach numbers. In fact, at this point the engine is acting like a ramjet because the compressor is not being powered. One should also note as well that the thrust is much lower and the *TSFC* is much higher than for lower Mach numbers. Thus, lower speeds are better suited for a turbojet. Also shown on this figure are general results for a ramjet. As can be seen, for high speeds the ramjet is preferred.

In Figure 3.21 a similar study is presented but now including a converging nozzle. As can be seen for both the turbojet and ramjet very significant decreases in thrust is seen for Mach numbers above about 1.0 for the turbojet and 2.0 for the ramjet. Again, for large Mach numbers the turbojet basically becomes a ramjet.

Example 3.5 | Ideal Turbojet with an Afterburner

The ideal turbojet as examined in Example 3.3 now has an afterburner with an afterburner exit total temperature of 3500 °R (1944 K). All other parameters match those in Example 3.3. Once again, find the developed thrust and *TSFC*.

Solution

Much of the work has already been performed for this analysis in Example 3.3. All calculations up through the turbine exit are still valid.

Afterburner

The additional fuel flow for the afterburner is given by

$$\dot{m}_{f_{ab}} = \frac{\dot{m}c_p[T_{t6} - T_{t5}]}{\Delta H} = \frac{165\dfrac{\text{lb}_\text{m}}{\text{s}} \times 0.2644\dfrac{\text{BTU}}{\text{lb}_\text{m}\text{-}^\circ\text{R}} \times [3500 - 2420]\ ^\circ\text{R}}{18{,}000\dfrac{\text{BTU}}{\text{lb}_\text{m}}} = 2.618\ \text{lb}_\text{m}/\text{s}\ (1.187\ \text{kg/s})$$

and thus the total fuel mass flow is

$$\dot{m}_{f_t} = \dot{m}_f + \dot{m}_{f_{ab}} = (4.485 + 2.618)\frac{\text{lb}_\text{m}}{\text{s}} = 7.103\frac{\text{lb}_\text{m}}{\text{s}}\left(3.221\frac{\text{kg}}{\text{s}}\right).$$

Nozzle

Because the inlet total pressure to the nozzle is the same as for the nonafterburning case, the nozzle exit Mach number is once again 2.122. However, since the nozzle inlet total temperature is much higher, the exit temperature is higher. That is, the nozzle is adiabatic ($T_{t8} = T_{t6} = 3500\ ^\circ\text{R}$) and so

$$T_8 = \frac{T_{t8}}{1 + \dfrac{\gamma - 1}{2}M_8^2} = \frac{3500}{1 + \dfrac{1.35 - 1}{2}2.112^2} = 1957\ ^\circ\text{R}\ (1087\ \text{K}).$$

As a result, the exit speed of sound and exit velocity are much higher and thus

$$u_8 = M_8 a_8 = M_8\sqrt{\gamma \mathcal{R} T_8} = 1.951\sqrt{1.35 \times 53.35\frac{\text{ft lb}_\text{f}}{\text{lb}_\text{m}\text{-}^\circ\text{R}} \times 32.17\frac{\text{lb}_\text{m}}{\text{sl}} \times 1957\ ^\circ\text{R} \times \frac{\text{sl ft}}{\text{s}^2\ \text{lb}_\text{f}}}$$

$$= 4519\ \text{fps}\ (1377\ \text{mps}),$$

$$\rho_8 = \frac{p_8}{\mathcal{R}T_8} = \frac{14.69\dfrac{\text{lb}_\text{f}}{\text{in.}^2} \times 144\dfrac{\text{in.}^2}{\text{ft}^2}}{53.35\dfrac{\text{ft-lb}_\text{f}}{\text{lb}_\text{m}\text{-}^\circ\text{R}} \times 1957\ ^\circ\text{R} \times 32.17\dfrac{\text{lb}_\text{m}}{\text{sl}}} = 0.0006299\ \text{sl/ft}^3\ (0.3247\ \text{kg/m}^3).$$

Therefore, the required nozzle exit area is

$$A_8 = \frac{\dot{m}}{\rho_8 u_8} = \frac{165\dfrac{\text{lb}_\text{m}}{\text{s}} \times 144\dfrac{\text{in.}^2}{\text{ft}^2}}{0.0006299\dfrac{\text{sl}}{\text{ft}^3} \times 4519\dfrac{\text{ft}}{\text{s}} \times 32.17\dfrac{\text{lb}_\text{m}}{\text{sl}}} = 259.4\ \text{in.}^2(0.1674\ \text{m}^2).$$

Thus, for the afterburner a larger nozzle diameter is needed. A fixed geometry cannot accommodate the flows with a higher exit temperature. Thus, a variable geometry would be needed. This topic will be considered in detail in later in this chapter and Chapter 10.

Thrust and TSFC

Because the exit pressure again matches the ambient pressure, the ideal thrust is

$$F = \dot{m}(u_e - u_a) = 165\frac{\text{lb}_\text{m}}{\text{s}}(4519 - 822.2)\frac{\text{ft}}{\text{s}}\frac{\text{sl}}{32.17\,\text{lb}_\text{m}}\frac{\text{lb}_\text{f}\text{-s}^2}{\text{sl-ft}} = 18{,}960\ \text{lb}_\text{f}\ (84{,}330\ \text{N}), \quad \text{<ANS}$$

$$TSFC = \frac{\dot{m}_{f_t}}{F} = \frac{7.103\dfrac{\text{lb}_\text{m}}{\text{s}} \times 3600\dfrac{\text{s}}{\text{h}}}{18{,}960\ \text{lb}_\text{f}} = 1.349\ \text{lb}_\text{m}/\text{h}/\text{lb}_\text{f}\ (0.1376\ \text{kg/h/N}). \quad \text{<ANS}$$

One can compare the results to Example 3.3 and see that the thrust has been increased by 25.8 percent. The *TSFC* has been increased by 25.0 percent. Thus, although the thrust is much higher, the fuel economy is relatively poor. This is typical trend of engines with afterburners. One should keep in mind, the primary purpose of afterburners is to achieve a sudden and large increase in thrust. For example, takeoff on short runways (aircraft carriers) or sudden dog-fights.

Example 3.6 | **Nonideal Turbojet with an Afterburner**

A turbojet with an afterburner flies at sea level at a Mach number of 0.75. It takes in 165 lb_m/s (74.83 kg/s) of air. The compressor operates with a pressure ratio of 15 and an efficiency of 88 percent. The fuel has a heating value of 18,000 BTU/lb_m (41,870 kJ/kg) and the burner total temperature is 3000 °R (1667 K). The burner has an efficiency of 91 percent and a total pressure ratio of 0.95, while the turbine has an efficiency of 85 percent. The afterburner exit total temperature is 3500 °R (1944 K). The afterburner has an efficiency of 89 percent and a total pressure ratio of 0.97. A converging nozzle is used and the nozzle efficiency is 96 percent. The total pressure recovery for the diffuser is 0.92 and the shaft efficiency is 99.5 percent. Find the developed thrust and *TSFC*. One should note that this is the same engine as in Example 3.5 with the exception of the nonideal effects. This is also the same engine as that in Example 3.4 with the addition of an afterburner.

Solution

The solution is the same up to the afterburner as for Example 3.4. Thus, only the steps after the turbine will be shown.

Afterburner

First, the total pressure ratio of the afterburner is given as 0.97, so $p_{t6} = \pi_{ab}\, p_{t5.5} = 0.97 \times 88.58$ psia $= 85.92$ psia (592.4 kPa) and the energy analysis is

$\Delta H \eta_{ab} \dot{m}_{f_{ab}} = (\dot{m} + \dot{m}_f\ )c_{p_{ab}}(T_{t6} - T_{t5.5}) + \dot{m}_{f_{ab}} c_{p_{ab}} T_{t6}.$

But $\eta_{ab} = 0.89$ and $\Delta H = 18{,}000$ BTU/lb$_m$ are given, so that solving for the afterburner fuel mass flow rate yields

$$\dot{m}_{f_{ab}} = (165 + 5.038)\frac{\text{lb}_m}{\text{s}} \times 0.3024\frac{\text{BTU}}{\text{lb}_m\text{-°R}} \frac{3500\ \text{°R} - 2402\ \text{°R}}{0.89 \times 18{,}000\frac{\text{BTU}}{\text{lb}_m} - 0.3024\frac{\text{BTU}}{\text{lb}_m\text{-°R}} \times 3500\ \text{°R}}$$

$$= 3.773\ \text{lb}_m/\text{s}\ (1.711\ \text{kg/s}),$$

where the burner specific heat is evaluated at the average afterburner temperature, which is not an iteration.

Now the total fuel flow rate is

$$\dot{m}_{ft} = \dot{m}_f + \dot{m}_{f_{ab}}$$

$$= 5.038 + 3.773 = 8.811\ \text{lb}_m/\text{s}\ (3.996\ \text{kg/s}).$$

Primary Nozzle

The inlet total pressure is $p_{t6} = 85.92$ psia (592.4 kPa). Since the engine has a fixed converging nozzle, one must first check to see if the nozzle is choked.

For choking, the nozzle exit pressure is

$$p_8^* = p_{t6}\left[1 + \frac{1 - \gamma_n}{\eta_n(1 + \gamma_n)}\right]^{\frac{\gamma_n}{\gamma_n - 1}};$$

thus for $p_{t6} = 85.92$

psia, $\eta_n = 0.96$ (given), and $\gamma_n = 1.2737$ (iteratively found for the average nozzle temperature). Note that is lower than for the nonafterburning case (Example 3.4) due to the significantly higher nozzle temperature:

$$p_8^* = 85.92\ \text{psia} \times \left[1 + \frac{1 - 1.2737}{0.96 \times (1 + 1.2737)}\right]^{\frac{1.2737}{1.2737 - 1}} = 46.06\ \text{psia}\ (317.6\ \text{kPa}),$$

but since $p_a = 14.69$ psia (101.3 kPa), which is well below the choking condition, $p_8{}^*$, the nozzle is choked and the exit Mach number is identically unity as in Example 3.4.

Therefore, the exit pressure is $p_8 = p_8{}^* = 46.06$ psia (317.6 kPa), and $T_8 = \dfrac{2T_{t6}}{1 + \gamma_n}$ for $M_8 = 1$.

Thus, for $T_{t6} = 3500$ °R (1944 K):

$$T_8 = \frac{2 \times 3500\ \text{°R}}{1 + 1.2737} = 3079\ \text{°R}\ (1711\ \text{K}).$$

Now the nozzle exit velocity is given by $u_8 = \sqrt{2\ c_{pn}\ (T_{t6} - T_8)}$, so that

$$u_8 = \sqrt{2 \times 0.3190\frac{\text{BTU}}{\text{lb}_m\text{-°R}} \times (3500\ \text{°R} - 3079\ \text{°R}) \times 32.17\frac{\text{lb}_m}{\text{sl}} \times 778.16\frac{\text{ft-lb}_f}{\text{BTU}} \times \frac{\text{sl-ft}}{\text{lb}_f\text{-s}^2}}$$

$$= 2594\ \text{fps}\ (790.6\ \text{mps}),$$

which is 19.8 percent higher than for the nonfterburning case (Example 3.4), due to the much higher temperatures.

Next, from continuity the exit area is given by $A_8 = \dfrac{\dot{m}_8}{\rho_8\, u_8}$, where $\dot{m}_8$ is the total mass flow rate at 8 and is given by $\dot{m}_8 = \dot{m}_{a8} + \dot{m}_{ft}$, and $\dot{m}_{a8}$ is the air mass flow rate at 8 and is equal to $\dot{m}$, the ingested flow, so $\dot{m}_{a8} = 165$ lb$_m$/s (74.83 kg/s).

The total mass flow rate at the exit is $\dot{m}_8 = 165 + 8.811 = 173.8\ \mathrm{lb_m/s}\ (78.82\ \mathrm{kg/s})$. Since both T_8 and p_8 are known, ρ_8 is found from the ideal gas equation:

$$\rho_8 = \frac{p_8}{\mathcal{R}T_8} = \frac{46.06\dfrac{\mathrm{lb_f}}{\mathrm{in.}^2} \times 144\dfrac{\mathrm{in.}^2}{\mathrm{ft}^2}}{53.35\dfrac{\mathrm{ft\text{-}lb_f}}{\mathrm{lb_m\text{-}^\circ R}} \times 3079\ ^\circ\mathrm{R} \times 32.17\dfrac{\mathrm{lb_m}}{\mathrm{sl}}} = 0.001255\ \mathrm{sl/ft^3}\ (0.6469\ \mathrm{kg/m^3}),$$

thus the nozzle exit area is

$$A_8 = \frac{173.8\dfrac{\mathrm{lb_m}}{\mathrm{s}} \times 144\dfrac{\mathrm{in.}^2}{\mathrm{ft}^2}}{0.001255\dfrac{\mathrm{sl}}{\mathrm{ft}^3} \times 2594\dfrac{\mathrm{ft}}{\mathrm{s}} \times 32.17\dfrac{\mathrm{lb_m}}{\mathrm{sl}}} = 238.9\ \mathrm{in.}^2\ (0.1541\ \mathrm{m}^2).$$

Total Thrust and TSFC

The thrust is given by

$$F = \dot{m}_8 u_8 - \dot{m}_{a8} u_a + A_8(p_8 - p_a),$$

so that

$$F = \frac{173.8\dfrac{\mathrm{lb_m}}{\mathrm{s}} \times 2594\dfrac{\mathrm{ft}}{\mathrm{s}}}{32.17\dfrac{\mathrm{lb_m}}{\mathrm{sl}}} - \frac{165\dfrac{\mathrm{lb_m}}{\mathrm{s}} \times 838.2\dfrac{\mathrm{ft}}{\mathrm{s}}}{32.17\dfrac{\mathrm{lb_m}}{\mathrm{sl}}} + 238.9\ \mathrm{in.}^2 \times \left(46.06\frac{\mathrm{lb_f}}{\mathrm{in.}^2} - 14.69\frac{\mathrm{lb_f}}{\mathrm{in.}^2}\right).$$

Thus $F = 9717$ (momentum) $\times$ 7493 (pressure) or $F = 17{,}210$ lb$_f$ (76,550 N).

Finally, $TSFC = \dot{m}_{ft}/F$, so that

$$TSFC = \frac{8.811\dfrac{\mathrm{lb_m}}{\mathrm{s}}}{17210\ \mathrm{lb_f}} \times 3600\frac{\mathrm{s}}{\mathrm{h}}$$

$$= 1.843\frac{\mathrm{lb_m}}{\mathrm{h\text{-}lb_f}}\left(0.1880\frac{\mathrm{kg}}{\mathrm{h\text{-}N}}\right).$$

By comparing these results with the nonafterburning case, one can see that the thrust has been increased by 31.3 percent and the *TSFC* has been increased by 33.3 percent. Thus, although the thrust is greatly increased, the fuel economy is much worse. Also, by comparing these results to those of the ideal case (Example 3.5), one finds that the thrust is 8.9 percent lower, but the *TSFC* is 36.6 percent higher due to the losses.

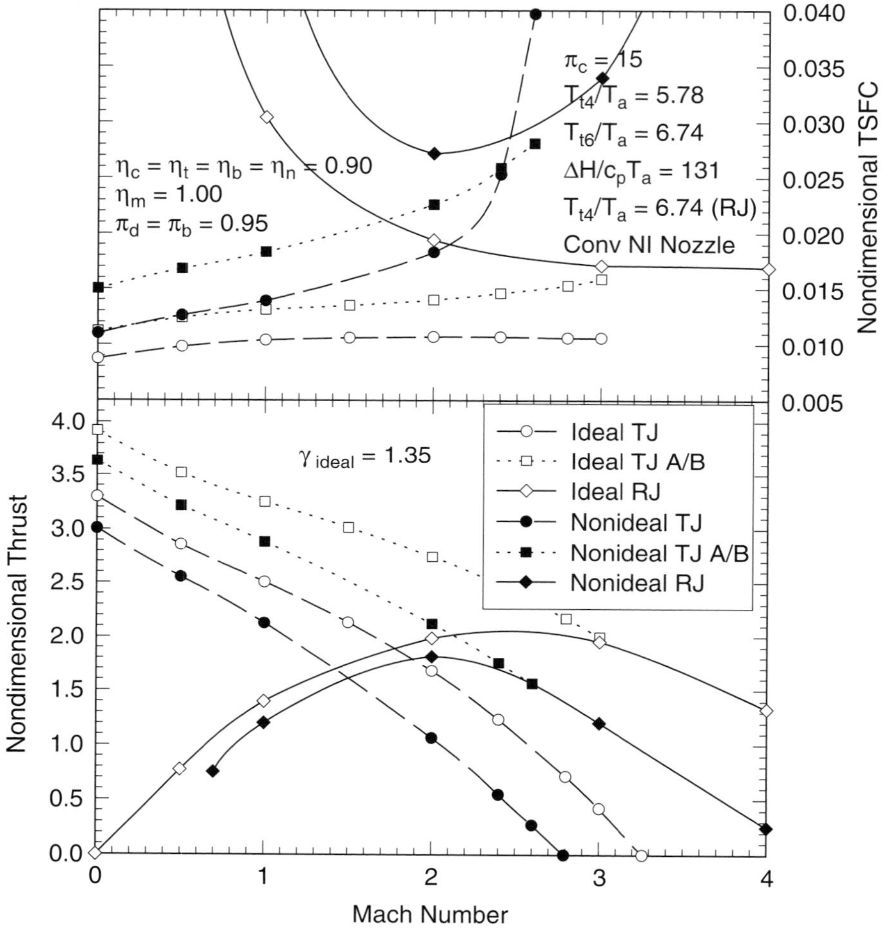

Figure 3.22 Comparison of thrust and *TSFC* vs. Mach number trends for a turbojet, turbojet with an afterburner, and ramjet.

Afterburning

Once again, various parametric studies can be performed to determine the overall performance of an afterburning turbojet with a fixed compressor pressure ratio both ideal and nonideal cases are shown. Two general trends are presented in Figure 3.22. The thrust is presented as a function of the Mach number and is compared with a turbojet with no afterburner and a ramjet. All engines have the same airflow rate. For this study the ramjet and afterburner temperatures are the same. One can see that the thrust for the afterburning case is always higher than for the nonafterburning case. Also, in Figure 3.22, it can be seen that the *TSFC* is likewise higher than for the nonafterburning case. This indicates that the extra fuel injected into the afterburner is not being used optimally. As indicated in Chapter 1 and in Example 3.6, however, this is generally not of importance when immediate extra thrust is required. In comparison with the ramjet, one finds that the turbojet thrust can be higher at low to moderate Mach numbers, but this will depend on how much afterburner fuel is used. At low Mach numbers, the *TSFC* for an afterburning turbojet is lower than for a ramjet but is once again less efficient at high Mach numbers. Note that at higher Mach numbers the thrust for the nonafterburning turbojet becomes zero ($M_a = 3.26$ ideally and $M_a = 2.79$ nonideally). This is due to the increased total temperature entering the compressor; all of the added enthalpy is extracted from the fluid by the turbine to drive the compressor. As a result, sufficient enthalpy is not delivered to the nozzle to generate any thrust. At these same Mach numbers the afterburning case becomes the same as the ramjet. All of the turbojet thrust is the added thrust from the afterburner and the engine is basically a ramjet. One should be careful and not directly compare the results from Figure 3.22 with those of Figure 3.20; in Figure 3.22 the compressor pressure ratio is fixed at 15, whereas in Figure 3.20 the optimum value of π_c is used, which is different than 15, except for one particular Mach number.

Turbofan: Fan Exhausted

The general diagram for a turbofan is shown in Figure 2.3. As indicated earlier in this chapter, this engine has up to three basic flow paths. Two simplifications of the general turbofan are covered for the ideal analysis: the fully exhausted fan and the ducted and mixed fan. Both nonafterburning and afterburning cases are considered. Later in this chapter, a hybrid fan (in which a portion of the secondary flow is mixed while the remainder is exhausted) is covered.

First, a fully exhausted turbofan is shown in Figure 3.4. For this configuration two flow paths exist. For one path, the air enters the diffuser, flows through the core, and exhausts through the primary nozzle. For the second path, air enters the diffuser, flows through the fan, and exits through the fan nozzle. Two separate subcases are considered here: without and with the afterburner. In Figure 3.4, an afterburner is present in the core between stations 5 and 6. Analyses for the two subcases are performed separately.

Nonafterburning For the nonafterburning case, stations 5 and 6 are one and the same and will be termed station 5, as shown in Figure 3.5. For all fan exhausted cases, the thrust is generated by two separate gas streams.

Example 3.7 | Ideal Fully Exhausted Turbofan without Afterburner

An ideal turbofan with an exhausted fan flies at sea level at a Mach number of 0.75. The primary flow is 74.83 kg/s (165 lb_m/s) and the bypass ratio is 1.20. The compressor pressure ratio is 15 whereas that of the fan is 3. The fuel has a heating value of 41,870 kJ/kg (18,000 BTU/lb_m) and the burner exit total temperature is 1667 K (3000 °R). Find the developed thrust and the *TSFC* if $\gamma = 1.35$.

Solution

If this example is compared with the previous examples for the ideal turbojet it can be seen that the only difference is the addition of the exhausted fan. All of the given conditions are identical. Once again a step-by-step solution will be presented because it allows the determination of internal differences between this example and the plain turbojet example.

Many of the values are the same as the previous turbojet example. These will be simply listed as follows.

$p_a = 101.3$ kPa (14.69 psia) $\quad T_a = 288.2$ K (518.7 °R)

$u_a = 250.6$ mps (822.2 fps) $\quad a_a = 334.1$ mps (1096 fps)

$T_{ta} = 316.5$ K (569.7 °R) $= T_{t2} \quad p_{ta} = 145.5$ kPa (21.11 psia) $= p_{t2}$

$p_{t3} = 2183$ kPa (316.4 psia) $= p_{t4} \quad T_{t3} = 638.7$ K (1150 °R)

Diffuser

As for the turbojet, $\rho_a = 1.225$ kg/m^3 (0.002376 sl/ft^3). Thus, if (not necessarily the case) the gas velocity at the diffuser inlet is the same as the jet velocity, the diffuser inlet area is:

$$A_{in} = \frac{\dot{m}(1+\alpha)}{\rho_a u_a} = \frac{74.83\dfrac{\text{kg}}{\text{s}} \times (1+1.20)}{1.225\dfrac{\text{kg}}{\text{m}^3} \times 250.6\dfrac{\text{m}}{\text{s}}} = 0.5363 \text{ m}^2 (831.3 \text{ in.}^2),$$

which yields an inlet diameter of 0.8456 m (33.29 in.), which is 52 percent larger than that for the simple turbojet.

Fan

From the fan pressure ratio the fan exit total pressure is

$p_{t7} = \pi_f p_{t2} = 3 \times 145.5 = 436.6$ kPa (63.32 psia).

Because the ideal fan is isentropic, the fan exit total temperature is

$$T_{t7} = T_{t2}\tau_f = T_{t2}[\pi_f]^{\frac{\gamma-1}{\gamma}} = 316.5[3]^{\frac{0.35}{1.35}} = 420.8\text{K } (757.4\,°\text{R}).$$

Fan Nozzle

Now, for the ideal fan nozzle the exit pressure matches the ambient pressure:

$p_9 = p_a = 101.3$ kPa (14.69 psi)

and, since the ideal nozzle is isentropic, the total pressure remains constant through the nozzle

$$p_{t9} = p_{t7} = 436.4 \text{ kPa (63.3 psia) and } p_{t9} = p_9\left[1 + \frac{\gamma-1}{2}M_9^2\right]^{\frac{\gamma}{\gamma-1}}.$$

Thus, solving for the exit Mach number:

$$M_9 = \sqrt{\left[\frac{2}{\gamma-1}\right]\left[\left[\frac{p_{t9}}{p_9}\right]^{\frac{\gamma-1}{\gamma}} - 1\right]} = \sqrt{\left[\frac{2}{1.35-1}\right]\left[\left[\frac{436.4}{101.3}\right]^{\frac{0.35}{1.35}} - 1\right]} = 1.622.$$

Consequently, the exit static temperature is

$$T_9 = \frac{T_{t9}}{1 + \frac{\gamma - 1}{2} M_9^2} = \frac{420.8}{1 + \frac{1.35 - 1}{2} 1.622^2} = 288.2 \text{ K } (518.7\ ^\circ\text{R}),$$

and the exit velocity is

$$u_9 = M_9 a_9 = M_9 \sqrt{\gamma \mathcal{R} T_9} = 1.622 \sqrt{1.35 \times 287.1 \frac{\text{J}}{\text{kg K}} \times 288.2 \text{ K} \times \frac{\text{N m}}{\text{J}} \times \frac{\text{kg m}}{\text{s}^2 \text{ N}}}$$

$$= 541.9 \text{ mps } (1778 \text{ fps}).$$

Turbine

For power balance on the shaft:

$$\dot{m} c_p (T_{t4} - T_{t5}) = \dot{m} c_p (T_{t3} - T_{t2}) + \alpha\ \dot{m} c_p (T_{t7} - T_{t2})$$

or $(T_{t4} - T_{t5}) = (T_{t3} - T_{t2}) + \alpha (T_{t7} - T_{t2})$;

thus $(1667 - T_{t5}) = (638.7 - 316.5) + 1.20(420.8 - 316.5)$.
Solving for the turbine exit total temperature yields: $T_{t5} = 1219$ K (2194 °R).

Note that because the turbine is also driving the fan, the turbine exit temperature is lower than that for the turbojet

Since the ideal turbine is isentropic it is possible to find the turbine exit total pressure

$$p_{t5} = p_{t4} \pi_t = p_{t4} [\tau_t]^{\frac{\gamma}{\gamma - 1}} = 2206 \left[\frac{1219}{1667}\right]^{\frac{1.35}{0.35}} = 2206 \times 0.2996 = 654.1 \text{ kPa } (94.87 \text{ psia}).$$

Primary Nozzle

Because the nozzle is isentropic: $p_{t8} = p_8 \left[1 + \frac{\gamma - 1}{2} M_8^2\right]^{\frac{\gamma}{\gamma - 1}}$

Or, solving for the exit Mach number yields

$$M_8 = \sqrt{\left[\frac{2}{\gamma - 1}\right] \left[\left[\frac{p_{t8}}{p_8}\right]^{\frac{\gamma - 1}{\gamma}} - 1\right]} = \sqrt{\left[\frac{2}{1.35 - 1}\right] \left[\left[\frac{654.1}{101.3}\right]^{\frac{0.35}{1.35}} - 1\right]} = 1.885.$$

Thus, the exit static temperature is

$$T_8 = \frac{T_{t8}}{1 + \left[\frac{\gamma - 1}{2}\right] M_8^2} = \frac{1151}{1 + \left[\frac{1.35 - 1}{2}\right] 1.885^2} = 752.0\text{K } (1354\ ^\circ\text{R})$$

and the exit gas velocity is

$$u_8 = M_8 a_a = M_8 \sqrt{\gamma \mathcal{R} T_8} = 1.885 \sqrt{1.35 \times 287.1 \frac{\text{J}}{\text{kg-K}} \times 752.0 \text{ K} \times \frac{\text{N-m}}{\text{J}} \times \frac{\text{kg-m}}{\text{s}^2\text{-N}}}$$

$$= 1017 \text{ mps } (3337 \text{ fps}).$$

Note here that u_8 and u_9 are not identical; thus, the thrust is not maximized.

Thrust and TSFC

Finally, because both nozzle exit pressures are atmospheric, there is not a pressure thrust term and the thrust is given by

$$F = \dot{m}(u_8 - u_a) + \dot{m}_s(u_9 - u_a) = \dot{m}(u_8 - u_a) + \alpha\,\dot{m}(u_9 - u_a)$$
$$= \left[74.83\frac{\text{kg}}{\text{s}}(1017 - 250.6)\frac{\text{m}}{\text{s}} + 1.2 \times 74.83\frac{\text{kg}}{\text{s}}(541.9 - 250.6)\frac{\text{m}}{\text{s}}\right] \times \frac{\text{Ns}^2}{\text{kg-m}}$$
$$= 57{,}400 \text{ N (primary)} + 26{,}170 \text{ N (fan)}$$
$$= 83{,}570 \text{ N } (18{,}790 \text{ lb}_\text{f}).$$ <ANS

This thrust is 18.1 percent more than for the turbojet, but the total air flow rate is 2.2 times that of the turbojet. Thus, the thrust to weight ratio will be higher for the turbojet. The nondimensional thrust is determined by using the total mass flow rate, $(1+\alpha)\dot{m}$, not the core flow rate as for the turbojet, $\dot{m}$:

$$\frac{F}{a_a\dot{m}(1+\alpha)} = \frac{83570 \text{ N} \times \dfrac{\text{kg-m}}{\text{s}^2\text{ N}}}{334.1\dfrac{\text{m}}{\text{s}} \times 74.83\dfrac{\text{kg}}{\text{s}} \times (1+1.20)} = 1.519.$$

This is considerably less than that for the turbojet (2.678), again showing that the thrust-to-weight ratio for the turbofan is less than that for the turbojet.

The fuel flow rate is exactly as for the turbojet, that is

$$\dot{m}_\text{f} = 2.035 \text{ kg/s } (4.487 \text{ lb}_\text{m}/\text{s}).$$

Thus,

$$TSFC = \frac{\dot{m}_\text{f}}{F} = \frac{2.035\dfrac{\text{kg}}{\text{s}} \times 3600\dfrac{\text{s}}{\text{h}}}{83{,}570 \text{ N}} = 0.08769 \text{ kg/h/N } (0.8597 \text{ lb}_\text{m}/\text{h}/\text{lb}_\text{f}),$$ <ANS

which is 24.6 percent less than for the turbojet. This is much more fuel efficient!

As an afterthought, to maximize the thrust one could find the optimum value of τ_f and then π_f from Eq. (3.2.13). These values are 1.81 and 9.87, respectively, but their determination is left as an exercise for the reader. This optimization yields a thrust of 89,490 N (20,120 lb_f) and a *TSFC* of 0.08186 kg/h/N (0.8025 $\text{lb}_\text{m}/\text{h}/\text{lb}_\text{f}$). Also, for this optimized case, $u_8 = u_9 = 794.0$ m/s (2605 ft/s).

Repeating the calculations for the nonideal case, using efficiencies and losses similar to those for the turbojet in Example 3.4, yields $F = 15{,}970$ lb_f (15.0 percent lower than ideal) and $TSFC =$ 1.136 $\text{lb}_\text{m}/\text{h}/\text{lb}_\text{f}$ (24.3 percent higher).

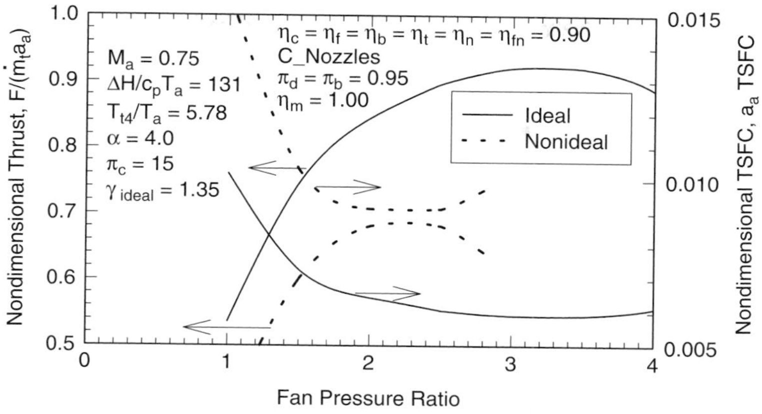

Figure 3.23 Effect of fan total pressure ratio on performance of an exhausted turbofan.

It is also possible to perform a series of parametric studies to compare engine types under different operating conditions.

For example, Figure 3.23 presents an ideal and nonideal study in which the thrust and *TSFC* are treated as functions of the fan total pressure ratio holding the compressor pressure ratio constant. For the ideal case the specific heat ratio is 1.35. For the nonideal case all thermodynamic efficiencies are 0.90; all of the pressure loss coefficients are 0.95 and the shaft efficiency is 1.00. The parameter used to nondimensionalize the thrust is the total flow rate, $\dot{m}_t = (1 + \alpha)\dot{m}$, as will be discussed below. As can be seen, a maximum thrust is realized as well as a minimum *TSFC*. The optimum pressure ratios are considerably different for the ideal and nonideal cases. The ideal optimum occurs at the total pressure ratio corresponding to the total temperature ratio given by Eq. (3.2.13).

Next, using this information, the dimensionless thrust is plotted versus the flight Mach number in Figure 3.24 using the core or primary mass flow rate as implied by Eq. (3.2.11). The fan pressure ratio has been optimized and both ideal and nonideal cases are presented. For the ideal case the specific heat ratio is 1.35. For the nonideal case all thermodynamic efficiencies are 0.90; all of the pressure loss coefficients are 0.95 and the shaft efficiency is 1.00. As can be seen on the right-hand axis, the thrust decreases with the Mach number but is higher than that of a turbojet. This is somewhat misleading, however. That is, the total airflow for a turbofan is considerably higher than the primary flow, which is used to nondimensionalize the thrust. Bypass ratios up to 10 are used in modern designs, which means that the total flow can be as much as 10 times the core flow. Thus, a better parameter to nondimensionalize the thrust is the total flow rate, $\dot{m}_t = (1 + \alpha)\dot{m}$. If this is used and the analysis is again performed, one obtains the left-hand axis in Figure 3.24. Now it can be seen that the dimensionless thrust is lower than that for the turbojet case. Thus, for the same total flow, which is represented by the physical size of the engine, the turbojet produces more thrust. Next, one should consider the *TSFC*, which is also characterized in Figure 3.24. In this respect the turbofan outperforms the turbojet. The turbofan produces more thrust per fuel flow than does the turbojet. This explains why most commercial craft, use turbofans with large bypass ratios. Fighter craft, on the other hand, use turbojets or turbofans with low bypass ratios to reduce engine size and weight.

Another trend study that can be performed is the effect of compressor pressure ratio on the overall performance. In Figure 3.25, such a study is shown. This is for a fixed Mach number, bypass ratio, fan pressure ratio, flow rate, and turbine inlet temperature. As for the simple turbojet, an optimum pressure ratio is found if

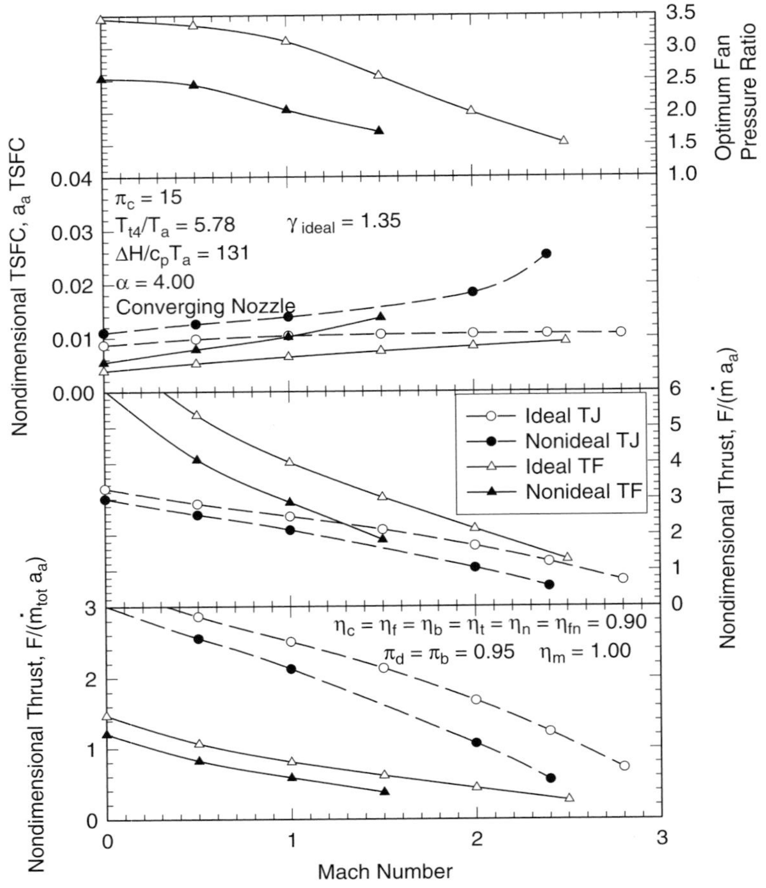

Figure 3.24 Dimensionless thrust (based on core flow), dimensionless thrust (based on total flow), and dimensionless *TSFC* vs. Mach number for an exhausted fan and turbojet.

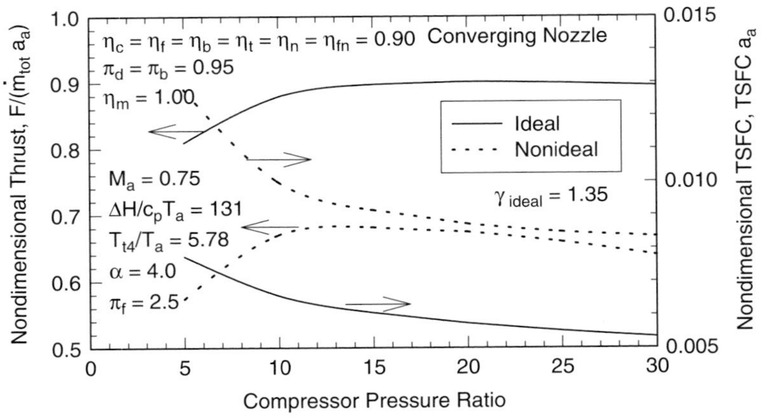

Figure 3.25 Effect of compressor total pressure ratio on performance of an exhausted turbofan.

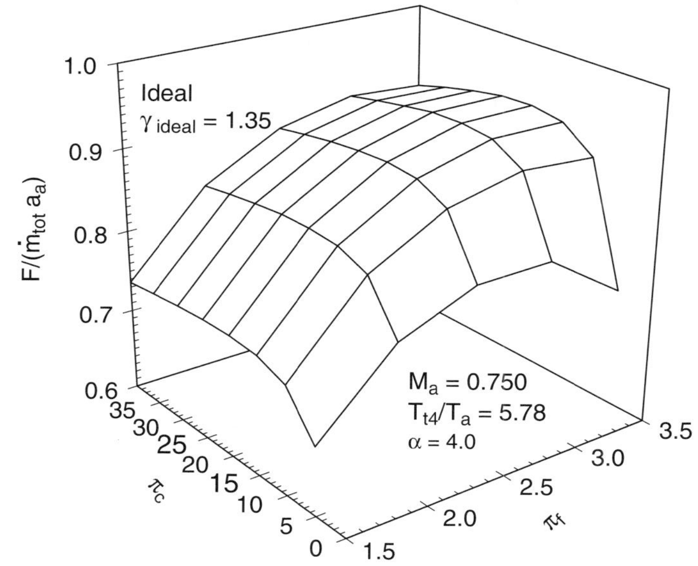

Figure 3.26 Effect of fan total pressure ratio and compressor total pressure ratio on dimensionless thrust of an ideal exhausted turbofan.

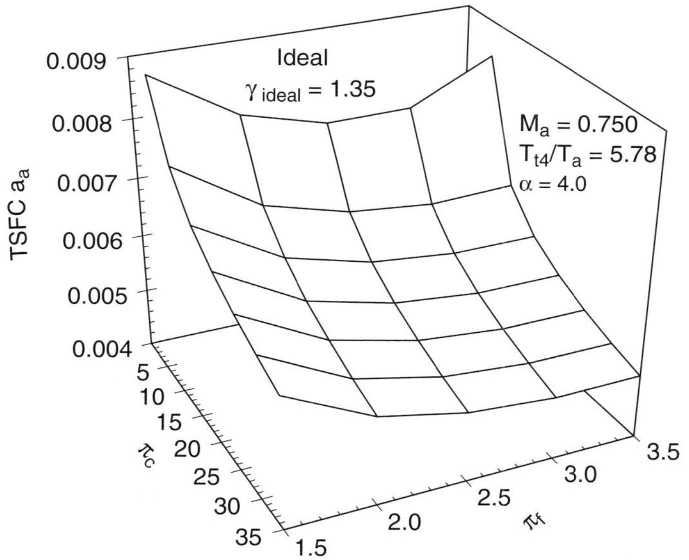

Figure 3.27 Effect of fan total pressure ratio and compressor total pressure ratio on dimensionless *TSFC* of an ideal exhausted turbofan.

the objective is to maximize the thrust. However, one can also see that the value of *TSFC* continuously decreases with increasing the compressor pressure ratio.

Figure 3.26 demonstrates the relationship of nondimensional thrust on both the compressor and fan pressure ratios for an ideal turbofan. As can be seen from the 3-D plot, a maximum occurs for one particular combination of π_c and π_f. The counterpart performance plot (nondimensional *TSFC*) is presented in Figure 3.27. The *TSFC* can be minimized with the right choice of π_f;

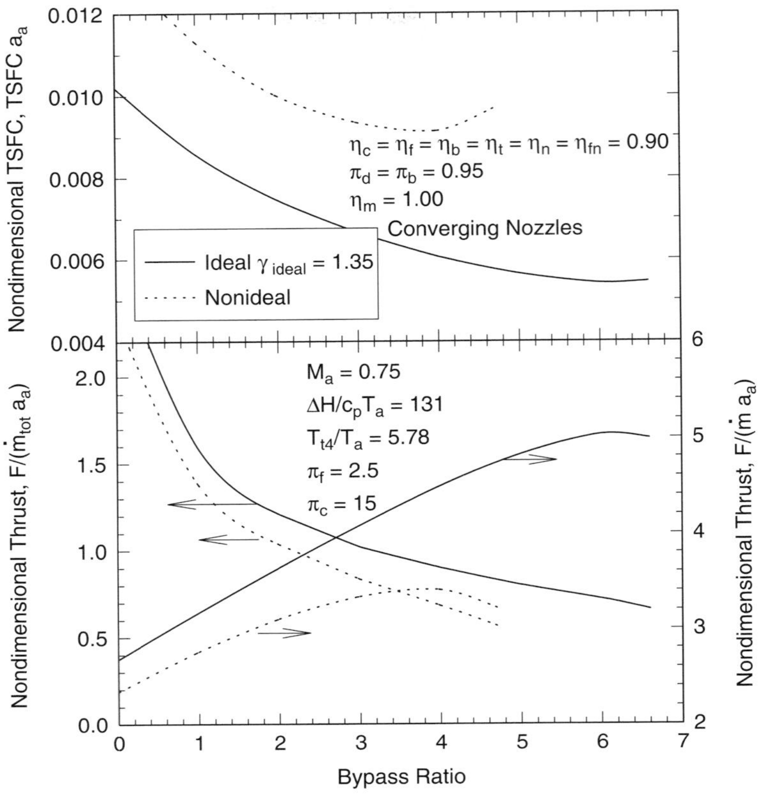

Figure 3.28 Effect of bypass ratio on dimensionless thrust and dimensionless *TSFC* of an exhausted turbofan.

however, it decreases monotonically in the π_c direction. Similar trends would be found for the nonideal case.

To expand on fan performance, one can study the effect of bypass ratio on the overall performance. Figure 3.28 illustrates the general dimensionless thrust and *TSFC* trends as functions of the bypass ratio. These are for fixed fan and compressor pressure ratios, Mach number, core flow rates, and turbine exit temperatures. When the thrust is nondimensionalized by the core flow rate, it can be seen from Figure 3.28 that an optimum thrust is attained for one particular bypass ratio. Note that a bypass ratio of zero represents a simple turbojet. When nondimensionalized by the total flow rate, the thrust always decreases with the bypass ratio. For this type of turbofan with the parameters fixed as stated, a maximum bypass ratio exists. If an attempt is made to operate above this, too much energy is required from the turbine to drive both the fan and the compressor and the engine will drop to a lower flow rate, different pressure ratios, or both. The *TSFC* trend is also shown in Figure 3.28. As one can see, a minimum *TSFC* is realized at the same bypass ratio that the maximum thrust (based on only the core mass flow rate) was obtained.

Before proceeding to afterburning cases, two complementing studies are presented. First, all of the ideal cases thus far have used a specific heat ratio of 1.35. To study the effect of the assumed specific heat ratio, Figure 3.29 is presented. For this figure a turbojet and an exhausted fan with fixed parameters as noted is shown for the nonideal cases. A constant specific heat ratio, which was varied from 1.30 to 1.40, is used for all components to see the effect on calculated performance. Both thrust and *TSFC* are studied. As can be seen, the *calculated* thrust

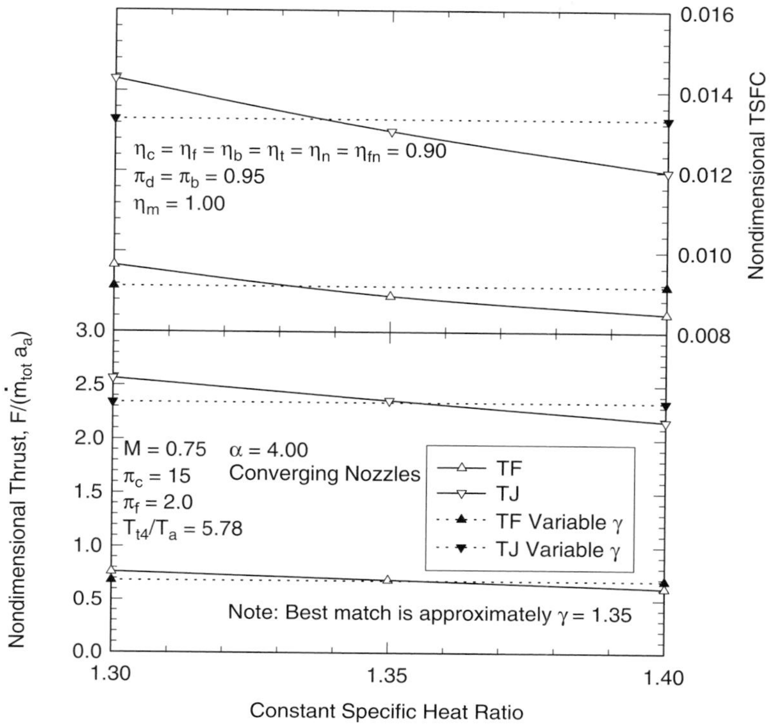

Figure 3.29 Effect of constant specific heat ratio on calculated performance for an exhausted turbofan.

decreases with increasing γ, as does the *TSFC*. Also depicted on this graph is the calculated thrust and *TSFC* with variable specific heats within the engine, namely as has been calculated in turbojet examples. These are straight lines and simply shown as a reference as this represents the best solution. As can be seen, the best constant specific heat solution is for $\gamma \cong 1.35$ to 1.36 when the thrust is considered, and $\gamma \cong 1.33$ to 1.34 when the *TSFC* is considered. So, the use of $\gamma \cong 1.35$ is reasonable for simple calculations.

Next, a study on the effects of efficiencies and pressure loss coefficients are presented for both a nonideal turbojet and exhausted turbofan in Figure 3.30. Engine parameters are fixed as noted but the thermodynamic efficiencies have been varied from 0.90 to 1.00 and the pressure loss coefficients have been varied from 0.95 to 1.00. Also, ideal results are presented for $\gamma = 1.35$ for both the turbojet and turbofan. As can be seen, the thrust is severely decreased for both the turbofan and turbojet as the efficiency and loss coefficients are reduced. Similarly, the *TSFC* is severely increased for both the turbofan and turbojet as the efficiency and loss coefficients are reduced.

Afterburning Next, the afterburning exhausted fan is considered. Figure 3.4 depicts this geometry. As can be seen, the afterburner is strictly in the core region and near the primary exhaust. In general, a second afterburner could be located near the fan exhaust. This geometry is seldom used, however, and will not be covered here.

A typical study is presented in Figure 3.31, where the thrust and *TSFC* for an ideal exhausted fan with an afterburner are depicted as functions of the Mach number for the optimum fan pressure ratio (maximum thrust). The fan pressure ratio is given in Figure 3.24. As should be expected, the thrust and *TSFC* are higher

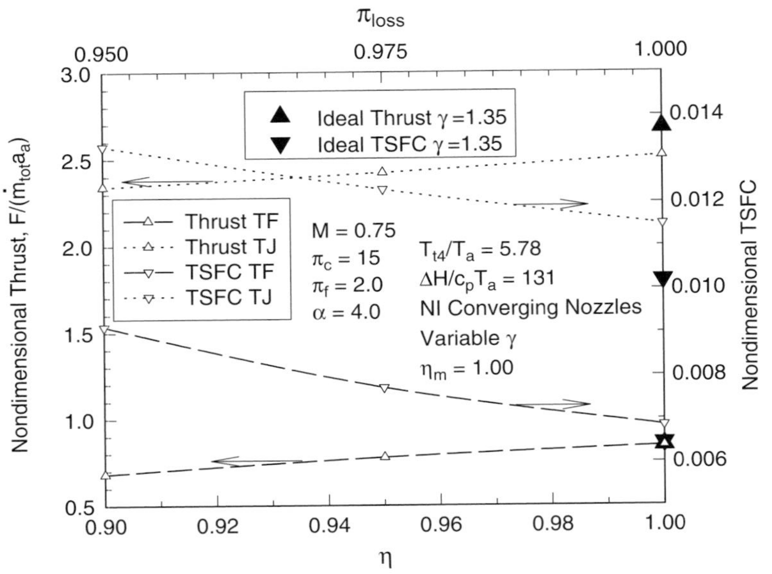

Figure 3.30 Effect of component efficiencies and pressure losses on performance for an exhausted turbofan.

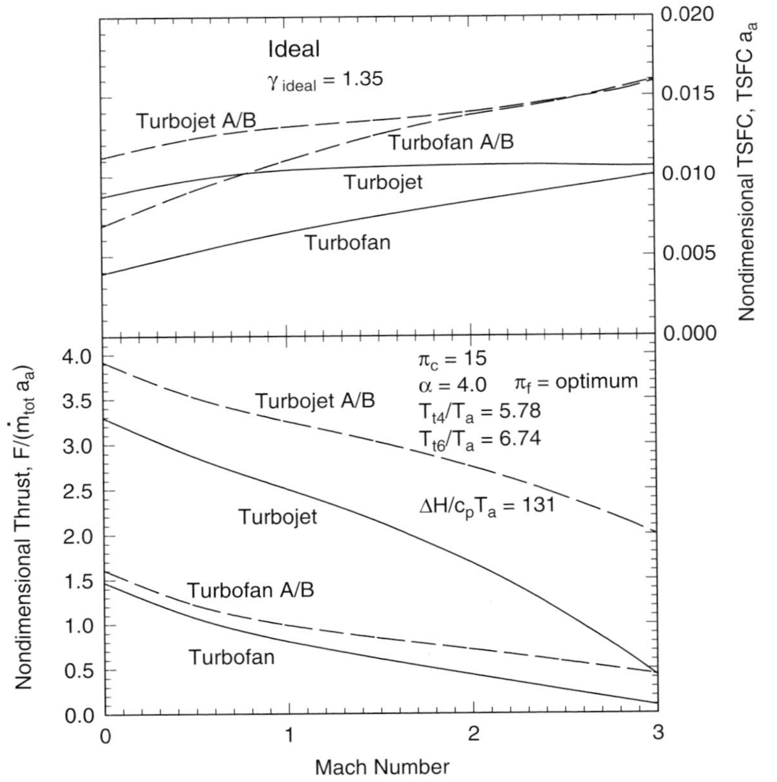

Figure 3.31 Dimensionless thrust (based on total flow) and dimensionless *TSFC* vs. Mach number for an ideal exhausted fan with an optimum total pressure ratio and turbojet with and without afterburners.

than for a nonafterburning case. Also, these trends are compared with those for a turbojet. As can be seen, the *TSFC* for the afterburning turbofan is lower than that for the afterburning turbojet at low Mach numbers. The dimensionless thrust (by total flow rate) is higher for a turbojet without afterburning than for a turbofan with an afterburner.

Turbofan: Fan Mixed

The general diagram for a turbofan is shown in Figure 2.3. The fully mixed type of fan is shown in Figure 3.16. For this configuration, the fan exit bypassed air passes through a bypass duct and is mixed with the turbine exhaust before station 5.5. Two separate subcases are considered here: without and with afterburner. In Figure 3.16, the afterburner exists between stations 5.5 and 6. Analyses for the two engines are performed separately in the following sections.

Nonafterburning For the nonafterburning case, stations 5.5 and 6 are one and the same and are termed station 6, as shown in Figure 3.17.

Example 3.8 Ideal Nonafterburning Fully Mixed Turbofan

An ideal turbofan with a mixed fan flies at sea level at a Mach number of 0.75. The primary flow is 165 lb_m/s (74.83 kg/s) and the bypass ratio is 1.20. The compressor pressure ratio is 15. The fuel has a heating value of 18,000 BTU/lb_m (41,870 kJ/kg) and the burner exit total temperature is 3000 °R (1667 K). The Mach numbers at the turbine and duct exits are the same. Find the developed thrust and the *TSFC* for $\gamma = 1.35$.

Solution

If this example is compared with the example for the turbojet (Example 3.3), it can readily be seen that the only difference is the addition of the mixed fan. All of the given conditions are identical. Once again, a step-by-step solution will be presented, for it allows internal differences between this example and the simple turbojet and the exhausted fan examples to be determined.

Many of the parameter values are the same as the previous turbojet example and are as follows.

$p_a = 14.69$ psia (101.3 kPa) $\quad T_a = 518.7$ °R (288.2 K)

$u_a = 822.2$ fps (250.6 mps) $\quad T_{ta} = 569.7$ °R (316.5 K) $= T_{t2}$

$a_a = 1096$ fps (334.1 mps)

$p_{ta} = 21.11$ psia (145.5 kPa) $= p_{t2}$ $\quad p_{t3} = 316.6$ psia (2183 kPa) $= p_{t4}$

$T_{t3} = 1150$ °R (638.7 K) $\quad \tau_c = 2.018$

Fan

First, the fan total pressure ratio, π_f, must be found. Obtaining this requires that the total temperature ratio, τ_f, be known. Ideally, Eq. (3.2.19) yields

$$\tau_f = \frac{\tau_c + \left[\frac{T_{ta}}{T_a}\right]\left[\frac{T_a}{T_{t4}}\right]\tau_c(1+\alpha-\tau_c)}{1+\left[\frac{T_{ta}}{T_a}\right]\left[\frac{T_a}{T_{t4}}\right]\tau_c\alpha} = \frac{2.018 + \left[\frac{569.7}{518.7}\right]\left[\frac{518.7}{3000}\right] \times 2.018 \times (1+1.20-2.018)}{1+\left[\frac{569.7}{518.7}\right]\left[\frac{518.7}{3000}\right] \times 2.018 \times 1.20}$$

$$= 1.430.$$

Thus, for the ideal (isentropic) fan, $\pi_f = [\tau_f]^{\frac{\gamma}{\gamma-1}} = [1.365]^{\frac{1.35}{0.35}} = 3.974$, which is somewhat higher than the exhausted fan case. For the nonideal case, this pressure ratio would be found iteratively as shown in a following example.

The total temperature exiting the fan is

$$T_{t7} = \tau_f \times T_{t2} = 1.430 \times 569.7 = 814.7\ °\text{R}\ (452.6\ \text{K}) = T_{t7.5}.$$

Also, the total pressure exiting the fan is

$$p_{t7} = \pi_f \times p_{t2} = 3.974 \times 21.11 = 83.89\ \text{psia}(578.4\ \text{kPa}) = p_{t7.5}.$$

Duct

Because the duct is ideal (adiabatic) the total temperature is constant, that is, $T_{t7.5} = T_{t7} = 814.7$ °R (452.6 K). Since the duct is ideal (isentropic), the total pressure is also constant,

$$p_{t7.5} = p_{t7} = 83.89\ \text{psia}\ (578.4\ \text{kPa}).$$

Turbine

The power balance on the shaft is determined as follows:

$$\dot{m}c_p(T_{t4} - T_{t5}) = \dot{m}c_p(T_{t3} - T_{t2}) + \alpha\dot{m}c_p(T_{t7} - T_{t2})$$

or $(T_{t4} - T_{t5}) = (T_{t3} - T_{t2}) + \alpha(T_{t7} - T_{t2})$.

Thus, $(3000 - T_{t5}) = (1249 - 569.7) + 1.20 \times (814.7 - 569.7)$,
and solving for the turbine exit total temperature yields $T_{t5} = 2126$ °R (1181 K).

For the isentropic turbine

$$p_{t5} = p_{t4}[\tau_t]^{\frac{\gamma}{\gamma-1}} = p_{t4}\left[\frac{T_{t5}}{T_{t4}}\right]^{\frac{\gamma}{\gamma-1}} 316.6\left[\frac{2126}{3000}\right]^{\frac{1.35}{0.35}} = 83.89\ \text{psia}\ (578.4\ \text{kPa}) = p_{t5.5} = p_{t8},$$

which one should note matches the value of $p_{t7.5}$ found in the preceding discussion of the duct (as it should).

Mixer

For the mixer the exit total temperature is found from a balance on the energy equation on the assumption that the exit temperature is uniform:

$$T_{t5.5} = \frac{\alpha T_{t7.5} + T_{t5}}{\alpha + 1} = \frac{1.20 \times 814.7 + 2126}{1.20 + 1} = 1410\ °\text{R}\ (783.5\ \text{K}).$$

The total pressure remains constant in an ideal mixer and thus

$p_{t5.5} = p_{t5} = 83.89$ psia (578.4 kPa).

Nozzle

For the ideal nozzle the exit total temperature is the same as the inlet value:

$T_{t8} = T_{t5.5} = 1410$ °R (783.3 K) and $p_{t8} = p_{t5.5} = 83.89$ psia (578.4 kPa).

For an ideal nozzle, $p_{t8} = p_8\left[1 + \frac{\gamma - 1}{2} M_8^2\right]^{\frac{\gamma}{\gamma-1}}$. Solving for the exit Mach number yields

$$M_8 = \sqrt{\left[\frac{2}{\gamma - 1}\right]\left[\left[\frac{p_{t8}}{p_8}\right]^{\frac{\gamma-1}{\gamma}} - 1\right]} = \sqrt{\left[\frac{2}{1.35 - 1}\right]\left[\left[\frac{83.89}{14.69}\right]^{\frac{0.35}{1.35}} - 1\right]} = 1.806.$$

Therefore, the exit temperature is

$$T_8 = \frac{T_{t8}}{1 + \left[\frac{\gamma - 1}{2}\right] M_8^2} = \frac{1410}{1 + \left[\frac{1.35 - 1}{2}\right] 1.806^2} = 898.1\ °\text{R}\ (498.9\ \text{K})$$

and the exit gas velocity is

$$u_8 = M_8 a_8 = M_8 \sqrt{\gamma \mathcal{R} T_8} = 1.806 \sqrt{1.35 \times 53.35 \frac{\text{ft-lb}_\text{f}}{\text{lb}_\text{m}\text{-°R}} \times 32.17 \frac{\text{lb}_\text{m}}{\text{sl}} \times 898.1\ °\text{R} \times \frac{\text{sl-ft}}{\text{s}^2\text{-lb}_\text{f}}}$$

$$= 2605\ \text{fps}\ (794.1\text{mps}).$$

Thrust and TSFC

Finally, since the nozzle exit pressure matches the ambient pressure, the thrust is given by:

$$F = \dot{m}_t(u_8 - u_a) = (1 + \alpha)\dot{m}(u_8 - u_a)$$

$$= 2.20 \times 165 \frac{\text{lb}_\text{m}}{\text{s}} \times (22{,}605 - 822.2)\frac{\text{ft}}{\text{s}} \times \left[\frac{\text{sl}}{32.17\ \text{lb}_\text{m}}\right]\left[\frac{\text{lb}_\text{f}}{\text{sl-ft}}\right]$$

$= 20{,}120\ \text{lb}_\text{f}$ (89,490 N). <ANS

This thrust is 7.1 percent higher than the fan exhausted case (Example 3.7).

The fuel flow rate is the same as for the turbojet once again:

$\dot{m}_f = 4.485\ \mathrm{lb_m/s}\ (2.034\ \mathrm{kg/s})$.

Thus

$$TSFC = \frac{\dot{m}_f}{F} = \frac{4.485\frac{\mathrm{lb_m}}{\mathrm{s}} \times 3600\frac{\mathrm{s}}{\mathrm{h}}}{20{,}120\ \mathrm{lb_f}} = 0.8025\ \mathrm{lb_m/h/lb_f}\ (0.08185\ \mathrm{kg/h/N}), \qquad \text{<ANS}$$

which is 7.7 percent lower than the exhausted fan case

Note that the thrust and *TSFC* match the optimized fan exhausted case (values noted at end of Example 3.7). This is not a coincidence and can be shown always to be the case for an ideal fan. Proving this is left as an exercise for the reader.

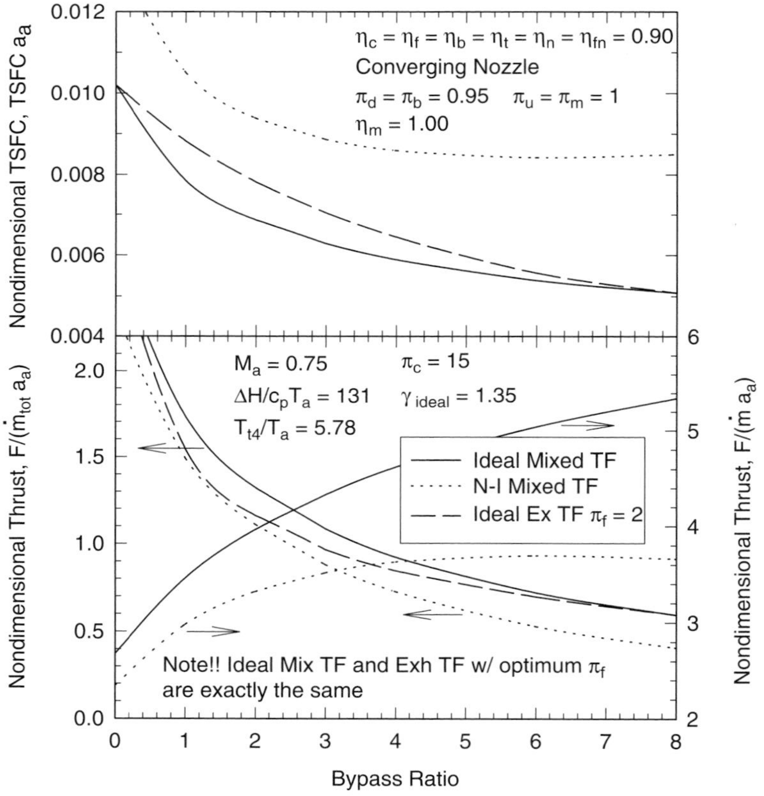

Figure 3.32 Effect of bypass ratio on performance of a mixed turbofan.

A representative trend study is presented in Figure 3.32. Here the trends of thrust and *TSFC* are shown for a given compressor pressure ratio, Mach number, turbine inlet temperature, and core flow rate as functions of bypass ratio. Unlike the exhausted fan case, both the dimensionless thrust (by $\dot{m}_t$) and *TSFC*

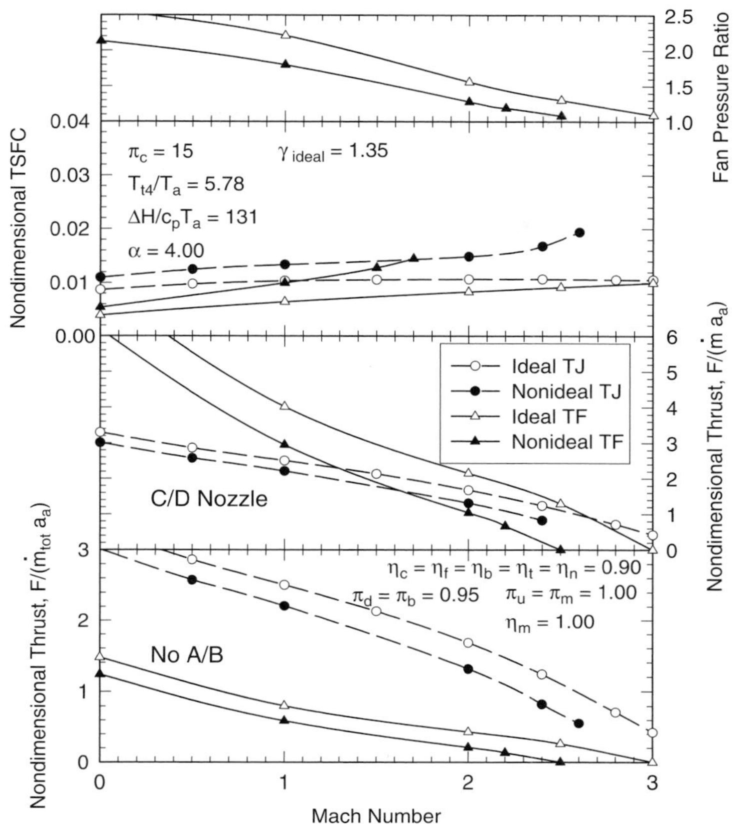

Figure 3.33 Dimensionless thrust (based on total flow) and dimensionless *TSFC* vs. Mach number for a mixed fan and turbojet without afterburners.

continuously decrease with increasing bypass ratio. Because the fan pressure ratio is not specified, a limiting or maximum bypass ratio value does not exist as it did for the exhausted fan case. As just noted, as the bypass ratio increases, the value of *TSFC* monotonically decreases, which is good from a fuel economy standpoint. However, as the bypass ratio increases, the value of $F/\dot{m}_t\, a_a$ decreases. Keeping in mind that the physical size of the engine increases as $\dot{m}_t$ increases, one should realize that, as the bypass ratio increases, the thrust-to-size or, more importantly, the thrust-to-weight ratio decreases. Thus, for a given needed thrust, if a large bypass ratio is used, a large and heavy engine will be required. However, if a low bypass ratio is used, a smaller, lighter engine can be used. As a result, for military (fighter) applications, where weight is a primary concern, low bypass ratios are used (with a mixed fan) (often less than unity). On the other hand, for commercial applications in which fuel economy is the primary concern, high bypass ratios up to 10 are used (and with the exhausted fan design).

The ideal version of the mixed fan operates with characteristics similar to those for the exhausted version of a turbofan. That is, the dimensionless thrust (by $\dot{m}_t$) is lower than that of a turbojet; however, the *TSFC* is also lower. Typical trends are shown in Figure 3.33. The characteristics of mixed and exhausted turbofans can be compared from Figures 3.33 and 3.31. One important conclusion is that ideal (and only ideal) nonafterburning models operate identically when (and only when) the exhausted fan operates with the optimum pressure ratio.

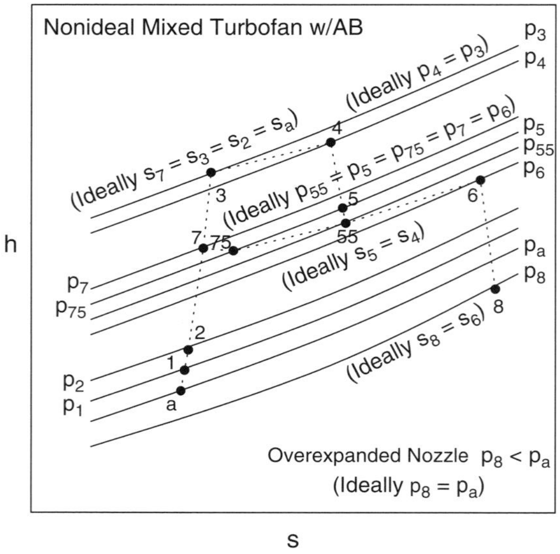

Figure 3.34 The h–s diagram for an afterburning mixed turbofan.

Afterburning The afterburning mixed fan is now considered. Figure 3.6 depicts this geometry. Many of the ideal equations from the nonafterburning case apply. The h–s diagram is shown in Figure 3.34. This diagram is considerably more complex than any other such diagrams to this point because of the increased number of processes. Diffusion process a to 2 is ideally isentropic but for the nonideal case the entropy increases. Compressor and fan processes 2 to 3 and 2 to 7 also are ideally isentropic but for the nonideal case the entropy increases. Combustion and afterburner processes 3 to 4 and 5.5 to 6 are ideally isobaric but nonideally the pressure drops. The duct flow and mixing processes 7 to 7.5 and 5+7.5 to 5.5 are ideally isobaric but nonideally the pressure drops. Turbine process 4 to 5 is ideally isentropic but for the nonideal case the entropy increases. Lastly the nozzle process 6 to 8 is also ideally isentropic but for the nonideal case the entropy increases. It is important to note that almost all afterburning fans use this mixed fan geometry.

As before, a series of parametric studies can be performed for this type of engine. The ideal

Example 3.9 | Ideal Fully Mixed Turbofan with an Afterburner

The ideal turbofan with a mixed fan in the previous example now has an afterburner. All other conditions are the same and the afterburner exit total temperature is 3500 °R (1944 K). The same fuel is used for both the primary and afterburners. Find the developed thrust and the *TSFC*.

Solution

If this example is compared with the previous example for the turbofan, the only difference that will be observed is the addition of the afterburner. All of the given conditions are identical. Once again a step-by-step solution will be presented as it allows one to determine internal differences between this example and the nonafterburning case.

Many of the values are the same as the previous nonafterburning example and are simply listed as follows.

$p_a = 14.69$ psia (101.3 kPa)

$T_a = 518.7$ °R (288.2 K)

$u_a = 822.2$ fps (250.6 mps)

$T_{ta} = 569.7$ °R (316.5 K) $= T_{t2}$

$p_{ta} = 21.11$ psia (145.5 kPa) $= p_{t2}$

$p_{t3} = 316.6$ psia (2183 kPa) $= p_{t4}$

$T_{t3} = 1150$ °R (638.7 K)

$\tau_c = 2.018$

$\tau_f = 1.430$

$\pi_f = 3.974$

$T_{t7.5} = 814.7$ °R (452.6 K)

$p_{t5} = p_{t5.5} = p_{t7.5} = 83.89$ psia (578.4 kPa)

$M_8 = 1.806$

$T_{t5} = 2126$ °R (1181 K)

$T_{t5.5} = 1411$ °R (783.9 K)

Afterburner

The total temperature at station 8 is the same as the afterburner exit (T_{t6}), which is given as follows:

$T_{t6} = T_{t8} = 3500$ °R (1944 K).

The total pressure at the exit of the afterburner is the same as the total pressure at the turbine exit:

$p_{t6} = p_{t5} = 83.89$ psia (578.4 kPa).

The primary fuel flow rate is the same as for the nonafterburning case and is

$\dot{m}_f = 4.485$ lb$_m$/s (2.034 kg/s),

but the afterburning fuel flow rate is needed and is given by

$$\dot{m}_{f_{ab}} = \frac{\dot{m}_t\, c_p \left[T_{t6} - T_{t5.5}\right]}{\Delta H} = \frac{(1+\alpha)\dot{m} c_p \left[T_{t6} - T_{t5.5}\right]}{\Delta H}$$

$$= \frac{2.20 \times 165 \dfrac{\text{lb}_\text{m}}{\text{s}} \times 0.2644 \dfrac{\text{BTU}}{\text{lb}_\text{m}\text{-}^\circ\text{R}} \times [3500 - 11411]\,^\circ\text{R}}{18{,}000 \dfrac{\text{BTU}}{\text{lb}_\text{m}}} = 11.14\ \text{lb}_\text{m}/\text{s}\ (5.052\ \text{kg/s}),$$

which is 2.48 times the primary burner rate. The afterburner is consuming large amounts of fuel and it is obvious that not much time will be required to drain the fuel tank. Thus, the total fuel flow rate is

$$\dot{m}_{f_t} = \dot{m}_f + \dot{m}_{f_{ab}} = 4.485 + 11.14 = 15.63 \text{ lb}_m/\text{s } (7.088 \text{ kg/s}).$$

Nozzle

Because the gas is expanding between the same pressures, the Mach number will be the same as the previous example. The exit temperature will be higher, as follows:

$$T_8 = \frac{T_{t8}}{1 + \left[\frac{\gamma - 1}{2}\right] M_8^2} = \frac{3500}{1 + \left[\frac{1.35 - 1}{2}\right] 1.806^2} = 2228\ ^\circ\text{R } (1238 \text{ K}).$$

The gas exit velocity is

$$u_8 = M_8 a_8 = M_8 \sqrt{\gamma \mathcal{R} T_8} = 1.806 \sqrt{1.35 \times 53.35 \frac{\text{ft-lb}_f}{\text{lb}_m\text{-}^\circ\text{R}} \times 32.17 \frac{\text{lb}_m}{\text{sl}} \times 2228\ ^\circ\text{R} \times \frac{\text{sl-ft}}{\text{s}^2\text{-lb}_f}}$$
$$= 4103 \text{ fps } (1251 \text{ mps}).$$

Thrust and TSFC

Finally, the nozzle exit pressure and atmospheric pressure are ideally the same and thus the thrust is given by

$$F = \dot{m}_t(u_8 - u_a) = (1 + \alpha)\dot{m}(u_8 - u_a)$$
$$= 2.20 \times 165 \frac{\text{lb}_m}{\text{s}} \times (4103 - 822.2) \frac{\text{ft}}{\text{s}} \times \left[\frac{\text{sl}}{32.17 \text{ lb}_m}\right]\left[\frac{\text{lb}_f}{\text{sl-ft}}\right] = 37{,}030 \text{ lb}_f \ (164{,}700 \text{ N}).$$

<ANS

This thrust is 84 percent higher than the nonafterburning case.

Thus, the *TSFC* is

$$TSFC = \frac{\dot{m}_{f_t}}{F} = \frac{15.63 \frac{\text{lb}_m}{\text{s}} \times 3600 \frac{\text{s}}{\text{h}}}{37{,}030 \text{ lb}_f} = 1.519 \text{ lb}_m/\text{h}/\text{lb}_f \ (0.1549 \text{ kg/h/N}),$$ <ANS

which is 89 percent higher than the nonafterburning case

By comparing this example with the afterburning turbojet example one can see that the thrust has been increased by 95 percent, but the *TSFC* has only been increased by 13 percent. This is a large increase in thrust and only a moderate decrease in fuel economy. This partially explains why low-bypass-ratio turbofans are used in military applications.

Example 3.10 | Nonideal Mixed Turbofan with an Afterburner

A turbofan flies at sea level at a Mach number of 0.75. It ingests 74.83 kg/s (165 lb_m/s) of air to the core. The compressor operates with a pressure ratio of 15 and an efficiency of 92 percent. The engine has a bypass ratio of 1.20. The efficiency of the fan is 91 percent. The fuel has a heating value of 41,870 kJ/kg (18,000 BTU/lb_m) and the burner total temperature is 1667 K (3000 °R). The burner has an efficiency of 91 percent and a total pressure ratio of 0.93, while the turbine has an efficiency of 89 percent. The duct has a total pressure ratio of 0.99 and the total pressure ratio of the mixer is 0.985. A variable CD nozzle (to match the exit pressure to the ambient pressure) is used for the primary nozzle and the efficiency is 94 percent. The total pressure recovery for the diffuser is 0.96 and the shaft efficiency is 99.5 percent. An afterburner is used and the exit temperature is 1944 K (3500 °R). The efficiency of the afterburner is 85 percent and the pressure ratio is 0.94. The Mach numbers at the turbine and duct exits are nearly the same. Find the developed thrust and *TSFC*.

Solution

Diffuser

First, the speed of sound is

$$a_a = \sqrt{\gamma_a \mathcal{R} T_a} = \sqrt{1.403 \times 287.1 \frac{\text{J}}{\text{kg-K}} \times 288.2\ \text{K} \times \frac{\text{N-m}}{\text{J}} \times \frac{\text{kg-m}}{\text{N-s}^2}}$$
$$= 340.7\text{mps (1118 fps)},$$

so the engine airspeed is $u_a = M_a\, a_a = 0.75 \times 340.7$ mps $= 255.5$ mps (838.2 fps).

Next, $\dfrac{T_{ta}}{T_a} = 1 + \dfrac{\gamma_d - 1}{2} M_a^2 = 1 + \dfrac{1.3997 - 1}{2} 0.75^2 = 1.1124$ for $\gamma_d = 1.3997$ and $M_a = 0.75$,

so that $T_{t2} = T_{ta} = 1.1124 = 288.2\ \text{K} = 320.6\ \text{K}$ (577.0 °R).

Also, $\dfrac{p_{ta}}{p_a} = \left[\dfrac{T_{ta}}{T_a}\right]^{\frac{\gamma_d}{\gamma_d - 1}} = [1.1124]^{\frac{1.3997}{1.3997-1}} = 1.4522.$

Next, the pressure recovery factor for the diffuser is 0.92, so that

$$\frac{p_{t2}}{p_a} = \frac{p_{t2}}{p_{ta}} \frac{p_{ta}}{p_a} = \pi_d \frac{p_{ta}}{p_a} = 0.96 \times 1.4522 = 1.3941,$$

and thus $p_{t2} = 1.3941 \times 101.3\ \text{kPa} = 141.3\ \text{kPa}$ (20.49 psia).

Compressor

The compressor total pressure ratio is given as 15 so
$p_{t3} = \pi_c\, p_{t2} = 15 \times 141.3\ \text{kPa} = 2118\ \text{kPa}$ (307.2 psia).
Next, one knows that for a nonideal compressor

$$\frac{p_{t3}}{p_{t2}} = \left\{1 + \eta_c\left[\frac{T_{t3}}{T_{t2}} - 1\right]\right\}^{\frac{\gamma}{\gamma-1}} = \pi_c = \{1 + \eta_c[\tau_c - 1]\}^{\frac{\gamma}{\gamma-1}}.$$

So solving for the total temperature ratio (the efficiency of the compressor is given as 92 percent) yields

$$\frac{T_{t3}}{T_{t2}} = \tau_c = \frac{\left[\frac{p_{t3}}{p_{t2}}\right]^{\frac{\gamma_c-1}{\gamma_c}} - 1}{\eta_c} + 1 = \frac{[15]^{\frac{1.3813-1}{1.3813}} - 1}{0.92} + 1 = 2.208 \text{ for } \gamma_c = 1.3805 \text{ (iterative)}.$$

Therefore, the exit temperature of the compressor is $T_{t3} = \tau_c\, T_{t2} = 2.208 \times 320.5\ \text{K} = 707.9\ \text{K}$ (1274 °R).

Fan

For a mixed nonideal turbofan geometry the fan total pressure ratio is unknown because the ideal equation (Eq. (3.2.19)) cannot be used. Thus, one needs to find π_f by iteration. In effect, the duct exit pressure (same as the turbine exit pressure) directly depends on the fan total pressure ratio. But this pressure ratio also affects the turbine power, which indirectly affects the turbine exit pressure. Thus, one can find the fan pressure ratio by a progressive iteration. Details of the "Regula Falsi" iteration are presented in Appendix I (Example I.2). Fortunately, the iteration process is quick to converge (seven iterations for this case) and easily programmed with a commercial math solver. In actuality, four or five iterations could have been used with minimal effects on the net results. Also, one should note that although precise values of the pressure ratio are desired for an engine design, it should not become the dominant feature of examples or chapter problems.

For the purposes of this example to demonstrate the thermodynamic calculations, one can first try $\pi_f = 3.049$, which will be checked later at the mixer:

$$p_{t7} = \pi_f\, p_{t2} = 3.049 \times 141.3\ \text{kPa} = 430.8\ \text{kPa (62.48 psia)}.$$

Next, one knows that

$$\frac{p_{t7}}{p_{t2}} = \left\{1 + \eta_f\left[\frac{T_{t7}}{T_{t2}} - 1\right]\right\}^{\frac{\gamma}{\gamma-1}} = \pi_f = \{1 + \eta_f[\tau_f - 1]\}^{\frac{\gamma}{\gamma-1}},$$

so solving for the total temperature ratio (the efficiency of the fan is given as 90 percent) yields:

$$\frac{T_{t7}}{T_{t2}} = \tau_f = \frac{\left[\frac{p_{t7}}{p_{t2}}\right]^{\frac{\gamma_f-1}{\gamma_f}} - 1}{\eta_f} + 1 = \frac{[3.049]^{\frac{1.393-1}{1.393}} - 1}{0.91} + 1 = 1.406 \text{ for } \gamma_f = 1.3930 \text{ (iterative)}.$$

Therefore, the exit temperature of the fan is

$T_{t7} = \tau_f T_{t2} = 1.406 \times 320.5 \text{ K} = 450.9 \text{ K } (811.6\ °\text{R}).$

Bypass Duct

Next, the pressure ratio for the bypass duct is $\pi_u = \dfrac{p_{t7.5}}{p_{t7}} = 0.99$, so that $p_{t7.5} = 0.99 \times 430.8 \text{ kPa} = 426.5 \text{ kPa } (61.85 \text{ psia})$.

Also, since the duct is adiabatic the total temperature is constant through the passage so $T_{t7.5} = T_{t7} = 450.9 \text{ K } (811.6\ °\text{R})$.

Primary Combustor

Next, the total pressure ratio for the burner is 0.93, so the burner exit total pressure is

$$p_{t4} = \pi_b\, p_{t3} = 0.93 \times 2119 \text{ kPa} = 1971 \text{ kPa } (285.9 \text{ psia})$$

and the energy balance yields $\Delta H \eta_b \dot{m}_f = \dot{m} c_{p_b}(T_{t4} - T_{t3}) + \dot{m}_f c_{p_b} T_{t4}$, where the burner efficiency is 91 percent and the heating value is 41,870 kJ/kg:

$$\dot{m}_f = 74.83 \frac{\text{kg}}{\text{s}} \times 1.170 \frac{\text{kJ}}{\text{kg-K}} \times \frac{(1667 - 707.9)\text{K}}{0.91 \times 41780 \dfrac{\text{kJ}}{\text{kg}} - 1.170 \dfrac{\text{kJ}}{\text{kg-K}} \times 1667 \text{ K}},$$

so that

$\dot{m}_f = 2.322 \text{ kg/s } (5.120 \text{ lb}_m\text{/s}).$

The specific heat is evaluated at the average temperature (not an iteration). One should note that $f = \dot{m}_f/\dot{m} = 0.03104$, which is small.

Turbine

Now the power balance for the shaft of the turbofan is

$$\dot{m} c_{p_c}(T_{t3} - T_{t2}) + \alpha \dot{m} c_{p_f}(T_{t7} - T_{t2}) = \eta_m\, \dot{m}_t c_{p_t}(T_{t4} - T_{t5}),$$

where the mechanical efficiency is 99.5 percent, or

$$74.83 \frac{\text{kg}}{\text{s}} \times 1.040 \frac{\text{kJ}}{\text{kg-K}} \times (707.9 \text{ K} - 320.5 \text{ K}) + 1.20 \times 74.83 \frac{\text{kg}}{\text{s}} \times 1.017 \frac{\text{kJ}}{\text{kg-K}}$$
$$\times (450.9 \text{ K} - 320.5 \text{ K}) = 0.995 \times \left(74.83 \frac{\text{kg}}{\text{s}} + 2.322 \frac{\text{kg}}{\text{s}}\right) \times 1.1232 \frac{\text{kJ}}{\text{kg-K}} \times (1667 \text{ K} - T_{t5});$$

thus, solving for the turbine exit temperature, $T_{t5} = 1219 \text{ K } (2194\ °\text{R})$. The value of the turbine specific heat is found iteratively.

Now, the turbine total temperature ratio is $\tau_t = \dfrac{T_{t5}}{T_{t4}} = \dfrac{1219 \text{ K}}{1667 \text{ K}} = 0.7313$.

Next, one knows for a nonideal turbine

$$\frac{p_{t5}}{p_{t4}} = \left\{1 - \frac{\left[1 - \frac{T_{t5}}{T_{t4}}\right]}{\eta_t}\right\}^{\frac{\gamma}{\gamma-1}} = \pi_t = \left\{1 - \frac{[1 - \tau_t]}{\eta_t}\right\}^{\frac{\gamma}{\gamma-1}},$$

but the turbine efficiency, η_t, is given as 89 percent, so that the turbine total pressure ratio is

$$\pi_t = \left\{1 - \frac{[1 - 0.7313]}{0.89}\right\}^{\frac{1.3065}{1.3065-1}} = 0.2164,$$

for $\gamma_t = 1.3065$ (iterative along with c_{p_t}).

Finally, the turbine exit total pressure is

$$p_{t5} = \pi_t\, p_{t4} = 0.2164 \times 1971\ \text{kPa} = 426.5\ \text{kPa}\ (61.86\ \text{psia}).$$

Mixer

One can now compare $p_{t7.5}$ and p_{t5} from above and see that they are the same. Also, since the Mach numbers at stations 5 and 7.5 are given to be approximately the same ($M_5 \cong M_{7.5}$) and since the total pressures are identical, the static pressures will match, so the initial guess on the fan pressure ratio, π_f, was excellent! If they had been different, a new value of π_f would have to be tried and all of the above work would have to be repeated as demonstrated in detail in Appendix I (Example I.2). It is important to note that if the Mach numbers were approximately known, we could have used the static pressures to match in the solution; that is, $p_5 = p_{7.5}$ instead of the total pressures. This is an option in the program JetEngineCycle.exe.

Next, one can find the temperature exiting the mixer $T_{t5.5}$ by

$$T_{t5.5} = \frac{(1+f)c_{p_{mc}} T_{t5} + \sigma\alpha c_{p_{mu}} T_{t7.5}}{(1+f)c_{p_{mc}} + \sigma\alpha c_{p_{mu}}}$$

$$= \frac{(1 + 0.03104 \times 1.1357 \frac{\text{kJ}}{\text{kg-K}} \times 1219\ \text{K} + 1.2 \times 1.062 \frac{\text{kJ}}{\text{kg-K}} \times 450.9\ \text{K}}{(1 + 0.03104 \times 1.1357 \frac{\text{kJ}}{\text{kg-K}} + 1.2 \times 1.062 \frac{\text{kJ}}{\text{kg-K}}},$$

where $c_{p_{mu}}$ is evaluated at the average temperature between stations 7.5 to 5.5 and $c_{p_{mc}}$ is from the average temperature between stations 5 to 5.5.

So, $T_{t5.5} = 818.9$ K (1474 °R). Also for this case, and with $M_5 = M_{7.5}$, $\pi_m = p_{t5.5}/p_{t5.2} = p_{t5.5}/p_{t5} = p_{t5.5}/p_{t7.5}$.

The total pressure ratio of the duct is 0.985, so that $p_{t5.5} = 0.985 \times 426.5$ kPa $= 420.1$ kPa (60.93 psia).

Afterburner

First, the total pressure ratio of the afterburner is given as 0.94, so $p_{t6} = \pi_{ab}\, p_{t5.5} = 0.94 \times 420.1$ kPa = 394.9 kPa (47.12 psi), and the energy analysis is

$$\Delta H \eta_{ab} \dot{m}_{f_{ab}} = (\dot{m} + \dot{m}_f) c_{p_{ab}} (T_{t6} - T_{t5.5}) + \dot{m}_{f_{ab}} c_{p_{ab}} T_{t6}.$$

The energy balance yields $\Delta H \eta_b \dot{m}_f = \dot{m} c_{p_b} (T_{t4} - T_{t3}) + \dot{m}_f c_{p_b} T_{t4}$, where the burner efficiency is 91 percent and the heating value is 41,870 kJ/kg, so that

$$\dot{m}_{f_{ab}} = (74.83 + 2.322 + 1.20 \times 74.83) \frac{\text{kg}}{\text{s}} \times 1.210 \frac{\text{kJ}}{\text{kg-K}}$$

$$\times \frac{(1944 - 818.9)\ \text{K}}{0.85 \times 41780 \dfrac{\text{kJ}}{\text{kg}} - 1.210 \dfrac{\text{kJ}}{\text{kg-K}} \times 1944\ \text{K}}$$

$$= 6.840\ \text{kg/s}\ (15.08\ \text{lb}_\text{m}/\text{s}).$$

The specific heat is evaluated at the average temperature (not an iteration). Now the total fuel flow rate is $\dot{m}_{f_t} = \dot{m}_f + \dot{m}_{f_{ab}}$ or $\dot{m}_{f_t} = 2.322 + 6.840 = 9.162 \dfrac{\text{kg}}{\text{s}} \left(21.19 \dfrac{\text{lb}_\text{m}}{\text{s}}\right)$.

Primary Nozzle

Since the afterburner is present, the inlet total temperature for the primary nozzle is

$T_{t6} = 1944$ K (3500 °R)

and the inlet total pressure is $p_{t6} = 394.9$ kPa (47.12 psia).

The nozzle is a variable CD nozzle, which is often used for afterburners, so the exit pressure matches the ambient pressure and one does not have to check for choking, namely, $p_8 = 101.3$ kPa (14.69 psia).

$$\text{Ideally, } T_8{}' = T_{t6} \left[\frac{p_8}{p_{t6}}\right]^{\frac{\gamma_n - 1}{\gamma_n}} = 1944 \left[\frac{101.3\ \text{kPa}}{394.9\ \text{kPa}}\right]^{\frac{1.2738-1}{1.2738}} = 1451\ \text{K}\ (2612\ °\text{R}),$$

where $\gamma_n = 1.3528$ (iterative).

The nozzle efficiency is 94 percent, thus the nozzle exit temperature is

$$T_8 = T_{t6} - \eta_n (T_{t6} - T_8{}') = 1944\ \text{K} - 0.94 \times (1944\ \text{K} - 1451\ \text{K}) = 1112\ \text{K}\ (2002\ °\text{R}).$$

Now $u_8 = \sqrt{2\, c_{p_n}\, (T_{t6} - T_8)}$,

so that

$$u_8 = \sqrt{2 \times 1.335 \frac{\text{kJ}}{\text{kg-K}} \times (1944\,\text{K} - 1481\,\text{K}) \times 1000 \frac{\text{J}}{\text{kJ}} \times \frac{\text{N-m}}{\text{J}} \frac{\text{kg-m}}{\text{N-s}^2}} = 1481\ \text{mps}\ (4859\ \text{fps}).$$

Next, the speed of sound at the exit is $a_8 = \sqrt{\gamma_8 \mathcal{R} T_8}$, so $a_8 = \sqrt{1.2738 \times 287.1 \dfrac{\text{J}}{\text{kg-K}} \times 1481\text{K} \times \dfrac{\text{N-m}}{\text{J}} \times \dfrac{\text{kg-m}}{\text{N-s}^2}} = 735.7 \text{ mps (1414 fps)}$

and thus $M_8 = u_8/a_8 = 1481 \text{ mps}/735.7 \text{ mps} = 1.512$.

Next, the exit area is given by $A_8 = \dfrac{\dot{m}_8}{\rho_8 u_8}$, where $\dot{m}_8$ is the total mass flow rate at 8 and is given by $\dot{m}_8 = \dot{m}_{a8} + \dot{m}_{ft}$ (air flow and fuel flow) and $\dot{m}_{a8}$ is the air mass flow rate at 8 and is given by $\dot{m}_{a8} = \dot{m} + \alpha\dot{m}$, namely the core air and the air from the bypass duct.

So, $\dot{m}_{a8} = 74.83 \text{ kg/s} + 1.2 \times 74.83 \text{ kg/s} = 164.6 \text{kg/s} \ (362.9 \text{ lb}_m/\text{s})$, such that the total mass flow is

$\dot{m}_8 = 164.6 \text{ kg/s} + 9.162 \text{ kg/s} = 173.8 \text{ kg/s} \ (383.2 \text{ lb}_m/\text{s})$.

Since both T_8 and p_8 are known, ρ_8 is found from the ideal gas equation:

$$\rho_8 = \frac{p_8}{\mathcal{R}T_8} = \frac{101.3 \text{ kPa} \times 1000 \dfrac{\text{Pa}}{\text{kPa}}}{287.1 \dfrac{\text{J}}{\text{kg-K}} \times 1112\text{K}} \times \frac{\text{N}}{\text{Pa-m}^2} \times \frac{\text{J}}{\text{N-m}} = 0.2385 \frac{\text{kg}}{\text{m}^3} \left(0.01488 \frac{\text{lb}_m}{\text{ft}^3}\right).$$

Total Thrust and TSFC

The thrust is given by

$F = \dot{m}_8 u_8 - \dot{m}_{a8} u_a + A_8(p_8 - p_a)$,

so that

$$F = 173.8 \frac{\text{kg}}{\text{s}} \times 1112 \frac{\text{m}}{\text{s}} \times \frac{\text{N-s}^2}{\text{kg-m}} - 164.6 \frac{\text{kg}}{\text{s}} \times 255.5 \frac{\text{m}}{\text{s}} \times \frac{\text{N-s}^2}{\text{kg-m}}$$

$$+ 0.6556 \, \text{m}^2 \times (101.3 \, \text{kPa} - 101.3 \, \text{kPa}) \times \frac{\text{N-m}^2}{\text{Pa}} \times \frac{1000\text{Pa}}{\text{kPa}}.$$

Thus, $F = 151{,}280$ (momentum) + 0 (pressure) N or F = 151,280 N (34,010 lb_f), so that the nondimensional thrust is

$$\overline{F} = \frac{F}{(1+\alpha)\dot{m}_a} = \frac{151,280 \text{ N}}{2.2 \times 74.83 \dfrac{\text{kg}}{\text{s}} \times 340.6 \dfrac{\text{m}}{\text{s}}} \times \frac{\dfrac{\text{kg-m}}{\text{s}^2}}{\text{N}} = 2.697.$$

Finally, $TSFC = \dot{m}_{f_t}/F$, so that

$$TSFC = \frac{9.162 \dfrac{\text{kg}}{\text{s}}}{151{,}280 \text{ N}} \times 3600 \frac{\text{s}}{\text{h}} = 0.2181 \frac{\text{kg}}{\text{h-N}} \left(2.138 \frac{\text{lb}_m}{\text{h-lb}_f}\right).$$

One can also compute the ideal performance for this engine. The ideal thrust is 164,690 N and the *TSFC* is 0.1549 kg/h/N. Thus, for the real case the thrust is reduced by 9 percent and the *TSFC* is increased by 41 percent because of the losses.

Extended Example 3.10.1 Nonideal Mixed Turbofan with an Afterburner: Match the Static Pressures at the Mixer

A turbofan flies at sea level at a Mach number of 0.75. It ingests 74.83 kg/s (165 lb_m/s) of air to the core. The compressor operates with a pressure ratio of 15 and an efficiency of 92 percent. The engine has a bypass ratio of 1.20. The efficiency of the fan is 91 percent. The fuel has a heating value of 41,870 kJ/kg (18,000 BTU/lb_m) and the burner total temperature is 1667 K (3000 °R). The burner has an efficiency of 91 percent and a total pressure ratio of 0.93, while the turbine has an efficiency of 89 percent. The duct has a total pressure ratio of 0.99 and the total pressure ratio of the mixer is 0.985. A variable CD nozzle (to match the exit pressure to the ambient pressure) is used for the primary nozzle, and the efficiency is 94 percent. The total pressure recovery for the diffuser is 0.96 and the shaft efficiency is 99.5 percent. An afterburner is used and the exit temperature is 1944 K (3500 °R). The efficiency of the afterburner is 85 percent and the pressure ratio is 0.94. The Mach numbers at the turbine exit is 0.50 and duct exit is 0.30. Find the developed thrust and *TSFC*. Note that this is a modification from Example 3.10. Namely, in that example total pressures were matched at the duct exit. Now we have some Mach number information so we can match static pressures at the duct exit.

Solution

This is similar to Example 3.10, but now the static pressures are matched at the mixer. Namely, $p_5 = p_{7.5}$ since the Mach numbers are known now ($p_{t5} = p_{t7.5}$). Since solution details are very similar to those in Example 3.10, only the results are presented.

Now, $\pi_f = 2.860$ (was 3.049), so that

$$\frac{p_{t7.5}}{p_{7.5}} = \left[1 + \frac{\gamma_u - 1}{2} M_{7.5}^2\right]^{\frac{\gamma_u}{\gamma_u - 1}}$$

$p_{t7.5} = 400.0$ kPa (58.01 psia) so $p_{7.5} = 376.0$ kPa (54.53 psia) for $\gamma_u = 1.389$ and $M_{7.5} = 0.30$,

$$\frac{p_{t5}}{p_5} = \left[1 + \frac{\gamma_t - 1}{2} M_5^2\right]^{\frac{\gamma_t}{\gamma_t - 1}}$$

$p_{t5} = 441.3$ kPa (64.00 psia) so $p_5 = 376.0$ kPa (54.53 psia) for $\gamma_t = 1.306$ and $M_5 = 0.50$.

So indeed static pressures are matched, $p_5 = p_{7.5}$,

$F = 153{,}380$ N (34,480 lb_f)

and $TSFC = 0.2152 \frac{kg}{h\text{-}N} \left(2.110 \frac{lb_m}{h\text{-}lb_f} \right)$.

These are about 1.4 percent different than as calculated by matching the total pressures in Example 3.10.

afterburning version of the mixed fan operates with characteristics similar to those for the exhausted version of a turbofan. That is, the dimensionless thrust (by $\dot{m}_t$) is lower than that of a turbojet; however, the *TSFC* is also lower. The addition of the afterburner increases the thrust and *TSFC* significantly, but both thrust and *TSFC* are below those for a similar turbojet with an afterburner. Typical trends are shown in Figure 3.35.

The characteristics of mixed and exhausted turbofans can be compared from Figures 3.35 and 3.31. A very important conclusion is that the ideal mixed turbofan has markedly more thrust than the ideal exhausted fan when the afterburner is lit. This finding is true because of the much larger airflow rate ($(1 + \alpha)\dot{m}$ and consequently a much larger fuel flow) present in the afterburner for the mixed configuration. However, on the basis of the same reasoning the *TSFC* is much larger for the mixed version. This partially explains why mixed turbofans are usually used in military applications. That is, military craft usually have afterburners and thus need an increase in thrust levels as large as possible when the afterburner is lit but also need to be fuel efficient when cruising. On the other hand, exhausted turbofans are usually used in commercial aircraft, and these do not usually have afterburners. Ideally, both types of turbofans without afterburners have been demonstrated in this chapter the capability to operate identically. However, as demonstrated in Figure 3.34 (h–s diagram), the ducts and mixers in mixed turbofans are quite nonisentropic and thus incur significant losses to the flow. As a result, if an afterburner is not to be used, an exhausted fan is more efficient than a mixed turbofan owing to reduced nonideal effects or losses.

Turbofan: Fan Mixed and Exhausted, Hybrid

The general diagram for a turbofan is shown in Figure 2.3. Thus far, fully exhausted and fully mixed turbofans have been covered. When a fan is present, and when at least a part of the flow is mixed into the turbine exit flow, a bypass duct is used to direct the bypassed air. For the ideal cases, all of the bypassed air is considered to be either all exhausted or all mixed back into the primary flow. In practice, a portion of the bypassed air is sometimes bled and mixed. Such a turbofan is shown in Figure 3.36, which includes an afterburner. A new parameter, termed the duct split ratio, is defined as the ratio of secondary air that is mixed to the total secondary air and is defined in Eq. (2.3.117). Deriving closed-form ideal cycle equations is

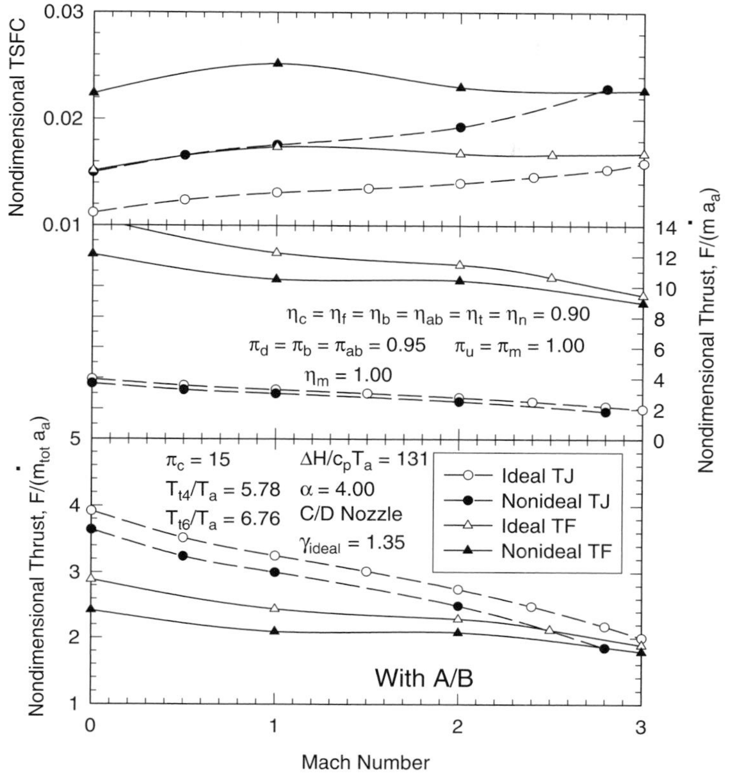

Figure 3.35 Dimensionless thrust (based on total flow) and dimensionless *TSFC* vs. Mach number for a mixed fan and turbojet with afterburners.

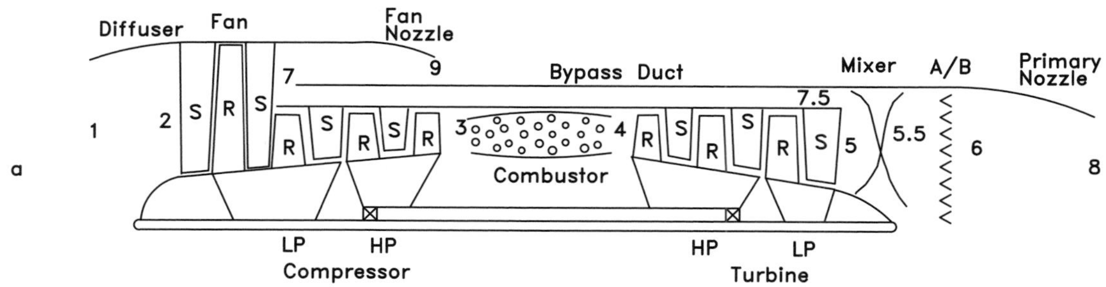

Figure 3.36 Turbofan with mixed and exhausted (hybrid) fan.

not possible for this geometry. However, one can still perform ideal analyses by hand or computer calculations or using the provided software JetEngineCycle.exe.

Example 3.11 Nonideal Nonafterburning Mixed plus Exhausted (Hybrid) Turbofan

A nonafterburning turbofan flies at sea level at a Mach number of 0.75. It ingests 74.83 kg/s (165 lb_m/s) of air to the core. The compressor operates with a pressure ratio of 15 and an efficiency of 88 percent. The engine has a bypass ratio of 3 and a split ratio of 0.25. The efficiency of the fan is 90 percent. The fuel has a heating value of 41,870 kJ/kg (18,000 BTU/lb_m) and the burner total temperature is 1667 K (3000 °R). The burner has an efficiency of 91 percent and a total pressure ratio of 0.95, while the turbine has an efficiency of 85 percent. The duct has a total pressure ratio of 0.98 and the total pressure ratio of the mixer is 0.97. A variable converging/diverging nozzle (to match the exit pressure to the ambient pressure) is used for the primary nozzle and the efficiency is 96 percent. A converging nozzle is used for the fan nozzle and the efficiency is 95 percent. The total pressure recovery for the diffuser is 0.92 and the shaft efficiency is 99.5 percent. The Mach numbers at the turbine and duct exits are nearly the same. Find the developed thrust and *TSFC*.

Solution

Diffuser

First, the speed of sound is

$$a_a = \sqrt{\gamma_a \mathcal{R} T_a} = \sqrt{1.403 \times 287.1 \frac{\text{J}}{\text{kg-K}} \times 288.2\ \text{K} \times \frac{\text{N-m}}{\text{J}} \times \frac{\text{kg-m}}{\text{N-s}^2}} = 340.7\ \text{mps}\ (1118\ \text{fps}),\ \text{so}$$

the engine airspeed is $u_a = M_a a_a = 0.75 \times 340.7$ mps $= 255.5$ mps (838.2 fps).

Next,

$$\frac{T_{ta}}{T_a} = 1 + \frac{\gamma_d - 1}{2} M_a^2 = 1 + \frac{1.3997 - 1}{2} 0.75^2 = 1.1124 \text{ for } \gamma_d = 1.3997 \text{ and } M_a = 0.75,$$

so that $T_{t2} = T_{ta} = 1.1124 \times 288.2\ \text{K} = 320.6\ \text{K}$ (577.0 °R), and also

$$\frac{p_{ta}}{p_a} = \left[\frac{T_{ta}}{T_a}\right]^{\frac{\gamma_d}{\gamma_d - 1}} = [1.1124]^{\frac{1.3997}{1.3997-1}} = 1.4522.$$

Next, the pressure recovery factor for the diffuser is 0.92, so that

$$\frac{p_{t2}}{p_a} = \frac{p_{t2}}{p_{ta}} \frac{p_{ta}}{p_a} = \pi_d \frac{p_{ta}}{p_a} = 0.92 \times 1.4522 = 1.3360$$

and thus $p_{t2} = 1.3360 \times 101.3\ \text{kPa} = 135.3\ \text{kPa}$ (19.63 psia).

Compressor

The compressor total pressure ratio is given as 15 so

$p_{t3} = \pi_c\, p_{t2} = 15 \times 135.3\ \text{kPa} = 2030\ \text{kPa}$ (294.4 psia).

Next, one knows that for a nonideal compressor

$$\frac{p_{t3}}{p_{t2}} = \left\{1 + \eta_c\left[\frac{T_{t3}}{T_{t2}} - 1\right]\right\}^{\frac{\gamma}{\gamma-1}} = \pi_c = \{1 + \eta_c[\tau_c - 1]\}^{\frac{\gamma}{\gamma-1}},$$

so solving for the total temperature ratio (the efficiency of the compressor is given as 88 percent) yields

$$\frac{T_{t3}}{T_{t2}} = \tau_c = \frac{\left[\frac{p_{t3}}{p_{t2}}\right]^{\frac{\gamma_c-1}{\gamma_c}} - 1}{\eta_c} + 1 = \frac{[15]^{\frac{1.3805-1}{1.3805}} - 1}{0.88} + 1 = 2.2607 \text{ for } \gamma_c = 1.3805 \text{ (iterative).}$$

Therefore, the exit temperature of the compressor is $T_{t3} = \tau_c\, T_{t2} = 2.2607 \times 320.6\text{ K} = 724.7\text{ K}$ (1305 °R).

Fan

Note: For a mixed turbofan geometry the fan total pressure ratio is unknown. Thus, one needs to find π_f by iteration. In effect, the duct exit pressure (same as the turbine exit pressure) directly depends on the fan total pressure ratio. But this pressure ratio also affects the turbine power, which indirectly affects the turbine exit pressure. Thus, one can find the fan pressure ratio by a progressive iteration. Details of the "Regula Falsi" iteration are presented in Appendix I (Example I.2). Fortunately, the iteration process is quick to converge (only four iterations for this case) and easily programmed with a commercial math solver. Also as before, one should note that although precise values of the pressure ratio are desired for an engine design, it should not become the dominant feature of examples or chapter problems.

For the purposes of this example to demonstrate the thermodynamic calculations, one can first try $\pi_f = 2.0144$, which will be checked later at the mixer:
$p_{t7} = \pi_f\, p_{t2} = 2.015 \times 135.3\text{ kPa} = 272.8\text{ kPa}$ (39.56 psia).

Next one knows that

$$\frac{p_{t7}}{p_{t2}} = \left\{1 + \eta_f\left[\frac{T_{t7}}{T_{t2}} - 1\right]\right\}^{\frac{\gamma}{\gamma-1}} = \pi_f = \{1 + \eta_f[\tau_f - 1]\}^{\frac{\gamma}{\gamma-1}},$$

so solving for the total temperature ratio (the efficiency of the fan is given as 90 percent) yields

$$\frac{T_{t7}}{T_{t2}} = \tau_f = \frac{\left[\frac{p_{t7}}{p_{t2}}\right]^{\frac{\gamma_f-1}{\gamma_f}} - 1}{\eta_f} + 1 = \frac{[2.015]^{\frac{1.396-1}{1.396}} - 1}{0.88} + 1 = 1.220 \text{ for } \gamma_f = 1.3960 \text{ (iterative).}$$

Therefore, the exit temperature of the fan is

$T_{t7} = \tau_f\, T_{t2} = 1.220 \times 320.6\text{ K} = 398.8\text{ K}$ (717.8 °R).

Bypass Duct

Next, the pressure ratio for the bypass duct is $\pi_u = \dfrac{p_{t7.5}}{p_{t7}} = 0.98$, so that $p_{t7.5} = 0.98 \times 272.8\text{ kPa} = 267.3\text{ kPa}$ (38.77 psia).

Also, since the duct is adiabatic the total temperature is constant through the passage, so $T_{t7.5} = T_{t7} = 398.8\text{ K}$ (717.8 °R).

Primary Combustor

Next, the total pressure ratio for the burner is 0.95 so the burner exit total pressure is

$p_{t4} = \pi_b \, p_{t3} = 0.95 \times 2030 \text{ kPa} = 1929 \text{ kPa } (279.7 \text{ psia})$

and the energy balance yields $\Delta H \eta_b \dot{m}_f = \dot{m} c_{p_b}(T_{t4} - T_{t3}) + \dot{m}_f c_{p_b} T_{t4}$, where the burner efficiency is 91 percent and the heating value is 41,870 kJ/kg, so that

$$\dot{m}_f = 74.83 \frac{\text{kg}}{\text{s}} \times 1.171 \frac{\text{kJ}}{\text{kg-K}} \times \frac{(1667 - 724.7)\text{K}}{0.91 \times 41780 \frac{\text{kJ}}{\text{kg}} - 1.171 \frac{\text{kJ}}{\text{kg-K}} \times 1667 \text{ K}}$$

$= 2.285$ kg/s (5.038 lb$_m$/s).

The specific heat is evaluated at the average temperature (not an iteration). Since no afterburner is present, this is also the total fuel flow rate ($\dot{m}_{ft}$). One should note that $f = \dot{m}_f / \dot{m} = 0.0305$, which is small.

Turbine

Now the power balance for the shaft of the turbofan is

$\dot{m} c_{p_c}(T_{t3} - T_{t2}) + \alpha \dot{m} c_{p_f}(T_{t7} - T_{t2}) = \eta_m \dot{m}_t c_{p_t}\,(T_{t4} - T_{t5})$, where the mechanical efficiency is 99.5 percent, or

$$74.83 \frac{\text{kg}}{\text{s}} \times 1.041 \frac{\text{kJ}}{\text{kg-K}} \times (724.7\,\text{K} - 320.6\,\text{K}) + 3 \times 74.83 \frac{\text{kg}}{\text{s}} \times 1.012 \frac{\text{kJ}}{\text{kg-K}} \times (398.8\,\text{K} - 320.6\,\text{K})$$
$$= 0.995 \times \left(74.83 \frac{\text{kg}}{\text{s}} + 2.285 \frac{\text{kg}}{\text{s}}\right) \times 1.1215 \frac{\text{kJ}}{\text{kg-K}} \times (1667\,\text{K} - T_{t5}).$$

Thus, solving for the turbine exit temperature $T_{t5} = 1138$ K (2048 °R). The value of the turbine specific heat is found iteratively.

Now the turbine total temperature ratio is $\tau_t = \dfrac{T_{t5}}{T_{t4}} = \dfrac{1138 \text{ K}}{1667 \text{ K}} = 0.6827.$

Next, one knows for a nonideal turbine:

$$\frac{p_{t5}}{p_{t4}} = \left\{ 1 - \frac{\left[1 - \frac{T_{t5}}{T_{t4}}\right]}{\eta_t} \right\}^{\frac{\gamma}{\gamma-1}} = \pi_t = \left\{ 1 - \frac{[1 - \tau_t]}{\eta_t} \right\}^{\frac{\gamma}{\gamma-1}},$$

but the turbine efficiency, η_t, is given as 85 percent, so the turbine total pressure ratio is

$$\pi_t = \left\{ 1 - \frac{[1 - 0.6827]}{0.85} \right\}^{\frac{1.3093}{1.3093-1}} = 0.1386 \text{ for } \gamma_t = 1.3093 \text{ (iterative with } c_{p_t}\text{)}.$$

Finally, the turbine exit total pressure is

$p_{t5} = \pi_t \, p_{t4} = 0.1386 \times 1929 \text{ kPa} = 267.3 \text{ kPa } (38.77 \text{ psia}).$

Mixer

Now one can compare $p_{t7.5}$ and p_{t5} and see that they are the same. Also, since the Mach numbers at stations 5 and 7.5 are given to be approximately the same ($M_5 \cong M_{7.5}$) and since the total pressures are identical, the static pressures will match, so the initial guess on the fan pressure ratio, π_f, was excellent! If they had been different, a new value of π_f would have to be tried and all of the above work would have to be repeated, as demonstrated in detail in Appendix I (Example I.2).

Next, one can find the temperature exiting the mixer $T_{t5.5}$ by

$$T_{t5.5} = \frac{(1+f)c_{p_{mc}} T_{t5} + \sigma\alpha c_{p_{mu}} T_{t7.5}}{(1+f)c_{p_{mc}} + \sigma\alpha c_{p_{mu}}}$$

$$= \frac{(1+0.0305) \times 1.1296 \dfrac{\text{kJ}}{\text{kg-K}} \times 1138\ \text{K} + 0.25 \times 3 \times 1.059 \dfrac{\text{kJ}}{\text{kg-K}} \times 398.8\ \text{K}}{(1+0.0305) \times 1.1296 \dfrac{\text{kJ}}{\text{kg-K}} + 0.25 \times 3 \times 1.059 \dfrac{\text{kJ}}{\text{kg-K}}},$$

where $c_{p_{mu}}$ is evaluated at the average temperature between stations 7.5 to 5.5 and $c_{p_{mc}}$ is from the average temperature between stations 5 to 5.5.

So, $T_{t5.5} = 838.4$ K (1509 °R). Also for this case, and with $M_5 = M_{7.5}$, $\pi_m = p_{t5.5}/p_{t5.2} = p_{t5.5}/p_{t5} = p_{t5.5}/p_{t7.5}$, the total pressure ratio of the duct is 0.97, so that $p_{t5.5} = 0.97 \times 267.3$ kPa $= 259.3$ kPa (37.61 psia).

Primary Nozzle

Since no afterburner is present the inlet total temperature for the primary nozzle is

$T_{t6} = T_{t5.5} = 838.4$ K (1509 °R)

and the inlet total pressure is $p_{t6} = p_{t5.5} = 259.3$ kPa (37.61 psia).

The nozzle is a variable CD nozzle so the exit pressure matches the ambient pressure and one does not have to check for choking, namely, $p_8 = 101.3$ kPa (14.69 psia).

Ideally $T_8{}' = T_{t6}\left[\dfrac{p_8}{p_{t6}}\right]^{\frac{\gamma_n-1}{\gamma_n}} = 838.4\left[\dfrac{101.3\ \text{kPa}}{259.3\ \text{kPa}}\right]^{\frac{1.3528-1}{1.3528}}$, where $\gamma_n = 1.3528$ (iterative)

$= 656.2$ K (1181 °R).

The nozzle efficiency is 96 percent. Thus the nozzle exit temperature is

$T_8 = T_{t6} - \eta_n (T_{t6} - T_8{}') = 838.4\ \text{K} - 0.96 \times (838.4\ \text{K} - 656.2\ \text{K}) = 663.5$ K (1194 °R).

Now $u_8 = \sqrt{2c_{p_n}(T_{t6} - T_8)}$, so that

$$u_8 = \sqrt{2 \times 1.100 \frac{\text{kJ}}{\text{kg-K}} \times (838.4\,\text{K} - 663.5\,\text{K}) \times 1000 \frac{\text{J}}{\text{kJ}} \times \frac{\text{N-m}}{\text{J}} \frac{\text{kg-m}}{\text{N-s}^2}} = 620.5\ \text{mps (2036 fps)}.$$

Next, the speed of sound at the exit is $a_8 = \sqrt{\gamma_8 \mathcal{R} T_8}$, so

$$a_8 = \sqrt{1.3528 \times 287.1 \frac{\text{J}}{\text{kg-K}} \times 663.5\ \text{K} \times \frac{\text{N-m}}{\text{J}} \times \frac{\text{kg-m}}{\text{N-s}^2}} = 507.5\ \text{mps}\ (1665\ \text{fps}).$$

Thus $M_8 = u_8/a_8 = 620.5\ \text{mps}/507.5\ \text{mps} = 1.222$.

Next, the exit area is given by $A_8 = \dfrac{\dot{m}_8}{\rho_8\ u_8}$,

where $\dot{m}_8$ is the total mass flow rate at 8 and is given by $\dot{m}_8 = \dot{m}_{a8} + \dot{m}_{ft}$ (air flow and fuel flow) and $\dot{m}_{a8}$ is the air mass flow rate at 8 and is given by $\dot{m}_{a8} = \dot{m} + \alpha\sigma\dot{m}$, namely the core air and the air from the bypass duct.

So, $\dot{m}_{a8} = 74.83\ \text{kg/s} + 3 \times 0.25 \times 74.83\ \text{kg/s} = 131.0\ \text{kg/s}\ (288.8\ \text{lb}_\text{m}/\text{s})$, such that the total mass flow is
$\dot{m}_8 = 131.0\ \text{kg/s} + 2.285\ \text{kg/s} = 133.2\ \text{kg/s}\ (293.7\ \text{lb}_\text{m}/\text{s})$.

Since both T_8 and p_8 are known, ρ_8 is found from the ideal gas equation:

$$\rho_8 = \frac{p_8}{\mathcal{R}T_8} = \frac{101.3\ \text{kPa} \times 1000 \dfrac{\text{Pa}}{\text{kPa}}}{287.1 \dfrac{\text{J}}{\text{kg-K}} \times 663.5\ \text{K}} \times \frac{\text{N}}{\text{Pa-m}^2} \times \frac{\text{J}}{\text{N-m}} r_8 = 0.5322\ \text{kg/m}^3\ (0.001032\ \text{sl/ft}^3).$$

Thus, the exit area is

$$A_8 = \frac{133.2 \dfrac{\text{kg}}{\text{s}}}{0.5322 \dfrac{\text{kg}}{\text{m}^3} \times 620.5 \dfrac{\text{m}}{\text{s}}} = 0.4037\ \text{m}^2 (625.7\text{in.}^2).$$

Fan Nozzle

Since the fan nozzle is a fixed converging nozzle, one must first check to see if it is choked.

For choking, the fan nozzle exit pressure is

$$p_9^* = p_{t7} \left[1 + \frac{1 - \gamma_{fn}}{\eta_{fn}(1 + \gamma_{fn})}\right]^{\frac{\gamma_{fn}}{\gamma_{fn}-1}}.$$

Thus, for $p_{t7} = 272.8$ kPa, $\eta_{fn} = 0.95$ (given), and $\gamma_{fn} = 1.3921$ (iterative):

$$p_9^* = 272.8\ \text{kPa} \times \left[1 + \frac{1 - 1.3921}{0.95 \times (1 + 1.3921)}\right]^{\frac{1.3921}{1.3921-1}} = 139.2\ \text{kPa}\ (20.19\ \text{psia}).$$

But since $p_a = 101.3$ kPa (14.69 psia), which is well below the exit pressure for choking, the nozzle is choked and the exit Mach number is identically unity.

Therefore $p_9 = p_9^* = 139.2$ kPa (20.19 psia) and

$$T_9 = \frac{2T_{t7}}{1 + \gamma_{fn}} \text{ for } M_9 = 1.00.$$

Thus, for $T_{t7} = 398.8$ K (717.8 °R), $T_9 = \dfrac{2 \times 398.8\ \text{K}}{1 + 1.3946} = 333.5\,\text{K}\,(600.3\ °\text{R})$.

Now $u_9 = \sqrt{2\, c_{p_{fn}}\,(T_{t7} - T_9)}$, so that

$$u_9 = \sqrt{2 \times 1.019 \frac{\text{kJ}}{\text{kg-K}} \times (398.8\ \text{K} - 333.5\ \text{K}) \times 1000 \frac{\text{J}}{\text{kJ}} \times \frac{\text{N-m}}{\text{J}} \frac{\text{kg-m}}{\text{N-s}^2}}$$

$$= 365.0\ \text{mps}\ (1198\ \text{fps}).$$

Alternatively, one could have found u_9 by first finding the speed of sound at the exit $a_9 = \sqrt{\gamma_9 \mathcal{R} T_9}$.

$$\text{So } a_9 = \sqrt{1.3921 \times 287.1 \frac{\text{J}}{\text{kg-K}} \times 333.5\ \text{K} \times \frac{\text{N-m}}{\text{J}} \times \frac{\text{kg-m}}{\text{N-s}^2}} = 365.0\ \text{mps}\ (1198\ \text{fps}).$$

But $M_9 = 1 = u_9/a_9$, so $u_9 = 365.0$ mps (1198 fps), as we found previously.

Next, the exit area is given by $A_9 = \dfrac{\dot{m}_9}{\rho_9\ u_9}$, where $\dot{m}_8$ is the total mass flow rate at 9 and is given by

$$\dot{m}_9 = \alpha(1 - \sigma)\dot{m} = 3 \times (1 - 0.25) \times 74.83\ \text{kg/s} = 168.4\ \text{kg/s}\ (371.3\ \text{lb}_\text{m}/\text{s}).$$

Since both T_9 and p_9 are known, ρ_9 is found from the ideal gas equation:

$$\rho_9 = \frac{p_9}{\mathcal{R}T_9} = \frac{139.2\ \text{kPa} \times 1000 \dfrac{\text{Pa}}{\text{kPa}}}{287.1 \dfrac{\text{J}}{\text{kg-K}} \times 333.5\ \text{K}} \times \frac{\text{N}}{\text{Pa-m}^2} \times \frac{\text{J}}{\text{N-m}} = 1.455 \text{kg/m}^3 (0.002823\ \text{sl/ft}^3);$$

thus the fan nozzle exit area is

$$A_9 = \frac{168.4 \dfrac{\text{kg}}{\text{s}}}{1.455 \dfrac{\text{kg}}{\text{m}^3} \times 365.0 \dfrac{\text{m}}{\text{s}}} = 0.3172 \text{m}^2 (491.7 \text{in.}^2).$$

Total Thrust and TSFC

The thrust is given by

$$F = \dot{m}_8 u_8 - \dot{m}_{a8} u_a + A_8(p_8 - p_a) + \dot{m}_9 u_9 - \dot{m}_9 u_a + A_9(p_9 - p_a),$$

so that

$$F = 133.2 \frac{\text{kg}}{\text{s}} \times 620.5 \frac{\text{m}}{\text{s}} \times \frac{\text{N-s}^2}{\text{kg-m}} - 131.0 \frac{\text{kg}}{\text{s}} \times 255.5 \frac{\text{m}}{\text{s}} \times \frac{\text{N-s}^2}{\text{kg-m}}$$

$$+ 0.4037\ \text{m}^2 \times (101.3\ \text{kPa} - 101.3\ \text{kPa}) \times \frac{\text{N-m}^2}{\text{Pa}} \times \frac{1000\ \text{Pa}}{\text{kPa}}$$

$$+ 168.4 \frac{\text{kg}}{\text{s}} \times 365.0 \frac{\text{m}}{\text{s}} \times \frac{\text{N-s}^2}{\text{kg-m}} - 168.4 \frac{\text{kg}}{\text{s}} \times 255.5 \frac{\text{m}}{\text{s}} \times \frac{\text{N-s}^2}{\text{kg-m}}$$

$$+ 0.3172\ \text{m}^2 \times (139.2\ \ \text{kPa} - 101.3\ \text{kPa}) \times \frac{\text{N-m}^2}{\text{Pa}} \times \frac{1000\ \text{Pa}}{\text{kPa}}.$$

Thus, $F = 49{,}233$ (momentum) + 0 (pressure) + 18,449 (fan momentum) + 12,023 (fan pressure), or $F = 79{,}704$ N (17,919 lb_f), so that the nondimensional thrust is

$$\overline{F} = \frac{F}{(1+\alpha)\dot{m}a_a} = \frac{79,704\ \text{N}}{4 \times 74.83\dfrac{\text{kg}}{\text{s}} \times 340.7\dfrac{\text{m}}{\text{s}}} \times \frac{\dfrac{\text{kg-m}}{\text{s}^2}}{\text{N}} = 0.7816.$$

Finally, $TSFC = \dot{m}_{f_t}/F$, so that

$$TSFC = \frac{2.285\dfrac{\text{kg}}{\text{s}}}{79,704\ \text{N}} \times 3600\frac{\text{s}}{\text{h}} = 0.1032\frac{\text{kg}}{\text{h-N}}\left(1.012\frac{\text{lb}_\text{m}}{\text{h-lb}_\text{f}}\right).$$

The nondimensional *TSFC* is

$$a_a \times TSFC = \frac{340.6\dfrac{\text{m}}{\text{s}} \times 0.1033\dfrac{\text{kg}}{\text{h-N}}}{3600\dfrac{\text{s}}{\text{h}}} \times \frac{\text{N}}{\dfrac{\text{kg-m}}{\text{s}^2}} = 0.009769.$$

One can also compute the ideal performance for this engine. The ideal thrust is 105,900 N and the *TSFC* is 0.0642 kg/h/N. Thus, for the nonideal case the thrust is reduced by 32.9 percent and the *TSFC* is increased by 33 percent because of the losses.

Extended Example 3.11.1 Nonideal Nonafterburning Mixed plus Exhausted (Hybrid) Turbofan: Effects of Bypass Ratio

Holding all other variables constant in the above example, the effect of the bypass ratio was considered in a complementary study. All of the calculations above were the same except for the parameter values. In Figure 3.37 the effects of changing the bypass ratio are shown. As the bypass ratio is increased, the nondimensional thrust is continually reduced, indicating that the thrust-to-weight ratio is decreasing. Also, as the bypass ratio is increased, the nondimensional *TSFC* reaches a minimum and then increases. Both of these trends were shown earlier for the ideal case of a fully exhausted turbofan ($\sigma = 0.0$). The optimum *TSFC* occurs at a value of the bypass ratio of approximately 5.5.

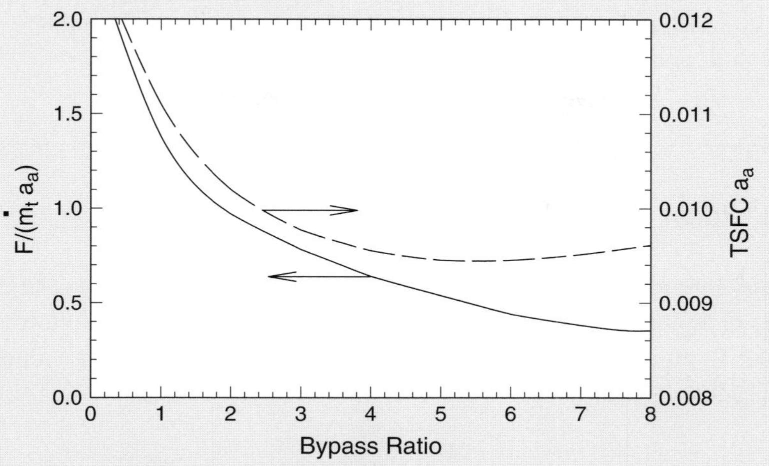

Figure 3.37 Dimensionless thrust (based on total flow) and dimensionless *TSFC* vs. bypass ratio for the hybrid turbofan in Extended Example 3.11.1

Turboprop

As indicated in Chapter 1, a turboprop is aimed at a different group of aircraft than the engines already covered. This geometry is different than that of the previous engines and is shown in Figure 3.8. Propellers tend to be relatively large; propeller tip speeds can be quite high. As a result, this engine type is primarily used on low-speed aircraft so that the tip speeds do not become prohibitively large. The behavior is somewhat similar to a very large bypass ratio turbofan; however, the flow through the propeller is not as well defined as in a fan. Also, the pressure ratio across the propeller is very nearly unity.
A considerable flow rate of air goes through the propeller to generate thrust. The remainder of the engine (stations 2 through 8) is fundamentally identical to a turbojet. A small portion of the flow from the propeller enters the compressor (station 2). The thrust from this engine type is developed from two sources: the propeller and the turbojet nozzle. Thus, the total power generated for thrust of a turboprop is given by

$$\mathcal{P}_{\mathrm{t}} = \mathcal{P}_{\mathrm{p}} + F u_{\mathrm{a}}, \tag{3.3.1}$$

where F is the thrust of a turbojet, which we have previously found, and $\mathcal{P}_{\mathrm{p}}$ is the propeller power. Next, a work coefficient for the engine can conveniently be defined as

$$C_{\mathrm{w_e}} \equiv \frac{\mathcal{P}_{\mathrm{t}}}{\dot{m} h_{\mathrm{a}}}, \tag{3.3.2}$$

where $\dot{m}$ is the airflow through the turbojet core. This now can be applied to Eq. (3.3.1):

$$C_{\mathrm{w_e}} = \frac{\mathcal{P}_{\mathrm{p}}}{\dot{m} c_{\mathrm{p}} T_{\mathrm{a}}} + \frac{u_{\mathrm{a}} F}{\dot{m} c_{\mathrm{p}} T_{\mathrm{a}}} = C_{\mathrm{w_p}} + \frac{u_{\mathrm{a}} F}{\dot{m} c_{\mathrm{p}} T_{\mathrm{a}}}, \tag{3.3.3}$$

where $C_{\mathrm{w_p}}$ is the propeller work coefficient previously given by Eq. (2.3.50).

For a less defined nature of the propeller flow (mass flow, pressure rise, and flow direction), a somewhat different approach to the analysis is needed. The thrust is a strong function of burner exit temperature, compressor pressure ratio, and the nozzle parameter. Of course, the Mach number and mass flow rate are strong influences for the turboprop just as they are for all other engines.

Example 3.12 | Ideal Turboprop

An ideal turboprop powers an aircraft at sea level at a Mach number of 0.7. The compressor has a pressure ratio of 6.5 and an airflow of 13.61 kg/s (30 lb_m/s). The burner exit total temperature is 1389 K (2500 °R) and the nozzle exit Mach number is 0.95. The heating value of the fuel is 43,960 kJ/kg (18,900 BTU/lb_m). Find the thrust and *TSFC* if $\gamma = 1.35$.

Solution

First, the known quantities are as follows.

$T_a = 288.2$ K (518.7 °R)	$p_a = 101.3$ kPa (14.69 psia)
$T_{t4} = 1389$ K (2500 °R)	$p_8 = 101.3$ kPa (14.69 psia)
$M_a = 0.7$	$M_e = 0.95$

Diffuser

The freestream velocity is

$$u_a = M_a a_a = M_a\sqrt{\gamma_a \mathcal{R} T_a} = 0.7\sqrt{1.35 \times 287.1\frac{\text{J}}{\text{kg-K}} \times 288.2\ \text{K} \times \frac{\text{N-m}}{\text{J}} \times \frac{\text{kg-m}}{\text{N-s}^2}}$$

$$= 233.9\ \text{mps}\ (767.4\ \text{fps})$$

and the total temperature at the inlet is

$$T_{ta} = T_a\left[1 + \frac{\gamma - 1}{2}M_a^2\right] = T_a\left[1 + \frac{1.35 - 1}{2}0.7^2\right] = 1.086 \times T_a = 312.9\ \text{K}\ (563.2\ °\text{R}),$$

which is also equal to T_{t2} because the process is adiabatic.

The total pressure at the inlet is $p_{ta} = p_a\left[1 + \frac{\gamma - 1}{2}M_a^2\right]^{\frac{\gamma}{\gamma-1}} = p_a\left[\frac{T_{ta}}{T_a}\right]^{\frac{\gamma}{\gamma-1}}$

and thus $p_{ta} = p_a[1.086]^{\frac{1.35}{0.35}} = 139.2$ kPa (20.19 psia).

Compressor

Since the compressor total pressure ratio, π_c, is given, the total pressure at the compressor exit is $p_{t3} = \pi_c \times p_{t2}$. But because the process is isentropic for an ideal diffuser $p_{t2} = p_{ta}$,

$p_{t3} = 6.5 \times 139.2 \text{ kPa} = 904.6 \text{ kPa } (131.2 \text{ psia}).$

For an isentropic compressor, $\tau_c = [\pi_c]^{\frac{\gamma-1}{\gamma}} = [6.5]^{\frac{0.35}{1.35}} = 1.625$, and thus the total temperature at the compressor exit is

$T_{t3} = t_c \times T_{t2} = 1.625 \times 312.9 \text{ K} = 508.3 \text{ K } (914.9\ °\text{R}).$

Burner

The fuel flow rate is given by

$$\dot{m}_f = \frac{\dot{m} c_p [T_{t4} - T_{t3}]}{\Delta H} = 13.61 \frac{\text{kg}}{\text{s}} \times 1.107 \frac{\text{kJ}}{\text{kg-K}} \times \frac{(1389 - 508.2) \text{ K}}{43960 \frac{\text{kJ}}{\text{kg}}}$$

$$= 0.3019 \text{ kg/s } (0.6657 \text{ lb}_m/\text{s}),$$

and the exit total pressure is the same as the inlet for an ideal burner:

$p_{t4} = p_{t3} = 904.6 \text{ kPa } (131.2 \text{ psia}).$

Nozzle

Because $\frac{p_{t8}}{p_8} = \left[1 + \frac{\gamma - 1}{2} M_8^2\right]^{\frac{\gamma}{\gamma-1}}$, the exit Mach number is known, and the nozzle exit pressure ideally matches atmospheric pressure, the total pressure at the exit can be found as follows:

$$p_{t8} = 101.3 \text{ kPa} \left[1 + \frac{1.35 - 1}{2} 0.95^2\right]^{\frac{1.35}{0.35}} = 178.3 \text{ kPa } (25.87 \text{ psia}).$$

For an ideal nozzle the total pressure is constant (isentropic flow) and thus $p_{t5} = p_{t8} = 178.3$ kPa (25.87 psia), which is also the exit pressure for the turbine.

Turbine

The total pressure ratio for the turbine can be found directly by

$$\pi_t = \frac{p_{t5}}{p_{t4}} = \frac{178.3 \text{ kPa}}{904.6 \text{ kPa}} = 0.1971$$

and thus the total temperature ratio for the isentropic case can be found:

$$\tau_t = [\pi_t]^{\frac{\gamma-1}{\gamma}} = [0.1971]^{\frac{0.35}{1.35}} = 0.6564.$$

The turbine total temperature is therefore

$T_{t5} = \tau_t \times T_{t4} = 0.6564 \times 1389 \text{ K} = 911.7 \text{ K } (1641\ °\text{R}).$

Nozzle

We look at the nozzle again, using turbine information.

From the turbine $T_{t6} = T_{t5} = 911.7$ K, and

$$T_8 = T_{t8}\left[\frac{p_8}{p_{t8}}\right]^{\frac{\gamma-1}{\gamma}} = 911.7\left[\frac{101.3}{178.4}\right]^{\frac{1.35-1}{1.35}} = 787.4 \text{ K}.$$

Also,

$$u_8 = M_8 a_8 = M_8\sqrt{\gamma_8 \mathcal{R} T_8} = 0.95\sqrt{1.35 \times 287.1\frac{\text{J}}{\text{kg-K}} \times 787.4\,\text{K} \times \frac{\text{N-m}}{\text{J}} \times \frac{\text{kg-m}}{\text{N-s}^2}} = 524.8\frac{\text{m}}{\text{s}}.$$

So only for the jet

$$F_{\text{jet}} = \dot{m}(u_8 - u_a) = 13.61(524.8 - 233.9) = 3959 \text{ N}$$

and $\dfrac{F_{\text{jet}}}{\dot{m}a_a} = \dfrac{3959}{13.61 \times 334.1} = 0.870.$

Also, $\dfrac{F_{\text{jet}}u_a}{\dot{m}c_p T_a} = \dfrac{3959 \times 233.9}{13.61 \times 1.107 \times 288.2}\,\text{N}\dfrac{\text{s}}{\text{kg}}\dfrac{\text{m}}{\text{s}}\dfrac{\text{kgK}}{\text{kJ}}\dfrac{1}{\text{K}}\dfrac{\text{kJ}}{1000\ \text{J}}\dfrac{\text{J}}{\text{Nm}} = 0.213.$

Propeller

The nozzle temperature parameter is

$$\tau_5 = \left[\frac{T_{ta}}{T_a}\right]\tau_c\tau_t = (1.086) \times (1.625) \times (0.6564) = 1.158$$

and the ideal work coefficient is (from Eq. (3.2.24)):

$$C_{W_e} = \left[\frac{T_{t4}}{T_a}\right]\left[1 - \frac{\tau_5}{\left[\frac{T_{ta}}{T_a}\right]\tau_c}\right] - \left[\frac{T_{ta}}{T_a}\right](\tau_c - 1) + (\gamma - 1)M_a^2\left[\sqrt{\frac{\left[\frac{T_{t4}}{T_a}\right]\left[\frac{T_a}{T_{ta}}\right]\left[\frac{\tau_5 - 1}{\tau_c}\right]}{\left[\frac{T_{ta}}{T_a}\right] - 1}} - 1\right].$$

Thus,

$$C_{W_e} = \left[\frac{1389}{288.2}\right]\left[1 - \frac{1.158}{1.086 \times 1.625}\right] - [1.086] \times (0.625) + (0.35)$$

$$\times\, 0.7^2\left[\sqrt{\frac{\left[\frac{1389}{288.2}\right]\left[\frac{1}{1.086}\right]\left[\frac{0.158}{1.625}\right]}{0.086}} - 1\right]$$

$$= 1.191 = C_{W_p} + \frac{u_a F_{\text{jet}}}{\dot{m}c_p T_a} = C_{W_p} + 0.213,$$

so solving for $C_{W_p} = 0.9783$ gives us the propeller work coefficient. This means that 81.4 percent of the thrust work or power comes from the propeller.

The total thrust power is therefore

$$\mathcal{P}_t = C_{W_e}\dot{m}\,h_a = C_{W_e}\dot{m}\,c_p T_a = 1.191 \times 13.61\frac{\text{kg}}{\text{s}} \times 1.107\frac{\text{kJ}}{\text{kg-K}} \times 288.2\ \text{K} \times 10^{-3}\frac{\text{MW}}{\text{kJ}}$$

$$= 5.171\,\text{MW}\ (6941\ \text{HP}).$$

The thrust is then

$$F_t = \frac{}{u_a} = \frac{5.171 \text{ MW}}{233.9 \dfrac{\text{m}}{\text{s}}} \times 10^6 \frac{\text{W}}{\text{MW}} \times \frac{\text{J}}{\text{W-s}} \times \frac{\text{N-m}}{\text{J}} = 22110 \text{ N } (4970 \text{ lb}_f)$$ <ANS

and thus the *SFC* is

$$SFC \equiv \frac{\dot{m}_f}{\mathcal{P}_t} = \frac{0.3019 \dfrac{\text{kg}}{\text{s}} \times 3600 \dfrac{\text{s}}{\text{h}}}{5.171 \text{ MW}} = 210.2 \frac{\text{kg}}{\text{MW-h}} \left(0.3456 \frac{\text{lb}_m}{\text{HPh}}\right).$$

The *TSFC* is

$$TSFC = \frac{\dot{m}_f}{F_t} = \frac{0.3019 \dfrac{\text{kg}}{\text{s}} \times 3600 \dfrac{\text{s}}{\text{h}}}{22,110 \text{ N}} = 0.04917 \frac{\text{kg}}{\text{N-h}} \left(0.4821 \frac{\text{lb}_m}{\text{h-lb}_f}\right).$$ <ANS

Extended Example 3.12.1 **Optimized Ideal Turboprop**

As a complementary study, one can maximize the thrust for Example 3.12. The optimum value of τ_5 is (from Eq. (3.2.28))

$$\tau_{5_{opt}} = 1 + \tau_c \left[\frac{T_{ta}}{T_a}\right]\left[\frac{T_a}{T_{t4}}\right]\left[\left[\frac{T_{ta}}{T_a}\right] - 1\right] = 1 + 1.625 \times [1.086] \times \left[\frac{288.2}{1389}\right] \times [0.086] = 1.0315.$$

This results in an engine work coefficient $C_{W_e} = 1.324$. The total power is thus $\mathcal{P}_t = 5.749$ MW (7717 hp). The total thrust is $F_t = 24{,}580$ N (5526 lb_f) and is all from the propeller; and the $TSFC = 0.04422$ kg/h/N (0.4335 $\text{lb}_m/\text{h}/\text{lb}_f$).

Also, for the ideal case

$$\frac{p_{t8}}{p_a} = \frac{p_{t5}}{p_a} = [\tau_5]^{\frac{\gamma}{\gamma-1}} = [1.0315]^{\frac{1.35}{0.35}} = 1.127,$$

which yields an exit Mach number of 0.424, a turbine total temperature ratio of 0.5845, a turbine exit total temperature of 811.9 K (1461 °R), a nozzle exit static temperature of 786.2 K (1415 °R), and an exhaust velocity of 234.1 mps (768.1 fps).

Example 3.13 | Nonideal Turboprop

A turboprop flies at sea level at a Mach number of 0.70. It intakes 13.61 kg/s (30 lb_m/s) of air. The compressor operates with a pressure ratio of 6.5 and an efficiency of 88 percent. The fuel has a heating value of 43,960 kJ/kg (18,900 BTU/lb_m) and the burner total temperature is 1389 K (2500 °R). The burner has an efficiency of 91 percent and a total pressure ratio of 0.95, while the turbine has an efficiency of 85 percent. A converging nozzle is used and the nozzle efficiency is 96 percent. The total pressure recovery for the diffuser is 0.92 and the shaft efficiency is 99.5 percent. The work coefficient for the propeller is 0.9783 and the propeller efficiency is 70 percent. Find the developed thrust and *TSFC*. One should note that this is the same engine as in Example 3.12 with the exception of the nonideal effects. The work coefficient for the propeller matches that for Example 3.12.

Solution

Diffuser

The standard conditions at sea level are $T_a = 288.2$ K (518.7 °R) and $p_a = 101.3$ kPa (14.69 psia). Next, the speed of sound is

$$a_a = \sqrt{\gamma_a \mathcal{R} T_a} = \sqrt{1.403 \times 287.1 \frac{\text{J}}{\text{kg-K}} \times 288.2\ \text{K} \times \frac{\text{N-m}}{\text{J}} \times \frac{\text{kg-m}}{\text{N-s}^2}} = 340.7\ \text{mps}\ (1118\ \text{fps}),$$

so the freestream airspeed is $u_a = M_a\, a_a = 0.70 \times 340.7$ mps $= 238.4$ mps (782.3 fps).

Next, $\frac{T_{ta}}{T_a} = 1 + \frac{\gamma_d - 1}{2} M_a^2 = 1 + \frac{1.4001 - 1}{2} \times 0.70^2 = 1.0980$ for $\gamma_d = 1.4001$ and $M_a = 0.70$, so that the exit temperature of the diffuser is $T_{t2} = T_{ta} = 1.0980 \times 288.2\text{K} = 316.4$ K (569.5 °R).

Also, $\frac{p_{ta}}{p_a} = \left[\frac{T_{ta}}{T_a}\right]^{\frac{\gamma_d}{\gamma_d - 1}} = [1.0980]^{\frac{1.4001}{1.4001-1}} = 1.3871.$

Next, the pressure recovery factor for the diffuser is 0.92, so $\frac{p_{t2}}{p_a} = \frac{p_{t2}}{p_{ta}} \frac{p_{ta}}{p_a} = \pi_d \frac{p_{ta}}{p_a} = 0.92 \times 1.3871 = 1.2762.$

Therefore, the diffuser exit total pressure is $p_{t2} = 1.2762 \times 101.3$ kPa $= 129.3$ kPa (18.75 psia).

Compressor

Next, the total pressure ratio of the compressor is given as $\pi_c = 6.5$. So the total pressure exiting the compressor is $p_{t3} = \pi_c\, p_{t2} = 6.5 \times 129.3$ kPa $= 840.3$ kPa (121.9 psia).

Ideally, $\tau_c' = [\pi_c]^{\frac{\gamma_c - 1}{\gamma_c}} = [6.5]^{\frac{1.3882-1}{1.3882}} = 1.6879$ for $\gamma_c = 1.3882$ (iterative); therefore, $T_{t3}' = \tau_c'\, T_{t2} = 1.6879 \times 316.4$ K $= 534.1$ K (961.3 °R), but the compressor efficiency is given and $\eta_c = 0.88 = \dfrac{T_{t3}' - T_{t2}}{T_{t3} - T_{t2}} = \dfrac{534.1\text{ K} - 316.4\text{ K}}{T_{t3} - 316.4\text{ K}}$.

So, solving for T_{t3} one finds the compressor total exit temperature is $T_{t3} = 563.7$ K (1015 °R).

Primary Combustor

Next, the total pressure ratio of the burner is given as 0.95. So the burner exit total pressure is $p_{t4} = \pi_b\, p_{t3} = 0.95 \times 840.2$ kPa $= 798.2$ kPa (115.8 psia), and the energy analysis is $\Delta H \eta_b \dot{m}_f = \dot{m} c_{p_b}(T_{t4} - T_{t3}) + \dot{m}_f c_{p_b} T_{t4}$.

But $\eta_b = 91$ perent and $\Delta H = 43{,}960$ kJ/kg are given, so that solving for the fuel mass flow rate yields

$$\dot{m}_f = 13.61\frac{\text{kg}}{\text{s}} \times 1.127\frac{\text{kJ}}{\text{kg-K}} \times \frac{(1389 - 563.7)\text{ K}}{0.91 \times 43960\dfrac{\text{kJ}}{\text{kg}} - 1.127\dfrac{\text{kJ}}{\text{kg-K}} \times 1389\text{ K}}$$
$$= 0.3293\text{ kg/s }(0.7261\text{ lb}_m/\text{s}),$$

where the burner specific heat is evaluated at the average burner total temperature, which for this case is not an iteration. Obviously no afterburner is present, so this is also the total fuel flow rate ($\dot{m}_{ft}$) one should also note that $\dot{m}_f/\dot{m} = 0.02420$, which, as for all previous examples, is small.

Turbine

Now the mechanical efficiency is given as 0.995 so the shaft power balance

$$\dot{m} c_{p_c}(T_{t3} - T_{t2}) + C_{w_p} \dot{m} c_{p_a} T_a = \eta_m \dot{m}_t c_{p_t}(T_{t4} - T_{t5})$$

becomes, knowing that the work coefficient for the propeller is 0.9783,

$$13.61\frac{\text{kg}}{\text{s}} \times 1.026\frac{\text{kJ}}{\text{kg-K}}(563.7\text{ K} - 316.4\text{ K}) + 0.9783 \times 13.61\frac{\text{kg}}{\text{s}} \times 0.9992\frac{\text{kJ}}{\text{kg-K}} \times 288.2\text{ K}$$
$$= 0.995 \times \left(13.61\frac{\text{kg}}{\text{s}} + 0.3293\frac{\text{kg}}{\text{s}}\right) \times 1.164\frac{\text{kJ}}{\text{kg-K}}(1389\text{ K} - T_{t5});$$

thus, solving for the turbine exit total temperature, $T_{t5} = 937.9$ K (1688 °R).

Again, the specific heat of the turbine is iteratively found for the average turbine temperature.

But the turbine efficiency is defined as $\eta_t = \dfrac{T_{t4} - T_{t5}}{T_{t4} - T_{t5}'}$ and given as 85 percent, so

$$0.85 = \frac{1389\text{ K} - 937.9\text{ K}}{1389\text{ K} - T_{t5}'};$$

thus, solving for the ideal exit total temperature, $T_{t5}' = 858.3$ K (1545 °R), and ideally

$$\tau_t' = \frac{T_{t5}'}{T_{t4}} = \frac{858.3}{1389} = 0.6179.$$

Also, $\tau_t' = [\pi_t]^{\frac{\gamma_t - 1}{\gamma_t}}$, so that the turbine total pressure ratio is

$\pi_t = [\tau_t']^{\frac{\gamma_t}{\gamma_t - 1}} = (0.6179)^{\frac{1.3269}{1.3269-1}}$,

$= 0.1417$ for $\gamma_t = 1.3269$ (evaluated at average turbine temperature).

Finally, the turbine exit total pressure is

$p_{t5} = \pi_t p_{t4} = 0.1469 \times 798.2 \text{ kPa} = 113.2 \text{ kPa (16.42 psia)},$

and since this is a simple turboprop, $T_{t5} = 937.9$ K (1688 °R) and $p_{t5} = 113.2$ kPa (16.42 psia).

Primary Nozzle

Since no afterburner is present the inlet total temperature for the nozzle is $T_{t5} = 937.9$ K (1688 °R) and the inlet total pressure is $p_{t5} = 113.2$ kPa (16.42 psia). As the engine has a fixed converging nozzle, one must first check to see if the nozzle is choked.

For choking, the nozzle exit pressure is

$$p_8^* = p_{t5}\left[1 + \frac{1 - \gamma_n}{\eta_n(1 + \gamma_n)}\right]^{\frac{\gamma_n}{\gamma_n - 1}}.$$

Thus, for $p_{t5} = 113.2$ kPa, $\eta_n = 0.96$ (given), and $\gamma_n = 1.3447$ (iteratively found):

$$p_8^* = 113.2 \text{ psia} \times \left[1 + \frac{1 - 1.3447}{0.96 \times (1 + 1.3447)}\right]^{\frac{1.3447}{1.3447-1}} = 59.17 \text{ kPa (8.582 psia)}.$$

But since $p_a = 101.3$ kPa (14.69 psia), which is well above the choking condition, p_8^*, the nozzle is not choked and so the exit pressure matches the ambient pressure. Therefore, $p_8 = p_a = 101.3$ kPa (14.69 psia).

Now ideally

$$T_8' = T_{t5}\left[\frac{p_8}{p_{t6}}\right]^{\frac{\gamma_n - 1}{\gamma_n}} = 930.3 \text{ K}\left[\frac{101.3 \text{ kPa}}{113.2.6 \text{ kPa}}\right]^{\frac{1.3447-1}{1.3447}} = 911.7 \text{ K (1641 °R)},$$

and thus $T_8 = T_{t5} - \eta_n (T_{t5} - T_8') = 937.9 \text{ K} - 0.96 (937.9 \text{ K} - 911.7 \text{ K}) = 912.8$ K (1643 °R).

Now the nozzle exit velocity is given by $u_8 = \sqrt{2c_{p_n}(T_{t5} - T_8)}$, so that the exit velocity is

$$u_8 = \sqrt{2 \times 1.120 \frac{\text{kJ}}{\text{kg-K}} \times (937.9 \text{ K} - 912.8 \text{ K}) \times 1000 \frac{\text{J}}{\text{kJ}} \times \frac{\text{N-m}}{\text{J}} \frac{\text{kg-m}}{\text{N-s}^2}} = 237.2 \text{ mps (778.3 fps)}.$$

Next, the speed of sound at the exit is $a_8 = \sqrt{\gamma_8 \mathcal{R} T_8}$, so

$$a_8 = \sqrt{1.3447 \times 287.1 \frac{\text{J}}{\text{kg-K}} \times 912.8 \text{ K} \times \frac{\text{N-m}}{\text{J}} \times \frac{\text{kg-m}}{\text{N-s}^2}} = 593.5 \text{ mps (1947 fps)}.$$

Thus $M_8 = \frac{u_8}{a_8} = \frac{237.2}{593.5} = 0.3996$, well into the subsonic range. Next, from the continuity equation the exit area is given by $A_8 = \frac{\dot{m}_8}{\rho_8 u_8}$, where $\dot{m}_8$ is the total mass flow rate at 8 and is given by $\dot{m}_8 = \dot{m}_{a8} + \dot{m}_{ft}$, and $\dot{m}_{a8}$ is the air mass flow rate at 8 and is equal to $\dot{m}$, the ingested flow.

So $\dot{m}_{a8} = 13.61$ kg/s (30 lb$_m$/s), such that the total mass flow rate at the exit is $\dot{m}_8 = 13.61 + 0.3293 = 13.93$ kg/s (30.73 lb$_m$/s).

Since both T_8 and p_8 are known, ρ_8 is found from the ideal gas equation:

$$\rho_8 = \frac{p_8}{\mathcal{R}T_8} = \frac{101.3\text{ kPa} \times 1000\dfrac{\text{Pa}}{\text{kPa}}}{287.1\dfrac{\text{J}}{\text{kg-K}} \times 912.8\text{ K}} \times \frac{\text{N}}{\text{Pa-m}^2} \times \frac{\text{J}}{\text{N-m}} = 0.3868\text{ kg/m}^3\ (0.0007503\text{ sl/ft}^3).$$

Thus, the nozzle exit area is

$$A_8 = \frac{13.93\dfrac{\text{kg}}{\text{s}}}{0.3868\dfrac{\text{kg}}{\text{m}^3} \times 237.2\dfrac{\text{m}}{\text{s}}} = 0.1520\text{ m}^2\ (235.6\text{ in.}^2).$$

Total Thrust and TSFC

The thrust is given by

$F = \dot{m}_8 u_8 - \dot{m}_{a8} u_a + A_8(p_8 - p_a) + C_{w_p}\eta_p c_{p_a} T_a \dot{m}/u_a,$

so that

$$F = 13.93\frac{\text{kg}}{\text{s}} \times 237.2\frac{\text{m}}{\text{s}} \times \frac{\text{N-s}^2}{\text{kg-m}} - 13.61\frac{\text{kg}}{\text{s}} \times 238.4\frac{\text{m}}{\text{s}} \times \frac{\text{N-s}^2}{\text{kg-m}}$$

$$+ 0.1520\text{ m}^2 \times (101.3\text{ kPa} - 101.3\text{ kPa}) \times \frac{\text{N-m}^2}{\text{Pa}} \times \frac{1000\text{ Pa}}{\text{kPa}}$$

$$+ \frac{0.9783 \times 0.70 \times 0.9992\dfrac{\text{kJ}}{\text{kg-K}} \times 288.2\text{ K} \times 13.61\dfrac{\text{kg}}{\text{s}} \times 1000\dfrac{\text{N-m}}{\text{kJ}}}{238.3\dfrac{\text{m}}{\text{s}}},$$

where the propeller efficiency is $\eta_p = 70$ percent.

Therefore, $F = 61.3$ N (momentum) + 0 N (pressure) + 11,260 N (propeller) or $F = 11{,}320$ N (2545 lb_f).

One should note that the momentum thrust is very small.

Finally, $TSFC = \dot{m}_{ft}/F$, so that

$$TSFC = \frac{0.3293\dfrac{\text{kg}}{\text{s}}}{11320\text{ N}} \times 3600\frac{\text{s}}{\text{h}} = 0.1048\frac{\text{kg}}{\text{h-N}}\left(1.027\frac{\text{lb}_\text{m}}{\text{h-lb}_\text{f}}\right).$$

By comparing these results with those of the ideal case (Example 3.12), one finds that the thrust is reduced by 49 percent and the *TSFC* is increased by 113 percent due to the losses.

Also, a set of more general studies can once again be performed. For example, the independent effect of both the compressor pressure ratio and the nozzle parameter π_5 on the ideal work coefficient (or thrust) can be studied. Such a general study is presented in Figure 3.38 for $\gamma = 1.35$, a fixed freestream Mach

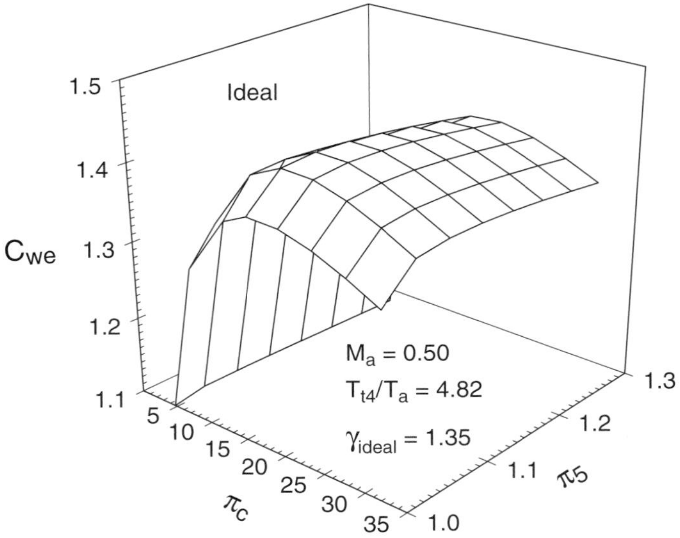

Figure 3.38 Effect of compressor total pressure ratio and nozzle parameter, π_5, on the work coefficient of an ideal turboprop.

number and temperature ratio T_{t4}/T_a. From this figure, it can be seen that optimum values of both π_c and π_5 exist if the thrust is to be maximized. That is, for a fixed value of π_5, one can optimize the compressor pressure ratio π_c or vice versa. Furthermore, by choosing optimum values of both π_5 and π_c, one can maximize the thrust. For this case, the optimum values are $\pi_{5_{opt}} \cong 1.08$, $\pi_{c_{opt}} \cong 16$, and $C_{W_e} \cong 1.428$. Similar trends are seen for nonideal cases.

3.3.3 Example Problems, Trend Studies, and Optimizations: Power Generation Gas Turbines

The next category of gas turbines to be analyzed is for power generation and not aircraft applications. Such a gas turbine receives worldwide applications. A simple power generation unit of general design is shown in Figure 3.9. The unit is very much like a turbojet engine, and one can see many similarities between this gas turbine and the aircraft engine; much of the cycle analysis is therefore identical. The gas generator portion or core of the unit is usually a derivative of a jet engine. However, an inlet bell, with a decreasing area along the flow path, replaces the diffuser, which has an increasing area, and an exhaust, in which the gas flow has a very low velocity, replaces the nozzle, which has a high gas velocity. Also, a power-extracting device is attached to the shaft. This device may be an electrical generator, ship propeller, a string of pumps, or one of several other power-extracting devices. Since the net power output is of primary concern (and not thrust), the cycle analysis is somewhat different.

A power generation gas turbine with two turbine shafts is shown in Figure 2.4 and a unit with three turbine shafts is shown in Figure 1.45a. In Figure 2.5 a diagram of a performance-enhancing power generation gas turbine is shown along with important station numbers. On this figure inlet cooling, intercooling, regeneration, and reheating are shown. Each and all improve the overall performance of such a unit. As a result, power generation units tend to be heavier than aircraft applications.

In this section, several examples are presented. These examples cover all of the important component concepts that may be encountered in nonideal analyses. Nonideal and ideal examples are compared. Trend studies are again presented to indicate that the basic trends are similar for both ideal components and nonideal components. Also, complementary or extended examples will be used for some examples to emphasize certain effects. Again, the examples and trend studies presented are an integral portion of this chapter and should not be treated as supplementary or complementary material.

Multiple Turbines

Example 3.14 Ideal Power Generation Gas Turbine: One Turbine

An ideal power generation gas turbine approximating a modern moderate-size unit operates at sea level. It ingests 66.67 kg/s (147 lb_m/s) of air. The compressor operates with a pressure ratio of 18. The fuel has a heating value of 42,800 kJ/kg (18,400 BTU/lb_m) and the burner exit total temperature is 1456 K (2620 °R). This is representative of a mid-range power generation unit such as the General Electric LM2500 (Figure 1.47). For this example, the free turbine and primary turbine are considered to be a single turbine. This can have a single shaft or two separate shafts. Namely, a portion of the turbine power drives the compressor and the net power drives the external load. Find the net output power, *SFC*, and net thermal efficiency for $\gamma = 1.35$. Station numbers are defined in Figure 2.5 and stations 5, 4.5c, and 4.5a are the same since only one turbine is present.

Solution

Inlet

First, flow is drawn into the gas turbine and accelerated to the inlet velocity from a stagnant condition far from the inlet. Because the velocity is zero far from the inlet, the ambient static and total pressures are equal ($p_{ta} = p_a = 101.3$ kPa) and the ambient static and total temperatures are equal ($T_{ta} = T_a = 288.2$ K). The process is also adiabatic, thus ideally the exit total temperature is $T_{t2} = T_{ta} = 288.2$ K (518.7 °R), and since the flow in the ideal inlet is isentropic, the exit total pressure is $p_{t2} = p_{ta} = 101.3$ kPa (14.69 psia).

Compressor

Next, the total pressure ratio of the compressor is given as $\pi_c = 18$.

The total pressure exiting the compressor is therefore

$p_{t3} = \pi_c p_{t2} = 18 \times 101.3 \text{ kPa} = 1823 \text{ kPa (264.4 psia)}$.

Ideally, the compressor total temperature ratio is

$$\tau_c = [\pi_c]^{\frac{\gamma-1}{\gamma}} = [18]^{\frac{1.35-1}{1.35}} = 2.116 \text{ for } \gamma = 1.35;$$

therefore, the compressor total exit temperature is

$$T_{t3} = \tau_c T_{t2} = 2.116 \times 288.2\text{K} = 609.7 \text{ K (1097 °R)}.$$

Combustor

The ideal burner exit pressure is $p_{t4} = p_{t3} = 1823$ kPa (264.4 psia), and the first law of thermodynamics yields

$\Delta H\, \dot{m}_f = \dot{m} c_{p_b}\, (T_{t4} - T_{t3});$

where the fuel heating value is 42,800 kJ/kg so that solving for the fuel mass flow rate yields

$$\dot{m}_f = 66.67 \frac{\text{kg}}{\text{s}} \times 1.107 \frac{\text{kJ}}{\text{kg-K}} \times \frac{(1456 - 609.7)\ \text{K}}{42800 \frac{\text{kJ}}{\text{kg}}} = 1.460\ \text{kg/s}(3.217\ \text{lb}_\text{m}/\text{s}).$$

Note that $\dot{m}_f / \dot{m} = 0.02189$ – again, small.

Exhaust

Next, the gas velocity at the exhaust exit is small, so that static and total pressures are equal ($p_{t8} = p_8$) Furthermore, since the velocity is small (subsonic), the exit and ambient pressures are the same ($p_8 = p_a$). Thus $p_{t8} = p_a = 101.3$ kPa.

So ideally (isentropic) the inlet total pressure to the exhaust is
$p_{t5} = p_{t8} = 101.3$ kPa (14.69 psia).

Single Turbine

Now, since the inlet total pressure to the exhaust is the same as the exit total pressure for the turbine, the turbine total pressure ratio is $\pi_t = p_{t5}/p_{t4} = 101.3\ \text{kPa}/1823\ \text{kPa} = 0.05556$.

Next, ideally the total temperature ratio is $\tau_t = [\pi_t]^{\frac{\gamma-1}{\gamma}} = [0.05556]^{\frac{1.35-1}{1.35}}$, so that $\tau_t = 0.4727$ for $\gamma = 1.35$.

Thus, the turbine exit total temperature is
$T_{t5} = \tau_t T_{t4} = 0.4727 \times 1456\ \text{K} = 688.2\ \text{K}\ (1238\ °\text{R})$.

Net Power and Thermal Efficiency

Now the power balance on the shaft is $\mathcal{P}_{net} = \dot{m} c_p (T_{t4} - T_{t5}) - \dot{m} c_p (T_{t3} - T_{t2})$.

The first term is the total power derived from the turbine and the second term is the power driving the compressor:

$$\mathcal{P}_{net} = \left[66.67 \frac{\text{kg}}{\text{s}} \times 1.107 \frac{\text{kJ}}{\text{kg-°R}} \times (1456 - 688.2)\text{K} - 66.67 \frac{\text{kg}}{\text{s}} \times 1.107 \frac{\text{kJ}}{\text{kg-°R}} \times (609.7 - 288.2)\text{K} \right] \times \frac{\text{kW}}{\text{kJ/s}}.$$

$\mathcal{P}_{net}$ is the net power from the gas turbine, which can be used for electric generation or other applications:
$\mathcal{P}_{net} = 32950\ \text{kW} = 32.95\ \text{MW}\ (44{,}160\ \text{hp})$.

Next, the overall thermal efficiency is $\eta_{th} = \frac{\mathcal{P}_{net}}{\mathcal{Q}_{in}}$, where the heat in is

$$\dot{Q}_{in} = \dot{m}_f \Delta H = 1.460 \frac{kg}{s} \times 42,800 \frac{kJ}{kg} \times \frac{kW}{kJ/s} = 62{,}490 \text{ kW} = 62.49 \text{ MW } (83{,}800 \text{ hp})$$

or $\eta_{th} = 32{,}930 \text{ kW}/62{,}490 \text{ kW} = 0.527$ or 52.7 percent.

Next, the heat rate is defined as $\mathcal{HR} = \dfrac{\dot{m}_f \Delta H}{\mathcal{P}_{net}}$, so

$$\mathcal{HR} = \frac{1.460 \frac{kg}{s} \times 42{,}800 \frac{kJ}{kg} \times 3600 \frac{s}{h}}{32{,}950 \text{ kW}} = 6827 \frac{kJ}{kW\text{-}h} \left(4825 \frac{BTU}{hp\text{-}h}\right),$$

and finally the specific fuel consumption is defined as $SFC = \dfrac{\dot{m}_f}{\mathcal{P}_{net}}$, so

$$SFC = \frac{1.460 \frac{kg}{s}}{32{,}950 \text{ kW}} \times 3600 \frac{s}{h} = 0.1595 \frac{kg}{kW\text{-}h} \left(0.2622 \frac{lb_m}{hp\text{-}h}\right).$$

Extended Example 3.14.1 | **Ideal Power Generation Gas Turbine: Optimum Compressor Pressure Ratio**

As a final exercise, optimize the above gas turbine with the compressor pressure ratio.

Solution

One can find the optimum operating total temperature ratio for the compressor from

$$\tau_{c_{opt}} = \sqrt{\frac{T_{t4}}{T_a}},$$

which yields $\tau_{c_{opt}} = 2.248$, which is close to the actual ratio of 2.116. This yields an optimum compressor total pressure ratio of 22.74, which is somewhat higher than the actual value. This optimum value yields a net power of 33,130 kW (44,390 hp) and an efficiency of 55.5 percent. On the basis of these characteristics, although not optimum, the gas turbine in Example 3.14 is still a fairly well-designed power unit.

Example 3.15 | **Ideal Power Generation Gas Turbine: Two Turbines**

An ideal power generation gas turbine approximating of a modern moderate size unit operates at sea level. It ingests 66.67 kg/s (147 lb_m/s) of air. The compressor operates with a pressure ratio of 18. The fuel has a heating value of 42,800 kJ/kg (18,400 BTU/lb_m) and the burner exit total temperature is 1456 K (2620 °R). The specifications are the same as for Example 3.14. However, for this case there is one separate primary turbine (LPT+HPT) and a power or free turbine. The primary turbine drives the compressor and the free turbine is for auxiliary use/power. Find the net output power, *SFC*, and net thermal efficiency for $\gamma = 1.35$. Station numbers are defined in Figure 2.5 and stations 4.5a and 4.5c are the same since only one primary turbine is present.

Solution

The solution is exactly the same as for Example 3.14 through the combustor, so these steps will not be repeated.

Turbine

All of the power to drive the compressor is from the turbine (excluding the power or free turbine):

$\dot{m}c_p(T_{t4} - T_{t45c}) = \dot{m}c_p(T_{t3} - T_{t2})$.

Thus, solving for T_{t45c} for the ideal case:

$T_{t45c} = T_{t4} - (T_{t3} - T_{t2}) = 1456 - (609.7 - 288.2) = 1135\,\mathrm{K}$ (2043 °R).

Next, for the pressure ratio of the turbine

$$\pi_t = [\tau_t]^{\frac{\gamma}{\gamma-1}} = \left[\frac{1135\,\mathrm{K}}{1456\,\mathrm{K}}\right]^{\frac{1.35}{1.35-1}} = 0.3820.$$

Thus $p_{t45c} = \pi_t p_{t4} = 0.3820 \times 1823$ kPa $= 696.5$ kPa (101.0 psi).

Exhaust

Next, the gas velocity at the exhaust exit is small, thus the static and total pressures are equal ($p_{t8} = p_8$). Furthermore, since the velocity is small (subsonic), the exit and ambient pressures are the same ($p_8 = p_a$). Thus, $p_{t8} = p_a = 101.3$ kPa.

Therefore, ideally (isentropic) the inlet total pressure to the exhaust is

$p_{t5} = p_{t8} = 101.3$ kPa (14.69 psia).

Power or Free Turbine

Because the inlet total pressure to the exhaust is the same as the exit total pressure for the free turbine, the power turbine total pressure ratio is

$\pi_{ft} = p_{t5}/p_{t45c} = 101.3$ kPa/696.5 kPa $= 0.1454$.

Ideally, the total temperature ratio is $\tau_{ft} = [\pi_{ft}]^{\frac{\gamma-1}{\gamma}} = [0.1454]^{\frac{1.35-1}{1.35}}$, and thus $\tau_{ft} = 0.6066$ for $\gamma = 1.35$. The power turbine exit total temperature is therefore

$T_{t5} = \tau_t\, T_{t45c} = 0.6066 \times 1135\,\mathrm{K} = 688.3\,\mathrm{K}$ (1239 °R).

Net Power and Thermal Efficiency

The power balance on the shaft yields

$$\mathcal{P}_{net} = \dot{m}c_p\,(T_{t45c} - T_{t5})$$
$$= \left[66.67\frac{kg}{s} \times 1.107\frac{kJ}{kg\text{-}K} \times (1135 - 688.3)\,K\right] \times \frac{kW}{kJ/s}$$

or the net power from the gas turbine which can be used for electric generation or other applications is

$\mathcal{P}_{net} = 32{,}950$ kW $= 32.95$ MW (44,160 hp).

The overall thermal efficiency is $\eta_{th} = \frac{\mathcal{P}_{net}}{\dot{Q}_{in}}$, where the heat in is

$$\dot{Q}_{in} = \dot{m}_f \Delta H = 1.460\frac{kg}{s} \times 42{,}800\frac{kJ}{kg} \times \frac{kW}{kJ/s} = 62450\,kW = 62.45\,MW\ (83830\,hp)$$

or $\eta_{th} = 32{,}950$ kW/62,450 kW $= 0.527$ or 52.7 percent.

The heat rate is defined as $\mathcal{HR} = \frac{\dot{m}_f \Delta H}{\mathcal{P}_{net}}$, so

$$\mathcal{HR} = \frac{1.460\frac{kg}{s} \times 42{,}800\frac{kJ}{kg} \times 3600\frac{s}{h}}{32{,}950\,kW} = 6823\frac{kJ}{kW\text{-}h}\left(4822\frac{BTU}{hp\text{-}h}\right),$$

and finally the specific fuel consumption is defined as $SFC = \frac{\dot{m}_f}{\mathcal{P}_{net}}$.
Thus

$$SFC = \frac{1.460\frac{kg}{s}}{32{,}950\,kW} \times 3600\frac{s}{h} = 0.1595\frac{kg}{kW\text{-}h}\left(0.2622\frac{lb_m}{hp\text{-}h}\right).$$

It should be no surprise that for the ideal case the results are exactly the same as those for Example 3.14

It also should be noted that these two ideal examples could be repeated with ideal separate low-pressure and high-pressure compressors with equal pressure ratios of $\sqrt{18} = 4.2426$, separate high-pressure and low-pressure turbines, and a separate power turbine, and the results would again be exactly the same.

Example 3.16 Nonideal Power Generation Gas Turbine: One Turbine

A power generation gas turbine approximating a modern and moderate-sized power generation unit operates at sea level. Again, this example is similar to Example 3.14 except for nonideal effects. It ingests 66.67 kg/s (147 lb_m/s) of air. The compressor operates with a pressure ratio of 18 and an efficiency of 88 percent. The fuel has a heating value of 42,800 kJ/kg (18,400 BTU/lb_m) and the burner exit total temperature is 1456 K (2620 °R). The burner has an efficiency of 96 percent and a total pressure ratio of 0.96, while the turbine has an efficiency of 91.5 percent. The exhaust has a pressure ratio of 0.93. The total pressure recovery for the inlet is 0.98 and the shaft efficiency is 98 percent. For this example, the free turbine and primary turbine are considered to be a single turbine. This can be a single shaft or two separate shafts. Namely, a portion of the turbine power drives the compressor and the net power drives the external load. Find the net output power, *SFC*, and net thermal efficiency. Compare with the ideal case, Example 3.14.

Solution

Inlet

First, flow is drawn into the gas turbine and accelerated to the inlet velocity from a stagnant condition far from the inlet. Since the velocity is zero far from the inlet, the ambient static and total pressures are equal ($p_{ta} = p_a = 101.3$ kPa) and the ambient static and total temperatures are equal ($T_{ta} = T_a = 288.2$ K). The process is also adiabatic, thus the exit total temperature is $T_{t2} = T_{ta} = 288.2$ K (518.7 °R).

Also the inlet pressure recovery factor is 0.98, so the exit total pressure of the inlet is $p_{t2} = \pi_i p_{ta} = 0.98 \times 101.3 \text{ kPa} = 99.27 \text{ kPa}$ (14.40 psia).

Compressor

Next, the total pressure ratio of the compressor is given as $\pi_c = 18$, so that the total pressure exiting the compressor is

$p_{t3} = \pi_c p_{t2} = 18 \times 99.27 \text{ kPa} = 1787 \text{ kPa}$ (259.1 psia).

Ideally, $\tau_c{}' = [\pi_c]^{\frac{\gamma_c - 1}{\gamma_c}} = [18]^{\frac{1.3836-1}{1.3836}} = 2.2285$ for $\gamma_c = 1.3836$ (iteratively found). Therefore, $T_{t3}{}' = \tau_c{}' T_{t2} = 2.2285 \times 288.2 \text{ K} = 642.2 \text{ K}$ (1156 °R), but the compressor efficiency is given and

$$\eta_c = 0.88 = \frac{T_{t3}{}' - T_{t2}}{T_{t3} - T_{t2}} = \frac{642.2 \text{ K} - 288.2 \text{ K}}{T_{t3} - 288.2 \text{ K}}$$

So solving for T_{t3} one finds the compressor total exit temperature is $T_{t3} = 690.5$ K (1243 °R).

Combustor

Next, the total pressure ratio for the burner is 0.96, so the exit pressure is $p_{t4} = \pi_b\, p_{t3} = 0.96 \times 1787 = 1715$ kPa (248.8 psia) and the first law of thermodynamics yields

$$\Delta H\ \eta_b\ \dot{m}_f = \dot{m}\ c_{p_b}\ (T_{t4} - T_{t3}) + \dot{m}_f\ c_{p_b}\ T_{t4},$$

where the burner efficiency is 96 percent and the heating value is 42,800 kJ/kg. So solving for the fuel mass flow rate yields

$$\dot{m}_f = 66.67\frac{\text{kg}}{\text{s}} \times 1.147\frac{\text{kJ}}{\text{kg-K}} \times \frac{(1456 - 690.5)\ \text{K}}{0.96 \times 42{,}800\frac{\text{kJ}}{\text{kg}} - 1.147\frac{\text{kJ}}{\text{kg-K}} \times 1456\ \text{K}}$$

$$= 1.485\ \text{kg/s}\ (3.271\ \text{lb}_m/\text{s}).$$

No reheat is present and this is also the total fuel flow rate ($\dot{m}_{ft}$). One should note that $\dot{m}_f/\dot{m} = 0.02226$ – once again, small.

Exhaust

Next, the gas velocity at the exhaust exit is very low, so that static and total pressures are equal ($p_{t8} = p_8$). Furthermore, since the velocity is small (subsonic) the exit and ambient pressures are the same ($p_8 = p_a$). So $p_{t8} = p_a = 101.3$ kPa.

Also the pressure recovery for the exhaust is specified as 0.93. So the inlet total pressure to the exhaust is $p_{t5} = p_{t8}/\pi_e = 101.3$ kPa/0.93 $= 108.9$ kPa (15.80 psia).

Single Turbine

Again, for this example the primary and free turbines are considered to be one unit. A portion of the derived power parasitically drives an external driven device, for example, a pump or electric generator. Since the inlet total pressure to the exhaust is the same as the exit total pressure for the turbine, the turbine total pressure ratio is $\pi_t = p_{t5}/p_{t4} = 108.9$ kPa/1715 kPa $= 0.06350$.

Next, ideally the total temperature ratio is $\tau_t{}' = [\pi_t]^{\frac{\gamma_t - 1}{\gamma_t}} = [0.06350]^{\frac{1.3297-1}{1.3297}}$, so that $\tau_t{}' = 0.5047$ for $\gamma_t = 1.3298$ (iterative).

Thus, $T_{t5}{}' = \tau_t{}' T_{t4} = 0.5047 \times 1456\ \text{K} = 734.8\ \text{K}\ (1322\ °\text{R})$.

But the turbine efficiency is $\eta_t = \dfrac{T_{t4} - T_{t5}}{T_{t4} - T_{t5}{}'}$ and is equal to 91.5 percent. So

$$0.915 = \frac{1456\,\text{K} - T_{t5}}{1456\ \text{K} - 734.8\,\text{K}};$$

thus, solving for the turbine exit total temperature, $T_{t5} = 796.1$ K (1433 °R).

Net Power and Thermal Efficiency

Now the power balance on the shaft is $\mathcal{P}_{net} = \eta_m \dot{m}_t c_{p_t} (T_{t4} - T_{t5}) - \dot{m} c_{p_c} (T_{t3} - T_{t2})$ or, for a mechanical efficiency of 98 percent:

$$\mathcal{P}_{net} = \left[0.98 \times (66.67 + 1.485)\frac{\text{kg}}{\text{s}} \times 1.157\frac{\text{kJ}}{\text{kg-K}} \times (1456 - 796.1)\ \text{K}\right] \times \frac{\text{kW}}{\text{kJ/s}}$$

$$- \left[66.67\frac{\text{kg}}{\text{s}} \times 1.035\frac{\text{kJ}}{\text{kg-K}} \times (690.5 - 288.2)\ \text{K}\right] \times \frac{\text{kW}}{\text{kJ/s}}.$$

The net power from the gas turbine which can be used for electric generation or other applications is
$\mathcal{P}_{net} = 23{,}240$ kW $= 23.24$ MW (31140 hp).

Next, the overall thermal efficiency is $\eta_{th} = \dfrac{\mathcal{P}_{net}}{\mathcal{Q}_{in}}$, where the heat in is

$$\dot{\mathcal{Q}}_{in} = \dot{m}_f \Delta H = 1.485\frac{\text{kg}}{\text{s}} \times 42{,}800\frac{\text{kJ}}{\text{kg}} \times \frac{\text{kW}}{\text{kJ/s}} = 63{,}560 \text{ kW} = 63.56 \text{ MW } (85180 \text{ hp})$$

or $\eta_{th} = 23{,}240$ kW/63,560 kW $= 0.366$ or 36.6 percent.

Next, the heat rate is defined as $\mathcal{HR} = \dfrac{\dot{m}_f \Delta H}{\mathcal{P}_{net}}$, so

$$\mathcal{HR} = \frac{1.485\frac{\text{kg}}{\text{s}} \times 42{,}800\frac{\text{kJ}}{\text{kg}} \times 3600\frac{\text{s}}{\text{h}}}{23{,}240 \text{ kW}} = 9845\frac{\text{kJ}}{\text{kW-h}}\left(6959\frac{\text{BTU}}{\text{hp-h}}\right).$$

Finally, the specific fuel consumption is defined as $SFC = \dfrac{\dot{m}_f}{\mathcal{P}_{net}}$, so

$$SFC = \frac{1.485\frac{\text{kg}}{\text{s}}}{23{,}240 \text{ kW}} \times 3600\frac{\text{s}}{\text{h}} = 0.2303\frac{\text{kg}}{\text{kW-h}}\left(0.378\frac{\text{lb}_m}{\text{hp-h}}\right).$$

By comparing these results to those of the ideal case (Example 3.14), one can see that the net power is decreased by 29.4 percent and the *SFC* is increased by 44.4 percent. Furthermore, the overall thermal efficiency for the ideal case was 52.7 percent compared with the current value of 36.6 percent. Obviously, the losses have significantly reduced the effectiveness of the gas turbine.

Example 3.17 | Nonideal Power Generation Gas Turbine: Two Turbines

A power generation gas turbine approximating a modern and moderate-sized power generation unit operates at sea level. This is similar to Example 3.16, but with two shafts now. It ingests 66.67 kg/s (147 lb_m/s) of air. The compressor operates with a pressure ratio of 18 and an efficiency of 88 percent. The fuel has a heating value of 42,800 kJ/kg (18,400 BTU/lb_m) and the burner exit total temperature is 1456 K (2620 °R). The burner has an efficiency of 96 percent and a total pressure ratio of 0.96. There are two sections of the turbine: the primary turbine, which drives the compressor, and the free turbine, which drives any auxiliary unit. The combined efficiency is 91.5 percent for the two turbines resulting in 89.7 percent for each of the two turbines (refer to Section 2.3.5). The exhaust has a pressure ratio of 0.93. The total pressure recovery for the inlet is 0.98 and the shaft efficiency is 98 percent. This unit has the same specifications as in Example 3.16. Again, station numbers are defined in Figure 2.5 and stations 4.5a and 4.5c are the same since only one primary turbine is present. Find the net output power, *SFC*, and net thermal efficiency.

Solution

The solution is exactly the same as in Example 3.16 through the combustor and will not be repeated here.

Primary Turbine

Starting with the primary turbine, all of the power to drive the compressor is from the turbine (excluding the power or free turbine):

$$\eta_m \dot{m}_t c_{p_t}(T_{t4} - T_{t45c}) = \dot{m} c_{p_c}(T_{t3} - T_{t2})$$

$$0.98 \times (66.67 + 1.485)\frac{\text{kg}}{\text{s}} \times 1.189\frac{\text{kJ}}{\text{kg-K}} \times (1456 - T_{t45a})\,\text{K}$$

$$= 66.67\frac{\text{kg}}{\text{s}} \times 1.035\frac{\text{kJ}}{\text{kg-K}} \times \ (690.5 - 288.2)\ \text{K}.$$

So $T_{t45c} = 1106$ K (1991 °R),

$$\eta_t = \frac{T_{t4} - T_{t45c}}{T_{t4} - T_{t45c}{}'} = \frac{1456 - 1106}{1456 - T_{t45c}{}'} = 0.897.$$

So $T_{t45c}{}' = 1066$ K (1919 °R),

$$\pi_t = [\tau_t{}']^{\frac{\gamma}{\gamma-1}} = \left[\frac{1066\ \text{K}}{1456\ \text{K}}\right]^{\frac{1.318}{1.318-1}} = 0.2749 \text{ for } \gamma_t = 1.318 \text{ (iterative)}.$$

Thus, $p_{t45c} = \pi_t \times p_{t4} = 0.2749 \times 1715$ kPa $= 471.6$ kPa (68.40 psi).

Exhaust

Next, the gas velocity at the exhaust exit is very low, so that static and total pressures are equal ($p_{t8} = p_8$). Furthermore, since the velocity is small (subsonic) the exit and ambient pressures are the same ($p_8 = p_a$). So $p_{t8} = p_a = 101.3$ kPa.

Also, the pressure recovery for the exhaust is specified as 0.93. So the inlet total pressure to the exhaust is $p_{t5} = p_{t8}/\pi_e = 101.3$ kPa/0.93 $= 108.9$ kPa (15.80 psia).

Free or Power Turbine

Now, since the inlet total pressure to the exhaust is the same as the exit total pressure for the free turbine, the turbine total pressure ratio is $\pi_t = p_{t5}/p_{t45c} = 108.9$ kPa/471.6 kPa $= 0.2310$.

Next, ideally the total temperature ratio is $\tau_t{}' = [\pi_t]^{\frac{\gamma_t-1}{\gamma_t}} = [0.2310]^{\frac{1.344-1}{1.344}}$, so that $\tau_t{}' = 0.6874$ for $\gamma_t = 1.3436$ (iterative).

Thus, $T_{t5}{}' = \tau_t{}'\, T_{t45c} = 0.6871 \times 1106\ \text{K} = 760.3\ \text{K}$ (1369 °R).

But the turbine efficiency is

$$\eta_{ft} = \frac{T_{t45c} - T_{t5}}{T_{t45c} - T_{t5}{}'}$$

and is equal to 89.7 percent, so

$$0.897 = \frac{1106\ \text{K} - T_{t5}}{1106\ \text{K} - 760.3\ \text{K}};$$

thus, solving for the turbine exit total temperature, $T_{t5} = 795.9\ \text{K}$ (1433 °R).

Net Power and Thermal Efficiency

The power balance on the free turbine shaft is

$$\mathcal{P}_{net} = \eta_m \dot{m}_{ft}\, c_p (T_{t45c} - T_{t5})$$

$$\mathcal{P}_{net} = 0.98 \times \left[(66.67 + 1.484)\frac{\text{kg}}{\text{s}} \times 1.122\frac{\text{kJ}}{\text{kg-K}} \times (1106 - 795.9)\ \text{K}\right] \times \frac{\text{kW}}{\text{kJ/s}},$$

or the net power from the gas turbine which can be used for electric generation or other applications is

$$\mathcal{P}_{net} = 23{,}250\ \text{kW} = 23.3\ \text{MW}(31180\ \text{hp}).$$

Next, the overall thermal efficiency is $\eta_{th} = \dfrac{\mathcal{P}_{net}}{\dot{Q}_{in}}$, where the heat in is

$$\dot{Q}_{in} = \dot{m}_f \Delta H = 1.484\frac{\text{kg}}{\text{s}} \times 42{,}800\frac{\text{kJ}}{\text{kg}} \times \frac{\text{kW}}{\text{kJ/s}} = 63{,}520\ \text{kW} = 63.52\ \text{MW}\ (85190\ \text{hp}).$$

Thus, $\eta_{th} = 23{,}250\ \text{kW}/63{,}520\ \text{kW} = 0.366$ or 36.6 percent.

Next, the heat rate is defined as $\mathcal{HR} = \dfrac{\dot{m}_f \Delta H}{\mathcal{P}_{net}}$, so

$$\mathcal{HR} = \frac{1.484\frac{\text{kg}}{\text{s}} \times 42800\frac{\text{kJ}}{\text{kg}} \times 3600\frac{\text{s}}{\text{h}}}{23{,}250\ \text{kW}} = 9836\frac{\text{kJ}}{\text{kW-h}}\left(6951\frac{\text{BTU}}{\text{hp-h}}\right).$$

Finally, the specific fuel consumption is defined as $SFC = \dfrac{\dot{m}_f}{\mathcal{P}_{net}}$, so

$$SFC = \frac{1.484\frac{\text{kg}}{\text{s}}}{23{,}250\ \text{kW}} \times 3600\frac{\text{s}}{\text{h}} = 0.2298\frac{\text{kg}}{\text{kW-h}}\left(0.3778\frac{\text{lb}_m}{\text{hp-h}}\right).$$

By comparing the results with those for Example 3.16, the net power for this example is 23,250 kW compared with 23,240 kW. The efficiency is 36.6 percent for both cases. The *SFC* for this case is 0.2298 kg/kW-h compared with 0.2303 kg/kW-h. Clearly there is little difference in all of the overall performance measures. The only small differences are due to the fact that for this case specific heats were evaluated for different turbine component temperatures and that the individual single-stage efficiencies for the two turbines are based on the assumption that the load or pressure drop for each turbine was the same. Regardless, for this study both analyses are valid.

Example 3.18 Nonideal Power Generation Gas Turbine: Three Turbines

A similar specification example as in the previous two examples is discussed here. For this study, two compressors (low-pressure compressor (LPC) and high-pressure compressor (HPC)) and three turbines (high-pressure turbine, low-pressure turbine and free turbine) are used for the analysis. The low-pressure turbine drives the low-pressure compressor and the high-pressure turbine drives the high-pressure compressor. The combined efficiency is 88 percent for the two compressors, resulting in 90.2 percent for each of the two compressors (refer to Section 2.3.3). The pressure ratio for the two compressors is $\sqrt{18} = 4.2426$. The combined efficiency is 91.5 percent for the three turbines resulting in 89.4 percent for each of the three turbines (refer to Section 2.3.5). Find the net output power, *SFC*, and net thermal efficiency.

Without going through the logical step-by-step process for the more complex example, only results are presented. In Example 3.19, the process of analyzing two compressors and three turbines will be covered with performance enhancers.

By comparing the results to those for Example 3.16 the net power for this example is 23,220 kW compared with 23,240 kW. The efficiency is 36.6 percent for both cases. The *SFC* for this case is 0.2298 kg/kW-h compared with 0.2303 kg/kW-h. Clearly there is little difference in all of the overall performance measures as well as comparing with Example 3.17. The only small differences are due to the fact that, for this case, specific heats were evaluated for different turbine and compressor component temperatures and that the individual single-stage efficiencies for the compressors and turbines are based on the assumption that the load pressure rations for each component was the same. Regardless, for this study all three analyses are valid.

Performance Enhancers

Example 3.19 Nonideal Power Generation Gas Turbine with Intercooling, Reheat Combustion, and Regeneration

A two-compressor and three-turbine power generation gas turbine (Figure 1.45a and Figure 2.5) operates at sea level. The unit has intercooling, regeneration, and reheat (Figure 2.5). It ingests 66.67 kg/s (147 lb_m/s) of air. The low-pressure compressor operates with a pressure ratio of 4.2426 and an efficiency of 90.2 percent. The high-pressure compressor also operates with a pressure ratio of 4.2426 and an efficiency of 90.2 percent, resulting in a total pressure ratio for both compressors of 18 and a combined efficiency of 88 percent. The fuel has a heating value of 42,800 kJ/kg (18,400 BTU/lb_m) and the primary burner and reheat burner exit total temperatures are 1456 K (2620 °R). The burners both have an efficiency of 96 percent and a total pressure ratio of 0.96. The three turbines (low pressure, high pressure, and free) have efficiencies of 89.4 percent, resulting in the total turbine efficiency of 91.5 percent. The exhaust/stack has a pressure ratio of 0.93. The total pressure recovery for the inlet is 0.98 and the shaft efficiency is 98 percent. The regenerator has an effectiveness of 75 percent and total pressure ratio of 0.99 in both directions. The intercooler has an effectiveness of 75 percent and total pressure ratio of 0.99 in both directions. The intercooler coolant flows at 22.7 kg/s (50 lb_m/s), enters at 289 K and has a specific heat of 4.2 kJ/kg-K. Inlet cooling is not used. Evaluate gas properties at the average stagnation temperatures. This is similar to the previous example, but with the added intercooler, reheat, and regenerator. Find the net output power, *SFC*, and net thermal efficiency. Compare to the simpler case in Example 3.18.

Solution

Inlet

As in the previous example, the velocity of the gas turbine is zero, so that $T_{t2} = T_{ta} = T_a =$ 288.2 K. Also, since the velocity of the gas turbine is zero $p_{ta} = p_a =$ 101.3 kPa, and the total pressure exiting the inlet is $p_{t2} = \pi_{in} \times p_{ta}$. So $p_{t2} = 0.98 \times 101.3 = 99.26$ kPa.

LP Compressor

Next, the pressure ratio for the low-pressure compressor is given as $\pi_{LPC} = 4.2426$, so that the total pressure at the exit of the low-pressure compressor is

$p_{t25a} = \pi_{LPC} \times p_{t2} = 4.2426 \times 99.3 = 421.1$ kPa.

Ideally, $\tau'_{LPC} = [\pi_{LPC}]^{\frac{\gamma_{LPC}-1}{\gamma_{LPC}}} = [4.2426]^{\frac{1.395-1}{1.395}} = 1.505$ for $\gamma_{LPC} = 1.395$. Note that γ_{LPC} is based on the average temperature of 369.0 K (iteratively found, typical).

Therefore, $T'_{t25a} = \tau'_{LPC} \times T_{t2} = 433.9$ *K* but the compressor efficiency is given and is

$$\eta_{LPC} = 0.902 = \frac{T_{t25a}' - T_{t2}}{T_{t25a} - T_{t2}} = \frac{433.9\text{ K} - 288.2\text{ K}}{T_{t25a} - 288.2\text{ K}}$$

Solving for the LPC exit total temperature yields $T_{t25a} = 449.7$ K.

Intercooler

The effectiveness of the intercooler, ε_{IC}, is given as 0.75, so that

$T_{t25b} = \varepsilon_{IC} \times (T_{t25c} - T_{t25a}) + T_{t25a}$.

The conditions of the coolant are given ($T_{t25c} = 289$ K) so the exit gas temperature and inlet temperature to the second compressor (HPC) is

$T_{t25b} = 0.75 \times (289 - 449.7) + 449.7 = 329.2$ K.

The exit coolant temperature is given by

$$T_{t25d} = T_{t25c} - (T_{t25a} - T_{t25b}) \times \frac{c_{p25ab} \times \dot{m}}{c_{pcoolant} \times \dot{m}_{coolant}}$$

$$= 289\text{ K} - (449.7\text{ K} - 329.2\text{ K}) \times \frac{1.0172 \times 66.67}{4.2 \times 22.7}$$

$$= 374.7\text{ K}.$$

One should note that this temperature has no bearing on the power generation.

Note that c_{p25} for the gas is based on the average static temperature of 389.4 K (iteratively found).

The pressure ratio for the inetercooler is given as, $\pi_{25ab} = 0.99$, so that entrance pressure to the second compressor (HPC) is

$$p_{t25b} = \pi_{25ab} \times p_{t25a}$$
$$= 0.99 \times 421.1 = 416.9 \text{ kPa}.$$

HP Compressor

Next, the pressure ratio for the high-pressure compressor is given as $\pi_{HPC} = 4.2426$ so that the total pressure at the exit is

$p_{t3} = \pi_{HPC} \times p_{t25b} = 4.2426 \times 416.9 = 1769$ kPa.

Ideally, $\tau'_{HPC} = [\pi_{HPC}]^{\frac{\gamma_{HPC}-1}{\gamma_{HPC}}} = [4.2426]^{\frac{1.390-1}{1.390}} = 1.500$ for $\gamma_{HPC} = 1.390$.

Note that γ_{HPC} is based on the average temperature of 420.4 K.

Therefore, $T'_{t3}{=}\pi'_{HPC} \times T_{t25b}{=}$ 493.8 K.

But the efficiency is

$$\eta_{HPC} = 0.902 = \frac{T_{t3}' - T_{t25b}}{T_{t3} - T_{t25b}} = \frac{493.8 \text{ K} - 329.2 \text{ K}}{T_{t3} - 329.2 \text{ K}}.$$

So, solving for the exit total temperature, T_{t3}, yields $T_{t3} = 511.7$ K.

Regenerator

To increase the temperature into the first turbine and thus reduce the needed fuel, a regenerator is used. The effectiveness is given as $\varepsilon_{RG} = 0.75$, so that

$T_{t3x} = \varepsilon_{RG} \times (T_{t5} - T_{t3}) + T_{t3}$.

But the temperature exiting the free turbine, T_{t5}, is unknown. So one must iterate on T_{t5}.

The first guess is $T_{t5} = 915.8$ K, which will be checked later after the free turbine calculations.

Now $T_{t3x} = 0.75 \times (915.8 \text{ K} - 511.7 \text{ K}) + 511.7 \text{ K} = 814.8$ K.

The exit temperature of the regenerator and entrance to the exhaust and exit stack is

$$T_{t5x} = T_{t5} - (T_{t3} - T_{t3x})\frac{c_{pRG3}}{c_{pRG5}} = 915.8 \text{ K} - (511.7 \text{ K} - 814.8 \text{ K}) \times \frac{1.0671}{1.0872}$$
$$= 623.8 \text{ K}.$$

NOTE. *It is very important to note here that T_{t5} is greater than T_{t3}. Had this not been the case a regenerator should or would not have been used.* This was discussed in Section 1.4.2.

Note that c_{pRG3} is based on the average total temperature of 663.2 K and c_{pRG5} is based on the estimated average temperature of 769.8 K.

The pressure ratio for the regenerator is given as, $\pi_{3-3x} = 0.99$, so that entrance pressure to first turbine is

$$p_{t3x} = \pi_{3-3x} \times p_{t3}$$
$$= 0.99 \times 1769 = 1751 \text{ kPa}.$$

Primary Combustor or Burner

The pressure ratio for the primary burner is given as, $\pi_{PB} = 0.96$, so that the entrance pressure to the high-pressure turbine is

$p_{t4} = \pi_{PB} \times p_{t3x} = 0.96 \times 1751 = 1681$ kPa.

The first law of thermodynamics yields

$\Delta H\ \eta_{PB}\ \dot{m}_{fPB} = \dot{m}c_{pPB}(T_{t4} - T_{t3x}) + \dot{m}_{fPB}\ c_{pPB}\ T_{t4},$

where the burner efficiency is 96 percent and the heating value is 42,800 kJ/kg. So solving for the primary fuel mass flow rate yields

$$\dot{m}_{fPB} = 66.67\frac{\text{kg}}{\text{s}} \times 1.159\frac{\text{kJ}}{\text{kg-K}} \times \frac{(1456 - 814.8)\ \text{K}}{0.96 \times 42800\frac{\text{kJ}}{\text{kg}} - 1.159\frac{\text{kJ}}{\text{kg-K}} \times 1456\ \text{K}} = 1.257\ \text{kg/s}.$$

Note that c_{pPB} is based on the average temperature of 1135 K. One should also note that $\dot{m}_{fPB}/\dot{m} = 0.01885$, which is small.

HP Turbine

Next, the high-pressure turbine is driving the high-pressure compressor so an energy balance on the shaft yields:

$$\dot{m}c_{pHPC}(T_{t3} - T_{t25b}) = \eta_m c_{pHPT}(\dot{m} + \dot{m}_{fPB})(T_{t4} - T_{t45a})$$

$$T_{t45a} = T_{t4} - \frac{\dot{m}}{\dot{m} + \dot{m}_{fPB}}\frac{c_{pHPC}}{c_{pHPT}}\frac{(T_{t3} - T_{t25b})}{\eta_{mech}},$$

$$\text{so } T_{t45a} = 1456\ \text{K} - \frac{66.67}{(66.67 + 1.257)}\frac{1.0227}{1.2094}\frac{182.5\ \text{K}}{0.98},$$

and so $T_{t45a} = 1301$ K.

Also, the high-pressure turbine efficiency is

$$\eta_{HPT} = \frac{T_{t4} - T_{t45a}}{T_{t4} - T_{t45a}{}'}$$

and the efficiency is given as 89.4 percent.

$$\text{So } T_{t45a}{}' = T_{t4} - \frac{T_{t4} - T_{t45a}}{\eta_{HPT}} = 1456\ \text{K} - \frac{154.5\ \text{K}}{0.894}$$

and thus $T_{t45a}{}' = 1283$ K.

Note that γ_{HPT} is based on the average temperature of 1378 K.

So the total pressure ratio for the HP turbine is

$$\pi_{HPT} = \left(\frac{T_{t45a}{}'}{T_{t4}}\right)^{\frac{\gamma_{HPT}}{\gamma_{HPT}-1}} = \left(\frac{1283}{1456}\right)^{\frac{1.3111}{0.3111}} = 0.5869,$$

and the exit total pressure is $p_{t45a} = \pi_{HPT} \times p_{t4} = 0.5869 \times 1681$,

$p_{t45a} = 986.6$ kPa.

Reheat Combustor

The pressure ratio, π_{RHB}, for the reheat burner is 0.96 so that the exit pressure is

$p_{t45b} = \pi_{RHB} \times p_{t45a} = 0.96 \times 986.6 = 947.2$ kPa,

and the first law of thermodynamics yields

$$\Delta H\,\eta_{RHB}\,\dot{m}_{fRHB} = (\dot{m} + \dot{m}_{fPB})\,c_{pRHB}\,(T_{t45b} - T_{t45a}) + \dot{m}_{fRHB}\,c_{pRHB}\,T_{t45b}.$$

Note that c_{pRHB} is based on the average temperature of 1378 K, so that the fuel flow rate for the reheat burner is

$$\dot{m}_{fRHB} = 67.93\frac{\text{kg}}{\text{s}} \times 1.2094\frac{\text{kJ}}{\text{kg-K}} \times \frac{(1456 - 1301)\ \text{K}}{0.96 \times 42{,}800\frac{\text{kJ}}{\text{kg}} - 1.2094\frac{\text{kJ}}{\text{kg-K}} \times 1456\ \text{K}} = 0.3229.$$

One should note that $\dot{m}_{fRHB}/\dot{m} = 0.00484$, and one should also note that

$$\frac{\dot{m}_{fTOTAL}}{\dot{m}} = \frac{1.257 + 0.3229}{66.67} = 0.02370.$$

LP Turbine

Next, the low-pressure turbine is driving the low-pressure compressor, so an energy balance on the shaft yields

$$\dot{m}c_{pLPC}(T_{t25a} - T_{t2}) = \eta_m c_{pLPT}(\dot{m} + \dot{m}_{fPB} + \dot{m}_{fRHB})(T_{t45b} - T_{t45c})$$

$$T_{t45c} = T_{t45b} - \frac{\dot{m}}{\dot{m} + \dot{m}_{fPB} + \dot{m}_{fRHB}} \times \frac{c_{pLPC}}{c_{pLPT}} \times \frac{(T_{t25a} - T_{t2})}{\eta_m},$$

so the exit total temperature for the LP turbine is

$$T_{t45c} = 1456\ \text{K} - \frac{66.67}{68.25} \times \frac{1.0135}{1.2115} \times \frac{161.5\ \text{K}}{0.98} = 1321\text{K}.$$

Also, the efficiency is

$$\eta_{LPT} = \frac{T_{t45b} - T_{t45c}}{T_{t45b} - T_{t45c}'}$$

and the efficiency is given as 89.4 percent,

so

$$T_{t45c}' = T_{t45b} - \frac{T_{t45b} - T_{t5}}{\eta_{LPT}} = 1456\ \text{K} - \frac{134.7\ \text{K}}{0.894}$$

and thus $T_{t45c}' = 1305$ K.

Note that γ_{LPT} is based on the average temperature of 1388 K.

The pressure ratio for the LP turbine is

$$\pi_{LPT} = \left(\frac{T_{t45c}'}{T_{t45b}}\right)^{\frac{\gamma_{LPT}}{\gamma_{LPT}-1}} = \left(\frac{1305}{1456}\right)^{\frac{1.3104}{0.3104}} = 0.6305,$$

so the exit total pressure is

$$p_{t45c} = \pi_{LPT} \times p_{t45b} = 0.6305 \times 947.2 = 597.1\ \text{kPa}.$$

Exhaust

Now, for the exhaust/stack and regenerator, and realizing that the exit of the stack is at atmospheric pressure, the exit pressure from the free turbine can be found:

$$p_{t5} = \frac{p_{ta}}{\pi_{EXH5x-5a}\ \pi_{REG5-5x}} = \frac{101.3\ \text{kPa}}{0.93 \times 0.99} = 110.0\ \text{kPa}.$$

Free or Power Turbine

Now the net output from the free turbine can be found.

First the pressure ratio for the free turbine is

$$\pi_{FT} = \frac{p_{t5}}{p_{t45c}} = \frac{110.0\ \text{kPa}}{597.1\ \text{kPa}} = 0.1842.$$

Ideally, $\tau'_{FT} = \pi_{FT}^{\frac{\gamma_{FT}-1}{\gamma_{FT}}} = 0.1842^{\frac{0.3304}{1.3304}}$, so $\tau'_{FT} = 0.6570$ for $\gamma_{FT} = 1.3304$.

Note that γ_{FT} is based on the average static temperature of 1118 K. Thus, $T'_{t5} = 867.8$ K but the efficiency is

$$\eta_{FT} = \frac{T_{t45c} - T_{t5}}{T_{t45c} - T'_{t5}} = \frac{1321\ \text{K} - T_{t5}}{1321\ \text{K} - 867.8\ \text{K}} = 0.894,$$

and thus $T_{t5} = 915.8$ K.

One should note that this value was estimated and used for the regenerator. Thus, no iteration is needed. WOW! The author made a spectacular estimate!

Net Power and Thermal Efficiency

Finally, using a power balance on the shaft

$$\mathcal{P}_{net} = \eta_m \dot{m}_{HPT} c_{pHPT}(T_{t4} - T_{t45a}) + \eta_m \dot{m}_{LPT} c_{pLPT}(T_{t45b} - T_{t45c}) + \eta_m \dot{m}_{FT} c_{pFT}(T_{t45c} - T_{t5})$$
$$- \dot{m}_{LPC} c_{pLPC}(T_{t25a} - T_{t2}) - \dot{m}_{HPC} c_{pHPC}(T_{t3} - T_{t25b}).$$

But from the calculations above one can realize that the power derived from the two compressors exactly drives the low-pressure and high-pressure turbines. So that the only net power derived is from the free turbine:

$$\mathcal{P}_{net} = \eta_m \dot{m}_{FT} c_{pFT}(T_{t5} - T_{t6})$$

$$= 0.98 \left[(66.67 + 1.2569 + 0.3229) \frac{\text{kg}}{\text{s}} \times 1.1556 \frac{\text{kJ}}{\text{kg-K}} \times (1321 - 915.8)\ \text{K} \right] \frac{\text{kW}}{\text{kJ/s}}$$
$$= 31{,}320\ \text{kW}\ (31.32\ \text{MW}).$$

The total fuel flow rate is $\dot{m}_{fPB} + \dot{m}_{fRH} = 1.2569 + 0.3229 = 1.5799$ kg/s.

The overall thermal efficiency is

$$\eta_{th} = \frac{\mathcal{P}_{net}}{\dot{Q}_{in}} = \frac{\mathcal{P}_{net}}{\dot{m}_{fTOT} \Delta H}$$

$$= \frac{\mathcal{P}_{net}}{\dot{Q}_{in}} = \frac{31{,}320\ \text{kW}}{1.5799 \frac{\text{kg}}{\text{s}} \times 42{,}800 \frac{\text{kJ}}{\text{kg}} \times \frac{\text{kW}}{\text{kJ/s}}} = 0.4631.$$

Next, the heat rate is defined: $\mathcal{HR} = \dfrac{\dot{m}_{\text{fTOT}} \times \Delta H}{\mathcal{P}_{\text{net}}}$, so

$$\mathcal{HR} = \frac{1.5799\dfrac{\text{kg}}{\text{s}} \times 42800\dfrac{\text{kJ}}{\text{kg}} \times 3600\dfrac{\text{s}}{\text{h}}}{31{,}320\ \text{kW}} = 7775\frac{\text{kJ}}{\text{kW-h}}.$$

Finally, the specific fuel consumption is $SFC = \dfrac{\dot{m}_{\text{fTOT}}}{P_{\text{net}}}$, so

$$SFC = \frac{1.5799\dfrac{\text{kg}}{\text{s}}}{31{,}320\ \text{kW}} \times 3600\frac{\text{s}}{\text{h}} = 0.1816\frac{\text{kg}}{\text{kW-h}}.$$

Comparison

In the previous example (Example 3.18), for the much simpler power generation unit (two compressors and three turbines without any performance improvements) the net power was 23,220 kW (now it is 31,320 kW). The overall efficiency was 36.6 percent (now it is 46.3 percent). The *SFC* was 0.2298 kg/kW-h (now it is 0.1816 kg/kW-h). All of these performance measures result in significant improvements. In the following table, the individual performance improvements for the intercooler, regenerator, and reheat are identified. The regenerator actually decreases the power output but improves both the efficiency and *SFC*. The intercooler increases the power output but lowers the efficiency. And the reheat increases the power output but lowers the efficiency and increases the *SFC*. From this table it is clear why combinations of the performance enhancers are often used instead of using only one.

	Simple	Regenerator	Intercooler	Reheat	All
Net power (kW)	23,220	22,860	27,680	28,850	31,320
Efficiency (%)	36.6	40.0	35.9	35.1	46.3
SFC (kg/kW-h)	0.230	0.210	0.234	0.240	0.182

A typical trend study is presented in Figure 3.39 for a temperature ratio, T_{t4}/T_a, of 5.78. In this figure, the net nondimensional power and efficiency are plotted as functions of the compressor total pressure ratio. For this study a single compressor and single turbine are considered. As predicted by Eq. (3.2.30), the ideal net power is maximized at a total pressure ratio value of 29.47. The nonideal pressure ratio is about 21.4. Very importantly, however, note that the thermal efficiency continues to increase as the compressor total pressure ratio increases past the optimum value based on power.

Another trend study is presented in Figure 3.40 for the same basis as Figure 3.39. Here the power and

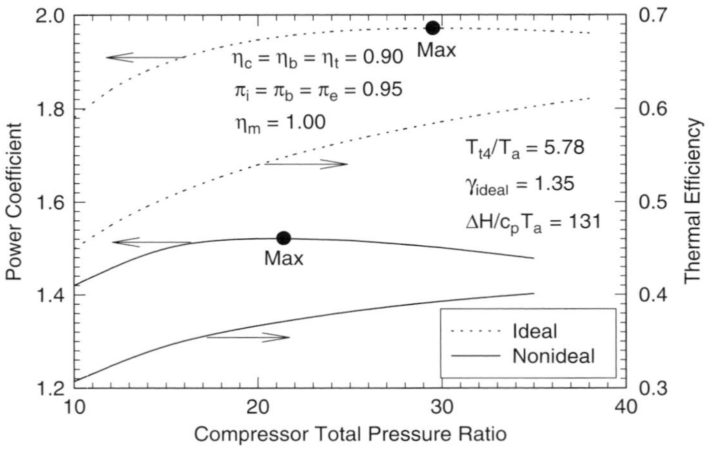

Figure 3.39 Effect of compressor total pressure ratio on net power and thermal efficiency of a power generation gas turbine.

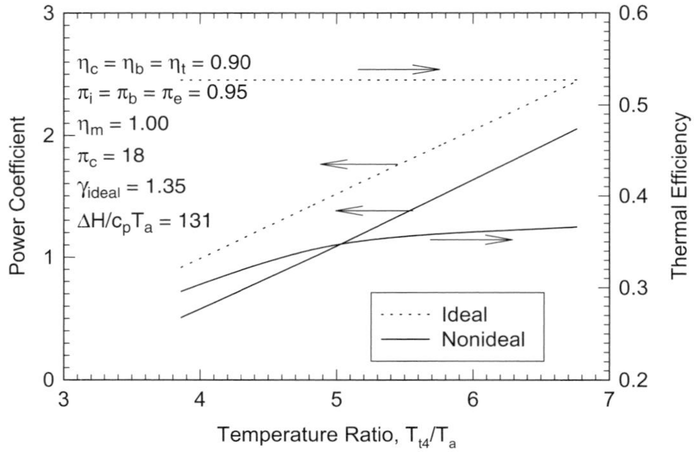

Figure 3.40 Effect of turbine inlet temperature on net power and thermal efficiency of a power generation gas turbine.

overall efficiency are shown to vary with the turbine inlet temperature. Ideally, the efficiency does not vary with T_{t4}. However, nonideally the efficiency increases with the temperature especially for the lower temperatures. This is due to the variations in specific heats. For both ideal and nonideal cases, the output power increases.

3.4 Use of Cycle Analysis in Preliminary Design

Thus far in this chapter, the concept of cycle analysis has been presented and used in examples as a method of directly analyzing engines with given parameters. In fact, the method in practice can be used to perform a preliminary design analysis of an advanced engine or power generation gas turbine or a unit that has just hit the drawing boards. For example, for a jet engine, a set of overall engine design conditions is first given to an engine manufacturer by a military group or commercial enterprise including, but not exclusively, thrusts, thrust-to-weight ratios, and *TSFC*s at different altitudes and Mach numbers. A group of system design engineers then uses an advanced cycle analysis to vary the engine types and component parameters parametrically to accomplish the overall design goals and to ensure that all of the components fit together. For example, the derived turbine power must match the compressor power requirements. The final sets of component parameters determined by this group then become the design goals of the more focused or specific component design groups – for example, compressor design or turbine design groups. To accomplish these component goals, detailed analyses and designs of the components are undertaken using methods presented in the next eight chapters. Such design methodology is a part of an "inverse" system design; that is, one starts with the overall design goals and works backward to determine the "inputs" or component characteristics that accomplish the goals. This is the first step in an aircraft engine or industrial gas turbine design. The software JetEngineCycle.exe or PowerGTCycle.exe are particularly useful in parametrically varying the design parameters without manually calculating step by step or writing independent software.

3.5 Summary

In this chapter the ideal and nonideal gas turbine cycle analyses were presented to facilitate calculation of the overall gas turbine performance. Ideal and nonideal components (inlet, diffuser, compressor, fan, propeller, combustor, turbine, bypass duct, bypass mixer, afterburner, nozzle, power gas turbine exhaust, regeneration, intercooling, and reheat) were first discussed in Chapter 2. In this chapter the components were integrated to find the overall performance of different gas turbines. For the nonideal components and for the compressor and turbine, two different efficiencies were defined. For the duct and mixer flow, stability was addressed. Simple models of losses for the components were chosen so that cycle analyses could be easily accomplished but would still enable the reader to make realistic predictions of engine performance. In addition, nonideal gas effects were included with the addition of variable specific heats for the different components.

Closed-form ideal cycle performance analyses were summarized for different gas turbine types: ramjet/scramjet, turbojet, turbofan with exhausted fan, turbofan with mixed fan, turboprop, and power generation gas turbine. Details of the derivations are included in Appendix H. A component station numbering identification scheme was also established that is used throughout the book. Ideal component equations were used with equations from Chapter 1 to determine overall performance. Ideally the specific heat ratio is constant through a gas turbine. Afterburners were included for the turbojet and turbofan. All of the components were assumed to operate without any losses, and the gas was assumed to behave ideally and with constant properties throughout the engine. Three performance parameters were demonstrated as being important comparative measures of a propulsive engine: thrust, nondimensional thrust, and *TSFC*. For the power generation units net power, *SFC*, and thermodynamic efficiency or heat rate are important measures of performance. From these equations the reader is able to observe the dependence of the overall performance of a unit on detail parameters conveniently and directly without having to resort to numerical parametric studies. For some cases the approach also allowed analytical optimization of performance characteristics.

In this chapter, components were assembled to determine the nonideal cycle performance of different gas turbine types. For nonideal cases, closed-form performance equations are not possible as they are for the ideal cases. Numerous quantitative jet engine and power gas turbine examples were presented in which the solutions were attained by stepping through the engines component by component both ideally and nonideally. All nonideal cases were iterative in that specific heats were not known a priori. Also, one of the problems (mixed turbofan) was iterative because the fan total pressure ratio was not known a priori. Another gas turbine (power generation with a regenerator) was iterative because the exit temperature from the turbine was unknown a priori. Although the examples required iteration methods, such iterations should not mask the purpose of the examples and should not dominate any problem solutions. Reasonable estimates can be made for any of the component properties for trend analyses and preliminary designs. The software JetEngineCycle.exe and PowerGTCycle.exe perform the above calculations and are particularly helpful when optimizing performance because they eliminate repetitive and detailed calculations.

Typical trend studies were also presented for different jet engine and power generation types. Also, some component variations were performed to identify the corresponding effects on engine performance. For example, some important trend studies showed that the thrust can be maximized for a turbojet or turbofan at fixed given conditions with a prudent choice of compressor pressure ratio, that decreasing the inlet temperature of the incoming air improves the *TSFC*, and that increasing the flight Mach number increases the *TSFC*. The trend studies also generally hold for nonideal cases. For a power generation unit, the output power was maximized for a particular value of compressor pressure ratio, but it was found that the overall thermal efficiency monotonically increased with increasing compressor pressure ratio.

Along with the trend studies, different engine types were compared. For example, a ramjet was shown to operate with a lower *TSFC* than a turbojet or a turbofan at very high Mach numbers. Also, afterburning in an engine was shown to increase the thrust significantly; however, the *TSFC* was also significantly increased. Thus, afterburning is good for needed short-term, sudden thrust increases, but degrades fuel economy. Turbofans were shown to operate with lower *TSFC*s than turbojets and better fuel economy; however, for the same thrust as a turbojet, the air mass flow rate is larger for a turbofan, thus increasing the required physical engine diameter or size (or weight). Also, as the bypass ratio of the turbofan was increased, the *TSFC* decreased but the engine size, including weight, increased for a given thrust level. For these reasons, military fighter craft are outfitted with low-bypass-ratio engines and commercial craft are designed with large-bypass-ratio engines. It was demonstrated that, ideally, both types of turbofans (exhausted and mixed) operate identically at the optimum conditions. If an afterburner is used with a turbofan, a mixed turbofan is usually utilized so that more airflow is available for afterburning than for an exhausted turbofan, thus increasing the possible thrust boost. The *TSFC* of a turboprop for a relatively slow aircraft is lower than for any of the other engines.

The objective of the quantitative examples was to step through each representative problem so that the individual steps could be understood and appreciated. This also gives the reader a "feel" for some of the different internal parameter values (pressures and temperatures) usually seen in gas turbines. For all jet engine cases, the nonideal thrusts were significantly lower than for the corresponding ideal cases. Inversely, the values of *TSFC* were significantly higher than in the ideal cases. Similarly, for the power generation units, the net power and efficiency of the nonideal cases were lower than the ideal values. However, the *SFC* was higher.

For the power generation studies three performance-enhancing components were included. Intercooling was shown to increase the power but decrease the efficiency. Regeneration was shown to decrease the power but increase the efficiency. Reheat was shown to increase the power but decrease the efficiency. When all three are used the power is increased and the efficiency is also increased.

List of Symbols

A	Area
a	Speed of sound
c_p	Specific heat at constant pressure
C_{w_p}, C_T, C_p C_{w_e}	Coefficients for propeller
f	Fuel ratio
F	Thrust
h	Specific enthalpy
ΔH	Heating value
$\mathcal{HR}$	Heat rate
$\dot{m}$	Mass flow rate
M	Mach number
p	Pressure
$\mathcal{P}$	Power
$\dot{Q}$	Heat transfer rate
$\mathcal{R}$	Ideal gas constant
s	Entropy
SFC	Specific fuel consumption
T	Temperature
$TSFC$	Thrust specific fuel consumption
u	Velocity
α	Bypass ratio
γ	Specific heat ratio
ε	Heat exchanger effectiveness
η	Efficiency
π	Total pressure ratio
ρ	Density
σ	Split ratio
τ	Total temperature ratio

Subscripts

a	Freestream
a	Due to air only
ab	Afterburner
b	Primary burner
c	Compressor
coolant	Coolant for intercooler
d	Diffuser
e	Exhaust
e	exit
f	Fan
f	Fuel
fn	Fan nozzle
FT	Free or power turbine
HPC	High-pressure compressor
HPT	High-pressure turbine
i	Inlet
in	Diffuser/inlet frontal plane
IC	Intercooler
LPC	Low-pressure compressor
LPT	Low-pressure turbine
m	Mechanical (shaft)
m	Mixer
mc	Mixer – core side
mu	Mixer – duct side
n	Nozzle
net	Net
opt	Optimum
p	Propeller
p	Polytropic
pc	Polytropic – compressor
pt	Polytropic – turbine
PB	Primary burner
RG	Regenerator
RH	Reheat
RHB	Reheat burner
s	Secondary (bypassed)
st	Stoichiometric
t	Total (stagnation)
t, TOT	Total (summation)
t	Turbine
T	Thrust
th	Overall or thermal
u	Duct
w	Work
1 ... 9	Positions in components

Superscripts

$'$	Ideal
*	Choked

Problems

3.1 The Mach number of a ramjet is 2.88, the external temperature is 222.2 K (400 °R), and the external pressure is 26.2 kPa (3.80 psia). The flow rate of air through the engine is 38.6 kg/s (85 lb_m/s). Because of the thermal limits of the materials, the burner exit total temperature is 1750 K (3150 °R). The heating value of the fuel is 41,640 kJ/kg (17,900 BTU/lb_m). The efficiencies of the burner and converging nozzle are 89 and 93 percent, respectively. The total pressure ratios for the diffuser and burner are 0.94 and 0.93, respectively. What are the developed thrust, fuel ratio and *TSFC* for this engine? Repeat for an ideal case with $\gamma = 1.40$.

3.2 An ideal ramjet engine is being designed for a Mach 3.2 aircraft at an altitude of 33,000 ft. The fuel has a heating value of 18,600 BTU/lb_m, and the burner exit total temperature is 3400 °R. A thrust of 9500 lb_f is needed. What is the required airflow? What is the resulting nozzle exit diameter? What is the resulting *TSFC* and dimensionless thrust? Assume $\gamma = 1.40$.

3.3 An ideal ramjet is to fly at 20,000 ft at a yet-to-be determined Mach number. The burner exit total temperature is to be 3200 °R and the engine will use 145 lb_m/s of air. The heating value of the fuel is 18,500 BTU/lb_m. At what Mach number will the *TSFC* be optimized? What is the optimum *TSFC*? What is the thrust and dimensionless thrust at this condition? Assume $\gamma = 1.40$.

3.4 An ideal ramjet is to fly stoichiometrically at 20,000 ft. The fuel ratio is 0.050 for the stoichiometric combustion of this fuel. The heating value is 18,500 BTU/lb_m. At what Mach number will the *TSFC* be optimized? What is the optimum *TSFC*? Assume $\gamma = 1.40$.

3.5 The airflow through an ideal ramjet is 156 lb_m/s. The aircraft flies at a Mach number of 2.5 at an altitude of 35,000 ft. The heating value of the fuel is 17,900 BTU/lb_m. Study how the thrust varies with the burner exit total temperature. That is, plot the thrust vs. T_{t4}. Also, plot the fuel ratio and *TSFC* vs. T_{t4}. Assume $\gamma = 1.40$.

3.6 A turbojet ingests 87.0 kg/s (192 lb_m/s) of air at an altitude of 6700 m (22,000 ft). It flies at a Mach number of 0.88, and the compressor pressure ratio is 17. The fuel has a heating value of 41640 kJ/kg (17,900 BTU/lb_m), and the burner exit total temperature is 1306 K (2350 °R). The efficiencies of the compressor, burner, turbine, shaft, and converging nozzle are 89, 95, 87, 99.7, and 97 percent, respectively. The total pressure ratios for the diffuser and burner are 0.93 and 0.93, respectively. Find the developed thrust and *TSFC*. Repeat for an ideal case with $\gamma = 1.40$.

3.7 An ideal turbojet powers a Mach 0.92 aircraft that requires 15,500 lbf of thrust at sea level. The compressor ratio is 16.4, and the exit total temperature from the combustor is 2450 °R. The heating value of the fuel is 18,100 BTU/lb_m.

Find the required airflow. How large will the nozzle exit diameter be? What is the resulting *TSFC* and dimensionless thrust? Assume $\gamma = 1.40$.

3.8 The Mach number of an ideal turbojet that flies at 27,000 ft is 0.82. The airflow is 205 lb_m/s, and the burner total temperature is 2550 °R. The heating value of the fuel is 17,700 BTU/lb_m. If the thrust is to be maximized, find the resulting compressor ratio and *TSFC*. How sensitive is the thrust to this pressure ratio; that is, if the pressure ratio is increased by 4 percent, how much will the thrust drop? How much will the *TSFC* drop? Comment. Assume $\gamma = 1.40$.

3.9 A turbojet with an afterburner flies at 6700 m (22,000 ft) with a Mach number of 0.88. It ingests 87.0 kg/s (192 lb_m/s) of air, and the compressor pressure ratio is 17. The fuel has a heating value of 41,460 kJ/kg (17,900 BTU/lb_m). The primary burner and afterburner exit total temperatures are 1306 K (2350 °R) and 1656 K (2980 °R), respectively. The efficiencies of the compressor, burner, turbine, shaft, and converging nozzle are 89, 95, 87, 99.7, and 97 percent, respectively. The total pressure ratios for the diffuser and burner are 0.93 and 0.93, respectively. The afterburner has an efficiency of 91 percent and a total pressure ratio of 0.97. (a) Find the developed thrust and *TSFC*. (b) Compare these values to those that would be obtained for the same engine without an afterburner. (c) Repeat for an ideal case with $\gamma = 1.40$.

3.10 An ideal turbojet with an afterburner is being designed for an aircraft that flies at sea level at Mach 0.92. The compressor ratio is 16.4. With the afterburner unlit, 15,500 lb_f of thrust are needed. The total temperature at the turbine inlet is 2450 °R. The heating value of the fuel is 18,100 BTU/lb_m. With the afterburner lit, 21,000 lb_f of thrust are needed. What is the required air mass flow? What is the required afterburner total exit temperature? What are the *TSFC* and dimensionless thrust for the engine with the afterburner on and off? Assume $\gamma = 1.40$.

3.11 A turbofan with the fan fully exhausted operates at 6096 m (20,000 ft) at Mach 0.82. The compressor pressure ratio is 16, and the fan pressure ratio is 2.2. The fuel has a heating value of 41,170 kJ/kg (17,700 BTU/lb_m), and the exit total temperature from the burner is 1361 K (2450 °R). The core airflow is 65.3 kg/s (144 lb_m/s), and the bypass ratio is 1.4. The compressor, burner, turbine, fan, shaft, primary variable CD nozzle (to match the exit pressure to the ambient pressure), and fan converging nozzle have efficiencies of 89, 94, 87, 88, 99.4, 97, and 95 percent, respectively. The diffuser and burner have total pressure ratios of 0.94 and 0.92, respectively. Find the developed thrust and the *TSFC*. Repeat for an ideal case with $\gamma = 1.40$.

3.12 An ideal turbofan with an exhausted fan is being designed to operate at 25,000 ft and at a Mach number of 0.78. It is to operate at maximum or optimum thrust. The compressor pressure ratio is to be 14.5. The developed thrust is to be 16,800 lb_f, and the bypass ratio is to be 4.4. The fuel has a heating value of 18,100 BTU/lb_m, and the exit total temperature from the combustor is to be 2600 °R. What is the fan pressure ratio? What is the core flow rate? What are the *TSFC* and dimensionless thrust? Assume $\gamma = 1.40$.

3.13 A turbofan with the fan fully exhausted operates at 4570 m (15,000 ft) at a Mach number of 0.93. The compressor and fan pressure ratios are 17 and 2.3, respectively. The core airflow rate is 64.9 kg/s (143 lb_m/s), and the bypass ratio is 1.1. The fuel has a heating value of 41,460 kJ/kg (17,900 BTU/lb_m), and the combustor exit total temperature is 1417 K (2550 °R). An afterburner is on the core and, when lit, results in a total temperature of 1778 K (3200 °R) in the nozzle. The compressor, burner, turbine, fan, shaft, converging primary nozzle, variable CD fan nozzle (to match the exit pressure to the ambient pressure), and afterburner have efficiencies of 90, 95, 84, 87, 99.2, 95, 97, and 90 percent, respectively. The diffuser, burner, and afterburner have total pressure ratios of 0.91, 0.94, and 0.97, respectively. Find the thrust and *TSFC* for both (a) nonafterburning and (b) afterburning conditions. (c) Compare the values found with those with the engine with no afterburner. (d) Repeat for an ideal case with $\gamma = 1.40$.

3.14 Two ideal engine types are being considered for a commercial application. Both have core airflow rates of 133 lb_m/s. Both use a fuel with a heating value of 17,800 BTU/lb_m, and both have turbine inlet total temperatures of 2469 °R. Both have a compressor pressure ratio of 13. The altitude will be 27,000 ft, and the Mach number will be 0.77. Engine "A" is a turbojet. Engine "B" is a turbofan with the fan exhausted and has a fan pressure ratio of 2.2 and a bypass ratio of 3.2. The total thrust required for the aircraft is 30,000 lb_f. How many of the two ideal engine types will be needed to power the aircraft? What will the *TSFC* and dimensionless thrust be of the two engine types? Assume $\gamma = 1.40$.

3.15 A turbofan with the fan fully mixed operates at 6700 m (22,000 ft) at Mach 0.89. The compressor pressure ratio is 13. The fuel has a heating value of 42,100 kJ/kg (18,100 BTU/lb_m), and the exit total temperature from the burner is 1375 k (2475 °R). The core airflow is 56.2 kg/s (124 lb_m/s), and the bypass ratio is 5.00. All of the aerodynamic efficiencies and recovery factors are 0.90 except for the shaft efficiency (99.4 percent), and the duct and mixer total pressure ratios (0.985 and 0.975, respectively). The nozzle is variable and CD (to match the exit pressure to the ambient pressure). Find the developed thrust and the *TSFC*. What is the fan pressure ratio? Repeat for an ideal case with $\gamma = 1.40$.

3.16 An ideal turbofan with a mixed fan is being designed to operate at 31,000 ft and at a Mach number of 0.86. The compressor pressure ratio is to be 16.5. The developed thrust is to be 18,800 lb_f, and the bypass ratio is to be 3.6. The fuel has a heating value of 18,300 BTU/lb_m, and the exit total temperature from the combustor is to be 2650 °R. What is the fan pressure ratio? What is the core flow rate? What is the *TSFC*? Assume $\gamma = 1.40$.

3.17 A turbofan with the fan fully mixed operates at 8382 m (27,500 ft) at Mach 0.93. The compressor pressure ratio is 17. The fuel has a heating value of 42,100 kJ/kg (18,100 BTU/lb_m), and the exit total temperature from the burner is 1239 K (2230 °R). The core airflow is 71.2 kg/s (157 lb_m/s), and the bypass ratio is 1.60. An afterburner is also used. When lit, the afterburner total exit temperature is 1833 K (3300 °R). The compressor, burner, turbine, fan, shaft, variable CD primary nozzle (to match the exit pressure to the ambient pressure), and afterburner have efficiencies of 88, 95, 86, 91, 99.3, 94, and 90 percent, respectively. The diffuser, burner, and afterburner have total pressure ratios of 0.93, 0.91, and 0.96, respectively. The duct and mixer pressure ratios are 0.99 and 0.98, respectively. Find the developed thrust and the *TSFC* for both when the afterburner is on and when it is off. What is the fan pressure ratio? Repeat for an ideal case with $\gamma = 1.40$.

3.18 An ideal turbofan (mixed) with an afterburner is being designed for an aircraft that flies at sea level at Mach 0.90. The compressor ratio is 14.6, and the bypass ratio is 1.35. Without the afterburner lit, 12,000 lb_f of thrust is needed. The total temperature at the turbine inlet is 2340 °R. The heating value of the fuel is 18,200 BTU/lb_m. With the afterburner lit, 27,000 lb_f of thrust are needed. What is the required air mass flow? What is the resulting fan pressure ratio? What is the required afterburner total exit temperature? What are the *TSFC*, dimensionless thrust, and exit diameter for the afterburner on and off? Assume $\gamma = 1.40$.

3.19 An ideal turboprop is used to propel an aircraft at 5000 ft. The craft flies with a Mach number of 0.4. The compressor pressure ratio is 4.8, and the airflow through the core is 24 lb_m/s. The burner exit total temperature is 2200 °R, and the exit Mach number is 0.7. The heating value of the fuel is 17,800 BTU/lb_m. Find the developed thrust, dimensionless thrust, and *TSFC*. Assume $\gamma = 1.40$.

3.20 An ideal turboprop is used to propel an aircraft at 5000 ft. The craft flies with a Mach number of 0.4. The compressor pressure ratio is 4.8, and the airflow through the core is 24 lb_m/s. The burner exit total temperature is 2200 °R, and the heating value of the fuel is 17,800 BTU/lb_m. Find the optimum developed thrust, dimensionless thrust, *TSFC*, and exit Mach number. Assume $\gamma = 1.40$.

3.21 An aircraft is to fly at 10,000 ft with a Mach number of 0.5. It is to be propelled with a turboprop and requires 7000 lb_f of thrust per engine. The compressor pressure ratio is 7.5, and the exit total temperature from the burner is 2340 °R. The fuel has a heating value of 17,800 BTU/lb_m. The ideal engine is to operate optimally at the preceding conditions. How much airflow is required? What are the resulting horsepower of the engine, dimensionless thrust, and *TSFC*? Assume $\gamma = 1.40$.

3.22 A turboprop is used to power an aircraft at 2133 m (7000 ft). It flies at Mach 0.45, and the compressor pressure ratio is 5.4. The airflow rate is 8.62 kg/s (19 lb_m/s). The turbine inlet total temperature is 1266 K (2278 °R), and the heating value of the fuel is 42,800 kJ/kg (18,400 BTU/lb$_m$). The work coefficient of the propeller is 0.85. The compressor, burner, turbine, shaft, converging primary nozzle, and propeller have efficiencies of 90, 95, 87, 99.6, 97, and 82 percent, respectively. The diffuser and burner have total pressure ratios of 0.93 and 0.93, respectively. Find the thrust, horsepower, exit Mach number, and *TSFC*. Repeat for an ideal case with $\gamma = 1.40$.

3.23 A transport craft with a nonideal engine is to fly at Mach 0.80 at sea level on a standard day. The burner exit total temperature is 1333 K (2400 °R), and the core airflow rate is 81.6 kg/s (180 lb$_m$/s). The fuel has a heating value of 41,640 kJ/kg (17,900 BTU/lb$_m$). Nonideal parameters for the different engine types are listed in the table.

$\pi_d = 0.94$	$\eta_c = 0.90$	$\pi_{ab} = 0.98$
$\eta_b = 0.93$	$\eta_f = 0.89$	$\eta_{ab} = 0.91$
$\pi_b = 0.95$	$\eta_t = 0.92$	$\pi_u = 0.99$
$\eta_m = 0.997$	$\pi_m = 0.98$	
Fan nozzle converging		Primary nozzle converging–diverging (CD)
$\eta_{fn} = 0.96$	$\eta_n = 0.98$	

(a) Find the thrust, dimensionless thrust, and *TSFC* for the following engines.

(1) Ramjet.
(2) Turbojet $\pi_c = 13$.
(3) Turbojet with afterburner $\pi_c = 13$; total afterburner exit temperature is 2111 K (3800 °R).
(4) Turbofan with mixed auxiliary air $\pi_c = 13$, $\alpha = 1.4$; what is π_f?
(5) Turbofan with the fan all exhausted $\pi_c = 13$, $\alpha = 1.4$; π_f is the same as for the mixed case.
(6) Turbofan with an afterburner and mixed auxiliary air $\pi_c = 13$, $\alpha = 1.4$; total afterburner exit temperature is 2111 K (3800 °R).

(b) Which engine would you choose and why?
(c) Repeat for the ideal case with $\gamma = 1.40$.

3.24 An ideal turbojet flies at sea level on a standard day. The airflow rate is 180 lb$_m$/s. The total pressure ratio for the compressor is 13, and the fuel has a heating value of 17,900 BTU/lb$_m$. Assume $\gamma = 1.40$.

(a) Consider the following cases and find the thrust, dimensionless thrust, and *TSFC*:

(1) Mach number is 0.80 and burner exit total temperature is 2400 °R,
(2) Mach number is 1.20 and burner exit total temperature is 2400 °R,
(3) Mach number is 0.80 and burner exit total temperature is 3000 °R.

(b) Comment on the effect of M_a and burner exit total temperature on performance.

3.25 The Mach number of an ideal ramjet is 2.80 at an external pressure of 2.71 psia and a temperature of 376.1 °R. The flow rate of air through the engine is unknown. Owing to thermal limits of the materials, the burner exit total temperature is 3300 °R. The heating value of the fuel is 18,100 BTU/lb$_m$. The developed thrust is 9127 lb$_f$. What are the mass flow rate, fuel ratio, dimensionless thrust, and *TSFC* for this engine? Assume $\gamma = 1.40$.

3.26 An ideal ramjet is to fly at an external pressure of 2.71 psia and a temperature of 376.1 °R at a yet-to-be-determined Mach number. The burner exit total temperature is to be 3300 °R, and the engine will use 130 lb$_m$/s of air. The heating value of the fuel is 18,100 BTU/lb$_m$. At what Mach number will the *TSFC* be optimized? What is the optimum *TSFC*? What are the fuel ratio, thrust, and dimensionless thrust at this condition? Assume $\gamma = 1.40$.

3.27 An ideal turbojet with an afterburner is being designed for an aircraft that flies at 15,000 ft at Mach

0.87. The compressor ratio is 18.2. With the afterburner unlit, 10,860 lb_f of thrust are needed. The total temperature at the turbine inlet is 2600 °R. The heating value of the fuel is 18,000 BTU/lb_m. With the afterburner lit, 15,740 lb_f of thrust are needed. What is the required air mass flow? What is the required afterburner total exit temperature? What are the *TSFC* and dimensionless thrust for the engine with the afterburner on and off? Assume $\gamma = 1.40$.

3.28 An ideal turbofan with an exhausted fan is being designed to operate at 32,000 ft and at a Mach number of 0.80. It is to operate at maximum or optimum thrust. The compressor pressure ratio is to be 19.0. The developed thrust is to be 20,660 lb_f, and the bypass ratio is to be 3.5. The fuel has a heating value of 17,850 BTU/lb_m, and the exit total temperature from the combustor is to be 2520 °R. What is the fan pressure ratio? What is the core flow rate? What is the *TSFC*? What is the dimensionless thrust? Assume $\gamma = 1.40$.

3.29 An ideal turbofan with the fan mixed operates at 29,200 ft at Mach 0.80. The compressor pressure ratio is 19.3. The fuel has a heating value of 17,750 BTU/lb_m, and the exit total temperature from the burner is 2460 °R. The core airflow is 164 lb_m/s, and the bypass ratio is 2.50. An afterburner is also used. When lit, the afterburner total exit temperature is 3350 °R and the aircraft travels at Mach 1.5. Find the developed thrust, dimensionless thrust, and *TSFC* for the engine with the afterburner on and off. What is the fan total pressure ratio? Assume $\gamma = 1.40$.

3.30 An ideal turboprop is used to propel an aircraft at 3500 ft. The craft flies with a Mach number of 0.44. The compressor pressure ratio is 4.3, and the airflow through the core is 26 lb_m/s. The burner exit total temperature is 2150 °R, and the propeller work coefficient is 0.97. The heating value of the fuel is 17,940 BTU/lb_m. Find the developed thrust, dimensionless thrust, and *TSFC*. Assume $\gamma = 1.40$.

3.31 An ideal turboprop is used to propel an aircraft at 3500 ft. The craft flies with a Mach number of 0.44. The compressor pressure ratio is 4.3, and the airflow through the core is 26 lb_m/s. The burner exit total temperature is 2150 °R.
The heating value of the fuel is 17,940 BTU/lb_m. Find the optimum developed thrust, dimensionless thrust, *TSFC*, and propeller work coefficient. Assume $\gamma = 1.40$.

3.32 You are to pick an engine for a transport aircraft flying at $M_a = 0.8$ at sea level on a standard day. The exit burner total temperature is 1667 K (3000 °R), and $\Delta H =$ 41870 kJ/kg (18,000 BTU/lb_m). The air mass flow rate in the core is 81.6 kg/s (180 lb_m/s). Use a nonideal cycle analysis and compare the results with those obtained from an ideal analysis with $\gamma = 1.40$. Nonideal parameters for the different engine types are listed in the table.

$\pi_d = 0.92$	$\eta_c = 0.91$	$\pi_{ab} = 0.98$
$\eta_b = 0.93$	$\eta_f = 0.90$	$\eta_{ab} = 0.92$
$\pi_b = 0.96$	$\eta_t = 0.93$	$\pi_{duct} = 0.99$
Nozzles all converging	$\eta_m = 0.996$	$\pi_{mixer} = 0.98$
$\eta_n = 0.97$	$\eta_{fn} = 0.96$	

(a) Find the dimensionless quantity $F/\dot{m}_t\, a_{at}$ and the dimensional quantities F and *TSFC* for the following engines.

(1) Ramjet.
(2) Turbojet $\pi_c = 16$.
(3) Turbojet with afterburner $\pi_c = 16$; total afterburner temperature is 2333 K (4200 °R).
(4) Turbofan with exhausted fan $\pi_c = 16$, $\pi_f = 4.0$, $\alpha = 1$.
(5) Turbofan with mixed secondary flow and with afterburner $\pi_c = 16$, $\alpha = 1$; total afterburner temperature is 2333 K (4200 °R). What is π_f?

(b) Which engine will you choose and why?
(c) Repeat for an ideal case with $\gamma = 1.40$.

3.33 You are to choose an engine for takeoff from an aircraft carrier on a standard day. The exit burner total temperature is 1444 K (2600 °R), and $\Delta H =$ 41,640 kJ/kg (17,900 BTU/lb_m). The air mass flow rate in the core is 104 kg/s (230 lb_m/s). Use a nonideal cycle analysis and compare the results with those from an ideal analysis with $\gamma = 1.40$. Nonideal parameters for the different engine types are listed in the table.

$\pi_d = 0.94$	$\eta_c = 0.92$	$\pi_{ab} = 0.98$
$\eta_b = 0.93$	$\eta_f = 0.91$	$\eta_{ab} = 0.92$
$\pi_b = 0.96$	$\eta_t = 0.94$	$\pi_u = 0.99$
Nozzles all converging	$\eta_m = 0.995$	$\pi_m = 0.98$
$\eta_n = 0.96$	$\eta_{fn} = 0.96$	

(a) Find the dimensionless quantity $F/\dot{m}_t\, a_a$ and the dimensional quantities F and *TSFC* for the following engines.

(1) Ramjet.
(2) Turbojet $\pi_c = 23.5$.
(3) Turbojet with afterburner $\pi_c = 23.5$; total afterburner temperature is 2167 K (3900 °R).
(4) Turbofan with exhausted fan $\pi_c = 23.5$, $\pi_f = 1.98$, $\alpha = 5.0$.
(5) Turbofan with mixed secondary flow and with afterburner $\pi_c = 23.5$, $\alpha = 0.7$; total afterburner temperature is 2167 K (3900 °R). What is π_f?

(b) Which engine will you choose and why?
(c) Repeat for an ideal case with $\gamma = 1.40$.

3.34 A ramjet cruises at 139 K (251 °R) and 2.23 kPa (0.324 psia) at a Mach number of 4.0. The flow rate of air through the engine is 86.2 kg/s (190 lb_m/s). The fuel ratio is 0.0457 and is used because of the thermal limits of the nozzle materials. The heating value of the fuel is 41,870 kJ/kg (18,000 BTU/lb_m). The efficiencies of the burner and converging nozzle are 91 and 95 percent, respectively. The total pressure ratios for the diffuser and burner are 0.92 and 0.88, respectively. The specific heat ratios for the ambient, diffuser, burner, and nozzle are 1.418, 1.377, 1.320, and 1.274, respectively. The specific heats at constant pressure are 0.975, 1.047, 1.185, and 1.335 kJ/kg-K (0.233, 0.250, 0.283, and 0.319 BTU/lb_m-°R), respectively. Find the burner exit total temperature, developed thrust, and *TSFC*. Repeat for an ideal case with $\gamma = 1.32$.

3.35 Draw a general h–s diagram for a turbofan engine, which has an afterburner for the core flow, with a totally exhausted fan (see Figure 3.4). Draw this to scales that show the correct relative magnitudes; that is, be careful to show reasonable relative pressures and enthalpies for the different components. Show all points (a, 1, 2, 3, 4, 5, 6, 7, 8, and 9) as well as static and total pressures. Show both the ideal and nonideal cases.

3.36 Draw a general h–s diagram for an afterburner-equipped turbofan engine with a totally mixed fan (see Figure 3.5). Draw this to scales that show the correct relative magnitudes; that is, be careful to show reasonable relative pressures and enthalpies for the different components. Show all points (a, 1, 2, 3, 4, 5, 5.5, 6, 7, 7.5, and 8) and show both static and total pressures. Show both ideal and nonideal cases.

3.37 A new commercial aircraft is to be designed to fly routinely at 10,970 m (36,000 ft) at an air speed of 252 m/s (826 ft/s). Owing to expected drag characteristics of the airframe, two engines are needed, and the required thrust is 189,000 N (42,500 lb_f) per engine. Fuel economy considerations dictate that the engines need to operate with *TSFC* values of 0.0632 kg/N-h (0.62 lb_m/h-lb_f). The limit on the exit temperature of the burner is 1639 K (2950 °R). Design an engine for this application. Make reasonable approximations for nonideal characteristics. Repeat for an ideal case with $\gamma = 1.40$.

3.38 A new military fighter aircraft is to be designed to fly routinely at a variety of altitudes. One particular condition is at 6096 m (20,000 ft) at an air speed of 285 m/s (935 ft/s). Owing to expected drag characteristics of the airframe, two engines are needed, and the required thrust is 73,400 N (16,500 lb_f) per engine. Engine weight considerations dictate that the engines need to operate with nondimensional thrusts based on total mass flow rate of 1.90 and with a *TSFC* value of 0.0938 kg/N-h (0.92 lb_m/h-lb_f). Moreover, since the aircraft is a fighter craft, short increases in thrust for rapid acceleration to supersonic speeds are required to a total of 140,100 N (31,500 lb_f) per engine. The limits on the exit temperatures of the primary burner and afterburner are 1667 K and 2000 K (3000 and 3600 °R), respectively. Design an engine for this application and these conditions. Make reasonable approximations for nonideal characteristics. Repeat for an ideal case with $\gamma = 1.40$.

3.39 For an ideal turbojet (no losses, properties are constant throughout, and the fuel ratio is negligible), start with Eq. (1.5.14) and analytically show that the maximum thrust is obtained when the exit pressure of the nozzle matches the ambient pressure. Consider the flight Mach number, altitude, core mass flow rate, compressor total pressure ratio, burner exit total temperature, and fuel properties to be known.

3.40 An ideal turbojet cruises at an altitude at which the conditions are 430 °R and 5.45 psia (local sound speed is 994 ft/s) and the Mach number is 0.85. The flow rate of air through the engine is 150 lb_m/s. The compressor total pressure ratio is 18.0. The fuel ratio is 0.0265. The turbine exit total temperature is 2228 °R. The heating value of the fuel is 17,800 BTU/lb_m. The average specific heat ratio for the engine is 1.340, and the average specific heat at constant pressure is 0.270 BTU/lb_m-°R. Find:

(a) the burner exit total temperature,
(b) the nozzle exit velocity.

3.41 An ideal turbojet ingests 200 lb_m/s of air at sea-level takeoff (SLTO). The compressor pressure ratio is 20. The fuel has a heating value of 17,800 BTU/lb_m, and the burner exit total temperature is 2700 °R. Find the developed thrust, dimensionless thrust, and *TSFC*. Assume $\gamma = 1.40$.

3.42 An ideal air standard Brayton cycle (power generation) operates with a maximum temperature of 1500 °C. The inlet air temperature and pressure to the compressor are 40 °C and 120 kPa, respectively, and the compressor pressure ratio is 18. Assume constant specific heats and $\gamma = 1.40$. The airflow rate is 20 kg/s. Find:

(a) the net power output,
(b) the system thermal efficiency.

3.43 A turbojet operates at 6706 m (22,000 ft) at Mach 0.89. The compressor pressure ratio is 13. The fuel has a heating value of 42,100 kJ/kg (18,100 BTU/lb_m), and the exit total temperature from the burner is 1375 K (2475 °R). The airflow is 56.25 kg/s (124 lb_m/s). All of the aerodynamic efficiencies and recovery factors are 0.90 except for the shaft efficiency (99.4 percent). The nozzle is variable and CD (to match the exit pressure to the ambient pressure). Find the developed thrust and the *TSFC*.

3.44 Consider the turbojet in Problem 3.43 and find what the performance would be if each component could be independently improved. That is, find the thrust and *TSFC* for the following independent conditions (where all other conditions are as in Problem 3.43): (a) compressor efficiency is 95 percent; (b) turbine efficiency is 95 percent; (c) burner efficiency is 95 percent; (d) burner total pressure ratio is 95 percent; (e) diffuser recovery factor is 0.95; and (f) nozzle efficiency is 95 percent. Where should the most money be spent in trying to improve the overall engine performance?

3.45 A turbofan with the fan fully mixed operates at 6706 m (22,000 ft) at Mach 0.89. The compressor pressure ratio is 13. The fuel has a heating value of 42,100 kJ/kg (18,100 BTU/lb_m), and the exit total temperature from the burner is 1375 K (2475 °R). The core airflow is 56.25 kg/s (124 lb_m/s), and the bypass ratio is 1.15. All of the aerodynamic efficiencies and recovery factors are 0.90 except for the shaft efficiency (99.4 percent), and the duct and mixer total pressure ratios (0.985 and 0.975, respectively). The nozzle is variable and CD (to match the exit pressure to the ambient pressure). Find the developed thrust and the *TSFC*. What is the fan pressure ratio?

3.46 Consider the turbofan in Problem 3.45 and find what the performance would be if each component could be independently improved. That is, find the thrust and *TSFC* for the following independent conditions (where all other conditions are as in Problem 3.45): (a) compressor efficiency is 95 percent; (b) fan efficiency is 95 percent; (c) turbine efficiency is 95 percent; (d) burner efficiency is 95 percent; (e) burner total pressure ratio is 95 percent; (f) diffuser recovery factor is 0.95; and (g) nozzle efficiency is 95percent. Where should the most money be spent in trying to improve the overall engine performance?

3.47 Consider the nonideal turbofan in Problem 3.15 and find what the performance would be if each component could be independently improved. That is, find the thrust and *TSFC* for the following independent conditions (where all other conditions are as in Problem 3.15): (a) compressor efficiency is 95 percent; (b) fan efficiency is 95 percent; (c) turbine efficiency is 95 percent; (d) burner efficiency is 95 percent; (e) burner total pressure ratio is 95 percent; (f) diffuser recovery factor is 0.95; and (g) nozzle efficiency is 95 percent. Where should the most money be spent in trying to improve the overall engine performance?

3.48 A turbofan with a split ratio of 0.5 operates at 8382 m (27,500 ft) at Mach 0.93. The compressor pressure ratio is 17. The fuel has a heating value of 42,100 kJ/kg (18,100 BTU/lb_m), and the exit total temperature from the burner is 1239 K (2230 °R). The core airflow is 71.21 kg/s (157 lb_m/s), and the bypass ratio is 1.60. An afterburner is also used. When lit, the afterburner total exit temperature is 1833 K (3300 °R). The compressor, burner, turbine, fan, shaft, variable CD primary nozzle, variable CD fan nozzle (both to match the exit pressure to the ambient pressure), and afterburner have efficiencies of 88, 95, 86, 91, 99.3, 94, 94, and 90 percent, respectively. The diffuser, burner, and afterburner have total pressure ratios of 0.93, 0.91, and 0.96, respectively. The duct and mixer pressure ratios are 0.99 and 0.98, respectively. Find the developed thrust and the *TSFC* for when the afterburner is on and when it is off. What is the fan pressure ratio?

3.49 A turbofan with the fan fully exhausted operates at 6706 m (22,000 ft) at Mach 0.89. The compressor pressure ratio is 13, and the fan pressure ratio is 1.6097. The fuel has a heating value of 42,100 kJ/kg (18,100 BTU/lb_m), and the exit total temperature from the burner is 1375 K (2475 °R). The core airflow is 56.25 kg/s (124 lb_m/s), and the bypass ratio is 5.00. All of the aerodynamic efficiencies and recovery factors are 0.90 except for the shaft efficiency (99.4 percent). The nozzles are both variable and CD (to match the exit pressure to the ambient pressure). Find the developed thrust and the *TSFC*.

3.50 Consider the turbofan in Problem 3.49 and find when the performance would be if each component could be independently improved. That is, find the thrust and *TSFC* for the following independent conditions (where all other conditions are as in Problem 3.49): (a) compressor efficiency is 95 percent; (b) fan efficiency is 95 percent; (c) turbine efficiency is 95 percent; (d) burner efficiency is 95 percent; (e) burner total pressure ratio is 95 percent; (f) diffuser recovery factor is 0.95; (g) primary nozzle efficiency is 95 percent; (h) fan nozzle efficiency is 95 percent. Where should the most money be spent in trying to improve the overall engine performance?

3.51 A turboprop is used to power an aircraft at 1158 m (3800 ft). It flies at Mach 0.38, and the compressor pressure ratio is 4.75. The airflow rate is 10.0 kg/s (22 lb_m/s). The turbine inlet total temperature is 1358 K (2445 °R), and the heating value of the fuel is 41,500 kJ/kg (17,850 BTU/lb_m). The work coefficient of the propeller is 0.89. The compressor, burner, turbine, shaft, converging primary nozzle, and propeller have efficiencies of 89, 94, 91, 99.6, 96 and 78 percent, respectively. The diffuser and burner have total pressure ratios of 0.94 and 0.92, respectively. Find the thrust, horsepower, exit Mach number, and *TSFC*.

3.52 A turbofan with the fan fully mixed operates at an external pressure of 2.99 psia and a temperature of

383.2 °R at Mach 0.87. The compressor pressure ratio is 19.5. The fuel has a heating value of 17,900 BTU/lb$_m$, and the exit total temperature from the burner is 2480 °R. The core airflow is 155 lb$_m$/s, and the bypass ratio is 3.50. The aerodynamic efficiencies of the compressor, fan, turbine, burner, and nozzle are 88, 87, 91, 92, and 95 percent, respectively. The recovery factor for the diffuser is 0.93, and the burner has a total pressure ratio of 0.94. The shaft efficiency is 99.7 percent, and the duct and mixer total pressure ratios are 0.99 and 0.99, respectively. The nozzle is variable and CD (to match the exit pressure to the ambient pressure). Find the developed thrust and the *TSFC*. What is the fan total pressure ratio?

3.53 A turbofan with the fan fully exhausted operates at an external pressure of 2.99 psia and a temperature of 383.2 °R at Mach 0.87. The compressor pressure ratio is 19.5, and the fan pressure ratio is 2.14. The fuel has a heating value of 17,900 BTU/lb$_m$, and the exit total temperature from the burner is 2480 °R. The core airflow is 155 lb$_m$/s, and the bypass ratio is 3.50. The aerodynamic efficiencies of the compressor, fan, turbine, burner, primary nozzle, and fan nozzle are 88, 87, 91, 92, 95, and 95 percent, respectively. The recovery factor for the diffuser is 0.93 and the burner has a total pressure ratio of 0.94. The shaft efficiency is 99.7 percent. The nozzles are both variable and CD (to match the exit pressure to the ambient pressure). Find the developed thrust and the *TSFC*.

3.54 A turbofan with the fan fully exhausted operates at an external pressure of 2.99 psia and a temperature of 383.2 °R at Mach 0.87. The compressor pressure ratio is 19.5. The fuel has a heating value of 17,900 BTU/lb$_m$, and the exit total temperature from the burner is 2480 °R. The core airflow is 155 lb$_m$/s, and the bypass ratio is 3.50. The aerodynamic efficiencies of the compressor, fan, turbine, burner, primary nozzle, and fan nozzle are 88, 87, 91, 92, 95, and 95 percent, respectively. The recovery factor for the diffuser is 0.93, and the burner has a total pressure ratio of 0.94. The shaft efficiency is 99.7 percent. The nozzles are both variable and CD (to match the exit pressure to the ambient pressure). What is the optimal fan total pressure ratio to maximize the thrust? Find the optimum developed thrust and the *TSFC*.

3.55 For a turbojet, the compressor operates with a pressure ratio of 14 and an efficiency of 0.88, and the inlet total temperature to the compressor is 700 °R. The total airflow is 100 lb$_m$/s. The burner has an efficiency of 0.85 and a pressure ratio of 0.93. The fuel has a heating value of 18,000 BTU/lb$_m$ and flows at 2.0 lb$_m$/s. The turbine pressure ratio is 0.164, and it has an efficiency of 0.83. The mechanical efficiency of the shaft system is 0.98. Assume $\gamma_c = 1.40$, $c_{p_b} = 0.29$ BTU/lb$_m$-°R, and $\gamma_t = 1.33$.

(a) What is the exit total temperature of the compressor?
(b) What is the exit total temperature of the burner?
(c) What is the exit total temperature of the turbine?
(d) What is the power (hp) delivered to the compressor from the turbine?

3.56 A ramjet cruises at an external pressure of 6.17 kPa (0.895 psia) and a temperature of 169.3 K (304.7 °R), and the Mach number is 5.0. The flow rate of air through the engine is 63.5 kg/s (140 lb$_m$/s). Because of the thermal limits of the materials, the burner exit total temperature is 1833 K (3300 °R). The heating value of the fuel is 43,030 kJ/kg (18,500 BTU/lb$_m$). The efficiencies of the burner and converging nozzle are 94 and 96 percent, respectively. The total pressure ratios for the diffuser and burner are 0.90 and 0.92, respectively. The specific heat ratios for the ambient, diffuser, burner, and nozzle are 1.415, 1.347, 1.312, and 1.281, respectively. The specific heats at constant pressure are 0.980, 1.110, 1.210, 1.310 kJ/kg-K (0.234, 0.266, 0.288, and 0.313 BTU/lb$_m$-°R), respectively.

(a) What are the developed thrust, fuel ratio, and *TSFC* for this engine?
(b) If the performance is poor, which component should be changed and how?

3.57 A high-bypass turbofan engine with the fan exhausted is to be analyzed. The core or primary flow has an afterburner. The fan does not have an afterburner. The fan has 2 stages and the compressor has 10; the turbine has 4 stages. The shaft operates at 10,500 rpm. The engine performance at 10,000 ft on a standard day is to be considered.

$\dot{m} = 75\ \text{lb}_\text{m}/\text{s}$ (primary)	$\eta_\text{b} = 0.93$
$\gamma_\text{d} = 1.40$	$T_{t7} = 690\ °\text{R}$
$\dot{m}_f = 1.74\ \text{lb}_\text{m}/\text{s}$	$\eta_\text{t} = 0.86$
$\gamma_\text{f} = 1.40$	$M_\text{a} = 0.66$
$\alpha = 3.2$	$\eta_\text{ab} = 0.88$
$\gamma_\text{c} = 1.38$	$\eta_\text{m} = 0.995$
$\Delta H = 17{,}700\ \text{BTU}/\text{lb}_\text{m}$	$p_8/p_a = 1.00$
$\gamma_\text{b} = 1.33$	$\eta_\text{n} = 0.97$
$p_\text{a} = 10.11$ psia	$\eta_\text{fn} = 0.92$
$\gamma_\text{t} = 1.33$	$\pi_\text{ab} = 0.98$
$T_\text{a} = 483\ °\text{R}$	$\pi_\text{b} = 0.95$
$\gamma_\text{ab} = 1.32$	$\pi_\text{d} = 0.92$
$\pi_\text{f} = 2.40$	Variable converging–diverging (CD) primary nozzle
$\gamma_\text{n} = 1.28$	
$\pi_\text{c} = 22.0$ (includes fan)	Converging fan nozzle
$\eta_\text{c} = 0.89$	Engine mass $= 2550\ \text{lb}_\text{m}$
$T_{t6} = 3200\ °\text{R}$	$N = 10{,}500$ rpm

(a) Sketch the engine and clearly indicate the station numbers.
(b) What is the exit total temperature of the compressor?
(c) What is the exit total temperature of the primary burner if the total temperature at the exit of the compressor is 1322 °R?
(d) If the total temperature out of the primary combustor is 2650 °R and the total temperature at the exit of the compressor is 1322 °R, what is the total pressure out of the turbine?
(e) If the total temperature into the afterburner is 1510 °R and the total pressure is 15.7 psia, what is the velocity exiting the primary nozzle?
(f) What is the pressure at the exit of the fan nozzle?
(g) What is the efficiency of the fan?

3.58 All of the following apply to the same turbojet.

(a) A compressor operates with a pressure ratio of 14 and an efficiency of 0.88. The inlet total temperature is 388.9 K (700 °R). What is the exit total temperature? Assume $\gamma = 1.40$.
(b) The airflow rate of a turbojet is 45.36 kg/s (100 lb_m/s). Air enters the burner at 611.1 K (1100 °R) (total) temperature. The burner has an efficiency of 0.85 and a pressure ratio of 0.93. The fuel has a heating value of 41,870 kJ/kg (18,000 B/lb_m) and flows at 0.907 kg/s (2.0 lb_m/s). Assume $c_\text{p} = 1.214$ kJ/kg-K (0.29 BTU/lb_m-°R). What is the exit total temperature?
(c) The turbine pressure ratio is 0.30, and the turbine has an efficiency of 0.83. The incoming total temperature is 1167 K (2100 °R), and the mechanical efficiency of the shaft system is 0.98. How much power is delivered to the compressor (kW or hp) if the airflow rate of a turbojet is 45.56 kg/s (100 lb_m/s)? Assume $c_\text{p} = 1.155$ kJ/kg-K (0.276 BTU/lb_m-°R) and $\gamma = 1.33$.

3.59 You are to analyze the following engines for an aircraft flying at $M_\text{a} = 0.8$ at sea level on a standard day. The exit burner total temperature is 3000 °R, and the compressor pressure ratio is 16. Assume all aerodynamic efficiencies and recovery factors are 0.90 and that $\pi_\text{b} = 0.90$. Assume mechanical efficiencies of 1.00. For each engine, indicate clearly on a diagram where each of the different nonideal effects takes place. Consider thrust and *TSFC* for the following engines with converging nozzles and with compressor flow rates of 150 lb_m/s and $\Delta H = 18{,}000$ BTU/lb_m. Compute the overall performance by computing the performance of each component and compare with the performance for the ideal case with $\gamma = 1.35$.

(a) Turbojet.
(b) Turbofan with mixed auxiliary flow, $\alpha = 1$. What is π_f?
(c) Turbofan with afterburner with mixed auxiliary flow, $\alpha = 1$, total afterburner temperature is 4200 °R, and $\pi_{ab} = 0.9$. What is π_f?

3.60 Consider the turbojet in Problem 3.59 and decide if the performances of different components could be independently improved. Again find the thrust and *TSFC* for the following independent conditions:
(a) $\eta_c = 0.95$, (b) $\eta_t = 0.95$, (c) $\eta_b = 0.95$, (d) $\pi_b = 0.95$, (e) $\pi_d = 0.95$, (f) $\eta_n = 0.95$.
Where should the most money be spent in trying to improve overall engine performance?

3.61 Consider the mixed turbofan with no afterburner in Problem 3.59 and decide if the performances of different components could be independently improved. Again find the thrust and *TSFC* for the following independent conditions:
(a) $\eta_c = 0.95$, (b) $\eta_t = 0.95$, (c) $\eta_f = 0.95$, (d) $\eta_b = 0.95$, (e) $\pi_b = 0.95$, (f) $\pi_d = 0.95$, (g) $\eta_n = 0.95$.
Where should the most money be spent in trying to improve overall engine performance?

3.62 A turbofan with the fan fully mixed operates at 8534 m (28,000 ft) (32.89 kPa and 232.7 K) (4.77 psia and 418.9 °R) at Mach 0.79. The compressor pressure ratio is 19.5. The fuel has a heating value of 41,400 kJ/kg (17,800 BTU/lb_m), and the exit total temperature from the burner is 1528 K (2750 °R). The core airflow is 68.95 kg/s (152 lb_m/s), and the bypass ratio is 3.4. The fuel flow to core airflow ratio is 0.0273. The compressor, burner, fan, shaft, and converging nozzle have efficiencies of 88, 94, 87, 99.6, and 95 percent, respectively. The diffuser, burner, duct, and mixer have total pressure ratios of 0.97, 0.93, 1.00, and 0.98, respectively. The specific heat ratios for the diffuser, compressor, fan, burner, turbine, mixer, and nozzle are 1.405, 1.387, 1.402, 1.333, 1.312, 1.384, and 1.383, respectively. The specific heats at constant pressure for the compressor, fan, turbine, burner, nozzle, and mixer are 1.030, 1.000, 1.185, 1.151, 1.038, and 1.034 kJ/kg-K (0.246, 0.239, 0.283, 0.275, 0.248, and 0.247 BTU/lb_m-°R), respectively. Solve the following five problems, which are independent of each other. Note that information given in each part does not necessarily apply to other parts.

(a) If the inlet total temperature to the compressor is 262.2 K (471.8 °R) and the inlet total pressure is 48.19 kPa (6.99 psia), what is the exit total temperature of the compressor?
(b) If the inlet total pressure to the nozzle is 104.6 kPa (15.17 psia) and the inlet total temperature is 495.6 K (892.0 °R), what is the thrust?
(c) If the fan total pressure ratio is 2.21 and the turbine efficiency increases by 0.03 (3 percent), will the fan total pressure ratio increase, remain constant, or decrease? Why?
(d) If the exit total temperature of the fan is 388.9 K (610.0 °R), the exit total pressure of the fan is 106.7 kPa (15.48 psia), and the exit total temperature of the turbine is 984.4 K (1772 °R), what are the total exit pressure and temperature from the mixer?
(e) If the exit total pressures from the fan and compressor are 106.7 kPa (15.48 psia) and 940.4 kPa (136.4 psia), respectively, and the exit total temperatures from the fan and compressor are 388.9 K (610.0 °R) and 647.7 K (1164 °R), respectively, what is the turbine efficiency?

3.63 You are to analyze a modern dual-spool turbofan engine with an afterburner.

The fan has three stages, operates at 9600 rpm, and is driven by a two-stage low-pressure turbine. The high-pressure compressor has 10 stages, operates at

14,650 rpm, and is driven by a two-stage high-pressure turbine. You are to analyze its performance at 20,000 ft on a standard day. You may assume frictional losses exist in the bypass duct that reduce the fan exit total pressure to that equal to the low-pressure turbine exit pressure. The following parameters are given.

$\dot{m}_t = 228\ \text{lb}_m/\text{s}$	$\gamma_d = 1.40$
$\alpha = 0.7$ (bypass ratio)	$\gamma_f = \gamma_{c\ high} = 1.40$
$\Delta H = 18{,}400\ \text{BTU/lb}_m$	$\gamma_b = \gamma_{ab} = 1.30$
$p_a = 6.76$ psia	$\gamma_{t\ low} = \gamma_{t\ high} = 1.30$
$T_a = 447\ °\text{R}$	$\gamma_n = 1.30$
$\pi_f = 2.10$	$\pi_d = 0.96$
$\pi_{c\ high} = 11.2$	$\eta_b = 0.88$
$T_{t4} = 2300\ °\text{R}$	$\pi_b = 0.93$
$T_{t3} = 1376\ °\text{R}$	$\eta_{t\ high} = 0.83$
$T_{t2.5} = 646\ °\text{R} = T_{t7}$	$\eta_{t\ low} = 0.78$
$M_a = 0.85$	$\eta_{ab} = 0.87$
$\pi_{ab} = 0.92$	
Variable CD nozzle	$p_8/p_a = 1.00$
Exhaust nozzle	$\eta_n = 0.97$
$N_{c\ high} = 14{,}650$ rpm	$N_{c\ low} = 9600$ rpm
Engine mass $= 3000\ \text{lb}_m$	

(a) If all stages in the high-pressure compressor have the same single-stage total pressure ratio, what is it?
(b) What is the efficiency of the high-pressure compressor?
(c) What is the primary fuel flow rate (lb_m/s)?
(d) What is the high-pressure turbine exit total pressure? What is the low-pressure turbine exit total pressure?

Note: All four parts can be worked on independently.

3.64 An engine similar in size to an older nonafterburning turbojet ingests 54.43 (120 lb_m/s) of air at sea level. When it flies at a Mach number of 0.00 (takeoff, maximum power), the compressor pressure ratio is 14.6. The fuel has a heating value of 41,400 kJ/kg (17,800 BTU/lb_m), and the burner exit total temperature is 1433 K (2580 °R). The efficiencies of the compressor, burner, turbine, shaft, and variable CD nozzle (to match the exit pressure to the ambient pressure) are 90, 95, 92, 100, and 96 percent, respectively. The total pressure ratios for the diffuser and burner are 0.97 and 0.95, respectively. The test stand thrust and *TSFC* were experimentally determined to be 49.82 kN (11,200 lb_f), and 0.0908 kg/n-N (0.89 $\text{lb}_m/\text{h-lb}_f$). Write an algorithm to calculate the specific heat and specific heat ratio for each component at the average component temperature (iterative) and then predict the developed thrust and *TSFC*. Estimates for the specific heat ratio for the inlet, compressor, burner, turbine and nozzle are 1.403, 1.386, 1.337, 1.318, and 1.329, respectively.

3.65 If all of the characteristics of the engine in Problem 3.64 at SLTO are held constant, except the compressor pressure ratio, vary this ratio to optimize the SLTO thrust.

3.66 A turbofan with the fan fully mixed operates at 32,000 ft (3.98 psia and 404.6 °R) at Mach 1.40. The compressor pressure ratio is 20.0. The fuel has a heating value of 17,800 BTU/lb_m, the exit total temperature from the burner is 2850 °R, and the fuel flow ratio is 0.0257. The core airflow is 175 lb_m/s, and the bypass ratio is 0.95. The turbine, burner, fan, shaft, and variable CD nozzle (to match the exit pressure to the ambient pressure) have efficiencies of 91, 93, 89, 100, and 95 percent, respectively. The diffuser, burner, duct, and mixer have total pressure ratios of 0.92, 0.90, 1.00, and 1.00, respectively. The specific heat ratios for the diffuser, compressor, fan, burner, turbine, mixer, and nozzle are 1.400, 1.378, 1.394, 1.326, 1.314, 1.384, and 1.357, respectively. The specific heats at constant pressure for the compressor, fan, turbine, burner, nozzle, and mixer are 0.250, 0.243, 0.287, 0.279, 0.260, and 0.247 BTU/lb_m-°R, respectively. Solve the following six parts, which are independent of each other. Note that information given in each part does not necessarily apply to other parts.

(a) If the inlet total temperature to the compressor is 563.4 °R, the exit total temperature is 1399.4 °R, and the inlet total pressure is 11.64 psia, what is the efficiency of the compressor?
(b) If the inlet total pressure to the nozzle is 35.72 psia and the inlet total temperature is 1416 °R, what is the thrust?
(c) If the fan total pressure ratio is 3.07 and the compressor efficiency increases by 0.06 (6 percent), will the fan total pressure ratio increase, remain constant, or decrease? Why?
(d) If the exit total temperature of the fan is 799.4 °R, the exit total pressure of the fan is 35.72 psia, and the exit total temperature of the turbine is 1955 °R, what is the total exit pressure from the mixer?
(e) If the exit total temperature of the fan is 799.4 °R, the exit total pressure of the fan is 35.72 psia, and the exit total temperature of the turbine is 1955 °R, what is the total exit temperature from the mixer?
(f) If the exit total pressure from the compressor is 232.8 psia, and the exit total temperatures from the fan and compressor are 799.4 °R and 1399 °R, respectively, what is the turbine exit total pressure?

3.67 A turbojet ingests 180 lb_m/s of air at an altitude of 27,000 ft. It flies at Mach 0.85, and the compressor pressure ratio is 15.2. The fuel has a heating value of 17,800 BTU/lb_m, and the fuel ratio is 0.026. The efficiencies of the compressor, burner, turbine, shaft, and converging nozzle are 89, 95, 91, 99.6, and 96 percent, respectively. The total pressure ratios for the diffuser and burner are 0.96 and 0.95, respectively. Find the developed thrust and *TSFC*.

3.68 A ground-based gas turbine with characteristics approximating those of a modern moderate-size unit and used for power generation ingests 147 lb_m/s of air at standard sea level. The compressor pressure ratio is 18, whereas the fuel has a heating value of 18,400 BTU/lb_m, and the burner exit total temperature is 2620 °R. The efficiencies of the compressor, burner, turbine, and shaft are 88, 96, 91.5, and 98 percent, respectively. The total pressure ratios for the inlet, burner, and exhaust are 0.98, 0.96, and 0.93, respectively. Find the overall thermal efficiency, developed net power output, heat rate (total input energy per unit time), and specific fuel consumption, *SFC*.

3.69 A turbofan with the fan fully mixed operates at SLTO. The compressor pressure ratio is 18. The fuel has a heating value of 17,800 BTU/lb_m, and the exit total temperature from the burner is 2850 °R. The core airflow is 150 lb_m/s, and the bypass ratio is 0.80. The aerodynamic efficiencies of the compressor, fan, turbine, burner, and nozzle are 89, 90, 92, 95, and 96 percent, respectively. The recovery factor for the diffuser is 0.95, and the burner has a total pressure ratio of 0.94. The shaft efficiency is 99.8 percent, and the duct and mixer total pressure ratios are 0.99 and 0.99, respectively. The nozzle is variable and CD (to match the exit pressure to the ambient pressure). Find the developed thrust and the *TSFC*. What is the fan total pressure ratio?

3.70 A turbofan with the fan fully exhausted operates at SLTO. The fan pressure ratio is 3.90. The fuel has a heating value of 17,800 BTU/lb_m, and the exit total temperature from the burner is 2850 °R. The core airflow is 150 lb_m/s, and the bypass ratio is 0.80. The aerodynamic efficiencies of the compressor, fan, turbine, burner, primary nozzle, and fan nozzle are 89, 90, 92, 95, 96, and 95 percent, respectively. The recovery factor for the diffuser is 0.95, and the burner has a total pressure ratio of 0.94. The shaft efficiency is 99.8 percent. The primary nozzle is variable and CD (to match the exit pressure to the ambient pressure), whereas the fan nozzle is fixed and converging. Vary the compressor pressure ratio and find the optimum developed thrust and the corresponding *TSFC*.

3.71 A turbofan with the fan fully exhausted operates at SLTO. The compressor pressure ratio is 18. The fuel has a heating value of 17,800 BTU/lb$_m$, and the exit total temperature from the burner is 2850 °R. The core airflow is 150 lb$_m$/s, and the bypass ratio is 0.80. The aerodynamic efficiencies of the compressor, fan, turbine, burner, primary nozzle, and fan nozzle are 89, 90, 92, 95, 96, and 95 percent, respectively. The recovery factor for the diffuser is 0.95, and the burner has a total pressure ratio of 0.94. The shaft efficiency is 99.8 percent. The primary nozzle is variable and CD (to match the exit pressure to the ambient pressure), whereas the fan nozzle is fixed and converging.

(a) Vary the fan pressure ratio and find the optimum developed thrust and the corresponding *TSFC*.
(b) For a fan pressure ratio equal to that in Problem 3.69, what are the thrust and *TSFC*? Why are these different from those found in Problem 3.69?

3.72 A turbofan with the fan fully exhausted operates at SLTO. The compressor pressure ratio is 18, and the fan pressure ratio is 3.90. The fuel has a heating value of 41,400 kJ/kg (17,800 BTU/lb$_m$), and the exit total temperature from the burner is 1583 K (2850 °R). The core airflow is 68.0 kg/s (150 lb$_m$/s). The aerodynamic efficiencies of the compressor, fan, turbine, burner, primary nozzle, and fan nozzle are 89, 90, 92, 95, 96, and 95 percent, respectively. The recovery factor for the diffuser is 0.95 and the burner has a total pressure ratio of 0.94. The shaft efficiency is 99.8 percent. The primary nozzle is variable and CD (to match the exit pressure to the ambient pressure), whereas the fan nozzle is fixed and converging. Vary the bypass ratio and find the optimum *TSFC* and corresponding thrust.

3.73 A turbojet flies at sea level at a Mach number of 0.75. It ingests 165 lb$_m$/s (74.83 kg/s) of air. The compressor operates with an efficiency of 88 percent. The fuel has a heating value of 17,800 BTU/lb$_m$ (41,400 kJ/kg), and the burner total temperature is 2500 °R (1389 K). The burner has an efficiency of 91 percent and a total pressure ratio of 0.95, whereas the turbine has an efficiency of 85 percent. A converging nozzle is used, and the nozzle efficiency is 96 percent. The total pressure recovery for the diffuser is 0.92, and the shaft efficiency is 99.5 percent. Find and plot the developed nondimensional thrust and *TSFC* for compressor pressure ratios of 10 to 20.

3.74 A turbofan flies at sea level at Mach 0.75. It ingests 74.83 kg/s (165 lb$_m$/s) of air to the core. The compressor operates with a pressure ratio of 15 and an efficiency of 88 percent. The engine has a split ratio of 0.25. The efficiency of the fan is 90 percent. The fuel has a heating value of 41,400 kJ/kg (17,800 BTU/lb$_m$), and the burner total temperature is 1389 K (2500 °R). The burner has an efficiency of 91 percent and a total pressure ratio of 0.95, whereas the turbine has an efficiency of 85 percent. The duct has a total pressure ratio of 0.98, and the total pressure ratio of the mixer is 0.97. A variable CD nozzle (to match the exit pressure to the ambient pressure) is used for the primary nozzle, and the efficiency is 96 percent. A converging nozzle is used for the fan nozzle, and the efficiency is 95 percent. The total pressure recovery for the diffuser is 0.92, and the shaft efficiency is 99.5 percent. The Mach numbers at the turbine and duct exits are nearly the same. Find and plot the developed nondimensional thrust and *TSFC* for bypass ratios of 1 to 6.

3.75 A low-bypass afterburning turbofan engine with the fan mixed is to be analyzed. The fan has 2 stages, and the compressor has 12; the turbine has 5 stages. The engine mass is 2950 lbm. The shafts operate at 8500 and 11,500 rpm. The engine performance at 37,000 ft on a standard day is to be considered. The following details are known.

$\dot{m} = 140\ \text{lb}_\text{m}/\text{s}$ (core)	$\gamma_d = 1.40$
$\dot{m}_{f_{ab}} = 9.121\ \text{lb}_\text{m}/\text{s}$	$\gamma_s = 1.39$
$\alpha = 0.7$	$\gamma_c = 1.38$
$\Delta H = 17{,}800\ \text{BTU}/\text{lb}_\text{m}$	$\gamma_b = 1.33$
$p_a = 3.14$ psia	$\gamma_t = 1.31$
$T_a = 387\ °\text{R}$	$\gamma_{ab} = 1.31$
$a_a = 968$ ft/s	$\gamma_n = 1.27$
$M_a = 1.40$	$\eta_c = 0.90$
$T_{t6} = 3500\ °\text{R}$	$\eta_b = 0.95$
$T_{t7} = 817\ °\text{R}$	$\eta_t = 0.93$
$T_{t4} = 2800\ °\text{R}$	$\eta_{ab} = 0.90$
$T_{t3} = 1272\ °\text{R}$	$\eta_m = 0.995$
$p_{t2} = 9.19$ psia	$p_8/p_a = 5.61$
$p_{t7} = 34.92$ psia	$\eta_n = 0.96$ (fixed converging)
Thrust (afterburner on) $= 23{,}250\ \text{lb}_\text{f}$	$\pi_d = 0.92$
Thrust (afterburner off) $= 11{,}120\ \text{lb}_\text{f}$	$\pi_{ab} = 0.96$
	$\pi_b = 0.93$

(a) What is the exit total pressure of the compressor?
(b) What is the fuel flow rate of the primary burner? What is the *TSFC* with the afterburner on? What is the *TSFC* with the afterburner off?
(c) What is the dimensionless thrust with the afterburner on?
(d) What is the exit total temperature of the turbine if the fuel flow rate is negligible?
(e) What is the efficiency of the fan?
(f) What type of application do you expect this engine to have? Why (list specific reasons)?

3.76 A turbojet flies at sea level at Mach 0.75. It ingests 165 lb_m/s (74.83 kg/s) of air. The compressor operates with a compressor pressure ratio of 15 and an efficiency of 88 percent. The fuel has a heating value of 17,800 BTU/lb_m (41,400 kJ/kg), and the burner total temperature is 2500 °R (1389 K). The burner has an efficiency of 91 percent and a total pressure ratio of 0.95, whereas the turbine has an efficiency of 85 percent. A converging nozzle is used, and the nozzle efficiency is 96 percent. The total pressure recovery for the diffuser is 0.92, and the shaft efficiency is 99.5 percent. Find the developed nondimensional thrust and *TSFC* for two independent conditions: (a) the compressor efficiency is increased by three points, and (b) the turbine efficiency is increased by three points. Which is more effective in improving performance?

3.77 A turbofan flies at sea level at a Mach number of 0.75. It ingests 74.83 kg/s (165 lb_m/s) of air to the core. The compressor operates with a pressure ratio of 15 and an efficiency of 88 percent. The engine has a bypass ratio of 3 and a split ratio of 0.25. The efficiency of the fan is 90 percent. The fuel has a heating value of 41,400 kJ/kg (17,800 BTU/lb_m), and the burner total temperature is 1389 K (2500 °R). The burner has an efficiency of 91 percent and a total pressure ratio of 0.95, whereas the turbine has an efficiency of 85 percent. The duct has a total pressure ratio of 0.98, and the total pressure ratio of the mixer is 0.97. A variable CD nozzle (to match the exit pressure to the ambient pressure) is used for the primary nozzle, and the efficiency is 96 percent. A converging nozzle is used for the fan nozzle, and the efficiency is 95 percent. The total pressure recovery for the diffuser is 0.92, and the shaft efficiency is 99.5 percent. The Mach numbers at the turbine and duct exits are nearly the same. Find the developed nondimensional thrust and *TSFC* for two independent conditions: (1) the compressor efficiency is increased by three points, and (2) the turbine efficiency is increased by three points. Which is more effective in improving performance?

3.78 A high-bypass afterburning turbofan engine with the fan exhausted is to be analyzed. The fan has 1 stage and the compressor has 14; the turbine has 4 stages. The primary airflow is 144 lb_m/s, and the bypass ratio is 4.40. The engine mass is 4540 lb_m. The primary burner fuel mass flow rate is 3.62 lbm/s. The inlet and exit total temperatures for the compressor are 529 and 1268 °R, respectively. The inlet total temperature for the fan is 529 °R. The exit total temperatures for the primary burner and turbine are 2770 and 1783 °R, respectively. The thrust without the afterburner is 15,760 lb_f, and the total temperature into the primary nozzle is 1783 °R. The thrust with the afterburner is 22,110 lb_f, and the total temperature into the primary nozzle is 3850 °R. The total pressure ratio for the primary burner and afterburners are 0.95 and 0.97, respectively. The efficiencies for the primary burner and after-burners are 97 and 90 percent, respectively. The heat of reaction of the fuel is 17,800 BTU/lb_m. Both nozzles are converging; the primary and fan nozzles have efficiencies of 97 and 96 percent, respectively. The inlet and exit total pressures to the turbine are 215.0 and 28.4 psia, respectively. The efficiencies of the fan and compressor are 89 and 90 percent, respectively. The recovery factor for the diffuser is 0.96, and the shaft mechanical efficiency is 99.7 percent. The shafts operate at 7400 and 10,900 rpm. The engine performance at a Mach number of 0.90 at 18,000 ft on a standard day ($p_a = 7.33$ psia, $T_a = 455$ °R, $a_a = 1047$ ft/s) is to be considered. The following specific heat ratios are known.

$\gamma_d = 1.40$	$c_{p_d} = 0.2390$ BTU/lb_m-°R
$\gamma_s = 1.40$	$c_{p_s} = 0.2401$ BTU/lb_m-°R
$\gamma_c = 1.38$	$c_{p_c} = 0.2477$ BTU/lb_m-°R
$\gamma_b = 1.33$	$c_{p_b} = 0.2762$ BTU/lb_m-°R
$\gamma_t = 1.32$	$c_{p_t} = 0.2832$ BTU/lb_m-°R
$\gamma_{ab} = 1.30$	$c_{p_{ab}} = 0.2985$ BTU/lb_m-°R
$\gamma_n = 1.26$	$c_{p_n} = 0.3301$ BTU/lb_m-°R
$\gamma_{fn} = 1.40$	$c_{p_{fn}} = 0.2412$ BTU/lb_m-°R

(a) What is the total pressure ratio of the fan (afterburner on)?
(b) What is the fuel flow rate of the afterburner when lit? What is the *TSFC* with the afterburner on?
(c) What is the dimensionless thrust with the afterburner off?
(d) What is the efficiency of the turbine (afterburner on)?
(e) What is the exit velocity from the primary nozzle (afterburner on)?

3.79 Draw a general *h–s* diagram for a turbojet engine that has a lit afterburner (see Figure 3.2). Draw this to scales that show the correct relative magnitudes; that is, be careful to show reasonable relative pressures and enthalpies for the different components. Show all points (a, 1, 2, 3, 4, 5, 6, and 8) as well as static and total pressures. Show only the nonideal case.

3.80 A high-bypass turbofan engine with the fan exhausted is to be analyzed. The core or primary flow has an afterburner. The fan does not have an afterburner. The fan has 2 stages, and the compressor has 10; the turbine has 4 stages. The single-spool shaft operates at 10,500 rpm. The engine performance at 10,000 ft on a standard day is to be considered.

Overall		
$\dot{m} = 75$ lb_m/s (core)	$T_a = 483$ °R	$p_a = 10.10$ psia
$M_a = 0.66$	$\eta_m = 0.995$	Thrust $= 8894$ lb_f
engine mass $= 2550$ lb_m	$N = 10{,}500$ rpm	
Diffuser		
$\pi_d = 0.92$	$\gamma_d = 1.40$	$c_{p_d} =$ 0.240 BTU/lb_m-°R
Compressor		
$\pi_c = 22.0$ (includes fan)	$\gamma_c = 1.38$	$c_{p_c} =$ 0.248 BTU/lb_m-°R
$\eta_c = 0.89$		
Fan		
$\pi_f = 2.40$	$\gamma_f = 1.40$	$c_{p_f} =$ 0.240 BTU/lb_m-°R
$\alpha = 3.2$	$T_{t7} = 687$ °R	

(cont.)

Burner		
$\dot{m}_f = 1.74\ lb_m/s$	$\gamma_b = 1.33$	$c_{p_b} =$ 0.275 BTU/lb_m-°R
$\Delta H =$ 17,700 BTU/lb_m	$\eta_b = 0.93$	$\pi_b = 0.95$
Turbine		
$\eta_t = 0.86$	$\gamma_t = 1.33$	$c_{p_t} =$ 0.278 BTU/lb_m-°R
Afterburner		
$\dot{m}_{f_{ab}} = 2.52\ lb_m/s$	$\gamma_{ab} = 1.32$	$c_{p_{ab}} = 0.285$ BTU/lb_m-°R
$T_{t6} = 3200$ °R	$\eta_{ab} = 0.88$	$\pi_{ab} = 0.98$
Primary nozzle		
Variable CD	$p_8/p_a = 1.00$	
$\pi_n = 0.97$	$\eta_n = 1.28$	$c_{p_n} =$ 0.310 BTU/lb_m-°R
Fan nozzle		
Converging		
$\eta_{fn} = 0.92$	$\gamma_{fn} = 1.39$	$c_{p_{fn}} =$ 0.243 BTU/lb_m-°R

(a) What is the exit total temperature of the compressor?
(b) What is the exit total temperature of the primary burner if the total temperature at the exit of the compressor is 1321 °R?
(c) If the total temperature out of the primary combustor is 2650 °R and the total temperature at the exit of the compressor is 1321 °R, what is the total pressure out of the turbine?
(d) What is the *TSFC*?
(e) What is the pressure at the exit of the fan nozzle?
(f) What is the efficiency of the fan?

All parts can be worked independently.

3.81 The following are from a mixed turbofan with a bypass ratio of 0.55.

(a) A compressor operates with an inlet total temperature of 518 °R and an exit total temperature of 1252 °R. The inlet total pressure is 12.5 psia, and the exit total pressure is 237.5 psia. The specific heat ratio is 1.3834.

(1) What is the compressor efficiency?
(2) If the input power remains constant, the inlet total conditions remain constant, and the efficiency is raised to 100 percent, what is the exit total pressure of the compressor?

(b) A turbine operates with an inlet total temperature of 2860 °R and an inlet total pressure of 230.4 psia. The compressor and fan have inlet total temperatures of 518 °R. The compressor and fan have exit total temperatures of 1252 and 815 °R, respectively. The core mass flow rate is 190 lb_m/s, and the fuel flow rate is 5.4 lbm/s. The turbine efficiency is 90 percent, and the shaft mechanical efficiency is 99.9 percent. The specific heat ratio of the compressor, fan, and turbine are 1.3834, 1.3949, and 1.3109, respectively. The specific heats at constant pressure for the compressor, fan, and turbine are 0.2474, 0.2422, and 0.2891 BTU/lb_m-°R, respectively.

(1) Find the exit total pressure of the turbine.

(c) At 10,000 ft, the net thrust with an afterburner lit is 34,240 lb_f. The primary burner fuel flow rate is 5.4 lb_m/s, and the afterburner fuel flow is 13.1 lb_m/s. The core flow rate is 190 lb_m/s. The inlet total temperature to the nozzle is 3860 °R, and the inlet total pressure to the nozzle is 54.0 psia.

(1) Find the nondimensional thrust.
(2) Find the *TSFC*.

3.82 An exhausted turbofan with a bypass ratio of 0.80 at SLTO has the following components. The core mass flow rate is 150 lb_m/s.

(a) The compressor operates with an inlet total temperature of 519 °R and an exit total temperature of 1235 °R. The inlet total pressure is 13.96 psia, and the efficiency is 89 percent. The specific heat ratio is 1.3838. What is the total exit total pressure from the compressor? If the input power remains constant, the inlet total conditions remain constant, and the efficiency is raised to 100 percent, what is the exit total pressure of the compressor?

(b) The turbine operates with an inlet total temperature of 2850 °R and an inlet total pressure of 236.1 psia. The compressor and fan have inlet total temperatures of 519 °R. The compressor and fan have exit total temperatures of 1235 and 790 °R, respectively. The fuel flow rate is 4.2 lb_m/s. The turbine exit total pressure is 53.9 psia, and the shaft mechanical efficiency is 99.8 percent. The specific heat ratio of the compressor, fan, and turbine are 1.3838, 1.3955, and 1.3119, respectively. The specific heats at constant pressure for the compressor, fan, and turbine are 0.2472, 0.2419, and 0.2884 BTU/lb_m-°R, respectively. Find the efficiency of the turbine. If the turbine efficiency is raised to 100 percent, what percent increase of power to the fan and compressor will result for the same turbine inlet total pressure and temperature and the same turbine exit total pressure?

(c) The thrust from the primary nozzle of an exhausted turbofan (with an afterburner off) is 13,212 lb_f. The primary burner fuel flow rate is 4.16 lb_m/s. The core flow rate is 150 lb_m/s. The inlet total temperature to the variable CD primary nozzle is 2074 °R, and the inlet total pressure to the primary nozzle is 53.9 psia; the efficiency is 96 percent, the specific heat ratio is 1.3278, and the exit velocity is 2757 ft/s. The inlet total temperature to the converging fan nozzle is 790 °R, and the inlet total pressure to the fan nozzle is 54.4 psia; the efficiency is 95 percent and the specific heat ratio is 1.3883. Find the nondimensional thrust. Find the *TSFC*.

3.83 An aircraft with an afterburning turbojet engine cruises at 866 ft/s with the ambient pressure at 5.45 psia and ambient temperature at 429.6 °R. The engine has the components listed below. The core mass flow rate is 210 lb_m/s.

(a) The compressor operates with an inlet total temperature of 492.3 °R and an exit total temperature of 1187 °R. The inlet total pressure is 8.22 psia, and the total pressure ratio is 19.0. The specific heat ratio is 1.3857. What is the efficiency of the compressor?

(b) The afterburner operates with an inlet total temperature of 2171 °R, an inlet total pressure of 50.95 psia, an efficiency of 92 percent, and a total pressure ratio of 0.99. The primary burner fuel flow rate is 5.59 lb_m/s, and the afterburner fuel flow rate is 6.29 lb_m/s. The specific heat at constant pressure for the afterburner is 0.3012 BTU/lb_m-°R, and the heating value of the fuel is 17,800 BTU/lb_m. Find the exit total temperature of the afterburner.

(c) The turbine operates with an inlet total temperature of 2750 °R and an inlet total pressure of 150.0 psia. The compressor has an inlet total temperature of 492.3 °R and an exit total temperature of 1187 °R. The core mass flow rate is 210 lb_m/s, and the fuel flow rate is 5.59 lb_m/s. The turbine exit total pressure is 51.0 psia, and the shaft mechanical efficiency is 99.8 percent. The specific heat ratios of the compressor, and turbine are 1.3857 and 1.3119, respectively. The specific heats at constant pressure for the compressor, and turbine are 0.2463 and 0.2883 BTU/lb_m-°R, respectively. Find the efficiency of the turbine.

3.84 An aircraft powered by a turbofan with the fan exhausted cruises at 993 ft/s with the ambient pressure at 10.10 psia and ambient temperature at 483 °R.

The engine has the components listed below. The core mass flow rate is 255 lb_m/s, and the primary burner fuel flow rate is 6.64 lb_m/s.

(a) The turbine operates with an inlet total temperature of 2950 °R and an exit total temperature of 2094 °R. The inlet total pressure is 411 psia, and the total pressure ratio is 0.205. The specific heat ratio is 1.310. What is the efficiency of the turbine?

(b) The primary combustor operates with an inlet total temperature of 1434 °R, an inlet total pressure of 419 psia, an efficiency of 96 percent, and a total pressure ratio of 0.98. The specific heat at constant pressure is 0.281 BTU/lb$_m$-°R, and the heating value of the fuel is 17,900 BTU/lb$_m$. Find the exit total temperature of the combustor.

(c) The turbine operates with an inlet total temperature of 2950 °R and an exit total temperature of 2094 °R; the inlet total pressure is 411 psia. The compressor has an inlet total temperature of 565 °R and an exit total temperature of 1434 °R; the inlet total pressure is 16.8 psia. The fan has an inlet total temperature of 565 °R, an inlet total pressure of 16.8 psia, and an exit total pressure of 37.7 psia. The bypass ratio is 0.95, and the shaft mechanical efficiency is 99.8 percent. The specific heat ratios of the fan, compressor, and turbine are 1.396, 1.378, and 1.310, respectively. The specific heats at constant pressure for the fan, compressor, and turbine are 0.242, 0.250, and 0.290 BTU/lb$_m$-°R, respectively. Find the efficiency of the fan.

3.85 A ground-based gas turbine with regeneration (Figure 1.50), which is used for electric power generation, ingests 132 lb$_m$/s of air at standard sea level. The compressor pressure ratio is 15.6, the fuel has a heating value of 18,000 BTU/lb$_m$, and the burner exit total temperature is 2477 °R. The efficiencies of the compressor, burner, turbine, and shaft are 87.7, 91, 92, and 99.5 percent, respectively. The total pressure ratios for the inlet, burner, and exhaust are 0.983, 0.967 and 0.979, respectively. The total pressure ratios for both paths (3–3x, and 5–5x) of the regenerator are 0.991 (owing to total pressure losses from friction), and the regenerator effectiveness is 0.75. The specific heat ratios for the compressor, burner, turbine, inlet, and exhaust are 1.385, 1.335, 1.334, 1.400, and 1.366, respectively. The specific heats for paths (3–3x, and 5–5x) of the regenerator are 0.257 and 0.258 BTU/lb$_m$-°R, respectively. Find:

(a) the overall thermal efficiency,
(b) the developed net power output,
(c) the specific fuel consumption, *SFC*.

For reference, if all component efficiencies and their effectiveness are unity and no irreversible pressure losses occur, the results are 56.1 percent, 36,660 hp, and 0.252 lb$_m$/hp-h.

3.86 Before reaching altitude, an aircraft powered by a turbofan with the fan totally mixed flies at 540 ft/s with the ambient pressure at 10.10 psia and ambient temperature at 483 °R. The core mass flow rate is 195 lb$_m$/s, and the primary burner fuel flow rate is 5.18 lb$_m$/s. The bypass ratio is 0.75.

(a) The turbine operates with an inlet total temperature of 2850 °R and an efficiency of 93 percent. The turbine inlet total pressure is 281.9 psia. The turbine specific heat ratio is 1.314, and the specific heat at constant pressure is 0.287 BTU/lb$_m$-°R. The compressor has inlet and exit total temperatures of 507 and 1308 °R, an efficiency of 91 percent, a specific heat ratio of 1.382, and the $c_p = 0.248$ BTU/lb$_m$-°R. The fan has inlet and exit total temperatures of 507 and 815 °R, a total pressure ratio of 4.66, a specific heat ratio of 1.395, and the $c_p = 0.242$ BTU/lb$_m$-°R. The shaft mechanical efficiency is 100 percent. What is the exit total temperature of the turbine?

(b) The afterburner operates with an inlet total temperature of 1509 °R, an inlet total pressure of 419 psia, an efficiency of 88 percent, and a total pressure ratio of 0.99. The afterburner fuel flow rate is 17.48 lb$_m$/s. The specific heat at constant pressure is 0.296 BTU/lb$_m$-°R, and the heating value of the fuel is 17,800 BTU/lb$_m$. Find the exit total temperature of the afterburner.

(c) The fan has inlet and exit total temperatures of 507 and 815 °R, a total pressure ratio of 4.66, a specific heat ratio of 1.395, and the $c_p = 0.242$ BTU/lb$_m$-°R. Find the efficiency of the fan.

(d) Analyze the fixed converging nozzle of the engine with the afterburner off. The inlet total temperature to the nozzle is 1509 °R, the inlet total pressure is 52.6 psia, the efficiency is 97 percent, the specific heat ratio is 1.353, and the specific heat at constant pressure is 0.263 BTU/lb_m-°R. The nozzle exit area is 498 in.^{2}. Find the exit Mach number of the nozzle, engine thrust, and *TSFC*.

3.87 A compressor is composed of *N* stages, each with a total pressure ratio of 1.25 and an efficiency of 90 percent. Plot the compressor efficiency and total pressure ratio for *N* from 1 to 16.

3.88 A ground-based gas turbine (Figure 1.11) used for electric power generation ingests 132 lb_m/s of air at standard sea level. The compressor pressure ratio is 15.6, the fuel has a heating value of 18,000 BTU/lb_m, and the burner exit total temperature is 2477 °R. The efficiencies of the compressor, burner, turbine, and shaft are 87.7, 91, 92, and 99.5 percent, respectively. The total pressure ratios for the inlet, burner, and exhaust are 0.983, 0.967, and 0.979, respectively. The specific heat ratios for the compressor, burner, turbine, inlet, and exhaust are 1.385, 1.338, 1.334, 1.403, and 1.366, respectively.

(a) Find the overall thermal efficiency, developed net power output, and specific fuel consumption, *SFC*.
(b) If the inlet air is precooled so that the inlet static temperature is dropped by 50 °R, find the overall thermal efficiency, developed net power output, and specific fuel consumption.

3.89 A fully mixed afterburning turbofan flies at 10,670 m (35,000 ft) at Mach 0.93. It ingests 43.10 kg/s (95 lb_m/s) of air to the core. The compressor operates with a pressure ratio of 24 and an efficiency of 91 percent. The engine has a bypass ratio of 0.66. The efficiency of the fan is 92 percent. The fuel has a heating value of 41400 kJ/kg (17,800 BTU/lb_m), and the burner total temperature is 1595 K (2870 °R). The burner has an efficiency of 97 percent and a total pressure ratio of 0.96, whereas the turbine has an efficiency of 93 percent. The duct has a total pressure ratio of 1.00, and the total pressure ratio of the mixer is 0.99. A variable CD nozzle (to match the exit pressure to the ambient pressure) is used, and its efficiency is 96 percent. The total pressure recovery for the diffuser is 0.96, and the shaft efficiency is 99.5 percent. The Mach numbers at the turbine and duct exits are nearly the same. The efficiency of the afterburner is 92 percent, and the total pressure ratio is 0.98. The exit total temperature from the afterburner is 2189 K (3940 °R). Find the developed thrust, nondimensional thrust, and *TSFC*.

3.90 A high-bypass turbofan engine with the fan exhausted is to be analyzed. The primary airflow is 90.72 kg/s (200 lb_m/s), and the bypass ratio is 6. The total pressure ratios for the fan and compressor are 2.2 and 36, respectively. The exit total temperature for the primary burner is 1678 K (3020 °R), and the total pressure ratio for the burner is 0.95. The efficiency for the burner is 95 percent, and the heat of reaction of the fuel is 41,400 kJ/kg (17,800 BTU/lb_m). Both nozzles are converging; the primary and fan nozzles have efficiencies of 96 and 96 percent, respectively. The efficiencies of the fan, compressor, and turbine are 92, 90, and 91 percent, respectively. The recovery factor for the diffuser is 0.97, and the shaft mechanical efficiency is 99.3 percent. The engine performance at Mach 0.30 at 0 m (0 ft) on a standard day is to be considered (just after takeoff). Find the developed thrust, nondimensional thrust, and *TSFC*.

3.91 The turbine of a turbofan with the fan mixed operates with an inlet total temperature of 2880 °R and an efficiency of 93 percent. The turbine inlet total pressure is 147.0 psia. The turbine specific heat ratio is 1.313, and the specific heat at constant pressure is 0.288 BTU/lb_m-°R. The core mass flow rate is 177 lb_m/s, the bypass ratio is 0.90, and the primary burner fuel flow rate is 5.00 lb_m/s. The aircraft flies at 919 ft/s with the ambient pressure at 3.14 psia and ambient temperature at 387 °R. The compressor has inlet and exit total temperatures of 458 and 1238 °R, a total pressure ratio of 29, a specific heat ratio of 1.385, and the $c_p = 0.246$ BTU/lb_m-°R. The fan has inlet and exit total temperatures of 458 and 761 °R, a total pressure ratio of 5.328, a specific heat ratio of 1.398, and the $c_p = 0.241$ BTU/lb_m-°R. The primary burner operates with an inlet total temperature of 1238 °R, an inlet total pressure of 153 psia, an efficiency of 95 percent, and a total pressure ratio of 0.96. The primary burner fuel flow rate is 5.00 lb_m/s, and the afterburner fuel flow rate is 16.78 lb_m/s. The burner-specific heat at constant pressure is 0.277 BTU/lb_m-°R, and the heating value of the fuel is 17,800 BTU/lb_m. The inlet total temperature to the fixed converging nozzle is 3975 °R, the inlet total pressure is 27.85 psia, the nozzle efficiency is 97 percent, the nozzle specific heat ratio is 1.258, the specific heat at constant pressure is 0.334 BTU/lb_m-°R, and the nozzle exit area is 1614 in.^{2}. The shaft mechanical efficiency is 100 percent.

(a) What is the exit total pressure of the turbine?
(b) Find the efficiency of the compressor.
(c) If the compressor has 14 stages, what is the average total pressure ratio for each stage?
(d) Find the exit total temperature of the burner.
(e) Find the exit Mach number of the nozzle.
(f) Find the thrust.
(g) Find the *TSFC*.

For comparison, if a CD nozzle were used matching the exit pressure to the ambient pressure, the thrust would be 44,113 lb_f and the *TSFC* would have a value of 1.778 lb_m/h-lb_f.

3.92 An air standard power generation Brayton cycle operates with a maximum temperature of 1606 K. The inlet air temperature and pressure to the compressor are 0 °C and 102 kPa, respectively, and the compressor pressure ratio is 21. The compressor, turbine, and burner have efficiencies of 91, 93, and 100 percent, respectively. The total pressure ratios for the inlet, burner, and exhaust are all unity. The airflow rate is 25 kg/s. The heating value of the fuel is 42,000 kJ/kg. Assume constant specific heats and $\gamma = 1.36$. Draw the T–s diagram and find:

(a) the net power output,
(b) the system thermal efficiency,
(c) the fuel flow rate.

3.93 A new advanced commercial aircraft is to be designed to routinely fly at 11,890 m (39,000 ft) at an air speed of 257.6 m/s (845 ft/s). Owing to expected drag characteristics of the airframe, three engines are required, and the required thrust is 71.17 kN (16,000 lb_f) per engine and fuel economies dictate that the engines are to operate with *TSFC* values of 0.060 kg/N-h (0.59 lb_m/h-lb_f). The material limit on the exit temperature of the burner is 1653 K (2975 °R). Design an engine for this application. Make reasonable approximations for nonideal characteristics.

3.94 A new advanced military fighter aircraft is to be designed to fly routinely at a variety of altitudes. One particular condition is at 7315 m (24,000 ft) at an air speed of 302 m/s (992 ft/s). Owing to expected drag characteristics of the airframe, two engines are required and the required thrust is 64.5 kN (14,500 lb_f) per engine, and engine weight considerations dictate that the engines are to operate with nondimensional thrusts of 2.00 and with a *TSFC* value of 0.094 kg/N-h (0.92 lb_m/h-lb_f). Furthermore, since the aircraft is a fighter aircraft, short increases in thrust for rapid acceleration to supersonic speeds are required for a total of 125 kN (28,100 lb_f) per engine. The material limits on the exit temperatures of the primary burner and afterburner are 1667 K and 2056 K (3000 °R and 3700 °R), respectively. Design an engine for this application and conditions. Make reasonable approximations for nonideal characteristics.

3.95 You are to analyze parts of a turbofan engine (with the fan fully exhausted).

The aircraft cruises at 818 ft/s with the ambient pressure at 4.36 psia and ambient temperature at 412 °R. The core air mass flow rate is 200 lb_m/s. The turbine operates with an inlet total temperature of 2870 °R and an exit total temperature of 1561 °R. The inlet total pressure is 158 psia and the exit total pressure is 8.66 psia. The compressor has an inlet total temperature of 468 °R and an exit total temperature of 1237 °R; the inlet total pressure is 6.58 psia. The fan has an inlet total temperature of 468 °R, an inlet total pressure of 6.58 psia, an exit total pressure of 14.5 psia, and an efficiency of 91 percent. The primary combustor operates with an inlet total temperature of

1237 °R, an inlet total pressure of 165 psia, an efficiency of 95 percent, and a total pressure ratio of 0.96. The primary burner fuel flow rate is 5.58 lb_m/s. The heating value of the fuel is 17,900 BTU/lb_m. The inlet total temperature to the fan nozzle is 600 °R, the inlet total pressure is 14.5 psia. The exit area is 3930 $in.^2$. The shaft mechanical efficiency is 100 percent. The specific heat ratios of the fan, compressor, turbine, and nozzle are 1.400, 1.385, 1.322, and 1.398, respectively. The specific heats at constant pressure for the fan, compressor, burner, turbine, and nozzle are 0.240, 0.247, 0.277, 0.281, and 0.241 BTU/lb_m-°R, respectively. Using intercomponent data, do the following.

(a) Calculate the efficiency of the turbine.
(b) Calculate the exit total temperature of the combustor.
(c) Calculate the exit total pressure of the combustor.
(d) Calculate the exit Mach number of the fan nozzle.
(e) Calculate the total mass flow rate through the fan nozzle.
(f) Calculate the bypass ratio.

3.96 A ground-based gas turbine with regeneration, intercooling, and reheat (Figure 1.50), which is used for electric power generation, ingests 140 lb_m/s of air at standard sea level. The low-pressure compressor pressure ratio is 4.5, the high-pressure compressor pressure ratio is also 4.5. The fuel has a heating value of 18,000 BTU/lb_m, and the burner and reheat burner exit total temperatures are both 2600 °R. The efficiencies of the LP compressor and HP compressor are 88.0 percent and 88.0 percent, respectively. The efficiencies of the burner and reheat burner are 91 percent and 91 percent, respectively. The efficiencies of the HP turbine, LP turbine and free turbine are 91 percent, 91 percent, and 91 percent, respectively. The efficiencies of the shafts are all 99.5 percent. The total pressure ratios for the inlet and exhaust are 0.980 and 0.970, respectively. The total pressure ratios for the primary and reheat burners are 0.94 and 0.94, respectively. The total pressure ratios for both paths (3–3x and 5–5x) of the regenerator are 0.985 (owing to total pressure losses from friction), and the regenerator effectiveness is 0.75. The total pressure ratios for both paths (25a–25b and 25c–25d) of the intercooler 0.985 (owing to total pressure losses from friction), and the intercooler effectiveness is 0.75. Water is the coolant for the intercooler and flows at 50 lb_m/s and enters at 500 °R. The specific heat ratios for the LP compressor, HP compressor, primary burner, reheat burner, HP turbine, LP turbine, free turbine, inlet, and exhaust are 1.394, 1.390, 1.330, 1.312, 1.312, 1.312, 1.332, 1.400, and 1.372, respectively. The specific heats for paths (3–3x, and 5–5x) of the regenerator are 0.2545 and 0.2588 BTU/lb_m-°R, respectively. The specific heats for paths 25a–25b, and 25c–25d of the intercooler are 0.2431 and 1.000 BTU/lb_m-°R, respectively. Find:

(a) the overall thermal efficiency,
(b) the developed net power output,
(c) the specific fuel consumption, *SFC*.

For reference, if all component efficiencies and their effectivness (rather hopeful) are unity and no irreversible pressure losses occur, the results are 67.2 percent, 56,810 hp, and 0.210 lb_m/hp-h.

3.97 A ground-based gas turbine with the possibility of using regeneration, intercooling, and reheat (Figure 1.50), which is used for electric power generation, ingests 140 lb_m/s of air at standard sea level. The low-pressure compressor pressure ratio is 4.5, the high-pressure compressor pressure ratio is also 4.5. The fuel has a heating value of 18,000 BTU/lb_m, and the burner and reheat burner exit total temperatures are both 2600 °R. The efficiencies of the LP compressor and HP compressor are 88.0 percent and 88.0 percent, respectively. The efficiencies of the burner and reheat burner are 91 percent and 91 percent, respectively. The efficiencies of the HP turbine, LP turbine and free turbine are 91 percent, 91 percent, and 91percent, respectively. The efficiencies of the shafts are all 99.5 percent. The total pressure ratios for the inlet and exhaust are 0.980 and 0.970, respectively. The total pressure ratios for the primary and reheat burners are 0.94 and 0.94, respectively. The total pressure ratios for both paths

(3–3x and 5–5x) of the regenerator are 0.985 (owing to total pressure losses from friction), and the regenerator effectiveness is 0.75. The total pressure ratios for both paths (25a–25b and 25c–25d) of the intercooler 0.985 (owing to total pressure losses from friction), and the intercooler effectiveness is 0.75. Water is the coolant for the intercooler and flows at 50 lb_m/s and enters at 500 °R. Consider four conditions: No enhancers, only intercooling, only regeneration, and only reheat. Note that this is similar to Problem 3.96. Find for each case:

(a) the overall thermal efficiency,
(b) the developed net power output,
(c) the specific fuel consumption.

3.98 A ground-based gas turbine with the possibility of using regeneration, which is used for electric power generation, ingests 140 lb_m/s of air at standard sea level. The value of the single pressure compressor pressure ratio is 20.25. The efficiency of the compressor is 85.75 percent. The fuel has a heating value of 18,000 BTU/lb_m, and the burner exit total temperatures is 2600 °R. The efficiency of the burner is 91 percent. The total pressure ratio for the burner is 0.94. The efficiency of the single turbine is 92.78 percent. The efficiency of the shaft is 99.5 percent. The total pressure ratios for the inlet and exhaust are 0.980 and 0.970, respectively. The total pressure ratios for both paths (3–3x and 5–5x) of the regenerator are 0.985 (owing to total pressure losses from friction), and the regenerator effectiveness is 0.75. Note that this is similar to Problem 3.96 except this has single turbine and compressor. Use a constant value of $\gamma = 1.35$, which is also different from Problem 3.96. Find for the cases of the unit with and without regeneration:

(a) the overall thermal efficiency,
(b) the developed net power output,
(c) the specific fuel consumption.

3.99 A ground-based gas turbine with the possibility of using regeneration, which is used for electric power generation, ingests 140 lb_m/s of air at standard sea level. The value of the single pressure compressor pressure ratio is to be optimized. The expected efficiency of the compressor is 85.75 percent. The fuel has a heating value of 18,000 BTU/lb_m, and the burner exit total temperatures is 2600 °R. The efficiency of the burner is 91 percent. The total pressure ratio for the burner is 0.94. The efficiency of the single turbine is 92.78 percent. The efficiency of the shaft is 99.5 percent. The total pressure ratios for the inlet and exhaust are 0.980 and 0.970, respectively. The total pressure ratios for both paths (3–3x and 5–5x) of the regenerator are 0.985 (owing to total pressure losses from friction), and the regenerator effectiveness is 0.75. Note that this is similar to Problem 3.98, except the power is to be optimized. Use a constant value of $\gamma = 1.35$. Find for the cases of the unit with and without regeneration:

(a) the optimum value of compressor pressure ratio to maximize power, along with
(b) the overall thermal efficiency,
(c) the developed net power output,
(d) the specific fuel consumption.

3.100 A ground-based gas turbine with regeneration, intercooling, and reheat (Figure 1.50), which is used for electric power generation, ingests 140 lb_m/s of air at standard sea level. The low-pressure compressor pressure ratio is to be optimized, the high-pressure compressor pressure ratio is also to be optimized and equal to that of the low-pressure compressor. The efficiencies of the LP compressor and HP compressor are 88.0 percent and 88.0 percent, respectively. The fuel has a heating value of 18,000 BTU/lb_m, and the burner and reheat burner exit total temperatures are both 2600 °R. The efficiencies of the burner and reheat burner are 91 percent and 91 percent, respectively. The efficiencies of the HP turbine, LP turbine and free turbine are 91 percent, 91 percent, and 91 percent, respectively. The efficiencies of the shafts are all 99.5 percent. The total pressure ratios for the inlet and exhaust are 0.980 and 0.970, respectively. The total pressure ratios for the primary and reheat burners are 0.94 and 0.94, respectively. The total pressure ratios for both paths (3–3x and 5–5x) of the regenerator are

0.985 (owing to total pressure losses from friction), and the regenerator effectiveness is 0.75. The total pressure ratios for both paths (25a–25b and 25c–25d) of the intercooler 0.985 (owing to total pressure losses from friction), and the intercooler effectiveness is 0.75. Water is the coolant for the intercooler and flows at 50 lb_m/s and enters at 500 °R. You may assume that the specific heat ratios for the LP compressor, HP compressor, primary burner, reheat burner, HP turbine, LP turbine, free turbine, inlet, and exhaust are 1.394, 1.390, 1.330, 1.312, 1.312, 1.312, 1.332, 1.400, and 1.372, respectively. The specific heats for paths (3–3x, and 5–5x) of the regenerator are 0.2545 and 0.2588 BTU/lb_m-°R, respectively. The specific heats for paths 25a–25b, and 25c–25d of the intercooler are 0.2431 and 1.000 BTU/lb_m-°R, respectively. This is similar to Problem 3.96 except for the optimization. Optimize the compressor pressure ratios for the three cases of: (1) no enhancers, (2) all enhancers, and (3) all enhancers except no reheat, and find:

(a) the overall thermal efficiency,
(b) the developed net power output,
(c) the specific fuel consumption, *SFC*.
(d) Comment on the use of the reheat and regeneration.

3.101 Find the efficiency of an ideal power generation gas turbine with intercooling in terms of T_{t5}, T_{t4}, T_{t3IC}, T_{t2}, $T_{t2.5a}$, $T_{t2.5b}$, and τ_{IC}, where $\tau_{IC} = T_{t2.5b}/T_{t2.5a}$. See Figure 1.51. Also find the efficiency of an ideal power generation gas turbine without intercooling in terms of T_{t5}, T_{t4}, T_{t3NoIC}, and T_{t2}. Note that T_{t3} will be different for the two cases ($T_{t3NoIC} > T_{t3IC}$). Compare the two cases using some typical numbers and show that although the net power is increased with intercooling the efficiency actually decreases (despite many published comments). Suggested representative values are: $T_{t5} = 1148$, $T_{t4} = 2620$, $T_{t3NoIC} = 1184$, $T_{t2} = 519$, $T_{t2.5b} = 500$, (all °R), and τ_{IC}=0.638. You will have to determine T_{t3IC}.

3.102 A turbojet flies at sea level at a Mach number of 0.75. It ingests 165 lb_m/s (74.83 kg/s) of air. The compressor operates with a pressure ratio of 15 and an efficiency of 88 percent. The fuel has a heating value of 17,800 BTU/lb_m (41,400 kJ/kg) and the burner total temperature is 2500 °R (1389 K). The burner has an efficiency of 91 percent and a total pressure ratio of 0.95, while the turbine has an efficiency of 85 percent. A converging nozzle is used and the nozzle efficiency is 96 percent. The total pressure recovery for the diffuser is 0.92 and the shaft efficiency is 99.5 percent. Find the developed thrust and *TSFC* for two cases.

(a) As you will find, the nozzle is choked so assume $M_8 = 1.000$.
(b) Find the nozzle exit Mach number due to the nozzle efficiency (96 percent) and use that and compare to results for $M_8 = 1.000$.

Part II
Rotating Components

Rolls-Royce RB211 compressor blades (© Rolls-Royce plc 2022. All rights reserved).

Nuovo Pignone PGT2 centrifugal compressor (courtesy of General Electric Gas Power).

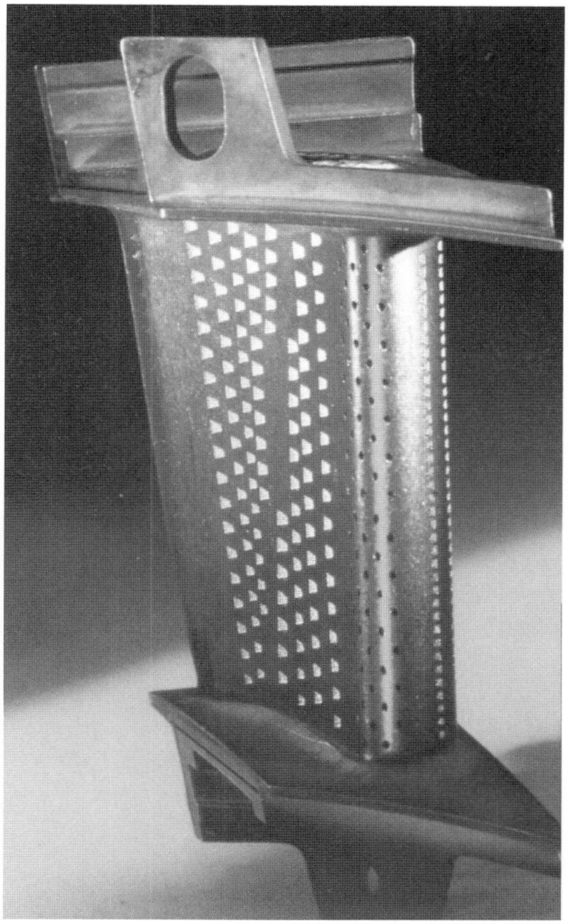

Pratt & Whitney/ Mitsubishi FT8 turbine blade (© 2022 Raytheon Technologies Corporation, Pratt & Whitney division. Reproduced with permission).

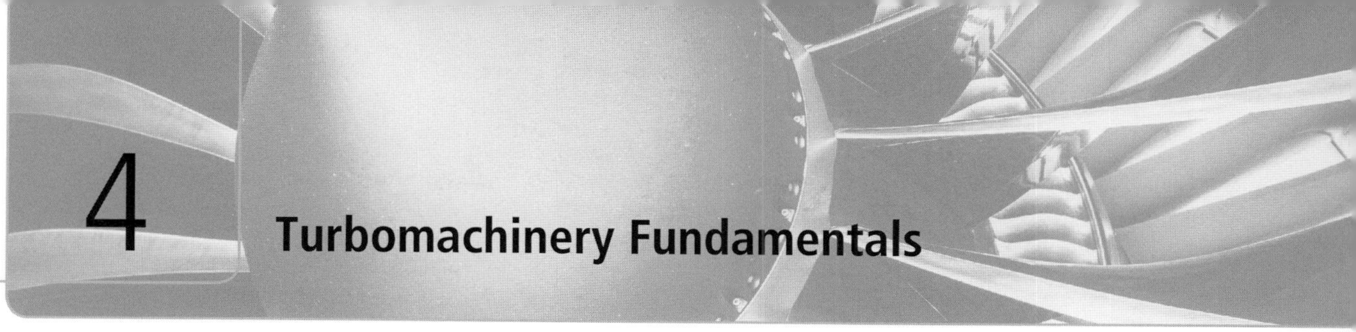

4 Turbomachinery Fundamentals

4.1 Introduction

Turbomachines are the heart of all production jet engines and power generation gas turbines. This book discusses six basic types of turbomachines directly: axial flow compressors, axial flow pumps, radial flow compressors, centrifugal pumps, axial flow gas turbines, and axial flow hydraulic turbines. Two other basic types that are used in practice are not covered in this text because of their limited application in propulsion and power generation: radial inflow gas turbines and radial inflow hydraulic turbines. The basic derivations of the equations for each machine type are covered in this chapter rather than in the chapters in which the machines are discussed. As will be shown, the resulting fundamental equations apply to all types of turbomachines, regardless of categorization. Application of the equations with complementing internal velocity information is, however, different for the different turbomachines. Advanced details can be found in texts, including Stodola (1927), Howell (1945a, 1945b), Shepherd (1956), Vavra (1974), Dixon (1975, 1998), Osborn (1977), Balje (1981), Whittle (1981), Wallis (1983), Turton (1984), Hill and Peterson (1992), Logan (1993), Japikse and Baines (1994), Cohen et al. (1996), Hah (1997), and Wilson and Korakianitis (2014). Cumpsty and Greitzer (2004) present an interesting review of turbomachinery development. Furthermore, Denton and Dawes (1998), Elmendorf et al. (1998), LeJambre et al. (1998), Rhie et al. (1998), Adamczyk (2000), and Denton (2010) show how modern computational fluid dynamic (CFD) tools can effectively be used for the complex 3-D analysis and design of turbomachines and also cover some of the limitations. In the following three chapters the equations developed in this chapter will be used to find the operating characteristics of compressors and turbines.

4.1.1 Objectives

After completing this chapter, the reader should be able to do the following.

- Understand the idea of a mean line single-stage analysis
- Understand the assumptions of a mean line single-stage analysis
- Develop fluid mechanic, thermodynamic, and gas dynamic equations to be used to find the performance of compressors and turbines from velocity data
- Understand the premises of pressure ratios, efficiency, and percent reaction
- Understand the definitions of "corrected" speed and "corrected" mass flow rate
- Understand the application of "corrected" speed and "corrected" mass flow rate

4.2 Single-Stage Energy Analysis

In this section, the equations are derived so that the internal fluid velocities (later referred to as velocity polygons or triangles) can be related to the pressure change, power, and other important turbomachine characteristics. This is done for a general single stage (rotor and stator for an axial machine or impeller and diffuser for a centrifugal machine). A well-documented and versatile control volume approach is used for the derivations. In Figure 4.1, a single stage is shown for an axial flow turbomachine, whereas Figure 4.2 shows a single stage for a radial flow turbomachine. Regardless of the type of machine, two different oblique cylindrical control volumes are drawn around the rotating element (rotor or impeller) in this stage – one stationary and one rotating. For the case of an axial flow machine, a second

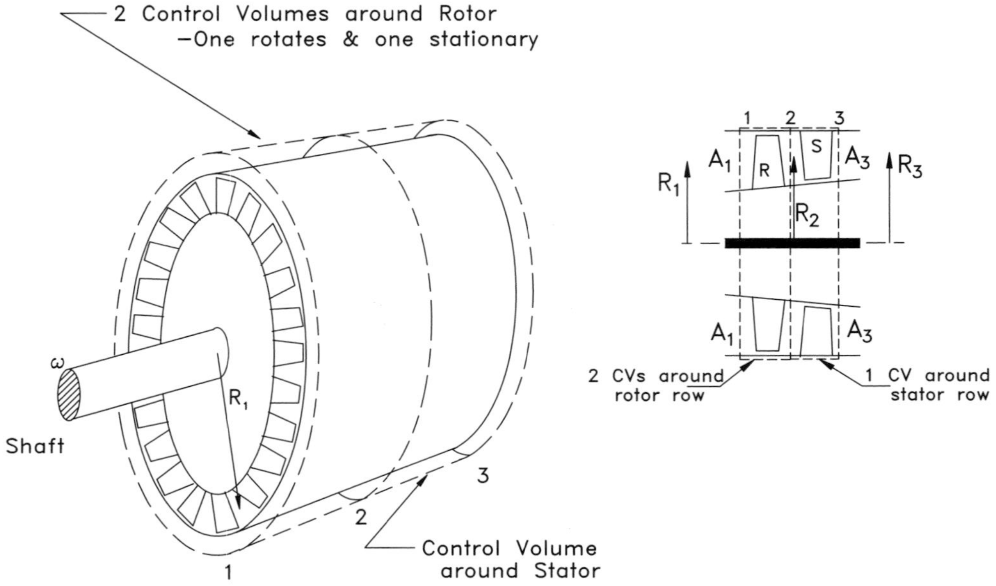

Figure 4.1 Control volume definition for single-stage axial flow turbomachine.

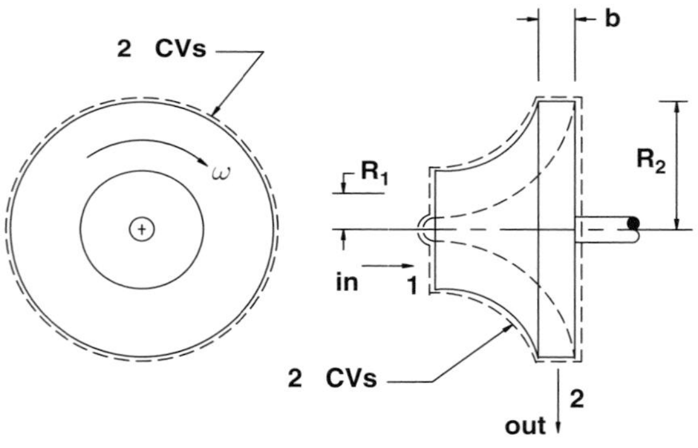

Figure 4.2 Control volume definition for single-stage radial flow turbomachine.

stationary control volume is drawn around the stator. All control volumes cut through the case and rotor shaft as shown. The objective is to relate the inlet and exit conditions to the property changes. As is shown in the following sections, the same control volume analysis applies to axial flow compressors, centrifugal compressors, and axial flow turbines which will be used in Chapters 5, 6, and 7. The same basic equations result from the analysis; however, the application of the equations for the three different machines is somewhat different.

4.2.1 Total Pressure Ratio

The three basic equations to be used are the continuity, moment of momentum, and energy equations. The first to be discussed is the general continuity equation applied to a control volume:

$$0 = \frac{\partial}{\partial t}\iiint_{CV} \rho d\mathcal{V} + \iint_{CS} \rho \vec{V} \cdot d\vec{A}. \tag{4.2.1}$$

The second equation is the moment of momentum, which is given by

$$\vec{r} \times \vec{F}_s + \iiint_{CV} (\vec{r} \times \vec{g})\rho d\mathcal{V} + \vec{T}_{sh} = \frac{\partial}{\partial t}\iiint_{CV} (\vec{r} \times \vec{V})\rho d\mathcal{V} + \iint_{CS} (\vec{r} \times \vec{V})\rho \vec{V} \cdot d\vec{A}, \tag{4.2.2}$$

where $\vec{F}_s$ is a surface force on a control volume, $\vec{g}$, is an applied body acceleration, and $\vec{T}_{sh}$ is a torque applied to a shaft. The final basic equation needed is the energy equation, which for a control volume is

$$\dot{Q} + \dot{W}_{\mathrm{sh}} + \dot{W}_{\mathrm{loss}} = \frac{\partial}{\partial t}\iiint_{CV}\left[u + \frac{V^2}{2} + gz\right]\rho d\mathcal{V}$$

$$+ \iint_{CS}\left[u + pv + \frac{V^2}{2} + gz\right]\rho\vec{V}\cdot d\vec{A}, \qquad (4.2.3)$$

where $\dot{Q}$ is the heat transfer rate, $\dot{W}_{\mathrm{sh}}$ is an applied or derived shaft power to or from the control volume, $\dot{W}_{\mathrm{loss}}$ is a power loss within the control volume, v is the specific volume, and u is the specific internal energy.

One must now make a few realistic assumptions so that these equations become manageable. First, the flow is assumed to be steady. Second, the control volumes are assumed to be free of any body forces (which will affect the potential energy) as well as surface forces. The control volumes are assumed to be adiabatic. Next, the power loss term is assumed to be negligible, although losses will eventually be included in the form of an efficiency. The flow in and out of the control volumes is assumed to be uniform as well. The flow is also be assumed to be planar 2-D, and a “mean line” or “meridional” analysis is used. That is, for an axial machine and for the inlet of a centrifugal machine a point located midway between the hub and tip will be used to evaluate the radii, fluid properties, and velocities. Also, in the case of an axial machine this implies that no radial component of velocity will be present. For the case of a radial flow machine, this implies that no radial component of velocity will be present in the inlet plane and that no axial component of velocity will be present in the exit plane. Lastly, the gas is assumed to be ideal. Although the assumptions may look restrictive, the analysis yields very realistic results for many applications. The list of assumptions is as follows.

(1) Steady	(4) Adiabatic	(7) Planar 2-D flow
(2) No body forces	(5) No power losses	(8) Ideal gas
(3) No surface forces	(6) Uniform flow	

Thus, by assumption (1), Eq. (4.2.1) becomes

$$0 = \iint_{CS}\rho\vec{V}\cdot d\vec{A} \qquad (4.2.4)$$

By the use of assumptions (1), (2), and (3), Eq. (4.2.2) becomes

$$\vec{T}_{\mathrm{sh}} = \iint_{CS}(\vec{r}\times\vec{V})\rho\,\vec{V}\cdot d\vec{A}, \qquad (4.2.5)$$

and using assumptions (1), (2), (4), and (5), Eq. (4.2.3) reduces to

$$\dot{W}_{\mathrm{sh}} = \iint_{CS}\left[u + pv + \frac{V^2}{2}\right]\rho\vec{V}\cdot d\vec{A} \qquad (4.2.6)$$

Now, by applying Eq. (4.2.4) with assumptions (6) and (7) to the stationary control volume around the rotor or impeller, one quickly finds that

$$\rho_1 c_{\mathrm{a}1} A_1 = \rho_2 c_{\mathrm{e}2} A_2, \qquad (4.2.7)$$

where A_1 is the annular flow area at the inlet, $c_{\mathrm{a}1}$ is the axial velocity at the inlet, and so forth. Note that the absolute exit velocity $c_{\mathrm{e}2}$ for the case of an axial flow machine is $c_{\mathrm{a}2}$, the axial velocity at the exit, whereas for the case of a radial machine it is $c_{\mathrm{r}2}$, the radial velocity at the exit. For an axial flow machine, A_2, is the annual area at the exit and for a radial machine it is the perimeter area at R_2. Equation (4.2.7) can also be written as

$$\dot{m}_1 = \dot{m}_2. \qquad (4.2.8)$$

Next, by applying Eq. (4.2.4) with the same assumptions to the stationary control volume around the stator (only for the case of the axial machines), one finds

$$\rho_2 c_{\mathrm{a}2} A_2 = \rho_3 c_{\mathrm{a}3} A_3 \qquad (4.2.9)$$

or

$$\dot{m}_2 = \dot{m}_3. \qquad (4.2.10)$$

Now one can apply the moment of momentum equation, Eq. (4.2.5), to the stationary control volume for the rotating element. Note that all of the torque delivered through the shaft is in the axial direction $(\hat{k})$. The quantity $(\vec{r}\times\vec{V})$ is also found under the integral. The vector $\vec{r}$ is normal to the axial direction, regardless of machine type. Furthermore, since the cross product must yield a component in only the axial direction, the only velocity component that can contribute to the torque is the absolute tangential component (c_{u}). Thus, one finds

$$\vec{T}_{\mathrm{sh}} = T_z\hat{k} = \hat{k}\iint_{CS} Rc_{\mathrm{u}}\rho\vec{V}\cdot d\vec{A}, \tag{4.2.11}$$

where T_z is the scalar of the shaft torque. Using the continuity equation, one finds that this reduces to

$$T_z\hat{k} = \hat{k}\iint_{CS} Rc_{\mathrm{u}}d\dot{m}. \tag{4.2.12}$$

Now, because the flow has been assumed to be uniform and 2-D, the integral across the inlet and exit planes of the rotor or impeller control volume can be evaluated as follows:

$$T_z = -R_1\, c_{\mathrm{u}1}\, \dot{m}_1 + R_2\, c_{\mathrm{u}2}\, \dot{m}_2. \tag{4.2.13}$$

For the case of a compressor or pump, this will be a positive quantity. However, for the case of a turbine this quantity will be negative, indicating that torque is derived from the fluid. Finally, using Eq. (4.2.8), one finds

$$T_z = \dot{m}\,[R_2\, c_{\mathrm{u}2} - R_1\, c_{\mathrm{u}1}]. \tag{4.2.14}$$

Thus, this relates the flow velocities to the torque applied to the shaft to transfer power to or from the stage. The applied or derived torque can easily be used to find the input or output power by

$$\dot{W}_{\mathrm{sh}} = T_z\omega, \tag{4.2.15}$$

where ω is the angular shaft speed. Also, the mean blade speed can be related to the shaft rotational speed by

$$U = R\omega, \tag{4.2.16}$$

where R is the distance to the center of the blade from the shaft centerline. Thus, by applying Eqs. (4.2.14), (4.2.15), and (4.2.16), one finds

$$\dot{W}_{\mathrm{sh}} = \dot{m}[U_2c_{\mathrm{u}2} - U_1c_{\mathrm{u}1}]. \tag{4.2.17}$$

Again, if the application is a turbine, the power will be negative, indicating that power is derived from the fluid. Equation (4.2.6) can be used for the stationary rotor or impeller control volume with assumptions (6) and (7) to yield

$$\dot{W}_{\mathrm{sh}} = \dot{m}_2\left(u_2 + p_2v_2 + {}^1\!/_2c_2^2\right) - \dot{m}_1\left(u_1 + p_1v_1 + {}^1\!/_2c_1^2\right), \tag{4.2.18}$$

where c is the absolute velocity. The enthalpy can be related to the internal energy by

$$h = u + pv. \tag{4.2.19}$$

Thus, by using Eqs. (4.2.8), (4.2.18), and (4.2.19), one obtains

$$\dot{W}_{\mathrm{sh}} = \dot{m}\left(h_2 + {}^1\!/_2c_2^2 - h_1 - {}^1\!/_2c_1^2\right). \tag{4.2.20}$$

Remembering the definition of total enthalpy from Eq. (H.2.5), however, one finds

$$\dot{W}_{\mathrm{sh}} = \dot{m}(h_{\mathrm{t}2} - h_{\mathrm{t}1}). \tag{4.2.21}$$

Thus, the transferred power is equal to the change in total enthalpy. In the case of a compressor, the power delivered to the rotor stage by the shaft is equal to the increase in the total enthalpy. For a turbine, the power removed from the rotor stage by the shaft is equal to the decrease in the total enthalpy. Up to this point in the derivation, the nonideal power loss has been neglected. Using a definition of efficiency similar to that applied to an entire turbomachine, one finds that

$$\dot{W}_{\mathrm{sh}} = \mathcal{E}\dot{m}(h_{\mathrm{t}2}{}' - h_{\mathrm{t}1}), \tag{4.2.22}$$

where $h_{\mathrm{t}2}{}'$ is the ideal total enthalpy at station 2 (for the same total pressure increase as in the nonideal case). Also, for a compressor, $\mathcal{E}$ is equal to $1/\eta_{12}$, and for a turbine $\mathcal{E}$ is equal to η_{12} (η_{12} is the aerodynamic efficiency of the rotor or impeller). Note that the actual (nonideal) work or power is more than for an ideal compressor but smaller than the ideal case of a turbine. For an ideal gas (assumption (8)), Eq. (4.2.22) becomes

$$\dot{W}_{\mathrm{sh}} = \mathcal{E}\dot{m}c_{\mathrm{p}}T_{\mathrm{t}1}\left[\frac{T_{\mathrm{t}2}{}'}{T_{\mathrm{t}1}} - 1\right]. \tag{4.2.23}$$

For isentropic flow, one can use Eq. (H.2.11) to find

$$\dot{W}_{\mathrm{sh}} = \mathcal{E}\,\dot{m}c_{\mathrm{p}}T_{\mathrm{t}1}\left[\left[\frac{p_{\mathrm{t}2}}{p_{\mathrm{t}1}}\right]^{\frac{\gamma-1}{\gamma}} - 1\right]. \tag{4.2.24}$$

Next, one can solve for $p_{\mathrm{t}2}/p_{\mathrm{t}1}$ and find

$$\frac{p_{\mathrm{t}2}}{p_{\mathrm{t}1}} = \left[\frac{\dot{W}_{\mathrm{sh}}}{\mathcal{E}c_{\mathrm{p}}T_{\mathrm{t}1}\dot{m}} + 1\right]^{\frac{\gamma}{\gamma-1}}. \tag{4.2.25}$$

Using the results from the moment of momentum equation (Eq. (4.2.17)), one obtains

$$\frac{p_{\mathrm{t}2}}{p_{\mathrm{t}1}} = \left[\frac{U_2c_{\mathrm{u}2} - U_1c_{\mathrm{u}1}}{\mathcal{E}c_{\mathrm{p}}T_{\mathrm{t}1}} + 1\right]^{\frac{\gamma}{\gamma-1}}, \tag{4.2.26}$$

which is greater than unity for the case of a compressor but less than unity for the case of a turbine. Thus, knowing the velocity information of a rotor and the efficiency, one can find the total pressure ratio. It is important to note that determining velocities such as U_2, c_{u2}, U_1, and c_{u1}, will be determined by velocity polygons or triangles in Chapters 5, 6, and 7. In this section, the equations are derived so that the internal fluid velocities can be related to the pressure change, power, and other important turbomachine characteristics and will be applied in those chapters. As a reminder, because of a subtle difference, for the case of a compressor, $\mathcal{E}$ is equal to $1/\eta_{12}$, and for a turbine $\mathcal{E}$ is equal to η_{12}. Also note that Eq. (4.2.26) applies to axial or radial flow machines. However, the application of the equation will be somewhat different because of the difference in velocity polygons in the two categories of turbomachines.

4.2.2 Percent Reaction

An important characteristic for an axial flow compressor or an axial flow turbine is the percent reaction. Such a parameter is not usually used for centrifugal turbomachines. This is a relation that approximates the relative loading of the rotor and stator. For a compressible flow, this is defined as the change in enthalpy across the rotor to the change in total enthalpy across the entire stage. In equation form, this is

$$\%\mathrm{R} \equiv \frac{h_2 - h_1}{h_{t3} - h_{t1}}. \tag{4.2.27}$$

Across the stator, however, no power is transferred to or from the flow, and so h_{t2} is the same as h_{t3}. Thus,

$$\%\mathrm{R} = \frac{h_2 - h_1}{h_{t2} - h_{t1}}. \tag{4.2.28}$$

From Eq. (H.2.5), Eq. (4.2.28) becomes

$$\%\mathrm{R} = \frac{1}{1 + {}^1\!/_2\left[\dfrac{c_2^2 - c_1^2}{h_2 - h_1}\right]}. \tag{4.2.29}$$

If a rotating control volume is used with the rotor so that no shaft power crosses the control volume, Eq. (4.2.6) becomes

$$0 = \left(u_2 + p_2 v_2 + {}^1\!/_2 w_2^2\right) - \left(u_1 + p_1 v_1 + {}^1\!/_2 w_1^2\right), \tag{4.2.30}$$

or with Eq. (4.2.19) it becomes

$$h_2 - h_1 = {}^1\!/_2[w_1^2 - w_2^2], \tag{4.2.31}$$

where once again w represents the relative velocity with respect to the rotor blade. Thus, from Eq. (4.2.29), this yields

$$\%\mathrm{R} = \frac{1}{1 + \left[\dfrac{c_2^2 - c_1^2}{w_1^2 - w_2^2}\right]}. \tag{4.2.32}$$

For compressible flow, therefore, the percent reaction of the stage can be directly related to the absolute and relative velocities at the inlet and exit of the rotor. The program Turbomachinery.exe was written to analyze four types of single-stage turbomachines: axial flow compressors, centrifugal compressors, axial flow turbines, and radial inflow turbines.

4.2.3 Incompressible Flow (Limiting Case)

For comparison, one may wish to find the pressure change of a turbomachine with an incompressible fluid – namely, $\rho_1 = \rho_2 = \rho$. This condition would apply to a pump or hydraulic turbine. Such incompressible flow turbomachines are not a part of jet engines or power generation gas turbines; however, this limiting case analysis provides significant insight as to the implications of the parameter percent reaction. For this reason, it is included herein. Equation (4.2.22) still applies but can be simplified. Using Eq. (4.2.18) with the incompressibility assumption and realizing that, for an incompressible flow, the internal energy will not increase, one finds

$$\dot{W}_{sh} = \mathcal{E}\,\dot{m}\left[\frac{p_2}{\rho} - \frac{p_1}{\rho} + \frac{c_2^2 - c_1^2}{2}\right]. \tag{4.2.33}$$

Equation (4.2.17) also applies. That is,

$$\dot{W}_{sh} = \dot{m}[U_2 c_{u2} - U_1 c_{u1}]. \tag{4.2.34}$$

Next, Eq. (4.2.30) can be compared with the Bernoulli equation $\left(pv + {}^{w^2}\!/_2 = constant\right)$. One can see that they are similar except for the $\Delta u = u_2 - u_1$ term. Thus, for reversible flow, u_2 is equal to u_1. As a result, for ideal flow,

$$p_2 - p_1 = \frac{\rho}{2}[w_1^2 - w_2^2]. \tag{4.2.35}$$

For incompressible flow, however, the total pressure is

$$p_t = p + {}^1/_2\rho c^2. \tag{4.2.36}$$

Therefore, using Eqs. (4.2.33), (4.2.34), and (4.2.36), one finds

$$p_{t2} - p_{t1} = \frac{\rho}{\mathcal{E}}[U_2 c_{u2} - U_1 c_{u1}], \tag{4.2.37}$$

where the efficiency parameter ($\mathcal{E}_c = 1/\eta_{12}$ and $\mathcal{E}_t = \eta_{12}$) accounts for nonideal effects. Also, using Eqs. (4.2.35) and (4.2.36) one finds

$$p_{t2} - p_{t1} = \frac{\rho}{2}(w_1^2 - w_2^2 + c_2^2 - c_1^2). \tag{4.2.38}$$

For comparison, it may be desirable to find percent reaction of an axial flow turbomachine with an incompressible fluid. Once again, the percent reaction is given by Eq. (4.2.27), which again becomes

$$\%R = \frac{1}{1 + \left[\dfrac{c_2^2 - c_1^2}{w_1^2 - w_2^2}\right]}. \tag{4.2.39}$$

One can compare this equation with that for compressible flow and see that they are identical. Also, using Eqs. (4.2.18) and (4.2.28), one finds for an incompressible fluid that

$$\%R = \frac{u_2 + \dfrac{p_2}{\rho} - u_1 - \dfrac{p_1}{\rho}}{u_{t3} + \dfrac{p_{t3}}{\rho} - u_{t1} - \dfrac{p_{t1}}{\rho}}. \tag{4.2.40}$$

Ideally, the static specific internal energies, u_2 and u_1, are identical, as shown above, and likewise the total internal energies u_{t3} and u_{t1}, are the same. Thus, the percent reaction for incompressible and ideal flow reduces to

$$\%R = \frac{p_2 - p_1}{p_{t3} - p_{t1}}. \tag{4.2.41}$$

This is a significant finding. For example, axial flow turbomachine stages have percent reactions of about 0.45 to 0.55. For a nonideal compressor, pump or turbine, the percent reaction indicates that approximately half of the enthalpy change occurs in the rotor and half in the stator. For an ideal pump or hydraulic turbine this also means that exactly half of the pressure change occurs in the rotor and half in the stator. Although not exactly the same, value for a compressible flow turbomachine, percent reaction is a practical measure of a good design. This implies that the force "loads" on the rotor and stator blades are about the same. For a compressor, if the pressure rises across the rotor blades and stator vanes are equal, the percent reaction tends to be somewhat higher owing to the compressibility of the flow and radial variations. For a compressible flow turbine, radial variations also influence the design reaction. For a compressor, if a large pressure rise in either the rotor blades or stator vanes occurs, separation or partial separation of the cascade is likely. As a result of the associated losses, the efficiency will be less than optimum. On the other hand, for a turbine, if a large pressure drop occurs in either the blades or vanes, choking or partial choking of the cascade is likely. Again, as will be shown for axial flow compressors in Section 5.11, the efficiency will be less than optimal as the percent reaction deviates greatly from 50 percent because of the associated losses. Furthermore, if higher loads than necessary are on either a set of rotor blades or stator vanes, the life expectancy of that component will be compromised.

Example 4.1 | Axial Flow Compressor Stage Pressure Ratio, Power, and Reaction

A stage approximating one of the last stages (rotor/stator) of a modern high-pressure compressor is to be analyzed. It rotates at 8500 rpm and compresses 250 lb_m/s (7.771 sl/s) of air. The inlet pressure and temperature are 250 psia and 740 °F (1200 °R), respectively. The average radius of the blades is 12 in and the inlet blade height is 1.2 in. The velocities are known: $c_1 = 739.9$ ft/s, $c_2 = 952.9$ ft/s, $w_1 = 977.1$ ft/s, $w_2 = 751.1$ ft/s, $c_{u2} = 638.2$ ft/s and $c_{u1} = 216.2$ ft/s. The specific heat ratio, γ, is 1.366 and the specific heat, c_p, is 0.2558 BTU/lb_m/ °R. The efficiency of the stage is 90 percent. The following details are to be found: the total pressure ratio for the stage, the required power for the stage, and the percent reaction for the stage.

Solution

Before proceeding, a few preliminary calculations are needed:

$$U = U_1 = U_2 = R\omega = \frac{12\text{in.}}{12\dfrac{\text{in.}}{\text{ft}}}\frac{8500\,\text{rev}}{\text{min}}\frac{2\pi\,\text{rad}}{\text{rev}}\frac{1}{60\dfrac{\text{s}}{\text{min}}} = 890.1\text{ fps}.$$

Finally,

$$a_1 = \sqrt{\gamma \mathcal{R} T_1} = \sqrt{1.366 \times 53.35\frac{\text{ft-lb}_\text{f}}{\text{lb}_\text{m}\text{-°R}} \times 32.17\frac{\text{lb}_\text{m}}{\text{sl}} \times 1200\text{ °R} \times \frac{\text{sl-ft}}{\text{s}^2 - \text{lb}_\text{f}}} = 1677\text{ fps},$$

so that $M_1 = \dfrac{c_1}{a_1} = \dfrac{739.9}{1677} = 0.4411.$

Next:

$$\frac{T_{t1}}{T_1} = 1 + \frac{\gamma - 1}{2}{M_1}^2 = 1 + \frac{0.366}{2}0.4411^2 = 1.036,$$

so that $T_{t1} = 1.036 \times 1200 = 1243\text{ °R} = 783\text{ °F}.$

From the above derivations for the total pressure ratio for a compressor can be found:

$$\frac{p_{t2}}{p_{t1}} = \left[\frac{\eta U(c_{u2} - c_{u1})}{c_p T_{t1}} + 1\right]^{\frac{\gamma}{\gamma-1}},$$

so that

$$\frac{p_{t2}}{p_{t1}} = \left[\frac{0.90 \times 890.1\dfrac{\text{ft}}{\text{s}}(638.2 - 216.3)\dfrac{\text{ft}}{\text{s}}}{0.2558\dfrac{\text{BTU}}{\text{lb}_\text{m}\text{- °R}} \times 1243\text{ °R} \times 778.16\dfrac{\text{ft-lb}_\text{f}}{\text{BTU}} \times 32.17\dfrac{\text{lb}_\text{m}}{\text{sl}}}\frac{\text{lb}_\text{f}\text{-s}^2}{\text{sl-ft}} + 1\right]^{\frac{1.366}{0.366}} = 1.168.$$

Next, one can find the input stage power by

$$\dot{W}_{sh} = \dot{m}U(c_{u2} - c_{u1}) = \frac{7.771\dfrac{\text{sl}}{\text{s}} \times 890.1\dfrac{\text{ft}}{\text{s}}(638.2 - 216.3)\dfrac{\text{ft}}{\text{s}}}{550\dfrac{\text{ft-lb}_\text{f}}{\text{s-hp}}}\left(\frac{\text{s}^2\text{-lb}_\text{f}}{\text{sl-ft}}\right) = 5306\text{ hp},$$ <ANS

and the percent reaction can be found by

$$\%\text{R} = \frac{1}{1 + \left[\dfrac{c_2^2 - c_1^2}{w_1^2 - w_2^2}\right]} = \frac{1}{1 + \left[\dfrac{952.9^2 - 739.9^2}{977.1^2 - 751.1^2}\right]} = 0.5200.$$ <ANS

It is noteworthy that the velocities were given for this example. As will be shown in details in the next three chapters one can find such velocities by using “velocity polygons.”

Example 4.2 Axial Flow Turbine Stage Power

A stage of approximately the same size as one of the first stages (rotor/stator) of modern high-pressure turbine is to be analyzed. It rotates at 8500 rpm and expands 250 lb_m/s (7.771 sl/s) of air. The inlet pressure and temperature are 250 psia and 2140 °F (2600 °R), respectively. The average radius of the blades is 16.8 in. and the inlet blade height is 2.04 in. The efficiency of the stage is 87 percent. The values of c_p and γ are 0.2890 BTU/lb_m-°R and 1.311, respectively. The total pressure ratio is 0.6588 and the percent reaction is 0.5087. The absolute velocity at the inlet is $c_1 = 1419$ fps. The following details are to be found: the change in absolute tangential velocity and the derived power from the stage. One should note that this turbine can be compared with the compressor in Example 4.1. Both turbomachines are from the same engine and both are operating at about the same pressure.

Solution

Before proceeding, a few preliminary calculations are needed.

$U = R\omega = (16.8/12)(8500 \times 2\pi/60) = 1246\,\text{fps}.$

Finally, $a_1 = \sqrt{\gamma \mathcal{R} T_1} = \sqrt{1.311 \times 53.35 \dfrac{\text{ft-lb}_\text{f}}{\text{lb}_\text{m}\text{-°R}} \times 32.17 \dfrac{\text{lb}_\text{m}}{\text{sl}} \times 2600\ \text{°R} \times \dfrac{\text{sl-ft}}{\text{s}^2\text{lb}_\text{f}}} = 2419\,\text{fps},$

and $M_1 = \dfrac{c_1}{a_1} = \dfrac{1419}{2419} = 0.5866.$

Also:

$$\frac{T_{t1}}{T_1} = 1 + \frac{\gamma - 1}{2} M_1^2 = 1 + \frac{0.311}{2} 0.5866^2 = 1.054$$

$T_{t1} = 1.054 \times 2600 = 2739\ \text{°R} = 2279\ \text{°F}.$

From the energy equation one can find

$$\frac{p_{t2}}{p_{t1}} = \left[\frac{(h_{t2} - h_{t1})}{\eta_{12} c_p T_{t1}} + 1\right]^{\frac{\gamma}{\gamma-1}}$$

$$= \left[\frac{(h_{t2} - h_{t1})}{0.87 \times 0.2890 \dfrac{\text{BTU}}{\text{lb}_\text{m}\text{-°R}} \times 2739\ \text{°R} \times 778.16 \dfrac{\text{ft-lb}_\text{f}}{\text{BTU}} \times 32.17 \dfrac{\text{lb}_\text{m}}{\text{sl}} \dfrac{\text{sl-ft}}{\text{lb}_\text{f}\text{-s}^2}} + 1\right]^{\frac{1.311}{0.311}} = 0.6588,$$

so solving for the change in total enthalpy:

$h_{t2} - h_{t1} = -1{,}625{,}000\,\text{ft}^2/\text{s}^2.$

One should notice the negative sign for Δh_t, indicating that power is derived from the fluid.

Also, $h_{t2} - h_{t1} = U(c_{u2} - c_{u1}) = 1246 \dfrac{\text{ft}}{\text{s}} (c_{u2} - c_{u1}),$

so $c_{u2} - c_{u1} = -1304\ \text{ft/s}.$ <ANS

Lastly, the power derived is:

$$\dot{W}_{sh} = \dot{m}(h_{t2} - h_{t1}) = 7.771\frac{\text{sl}}{\text{s}} \times \left(-1{,}625{,}000\frac{\text{ft}^2}{\text{s}^2}\right)\frac{1}{550}\frac{\text{hp-s}}{\text{ft-lb}_f} = -22,962 \text{ hp.}$$ <ANS

For comparison, the power for one of the last stages of the high-pressure compressor (Example 4.1) is only 5306 hp. Thus, all of the variables of interest have been found.

Example 4.3 | Radial Flow Compressor Stage Pressure Ratio and Power

A stage of a radial compressor is to be analyzed. It rotates at 13,000 rpm and compresses 25 kg/s of air. The inlet pressure and temperature are 300 and 250 K, respectively. The average radius of the blades at the inlet is 100 mm, respectively and the exit radius is 250 mm and the exit blade height is 20 mm. The efficiency of the stage is 90 percent. The values of c_p and γ are 0.998 kJ/kg-K and 1.404, respectively. Velocities are: $c_1 = 123.2$ mps, $c_2 = 345.5$ mps, $w_1 = 205.9$ mps, $w_2 = 123.4$ mps, $c_{u1} = -31.9$ mps, and $c_{u2} = 323.2$ mps. The following details are to be found: the total pressure ratio and the required power for the stage.

Solution

Before proceeding, a few preliminary calculations are needed.

The average (mean line) blade speed at the inlet is

$U_1 = R_1\omega = (100)(13{,}000 \times 2\pi/60) \doteq 136{,}100$ mm/s $= 136.1$ mps,

the average blade speed at the exit is

$U_2 = R_2\omega = (250)(13{,}000 \times 2\pi/60) = 340{,}300$ mm/s $= 340.3$ mps

and $a_1 = \sqrt{\gamma R T_1} = \sqrt{1.404 \times 287.1\frac{\text{J}}{\text{kg-K}} \times 250\,\text{K} \times \frac{\text{N-m}}{\text{J}} \times \frac{\text{kg-m}}{\text{s}^2\text{-N}}} = 317.3$ mps.

Finally, $M_1 = \frac{c_1}{a_1} = \frac{123.2}{317.3} = 0.3881.$ <ANS

Next, $\frac{T_{t1}}{T_1} = 1 + \frac{\gamma - 1}{2}M_1^2 = 1 + \frac{0.404 \times 0.3881^2}{2} = 1.030,$

$T_{t1} = 1.030 \times 250 = 257.6$ K,

and $c_{u2} = 323.2$ mps is given.

Next, the moment of momentum equation can be used to yield

$h_{t2} - h_{t1} = U_2 c_{u2} - U_1 c_{u1} = 340.3 \times 323.2 + 136.1 \times 31.9 = 114{,}300\ \text{m}^2/\text{s}^2.$

From the energy equation one can find

$$\frac{p_{t2}}{p_{t1}} = \left[\frac{\eta_{12}[h_{t2} - h_{t1}]}{c_p T_{t1}} + 1\right]^{\frac{\gamma}{\gamma-1}} = \left[\frac{0.90 \times 114{,}300\frac{\text{m}^2}{\text{s}^2}}{0.998\frac{\text{kJ}}{\text{kg-K}} \times 257.6\,\text{K}} \times \frac{\text{N}}{\text{kg-}\frac{\text{m}}{\text{s}^2}} \times \frac{\text{J}}{\text{N-m}} \times \frac{\text{kJ}}{1000\,\text{J}} + 1\right]^{\frac{1.403}{0.403}}$$

$= 3.224.$

One can note that this pressure ratio is much larger than that of the axial stage in the previous example.

Finally, one can find the input stage power by

$$\dot{W}_{sh} = \dot{m}(h_{t2} - h_{t1}) = 25\frac{\text{kg}}{\text{s}} \times 114{,}300\frac{\text{m}^2}{\text{s}^2} \times \frac{\text{N}}{\text{kg-}\frac{\text{m}}{\text{s}^2}} \times \frac{\text{J}}{\text{N-m}} \times \frac{\text{W}}{\text{J/s}} = 2{,}859{,}000\ \text{W} = 2.859\ \text{MW}$$

Thus, all of the variables of interest have been found.

4.3 Similitude

4.3.1 Dimensional Analysis: Compressible Flow

Experimental performance curves or "maps" are usually used to provide a working, versatile set of data for a compressor engineer or compressible flow turbine engineer. The data should be in the most general form possible so that air speed, altitude, and other variables have minimal effects. These data provide an engineer with a quick and accurate view of the overall performance of the machine at its operating and an understanding of what can be expected of the machine if the flow conditions are changed. Of primary concern are the total pressure ratio and efficiency of a compressor or turbine. The pressure ratio is p_{te}/p_{ti}, where the subscript e represents the outlet condition and the subscript i denotes the inlet condition. Thus, for a compressor, $p_{te}/p_{ti} = p_{t3}/p_{t2}$, and for a turbine, $p_{te}/p_{ti} = p_{t5}/p_{t4}$. Experimentally, the pressure ratio and efficiency are functions of many variables, and the functional relationship for the pressure ratio can be written as

$$p_{te}/p_{ti} = f\{\dot{m}, T_{ti}, A, p_{ti}, U, \gamma, \mathcal{R}\}. \tag{4.3.1}$$

One can use dimensionless parameters to reduce the number of independent variables. Using dimensional analysis/similitude or the Buckingham Pi theorem, one finds for a given values of γ and $\mathcal{R}$ that

$$p_{te}/p_{ti} = f\left\{\dot{m}\frac{\sqrt{\mathcal{R}T_{ti}}}{Ap_{ti}}, \frac{U}{\sqrt{\gamma\mathcal{R}T_{ti}}}\right\}. \tag{4.3.2}$$

This form is rarely used to document the performance of a compressible flow turbomachine, however. Usually, one finds a modification of this because $\mathcal{R}$ and γ are constant for a given gas and area, A, and radius, R, are constant for a given compressor or turbine. Also, $U = R\omega = 2\pi RN/60$. Thus, the only controlled variable parameters for a given turbomachine are N, $\dot{m}$, p_{ti}, and T_{ti}. As a result, we can write for a given machine:

$$p_{te}/p_{ti} = f\left\{\dot{m}\frac{\sqrt{\theta_{ti}}}{\delta_{ti}}, \frac{N}{\sqrt{\theta_{ti}}}\right\}, \tag{4.3.3}$$

where

$$\delta_{ti} = p_{ti}/p_{stp} \tag{4.3.4}$$

and

$$\theta_{ti} = T_{ti}/T_{stp}, \tag{4.3.5}$$

and where the subscript stp refers to the standard conditions and the subscript ti refers to the inlet total condition to the component. Also, of concern to a design or applications engineer is the efficiency of a turbomachine. Similarly, for the efficiency one finds

$$\eta = g\left\{\dot{m}\frac{\sqrt{\theta_{ti}}}{\delta_{ti}}, \frac{N}{\sqrt{\theta_{ti}}}\right\}, \tag{4.3.6}$$

and two new parameters can now be defined:

$$\dot{m}_{ci} \equiv \dot{m}\frac{\sqrt{\theta_{ti}}}{\delta_{ti}} \tag{4.3.7}$$

and

$$N_{ci} \equiv \frac{N}{\sqrt{\theta_{ti}}}. \tag{4.3.8}$$

Note that these two independent parameters are not truly dimensionless. The first is called the "corrected" mass flow and has dimensions of mass flow, whereas the second is called the "corrected" speed and has dimensions of rotational speed. As a result,

$$p_{te}/p_{ti} = f\{\dot{m}_{ci}, N_{ci}\} \tag{4.3.9}$$

and

$$\eta = g\{\dot{m}_{ci}, N_{ci}\}. \tag{4.3.10}$$

The functions f and g can be determined from modeling, experimental data, or a combination of empiricism and modeling. The corrected mass flow and corrected speed are also often referred to as a percent of the design conditions. For example, 100 percent speed and 100 percent mass flow would be the design condition or point of maximum efficiency. These equations can be in the form of analytical equations, curve fits, tables, or graphs.

Thus, since the two parameters are not dimensionless, the resulting functions cannot be used to correlate or compare different compressible flow turbomachines. One can use the functions for one particular turbomachine, however, as is the usual case. The functions f and g retain most of the nondimensional characteristics. Therefore, raw data can be obtained under one set of conditions (e.g., at STP) but applied over a wider range of conditions if "corrected" parameters are used. That is, the map is applicable regardless of altitude, speed, atmospheric conditions, and so on. Thus, obtaining raw data is not necessary for all possible conditions.

Example 4.4 | Compressor Similitude from Test Stand Data to Flight Conditions

A compressor is tested on a dedicated test stand at 500 ft, $p = 14.43$ psia, on a cold day, $T = 495.0$ °R. Compressor testing will be covered in Chapter 5. For one test condition the measured compressor efficiency is 88 percent and the total pressure ratio is 18. For this test condition the compressor runs at 8000 rpm and the mass flow rate is 150 lb_m/s. If the engine is flown at 30,000 ft at Mach 0.800, what is the engine rotational speed and flow rate for compressor similitude?

Solution

First note that for similitude

$$p_{te}/p_{ti} = f\left\{\dot{m}\frac{\sqrt{\theta_{ti}}}{\delta_{ti}}, \frac{N}{\sqrt{\theta_{ti}}}\right\}$$

$$\eta = g\left\{\dot{m}\frac{\sqrt{\theta_{ti}}}{\delta_{ti}}, \frac{N}{\sqrt{\theta_{ti}}}\right\},$$

where

$$\theta_{ti} = T_{ti}/T_{stp},$$

and where the subscript stp refers to the standard conditions and the subscript ti refers to the inlet total condition to the component or engine. At sea level and stp, $p = 14.70$ psia and T = 518.7 °R. At 30,000 ft, $p = 4.373$ psia and $T = 411.9$ °R. The specific heat ratio, γ, is about 1.399 at both altitudes.

Thus, at the test conditions since the compressor is on a test stand, the Mach number is 0.000:

$$\frac{T_{t1}}{T_1} = 1 + \frac{\gamma - 1}{2} M_1^2 = 1 + \frac{0.399 \times 0.000^2}{2} = 1.000$$

$$\text{and } \frac{p_{t1}}{p_1} = \left(\frac{T_{t1}}{T_1}\right)^{\frac{\gamma}{\gamma-1}} = (1.000)^{\frac{1.399}{.399}} = 1.000.$$

So at the testing condition, $p_{t1} = 14.43$ psia and $T_{t1} = 495.0$ °R.

Also, at 30,000 ft and $M = 0.800$:

$$\frac{T_{t1}}{T_1} = 1 + \frac{\gamma - 1}{2} M_1^2 = 1 + \frac{0.399 \times 0.80^2}{2} = 1.128$$

$$\text{and } \frac{p_{t1}}{p_1} = \left(\frac{T_{t1}}{T_1}\right)^{\frac{\gamma}{\gamma-1}} = (1.128)^{\frac{1.399}{.399}} = 1.525.$$

So at 30,000 ft, $p_{t1} = 6.669$ psia and $T_{t1} = 464.6$ °R.

So at the testing condition

$$\dot{m}_c = \dot{m}\frac{\sqrt{\theta_{ti}}}{\delta_{ti}} = 150\frac{\sqrt{\dfrac{495.0}{518.7}}}{\dfrac{14.43}{14.70}} = 149.3\frac{\text{lb}_\text{m}}{\text{s}}$$

and

$$N_c = \frac{N}{\sqrt{\theta_{ti}}} = \frac{8000}{\sqrt{\dfrac{495.0}{518.7}}} = 8189 \text{ rpm}.$$

At 30,000 ft for similitude:

$$\dot{m}_c = \dot{m}\frac{\sqrt{\theta_{ti}}}{\delta_{ti}} = \dot{m}\frac{\sqrt{\dfrac{464.6}{518.7}}}{\dfrac{6.669}{14.70}} = 149.3\frac{\text{lb}_\text{m}}{\text{s}},$$

so solving for the true mass flow rate,

$$\dot{m} = 71.57\frac{\text{lb}_\text{m}}{\text{s}}$$

$$\text{and } N_c = \frac{N}{\sqrt{\theta_{ti}}} = \frac{N}{\sqrt{\dfrac{464.6}{518.7}}} = 8189 \text{ rpm},$$

and solving for the true rotational speed,

$N = 7750$ rpm.

Thus, the pressure ratio and efficiency will be 18 and 88 percent for this flow rate and rotational speed at Mach 0.800 at 30,000 ft.

4.4 Summary

In this chapter the basic control volume and dimensional analysis equations for fluid mechanics and gas dynamics were applied to turbomachines. Single stages for axial flow and radial flow turbomachines were analyzed. Working equations were developed in which internal velocity information can be used to find the important parameters total pressure ratio and power generated or consumed by a single stage. Other important parameters defined and developed include the percent reaction, efficiency, "corrected" mass flow rate, and "corrected" rotational speed. All of these parameters will be applied extensively to axial flow compressors, radial flow compressors, and axial flow turbines in the following three chapters. Specific working equations will be individually developed for all three types of turbomachines. Implications and design characteristics for these parameters will be further discussed for each of these turbomachines.

List of Symbols

A	Area
c	Absolute velocity
c_p	Specific heat at constant pressure
$\mathcal{E}$	Efficiency parameter
F_b	Body force
F_s	Surface force
g	Gravitational constant
h	Specific enthalpy
$\dot{m}$	Mass flow rate
N	Rotational speed
p	Pressure
$\dot{Q}$	Heat transfer rate
$\mathcal{R}$	Ideal gas constant
r, R	Radius
%R	Percent reaction
T	Temperature
T	Torque
U	Blade velocity
u	Specific internal energy
v	Specific volume
V	Velocity
$\mathcal{V}$	Volume
w	Relative velocity
$\dot{W}$	Power
γ	Specific heat ratio
δ	Ratio of pressure to standard pressure
η	Efficiency
θ	Ratio of temperature to standard temperature
ρ	Density
ω	Rotational speed

Subscripts

a	Axial
c	Compressor
c	Corrected
e	Exit
i	Inlet
loss	Power loss
r	Radial
sh	Applied to shaft
stp	Standard conditions
t	Total (stagnation)
t	Turbine
u	Tangential
z	Axial
1	Inlet to rotor
2	Exit of rotor and inlet to stator
3	Exit of stator

Superscripts

′ Ideal

Problems

4.1 A stage approximating one of the last stages (rotor/stator) of a high-pressure compressor is to be analyzed. It rotates at 9000 rpm and compresses 300 lb_m/s of air. The inlet pressure and temperature are 260 psia and 1300 °R, respectively. The average radius of the blades is 14.4 in. and the inlet blade height is 1.2 in. The velocities are known: $c_1 = 794.9$ ft/s, $c_2 = 1037.9$ ft/s, $w_1 = 1112.4$ ft/s, $w_2 = 838.7$ ft/s, $c_{u2} = 730.7$ ft/s, and $c_{u1} = 297.8$ ft/s. The specific heat ratio, γ, is 1.361 and the specific heat, c_p, is 0.2586 BTU/lb_m/°R. The efficiency of the stage is 88 percent. The following details are to be found: the total pressure ratio for the stage, the required power for the stage, and the percent reaction for the stage.

4.2 A stage of approximately the same size as one of the first stages (rotor/stator) of a high-pressure turbine is to be analyzed. It rotates at 9000 rpm and expands 300 lb_m/s of air. The inlet pressure and temperature are 280 psia and 2800 °R, respectively. The average radius of the blades is 18 in and the inlet blade height is 1.8 in. The efficiency of the stage is 88 percent. The values of c_p and γ are 0.2926 BTU/lb_m-°R and 1.306, respectively. The total pressure ratio is 0.5723 and the percent reaction is 0.5856. The absolute velocity at the inlet is $c_1 = 1621$ fps. The following details are to be found: the change in absolute tangential velocity and the derived power from the stage.

4.3 A stage of a radial compressor is to be analyzed. It rotates at 12,000 rpm and compresses 28 kg/s of air. The inlet pressure and temperature are 270 and 220 K, respectively. The average radius of the blades at the inlet is 115 mm, and the exit radius is 260 mm and the exit blade height is 20 mm. The efficiency of the stage is 88 percent. The values of c_p and γ are 0.993 kJ/kg-K and 1.407, respectively. Velocities are: $c_1 = 104$ mps, $c_2 = 326.5$ mps, $w_1 = 198.9$ mps, $w_2 = 89.1$ mps, $c_{u1} = -27.0$ mps, and $c_{u2} = 314.3$ mps. The following details are to be found: the total pressure ratio and the required power for the stage.

4.4 A compressor is tested on a dedicated test stand at 200 ft, $p = 14.59$ psia, on a hot day, $T = 530.0$ °R. For one test condition the measured compressor efficiency is 90 percent and the total pressure ratio is 30. For this test condition the compressor runs at 9400 rpm and the mass flow rate is 300 lb_m/s. If the engine is flown at 35,000 ft at Mach 0.850, what is the engine rotational speed and flow rate for compressor similitude? What are the efficiency and total pressure ratio for this condition?

4.5 A compressor is tested on a dedicated test stand at 50 m, $p = 100.73$ kPa psia, on a hot day, $T = 290.0$ K. For one test condition the measured compressor efficiency is 92 percent and the total pressure ratio is 25. For this test condition the compressor runs at 9000 rpm and the mass flow rate is 150 kg/s. If the engine is flown at 10,000 m at Mach 0.800, what is the engine rotational speed and flow rate for compressor similitude? What are the efficiency and total pressure ratio for this condition?

5 Axial Flow Compressors and Fans

5.1 Introduction

As discussed in previous chapters, a fan or compressor is the first rotating component that the fluid encounters. A cross-sectional view of a compressor for a simple, single-shaft turbojet is shown in Figure 5.1. The basic function of a compressor is to impart kinetic energy to the working fluid (air) by means of some rotating blades and then to convert the increase in energy to an increase in total pressure, which is needed by the combustor. The limits of operation of an engine are often dictated by a compressor, as is discussed in this chapter. Furthermore, the design of an efficient axial flow fan or compressor remains such a complex process that the success or failure of an engine often revolves around the design of a compressor. Many fundamental and advanced design details are available in Howell (1945a, 1945b), Horlock (1958), Hawthorne (1964), Johnsen and Bullock (1965), Cumpsty (1988), and Wilson and Korakianitis (2014). Denton (1998), Elmendorf et al. (1998), LeJambre et al. (1998), Rhie et al. (1998), Adamczyk (2000), and Denton and Dawes (2010), demonstrate how modern computational fluid dynamic (CFD) tools can be used for the complex 3-D analysis and design of compressors.

Compressors were the main stumbling block of the early engines and the primary cause of the delays in the development of jet engines for World War II. Meher-Homji (1996, 1997a, 1999), Dunham (2000), Bakken, Syverud, Jordal, and Veer (2004), and Bakken (2018) present interesting historical perspectives and technical information on early engine and compressor development. The Germans were the first to apply axial flow compressors successfully. The first Norwegian and British engines incorporated centrifugal compressors, which is the topic of Chapter 6. Centrifugal compressors are comparatively inefficient with some room for improvement. Because of this, it became apparent during the War that any large-thrust engine would require an efficient compressor. Since that time, considerable research and development have occurred, and thus current axial flow compressors are reliable and efficient components.

The development of compressors was very rapid in the 1950s through the 1990s due to better computing and testing facilities, and CFD has had a great impact since then. In some ways the development has slowed somewhat, but yet many new and important advancements have been made in the last decade. The GE GE9X (Figure 1.11) is a very good example of some of the new developments. It powers the Boeing 777X with 110,000 lb_f thrust. The overall pressure ratio is 60, compared with early engines with ratios of 3. The efficiency of the compressor was increased by 2 percent due to advanced aerodynamic design capabilities. The first-stage fan blades are representative of many new large fans, in that the blade shapes have previously unconventional shapes, longer chords, and twists than previous designs. Carbon fiber composites are used for these fan blades, with steel leading edges and glass fiber trailing edges. As a result, the fan is lighter and can rotate faster to better match the LP turbine. The first five compressor stages are single piece bladed disks ("blisks"). Three-dimensional "printing" also reduces such complex manufacturing costs; this is a technique that is also being developed by other manufacturers. The Pratt & Whitney 7200 is an Engine Alliance (50 percent P & W and 50 percent GE) engine derived from the PW4000 and GE90 engines. It powers the Airbus A380 with 70,000 lb_f thrust. It has a total pressure ratio of 36 and has hollow titanium fan blades. The Rolls-Royce Trent XWB (Figure 1.28) is a three-spool engine that powers the Airbus A350. It develops thrusts up to 97,000 lb_f

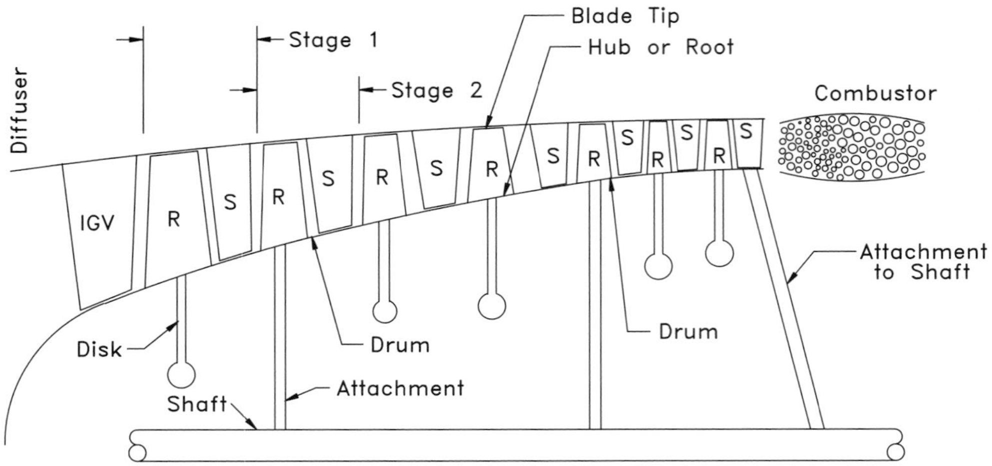

Figure 5.1 Single-shaft compressor.

and has a pressure ratio of 50. The fan blades are hollow titanium. Some of the compressor stages are single-piece bladed disks.

The compressor is vital to the operation of the engine because it readies the air for the combustor. Work is done on the fluid by the compressor, which is derived from the turbine. The total pressure is thus increased across the compressor, which is necessary for the combustion process, as is discussed in Chapter 9, and the production of thrust. Typically, for most engines the total pressure ratio across the compressor is from 5 to 40 as determined by the particular engine. As indicated in Chapter 1, by examining Eq. (1.2.4), one can see that, as the compressor pressure ratio increases, the thermodynamic efficiency increases for a closed cycle and likewise open cycle. Also, in general, the *TSFC* of a jet engine decreases with increasing compressor pressure ratio. In short, as the pressure ratio increases, the required fuel flow decreases and the extracted power increases.

In addition to the function just described in the previous paragraph, the compressor delivers high-pressure air to serve other functions. For example, a small portion of the air is bled to provide some cockpit and electronics environmental control. Also, a small portion is bled to provide pressurized air for inlet anti-icing. Finally, some of the high-pressure "cool" air is directed to the turbine and used to reduce the temperature of the hot turbine blades, as is discussed in Chapter 7.

5.1.1 Objectives

After completing this chapter, the reader should be able to do the following.

- Understand the basic geometry of a compressor and fan
- Understand the difference of rotor blades and stator vanes
- Understand the pressure changes in a single stage
- Understand and be able to develop a velocity triangle or polygon
- Understand the different flow and blade/vane angles in a stage
- Understand the influences of blade/vane geometry on pressure and velocities
- Be able to apply equations from Chapter 4 along with velocity polygons to find single-stage performance
- Understand the implications of percent reaction

- Understand the development and application of two compressor "maps"
- Understand surge and the control of
- Understand the limits of stage pressure increases
- Appreciate the compressor and turbine geometries and characteristics of a large range of gas turbines
- Understand the reason for and application of variable stators
- Understand the reason for and application of two or three spools
- Understand and appreciate that a percent reaction of 0.50 not only approximately equalizes pressure the rise in rotor and stator but it also approximately optimizes the stage efficiency
- Appreciate the 3-D effects (hub to tip) of flow using radial equilibrium (optional)
- Understand the application of the streamline analysis method to predict 3-D flow (optional)
- Understand the idea of using cascade lift and drag data, percent reaction, and flow rate to predict stage efficiency (optional)

5.2 Geometry

A compressor comprises a series of stages that all share the work load. The total pressure ratio across a single stage is typically from 1.15 to 1.28. The typical number of stages in an axial compressor is from 5 to 20. A stage is composed of two blade components (Figure 5.1). A set (or cascade) of rotating blades are attached to a "disk" at the "hub" or "root." The region of the blade with the largest radius is called the "tip." The tip diameter and housing diameter are approximately constant through a compressor. This first blade component is called the set of rotor blades. These blades do the work on the fluid. The second blade component is a set of stationary blades, which are attached to the case at the tips of the blades. These are termed stator blades or, more commonly, vanes. These vanes take the fluid from the previous set of rotor blades and ready the fluid velocity and direction for incidence-free entrance into the next or following set of rotor blades. Thus, the stator vanes do not input any energy into the fluid but are necessary for the operation of the compressor. A generalized three-dimensional view of the rotor and stator blades is shown in Figure 5.2; the blades, vanes, and disks are clearly visible. Fan blades are often constructed from titanium alloys, whereas the compressor blades operating in the "cool" region are often made from titanium alloys and the compressor blades operating in the "warm" region are often constructed from nickel alloys. Also see Figures 2.1, 2.15, and 2.21 and the image at the opening of Part I (GE90-115B) for views of the compressor and fan blading.

As noted, the blades are attached to a "disk," which is sometimes attached directly to the shaft. Disks are used for structural and dynamic stabilizing integrity. Each cascade of blades is attached to a different disk. Disk diameters range from 6 in. for small engines to over 40 in. for large turbofans with fan diameters over 90 in.

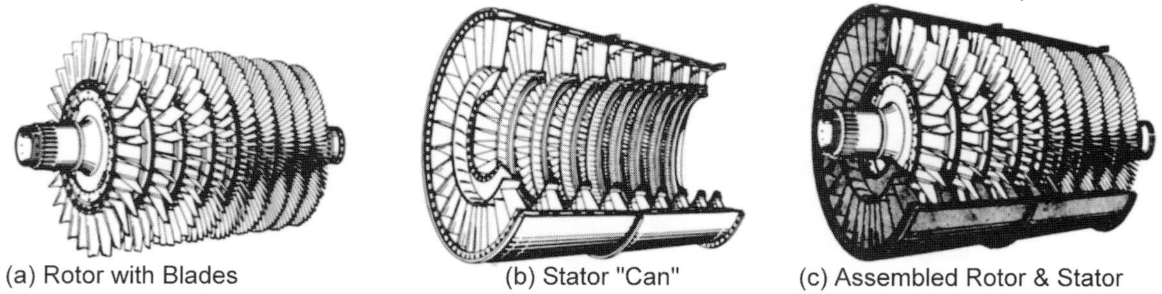

Figure 5.2 The 3-D view of rotor and stator blades and disks (© 2022 Raytheon Technologies Corporation, Pratt & Whitney division. Reproduced with permission).

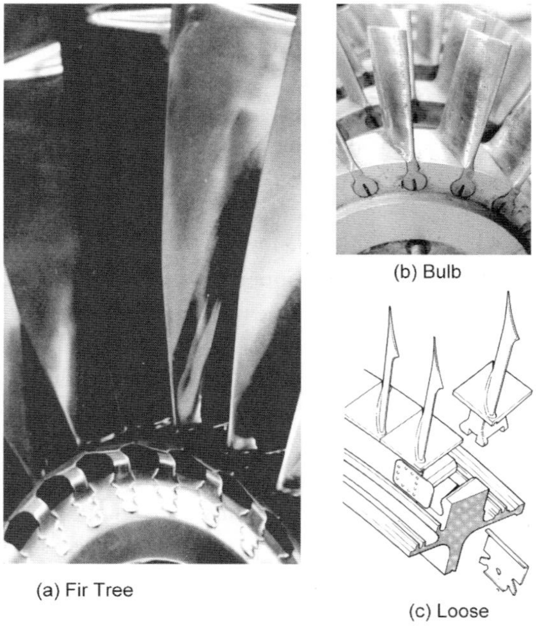

Figure 5.3 Attachment of a rotor blade to disk ((a) and (c) © Rolls-Royce plc 2022. All rights reserved; (b) photo by R. Flack).

Several blade attachment methods are used. Three such arrangements are shown in Figure 5.3. The first two are older methods using tight fits into the disk and are called a "fir tree" from the Rolls-Royce Pegasus and a bulb from the Westinghouse J30. The third method from Rolls-Royce (Figure 5.3c) is currently used on many compressors. In this case the blades are loose fits into the disk and held in place axially with a cover plate. Upon installation the blades are still loose (a few hundredths of a millimeter or less). However, during compressor operation, the blades are held tightly in place as the result of thermal growth and centrifugal forces. This attachment method allows the blades to be removed easily for replacement in case of blade damage. Figure 5.4 is a photograph of the Rolls-Royce RB211 high-pressure (HP) compressor.

As is discussed in Section 5.11.2 and Appendix L.2.2, tip losses can considerably compromise compressor efficiency as a result of flow moving from the pressure sides of the vanes or blades to the suction sides. Thus, some of the high-pressure fluid is lost. For the case of the rotor blades, narrowing the gap between

Figure 5.4 Rolls-Royce RB211 HP compressor (© Rolls-Royce plc 2022. All rights reserved).

the blade tip and housing reduces the flow (and associated losses). Of course any shaft vibrations will require the tip clearance to be finite, or the blade can rub the housing and fail. Two methods are used to minimize the gap. First, the blade tips are sometimes made of an ablating material that conforms to the housing. Consequently, as the compressor rotates the blades "wear in" their own required clearances. Second, the housing temperature is sensed, and cooling can be controlled automatically to change the housing temperature and thus the housing diameter due to thermal growth of the housing. As a result, the rotor-blade-tip-to-housing clearance is dynamically changed.

At the point before the fluid enters the first stage, many engines have another set of nonrotating blades, which are a set of airfoils termed the inlet guide vanes (IGV). These are similar in operation to the stator vanes, because this set of vanes readies the fluid direction and velocity for the first set of rotor blades. However, the design of such vanes is quite different from that of the stator vanes. These vanes turn the incoming air, which is predominately in the axial direction, in a direction that makes the flow into the first rotor blades incident free. Also, sometimes an engine has a set of stator vanes after the last stage that readies the flow for entrance to the combustor. These vanes are termed the exit guide vanes (EGV). The geometry for this set of vanes is also different from that of a normal set of stator vanes and is typically designed to add swirl to the flow, which aids in mixing within the combustor. Also, a set of exit guide vanes is often used for a turbofan at the exit of the fan to ready the flow either for the fan nozzle or bypass duct. These exit guide vanes straighten the flow and reduce the absolute velocity.

As discussed in Chapters 1, 2, and 3 many modern engines operate with two shafts and a fan. Such a design is shown in Figure 5.5. As illustrated, the fluid first enters the fan and then the low-pressure (LP) compressor, which is usually on the same shaft as the fan. The air next enters the high-pressure (HP) compressor, which is on a higher-

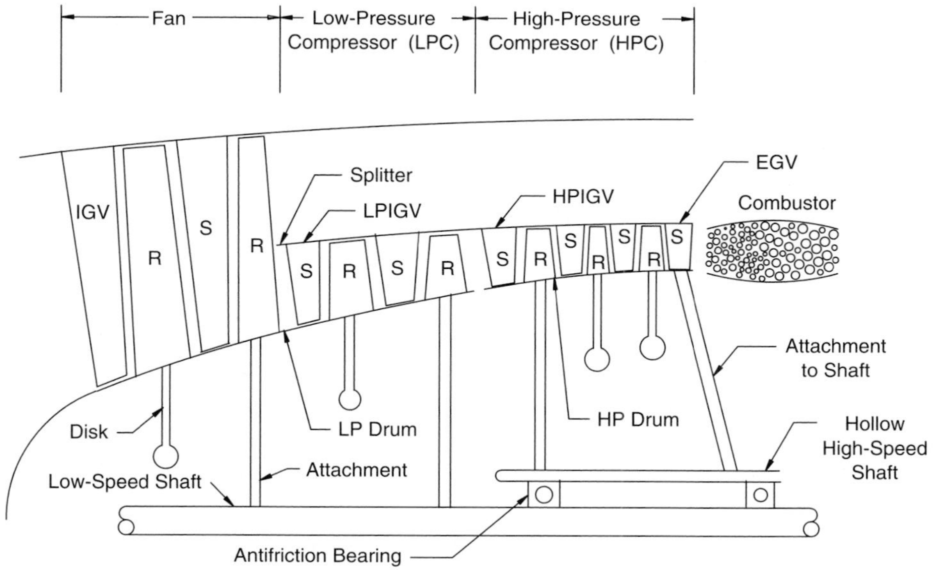

Figure 5.5 Twin-spool turbofan.

speed shaft. One set of inlet guide vanes is used for the fan or low-pressure compressor, and a second set is used for the high-pressure compressor. The blades of a fan are usually quite large, and blade heights decrease in size as the pressure increases down the engine. The actual blade shapes and designs are somewhat different for a fan, low-pressure compressor, and high-pressure compressor. In particular, fan blades are quite long with a large tip-to-hub radius ratio and, as will be discussed later, this results in blades having more curvature at the hub and less at the tip and with significant blade twist from hub to tip. Typical total pressure ratios for fans are 1.3 to 1.5 per stage. As a result, designing fan blades is more complicated. However, the fundamental thermodynamics, fluid mechanics, and design methodology are the same for the fan and two compressors. Thus, for the remainder of this chapter no distinctions are made in the analysis of the three stage types.

If one were to unwrap the blades around the periphery of the compressor and consider its geometry to be planar 2-D when viewed from the top, a series of "cascades" would be seen as shown in Figure 5.6. The fluid first enters the inlet guide vanes. Next, it enters the first rotating passage. In Figure 5.6, the rotor blades are shown to be moving with linear velocity U, which is found from $R\omega$, where ω is the angular speed and R is the mean radius of the passage. The blades and vanes are in fact airfoils, and the pressure (+ +) and suction surfaces (– –) can be defined as shown in the figure. After passing through the cascade of rotor blades, the fluid enters the cascade of stator vanes. The fluid is turned by these stationary vanes and readied to enter

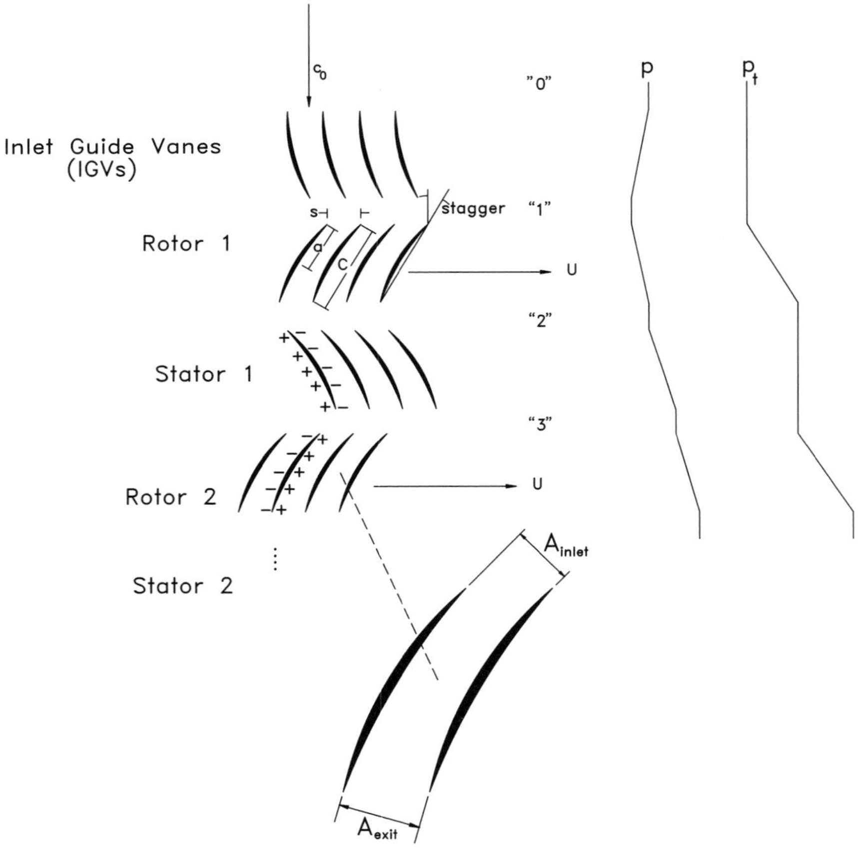

Figure 5.6 Cascade view of blades/vanes with pressure profiles and areas in a blade passage.

the second stage, beginning with the second rotor blades, ideally incidence free. In general, the second stage has a slightly different design than the first stage (all of the stages are different). The process is repeated for each stage. In Figure 5.6, compressor component stations 0 through 3 are defined. It is very important not to confuse these with engine station designations.

An important parameter in compressor design is the solidity, which is defined as C/s, where s is the blade spacing, or pitch, and C is the chord. The solidity is the inverse of the pitch-to-chord ratio s/C. Both parameters are defined in Figure 5.6. If the solidity becomes too large, frictional effects, which decrease the efficiency and total pressure ratio, become large because the boundary layers dominate the passage flow. However, if the solidity becomes too small, sufficient flow guidance is not attained (this phenomenon is termed "slip") and the flow thus does not adequately follow the blade shape. Separation can also become a major problem. For the low-solidity case due to slip, less power is added to the flow than desired; as a result, the compressor does not operate at the necessary pressure ratio. Because of accompanying separation (and losses), the maximum efficiency is not realized. Thus, a compromise is needed. In Section 5.11 a method is presented to optimize the single-stage aerodynamic performance or maximize the efficiency. Typical values of solidity for a compressor are approximately 1. The selection of solidity is partially responsible for the resulting number of blades in a cascade. For example, for the Pratt & Whitney JT9D turbofan, 46 blades are used on the fan rotor stage, and the number of blades on the compressor stages ranges from 60 to 154. For comparison, the exit guide vanes for the fan (which has an extremely large chord) have only nine blades.

It is very important that the number of stator vanes and rotor blades for a stage or nearby stages be different. If they were equal, a resonance due to fluid dynamic blade interactions could be generated, resulting in large blade, disk, and shaft vibrations accompanied by noise. This would reduce the life of the blades and compromise engine safety. Often, and if possible, blade or vane numbers are selected as prime numbers, but they are always chosen such that common multiple resonances are not excited. Cumpsty (1977) reviews this topic in detail.

Another important parameter for blade performance is the ratio of the distance to the maximum camber to the chord length of the blade, a. This distance is also shown in Figure 5.6. This parameter strongly influences the lift and drag characteristics of blades, which in turn have marked effects on the efficiency and pressure ratio of a stage.

If one examines the cross-sectional flow areas in the passages between the blades (area normal to the mean flow direction), it will be seen that the areas increase from the inlet to the exit of the rotor, as shown in Figure 5.6, and also increase from the inlet to the exit of the stator. Thus, both the rotor and stator blade rows act like diffuser sections.

Also shown in Figure 5.6 are axial pressure profiles down the compressor. Both the total and static pressures are indicated. As can be seen, the static pressure decreases across the inlet guide vanes but increases across all of the rotor and stator blades, which is consistent with the observation that rotor and stator blade rows act like diffusers. The total pressure remains approximately constant across the inlet guide vanes and stator vanes but increases across the rotor blades, where energy is put into the flow and the enthalpy increases. For the remainder of this chapter, pressure changes are related to the geometry of the compressor. That is, for a given geometry of blades and rotational speed, what are the static and total pressure rises and ratios?

A few design concerns should be identified at this point. First, as a fluid particle traverses a compressor, the pressure increases. As a result, the density increases. A common design practice is to keep the axial velocity component approximately constant through a compressor. Therefore, to compensate for the increase in fluid density, the blade heights must decrease through a compressor. Next, a compressor stage cannot be expected to increase the pressure without limits. When pressure increases in a stage, it acts like a diffuser. If the pressure rise becomes too large, the flow in the stage separates, causing a gross reduction in efficiency and a phenomenon called surge. Such a condition reduces the thrust to zero and also causes large oscillatory pressure fluctuations in the compressor, which can cause case mechanical failure. Finally, flow compressibility induces limits. For example, supersonic flow over the blade surfaces generates shock waves. Such a formation both reduces the total pressure of the gas (thus reducing the compressor efficiency) and also leads to flow separation, which also reduces the efficiency. These concerns are addressed more throughout the chapter.

5.3 Velocity Polygons or Triangles

Because energy is being added to the flow (which results in a pressure rise) and a part of the compressor is moving, one must be able to understand the complex flow patterns in a compressor. As the theoretical analysis in Chapter 4 and in the next section demonstrates, the pressure rise is directly dependent on the fluid velocity magnitudes and directions in a compressor. The purpose of this section is to introduce a means of interpreting the fluid velocities for use with the governing equations. That is, since one component rotates (rotor) and one component is stationary (stator), an analyst must be able to relate the velocities between the two so that the components are compatible. The rotor blades have a set of coordinates that move with the blades, and the stator vanes have a set of coordinates that are stationary with the blades. Converting velocities from one frame to the other is accomplished through a vector analysis termed velocity polygons or sometimes velocity triangles.

First consider the inlet guide vanes, as shown in Figure 5.7a. The inlet flow to the inlet guide vanes is typically aligned with the axis of the engine. The axial flow velocity relative to the engine frame (referred to as absolute velocity for the remainder of the book) at the IGV inlet is c_0. The blade inlet angle is usually aligned with the axis. As the fluid passes through the IGV, it is turned; thus, the absolute flow velocity (or velocity in the nonrotating frame) at the IGV exit (and the rotor inlet) is c_1, and it has a flow angle α_1 relative to the axis of the engine. The IGV exit blade angle is α_1'. If the flow is exactly parallel to the blade at the exit, α_1 and α_1' are identical. This is usually not true in an actual design, however, because the fluid does not exactly follow the blade/vane because of pressure gradients and fluid momentum. This is a term called slip.

The inlet to the rotor stage, as shown in Figure 5.7b, can be considered next. However, the rotor blades are rotating around the engine centerline or are moving in the planar 2-D plane with absolute velocity U_1 in the tangential direction. The objective is therefore to relate the velocities in the stationary frame to those in the rotating frame and vice versa. Thus, to find the velocity of the fluid relative to the rotating blades, one must subtract the blade velocity vector from the absolute flow vector ($V_{abs} = V_{rel} + V_{ref\ frame}$) or ($c = w + U$). This is graphically performed in Figure 5.7b. The resulting velocity vector in the rotating frame is w_1 and has a flow angle β_1 relative to the axial direction. It is important to note that the rotor blades have inlet and exit angles (relative to the axial direction) of β_1' and β_2', respectively. Once again, if the relative flow direction matches the blade angles exactly, the values β_1' and β_1 are the same, and the incidence angle (the difference between β_1' and β_1 at the leading edge) is zero. However, this usually does not occur in practice – especially at off-design conditions. The difference at the trailing edge β_2 and β_2', is called the deviation.

When drawing the polygons, one should always draw the triangles exactly to scale. This not only allows "rough" checking of algebraic computations but also allows makes it possible to observe the relative magnitudes of the different velocity vectors and the "health" of a compressor stage, as is discussed in Sections 5.4.2 and 5.4.3 with the percent reaction. When an undesirable situation is observed, it is much easier to see how to correct the condition by changing geometric parameters and visualizing them with a scale diagram than by means of equations.

Next, one can break the vectors into components, as indicated in Figure 5.7b. As shown, the absolute velocity c_1 has a component in the tangential direction that is equal to c_{u1} and a component in the axial direction that is c_{a1}. Also, the relative velocity w_1 has components w_{u1} and w_{a1} in the tangential and axial directions, respectively.

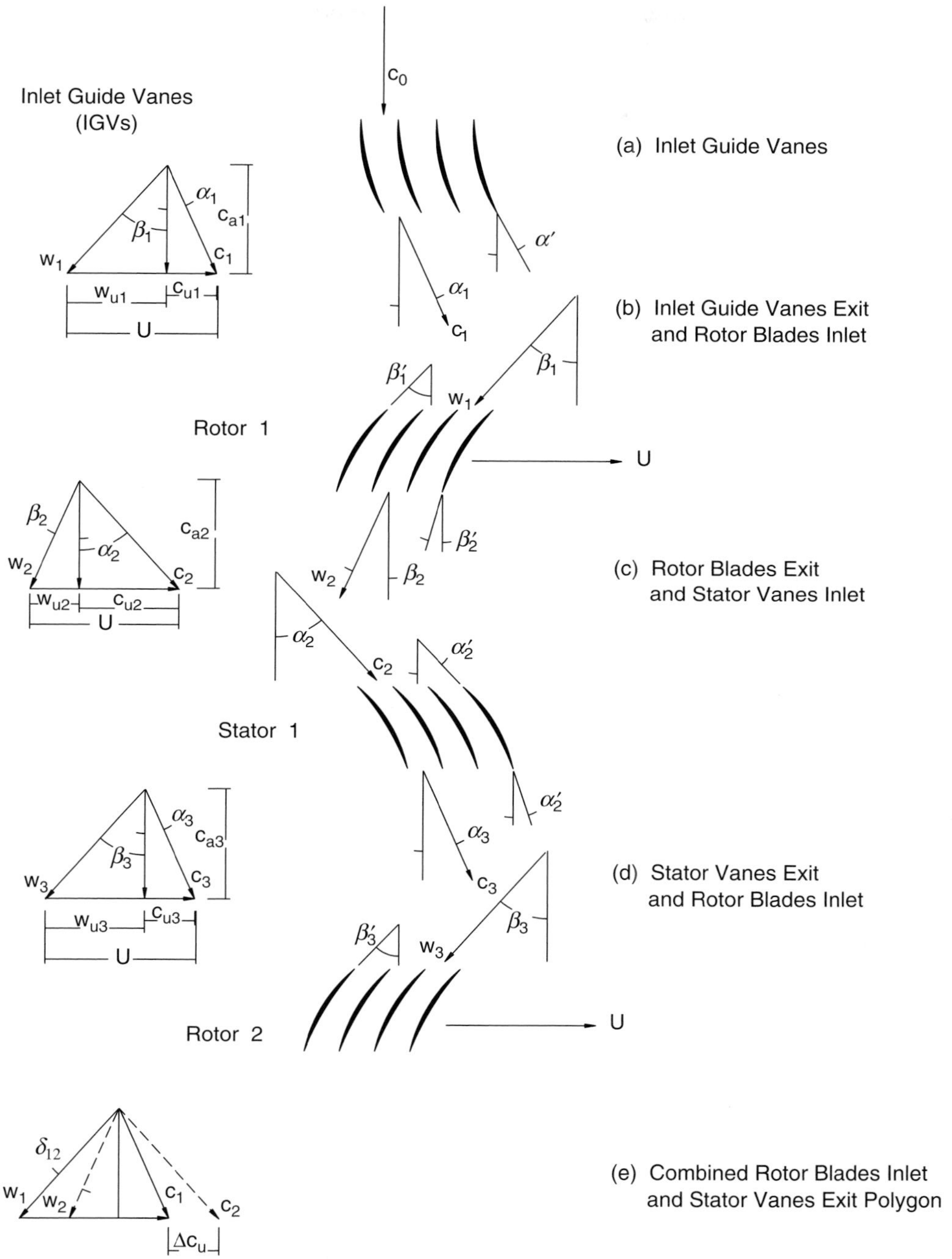

Figure 5.7 Definitions of blade geometries and velocity polygons for compressors.

Figure 5.7c shows the velocity polygon for the rotor exit and stator inlet. The exit relative velocity (to the rotor blades) is w_2, and this velocity is at an angle of β_2 relative to the axial direction. The blade velocity is U_2 and can be vectorially added to w_2 to find the absolute exit velocity c_2. Note that U_2 and U_1 may be slightly different because the radii at the rotor inlet and exit may be slightly different. The absolute flow angle of c_2 is α_2 relative to the engine axis. Once again, the rotor blade exit angle is β_2', which may be different from the flow angle β_2. Furthermore, the stator vane inlet angle is α_2' and will in general be different from α_2. Also shown in Figure 5.7c are the components of both w_2 and c_2 in the tangential and axial directions.

In Figure 5.7d, the stator exit triangle is shown (which is also the inlet velocity polygon to the next set of rotor blades). The absolute velocity is c_3, which makes an angle of α_3 to the axial direction. Once again, the absolute flow angle α_3 will in general be different from the absolute stator vane angle α_3' The next rotor blade velocity at the inlet is U_3, and the relative velocity (to the next rotor blades) is w_3, which makes an angle of β_3 to the axis. The method just described can be used for all of the downstream stages.

Next, as shown in Figure 5.7e, the velocity polygons for the rotor entrance and exit are often combined to form one polygon. As is shown in the Section 5.4, the power and pressure ratio of the stage are directly related to the change in absolute tangential velocity across a set of rotor blades if the radius is approximately constant ($\Delta c_u = c_{u1} - c_{u2}$). Thus, combining the two triangles makes the change easy to observe graphically. Furthermore, by examining the difference of the two vectors w_1 and w_2, one can see that the rotor flow turning angle or deflection is $\beta_2 - \beta_1$ or δ_{12}. Once again, if the vectors are to scale, it is relatively easy to see how to alter the performance of a stage by changing geometric parameters.

Finally, in Figure 5.8 a side view of the first stage of a typical compressor is shown. Defined are the blade heights (t) and the average radii of the blades (R).

Before one proceeds, a consistent sign notation is needed. First, all positive tangential velocities are defined to the right. Thus, rotor blades always move to the right with positive velocity U. Second, all angles are measured relative to the axial direction and positive angles are considered to be counterclockwise.

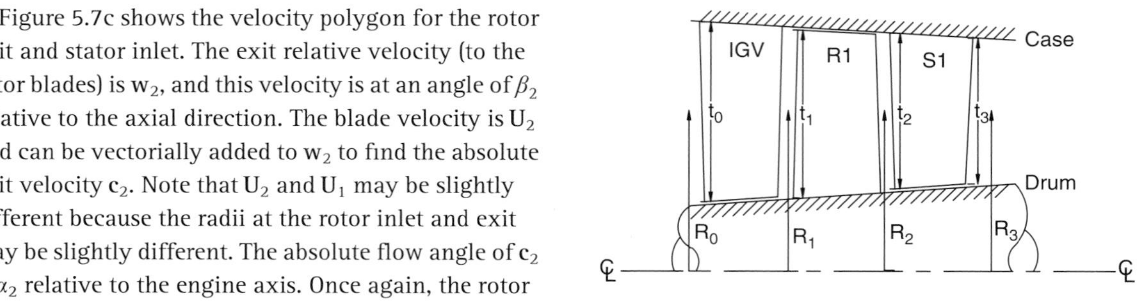

Figure 5.8 Side view of first stage.

Note that the velocity polygons are similar for an incompressible axial turbomachine (i.e., pumps). Thus, the same type of analysis is used to understand the flow velocities and relate them to power requirements, pressure rises, and so on. One basic difference, however, is that pump manufacturers typically do not measure the flow angles relative to the axial direction. They instead use the tangential direction as a reference. Both reference conditions lead to identical velocity polygons and other results. One needs to be aware of the different conventions and angle definitions when using other reference material, however.

In Table 5.1, the trends of the velocity magnitudes, flow passage areas or cross-sectional areas normal to the mean flow, and pressures across the different components are shown. As can be seen, the static pressure increases across both the stator and rotor blades because the relative velocities across both decrease. On the other hand, as the absolute and relative velocities across the IGVs increase, the static pressure decreases. Details of, and reasons for, these trends are covered in the next section.

5.4 Single-Stage Energy Analysis

In this section, the equations used in single-stage energy analysis are summarized so that the velocities

Table 5.1 **Axial flow compressor component trends.**

	Absolute velocity	Relative velocity	Area	p	p_t
IGV	Increases	Increases	Decreases	Decreases	Constant (ideally)
Rotor	Increases	Decreases	Increases	Increases	Increases
Stator	Decreases	Decreases	Increases	Increases	Constant (ideally)

from polygons can be related to the pressure rise and other important compressor characteristics. In Chapter 4, this derivation was performed for a single-stage (rotor and stator) using a two-dimensional planar "mean line" control volume approach. That is, for an axial flow compressor, a point located midway between the hub and tip is used to evaluate the radii, fluid properties, and velocities. Figure 4.1 illustrates a single stage. The objective of this section is to relate the inlet and exit conditions to the property changes.

5.4.1 Total Pressure Ratio

The three basic equations used in Chapter 4 are those for continuity, moment of momentum, and energy equations. For the power input to the shaft, one finds (Eq. (4.2.17))

$$\dot{W}_{sh} = \dot{m}[U_2 c_{u2} - U_1 c_{u1}]. \tag{5.4.1}$$

Next, one can solve for the total pressure ratio of the stage and find (Eq. (4.2.26))

$$\frac{p_{t2}}{p_{t1}} = \left[\eta_{12}\frac{U_2 c_{u2} - U_1 c_{u1}}{c_p T_{t1}} + 1\right]^{\frac{\gamma}{\gamma-1}}, \tag{5.4.2}$$

which is greater than unity. Note that if $R_2 = R_1 = R\,(U_2 = U_1 = U)$, p_{t2}/p_{t1} is proportional to $U\,\Delta c_u$ to a power. Furthermore, by referring to Figure 5.7e, one can see that, as the rotor flow turning angle δ_{12} increases, so does Δc_u. Therefore, increasing the pressure ratio of a stage is accomplished by increasing the rotor turning angle. Note also that U is proportional to R and, as U increases, the total pressure ratio increases. Thus, as the radius becomes larger, so does the total pressure ratio. Therefore, once the velocity information of a rotor is known from the velocity polygons and the efficiency, it is possible to find the total pressure rise.

5.4.2 Percent Reaction

An important characteristic for a compressor is termed the percent reaction. This is a relation that approximates the relative loading of the rotor and stator based on the enthalpy rise. From Chapter 4 (Eq. (4.2.32)),

$$\%R = \frac{1}{1 + \left[\dfrac{c_2^2 - c_1^2}{w_1^2 - w_2^2}\right]}. \tag{5.4.3}$$

Therefore, for compressible flow the percent reaction of the stage can be related to the absolute and relative velocities at the inlet and exit of the rotor. Implications and physical insight of the percent reaction are discussed below. An example of a single stage with all of the calculation details will follow. One should note that the software Turbomachinery.exe performs the same calculations.

5.4.3 Incompressible Flow (Limiting Case)

For comparison, one can find the pressure rise and percent reaction of a turbomachine with an incompressible fluid (i.e., an axial flow pump). This limiting case is presented because, as will be shown below, this comparison lends to great insight as to the significance of percent reaction and how it relates to the pressure rise. Although gas turbines are compressible turbomachines, for this reason incompressible flow is discussed herein. More pump information can be found in Stepanoff (1957) and Brennen (1994), who give great details on axial flow pump performance. However, Chapter 4 presents (Eqs. (4.2.34), (4.2.37), and (4.2.39)):

$$\dot{W}_{sh} = \dot{m}[U_2 c_{u2} - U_1 c_{u1}] \quad (5.4.4)$$

$$p_{t2} - p_{t1} = \rho\eta_{12}[U_2 c_{u2} - U_1 c_{u1}] \quad (5.4.5)$$

$$\%R = \frac{1}{1 + \left[\frac{c_2^2 - c_1^2}{w_1^2 - w_2^2}\right]}. \quad (5.4.6)$$

Comparing Eq. (5.4.6) with that for compressible flow reveals that they are identical. Also, the percent reaction for incompressible and ideal flow (no losses) reduces to Eq. (4.2.41), an equation with a practical implication:

$$\%R = \frac{p_2 - p_1}{p_{t3} - p_{t1}}. \quad (5.4.7)$$

For example, typical incompressible turbomachine stages with short blades have percent reactions of 0.50 to 0.55. For a nonideal pump, this dictates that approximately half of the enthalpy increase occurs in the rotor and half in the stator. For an ideal pump, this also means that half of the static pressure increase occurs in the rotor and half in the stator. This implies that the force or "load" distributions on the rotor and stator blades are about the same, and thus neither the stator vanes nor rotor blades are subjected to higher-than-needed stresses, which can lead to catastrophic failures. If, however, the percent reaction differs greatly from 0.5, a large pressure rise occurs in either the stator ($\%R < 0.5$) or the rotor ($\%R > 0.5$). If this occurs, separation or partial separation of the cascade is possible or likely. As a result of the associated losses, the efficiency will be less than optimum. Such an optimization is covered in Section 5.11 using cascade lift and drag data for different blade shapes in detail. However, in summary, the percent reaction and efficiency are NOT independent variables as is shown in Section 5.11. For an axial flow pump the percent reaction should be close to 0.50 for two independent reasons: (1) the stage efficiency will be maximized and (2) force loads will be distributed so that blades and vanes are less susceptible to material failure.

Owing to compressibility of the gas and losses in a compressor, the percent reaction tends to be higher than 0.5 in modern compressors. That is, as the stage total pressure rise increases to maintain an even pressure rise in the rotor and stator, the percent reaction tends to be somewhat greater than 0.5, namely 0.55 to 0.60. So although the convenient incompressible analysis presented in Section 5.11 can be used as a starting point, one should keep in mind that the overriding idea is to control the separation and maximize the efficiency. Also, as the flow becomes more 3-D (e.g., as the tip-to-hub diameter increases for front stages), maintaining a constant percent reaction in the radial direction becomes more difficult, as discussed in Section 5.10. Even worse, if a stage were designed with a mean line reaction of 50 percent, some hub regions would possibly experience negative values of reaction due to radial flow, which is obviously unacceptable. As a result, some stages are designed so that the mean line reaction is significantly above 50 percent; this results in tip regions well above 50 percent with the hub regions above 0 percent. Fortunately, as will be shown later, the efficiency does not decrease rapidly as one deviates from the optimal 50 percent reaction.

5.4.4 Relationships of Velocity Polygons to Percent Reaction and Pressure Ratio

As indicated by Eqs. (5.4.3) and (5.4.6), the velocities dictate the percent reaction. In Figure 5.9, three special cases are shown. In particular, sets of velocity polygons (rotor inlet and exit) for 50 percent, 0 percent, and 100 percent reaction compressor stages are shown for a constant axial velocity. Of particular interest is the set of polygons for a 50 percent stage. As can be seen, for this case the inlet and exit polygons are similar triangles but reversed. Once again, if the polygons are drawn to scale, one can easily observe if the percent reaction is close to 50 percent simply by examining the symmetry of the triangles. Also, one can use polygons to make preliminary predictions on performance trends due to geometry changes. For example, for a compressor stage that originally operates close to a 50 percent reaction, if δ_{12} is increased the percent reaction is decreased if other parameters are constant (Figure 5.9b). Similarly, if the absolute inlet angle is decreased while holding other parameters constant, the percent reaction increases (Figure 5.9c).

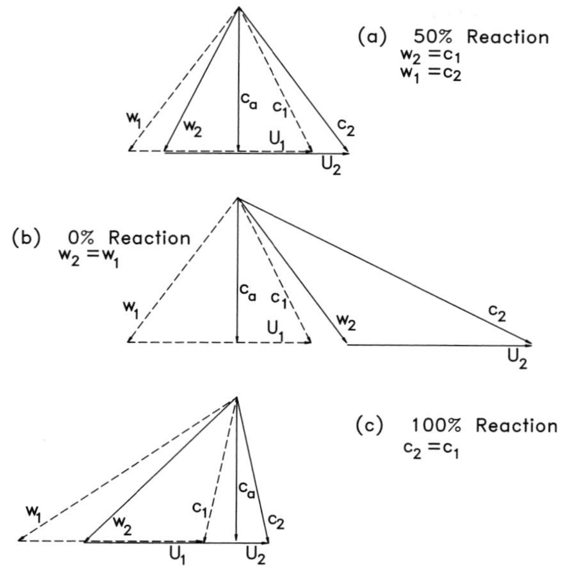

Figure 5.9 Three special cases of velocity polygons.

Table 5.2 **Geometric and operating condition effects on compressor parameters.**

Decrease in	Percent reaction, %R	Total pressure ratio, p_{t3}/p_{t1}
Absolute inlet angle, α_1	Increase	Increase
Rotor flow turning angle, δ_{12}	Increase	Decrease
Rotational speed, N	Decrease	Decrease
Axial velocity component, c_a	Increase	Increase

Moreover, for general trends, one can examine both the governing equations and velocity polygons for individual variations in the absolute inlet angle α_1, the rotor flow turning angle δ_{12}, the rotational speed N, and mass flow rate (or axial velocity component) $\dot{m}$ (or c_a). The resulting effects on the stage total pressure ratio and percent reaction can be determined.
A matrix of general trends is presented in Table 5.2.

Example 5.1 | Axial Compressor Stage

A stage approximating the size one of the last stages (rotor and stator) of a high-pressure compressor is to be analyzed. It rotates at 8000 rpm and compresses 280 lb_m/s (8.704 sl/s) of air. The inlet pressure and temperature are 272 psia and 1310 °R, respectively. The average radius of the blades is 13.2 in., and the inlet blade height is 1.24 in. The absolute inlet flow angle to the rotor is the same as the stator exit flow angle (15°), and the rotor flow turning angle (δ_{12}) is 25°. The stage has been designed so that the blade height varies and the axial velocity remains constant through it. The efficiency of the stage is 90 percent. The values of c_p and γ are 0.2587 BTU/lb_m-°R and 1.361, respectively, *which are based on the resulting value* of T_2 or average static temperature of the stage. The following details are to be found: blade heights at the rotor and stator exits, the static and total pressures at the rotor and stator exits, the stator turning angle, the Mach numbers at the rotor and stator exits, the required power for the stage, and the percent reaction for the stage.

Solution Note

Before proceeding and because the solution is long, an outline for the solution is appropriate.

(1) First, some preliminary calculations are presented.
(2) Next, the polygon of the velocities at the rotor inlet is "drawn." Various angles and velocities are found.

(3) Next, the triangle of the velocities at the rotor exit/stator inlet is "drawn" using the above and problem specifications. Various angles and velocities are found.
(4) Next, using this information the power, pressure ratio, and percent reaction are found.
(5) Lastly, the triangle of the velocities at the stator exit is "drawn" using the above and problem specifications, which would be needed for the following stage.
(6) Intermediate steps also allow for the determination of such important characteristics such as Mach numbers and pressure coefficients.

Solution

To obtain the solution, velocity polygons will be used. Before proceeding, a few preliminary calculations are needed:

$$U = R\omega = \frac{13.2\text{in.}}{12\dfrac{\text{in.}}{\text{ft}}}\frac{8000\text{ rev}}{\text{min}}\frac{2\pi\text{ rad}}{\text{rev}}\frac{1}{60\dfrac{\text{s}}{\text{min}}} = 921.5\text{ ft/s}$$

$$A_1 = \pi D_1 t_1 = \pi\frac{2\times 13.2\text{ in.}}{12\dfrac{\text{in.}}{\text{ft}}}\frac{1.24\text{ in.}}{12\dfrac{\text{in.}}{\text{ft}}} = 0.7142\text{ ft}^2.$$

For $p_1 = 272$ psia and $T_1 = 1310$ °R $= 850$ °F, the ideal gas equation yields

$$\rho_1 = \frac{p_1}{\mathcal{R}T_1} = \frac{272\dfrac{\text{lb}_\text{f}}{\text{in.}^2}\times 144\dfrac{\text{in.}^2}{\text{ft}^2}}{53.35\dfrac{\text{ft-lb}_\text{f}}{\text{lb}_\text{m}\text{-°R}}\times 1310\text{ °R}\times 32.17\dfrac{\text{lb}_\text{m}}{\text{sl}}} = 0.01742\frac{\text{sl}}{\text{ft}^3}.$$

Next,

$$c_{a1} = \frac{\dot{m}}{\rho_1 A_1} = \frac{8.704\dfrac{\text{sl}}{\text{s}}}{0.01742\dfrac{\text{sl}}{\text{ft}^3}\times 0.7142\text{ ft}^2} = 699.6\text{ ft/s}$$

Finally,

$$a_1 = \sqrt{\gamma\mathcal{R}T_1}$$

$$= \sqrt{1.361\times 53.35\frac{\text{ft-lb}_\text{f}}{\text{lb}_\text{m}\text{-°R}}\times 32.17\frac{\text{lb}_\text{m}}{\text{sl}}\times 1310\text{ °R}\times\frac{\text{sl-ft}}{\text{s}^2\text{lb}_\text{f}}} = 1749\text{ ft/s}.$$

Rotor Inlet

Next, one must refer to Figure 5.10a, which is the velocity polygon (to scale) for the rotor inlet. First,

$$c_1 = \frac{c_{a1}}{\cos\alpha_1} = \frac{699.6}{\cos(15°)} = 724.2\text{ ft/s}.$$

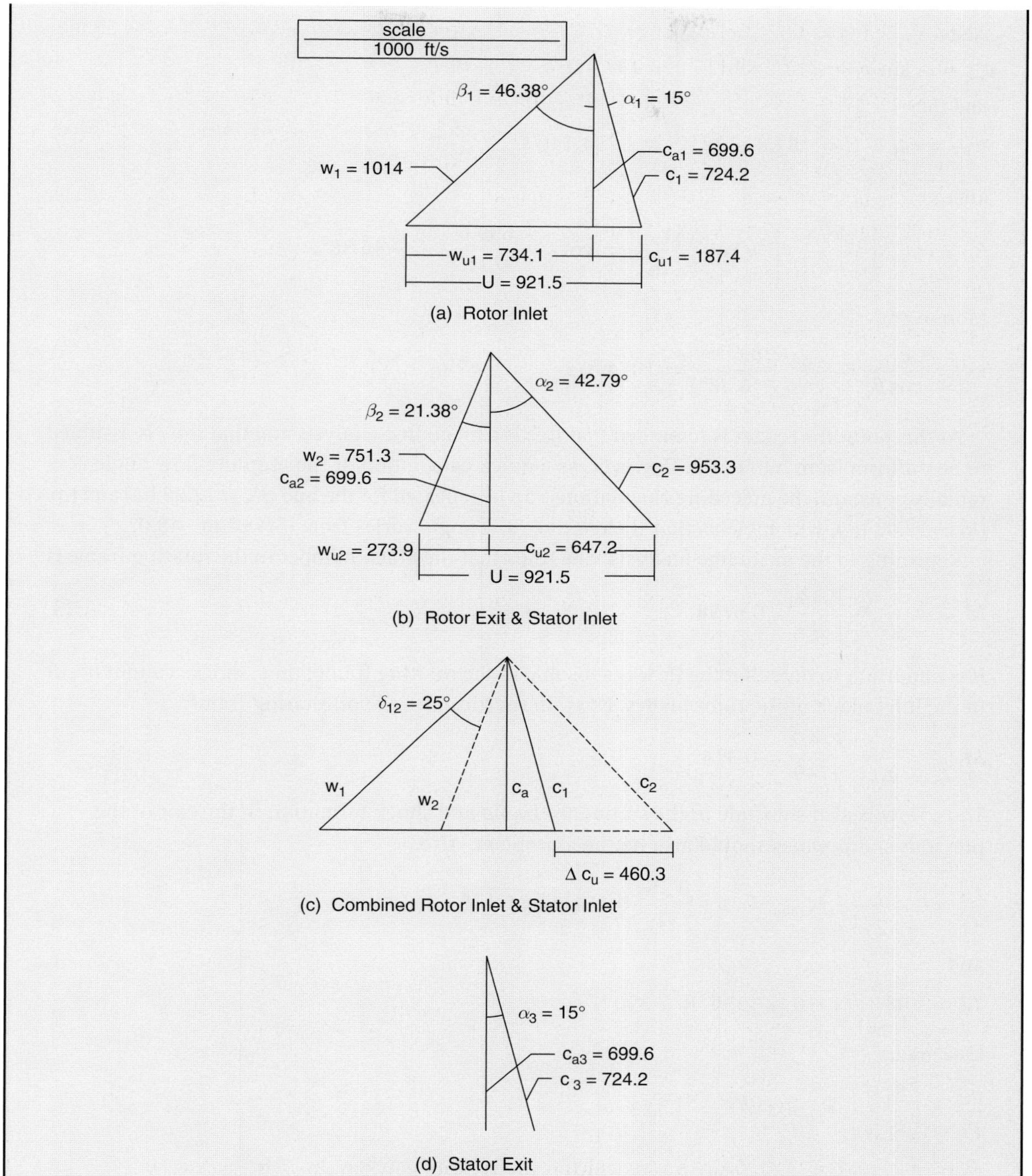

Figure 5.10 Velocity polygons for Example 5.1.

Also,

$$c_{u1} = c_1 \sin\alpha_1 = 724.2 \sin(15°) = 187.4 \text{ ft/s},$$

and thus

$$w_{u1} = c_{u1} - U = 187.4 - 921.5 = -734.1 \text{ ft/s}$$

and

$$\beta_1 = \cot^{-1}\left[\frac{c_{a1}}{c_{u1} - U_1}\right] = \cot^{-1}\left[\frac{c_{a1}}{w_{u1}}\right] = \cot^{-1}\left[\frac{699.6}{-734.1}\right] = -46.38°.$$

Moreover,

$$w_1 = \frac{c_{a1}}{\cos\beta_1} = \frac{699.6}{\cos(-46.38°)} = 1014 \text{ ft/s}.$$

At this point the reader is reminded that this is a mean line analysis and that flow is assumed to be uniform from hub to tip. However, as a quick calculation, if the absolute flow angle is radially constant, the preceding calculations can be repeated for the hub ($R_h = 12.58$ in.) and tip ($R_t = 13.82$ in.), which yields that the relative flow angle varies from −44.6° to −48.0°.

Returning to the mean line analysis, one finds that the Mach number in the rotating frame is

$$M_{1rel} = \frac{w_1}{a_1} = \frac{1014}{1749} = 0.5798.$$ <ANS

It is important to note that the flow is subsonic in the rotating frame; thus, shocks cannot occur in the inlet plane of the rotor blades. Next, in the absolute or nonrotating frame,

$$M_{1abs} = \frac{c_1}{a_1} = \frac{724.2}{1749} = 0.4141.$$

Thus, flow is also subsonic in the stationary frame and shock formation in the exit of the previous stator vanes should also not be a problem. Also,

$$\frac{T_{t1}}{T_1} = 1 + \frac{\gamma - 1}{2} M_{1abs}{}^2 = 1 + \frac{0.361}{2} 0.4141^2 = 1.031$$

and

$$T_{t1} = 1.031 \times 1310 = 1350\,°\text{R} = 890\,°\text{F}.$$

Moreover,

$$\frac{p_{t1}}{p_1} = \left[\frac{T_{t1}}{T_1}\right]^{\frac{\gamma}{\gamma-1}} = [1.031]^{\frac{1.361}{0.361}} = 1.122$$

$$p_{t1} = 1.122 \times 272 = 305.1 \text{ psia}.$$

Rotor Exit–Stator Inlet

For the rotor exit and stator inlet velocity polygon, refer to Figure 5.10b. Since the rotor flow turning angle is known, it is possible to find the relative exit angle by

$\beta_2 = \beta_1 + \delta_{12} = -46.38° + 25° = -21.38°$

Also, because the axial velocity is constant through the stage,

$c_{a2} = c_{a1} = 699.6\ \text{ft/s}.$

Next, $w_2 = \dfrac{c_{a2}}{\cos\beta_2} = \dfrac{699.6}{\cos(-21.38°)} = 751.3\ \text{ft/s}.$

Also, $w_{u2} = c_{a2}\tan\beta_2 = 699.6\tan(-21.38°) = -273.9\ \text{ft/s},$

and $c_{u2} = U + w_{u2} = 921.5 - 273.9 = 647.7\ \text{ft/s};$

thus,

$$\alpha_2 = \cot^{-1}\left[\frac{c_{a2}}{U + w_{u2}}\right] = \cot^{-1}\left[\frac{c_{a2}}{c_{u2}}\right] = \cot^{-1}\left[\frac{699.6}{647.7}\right] = 42.79°,$$

and $c_2 = \dfrac{c_{a2}}{\cos\alpha_2} = \dfrac{699.6}{\cos(42.79°)} = 953.3\ \text{ft/s}.$

Next, the moment of momentum equation can be used with results from the velocity polygons to yield

$h_{t2} - h_{t1} = U(c_{u2} - c_{u1}) = 921.5 \times (647.7 - 187.4) = 424{,}107\ \text{ft}^2/\text{s}^2.$

From the energy equation, one can find

$$\frac{p_{t2}}{p_{t1}} = \left[\frac{\eta_{12}(h_{t2} - h_{t1})}{c_p T_{t1}} + 1\right]^{\frac{\gamma}{\gamma-1}}.$$

Thus,

$$\frac{p_{t2}}{p_{t1}} = \left[\frac{0.90 \times 424{,}107\,\dfrac{\text{ft}^2}{\text{s}^2}}{0.2587\,\dfrac{\text{BTU}}{\text{lb}_\text{m}\text{-}°\text{R}} \times 1350\,°\text{R} \times 778.16\,\dfrac{\text{ft-lb}_\text{f}}{\text{BTU}} \times 32.17\,\dfrac{\text{lb}_\text{m}}{\text{sl}}}\,\frac{\text{lb}_\text{f}\text{-s}^2}{\text{sl-ft}} + 1\right]^{\frac{1.361}{0.361}},$$

$$= 1.175, \qquad \text{<ANS}$$

and so $p_{t2} = 1.175 \times 305.1 = 358.5$ psia.

Also,

$$T_{t2} - T_{t1} = \frac{h_{t2} - h_{t1}}{c_p} = \frac{424{,}107\,\dfrac{\text{ft}^2}{\text{s}^2}}{0.2587\,\dfrac{\text{BTU}}{\text{lb}_\text{m}\text{-}°\text{R}} \times 778.16\,\dfrac{\text{ft-lb}_\text{f}}{\text{BTU}} \times 32.17\,\dfrac{\text{lb}_\text{m}}{\text{sl}}}\,\frac{\text{lb}_\text{f}\text{-s}^2}{\text{sl-ft}},$$

$$= 65.5\ °\text{R};$$

thus, $T_{t2} = 1416\ °\text{R} = 956\ °\text{F}.$

Moreover,

$$T_2 = T_{t2} - \frac{c_2^2}{2c_p}$$

$$= 1416\ °\text{R} - \frac{953.2^2\,\frac{\text{ft}^2}{\text{s}^2}}{2 \times 0.2587\,\frac{\text{BTU}}{\text{lb}_\text{m}\text{-°R}} \times 778.16\,\frac{\text{ft-lb}_\text{f}}{\text{BTU}} \times 32.17\,\frac{\text{lb}_\text{m}}{\text{sl}}}\,\frac{\text{lb}_\text{f}\text{-s}^2}{\text{sl-ft}}$$

$$= 1346\ °\text{R} = 886\ °\text{F}.$$

Next,

$$a_2 = \sqrt{\gamma \mathcal{R} T_2} = \sqrt{1.361 \times 53.35\,\frac{\text{ft-lb}_\text{f}}{\text{lb}_\text{m}\text{-°R}} \times 32.17\,\frac{\text{lb}_\text{m}}{\text{sl}} \times 1346\ °\text{R} \times \frac{\text{sl-ft}}{\text{s}^2\text{-lb}_\text{f}}}$$
$$= 1773\ \text{ft/s}.$$

Thus, the Mach number in the rotating frame is

$$M_{2\text{rel}} = \frac{w_2}{a_2} = \frac{751.3}{1773} = 0.4237.$$ <ANS

It is once again important to note that the flow is thus subsonic in the rotating frame, and shock formation in the exit of the rotor should not be a problem. Next, in the absolute frame,

$$M_{2\text{abs}} = \frac{c_2}{a_2} = \frac{953.3}{1773} = 0.5378.$$ <ANS

Thus, the flow is also subsonic in the stationary frame, and so shocks cannot occur in the inlet plane of the following stator blades. Using the absolute Mach number yields

$$\frac{p_{t2}}{p_2} = \left[1 + \frac{\gamma - 1}{2} M_{2\text{abs}}{}^2\right]^{\frac{\gamma}{\gamma-1}} = \left[1 + \frac{0.361}{2} 0.5378^2\right]^{\frac{1.361}{0.361}} = 1.211,$$

and thus

$$p_2 = \frac{358.5}{1.211} = 295.9\ \text{psia}.$$ <ANS

Next, from the ideal gas law for $p_2 = 295.9$ psia and $T_2 = 1346$ °R one obtains

$$\rho_2 = \frac{p_2}{\mathcal{R} T_2} = \frac{295.9\,\frac{\text{lb}_\text{f}}{\text{in.}^2} \times 144\,\frac{\text{in.}^2}{\text{ft}^2}}{53.35\,\frac{\text{ft-lb}_\text{f}}{\text{lb}_\text{m}\text{-°R}} \times 1346\ °\text{R} \times 32.17\,\frac{\text{lb}_\text{m}}{\text{sl}}} = 0.01845\,\frac{\text{sl}}{\text{ft}^3},$$

and because $\dot{m} = \rho_2 c_{a2} A_2$,

$$A_2 = \frac{\dot{m}}{\rho_2 c_{a2}} = \frac{8.704\,\frac{\text{sl}}{\text{s}}}{0.01845\,\frac{\text{sl}}{\text{ft}^3} \times 699.6\,\frac{\text{ft}}{\text{s}}} = 0.6744\ \text{ft}^2.$$

Also $A_2 = \pi D_2 t_2$;

thus,

$$t_2 = \frac{A_2}{\pi D_2} = \frac{0.6744\,\text{ft}^2 \times 144\dfrac{\text{in.}^2}{\text{ft}^2}}{\pi \times 2 \times 13.2\,\text{in.}} = 1.171\,\text{in.}$$ <ANS

Note that this is smaller than the rotor inlet height because of the compressibility of the fluid.

One can define a pressure coefficient for the rotor (which is covered in detail in Section 5.6) as

$$C_{\text{pr}} = \frac{p_2 - p_1}{\dfrac{1}{2}\rho_1 w_1^2} = \frac{(295.9 - 272.0)\dfrac{\text{lb}_\text{f}}{\text{in.}^2} \times \dfrac{\text{sl-ft}}{\text{s}^2\text{-lb}_\text{f}} \times \dfrac{144\,\text{in.}^2}{\text{ft}^2}}{\dfrac{1}{2} \times 0.01742\dfrac{\text{sl}}{\text{ft}^3} \times 1014\dfrac{\text{ft}^2}{\text{s}^2}} = 0.384.$$

Implications of this parameter are discussed in Section 5.6. Next, one can find the input stage power by

$$\dot{W}_{\text{sh}} = \dot{m}(h_{\text{t}2} - h_{\text{t}1}) = \frac{8.704\dfrac{\text{sl}}{\text{s}} \times \left(424{,}107\dfrac{\text{ft}^2}{\text{s}^2}\right)}{550\dfrac{\text{ft-lb}_\text{f}}{\text{s-hp}}}\left(\frac{\text{s}^2\text{-lb}_\text{f}}{\text{sl-ft}}\right)$$

$$= 6712\ \text{hp}.$$ <ANS

The percent reaction can be found by

$$\%\text{R} = \frac{1}{1 + \left[\dfrac{c_2^2 - c_1^2}{w_1^2 - w_2^2}\right]} = \frac{1}{1 + \left[\dfrac{953.3^2 - 724.2^2}{1014^2 - 751.3^2}\right]} = 0.5469.$$

On initial inspection, this is a reasonable value for the percent reaction, for it is slightly above 0.5. One can also combine the polygons for the rotor inlet and exit as shown in Figure 5.10c. The rotor turning angle is easily seen as well as the change in absolute tangential velocity (Δc_u). As indicated in Section 5.4.1 and Table 5.2, if the rotor turning is increased, so is the change in absolute tangential velocity. Also, it is apparent that the two polygons in Figure 5.10c are approximately symmetric, indicating that the percent reaction should be about 0.5, which it is. One should, however, compare the pressure rises for the rotor and stator later.

Stator Exit

Next, refer to Figure 5.10d for the stator exit velocity polygon. Once again, the axial velocity is constant through the stage, and so

$$c_{a3} = c_{a2} = c_{a1} = 699.6\,\text{ft/s}.$$

Also, the absolute exit angle is given, $\alpha_3 = 15°$; thus,

$$c_3 = \frac{c_{a3}}{\cos\alpha_3} = \frac{699.6}{\cos(15°)} = 724.2\ \text{ft/s}.$$

The turning angle of the stator can be found by

$$\delta_{23} = \alpha_3 = \alpha_2 = 15 - 42.79 = -27.79°.$$

Thus, the magnitude of the turning angle for the stator vane is approximately that of the rotor blade. Next, for the stator,

$$T_{t3} = T_{t2} = 1416\ °\text{R},$$

and

$$T_3 = T_{t3} - \frac{c_3^2}{2c_p}$$

$$= 1416\ °\text{R} - \frac{724.2^2 \dfrac{\text{ft}^2}{\text{s}^2}}{2 \times 0.2587 \dfrac{\text{BTU}}{\text{lb}_\text{m}\text{-}°\text{R}} \times 778.16 \dfrac{\text{ft-lb}_\text{f}}{\text{BTU}} \times 32.17 \dfrac{\text{lb}_\text{m}}{\text{sl}} \dfrac{\text{lb}_\text{f}\text{-s}^2}{\text{sl-ft}}}$$

$$= 1375\ °\text{R}.$$

Next,

$$a_3 = \sqrt{\gamma \mathcal{R} T_3} = \sqrt{1.361 \times 53.35 \frac{\text{ft-lb}_\text{f}}{\text{lb}_\text{m}\text{-}°\text{R}} \times 32.17 \frac{\text{lb}_\text{m}}{\text{sl}} \times 1375\ °\text{R} \times \frac{\text{sl-ft}}{\text{s}^2\text{-lb}_\text{f}}}$$

$$= 1792\ \text{ft/s},$$

and so

$$M_{3\text{abs}} = \frac{c_3}{a_3} = \frac{742.2}{1792} = 0.4041.$$ <ANS

Therefore, the flow is once again subsonic in the stationary frame, and so shocks cannot occur in the exit plane of the stator blades.

In addition,

$$\frac{p_{t3}}{p_3} = \left[1 + \frac{\gamma - 1}{2} M_{3\text{abs}}{}^2\right]^{\frac{\gamma}{\gamma-1}} = \left[1 + \frac{0.361}{2} 0.4041^2\right]^{\frac{1.361}{0.361}} = 1.116.$$

Also, for the stator, $p_{t3} = p_{t2} = 358.5$ psia; thus,

$$p_3 = \frac{358.5}{1.116} = 321.3\ \text{psia}.$$ <ANS

From the ideal gas law, for $T_3 = 1375$ °R and $p_3 = 321.3$ psia,

$$\rho_3 = \frac{p_3}{\mathcal{R} T_3} = \frac{321.3 \dfrac{\text{lb}_\text{f}}{\text{in.}^2} \times 144 \dfrac{\text{in.}^2}{\text{ft}^2}}{53.35 \dfrac{\text{ft-lb}_\text{f}}{\text{lb}_\text{m}\text{-}°\text{R}} \times 1375\ °\text{R} \times 32.17 \dfrac{\text{lb}_\text{m}}{\text{sl}}} = 0.01960 \frac{\text{sl}}{\text{ft}^3}.$$

One can define a pressure coefficient for the stator (which is covered in detail in Section 5.6) as

$$C_{ps} = \frac{p_3 - p_2}{\frac{1}{2}\rho_2 c_2^2}$$

$$C_{ps} = \frac{(321.3 - 295.9)\frac{\text{lb}_f}{\text{in.}^2} \times \frac{\text{sl-ft}}{\text{s}^2\text{-lb}_f} \times \frac{144\ \text{in.}^2}{\text{ft}^2}}{\frac{1}{2} \times 0.01845\frac{\text{sl}}{\text{ft}^3} \times +953.3^2\frac{\text{ft}^2}{\text{s}^2}} = 0.436.$$

As for the rotor, implications of this parameter are discussed in Section 5.6. Also, since

$$\dot{m} = \rho_3 c_{a3} A_3,$$

$$A_3 = \frac{\dot{m}}{\rho_3 c_{a3}} = \frac{8.704\frac{\text{sl}}{\text{s}}}{0.01960\frac{\text{sl}}{\text{ft}^3} \times 699.6\frac{\text{ft}}{\text{s}}} = 0.6348\ \text{ft}^2.$$

In addition, $A_3 = \pi D_3\, t_3$, and so

$$t_3 = \frac{A_3}{\pi D_3} = \frac{0.6348\ \text{ft}^2 \times 144\frac{\text{in.}^2}{\text{ft}^2}}{\pi \times 2 \times 13.2\ \text{in.}} = 1.102\ \text{in.}$$ <ANS

Once again, note that this is smaller than the stator inlet height because of the compressibility of the fluid. Note also that the static pressure rises across the rotor and stator are 23.9 and 25.4 psia, respectively. Thus, they are approximately the same. One should remember that the calculated percent reaction was 0.547, which is in keeping with the pressure distribution for nonideal compressible flow.

Thus, all of the variables of interest have been found. For the sake of comparison, one can find similar parameters for an approximation to the fifth engine stage or first high-pressure stage, of the same engine. For this stage, the radius is 14.9 in. and the inlet blade height is 4.70 in. (much larger than the last stage). Also, the inlet pressure is 32 psia, whereas the inlet temperature is 220 °F. The stage efficiency is again 0.90. The inlet absolute flow angle to the rotor is 23° and the rotor blade turning angle is 15°, which is lower than that of the last stage. The resulting values of c_p and γ are 0.2432 BTU/lb$_m$-°R and 1.393, respectively. The resulting axial velocity is 721.4 ft/s, which is close to that of the last stage. The relative inlet angle to the rotor blade is 45.49°, and the absolute inlet angle to the stator is 40.46°. The percent reaction is 0.5570, which is again reasonable, and the total pressure ratio is 1.251, which is higher than that of the last stage. Details of the calculations are left as an exercise to the reader.

Also, in this example the axial component of velocity was given as being constant with axial position, and the blade heights were calculated. However, if the blade heights would have been given, the axial component of velocity would have needed to be determined. The solution would have been iterative, using the same procedure as given.

Repeating the above for the ideal case ($\eta = 1.00$) yields a total pressure ratio of 1.196, which is significantly higher than the nonideal case. Also %R = 0.547. And, $p_2 - p_1 = 29.1$ psia (rotor) and $p_3 - p_2 = 25.9$ psia (stator), indicating a percent reaction greater than 0.50.

Example 5.2 Axial Pump Stage

Solve Example 5.1 if an incompressible flow with a density of that at the inlet ($\rho = 0.01742$ sl/ft^3) is assumed. Namely, the compressor now is actuality a pump. Compare to the compressible case.

Solution

Many of the results can be used from Example 5.1. That is,

$$U = R\omega = \frac{13.2\text{in.}}{12\frac{\text{in}}{\text{ft}}}\frac{8000\text{ rev}}{\text{min}}\frac{2\pi\text{ rad}}{\text{rev}}\frac{1}{60\frac{\text{s}}{\text{min}}} = 921.5\text{ ft/s},$$

$$A_1 = \pi D_1 t_1 = \pi\frac{2\times 13.2\text{ in.}}{12\frac{\text{in.}}{\text{ft}}}\frac{1.24\text{ in.}}{12\frac{\text{in.}}{\text{ft}}} = 0.7142\text{ ft}^2,$$

and

$$c_{a1} = \frac{\dot{m}}{\rho_1 A_1} = \frac{8.704\frac{\text{sl}}{\text{s}}}{0.01742\frac{\text{sl}}{\text{ft}^3}\times 0.7142\text{ ft}^2} = 699.6\text{ ft/s}.$$

Thus, since the axial and rotational velocities and the angles are the same as for the compressible case, the velocity polygons are the same.

Rotor Inlet

Refer to Figure 5.10a, which is the velocity polygon for the rotor inlet. First,

$$c_1 = \frac{c_{a1}}{\cos\alpha_1} = \frac{699.6}{\cos(15°)} = 724.2\text{ft/s}.$$

Also,

$$c_{u1} = c_1\sin\alpha_1 = 724.2\sin(15°) = 187.4\text{ ft/s};$$

thus,

$$w_{u1} = c_{u1} - U = 187.4 - 921.5 = -734.1\text{ ft/s}$$

and

$$\beta_1 = \cot^{-1}\left[\frac{c_{a1}}{c_{u1} - U_1}\right] = \cot^{-1}\left[\frac{c_{a1}}{w_{u1}}\right] = \cot^{-1}\left[\frac{699.6}{-734.1}\right] = -46.38°.$$

Moreover,

$$w_1 = \frac{c_{a1}}{\cos\beta_1} = \frac{699.6}{\cos(-46.38°)} = 1014\text{ ft/s}.$$

Also, for the incompressible case the total pressure is

$$p_{t1} = p_1 + \frac{\rho c_1^2}{2} = 272\ \text{psia} + \frac{0.01742\dfrac{\text{sl}}{\text{ft}^3} \times 724.2^2\dfrac{\text{ft}^2}{\text{s}^2}}{2 \times 12\dfrac{\text{in.}}{\text{ft}} \times 12\dfrac{\text{in.}}{\text{ft}}} = 303.7\ \text{psia}.$$

Rotor Exit and Stator Inlet

For the rotor exit and stator inlet velocity polygon, refer to Figure 5.10b. Because the rotor flow turning angle is known, one can find the relative exit angle by

$\beta_2 = \beta_1 + \delta_{12} = -46.38 + 25 = -21.38°.$

Also, since the axial velocity is constant through the stage,

$c_{a2} = c_{a1} = 699.6$ ft/s.

Next,

$$w_2 = \frac{c_{a2}}{\cos\beta_2} = \frac{699.6}{\cos(-21.38°)} = 751.3\ \text{ft/s}.$$

Also, $w_{u2} = c_{a2}\tan\beta_2 = 699.6\tan(-21.38°) = -273.9$ ft/s, and
$c_{u2} = U + w_{u2} = 921.5 - 273.9 = 647.7$ ft/s;

thus,

$$\alpha_2 = \cot^{-1}\left[\frac{c_{a2}}{U + w_{u2}}\right] = \cot^{-1}\left[\frac{c_{a2}}{c_{u2}}\right] = \cot^{-1}\left[\frac{699.6}{647.7}\right] = 42.79°$$

and

$$c_2 = \frac{c_{a2}}{\cos\alpha_2} = \frac{699.6}{\cos(42.79°)} = 953.3\text{ft/s}.$$

Next, for the incompressible case,

$$\begin{aligned} p_{t2} - p_{t1} &= \rho\eta_{12}U[c_{u2} - c_{u1}] \\ &= 0.01742\frac{\text{sl}}{\text{ft}^3} \times 0.90 \times 921.5\frac{\text{ft}}{\text{s}} \times [647.7 - 187.4]\frac{\text{ft}}{\text{s}} \times \frac{1}{144\dfrac{\text{in.}^2}{\text{ft}^2}} \\ &= 46.2\ \text{psia}, \end{aligned}$$

and so $p_{t2} = 303.72 + 46.18 = 349.9$ psia.

Therefore,

$$\frac{p_{t2}}{p_{t1}} = \frac{349.9}{303.7} = 1.152,$$

which is 1.96 percent lower than found in Example 5.1 owing to the incompressibility of the fluid.

Next,

$$p_2 = p_{t2} - \frac{\rho c_2^2}{2} = 349.9\,\text{psia} - \frac{0.01742\frac{\text{sl}}{\text{ft}^3} \times 953.3^2\frac{\text{ft}^2}{\text{s}^2}}{2 \times 12\frac{\text{in.}}{\text{ft}} \times 12\frac{\text{in.}}{\text{ft}}} = 294.9\,\text{psia}.$$

Also, the required power is

$$\dot{W}_{\text{sh}} = \dot{m}U(c_{u2} - c_{u1}) = \frac{8.704\frac{\text{sl}}{\text{s}} \times 921.5\frac{\text{ft}}{\text{s}} \times (647.7 - 187.4)\frac{\text{ft}}{\text{s}}}{550\frac{\text{ft-lb}_\text{f}}{\text{s-hp}}} = 6712\ \text{hp},$$

which is identical to that in Example 5.1.

Finally,

$$\%\text{R} = \frac{1}{1 + \left[\frac{c_2^2 - c_1^2}{w_1^2 - w_2^2}\right]} = \frac{1}{1 + \left[\frac{953.3^2 - 724.2^2}{1014^2 - 751.3^2}\right]} = 0.5469,$$

which is also identical to the percent reaction in Example 5.1.
Also, since the flow is incompressible, the blade heights remain constant through the stage; thus, $t_2 = 1.24$ in.

Stator Exit

Refer to Figure 5.10c for the stator exit velocity polygon.

Once again, the axial velocity is constant through the stage, and thus

$$c_{a3} = c_{a2} = c_{a1} = 699.6\ \text{ft/s}.$$

Also, the absolute exit angle is given, $\alpha_3 = 15°$, and so

$$c_3 = \frac{c_{a3}}{\cos\alpha_3} = \frac{699.6}{\cos(15°)} = 724.2\ \text{ft/s}.$$

Since $p_{t3} = p_{t2}$, the exit static pressure is

$$p_3 = p_{t2} - \frac{\rho c_3^2}{2} = 349.9\,\text{psia} - \frac{0.01742\frac{\text{sl}}{\text{ft}^3} \times 724.2^2\frac{\text{ft}^2}{\text{s}^2}}{2 \times 12\frac{\text{in.}}{\text{ft}} \times 12\frac{\text{in.}}{\text{ft}}} = 318.2\,\text{psia}.$$

One can note that $p_2 - p_1 = 22.9$ psia and $p_3 - p_2 = 23.3$ psia, or they are nearly equal, which is in agreement with the percent reaction.

The turning angle of the stator can lastly be found by

$$\delta_{23} = \alpha_3 - \alpha_2 = 15 - 42.79 = -27.29°\cdot$$

Thus, the turning angle for the stator vane is approximately that of the rotor blade. Again, because the flow is incompressible, the blade height remains constant throughout the stage; thus, $t_3 = 1.24$ in.

5.5 Performance Maps

5.5.1 Dimensional Analysis

As a working medium for a compressor engineer, performance curves or "maps" are usually used to present experimental data. These enable an engineer to make a quick, comprehensive, and accurate assessment of the general conditions at which the turbomachine is operating and to understand what can be expected of the machine if the flow conditions are changed. In Chapter 4, a dimensional analysis of the different parameters is performed, and similitude is shown. Of primary concern is the total pressure ratio of a compressor, namely, p_{t3}/p_{t2}. From Chapter 4 (Eqs. (4.3.3), (4.3.7), and (4.3.8)),

$$\pi_c = p_{t3}/p_{t2} = f\left\{\dot{m}\frac{\sqrt{\theta_{t2}}}{\delta_{t2}}, \frac{N}{\sqrt{\theta_{t2}}}\right\} = f\{\dot{m}_{c2}, N_{c2}\}, \tag{5.5.1}$$

where

$$\delta_{t2} = p_{t2}/p_{stp} \tag{5.5.2}$$

and

$$\theta_{t2} = T_{t2}/T_{stp}, \tag{5.5.3}$$

where the subscript stp refers to standard conditions, the subscript t2 denotes the inlet total condition to the compressor (not to be confused with the exit total condition of a rotor cascade), and the subscript t3 signifies the exit total condition out of the compressor. Similarly, for the efficiency, one finds (Eqs. (4.3.6), (4.3.7), and (4.3.8)):

$$\eta_c = g\left\{\dot{m}\frac{\sqrt{\theta_{t2}}}{\delta_{t2}}, \frac{N}{\sqrt{\theta_{t2}}}\right\} = g\{\dot{m}_{c2}, N_{c2}\}. \tag{5.5.4}$$

The two functions f and g are empirically determined from experimental data. These functions are not necessarily analytical (closed-form algebraic expressions) (and in fact rarely are) but are most commonly graphical presentations or tabular data. The two independent parameters ($\dot{m}_{c2}$ and N_{c2}) are not truly dimensionless. The first is called the "corrected" mass flow and has dimensions of mass flow, whereas the second is called the "corrected" speed and has dimensions of rotational speed. Thus, because they are not dimensionless, the resulting function cannot be used to correlate or compare different engines. The function for one particular gas turbine can be used under different operating conditions, however, as is the usual case. That is, the compressor can be tested under one set of conditions but applied to markedly different conditions so long as the corrected speed and mass flow rates are identically matched. For example, for a jet engine the map is applicable regardless of altitude, aircraft speed, atmospheric conditions, and so on.

5.5.2 Trends

One should also know how the developed pressure ratio varies with the two independent variables – namely, the trends. First, consider how p_{t3}/p_{t2} varies with $\dot{m}$ if all other parameters are held constant. Figure 5.11 is applicable to this study. In Figure 5.11a, the inlet and exit velocity polygons for one stage are shown for the design condition. If the mass flow is decreased, the axial component of velocity is decreased, as shown in Figure 5.11b. However, for the same inlet flow angle and rotor turning angle, the difference $\Delta c_u = c_{u2} - c_{u1}$ is increased. Thus, from Eq. (5.4.2) it can be seen that, as the flow rate decreases, the pressure ratio increases.

The next consideration is how the pressure ratio will change as the rotational speed is changed with all other parameters are held constant. For example, in Figure 5.11a, the inlet and exit velocity polygons are presented for the design condition. In Figure 5.11c, the velocity polygons are presented for a decreased rotational speed with the inlet angle and rotor turning angle the same as for the design condition. Now, $\Delta c_u = c_{u2} - c_{u1}$ has been decreased. Also, U has been decreased. Thus, one can examine Eq. (5.4.2) again and see that the pressure ratio decreases with decreasing rotational speed.

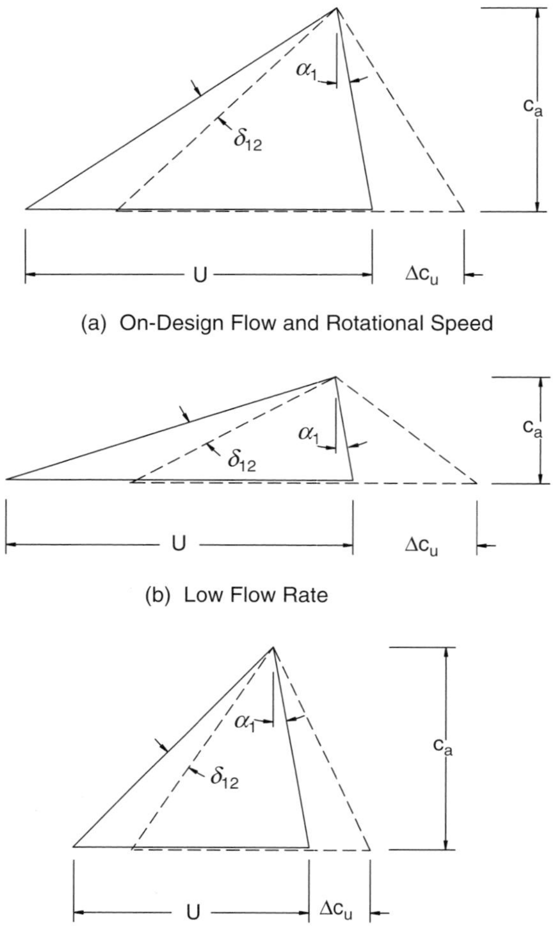

Figure 5.11 Trends of compressor velocity polygons.

5.5.3 Experimental Data

As indicated in Section 5.5.1, the two compressor parameters that are most important are the total pressure ratio and the efficiency. To determine these parameters, the compressor must be tested independently of the other engine components in a compressor test facility (Figure 5.12). In such a facility, the compressor is usually driven by an auxiliary gas turbine. The rotational speed is independently controlled through a gearbox, and the flow rate is independently controlled by a large valve system. Typically, the test rig is run at a series of constant rotational speeds. At each speed the flow rate is incrementally decreased, the system is allowed to reach steady state, are pressure and temperature measurements are taken at every mass flow rate until the compressor pressure ratio becomes so large that the compressor has massive flow reversal or "surges." Surge results when one or more rows of blades or vanes stall, namely flow separates. The unstable behavior in one stage expands to other stages until all or most of the stages have separated. This unstable flow reversal is accompanied by tremendous vibration forces on the blades and disks as well as a possible burner "flameout" in an engine. Under severe surge an engine can experience backflow out of the front of a gas turbine with combustion occurring in the compressor and inlet. Clearly this is a dangerous situation causing severe damage to the gas turbine with possible catastrophic results.

The compressor stages are all instrumented with a series of total and static pressure probes and total and static temperature thermocouples. At the inlet and exit of each stage, a pressure and temperature "rake" (a series of radially located probes) is installed. Thus, for the inlet and exit of each stage, a total pressure and temperature profile from hub to tip is measured, giving the design engineer great detail about the internal performance. By averaging (weighted by mass flow rate) the total pressures at the inlet to, and exit of, the compressor, the total pressure ratio of the compressor can be found for each speed and flow rate condition. Next, by averaging (again weighted by mass flow rate) the total temperature at the inlet to, and exit of, the compressor, the total temperature ratio of the compressor can be determined. Finally, from the experimental data the overall compressor efficiency can be found by using Eq. (2.3.30) as follows:

$$\eta_c = \frac{\left[\frac{p_{t3}}{p_{t2}}\right]^{\frac{\gamma-1}{\gamma}} - 1}{\left[\frac{T_{t3}}{T_{t2}}\right] - 1}. \tag{5.5.5}$$

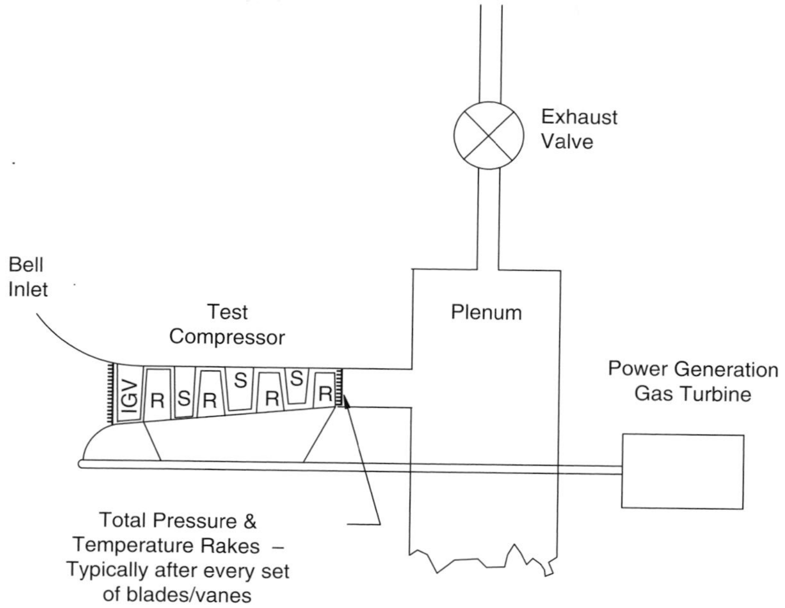

Figure 5.12 Compressor rig test facility.

While a constant speed line is traversed, as the mass flow decreases, the efficiency increases to a maximum and then decreases as the compressor nears surge. The maximum efficiency on this line becomes a part of the desired operating line.

All of the preceding discussions in this section have been confined to analysis of the overall compressor data. Note that the interstage total pressure and temperature profile data are also used to find the efficiency and total pressure ratio for only the fan and the low- and high-pressure compressors. Furthermore, data from individual stages are examined using an equation similar to Eq. (5.5.5) and reduced to give total pressure ratios and efficiencies of individual stages. Even furthermore, data from the hub, midstream, and tip regions of all stages are analyzed so that performance of the radial regions of stages can be investigated. Thus, not only are the overall conditions of the compressor found, but the efficiency and total pressure ratio of the subcomponents, each stage, and radial regions of all stages are determined. These specific and local details obviously allow an engineer to determine the inefficient regions of a compressor so that local blade shape changes can be made and, thus, improvements to the overall improvement of the whole compressor can be achieved.

5.5.4 Mapping Conventions

Finally, in Figure 5.13 a typical compressor map, which is generated from the data taken during compressor stand tests, is presented. The pressure ratio is plotted vs. the corrected mass flow, and the corrected speed is a second parameter. Thus, a series of corrected speed lines are also included in the figure, and these are often presented as a percentage of the nominal or design rotational speed. Also shown in this figure is the so-called surge line. Surge arises when the pressure ratio becomes so large across the compressor that the pressure gradient overwhelms the flow momentum and causes massive flow reversal in the engine. Locally, massive flow separation occurs on the

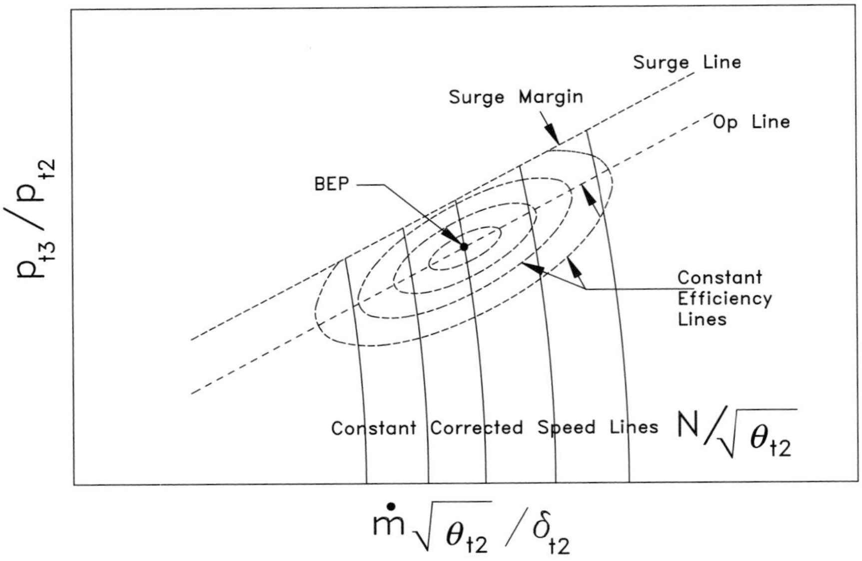

Figure 5.13 Typical compressor map with efficiency islands.

different blades in the engine at this condition, and the performance goes to zero. This is a very dangerous condition for a jet engine to operate in because of excessive vibrations and the zero overall engine thrust.

A second set of complementary compressor data is plotted on this map so that curves are fitted between points of constant efficiency on the different speed lines. Lines on which the efficiency is constant are shown (dotted) in this figure and are called efficiency "islands." The typical design point of an engine is near the peak or best efficiency point (BEP). Thus, as one changes conditions from the design point, the efficiency decreases. Also shown in Figure 5.13 is the operating line. This line passes through the design point and represents the different conditions at which an engine typically operates. In this figure, one can easily see that, along a constant speed line, as the mass flow decreases, the efficiency increases to a maximum on the operating line and then decreases as the compressor nears surge. The surge margin is the remaining performance information presented shown in this figure. This is the margin between the operating line and the surge line and is regarded a safety margin. In designing a compressor, one encounters a tradeoff because the objective is usually to design for a large pressure ratio at the operating condition. On the other hand, a small surge margin is undesirable because the compressor could surge as the result of a small perturbation – for example, a wind gust or sudden yaw, acceleration for an engine or sudden change in load condition for a power unit. Typical surge margins for modern gas turbines range from 15 to 20 percent. Overall, once a compressor map is found, if the corrected mass flow and corrected speed are known, the total pressure ratio and efficiency can be found regardless of other conditions.

In Figure 5.14, a second method of mapping the compressor performance is shown. This has exactly the same data as Figure 5.13 but in a different format. For this format the pressure ratio data are the same format as in Figure 5.13. However, the efficiency data are plotted separately and are a function of the corrected mass flow with the corrected speed as another variable.

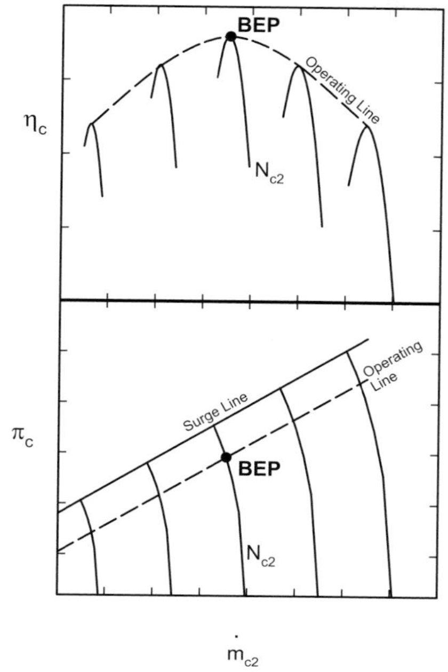

Figure 5.14 Typical compressor map with efficiency plot.

This format actually presents the efficiency data is a more accurate manner. However, the presentation of Figure 5.13 is a better graphical presentation as it gives a better "feel" of efficiency and pressure ratio combined.

As discussed in this section, maps are considered for the entire compressor as well as individual maps for the fan and low- and high-pressure compressors. There will be a map for each stage of the engine and, in most instances, for the hub, pitch, and tip regions of each stage. Maps and the included data are very important in the development and application of a gas turbine design. Three important concepts are a part of the maps. First, a compressor must have the pressure ratio and efficiency for the design condition applied at the initial stages of design. Second, as will be seen in Chapter 12, such data are used with turbine and other maps and data to insure the components are properly "matched." Matching implies fundamentals such as the best efficiency should be derived from each component at each operating point. Third, the surge line will become very important in the final application of the compressor and other components. Such surge and surge margin data will become a benchmark and a trigger when a gas turbine is approaching surge, such that proper modifications to the operation can be modified. This will be further discussed in Section 5.5.5.

In Table 5.3, some characteristic fan, compressor, and turbine parameters are shown for selected modern engines and power units. The reader should be attentive to the pressure ratios, bypass ratios, and number of stages. Table 5.3 was integrated from engine data in *Aviation Week & Space Technology*; *Flight International*; information from manufacturers, users, and the military; brochures; and web pages. Although every effort has been made to ensure maintain accuracy by cross-checking, the information is for reference and is not intended for technical applications. Furthermore, specifications for gas turbines are forever changing due to improvements in design. Also, many of the gas turbines have different "builds" for different applications and for those cases typical characteristics are presented.

5.5.5 Surge Control

Because surge is a dangerous condition and the operating lines for most compressors are near surge and at the edge of the surge margin, active control is often used to ensure safe operation. Such automatic control is usually in three forms and all rely on an early detection system that monitors the pressure profiles on vanes in selected stages to detect any flow separation as well as with comparison to the surge data in the different maps. Such controls operate on a millisecond timescale. First, if a detection system senses preliminary stall for a stage, air can be bled from the compressor to increase the mass flow rate. This is usually done with bleed vanes on the last stages of a high-pressure compressor for a single-spool

Table 5.3 **Representative compressor and turbine characteristics at maximum power.**

Engine	Manufacturer	Type & no. of compressive stages (F+LP+IP+HP)	Overall total pressure	Bypass ratio	LP+(IP)+HP turbine stages	Gas turbine type
CF6-50C2B	GE	1F, 3A, 14A	31.1	4.6	4+2	TF
CF6-80C2A5	GE	1F, 4A, 14A	31.5	5.3	5+2	TF
F101-102	GE	2F, 9A	26.8	1.9	1+2	TF
F110-100	GE	3F, 9A	30.4	0.8	1+2	TF
CF34-3A	GE	1F, 14A	21	6.2	4+2	TF
F404	GE	3F, 7A	26	0.33	1+1	TF
GE90-115B	GE	1F, 4A, 9A	42	9	6+2	TF
GE9X	GE	1F, 3A,11A	60	9.9	6+2	TF
J79-8	GE	17A	12.9	NA	3	TJ
J85-17AB	GE	8A	6.9	NA	2	TJ
LM2500-RC	GE	16	33.7	NA	2+6 Free	PGT
LM6000-PG	GE/MTU	5A, 14A	32.3	NA	5+2	PGT
CFM56-5B1	CFM	1F, 4A, 9A	32	5.5	4+1	TF
GP 7200	EA	1F, 5A, 9A	36	8.8	6+2	TF
TF30-P-100	PWA	3F, 13A	21.8	0.7	3+1	TF
F100-PW-232	PWA	3F, 10A	35	0.34	2+2	TF
JT8D	PWA	2F, 11A	18.2	1.7	3+1	TF
JT9D	PWA	1F, 15A	26.7	4.8	4+2	TF
PW2037	PWA	1F, 5A, 11A	27.4	6	5+2	TF
PW4056	PWA	1F, 4A, 11A	30	4.9	4+2	TF
PW4098	PWA	1F, 7A, 11A	42.8	6.4	7+2	TF
F117	PWA	1F, 5A, 11A	30.8	5.9	5+2	TF
J52	PWA	12A	13.6	NA	2	TJ
J58	PWA	9A	8.8	NA	2	TJ
J75	PWA	8A, 7A	12	NA	2+1	TJ

Table 5.3 **(cont.)**

Engine	Manufacturer	Type & no. of compressive stages (F+LP+IP+HP)	Overall total pressure	Bypass ratio	LP+(IP)+HP turbine stages	Gas turbine type
FT8	Mitsubishi	8A,7A	23	NA	2+1+4 Free	PGT
Spey-101	RR	4F, 12A	16.5	0.7	2+2	TF
RB211-535C	RR	1F, 7A, 6A	34.5	4.3	3+1+1	TF
Trent 895	RR	1F, 8A, 6A	41.6	6.5	5+1+1	TF
Trent 1000	RR	1F, 8A, 6A	50	10.1	6+1+1	TF
Trent XWB	RR	1F, 8A, 6A	50	9.6	6+2+1	TF
Tyne	RR	6A, 9A	13.5	NA	3+1	TP
Viper 522	RR	8A	5.6	NA	1	TJ
T56A-15	RR	14A	9.5	NA	4	TP
RB211-H63	RR	8A, 6A	25.1	NA	1+1+2 Free	PGT
Coberra 6462	Cooper-Rolls	7A, 6A	20.1	NA	1+1+2 Free	PGT
ALF502	Honeywell	3F, 7A, 1R	13.8	5.6	2+2	TF
TFE731-2	Honeywell	1F, 4A, 1R	13	2.7	3+1	TF
T55-L-714A	Lycoming	7A, 1R	9.3	NA	2+2 Free	HC
GT26	GE/Alstom	22A	32	NA	4+1	PGT

F = fan stages.
A = axial compressor stages.
1F, 4A, 9A = 1 fan + 4 axial stages on low speed shaft, 9 axial stages on high-speed shaft, etc.
R = radial stages.

engine and sometimes also on the last stages of the low-pressure compressor for a two-spool engine. Thus, on the map, as the mass flow increases, the compressor moves along a constant speed line away from the surge line and reduces the pressure ratio. The second form of active control used if a detection system senses preliminary stall for a stage "entails using variable" stators, which are discussed in Section 5.7, that can be turned so that the incidence angles are decreased, thus reducing stall. Third, since surge is often or usually associated with sudden increases in needed fuel, the fuel rate can be reduced. Often, a combination of the three methods is used.

A related instability is rotating stall. As its name implies, stall occurs in one or more blade passages. The stall results from massive separation on the blade because of too much incidence or too large of a pressure rise. For rotating stall not all blade passages are stalled, but the stall propagates circumferentially from one blade passage to another or several passages to adjacent passages. This instability occurs at a frequency lower than the rotational speed and can cause massive vibrations which can result in shaft, bearing, and/or blade failures.

A historical review of surge and rotating stall is presented by Champ, Massarado, Ferrari, and Silvestri (2019) and an expanded tutorial for centrifugal compressors is presented by Kurz, White, Brun, and Winkelman (2016).

5.6 Limits on Stage Pressure Ratio

As indicated in Chapter 1, the overall thermodynamic efficiency of a machine is improved if the compressor pressure ratio is increased. Thus, it is desirable to have as high a pressure ratio as possible across each stage without separation, which leads to lower efficiencies and stall. In Section 5.4, equations were derived whereby the pressure ratio can be found for given blade and flow geometries. If one examines these equations, it becomes apparent that there are essentially no limits on the possible developed pressure ratios. Yet, modern compressors have maximum stage total pressure ratios of about 1.25 to 1.30. In this section, the practical stage limits are examined.

First, keep in mind that both the rotor and stator blade passages are essentially diffusers. Both have flow passage areas that increase as the flow passes through, and the static pressures increase. Thus, in both cases an adverse pressure gradient is present that tends to cause separation of the boundary layer. It is this separation that limits the operation of a compressor through surge or stall. One can use a relatively simple model postulated by Hill and Peterson (1992) using pressure coefficients. The pressure coefficient is defined by

$$C_{\mathrm{p}} \equiv \frac{\Delta p}{\frac{1}{2}\rho_{\mathrm{i}} w_{\mathrm{i}}^2}, \tag{5.6.1}$$

where w_{i} is the inlet relative velocity to the diffuser. Typically, for a simple diffuser having walls with limited curvature and ideal incidence-free inlet conditions (Figure 5.15a), C_{p} is less than or approximately equal to 0.6 for flow without separation. If the pressure coefficient becomes greater than about 0.6, the flow separates. Once the flow separates, the effective area ratio is not very large, and thus the pressure ratio of the stage is minimal. Also, the efficiency of the stage is greatly reduced. And finally, a stage separation can lead to compressor surge. If the walls have significant curvature or if the flow enters the diffuser in a skewed direction, or with incidence (Figure 5.15b), this upper limit on the pressure coefficient will be reduced significantly to perhaps 0.1 to 0.4.

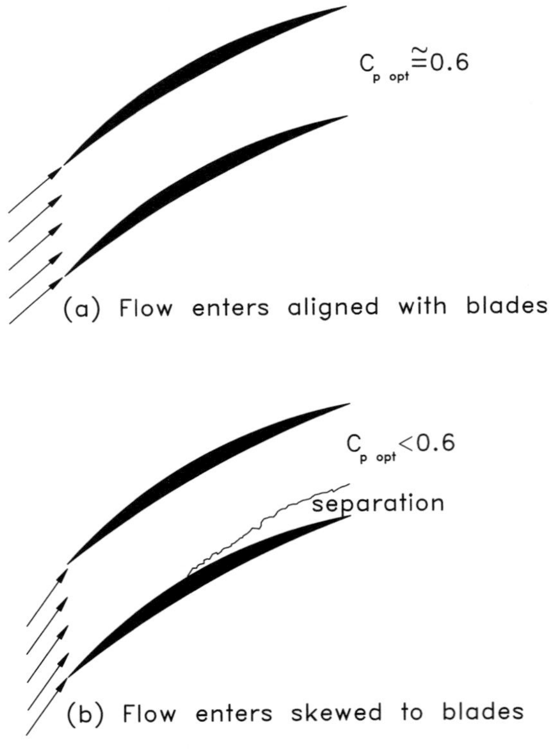

Figure 5.15 Simple blade passage.

Now, referring to Figures 5.6 and 5.7, one can apply Eq. (5.6.1) and define a pressure coefficient for the rotor. Because the rotor is moving, the relative velocity must be used; thus,

$$C_{pr} = \frac{p_2 - p_1}{\frac{1}{2}\rho_1 w_1^2}. \tag{5.6.2}$$

For the stator, which is stationary, the absolute velocity is used; thus,

$$C_{ps} = \frac{p_3 - p_2}{\frac{1}{2}\rho_2 c_2^2}, \tag{5.6.3}$$

where the velocity subscripts 1 and 2 refer to the stage inlet and midstage properties, respectively. Next, for the rotor one can consider the square of the relative inlet Mach number as

$$M_{1rel}^2 = \frac{w_1^2}{a_1^2} = \frac{\rho_1 w_1^2}{\gamma p_1}. \tag{5.6.4}$$

Therefore, from Eq. (5.6.2), one finds

$$\frac{p_2}{p_1} = \frac{1}{2}\frac{\rho_1 w_1^2 C_{pr}}{p_1} + 1, \tag{5.6.5}$$

and from Eq. (5.6.4),

$$\frac{p_2}{p_1} = 1 + \frac{1}{2}C_{pr}\gamma M_{1rel}^2. \tag{5.6.6}$$

Next, the square of the inlet Mach number as experienced by the stator is

$$M_{2abs}^2 = \frac{c_2^2}{a_2^2} = \frac{\rho_2 c_2^2}{\gamma p_2}. \tag{5.6.7}$$

Finally, from Eqs. (5.6.3) and (5.6.7) one finds

$$\frac{p_3}{p_2} = 1 + \frac{1}{2}C_{ps}\gamma M_{2abs}^2. \tag{5.6.8}$$

One very important fact is obvious. Since the maximum allowable C_p is approximately fixed, if a designer wants to increase the pressure ratio limit of a rotor, stator passage, or both for a given pressure coefficient (without stall), it is necessary to increase the inlet Mach number.

Example 5.3 Axial Compressor Stage: Influence of Pressure Coefficient and Mach Numbers

For the stage geometry in Example 5.1, find the approximate optimum limit on p_3/p_1. The pressure coefficients can be assumed to be 0.6 (the upper limit) for both the rotor and stator. Note that, in Example 5.1, the rotor and stator pressure coefficients were found to be 0.384 and 0.436, respectively. Thus, as compared with the maximum or optimum limit of approximately 0.6, a reasonable safety margin was used, which in turn yields the surge margin.

Solution

The specific heat ratio is $\gamma = 1.361$, and the relative inlet Mach number to the rotor is

$$M_{1rel} = \frac{w_1}{a_1} = \frac{1014}{1749} = 0.5798.$$

The inlet Mach number to the stator was already found and is as follows:

$$M_{2abs} = 0.5378.$$

Thus, for the maximum pressure coefficient,

$$\left.\frac{p_2}{p_1}\right|_{\text{limit}} = 1 + \frac{1}{2} C_{\text{pr}} \gamma M_{1\text{rel}}^2 = 1 + \frac{1}{2} \times 0.6 \times 1.361 \times 0.5798^2 = 1.137.$$

Also

$$\left.\frac{p_3}{p_2}\right|_{\text{limit}} = 1 + \frac{1}{2} C_{\text{ps}} \gamma M_{2\text{abs}}^2 = 1 + \frac{1}{2} \times 0.6 \times 1.361 \times 0.5798^2 = 1.118,$$

and so

$$\left.\frac{p_3}{p_1}\right|_{\text{limit}} = \left.\frac{p_2}{p_1}\right|_{\text{limit}} \times \left.\frac{p_3}{p_2}\right|_{\text{limit}} = 1.137 \times 1.118 = 1.271.$$

The true pressure ratio from Example 5.1 is

$$\frac{p_3}{p_1} = \frac{321.3}{272} = 1.181.$$

Thus, the stage is not running at the maximum limit. It likely is running at the limit with a reasonable surge margin for the particular geometry determined from tests, however.

For comparison one can consider the effect if the inlet relative Mach numbers were increased to 0.8. Repeating the preceding calculations, one finds that

$$\left.\frac{p_3}{p_1}\right|_{\text{limit}} = 1.591.$$

Thus, the limit is much higher for the higher Mach numbers. However, other losses that accompany higher Mach numbers in a design must be considered.

Therefore, as shown previously, increasing the inlet Mach numbers to the transonic regime greatly increases the potential pressure ratio. However, increasing the Mach number into the transonic regime also decreases the efficiency of the stage. Typically, when a stage is operating transonically, several localized supersonic regions occur. These regions are accompanied by shocks, which decrease the total pressure and thus reduce the efficiency. Also, shocks can lead to separation. Therefore, a tradeoff is realized. If the Mach number is high, a large pressure ratio can be attained; however, usually the efficiency is low. As a result, although a few modern compressors operate near the transonic regime, most operate with inlet Mach numbers of approximately 0.6 to 0.9. Also, most compressors do not operate with the optimum value of C_p. One must remember that separation occurs at a pressure coefficient of 0.6 for only very well controlled conditions. When time-varying or periodic inlet conditions (especially velocities and velocity directions) are present at a compressor stage as the result of the blade jet and wakes of the previous stage, the critical pressure coefficient is reduced considerably from 0.6 because the incidence angle will be periodic as well.

5.7 Variable Stators

5.7.1 Theoretical Reasons

Currently, most modern jet engines and power generation gas turbines have at least some stages with variable stators. This means that the stator vanes are turned or rotated about the blade radial axis as determined by the operating conditions. Basically, the idea is to change the vane angles so that the leading and trailing vane angles match the flow angles as close as possible, thus providing a flow with minimal incidence to both the rotor blades and stator vanes. This is done to improve off-design performance and stall control. Stator vanes are chosen because they are not moving and can be rotated relatively easily. Rotating the rotor blades on the shaft would be a difficult task to accomplish mechanically. The concept is best demonstrated with an example.

Example 5.4 Axial Compressor Stage: Off-Design Implications, Variable Stator Vanes

The same stage geometry as in Example 5.1 is to be studied for an operating flow rate of 50 percent of the design condition (140 lb_m/s). One can assume that all conditions are the same for the purpose of demonstration and that, in the previous example, the blade angles exactly matched the flow angles and that all stages have similar geometries. Using a simplified analysis, determine what angle the stator will have to be turned so that the blades best match the flow. As will be seen, the incidence angle cannot be forced to zero for both the rotor and stator, and so a tradeoff will be necessary as determined through a design process.

Solution

Once again, velocity polygons will be necessary for the solutions, and one can refer to Figure 5.16, which is again to scale. First, recall that, at design, the flow angle is,

$$\beta_1 = -46.38°$$

thus, the blade angle is $\beta'_1 = -46.38°$.

Now, if $\alpha_1 = 15°$, one can draw a new velocity polygon (Figure 5.16a) and find that the new flow angle is

$$\beta_1 = -67.09°.$$

Thus, the rotor blade inlet incidence angle ($\iota_{\beta 1}$) is shown in Figure 5.16b and is

$$\iota_{\beta 1} = \beta_1 - \beta_1{}' = -67.09° - (-46.38°) = -20.71°.$$

This angle is very large and would lead to a premature flow separation from the rotor blade suction surface and result in very poor performance (pressure rise and efficiency). Next, one can consider changing the upstream stator exit angle so that the flow angle matches the blade angle. For this case (Figure 5.16c), one has

$$\beta_1 = -46.38°,$$

but now, working backward on the velocity polygon, one finds the flow angle is

$$\alpha_1 = 57.76° (\text{vs. } 15° \text{ previously}).$$

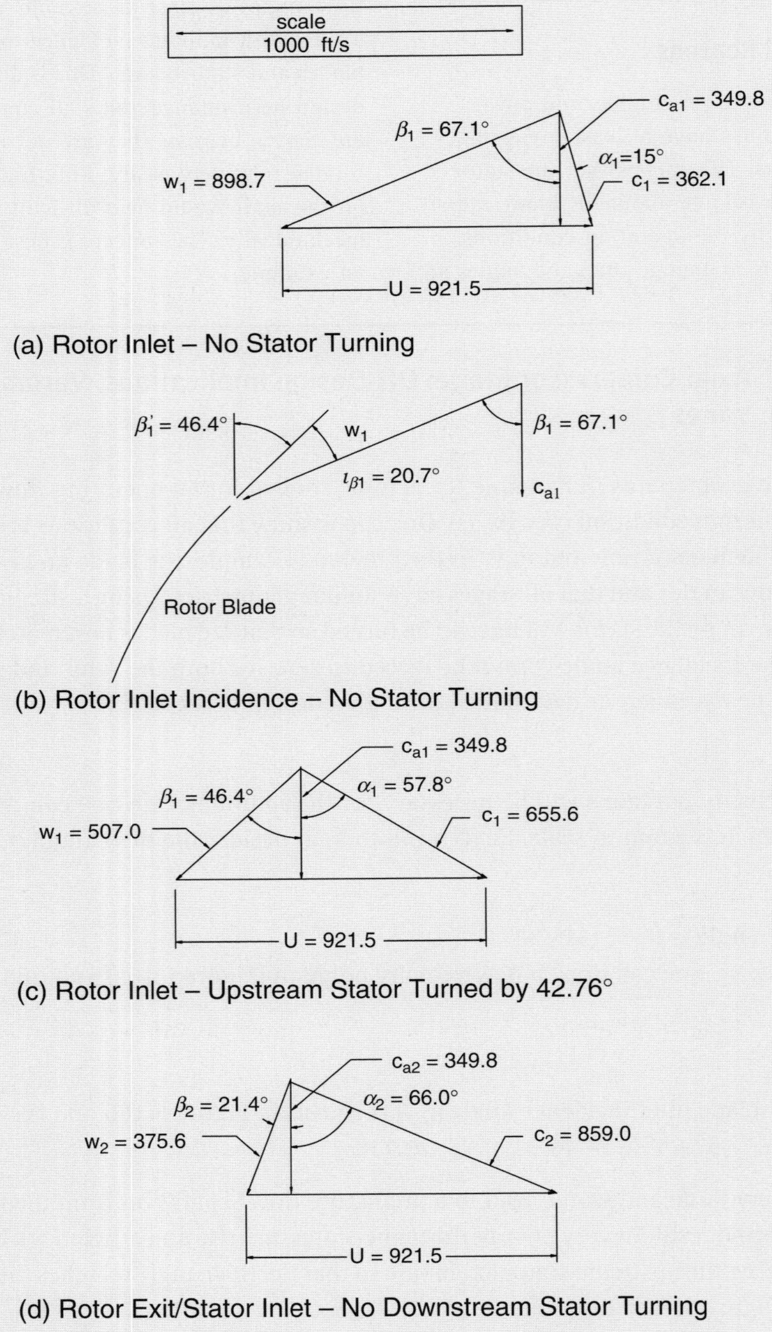

Figure 5.16 Rotor and stator polygons and resulting vane positions for Example 5.4.

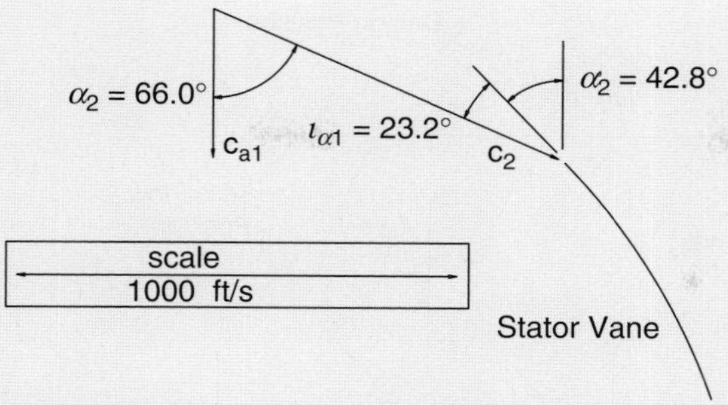

(e) Stator Inlet Incidence – No Downstream Stator Turning

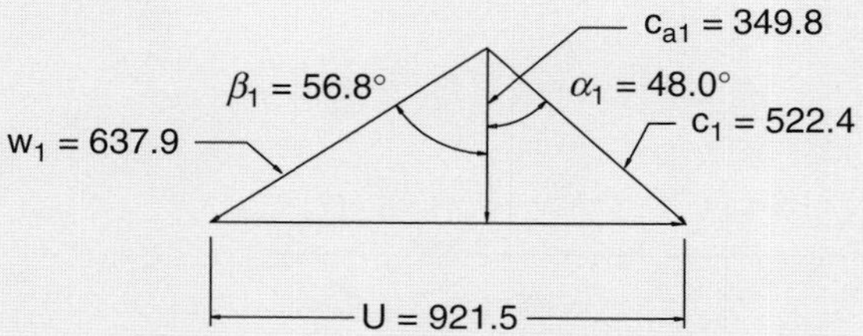

(f) Rotor Inlet – Upstream Stator Turned by 32.97°

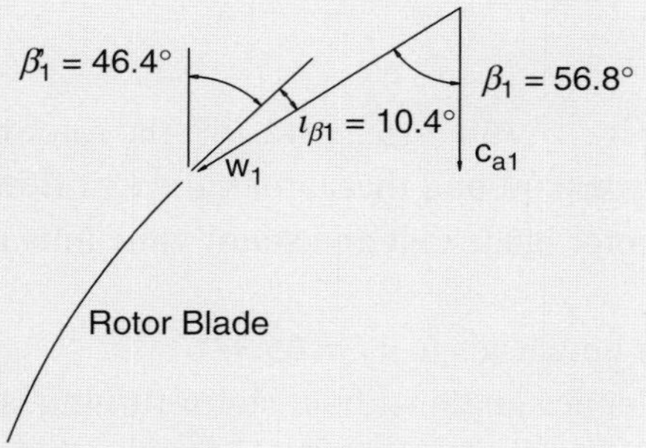

(g) Rotor Inlet Incidence – Upstream Stator Turned by 32.97°

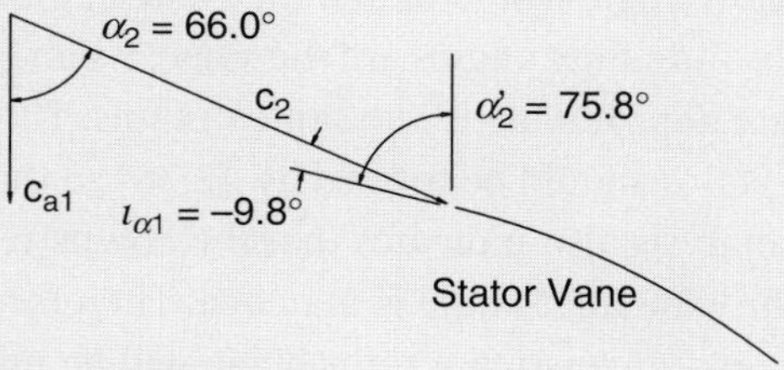

(h) Stator Inlet Incidence – Downstream Stator Turned by 32.97°

Figure 5.16 (cont.)

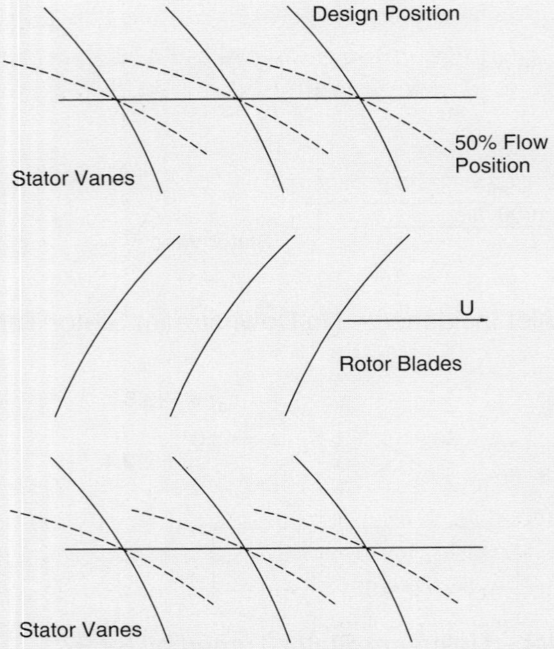

Figure 5.16 (cont.)

Thus, if one were to rotate the previous stator by 57.76° – 15° = 42.76°, the rotor blade inlet would be matched if all conditions remained constant and the stator vane exit flow angle matched the stator vane exit angle. Now the rotor blade exit and stator vane inlet angles (Figure 5.16d) should be considered.

As for the design case, $\beta_2 = -21.38°$, which now leads to $\alpha_2 = 65.97°$.

Thus, the downstream stator vane inlet incidence angle without stator turning is shown in Figure 5.16e and is

$\imath_{\alpha 2} = \alpha_2 - \alpha_2{}' = 65.97° - 42.79° = 23.18°$.

Once again, the incidence angle is large and would lead to reduced performance. If one were to rotate the downstream stator by 23.18°, the flow angle would match the blade angle ($\alpha_2 = \alpha_2'$). Since the stage inlet and stage exit design absolute flow angles are the same, it can be assumed that the upstream and downstream stators are approximately the same designs. Therefore, the current analysis indicates that the upstream stator should be rotated by 42.76° so that the rotor blade inlet angle is matched; however, the analysis also indicates that the downstream stator should be rotated by 23.18° so that the stator vane inlet angle is matched. Therefore, if the downstream stator is rotated by 23.18°, the next downstream rotor blade will be mismatched because, to match this rotor blade, the stator will have to be rotated by approximately 42.76°. Thus, it will unfortunately be impossible to match both the rotor and stator inlet angles by simply rotating the stator vanes. At this point some engineering judgment must be exercised to decide if the off-design condition with fixed stators can at least be improved upon. One can consider rotating the stators by an average of the two specified rotations, that is:

$(23.18° + 42.76°)/2 = 32.97°$.

In Figure 5.16f, one can see that now

$\alpha_1 = 15° + 32.97° = 47.97°$

and thus the flow angle is $\beta_1 = -56.75°$.

However, the blade angle is $\beta_1{}' = -46.38°$. The inlet incidence angle is thus

$\imath_{\beta 1} = -56.75° - (-46.38°) = -10.37°$,

as shown in Figure 5.16g.

This is better than the incidence angle without stator turning that was found above (−20.71°). Consequently, the new inlet stator vane angle is $\alpha_2{}' = 42.79° + 32.97 = 75.76°$.

If the exit absolute angle from the rotor remains the same ($\alpha_2 = 65.97°$), the incidence angle of the stator inlet is $\imath_{\alpha 2} = 65.97° - 75.76° = -9.79°$, as shown in Figure 5.16h. This is also better than the incidence angle without stator turning (23.18°) that was found earlier in this section.

Thus, as a result of the stator turning, the incidence angles to both the rotor and stator inlets have been reduced by more than a factor of 2. In Figure 5.16i, the resulting schematic of the stage, including the previous stator, is shown indicating the variable stator locations.

5.7.2 Turning Mechanism

Turning of the stator vanes is essentially accomplished through a mechanical linkage. A schematic of the mechanism is shown in Figures 5.17a and 5.17b, and photographs of an application are shown in Figure 5.17c for a Ruston power generation gas turbine. Also the GE LM2500 power generation unit is seen to have variable stators in Figure 1.47. Basically, each vane is rigidly attached to a linkage and pivot but is free to pivot around a point. The linkages of a given row of vanes are attached to an annular ring by pegs and holes. Thus, as the ring rotates, the vanes of the row are turned by equal amounts. The rings of the different stator rows are attached to a servo system, or, for some cases, a separate servo system is used for each row. Thus, as the servo is activated as shown, all of the vanes of all of the variable stages are rotated. By making the linkage lengths different for the different stator rows or by having a separate servo system for each ring, the stators of the different rows can be rotated by different amounts. Thus, each stator row can be rotated by a different angle as determined by the particular design and conditions.

5.8 Twin Spools

5.8.1 Theoretical Reasons

All turbofans and most modern turbojets operate with at least two shafts. Such a design has previously been defined as a twin-spool engine. Modern power generation gas turbines have either one or two shafts. The low-pressure section runs at a markedly lower rotational speed than does the high-pressure section. The purpose of this section is to demonstrate that two or more shafts are used in modern engines for two

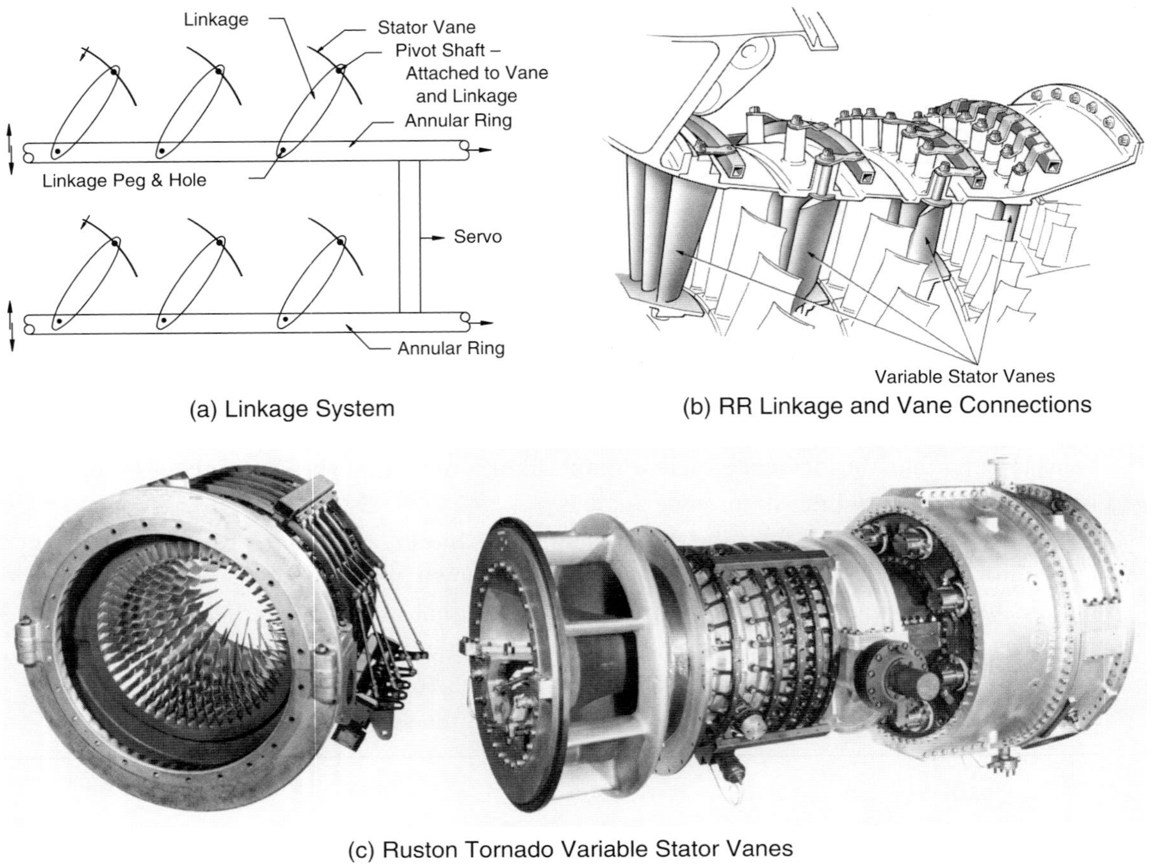

Figure 5.17 (a) Linkage systems for variable stators. (b) Cutaway view of variable stator linkage system (© Rolls-Royce plc 2022. All rights reserved). (c) Front and side view of Ruston Tornado showing servos (© Siemens Energy, 1985).

basic reasons: off-design operation and power distribution.

Off-Design Operation

Like variable stators, one basic purpose of twin spools is to improve off-design performance. If an engine were to operate at one flow rate, one speed, and so on, it would be possible to design all of the blade angles so that the flow matched every stage without any incidence. However, engines must be able to operate over a range of flow rates and speeds efficiently (takeoff, cruise, dog fight, landing, etc.). Thus, the incidence angles on the different stages will change with operating conditions.

Consider a typical compressor map for a single-spool gas turbine, as shown in Figure 5.18a. In this figure, the design point is designated as point A. If the speed of the engine were reduced, its operation would be at some point B. Typically, the air mass flow rate experimentally decreases faster than the rotational speed of the gas turbine.

Consider now the two rotor inlet velocity polygons for the first stage of the engine, as shown in Figure 5.18b. First, one should examine the design polygon (solid line). Next, as the rotational speed is decreased, so is the mass flow rate and thus so is the inlet axial velocity component (c_a). The absolute flow angle is the same for both cases. Because c_a is reduced faster than U, the

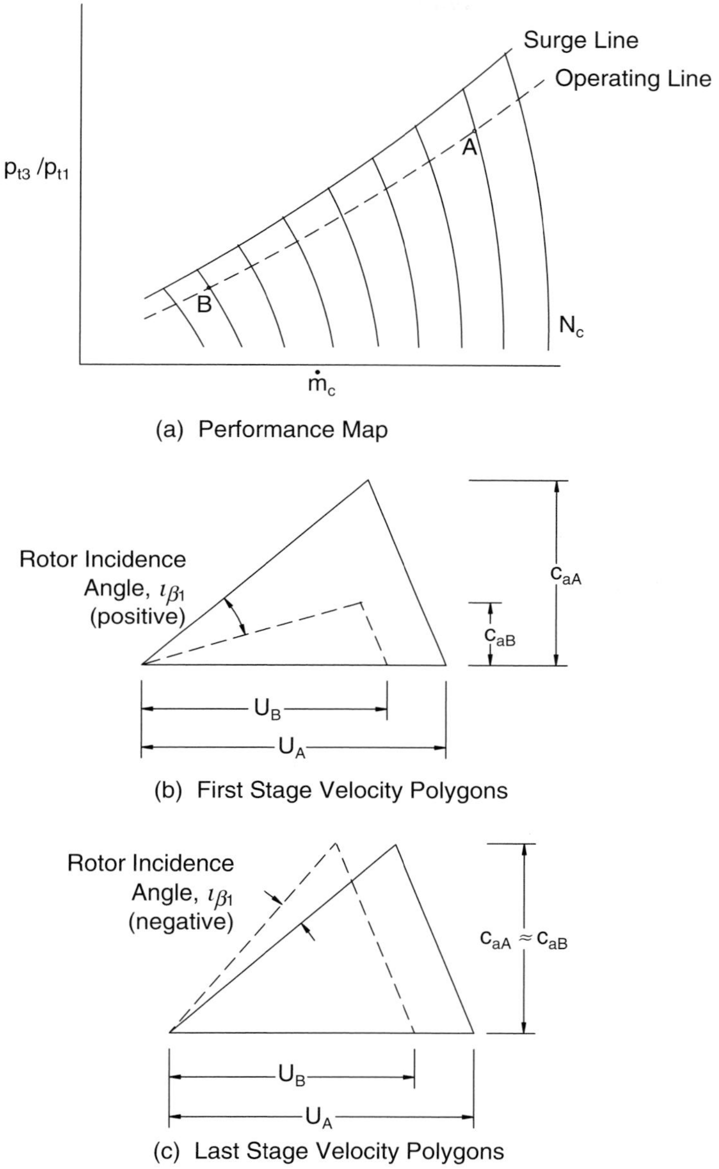

Figure 5.18 Velocity polygons for off-design operation.

polygons are not similar. Thus, a positive incidence angle arises for the rotor inlet. Thus, for this flow rate it would be desirable to reduce U (and thus N) to force the incidence to zero for an early stage.

Next, the two velocity polygons for one of the last stages as shown in Figure 5.18c (solid line is the design case) should be examined. For the latter stages, the pressure is not as great for the off-design condition as for the design case. Thus, the density is also lower for the off-design condition. As a result, although the mass flow rate may be significantly lower, the axial component of velocity remains approximately constant (in fact it can increase slightly). Again, the absolute flow angle is the same for both cases. Now the velocity polygons indicate that a negative incidence angle results for the rotor inlet. Thus, to force this incidence angle to zero, it would be desirable to increase the rotational speed for the latter stage – this is opposite to the effect from the early stage.

Therefore, to operate at off-design conditions, one must be able to change the rotational speed of the compressor with opposite trends from the front to the rear. Ideally, one would like to operate each stage with a different rotational speed because, for this situation, the incidence angle for each stage would be reduced to zero. However, the mechanical complexities would outweigh the aerodynamic gains. Thus, as a tradeoff for modern engines, two shafts are typically used. Incidence angles are still realized by some stages at off-design conditions; however, they are much smaller than would be seen with a single shaft.

Power

Power considerations constitute a second reason for using twin shafts. That is, if one recalls Eq. (4.2.24),

$$\dot{W}_{\mathrm{sh}} = \frac{\dot{m} c_{\mathrm{p}} T_{\mathrm{t}}}{\eta_{12}} \left[\left[\frac{p_{\mathrm{t2}}}{p_{\mathrm{t1}}} \right]^{\frac{\gamma-1}{\gamma}} - 1 \right]. \tag{5.8.1}$$

Inasmuch as the total pressure ratios are approximately the same for all stages of a compressor, the required power to be delivered to the last or high-pressure stages must be higher because the total temperature is higher for these stages. Next, recalling Eq. (5.4.1), one obtains

$$\dot{W}_{\mathrm{sh}} = \dot{m}[U_2 c_{\mathrm{u2}} - U_1 c_{\mathrm{u1}}]. \tag{5.8.2}$$

Now given that $c_{\mathrm{u2}} - c_{\mathrm{u1}}$ is limited by the turning angle of the rotor blades and that all stages have approximately the same turning angles, the only practical means to increase the input power is to increase the rotational speed U. Thus, once again, the high-pressure sections must rotate at a higher speed than the low-pressure stages.

5.8.2 Mechanical Implementation

In Figure 5.5, the geometry for a twin-spool compressor is shown. The high-pressure shaft is hollow. A set of antifriction or rolling element bearings separates the high- and low-speed shafts. The gain in aerodynamic performance is not without difficulties, however. For example, with the addition of more bearings, the mechanical or shaft efficiency is reduced. Also, because the high-speed shaft is hollow, it is more flexible. Thus, the shaft may demonstrate several resonances or critical speeds in the operating range. Care must be taken in the design so that the resonances do not occur at design points and enough damping is in the engine so that the resonances are not excited upon engine transients.

5.8.3 Three Spools

Although most modern jet engines operate with two spools, a few successful engines have three shafts, as shown in Figure 5.19. In such an engine, three compressors and turbines are used (low-, intermediate-,

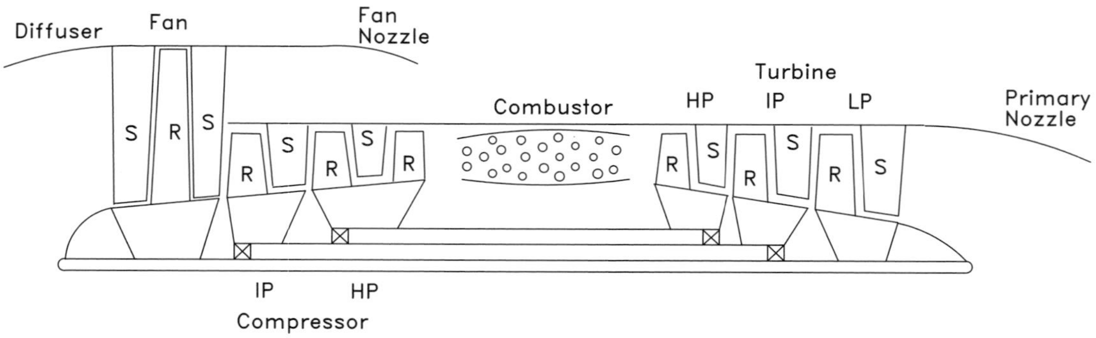

Figure 5.19 Three-spool turbofan.

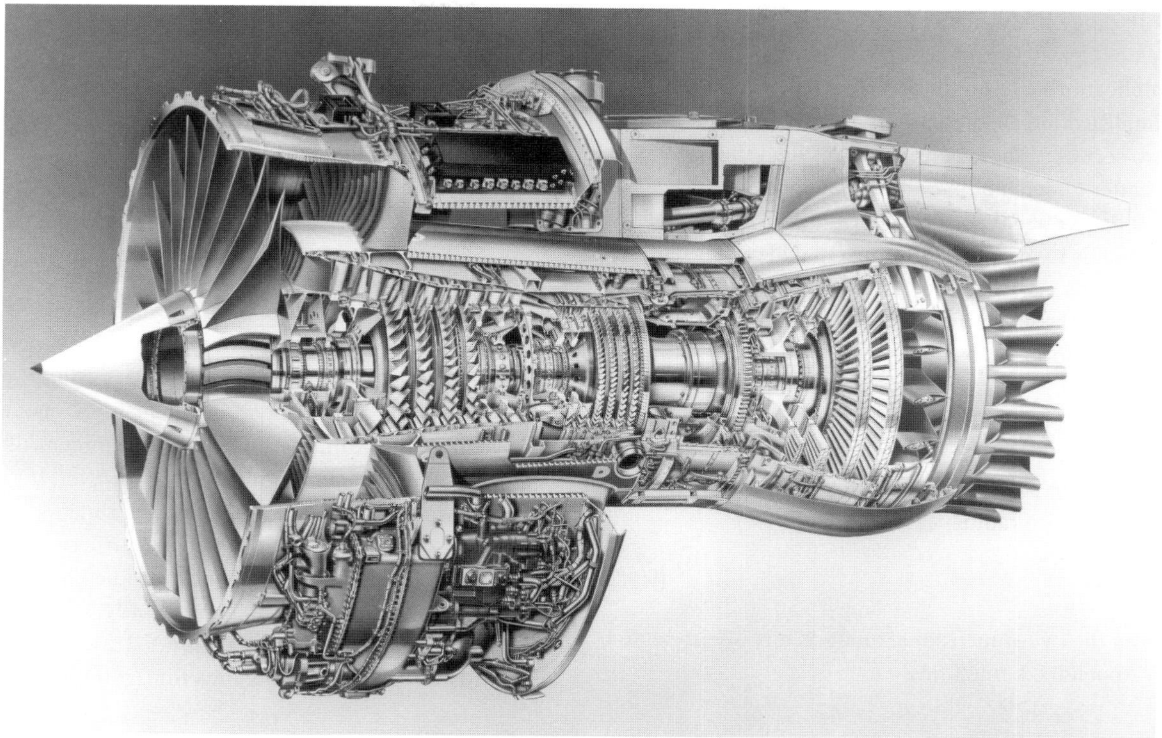

Figure 5.20 Rolls-Royce RB211 three-spool turbofan ().

and high-pressure). An example is the Rolls-Royce RB 211, as shown in Figures 1.26 and 5.20. The RB211 is used to power the Lockheed L 1011. For this engine, a single-stage fan operates at 3530 rpm, an intermediate pressure compressor operates at 5100 rpm, and the high-pressure stages run at 9390 rpm. Such an engine is aerodynamically superior to twin-spool engines but is relatively complex because of the increased number of bearings and resonances in the operating speed range. As a result, only a few such designs have been used successfully and this market is dominated by Rolls-Royce.

5.9 Radial Equilibrium Summary (from Appendix J) (Advanced Topic)

So far in this chapter, only a mean line analysis of the flow has been considered. In this section, three-dimensionality will be addressed – that is, hub-to-tip variations. A possible phenomenon occurs at the interface between each rotor and stator that is sometimes considered in the design of stages. At these interblade stations, an equilibrium condition is often assumed for which the flow is to be only in the axial and tangential directions (no radial flow). Namely, at the interfaces the sum of the surface forces on each fluid element is assumed to be equal to the centrifugal force. Between the interfaces, such conditions are not imposed, and thus radial flow does exist within the blade rows. It is this so-called radial equilibrium condition that allows an analysis of the three-dimensionality of the flow.

Through a detailed differential analysis of a force balance on a fluid element in Appendix J, one obtains:

$$\frac{dp}{dr} = \frac{\rho}{r} c_u^2. \tag{5.9.1}$$

This is the requirement for radial equilibrium. The radial pressure gradient is shown to be a function of both tangential velocity and radius. Several different velocity profiles can be fitted to this condition, and the implications can be studied. Two representative and simplification considerations are presented in the next two sections. These considerations lend themselves to a better understanding of the complexity of the 3-D flows due to radial equilibrium. The purpose of this section is to primarily introduce the idea that radial variations occur in blades, vanes, and passages between them and identify the trends that result from this equilibrium principle.

5.9.1 Free Vortex

One goal of a design may be to keep the total pressure constant with radius:

$$\frac{\partial p_t}{\partial r} = 0. \tag{5.9.2}$$

From the two-dimensional analysis already presented, in Appendix J, one finds

$$c_u = \frac{\kappa}{r}, \tag{5.9.3}$$

where κ is a constant.

If the Bernoulli equation is used to identify the total pressure, in Appendix J we find:

$$\rho c_a \frac{\partial c_a}{\partial r} = 0. \tag{5.9.4}$$

Consequently, for this special case, the axial component of velocity is not a function of radius; hence, c_a is exactly constant with radius.

Finally, some of the implications of this condition need to be addressed. If the radius is nearly constant from blade hub to tip, the tangential component of velocity will not change significantly. However, if the radius change is large (for example, a fan blade), significant variations in c_u can occur. Furthermore, the axial component has been shown to be constant with radius. Thus, if c_u varies significantly, so will blade angles. As an example, Figure 5.21, which represents three velocity polygons for a rotor blade, is

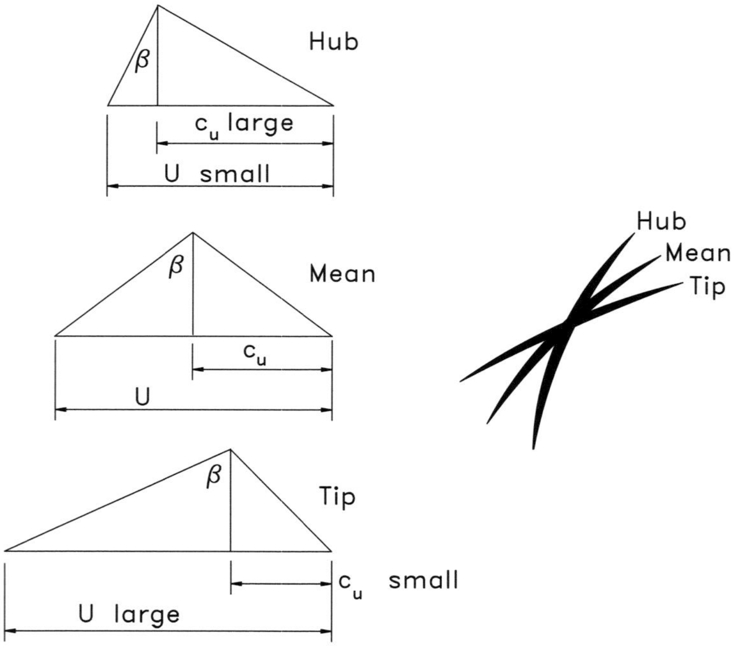

Figure 5.21 Velocity polygons and blade shapes for free vortex specification.

presented. At the blade hub, the blade speed U is relatively small, but c_u is relatively large. At the tip, the blade speed is relatively large, but now c_u is relatively small. From the diagrams one can easily see that the blade angle β changes markedly from hub to tip. Thus, radial equilibrium results in highly twisted blades as well as large changes in the blade curvature (most curved at the hub and flatter at the tip). Furthermore, the condition can lead to large (and undesirable) variations in the percent reaction from hub to tip.

5.9.2 "Constant" Reaction

As implied in the previous section, another consideration may be to force the reaction to be constant from hub to tip (or approximately so). With this idea in mind, a traditional proposal for radial variations for the tangential velocity is

$$c_{u1} = ar - b/r \tag{5.9.5}$$

for the velocity after the stator and

$$c_{u2} = ar + b/r \tag{5.9.6}$$

for the velocity after the rotor, where a and b are constants. Next, it can be demonstrated that, for incompressible flow without losses, if the axial velocity is approximately constant across the rotor ($c_{a1} \cong c_{a2}$),

$$\%R \cong 1 - \frac{a}{\omega}. \tag{5.9.7}$$

Next it is shown in Appendix J.3, at any value of r, c_{a1} (and c_{a3}) does not exactly equal c_{a2}, as assumed earlier in this section. As a result, the percent reaction is not exactly constant with radius. However, for many cases of interest, the approximation is close and allows for the a priori specification and understanding of a velocity profile.

Once again, if the implications of this condition are examined, one finds that the blades are moderately twisted. The geometries are not as awkward as for a free-vortex distribution but are not as simple as for a mean line analysis. Two typical sets of angle profiles are presented in Figure 5.22 for designs based on the free vortex and constant reaction specifications; for

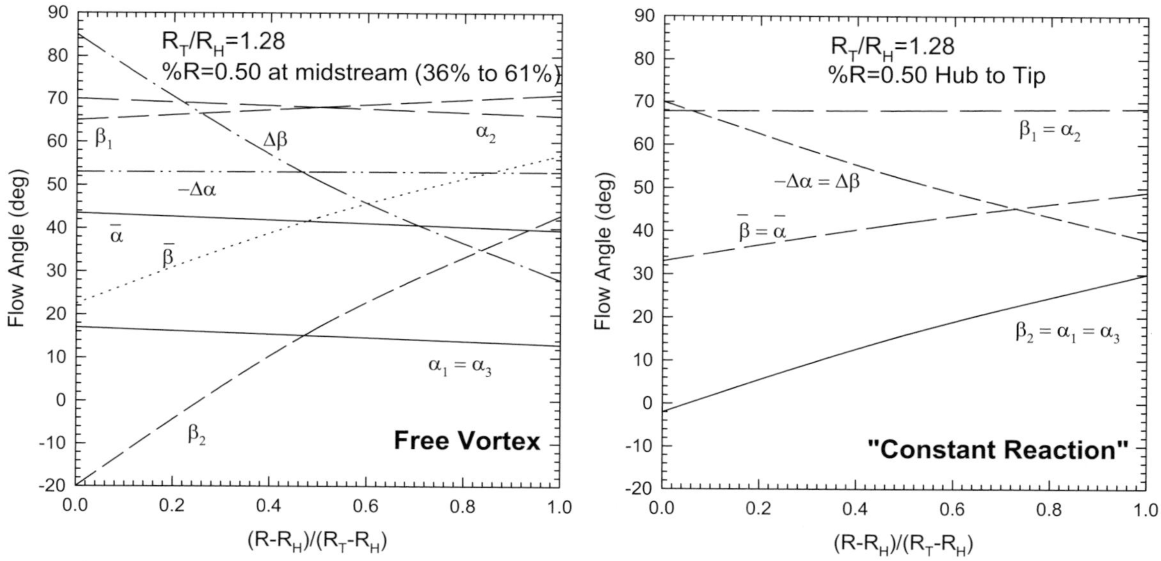

Figure 5.22 Typical flow angle distributions for free vortex and "constant reaction" specifications.

these geometries the tip-to-hub diameter ratio is 1.28 and %R = 0.50 at the midstream. In these figures, the variation of different angles from hub, $(R - R_H)/(R_T - R_H) = 0$, to tip, $(R - R_H)/(R_T - R_H) = 1$, are presented. As can be seen, the rotor blade angles change significantly from hub to tip for the free vortex case; in fact, β_2 changes sign. Also note that much more rotor flow turning ($\Delta\beta$) is near the hub than near the tip. The average rotor flow angle ($\bar{\beta}$) changes markedly from hub to tip, indicating severe blade twist. The stator angles are similarly behaved in stator turning ($\Delta\alpha$) compared with the rotor turning. Because of the different rotor and stator flow turning angles, the percent reaction changes from 36 to 61 percent from hub to tip for the free vortex. For the constant reaction case, when compared to the free vortex case, the rotor blade angles do not change as drastically from blade hub to tip (i.e., the blades are not as twisted), but the stator vanes change more.

The percent reaction for the "constant reaction" case is in fact essentially constant within 0.001 at 50 percent. Significant changes in rotor flow turning are still present from hub to tip, but there is less variation than in the free vortex case. As one might conclude from these two cases, in reality no one such simple postulation is used in blade designs. But these cases provide insight and do show the trends of hub to tip variations. By examining Figures 5.3a, 5.3b, and 5.21, and the GE90-115B image at the opening of Part I, one can easily see the relative changes in blade curvature and twist. The two specifications in this section are primarily intended as learning demonstrations to exhibit the blade angle twist and curvature trends. Years ago, they could also have been used as starting profiles that were attempted before refinements are made to an iterative design.

5.10 Streamline Analysis Method Summary (from Appendix K) (Advanced Topic)

For the most part in the chapter, the analyses of flows in the compressor have been assumed to be meanline 2-D and any radial distributions or velocity components have been assumed to be negligible in calculations. Section 5.9 indicates that three-dimensionality is indeed present, where radial equilibrium was covered and the implications for three-dimensionality were discussed. As implied, because the ratios of blade height to hub diameter for many modern machines are not small and centrifugal forces are large, 3-D flows result. As is demonstrated in this section, significant radial variations in parameter profiles will result when the tip radius is approximately 1.1 times the hub radius or greater. The variations become more pronounced as the rotational speed increases as other conditions are imposed. For example, in Figure 5.3a (Rolls-Royce Pegasus), Figure 1.28 (Rolls-Royce XWB), Figure 1.11 (GE9X), and in the opening image of Part I (GE 90-115B), significant blade twist is evident; that is, the average blade angle varies from hub to tip.

Several numerical "throughflow" techniques have been developed that can be used to predict the flow's three-dimensionality. Furthermore, CFD is certainly an important and essential tool used in the design of compressors and fans, as discussed early in this chapter and considered by Elmendorf et al. (1998), LeJambre et al. (1998), Rhie et al. (1998), and Adamczyk (2000). All of these approaches are for compressible flow and are robust methods intended for sophisticated predictions. Unfortunately, these methods require significant background and have a significant learning curve. They are thus not appropriate to learning the fundamentals of 3-D flows in turbomachines.

This section summarizes a streamline analysis method. Basically, different streamline paths are analyzed by a control volume approach through a compressor passage. The method is for flow in axial flow fans, compressors, or turbines and is primarily intended for educational purposes; however, it can also be used for preliminary designs and applied in the inverse mode for the analysis of experimental pressure data. Streamline analysis can be used to predict 3-D trends resulting from blade curvatures and twists and centrifugal forces, thus contributing valuable insight into the flow's three-dimensionality. This method provides a tool through which a "feel" for the magnitudes of the resulting three-dimensionality of the flow fields can easily be developed as blade configurations are varied. The technique converges relatively quickly and can be written with commercial math solvers. The software SLA.exe was developed to accommodate a few select boundary condition types.

Basically, the flow passage is divided into a finite number of annuli, and the equations are solved for each blade row independently. As is demonstrated, different conditions can be considered with the method. For example, free vortex or irrotational flow can be assumed throughout, and the flow angles are then calculated. As discussed in Section 5.9, this condition does not result in radial flow and therefore does not require an iterative solution. Also, for a more complex condition, the flow angles or other parameters can be specified. This case has an iterative solution whereby radial migration of the flow is predicted. Radial efficiency distributions for each blade row can also be specified for the second condition. Other conditions are discussed.

Steady incompressible adiabatic flow is assumed to simplify the analysis and to accelerate the numerical convergence of the solution. Moderate accuracy can be obtained using an average density – particularly for one- or two-stage analyses. Moreover, the incompressible assumption does not mask the fundamental 3-D trends predicted from the method, which can also be extended to include compressibility effects. Flack (1987) gives more details and other applications. In this section the fundamental concepts ideas are presented along with some example problems. Details of the equations and solution methods are presented in Appendix K.

5.10.1 Flow Geometry

In Figure 5.23 a stage in an axial flow turbomachine is shown with varying cross-sectional area, although more stages can be stacked behind. The analysis begins at the inlet to the previous stator (which can be an inlet guide vane, IGV). Each stage is divided into N annuli. In Figure 5.23, the annuli number (i) and the axial position number (j) are defined. For the sake of simplicity, only four annuli are shown. Each annulus represents a bundle of streamlines, and a different control volume is applied to each annulus. In general, the annuli will not be constant in area between axial stations. This is due to the resulting radial flow and changing cross section of the stage. For the control volume analysis, the flow is assumed to be uniform, both radially and circumferentially, within each annulus, and the velocities are evaluated at the radial center of the

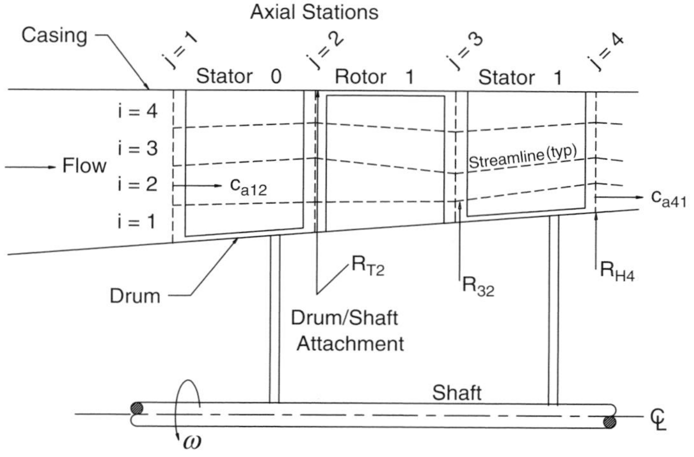

Figure 5.23 Axial flow turbomachine for streamline analysis.

annulus. Thus, circumferential blade-to-blade variations are not predicted as is the case in a CFD analysis. The exiting conditions for stator 0 are used as the inlet conditions for rotor 1, and so on. The radii of all of the axial and radial stations can in general be different. The radius R_{32} (typical) represents the outer radius of the first streamline bundle and the inner radius of the second streamline bundle at the third axial station. Also, the axial component of velocity at axial station 4 and annulus 1 is shown, c_{a41}.

5.10.2 Working Equations

The equations for each annulus are described in this section. The equations are independently derived for each annulus by a control volume analysis similar to the mean line flow analysis for incompressible flow described in Section 5.4. However, equations for all of the annuli must be solved simultaneously for each stage. It is important to note that the control volumes can be drawn in general. However, the exact positions and sizes (radii) of the control volumes are actually found as a part of the solution.

Geometric Equations

The velocity of the blades at the radial center of the element is U_{ji}. Shown in Figure 5.7 are the absolute flow velocity (c) and the flow velocity relative to the blades (w). Also shown is the normal component c_a. Velocity polygons, as in Figure 5.7, are constructed for each annulus and axial position. As in previous sections, α and β are the absolute and relative flow angles, respectively. Important relationships that result from the velocity polygons for each annulus are developed and listed in Appendix K.

Control Volume Equations

Next, the conservation of mass equation, Eq. (4.2.1), is considered and applied to each control volume. Because each control volume follows the streamlines, no flow crosses the annuli, which accounts for the name of the method. Namely, flow never crosses a streamline. As a result, the only mass flux occurs at the axial planes. The moment of momentum equation, Eq. (4.2.2), is also applied to each control volume. Thus, summing all annular torques yields the total torque for each stage. The energy equation, Eq. (4.2.3), is also applied to each control volume. Furthermore, the torque and rotor power for each annulus are related by the rotational speed. When stator blades are analyzed for the power, the power is of course zero. The last consideration occurs at the interface between each rotor and stator. At these stations, radial equilibrium (which is covered in Section 5.9) is assumed. It is again important to realize the sizes of the control volumes are not specified but are found as a part of the solution!

Closure Equations

Before one attempts a solution, counting of the variables is appropriate. As outlined in Appendix K, there are $15N + 2$ variables and $12N + 2$ equations for each blade row. Clearly, to obtain a solution, one must specify $3N$ variables or $3N$ more equations are needed for each blade row. Various specified conditions are considered in detail in the following sections. In all cases, all flow conditions into the inlet of the previous stator are specified (N variables). Some different cases to provide the other $2N$ conditions are discussed in detail in the following sections. Other cases can certainly be used. Specific details are presented in Appendix K.

Potential Flow or Free Vortex Flow The flow is assumed to be irrotational as well as inviscid. Also, as shown in Section 5.9.1, this condition, along with radial equilibrium, results in a constant axial velocity at each axial plane and no radial flow between the axial stations. Lastly, for this condition the absolute flow angle at some radius (usually at the center of each axial plane) is usually specified. As an equally valid alternative specification for this case, the relative flow angle at some radius could be specified. This case is relatively simple and does not require an iteration technique.

Specified Flow Angle Distribution For this case, a set of boundary conditions is for specified flow angle variations (hub to tip). Absolute flow angles and efficiencies are chosen to be input at specified radial positions and at all axial stations. Thus, $2N$ variables (flow angles and efficiencies) are specified, and a solution is possible. Although flow angles and efficiencies are specified at certain radii, these do not necessarily correspond to annuli centroids and can be a curve fit to a radial distribution or table for interpolation at different radii. Alternatively, and equally valid, the relative flow angles and efficiencies could be chosen to be input at all axial stations and at specified radial positions. The solution to this case requires a two-nested loop iterative solution.

Specified Pressure Distribution If limited experimental data are available, radial profiles of static and total pressure may be known. Thus, for this case, the second set of boundary conditions is for specified static and total pressure variations (hub to tip) and a solution is possible providing angle, efficiency, and other parameter radial distributions to complement the data. Therefore, the static pressure and total pressure are input at specified radial positions and at all axial stations. Again, $2N$ variables and a solution is possible. As in the previous section, the radii do not necessarily correspond to annuli centroids. The solution to this case also requires a two-nested loop iterative solution.

Other Specified Conditions As indicated previously, in general there are $15N + 2$ variables and $12N + 2$ equations. One can designate conditions other than specified angles or pressures that would allow a solution. Many other possibilities could be used to specify closure equations.

Solution Method The equations discussed in Section 5.10.2 and its subsections can be solved using any of the commercial math solvers. To find a solution, one starts at $j = 2$ (if conditions at $j = 1$ are known) and analyzes each axial station separately, repeatedly marching in the axial direction. For the most solutions (other than the free vortex), a two-nested loop iterative solution is usually required. Two solution approaches for specified angles plus efficiencies and specified static pressures and total pressures are outlined in Appendix K. The Regula Falsi method (Appendix I) can be used to accelerate the converging iteration for two loops. As discussed at the beginning of this section, the radial size and position of the control volumes are found as a part of the solution. Also, initial estimates can have a strong influence on convergence of such schemes. Three examples of the streamline analysis are presented in the section.

Example 5.5 | Streamline Analysis of a Single Stage: Irrotational Flow

A single stage of a compressor with an IGV and free vortex or potential flow is analyzed ($\eta = 1$). The compressor geometry and operating conditions approximate those on the first low-pressure compressor stage on a small engine. The tip and hub diameters, which are axially constant, are 0.391 and 0.320 m (1.283 and 1.050 ft), respectively, and the compressor operates at 12,000 rpm. The resulting tip-to-hub diameter ratio is 1.222, which indicates that predictable radial flow and variations should exist. Air enters at 97.9 kPa (14.2 psia), the density is 1.21 kg/m^3 (0.00235 sl/ft^3), and the flow rate is 11.3 kg/s (25 lb$_m$/s). The midstream absolute flow angles at the exit of the IGV, rotor blades, and stator blades are 11°, 34.2°, and 11°, respectively. The flow angles, blade curvatures, blade twists, percent reaction, and pressure profiles at the exit of the rotor and stator are to be found for nine annuli.

Results

Results for this case are shown in Figures 5.24 and 5.25, in which characteristics are plotted from hub to tip (as a function of the dimensionless radial position, $\hat{R} = (R - R_H)/(R_T - R_H)$). Because the flow is irrotational, the axial component of velocity is constant at all radial and axial positions and equal to 236.1 m/s (774.6 ft/s), and the streamlines are all axial and parallel.

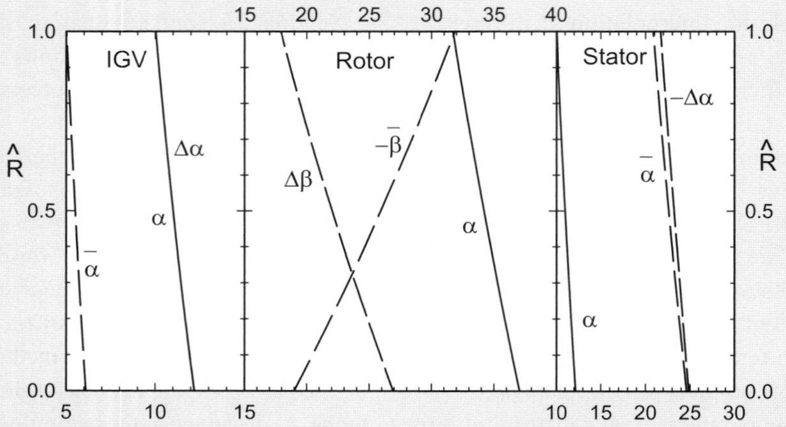

Figure 5.24 Exit absolute flow angles, average flow angles, and flow turning angles (Example 5.5).

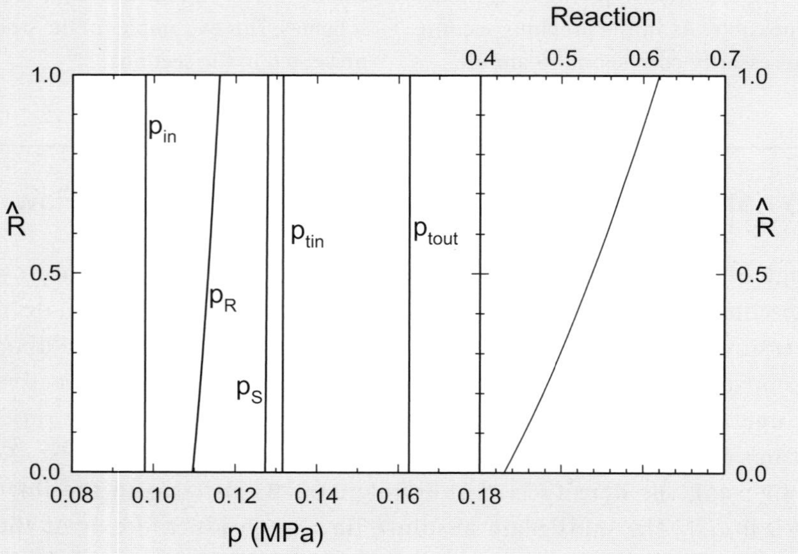

Figure 5.25 Exit pressure and rotor reaction profiles (Example 5.5).

In Figure 5.24, the absolute flow angles at the exits of the IGV, rotor, and stator are shown. As can be seen, the angles vary by approximately 2° across each passage for the IGV and stator and approximately 5° for the rotor.

The average flow angle $\left(\bar{\beta} = \frac{\beta_{\text{exit}} + \beta_{\text{entrance}}}{2}\right)$, averaged from leading edge to exit, is shown in Figure 5.24. This angle varies by only 1° for the IGV. However, the rotor flow angle varies by approximately 13° across the passage. This indicates that the blade is moderately twisted. The stator vane is less twisted, and the average flow angle varies by approximately 4°.

The turning angles of the blades are also presented in Figure 5.24 (namely, $\Delta\beta = \beta_{\text{exit}} - \beta_{\text{entrance}}$). The IGV turning angle varies by approximately 2° from hub to tip. The rotor turning angle varies by 9° across the passage. This figure indicates that, at the hub, the blade is moderately curved, whereas at the tip the blade is relatively flat. Also, the turning angle for the stator vane varies by only 3°.

The performance of this geometry is shown in Figure 5.25. The static and total pressure profiles at the exits of the rotor and stator are presented along with the rotor reaction profile. The static pressure varies by 3 percent across the exit of the rotor blades. However, at the exit of the stator the pressure varies by only 0.3 percent. The exit total pressure profile is exactly constant across the passage, as it should be for free vortex flow. The rotor reaction varies from 0.43 to 0.62 from hub to tip. Thus, this design is least efficient near the hub owing to the low reaction. The total power required is 290 kW (389 hp).

The three-dimensionality of the flow is obvious from this example. The absolute flow angles are mild functions of the radius, whereas the rotor blade angles, pressure profiles, and reaction profiles are stronger functions of the radius.

For comparison with the nine-annuli case, this geometry was also solved for one annulus and also with an isentropic, 2-D, mean line compressible flow analysis. For the three methods, the total power and the mean line pressures were exactly the same. The total pressure from streamline analysis at the exit of the stage was found to be within 1 percent of the value calculated using the compressible mean line flow analysis.

Extended Example 5.5.1 Streamline Analysis of a Single Stage: Variation of Hub-to-Tip Diameter Ratios

The next four cases to be considered are also single-stage compressors with potential flow and are extensions of Example 5.5. The mean diameter is 0.356 for all four cases, and the compressor again operates at 12,000 rpm. The hub and tip diameters are parametrically varied so that the tip-to-hub diameter ratio increases from 1.22 to 2.11. The axial velocity is the same as in Example 5.5; as a result, the mass flow rate increases as the tip-to-hub diameter increases. The average density is 1.21 kg/m^3 (0.00235 sl/ft^3). An inlet guide vane is used, and the mean line flow angles at the exit of the IGV, rotor blades, and stator blades are 11°, 34.2°, and 11°, respectively. Of particular interest is the percent reaction profile.

Results

The results are shown in Figure 5.26. As can be seen, all of the geometries operate with a mean line value of reaction of 54 percent. Near the tip, the reaction increases as the diameter ratio increase. Most importantly, near the hub, as the diameter ratio increases, the reaction rapidly decreases. In fact, for $D_T/D_H = 2.11$, the reaction is negative at the hub. This is obviously unacceptable for a compressor because it implies a pressure drop.

As a result, for cases with low or negative values of reaction, the compressor must be designed so that the mean line reaction will be large enough that such problems are not encountered at the hub. This example is therefore extended one step further. It can be noted that, for the highest diameter ratio case, the flow angles were adjusted so that the reaction increased but the total pressure rise remained constant. The mean line flow angles at the exit of the IGV, rotor blades, and stator blades are now 2°, 27.5°, and 2°, respectively. Results are also shown in Figure 5.26. Reactions at the mean and tip are significantly larger than 50 percent for this case, and tip reactions may approach unity for some cases but the percent reaction at the hub is now greater than zero. This is a significant observation – particularly for front stages of a compressor or fans, where the diameter ratio is much larger than 1.

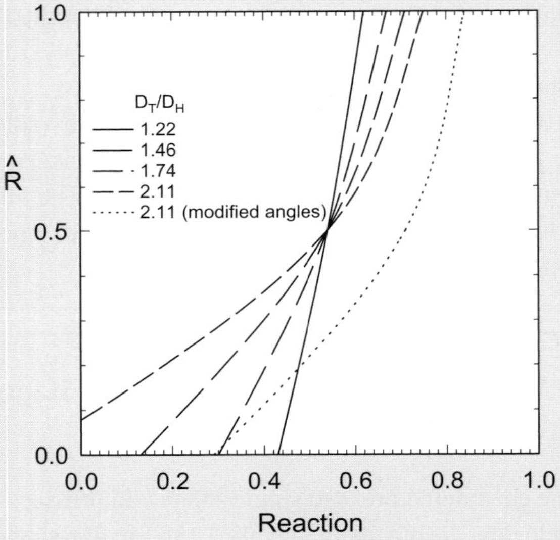

Figure 5.26 Reaction profiles for Extended Example 5.5.1: effect of diameter ratio.

Example 5.6 Streamline Analysis of a Single Stage: Specified Flow Angles

In Example 5.5, a compressor with irrotational flow was studied. The second case is the same turbomachine, but now the absolute flow angles across the exit of each blade row are specified and are constant from hub to tip. The same flow rate and speed are studied, and the flow angles are presented in Figure 5.27. Also in Figure 5.27 is the specified stage efficiency. Efficiencies of 87 percent are used at the hub and tip and 90 percent at the midstream. This is a limiting case to demonstrate the changes in profiles from the free vortex case.

Results

Results are presented in Figures 5.27 through 5.29. Figure 5.28 presents nine streamlines. As can be seen in the rotor cascades, a slight, net inward radial flow results from the radial equilibrium condition. In the stator cascades, a net outward radial flow results. For the first case (Example 5.5), one should remember that no radial flow existed. Thus, the present case exemplifies the types of radial flows that can result. Note that, owing to the radial equilibrium condition, either a net inward or outward flow can be realized as determined by the specified flow angles.

Axial velocity profiles are also presented in Figure 5.28. The profile for the IGV exit varies by 0.4 percent from the mean, whereas the rotor exit profile varies by 2 percent. The stator exit profile varies by 5 percent from the mean. As can be seen for the stator, the highest velocities are near the tip, which also indicates a net flow outward.

Average and turning flow angles are also presented in Figures 5.27. The blade shapes are different from those in the preceding case, and the magnitude of the twists for this case is slightly smaller. For the IGV, rotor, and stator blades, twists (radial change in $\overline{\alpha}$ or $\overline{\beta}$) of 0°, 9°, and 0° from hub to tip are present. Similarly, the amount of change in the blade curvatures from hub to tip is slightly less. For the IGV, rotor, and stator cascades the turning angles ($\Delta\alpha$ or $\Delta\beta$) vary by 0°, 4°, and 0° across the passages.

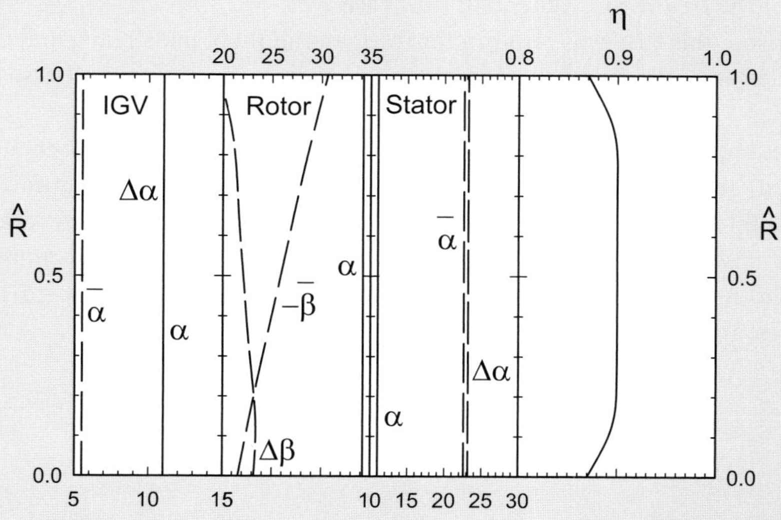

Figure 5.27 Specified exit absolute flow angles, average flow angles, flow turning angles, and stage efficiency (Example 5.6).

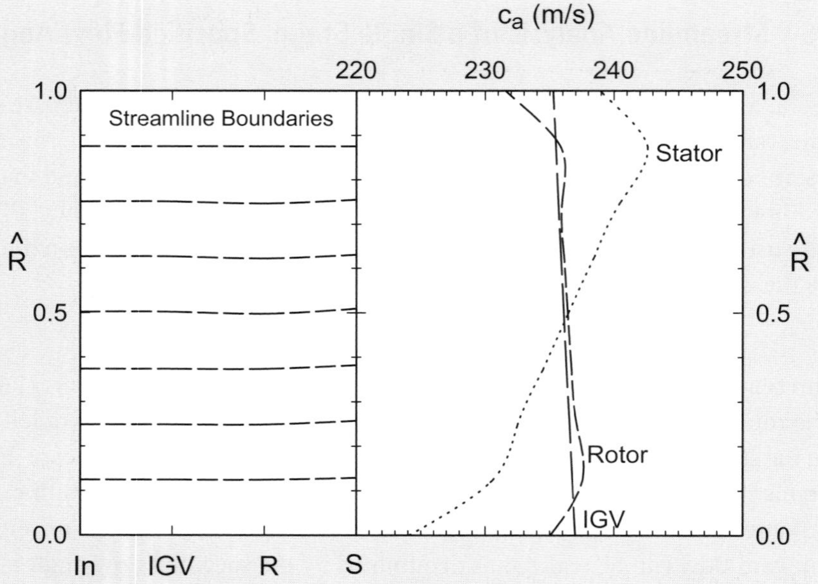

Figure 5.28 Streamlines through stage and exit velocity profiles (Example 5.6).

Lastly, the performance for this case is illustrated in Figure 5.29. Once again, the static pressure profile from the rotor stage is skewed, whereas the profile from the stator stage is flat within 1 percent. The total pressure profile varies by 3 percent from the mean and is lowest near the hub. This low total pressure area is the result of the low efficiency in this region. The rotor reaction varies from 0.51 to 0.62. The reaction is between 0.50 and 0.60 for over 90 percent of blade span, however, which indicates that this blade row is operating relatively efficiently.

For this case, 291 kW (390 hp) is required to power the compressor. This power is minimally higher than for the free vortex (potential) flow case.

For comparison, this case was also run for three annuli (hub, midstream, and tip). The axial pressure profiles and other results were found to vary from the nine-annuli case at most by 0.3 percent.

This last example was obtained by varying the flow angles and observing how the performance of the machine varied. It is not meant to represent an optimized geometry but is merely intended to illustrate the type of study that can be performed. It demonstrates how one can iterate on geometries to improve the performance characteristics. As indicated earlier, the 3-D effects increase as the tip-to-hub diameter ratio increases and other parameters change.

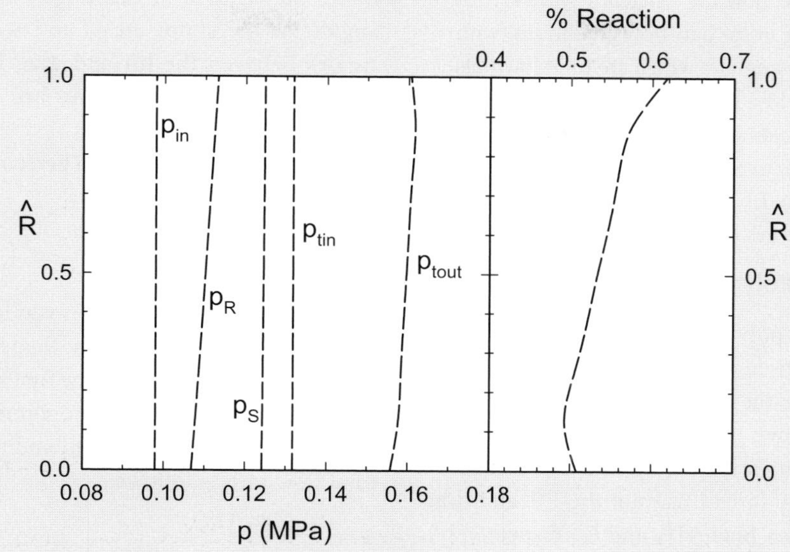

Figure 5.29 Exit pressure and rotor reaction profiles (Example 5.6).

5.11 Performance Prediction of a Compressor Stage Summary (from Appendix L) (Advanced Topic)

Thus far, predictions or understanding of the basic principles of stage performance (primarily efficiency) for given geometries have not been addressed. That is, methods of relating changes in operating conditions (e.g., flow rate) to changes in efficiency have not been covered. One should remember that a compressor stage or cascade is made up of a series of airfoils. Like wings, airfoils have lift and drag characteristics. For the case of a cascade, a large lift implies a large torque, which has been shown to result in a large pressure ratio. On the other hand, drag implies frictional losses, which result in a loss in efficiency and total pressure. Furthermore, both the lift and drag are (strong) functions of leading-edge incidence angle. Sizable amounts of detailed cascade and blade data are available for the varied series of NACA and NASA blade shapes. Abbott and von Doenhoff (1959) present a considerable amount of pressure, lift, and drag data for many blades shapes, but this still is only a small proportion of the data available in many sources. These data are in terms of lift and drag coefficients rather than efficiency information. A compressor designer is interested in efficiency, however. Thus, the purpose of this section is to present a method to relate the lift and drag characteristics directly to efficiency. The method is a combination of theoretical considerations and empiricism. Incompressible flow is assumed to simplify the mathematics. The method can be used to select the blade characteristics – for example, angles, solidity, pitch, chord, and so on – to maximize the efficiency. This general approach was used in early compressor design and was developed by Vavra (1974) in 1960, Logan (1993), and several others; the author does not want to imply it is a modern technology. As indicated earlier, CFD is certainly an important tool in the design of modern compressors and fans. These CFD codes are robust methods and are thus not appropriate for teaching the fundamental characteristics of efficiency variations in turbomachines. Therefore, most importantly, the method in this section gives great insight into how the stage performance and blade characteristics are fundamentally related – for example, how the efficiency will vary for a basic design choice or for part flow conditions (with increasing incidence angle) – which are fundamentals that engineers often lose sight of in CFD analyses. The analysis is summarized herein and expanded upon, including application examples. Derivation details are included in Appendix L.

5.11.1 Velocity Polygons

Although velocity polygons have been covered already, they are discussed again in a different light in this section. For the rotor, the relevant velocities for a blade are shown in Figure 5.30. For the stator, the relevant velocities for a vane are also shown in Figure 5.30. Both figures illustrate the lift and drag forces acting on the blades (F_l and F_d, respectively). As usual, the inlet and exit flow angles to the rotor blade are β_1 and β_2 respectively, and the flow deflection is δ_{12}. The average flow angle is defined as β_m, which is approximately the stagger, and the forces are relative to this angle. Inlet and exit flow angles for the stator are α_2 and α_3, respectively. Angles between the lift and drag loads for both rotor blades and stator vanes are ε, and they are shown in Figure 5.30.

Next, recall the definition of percent reaction results in

$$\%R = \frac{1}{1 + \left[\dfrac{c_2^2 - c_1^2}{w_1^2 - w_2^2}\right]}. \tag{5.11.1}$$

The blade velocity, U, is conventionally used to nondimensionalize velocities. Next, two new definitions are introduced. The first is the conventional flow coefficient, which is a nondimensional velocity based on the axial component and thus represents the total flow rate as follows:

$$\phi = \frac{c_a}{U} = \frac{w_a}{U}. \tag{5.11.2}$$

The second is the deflection coefficient (nondimensional deflection), which represents the blade turning. For a rotor blade,

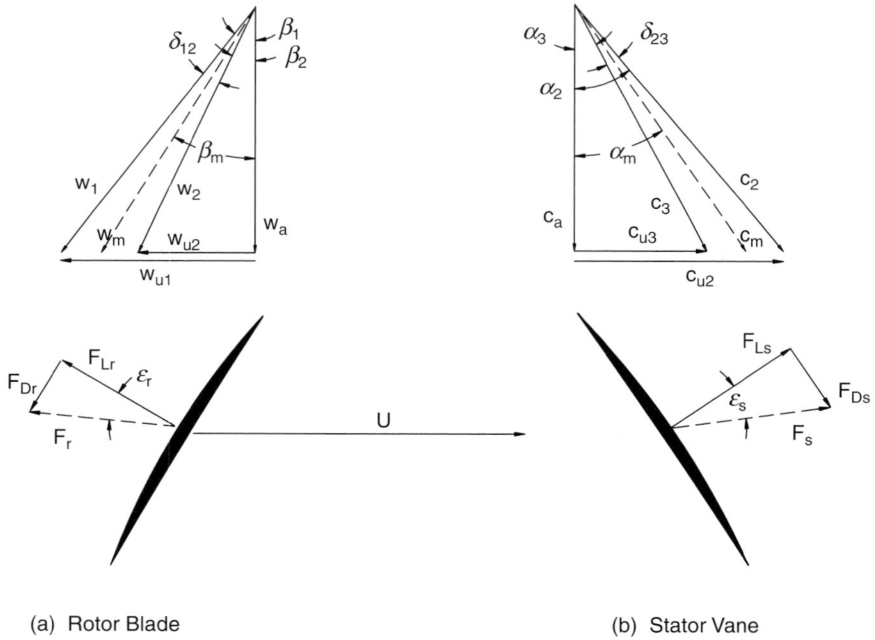

Figure 5.30 Velocity polygons and force vectors.

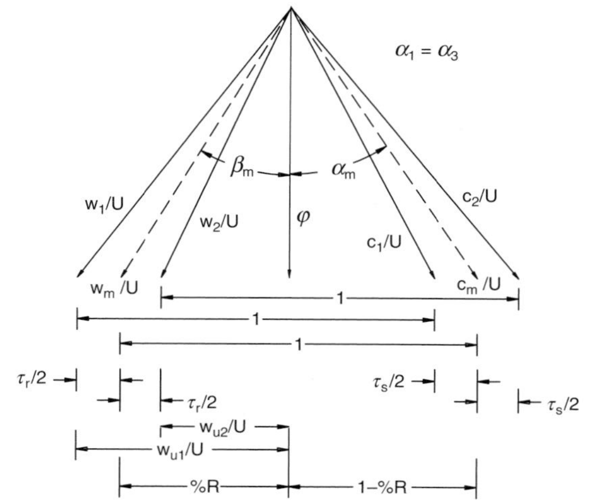

Figure 5.31 Nondimensional velocity polygons for a compressor stage.

$$\tau_r = \frac{w_{u2} - w_{u1}}{U}, \tag{5.11.3}$$

and, for a stator vane,

$$\tau_s = \frac{c_{u2} - c_{u3}}{U}; \tag{5.11.4}$$

thus, one can now nondimensionalize the entire set of velocity polygons in Figure 5.30 by U to generate Figure 5.31. In Figure 5.31, one can see the different physical quantities %R, ϕ, and τ and thus gain an understanding of the relevance of velocity polygons to compressor stage performance. For example, if α_1 is equal to α_3 and %R $= 0.50$, then τ_r is equal to τ_s and once again the polygon is symmetric, as discussed in Section 5.4.4. Also shown on Figure 5.31 are the average angles β_m and α_m.

5.11.2 Lift and Drag Forces and Control Volume Equations

Next, the lift and drag forces and the angle between the two (ε_r and ε_s shown on Figure 5.30) are considered:

$$\tan \varepsilon_r = \frac{F_{dr}}{F_{lr}} = \frac{C_{dr}}{C_{lr}} \tag{5.11.5}$$

$$\tan \varepsilon_s = -\frac{F_{ds}}{F_{ls}} = -\frac{C_{ds}}{C_{ls}}, \tag{5.11.6}$$

where C_l and C_d are the lift and drag coefficients. These can be related to the pressure difference using a momentum control volume approach in a blade or vane passage. It should be noted that finite length vanes/blades will have more drag due to tip leakage and is covered in Appendix L.3. And finally, the power can be related to the pressures by a control volume approach with the energy equation. As a part of this energy balance the efficiency is again defined as

$$\eta = \frac{\Delta p_t}{\Delta p'_t}. \tag{5.11.7}$$

Therefore, using the forces and control volume equations (development and details are in Appendix L) we find

$$\eta = \phi \left\{ \left[\frac{\%R - \phi\varepsilon_r}{\phi + \varepsilon_r \%R} \right] + \left[\frac{1 - \%R + \phi\varepsilon_s}{\phi - \varepsilon_s(1 - \%R)} \right] \right\}. \tag{5.11.8}$$

The stage efficiency is consequently (and very simply) a function of only four variables: percent reaction, flow coefficient, the angle between the drag and lift for the rotor, and the angle between the drag and lift for the stator. Thus, if it is possible to evaluate the lift and drag for a given condition, the efficiency can be estimated. One observation is that, as the drag on blades and vanes increases, the stage efficiency decreases. Furthermore, velocity polygons for different operating conditions can be used to evaluate the flow angles (and thus yield the vane and blade incidence angles), and if information or data for the lift and drag on a cascade (as a function of incoming incidence angle) are available, the efficiency can be estimated.

5.11.3 Effects of Parameters

One may wish to see the effects of the different parameters on the overall performance. It is desirable for the efficiency to be as high as possible. Thus, from Eq. (5.11.8), Vavra (1974) showed that one can differentiate η with respect to %R and set to zero to maximize the efficiency.

$$\%R_{opt} = \frac{1 - \frac{\phi}{\varepsilon_s}\left[1 - \sqrt{\frac{1+\varepsilon_s^2}{1+\varepsilon_r^2}}\right]}{1 - \frac{\varepsilon_r}{\varepsilon_s}\sqrt{\frac{1+\varepsilon_s^2}{1+\varepsilon_r^2}}}. \tag{5.11.9}$$

If both ε_r and ε_s are small,

$$\%R_{opt} \cong \frac{1}{1 - \frac{\varepsilon_r}{\varepsilon_s}}, \tag{5.11.10}$$

and if ε_r and ε_s are the same magnitude ($\varepsilon_r = -\varepsilon_s = \varepsilon$),

$$\%R_{opt} = 0.50. \tag{5.11.11}$$

This is an extremely interesting and important finding. When the percent reaction is 50 percent not only are the forces equal on the rotor and stator blades, but the efficiency is at a maximum. Thus, most axial flow pumps are designed with percent reactions of approximately 50 percent. Note that incompressible flow was assumed for this analysis. For the same performance phenomena to be realized in compressors, compressor stages tend to have percent reactions somewhat higher than 0.5 (especially for highly loaded stages) due to flow compressibility and long blades. The efficiency at this value of %R is

$$\eta_{\%R=0.50} = 2\phi\left[\frac{1-2\varepsilon\phi}{\varepsilon+2\phi}\right]. \tag{5.11.12}$$

The optimum value of flow coefficient (maximum efficiency) can be next be found by differentiating Eq. (5.11.12) with respect to ϕ and setting to zero, which yields

$$\phi opt_{\%R=0.50} = \frac{1}{2}\left[\sqrt{1+\varepsilon^2} - \varepsilon\right], \tag{5.11.13}$$

and if ε is small:

$$\phi opt_{\%R=0.50} \cong \frac{1}{2}[1-\varepsilon]. \tag{5.11.14}$$

From the optimized values of %R and ϕ one can find the optimum flow angles. As a limiting case, note that, as the drag approaches zero, the optimum value of ϕ approaches 0.50. Thus, the flow angles β_m and α_m approach 45°. Finally, the maximum value of efficiency under the general conditions is

$$\eta \max_{\%R=0.50} = 1 + 2\varepsilon^2 - 2\varepsilon\sqrt{1+\varepsilon^2}, \tag{5.11.15}$$

which, for small ε, becomes

$$\eta \max_{\%R=0.50} \cong 1 - 2\varepsilon[1-\varepsilon]. \tag{5.11.16}$$

Obviously, as the drag approaches zero as do the losses, the efficiency approaches unity. Figure 5.32 indicates the order of magnitudes and trends one can expect. For this figure, percent reactions of 50 percent and 60 percent (done numerically) are used with ε varying from 0.0 to 0.10. Both the optimum flow coefficient and resulting efficiency are shown. As can be seen, the optimum flow coefficient decreases with ε and increases with %R. The efficiency decreases with increasing ε (or drag), as expected. However, for the optimum flow coefficient, the efficiency has minimal dependence on %R, which is an interesting and important finding. Thus, although the optimum reaction is 50 percent, the value can be increased without major penalties on efficiency.

It is also possible to acquire a better overall understanding of the influences of the different parameters through the following analysis. By using Eq. (5.11.8), which is the most general form of the

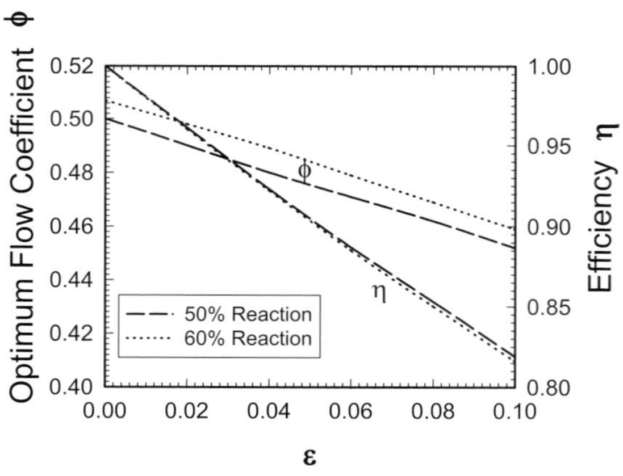

Figure 5.32 Dependence of optimum flow coefficient and efficiency on ε.

equation, one can combine terms and ignore second-order terms, which are very small. If one assumes that $\varepsilon_r = -\varepsilon_s = \varepsilon$, this results in

$$\eta = \frac{\phi + 2\varepsilon(\%R - \%R^2 - \phi^2)}{\phi + \varepsilon}. \quad (5.11.17)$$

A general 3-D depiction of the dependence of the efficiency on the flow coefficient, reaction, and drag to lift angle is presented in Figure 5.33. For clarity, only one value of efficiency (90 percent) is depicted. A series of conoids result – one for each value of η. Vavra (1974) has shown that the loci of constant efficiency are indeed circles. As has been shown in Section 5.7 as the flow rate deviates from the design value, the flow angle does not match the blade angle (incidence). Thus, as the flow rate deviates from the design value, the drag increases, thus possibly increasing ε (lift may increase also). As a result, as one deviates from the design point, the efficiency changes because of two different effects. That is, as the flow rate is changed from the design value, movement is in both the ϕ and ε directions. Furthermore, with each parameter treated separately, if %R and ϕ are held constant, as ε decreases, the efficiency always increases.

Other trends for a particular value of ε (0.03 shown here) yields Figure 5.34. As can be seen in Figure 5.34, the maximum efficiency for $\varepsilon = 0.03$ is 0.942 and occurs at %R = 0.50 and $\phi = 0.485$. As the flow coefficient or percent reaction deviates from the optimum values, the efficiency drops off. Note particularly that, for low flow rates, as the flow rate drops, the efficiency decreases very quickly. However, for large flow rates, as the flow rate increases the efficiency again decreases – but not rapidly. For small values of flow coefficient, as the reaction increases from 0, the efficiency rapidly increases and then decreases. For large flow coefficients, however, as the reaction changes, changes in efficiency are not large.

5.11.4 Empiricism Using Cascade Data

Next, 2-D cascade data are discussed as an empirical method that integrates the previous analyses to predict the efficiency performance of a compressor stage. For example, for a given cascade of stator vanes, as in Figures 5.7c and 5.7d, if the flow enters at angle α_2, the flow will exit at

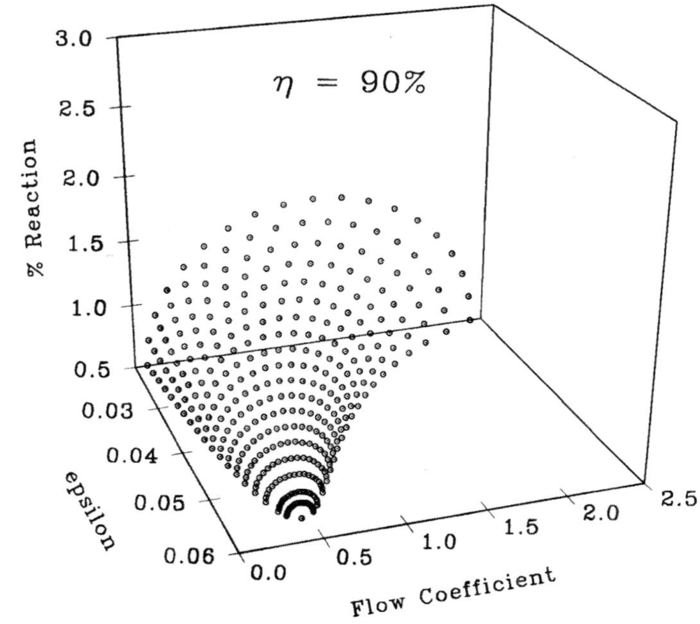

Figure 5.33 Constant single-stage efficiency conoid for $\eta = 90$ percent.

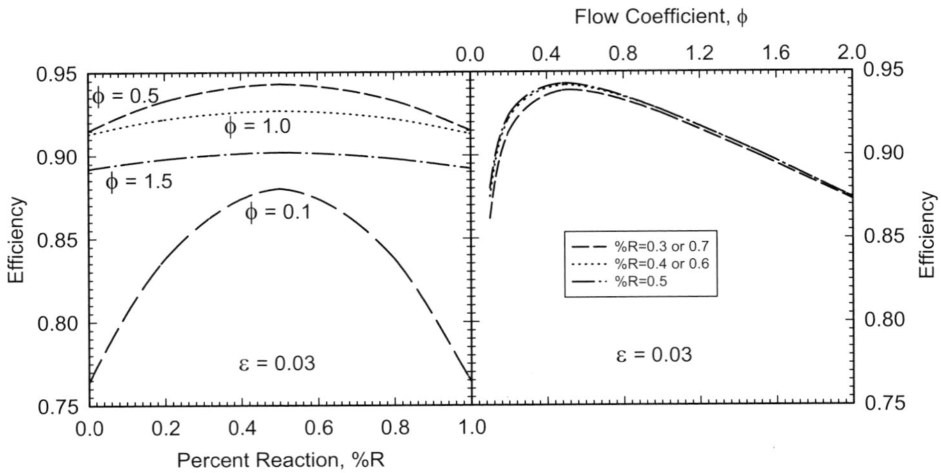

Figure 5.34 Dependence of the single-stage efficiency on flow coefficient and percent reaction for $\varepsilon_r = -\varepsilon_s = 0.03$.

angle α_3. The exit angle will be determined by several variables, including the vane shapes, pitch-to-chord ratio (s/C), incidence angle ($\alpha_2 - \alpha_2'$), and stagger angle (Figure 5.6). For a given set of vanes (fixed shape and fixed s/C), one can experimentally correlate the functional relationship of α_2 and α_3. Similar correlations can be generated for a cascade of rotor blades. Note that the difference in the inlet and exit flow angles is the flow deflection – that is, how much the flow is turned. These data and similar data for other values of solidity can then be correlated with generate defection data. The deflection, δ, or the flow turning angle is then ($\alpha_3 - \alpha_2$) for a stator and ($\beta_2 - \beta_1$) for a rotor. The flow deflection is in general a function of exit angle for a given family of blades and given stagger. Such a correlation can be used, for example, to determine the necessary solidity for a given deflection and exit flow angle. Horlock (1958) and others present a large amount of such detailed empirical data. Use of such detailed data is presented in the following example.

Example 5.7 Single-Stage Efficiency Estimation

A stage with a relative rotor inlet blade angle of $-42°$ and a relative rotor exit blade angle of $-8°$ is to be analyzed. The stator inlet blade angle is 42° and the stator exit blade angle is 8°. Cascade data are presented in Figure 5.35 for this particular stagger. On the basis of the sign definitions adopted in this book, for the rotor blades the inlet and exit flow angles are $-\beta_1$ and $-\beta_2$ and for the stator they are α_2 and α_3, respectively. The maximum camber is at 40 percent of the chord, the pitch-to-chord ratio is 1.0, and the pitch-to-blade height is 0.75. The percent reaction is 50 percent, and the incidence to the rotor is +2.3°. Find the flow coefficient and predict the flow angles and efficiency. As the flow coefficient changes, consider other incidence angles for a range of incidence angles from $-12°$ to $+12°$and for a constant percent reaction predict efficiency as a function of flow coefficient.

Solution

First, one can determine the blade deflection angles or ideal deflections:

$$\delta'_{12} = \beta'_2 - \beta'_1 = 34°$$

$$\delta'_{23} = \alpha'_2 - \alpha'_1 = 34°.$$

Second, the flow deflection is determined. One must consider the incidence and inlet flow angles. For the rotor,

$$\imath_1 = 2.3°$$

Therefore,

$$\beta_1 = \beta'_1 - \imath_1 = -44.3°,$$

and thus from Figure 5.35 the rotor exit flow angle is

$$\beta_2 = -15.8°.$$

Ideally, the exit angle is −8°, the difference being the deviation. Also, for reference,

$$\beta_2 = \beta_1 + \delta_{12},$$

and so the actual deflection is

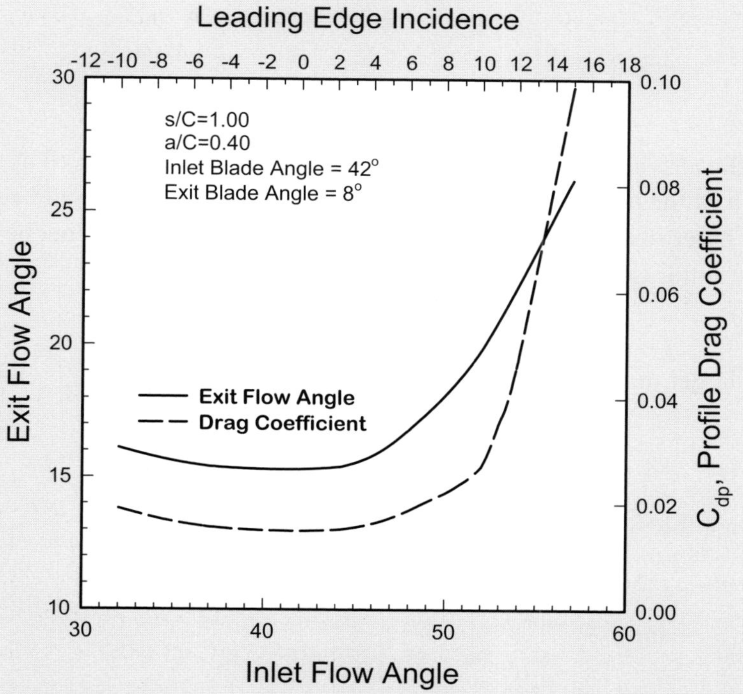

Figure 5.35 Cascade data for Example 5.7.

$\delta_{12} = 28.5°$ (less than ideal).

Next,

$$\tan \beta_m = \frac{1}{2}(\tan \beta_1 + \tan \beta_2),$$

and so $\beta_m = -32.2°$.
Finally,

$$-\tan \beta_m = 0.6294 = \frac{\%R}{\phi};$$

thus,

$$\phi = \frac{\%R}{-\tan \beta_m} = \frac{0.50}{0.6294} = 0.794.$$

For the stator, one can examine the polygons in Figure 5.31 and see that

$$\tan \alpha_2 = \frac{1 + \phi \tan \beta_2}{\phi}.$$

Therefore, solving for α_2 yields

$$\alpha_2 = 44.3°,$$

but

$$\alpha_2 = \alpha_2' - \imath_2;$$

thus,

$$\imath_2 = -2.3°.$$

Note also that the absolute values of β_1 and α_2 are exactly the same, as well as the incidences, which is a result of the percent reactions being 50 percent and the blade and vane angles are symmetric. For other cases of %R, the angles and incidences will, in general, be different. From Figure 5.35, the stator exit flow angle is

$$\alpha_3 = 15.8°,$$

which is again larger than ideal. As reference,

$$\alpha_3 = \alpha_2 + \delta_{23};$$

thus,

$$\delta_{23} = -28.5°\text{(again less than ideal)}.$$

Next,

$$\tan \alpha_m = \frac{1}{2}(\tan \alpha_2 + \tan \alpha_3)$$

and so $\alpha_m = 32.2°$.

The lift and drag coefficients must also be considered. For the profile drag of the rotor blades, using Figure 5.35 yields

$C_{dr_p} = 0.015.$

Next, the lift data are needed. Because it was not given, it can be estimated using Eq. (L.3.26) from Appendix L:

$$\begin{aligned} C_{lr} &= 2\frac{S_r}{C}\cos\beta_m[\tan\beta_2 - \tan\beta_1] + C_{dr_p}\tan\beta_m \\ &= 2 \times 1.0 \times \cos(-32.2°) \times [\tan(-15.8°) - \tan(-44.3°)] + 0.015 \times \tan(-32.2°) \\ &= 1.164. \end{aligned}$$

Secondary flow losses can be estimated using Eq. (L.3.34),

$C_{dr_s} = 0.018 C_{lr}^2 = 0.025,$

and annulus flow losses can be obtained from Eq. (L.3.35),

$$C_{dr_a} = 0.020\frac{S_r}{h} = 0.015;$$

thus, the total drag is, from Eq. (L.3.36),

$C_{dr_t} = C_{dr_p} + C_{dr_s} + C_{dr_a} = 0.055.$

Finally, the angle between drag and lift is defined by

$$\tan\varepsilon_r = \frac{C_{dr_t}}{C_{lr}} = 0.0473$$

and is positive because of the definition of angles and since ε_r is small, $\varepsilon_r = 0.0473$ radians. Similarly, for the profile drag of the stator, using Figure 5.35,

$C_{ds_p} = 0.015.$

Thus, using Eq. (L.3.33),

$$C_{ls} = 2\frac{S_s}{C}\cos\alpha_m[\tan\alpha_2 - \tan\alpha_3] - C_{ds_p}\tan\alpha_m,$$

results in

$C_{ls} = 1.164.$

Secondary flow losses can be estimated using (Eq. L.3.37),

$C_{ds_s} = 0.018 C_{ls}^2 = 0.025,$

and annulus flow losses can be obtained from Eq. (L.3.38),

$$C_{ds_a} = 0.020\frac{S_s}{h} = 0.015;$$

thus, the total drag is from Eq. (L.3.39),

$C_{ds_t} = C_{ds_p} + C_{ds_s} + C_{ds_a} = 0.055.$

Finally, the angle between drag and lift is defined by

$$\tan \varepsilon_s = \frac{C_{ds_t}}{C_{ls}} = -0.0473$$

and is negative because of the definition of angles and since ε_s is small, $\varepsilon_s = -0.0473$ radians.

Finally, the efficiency is predicted by Eq. (L.5.15):

$$\eta = \phi\left\{\left[\frac{\%R - \phi\varepsilon_r}{\phi + \varepsilon_r\%R}\right] + \left[\frac{1 - \%R + \phi\varepsilon_s}{\phi - \varepsilon_s(1 - \%R)}\right]\right\}$$

$$= 0.794\left\{\left[\frac{0.50 - 0.794 \times 0.0473}{0.794 + 0.0473 \times 0.50}\right] + \left[\frac{1 - 0.50 - 0.794 \times 0.0473}{0.794 - (-0.0473) \times (1 - 0.50)}\right]\right\}$$

$$= 0.898.$$

Thus, for the given blades and vanes and the given incidence angle, the flow coefficient is 0.794 and the efficiency is 89.8 percent.

Extended Example 5.7.1 | Single-Stage Efficiency Estimation: Incidence Angle Varied

Next, the process is repeated for the above example for a range of incidence angles from $-12°$ to $+12°$. Because the calculation procedure is exactly the same, details will not be repeated. Only the final results are shown. The maximum efficiency occurs at an incidence angle of 5.5° – not zero. Although the minimum drag is at zero incidence, the lift increases significantly as the incidence increases, thus reducing ε, at least for small positive incidence angles. The efficiency and angle between the lift and drag (ε) are plotted in Figure 5.36. The optimum flow coefficient is 0.72. The minimum value of ε is 0.046, which corresponds to the maximum efficiency. However, from Figure 5.32 one can see that the optimum value of ϕ is below 0.5 for a constant value of ε between 0.04 and 0.05. From Figure 5.36, it is apparent that, as the flow coefficient decreases below the optimum (incidence increases), the value of ε increases sharply, implying a reduction in efficiency. Thus, as observed in Section 5.11.4, the efficiency is really a function of variations of both ϕ and ε. For this particular case, reducing ϕ (independent of ε) from 0.72 would increase the efficiency. However, because this change also increases ε, the efficiency decreases. Also for this example, 2-D cascade data for the given blade shapes for deflection and drag were used and lift was calculated. However, for an accurate prediction of stage performance, including efficiency, CFD would certainly be the best method.

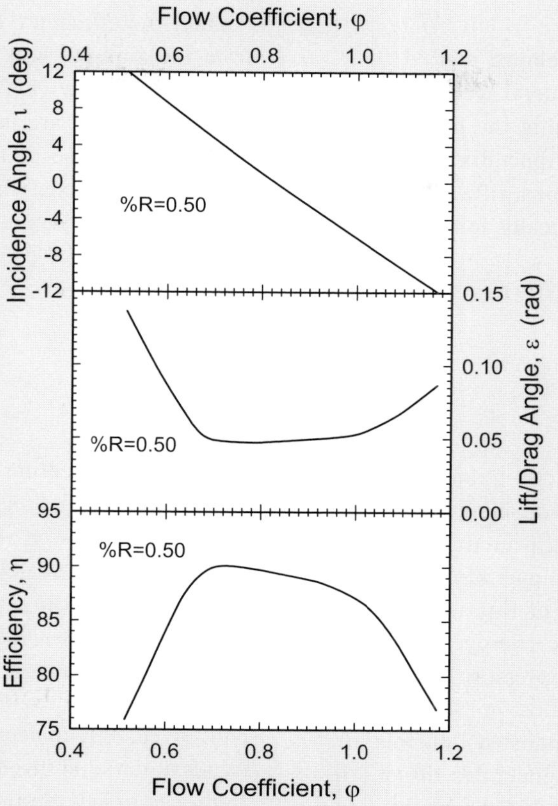

Figure 5.36 Dependence of the lift/drag angle and stage efficiency on incidence angle and flow coefficient for Extended Example 5.7.1.

5.11.5 Further Empiricism

Empiricism has shown that much of the data for different blade or vane shapes and conditions can be correlated with a single figure. Thus, if particular data are not available, general correlated data can be used to make predictions. Details of correlated data are included in Appendix L.8. For these correlations, nominal parameters are defined and used for the correlations. Using these definitions and extended data, which include the stall condition, results in the data empirically (and conveniently) reducing to two figures (Figures L.6 and L.7). Data in these figures, including drag data, are conveniently not functions of the inlet angle or blade shapes.

The generalized correlations presented in Figure L.7 give reasonable to very good results. It goes without saying, however, that if deflection data are available for specific blades and vanes – for example, specific data as in Figure 5.35 and as used in Example 5.7, experimental drag and lift data for a specific blade row, or all of these data – such data should be used in lieu of general correlations.

5.11.6 Implementation of the General Method

The previous five subsections are integrated, in this section, and the implementation of the method to predict the performance of a single stage using these results is discussed. For the purposes of discussion, the blade angles are assumed to be known as a part of the design process or because the stage has already been built. Blade shapes, heights, chord lengths, pitch, distance to maximum camber, and flow coefficient are also assumed to be known.

Howell (1942, 1945a) developed and demonstrated the overall general technique. The method starts with determining the nominal parameters. As outlined in Appendix L.9, incidence angles, flow angles, deflections, lift, and drag coefficients are all systematically found using particular data or the generalized correlated data. Finally, the efficiency of the stage can be found from these angles and data as well as the percent reaction and flow coefficient. Step-by-step details of the procedure are included in Appendix L.9. Solutions can also be obtained using the software SSAxCompPerf.exe, which incorporates all of the concepts in this section.

5.12 Summary

The thermodynamics of strictly axial flow compressors and fans was considered in this chapter. The basic operating principle of a compressor or fan is to impart kinetic energy to the incoming air by means of rotating blades and then convert the increase in energy to an increase in total pressure. For a compressor, this high-pressure air enters the combustor, which requires such conditions. For an exhausted fan, this air enters a fan nozzle, which generates thrust. The basic geometries were discussed, and hardware parts were identified; the important geometric parameters that affect performance were also defined. The air was seen to flow alternately between rows of rotating rotor blades and stationary stator vanes. Velocity triangles or polygons, which are a means of changing from a rotating to a stationary reference frame and vice versa, were used to analyze such turbomachines. Although the geometries (especially blade shape) and operating performance can be quite different for fans and compressors, the fundamental thermodynamics and fluid mechanics are the same.

First, for analysis of the compressors, 2-D flow or mean flow was assumed and the radial components of velocity were assumed to be small. On the basis of these geometric parameters, tools were developed using a control volume approach through which the resulting performance of a stage can be predicted. Continuity, momentum, and energy equations were used to derive equations from which the total pressure ratio, required power, percent reaction, and other parameters could be determined. The software Turbomachinery.exe performs the above calculations and is particularly helpful when optimizing performance because it eliminates repetitive and detailed calculations. Incompressible cases were also considered, only to demonstrate some basic important trends. Rotational speed and blade and vane turning were seen to have strong influences on performance. Although the total pressure ratio of a stage is very important, percent reaction, which will usually be somewhat above 50 percent owing to radial variations and compressibility, was identified as a measure of the relative loading of the rotor and stator blade rows and a measure of resulting stage efficiency. Also, the limiting factors on stage pressure ratio before flow separation from the blades and vanes occurs (which leads to surge) were discussed. These can usually be controlled by prudent designs in which the operating Mach number and the pressure coefficients based on geometry of the blade and vane rows are carefully considered. Also, performance maps were described by which the important operating characteristics of a compressor can accurately and concisely be presented. Although the maps are dimensional they are based on a dimensional analysis of the turbomachine characteristics. Methods of experimentally deriving stage, compressor, and fan maps were discussed. Four of the most important output characteristics on a map are the pressure ratio, efficiency, surge line, and operating line. From a map, for example, if the corrected speed and flow rates are known, the efficiency and pressure ratio can be found. Methods of improving off-design performance were covered such as variable stators and multispool machines. These more complex geometries basically allow the blade rows to operate with low incidence angles over a wide range of conditions, which directly improves the efficiency over the range. Methods of controlling surge were also identified.

A more complex (advanced) streamline analysis method was also used to predict and account for the 3-D flows resulting from radial equilibrium. Such an analysis becomes more important as longer blades and vanes are considered. Significant hub-to-tip blade twist and turning variations were seen to be required for longer blades. Finally, because a cascade is made up of a series of airfoils, a method of using lift and drag coefficients of blade cascades and other empirical data was demonstrated for incompressible flow to predict and optimize stage efficiency, including such effects as blade incidence angles. A percent reaction of approximately 50 percent was shown to optimize the efficiency of a stage, although the efficiency does not rapidly deteriorate as the reaction deviates from this value. For incompressible flow, the method gives great insight into how the stage performance and fundamental airfoil (vane or blade) characteristics are related and can be used to maximize the efficiency, for example, with the proper choice of blade shapes and solidity or as the leading-edge incidence increases. Throughout the chapter, design guidelines were presented to make it possible to predict or determine if a stage is operating reasonably.

List of Symbols

A	Area
a	Speed of sound
a	Leading edge to maximum camber length
a	"Constant reaction" coefficient
b	"Constant reaction" coefficient
c	Absolute velocity
c_p	Specific heat at constant pressure
C	Chord length
C_d	Drag coefficient
C_l	Lift coefficient
C_p	Pressure coefficient
D	Diameter
F	Force
g	Gravitational constant
h	Specific enthalpy
i	Radial counter (streamline analysis)
j	Axial counter (streamline analysis)
$\dot{m}$	Mass flow rate
M	Mach number
N	Rotational speed
p	Pressure
Q	Volumetric flow rate
$\mathcal{R}$	Ideal gas constant
R	Radius
%R	Percent reaction
s	Blade spacing or pitch
T	Temperature
t	Blade height
u	Specific internal energy
U	Blade velocity
w	Relative velocity
$\dot{W}$	Power
α	Absolute (stator) angle
β	Relative (rotor) angle
γ	Specific heat ratio
δ	Flow turning angle, deflection
δ	Ratio of pressure to standard pressure

ε Angle between lift and drag forces
κ Free vortex constant
η Efficiency
θ Ratio of temperature to standard temperature
ι Incidence angle
ξ Deviation
ρ Density
τ Deflection coefficient
ϕ Flow coefficient
ω Rotational speed

Subscripts

a Axial
abs In absolute frame (stator)
c Corrected
d Drag
H Hub
i Radial counter (streamline analysis)
i Inlet to a diffuser
j Axial counter (streamline analysis)
l Lift
m Mean or average
max Maximum
opt Optimum
r Rotor
rel In rotating frame (rotor)
s Stator
sh Applied to shaft
stall Separation/stall
stp Standard conditions
t Total (stagnation)
T Tip
u Tangential
0 Upstream of IGV
1, 2, 3 Positions in stage
2 Inlet to compressor
3 Exit of compressor

Superscripts

$\bar{}$ Average
$'$ For the blade
$'$ Ideal
$\wedge$ Normalized radius

Problems

5.1 A single stage of a compressor is to be analyzed. It rotates at 12,400 rpm and compresses 54.43 kg/s (120 lb_m/s) of air. The inlet pressure and temperature are 0.9653 MPa and 500 K (140 psia and 900 °R), respectively. The average radius of the blades is 203.2 mm (8.0 in.), and the inlet blade height is 29.21 mm (1.15 in). The absolute inlet angle to the rotor is the same as the stator exit angle (18°), and the rotor flow turning angle (δ_{12}) is 17°. The compressor stage has been designed so that the blade height varies and the axial velocity remains constant through the stage. The efficiency of the stage is 88 percent. The values of c_p and γ are 1.039 kJ/kg-K (0.2483 BTU/lb_m-°R) and 1.381, respectively.

Find the following:

(a) the blade heights at the rotor and stator exits,
(b) the static pressure at the rotor and stator exits,
(c) the total pressure ratio for the stage,
(d) the stator turning angle,
(e) the required power for the stage,
(f) the percent reaction for the stage.

5.2 A single stage of a compressor is to be analyzed. It rotates at 12,400 rpm and compresses 54.43 kg/s (120 lb_m/s) of air. The inlet pressure and temperature are 0.9653 MPa and 500 K (140 psia and 900 °R), respectively. The average radius of the blades is 203.2 mm (8.0 in.), and the inlet blade height is 29.21 mm (1.15 in.). The absolute inlet angle to the rotor is the same as the stator exit angle (18°), and the percent reaction is 0.500. The compressor stage has been designed so that the blade height varies and the axial velocity remains constant through the stage. The efficiency of the stage is 88 percent. The values of c_p and γ are 1.040 kJ/kg-K (0.2484 BTU/lb$_m$-°R) and 1.381, respectively.

Find the following:

(a) the blade heights at the rotor and stator exits,
(b) the static pressure at the rotor and stator exits,
(c) the total pressure ratio for the stage,
(d) the stator turning angle,
(e) the rotor turning angle,
(f) the required power for the stage.

5.3 A single stage of a compressor is to be analyzed. It rotates at 12,400 rpm and compresses 54.43 kg/s (120 lb$_m$/s) of air. The inlet pressure and temperature are 0.9653 MPa and 500 K (140 psia and 900 °R), respectively. The average radius of the blades is 203.2 mm (8.0 in.), and the inlet blade height is 29.21 mm (1.15 in). The absolute inlet angle to the rotor is the same as the stator exit angle (18°), and the total pressure ratio is 1.200. The compressor stage has been designed so that the blade height varies and the axial velocity remains constant through the stage. The efficiency of the stage is 88 percent. The values of c_p and γ are 1.040 kJ/kg-K (0.2484 BTU/lb$_m$-°R) and 1.381, respectively.

Find the following:

(a) the blade heights at the rotor and stator exits,
(b) the static pressure at the rotor and stator exits,
(c) the stator turning angle,
(d) the rotor turning angle,
(e) the required power for the stage.

5.4 A stage approximating the size the first stage (rotor and stator) of an older low-pressure compressor operating at Standard temperature and pressure is to be analyzed. It rotates at 7500 rpm and compresses 153 lb$_m$/s of air. The inlet pressure and temperature are 31.4 psia and 655 °R, respectively. The average radius of the blades is 14.4 in., and the inlet area is 507 in.^{2}. The absolute inlet angle to the rotor is the same as the stator exit angle (37°), and the rotor flow turning angle (δ_{12}) is 13°. The stage has been designed so that the blade height varies and the axial velocity remains constant through the stage. The efficiency of the stage is 88 percent. The values of c_p and γ are 0.2425 BTU/lb$_m$-°R and 1.394, respectively.

(a) If the flow is compressible, find the following: (i) blade heights at the rotor and stator exits, (ii) the static pressure at the rotor and stator exits, (iii) the total pressure ratio for the stage, (iv) the stator turning angle, (v) the required power for the stage, and (vi) the percent reaction for the stage.
(b) If the flow can be assumed to be incompressible and the density is 0.004 sl/ft^3, find the following: (i) blade heights at the rotor and stator exits, (ii) the static pressure at the rotor and stator exits, (iii) the total pressure ratio for the stage, (iv) the stator turning angle, (v) the required power for the stage, and (vi) the percent reaction for the stage.
(c) (i) If the flow rate is dropped to 123 lb$_m$/s, what angle would the LPIGVs have to be turned so that the flow matches the rotor blade inlet angle (compressible flow)? (ii) To what angle would the stators have to be turned so that the flow out of the rotor would match the inlet angle of the stator blades?

5.5 A stage approximating the first stage (rotor and stator) of a high-pressure compressor operating at standard temperature and pressure is to be analyzed. It rotates at 10,500 rpm and compresses 54.43 kg/s (120 lb$_m$/s) of air. The inlet pressure and temperature are 0.427 MPa and 413.9 K (62.0 psia and 745 °R), respectively. The average radius of the blades is 269.2 mm (10.6 in.), and the inlet blade height is 62.74 mm (2.47 in). The absolute inlet angle to the rotor is the same as the stator exit angle (29°), and the rotor flow turning angle (δ_{12}) is 21°. The stage has been designed so that the blade height varies and the axial velocity remains constant through the stage. The efficiency of the stage is 86 percent. The values of c_p and γ are 1.025 kJ/kg-K (0.2448 BTU/lb_m-°R) and 1.389, respectively.

(a) If the flow is compressible, find the following: (i) blade heights at the rotor and stator exits, (ii) the static pressure at the rotor and stator exits, (iii) the total pressure ratio for the stage, (iv) the stator turning angle, (v) the required power for the stage, (vi) the percent reaction for the stage, and (vii) the pressure coefficients of the rotor and stator.
(b) If the flow can be assumed to be incompressible and the density is 3.600 kg/m^3 (0.00698 slg/ft^3), find the following: (i) blade heights at the rotor and stator exits, (ii) the static pressure at the rotor and stator exits, (iii) the total pressure ratio for the stage, (iv) the stator turning angle, (v) the required power for the stage, (vi) the percent reaction for the stage, and (vii) the rotor and stator pressure coefficients.
(c) (i) If the flow rate is dropped to 35.38 kg/s (78 lb_m/s) (65 percent flow rate), to what angle would the HPIGVs have to be turned so that the flow would match the rotor blade inlet angle (compressible flow)? (ii) To what angle would the stators have to be turned so that the flow out of the rotor would match the inlet angle of the stator blades?

5.6 A stage approximating the size one of the last stages (rotor and stator) of an older high-pressure compressor is to be analyzed. It rotates at 8000 rpm and compresses 280 lb_m/s of air. At midspan the inlet pressure and density are 272 psia and 0.01742 sl/ft^3, respectively. The average radius of the blades is 13.2 in., and the blade height is 1.24 in. At midspan the absolute inlet flow angle to the rotor is the same as the stator exit flow angle (15°), and the absolute exit rotor flow angle is 42.79°. The following details are to be found as functions of radius for incompressible flow with radial equilibrium: the static and total pressures and axial velocities at the rotor and stator exits, the rotor and stator turning angles, the required power for the stage, and the percent reaction for the stage. Solve for the following conditions:

(a) if free vortex flow is assumed (use one annulus),
(b) if free vortex flow is assumed (use three annuli),
(c) if the absolute flow angles and efficiencies (90 percent) are assumed to be constant with radius (use three annuli).

5.7 If the drag-to-lift ratio for a rotor is 0.042 and the drag-to-lift ratio for a stator is 0.038, what is the maximum efficiency that can be derived from the stage? At what flow coefficient does this maximum occur?

5.8 The efficiency of a stage with a design flow coefficient of 0.48 is to be predicted. The stator turning angle is 22.00°, and the rotor turning angle is 19.00°. Both the stator and rotor are given profiles, and lift and drag coefficient data for single airfoils are shown in Figure 5.P.1. The Reynolds number is very large. The spacing of the blades is large enough so that one can treat the cascades as single blades. The stage is designed so that at design the incidence angles to the rotor and stator are 4.00° and 6.00°, respectively. One can assume that this is a middle stage and that similar stages are upstream and downstream. To simplify the calculations, one can also assume that the turning angles do not change with the incidence angles. The efficiency is to be estimated at flow rates of (a) 100, (b) 130, and (c) 70 percent.

5.9 A stage approximating a middle stage (rotor and stator) of a high-pressure compressor is to be designed. The rotor rotates at 12,000 rpm, and the compressor compresses 45.36 kg/s (100 lb_m/s) of air. The inlet pressure and temperature to the stage are 0.8374 MPa and 555.6 K (120 psia and 1000 °R), respectively. The average radius of the blades is 304.8 mm (12.0 in). The efficiency of the stage is 86 percent. Clearly state the design objectives and proceed to design the stage. When finished, check to see if the design objectives were met and draw the blade geometries.

5.10 An axial flow compressor is to be analyzed. One should consider one set of rotor blades that

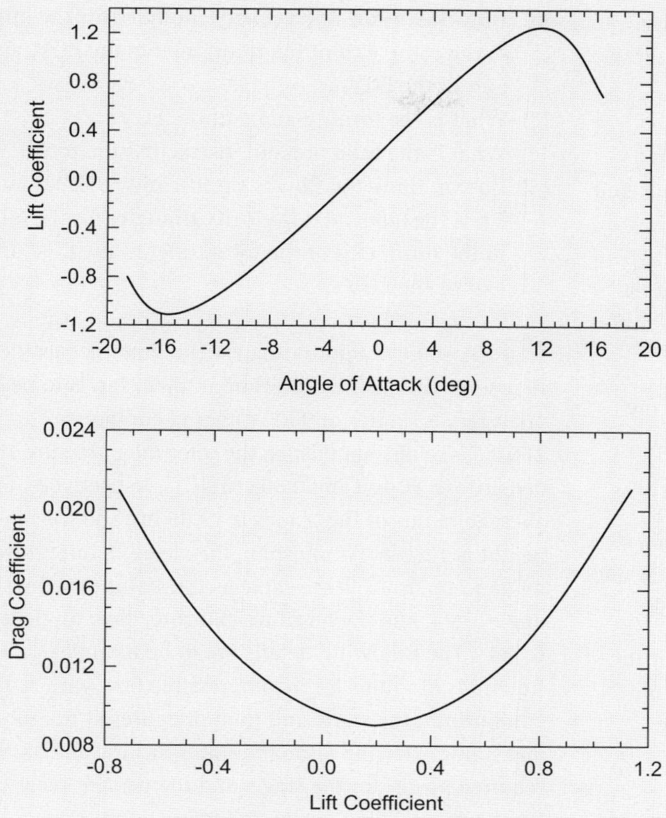

Figure 5.P.1 Lift and drag coefficients for blades approximating those of NACA4212 blades for Problem 5.8.

rotate at 9000 rpm. The average diameter is 20 in. and the axial velocity is 750 ft/s throughout the compressor. The previous exit stator angle is 13°. The rotor blade turning angle is 26°. Find the percent reaction.

5.11 In an axial compressor, if the axial component of velocity remains constant through a stage, the Mach number decreases across that stage.

(a) Why?
(b) If this is allowed to happen through many stages, why will this result in a poor design?

5.12 One of the first stages (rotor and stator) of a high-pressure axial compressor is to be analyzed. It rotates at 9600 rpm and compresses 70.31 kg/s (155 lb_m/s) of air. The inlet pressure and temperature are 0.3447 MPa and 411.1 K (50 psia and 740 °R), respectively. The average radius of the blades is 281.9 mm (11.1 in.) and the inlet blade height is 54.61 mm (2.15 in.). The absolute inlet flow angle to the rotor is the same as the stator exit flow angle (18°), and the rotor flow turning angle is 22° The stage has been designed so that the blade height varies and the axial velocity increases by 2 percent across each blade row of the stage; thus, the Mach number remains about constant. The efficiency of the stage is 87 percent. The values of c_p and γ are 1.024 kJ/kg-K (0.2446 BTU/lb_m) and 1.389, respectively. Find

(a) the total pressure at the stator exit,
(b) the required power for the stage,
(c) the percent reaction for the stage.

5.13 You are to analyze the velocity polygons of a single stage of an axial compressor. The mass flow rate is 72.57 kg/s (160 lb_m/s). The inlet pressure is 1.034 MPa (150 psia) and the inlet temperature is 555.6 K (1000 °R). The inlet density is 6.487 kg/m^3 (0.405 lb_m/ft^3). The hub radius is 127.0 mm (5.0 in.), and the tip radius is 190.5 mm (7.5 in). The absolute flow angles into the rotor and out of the stator are 20°. The turning angle of the rotor is 22°. The efficiency of the stage is 90 percent. The shaft speed is 14,000 rpm, and the axial velocity remains constant in the stage.

(a) Draw the velocity polygons into and out of the rotor. You may assume 2-D flow. Indicate all important velocity and angle magnitudes.
(b) Evaluate the percent reaction.
(c) For a 3-D analysis, qualitatively sketch the rotor exit axial velocity profile if the flow is (i) free vortex, and (ii) if the absolute flow angle is constant with *r*.

5.14 You are to analyze a proposed design for the first stage of a fan similar in size to that of an older engine at sea level. The total air mass flow rate is 260 lb_m/s, and it runs at 9300 rpm. The inlet pressure to the fan rotor is 21 psia, and the temperature is 540 °R. The inlet hub and tip radii are 6.8 in. and 18.9 in., respectively. An inlet guide vane is used so that the inlet flow has an angle of 35° relative to the axial component (α_1). The turning angle of the rotor blades is 20°. Finally, the exit angle of the first row of stator blades is also 35° (α_3). You may assume constant axial velocities along the stage and 100 percent efficiency. Find (a) α_2, (b)%R, and (c) p_{t3}/p_{t1}. (d) Comment on the likelihood of success.

5.15 The total mass flow rate through the cowl of an engine similar in size to a commercial engine at sea level and standard temperature and pressure is 1560 lb_m/s. The single-stage fan does not have inlet guide vanes but does have exit guide vanes (EGVs). EGVs are essentially stators. The absolute exit angle for the EGVs is 0°. The outer radius of the blades is 48 in., and the inner radius is 20 in. The rotational speed is 3650 rpm. The turning angle of the rotor is 11°, and the stage efficiency is 89 percent. You may assume the axial velocity remains constant.

(a) Draw the mean line velocity triangles for the inlet to the rotor, exit of the rotor, inlet to the EGV, and exit of the EGV.
(b) What is the turning angle of the EGV?
(c) What is the total pressure rise across the rotor?
(d) Do you think the flow is significantly 3-D? Why?
(e) Draw the hub and tip velocity triangles for the inlet to the rotor, exit of the rotor, inlet to the EGV and exit of the EGV.

5.16 A stage of approximately the same size as the first stage (IGV/rotor/stator) of an older fan is to be analyzed. It rotates at 9300 rpm and compresses 260 lb_m/s of air. At midspan the rotor inlet pressure and density are 21 psia and 0.00326 sl/ft^3, respectively. The average radius of the blades is 12.85 in., and the blade height is 12.1 in. At midspan, the absolute inlet flow angle to the rotor is the same as the stator exit flow angle (35°), and the absolute exit rotor flow angle is 61.64°. The following details are to be found as functions of radius for incompressible flow with radial equilibrium: the static and total pressures at the rotor and stator exits, the rotor and stator turning angles, the required power for the stage, and the percent reaction for the stage. Solve for the following conditions:

(a) if free vortex flow is assumed (use one annulus),
(b) if free vortex flow is assumed (use three annuli),
(c) if the absolute flow angles and efficiencies (90 percent) are assumed to be constant with radius (use three annuli).

5.17 A stage of approximately the same size as a middle stage (rotor and stator) of a modern high-pressure compressor is to be designed. The rotor is to rotate at 14,500 rpm, and the compressor is to compress 135 lb_m/s of core air. The inlet pressure and temperature to the stage are estimated to be 190 psia and 1150 °R, respectively. The average radius of the blades and vanes of this stage is to be 8.50 in. The efficiency of the stage is approximately 87 percent. The compressor is to have 13 stages, and the pressure ratio for the entire compressor is to be 24.0. Clearly state the design objectives and proceed to design the stage. When finished, check to see if the design objectives were met and draw the blade geometries.

5.18 An axial flow compressor stage rotates at 12,000 rpm and has an average radius of 9 in., with an inlet blade height of 1.1 in. The absolute inlet angle to the rotor blades is the same as the absolute exit angle from the stator blades (40°) relative to the axial direction. The rotor turning angle is 33°, and the stage efficiency is 89 percent. The axial velocity component remains constant at 724 ft/s. The inlet temperature and pressure are 950 °R and 135 psia, and the total pressure ratio of the stage is 1.227. The specific heat at constant pressure is 0.249 BTU/lb_m-°R, and the specific heat ratio is 1.380.

(a) Accurately draw the velocity polygons.
(b) Find the percent reaction.
(c) Find the stator turning angle.
(d) Is this a good design? Why or why not? Be specific.
(e) If the design is not good, what would you do to improve it? Be specific.

5.19 A single stage of an axial flow compressor stage rotates at 9500 rpm and has an average radius of 208.3 mm (8.20 in.) with an inlet blade height of 30.48 mm (1.20 in.). The absolute inlet angle to the rotor blades is the same as the absolute exit angle from the stator blades (0°), relative to the axial direction. The rotor turning angle is 20°, and the stage efficiency is 87 percent. The mass flow rate is 45.36 kg/s (100 lb_m/s), and the axial velocity component remains constant at 263.0 m/s (863 ft/s). The inlet temperature and pressure are 500 K and 0.6205 MPa (900 °R and 90 psia), respectively, and the total pressure ratio of the stage is 1.149. The specific heat at constant pressure is 1.038 kJ/kg-K (0.248 BTU/lb_m-°R), and the specific heat ratio is 1.381.

(a) Accurately draw the velocity polygons.
(b) Find the percent reaction.
(c) Find the stator turning angle.
(d) Is this a good design? Why or why not? Be specific.
(e) If the design is not good, what would you do to improve it? Be specific.

5.20 Initially an axial flow compressor stage operates with a 50 percent reaction. However, in a later test the rotational speed is increased. If the absolute exit angles from the previous stator and stator of the stage under analysis remain constant, the relative rotor turning angle remains the same, and the mass flow rate remains constant, what happens to the reaction? Explain with polygons.

5.21 A stage of approximately the same size as the last stage (rotor and stator) of a modern military high-pressure compressor is to be designed. The rotor is to rotate at 14,500 rpm, and thrust requirements dictate that the high-pressure compressor is to compress 146 lb_m/s of core air. The inlet pressure and temperature to the stage are estimated to be 480 psia and 1410 °R, respectively. The average radius of the blades and vanes of this stage is to be 8.0 in. The efficiency of the stage is approximately 89 percent. The compressor is to have 13 stages, and the pressure ratio for the entire compressor is to be 32.0. Clearly state the design objectives and proceed to design the stage. When finished, check to see if the design objectives were met and draw the blade geometries.

5.22 A single stage of an axial flow compressor stage rotates at 11,500 rpm and has an average radius of 186.7 mm (7.35 in.) with an inlet blade height of 44.45 mm (1.75 in.). The absolute inlet angle to the rotor blades is the same as the absolute exit angle from the stator blades and is 20° relative to the axial direction. The rotor turning angle is 28°, and the stage efficiency is 88 percent. The mass flow rate is 36.29 kg/s (80 lb_m/s); the axial velocity component enters the stage at 275.5 m/s (904 ft/s) and increases by 3 percent across the rotor and another 3 percent across the stator. The inlet temperature and pressure are 427.8 K and 0.3103 MPa (770 °R and 45 psia), respectively. The specific heat at constant pressure is 1.026 kJ/kg-K (0.245 BTU/lb_m-°R), and the specific heat ratio is 1.389. The stage consumes 1.165 MW (1562 hp) of power.

(a) Accurately draw the velocity polygons.
(b) Find the percent reaction.
(c) Find the total pressure ratio.
(d) Find the stator turning angle.

(e) Is this a good design? Why or why not? Be specific.
(f) If not good, what would you do to improve it? Be specific.

5.23 A single stage of an axial flow compressor stage rotates at 11,500 rpm and has an average radius of 186.7 mm (7.35 in.) with an inlet blade height of 44.45 mm (1.75 in.). The absolute inlet angle to the rotor blades is the same as the absolute exit angle from the stator blades and is 20° relative to the axial direction. The rotor turning angle is 24°, and the stage efficiency is 88 percent. The mass flow rate is 36.29 k-g/s (80 lb_m/s); the axial velocity component enters the stage at 275.5 m/s (904 ft/s) and increases by 3 percent across the rotor and another 3 percent across the stator. The inlet temperature and pressure are 427.8 K and 0.3103 MPa (770 °R and 45 psia), respectively. The specific heat at constant pressure is 1.026 kJ/kg-K (0.245 BTU/lb_m-°R), and the specific heat ratio is 1.389. The stage consumes 1.003 MW (1345 hp) of power.

(a) Accurately draw the velocity polygons.
(b) Find the percent reaction.
(c) Find the total pressure ratio.
(d) Find the stator turning angle.
(e) Is this a good design? Why or why not? Be specific.
(f) If the design is not good, what would you do to improve it? Be specific.

5.24 You are to analyze the first stage of the fan on an engine similar in size to an older military engine at sea level. The total air mass flow rate is 260 lb_m/s, and it runs at 9300 rpm. The inlet pressure to the fan rotor is 21 psia, and the temperature is 540 °R. The inlet hub and tip radii are 6.8 in. and 18.9 in., respectively. An inlet guide vane is used so that the inlet flow has an angle of 45° relative to the axial component (α_1). The turning angle of the rotor blades is 12°. Finally, the exit angle of the first row of stator blades is also 45° (α_3). You may assume constant axial velocities along the stage and 90 percent efficiency. Find (a) α_2, (b) %R, and (c) p_{t3}/p_{t1}.

5.25 A stage approximating the first stage (rotor and stator) of a high-pressure compressor is to be analyzed. It rotates at 10,500 rpm and compresses 120 lb_m/s of air. At midspan the inlet pressure and density are 62.0 psia and 0.00698 sl/ft^3, respectively. The diameters of the hub and tip are 18.756 and 23.676 in., respectively. At midspan, the absolute inlet flow angle to the rotor is the same as the stator exit flow angle (29°), and the absolute exit rotor flow angle is 53.60°. The total pressure is uniform at the inlet. The following details are to be found and plotted as functions of radius for incompressible flow with radial equilibrium at all three axial stations: the static and total pressures and axial velocities at the rotor and stator exits, the rotor and stator turning angles, the required power for the stage, and the percent reaction for the stage. Solve for the following conditions:

(a) if free vortex flow is assumed (use one annulus),
(b) if free vortex flow is assumed (use three annuli),
(c) if the absolute flow angles and efficiencies (86 percent) are assumed to be constant with radius (use three annuli).

5.26 A stage approximating the first stage (rotor and stator) of a high-pressure compressor is to be analyzed. It rotates at 10,500 rpm and compresses 120 lb_m/s of air. The density is 0.00698 sl/ft^3, and the fluid can be considered to be incompressible. The diameters of the hub and tip are 18.756 and 23.676 in., respectively. The efficiency is constant with diameter and is equal to 86 percent. Pressure and axial velocities are presented in Table 5.P.1 for the stage. Find the absolute flow angle distribution (including absolute and relative angles) at the rotor inlet, rotor exit, and stator exit if radial equilibrium is assumed at all three axial stations. Find the percent reaction for each annulus. Plot the angles and percent reaction versus radius.

5.27 An entire compressor was tested, and internal total pressures and temperatures were monitored. Flow rates were controlled. Experimental data for a single stage of the

Table 5.P.1 Pressure and axial velocities for Problem 5.26.

Rotor inlet			
Dia. (ft)	c_a (ft/s)	p (psia)	p_t (psia)
1.563	496.86	59.93	67.33
1.762	471.73	60.30	67.33
1.973	443.96	60.69	67.33
Rotor exit			
Dia. (ft)	c_a (ft/s)	p (psia)	p_t (psia)
1.563	531.31	64.84	81.82
1.756	474.25	67.18	82.53
1.973	419.67	69.48	83.57
Stator exit			
Dia. (ft)	c_a (ft/s)	p (psia)	p_t (psia)
1.563	468.01	75.25	81.82
1.769	467.41	75.61	82.53
1.973	473.45	76.02	83.57

compressor are presented in Table 5.P.2. Construct the corrected map for these data. Include both the total pressure ratio and efficiency islands.

5.28 A single stage of an axial flow compressor stage rotates at 9000 rpm and has an average radius of 11.0 in. (so that the blade speed is 864 ft/s) with an inlet blade height of 0.80 in. Flow enters the rotor blades axially in the fixed reference frame and also exits the stator blades axially. The rotor flow turning angle is 20°, and the stage efficiency is 90 percent. The mass flow rate is 175 lb_m/s; the axial velocity component remains constant through the stage (938 ft/s). The inlet static temperature and pressure are 1000 °R and 180 psia, respectively. The stage consumes 4038 hp of power. The specific heat at constant pressure is 0.251 BTU/lb_m-°R, and the specific heat ratio is 1.375. You are asked to evaluate the design as follows.

(a) Accurately draw the velocity polygons.
(b) Find the percent reaction.
(c) Find the total pressure ratio.
(d) Find the stator turning angle.
(e) Is this a good design? Why or why not? Be specific.
(f) If not good, what would you do to improve it? Be specific.

Table 5.P.2 Experimental data for a single stage of the compressor for Problem 5.27.

$\dot{m}$ (lbm/s)	p_{t1} (psia)	T_{t1} (°R)	p_{t3} (psia)	T_{t3} (°R)	N (rpm)
126.5	20.0	600	20.0	600	4301
127.9	22.9	624	24.5	640	4385
119.6	24.3	634	26.8	657	4422
107.6	25.1	640	28.1	670	4442
97.7 97.6[a]	25.4	642	28.6	676	4449
273.5	30.0	650	30.0	650	6715
293.6	35.3	681	38.3	706	6873

Table 5.P.2 **(cont.)**

286.4	41.1	711	48.0	748	7023
269.1	42.3	717	50.2	763	7052
249.3 249.2[a]	43.2	721	51.8	778	7073
465.3	40.0	710	40.0	710	9357
526.0	48.4	750	53.2	784	9615
564.3	56.6	784	67.4	840	9834
554.0	63.6	811	80.2	884	9998
526.1 526.0[a]	67.2	823	87.1	919	10,077

[a] Surge.

5.29 Initially an axial flow compressor stage operates with a 50 percent reaction. However, in a later test the mass flow is decreased. If the absolute inlet angle to the rotor and the absolute exit angle from the stator remain constant, and the relative rotor turning angle remains the same, what happens to the reaction if the rotational speed is the same? Explain with polygons.

5.30 A stage approximating the first stage (rotor and stator) of a high-pressure compressor is to be analyzed. It rotates at 10,500 rpm and compresses 54.45 kg/s (120 lb_m/s) of air. The inlet pressure and temperature are 0.4275 MPa and 413.9 K (62.0 psia and 745 °R), respectively. The average radius of the blades is 269.2 mm (10.6 in.), and the inlet blade height is 62.74 mm (2.47 in.). The absolute inlet angle to the rotor is the same as the stator exit angle (29°), and the rotor flow turning angle (δ_{12}) is 21°. The stage has been designed so that the blade height varies and the axial velocity remains constant through the stage. The efficiency of the stage is 86 percent.

(a) Find the following: (i) flow angles, (ii) blade heights at the rotor and stator exits, (iii) the static pressure at the rotor and stator exits, (iv) the total pressure ratio for the stage, (v) the stator turning angle, (vi) the required power for the stage, (vii) the percent reaction for the stage, and (viii) the pressure coefficients of the rotor and stator. Consider this to be the on-design condition.

(b) If the rotational speed is reduced to 8500 rpm for an off-design condition, what mass flow rate will result in an incidence-free flow into the rotor? Use the same rotor inlet angle and blade heights you found in part (a). Again find the total pressure ratio for the stage if the efficiency remains constant.

(c) If the rotational speed is increased to 12,500 rpm for an off-design condition, what mass flow rate will result in an incidence-free flow into the rotor? Again, use the same rotor inlet angle and blade heights you found in part (a). Again find the total pressure ratio for the stage if the efficiency remains constant. Plot the total pressure ratio versus mass flow rate for parts (a), (b), and (c). What does this line represent?

(d) If the mass flow rate is reduced to 45.36 kg/s (100 lb_m/s), the speed remains 10,500 rpm for another off-design condition, the absolute inlet and exit angles (α_1 and α_3) remain the same, the relative rotor exit angle (β_2) remains the same, blade heights remain the same, and the efficiency drops to 70 percent owing to increased incidence, again find the total pressure ratio for the stage.

(e) If the mass flow rate is increased to 63.50 kg/s (140 lb_m/s), the speed remains 10,500 rpm for another off-design condition, the absolute inlet and exit angles (α_1 and α_3) remain the same, the relative rotor exit angle (β_2) remains the same, blade heights remain the same, and the efficiency drops to 70 percent, again find the total pressure ratio for the stage. Plot the total pressure ratio vs. mass flow rate for parts (a), (d), and (e). What does this line represent?

5.31 You are given a 50 percent reaction stage with a relative rotor inlet blade angle of −45° and a relative rotor exit blade angle of −5°. The stator inlet blade angle is 45°, and the stator exit blade angle is 5°. The maximum camber is at 33 percent of the chord, the pitch-to-chord ratio is 1.2, and the pitch-to-blade height is 0.8. The flow coefficient ranges from 0.45 to 1.25. Find the flow angles and predict the efficiency and incidence angles as functions of the flow coefficient.

5.32 You are given a stage with a relative rotor inlet blade angle of −45° and a relative rotor exit blade angle of −5°. The stator inlet blade angle is 45°, and the stator exit blade angle is 5°. The maximum camber is at 33 percent of the chord, the pitch-to-chord ratio is 1.2, and the pitch-to-blade height is 0.8. The flow coefficient ranges from 0.45 to 1.25, but the incidence to the rotor remains constant at +1.5°. Find the flow angles and predict the efficiency and percent reaction as functions of the flow coefficient.

5.33 Initially, an axial flow compressor operates with a 50 percent reaction. However, in a later test the rotor blades are replaced and thus the relative flow turning in the rotor is less. If the absolute inlet angle to the rotor, the absolute exit angle from the stator, the mass flow rate, and the rotational speed remain the same, what happens to the total pressure ratio and reaction (increase, decrease, or remain the same)? Explain with polygons.

5.34 Initially a stage of an axial flow compressor operates with a 50 percent reaction. However, in a later test the rotational speed is decreased. If the absolute inlet angle to the rotor remains constant, the relative rotor turning angle remains the same, and the axial velocity component at the inlet to the stage remains the same (explain with velocity polygons and a diagram of the rotor blade),

(a) What happens to the reaction?
(b) What happens to the incidence to the rotor blades?

5.35 A single stage of an axial flow compressor stage rotates at 8000 rpm and has an average radius of 10.5 in. (and thus the blade speed is 733 ft/s) with an inlet blade height of 1.30 in. Flow enters the rotor blades axially in the fixed reference frame and also exits the stator vanes axially. The rotor flow turning angle is 38°, and the stage efficiency is 91 percent. The mass flow rate is 220 lb_m/s; the axial velocity component remains constant through the stage (1000 ft/s). The inlet static temperature and pressure are 1520 °R and 208 psia, respectively, and the speed of sound at the inlet is 1876 ft/s. The stage consumes 6960 hp of power. The specific heat at constant pressure is 0.264 BTU/lb_m-°R, and the specific heat ratio is 1.350. You are asked to evaluate the design.

(a) Accurately draw or plot the velocity polygons.
(b) Find the percent reaction.
(c) Find the total pressure ratio.
(d) On the basis of the rotor blade performance (do not calculate stator vane parameters), is this a good design? Why or why not? Be specific.

5.36 Initially an axial flow compressor stage operates with a 50 percent reaction. However, in a later test the rotor blades are replaced so that the relative flow turning in the rotor increases. If the absolute inlet angle to the rotor, the absolute exit angle from the stator, the mass flow rate, and the rotational speed remain the same, what happens to the total pressure ratio? Explain with polygons.

5.37 A stage approximating a middle stage (rotor and stator) of a high-performance, high-pressure compressor is to be designed. The rotor is to rotate at 14,500 rpm, and the compressor is to compress 66.21 kg/s (146 lb_m/s) of core air. The inlet pressure and temperature to the stage are estimated to be 3.310 MPa and 783.3 K (480 psia and 1410 °R), respectively. The radius of the blades and vanes at the case (i.e., rotor blade tip) of this stage is to be 209.6 mm (8.25 in.). The efficiency of the stage is approximately 89 percent. The compressor is to have 12 stages, and the pressure ratio for the entire compressor is to be 37.0. Clearly state the design objectives and proceed to design the stage. When finished, check to see if the design objectives were met and draw the blade geometries.

5.38 An entire compressor (low-pressure and high-pressure) was tested, and internal total pressures and temperatures were monitored. Flow rates and rotational speeds were controlled. Experimental data for the high-pressure compressor are presented in Table 5.P.3. Construct the corrected map of the HPC from these data. Include both the total pressure ratio and efficiency islands.

Table 5.P.3 Experimental data for the high-pressure compressor for Problem 5.38.

$\dot{m}$ (lb_m/s)	$p_{t2.5}$ (psia)	$T_{t2.5}$ (°R)	p_{t3} (psia)	T_{t3} (°R)	N (rpm)
159.2	15.0	480	15.0	480	5770
169.2	16.2	491	35.7	668	5836
179.0	17.6	503	88.1	884	5904
178.8	18.5	509	147.7	1014	5944
175.2 175.1[a]	19.0	513	199.3	1216	5967
217.5	16.0	520	16.0	520	8008
248.3	19.0	546	106.4	1022	8207
250.8	19.7	552	157.6	1098	8249
250.1	20.1	555	199.2	1126	8274
245.9 245.7[a]	20.6	559	259.7	1408	8303
297.4	18.0	550	18.0	550	10,294
332.1	20.6	571	78.2	967	10,493
349.2	22.0	583	165.1	1204	10,595
351.3	23.0	590	271.9	1319	10,664
342.3 342.1[a]	23.6	594	356.6	1589	10,701

[a] Surge.

5.39 Initially, an axial flow compressor stage operates with a 50 percent reaction. However, in a later test the rotor and stator blades are replaced so that the inlet relative flow angle in the rotor is 0° (i.e., the relative flow is purely axial). If the flow turning in the rotor, rotational speed, and mass flow rate are the same for both cases, answer the following.

(a) Qualitatively what happens to the percent reaction?
(b) Qualitatively what happens to the total pressure ratio?

Explain with accurate polygons.

5.40 A single stage of an axial flow compressor stage rotates at 7000 rpm and has an average radius of 335.3 mm (13.2 in.) (so that the blade speed is 245.7 m/s or 806 ft/s) with an inlet blade height of 25.40 mm (1.00 in.). Flow enters the rotor blades at +7° in the fixed reference frame and also exits the stator vanes at the same angle. The rotor flow turning angle is 36°, and the stage efficiency is 89 percent. The mass flow rate is 111.1 kg/s (245 lb_m/s); the axial velocity component at the rotor inlet is 245.4 m/s (805 ft/s) and increases by 2 percent across each blade and vane row. The inlet static temperature and pressure are 638.9 K and 1.551 MPa (1150 °R and 225 psia), respectively; the speed of sound at the inlet is 500.8 m/s (1643 ft/s). The inlet total temperature and pressure in the stationary frame are 677.8 K and 1.827 MPa (1202 °R and 265 psia), respectively. The stage consumes 5.256 MW (7049 hp) of power. The specific heat at constant pressure is 1.067 kJ/kg-K (0.255 BTU/lb_m-°R), and the specific heat ratio is 1.368. You are asked to evaluate the design.

(a) Accurately draw or plot the velocity polygons.
(b) Find the percent reaction.
(c) Find the total pressure ratio.
(d) On the basis of the percent reaction and total pressure ratio, is this a good design? Why or why not? What other parameters would you also check and what should they be? (Do not calculate them.)

5.41 Initially an axial flow compressor operates with a 50 percent reaction. However, in a later test the rotor and stator blades are replaced so that the inlet absolute flow angle for the rotor is 0° (i.e., the absolute flow is purely axial). If the flow turning in the rotor, rotational speed, and mass flow rate are the same for both cases, answer the following.

(a) Qualitatively what happens to the percent reaction?
(b) Qualitatively what happens to the total pressure ratio?

Explain with accurate velocity polygons.

5.42 Consider a new axial flow compressor single stage for which you are making estimates of efficiency. From previously determined experimental airfoil data, the angles between the drag and lift for the rotor blades and stator vanes are 1.43° and 1.78°, respectively. Estimate the dependence of efficiency on the flow coefficient and plot for a range of 0.45 to 0.85. Consider four different percent reactions from 0.45 to 0.75; that is, independently vary the percent reaction and plot as separate lines. Comment on the dependence and sensitivity of efficiency on each parameter.

5.43 A single stage of an axial flow compressor stage rotates at 10,000 rpm and has an average radius of 10.0 in. with an inlet blade height of 1.70 in. The absolute inlet angle to the rotor blades is the same as the absolute exit angle from the stator blades and is 10° relative to the axial direction. The rotor turning angle is 25°, and the stage efficiency is initially estimated to be 90 percent. The mass flow rate is 133 lb_m/s; the axial velocity component enters the stage at 854.1 ft/s and increases by 2 percent across the rotor and another 2 percent across the stator. The inlet temperature and pressure are 900 °R and 70 psia, respectively; the density at the inlet is 0.00653 sl/ft^3. The speed of sound at the inlet is 1460 ft/s, the blades move tangentially at 872.7 ft/s, and the Mach number in the absolute frame at the rotor exit or stator inlet is 0.725. The specific heat at constant pressure is 0.249 BTU/lb_m-°R, and the specific heat ratio is 1.381. The stage consumes 3183 hp of power.

(a) Accurately draw the velocity polygons.
(b) Find the percent reaction.
(c) Find the pressure coefficient for the rotor blades.
(d) Find the total pressure ratio.

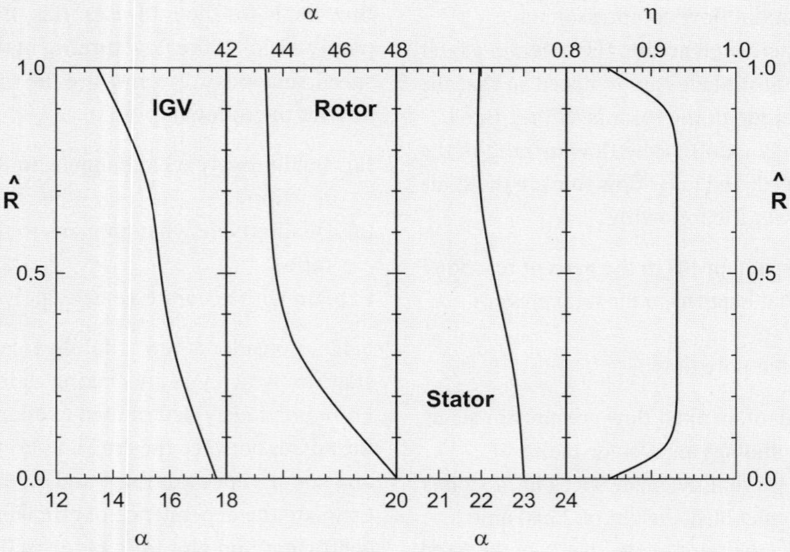

Figure 5.P.2 Specified flow angles and stage efficiency for Problem 5.46.

(e) Find the stator turning angle.
(f) If the lift and drag coefficients of the stator vanes are 0.72 and 0.0338, respectively, and the lift and drag coefficients of the rotor blades are 0.70 and 0.028, respectively, using the velocity information and performance data that you just calculated, better estimate the stage efficiency (do not repeat any of the preceding calculations).

5.44 Starting with Eq. (5.11.17) for ($\varepsilon = \varepsilon_r = -\varepsilon_s$), show that the loci of constant efficiency are circles in the %R and ϕ plane. Find the radii of circles of constant efficiency and the center of the circles in the plane.

5.45 A single-stage compressor with potential flow is to be analyzed. The compressor size and operating conditions approximate those on the first low-pressure compressor stage on an older commercial turbofan engine. The tip and hub diameters are 1.228 and 0.968 m (4.029 and 3.176 ft), respectively and the compressor operates at 3650 rpm. The resulting tip to hub diameter ratio is 1.269, which indicates that significant radial flow and variations should exist. Air entering at 0.137 MPa (19.9 psia) is compressed at a flow rate of 109 kg/s (240 lb_m/s). The average density is 1.70 kg/m^3 (0.0033 sl/ft^3). An inlet guide vane is used, and the absolute flow angles at midstream at the exit of the IGV, rotor blades, and stator blades are 21°, 46°, and 21°, respectively. Because potential flow is assumed, the efficiency is unity everywhere. Find the radial variations of the flow angles, blade curvatures, blade twists, reaction, and pressure profiles at the exit of the rotor and stator.

5.46 In Problem 5.45, a low-pressure compressor with irrotational flow was studied. For this problem the absolute flow angles across the exit of each blade row are specified for the same geometry. The same flow rate and speed are studied, and the flow angles are presented in Figure 5.P.2. Also in the figure is the specified stage efficiency. Efficiencies of 85 percent are used at the hub and tip and 93 percent at the midstream. Find the streamlines, radial variations of the flow angles, blade curvatures, blade twists, reaction, and pressure profiles at the exit of the rotor and stator.

5.47 A single (first) stage of a small compressor is to be analyzed. It rotates at 12,000 rpm and compresses 25 lb_m/s of air. The inlet pressure and temperature are 14.2 psia and 507 °R, respectively. The average radius of the blades is 7.0 in., and the inlet blade height is

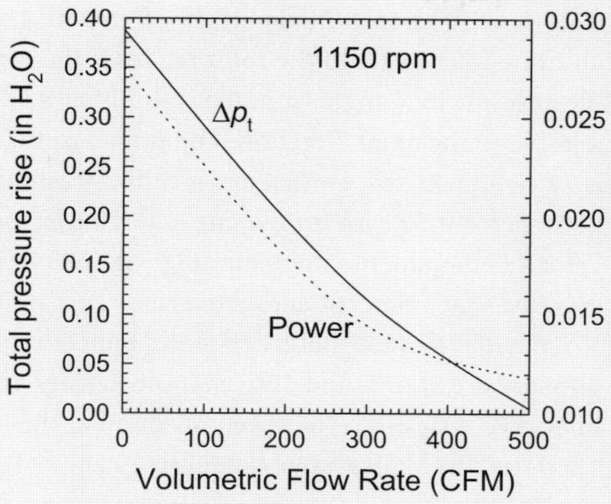

Figure 5.P.3 Total pressure rise and power data for Problem 5.48.

1.40 in. The absolute inlet angle to the rotor is the same as the stator exit angle (11°), and the rotor flow turning angle (δ_{12}) is 22°. It has been designed so that the blade height varies and the axial velocity remains constant through the stage. The efficiency of the stage is 90 percent. One should note that this is similar to Example 5.5.

Find the following:

(a) blade heights at the rotor and stator exits,
(b) the static pressure at the rotor and stator exits,
(c) the total pressure ratio for the stage,
(d) the stator turning angle,
(e) the required power for the stage,
(f) the percent reaction for the stage.

5.48 Low-speed fans are often used to move cool air through electronic and other systems to ensure that critical temperatures are not exceeded. Although simple, these fans are low-speed axial turbomachines with essentially incompressible flow. Performance characteristics of one such fan at standard temperature and pressure are presented here in Figure 5.P.3. The fan has an outer radius of 5.00 in. and an inner radius of 1.30 in. It has five blades and rotates at 1150 rpm. The average pitch of the blades is 57° relative to the axial direction (33° relative to the tangential direction).

(a) Using the Δp vs. Q curve and power vs. Q curve, find the efficiency vs. Q curve. Where is the best efficiency point (BEP)?
(b) At BEP, using aerodynamics, the efficiency vs. Q curve, and the Δp vs. Q curve, assume that the inlet flow is purely axial and iterate on the flow angles to match the experimental total pressure rise. Compare the average pitch of the flow to the blade pitch. Predict the power at BEP.
(c) At BEP, using aerodynamics, the efficiency vs. Q curve, and the Δp vs. Q curve, iterate on the inlet flow angle (preswirl) using the average pitch of the flow equal to the blade pitch so that the experimental total pressure rise is matched. Comment on the preswirl. Predict the power at BEP.

5.49 One of the first stages (rotor and stator) of a high-pressure axial compressor is to be analyzed. It rotates at 9600 rpm and compresses 70.31 (155 lb_m/s of air). The inlet pressure and temperature are 0.3447 MPa and 411.1 K (50 psia and 740 °R), respectively. The average radius of the blades is 281.9 mm (11.1 in.), the inlet blade height is 54.61 mm (2.15 in.), the rotor exit blade height is 50.29 mm (1.98 in.), and the stator exit blade height is 44.96 mm (1.77 in). The absolute inlet flow angle to the rotor is the same as the stator exit flow angle (1 °), and the rotor flow turning angle is 22°. The efficiency of the stage is 87 percent. The values of c_p and

γ are 1.024 kJ/kg-K (0.2446 BTU/lb$_m$-°R) and 1.389, respectively. Find:

(a) the total pressure at the stator exit,
(b) the required power for the stage,
(c) the percent reaction for the stage,
(d) the stator exit axial velocity.

5.50 Find the corrected mass flow rate and corrected speed for the stage in Problem 5.1. If the stage is run at 10,000 rpm, and the flow rate is 47.63 kg/s (105 lb$_m$/s), find the inlet pressure and inlet temperature so that the corrected parameters are matched. Use these conditions to find the resulting flow angles and total pressure ratio and compare with those from Problem 5.1. Comment.

5.51 A single stage of a small axial flow compressor stage rotates at 16,000 rpm and has an average radius of 6.0 in. with an inlet blade height of 2.00 in. The absolute inlet angle to the rotor blades is the same as the absolute exit angle from the stator vanes and is 7° relative to the axial direction. The absolute inlet angle to the stator vanes is 32.5° relative to the axial direction, and the stage efficiency is estimated to be 92 percent. The total pressure ratio for the stage is 1.289. The mass flow rate is 75 lb$_m$/s; the axial velocity component enters the stage at 956 ft/s and increases by 1 percent across the rotor and another 1 percent across the stator. The inlet temperature and pressure are 775 °R and 43 psia; the density at the inlet is 0.00465 sl/ft^3. The speed of sound at the inlet is 1358 ft/s, the blades move tangentially at 838 ft/s, and the Mach number in the absolute frame at the rotor exit or stator inlet is 0.824. The specific heat at constant pressure is 0.246 BTU/lb$_m$-°R, and the specific heat ratio is 1.387.

(a) Accurately draw (to scale) the velocity polygons.
(b) Find the rotor turning angle.
(c) Find the percent reaction.
(d) Find the required power.
(e) Find the pressure coefficient for the rotor blades.

5.52 A single stage of an axial flow compressor stage rotates at 9500 rpm and has an average radius of 10.0 in. with an inlet blade height of 1.00 in. The absolute inlet angle to the rotor blades is the same as the absolute exit angle from the stator blades and is 7° relative to the axial direction. The rotor turning angle is 27° while the stage efficiency is initially estimated to be 91 percent. The mass flow rate is 170 lb$_m$/s; the axial velocity component enters the stage at 1083 ft/s and decreases by 1 percent across the rotor and another 1 percent across the stator. The inlet temperature and pressure are 900 °R and 120 psia; the density at the inlet is 0.0112 sl/ft^3. The speed of sound at the inlet is 1460 ft/s; the blades move tangentially at 829 ft/s; and the Mach number in the absolute frame at the rotor exit or stator inlet is 0.866. The stage consumes 4686 hp of power. The specific heat at constant pressure is 0.249 BTU/lb$_m$-°R and the specific heat ratio is 1.381.

(a) Accurately draw the velocity polygons.
(b) Find the percent reaction.
(c) Find the pressure coefficient for the rotor blades.
(d) Find the total pressure ratio.
(e) Find the stator turning angle.

5.53 A single stage of a compressor is to be analyzed. It rotates at 14,000 rpm and compresses 60 lb$_m$/s of air. The inlet pressure and temperature are 100 psia and 800 °R, respectively. The average radius of the blades is 6.0 in., and the inlet blade height is 0.96 in. The absolute inlet angle to the rotor is the same as the stator exit angle (15°), and the rotor flow turning angle (δ_{12}) is 24°. The compressor stage has been designed so that the blade height varies and the axial velocity remains constant through the stage. The efficiency of the stage is 90 percent. The values of c_p and γ are 0.2459 BTU/lb$_m$-°R and 1.387, respectively.

Find the following:

(a) the blade heights at the rotor and stator exits,
(b) the static pressure at the rotor and stator exits,
(c) the total pressure ratio for the stage,
(d) the stator turning angle,
(e) the required power for the stage,
(f) the percent reaction for the stage.

6 Centrifugal Compressors

6.1 Introduction

As discussed in Chapters 2 and 4, the basic operating principle of a compressor is to impart kinetic energy to the working fluid by the means of some rotating blades and then to convert the increase in energy to an increase in total pressure. Axial flow compressors are covered in Chapter 5. These compressors are used on large engines and gas turbines. However, for small engines – particularly turboshafts and turboprops – centrifugal (or radial) compressors are used.

These compressors have higher single-stage pressure ratios than axial compressors (typically 2 to 4 compared with ~1.25). As a result, centrifugal compressors have lower cross-sectional flow areas per mass flow rate than do axial compressors. They also have larger diameters but shorter lengths per unit mass flow rate than do axial compressors. However, the rotating element (impeller) of the compressor is an integral unit of blades and a disk, and thus if one blade is damaged the entire unit is replaced. The weights of centrifugal units are approximately the same as for axial compressors used for the same application. They, however, demonstrate the characteristic of lower flow rates. As a result, they are physically small units with low flow rates, which makes them ideal for helicopter and small aircraft applications. Because the flow is turned by 90° (from axial flow at the impeller inlet to radial flow at the impeller exit), however, they are slightly less efficient than axial machines – particularly for multistage compressors (typically by 5 percent), and thus they are not used in applications in which efficiency is at a premium. Given the bulk and one-piece construction of an impeller, centrifugal compressors are much less susceptible to foreign object damage than are axial machines and are typically more durable. Finally, because centrifugal impellers are often cast in one piece and axial flow compressor blades must be individually cast and then assembled, the costs of centrifugal compressors are usually less than for axial compressors.

One application of such compressors is the Rolls-Royce Dart shown in Figure 6.1, which has an overall pressure ratio of 5.6. The design mass flow rate for this engine is 10.7 kg/s. This engine has two such centrifugal compressor "wheels" and is driven by three axial flow turbine stages. This is a turboprop engine and is used to power the Fairchild F-27 and other aircraft. The reduction gear nest is also depicted.

Another application is the Honeywell T55 turboshaft shown in Figure 6.2, which has an overall pressure ratio of 9.3. This dual-shaft engine has seven low-pressure axial flow compressor stages before the single high-pressure centrifugal compressor stage. The first axial flow turbine stages (HPT) drive all compressor stages and the last two stages (LPT) are the output power. Note that, to shorten the length of the engine, the flow in the annular combustor actually flows in the opposite axial direction to that of the flow in the axial stages (called counterflow), as with the early Whittle engines. Air enters the combustor near the rear of the engine. This engine is used on several helicopters.

6.1.1 Objectives

After completing this chapter, the reader should be able to do the following.

- Understand the basic geometry of a centrifugal compressor
- Understand the difference and roles of the impeller and diffuser
- Understand the pressure changes in a single stage

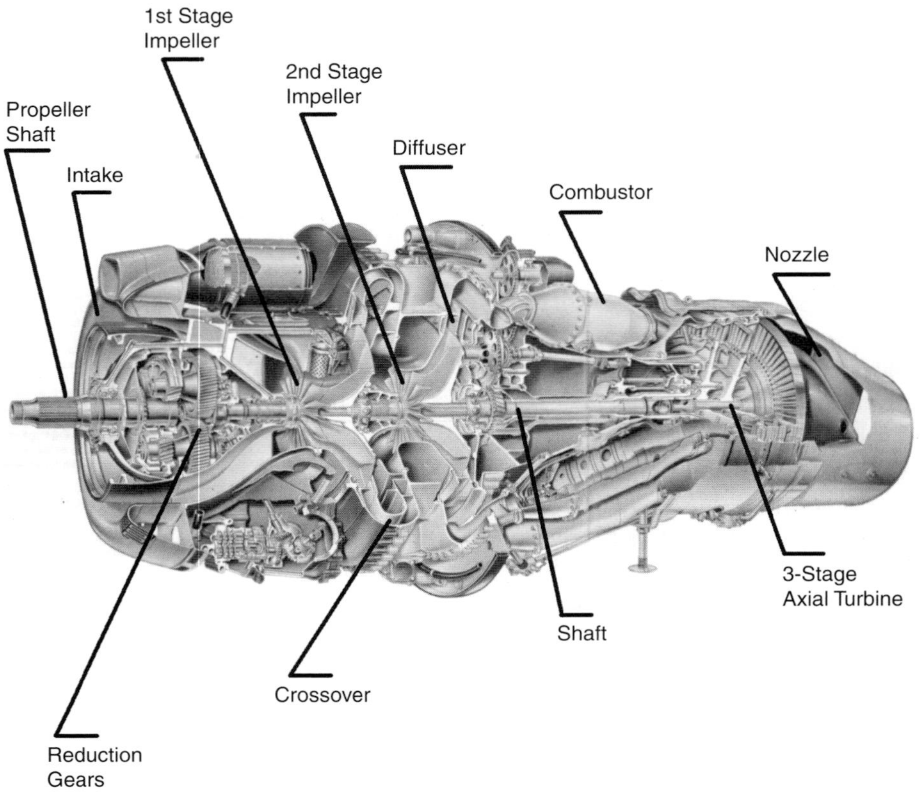

Figure 6.1 Rolls-Royce Dart (© Rolls-Royce plc 2022. All rights reserved).

- Understand and be able to develop a velocity triangle or polygon
- Understand the different flow and blade angles in a stage
- Understand the influences of impeller diameter on the pressure increase and velocities
- Understand the influences of blade geometry on pressure and velocities
- Be able to apply equations from Chapter 4 along with velocity polygons to find single-stage performance
- Understand the development and application of two compressor "maps"
- Understand the basic concepts of losses and slip for a centrifugal machine
- Understand the basic design parameters
- Understand the role diffusers and crossovers have for a centrifugal unit
- Appreciate the differences performance and other characteristics in centrifugal and axial flow compressors

6.2 Geometry

6.2.1 Pure Centrifugal/Radial Exit Impeller

The geometry of a centrifugal compressor is significantly different than that of an axial compressor. The basic geometry of a single stage is shown in Figure 6.3a. Flow enters the rotating element (impeller) predominantly in the axial direction with no

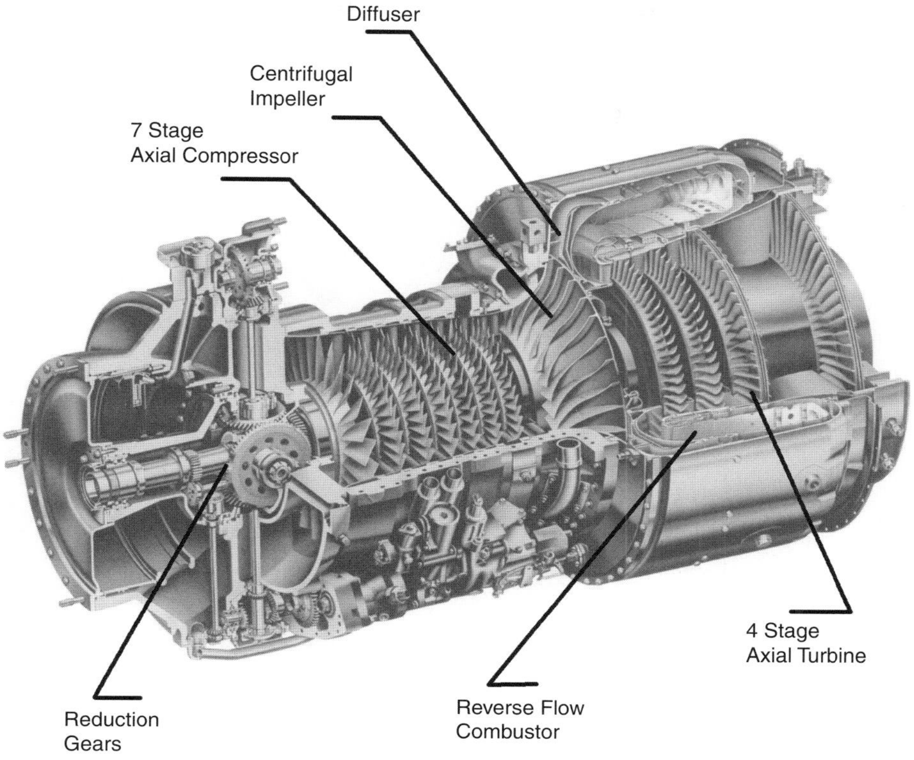

Figure 6.2 TF55 turboshaft (courtesy of Honeywell).

radial component of velocity. The entrance region to the impeller is called the inducer, and the impeller rotates with angular velocity ω. Thus, in the inlet section the blades move with linear velocity $R_1\ \omega$, where R_1 is the average of the hub and eye radii, R_{hub} and R_{eye} (Figure 6.3b). If the flow enters the inducer strictly in the axial direction (no tangential component of velocity), the flow is said to be swirl free or to be without preswirl. However, if the flow at the inlet has both axial and tangential components of velocity, the flow is said to have preswirl. Photographs of two impellers are shown in Figure 6.3c. One is a single compressor called a single faced impeller. Splitter blades are short blades near the impeller exit and are seen on the photograph of this impeller and on Figure 6.3a. Splitter blades will be discussed further in Section 6.6. The second is two impellers back-to-back with a single exit called a double-faced or dual-inlet impeller. This design results in a shorter but more complex compressor. Also, such a design results in a smaller impeller diameter since the flow rate for each inlet is smaller. However, this design usually has an increased rotational speed due to the smaller diameter.

By means of centrifugal forces acting on the fluid, the flow is turned so that it exits the impeller radially and tangentially – usually with no or minimal axial component of velocity. The blade velocity at the exit is $R_2\ \omega$. As is true for an axial compressor stage, the impeller imparts energy to the fluid, which increases the total pressure; the centrifugal action of the impeller is used to increase the pressure. Because the gas exits the impeller at a larger radius than it entered, the absolute velocity of the gas is higher at the exit. After the fluid exits the impeller, it enters the diffuser, which slows the fluid and directs it to the outlet or discharge region; that is, the high-velocity,

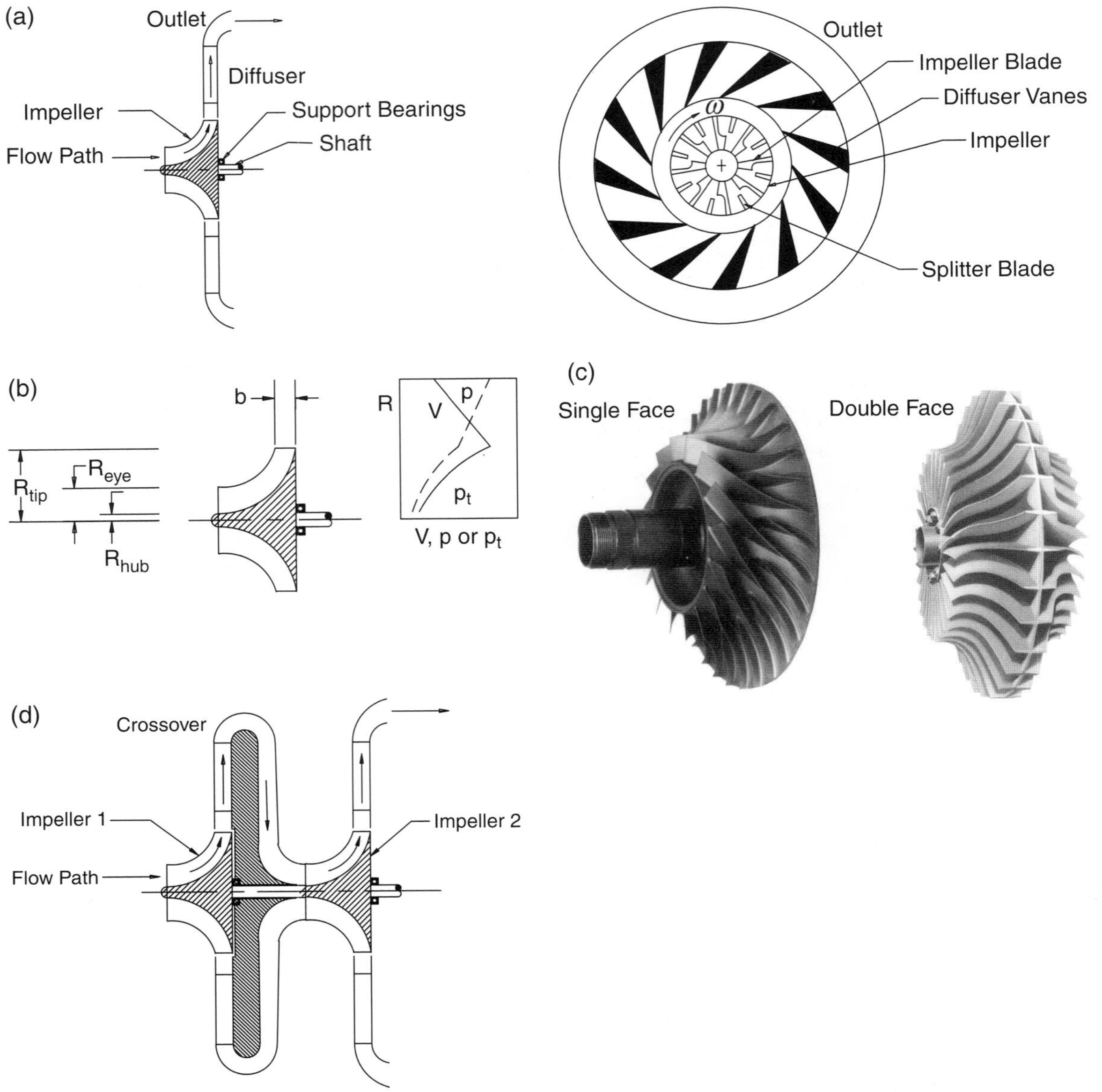

Figure 6.3 (a) Geometry of a radial compressor with a single discharge. (b) Velocity and pressure trends in impeller. (c) Photograph of a Rolls-Royce centrifugal impeller (© Rolls-Royce plc 2022. All rights reserved). (d) Geometry of a two-stage radial compressor. (e) Exploded view of a radial compressor (© 2022 Raytheon Technologies Corporation, Pratt & Whitney division. Reproduced with permission). (f) Three types of impeller blade designs.

high-kinetic-energy gas is converted to a low-velocity, high-pressure gas. Two types of diffusers are used. One is the vaneless diffuser, and the second is a vaned diffuser. The vaned diffuser has a series of blades in the circumferential–radial direction at the impeller exit, which provide guidance to the fluid. After the fluid exits the diffuser, it enters the manifold, collector, or scroll in the axial direction.

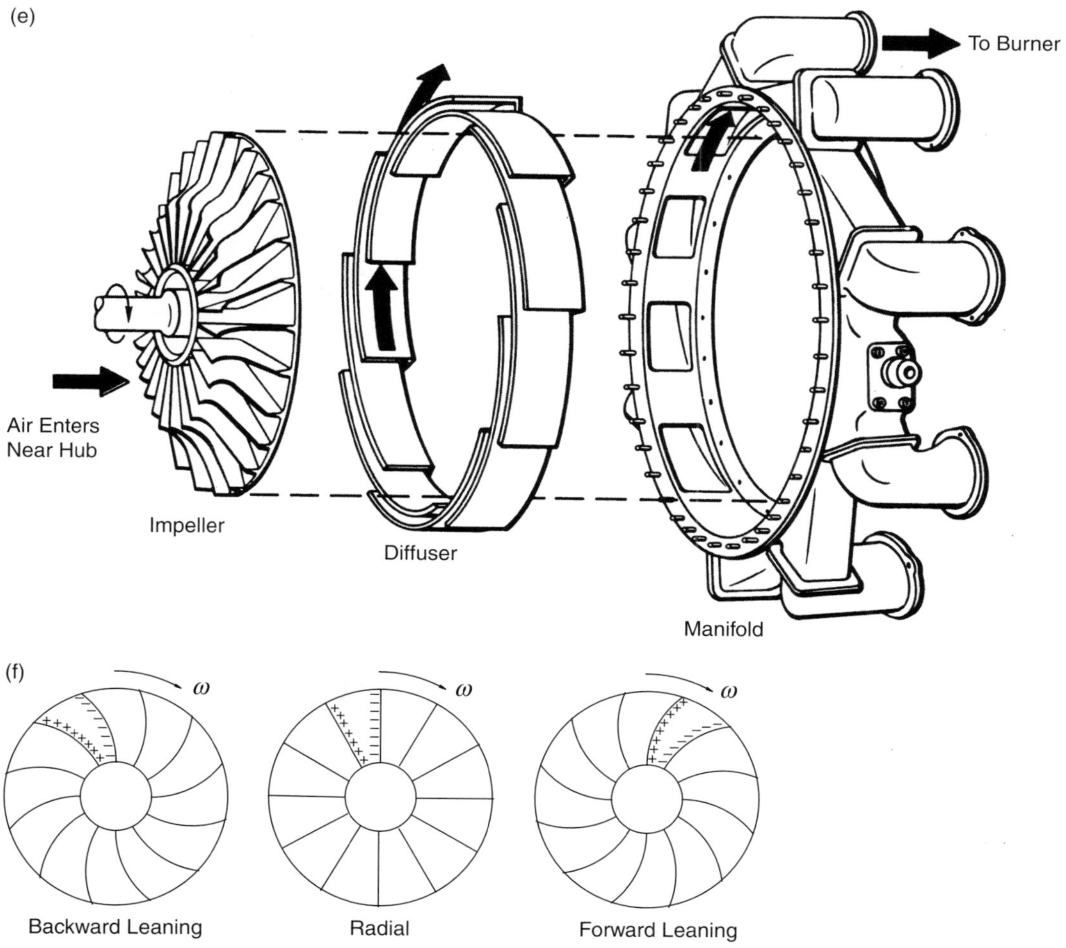

Figure 6.3 (cont.)

This region directs the flow to the outlet, exit, or discharge. Figure 6.3a shows a single annular discharge outlet. Typically, a jet engine application will not have a single discharge but will usually have a continuous annular exit or manifold that exhausts directly into the combustor. In Figure 6.3d, a multiple-stage centrifugal compressor is shown. For a multiple-stage machine, a stationary "crossover" is needed to direct the exit flow from the first stage to the inlet of the second stage. In Figure 6.3e, an exploded view of a single-stage compressor is shown for the flow through the three components.

Also, Figure 6.3b illustrates the trends of velocity and static pressure as functions of radius in a centrifugal compressor stage. As can be seen, the velocity increases in the impeller but decreases in the diffuser and scroll. On the other hand, the static pressure increases in both. The total pressure increases in the impeller but remains approximately constant (except for losses) in the diffuser and scroll.

Three basic types of blade geometries are used for the exit region of the impeller. These are called backward-facing (or backward-swept), forward-facing (or forward-swept), or purely radial blades. The geometries are shown in Figure 6.3f. As can be seen, the flow in a backward-swept blade passage exits from the impeller with a negative tangential velocity relative to the impeller rotation, whereas the flow in a forward-swept blade passage exits with a positive magnitude. The flow in a purely radial passage exits from the impeller with no relative tangential velocity. Backward-swept blades are used for hydraulic pumps and small compressors, radial blades are used for large compressors, and forward-swept blades are more of an academic interest. Advantages to the different designs will be discussed in Section 6.6.

Because the impeller is rotating (at sometimes very high speeds), the developed pressure field between the blades is not uniform. In fact, because the blades are "pushing" the fluid on one side and "pulling" it on the backside, the two surfaces typically have higher and lower pressures than the average, respectively. Thus, one side of the blade is called the pressure surface and the opposite side is called the suction surface, as is the case for a conventional airfoil. Shown in Figure 6.3f are the definitions of these two surfaces. Note that this pressure distribution is only typical. Under off-design conditions the opposite trend can occur – at least at some radii; that is, the pressure surface will develop a lower-than-average pressure and vice versa. Nevertheless, the definition of pressure and suction surface remains the same as shown in Figure 6.3f.

Thus, the impeller acts similarly to the rotor blades in an axial machine, and the diffuser has the same function as the stator vanes. Both the total and static pressures are increased across the impeller, and the static pressure is increased across the diffuser and collector. In the remainder of this chapter, it is the intention to relate the pressure changes to the geometry and flows of the compressor. That is, for a given geometry and rotational speed, what are the static and total pressure rises and ratios?

6.2.2 Mixed-Flow Impeller

In Section 6.2.1 a purely radial exit impeller was covered. Another impeller design is called the mixed-flow impeller and is shown in Figure 6.4. For this type of impeller, flow exits with both radial and axial components of velocity. The angle between the radial direction and flow or geometry is β_x. This geometry is seldom used for gas turbines but receives wider spread use in the pump industry. For the remainder of this chapter the radial impeller will be covered. However, it is worth noting for reference the definition of a mixed-flow impeller.

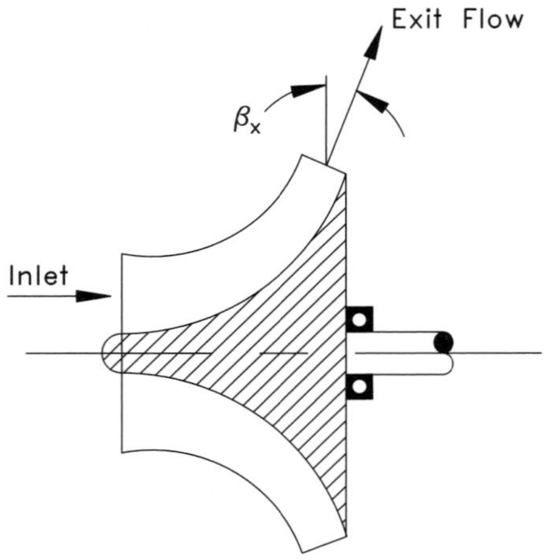

Figure 6.4 Mixed-flow impeller.

6.3 Velocity Polygons or Triangles

Because energy is being added to the flow (which results in a pressure rise) and a part of the compressor is moving, it is important to understand the complex flow patterns. As is the case for an axial flow compressor, the pressure rise is directly dependent on the velocity magnitudes and directions in a compressor. This section defines a method for interpreting the fluid velocities. That is, because one component rotates and the other is stationary, it must be possible to relate the velocities so that the components are compatible. As will be shown in this section, the velocity polygons or triangles are different for a centrifugal compressor than for an axial compressor.

First, consider the inlet to the impeller for a "swirl-free" condition, as shown in Figure 6.5. "Swirl free" is

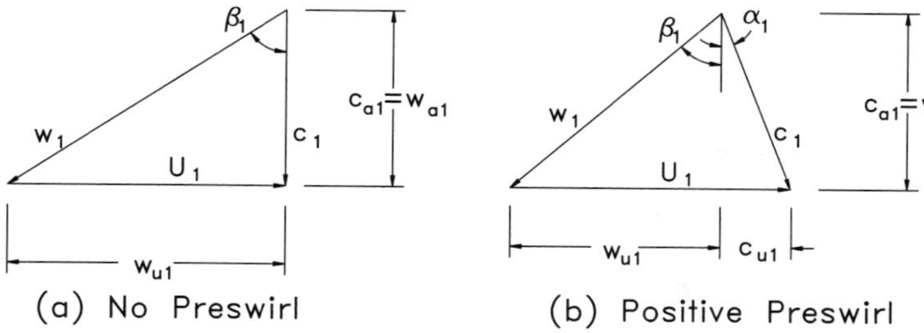

Figure 6.5 Velocity polygon at impeller inlet.

the condition in which there is no tangential component of velocity and the velocity vector is purely axial. Because the flow is axial and the fluid does not have a radial component of velocity, the polygons for the impeller inlet will be similar to those for an axial machine. Once again, the impeller blades are rotating around the engine centerline or are moving in the planar 2-D plane with absolute velocity $\mathbf{U}_1$ in the tangential direction. For this first case, the inlet absolute velocity is $\mathbf{c}_1$ and is aligned with the machine axis. Thus, to find the velocity of the fluid relative to the rotating blades, one must subtract the blade velocity vector from the absolute flow vector. This is graphically performed in Figure 6.5. The resulting vector is $\mathbf{w}_1$ and has a flow angle β_1 relative to the axial direction. It is important to note that the impeller blades have an inlet angle (relative to the axial direction) of β'_1. If the relative flow direction matches the blade angles exactly, the values β_1 and β'_1 are the same and the incidence angle is zero. However, this usually does not occur in practice – especially at off-design conditions.

The vectors can be revolved into different components as indicated in Figure 6.5. As shown, the absolute velocity $\mathbf{c}_1$, has no component in the tangential direction for the swirl-free case. The relative velocity $\mathbf{w}_1$ has components w_{u1} and w_{a1} in the tangential and axial directions, respectively.

The inlet to the impeller can next be considered for flow with positive preswirl, as also shown in Figure 6.5. Again the inlet absolute velocity is $\mathbf{c}_1$; however, it has a flow angle of α_1 relative to the machine axis, which for this case is positive. To find the velocity of the fluid relative to the rotating blades, one must again subtract the blade velocity vector $\mathbf{U}_1$, from the absolute flow vector. This is performed in Figure 6.5. The resulting vector is $\mathbf{w}_1$ and has a flow angle β_1 relative to the axial direction. Again, the impeller blades have an inlet angle of β'_1.

One can also break the vectors into components as indicated in Figure 6.5. As shown, the absolute velocity $\mathbf{c}_1$ has a component in the tangential direction that is equal to c_{u1} and a component in the axial direction that is c_{a1}. Also, the relative velocity $\mathbf{w}_1$ has components w_{u1} and w_{a1} in the tangential and axial directions, respectively. It is worth noting that negative preswirl can also be encountered.

In Figure 6.6, the velocity polygon is shown for the impeller exit (and diffuser inlet) for backward-swept blades and no axial component. The exit relative velocity (to the impeller blades) is $\mathbf{w}_2$ and this velocity is at an angle of β_2 relative to the radial (not axial as for the inlet) direction. The blade velocity is $\mathbf{U}_2$ and can be vectorially added to $\mathbf{w}_2$ to find the absolute exit velocity $\mathbf{c}_2$. Note that $\mathbf{U}_2$ and $\mathbf{U}_1$ are markedly different because the radii at the rotor inlet and exit are significantly different. The absolute flow angle of $\mathbf{c}_2$ is α_2 relative to the radial direction. Once again, the impeller blade exit angle is β'_2 which may be different from the flow angle β_2 owing to "slip." Furthermore, the diffuser vane inlet angle is α'_2 and will in general be different from α_2. Also shown in Figure 6.6 are the components of both $\mathbf{w}_2$ and $\mathbf{c}_2$ in the tangential and radial directions, which are w_{r2} and c_{r2}, respectively.

Table 6.1 **Centrifugal compressor component trends.**

	Absolute velocity	Relative velocity	Area	Static pressure	Total pressure
Impeller	Increases	Decreases	Increases	Increases	Increases
Diffuser	Decreases	Decreases	Increases	Increases	Constant (ideal)

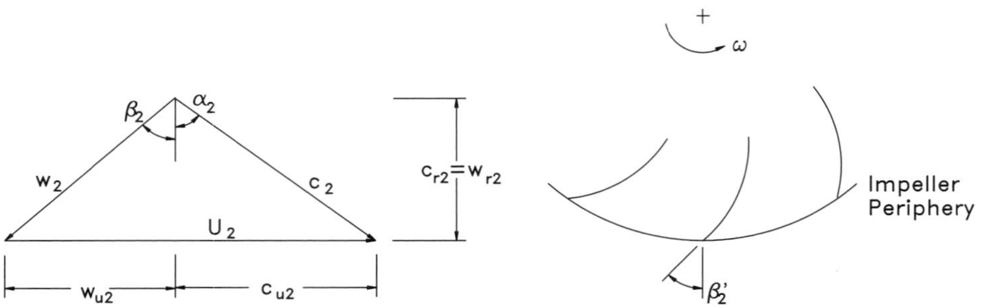

Figure 6.6 Flow and blade angles at impeller exit.

Once again, a consistent sign notation is needed, and the notation will be the same as that used for axial flow compressors. First, all positive tangential velocities will be to the right. Thus, rotor blades will always move to the right with positive velocity U owing to the counterclockwise rotation of the impeller. Second, all angles will be measured relative to the axial direction, and positive angles will be considered to be counterclockwise.

It is noteworthy that the velocity polygons are similar for an incompressible centrifugal turbomachine – namely, pumps. Thus, the same type of analysis is used to understand the flow velocities and relate them to power requirements, pressure rises, and so on. One basic difference, however, is that pump manufacturers typically do not measure the flow angles relative to the axial or radial directions. They instead use the tangential direction as a reference. Both reference conditions lead to identical velocity polygons and other results. One needs to be aware of the different methods and sign conventions when using other texts, however.

In Table 6.1 the trends of the velocity magnitudes, flow cross-sectional areas, and pressures across the different components are shown. As can be seen, the static pressure increases across both the diffuser and impeller blades because the relative velocities across both decrease. Details and predictions of these trends are covered in Section 6.4.

6.4 Single-Stage Energy Analysis

In this section, the equations are summarized so that the velocities from polygons can be related to the pressure rise and other important compressor characteristics. In Chapter 4, these equations were found for a single-stage (impeller and diffuser) using a "mean line" control volume approach. That is, conditions at the center of the blade passages are used for the analysis. In Figure 4.2, a single stage is shown. The objective of this section is to relate the inlet and exit conditions to the property changes.

6.4.1 Total Pressure Ratio

As is the case for axial flow compressors, the continuity, moment of momentum, and energy equations are utilized. For the shaft power (Eq. (4.2.17)),

$$\dot{W}_{sh} = \dot{m}[U_2 c_{u2} - U_1 c_{u1}], \tag{6.4.1}$$

and for the total pressure ratio (Eq. (4.2.26)),

$$\frac{p_{t2}}{p_{t1}} = \left[\frac{\eta_{12}[U_2 c_{u2} - U_1 c_{u1}]}{c_p T_{t1}} + 1\right]^{\frac{\gamma}{\gamma-1}}, \tag{6.4.2}$$

which is significantly greater than unity. Thus, knowing the velocity information of an impeller and the efficiency, one can find the total pressure rise.

The results for a centrifugal compressor can now be compared with those for an axial flow compressor. For example, Eq. (6.4.2) is identical to Eq. (5.4.2). That is, the equations that result apply to both radial and axial flow machines. However, the velocity polygons and application of the equations are different for the two categories of machines. For example, for an axial flow machine, U_1 and U_2 are approximately the same (R_2 and R_1 are about the same). However, for a radial machine U_2 is much larger than U_1 because R_2 is much larger than R_1, which is the average of the hub and eye radii. Thus, a larger pressure ratio is possible for a radial than for an axial flow machine. Percent reaction is not normally used as a design parameter for a centrifugal compressor. It should be noted that these equations also hold for mixed-flow geometries. However, the vector analysis becomes somewhat more complicated for the exit flow since both radial and axial components of velocity are present. Regardless, this is a relatively straightforward calculation. An example of a single stage with all of the calculation details will follow. One should note that the software Turbomachinery.exe performs the same calculations and can be used for a series of repetitious analyses.

6.4.2 Incompressible Flow (Limiting Case)

For comparison, one may wish to find the pressure rise of a turbomachine with an incompressible fluid – namely, a centrifugal pump. This type of machine has many and varied industrial and other applications and is covered in such references as Shepherd (1956), Stepanoff (1957), Pfleiderer (1961), Karassik, Krutzsch, Fraser, and Messina (1976), Miner, Beaudoin, and Flack (1989), and Brennen (1994). It is worth including herein because some slip correlations and losses for pumps also have implications and insight for compressors. Although this section does not have direct applications to gas turbines, it does provide valuable insight into some of the fundamentals of centrifugal flow turbomachines. From Chapter 4 (Eqs. (4.2.34) and (4.2.37)), one obtains

$$\dot{W}_{sh} = \dot{m}[U_2 c_{u2} - U_1 c_{u1}] \tag{6.4.3}$$

and

$$p_{t2} - p_{t1} = \rho\eta_{12}[U_2 c_{u2} - U_1 c_{u1}]. \tag{6.4.4}$$

From fluid statics (no fluid velocity), one can recall that

$$\frac{dp}{dz} = -\rho g. \tag{6.4.5}$$

Thus, integrating yields that the height of a liquid generated by a pressure difference is

$$H = \frac{\Delta p_t}{\rho g}. \tag{6.4.6}$$

This is termed the head of a pump. It is the height of the working fluid that will be measured by, for example, with a U-tube manometer for the developed total pressure rise of the pump. Thus, from Eqs. (6.4.4) and (6.4.6), for an ideal pump ($\eta_{12} = 1$),

$$\frac{p_{t2} - p_{t1}}{\rho g} = \frac{U_2 c_{u2} - U_1 c_{u1}}{g} = H_i, \tag{6.4.7}$$

where H_i is the ideal developed head. Finally, using Eqs. (6.4.4) and (6.4.6), one finds the following for the nonideal head:

$$H = \eta_{12}\frac{U_2 c_{u2} - U_1 c_{u1}}{g} = \eta_{12} H_i. \tag{6.4.8}$$

Thus, for a pump the actual head is simply the ideal head multiplied by the efficiency. For pumps, peak efficiencies typically range 50 to 85 percent, as governed by the size and application. The effects of efficiency on pressure rise are quite different and more complicated for a compressor, as shown by Eq. (6.4.2).

6.4.3 Slip

Another nonideal effect occurs in all turbomachinery, and considerable effort has been expended to understand this effect in centrifugal pumps and compressors. This effect is termed "slip" and is a measure of how well the flow follows the blades. Ideally, the exit flow angle is

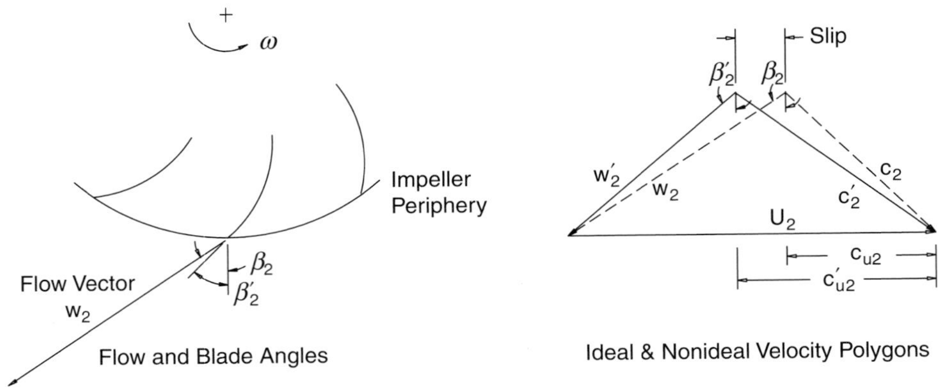

Figure 6.7 Ideal and nonideal velocity polygons at impeller exit.

identical to the blade angle. Because of pressure gradients within the passages (between the pressure and suction surfaces, Figure 6.3f) and the tangential momentum of the fluid, these angles will be different.

First consider Figure 6.7. The blade angle is β_2' and the flow angle is β_2, as discussed earlier in Section 6.3. If the two are identical, the "slip" is zero. Next, one can examine the ideal and nonideal velocity polygons at the impeller exit in Figure 6.7. The slip is defined as the difference in ideal and nonideal absolute tangential velocities. The slip factor μ is defined as

$$\mu \equiv \frac{c_{u2}}{c_{u2}'} \tag{6.4.9}$$

and is always less than or equal to unity. From the power equation for either a pump or compressor (Eq. (6.4.1)), one obtains

$$\dot{W}_{sh} = \dot{m}[U_2 c_{u2}' - U_1 c_{u1}]. \tag{6.4.10}$$

Note specifically that the actual power needed to drive the pump or compressor is less than the ideal power due to slip, which is accompanied by a decrease in total pressure rise. This is a major difference between slip and loss. A loss increases the required power for the stage for a given total pressure rise.

Compressor

This slip concept can be applied to compressors. Once again,

$$\dot{W}_{sh} = \dot{m}[h_{t2} - h_{t1}]. \tag{6.4.11}$$

Thus, owing to slip, the increase in total enthalpy and total temperature is less. This results in a total pressure ratio of

$$\frac{p_{t2}}{p_{t1}} = \left[\frac{\eta_{12}[U_2 \mu c_{u2}' - U_1 c_{u1}]}{c_p T_{t1}} + 1\right]^{\frac{\gamma}{\gamma-1}}. \tag{6.4.12}$$

One can see that the actual pressure ratio is less than the ideal one because of both the efficiency and slip. Once again, however, note that the efficiency represents a loss that causes extra power to be used, whereas slip does not.

Two limiting cases can be studied. First, if no preswirl is present, ($c_{u1} = 0$), then Eq. (6.4.12) reduces to

$$\frac{p_{t2}}{p_{t1}} = \left[\frac{\eta_{12} U_2 \mu c_{u2}'}{c_p T_{t1}} + 1\right]^{\frac{\gamma}{\gamma-1}} \tag{6.4.13}$$

and the power is

$$\dot{W}_{sh} = \dot{m} U_2 \mu c_{u2}'. \tag{6.4.14}$$

If the case of no preswirl and purely radial exit blades ($\beta_2' = 0$) is considered, the pressure ratio and required power are

$$\frac{p_{t2}}{p_{t1}} = \left[\frac{\eta_{12} \mu U_2^2}{c_p T_{t1}} + 1\right]^{\frac{\gamma}{\gamma-1}} \tag{6.4.15}$$

and

$$\dot{W}_{sh} = \dot{m}\mu U_2^2. \quad (6.4.16)$$

For both of these simplifying cases, it can be seen that the power is proportional to the slip factor. Also, one can see that the pressure ratio is affected in the same way for both the slip factor and efficiency.

Hydraulic Pump (Limiting Case)

Next, the effects of slip on pump performance can be examined. From the definition of slip factor, Eq. (6.4.8) becomes

$$H = \frac{p_{t2} - p_{t1}}{\rho g} = \frac{\eta_{12}}{g}\left[\mu U_2 c'_{u2} - U_1 c_{u1}\right]. \quad (6.4.17)$$

From the exit velocity polygon (Figure 6.6), one can readily see that, ideally,

$$c'_{u2} = U_2 - w'_{u2} = U_2 - c_{r2} \tan\beta'_2. \quad (6.4.18)$$

Now the head becomes

$$H = \frac{\eta_{12}}{g}\left[\mu U_2^2 - \mu U_2 c_{r2} \tan\beta'_2 - U_1 c_{u1}\right]. \quad (6.4.19)$$

The magnitude of c_{r2} is proportional to the mass flow rate, or in fact

$$c_{r2} = \frac{\dot{m}}{\pi\rho D_2 b}, \quad (6.4.20)$$

where D_2 and b are the impeller diameter and passage height at the exit, respectively. The head is therefore

$$H = \frac{\eta_{12}}{g}\left[\mu U_2^2 - \frac{\mu U_2 \tan\beta'_2}{\pi\rho D_2 b}\dot{m} - U_1 c_{u1}\right]. \quad (6.4.21)$$

Next, a limiting case is considered. If flow enters the pump with no preswirl, no losses exist ($\eta_{12} = 1$), and the fluid perfectly follows the blades ($\mu = 1$), the so-called Euler head results

$$H_e = \frac{U_2^2}{g} - \frac{U_2 \tan\beta'_2}{\pi\rho g D_2 b}\dot{m}. \quad (6.4.22)$$

Thus, the Euler head varies linearly with flow rate. In Figure 6.8, a typical variation is shown for one rotational speed. Two limiting conditions can be

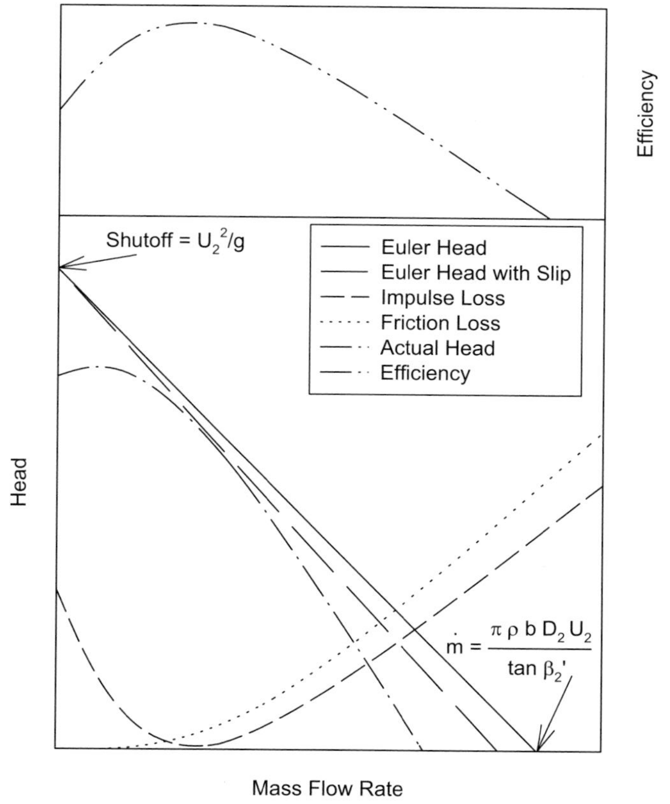

Figure 6.8 Ideal and nonideal pump head capacity curves.

considered. First, under the no-flow condition ($\dot{m} = 0$), the Euler head is U_2^2/g. Next, under the condition with no developed head ($H_e = 0$), the mass flow rate is $\pi\rho D_2 b U_2/\tan\beta'_2$. Thus, the Euler head curve is easy to determine. Also shown in Figure 6.8 is the Euler head with the slip included. At shut-off, the mass flow rate is zero; as a result, the slip factor is unity. Also shown are two important losses. The first is the friction loss due to boundary layers, separation, and turbulence, which increases monotonically with flow rate. The second is the so-called impulse loss (or "shock" loss – although for a pump it has nothing to do with the compressible shock wave), which is a result of incident losses at the blade's leading edge. At one mass flow rate, the incidence angle is nearly zero; consequently, the losses are near zero. By subtracting the two losses from the Euler head with slip, one finds the actual head capacity curve. As can be seen, the actual is always lower than the Euler curve. Also shown is the variation in efficiency. The efficiency reaches a maximum when losses are at the minimum and becomes zero when the head reaches zero.

As a final limiting case, one can consider if the impeller vanes under discussion are strictly radial at the exit ($\beta'_2 = 0$). For this case, the Euler head is

$$H_e = \frac{U_2^2}{g}, \tag{6.4.23}$$

which is obviously not a function of the flow rate. Therefore, for the case of no preswirl and strictly radial vanes, Eq. (6.4.21) reduces to

$$H = \eta_{12}\mu\frac{U_2^2}{g} \tag{6.4.24}$$

or

$$H = \eta_{12}\mu H_e. \tag{6.4.25}$$

Thus, the product of efficiency and slip factor appears to be one characteristic for this particular case. In general, however, this is not true. As noted above, although this analysis is for incompressible flow, one can use this information to understand for any radial machine how losses and slip can affect performance.

Slip Correlations

As indicated in the opening statements of this section, major efforts have been made to understand the dependence of slip factor on the different conditions. Several empirical correlations have been postulated to better serve engineers in the design stage of a centrifugal machine, as reviewed by Wiesner (1967). In general, as the number of vanes increases, the slip factor increases; that is, better control of the flow is achieved with more blades. Also, because of the momentum of the fluid, as the blade geometry varies from forward swept to purely radial to backward swept, the slip factor increases; thus, for a backward-swept geometry the flow follows the most natural or convenient flow path to the exit of the impeller. Although increasing the number of blades improves the slip factor, wall friction increases, reducing the efficiency, so a tradeoff is experienced. Also, although backward-swept blades increase the slip factor, the total pressure rise is reduced with this geometry.

Four correlations based on empiricism are recognized here. Stodola (1927) postulated that

$$\mu = 1 - \frac{\pi\cos\beta_2'}{Z}, \tag{6.4.26}$$

where Z is the number of vanes. Busemann (1928) correlated data graphically, which in general was higher than predicted by Stodola. Next, Stanitz (1952) proposed that

$$\mu = 1 - \frac{0.63\pi}{Z\left[1 - \frac{c_{r2}}{U_2}\tan\beta'_2\right]}. \tag{6.4.27}$$

Lastly, Pfleiderer (1961) presented the following:

$$\mu = \frac{1}{1 + \widehat{a}\left[1 + \frac{90 - \beta_2'}{60}\right]\frac{R_2^2}{ZS}}, \tag{6.4.28}$$

where

$$S = \int_{R_1}^{R_2} r\,dx \tag{6.4.29}$$

and x is the meridional distance the fluid traverses. Also, in Eq. (6.4.28), the factor $\widehat{a}$ is empirically determined. For a single volute geometry (pump), $\widehat{a}$ ranges between 0.65 and 0.85. For a vaned diffuser compressor, $\widehat{a}$ is approximately 0.6. For a vaneless diffuser geometry, $\widehat{a}$ is between 0.85 and 1.0.

6.4.4 Relationships of Velocity Polygons to Pressure Ratio

Both the governing equations and velocity polygons can be examined for general trends for variations in the absolute inlet angle α_1, the impeller exit angle β_2, the rotational speed N, mass flow rate (or axial velocity component) $\dot{m}$ (or c_a), and radius ratio R_2/R_1. The resulting effects on the stage total pressure ratio can be determined. A matrix of general trends is presented in Table 6.2.

Table 6.2 **Geometric/operating condition effects on compressor pressure ratio.**

Decrease in	Total pressure ratio, p_{t3}/p_{t1}
Preswirl inlet angle, α_1	Increase
Exit angle, β_2	Decrease
Rotational speed, N	Decrease
Axial velocity component, c_a	Increase
Radius ratio, R_2/R_1	Decrease

Example 6.1 Radial Compressor Stage

A stage of a radial compressor is to be analyzed. It rotates at 12,000 rpm and compresses 29.21 kg/s (2.002 sl/s) of air. The inlet pressure and temperature are 337.9 kPa (49 psia) and 282.2 K (508 °R), respectively. The hub and eye radii of the blades at the inlet are 63.5 and 152.4 mm, respectively, the exit radius is 304.8 mm, and the exit blade height is 25.4 mm. Flow enters the inducer with −20° of preswirl, and the impeller has backward-swept blades with a relative exit flow direction (β_2) of 10°. The efficiency of the stage is 90 percent. The values of c_p and γ are 1.005 kJ/kg-K and 1.399, respectively, *which are based on the resulting value* of the average static temperature. The following details are to be found: relative flow angle at the inlet, the static and total pressures at the impeller exit, the Mach numbers at the impeller inlet and exit, and the required power for the stage.

Solution Note

As noted for Example 5.1, before proceeding and because the solution is long, an outline for the solution is appropriate.

(1) First, some preliminary calculations are presented.
(2) Next, the polygon of the velocities at the impeller inlet is "drawn." Various angles and velocities are found.
(3) Next, the triangle of the velocities at the impeller exit/diffuser inlet is "drawn" using the above and problem specifications. Various angles and velocities are found.
(4) Next, using this information the power and pressure ratio are found.
(5) Intermediate steps also allow for the determination of such important characteristics such as Mach numbers and pressure coefficients.

Solution

Velocity polygons will be used to obtain the solution. Before proceeding, a few preliminary calculations are needed.

The average (mean line) blade speed at the inlet is

$$U_1 = R_1\omega = [(63.5 + 152.4)/2](12{,}000 \times 2\pi/60)$$
$$= 135{,}700\,\text{mm/s} = 135.7\,\text{m/s},$$

and the average blade speed at the exit is

$$U_2 = R_2\omega = (304.8)(12{,}000 \times 2\pi/60) = 383{,}000\,\text{mm/s} = 383.0\,\text{m/s}.$$

The cross-sectional flow areas at the inlet and exit are

$$A_1 = \pi(R_t^2 - R_h^2) = \pi(152.4^2 - 63.5^2)/1000^2 = 0.06030\,\text{m}^2$$
$$A_2 = 2\pi R_2 b = 2\pi \times 304.8 \times 25.4/1000^2 = 0.04864\,\text{m}^2.$$

For $p_1 = 337.9$ kPa and $T_1 = 282.2$ K, the ideal gas equation yields

$$\rho_1 = 4.172\,\text{kg/m}^3.$$

Next, $c_{a1} = \dfrac{\dot{m}}{\rho_1 A_1} = \dfrac{29.21}{4.172 \times 0.06030} = 116.1\,\text{m/s}.$

Finally, $a_1 = \sqrt{\gamma \mathcal{R} T_1} = \sqrt{1.399 \times 287.1 \times 282.2} = 336.7\,\text{m/s}.$

Impeller Inlet

Refer to Figure 6.9, which is the velocity polygon (to scale) for the impeller inlet.

First, $c_1 = \dfrac{c_{a1}}{\cos\alpha_1} = \dfrac{116.1}{\cos(-20^\circ)} = 123.6\,\text{m/s}.$

Also,

$$c_{u1} = c_1 \sin\alpha_1 = 123.6 \sin(-20^\circ) = -42.3\,\text{m/s},$$

and thus

$$\beta_1 = \cot^{-1}\left[\frac{c_{a1}}{c_{u1} - U_1}\right] = \cot^{-1}\left[\frac{116.1}{-42.3 - 135.7}\right] = -56.89^\circ \qquad \text{<ANS}$$

and $w_1 = \dfrac{c_{a1}}{\cos\beta_1} = \dfrac{116.1}{\cos(-56.89^\circ)} = 212.5\,\text{mps}.$

Note that the relative flow angle varies from -46.4° to -63.6° (quite significant) from the hub to eye.

Next, $M_{1\text{rel}} = \dfrac{w_1}{a_1} = \dfrac{212.5}{336.8} = 0.6312.$ <ANS

It is important to note that the flow is subsonic in the rotating frame. Thus, it is not possible for shock waves to be present within the impeller inlet.

Now, $M_{1\text{abs}} = \dfrac{c_1}{a_1} = \dfrac{123.6}{336.8} = 0.3671.$ <ANS

It is also noteworthy that the flow is subsonic in the stationary frame. Thus, the presence of shock waves in the duct before the impeller inlet is also not possible.

Next, $\dfrac{T_{t1}}{T_1} = 1 + \dfrac{\gamma - 1}{2} M_{1abs}^2 = 1 + \dfrac{0.399}{2} \times 0.3671^2 = 1.027;$

$T_{t1} = 1.027 \times 282.2 = 289.8\ \text{K}.$

Also, $\dfrac{p_{t1}}{p_1} = \left[\dfrac{T_{t1}}{T_1}\right]^{\frac{\gamma}{\gamma-1}} = [1.027]^{\frac{1.399}{0.399}} = 1.098;$

$p_{t1} = 1.098 \times 337.9 = 370.8\ \text{kPa}.$

Impeller Exit and Diffuser Inlet

For the impeller exit and diffuser inlet velocity polygon, refer to Figure 6.9. First, note that, in contrast to the inlet polygon and unlike an axial machine, U_2 is much larger than U_1. Next, the relative flow angle is known ($\beta_2 = -10°$); the negative sign implies backward swept. However, the other exit conditions are unknown. Thus, the problem will have to be solved by iteration. First, the exit density will be guessed (and will be checked later): $\rho_2 = 7.036\ \text{kg/m}^3$.

Thus,

$$c_{r2} = \frac{\dot{m}}{\rho_2 A_2} = \frac{29.21}{7.036 \times 0.04864} = 85.4\ \text{m/s}.$$

Next, $w_2 = \dfrac{c_{r2}}{\cos\beta_2} = \dfrac{85.4}{\cos(-10^\circ)} = 86.7\ \text{m/s}$

and $w_{u2} = c_{r2}\tan\beta_2 = 85.4\tan(-10°) = -15.1\ \text{m/s};$

thus, $\alpha_2 = \cot^{-1}\left[\dfrac{c_{r2}}{U_2 + w_{u2}}\right] = \cot^{-1}\left[\dfrac{85.4}{383.0 - 15.1}\right] = 76.93^\circ,$

$$c_2 = \frac{c_{r2}}{\cos\alpha_2} = \frac{85.4}{\cos(76.93°)} = 377.6\ \text{m/s},$$

and $c_{u2} = U_2 + w_{u2} = 383.0 - 15.1 = 367.9\ \text{m/s}.$

Next, the moment of momentum equation can be used to yield

$$\begin{aligned} h_{t2} - h_{t1} &= U_2 c_{u2} - U_1 c_{u1} = 383.0 \times 367.9 - 135.7 \times (-42.3) \\ &= 146{,}650\ \text{m}^2/\text{s}^2. \end{aligned}$$

From the energy equation, one can find

$$\frac{p_{t2}}{p_{t1}} = \left[\frac{\eta_{12}[h_{t2} - h_{t1}]}{c_p T_{t1}} + 1\right]^{\frac{\gamma}{\gamma-1}} = \left[\frac{0.90 \times 146{,}650}{1.005 \times 289.8 \times 1000} + 1\right]^{\frac{1.399}{0.399}} = 3.705,$$

and thus $p_{t2} = 3.705 \times 370.8 = 1373\ \text{kPa}\ (199\ \text{psia})$. <ANS

One should note that this pressure ratio is much higher than a single-stage axial compressor pressure ratio.

Also, $T_{t2} - T_{t1} = \dfrac{h_{t2} - h_{t1}}{c_p} = \dfrac{146{,}650}{1.005 \times 1000} = 145.9\ \text{K};$

therefore $T_{t2} = 435.7\ \text{K}.$

Now $T_2 = T_{t2} - \dfrac{c_2^{\,2}}{2c_p} = 435.7 - \dfrac{377.6^2}{2 \times 1.005 \times 1000} = 364.8\ \text{K}.$

Next, $a_2 = \sqrt{\gamma \mathcal{R} T_2} = \sqrt{1.399 \times 287.1 \times 364.8} = 382.8\ \text{m/s}$

and $M_{2\text{abs}} = \dfrac{c_2}{a_2} = \dfrac{377.6}{382.8} = 0.9865.$ <ANS

It is important to note that the flow is barely subsonic in the stationary frame. Thus, because of 2-D and 3-D effects, regions of supersonic flow may exist and some shock activity is therefore probable. Such Mach numbers are typical of centrifugal compressors near the exit because of the large radius and high rotational speeds. Moreover,

$$M_{2\text{rel}} = \frac{w_2}{a_2} = \frac{86.7}{382.8} = 0.2265.$$ <ANS

Note, however, that the flow is once again subsonic in the rotating frame. Thus, the presence of shock waves within the impeller exit is not possible.

$$\text{Now } \frac{p_{t2}}{p_2} = \left[1 + \frac{\gamma - 1}{2} M_{2\text{abs}}^2\right]^{\frac{\gamma}{\gamma-1}} = \left[1 + \frac{0.399}{2} \times 0.9867^2\right]^{\frac{1.399}{0.399}} = 1.863,$$

and so $p_2 = \dfrac{1373 \text{ kPa}}{1.863} = 736.6\,\text{kPa}\ (106.8\,\text{psia}).$ <ANS

The pressure coefficient is

$$C_{pr} = \frac{p_2 - p_1}{\frac{1}{2}\rho_1 w_1^2}$$

$$= \frac{(736.6 - 337.9)\,\text{kPa}}{\frac{1}{2} \times 4.172 \frac{\text{kg}}{\text{m}^2} \times 212.5^2 \frac{\text{m}^2}{\text{s}^2}} \times 1000 \frac{\text{Pa}}{\text{kPa}} \times \frac{\frac{\text{N}}{\text{m}^2}}{\text{Pa}} \times \frac{\text{kg-m}}{\text{N-s}^2} = 4.23.$$

Compared with an axial flow stage, this pressure coefficient is huge. Why is such a large pressure coefficient possible without surge? For an axial stage, surge occurs when the pressure gradient is large enough to overcome the fluid particle momentum such that the particle stops. In a radial stage, however, the fluid has the added centrifugal force in addition to momentum to aid in overcoming stall. As a result, such a stage can operate at a much higher pressure ratio without surge.

Next, from the ideal gas law for $p_2 = 736.6\,\text{kPa}$ and $T_2 = 364.8\,\text{K}$,

$\rho_2 = 7.036\,\text{kg/m}^3$,

which closely matches the initial guess. Thus, further iteration is not necessary. As noted in earlier chapters, iteration methods are presented in Appendix I, which would help in finding ρ_2. Finally, the input stage power can be found by

$$\begin{aligned} \dot{W}_{sh} &= \dot{m}(h_{t2} - h_{t1}) = 29.21 \times 146{,}650 = 4{,}284{,}000\,\text{W} \\ &= 4.284\,\text{MW}\ (5746\,\text{hp}). \end{aligned}$$ <ANS

Thus, all of the variables of interest have been found.

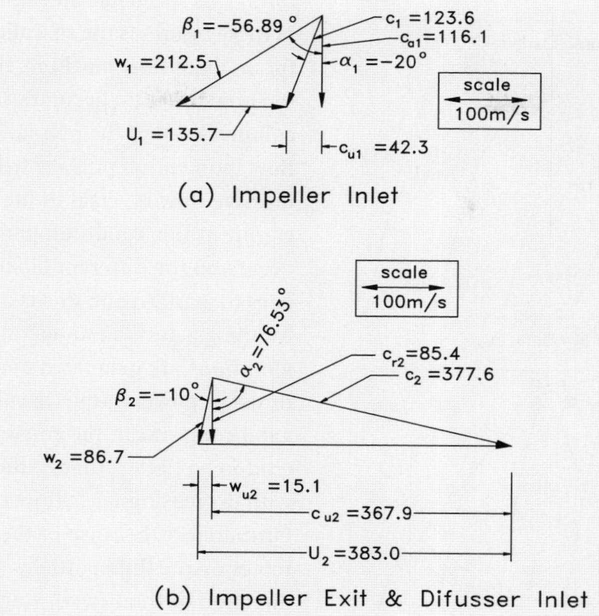

Figure 6.9 Velocity polygons for impeller inlet and exit (Example 6.1).

6.5 Performance Maps

6.5.1 Dimensional Analysis

As is the case for an axial flow compressor, radial machine performance curves or "maps" are usually used to provide an experimental working medium for a compressor engineer. Again, these provide an engineer with a quick and accurate view of the conditions at which the machine is operating and what can be expected of the machine if the flow conditions are changed. Of primary concern is the total pressure ratio of a compressor; namely, p_{t3}/p_{t2}. Once again, from Chapter 4 (Eqs. (4.3.3), (4.3.7), and (4.3.8)),

$$p_{t3}/p_{t2} = f\left\{\left[\frac{\dot{m}\sqrt{\theta_{t2}}}{\delta_{t2}}\right], \frac{N}{\sqrt{\theta_{t2}}}\right\} = f\{\dot{m}_{c2}, N_{c2}\}, \tag{6.5.1}$$

where

$$\delta_{t2} = p_{t2}/p_{stp}, \tag{6.5.2}$$

$$\theta_{t2} = T_{t2}/T_{stp}, \tag{6.5.3}$$

and the subscripts stp and t2 refer to the standard conditions and the inlet total condition to the compressor (which should not be confused with the exit total condition of a stage), respectively. Similarly, for the efficiency one finds (Eqs. (4.3.6), (4.3.7), and (4.3.8)) that

$$\eta = g\left\{\left[\frac{\dot{m}\sqrt{\theta_{t2}}}{\delta_{t2}}\right], \frac{N}{\sqrt{\theta_{t2}}}\right\} = g\{\dot{m}_{c2}, N_{c2}\}. \tag{6.5.4}$$

Again, $\dot{m}_{c2}$ and N_{c2} are not truly dimensionless. The first is called the "corrected" mass flow rate and has dimensions of mass flow, whereas the second is called the "corrected" speed and has dimensions of rotational speed. Thus, because they are not dimensionless, the resulting function cannot be used to correlate or compare different engines. The function for one particular engine can be used regardless of the external operating conditions, however, as is the usual case.

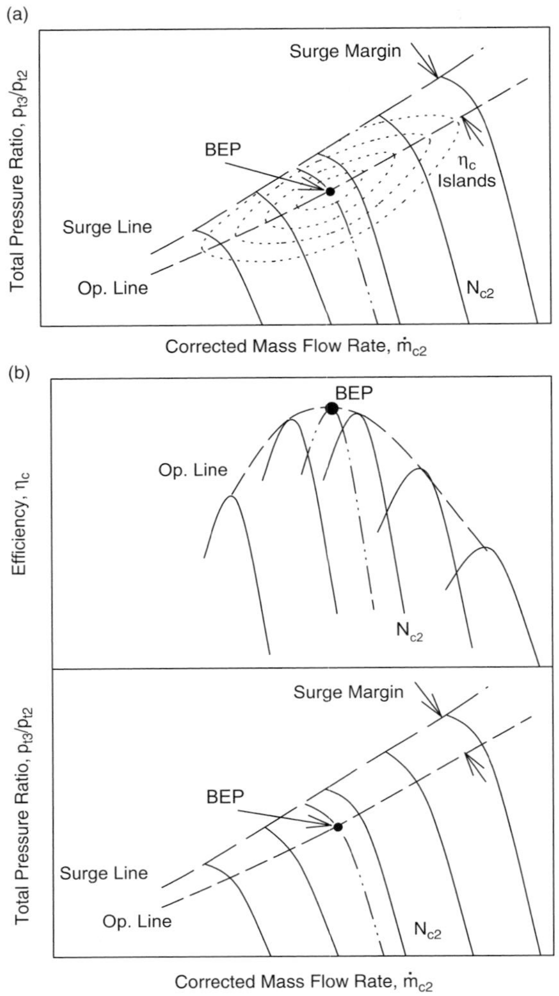

Figure 6.10 (a) Typical compressor map with efficiency islands. (b) Typical compressor map with efficiency plot.

6.5.2 Mapping Conventions

Finally, in Figures 6.10a and 6.10b, typical centrifugal compressor maps are presented. These formats are identical to axial flow compressor maps and are found experimentally with a similar test rig and apparatus. In both figures the pressure ratio is plotted versus the corrected mass flow with the corrected speed as a second parameter. Five constant speed lines are shown. The optimum speed line is shown with dotted lines. This speed has the highest efficiency. Also shown in these figures is the so-called surge line. As is the case for an axial flow machine, this condition occurs when the pressure ratio becomes so large across the compressor that the pressure gradient overwhelms the flow momentum plus centrifugal forces and causes massive flow reversal in the engine. Locally, in the engine at this condition massive flow separation occurs on the different blades and the performance goes to zero. Again, this is a very dangerous condition for the gas turbine to operate in because excessive vibrations are generated and the overall gas turbine thrust or power output becomes zero. In contrast to an axial compressor, the constant corrected speed lines tend to be flatter; that is, the increase in pressure ratio with decreasing mass flow rate is not as great. Furthermore, because of the additional centrifugal forces on the fluid particles (which are obviously in the direction of fluid flow), stage pressure ratios can be much higher than for an axial compressor stage before the onset of surge. In addition, the operating line is shown. This line passes through the design point and represents the different conditions at which an engine typically operates. The surge margin is the remaining performance information presented shown in these figures. This is the margin between the operating line and the surge line and is regarded a safety margin. In designing a compressor, one encounters a tradeoff because the objective is usually to design for a large pressure ratio at the operating condition. On the other hand, a small surge margin is undesirable because the compressor could surge as the result of a small operating perturbation. As for an axial compressor, once a compressor map is found, if the corrected mass flow and corrected speed are known, the total pressure ratio and efficiency can be found regardless of other conditions.

Also shown in Figure 6.10a are lines or "islands" of constant efficiency, shown with dotted lines. The typical design point of an engine is near the peak efficiency point. Thus, as one changes conditions from the design point, the efficiency decreases. In this figure, one can easily see that, along a constant speed line, as the mass flow decreases, the efficiency

Table 6.3 **Typical compressor and turbine characteristics at maximum power for gas turbines with centrifugal compressors.**

Gas turbine	Manufacturer	Type and no. of compressive stages	Overall pressure ratio	No. of turbine stages (LP + HP)
TPE331-1	Honeywell	2 R	8.3	3
ALF502	Honeywell	3F, 7A, 1 R	13.8	2 + 2
TFE731-3	Honeywell	1F, 4A, 1 R	14.6	3 + 1
T53A	Honeywell	5A, 1 R	7.0	2 + 2
T55-L-712	Honeywell	7A, 1 R	8.2	2 + 2
J69-T-29	Teledyne	1A, 1 R	5.3	1
PT6A-34	PWAC	3A, 1 R	7.0	1 + 1
T63-A-720	RR	6A, 1 R	7.3	2 + 2
250-B17C	RR	6A, 1 R	7.2	2 + 2
Dart	RR	2 R	5.6	3
T700-GE-700	GE	5A, 1 R	17.0	2 + 2
FJ44	Williams	1F, 1A, 1 R	12.7	2 + 1

F = fan stages; A = axial compressor stages; R = radial compressor stages.

increases to a maximum on the operating line and then decreases as the compressor nears surge.

In Figure 6.10b, a second method of mapping the compressor performance is shown. This has exactly the same data as Figure 6.10a but is in a different format. For this format the pressure ratio data is the same format as in Figure 6.10a. However, the efficiency data is plotted separately and is a function of the corrected mass flow with the corrected speed as another variable. This format actually presents the efficiency data is a more accurate manner. However, the presentation of Figure 6.10a is a better graphical presentation as it gives a better “feel” of efficiency and pressure ratio combined.

Lastly, Table 6.3 presents some characteristic parameters for selected engines. In particular, the reader should pay attention to the overall pressure ratios and number of stages and compare them with the axial flow compressor characteristics shown in Table 5.3. Table 6.3 was assembled from engine data in *Aviation Week & Space Technology*, *Flight International*, engine and airframe manufacturers' materials, brochures, web pages, and military information. Although every effort has been made to maintain accuracy by cross-checking, information is meant for reference only and is not intended for technical applications. Furthermore, specifications for gas turbines are forever changing due to improvements in design. Also, several of the engines have different “builds” for different applications and for those cases typical characteristics are presented.

6.6 Impeller Design Geometries

Although the “mean line” analysis presented in Section 6.4 can be used as a starting point for an impeller and diffuser design, one must consider other more advanced constraints and parameters as outlined by Aungier (2000) and the status of predictions as summarized by Adler (1980). Some considerations are addressed in this section.

6.6.1 Eye Diameter

The relative Mach number at the impeller inlet should be subsonic. Thus, the eye diameter must be chosen appropriately. If the eye becomes too large at R_{eye}, the blade velocity becomes large and, as a result, the relative velocity becomes large. On the other hand, if the eye is too small (resulting in a small cross-sectional area), for a given mass flow rate the axial velocity becomes large, which again results in a large relative velocity. In fact, one can show (Problem 6.13) that the eye diameter can be chosen to minimize the relative velocity.

6.6.2 Basic Blade Shapes

As discussed in Section 6.2, three basic blade shapes are used for a compressor impeller: backward-curved, forward-curved, and purely radial. These are shown in Figure 6.3f. For a given size compressor, the trends of the resulting maps are summarized in Figure 6.11 for one speed line. As can be seen, the forward-facing blades result in higher developed pressure ratios than do the other two geometries. However, for the three geometries, the forward-facing blades demonstrate the lowest efficiency and the greatest slip. On the other hand, the backward-facing blades demonstrate the highest efficiency and least slip of the three geometries but lowest pressure ratios.

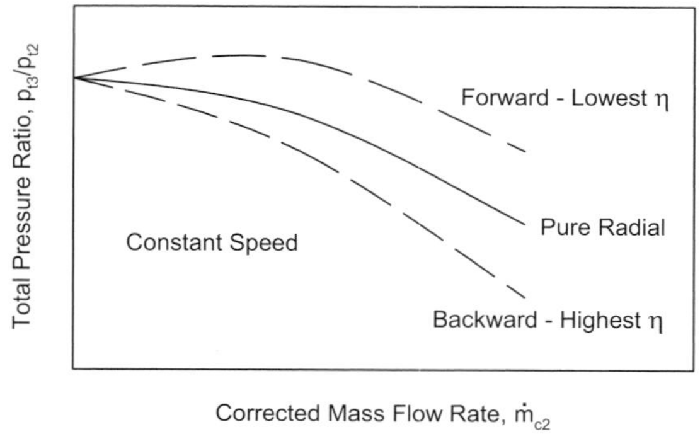

Figure 6.11 Trends of performance curve for different blade geometries.

6.6.3 Blade Stresses

In general, blade stresses increase as the square of the rotational speed. Of the three blade geometries, the purely radial blades demonstrate the lowest stress levels. This is because, at operating speeds, centripetal forces are trying to straighten the blades for the other two geometries. For aircraft and ground applications, the operating speeds are high (6000 to 20,000 rpm), and the resulting stresses for backward- or forward-facing blades would be prohibitively high. As a result, for aircraft engines and power generation gas turbines, purely radial blades are almost always used.

For centrifugal pumps, lower speeds (1800 to 3600 rpm) are used for comparison so that the working liquids do not cavitate. For these applications, backward-facing blades are usually utilized because of the high efficiencies. The only applications for forward-facing blades are designs that require the highest developed pressure rise possible in the smallest packages and in which thermodynamic efficiency is of lesser importance.

6.6.4 Number of Blades

The function of the blades is to guide the fluid through the impeller. If too few blades are used, the flow is more chaotic, and the fluid does not follow the desired path due to increased slip. As a result, more than the optimum required work is needed to obtain a given pressure rise. However, if too many blades are used, boundary layer and frictional effects become dominant and more important because of the finite thicknesses of the blades and resulting smaller flow passages. The friction causes pressure losses, and again more work is needed than is optimally required. This can be summarized by Figure 6.12. For a given speed, the product of slip factor × efficiency of the compressor stage increases owing to the rise in slip factor with an increasing number of blades until an optimum is reached. After this point, as more blades are added, the slip factor × efficiency product decreases as the result of declining efficiency.

The use of shortened blades, which are called splitter blades, is a frequent tradeoff. These blades do not traverse the entire flow path in the impeller. They usually begin at about half the meridional distance between the hub (at the inlet) and tip (exit) and continue to the tip. In this region, the radius is larger than near the hub, resulting in a larger spacing between primary blades; consequently, having more blades does not constrict the flow path too much. Thus, with splitter blades, more control of the flow is exercised near the exit (increasing the slip factor), but small flow passages are not encountered at the inlet. Splitter blades can be seen in Figures 6.3a, 6.3c (single face), and also in Figure 1.38 (TPE331) for the first compressor.

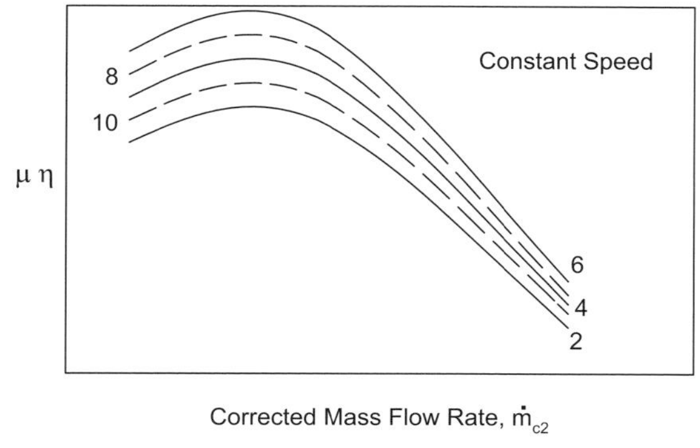

Figure 6.12 Trends of efficiency–slip factor variation with increasing blade number.

6.6.5 Blade Design

The overall design of a centrifugal compressor blade is more complex than that of an axial flow compressor blade. For an axial flow machine, the flow is largely 2-D and conventional blade shapes are mainly used (specified NASA, NACA, etc., shapes). However, the flow field is primarily 3-D for centrifugal machines. As a result, because of the blade curvature and twisting in all three planes, obtaining the optimum blade shape for a given application is a more difficult process for a centrifugal machine than for an axial compressor. In fact, the design of centrifugal machines is much less straightforward than for axial flow compressors, and the design process therefore becomes somewhat of an art. More iterations during design and testing are often required for a centrifugal compressor than for an axial flow compressor.

6.7 Diffusers and Crossovers

6.7.1 Vaned Diffusers and Vaneless Diffusers

Once the flow exits the impeller, it enters the diffuser. The purpose of a diffuser is similar to that of a stator cascade in an axial compressor. That is, the static pressure is further increased in the diffuser. Two types of diffusers are used: vaneless and vaned. In Figure 6.13, two variations of vaned diffuser schematics are shown. In Figure 6.14, a photograph is presented for a Nuovo Pignone/GE power generation unit. For these vaned geometries, a series of stationary vanes are used to ensure that the flow swirling and decelerating will not be chaotic. In general in more modern diffusers, wedge-shaped vanes are often used;

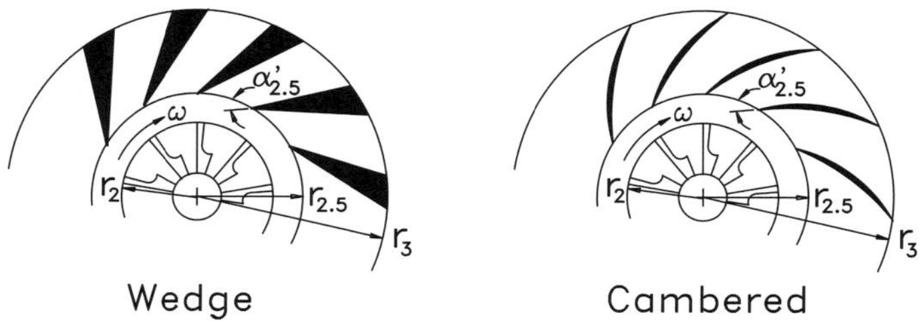

Figure 6.13 Two vaned diffuser geometries.

Figure 6.14 Nuovo Pignone/GE vaned diffuser (courtesy of General Electric Gas Power).

in older geometries, curved or airfoil-shaped vanes (cambered) were used. As shown in Figure 6.14, the blades are initially in the radial plane but at the outer radius of the diffuser wrap into the axial direction, thus controlling the three-dimensionality of the flow. In general the wedge geometry has a better efficiency than a cambered version. However, in general, this geometry has less range than the cambered version for off-design operation.

As noted earlier, owing to the large radius of the impellers and high rotational speeds, very high velocities entering the diffuser are encountered. These can be high subsonic velocities, but Mach numbers of 1.3 are not uncommon especially for compressors with high-pressure ratios. As a result, the inlet geometry, blade angle, wedge angle, or camber leading-edge angle becomes very important to minimize the resulting shock wave and ensuing pressure losses. Much more discussion on oblique shock waves is included in Appendix C.8 and Chapter 8.

As can be seen, the radius of the diffuser inlet ($r_{2.5}$) may be significantly larger than the exit of the impeller (r_2), and thus the flow conditions change between the two components. For this reason, the new position number 2.5 is used. If the two radii are essentially the same, only station number 2 is used. The distance between the impeller exit and diffuser inlet is an important design parameter. If the distance is too small, considerable vibration and noise can occur owing to the interaction of the flow between the impeller blades and diffuser vanes, as discussed by Caruthers and Kurosaka (1982). If the distance is too large, the effect of the vanes is diminished. As is true for an axial flow rotor and stator, the number of blades and vanes must be different and cannot have common multiples or significant resonance will be generated. Such resonances cause significant noise and vibration problems (with shortened blade or vane lives).
A vaneless diffuser does not contain these vanes and is a simple collection or scroll region.

Both vaned and vaneless diffusers have their advantages. For example, a vaned diffuser usually has fixed vanes; the angles do not change as a function of operating conditions as do variable stators. As a result, the inlet vane angle ($\alpha'_{2.5}$; see Figure 6.13) matches the flow angle for only one flow rate/rotational speed condition. Thus, for off-design operation, nonzero incidence angles result, flow separation occurs, and significant losses are incurred, thereby decreasing the efficiency. On the other hand, a vaneless diffuser does not induce these separation or frictional losses when operating at off-design conditions. Thus, at off-design conditions the efficiency is best for a vaneless diffuser. However, the flowfield is more chaotic at design conditions with a vaneless diffuser than for a vaned diffuser. Thus, at design conditions the efficiency is highest for a vaned diffuser. Another advantage of a vaneless diffuser is that no noise-generating impeller blade and diffuser vane interactions are present. Lastly, vaned diffusers tend to be heavier because of the extra material.

In Figure 6.15, typical map trends are shown for two similar-sized compressors; one has a vaned diffuser whereas the other has a vaneless diffuser. As can be seen, the highest pressure ratios are obtained for the vaned diffuser. Thus, the pressure rise per size of compressor is highest for a vaned diffuser. One must

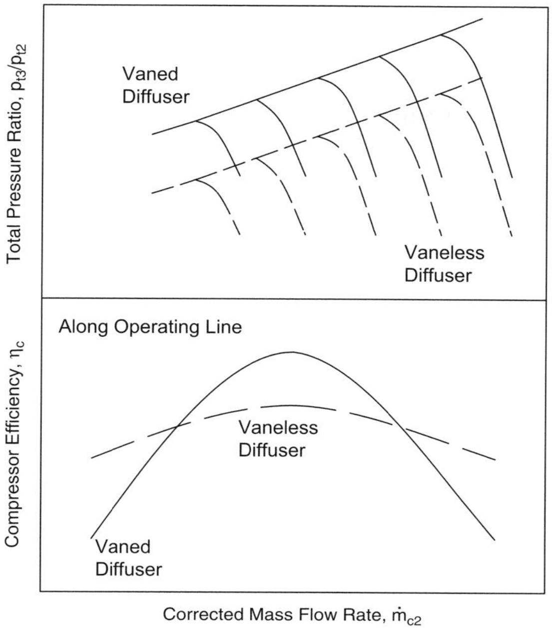

Figure 6.15 Performance and efficiency trends of vaneless and vaned diffusers.

remember, however, that the power input may be significantly higher for such a machine.

Also in Figure 6.15, the efficiency along the operating lines for two similar centrifugal compressors is shown. The effect of vaned vs. vaneless diffuser design is easily seen. That is, at the peak efficiency, the vaned diffuser yields the largest efficiency. However, as the corrected mass flow rate deviates from the maximum efficiency point, the efficiency drops much more rapidly for the vaned diffuser design because of the resulting incidence losses on the vanes. On the other hand, for the vaneless diffuser design the efficiency decreases much more gradually as the mass flow is changed from the maximum efficiency point. Thus, if a compressor is to be operated over a wide range of flow conditions, a vaneless diffuser is typically used. However, if the compressor is to be run at basically one flow rate, a vaned diffuser is typically utilized.

Generally, for a centrifugal compressor the percent reaction is not normally used as a performance measure. Nonetheless, the design of a centrifugal compressor is similar to that of an axial machine as far as loading the components is concerned. That is, the impeller and diffuser of a well-designed centrifugal compressor will be about equally loaded. For example, the static pressure rise across the impeller will be nearly the same as that of the diffuser. Thus, the static pressure rise across the impeller will be around half of that for the stage.

Lastly, as noted above, the exit Mach number of a centrifugal compressor is high (often supersonic) owing to the high rotational speed and large diameter of the wheel. However, with a properly designed diffuser, either vaned or vaneless, this difficulty can be overcome. A design guideline that is commonly used in evaluating an impeller design is that the impeller exit, Mach number in the stationary frame based on only the radial component of velocity should be less than unity. When supersonic flow exists at the impeller exit, the flow passage of the diffuser must be designed to be converging–diverging so that the flow is decelerated to a subsonic condition before entering the crossover or combustor.

6.7.2 Crossovers

When multiple centrifugal impellers are used, a "crossover" is used. This stationary component is shown in Figures 6.1, 6.3.d, and 6.16. Namely, after exiting the diffuser, flow exits the impeller in the radial direction, then is turned to the axial direction, and then turned inward toward the centerline in the reverse radial direction, and finally turned in the axial direction and is ready to enter the next impeller. In Figure 6.16 two geometries are shown. The simplest has basically strictly radial inflow in the return section. A more complex design also has a negative axial component in the return section. This design reduces the axial length. In both cases the flow is turned by sections that have approximately constant radii, as shown in Figure 6.16. Clearly the radii must be chosen to be large enough to reduce separation. All sections must be chosen to be wide enough such that wall friction is minimized. Aungier (2000) covers many aspects of advanced diffuser and crossover design in great detail and is a very good reference.

Table 6.4 **Centrifugal compressor characteristics compared to those of an axial flow compressor.**

Governing equations	↔ Same
Pressure ratio	↑ Higher
Net size	↓ Lower
Weight	↓ Lower
Frontal area	↑ Higher
Flow rate	↓ Lower
Net power	↓ Lower
Pressure ratio at surge	↑ Higher
Manufacturing cost	↓ Lower
Number of parts	↓ Lower
Durability	↑ Higher
Efficiency	↓ Lower
Operating range	↑ Higher
Maximum Mach number	↑ Higher

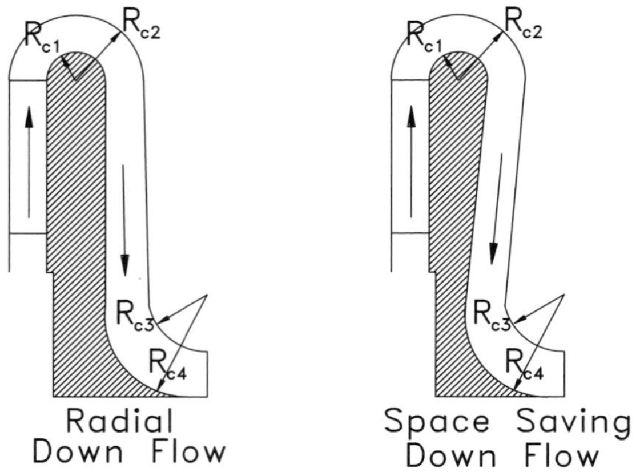

Figure 6.16 Complex crossover section.

6.8 Comparison of Centrifugal and Axial Flow Compressors

Throughout this chapter comparisons were made to axial flow compressors. At this point these will be summarized and complimented with other comparisons. First, the governing equations are the same for both centrifugal and axial compressors. However, the velocity polygons are treated much differently because of the large change in radius from inlet to exit for the centrifugal units. The pressure ratio per stage is much higher for a centrifugal unit. For a centrifugal compressor this can be as high as 3 or 4 whereas for an axial compressor it is only about 1.25. A centrifugal compressor will surge at a much higher pressure ratio due to the centrifugal forces on fluid particles. Maximum flow rates are much lower for the centrifugal compressor. As a result, the maximum power or thrust is much lower for a gas turbine with centrifugal compressors. With lower flow rates the net size and weight of gas turbines with centrifugal compressors is lower. However, because of the large wheel diameters the frontal area is higher. These characteristics lead to applications for smaller aircraft or power applications where minimizing size is important with lower net output. The centrifugal compressors have fewer parts. Also, manufacturing compressor wheels is less expensive than for an axial machine where multiple blades and attachment is required. Since the compressor wheels are one piece, they are more durable (like debris ingested) than an axial compressor stage. Because of the large radius ratio of a centrifugal compressor wheel, the exit Mach number is often supersonic, which makes the design of the diffuser very important. The efficiency of a centrifugal unit is lower than that of an axial compressor. However, the operating range of a centrifugal compressor is wider than the axial counterpart. Namely, the efficiency does not drop as dramatically as the conditions deviate from the design point. This also is evident on the maps as the pressure rise vs. mass flow rate tend to be flatter for the centrifugal compressor. To summarize these points Table 6.4 is presented.

6.9 Summary

This chapter has addressed the thermodynamics of a strictly centrifugal compressor. As with an axial compressor, the basic idea is to impart kinetic energy to the incoming fluid and convert the increase in energy to an increase in total pressure. Centrifugal compressors are used on smaller engines with lower flow rates but higher single-stage pressure ratios than their axial counterparts. Geometries and hardware were discussed, and the important geometric parameters were defined. For this geometry, the air was seen to enter a rotating impeller or "wheel" axially at a small radius and exit radially at a large radius. Single-faced and double-faced impellers were discussed. To analyze the turbomachines, velocity triangles or polygons were again used. To analyze the compressors, 2-D flow was again assumed – separately at the inlet and exit. At the inlet of the impeller, the radial components of velocity were assumed to be small and the polygons were similar to those for an axial flow compressor, whereas at the impeller exit the axial components of velocity were assumed to be negligible. Mixed-flow compressor impellers (axial and radial flow at the exit) were identified but not covered in detail.

Knowing the geometric and flow parameters made it possible to develop tools using the continuity, momentum, and energy equations with a control volume approach through which the resulting total pressure ratio and required power of a stage can be predicted. The software Turbomachinery.exe performs the above calculations and is particularly helpful when optimizing performance because it eliminates repetitive and detailed calculations. The same basic equations were found to apply equally to a centrifugal machine and an axial compressor; however, application of the equations to the geometries was significantly different. Because of the large difference in inlet and exit radii of the impeller, centrifugal compressors were found to generate much larger single-stage pressure ratios than axial flow compressors. Incompressible flow (centrifugal hydraulic pumps) was also covered only to emphasize the fundamentals. The concept of slip was introduced, which accounts for the fact that the fluid flow does not exactly follow the blades. Slip factor correlations were presented. Also, performance maps (again derived from dimensional analysis) were described by which the important operating characteristics of a compressor can accurately and concisely be presented, and these had the same form as those used for an axial flow compressor. As is true for an axial compressor, four of the most important output characteristics are the pressure ratio, efficiency, operating line, and surge line. For example, if the speed and flow rates are known, from a map the efficiency and pressure ratio can be found. The shapes and slopes of the speed lines on the maps are somewhat different for centrifugal and axial flow compressors. Surge at the high single-stage pressure ratios is not as great a problem as it is for axial flow compressors; this is because centrifugal forces on the fluid particles tend to help the fluid move toward the exit despite the adverse pressure gradient in the passages. Also, the different types of impeller exit blade shapes were discussed: forward-facing, backward-facing, and strictly radial blades. Owing to centrifugal forces and stress considerations, large high-speed impellers almost always have strictly radial blades. Also, an optimal number of blades can be found to offset large losses, which occur with many blades (small flow areas), and a low slip factor (large slip), which occurs for a small number of blades. Adding short splitter blades near the impeller exit can also decrease slip without adding too many losses.

Finally, the different operating characteristics of vaneless and vaned diffusers were discussed. Two types of vaned diffusers were identified: wedge and cambered. Wedge vanes yield better efficiency but have a shorter range. Also, a vaned diffuser exhibits a better efficiency at a single, design point, but the efficiency of a vaneless diffuser does not vary as much over a range of conditions.

Diffusers must be designed recognizing that supersonic flow is often encountered at the inlet, thus not only must shocks must be considered but the passage shape must decelerate the flow. Lastly, centrifugal compressor characteristics were compared to those of axial flow units. Design guidelines for centrifugal compressors were again presented throughout the chapter.

List of Symbols

A	Area
a	Speed of sound
$\widehat{a}$	Parameter in slip correlation
b	Blade height
c	Absolute velocity
c_p	Specific heat at constant pressure
C_p	Pressure coefficient
D	Diameter
g	Gravitational constant
H	Head
h	Specific enthalpy
$\dot{m}$	Mass flow rate
M	Mach number
N	Rotational speed
p	Pressure
$\mathcal{R}$	Ideal gas constant
R	Radius
S	Parameter in slip correlation
T	Temperature
U	Blade velocity
w	Relative velocity
$\dot{W}$	Power
x	Meridional direction
Z	Number of blades
α	Absolute angle
β	Relative (rotor) angle
γ	Specific heat ratio
δ	Ratio of pressure to standard pressure
η	Efficiency
θ	Ratio of temperature to standard temperature
μ	Slip factor
ρ	Density
ω	Rotational speed

Subscripts

a	Axial
abs	In absolute frame
c	Corrected
e	Euler
r	Radial
rel	In rotating frame
sh	Applied to shaft
stp	Standard conditions
t	Total (stagnation)
u	Tangential
1,2,2.5,3	Positions in stage
2	Inlet to compressor
3	Exit of compressor

Superscripts

$'$	For the blade or ideal

Problems

6.1 A stage of a radial compressor is to be analyzed. It rotates at 16,300 rpm and compresses 50.0 lb_m/s of air. The inlet pressure and temperature are 40 psia and 500 °R, respectively. The hub and tip radii of the blades at the inlet are 2.0 and 4.5 in., respectively, and the exit radius is 8.0 in. and the exit blade height is 0.85 in. The slip factor is unity. Flow enters the inducer with no preswirl, and the impeller has straight radial blades. The efficiency of the stage is 92 percent. The values of c_p and γ are 0.2400 BTU/lb$_m$- °R and 1.401, respectively. Find the following:

(a) relative flow angle at the inlet (hub, mean, eye),
(b) the static pressure at the impeller exit,
(c) the total pressure ratio for the stage (describe how this compares with an axial compressor stage),
(d) the Mach numbers at the impeller inlet and exit,
(e) the required power for the stage.

6.2 A stage of a radial compressor is to be analyzed. It rotates at 16,300 rpm and compresses 50.0 lb$_m$/s of air. The inlet pressure and temperature are 40 psia and 500 °R, respectively. The hub and tip radii of the blades at the inlet are 2.0 and 4.5 in., respectively, the exit radius is 8.0 in., and the exit blade height is 0.85 in. The slip factor is unity. Flow enters the inducer with −10° of preswirl, and the impeller has forward-swept blades at 10°. The efficiency of the stage is 92 percent. The values of c_p and γ are 0.2400 BTU/lb$_m$-°R and 1.401, respectively. Find the following:

(a) relative flow angle at the inlet (hub, mean, eye),
(b) the static pressure at the impeller exit,
(c) the total pressure ratio for the stage (describe how this compares with an axial compressor stage),
(d) the Mach numbers at the impeller inlet and exit,
(e) the required power for the stage.

6.3 A stage of a radial compressor is to be analyzed. It rotates at 12,300 rpm and compresses 31.75 kg/s (70.0 lb$_m$/s) of air. The inlet pressure and temperature are 241.3 kPa and 305.6 K (35 psia and 550 °R), respectively. The hub and tip radii of the blades at the inlet are 76.3 and 139.7 mm (3.0 and 5.5 in.), respectively, the exit radius is 279.4 mm (11.0 in.), and the exit blade height is 25.4 mm (1.00 in). The slip factor is unity. Flow enters the inducer with no preswirl, and the impeller has straight radial blades. The efficiency of the stage is 88 percent. The values of c_p and γ are 1.100 kJ/kg-K (0.2413 BTU/lb$_m$-°R) and 1.397, respectively. Find the following:

(a) relative flow angle at the inlet (hub, mean, eye),
(b) the static pressure at the impeller exit,
(c) the total pressure ratio for the stage (describe how this compares with an axial compressor stage),
(d) the Mach numbers at the impeller inlet and exit,
(e) the required power for the stage.

6.4 A stage of a radial compressor is to be analyzed. It rotates at 12,300 rpm and compresses 31.75 kg/s (70.0 lb$_m$/s) of air. The inlet pressure and temperature are 241.3 kPa and 305.5 K (35 psia and 550 °R), respectively. The hub and tip radii of the blades at the inlet are 76.2 and 139.7 mm (3.0 and 5.5 in.), respectively, the exit radius is 279.4 mm (11.0 in.), and the exit blade height is 25.4 mm (1.00 in). The slip factor is unity. Flow enters the inducer with +10.0° of preswirl, and the impeller has straight radial blades. The efficiency of the stage is 88 percent. The values of c_p and γ are 1.010 kJ/kg-K (0.2412 BTU/lb$_m$-°R) and 1.397, respectively. Find the following:

(a) relative flow angle at the inlet (hub, mean, eye),
(b) the static pressure at the impeller exit,
(c) the total pressure ratio for the stage (describe how this compares with an axial compressor stage),
(d) the Mach numbers at the impeller inlet and exit,
(e) the required power for the stage.

6.5 A stage of a radial compressor is to be designed. It compresses 6.0 lb$_m$/s of air, and the inlet pressure and temperature are 15 psia and 520 °R, respectively. Because of geometric constraints, the hub radius at the inlet is 1.0 in., and the exit radius is 3.0 in. The slip factor is unity. Flow enters the inducer with no preswirl. The efficiency of the stage is 88 percent. The values of c_p and γ are 0.241 BTU/lb$_m$-°R and 1.40, respectively. Clearly state the design objectives and proceed to design the stage. When finished, check to see if the design objectives were met and draw the impeller geometry.

6.6 A centrifugal compressor has an outside radius of 15 in. and a blade height (passage width) of 2 in. at the exit. The exit blade angle is 7° (backward swept) relative to the radial direction. The slip factor is unity. The inlet has radii of 1.2 and 8.4 in. The inlet swirl angle is 10° (positive) relative to the axial direction. The shaft runs at 9500 rpm, and the mass flow rate of air is 100 lb_m/s. Experimental measurements indicate the inlet and exit static pressures are 50 and 79.6 psia, respectively, whereas the inlet and exit static temperatures are 750 and 874.9 °R, respectively. Assume $\gamma = 1.387$. Find the following:

(a) the inlet blade angle (hub, mean, eye),
(b) the efficiency,
(c) the power.

6.7 Initially, a radial compressor impeller operates with purely radial exit blades and no preswirl.

(a) In a later test, negative preswirl is added before the inducer. What happens to the pressure rise across the impeller for the same flow rate and rotational speed? Explain with velocity polygons.
(b) In another later test, a similar impeller is used with slightly backward-facing blades (again no preswirl). What happens to the pressure rise across the impeller for the same flow rate and rotational speed (compared with the original test)? Explain with velocity polygons.

6.8 A stage of a radial compressor is to be analyzed. It rotates at 14,000 rpm and compresses 13.60 kg/s (30.0 lb_m/s) of air. The inlet pressure and temperature are 262.0 kPa and 286.1 K (38 psia and 515 °R), respectively. The hub and tip radii of the blades at the inlet are 48.26 and 101.67 mm (1.9 and 4.0 in.), respectively, the exit radius is 177.8 mm (7.0 in.), and the exit blade height is 15.24 mm (0.60 in). Flow enters the inducer with −5° of preswirl, and the impeller has forward-swept blades at 5°. The slip factor is unity. The efficiency of the stage is 91 percent. The values of c_p and γ are 1.002 kJ/kg-K (0.2392 BTU/lb_m-°R) and 1.402, respectively. If needed, you may assume that the impeller exit density is 3.763 kg/m^3 (0.0073 sl/ft^3) but indicate where this is used and what you will do to justify the assumption. Find the following:

(a) the relative flow angle at the inlet (hub, mean, eye),
(b) the static pressure at the impeller exit,
(c) the total pressure ratio for the stage (describe how this compares with an axial compressor stage),
(d) the Mach numbers at the impeller inlet and exit,
(e) the required power for the stage.

6.9 You are given a single-stage centrifugal air compressor with a mass flow rate of 24.95 kg/s (55 lb_m/s) and a rotational speed of 14,000 rpm. The blades are purely radial at the impeller exit to minimize stresses, and the efficiency is 82 percent. Flow enters the impeller eye with a swirl component (c_u), which is 7.75 percent of the total incoming velocity (c). The exit radius of the impeller is 254.0 mm (10 in.), whereas the inlet tip and hub radii are 127.0 and 50.8 mm (5 and 2 in.), respectively. The blade height is 22.86 mm (0.9 in.) at the impeller exit. The slip factor is unity. The incoming total temperature is 388.9 K (700 °R), and the incoming total pressure is 449.5 kPa (65.2 psia). For the problem solution, an initial guess of 4.805 kg/m^3 (0.300 lb_m/ft^3) for the exit density is suggested. Find the following:

(a) p_{t2}/p_{t1} (describe how this compares with a single-stage axial compressor),
(b) the power,
(c) the inlet blade angle.

6.10 You are given a single-stage centrifugal air compressor with a mass flow rate of 18.14 kg/s (40 lb_m/s) and a rotational speed of 13,000 rpm. The blades are purely radial at the impeller exit to minimize stresses, and the efficiency is 87 percent. Flow enters the impeller eye with no preswirl. The exit radius of the impeller is 304.8 mm (12 in.), and the inlet tip and hub radii are 114.3 and 63.5 mm (4.5 and 2.5 in.), respectively. The blade height is 25.4 mm (1.0 in.) at the impeller exit. The slip factor is unity. In the absolute reference frame, the incoming total temperature is 440.6 K (793 °R), and the incoming total pressure is 421.3 kPa (61.1 psia). The incoming static temperature is 416.7 K (750 °R), and the incoming static pressure is 344.7 kPa (50.0 psia). The inlet axial velocity is 221 m/s (728 ft/s). The specific heat at constant pressure is 1.030 kJ/kg-K (0.246 BTU/lb_m-°R), and the specific heat ratio is 1.386. The exit density is 4.510 kg/m^3 (0.00875 sl/ft^3). Find the following:

(a) p_{t2}/p_{t1} (describe how this compares with a single-stage axial compressor),
(b) the power,
(c) the inlet blade angle to the impeller (hub, mean, eye).

6.11 In 1975, a series of tests were performed by D. Eckardt (1975) on a small, single-stage centrifugal air compressor. You are to analyze this wheel for the following conditions: a mass flow rate of 11.7 lb_m/s and a rotational speed of 14,000 rpm. The blades are purely radial at the impeller exit to minimize stresses. Flow enters the impeller eye with no preswirl. The exit radius of the impeller is 7.87 in., and the inlet tip and hub radii are 5.52 and 1.77 in., respectively. The blade height is 1.02 in. at the impeller exit. The slip factor is unity. The incoming temperature is 513.2 °R, and the incoming pressure is 14.13 psia. For the problem solution, an initial guess of 0.003 sl/ft^3 for the exit density is suggested. Find the following:

(a) the corrected mass flow rate,
(b) the corrected speed,
(c) the efficiency.

Using this efficiency, predict the following:

(d) p_{t2}/p_{t1} (describe how this compares with the measured value as well as how this compares with the measured value for a single-stage axial compressor),
(e) the power,
(f) the inlet blade angle (hub, mean, eye).

6.12 For a centrifugal compressor, as the preswirl increases from a negative to a positive value, what happens to the developed pressure rise? Assume all parameters remain constant except the preswirl. Explain with velocity polygons and equations.

6.13 Air flow with density ρ enters a centrifugal impeller at a mass flow rate of $\dot{m}$ with no preswirl. The impeller rotates with angular velocity ω and has a hub radius of R_h.

(a) Find the inlet tip (eye) radius for which the relative velocity at the same radius is minimized.
(b) Numerically evaluate the condition (radius and Mach number) for a mass flow rate of 5 lb_m/s, a density of 0.002327 sl/ft^3, an inlet hub radius of 1.00 in., and a rotational speed of 25,000 rpm. The inlet temperature is 530 °R.
(c) How much does the relative Mach number at the inlet tip radius depend on the inlet tip radius? That is, for part (b), find the relative Mach number at the inlet tip radius if the tip radius is doubled or halved from the optimum value.

6.14 A centrifugal compressor with no inlet preswirl originally has an impeller with four radial blades and a slip factor of 0.70. A second impeller will be used to replace the original impeller. The new impeller has eight blades and eight splitter blades. The slip factor of the second impeller is 0.98. The efficiencies of the two impellers are identical. Sketch the two impellers. Which of the two impellers will produce the largest pressure rise? Explain with velocity polygons.

6.15 You are given a single-stage centrifugal air compressor with a mass flow rate of 19 lb_m/s and a rotational speed of 14,500 rpm. The blades are backward swept at 15° (relative to the radial direction) at the impeller exit. The efficiency is 89 percent, and you can consider the slip factor to be unity. Flow enters the impeller eye with +10° of preswirl. The exit radius of the impeller is 9.0 in., whereas the inlet tip and hub radii are 5.0 and 3.0 in., respectively. The blade height is 1.0 in. at the impeller exit. In the absolute reference frame, the incoming total temperature is 552 °R, and the incoming total pressure is 18.3 psia. The incoming static temperature is 500 °R, and the incoming static pressure is 13.0 psia. The inlet axial velocity is 776 ft/s. The average blade speed at the inlet is 506 ft/s, and the blade speed at the exit is 1139 ft/s. The specific heat at constant pressure is 0.240 BTU/lb_m-°R, and the specific heat ratio is 1.400. The exit density is 0.00356 sl/ft^3. Find the following:

(a) the velocity polygons (to scale),
(b) the inlet blade angle of the impeller assuming no incidence (hub, mean, eye),
(c) the power required,
(d) p_{t2}/p_{t1}.

6.16 You are given a single-stage centrifugal air compressor that was designed to operate with a mass flow rate of 19 lb_m/s and a rotational speed of 14,500 rpm. The blades are backward swept at 15° (relative to the radial direction) at the impeller exit. The design efficiency is 89 percent, and the slip factor is unity. Flow enters the impeller eye with +10° of preswirl. The exit radius of the impeller is 9.0 in., and the inlet tip and hub radii are 5.0 and 3.0 in., respectively. The blade height is 1.0 in. at the impeller exit. In the absolute reference frame, the incoming total temperature is 552 °R and the incoming total pressure is 18.3 psia. If the compressor is run at speeds of 13,500 and 15,500 rpm, the maximum efficiency is 85 percent. Find the following:

(a) the inlet blade angle of the impeller if it is assumed there is no incidence at the design condition,
(b) the power required at the design condition,
(c) the mass flow rates at 13,500 and 15,500 rpm if the flow is incidence free,
(d) p_{t2}/p_{t1} at all three speeds.
(e) Plot p_{t2}/p_t vs. mass flow rate. What does this curve represent?

6.17 You are given a single-stage centrifugal air compressor designed to operate with air and a mass flow rate of 19 lb_m/s and a rotational speed of 14,500 rpm. The blades are backward swept at 15° (relative to the radial direction) at the impeller exit. The design efficiency is 89 percent, and the slip factor is unity. Flow enters the impeller eye with +10° of preswirl. The exit radius of the impeller is 9.0 in., and the inlet tip and hub radii are 5.0 and 3.0 in., respectively. The blade height is 1.0 in. at the impeller exit. In the absolute reference frame, the incoming total temperature is 552 °R, and the incoming total pressure is 18.3 psia. Find the following:

(a) the inlet blade angle of the impeller on the assumption there is no incidence at the design condition,
(b) the power required at the design condition,
(c) p_{t2}/p_{t1}.
(d) If the air is replaced with helium, what are the design mass flow rate, power, and p_{t2}/p_{t1} ?

6.18 You are given a single-stage centrifugal air compressor designed to operate with a mass flow rate of 19 lb_m/s and a rotational speed of 14,500 rpm. The blades are backward swept at 15° (relative to the radial direction) at the impeller exit. The design efficiency is 89 percent, and the slip factor is unity. Flow enters the impeller eye with +10° of preswirl. The exit radius of the impeller is 9.0 in., and the inlet tip and hub radii are 5.0 and 3.0 in., respectively. The blade height is 1.0 in. at the impeller exit. In the absolute reference frame, the incoming total temperature is 552 °R, and the incoming total pressure is 18.3 psia. If the compressor is run at mass flow rates of 17.5 and 20.5 lb_m/s and a rotational speed of 14,500 rpm, the efficiency drops to 75 percent. Find the following:

(a) the inlet blade angle of the impeller if it is assumed there is no incidence at the design condition (hub, mean, eye),
(b) the power required at the design condition,
(c) p_{t2}/p_{t1} at all three mass flow rates if the efficiency remains at the design value,
(d) p_{t2}/p_{t1} at all three mass flow rates if the efficiency drops from the design value as indicated.
(e) Plot p_{t2}/p_{t1} vs. mass flow rate for both efficiency cases. What do these curves represent?

6.19 Initially the impeller of a centrifugal flow compressor, which has positive preswirl and purely radial blades at the exit, operates incidence free at rotational speed $N1$. In a later test, the rotational speed is increased to $2 \times N1$. For the second test, the flow at the impeller inlet is to remain incidence free.

(a) Estimate quantitatively by how much the mass flow rate will have to be adjusted (i.e., increase or decrease and by how much).
(b) Estimate what happens quantitatively to the required power (i.e., increase or decrease and by how much).

Explain with scaled velocity polygons and equations.

6.20 Initially, a radial compressor impeller operates with purely radial exit blades and positive preswirl.

(a) In a later test, positive preswirl is removed so that the inlet flow is swirl free before the inducer. What happens to the pressure rise across the impeller for the same flow rate and rotational speed? Explain with velocity polygons.
(b) In another later test, a similar impeller is used with backward-facing blades (with the same original positive preswirl). What happens to the pressure rise across the impeller for the same flow rate and rotational speed (compared with the original test)? Explain with velocity polygons.

6.21 A stage of a radial compressor is to be designed for an engine. The engine has five axial compressor stages and one centrifugal compressor stage; the total pressure ratio for the entire compressor is 7.0. The centrifugal compressor stage compresses 12.0 lb_m/s of air, and the inlet pressure and temperature are 30 psia and 650 °R, respectively. Because of geometric constraints, the hub radius at the inlet is 3.0 in. and the exit radius is 8.0 in. Flow enters the inducer with no preswirl. The efficiency of the stage is 90 percent. Clearly state the design objectives and proceed to design the stage. When finished, check to see if the design objectives were met and draw the impeller geometry.

6.22 You are given a single-stage centrifugal air compressor with a mass flow rate of 22.68 kg/s (50 lb_m/s) and a rotational speed of 11,000 rpm. At the impeller exit the blades are radial but the flow exits the impeller with backward sweep of 20° (relative to the radial direction). The efficiency is 88 percent. Flow enters the impeller eye with +5° of preswirl. The exit radius of the impeller is 355.6 mm (14.0 in.), and the inlet tip and hub radii are 177.8 and 76.2 mm (7.0 and 3.0 in.), respectively. The blade height is 25.4 mm (1.0 in.) at the impeller exit. In the absolute reference frame, the incoming total temperature is 352.2 K (634 °R) and the incoming total pressure is 167.5 kPa (24.3 psia). The incoming static temperature is 333.3 K (600 °R), and the incoming static pressure is 137.9 kPa (20.0 psia). The inlet axial velocity is 194.2 m/s (637 ft/s), and the exit radial velocity is 180.4 m/s (592 ft/s). The input power is 3138 kW (4209 hp). The average blade speed at the inlet is 146.3 m/s (480 ft/s), and the blade speed at the exit is 409.7 m/s (1344 ft/s). The specific heat at constant pressure is 1.013 kJ/kg-K (0.242 BTU/lb_m-°R), and the specific heat ratio is 1.3945. The inlet density is 1.443 kg/m^3 (0.00280 sl/ft^3), and the exit density is 2.216 kg/m^3 (0.00430 sl/ft^3). Find the following:

(a) the slip factor,
(b) the total pressure ratio.

6.23 A single-stage centrifugal air compressor with a mass flow rate of 12.70 kg/s (28 lb_m/s) rotates at 16,400 rpm. At the impeller exit, the flow exits the impeller purely radially (in the rotating frame) with a velocity of 180.4 m/s (592 ft/s). The efficiency is 91 percent. Flow enters the impeller eye with +4° of preswirl. The exit radius of the impeller is 195.6 mm (7.7 in.), and the inlet tip and hub radii are 101.6 and 17.78 mm (4.0 and 0.7 in.), respectively. The blade height is 22.86 mm (0.90 in.) at the impeller exit. In the absolute reference frame, the incoming total temperature is 372.8 K (671 °R), and the incoming total pressure is 195.8 kPa (28.4 psia). The incoming static temperature is 322.2 K (580 °R), and the incoming static pressure is 117.2 kPa (17.0 psia). The inlet axial velocity is 318.2 m/s (1046 ft/s). The input power is 1404 kW (1883 hp). The average blade speed at the inlet is 102.4 m/s (336 ft/s), and the blade speed at the exit is 335 m/s (1102 ft/s). The specific heat at constant pressure is 1.013 kJ/kg-K (0.242 BTU/lb_m-°R), and the specific heat ratio is 1.3956. The inlet density is 1.268 kg/m^3 (0.00246 sl/ft^3). Find the total pressure ratio.

6.24 Flow exits the impeller of a centrifugal compressor with a relative backward sweep of 15°. The blades are purely radial and move with a tangential velocity of 1200 ft/s. The radial component of the flow at the impeller exit is 600 ft/s. Find the slip factor.

6.25 Find the corrected mass flow rate and corrected speed for the stage in Problem 6.1. If the stage is run at 10,000 rpm and the flow rate is 40 lb_m/s, find the inlet pressure and temperature so that the corrected parameters are matched. Use these conditions to find the resulting flow angles and total pressure ratio, and compare with those from Problem 6.1. Comment.

6.26 A stage of a radial compressor is to be analyzed. It rotates at 15,000 rpm and compresses 40.0 lb_m/s of air. The inlet pressure and temperature are 35 psia and 500 °R, respectively. The hub and tip radii of the blades at the inlet are 2.4 and 4.8 in., respectively, the exit radius is 8.4 in., and the exit blade height is 0.96 in. The slip factor is unity. Flow enters the inducer with no preswirl, and the impeller has forward-facing exit blades at 5°. The efficiency of the stage is 90 percent. The values of c_p and γ are 0.2400 BTU/lb_m-°R and 1.401, respectively. Find the following:

(a) the relative flow angle at the inlet (hub, mean, eye),
(b) the static pressure at the impeller exit,
(c) the total pressure ratio for the stage (describe how this compares with an axial compressor stage),
(d) the Mach numbers at the impeller inlet and exit,
(e) the required power for the stage.

7 Axial Flow Turbines

7.1 Introduction

As discussed in Chapters 1–4, the purpose of the turbine is to extract energy from the fluid to drive the compressive devices. The actual operation of the turbine is in some respects similar to, but opposite that, of the compressor. That is, energy is extracted from the fluid and the pressure and temperature drop through the turbine. Typically, 70–80 percent of the enthalpy increase from the burner is used by the turbine to drive the compressor. For jet propulsion the remainder is used to generate thrust in the nozzle, while for a power generation unit the remainder is used to generate auxiliary power.

Two major differences are apparent between compressors and turbines, however. First, the pressure decreases through a turbine. In compressors, an adverse pressure gradient is present, which has the potential to cause blade stall. Such problems do not occur in a turbine, and turbine efficiency is typically higher than compressor efficiency. Also, because of this favorable pressure gradient, fewer stages are required for a turbine than for a compressor. As a result, the aerodynamic loading per stage is higher for a turbine than for a compressor. However, the inlet temperature is very high and limits the operation of the turbine. Thus, aerodynamically, the turbine is somewhat easier to design but is structurally more difficult.

Historically, more emphasis was initially placed on compressor design than on the design of turbines. For simple jet engines, one can show (Problem 3.44) that the overall engine performance is enhanced more by improving compressor performance than by improving turbine performance. However, for high-bypass turbofans, one can show that the opposite trend is true (Problem 3.47). For moderate-bypass ratio fans, the two components have approximately the same effect (Problem 3.46). Thus, for modern engines considerable emphasis has been placed on turbine aerodynamic design. Furthermore, many of the modern engines are limited by high temperatures from a materials viewpoint at the turbine inlet. Of all of the components in an engine, the turbine inlet blades are most likely to fail because of the combination of the high temperatures, high rotational speeds, and high aerodynamic loads. Thus, in recent years considerable emphasis has been placed on the material development for, and the cooling of, turbine blades.

As noted earlier, advances have been made over the past decade with compressor design. The GE GE9X (Figure 1.11) is a very good example of some of the new developments. Along with compressor advances, turbines are more advanced as well. Parts of the high-pressure turbine are made from ceramic matrix composites, reducing the needed cooling. Low-pressure turbine blades are manufactured from titanium aluminide, which are stronger and lighter than conventional nickel compounds, again allowing this component to rotate at higher speeds. The Rolls-Royce Trent XWB (Figure 1.28) is a three-spool engine that has turbine blades that are coated with a low-conductivity ceramic.

Usually, axial flow compressors are generally used in larger jet engines or power generation units, although in some cases radial compressors are used for small gas turbines with small mass flow rates. However, because the temperature is so high in a turbine, the gas density is low. Thus, larger flow areas are required to keep velocities at reasonable levels (i.e., subsonic) in a turbine than in a compressor. Because of this, a radial flow turbine would require a very large passage depth at the small radii. As a result, either a large overall engine diameter or engine length would be required. Furthermore, although a radial machine is capable of higher pressure ratios than an axial machine, developing high-pressure ratio axial flow turbines is not difficult because of the favorable pressure gradient. Therefore, in almost all gas turbines, axial flow turbines are used. This book covers only such turbines, although radial inflow turbines have applications in other industries.

7.1.1 Objectives

After completing this chapter, the reader should be able to do the following.

- Understand the basic geometry of a turbine
- Understand the difference of rotor blades and stator vanes
- Appreciate the compressor and turbine geometries and characteristics of a large range of gas turbines
- Understand the pressure changes in a single stage
- Understand and be able to develop a velocity triangle or polygon
- Understand the different flow and blade/vane angles in a stage
- Understand the influences of blade/vane geometry on pressure and velocities
- Be able to apply equations from Chapter 4 along with velocity polygons to find single-stage performance
- Understand the implications of percent reaction
- Understand the development and application of two turbine "maps"
- Understand stage choking
- Understand the limits of stage temperatures and failures
- Understand the different and historical methods of dealing with extreme temperatures
- Understand the application of the streamline analysis method to predict 3-D flow (optional)

7.2 Geometry

7.2.1 Configuration

Like a compressor, a turbine comprises a series of stages, all of which derive energy from the fluid. A stage again has two components: a set of rotor blades, which are attached to the disks and shaft, and the stator vanes, which are attached to the engine case. The rotor blades extract energy from the fluid, and the stator vanes ready the flow for the following set of rotor blades.

Many modern engines operate with two spools. Such a design is shown in Figure 7.1. The high-temperature gas leaves the combustor and enters the high-pressure turbine. This component is on the high-speed shaft. A set of inlet guide vanes is sometimes used to ready the flow for the first set of rotor blades. After exiting the high-pressure turbine, the gas enters the low-pressure turbine. This component is on the low-speed shaft. The energy from the low-pressure turbine drives the low-pressure compressor (and fan if present). The high-pressure turbine drives the high-pressure compressor. Because the gas pressure is dropping down the turbine, the gas density decreases down the turbine axis. Thus, the flow areas and blade heights increase down the turbine. Often a set of nonrotating blades is used at the exit of the turbine to ready the flow for the afterburner (if present), the nozzle, the free turbine, or exhaust. The blades are termed the exit guide vanes (EGV) and differ in design from normal stator vanes. For a jet engine if an afterburner is present, the vanes impart swirl to the flow to encourage mixing. If an afterburner is not present, the vanes straighten the flow for entrance into the nozzle. For a power unit, the vanes prepare the flow of the free turbine or straighten the flow for the exhaust system. Finally, although the actual designs for high- and low-pressure stages are different, the same basic thermodynamics, gas dynamics, and design methodology apply to both. For example, larger variations in blade angle and twist are found for a low-pressure turbine than for a high-pressure turbine owing to the larger radius ratio (similar to that for an axial flow compressor). Thus, for the remainder of this chapter, distinctions are not made between the two components.

If one were to look at a turbine from the top and consider the geometry to be planar 2-D by unwrapping the geometry, a series of cascades would be seen as shown in Figure 7.2. The fluid

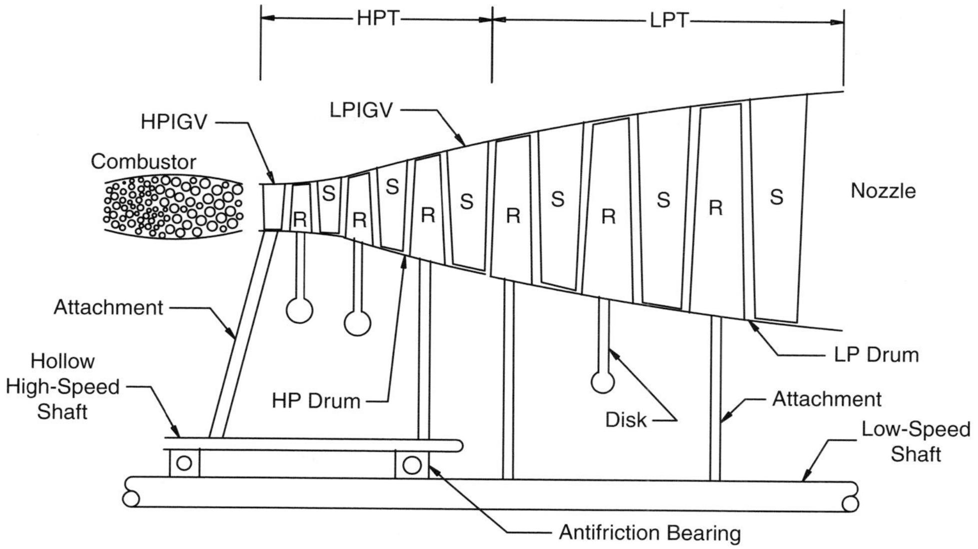

Figure 7.1 Twin-spool turbine.

first enters the inlet guide vanes (if used). Next, it enters the first rotating passage. In Figure 7.2, the rotor blades are shown to be moving with linear velocity *U*, which is found from $R\omega$, where ω is the angular speed and *R* is the mean radius of the passage. After passing through the rotor blades, the fluid enters the stator vanes. The fluid is turned by these stationary vanes and readied to enter the second stage beginning with the second rotor blades. In general, the second stage is slightly different in design than the first stage (all of the stages are different). The process is repeated for each stage. In Figure 7.2, turbine component stations 0 through 3 are defined. It is very important not to confuse these with engine station numbers.

As is the case for compressors, an important parameter is the solidity (C/s), which is the inverse of the pitch-to-chord ratio s/C. Again, if the solidity is too large, the frictional losses will be too large. If the solidity is too small, the flow will not follow the blades (slip), and less than ideal power will be derived from the hot gas. Again, for a turbine, typical solidities are unity. For example, for the Pratt & Whitney JT9D, the number of turbine blades on the stages ranges from 66 to 138. For comparison, the turbine exit guide vane (which has a very long chord) has 15 blades.

Also shown in Figure 7.2 are axial pressure profiles down the turbine. Both the total and static pressures are shown. As can be seen, the static pressure decreases across the inlet guide vanes and also decreases across all of the rotor and stator blades. The total pressure ideally remains approximately constant across the inlet guide vanes and stator vanes but decreases across the rotor blades. Consistent with the approach to compressors, it is the objective for the remainder of this chapter to relate the pressure changes to the geometry of the turbine; that is, to relate the static and total pressure rises and ratios to a given geometry of blades and rotational speed.

Next, a few of the design considerations are discussed. First, as indicated, the temperature is of great concern. Such high temperatures coupled with high rotational speeds lead to high centrifugal stresses in the blades. Thus, not only are the aerodynamics important in the design of blade shapes but so are the developed stresses. As is discussed later in this chapter,

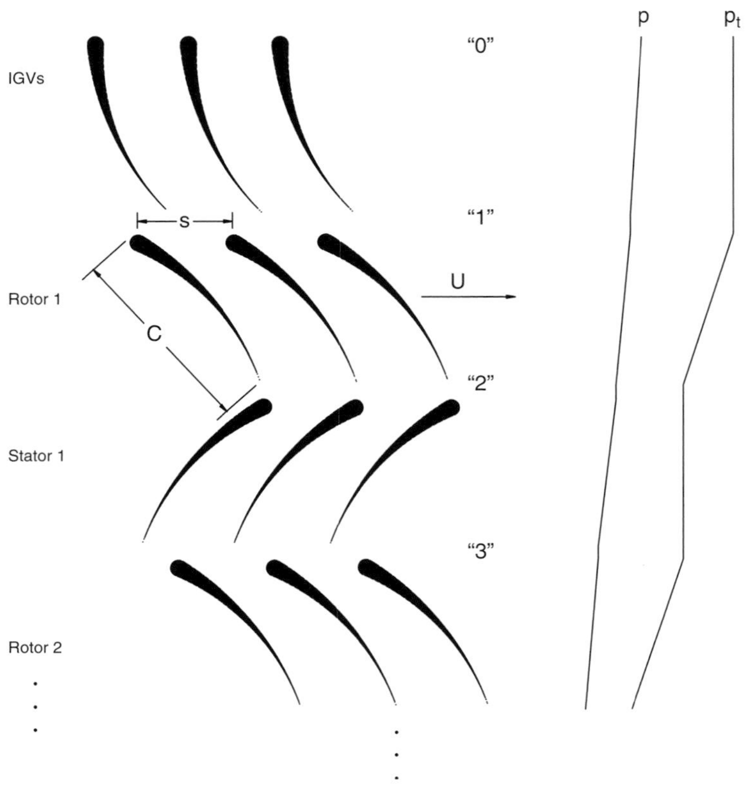

Figure 7.2 Cascade view of blades.

blade cooling is often used to combat this problem. Second, high efficiencies are of course desirable. However, airflow leakage around the blade tips becomes more important than is the case for a compressor because of the higher pressure drops across the stages and higher temperatures. The clearance between the blade tip and case varies with operating conditions because of the thermal growth of the blades. A large clearance is not desirable because the turbine efficiency drops about 1 percent with each 1 percent of flow across the tip. That is, expanding gas bypasses the turbine stage by flowing from the rotor-blade pressure surface to the suction surface in the tip clearance, but no useful work is derived from the flow. However, if the clearance is too small, thermal growth or vibrations can cause a blade rub, which can result in total destruction of the engine. Third, losses also occur owing to viscous shear and shocks. For example, high blade loading produces high velocities on the suction surfaces of the blades; with high velocities, flows can become locally supersonic and form shocks. These shocks drop the total pressure nonisentropically as well as set up separation regions, which further decrease the total pressure. The end result is to reduce the stage efficiency. Finally, flow must be properly directed into the turbine. A uniform flow (velocities, pressures, and temperatures) of gas is desired. For example, if more flow is directed near the blade root, the amount of torque (and thus power) derived from the fluid will be reduced. Also, if hot areas of flow exist at the turbine inlet, early blade failure can result; again such failures can be catastrophic. These and other concerns are discussed further in the chapter. More details can be found in Hawthorne (1964), Horlock (1966), Glassman (1972a, 1972b, 1973, 1975), Kercher (1998, 2000), Adamczyk (2000), Fielding (2000), Han, Dutta, and Ekkad (2000), Dunn (2001), and Shih and Sultanian (2001).

7.2.2 Comparison with Axial Flow Compressors

Passage Areas between Blades

Figure 7.3 illustrates the passage shapes of a turbine and compressor. For a turbine cascade, the exit area is less than the inlet area ($A_{exit} < A_{inlet}$). Consequently, the passage acts like a nozzle. For comparison, however, the exit area for a compressor cascade is greater than that of the inlet. Thus, the passage acts like a diffuser, as previously discussed in Chapter 5.

Pressures

The purpose of a turbine is to extract energy from the flow. As a result, the total and static pressures drop through a turbine; that is, the flows are expanding. For comparison, the pressures increase in a compressor. Also, the flows in turbines are usually subsonic. Thus, the trends agree with those for the areas as discussed in the paragraph above. Typical pressure ratios across (inlet to exit) a turbine stage are two.

Turning Angles

For a turbine the inlet pressure to a stage is greater than the exit pressure. Thus, separation will not occur on the blades because of an adverse pressure gradient. As a result, more expansion can be accomplished by a turbine than compression can be accomplished by a compressor. This means that ($\{A_{inlet}/A_{exit}\}_{turbine} > \{A_{exit}/A_{inlet}\}_{compressor}$) for a cascade. Thus, on the basis of Figure 7.3, the turning angle is greater for a turbine cascade than for a compressor cascade. Because each turbine stage has more flow turning than a compressor stage, more energy is extracted per stage than in a compressor.

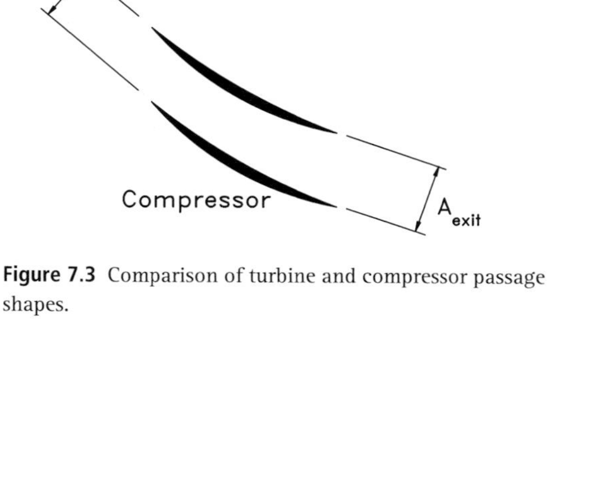

Figure 7.3 Comparison of turbine and compressor passage shapes.

Number of Stages

As just noted, more energy is extracted per stage than in a compressor. Since all of the turbine power is used to drive the compressive devices for a jet engine, the number of turbine stages is less than the number of compressor stages. Power generation gas turbines have additional stages for the external load, but still the total number of stages is less than for the compressor. The typical ratio of the number of compressor stages to turbine stages is from 2 to 6.

Temperatures

The typical inlet temperature to a turbine is currently around 3000 °R, which is now limited from a practical standpoint by the materials properties. For comparison, the hottest temperature realized by a compressor is at the exit and is typically 1400 °R. As discussed later in Section 7.6, considerable efforts have been undertaken to increase the turbine inlet temperature limit. One method entails using better materials. A second method involves blade cooling, as shown in one typical arrangement depicted in Figure 7.4. Cool air is pumped through a hollow core in the turbine blades to cool them. Sometimes small holes are machined into the blade so that the cool air forms a protective boundary layer between the blade and the hot gas exiting from the combustor. Because of the cooler operating temperatures, compressors do not need such complex geometries.

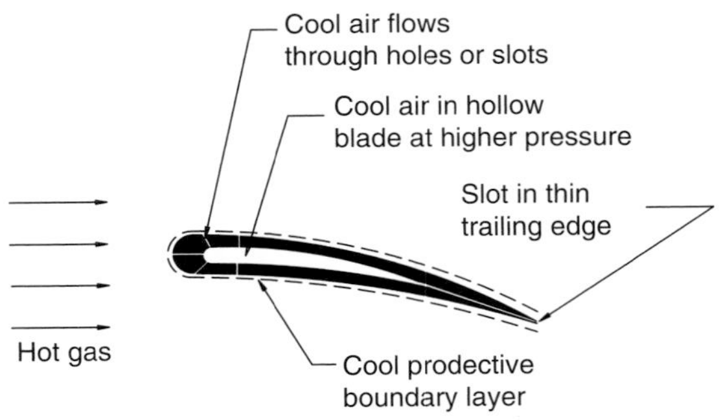

Figure 7.4 Blade cooling.

Radii and Blade Height

The air temperature is much higher in a turbine than in a compressor. However, the pressures are comparable. Thus, the density in a turbine is lower than the density in a compressor. The turbine area must be larger than the compressor area to pass the same flow rate.

In estimating the relative sizes, the compressor air flow rate is given by $\dot{m}_c = \rho_c c_{ac} A_c$, whereas that for the turbine is $\dot{m}_t = \rho_t c_{at} A_t$, which is approximately equal to $\dot{m}_c$. The density can be found from the ideal gas equation $\rho = p/\mathcal{R}T$. One can now consider the axial Mach number, $M_a = c_a/a$ and realize that $a = \sqrt{\gamma \mathcal{R} T}$. Thus, if γ_c is approximately equal to γ_t, p_c is approximately equal to p_t and the turbine and compressor Mach numbers are about the same (which is typical). The turbine cross-sectional area is $A_t \approx A_c\sqrt{T_t/T_c}$. Thus, turbine passage areas are typically larger than compressor passage areas by 30 to 50 percent. This is accomplished by making the blade heights larger for a turbine than for a compressor. Typically, owing to the increased blade heights, the hub radius is smaller for the turbine than for a compressor, but not always.

Blade Thickness

The thickness of a turbine blade is typically greater than the thickness of a compressor blade. This is the result of three factors. First, the inlet or exit area ratio is larger for a turbine cascade. Second, blade cooling is often used, which, because of the hollow region, requires a thicker blade. And third, to improve the structural soundness or reliability of the blade, the blade must be thicker. Care must be taken in the design of such blades to minimize the "profile" losses and to avoid blocking the flow too much.

Shrouding

Unlike compressors, some turbines have "shrouds." These are circumferential bands that wrap around the outside diameter of the blade tips and lock them together, as shown in Figure 7.5 (turbine blades are often attached to the disk using "fir trees," as was discussed for some axial flow compressors). Shrouds are used on turbines for four reasons. First, because of the high operating temperatures, the blades are weakened and thus need additional support so they do not fail in the presence of the steady-state forces (or loads) resulting from flow turning. Second, similarly, they need extra support to control blade vibrations, which can lead to blade failure. Third, because more flow turning is used in turbines than in compressors, larger loads occur on a turbine blade than on a compressor blade; once again more structural support is needed. Fourth, again because more flow turning is used, larger pressure changes occur in a turbine stage than in a compressor stage. As a result, flow tends to "leak" around the tip of a turbine blade from the pressure surface to the suction surface, which is similar to the wing-tip vortices on an aircraft wing. These secondary flows reduce the aerodynamic efficiency of a turbine stage; shrouds block these leakages and thus improve the efficiency. Shrouds are used less in modern turbines than in the past by reducing the clearance between the blades and the casings.

Number of Blades per Stage

Although the blades of a turbine are typically thicker and longer, the solidity of both compressors and turbines is about the same. Thus, the blade spacing is larger for a turbine than for a compressor. However, the hub radius of the blades is typically larger. As a result, approximately the same number of blades is used on a turbine or compressor cascade. As is the case for a compressor, enough blades are needed to

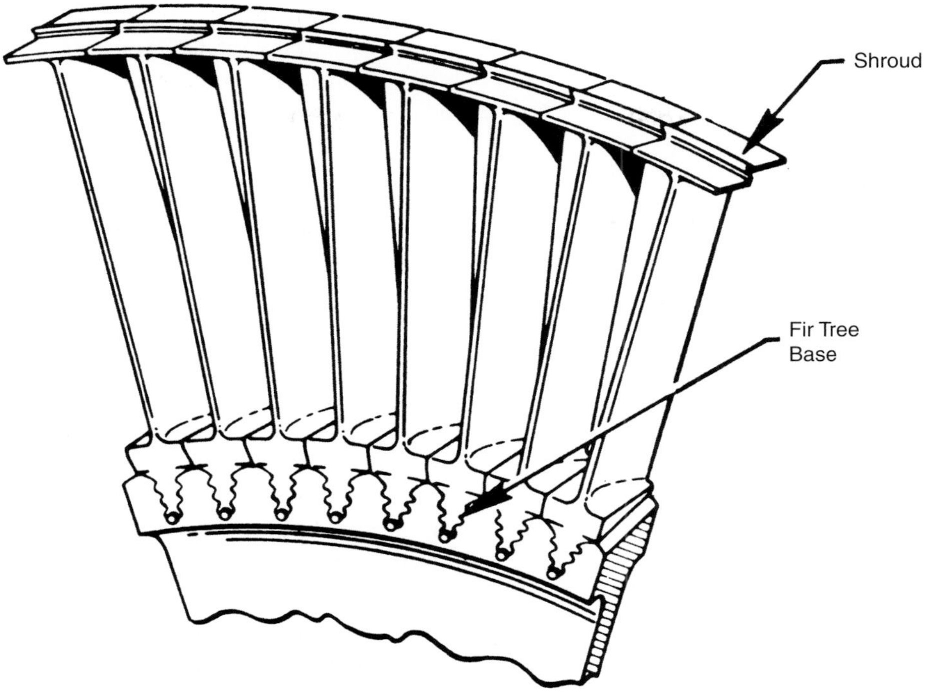

Figure 7.5 Shrouded turbine rotor blades (© 2022 Raytheon Technologies Corporation, Pratt & Whitney division. Reproduced with permission).

minimize separation and direct the flow in the desired direction, but not so many blades should be used that frictional losses become intolerable. Thus, efficiency is maximized by selecting the optimum number of blades. Ainley and Mathieson (1951) and Traupel (1958) postulated the optimum pitch-to-chord ratio correlations that are still often used today. The recommended values are correlated in Figure 7.6 in a format similar to that used for presenting compressor values and are dependent on the exit and deflection flow angles as is true for the compressor.

As is the case for an axial compressor, choosing different numbers of stator vanes and rotor blades for a stage or nearby stages is very important. If the numbers of vanes and blades are equal, a resonance due to fluid dynamic blade interactions can be generated and large blade, disk, and shaft vibrations and noise can result. This reduces the life of the blades and safety of the engine. Blade or vane numbers are often selected as prime numbers (but less so than in the past), but they are always chosen so that common multiple resonances are not excited. Dring, Joslyn, Hardin, and Wagner (1982) review details of this topic.

Efficiency

Overall, the efficiency is slightly larger for a turbine than for a compressor. This results from two opposite effects. Because of the favorable pressure gradient, one would expect the efficiency to be improved more than would be the case for a compressor. However, as a result of the blade cooling and injected cool air (including added turbulence, frictional losses, pressure drops, and enthalpy reduction), a decrease in efficiency would be anticipated. The net effect of the conditions is to improve the efficiency slightly as compared with a compressor.

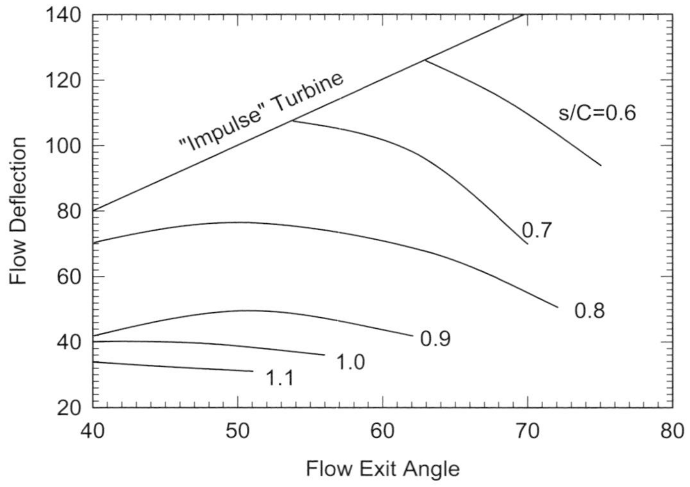

Figure 7.6 Optimum pitch-to-chord ratio (correlated from data of Ainley and Mathleson (1951) and Traupel (1958)).

7.3 Velocity Polygons or Triangles

Because energy is being derived from the flow (which results in a pressure drop) and part of the turbine is moving, it is important to be able to understand the complex flow patterns in a turbine. As is the case for a compressor and as is shown in Section 7.4, the pressure drop is directly dependent on the velocity magnitudes and directions in a turbine. The purpose of this section is to present a means for interpreting fluid velocities. That is, since one component rotates and one component is stationary, the velocities must again be related so that the components are compatible.

First consider the inlet guide vanes shown in Figure 7.7a. The exit flow from the combustor or the inlet flow to the inlet guide vanes is typically aligned with the axis of the engine. The flow velocity relative to the stationary frame (referred to as absolute velocity for the remainder of the book) at the IGV inlet is c_0. The blade inlet angle is usually aligned with the axis. As the fluid passes through the IGV, it is turned so that the absolute flow velocity at the IGV exit (and the rotor inlet) is c_1, and it has a flow angle α_1 relative to the axis of the engine. The IGV exit blade angle is α'_1 If the flow is exactly parallel to the blade at the exit, α_1 and α'_1 are identical. This is usually not true in an actual design, however, owing to slip.

Next, consider the inlet to the rotor stage as shown in Figure 7.7b. However, the rotor blades are rotating around the engine centerline or are moving in the planar 2-D plane with absolute velocity U_1 in the tangential direction. Thus, to find the velocity of the fluid relative to the rotating blades, one must subtract the blade velocity vector from the absolute flow vector. This is performed in Figure 7.7b. The resulting vector is w_1, and it has a flow angle β_1 relative to the axial direction. It is important to note that the rotor blades have inlet and exit angles (relative to the axial direction) of β'_1 and β'_2, respectively. Once again, if the relative flow direction matches the blade angles exactly, the values β'_1 and β_1 are the same and the incidence angle is zero. However, this usually does not occur in practice – especially at off-design conditions.

When drawing the polygons, one should always draw the triangles to scale. This not only allows "rough" checking of algebraic computations but provides the capability of observing the relative magnitudes of the different velocity vectors and the viability of a turbine stage design, as will be discussed in Section 7.4, for example, with

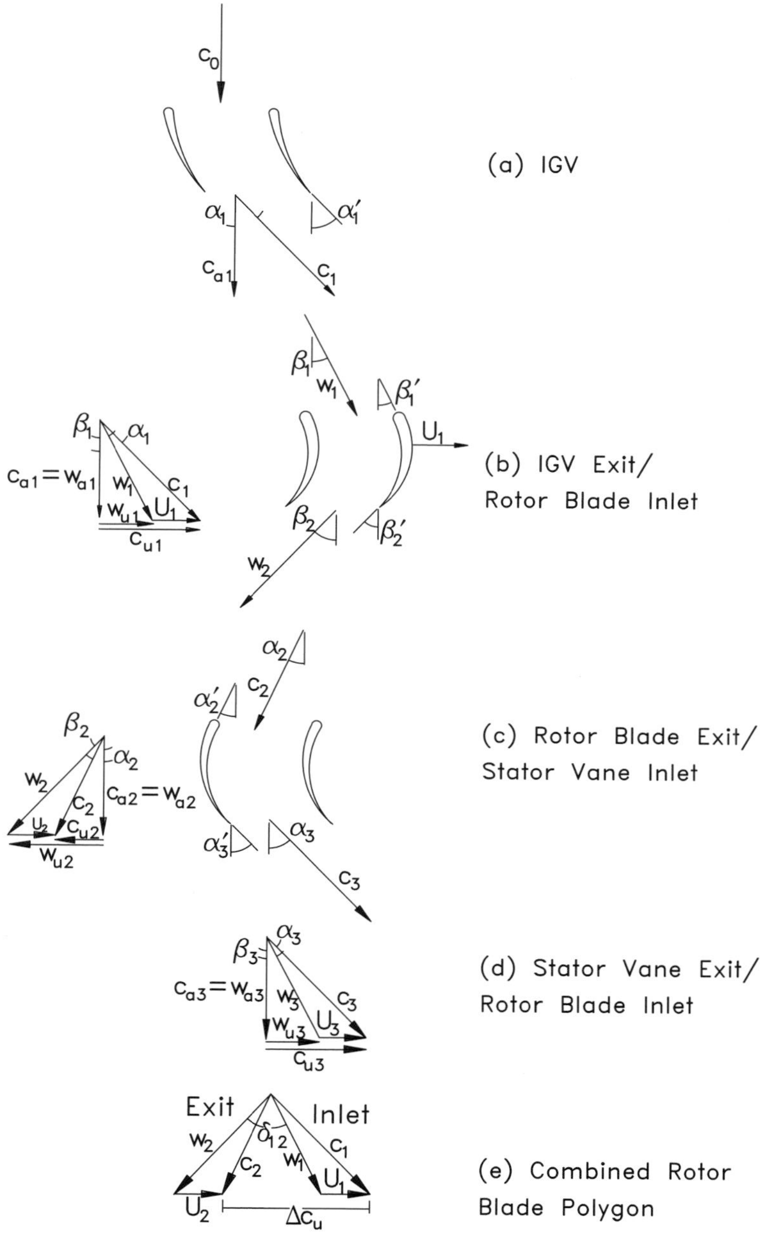

Figure 7.7 Definition of IGV, rotor blades, stator vanes, geometry, and velocity polygons.

the percent reaction. When a compromising situation is observed, it is easier to see how to change the condition by altering geometric parameters by using a scale diagram rather than equations.

One can break the vectors into components as indicated in Figure 7.7b. As shown, the absolute velocity $\mathbf{c}_1$ has a component in the tangential direction that is equal to c_{u1} and a component in the axial direction that is c_{a1}. Also, the relative velocity, $\mathbf{w}_1$ has components w_{u1} and w_{a1} in the tangential and axial directions, respectively.

Figure 7.7c shows the velocity polygon for the rotor exit and stator inlet. The exit relative velocity (to the rotor blades) is $\mathbf{w}_2$, and this velocity is at an angle of β_2 relative to the axial direction. The blade velocity is U_2 and can be vectorially added to $\mathbf{w}_2$ to find the absolute exit velocity $\mathbf{c}_2$. Note that U_2 and U_1 may be slightly different because the radii at the rotor inlet and exit may be slightly different. The absolute flow angle of $\mathbf{c}_2$ is α_2 relative to the engine axis. Once again, the rotor blade exit angle is β'_2, which may be different from the relative flow angle β_2 owing to slip. Furthermore, the stator vane inlet angle is α'_2 and will in general be different from α_2, indicating nonzero incidence. Also shown in Figure 7.7c are the components of both $\mathbf{w}_2$ and $\mathbf{c}_2$ in the tangential and axial directions.

The stator exit triangle, which is also the inlet velocity polygon to the next rotor blades, is shown in Figure 7.7d. The absolute velocity is $\mathbf{c}_3$, which makes an angle of α_3 to the axial direction. Once again, the flow absolute flow angle α_3 will, in general, be different from the absolute stator vane angle α'_3. The next rotor blade velocity at the inlet is U_3, and the relative velocity (to the next rotor blades) is $\mathbf{w}_3$, which makes a relative angle of β_3 to the axis. The methodology described here can be used for all of the downstream stages.

As shown in Figure 7.7e, the polygons for the rotor entrance and exit are often combined to form one polygon. As is true for a compressor and demonstrated in Section 7.4, the power of the stage is directly related to the change in absolute tangential velocity if the radius is approximately constant ($\Delta c_u = c_{u1} - c_{u2}$). Thus, by combining the two triangles, one can easily observe the change graphically.

Finally, a side view of a turbine first stage is presented in Figure 7.8. In this figure, the average blade radii (R) and blade heights (t) are defined.

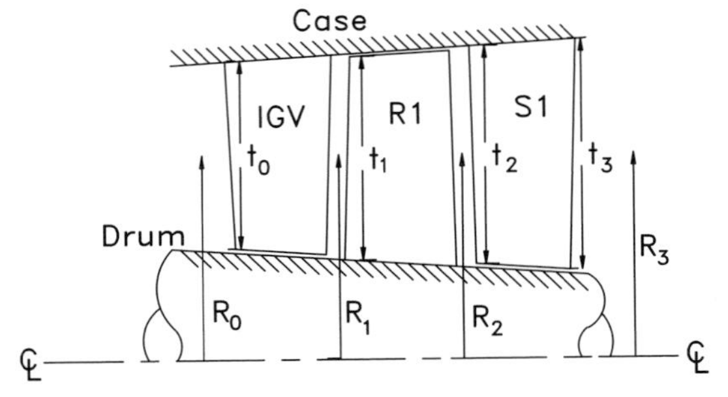

Figure 7.8 Side view of first stage.

Once again, a consistent sign notation is needed, and this notation will be the same as that used for axial flow and centrifugal compressors. First, all positive tangential velocities are to the right. Thus, rotor blades always move to the right with positive velocity U. Second, all angles are measured relative to the axial direction, and positive angles are considered to be counterclockwise.

In Table 7.1, the trends of the velocity magnitudes, flow cross-sectional areas, and pressures across the different components are shown. As can be seen, the static pressure decreases across both the stator and rotor blades because the relative velocities across both increase. Also, as the absolute and relative velocities across the IGVs increase, the static pressure decreases. Details of these trends are discussed in Section 7.4.

7.4 Single-Stage Energy Analysis

This section summarizes the equations so that the velocities can be related to the derived power, pressure drop, and other important turbine characteristics. The derivation is performed using a "mean line" control volume approach described in detail in Chapter 4 for a single stage (rotor and stator). Figure 4.1 shows a single stage. The objective is to relate the inlet and exit conditions to the property changes.

Table 7.1 **Axial flow turbine component trends.**

	Absolute velocity	Relative velocity	Area	p	p_t
IGV	Increases	Increases	Decreases	Decreases	Constant (ideal)
Rotor	Decreases	Increases	Decreases	Decreases	Decreases
Stator	Increases	Increases	Decreases	Decreases	Constant (ideal)

7.4.1 Total Pressure Ratio

In Chapter 4, the continuity, moment of momentum, and energy equations are used. For the delivered shaft power, they result in (Eq. (4.2.17))

$$\dot{W}_{sh} = \dot{m}[U_2 c_{u2} - U_1 c_{u1}], \tag{7.4.1}$$

and for the total pressure drop they result in (Eq. (4.2.26))

$$\frac{p_{t2}}{p_{t1}} = \left[\frac{[U_2 c_{u2} - U_1 c_{u1}]}{\eta_{12} c_p T_{t1}} + 1\right]^{\frac{\gamma}{\gamma-1}}. \tag{7.4.2}$$

Thus, knowing the velocity information of a rotor from polygons and the efficiency, one can find the total pressure drop. A few subtle differences exist between the equations for the compressor and turbine. For example, the power or work output from the turbine stage is less than the ideal value, but it is more for a compressor. Also, the pressure ratio (p_{t2}/p_{t1}) is less than unity for a turbine but is greater than 1 for a compressor.

7.4.2 Percent Reaction

An important characteristic for an axial flow turbine is the percent reaction. This relation again approximates the relative loading of the rotor and stator and is given by (Eq. (4.2.32))

$$\%R = \frac{1}{1 + \left[\frac{c_2^2 - c_1^2}{w_1^2 - w_2^2}\right]}. \tag{7.4.3}$$

Thus, for compressible flow the percent reaction of the stage can be related to the absolute and relative velocities at the inlet and exit of the rotor. An example of a single stage with all of the calculation details will follow. One should note that the software Turbomachinery.exe performs the same calculations.

7.4.3 Incompressible Flow (Limiting Case)

For comparison, one may wish to find the delivered power, pressure drop, and percent reaction of a turbine with an incompressible fluid. This condition would apply to a hydraulic turbine. As for an axial compressor, Chapter 5, incompressible flow is discussed only because it provides valuable insight as to the implications of percent reaction. From Chapter 4 (Eqs. (4.2.34), (4.2.37), and (4.2.39)),

$$\dot{W}_{sh} = \dot{m}[U_2 c_{u2} - U_1 c_{u1}] \tag{7.4.4}$$

$$p_{t2} - p_{t1} = \frac{\rho}{\eta_{12}}[U_2 c_{u2} - U_1 c_{u1}] \tag{7.4.5}$$

$$\%R = \frac{1}{1 + \left[\frac{c_2^2 - c_1^2}{w_1^2 - w_2^2}\right]}. \tag{7.4.6}$$

The percent reaction for incompressible and ideal flow is given by (Eq. (4.2.41))

$$\%R = \frac{p_2 - p_1}{p_{t3} - p_{t1}}. \tag{7.4.7}$$

For example, typical axial flow turbine stages have percent reactions of about 0.50. For a nonideal hydraulic turbine, the percent reaction dictates that approximately half of the enthalpy drop occurs in the rotor and half in the stator. For an ideal hydraulic turbine, this also means that half of the pressure drop occurs in the rotor and half in the stator. This implies that the force "loads" on the rotor and stator blades are about the same. For a compressible flow turbine, however, it merely means that half of the enthalpy increase occurs in the rotor and half in the stator. The pressure drops may not be exactly the same – even for the ideal case – but they are approximately the same. Regardless, if the percent reaction differs greatly from 0.5, a large pressure drop occurs in either the stator (%R < 0.5) or the rotor (%R > 0.5). For a turbine, if a large pressure drop occurs, choking or partial choking of the cascade is likely. The resulting efficiency will be less than optimum because of the associated losses. Furthermore, as is observed for a compressor, when the blade heights are significant relative to the radius, the reaction increases from hub to tip. For a turbine, if the percent reaction is less than zero, the stage operates as a compressor. Thus, the mean line reaction must be chosen such that the reaction at the hub is greater than zero.

7.4.4 Relationships of Velocity Polygons to Percent Reaction and Performance

As occurs for a compressor stage, the velocities dictate the percent reaction. Three special cases are shown in Figure 7.9. In particular, sets of velocity polygons (rotor inlet and exit) for 100 percent, 0 percent, and 50 percent reaction turbine stages are shown for the same axial velocity. Of particular interest is the set of polygons for a 50 percent reaction (Figure 7.9a). For this case, the inlet and exit polygons are similar triangles but reversed. If the polygons are drawn to scale, one can easily observe if the percent reaction is close to 50 percent simply by examining the symmetry of the triangles. Also, the polygons can be used to make preliminary predictions on performance trends due to geometry changes. For example, one can see from Figure 7.9c that, for a turbine stage that originally operated close to a 50 percent reaction, if $|\delta_{12}|$ is

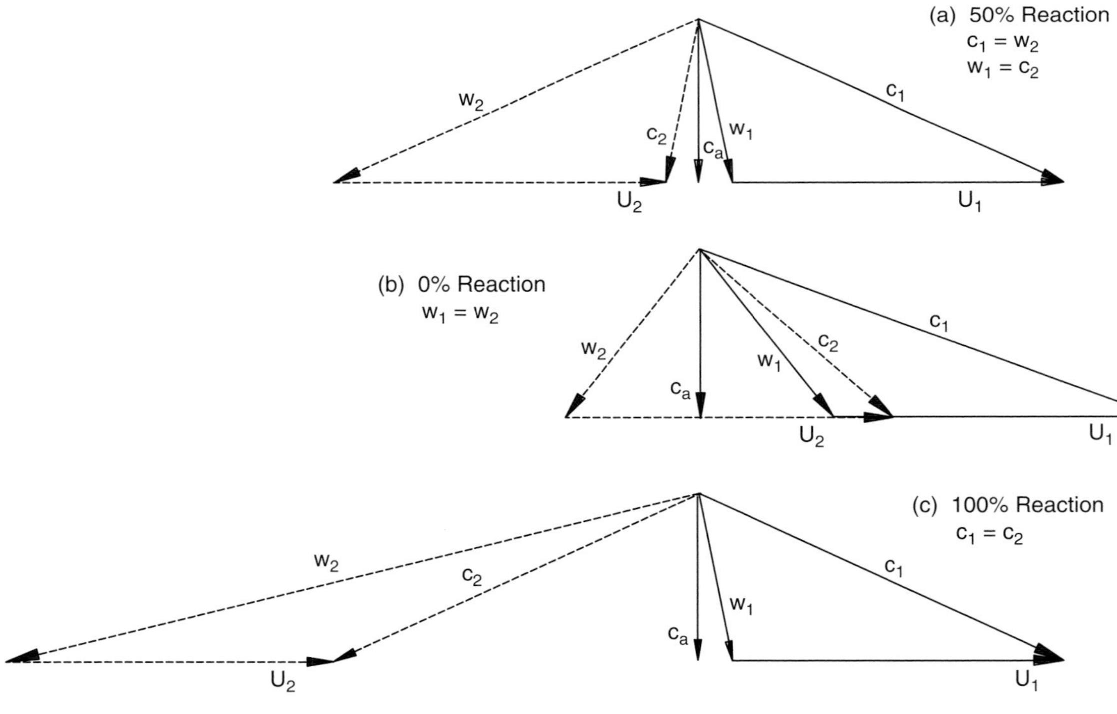

Figure 7.9 Three special cases of velocity polygons.

Table 7.2 **Geometric and operating condition effects on turbine parameters**

Decrease in	Percent reaction, %R	Total pressure ratio, p_{t3}/p_{t1}	Power output
Absolute inlet angle, α_1	Increase	Decrease	Increase
Rotor flow turning angle, $\lvert\delta_{12}\rvert$	Decrease	Increase	Decrease
Rotational speed, N	Decrease	Increase	Decrease
Axial velocity component, c_a	Increase	Decrease	Increase

increased and other parameters are constant (see Figure 7.7e), the percent reaction increases – eventually going to 100 percent. Such a 100 percent reaction turbine is often called a "reaction turbine," and all of the pressure drop occurs in the rotor. Also, one can see that, if α_1 is increased (with other parameters held constant – see Figure 7.7b), the percent reaction decreases – eventually going to 0 percent (Figure 7.9b). This is often called an "impulse turbine," and all of the static pressure drop takes place in the stator.

Moreover, for general trends, one can examine both the governing equations and velocity polygons for variations in the absolute inlet angle α_1, the rotor flow turning angle δ_{12}, the rotational speed N, and the mass flow rate (or axial velocity component) $\dot{m}$ (or c_a). The resulting effects on the stage total pressure ratio and percent reaction can be determined. A matrix of general trends is presented in Table 7.2.

Example 7.1 | Axial Turbine Stage

A stage of approximately the same size as one of the first stages (rotor and stator) of a high-pressure turbine is to be analyzed. It rotates at 8000 rpm and expands 280 lb_m/s (8.704 sl/s) of air. The inlet pressure and temperature are 276 psia and 2240 °F, respectively. The average radius of the blades is 17.5 in., and the inlet blade height is 2.12 in. The absolute inlet flow angle to the rotor is the same as the stator exit flow angle (65°), and the rotor flow turning angle (δ_{12}) is 75°. The stage has been designed so that the blade height varies and the axial velocity remains constant through the stage. The efficiency of the stage is 85 percent. The values of c_p and γ are 0.2920 BTU/lb_m-°R and 1.307, respectively, *which are based on the resulting value* of T_2. The following details are to be found: blade heights at the rotor and stator exits, the static and total pressures at the rotor and stator exits, the stator turning angle, the Mach numbers at the rotor and stator exits, the derived power from the stage, and the percent reaction for the stage. Note that this turbine can be compared with the compressor in Example 5.1. Both turbomachines are from the same engine and are operating at about the same pressure.

Solution Note

Before proceeding, and because the solution is long, an outline for the solution is appropriate.

(1) First, some preliminary calculations are presented.
(2) Next, the polygon of the velocities at the rotor inlet is "drawn." Various angles and velocities are found.
(3) Next, the triangle of the velocities at the rotor exit/stator inlet is "drawn" using the above and problem specifications. Various angles and velocities are found.
(4) Next, using this information the power, pressure ratio and percent reaction are found.
(5) Lastly, the triangle of the velocities at the stator exit is "drawn" using the above and problem specifications, which would be needed for the following stage.
(6) Intermediate steps also allow for the determination of such important characteristics such as Mach numbers.

Solution

Velocity polygons will be used to obtain the solution. Before proceeding, a few preliminary calculations are needed:

$U = R\omega = (17.5/12)(8000 \times 2\pi/60) = 1222\ \mathrm{ft/s}$

$A_1 = \pi D_1 t_1 = \pi(2 \times 17.5)(2.12)/144 = 1.619\ \mathrm{ft}^2.$

For $p_1 = 276$ psia and $T_1 = 2240\ °\mathrm{F} = 2700\ °\mathrm{R}$ the ideal gas equation yields

$\rho 1 = 0.008577\ \mathrm{sl/ft}^3.$

$$\text{Next, } c_{a1} = \frac{\dot{m}}{\rho_1 A_1} = \frac{8.704}{0.008577 \times 1.619} = 626.9\ \mathrm{ft/s}.$$

$$\text{Finally, } a_1 = \sqrt{\gamma \mathcal{R} T_1} = \sqrt{1.307 \times 53.35 \times 2700 \times 32.17} = 2461\ \mathrm{ft/s}.$$

Rotor Inlet

Next, refer to Figure 7.10a, which is the velocity polygon for the rotor inlet.

$$\text{First, } c_1 = \frac{c_{a1}}{\cos\ \alpha_1} = \frac{626.9}{\cos(65°)} = 1483\ \mathrm{ft/s}.$$

Also, $c_{u1} = c_1 \sin\ \alpha_1 = 1483\ \sin(65°) = 1344\ \mathrm{ft/s}$;

thus,

$$\beta_1 = \cot^{-1}\left[\frac{c_{a1}}{c_{u1} - U_1}\right] = \cot^{-1}\left[\frac{626.9}{1344 - 1222}\right] = 11.07°$$

$$\text{and } w_1 = \frac{c_{a1}}{\cos\ \beta_1} = \frac{626.9}{\cos(11.07°)} = 638.8\ \mathrm{ft/s}.$$

Thus, the Mach number in the rotating frame is

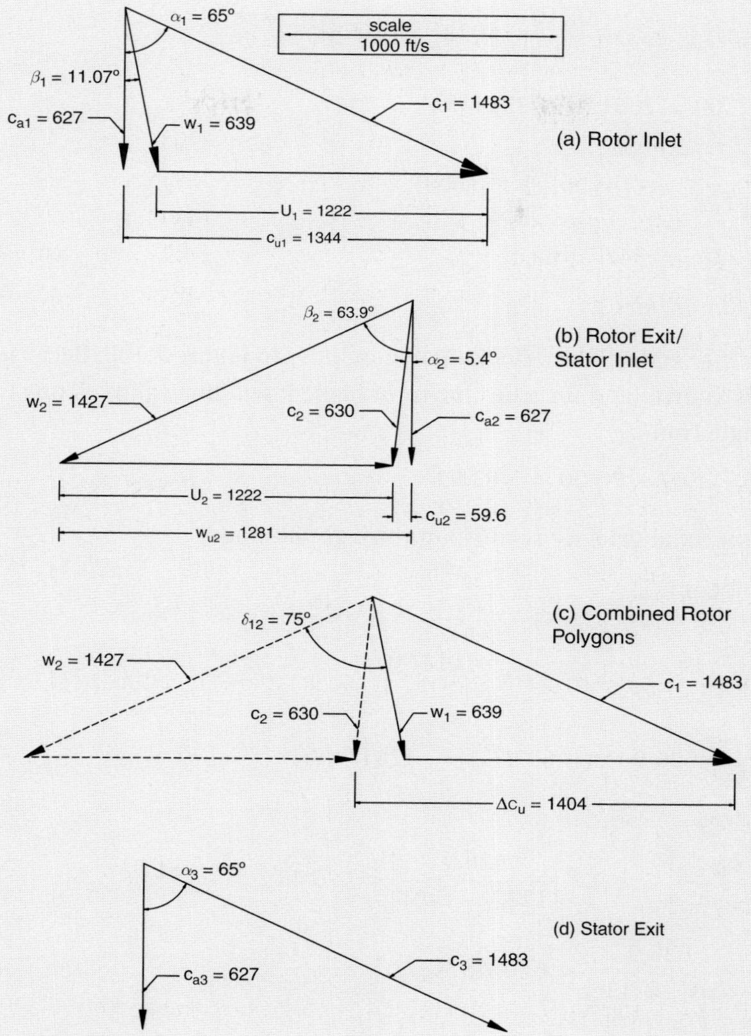

Figure 7.10 Velocity polygons for Example 7.1.

$$M_{1\text{rel}} = \frac{w_1}{a_1} = \frac{638.8}{2461} = 0.2596.$$

Flow is thus subsonic in the rotating frame, and shock formation in the inlet of the rotor should not be a problem. Next, in the absolute frame,

$$M_{1\text{abs}} = \frac{c_1}{a_1} = \frac{1483}{2461} = 0.6028.$$

Flow therefore is also subsonic in the stationary frame, and shock formation in the exit of the previous stator should also not be a problem. Additionally,

$$\frac{T_{t1}}{T_1} = 1 + \frac{\gamma - 1}{2} M_{1abs}^2 = 1 + \frac{0.307}{2} 0.6028^2 = 1.056$$

$T_{t1} = 1.056 \times 2700 = 2851\,°R = 2391\,°F.$

Finally, $\frac{p_{t1}}{p_1} = \left[\frac{T_{t1}}{T_1}\right]^{\frac{\gamma}{\gamma-1}} = [1.056]^{\frac{1.307}{0.307}} = 1.260$

$p_{t1} = 1.260 \times 276 = 347.7$ psia.

Rotor Exit and Stator Inlet

For the rotor exit or stator inlet velocity polygon, refer to Figure 7.10b. Because the rotor flow turning angle is known (and for a turbine rotor blade it will be negative), one can find the relative exit angle from:

$\beta_2 = \beta_1 + \delta_{12} = 11.07 - 75.00 = -63.93°.$

Also, because the axial velocity is constant through the stage,

$c_{a2} = c_{a1} = 626.9$ ft/s.

Next, $w_2 = \frac{c_{a2}}{\cos \beta_2} = \frac{626.9}{\cos(-63.93°)} = 1427$ ft/s

and

$w_{u2} = c_{a2} \tan \beta_2 = 626.9 \tan(-63.93°) = -1281$ ft/s;

thus,

$$\alpha_2 = \cot^{-1}\left[\frac{c_{a2}}{U + w_{u2}}\right] = \cot^{-1}\left[\frac{626.9}{1222 - 1281}\right] = -5.43°,$$

$$c_2 = \frac{c_{a2}}{\cos \alpha_2} = \frac{626.9}{\cos(-5.43°)} = 629.7 \text{ ft/s},$$

and $c_{u2} = U + w_{u2} = 1222 - 1281 = -59.6$ ft/s.
Next, the moment of momentum equation can be used to yield

$$h_{t2} - h_{t1} = U(c_{u2} - c_{u1}) = 1222(-1404)$$
$$= -1{,}715{,}329 \text{ ft}^2/\text{s}^2.$$

The quantity $\Delta c_u = (c_{u2} - c_{u1}) = -1404$ ft/s is shown in Figure 7.10c. Also, note the negative sign for Δh_t indicating that power is derived from the fluid. From the energy equation, one can find

$$\frac{p_{t2}}{p_{t1}} = \left[\frac{(h_{t2} - h_{t1})}{\eta_{12} c_p T_{t1}} + 1\right]^{\frac{\gamma}{\gamma - 1}}$$

$$= \left[\frac{-1{,}715{,}329}{0.85 \times 0.2920 \times 2851 \times 778.16 \times 32.17} + 1\right]^{\frac{1.307}{0.307}} = 0.6480,$$

and thus $p_{t2} = 0.6480 \times 347.7 = 225.3$ psia. <ANS

Note that the total pressure change in this turbine is much larger than the change in total pressure in the compressor in Example 5.1.

Also, $T_{t2} - T_{t1} = \dfrac{h_{t2} - h_{t1}}{c_p} = \dfrac{-1{,}715{,}329}{0.2920 \times 778.16 \times 32.17} = -234.7\,°\text{R}$, and so

$T_{t2} = 2616\,°\text{R} = 2156\,°\text{F}$.

Now $T_2 = T_{t2} - \dfrac{c_2^2}{2c_p} = 2616 - \dfrac{629.7^2}{2 \times 0.2920 \times 778.16 \times 32.17} = 2589\,°\text{R} = 2129\,°\text{F}$.

Next, $a_2 = \sqrt{\gamma \mathcal{R} T_2} = \sqrt{1.307 \times 53.35 \times 2589 \times 32.17} = 2410\,\text{ft/s}$.

Thus, the Mach number in the rotating frame is

$M_{2\text{rel}} = \dfrac{w_2}{a_2} = \dfrac{1427}{2410} = 0.5921.$ <ANS

Flow is thus subsonic in the rotating frame, and shock formation in the exit of the rotor should not be a problem. Next, in the absolute frame,

$M_{2\text{abs}} = \dfrac{c_2}{a_2} = \dfrac{629.7}{2410} = 0.2613.$ <ANS

Flow is therefore also subsonic in the stationary frame, and shock formation in the inlet of the stator should not be a problem either. Using the absolute Mach number yields

$$\frac{p_{t2}}{p_2} = \left[1 + \frac{\gamma - 1}{2} M_{2\text{abs}}^2\right]^{\frac{\gamma}{\gamma-1}} = \left[1 + \frac{0.307}{2} 0.2613^2\right]^{\frac{1.307}{0.307}} = 1.045$$

and so $p_2 = 225.3/1.045 = 215.5\,\text{psia}$. <ANS

Next, from the ideal gas law for $p_2 = 215.5\,\text{psia}$ and $T_2 = 2589\,°\text{R}$,
$\rho_2 = 0.006986\,\text{sl/ft}^3$,
and, since $\dot{m} = \rho_2 c_{a2} A_2$,

$$A_2 = \frac{\dot{m}}{\rho_2 c_{a2}} = \frac{8.704}{0.006986 \times 626.9} = 1.987\,\text{ft}^2.$$

Also, $A_2 = \pi D_2 t_2$,

and so $t_2 = \dfrac{A_2}{\pi D_2} = \dfrac{1.987 \times 144}{\pi \times 2 \times 17.5} = 2.603\,\text{in}.$ <ANS

Note that this is larger than the rotor inlet height because of the compressibility of the fluid. Next, the input stage power can be found from

$\dot{W}_{sh} = \dot{m}(h_{t2} - h_{t1}) = 8.704 \times (-1{,}715{,}329)/550 = -27{,}145\,\text{hp}.$ <ANS

For comparison, the power for one of the last stages of the high-pressure compressor (Example 5.1) is only 6712 hp. The percent reaction can be found from

$$\%R = \frac{1}{1 + \left[\dfrac{c_2^2 - c_1^2}{w_1^2 - w_2^2}\right]} = \frac{1}{1 + \left[\dfrac{629.7^2 - 1483^2}{638.9^2 - 1427^2}\right]} = 0.4742.$$ <ANS

This is a reasonable value for the percent reaction.

Stator Exit

Refer to Figure 7.10d for the stator exit velocity polygon. Once again, the axial velocity is constant through the stage, and so

$c_{a3} = c_{a2} = c_{a1} = 626.9\,\text{ft/s}.$

Also, the absolute exit angle is given, $\alpha_3 = 65°$, and thus

$$c_3 = \frac{c_{a3}}{\cos\alpha_3} = \frac{626.9}{\cos(65°)} = 1483\,\text{ft/s}.$$

The turning angle of the stator can be found from

$\delta_{23} = \alpha_3 - \alpha_2 = 65 + 5.43 = 70.43°.$

Thus, the turning angle for the stator vane is approximately that of the rotor blade as one would expect for a percent reaction of about 0.50. For the stator,

$T_{t3} = T_{t2} = 2616\,°\text{R}.$

$$T_3 = T_{t3} - \frac{c_3^2}{2c_p} = 2616 - \frac{1483^2}{2 \times 0.2920 \times 778.16 \times 32.17} = 2465\,°\text{R}.$$

Furthermore, $a_3 = \sqrt{\gamma \mathcal{R} T_3} = \sqrt{1.307 \times 53.35 \times 2465 \times 32.17} = 2352\,\text{ft/s}.$

Thus, the Mach number in the absolute frame is

$$M_{3\text{abs}} = \frac{c_3}{a_3} = \frac{1483}{2352} = 0.6308. \qquad \text{<ANS}$$

Flow is therefore also subsonic in the stationary frame, and shock formation in the exit of the stator should not be a problem. Since the absolute Mach number is known,

$$\frac{p_{t3}}{p_3} = \left[1 + \frac{\gamma - 1}{2} M_{3\text{abs}}^2\right]^{\frac{\gamma}{\gamma-1}} = \left[1 + \frac{0.307}{2} 0.6308^2\right]^{\frac{1.307}{0.307}} = 1.287.$$

Also, for the stator, $p_{t3} = p_{t2} = 225.3$ psia, and so

$p_3 = 225.3/1.287 = 175.1$ psia. <ANS

From the ideal gas law, for $T_3 = 2465\,°\text{R}$ and $p_3 = 175.1$ psia,

$\rho_3 = 0.005958\,\text{sl/ft}^3,$

and so $\dot{m} = \rho_3 c_{a3} A_3$

$$A_3 = \frac{\dot{m}}{\rho_3 c_{a3}} = \frac{8.704}{0.005958 \times 626.9} = 2.330\,\text{ft}^2.$$

Moreover, $A_3 = \pi D_3 t_3$ and so

$$t_3 = \frac{A_3}{\pi D_3} = \frac{2.330 \times 144}{\pi \times 2 \times 17.5} = 3.052\text{ in}. \qquad \text{<ANS}$$

Once again, note that this is larger than the stator inlet height because of the compressibility of the fluid. Also observe that all of the blade height changes are much larger for this turbine than for the compressor in Example 5.1. Thus, all of the variables of interest have been found.

7.5 Performance Maps

7.5.1 Dimensional Analysis

Experimental performance curves or "maps" are usually used that are similar to those for a compressor to provide a working medium for a turbine engineer. These provide an engineer with a quick and accurate view of the conditions at which the machine is operating and what can be expected of the turbine if the flow conditions are changed. Similitude is derived in Chapter 4 for turbines, and results are presented in this section. Of primary concern is the total pressure ratio of a turbine; namely, p_{t5}/p_{t4} or p_{t4}/p_{t5} (Eqs. (4.3.3), (4.3.7), and (4.3.8)):

$$p_{t4}/p_{t5} = f\left\{\frac{\dot{m}\sqrt{\theta_{t4}}}{\delta_{t4}}, \frac{N}{\sqrt{\theta_{t4}}}\right\} = f\{\dot{m}_{c4}, N_{c4}\}, \tag{7.5.1}$$

where

$$\delta_{t4} = p_{t4}/p_{stp} \tag{7.5.2}$$

and

$$\theta_{t4} = T_{t4}/T_{stp}, \tag{7.5.3}$$

and the subscript stp refers to the standard conditions and t4 refers to the inlet total conditions. Similarly, for the efficiency, one finds (Eqs. (4.3.6), (4.3.7), and (4.3.8))

$$\eta_t = g\left\{\frac{\dot{m}\sqrt{\theta_{t4}}}{\delta_{t4}}, \frac{N}{\sqrt{\theta_{t4}}}\right\} = g\{\dot{m}_{c4}, N_{c4}\}. \tag{7.5.4}$$

Again, note that the two independent parameters ($\dot{m}_{c4}$ and N_{c4}) are not truly dimensionless. The first is called the "corrected" mass flow and has dimensions of mass flow, whereas the second is called the "corrected" speed and has dimensions of rotational speed. Note that for a given engine at a given operating condition, the "corrected" values for a turbine and compressor are markedly different because of the significantly different reference or inlet conditions.

7.5.2 Mapping Conventions

Figure 7.11a presents a typical turbine map. Data are taken in a dedicated turbine test facility similar to a compressor facility. However, for the case of a turbine, high-pressure ("blowdown") air is expanded through the turbine and energy is derived from the turbine. Thus, a load is placed on the shaft. Static and total pressures and temperatures are measured between stages as well as the flow rate and derived power. The pressure and temperature ratios are found and the efficiency is calculated as follows:

$$\eta_t = \frac{1 - \left[\dfrac{T_{t5}}{T_{t4}}\right]}{1 - \left[\dfrac{p_{t5}}{p_{t4}}\right]^{\frac{\gamma-1}{\gamma}}}. \tag{7.5.5}$$

The pressure ratio is plotted vs. the corrected mass flow with the corrected speed as a second parameter. Also shown in this figure is the so-called choke line. This condition occurs when the pressure ratio grows so large across the turbine that the flow becomes sonic at one or more of the stage exits and the flow is thus choked. This condition basically limits the airflow rate

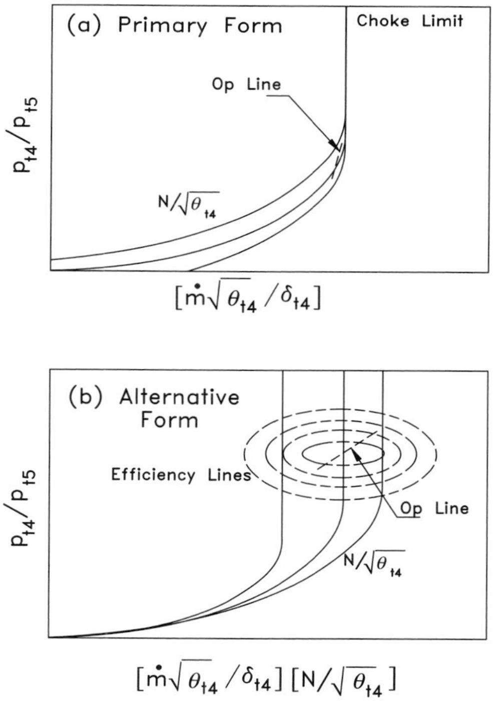

Figure 7.11 Typical turbine map – primary and alternative forms.

through the entire engine. Note that, because of the favorable pressure gradient in a turbine, massive stall and, therefore, surge are not present.

One problem with the presentation in Figure 7.11a is the congestion of data near the choke line (and this is where most turbines operate). An alternative form of presentation is given in Figure 7.11b. Here the corrected mass flow is multiplied by the corrected speed so that each speed line has a different choke line. Also shown in this figure are lines of constant efficiency. The typical design point of an engine is near the peak efficiency point. Thus, as one changes conditions from the design point, the efficiency decreases. Finally, shown in Figures 7.11a and 7.11b is the operating line. This line passes through the design point and represents the different conditions at which an engine typically operates.

Lastly, up to this point maps have been considered for the entire turbine. Typically, as is the case for a compressor, the same techniques are applied to a variety of combinations during the development stage. For example, one would have a map for only the low-pressure turbine and another for only the high-pressure turbine. Each stage of the engine will also have a map as well as the hub, pitch, and tip regions of each stage to enable the design of particular regions of a turbine to be improved. The maximum efficiencies are typically 88 percent to 90 percent for modern engines. Tables 5.3 and 6.3 list characteristics of some turbines from modern engines that complement the compressor information. One interesting note from these tables is that, for high-bypass turbofans (large mass flow rates), many low-pressure turbine stages are required to deliver the needed power to the fan even though the pressure ratios for the fans are relatively low.

7.6 Thermal Limits of Blades and Vanes

As indicated in Chapter 3, the thermodynamic efficiency and net power output of the overall engine can be improved by increasing the turbine operating temperature. However, the material composition of the turbine limits the safe operation of the engine. Excessive temperatures can lead to catastrophic blade failures, as shown in Figure 7.12 (also note the fir tree attachment to the disk). With failures such as these, blades can be thrown through the engine case and into the aircraft.

An historical view of the maximum turbine temperature limit is shown in Figure 7.13. As can be seen, the limit has steadily increased over the years. Until about 1960, increases in the temperature limit were due strictly to enhancements in material characteristics. However, around 1960 the slope of the line in the figure increases significantly. At that time, air cooling of the blades came into more widespread use. Since the early 1960s, improvements have been realized owing both to advances in heat transfer enhancement and material characteristics. However, the current limit is still well below the typical stoichiometric temperature of modern fuels. Currently, considerable research is being conducted in materials, including ceramics and composites, and better heat transfer mechanisms to increase the limit further. "Blade cooling" is often used to reduce the temperature of the blades. Three different types of cooling are used. All three bleed "cool," high-pressure air from the compressor. This air is directed into both stator vanes and rotor blades, as shown in Figure 7.14. Furthermore, thin thermal coatings of low-conductivity materials on blade and vane surfaces have further significantly enhanced the thermal limits of turbine blades over the past two decades, especially when combined with blade cooling. More details about the prediction and measurement of these effects can be found in Kercher (1998, 2000), Han et al. (2000), Dunn (2001), and Shih and Sultanian (2001).

7.6.1 Blade Cooling

The first type of blade cooling to be discussed is internal or convection cooling. For this type of cooling, cool air is pumped through the inside of the blades and the material is cooled by forced convection on the inside surface and by conduction through the blade. A schematic for this method is shown in Figure 7.15a.

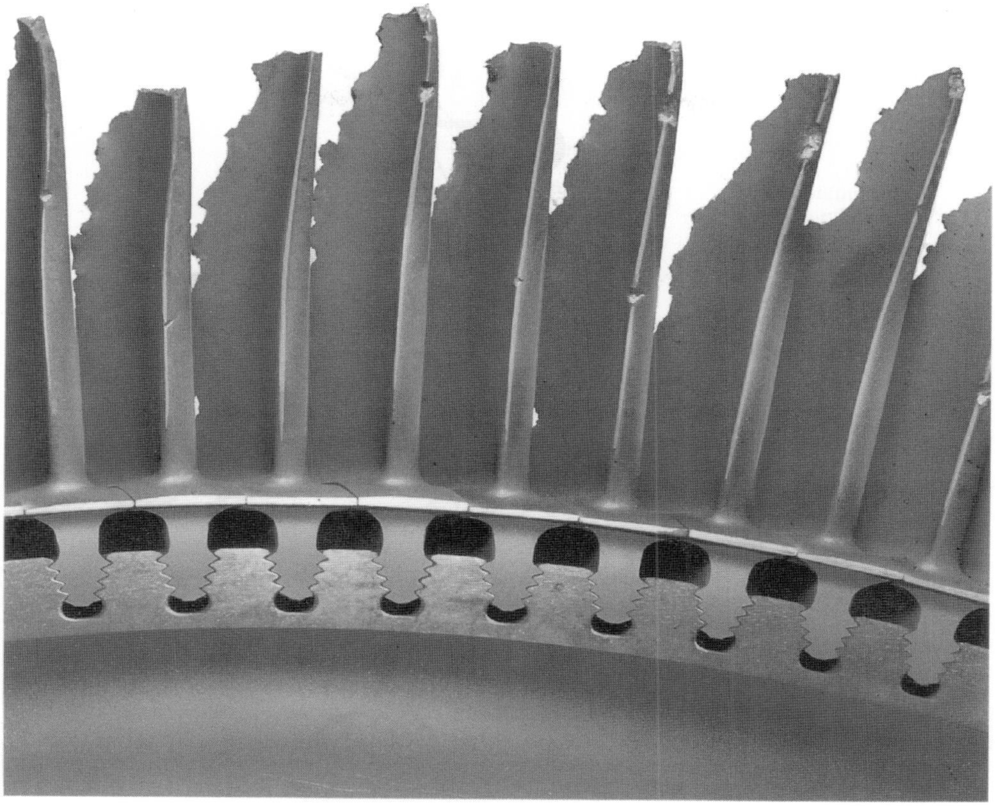

Figure 7.12 Turbine rotor blade failure due to excessive temperatures (© Rolls-Royce plc 2022. All rights reserved).

Overall, for this type of cooling, high-velocity cool air scrubbing the inside surface is desirable. Also, a large surface area on the inside is preferable; thus, for many designs, roughened internal "microfins" are utilized, as can be seen in Figure 7.15b. The internal flow paths are also shown in Figure 7.15b. For this blade, cool air is pumped into the blade at the root and makes multiple passes before exiting through the tail or trailing edge.

The second type of cooling is termed film cooling. In this method, air is pumped out of the blade (Figures 7.4 and 7.16) at different locations along the blade – but particularly at the leading edge. This cool air provides a cool protective boundary layer along the blade so that the high-temperature gas does not come in contact with the blade. Many arrangements have been chosen for the injection holes. Overall, for this type of cooling enough air is needed to cool the blades; however, the turbine efficiency decreases significantly with increasing bleed air. Thus, bleeding too much air is detrimental.

The third type of blade cooling is transpiration cooling. In this approach, a wire cloth or wire mesh is used for the exterior of the blade and air is leaked uniformly through it as shown in Figure 7.17. This type tends both to cool the surface and provide a protective layer. However, once again, too much cooling will tend to decrease the turbine efficiency markedly.

Also, combinations of these three types of cooling are often used. For example, in Figure 7.18a combination of internal and film cooling is used. Convection cooling is used near the leading edge (called impingement cooling), and air is injected

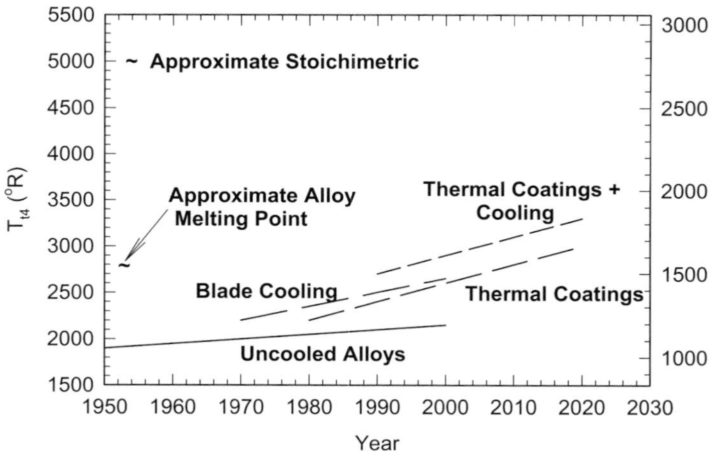

Figure 7.13 Turbine inlet temperature history.

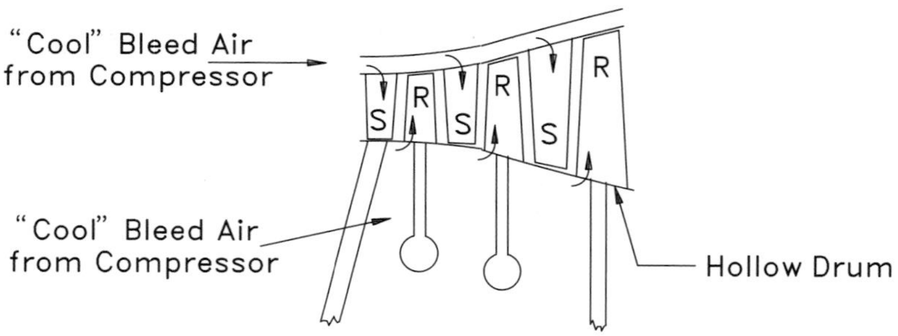

Figure 7.14 Cooling of turbine components using bleed "cool" air from compressor.

into the freestream at a few locations along the blade. One can also note the shrouding and fir tree on this blade.

Overall, blade cooling is a very complex and expensive solution to increasing the turbine inlet temperature. Manufacturing of blades with internal passages is not an easy or inexpensive process. Also, the hollow blades are weaker than their solid counterparts. Furthermore, ducting the cool air from the compressor to the turbine without losses is a complicated design problem. For these reasons, any large improvements in the future in turbine operating temperature will probably result from advances in blade material.

7.6.2 Blade and Vane Materials

Although blade cooling can be used to reduce the temperatures, heat transfer mechanisms limit the blade temperatures that can be obtained in such hostile environments. Thus, considerable effort has been made in recent years to improve the material characteristics. Nickel-based alloys containing chromium–cobalt or a titanium aluminide (a lighter material) are common materials for modern turbine blades. Turbine disks are typically made from a nickel-based alloy, a cobalt–nickel alloy, or a titanium aluminide.

Three types of blade structure are usually used. First, a conventionally cast blade is a myriad of crystals. Such a blade has multidirectional mechanical

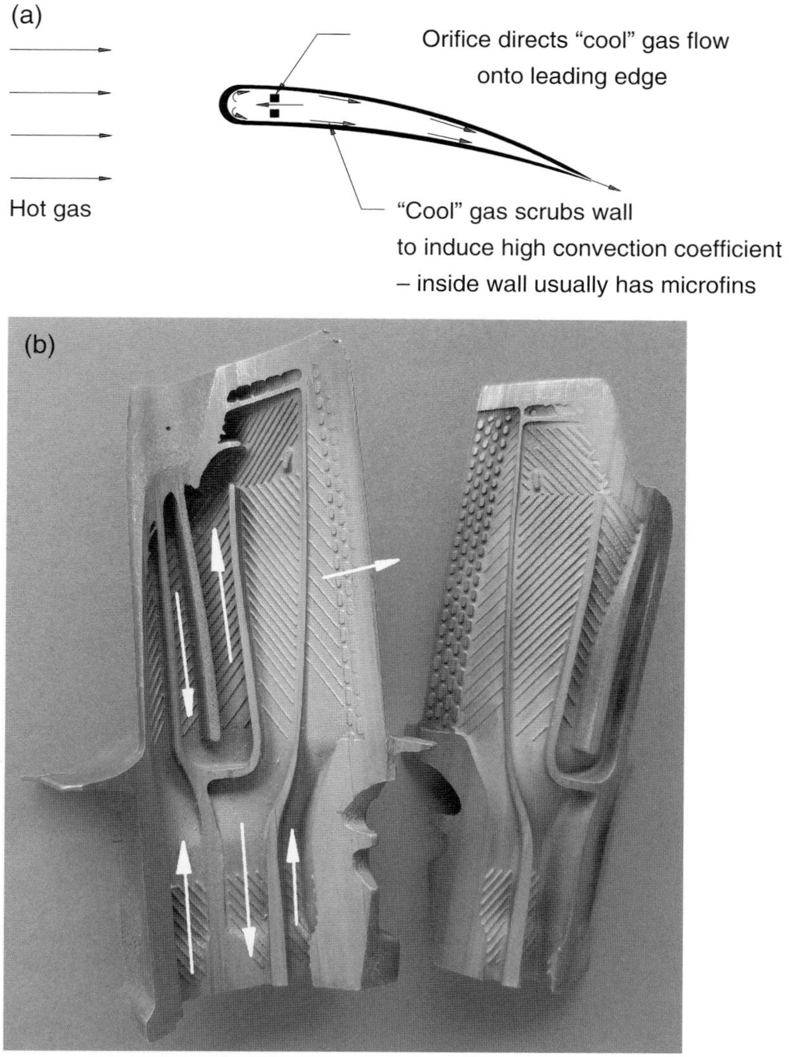

Figure 7.15 (a) Impingement-cooled leading edge; convection-cooled pressure and suction surfaces; chordwise fins in midchord region. (b) Split Pratt & Whitney rotor blade showing cooling passages and flow paths (photo by R. Flack; split blade © 2022 Raytheon Technologies Corporation, Pratt & Whitney division. Reproduced with permission).

properties – that is, the blades are strong in all directions. However, because of the boundaries between the crystals, the strength is compromised and blade failures usually occur at these boundaries. Such failures are typically due to long-term high stress levels accompanied by creep caused by the centrifugal and fluid forces, fatigue due to the high-frequency blade-pass frequencies, or corrosion resulting from the high temperatures and combustion products. Yet, these blades are relatively inexpensive (compared with the process described in Section 7.6.3) and find applications in sections in which the temperatures are not high.

A second type that is used is the directionally solidified blade. During the manufacturing process, the blade is cooled so that it comprises many long or columnar crystals. The blade has a dominant axis along which the blade demonstrates excellent strength properties. For rotating blades, this axis is chosen along the blade length owing to the high tensile

Figure 7.16 Film cooling on a Rolls-Royce nozzle inlet guide (© Rolls-Royce plc 2022. All rights reserved).

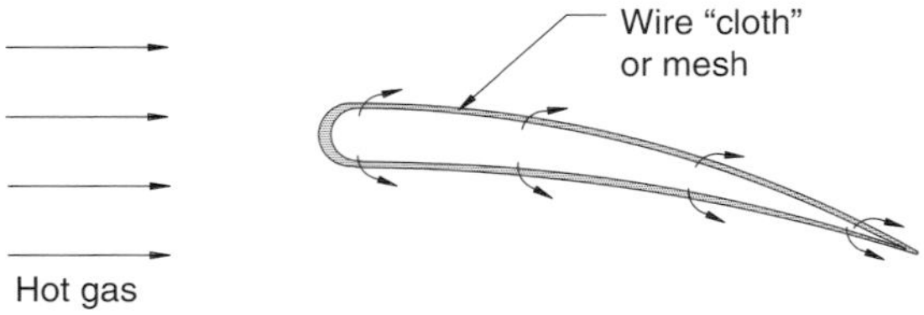

Figure 7.17 Transpiration cooling.

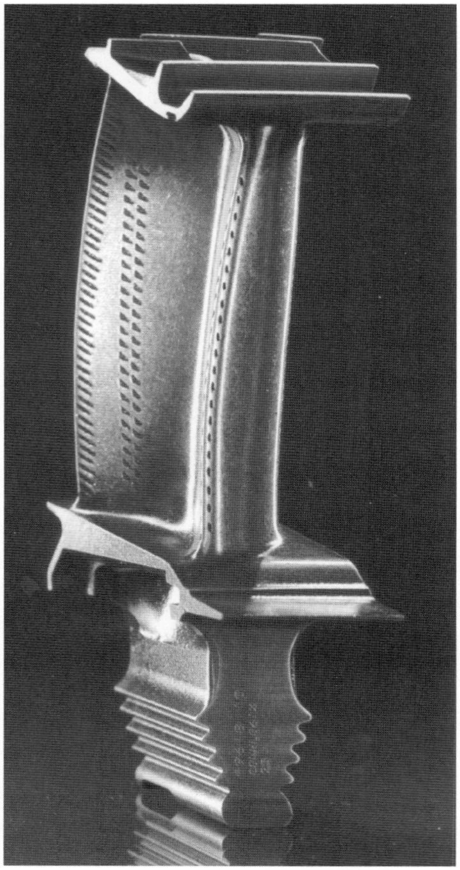

Figure 7.18 Combined film and convection cooling on a Rolls-Royce turbine rotor blade (© Rolls-Royce plc 2022. All rights reserved).

stresses that result from the centrifugal forces. This type of blade is usually used in the first stages of a turbine because of the extreme temperatures. A photograph of such a blade is presented in Figure 7.19

A third type is the "single crystal" blade, which, as its name implies, is a blade that, during the manufacturing process, cools as a single crystal. This blade type does not have any crystal boundaries and has multidirectional mechanical properties, making it ideal for turbine applications.

As noted in Sections 1.3.5 and 7.1, ceramics and composites are also currently being developed for turbine applications, and ceramic matrix blades are in fact used in the GE9X high-pressure section. Previously ceramic blades were susceptible to breakage due to the brittle nature but are improving dramatically.

7.6.3 Blade and Vane Coatings

To further minimize the effects of high temperatures on the blades, coating the blades with a layer of low-thermal-conductivity material is used. Many of the applications use aluminides. Ceramic coatings are under development and are used in the Rolls-Royce Trent XWB. The thicknesses of such coatings are usually on the order of 500 μm (0.02 in.) or less.

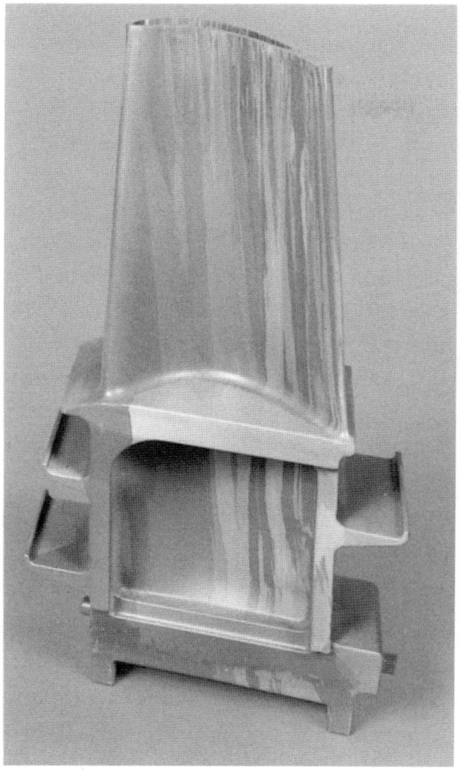

Figure 7.19 Howmet directionally solidified blade for a General Electric gas turbine (courtesy General Electric Gas Power).

7.6.4 Blade and Vane Manufacture

The manufacture of intricate turbine blades, such as shown in Figure 7.15b, is a very complex process. It is accomplished by the practice known as the investment or lost-wax process. It is about a 14-step process as follows.

(1) A two-piece die is made that is an exact (very accurate) "negative" of the eventual blade shape. This die will be used thousands of times.
(2) If the eventual blade is hollow, a ceramic core in the exact shape of the internal passages is precisely placed in the die. The ceramic core is of course a "negative" of the passages.
(3) The die is filled with a hot paraffin-based wax (liquid), and the wax is allowed to cool and harden. Waxes are chosen so that they do not shrink upon cooling.
(4) The die is separated and the wax piece is removed. The wax is now an exact replica of the eventual metal blade. If the eventual blade is hollow, the ceramic core is still in the wax replica.
(5) The wax blade is coated (by dipping, spraying, or both) with a slurry and then stucco with multilayers. A silica, alumina, or other ceramic "flour," or a combination of these, are typically used to create the hard stucco shell.
(6) The wax inside of the stucco shell is melted and escapes through an exit hole. If the blade is to be hollow, the ceramic core remains accurately in place in the stucco shell.
(7) The stucco shell, which is heat resistant, is filled with the blade material and the blade material is allowed to cool.
(8) The stucco shell is removed from the blade by air or sand blasted.
(9) If a hollow core is present, the ceramic core is removed by immersion in a caustic solution that dissolves the internal ceramic core.
(10) Finishing or trimming is accomplished by removing any metal used for holding the blades in place.
(11) The blades are inspected by X-ray and fluorescing surface die for internal and surface defects. If minor imperfections are found, they are repaired. If major imperfections are found, the blade is discarded.
(12) Some blades are coated with a very thin film of a poor heat conductor.
(13) Any surface holes for film cooling are "drilled" using precise, electrochemical, electrodischarge, waterjet, or laser machining.
(14) Some machining may be needed on the root (e.g., a fir tree) for placement on the wheel or drum. After this expensive and long multistep process, the blades are ready for installation.

7.7 Streamline Analysis Method (from Appendix K) (Advanced Topic)

In the thermodynamic and gas dynamic analyses in Section 7.4 the flows in the turbine are assumed to be 2-D (axial and tangential), and any radial distributions or velocity components are ignored. However, because the ratios of blade height to hub diameter for most of these machines are not small and centrifugal forces are large, 3-D flows result; that is, radial flows can result. As is the case for axial compressors, significant radial variations in parameter profiles will result when the tip radius is approximately 1.1 times the hub radius or greater.

As done for axial compressors, a streamline analysis method can be used to predict the 3-D behavior of the flow due to the centrifugal forces acting on the fluid elements. The control volume forms of the continuity, moment of momentum, and energy equations are again applied to streamline bundles, and radial equilibrium is used along with the appropriate boundary conditions.

The incompressible method is nearly identical to that covered in Chapter 5 by compressor examples and in detail in Appendix K. As a result, this will not be considered in detail in this chapter. One difference is in the energy equation and the definition of efficiency. The remainder of the equations are identical for axial flow turbines and compressors. The application of the equations and solution methods is the same as before.

7.8 Summary

The thermodynamics of a strictly axial flow turbine were considered in this chapter. The purpose of the turbine is to extract energy (enthalpy) from the fluid to drive the compressive devices; thus, the actual operation of a turbine is the opposite of that of a compressor. For propulsion gas turbines, the remaining enthalpy is used to generate thrust in the nozzle. For power generation units, the remaining enthalpy is used to generate auxiliary power, which can be used for electric generation or powering other equipment, such as a propeller or pump train. As in an axial flow compressor, the air was seen to flow alternately between rows of rotating rotor blades and stationary stator vanes. Geometries and hardware were discussed, and the important geometric parameters were defined. Comparisons were made with axial compressors, and several major differences in blade and other designs and operating conditions were discussed, most of which result from the pressure drop through the turbine (as compared with a pressure rise for a compressor) and the high inlet temperatures. Velocity triangles or polygons were once again used to analyze the turbomachines. In the initial analysis of turbines, 2-D flow was postulated, and the radial components of velocity were assumed to be small. From the known geometric parameters, working equations were derived through which to predict the resulting derived power, pressure ratios, and percent reactions using a control volume approach (continuity, momentum, and energy equations). The fundamental equations are very similar to those for an axial flow compressor, but their application to turbines is significantly different. The software Turbomachinery.exe performs the above calculations and is particularly helpful when optimizing performance because it eliminates repetitive and detailed calculations. Rotational speed and blade and vane turning were seen to be very important to the overall performance. The total power of a stage is a major consideration since this is used to drive the compressive devices, and percent reaction was identified as a measure of relative loading of the rotor and stator blade rows, which should be about 50 percent. Because of the favorable pressure gradient (pressure drop), separation does not pose the same problem for a turbine that it does for a compressor.

Axial flow turbine stages were found to operate with much larger blade turning and pressure differences than axial flow compressors; thus, fewer turbine stages are needed in an engine. However, because of the pressure drop, a turbine stage can choke. Performance maps (derived from dimensional analysis) were also described by which the important operating characteristics of a turbine can accurately and concisely be presented using similarity parameters. The data for a turbine map are best presented in a slightly different format than for a compressor. Four of the most important output characteristics are the pressure ratio, efficiency, operating line, and choke line. From a map, for example, if the speed and flow rates are known, the efficiency and pressure ratio can be found. From these maps, choking was seen to be a serious problem in turbines and needs to be carefully addressed in stage design.

Next, over the past decades turbine inlet temperatures have increased markedly, thus potentially reducing the integrity of the turbine blades. Developments in new materials and methods of cooling the turbine blades were covered: convection, film, transpiration cooling, and low-conductivity coatings. Blade materials and material structures have also improved dramatically over the past decades, allowing higher burner exit temperatures and better gas turbine performance. The manufacturing of turbine blades is complex and costly, particularly for hollow blades, and was discussed. Lastly, more complex streamline analysis methods for single stages were discussed to account for the 3-D flows resulting from radial equilibrium. These analysis approaches disclose that radial variations result, as they do for axial flow compressors, when the blade heights are significant. Throughout the chapter, turbine design guidelines were presented to facilitate prediction or determination of the reasonable stage operation.

List of Symbols

A	Area
a	Speed of sound
c	Absolute velocity
c_p	Specific heat at constant pressure
C	Chord length
D	Diameter
g	Gravitational constant
h	Specific enthalpy
$\dot{m}$	Mass flow rate
M	Mach number
N	Rotational speed
p	Pressure
Q	Volumetric flow rate
$\mathcal{R}$	Ideal gas constant
R	Radius
%R	Percent reaction
s	Blade spacing
T	Temperature
t	Blade height
U	Blade velocity
w	Relative velocity
$\dot{W}$	Power
α	Absolute (stator) angle
β	Relative (rotor) angle
γ	Specific heat ratio
δ	Flow turning angle
δ	Ratio of pressure to standard pressure
η	Efficiency
θ	Ratio of temperature to standard temperature
ρ	Density
ω	Rotational speed

Subscripts

a	Axial
abs	In absolute frame (stator)

c	Compressor	0	Upstream of IGV
c	Corrected	1,2,3	Positions
rel	In rotating frame (rotor)	4	Inlet to turbine
sh	Applied to shaft	5	Exit of turbine
stp	Standard conditions		
t	Total (stagnation)		
t	Turbine		
u	Tangential		

Superscripts

′	For the blade

Problems

7.1 A single stage (rotor and stator) of a high-pressure turbine is to be analyzed. It rotates at 11,000 rpm and expands 105 lb_m/s of air. The inlet pressure and temperature are 207 psia and 1700 °R, respectively. The average radius of the blades is 10.0 in., and the inlet blade height is 1.25 in. The absolute inlet flow angle to the rotor is the same as the stator exit flow angle (55°), and the rotor flow turning angle (δ_{12}) is 45°. The stage has been designed so that the blade height varies and the axial velocity remains constant through the stage. The efficiency of the stage is 83 percent. The values of c_p and γ are 0.2662 BTU/lbm-°R and 1.347, respectively. Find the following:

(a) the blade heights at the rotor and stator exits,
(b) the static pressure at the rotor and stator exits,
(c) the total pressure ratio across the stage,
(d) the stator flow turning angle,
(e) the derived power from the stage,
(f) the percent reaction for the stage.

7.2 A single stage (rotor and stator) of a high-pressure turbine is to be analyzed. It rotates at 13,200 rpm and expands 45.36 kg/s (100 lb_m/s of air). The inlet pressure and temperature are 1765 kPa and 1167 K (256 psia and 2100 °R), respectively. The average radius of the blades is 279.4 mm (11.0 in.), and the inlet blade height is 27.94 mm (1.10 in.). The absolute inlet flow angle to the rotor is the same as the stator exit flow angle (65°), and the total pressure ratio is 0.65. The stage has been designed so that the blade height varies and the axial velocity remains constant through the stage. The efficiency of the stage is 87 percent. The values of c_p and γ are 1.156 kJ/kg-K (0.2761 BTU/lb_m-°R) and 1.330, respectively. Find the following:

(a) the blade heights at the rotor and stator exits,
(b) the static pressure at the rotor and stator exits,
(c) the rotor flow and stator flow turning angles,
(d) the derived power from the stage,
(e) the percent reaction for the stage.

7.3 A single stage (rotor and stator) of a high-pressure turbine is to be analyzed. It rotates at 10,000 rpm and expands 45.36 kg/s (100 lb_m/s) of air. The inlet pressure and temperature are 1765 kPa and 1167 K (256 psia and 2100 °R), respectively. The average radius of the blades is 279.4 mm (11.0 in.), and the inlet blade height is 27.945 mm (1.10 in.). The absolute inlet flow angle to the rotor is the same as the stator exit flow angle (50°), and percent reaction is 0.50. The stage has been designed so that the blade height varies and the axial velocity remains constant through the stage. The efficiency of the stage is 87 percent. The values of c_p and γ are 1.162 kJ/kg-K (0.2776 BTU/lb_m-°R) and 1.328, respectively. Find the following:

(a) the blade heights at the rotor and stator exits,
(b) the static pressure at the rotor and stator exits,
(c) the rotor flow and stator flow turning angles,

(d) the derived power from the stage,
(e) the total pressure ratio across the stage.

7.4 An axial flow turbine stage rotates at 10,000 rpm with a mean radius of 11 in. and a blade height of 1.10 in. The efficiency of the stage is 87 percent, and 100 lb_m/s of air pass through the stage. The absolute flow angle into the rotor is 50°, which is the same as the absolute flow angle out of the stator. The rotor turning angle is 25°. Assume $\gamma = 1.30$. The inlet static pressure is 256 psia, and the inlet static temperature is 2100 °R.

(a) What is the percent reaction?
(b) What is the total pressure ratio?
(c) What is the power derived from the fluid (hp)?
(d) What is the stator blade turning angle?

7.5 A single stage of an axial flow turbine is to be analyzed. The shaft rotates at 11,000 rpm, and the average radius is 10 in. The inlet blade height is 1.25 in., and the inlet pressure and temperature are 207 psia and 1700 °R, respectively. The specific heat at constant pressure is 0.266 BTU/lb_m- °R, and the specific heat ratio is 1.347. The absolute angle at the rotor inlet is 55° (relative to the axial direction), and the rotor turning angle is 45°. The mass flow rate through the stage is 105 lb_m/s. The axial velocity remains constant at 586 ft/s. The total pressure ratio of the stage is 0.737, and the percent reaction is 0.532.

(a) Find the stage efficiency.
(b) Is this a good design? Why or why not? Be specific.
(c) If the design is not good, what would you do to improve it? Be specific.

7.6 A single stage (rotor and stator) of a high-pressure turbine is to be analyzed. It rotates at 8000 rpm and expands 90.72 kg/s (200 lb_m/s) of air. The inlet pressure and temperature are 1531 kPa and 1000 K (222 psia and 1800 °R), respectively. The average radius of the blades is 330.2 mm (13.0 in.), and the inlet blade height is 30.48 mm (1.20 in.). The absolute inlet flow angle to the rotor is the same as the stator exit flow angle (50°), and the rotor flow turning angle (δ_{12}) is 58°. The stage has been designed so that the blade height varies and the axial velocity changes by +3.0 percent across ach blade row (not stage). The total pressure ratio of the stage is 0.688. The values of c_p and γ are 1.122 kJ/kg-K (0.268 BTU/lb_m-°R) and 1.343, respectively. Find the following:

(a) blade heights at the rotor and stator exits,
(b) the static pressure at the rotor and stator exits,
(c) efficiency of the stage,
(d) the stator flow turning angle,
(e) the derived power from the stage,
(f) the percent reaction for the stage.

7.7 Initially an axial flow turbine stage operates with a 50 percent reaction. However, in a later test the mass flow rate decreases. If the absolute inlet angle to the rotor and the absolute exit angle from the stator remain constant, the relative rotor turning angle remains the same, the inlet conditions remain constant, and the rotational speed remains the same, what happens to the reaction? Explain with polygons.

7.8 Initially, an axial flow turbine stage operates with a 50 percent reaction. However, in a later test the rotational speed decreases. If the absolute inlet angle to the rotor and the absolute exit angle from the stator remain constant, the relative rotor turning angle remains the same, and the inlet axial velocity component remains the same, what happens to the reaction? Explain with polygons.

7.9 A single stage (rotor and stator) of a low-pressure turbine is to be analyzed. It rotates at 7000 rpm and expands 250 lb_m/s of air. The inlet pressure, temperature, and density are 120 psia, 1200 °R, and 0.00839 sl/ft^3, respectively. The inlet total pressure and total temperature are 179.3 psia and 1338 °R, respectively. The resulting axial velocity at the inlet is 758.1 ft/s; the speed of sound at the inlet is 1681 ft/s. The average radius of the blades is 14.0 in., and the inlet blade height is 2.00 in. The resulting tangential blade velocities are 855.2 ft/s. The absolute inlet flow angle to the rotor is the same as the stator exit flow angle (+55°), and the rotor flow turning angle ($|\delta_{12}|$) is 73°. The stage has been designed so that the blade height varies and the axial velocity decreases by 2.5 percent across each blade row (not stage). The total pressure ratio of the stage is 0.5573. The values of c_p and γ are 0.2528 BTU/lb$_m$-°R and 1.372, respectively. Find the following:

(a) the sign of δ_{12},
(b) the velocity polygons,
(c) the efficiency of the stage,
(d) the stator flow turning angle,
(e) the percent reaction for the stage,
(f) the derived power from the stage.

7.10 Initially an axial flow turbine stage operates with a 50 percent reaction. However, in a later test the rotor blades are replaced so that the relative flow turning in the rotor is less. If the absolute inlet angle to the rotor, the absolute exit angle from the stator, the mass flow rate, and the rotational speed remain the same, what happens to (explain with polygons and equations):

(a) the total pressure ratio of the exit to inlet (increases, decreases, or remains the same),
(b) percent reaction (increases, decreases, or remains the same)?

7.11 The shaft of a single stage (rotor and stator) of a low-pressure turbine rotates at 7600 rpm. The turbine expands 195 lb$_m$/s of air. The average radius of the blades is 12.0 in., and the inlet blade height is 1.60 in. The resulting tangential blade velocities are 795.9 ft/s. The inlet total pressure and total temperature are 217.2 psia and 1327 °R, respectively. The inlet static pressure, temperature, and density are 150 psia, 1200 ° R, and 0.01049 sl/ft^3, respectively. The speed of sound at the inlet is 1681 ft/s. The resulting axial velocity at the inlet is 689.9 ft/s; the turbine has been designed so that the axial velocity remains constant across each blade row. The absolute inlet flow angle to the rotor is the same as the stator exit flow angle (+57°), and the rotor flow turning angle ($|\delta_{12}|$) is 80°. The total pressure ratio of the stage is 0.5645. The values of c_{ab} and γ are 0.2527 BTU/lb$_m$-°R and 1.372, respectively. Find the following:

(a) the velocity polygons (to scale),
(b) the sign of δ_{12},
(c) the efficiency of the stage,
(d) the stator flow turning angle,
(e) the percent reaction for the stage.

7.12 The shaft of a single stage (rotor and stator) of a low-pressure turbine rotates at 9200 rpm. The turbine expands 80.29 kg/s (177 lb$_m$/s) of air. The average radius of the blades is 228.6 mm (9.0 in.), and the inlet blade height is 27.94 mm (1.10 in.). The resulting tangential blade velocities are 220.2 m/s (722.6 ft/s). The inlet static pressure, temperature, and density are 1655 kPa, 666.7 K, and 8.660 kg/m^3 (240 psia, 1200 °R, and 0.0168 sl/ft^3), respectively. The speed of sound at the inlet is 512.1 m/s (1680 ft/s). The inlet total pressure and total temperature are 2167 kPa and 717.1 K (314.3 psia and 1291 °R), respectively. The resulting axial velocity at the inlet is 231.1 m/s (759.0 ft/s); the turbine has been designed so that the axial velocity remains constant across each blade row. The absolute inlet flow angle to the rotor is the same as the stator exit flow angle (+45°), and the rotor flow turning angle ($|\delta_{12}|$) is 50°. The efficiency of the stage is 93 percent. The values of c_p and γ are 1.062 kJ/kg-K (0.2538 BTU/lb$_m$-°R) and 1.370, respectively. Find the following:

(a) the accurate velocity polygons (to scale),
(b) the sign of δ_{12},
(c) the total pressure ratio of the stage,
(d) the percent reaction for the stage,
(e) the optimum pitch-to-chord ratio for the stator vanes (estimate).

7.13 Find the corrected mass flow rate and corrected speed for the stage in Problem 7.1. If the stage is run at 9000 rpm and the flow rate is 90 lb$_m$/s, find the inlet pressure and inlet temperature so that the corrected parameters are matched. Use these conditions to find the resulting flow angles and total pressure ratio and compare them with those from Problem 7.1. Comment.

7.14 A single stage (rotor and stator) of a high-pressure turbine is to be analyzed. It rotates at 12,000 rpm and expands 90 lb_m/s of air. The inlet pressure and temperature are 150 psia and 2000 °R, respectively. The average radius of the blades is 9.60 in., and the inlet blade height is 1.20 in. The absolute inlet flow angle to the rotor is the same as the stator exit flow angle (45°), and the rotor flow turning angle (δ_{12}) is 40°. The stage has been designed so that the blade height varies and the axial velocity remains constant through the stage. The efficiency of the stage is 88 percent. The values of c_p and γ are 0.2739 BTU/lb_m-°R and 1.339, respectively. Find the following:

(a) the blade heights at the rotor and stator exits,
(b) the static pressure at the rotor and stator exits,
(c) the total pressure ratio across the stage,
(d) the stator flow turning angle,
(e) the derived power from the stage,
(f) the percent reaction for the stage.

Part III
Nonrotating Components

Rolls Royce RB211-524 inlet (© Rolls-Royce plc 2022. All rights reserved).

Pratt & Whitney F100-229A nozzle (©; 2022 Raytheon Technologies Corporation, Pratt & Whitney division. Reproduced with permission).

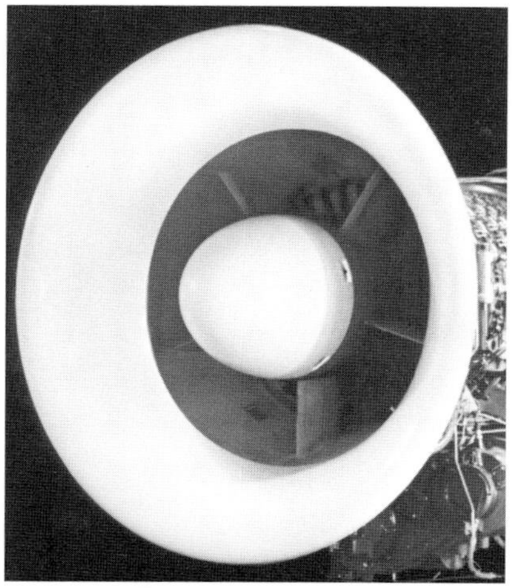

GE LM2500 inlet (courtesy of General Electric Gas Power).

Ruston Tornado burner (© Siemens Energy, 1985).

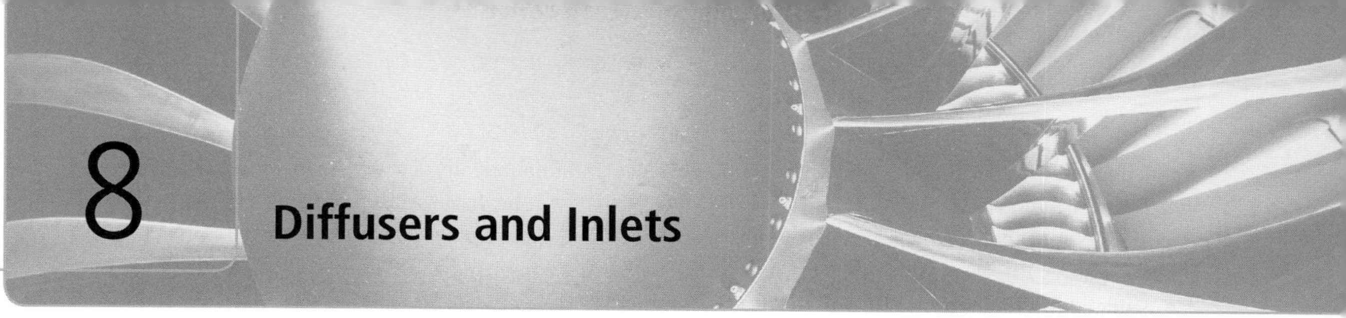

8 Diffusers and Inlets

8.1 Introduction

The purposes of the diffuser or inlet are first to bring air smoothly into the gas turbine. Second, for a propulsion diffuser to slow the fluid and to increase the pressure, or for a power generation inlet to increase the fluid speed and to decrease the pressure. And third, to deliver a uniform flow to the compressor. As indicated by studies for cycle analyses in Chapter 3, gas turbine performance improves with increasing pressure to the burner. The first component the air encounters is the diffuser or inlet, and the second component is the compressor. Thus, if the diffuser or inlet incurs a large total pressure loss, the total pressure into the burner will be reduced by the compressor total pressure ratio times this loss. For example, if 2 psia are lost in the diffuser or inlet, for a large engine this can result in 50 psia less in the burner.

For a propulsion diffuser, the losses result from several processes and affect engine performance in several ways. First, losses in total pressure occur outside of the propulsion diffuser primarily as the result of shock interactions. The flow outside the propulsion diffuser can also actually affect the flow around the entire engine and add a drag to the aircraft. Second, ideally the flow inside the diffuser is uniform and the streamlines are smooth. However, because of the unfavorable pressure gradient, the boundary layer on the walls tends to grow and separate. This has three subeffects. First, as the result of the wall shear, the flow is not isentropic; thus, the total pressure is further reduced. Second, because of the separation, the flow area is decreased. The flow velocity therefore remains high and the static pressure is not increased as much as in the ideal situation. Lastly, because of the separation, the flow is highly nonuniform as it enters the compressor. This, as a result, reduces the compressor efficiency and stability. Thus, it is very important for the diffuser to operate efficiently both internally and externally.

For a bell inlet there is minimal loss outside of the inlet. Ideally the flow inside the inlet is uniform and the streamlines are smooth. However, because of the boundary layer the flow is not uniform. Also, a strong crosswind may cause separation inside the inlet but this is somewhat avoided by the well curved shape of the bell inlet. These have two effects. First, as the result of the wall shear, the flow is not isentropic; thus, the total pressure is reduced. Second, because of the boundary layer or possible separation due to crosswinds, the flow area is slightly decreased, and the flow is not uniform as it enters the compressor. This, as a result, reduces the compressor efficiency and stability. Thus, it is very important for the inlet to operate efficiently both internally and externally.

Computational fluid dynamic (CFD) three-dimensional (3-D) and unsteady methods have been developed extensively over the past decades and are extensively used in the design of all gas turbine components. However, they are not exclusively used because experimental data are required to verify results. As a result, empiricism and experimental testing are used to complement CFD. Furthermore, the use of CFD is beyond the scope of this book. Thus, empiricism and one-dimensional (1-D) and two-dimensional (2-D) models will be presented instead to demonstrate trends as well as a first-pass (but reasonable) calculation method consistent with the effort required by the included problems.

8.1.1 Objectives

After completing this chapter, the reader should be able to do the following.

- Understand the principles and definitions of the diffuser recovery factor
- Understand the basic geometries of subsonic vs. supersonic diffusers
- Understand the external flow patterns of subsonic diffusers
- Understand the limits on pressure rises in subsonic diffusers
- Understand the losses in subsonic diffusers, including pressure losses
- Understand the basic properties of normal, planar oblique, and conical oblique shocks, including pressure losses
- Understand the considerations of mass flow patterns and different operating conditions of supersonic diffusers, including spillage
- Be able to predict the recovery factor for different diffusers
- Understand the concepts of additive drag for supersonic diffusers
- Understand what is meant by "starting" a supersonic diffuser
- Understand the basic geometry for a bell inlet
- Understand the principles and definitions of the inlet recovery factor
- Understand the losses in bell inlets, including pressure losses
- Understand the basic differences in the operation of bell inlets compared with diffusers
- Be able to predict the recovery factor for an inlet
- Understand the purposes of inlet cooling for both propulsion and power generation units
- Understand the principles and working concepts of diffuser and inlet maps

8.2 Subsonic Diffusers for Propulsion

8.2.1 Recovery Factor

Ideally, the diffuser is assumed to operate isentropically both internally and externally. As discussed in Chapters 2 and 3, this is not the case. Losses can occur both before the air reaches the diffuser and also once it enters the diffuser. The diffuser essentially operates adiabatically. The losses inside and outside the diffuser will be discussed separately.

The concept of total pressure recovery is introduced to quantify these losses. This is defined as

$$\pi_{\mathrm{d}} = \frac{p_{\mathrm{t2}}}{p_{\mathrm{ta}}}, \tag{8.2.1}$$

where stations a, 1, and 2 are defined in Figure 2.3. Note that this recovery factor represents the loss from the freestream to the diffuser exit. This can be expanded to

$$\pi_{\mathrm{d}} = \left(\frac{p_{\mathrm{t2}}}{p_{\mathrm{t1}}}\right) \times \left(\frac{p_{\mathrm{t1}}}{p_{\mathrm{ta}}}\right). \tag{8.2.2}$$

The first term, or loss, is often defined to identify internal diffuser losses. This represents the losses only within the diffuser due to boundary layers, turbulence, and separation. This is the internal diffuser pressure recovery factor. Losses can also occur before the flow enters the diffuser, which is the second term, and which represents the loss from the far freestream to the diffuser inlet. Equation (8.2.2) will be redefined as

$$\pi_{\mathrm{d}} = \left(\frac{p_{\mathrm{t2}}}{p_{\mathrm{t1}}}\right) \times \left(\frac{p_{\mathrm{t1}}}{p_{\mathrm{ta}}}\right) = \pi_{\mathrm{d\text{-}int}} \times \pi_{\mathrm{d\text{-}ext}}. \tag{8.2.3}$$

For subsonic flow, the quantity $\pi_{\mathrm{d\text{-}ext}}$ is approximately unity. Determining $\pi_{\mathrm{d\text{-}int}}$ is discussed in the following sections below. Consequently, by specifying either π_{d} or both $\pi_{\mathrm{d\text{-}int}}$ and $\pi_{\mathrm{d\text{-}ext}}$ for subsonic flow, the losses associated with the diffuser can be included in an analysis. For most of this text π_{d} will be primarily used.

8.2.2 External Flow Patterns

A diffuser must be designed for a variety of flight conditions. That is, it must be able to collect the needed airflow and ready the flow for the compressor for different Mach numbers, altitudes, engine mass flows, and so on. For a subsonic diffuser, the design is not nearly as critical as for a supersonic diffuser. This is because the governing equations of the flow are elliptic, and small pressure fluctuations are signaled upstream of the diffuser to adjust flow patterns into the diffuser as the needs change.

For example, streamlines for two flow conditions are shown in Figure 8.1. First, the streamline for high

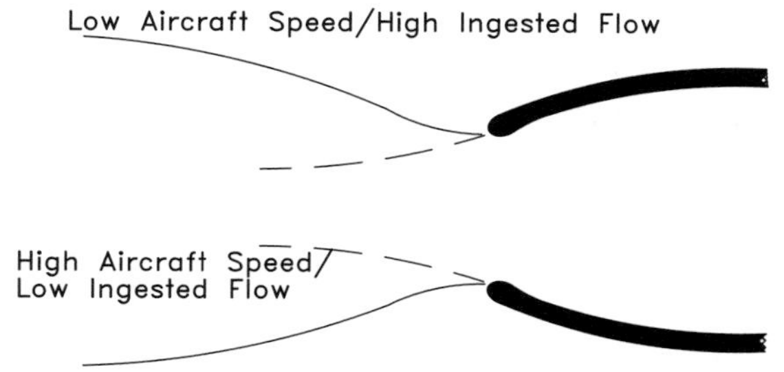

Figure 8.1 Inlet flow pattern for two conditions.

aircraft speed, or low ingested air flow rate, or both, is shown. Far away from the inlet, the encompassed area of the airflow is smaller than the diffuser inlet area. Thus, as it approaches the inlet, the flow decelerates and the pressure increases. To allow for streamline curvature at the intersection of the inlet and for crossflows, the inlet is well rounded at the leading edge so that air enters with minimal separation.

Next, the streamline for low aircraft speed, high airflow rate, or both, is shown. This condition would occur for an aircraft on the runway or an engine on the test stand. Far away from the inlet, the encompassed area of the airflow is now larger than the diffuser inlet area. Thus, as the flow approaches the inlet, the flow accelerates and the pressure decreases. A well-rounded inlet at the leading edge is even more important for this condition as the streamlines change direction (converging to diverging) at the front of the diffuser. Predictions from CFD aid significantly in the design of the lip of the cowl.

8.2.3 Limits on Pressure Rise

As indicated in Chapters 1–3, the developed thrust of the engine will be increased if the pressure into the nozzle is increased. Thus, it is desirable to increase the pressure out of the diffuser as much as possible as well as to reduce the velocities into the compressor. As a result, an adverse pressure gradient is present that tends to cause separation of the boundary layer. It is this separation that is one of the limits of diffuser operation. To estimate the limit, Hill and Peterson (1992) considered the general pressure coefficient defined by

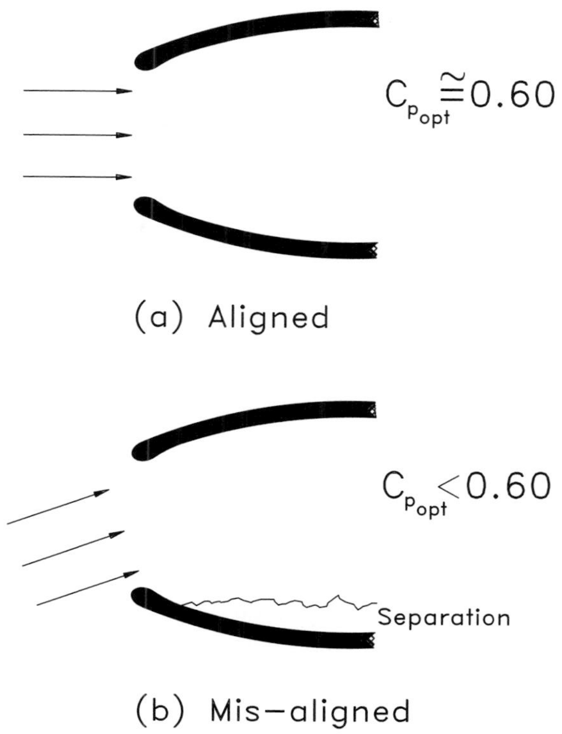

Figure 8.2 Separation in diffuser.

$$C_p \equiv \frac{\Delta p}{\frac{1}{2}\rho u^2}, \tag{8.2.4}$$

where u is the inlet relative velocity to the component. Typically, for a simple diffuser having walls with limited curvature and ideal inlet conditions (Figure 8.2a), C_p is less than or approximately equal to 0.6 for flow without separation. If the pressure coefficient becomes greater than about 0.6, the flow separates. If the walls have a pronounced curvature, or if the flow enters the diffuser in a skewed direction (Figure 8.2b) or the flow direction fluctuates, this upper limit on the pressure coefficient will be reduced appreciably to perhaps 0.1 to 0.2. Again, a well-rounded leading edge improves performance.

A second empirical method of predicting separation (which complements the preceding results) is shown in Figure 8.3. Here the limiting expansion angle of a general simple diffuser is shown as a function of diffuser length-to-width ratio. Such data can be

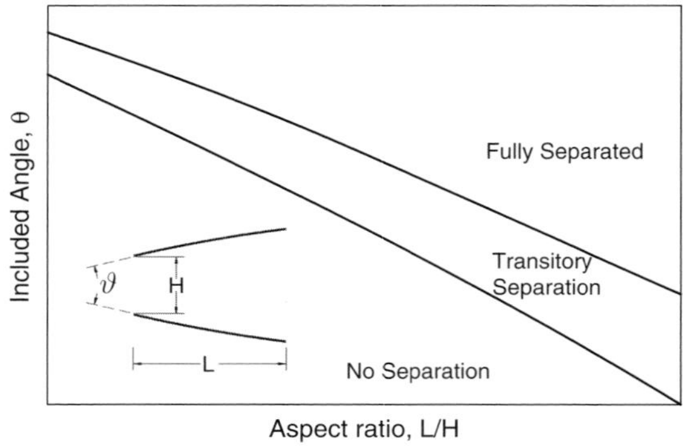

Figure 8.3 General empirical data trends for separation in a simple diffuser.

achieved for planar, axisymmetric, or other elliptical diffusers with a variety of wall shapes (straight, curved, etc.). The particular empiricism will depend on the diffuser of interest and is usually found with a well-controlled flow into the diffuser. For a given aspect ratio (L/H), if the included angle is too large, the diffuser area is enlarging too quickly and separation will occur. On the other hand, for a given expansion angle ϑ, if the aspect ratio is too large, separation will occur. In general, a region exists for which separation can be transitory – that is, the size of the separated region fluctuates.

Fox and Kline (1962) present data such for a simple 2-D planar diffuser with straight walls. Curve fitting their data yields two equations that approximate the two curves:

$$\ln \vartheta_{\min} = 3.28 - 0.46\ \ln(L/H) - 0.031(\ln(L/H))^2 \tag{8.2.5a}$$

$$\ln \vartheta_{\max} = 3.39 - 0.38\ \ln(L/H) - 0.020(\ln(L/H))^2, \tag{8.2.5b}$$

where ϑ is in degrees. For example, if $L/H = 4$, separation will be expected to initiate for an included angle between about 13° and 17° for a simple straight-walled planar diffuser with a well-controlled inlet.

Next, a pressure coefficient can be defined specifically for a diffuser as

$$C_p = \frac{p_2 - p_1}{\frac{1}{2}\rho_1 u_1^2}. \tag{8.2.6}$$

For the diffuser, one can consider the square of the inlet Mach number as

$$M_1^2 = \frac{u_1^2}{a_1^2} = \frac{\rho_1 u_1^2}{\gamma p_1}. \tag{8.2.7}$$

Thus, from Eq. (8.2.6), one finds

$$\frac{p_2}{p_1} = \frac{1}{2}\frac{\rho_1 u_1^2 C_p}{p_1} + 1, \tag{8.2.8}$$

and from Eq. (8.2.7) the following expression results:

$$\frac{p_2}{p_1} = 1 + \frac{1}{2}\gamma C_p M_1^2. \tag{8.2.9}$$

Namely, as the Mach number increases a larger pressure rise can be expected, without an increasing concern for separation, due to the added momentum of the fluid.

To appreciate the order of magnitude of the area ratios at hand, note that, for ideal incompressible flow, Eq. (8.2.6) will yield

$$C_p = 1 - \left[\frac{A_1}{A_2}\right]^2; \tag{8.2.10}$$

thus, for a pressure coefficient limit of 0.6 the exit to inlet area ratio becomes 1.581, which is much larger than would be present on an engine inlet. Note in particular that Eq. (8.2.10) does not apply to compressible flow.

Example 8.1 Isentropic Subsonic Diffuser Flow with Maximum Pressure Coefficient

Flow enters an ideal diffuser with a Mach number of 0.8, an inlet pressure of 13.12 psia, and an inlet diameter of 40 in. If the diffuser operates with the optimum (or maximum or limiting pressure) coefficient of 0.6, what is the resulting exit Mach number and diffuser area ratio? Assume isentropic flow and $\gamma = 1.40$.

Solution

For the given inlet Mach number, general one-dimensional isentropic flow (Appendix C) and the tables of Appendix D or Isentropic.exe reveal that the static-to-total pressure ratio is

$p_1/p_{t1} = 0.6560$

and $A_1/A^* = 1.0382$;

thus, the total pressure at the inlet is $p_{t1} = 13.12/0.6560 = 20.00$ psia, which for the case of isentropic flow is also the pressure at the exit, p_{t2}. From Eq. (8.2.9), one finds that

$$\frac{p_2}{p_1} = 1 + \frac{1}{2}\gamma C_p M_1^2,$$

and so for $C_p = 0.60$ the ratio of static pressures is

$$\frac{p_2}{p_1} = 1 + \frac{1}{2} \times 1.40 \times 0.60 \times 0.80^2 = 1.2688;$$

thus, the exit static pressure is $p_2 = 1.2688 \times 13.12 = 16.65$ psia. The static-to-total pressure ratio at the exit is $p_2/p_{t2} = 16.65/20.00 = 0.8323$. From the isentropic flow tables for this pressure ratio, the exit Mach number is

$M_2 = 0.5188$

and $A_2/A^* = 1.3055$.

Thus, the area ratio is $A_2/A_1 = 1.3055/1.0382 = 1.257$, which is different than it would be if incompressible flow were assumed. Note that if $L/H = 2$, this leads to an included angle of 6.9°, which, if one can use data from Fox and Kline (1962) (developed for planar flow), is well below the separation limit.

8.2.4 Fanno Line Flow

Boundary layers and other viscous flows are the primary means by which total pressure losses occur within diffusers. This will be analyzed based on Fanno line flow, which is flow with friction but no heat addition or loss. Fanno line flow is a constant-area process that does not exactly represent a diffuser. However, when the exit-to-inlet area ratio is near unity and the flow consequently does not separate, the Fanno line analysis can be used to make a reasonable prediction of the total pressure losses. The flow is analyzed in Appendix C with a control volume approach. The analysis results in 11 equations. Seventeen variables are present, and so six must be specified. For example, if inlet conditions M_1, T_1, and p_{t1}, and geometry parameters L, D, and f (Fanning friction factor) are specified, exit conditions, including the total pressure p_{t2}, can be found.

Example 8.2 | Fanno Line Flow

Flow with an incoming Mach number of 0.8 and an inlet total pressure of 20 psia enters a diffuser 24 in. long with a 40 in. average diameter and a Fanning friction factor of 0.020. What is the total pressure ratio if the specific heat ratio is 1.40?

Solution

From the tables in Appendix E or Fanno.exe for a Mach number of 0.80 it is possible to find the following nondimensional length-to-diameter and pressure parameters:

$$\left.\frac{4fL^*}{D}\right]_1 = 0.0723$$

and

$$\frac{p_{t1}}{p_t^*} = 1.0382;$$

thus, one can find the total pressure at the sonic condition, which is a *reference* condition (which means it usually does not occur in the actual conditions), as follows:

$$p_t^* = 20/1.0382 = 19.26\,\text{psia}.$$

From the duct geometry, it is possible to find the parameter

$$\left.\frac{4fL}{D}\right]_{1-2} = \frac{4 \times 0.020 \times 24}{40} = 0.048;$$

thus, the corresponding length-to-diameter parameter at the exit can be found as follows:

$$\left.\frac{4fL^*}{D}\right]_2 = \left.\frac{4fL^*}{D}\right]_1 - \left.\frac{4fL}{D}\right]_{1-2} = 0.0723 - 0.048 = 0.0243.$$

The exit Mach number can then be found from the tables; namely,

$$M_2 = 0.874.$$

Thus, as a result of friction the Mach number increased about 9 percent. From the tables or Fanno.exe, at this Mach number

$$\frac{p_{t2}}{p_t^*} = 1.0143,$$

and so the exit total pressure can be found: $p_{t2} = 1.0143 \times 19.26 = 19.54$ psia. The pressure ratio for the inlet for Fanno line flow is therefore

$$\pi_{\text{d-int}} = \frac{p_{t2}}{p_{t1}} = \frac{19.54}{20} = 0.977.$$

8.2.5 Combined Area Changes and Friction

In the preceding two sections, area changes and friction were treated separately. However, in a diffuser the two effects occur simultaneously. Thus, an analysis is needed in which both are included. The technique that will be used here is the so-called generalized 1-D flow method; it is described in Appendix C and is a differential equation analysis from which influence coefficients are derived. These influence coefficients relate changes of one variable on another. Applying the method entails numerically integrating the differential equations along the length of the diffuser, which incorporate both friction and area changes.

Example 8.3 Diffuser with Combined Area Change and Friction

Flow with an incoming Mach number of 0.8 and an inlet total pressure of 15 psia enters a diffuser channel 36 in. long and with a 36 in. diameter (with an inlet area of 1018 in.^{2}). The flow has a Fanning friction factor of 0.020 and an exit-to-inlet area ratio of 1.50. The total temperature remains constant at 500 °R. What is the total pressure ratio if the specific heat ratio is 1.40?

Solution

This problem will be solved by numerically integrating Eqs. (C.14.3) and (C.14.4). One hundred steps will be used, and area change and friction will be assumed to be uniform along the length of the diffuser. The exit area is 1527 in.^{2}; thus, the area increases by 509 in.^{2}. Therefore, $\Delta A = 5.09$ in.^{2}, and $\Delta x = 0.36$ in. for each step. A forward difference method is used. Thus, for the first step from Eq. (C.14.1) (the first term is due to area changes, and the second term results from friction),

$$\Delta(M^2) = -2M^2\left[\frac{\left(1+\frac{\gamma-1}{2}M^2\right)}{1-M^2}\right]\left[\frac{\Delta A}{A}\right] + 4f\gamma M^4\left[\frac{\left(1+\frac{\gamma-1}{2}M^2\right)}{1-M^2}\right]\left[\frac{\Delta x}{D}\right]$$

so $M_i^2 = M_{i-1}^2 + \Delta(M^2)$

$$M^2 = 0.8^2 - 2\times 0.8^2\left[\frac{\left(1+\frac{1.40-1}{2}0.8^2\right)}{1-0.8^2}\right]\left[\frac{5.09}{1018}\right]$$

$$+4\times 0.020\times 1.40\times 0.8^4\left[\frac{\left(1+\frac{1.40-1}{2}0.8^2\right)}{1-0.8^2}\right]\left[\frac{0.36}{36}\right] = 0.6214,$$

or $M = 0.7883$.

From Eq. (C.14.2), the total pressure loss is only due to friction:

$$\Delta p_t = -\frac{1}{2}p_t 4f\gamma M^2\left[\frac{\Delta x}{D}\right] = -2p_t f\gamma M^2\left[\frac{\Delta x}{D}\right],$$

and so $p_{ti} = p_{t\,i-1} + \Delta(p_t)$

$$p_t = 15 - 2\times 15\times 0.020\times 1.40\times 0.8^2\times\frac{0.36}{36} = 14.995.$$

This is repeated for all 100 steps, where the previous step is used for the initial conditions for the following step. The area and diameter are of course updated from step to step. Thus, one finds after 100 steps that

$M_2 = 0.4160$

$p_{t2} = 14.76$ psia,

and thus

$$\pi_{\text{d-int}} = \frac{p_{t2}}{p_{t1}} = \frac{14.76}{15} = 0.984.$$

Repeating the above for 15,000 steps, not 100, yields

$$\pi_{\text{d-int}} = \frac{p_{t2}}{p_{t1}} = \frac{14.76}{15} = 0.984,$$

namely the same. One should note that General1D.exe can be used for this calculation.

For comparison, one can approximate the flow by separately analyzing the flow with area change (isentropic flow) and Fanno line flow and superimposing the two results. For example, when the simple area change is considered for the area ratio ($A_2/A_1 = 1.50$) in this example, the exit Mach number is found to be 0.4105 for isentropic flow and the total pressure does not change ($p_{t2} = p_{t1}$). Thus, the average Mach number is 0.6053 and the average diameter is 40 in. Consequently, for Fanno line flow

$$\left.\frac{4fL}{D}\right]_{1-2} = \frac{4 \times 0.020 \times 36}{40} = 0.072.$$

Using this in an analysis similar to that of Example 8.2 and the average Mach number (0.6053) as the inlet condition, one finds an exit Mach number due to only friction of 0.6257, and, most importantly,

$$\pi_{\text{fanno}} = \frac{p_{t2}}{p_{t1}} = 0.981.$$

So, superimposing, $\pi_{\text{d-int}} = \pi_{\text{fanno}} \times \pi_{\text{isentropic}} = 0.981 \times 1.00 = 0.981$.

Thus, for this case, the approximate superposition analysis is quite close to the more accurate method, $\pi_{\text{d}} = 0.984$. In practice, a CFD method is used to calculate the total pressure loss and resulting exiting 2-D or 3-D velocity profile. As already noted, however, using CFD is beyond the scope of this book.

8.3 Supersonic Diffusers for Propulsion

8.3.1 Recovery Factor

As discussed in Section 8.2.1 losses can occur both outside of the diffuser and inside of the diffuser. For supersonic operation, additional losses can occur owing to resulting shocks. As the freestream Mach number increases past unity, this loss can increase quickly if design precautions are not taken as will be discussed. This is due to shocks outside of, or in the plane of, the diffuser inlet. This loss is *conventionally* defined by $\eta_r \equiv \left(\frac{p_{t1}}{p_{ta}}\right) \equiv \pi_{\text{d-ext}}$ and is the loss from the free-stream to the diffuser inlet. Thus, $\pi_{\text{d}} = \left(\frac{p_{t2}}{p_{t1}}\right) \times \left(\frac{p_{t1}}{p_{ta}}\right) = \pi_{\text{d-int}} \times \eta_r$, which replaces Eq. (8.2.2) for the supersonic case, and where η_r represents the shock loss, and $\pi_{\text{d-int}}$ is the internal diffuser recovery factor.

Determining this shock loss is a purpose of this section. The internal diffuser recovery factor is again due to nonisentropic losses within the diffuser. Because of shocks both $\pi_{d\text{-int}}$ and especially η_r can be significantly less than unity for the supersonic cases. Consequently, by specifying either π_d or both $\pi_{d\text{-int}}$ and η_r for supersonic cases, the losses associated with the diffuser can be included in an analysis. For most of this text π_d will be primarily used.

8.3.2 Shocks

Fundamentals

Shocks are the primary means by which total pressure losses occur outside of a propulsion diffuser. Normal shocks are a 1-D flow and lead to large total pressure losses. This flow is analyzed based on flow with no friction and no heat addition or loss. The normal shock process is also a constant area process. Normal shocks are discussed and derived in detail in Appendix C.7. For the normal shock process, the h–s diagram and the continuity, linear momentum, and energy equations are used. Thus, the process is in fact a superposition of Fanno line and Rayleigh line processes.

Oblique planar shocks result from supersonic flow over a 2-D wedge and incur lower losses in total pressure than do normal shocks, as will be demonstrated in following examples and figures. Although the flow is truly 2-D, it has many characteristics of 1-D flow. Flow before the shock is uniformly parallel to the axis, and behind the planar shock, flow is uniformly parallel to the wedge. For the analysis the velocities are broken into the components parallel and normal to the resulting shock. Equations and solutions are derived and discussed in Appendix C.8.

Only resulting equations and figures are presented in this chapter. If the reader is not familiar with the fundamentals of normal and oblique shocks, he/she is encouraged to read Appendix C in detail.

Normal Shocks

When an aircraft flies supersonically, shocks will usually occur exterior to, or near, the inlet plane of the

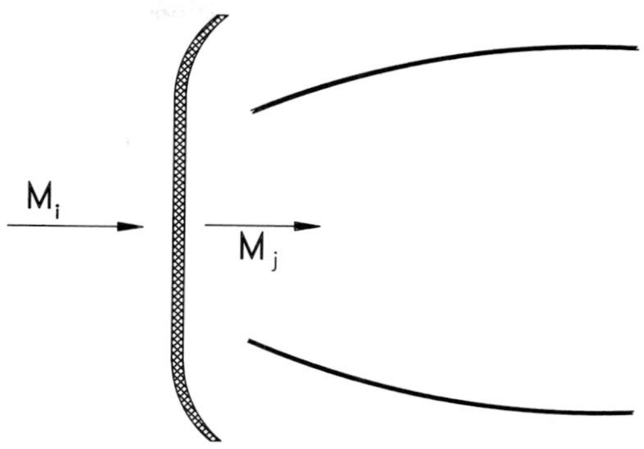

Figure 8.4 Standing normal shock at diffuser inlet.

diffuser. The strongest possible shock is the normal shock. This type of shock is covered extensively in fluid mechanics and gas dynamics texts; because of this, the development of the analysis is reviewed in Appendix C.7. In Figure 8.4, a standing normal shock is shown. Fundamentals of normal shocks area also reviewed in Section 2.2.7, so the details will not be reviewed here. The inlet conditions to the shock are subscripted with i, whereas the exit conditions are noted with j.

For an inlet Mach number M_i, the exit Mach number is determined by

$$M_j^2 = \frac{M_i^2 + \dfrac{2}{\gamma - 1}}{\dfrac{2\gamma}{\gamma - 1} M_i^2 - 1}, \tag{8.3.1}$$

and the total properties are related by

$$T_{tj} = T_{ti} \tag{8.3.2}$$

$$\frac{p_{tj}}{p_{ti}} = \frac{\left[\dfrac{\dfrac{\gamma+1}{2} M_i^2}{1 + \dfrac{\gamma-1}{2} M_i^2}\right]^{\frac{\gamma}{\gamma-1}}}{\left[\dfrac{2\gamma}{\gamma+1} M_i^2 - \dfrac{\gamma-1}{\gamma+1}\right]^{\frac{1}{\gamma-1}}}. \tag{8.3.3}$$

As an aid in problem solving, these and other quantities are presented in tables in Appendix G for two values of γ. Note that the exit Mach number is always less than unity. NormalShock.exe can also be used. The software Shock.exe can also be used for individual calculations. Readers are encouraged to use this software because it is

Example 8.4 Aircraft with a Standing Normal Shock

A diffuser on a Mach 2 aircraft operates with a standing normal shock outside of the inlet at standard conditions (STP). If the internal diffuser recovery factor is $\pi_{\text{d-int}} = 0.90$, what is the diffuser exit total pressure and the total pressure recovery from the freestream to the diffuser exit?

Solution

From the tables for $\gamma = 1.40$ and $M_i = 2.00$, or from the equations or NormalShock.exe one finds that the exit Mach number from the normal shock is

$$M_j^2 = \frac{M_i^2 + \dfrac{2}{\gamma - 1}}{\dfrac{2\gamma}{\gamma - 1}M_i^2 - 1} = 0.3333$$

$M_j = 0.5774.$

It can be found from the tables or equations that the total pressure ratio is

$$\frac{p_{tj}}{p_{ti}} = \frac{\left[\dfrac{\dfrac{\gamma+1}{2}M_i^2}{1 + \dfrac{\gamma-1}{2}M_i^2}\right]^{\frac{\gamma}{\gamma-1}}}{\left[\dfrac{2\gamma}{\gamma+1}M_i^2 - \dfrac{\gamma-1}{\gamma+1}\right]^{\frac{1}{\gamma-1}}} = 0.7209,$$

which is also $\eta_r = \dfrac{p_{t1}}{p_{ta}}$, and the static pressure ratio is

$$\frac{p_j}{p_i} = \frac{2\gamma}{\gamma+1}M_i^2 - \frac{\gamma-1}{\gamma+1} = 4.500.$$

Next, from Eq. (2.3.5) one finds for $p_a = 101.3$ kPa (14.69 psia) that

$$p_{ta} = p_a\left[1 + \frac{\gamma-1}{2}M_a^2\right]^{\frac{\gamma}{\gamma-1}} = 11.3\text{kPa}\left[1 + \frac{1.40-1}{2}2^2\right]^{\frac{1.40}{1.40-1}}$$

$$= 792.6\,\text{kPa}\ (115.0\,\text{psia}).$$

Thus, the total pressure at the entrance to the diffuser is
$p_{t1} = 0.7209 \times 792.6 = 571.3\,\text{kPa}\ (82.86\,\text{psia})$.
Therefore, a loss of 221.3 kPa (32.09 psi) in total pressure is realized as a result of the normal shock.

The exit total pressure from the diffuser is

$p_{t2} = \pi_{\text{d-int}}p_{t1} = 0.90 \times 571.3 = 514.2\,\text{kPa}\ (74.57\,\text{psia}).$ <ANS

Therefore, the overall recovery factor is

$$\pi_d = \frac{p_{t2}}{p_{ta}} = \frac{514.2}{792.6} = 0.649. \qquad \text{<ANS}$$

Thus, the loss is very high and methods should be found to reduce it.

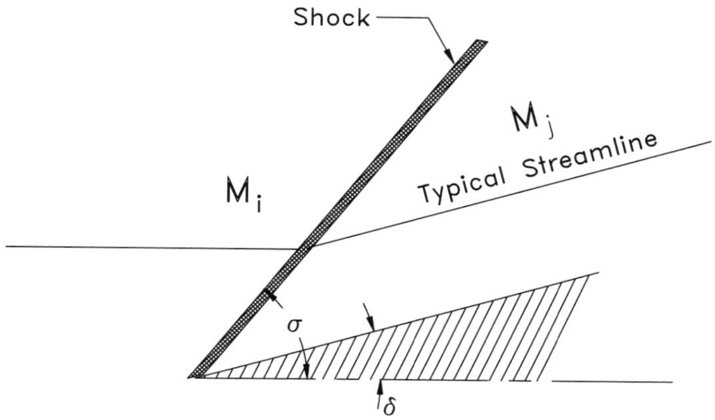

Figure 8.5 Oblique shock.

easier and more accurate than interpolating tables. Also, several equations are listed in Appendix C to relate exit and inlet conditions to the Mach numbers.

Oblique Planar Shocks

To reduce the losses encountered with normal shocks, supersonic diffusers are designed so that oblique shocks occur. Figure 8.5 shows an oblique shock. The losses that are encountered with oblique shocks are significantly less than with normal shocks, and the losses in total pressure are much lower. In actuality, many oblique shocks in aircraft applications are conical. However, planar 2-D shocks are first considered for two reasons. First, the required mathematics is much simpler for planar than for conical shapes, and thus the fundamentals can be better understood. Flow behind the planar shock is uniformly parallel to the wedge. Second, some applications exist in which oblique planar shocks are present; for example, when an inlet is attached to the fuselage of the aircraft, the inlet is more or less rectangular, resulting in a planar shock. Results for conical shocks are covered in the next section. Further details can be found in several gas dynamics texts, including Shapiro (1953), and the mathematics are reviewed in Appendix C. Fundamentals of planar oblique shocks area also reviewed in Section 2.2.8, so the details will not be reviewed here. Again, the inlet conditions to the shock are subscripted with i, whereas the exit conditions are noted with j.

The six resulting important equations are repeated here:

$$\frac{\rho_j}{\rho_i} = \frac{\left[\dfrac{\gamma+1}{\gamma-1}\right]\dfrac{p_j}{p_i} + 1}{\left[\dfrac{\gamma+1}{\gamma-1}\right] + \dfrac{p_j}{p_i}} \tag{8.3.4}$$

$$\frac{p_j}{p_i} = \frac{\left[\dfrac{\gamma+1}{\gamma-1}\right]\dfrac{\rho_j}{\rho_i} - 1}{\left[\dfrac{\gamma+1}{\gamma-1}\right] - \dfrac{\rho_j}{\rho_i}}. \tag{8.3.5}$$

It is noteworthy that the two equations above are named the Rankine–Hugoniot equations and relate pressure and density ratios across shocks.

$$\frac{\rho_j}{\rho_i} = \frac{\tan\sigma}{\tan(\sigma-\delta)} \tag{8.3.6}$$

$$\frac{p_j}{p_i} = 1+\gamma M_i^2 \sin^2\sigma\left[1-\frac{\rho_i}{\rho_j}\right] \tag{8.3.7}$$

$$\frac{p_i}{p_j} = 1+\gamma M_j^2 \sin^2(\sigma-\delta)\left[1-\frac{\rho_j}{\rho_i}\right] \tag{8.3.8}$$

$$\frac{p_{tj}}{p_{ti}} = \frac{p_j}{p_i}\left[\frac{1+\dfrac{\gamma-1}{2}M_j^2}{1+\dfrac{\gamma-1}{2}M_i^2}\right]^{\frac{\gamma}{\gamma-1}}. \tag{8.3.9}$$

Equations (8.3.5)–(8.3.9) contain seven independent variables: σ, δ, M_i, M_j, p_j/p_i, p_{tj}/p_{ti}, and ρ_j/ρ_i. Therefore, if two are specified, the remaining five can be determined. For example, for a given incoming Mach number (M_i) and a given turning angle (δ), one can find the shock angle, exit Mach number, total pressure ratio, and all of the other variables. These equations can easily be programmed in a commercial math solver. Figure 8.6 presents graphs for $\gamma = 1.40$ to aid in the theoretical solution. This figure is, in a sense, a large number of general solutions to the five equations of interest. It is important to note that, for every set of conditions, two solutions exist. The first is the "strong" shock in which the exit Mach number is subsonic and that results in a large loss in stagnation pressure. The second is the "weak" shock in which the exit Mach number is supersonic or high subsonic for a few cases. The weak shock occurs when a low pressure is incurred downstream of the shock and results in a smaller total pressure loss. For most jet-engine applications, weak oblique shocks occur. One limit is for $\delta = 0°$ for which the strong solutions reduce to the normal shock solutions. A second limit is for $\delta = 0°$ for which the weak solutions reduce to the Mach line solutions. Another important limit is reached when the turning is too large for a given Mach number (or the Mach number is too small for a given turning angle). When this limit is exceeded, an oblique shock solution is not possible. Physically, the oblique shock is no longer attached and "pops" off of the wedge, resulting in a standing normal shock in front of the inlet. These limiting lines are

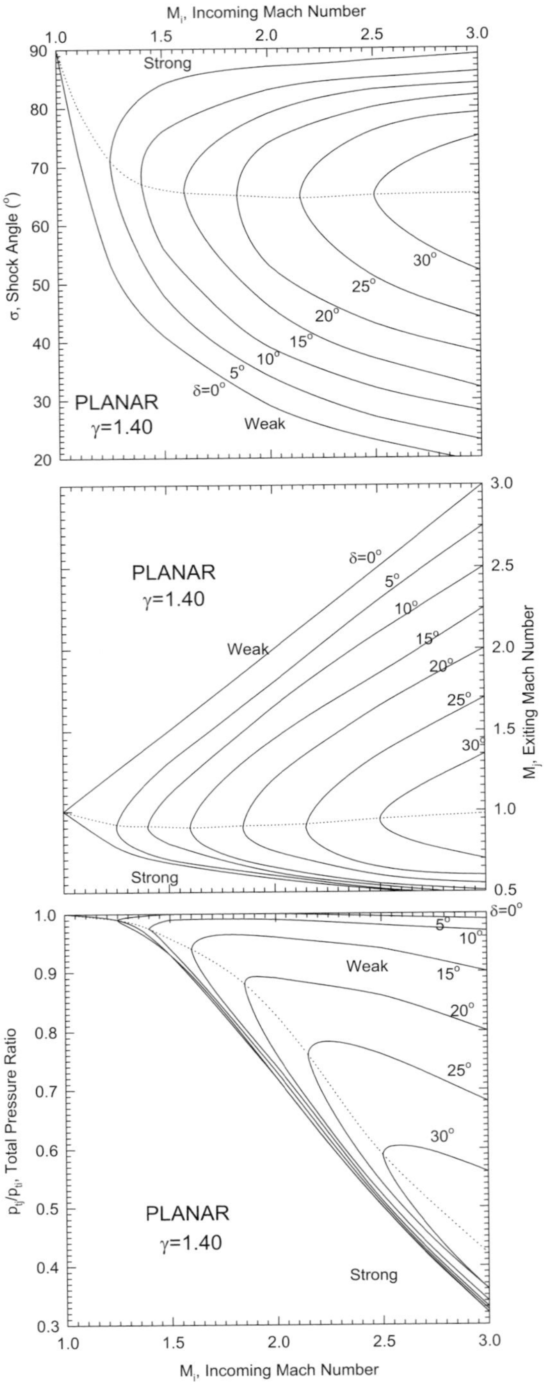

Figure 8.6 Oblique planar shocks – shock angle, exit Mach number, and total pressure ratio as functions of incoming Mach number. The shocks can be duplicated with Shock.exe.

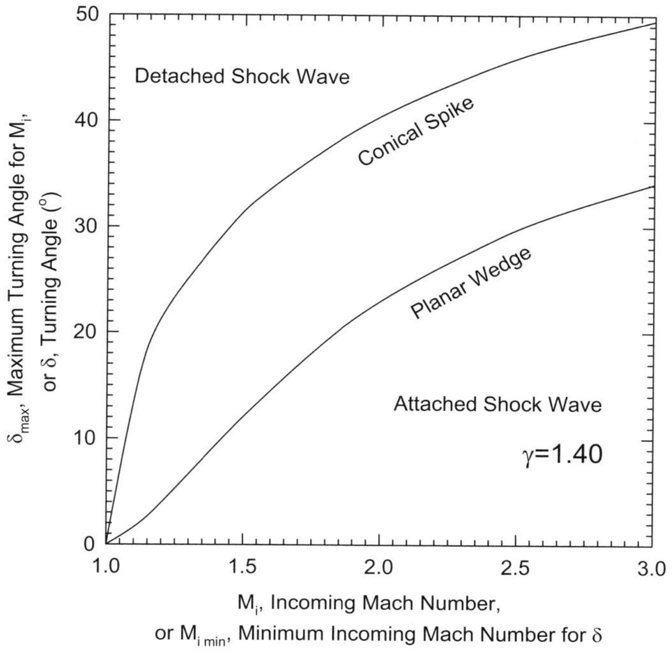

Figure 8.7 Oblique planar and conical shocks: conditions for shock detachment.

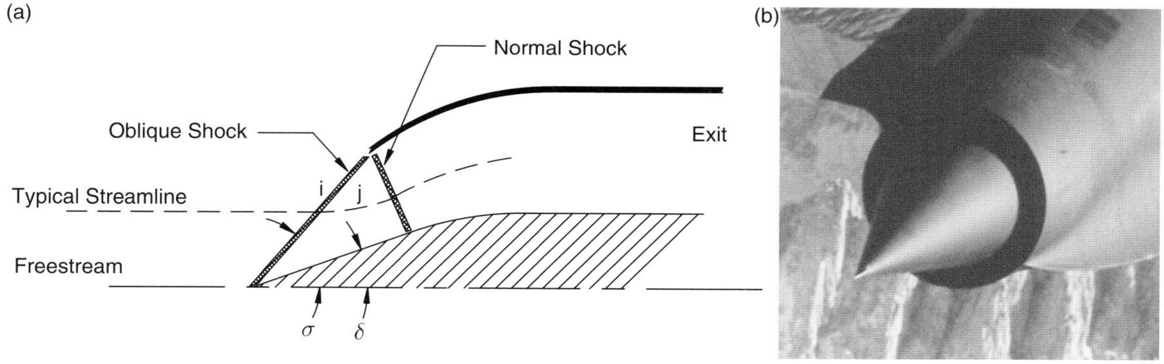

Figure 8.8 (a) Diffuser inlet with one conical oblique shock and one normal shock. (b) Spike on Blackbird (courtesy of NASA).

shown in Figure 8.6 and are also summarized in Figure 8.7. Solutions can be obtained using the software Shock.exe. Readers are encouraged to use this software because it is easier than using the equations in Appendix C.8.

Oblique Conical Shocks

In Figure 8.8a, a typical simple oblique shock system in the design condition is shown for a supersonic inlet. First, a conical (axisymmetric) ramp is used to generate an oblique shock, which

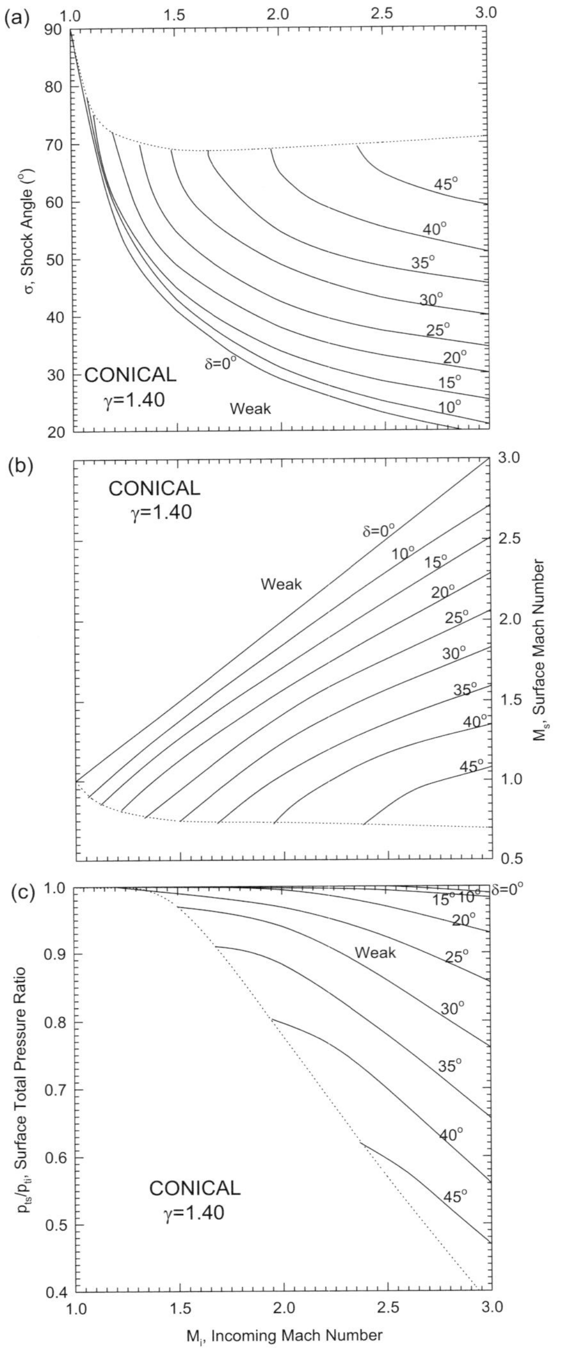

Figure 8.9 Oblique conical shocks – shock angle, exit Mach number, and total pressure ratio as functions of incoming Mach number. The shocks can be duplicated with Shock.exe.

decelerates the flow to a less supersonic condition. An example is the spike on the Blackbird shown in Figure 8.8b. Next, a normal shock further decelerates the flow to a subsonic condition for the internal flow in the diffuser. For the case of axisymmetric oblique shocks, the flow is not uniformly parallel to the spike, and streamlines curve toward the cone. One can develop the equations for oblique conical shocks in much the same way as those for oblique planar shocks. However, this is a more complicated procedure and requires numerical integration of some equations owing to the coordinate system. It suffices to say that the number of variables and equations remains the same. Details can be found in several gas dynamics texts, including Zucrow and Hoffman (1976). Again, the equations can be programmed, and theoretical solutions to these equations are presented for $\gamma = 1.40$ in the form of graphs in Figure 8.9. Note, however, that, unlike oblique planar shocks, oblique conical shocks only have one solution (weak). Also, the flowfield is 2-D and the pressure and Mach number are not uniform behind the shock as in the case of a planar oblique shock, but they are functions of radius because the differential flow area changes with radius. Solutions for only the spike surface are presented here. As for planar shocks, when the turning is too large for a given Mach number (or the Mach number is too small for a given turning angle), an oblique shock solution is not possible and a standing normal shock in front of the inlet results. These limiting lines are shown in Figure 8.9 and are also summarized in Figure 8.7. As can be seen, more turning is allowed for the conical shock than for the oblique shock before detachment. Solutions for normal, planar oblique, and conical oblique shocks can also be obtained using the software Shock.exe, which can be used more effectively and accurately than reading the figures.

To reduce the losses even further, one can use a series of oblique shocks, as indicated in Figure 8.10.

Example 8.5 | Aircraft with One Oblique Shock and One Normal Shock

A diffuser on a Mach 2 aircraft operates with one oblique conical and one normal shock outside of the inlet at STP. If the turning angle is 20° and the internal diffuser recovery factor ($\pi_{d\text{-}int}$) is 0.90, what is the diffuser exit total pressure and the total pressure recovery from the freestream to the diffuser exit on the spike?

Note that this is the same problem as found in Example 8.4 with the addition of the oblique shock.

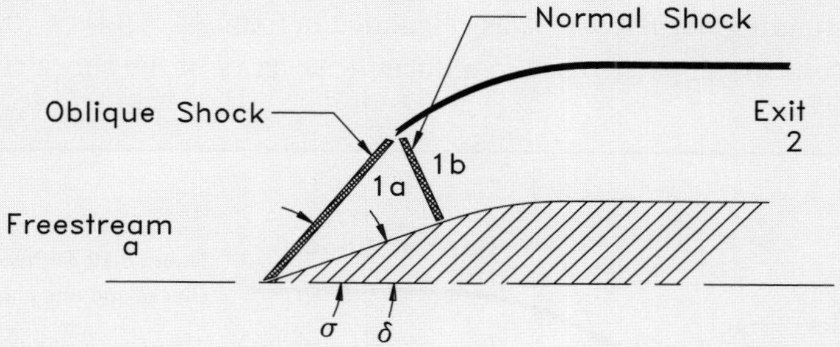

Solution

Referring to the figure shown in this example, one finds for the notation that for a freestream Mach number of $M_a = 2.00$ and a turning angle of $\delta = 20°$ from Figure 8.9b, the Mach number behind the oblique shock and on the spike surface is

$M_{1a} = 1.568,$

and from Figure 8.9c or Shock.exe the total pressure ratio for the oblique shock on the surface is

$p_{t1a}/p_{ta} = 0.9901.$

Also, from Figure 8.9a or Shock.exe the shock angle is

$\sigma = 37.80°.$

Although the latter quantity is not used in the solution, the diffuser should be designed for this shock angle so that the diffuser cowl intersects the shock as shown. If it does not, "spillage" and additional drag can occur, which will be covered in Section 8.3.2. Next, to analyze the normal shock for an incoming Mach number of $M_{1a} = 1.568$, one finds from Appendix G or NormalShock.exe that the Mach number behind the normal shock is

$M_{1b} = 0.6784,$

and the total pressure ratio for the normal shock is $p_{t1b}/p_{t1a} = 0.9070$; thus, combining results yields

$p_{t1b}/p_{ta} = p_{t1a}/p_{ta} \times p_{t1b}/p_{t1a} = 0.9901 \times 0.9070 = 0.8980,$

which is the same as η_r.

Therefore, from Example 8.4, the freestream total pressure is $p_{ta} = 792.6$ kPa (115.0 psia), and thus the total pressure at the inlet plane of the diffuser is

$p_{t1b} = 0.8980 \times 792.6 = 711.7\,\text{kPa}\ (103.2\,\text{psia}).$

Because $p_{t2} = \pi_{d\text{-int}}\, p_{t1b}$, the total pressure at the exit of the diffuser is

$p_{t2} = 0.90 \times 711.7 = 640.5\,\text{kPa}\ (92.89\,\text{psia}).$ <ANS

Finally, the total recovery of the diffuser is

$$\pi_d = \frac{p_{t2}}{p_{ta}} = \frac{640.5}{792.6} = 0.808.$$ <ANS

Although still low, which indicates a significant total pressure loss, this value is much higher than for a single standing normal shock, as calculated in Example 8.4 ($\pi_d = 0.649$). It is easy to understand why conical ramps are used to reduce the severity of the shock structure.

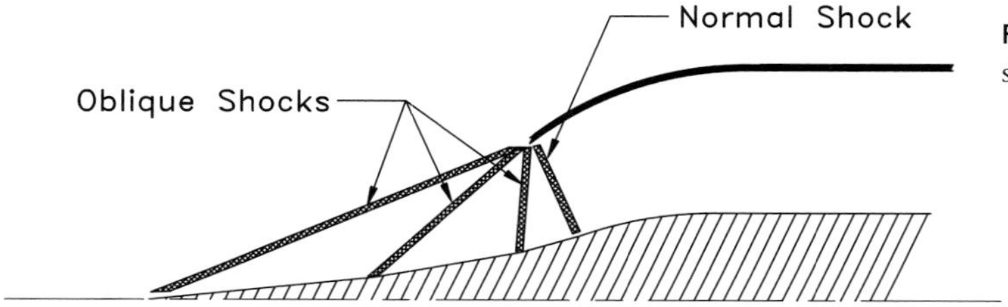

Figure 8.10 Diffuser inlet with three oblique shocks and one planar normal shock.

This has the effect of decreasing the incoming Mach number to the normal shock and thus reducing the losses. Unfortunately, this also has the effect of adding to the complexity of the geometric design, increasing the total turning angle and length (and possibly weight). Optimally designing such an inlet so that the total pressure ratio of the unit is minimized requires that each of the losses across the individual shocks be the same. For example, if two oblique and one normal shock are used, they all should have the same total pressure ratio. Figure 8.11 summarizes the typical effects. In this figure, the value of η_r and total turning angle are presented as a function of the freestream Mach number for four different designs ranging from a single normal shock to three oblique shocks with a normal shock. One can also take advantage of oblique shocks that result in high but subsonic exit Mach numbers. As can be seen, as more oblique shocks are used, the loss diminishes – particularly for large Mach numbers. It is important to realize that the optimized spike shape will be different for different Mach numbers.

8.3.3 Internal Area Considerations

Mass Flow or Area Ratio

A fixed-area diffuser will only operate at peak efficiency at one condition. However, it will usually be run at many different conditions. As an aid to the understanding of off-design performance and a help in determining the required area variation to extend the design range, the so-called mass flow or area ratio is used. This parameter will assist in defining the different off-design regimes in which a diffuser may operate.

At the design condition, a supersonic diffuser will operate as shown in Figure 8.8a. As depicted, the oblique

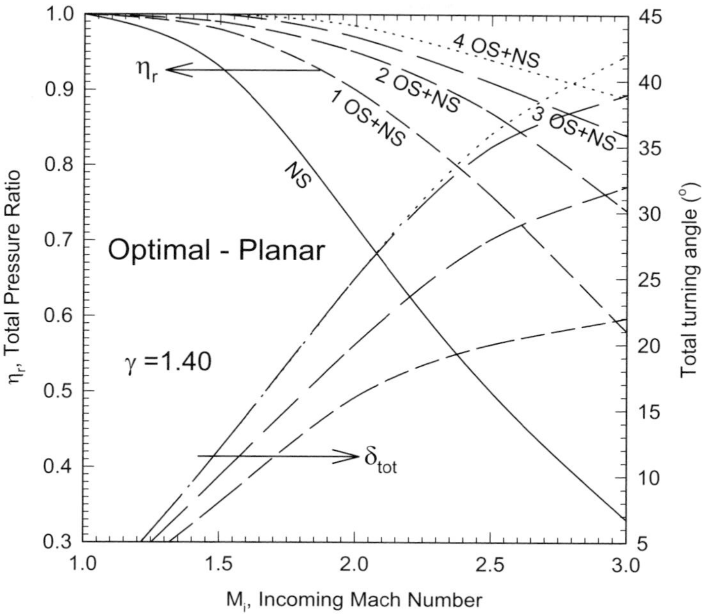

Figure 8.11 External total pressure ratio as a function of Mach number for different optimal diffuser planar shock designs.

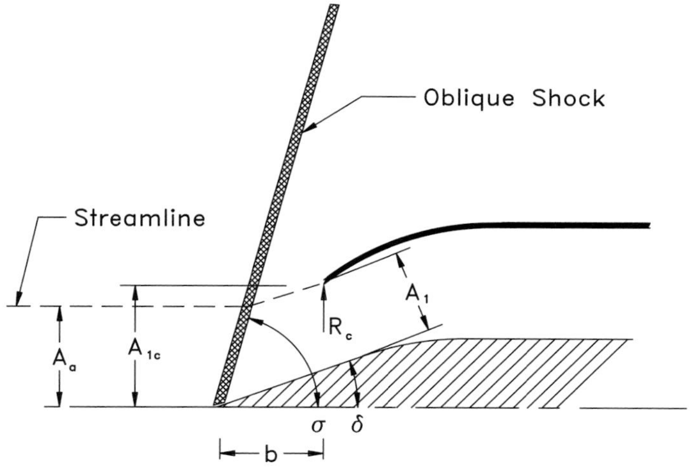

Figure 8.12 Definition of areas for mass flow ratio.

shock intersects the diffuser cowl and does not extend any farther. For this condition, all of the air that crosses the oblique shock enters the engine. One off-design condition is shown in Figure 8.12. As can be seen, for this condition the oblique shock extends outside the diffuser. Several definitions are made in this figure. For example, the area A_{1c} is the projected area based on the outside diameter of the cowl, area A_a is defined by the streamline that intersects the cowl, and area A_1 is the cross-sectional area normal to the flow at the inlet plane of the diffuser.

A reference parameter with the dimensions of mass flow rate is defined as

$$\dot{m}_i = \rho_a u_a A_{1c}, \tag{8.3.10}$$

and the true ingested mass flow rate is of course

$$\dot{m}_1 = \rho_1 u_1 A_1; \tag{8.3.11}$$

thus, the mass flow ratio is defined as

$$MFR \equiv \frac{\dot{m}_1}{\dot{m}_i}, \tag{8.3.12}$$

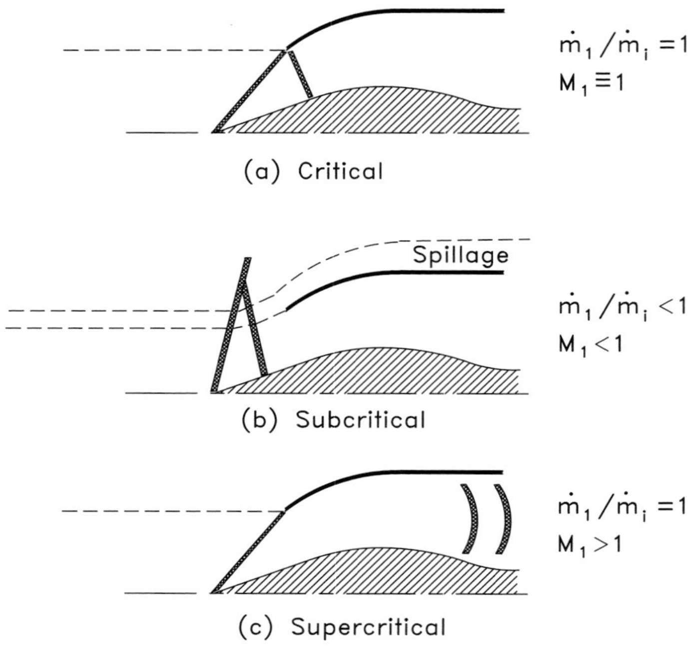

Figure 8.13 Supersonic operation conditions.

and so

$$MFR = \frac{\rho_1 u_1 A_1}{\rho_a u_a A_{ic}}. \tag{8.3.13}$$

The mass flow that enters the engine is, however, also given by (from Figure 8.12):

$$\dot{m}_a = \rho_a u_a A_a, \tag{8.3.14}$$

which is also $\dot{m}_1$. So from Eqs. (8.3.12) and (8.3.14) one finds

$$\frac{\dot{m}_1}{\dot{m}_i} = MFR = \frac{A_a}{A_{ic}}, \tag{8.3.15}$$

thus giving rise to the other name, area ratio. Note that the quantity A_a will not always be easy to define for a diffuser because it is a function of the operating and flight conditions. The mass flow ratio represents the inlet flow rate normalized to the maximum inlet flow rate. The value of *MFR* is thus between zero and unity.

Modes of Operation

As indicated in the previous section, a fixed-area diffuser operates most effectively for any given condition at only one flow rate. Three modes of operation exist: the flow rate is at the design rate, the flow rate is too small, and the flow rate is too great. The first to be discussed is the critical condition, which is by definition the same as the design condition. This is shown in Figure 8.8 and duplicated in Figure 8.13a. For this case the mass flow rates $\dot{m}_1$ and $\dot{m}_i$ are the same and the terminal shock occurs just inside the cowl lip; moreover, no "spillage" occurs. For this case internal shocks may not be present (not shown) or a single normal shock may sit in the inlet plane as shown. For the latter case the Mach number at the diffuser inlet plane is defined as unity, because the Mach number is ill-defined at exactly the inlet plane (i.e., a normal shock stands in the inlet).

If the flow rate is decreased, the pressure in the diffuser will be increased owing to the decreased Mach number in the diffuser. As a result, the normal shock will be pushed outside of the inlet plane. It will be stronger and will result in a larger total pressure loss. The normal shock will also join with the oblique shock to form a "λ-shock." Note that, ideally (i.e., at design), air encompassed by A_{1c} (far from the inlet) enters the diffuser. However, for this condition some of the air within A_{1c} (far from the diffuser, Figure 8.12) will be "spilled." The axial position of the spike can affect the resulting area ratio. This condition is undesirable for a variety of reasons. One reason is that the engine is compressing air using the shock (and thus working on the fluid) outside of the diffuser that is not being used, thereby wasting power. As discussed in the next section, this spillage of high-pressure air results in additive drag. For this condition, the Mach number in the inlet plane is less than unity. This overall condition is called the subcritical condition, as shown in Figure 8.13b. Hill and Peterson (1992) further subcategorize the condition into high and low subcritical.

As the flow increases beyond the critical condition, the pressure in the diffuser decreases. Thus, the normal shock moves into the diffuser. The diffuser is now at least in part acting like a supersonic nozzle. Shocks occur in the diverging section of the diffuser at high Mach numbers and, as a result, more total pressure is lost. This condition is the supercritical condition and is shown in Figure 8.13c. Now the Mach number in the inlet plane is greater than unity. However, because the external flow is supersonic, the mass flows $\dot{m}_1$ and $\dot{m}_i$ are still the same. No spillage occurs for this condition.

Off-Design Operation of the Diffuser

When a diffuser is operated at off-design conditions (i.e., at conditions different than those for which it was designed), the areas should be varied so that it operates efficiently. This is often accomplished by the axial location of the wedge or spike. The objective of this section is to develop a method by which the inlet area can be determined for given conditions and a given spike (or wedge).

First, one can examine Figure 8.12 and see immediately that, for the case of a single planar oblique shock,

$$\frac{A_a}{A_1} = \frac{\sin\sigma}{\sin(\sigma-\delta)}. \tag{8.3.16a}$$

Note that this ratio of areas is greater than unity. For a conical shock, this ratio can be found by CFD or by the numerical solution to the "Taylor–Maccoll" flow around a cone as described by Maccoll (1937) and Zucrow and Hoffman (1976):

$$\frac{A_a}{A_1} = f\left(\sigma, \delta, \frac{b}{R_c}\right). \tag{8.3.16b}$$

Note that the ratio of b/R_c (spike position to cowl radius) affects this ratio for a conical shock because of the streamline curvature but does not for the planar shock.

The engine mass flow rate is given by

$$\dot{m}_1 = \rho_1 u_1 A_1 \tag{8.3.17}$$

or by

$$\dot{m}_1 = \rho_a u_a A_a. \tag{8.3.18}$$

As indicated earlier, the quantity A_a is difficult to determine for conditions other than design conditions. Similarly, ρ_1 and u_1 are difficult to determine because they vary with flight and engine conditions. However, from Eq. (8.3.18), one finds

$$\dot{m}_1 = \rho_a u_a A_1 \left[\frac{A_a}{A_1}\right]. \tag{8.3.19}$$

Similarly,

$$\dot{m}_1 = \rho_a u_a A_{1c} \left[\frac{A_a}{A_{1c}}\right] = MFR\ \dot{m}_1. \tag{8.3.20}$$

One can now see more usefulness to the reference mass flow rate $\dot{m}_i$, for it is an easily definable quantity. The problem of finding the area variation has not been solved yet, however. First, from solution of the oblique shock equations one can find the variation of shock angle σ with the Mach number for a given turning angle. Next a curve of the required variation of the area ratio A_a/A_1 with Mach number can be generated from the shock angle variation and Eq.(8.3.16a). Next, other information on the engine performance must be used. For the operation of the engine as a system, as is covered in Chapter 12 (i.e., the matching of the inlet, compressor, burner, turbine, nozzle, etc.), one finds an engine and aircraft specification that results in the proper thrust for given conditions. These specifications result in the specification of engine mass flow rate as a function of the freestream Mach number. Lastly, one finds from Eq. (8.3.19) that

$$A_1 = \frac{\dot{m}_1}{\rho_a u_a \left[\frac{A_a}{A_1}\right]}, \tag{8.3.21}$$

and so for any given freestream Mach number one can find $\dot{m}_1$ from the specifications, A_a/A_1 from the shock angle and Eq. (8.3.16a), and ρ_a and u_a for a given altitude to generate the relation of required inlet area as a function of Mach number. Thus, it is now possible to define how the inlet area of the diffuser must vary with Mach number – that is, how the axial wedge or spike position varies with Mach number. For the case of a planar shock, the process is straightforward. For the case of the conical shock, however, an iteration loop is needed because Eq. (8.3.16b) is a function of spike position, which is also the desired end result.

Example 8.6 | Aircraft with One Planar Oblique Shock and Variable Inlet Area

An engine is used on a Mach 2.5 aircraft at sea level and STP ($T_a = 288.2$ K or 518.7 °R and $p_a = 101.3$ kPa or 14.69 psia). The craft routinely flies between Mach numbers of 1.5 and 2.5 and at Mach numbers of 1.5, 2.0, and 2.5. At this altitude the required mass flow rate of the engine resulting from a component matching analysis is 36.28, 54.42, and 64.40 kg/s (80, 120, and 142 lb_m/s), respectively. A planar wedge with an angle of 10.0° and a weak shock is used. What is the required inlet area variation? Assume $\gamma = 1.40$.

Solution

First, from Figure 8.6 or Shock.exe one finds, for an incoming Mach number of $M_1 = 2.00$ and a turning angle of $\delta = 10.00°$, that the shock angle is

$\sigma = 39.31°$;

thus,

$$\frac{A_a}{A_1} = \frac{\sin\sigma}{\sin(\sigma - \delta)} = \frac{\sin 39.31°}{\sin(39.31° - 10.00°)} = 1.294.$$

Next,

$a_a = \sqrt{\gamma \mathcal{R} T_a}$, and so at this altitude at STP,

$$a_2 = \sqrt{1.4 \times 287.1 \frac{\text{J}}{\text{kg-K}} \times 288.2 \text{ K} \times \frac{\text{N-m}}{\text{J}} \times \frac{\text{kg-m}}{\text{N-s}^2}}$$

$$= 340.3 \text{ m/s } (1116.4 \text{ ft/s})$$

and

$$\rho_a = \frac{p_a}{\mathcal{R}T_a} = \frac{101.3 \text{ kPa} \times 1000 \frac{\text{N/m}^2}{\text{kPa}}}{287.1 \frac{\text{J}}{\text{kg-K}} \times 288.2 \text{ K} \times \frac{\text{N-m}}{\text{J}}}$$

$$= 1.225 \text{ kg/m}^3 \ (0.00238 \text{ sl/ft}^3);$$

thus,

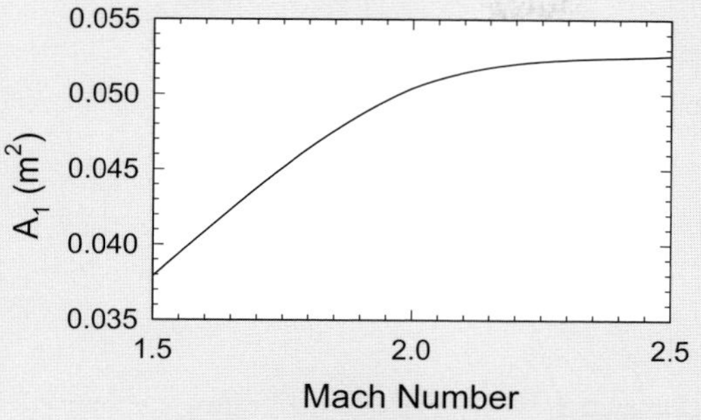

Figure 8.14 Required area variation of inlet for Example 8.6.

$$u_a = M_a a_a = 2.00 \times 340.3 = 680.6\,\text{m/s}\ (2232.8\,\text{ft/s}).$$

Finally, from Eq. (8.3.21), one obtains

$$A_1 = \frac{\dot{m}_1}{\rho_a u_a \left[\dfrac{A_a}{A_1}\right]} = \frac{54.42\dfrac{\text{kg}}{\text{s}}}{1.225\dfrac{\text{kg}}{\text{m}^3} \times 680.6\dfrac{\text{m}}{\text{s}} \times 1.294}$$

$$= 0.0504\,\text{m}^2\ (0.543\,\text{ft}^2 \text{ or } 78.2\,\text{in.}^2).$$

Repeating the calculations for the other two Mach numbers yields the area variation shown in Figure 8.14.

8.3.4 Additive Drag

Figure 8.15 displays a general turbojet. In the earlier analysis in Chapter 1, the flows both inside and outside the engine inlet and exit planes were assumed to be uniform.

However, owing to diffuser spillage and boundary layers on the exterior of the engine, this is not true. These interactions lead to a reduction of thrust called additive drag. At this point, calculation of engine internal processes are not undertaken, although the results are important. A control volume, which is affixed to the engine, is drawn around the engine as shown. As before, the engine moves in air to the left with velocity u_a. The cylindrical control volume cuts through the engine strut. Fuel is also injected into the engine. One surface is at the exit plane of the engine, and the front plane is far from the inlet.

Writing the general momentum equation in rectangular coordinates yields

$$\sum \mathrm{F_s} + \sum \mathrm{F_b} = \frac{\partial}{\partial t}\int_{CV} \mathrm{V}\rho\, d\mathcal{V} + \int_{CV} \mathrm{V}\rho\mathrm{V} \bullet d\mathrm{A}. \tag{8.3.22}$$

However, if the flow is considered to be steady, the first term on the right-hand side ($\partial/\partial t$ or unsteady term) disappears. Next, the thrust (x) direction should be considered. The surface forces are due to pressures on the control volume and to the cutting of the strut F, which is the force holding the engine in place. Thus, the force F represents the thrust. Because the front surface is far from the engine, the pressure is uniform and at atmospheric conditions, p_a, at this surface. At the front surface, the air flows at velocity u_a across the area $A_a + A_1$. Also, at the engine exit, the pressure in the engine

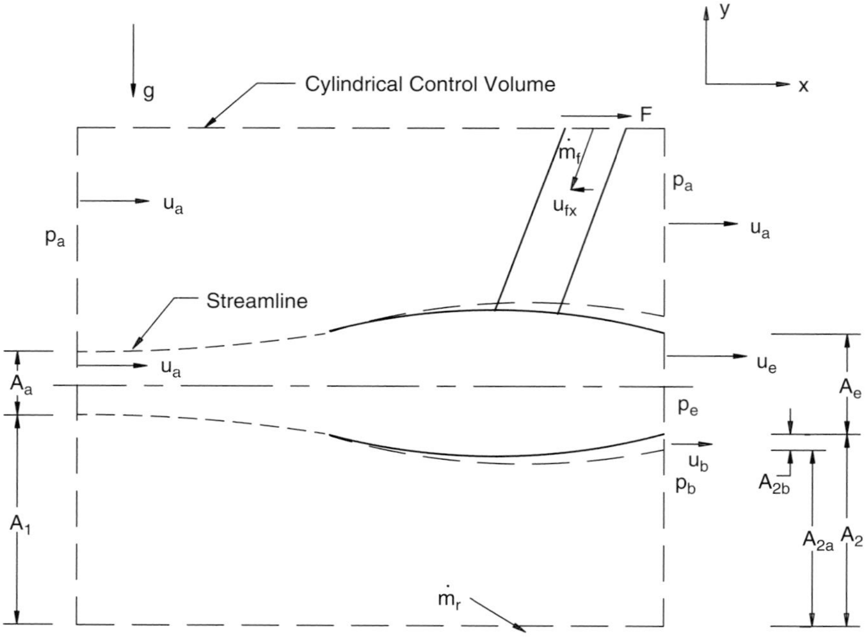

Figure 8.15 Control volume definition for a turbojet with additive drag.

is p_e, the pressure in the boundary layer is p_b, and the pressure far outside the engine is atmospheric pressure p_a. The velocity of the engine exhaust is u_e. The area of the engine exhaust is A_e, and the area of the boundary layer at the exit plane is A_{2b}. Air crosses the exit areas A_{2b} and A_{2a} with velocities u_b and u_a, respectively. The fuel is injected at a mass flow rate of $\dot{m}_f$ with velocity $-u_{fx}$ in the x direction, but the fuel flow does not significantly contribute to the flux portion of the momentum equation. A mass flow term $\dot{m}_r$ exits from the sides of the control volume. Thus, evaluating each of the terms and rewriting the momentum equation yields

$$\begin{aligned} F + p_a A_a + p_a A_1 - p_e A_e - p_a A_{2a} - p_b A_{2b} + F_{bx} \\ = -u_a \rho_a A_a u_a + u_e \rho_e A_e u_e - u_a \rho_a A_1 u_a + u_a \rho_a A_{2a} u_a \\ + u_b \rho_b A_{2b} u_b + \dot{m}_r u_a - \dot{m}_f u_{fx}. \end{aligned} \tag{8.3.23}$$

Next, one should realize that no body forces exist in the x direction ($F_{bx} = 0$). Also, uniform flow is assumed across areas A_e, A_{2b}, A_{2a}, A_1, and A_a. The mass fluxes across these inlet and exit surfaces, which are the gas flows into and out of the engine, respectively, are once again

$$\dot{m}_a = \rho_a u_a A_a \tag{8.3.24}$$

and

$$\dot{m}_e = \rho_e u_e A_e. \tag{8.3.25}$$

One can also define

$$\dot{m}_b = \rho_b u_b A_{2b}. \tag{8.3.26}$$

Next, by inspecting the control volume, one can see the areas are related by

$$A_1 + A_a = A_2 + A_e \tag{8.3.27}$$

and

$$A_2 = A_{2a} + A_{2b}. \tag{8.3.28}$$

Because the flow is steady, the continuity equation yields

$$\dot{m}_e = \dot{m}_a + \dot{m}_f \tag{8.3.29}$$

and

$$\dot{m}_e + \rho_a u_a A_{2a} + \rho_b u_b A_{2b} + \dot{m}_r = \dot{m}_a + \rho_a u_a A_1 + \dot{m}_f; \tag{8.3.30}$$

thus, using Eq. (8.3.29) yields

$$\rho_a u_a A_{2a} + \rho_b u_b A_{2b} + \dot{m}_r = \rho_a u_a A_1, \tag{8.3.31}$$

and therefore

$$\dot{m}_r = \rho_a u_a (A_1 - A_{2a}) - \dot{m}_b. \tag{8.3.32}$$

Now, using Eqs. (8.3.23) and (8.3.32), one finds

$$\begin{aligned} F + p_a A_a + p_a A_1 - p_e A_e - p_a A_{2a} - p_b A_{2b} \\ = u_e \rho_e A_e u_e - u_a \rho_a A_a u_a - u_a \rho_a A_1 u_a + u_a \rho_a A_{2a} u_a \\ + u_b \rho_b A_{2b} u_b + \rho_a u^2 (A_1 - A_{2a}) - \dot{m}_b u_a - \dot{m}_f u_{f.x}, \end{aligned} \tag{8.3.33}$$

or, using Eq. (8.3.26) and adding and subtracting like terms, yields

$$\begin{aligned} F = -p_a A_a - p_a A_1 + p_e A_e + p_a A_{2a} + p_b A_{2b} + p_a A_{2b} \\ - p_a A_{2b} + \dot{m}_e u_e - \dot{m}_a u_a + \dot{m}_b (u_b - u_a) - \dot{m}_f u_{f.x}. \end{aligned} \tag{8.3.34}$$

Thus, using Eq. (8.3.28) results in

$$\begin{aligned} F = -p_a A_a - p_a A_1 + p_e A_e + p_a A_2 + A_{2b}(p_b - p_a) \\ + \dot{m}_e u_e - \dot{m}_a u_a + \dot{m}_b (u_b - u_a) - \dot{m}_f u_{f.x}, \end{aligned} \tag{8.3.35}$$

and so using Eq. (8.3.27) yields

$$\begin{aligned} F = (p_e - p_a) A_e + A_{2b}(p_b - p_a) + \dot{m}_e u_e - \dot{m}_a u_a \\ + \dot{m}_b (u_b - u_a) - \dot{m}_f u_{f.x}. \end{aligned} \tag{8.3.36}$$

One can next reexamine Eq. (1.5.10) for the ideal case with no additive drag and recognize that the thrust is

$$F' = (\dot{m}_e u_e - \dot{m}_a u_a) + A_e (p_e - p_a) - \dot{m}_f u_{f.x}. \tag{1.5.10}$$

From Eq. (8.3.36), it is possible to see that, for the current case,

$$F = F' + A_{2b}(p_b - p_a) + \dot{m}_b (u_b - u_a), \tag{8.3.37}$$

or, by defining an additive drag, as

$$D_a = A_{2b}(p_a - p_b) + \dot{m}_b (u_a - u_b), \tag{8.3.38}$$

one finds

$$F = F' - D_a. \tag{8.3.39}$$

Thus, the current analysis yields the same thrust as in the ideal case considered in Chapter 1 with the exception of a thrust reduction due to the additive drag. Unfortunately, evaluating the additive drag is difficult. Obtaining the data required to evaluate the drag is a time-consuming process. Estimating these pressures and velocities for an engine on the drawing board can be accomplished by CFD. Thus, for the sake of practicality, dimensionless wind tunnel studies have been performed, correlated, and published for a variety of engines with different diffusers. The drags are then presented as drag coefficients; that is,

$$C_{da} = \frac{D_a}{\frac{1}{2}\gamma M_a^2 p_a A_{in}}, \tag{8.3.40}$$

where A_{in} is the frontal area of the inlet.

Note that the preceding analysis was performed for a turbojet engine. However, if one were to repeat the analysis for any type of engine, the thrust would be found equal to that for the ideal case with a reduction due to additive drag. For any type of engine, the actual drag will be evaluated based on CFD predictions in conjunction with wind tunnel testing of that particular geometry.

In general the additive drag coefficient for a given inlet and cowl will be the function:

$$C_{da} = f\left(\frac{\dot{m}_1}{\dot{m}_i}, M_a, \delta\right). \tag{8.3.41}$$

For an inlet with a spike or wedge, M_a and δ predetermine the shock structure. In general, the drag coefficient decreases with increasing mass flow ratio because of reduced spillage. At a mass flow ratio of unity, spillage does not occur and the drag coefficient is near zero. Also, as would be expected, the drag is less for a diffuser with an oblique shock than for a diffuser with a normal shock. For a normal shock, the drag coefficient increases with increasing Mach number.

8.3.5 "Starting" an Inlet

A few supersonic engines are designed to operate at least part of the time with an inlet without shocks – that is, a monotonically decelerating inlet using "internal compression" from supersonic flow to sonic to subsonic due to the CD geometry. Such an inlet has the advantage of minimal total pressure losses due to the lack of shocks. However, the idea of "starting" the diffuser becomes a problem. "Starting" can also be a problem for inlets designed to operate with oblique

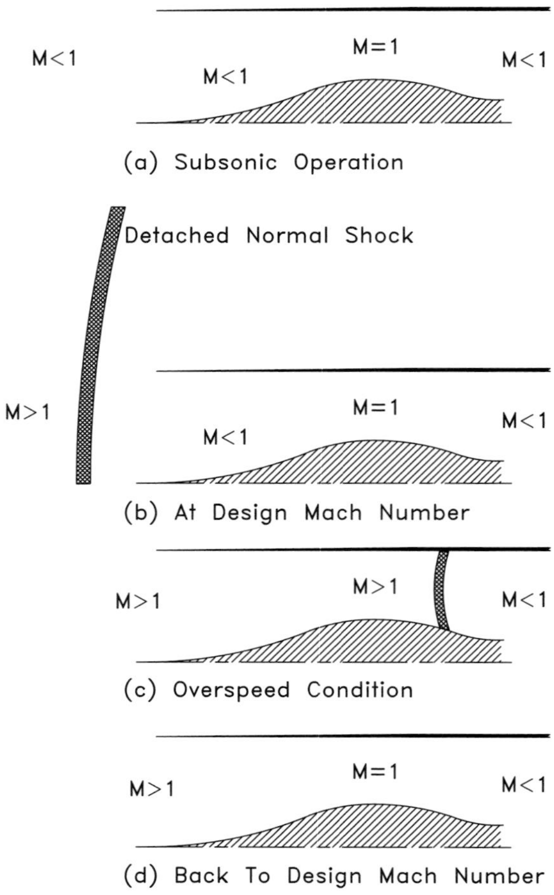

Figure 8.16 Transient shock development for an accelerating supersonic aircraft with a fixed-geometry diffuser.

shocks to ensure proper sizing of the inlet and location of the shocks. For either case, the aircraft must accelerate from takeoff to the nominal flight Mach number with the inlet operating efficiently. Starting is simply defined as having the aircraft reach the desired speed and having the inlet operate at the desired design condition, including proper location of or lack of any shocks. The problem encountered is similar to starting a supersonic wind tunnel. Two methods can be used to start an inlet as described in the following paragraphs.

For this concept, refer to Figure 8.16. For the following scenario the aircraft is considered to have a fixed-area diffuser and is eventually to operate supersonically at Mach number M_d and without any shocks, but initially it will be at rest. The diffuser has fixed inlet, minimum, and exit areas, which are designed for one particular freestream Mach number M_d so that the flow enters supersonically, decelerates to the sonic condition at the throat using internal compression. and then decelerates further in the diverging section. Zucrow and Hoffman (1976) and Hill and Peterson (1992) discuss the procedure in greater detail and with analyses.

First, however, consider the aircraft to be moving very slowly (e.g., at takeoff) – that is, well into the subsonic regime. For this condition, the flow enters the diffuser and accelerates into the minimum area and decelerates in the diverging area, but the flow remains subsonic throughout. Next, as the aircraft speed increases but remains subsonic, the Mach number increases in the minimum area until it eventually chokes at the condition shown in Figure 8.16a. In the diverging region, the flow decelerates subsonically. For this condition, the diffuser is actually acting like a choked nozzle, which is the opposite of the desired effect. Then, as the aircraft speed increases further, a standing normal shock develops outside of the diffuser. Initially, this shock is weak because of a low supersonic Mach number but eventually becomes strong with further increases in aircraft speed until the design speed M_d is attained (Figure 8.16b). The diffuser continues to operate like a choked nozzle with the strong standing shock, and the inlet is not yet operating efficiently. Next, the aircraft must be overspeeded to a Mach number M_s so that the shock moves toward the inlet and is suddenly swallowed (M_s is significantly greater than M_d) into the diverging part of the diffuser (Figure 8.16c). At this condition the flow is supersonic in the diffuser inlet, decelerates to the minimum area but remains supersonic, and accelerates supersonically up to the normal shock, where it decelerates to a subsonic condition. The standing shock is still undesirable because it results in large losses in the diffuser. As a final step, the aircraft Mach number is reduced back to M_d, where the flow is supersonic in the converging section, sonic at the throat (i.e., it is an infinitely weak shock), and subsonic in the diverging section (Figure 8.16d). At this point the diffuser is operating at maximum efficiency. It is important to note that the area ratio was selected for the given nominal flight Mach number M_d and does not apply to other Mach numbers.

A second, more mechanically complex method can be used to start a shockless inlet, but it does not require overspeeding. For example, consider an aircraft at operating speed for which the geometry is held constant (Figure 8.16b). This condition is also shown in Figure 8.17a. Note that for this condition the diffuser is acting like a choked nozzle, and thus the Mach number in the minimum area is unity. Next, if the minimum area is increased, the shock will be swallowed or move into the diverging region, as shown in Figure 8.17b. Now the Mach number in the minimum area is supersonic but less so than in the inlet plane. Finally, if the minimum area is decreased to the proper value, the shock will move back to the throat and become infinitely weak – that is, the Mach number is unity (Figure 8.17c). Thus, now the Mach number in the inlet plane is supersonic and the flow decelerates monotonically through the diffuser. Such a method of "starting" is also used for supersonic wind tunnels. The RR/Snecma Olympus 593 (Concorde) used a variation of this approach. For an aircraft, another method of changing the areas is by an axially moving plug, as shown in Figure 8.18. By proper design of the plug and internal cowl shapes and by controlling the axial placement of the plug, the minimum and inlet areas are varied. In Figure 8.18, two plug placements are shown, depicting how the inlet and minimum areas change with axial plug position. In general, the position and area can be changed during acceleration at different altitudes and Mach numbers, increasing the envelope of efficient operation over a fixed geometry inlet.

Furthermore, a plug has the added advantage of affecting the shock positions and thus the mass flow ratio for inlets designed to operate with shocks. For example, if an aircraft has attained the design speed and if a detached normal shock is present instead of the desired oblique shock, the plug can be moved so that the area changes, thus swallowing the normal shock and setting up the oblique shock to intersect the cowl. In Figure 8.19, a supersonic inlet designed to operate with one oblique shock and internal deceleration is shown. First, the figure

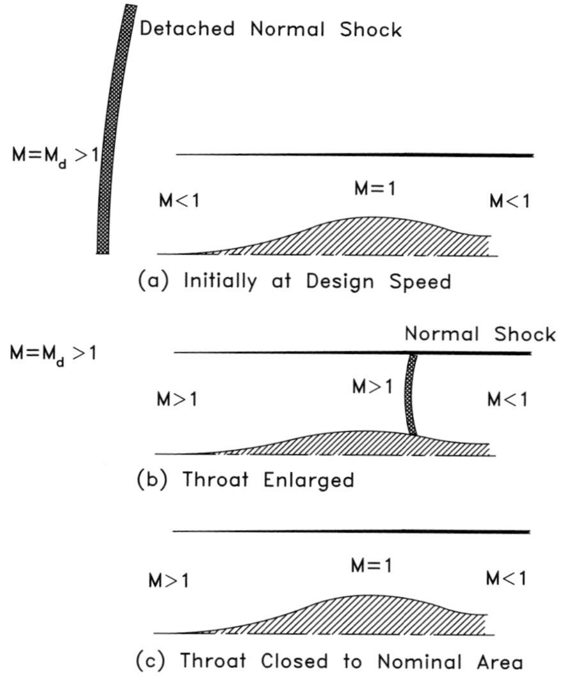

Figure 8.17 Transient shock development for a supersonic aircraft at the design Mach number with a variable geometry diffuser.

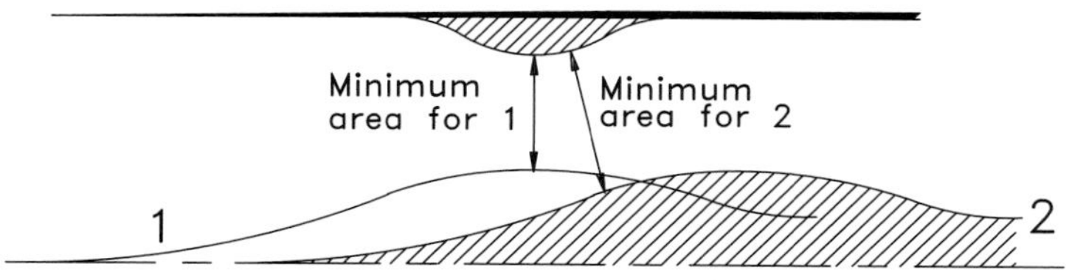

Figure 8.18 Variable diffuser geometry using a plug.

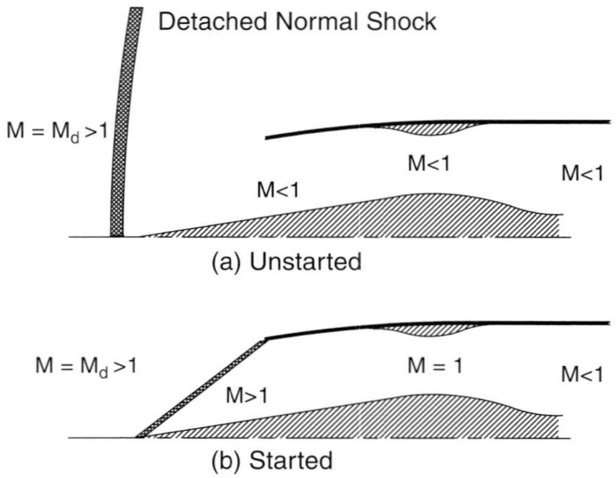

Figure 8.19 Inlet with an oblique shock.

presents the unstarted case, which includes a detached normal shock. Second, the started inlet is shown.

8.4 "Bell" Inlet for Power Generation Gas Turbines

8.4.1 Recovery Factor

The purpose of the bell inlet is to draw air into the power generation gas turbine. As previously presented, Figure 2.11 shows the inlet of the Pratt & Whiney FT8 and the inlet for the Coberra 6000 is presented in Figure 2.47. As compared with a jet engine diffuser, the flow area decreases and thus the flow velocity increases. There is a boundary layer on the internal walls and turbulence in the atmosphere, as shown on Figure 2.12. Owing to the wall shear, boundary layer turbulence, and freestream turbulence, the flow is nonisentropic because of viscous effects; thus, the total pressure is reduced. Also, the flow may not be uniform. As a result, compressor performance is compromised. One major difference in a bell inlet compared with a propulsion diffuser is that the pressure decreases through the process. As a result, separation is not a problem as for a diffuser unless strong crosswinds prevail for the inlet. Also, clearly, shocks never exist for a bell inlet.

The overall operation has been presented in Chapter 2. The flow accelerates from a to 1, as the static pressure decreases from a to 1 and also the enthalpy or temperature also decreases. Mach numbers at station 1 are typically low, on the order of 0.1. The flow velocity continues to increase from 1 to 2 due to the decreasing area. Thus, the pressure and temperature continue to decrease. The freestream air is at pressure p_a and can be decelerated isentropically to total enthalpy, h_{ta}, and total pressure, $p_{ta.}$ However, for the inlet station a, the static pressure, p_a, and total pressure, p_{ta}, are the same since the freestream velocity is zero. Ideally, this results in a total exit pressure $p'_{t2} = p_{ta}$. However, owing to losses in the inlet, the entropy increases from station a to station 2. Because the process is very nearly adiabatic (this assumption will not be relaxed), the total enthalpy (and total temperature) does not change from a to 2; thus, $h_{t2} = h_{ta}$ and $T_{t2} = T_{ta}$. As a result of losses, however, the total pressure at station 2 is p_{t2}, which is less than p'_{t2}.

However, as discussed earlier, due to losses, the total pressure decreases both outside and inside the inlet. The concept of total pressure recovery is introduced to quantify this loss. This is defined similarly to that for the engine diffuser as

$$\pi_i = \frac{p_{t2}}{p_{ta}}. \tag{8.4.1}$$

Note that this recovery factor represents the loss from the freestream to the bell exit and can be expanded to

$$\pi_i = \left(\frac{p_{t2}}{p_{t1}}\right) \times \left(\frac{p_{t1}}{p_{ta}}\right). \tag{8.4.2}$$

The first term, or loss, is often defined to identify internal bell inlet losses. This is the inlet pressure recovery factor and is defined similar to that of the diffuser as

$$\pi_{i\text{-int}} = \frac{p_{t2}}{p_{t1}}. \tag{8.4.3}$$

This expression represents losses only within the bell. Losses can also occur before the flow enters the inlet, mostly due to turbulence in the freestream. This loss is the second term in Eq. (8.4.2) and is defined by

$$\pi_{\text{i-ext}} = \frac{p_{t1}}{p_{ta}}, \tag{8.4.4}$$

which represents the loss from the far freestream to the inlet plane. This quantity is usually unity or very close to unity. Thus, from Eqs. (8.4.2), (8.4.3), and (8.4.4) one finds that

$$\pi_i = \pi_{\text{i-int}} \times \pi_{\text{i-ext}}. \tag{8.4.5}$$

As a result, by specifying either π_i or both $\pi_{\text{i-int}}$ and $\pi_{\text{i-ext}}$, the losses associated with the bell inlet can be included in the analysis.

8.4.2 Isentropic Flow with Area Change

The flow cross-sectional area decreases axially and ideally the flow is without friction or heat transfer as discussed in Chapter 2 and Appendix C. The nozzle, diffuser, inlet, and exhaust are the most obvious components in which the area changes. As the area changes, so do the Mach number, velocity, and other properties. Uniform flow enters a component with characteristics and properties at 1 (e.g., p_1, T_1, V_1, M_1, etc.) and leaves uniformly with all properties at 2 (e.g., p_2, T_2, V_2, M_2, etc.). For the isentropic process, the continuity and energy equations are used.

The flow is analyzed with a control volume approach in Appendix C.4. For an ideal gas the energy equation, continuity equation, and isentropic relationships of the properties are used. In Appendix C.4, 11 equations are summarized for the flow conditions. Sixteen variables are present, so five must be specified. For example, if M_1, T_1, p_1, A_1, and A_2 are specified, one can find M_2. For subsonic internal flow the Mach number increases as the area decreases (the usual case for an inlet).

To aid in the solution of these equations, abbreviated tables are presented in Appendix D for two values of γ. In these tables T/T_t, p/p_t, ρ/ρ_t, and A/A^* are presented vs. the Mach number. The superscript * indicates reference conditions at the sonic condition. Also, the software Isentropic.exe can be used to generate larger tables or specific individual results.

Example 8.7 | Isentropic Inlet Flow

Flow enters an inlet at Mach 0.0833. The inlet area is 20.75 ft^2 and the exit area is 5.187 ft^2. What is the exit Mach number for isentropic flow if $\gamma = 1.40$?

Solution

For $M_1 = 0.0833$ the ratio $\left.\dfrac{A}{A^*}\right]_1$ is 6.9762 from tables (or Isentropic.exe).

Therefore, one can find the area at the sonic condition, which is only a reference condition (which might not occur in the actual conditions):

$A_1^* = 20.75/6.9762 = 2.974$ ft^2.

But since the flow is isentropic, $A_1^* = A_2^*$.

Therefore, $\left.\dfrac{A}{A^*}\right]_2 = \dfrac{5.187}{2.974} = 1.744.$

So now, interpolating tables (or using Isentropic.exe) for this ratio yields $M_2 = 0.3580$.

8.4.3 Fanno Line Flow

Boundary layers and other viscous flows are the primary means by which total pressure losses occur within diffusers. This loss can be represented by Fanno line flow, which is flow with friction but no heat addition or loss. Fanno line flow is a constant area process that does not exactly represent an inlet. However, for an inlet the area decreases significantly as seen on Figures 2.11 and 2.47, typically by a ratio of 0.2 to 0.3. Only if the exit-to-inlet area ratio is near unity can the Fanno line analysis can be used to make a reasonable prediction of the total pressure losses. The flow is analyzed in Appendix C.5 with a control volume approach. An example is presented in this chapter (Example 8.2). The analysis results in 11 equations. Seventeen variables are present, and so six must be specified. For example, if inlet conditions M_1, T_1, and p_{t1}, and geometry parameters L, D, and f (Fanning friction factor) are specified, exit conditions, including the total pressure p_{t2}, can be found.

Example 8.8 | Fanno Line Inlet Flow

Flow with an averege Mach number of 0.2207 (the average of 0.0833 and 0.3580 from Example 8.7) and an inlet total pressure of 14.69 psia enters an inlet 2.14 ft long with a 3.86 ft average diameter and a Fanning friction factor of 0.020. What is the total pressure ratio if the specific heat ratio is 1.40?

Solution

From the tables in Appendix E or Fanno.exe for a Mach number of 0.2207 it is possible to find the following nondimensional length-to-diameter and pressure parameters:

$$\left.\frac{4fL^*}{D}\right]_1 = 11.508$$

and

$$\frac{p_{t1}}{p_t^*} = 2.6995;$$

thus, one can find the total pressure at the sonic condition, which is a reference condition (which means it usually does not occur in the actual conditions), as follows:

$p_t^* = 14.69/2.6995 = 5.442$ psia.

From the duct geometry, it is possible to find the parameter

$$\left.\frac{4fL}{D}\right]_{1-2} = \frac{4 \times 0.020 \times 2.14}{3.86} = 0.0444;$$

thus, the corresponding length-to-diameter parameter at the exit can be found as follows:

$$\left.\frac{4fL^*}{D}\right]_2 = \left.\frac{4fL^*}{D}\right]_1 - \left.\frac{4fL}{D}\right]_{1-2} = 11.508 - 0.0444 = 11.464.$$

The second average Mach number can then be found from the tables or Fanno.exe; namely,

$M_2 = 0.2211.$

Thus, as a result of friction the Mach number increased about 0.2 percent. From the tables, at this Mach number

$$\frac{p_{t2}}{p_t^*} = 2.6955$$

and so the exit total pressure can be found: $p_{t2} = 2.6955 \times 5.442 = 14.67$ psia. The pressure ratio for the inlet for Fanno line flow is therefore

$$\pi_i = \frac{p_{t2}}{p_{t1}} = \frac{14.67}{14.69} = 0.9986.$$

8.4.4 Combined Area Changes and Friction

In the preceding sections, area changes and friction were treated separately. However, in a diffuser the two effects occur simultaneously. Thus, an analysis is needed in which both are included. The technique that will be used here is the so-called generalized 1-D flow method; it is described in Appendix C.14 and is a differential equation analysis from which influence coefficients are derived. These influence coefficients relate changes of one variable on another. Applying the method entails numerically integrating the differential equations along the length of the inlet, which incorporate both friction and area changes. Also see General1D.exe.

Example 8.9 Combined Area Decrease with Friction

Flow is ingested into an inlet at 147 lb_m/s and an inlet total pressure of 14.69 psia. It enters an inlet 2.14 ft long and with a 5.14 ft diameter (with an inlet area of 20.75 ft^2). The flow has a Fanning friction factor of 0.020 and a converging exit-to-inlet area ratio of 0.25. The total temperature remains constant at 518.7 °R. What is the total pressure ratio if the specific heat ratio is 1.40?

Solution

First the Mach number is needed at the inlet plane. We use

$$\dot{m} = \rho_1 V_1 A_1.$$

The freestream conditions are known but the inlet conditions depend on the Mach number. The freestream total density is from the ideal gas equation and equal to 0.07644 lb_m/ft^3. The freestream sound speed is $a_a = 1116$ ft/s. But none of the static conditions are known since the incoming Mach number is not known. So an incoming Mach number of 0.0833 is guessed. Using isentropic flow tables or Isentropic.exe the density at the inlet plane is

$$\rho_1 = \frac{\rho_1}{\rho_t} \times \rho_t = 0.9965 \times 0.07644 = 0.07617 \text{ lb}_m/\text{ft}^3.$$

So

$$\dot{m} = \rho_1 V_1 A_1 = 147 \text{ lb}_m/\text{s} = 0.07617 \frac{\text{lb}_m}{\text{ft}^3} \times V_1 \times 20.75 \text{ ft}^2.$$

So $V_1 = 93.00$ ft/s and $M_1 = V_1/a_1$, and

$$T_1 = \frac{T_1}{T_t} \times T_t = 0.9986 \times 518.7 = 518.0\,°\text{R}.$$

So $a_1 = 1116$ ft/s and $M_1 = 0.0833$.

So the initial guess was good.

This problem will be solved by numerically integrating Eqs. (C.14.3) and (C.14.4); 15,000 steps will be used, and area change and friction will be assumed to be uniform along the length of the diffuser. The exit area is 5.19 ft^2 ; thus, the area decreases by 15.56 ft^2. Therefore, $\Delta A = -0.00103$ ft^2, and $\Delta x = 0.00014267$ ft for each step. A forward difference method is used. Thus, for the first step from Eq. (C.14.1) (the first term is due to area changes, and the second term results from friction),

$$\Delta(M^2) = -2M^2 \left[\frac{\left(1 + \frac{\gamma - 1}{2} M^2\right)}{1 - M^2}\right] \left[\frac{\Delta A}{A}\right] + 4f\gamma M^4 \left[\frac{\left(1 + \frac{\gamma - 1}{2} M^2\right)}{1 - M^2}\right] \left[\frac{\Delta x}{D}\right]$$

so $M_i = M_{i-1}^2 + \Delta(M^2)$.

From Eq. (C.14.2), the total pressure loss is only due to friction:

$$\Delta p_t = -\frac{1}{2} p_t 4f\gamma M^2 \left[\frac{\Delta x}{D}\right] = -2p_t f\gamma M^2 \left[\frac{\Delta x}{D}\right],$$

and so $p_{ti} = p_{ti-1} + \Delta(p_t)$.

This is repeated for all 15,000 steps similar to that done in Example 8.3, where the previous step is used for the initial conditions for the following step. The area and diameter are of course updated from step to step. Thus, one finds after 15,000 steps that

$M_2 = 0.3580$
$p_{t2} = 14.67$ psia,

and thus

$$\pi_i = \frac{p_{t2}}{p_{t1}} = \frac{14.67}{14.69} = 0.9989,$$

which is in agreement with Examples 8.7 and 8.8 combined. Thus, for this case the approximate superposition analysis is quite close to the more accurate method. In practice, a CFD method is used to calculate the total pressure loss and resulting exiting 2-D or 3-D velocity profile. As already noted, however, using CFD is beyond the scope of this book.

To complement this example, other inlet Mach numbers were considered to understand the trends on the recovery factor. Mach numbers from 0 to 0.146, where the inlet choked, were considered. Results are presented in Figure 8.20. As can be seen, as the Mach number increases, the recovery factor decreases at an increasing rate.

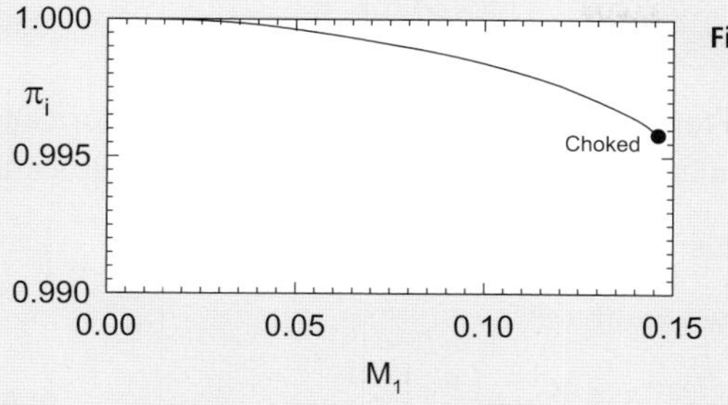

Figure 8.20 Pressure ratio as a function of Mach number for Example 8.9.

8.5 Inlet Cooling

Also, on occasion, inlet cooling is used to lower the total temperature at the inlet or diffuser for power generation units and propulsion applications. This is often done in hot climates with low humidity where liquid water or a water–alcohol mixture is sprayed into the inlet. Small holes or "jet nozzles" are located on the inside of the inlet or diffuser. The liquid evaporates and the evaporation causes the total temperature to drop because heat is absorbed by the water to vaporize or change the phase of the liquid. This has the same fundamental effect as intercooling as discussed in Chapter 1. Namely, by decreasing the inlet temperature less energy is needed by the compressor to develop the same exit pressure from the compressor. The result is for power generation units the net power is increased. For jet engines, the result is increased thrust. For propulsion this is only used for takeoff in hot and dry conditions. At cruise, the external temperatures are already low. A second advantage of inlet cooling for power generation is that the stream of liquid is sometimes used to clean the compressor when the unit is operated in hostile or dirty conditions.

8.6 Performance Maps

8.6.1 Propulsion Diffusers

Performance curves or "maps," which are developed for all of the components, are usually used to provide a working medium or methodology. These maps are used to condense a wealth of data and information as much as possible and to make the data more versatile for use over a wide range of operating conditions. These maps provide a quick and accurate view of the conditions at which the inlet is operating and of what can be expected of the diffuser (and what goes to the compressor) if the flow conditions are changed. As has been seen for subsonic flow, the recovery is a function of the inlet Mach number due to friction, which is in turn a function of the freestream Mach number. For supersonic operation, the recovery factor is a strong function of the freestream Mach number due to shocks and to the inlet Mach number due to friction. Thus, the total pressure ratio for a diffuser can be written in the general form

$$\pi_d = f(M_a), \tag{8.6.1}$$

where f can be determined from shock and Fanno line analyses, generalized 1-D flow analysis, CFD analysis, or experimental data, and can be found based on geometrical considerations. For example, for supersonic conditions for a given Mach number, the shock pattern can be found and the value of π_d can be calculated. This will not necessarily be done through a closed-form equation, for graphical or tabular means can be used as well. As has been seen through analyses for either subsonic or supersonic operation, as the Mach number increases, the recovery factor decreases.

More simply, diffuser designers have found an approximate and simple (but moderately accurate) empirical correlation for well-designed diffusers. The recovery factor can be modeled as

$$\pi_d = \pi_{d_d}\left[1 - d(M_a - 1)^{1.35}\right], \quad (8.6.2)$$

where π_{d_d} is the maximum pressure recovery and the constant d is 0.0 for subsonic operation and is typically 0.075 for supersonic operation. Of course, if data or analyses are available, they should be used in lieu of Eq. (8.6.2).

8.6.2 Power Generation Bell Inlets

Dimensional Analysis

These maps are used to simplify a substantial amount of data and make these data most versatile for use over a wide range of operation. These maps provide a quick and accurate view of the conditions at which the inlet is operating and what can be expected if the flow conditions are changed. Of primary concern is the mass flow rate and total pressure ratio of an inlet. The mass flow rate and pressure ratio are a function of many variables, and the functional relationship for an inlet can be written as

$$\dot{m}_1 = f\{T_{ta}, A_1, A_2, L, p_{ta}, p_{t2}, \gamma, \mathcal{R}\}. \quad (8.6.3)$$

One can use dimensionless parameters to reduce the number of independent variables. Using the Buckingham Pi theorem, one finds for a given value of γ that

$$\dot{m}_1 = \frac{\sqrt{\mathcal{R}T_{t1}}}{A_1 p_{t1}} = f\left\{\frac{p_{t2}}{p_{ta}}, \frac{A_2}{A_1}, \frac{L^2}{A_1}\right\}. \quad (8.6.4)$$

This form is rarely used to document the performance of an inlet, however. Usually, one finds a modification of this; that is, for a given inlet $\left\{\frac{A_2}{A_1}\right\}$ and $\left\{\frac{L^2}{A_1}\right\}$ are known and fixed, so

$$\dot{m}_1 \frac{\sqrt{\theta_{t1}}}{\delta_{t1}} = \dot{m}_{c1} = f\left\{\frac{p_{t2}}{p_{ta}}\right\}, \quad (8.6.5)$$

or inversely

$$\pi_i = \frac{p_{t2}}{p_{ta}} = f\left\{\dot{m}_1 \frac{\sqrt{\theta_{t1}}}{\delta_{t1}}\right\} = f\{\dot{m}_{c1}\}, \quad (8.6.6)$$

where, as before,

$$\delta_{t1} = \frac{p_{t1}}{p_{stp}} \quad (8.6.7)$$

and

$$\theta_{t1} = \frac{T_{t1}}{T_{stp}}, \quad (8.6.8)$$

and where the subscript stp refers to the standard conditions. For reference and from Appendix B, p_{stp} is equal to 14.69 psia or 101.33 kPa, and T_{stp} is equal to 518.7 °R or 288.2 K. Note that the two parameters in Eq. (8.6.6) are not truly dimensionless but are prudently chosen so that the function f retains the important characteristics of nondimensionalization. As for compressors and turbines, the term inside the brackets is called the "corrected" mass flow and has dimensions of mass flow, whereas the pressure ratio (left-hand side) is dimensionless and is the recovery factor. Thus, since they both are not dimensionless and, in particular, since the parameters A_1, $\left\{\frac{A_2}{A_1}\right\}$, and $\left\{\frac{L^2}{A_1}\right\}$ have been eliminated, the resulting function cannot be used to correlate or compare different inlets. The function can be used to characterize the data over a range of conditions for one particular inlet, however, as is the usual case. Thus, an inlet can be tested under one set of conditions, and the results can be "corrected" so that results are applicable to other conditions so long as the parameter $\dot{m}_{c1}$ is matched.

Trends

In Figure 8.21, the typical tends for the recovery factor for a power generation inlet are shown. As can be seen as the corrected mass flow rate increases the recovery factor decreases. The trend is dominated by the frictional loss in the inlet as predicted by Fanno line flow with a decreasing area. Namely as the mass flow increases for a given inlet so does the velocity. As the velocity increases so do the losses. This was seen in Example 8.9 and Figure 8.20.

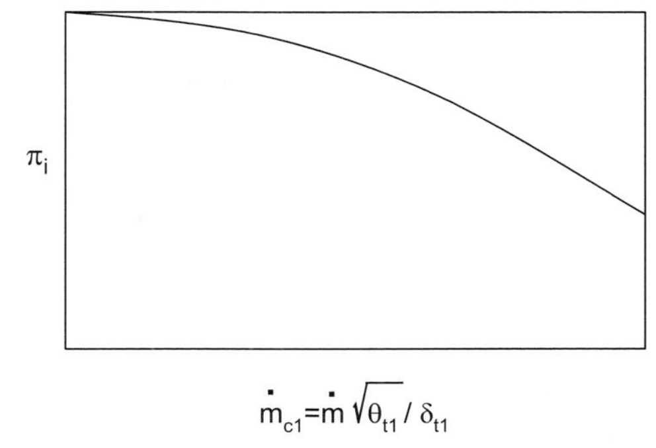

Figure 8.21 Typical map trends for a power generation inlet.

8.7 Summary

In this chapter, the basic operating characteristics of propulsion diffusers and power generation inlets were covered. The basic goal of the propulsion diffuser is to bring the flow of air smoothly into the engine, slow the flow and increase the pressure, and deliver the air to the compressor uniformly. Two losses were found to exist. The first is due to the flow up to the inlet plane of the diffuser. This loss is greatest for a supersonic diffuser because of shocks and is minimal for subsonic aircraft. The second loss occurs in the internal flow in the diffuser between the diffuser inlet and inlet to the compressor. This loss is primarily caused by friction resulting from the boundary layer, turbulence, and associated separation zones. Limitations on the attainable pressure rise were addressed, and guidelines for minimizing separation areas were given. The frictional losses can first be approximated by Fanno line flow; more advanced predictions can be made using generalized 1-D frictional flow with area changes. Normal, planar oblique, and conical oblique shocks were all discussed and were seen to be functions of the Mach number and inlet geometry, which makes it possible for external loss predictions to be made. Oblique shocks or a series of gradual oblique shocks can be used to minimize losses below the level that result from to normal shocks. Limits on flow turning and attached oblique shocks were discussed; if the Mach number is too low or the turning is too large, a standing normal shock (with large losses) was seen to result. For supersonic aircraft, the mass flow ratio was defined for use in developing a method to predict the off-design geometry requirements. Two distinct modes of off-design operation, which change with aircraft speed and engine mass flow rates, were discussed for supersonic diffusers. For subcritical operation, mass flow spillage occurs – that is, a portion of the airflow that ideally enters the engine spills around the inlet. For supercritical operation, the flow is supersonic in the diffuser and the resulting shocks cause a higher level of internal losses than would be experienced during subsonic operation. Additive drag on the engine cowl was found to detract directly from engine thrust because of mass flow spillage. Lastly, the concept of "starting" a supersonic inlet was discussed. Starting implies that incoming supersonic flow gradually decreases to subsonic flow without strong shocks. To start a diffuser properly one must either overspeed the aircraft until the shocks are "swallowed" or vary the inlet geometry to swallow the shocks.

The basic goal of the power generation inlet is to bring the flow of air smoothly into the gas turbine, increase the velocity of the flow and decrease the pressure, and deliver the air to the compressor uniformly. Two losses were found to exist. The first is due to the flow up to the inlet plane of the inlet. This loss is minimal for the gas turbine inlet. The second loss occurs in the internal flow in the inlet between the inlet and front of the compressor. This loss is primarily caused by friction resulting from the boundary layer, turbulence, and very limited separation zones because of crosswinds. Since the pressure is decreasing, separation is not as critical as for propulsion diffusers. The frictional losses can first be approximated by Fanno line flow; more advanced predictions can be made using generalized 1-D frictional flow with area changes.

Inlet cooling can be used in the diffuser and inlet to improve performance in dry and hot conditions. Namely, increased thrust on takeoff for propulsion applications and increased output power for power generation units. Lastly, maps were discussed for both diffusers and inlets, and models were presented.

List of Symbols

A	Area
a	Speed of sound
b	Spike or wedge position
C_d	Drag coefficient
C_p	Pressure coefficient
d	Empirical factor in supersonic recovery factor (Eq. (8.6.2))
D	Diameter
D_a	Additive drag
F	Force
f	Fanning friction factor
g	Gravitational constant
h	Specific enthalpy
H	Height
L	Length
M	Mach number
MFR	Mass flow ratio
p	Pressure
R	Radius
$\mathcal{R}$	Ideal gas constant
s	Entropy
T	Temperature
u	Specific internal energy
u	Velocity
V	Velocity
γ	Specific heat ratio
δ	Flow turning angle
δ	Pressure divided by standard pressure
η	Pressure recovery due to shocks
θ	Temperature divided by standard temperature
ϑ	Diffuser angle
π	Total pressure ratio
ρ	Density
σ	Shock angle

Subscripts

a	Axial
a	Additive drag
a	Ambient
b	Boundary layer
b	Body force
c	Cowl
c	"Corrected"
1c	Projected area
d	Diffuser (propulsion gas turbine)
d	Design
d-ext	Outside of diffuser
d-int	Internal to diffuser
d_d	Maximum supersonic recovery factor (Eq. (8.6.2))
e	Exit
f	Fuel
i	Inlet (power gas turbine)

i	Before shock
in	Inlet or diffuser
i-ext	Outside of inlet
i-int	Internal to inlet
j	After shock
max	Maximum before detachment
max	See Eq. (8.2.5b)
min	See Eq. (8.2.5a)
n	Normal to
r	Recovery
r	Rejected
s	Surface
s	Overspeed
stp	Standard pressure and temperature
t	Total (stagnation)
t	Tangent to
x	x direction
1	Inlet to diffuser/inlet
2	Exit of diffuser/inlet
1, 2	Areas of control volume

Superscripts

′	Ideal
*	Choked

Problems

8.1 A supersonic inlet is designed with a 2-D conical spike (with a half-cone angle of 25°) to operate at a flight Mach number of 2.6. The standing oblique shock is attached to the spike and cowl, and a converging inlet section with a throat of area A_m is used to decelerate the flow using internal compression. Assume $\gamma = 1.4$ and $\pi_{d\text{-int}} = 1.0$.

(a) Estimate π_d on the assumption that the inlet starts (i.e., the normal shock is swallowed; see Figure 8.19b). Also, find the required A_m/A_1 and sketch the geometry to scale.
(b) Find π_d on the assumption the inlet does not start and has a standing normal shock located in front of the spike (Figure 8.19a).

8.2 A supersonic inlet is designed with a 2-D conical spike (with two half-cone angles of 12° and 25° relative to the axial centerline, respectively; i.e., the turning angles are 12° and 13°, respectively). The inlet is to operate at a flight Mach number of 2.6. The two standing oblique shocks are attached to the spike and cowl, and a converging inlet section with a throat of area A_m is used to decelerate the flow using internal compression. Assume $\gamma = 1.4$ and $\pi_{d\text{-int}} = 1.0$.

(a) Estimate π_d on the assumption the inlet starts (i.e., the normal shock is swallowed; see Figure 8.19b). Also, find the required A_m/A_1 and sketch the geometry to scale.
(b) Find π_d on the assumption the inlet does not start and has a standing normal shock located in front of the spike (Figure 8.19a).

8.3 Analyze a turbojet engine for an aircraft flying at $M_a = 2.6$ at sea level on a standard day. The exit burner total temperature is 3200 °R, and the compressor pressure ratio is 16. Assume all aerodynamic coefficients and recovery factors are 0.90 and that $\pi_b = 0.90$ except for the diffuser, which has an internal recovery factor of 0.97. Assume a mechanical efficiency of 1.00. Consider thrust and *TSFC* for the engine with the following three inlets having a CD nozzle, a compressor flow rate of 150 lb_m/s, and $\Delta H = 18{,}000$ BTU/lb_m.

(a) Inlet has a standing normal shock.
(b) Inlet has a conical cone with one ramp at 25° to the centerline.

(c) Inlet has a cone with two ramps. The first is 12° to the centerline and the second is 13° to the first (25° to the centerline).

(d) Compare the results with those that would be obtained in a subsonic case with $M_a = 0.8$ and a CD nozzle.

8.4 Find the external recovery factor and the Mach number after the shock of a supersonic diffuser as functions of the freestream Mach number for a normal shock if $\gamma = 1.40$. Plot the results.

8.5 Find the external recovery factor and the Mach number after the shock of a supersonic diffuser having a spike with a cone half-angle of 15° as functions of freestream Mach number if $\gamma = 1.40$. Plot the results.

8.6 Flow with an incoming Mach number of 0.72 and an inlet total pressure of 82.74 kPa (12 psia) enters a straight duct 1.14 m (45 in.) long and 1.14 m (45 in.) in diameter with a Fanning friction factor of 0.015. What is the total pressure ratio if the specific heat ratio is 1.40?

8.7 Flow with an incoming Mach number of 0.85 and an inlet total pressure of 82.74 kPa (12 psia) enters a diffuser channel 1.14 m (45 in.) long and 0.914 m (36 in.) in diameter with an inlet area of 0.657 m^2 (1018 $in.^2$). The flow has an area ratio of 1.60. The total temperature remains constant at 266.7 k (480 °R). What is the exit Mach number if there are no frictional losses and if the specific heat ratio is 1.40? What is the exit pressure and pressure coefficient? Comment.

8.8 Flow with an incoming Mach number of 0.85 and an inlet total pressure of 82.74 kPa (12 psia) enters a diffuser channel 1.14 m (45 in.) long and 0.914 m (36 in.) in diameter with an inlet area of 0.657 m^2 (1018 $in.^2$). The flow has a Fanning friction factor of 0.015 and an area ratio of 1.60. The total temperature remains constant at 266.7 K (480 °R). What is the total pressure ratio if the specific heat ratio is 1.40?

8.9 A supersonic inlet is designed with a 2-D conical spike (with a half-cone angle of 20°) to operate at a flight Mach number of 1.9. The standing oblique shock is attached to the spike and cowl, and a converging inlet section with a throat of area A_m is used to decelerate the flow using internal compression. Assume $\gamma = 1.4$ and $\pi_{d\text{-int}} = 0.97$.

(a) Estimate π_d on the assumption the inlet starts (i.e., the normal shock is swallowed; see Figure 8.19b). Also, find the required A_m/A_1 and sketch the geometry to scale.

(b) Find π_d on the assumption the inlet does not start and has a standing normal shock located in front of the spike (Figure 8.19a).

8.10 A supersonic inlet is designed with a 2-D conical spike (with two half-cone angles 10° and 20° relative to the axial centerline, respectively; i.e., the turning angles are 10° and 10°, respectively). The inlet is to operate at a flight Mach number of 1.9. The two standing oblique shocks are attached to the spike and cowl, and a converging inlet section with a throat of area A_m is used to decelerate the flow through internal compression. Assume $\gamma = 1.4$ and $\pi_{d\text{-int}} = 0.97$.

(a) Estimate π_d on the assumption the inlet starts (i.e., the normal shock is swallowed). Also, find the required A_m/A_1 and sketch the geometry to scale.

(b) Find π_d on the assumption the inlet does not start and has a standing normal shock located in front of the spike.

8.11 Analyze a turbofan engine (fully mixed) with a bypass ratio of 2.2 for an aircraft flying at $M_a = 1.9$ at 6706 m (22,000 ft) on a standard day. The burner exit total temperature is 1556 K (2800 °R), and the compressor pressure ratio is 19. Assume all aerodynamic coefficients and recovery factors are 0.90 and $\pi_b = 0.90$ except for the diffuser, which has an internal recovery factor of 0.97. Assume a mechanical efficiency of 0.998, a mixer total pressure ratio of 0.99, and a duct total pressure ratio of 0.99. Find the thrust and *TSFC* for the engine with a converging nozzle, a compressor flow rate of 63.5 kg/s (140 lb_m/s), and $\Delta H = 41{,}400$ kJ/kg (17,800 BTU/lb_m) for the following three inlets.

(a) Inlet has a standing normal shock.
(b) Inlet has a conical cone with one ramp at 20° to the centerline.
(c) Inlet has a cone with two ramps. The first is 10° to the centerline, and the second is 10° to the first (20° to the centerline).

8.12 Analyze a turbofan engine (fully exhausted) with a bypass ratio of 2.2 for an aircraft flying at $M_a = 1.9$ at 6706 m (22,000 ft) on a standard day. The burner exit total temperature is 1556 K (2800 °R), and the compressor pressure ratio is 19. Assume all aerodynamic coefficients and recovery factors are 0.90 and $\pi_b = 0.90$ except for the diffuser, which has an internal recovery factor of 0.97. Assume a mechanical efficiency of 0.998 and a fan total pressure ratio of 1.90. Find the thrust and *TSFC* for the engine with converging nozzles, a compressor flow rate of 63.5 kg/s (140 lb_m/s), and $\Delta H = 41{,}400$ kJ/kg (17,800 BTU/lb_m) for the following three inlets.

(a) Inlet has a standing normal shock.
(b) Inlet has a conical cone with one ramp at 20° to the centerline.
(c) Inlet has a cone with two ramps. The first is 10° to the centerline and the second is 10° to the first (20° to the centerline).

8.13 A diffuser with a spike is used on a supersonic aircraft. The freestream Mach number is 2.0, and the cone half-angle is 25°. Inside the diffuser (which is designed to be shockless), there is a recovery factor of 0.95. What is the total recovery factor (freestream to diffuser exit) of the diffuser? Assume the diffuser has started.

8.14 A standing normal shock occurs on an aircraft flying at Mach 1.50. The internal recovery factor of the diffuser is 0.98, and the specific heat ratio is 1.40. Find the total recovery factor for the diffuser.

8.15 An aircraft moves at Mach 1.6 and has a diffuser with a conical spike. The cone has an angle of 15° relative to the centerline. You can assume that the diffuser has started (swallows the shock and is shockless inside the diffuser). If the internal total pressure ratio is 0.98, what is the overall pressure recovery factor? What is the inlet Mach number at the diffuser inlet? Assume a specific heat ratio of 1.40.

8.16 A supersonic inlet is designed with a 2-D conical spike (with a half-cone angle of 24°) to operate at a flight Mach number of 2.2. The standing oblique shock is attached to the spike and cowl, and a converging inlet section with a throat of area A_m is used to decelerate the flow through internal compression. Assume $\gamma = 1.40$ and $\pi_{d\text{-int}} = 0.98$.

(a) Estimate π_d on the assumption the inlet starts (i.e., the normal shock is swallowed; see Figure 8.19b). Also, find the required A_m/A_1 and sketch the geometry to scale.
(b) Find π_d on the assumption the inlet does not start and has a standing normal shock located in front of the spike (Figure 8.19a).

8.17 A supersonic inlet is designed with a 2-D conical spike (with two half-cone angles 12° and 24° relative to the axial centerline, respectively; i.e., the turning angles are 12° and 12°, respectively). The inlet is to operate at a flight Mach number of 2.2. The two standing oblique shocks are attached to the spike and cowl, and a converging inlet section with a throat of area A_m is used to decelerate the flow using internal compression. Assume $\gamma = 1.4$ and $\pi_{d\text{-int}} = 0.98$.

(a) Estimate π_d on the assumption the inlet starts (i.e., the normal shock is swallowed; see Figure 8.19b). Also, find the required A_m/A_1 and sketch the geometry to scale.
(b) Find π_d on the assumption the inlet does not start and has a standing normal shock located in front of the spike (Figure 8.19a).

8.18 An engine similar in size to an older military nonafterburning turbojet ingests 120 lb_m/s of air at sea level. When it flies at a Mach number of 2.2, the compressor pressure ratio is 14.6. The fuel has a heating value of 17,800 BTU/lb_m, and the burner exit total temperature is 2580 °R. The efficiencies of the compressor, burner, turbine, shaft, and CD nozzle are 90, 95, 92, 100, and 96 percent, respectively. The total pressure ratio for the burner is 0.95. The diffuser has an internal pressure recovery factor of 0.98. Find the thrust and *TSFC* for the engine for the following three inlets.

(a) Inlet has a standing normal shock.
(b) Inlet has a conical cone with one ramp at 24° to the centerline.
(c) Inlet has a cone with two ramps. The first is 12° to the centerline, and the second is 12° to the first (24° to the centerline).

8.19 Design a diffuser to attain the maximum pressure rise if the incoming Mach number is 0.80. That is, find the aspect ratio (L/D_1), area ratio (A_2/A_1), pressure ratio, and exit Mach number.

8.20 An aircraft moves at Mach 2.5 and has a diffuser with a conical spike. The cone has an angle of 50° relative to the centerline. If the internal total pressure ratio is 0.97, what is the overall pressure recovery factor? What is the inlet Mach number at the diffuser inlet? Assume a specific heat ratio of 1.40.

8.21 Flow enters a long converging inlet for a gas turbine at Mach 0.30. The mass flow rate is 100 lb_m/s, and the inlet total pressure and temperature are 14 psia and 500 °R, respectively. The L/D ratio is 20, and the area ratio is 0.70. The Fanning friction coefficient is 0.015, and the specific heat ratio is 1.40. The air warms slightly to a total temperature of 530 °R. Use a generalized 1-D analysis and find the internal recovery factor of the inlet.

8.22 An aircraft moves over a range of Mach numbers and has a diffuser with a movable conical spike. The cone has a fixed angle of 20° relative to the centerline. What is the minimum flight Mach number that will allow the diffuser to operate efficiently? If the internal total pressure ratio is 0.98, what is the overall pressure recovery factor? Accurately sketch the diffuser. Should the diffuser have a minimum area for this minimum flight condition? Assume a specific heat ratio of 1.40.

8.23 A rectangular supersonic inlet is constructed with a 2-D planar ramp. Following the oblique shock, the inlet is designed to be shockless at the design Mach number of 2.00. The engine mass flow rate is 100 lb_m/s at sea level. Assume $\gamma = 1.40$ and an internal recovery coefficient of 0.97.

(a) What is the maximum deflection angle before the shock detaches?
(b) Obtain the total recovery coefficient for the diffuser for the maximum deflection angle.
(c) Sketch the inlet.
(d) What is the required inlet area of the diffuser for the maximum deflection angle?

8.24 An annular supersonic inlet is constructed with a 2-D cone. Following the oblique shock, the inlet is designed to operate without any shocks at a design freestream Mach number of 2.00. The engine mass flow rate is 145 lb_m/s at sea level. Assume $\gamma = 1.40$ and an internal recovery coefficient of 0.96.

(a) What is the maximum deflection (half-cone) angle before the oblique shock detaches?
(b) Obtain the total recovery coefficient for the diffuser for the maximum deflection angle.
(c) Sketch the inlet showing important angles.

8.25 A planar 2-D ramp is used with a rectangular supersonic inlet. The aircraft flies at a Mach number of 1.70 ($\gamma = 1.40$). A shock angle of 40° results.

(a) What is the ramp angle, exit Mach number, and total pressure ratio of the shock?
(b) To what angle could the ramp be increased and still maintain an attached shock? For this condition, what is the exit Mach number and total pressure ratio of the shock?

8.26 A conical 2-D spike is used with a supersonic inlet. The aircraft flies at a Mach number of 2.60 ($\gamma = 1.40$). A shock angle of 35° results.

(a) What is the ramp angle, exit Mach number, and total pressure ratio of the shock?
(b) To what angle could the ramp be increased and still maintain an attached shock? For this condition, what is the exit Mach number and total pressure ratio of the shock?

8.27 Find the external recovery factor and the Mach number after the shock of a supersonic diffuser with a wedge having a wedge angle of 15° as functions of freestream Mach number if $\gamma = 1.40$. Plot the results.

8.28 A conical 2-D spike is used with an inlet for an aircraft that flies at a Mach number of 1.60 ($\gamma = 1.40$) at an altitude with a static pressure of 5 psia. A half-cone angle of 36° is used. The internal pressure recovery is 0.98.

(a) What are the Mach number and total pressure behind the shock?
(b) What is the overall pressure recovery of the inlet?

8.29 Find the external recovery factor and the Mach number after a single oblique shock and normal shock of a supersonic diffuser with a spike having a cone half-angle of 25° as functions of freestream Mach number if $\gamma = 1.40$. Plot the results.

8.30 A planar 2-D wedge is used with an inlet for an aircraft that flies at a Mach number of 2.00 ($\gamma = 1.40$) at an altitude with a static pressure of 7 psia. A half-wedge angle of 21° is used. The internal pressure recovery is 0.975.

(a) What is the Mach number behind the shock?
(b) What is the shock angle?
(d) For this wedge angle, what is the minimum freestream Mach number before the shock detaches?
(c) What is the total pressure at the exit of the diffuser?

8.31 A planar 2-D wedge and resulting oblique shock is used with a normal shock in an inlet for an aircraft that flies at a Mach number of 2.50 ($\gamma = 1.40$) at an altitude with a static pressure of 5 psia. A half-wedge angle of 17° is used. The internal pressure recovery is 0.980.

(a) What is the Mach number behind the normal shock?
(b) What is the oblique shock angle?
(c) What is the total pressure at the exit of the diffuser?
(d) For this wedge angle, what is the minimum freestream Mach number before the shock detaches?

8.32 A planar 2-D wedge with a half-wedge angle of 18° is used in an inlet for an aircraft. When a resulting oblique shock occurs, a normal shock follows, which reduces the Mach number below unity. Plot the external recovery factor for a Mach number range of 1.0 to 3.0

8.33 Flow with an incoming Mach number of 0.80 and an inlet total pressure of 82.64 kPa (12 psia) enters a diffuser channel 1.14 m (45 in.) long having a 0.914 m (36 in.) diameter. The diffuser has an area ratio of 1.20. The total temperature remains constant at 266.7 K (480 °R). The flow has a Fanning friction factor of 0.015 and the specific heat ratio is 1.40.

(a) What is the total pressure ratio?
(b) What is the pressure coefficient? Comment.

8.34 A solid cone and resulting conical oblique shock are used in an inlet of an aircraft that flies at Mach 2.00 ($\gamma = 1.40$) at an altitude of 9144 m (30,000 ft).

(a) What cone angle will yield the smallest Mach number behind the oblique shock?
(b) What is the resulting Mach number behind the shock on the cone surface?
(c) What is the oblique shock angle?
(d) What is the total pressure ratio?
(e) For this cone angle, what is the minimum freestream Mach number before the oblique shock detaches?

8.35 A planar 2-D wedge is used for an inlet of an aircraft that flies at a Mach number of 2.00 ($\gamma = 1.40$) at an altitude with a static pressure of 7 psia. A half-wedge angle of 28° is used. The internal pressure recovery is 0.980. What is the total pressure at the exit of the diffuser?

8.36 A conical 2-D spike is used for an inlet of an aircraft that flies at a Mach number of 2.00 ($\gamma = 1.40$) at an altitude with a static pressure of 7 psia. A half-cone angle of 45° is used. The internal pressure recovery is 0.980. What is the total pressure at the exit of the diffuser?

8.37 Flow is ingested into a bell inlet at 120 lb_m/s and an inlet total pressure of 14.50 psia. It enters an inlet 1.50 ft long and with a 4.5 ft diameter. The flow has a Fanning friction factor of 0.017 and a converging exit-to-inlet area ratio of 0.30. The total temperature remains constant at 516.0 °R. What is the total pressure ratio if the specific heat ratio is 1.40?

8.38 Flow is ingested into a bell inlet at 59 kg/s and an inlet total pressure of 100 kPa. It enters an inlet 0.4 m long and with a 2 m diameter. The flow has a Fanning friction factor of 0.016 and a converging exit-to-inlet area ratio of 0.35. The total temperature remains constant at 290 K. What is the total pressure ratio if the specific heat ratio is 1.40?

8.39 Flow is ingested into a bell inlet with inlet total pressure of 14.50 psia. It enters an inlet 1.50 ft long and with a 4.5 ft diameter. The flow has a Fanning friction factor of 0.017 and a converging exit-to-inlet area ratio of 0.30. The total temperature remains constant at 516.0 °R. What is the inlet Mach number and flow rate if the exit Mach number is 0.60? What is the total pressure ratio if the specific heat ratio is 1.40?

9 Combustors, Reheat Burners, and Afterburners

9.1 Introduction

The burner, reheat burner, and afterburner are the only components through which energy is added to the gas turbine. That is, in these components the total temperature of the gas increases. For the jet engine primary burner, a part of this energy is used by the turbine to drive the compressor, and the other part is left to generate a high-velocity gas from the nozzle, generating thrust. For the afterburner, all of this energy increase is used to generate an increase of the fluid enthalpy and consequently a higher velocity gas from the nozzle, generating more thrust. For a power generation unit primary burner, a part of this energy is used by the turbine to drive the compressor, and the other part is left to generate power in second turbine. For a reheat combustor, more energy is added to the turbine train. As a result of these direct impacts, efficient operation of these components is necessary for the overall efficiency of the gas turbine. However, several complex considerations must be realized in the design of either of these components, and some of the more advanced topics are summarized by Peters (1988) and the more recent exhaustive text of Lefebvre and Ballal (2010). Furthermore, Malecki et al. (2001) demonstrate the application of CFD predictions to modern combustor design.

The following are essential considerations in the design of a burner.

(1) A major objective is complete combustion so that fuel will not be wasted.
(2) Minimal total pressure loss is another important design goal. As will be discussed in Section 9.2, these first two objectives are in direct conflict.
(3) All of the primary combustion processes must take place in the combustor and not the turbine, or the turbine life will be reduced and emissions can result.
(4) Minimal deposits are desired because large deposits indicate burning inefficiencies and further compromise the burning efficiency, reduce the pressure, and cause hot spots.
(5) For obvious reasons, one wants easy ignition and relighting of the fuel as well as a high entrance pressure (exit of the high-pressure compressor).
(6) Burners, reheaters, and afterburners should have long lives and should not fail, because burner failure can lead to a gas turbine failure or explosion.
(7) The exit temperature should be uniform both in space and time because extremes should not be introduced to the turbine in the case of a primary combustor or to the nozzle in the case of an afterburner.
(8) No internal hot spots should be present because these can lead to burner failure.
(9) The flame should be stable over a range of mass flows and other operating conditions. An unstable flame can lead to fluctuating exit conditions, which can adversely affect thrusts, power, and loading of the other components and ultimately lead to their failure.
(10) The flame should not be prone to "flame out" because this can result in total loss of power or thrust.
(11) The aircraft burner or afterburner should have minimal volume because weight is of primary concern – especially with military craft.

(12) Overall, the design of a combustor is not as straightforward as the design of a compressor or turbine and requires more empiricism and testing.

9.1.1 Objectives

After completing this chapter, the reader should be able to do the following.

- Understand the basic geometry of a primary combustor, reheat burner, and an afterburner
- Understand the concepts and fundamentals of flame speed and instabilities
- Understand burner ignition and gas turbine starting
- Understand the basic chemistry and chemical reaction of a burner
- Understand the implications of a stoichiometric process
- Understand the basics of disassociation, equilibrium, and emissions
- Be able to determine the adiabatic flame temperature of a reaction
- Understand the relation of the chemistry and the previously used heating value
- Appreciate the conflicts of efficient burning and associated pressure losses
- Be able to predict pressure losses due to combustion, friction, and flameholders, and combinations thereof
- Understand the concepts and the application of maps
- Understand the different fuel types and their properties

9.2 Geometries

Several design criteria need to be considered, as follows.

(1) A good mixture ratio of fuel to air is required along the burner axis. Although the stoichiometric ratio for most fuels is typically 0.06 to 0.07 and the fuel-to-air ratio for the gas turbine is typically 0.015 to 0.03, in the primary burning zone the mixture is fuel rich and this ratio is typically 0.08.

(2) The temperature of the reactants should be heated to above the ignition temperature of the fuel before entering the burner, preferably in a gas state.

(3) The proper temperature in the burner must be achieved to sustain the burn.

(4) Good turbulence is required so that the fuel will be well mixed into the air.

(5) A low pressure loss is desired so the gas will enter the turbine with a high pressure. However, high turbulence results in a large pressure loss. Thus, although a high turbulence level is desired, it should not be any higher than necessary to ensure a good mix.

(6) Time is needed for each fuel particle to burn completely. If the gas velocity is higher than the flame speed, the burner will "flame out." The flame speed is the velocity at which the flame will move in a quiescent and uniform mixture of fuel and air. Typically, laminar flame speeds are 1 to 5 ft/s and turbulent flame speeds are 60 to 100 ft/s. Therefore, the size of the burner should be large enough so that the velocities will remain low. If large sizes cannot be maintained, flameholders (flame stabilizers) are often used to set up standing recirculation wakes. Thus, low-velocity zones are realized; however, they are at the expense of increased pressure losses.

(7) The exit temperatures of a primary or reheat burner and aircraft afterburner are significantly different. The allowable turbine inlet temperature limits the exit temperature of the primary burner. On the other hand, the allowable nozzle inlet temperature limits the exit temperature of the afterburner. Because the turbine blades are rotating, the stress levels are much higher than the nozzle stress levels. As a result, the turbine inlet temperatures must be much lower than the aircraft nozzle temperatures.

9.2.1 Primary Combustors

Three types of main burners are used: can (or tubular), annular, and cannular (or can–annular). The three general geometries are compared in Figure 9.1. Typically, cans were used on early engines and annular and cannular burners are used on modern gas turbines. Each will be discussed separately.

In Figures 9.2 and 9.3, cross sections of a can burner and assembly are shown, and in Figure 9.4 a schematic is presented. Several of these self-contained cans are evenly spaced and situated circumferentially around the engine. Typically, 7 to 14 cans are used. As can be seen from Figure 9.4, "primary" air first enters from the

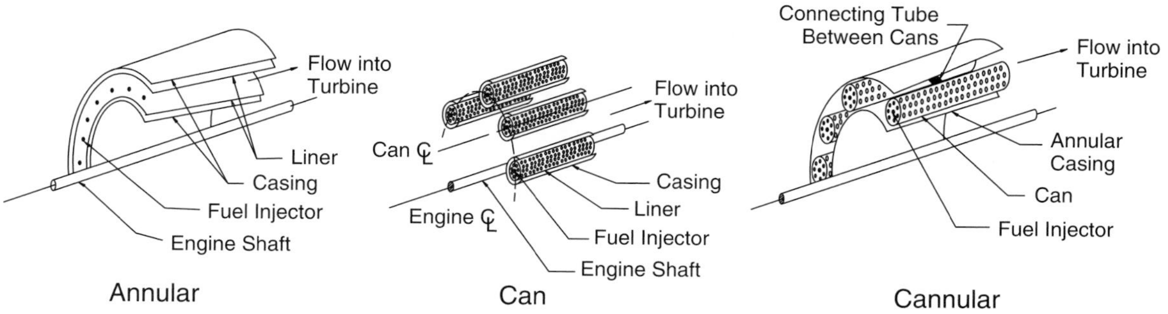

Figure 9.1 Three basic types of primary burners.

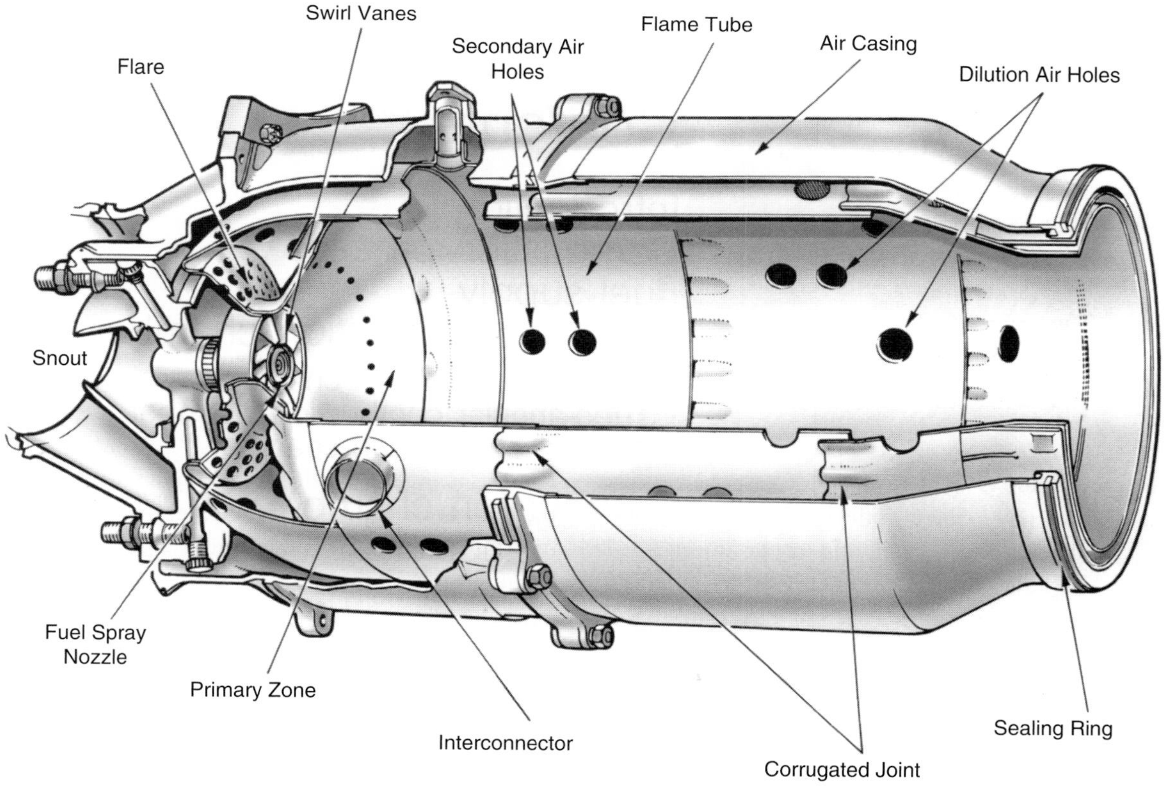

Figure 9.2 An older model can burner ().

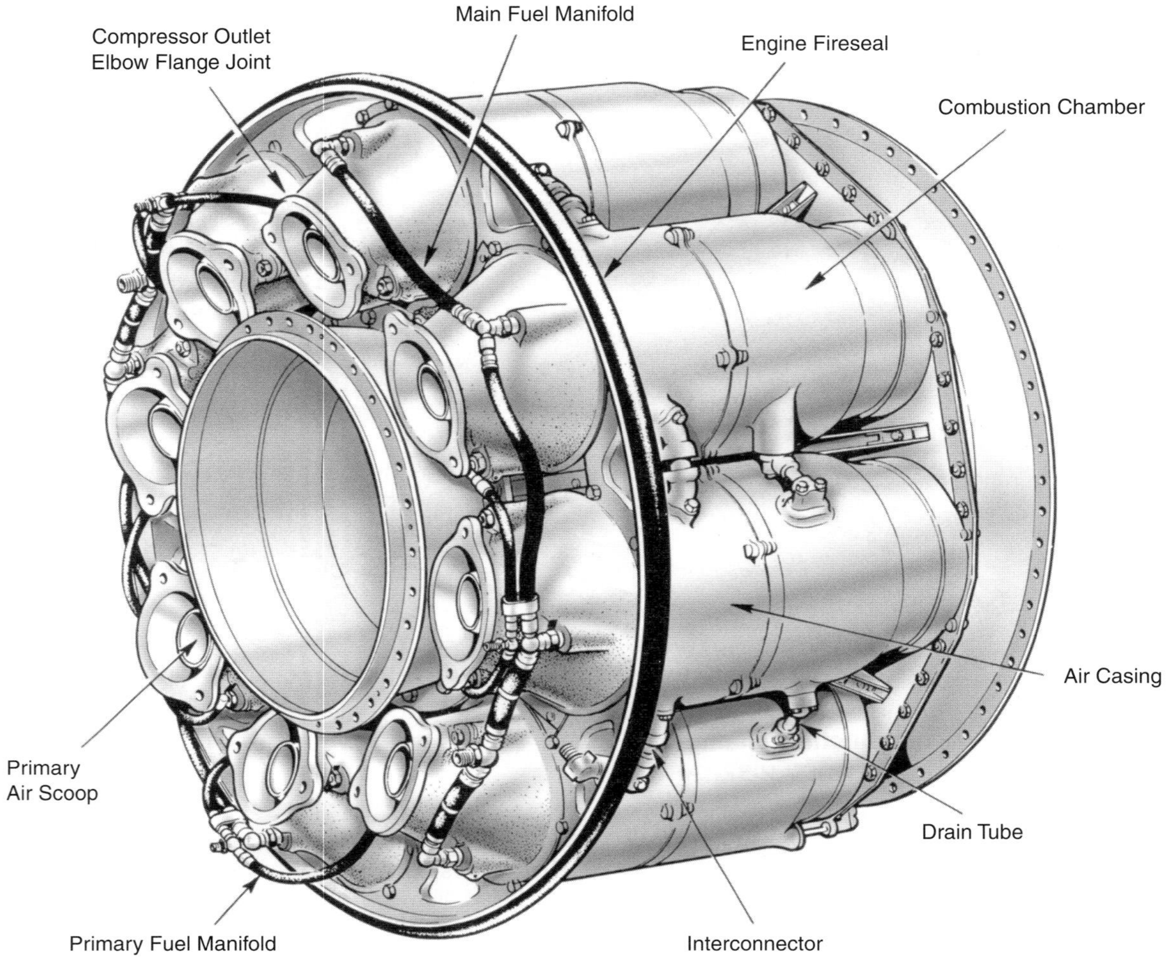

Figure 9.3 A 3-D view of an assembled set of can burners (© Rolls-Royce plc 2022. All rights reserved).

compressor and is decelerated to decrease the velocity and then swirled to increase the turbulence level of the flow, which enhances the mixing. Fuel is injected at a relatively low velocity. "Secondary" air is bounded by the outer liner (which also acts as a heat shield) and gradually fed into the combustion zone through holes or slits in the inner liner. Slits are used to form a cool boundary layer on the liner to maintain a tolerable liner temperature; holes are used to force air to the center of the burner so that the combustion will occur away from the walls. Usually, a combination of holes and slits is used. Typically, smaller holes are used near the front of the inner liner and larger holes at the rear of the liner so that air will be uniformly fed into the burning zone along the axis of the burner. Air finally exits at a relatively high velocity into the turbine.

Figure 9.5 shows a cutaway of an annular burner. Air enters the burning zone both from the inner and outer diameters, and hole sizes are again used to control the relative air flow rates along the axis, as seen in Figure 1.10. The Honeywell T55 turboshaft, previously shown in Figure 6.2, has an annular burner.

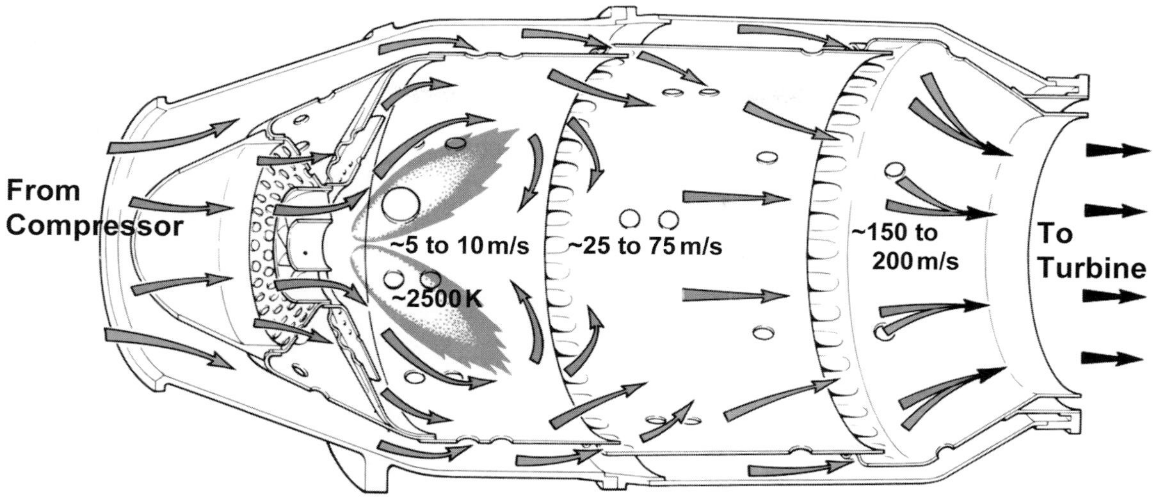

Figure 9.4 Flow patterns in a can burner (adapted from © Rolls-Royce plc 2022. All rights reserved).

To shorten the length of the engine, the burner is actually outside or around the periphery of the turbine. The flow in the combustor actually flows in the opposite axial direction to that of the flow in the axial stages (called counterflow), as with the early Whittle engines. Air enters the combustor near the rear of the engine.

As implied in the name, a cannular burner is a combined version of the can and annular geometries. Such a geometry is shown in Figure 9.6. Basically, a series of cans are circumferentially connected so that they interact more so than a series of cans. For example, any circumferential pressure variations are minimized. Connecting tubes, which directly connect the cans and allow for flow between cans, are often used. Furthermore, as shown in Figures 9.1 and 9.6, the can casings have holes that allow moderate circumferential flow.

One can compare the relative advantages of the three types of burners. With a can burner, the fuel-to-air ratio can easily be controlled circumferentially. If a particular burner can is suspected of failing during use, one can economically inspect and replace only that component. During testing, only one can needs to be tested, thus reducing the required flow rates and therefore test chamber size and cost. However, when fully assembled, can burners tend to be larger and heavier than the other types. Also, pressure drops are higher (about 7 percent), and each can must contain its own igniter; moreover, the flow into the turbine tends to be less uniform in temperature than for the other two geometries. As a result of these disadvantages, can burners are rarely used on modern large engines. They are of small diameter and have relatively long lengths and good survivability characteristics. They are used extensively in small turboshaft engines with centrifugal compressors.

Cannular burners tend to have smaller cross-sectional areas and thus weigh less. Fewer ignition systems are required because flow can propagate circumferentially. Circumferentially controlling the temperatures is more difficult with cannular burners, but they tend naturally to produce uniform exit temperatures. Also, if one component fails, replacing only that component can be more expensive than for a can burner. Finally, owing to the method by which

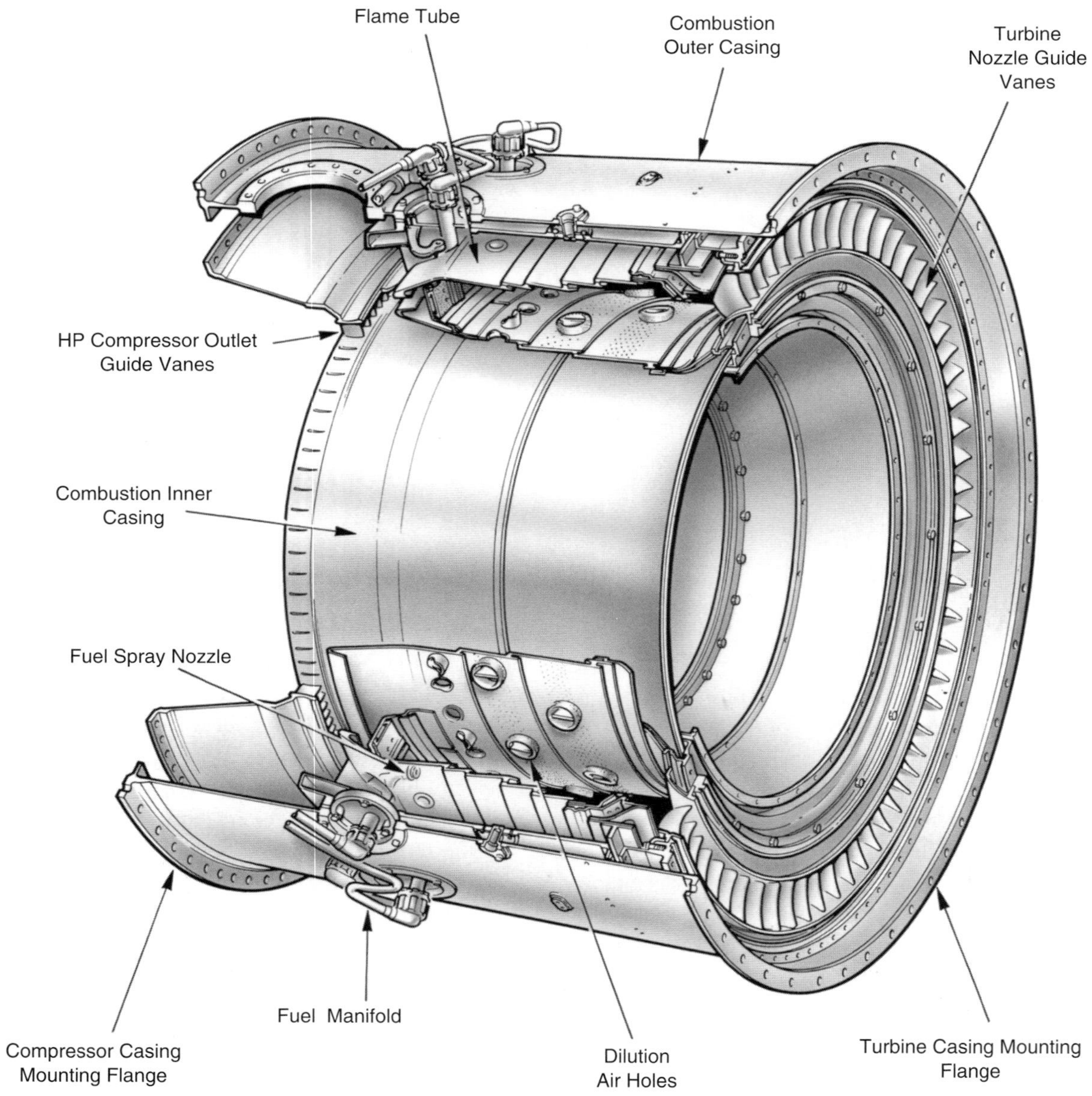

Figure 9.5 A 3-D view of an annular burner (© Rolls-Royce plc 2022. All rights reserved).

the burn areas are connected, thermal growth can cause thermal stresses to be generated. Pressure drops are lower than for can burners (approximately 6 percent). In general, however, cannular burners capture the advantages of the other two types and are used on both turbojets and turbofans or power units.

Annular burners are the simplest, the most compact, and the lightest. They demonstrate the lowest pressure drop (about 5 percent) and feature

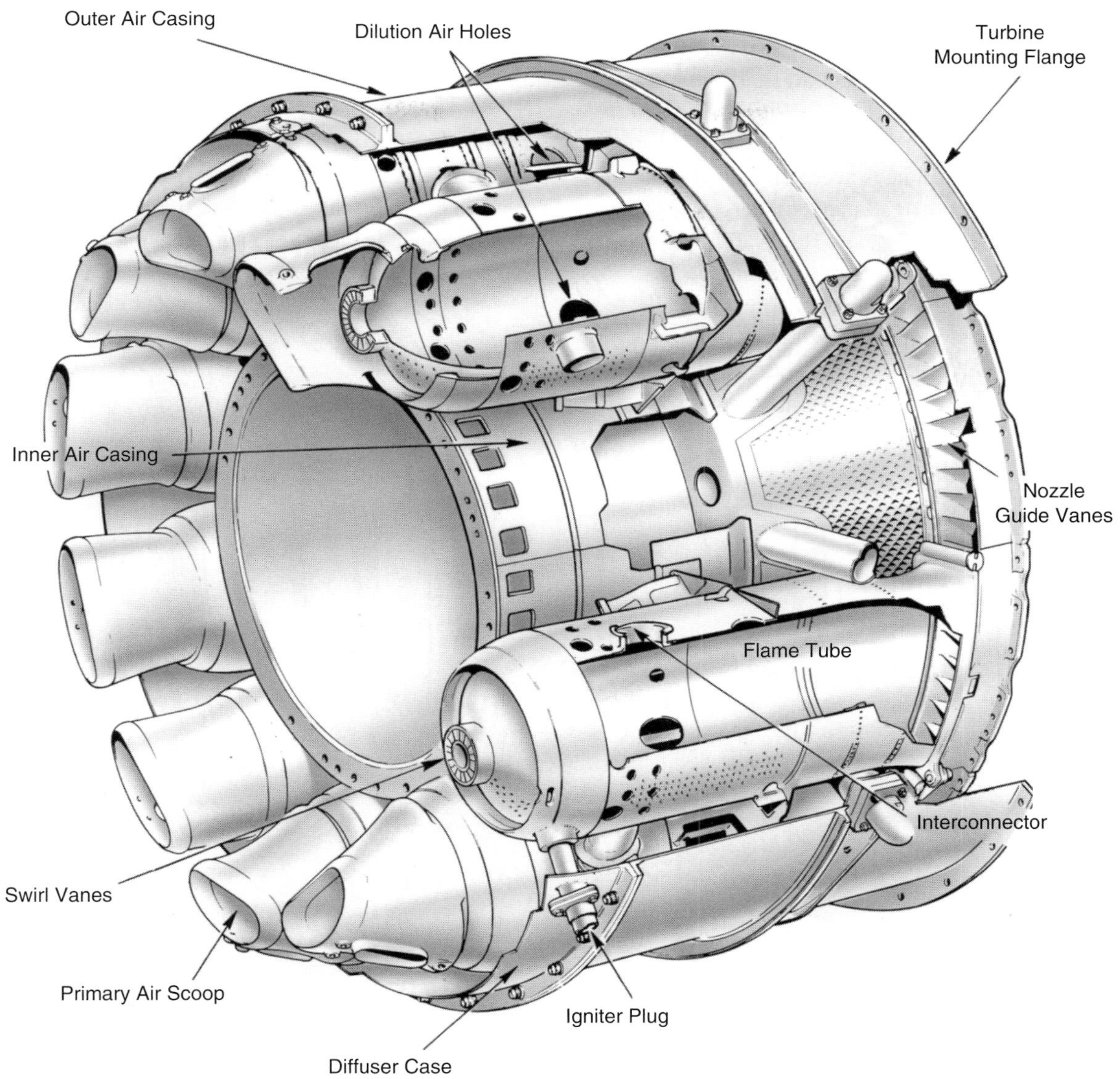

Figure 9.6 A 3-D view of a cannular (can-annular) burner (© Rolls-Royce plc 2022. All rights reserved).

good mixing and high efficiencies. Controlling the circumferential variation of temperature is most difficult with annular burners, but they also tend naturally to produce uniform exit temperatures. Mechanically, they also are not quite as rugged as the other types but have improved dramatically over the past few years as the result of improved materials. These burners are the hardest to service; often if an annular burner needs to be replaced, the entire engine must be removed from an aircraft and disassembled.

Hardware and technology called dry low emissions (DLE) are used on some ground- or marine-based gas turbines. This hardware improves overall efficiency and reduces emissions, particularly, NO_x. These devices result in the air/fuel to premix more efficiently before entering the combustor. These are most often used to retrofit existing combustors and require significant changes in the gas turbine as more space between the compressor and turbine is needed for the retrofit. The GE LM2500 is a good example of this retrofit which becomes the GE LM2500+G4. This is attained by using a triple annular configuration replacing the conventional annular combustor. The burn is thus more uniform throughout the entire configuration. This is shown in Figure 9.7. The Ansoldo Energia GT26 (Figure 2.32) uses two combustion chambers to improve efficiency, decrease emissions, and increase power. Both are annular and the first is before the turbine and partially surrounds the compressor. The second is after the first turbine stage. The first can be considered to be the primary burner and the second to be the reheat burner. This arrangement can also somewhat be considered a type of DLE since the fuel is more gradually added to the combustion process with the idea of decreasing emissions.

Most primary burners are constructed of nickel-based alloys; ceramic composites are developing as good candidates for near future materials. For example, ceramic matrix composites are used for parts

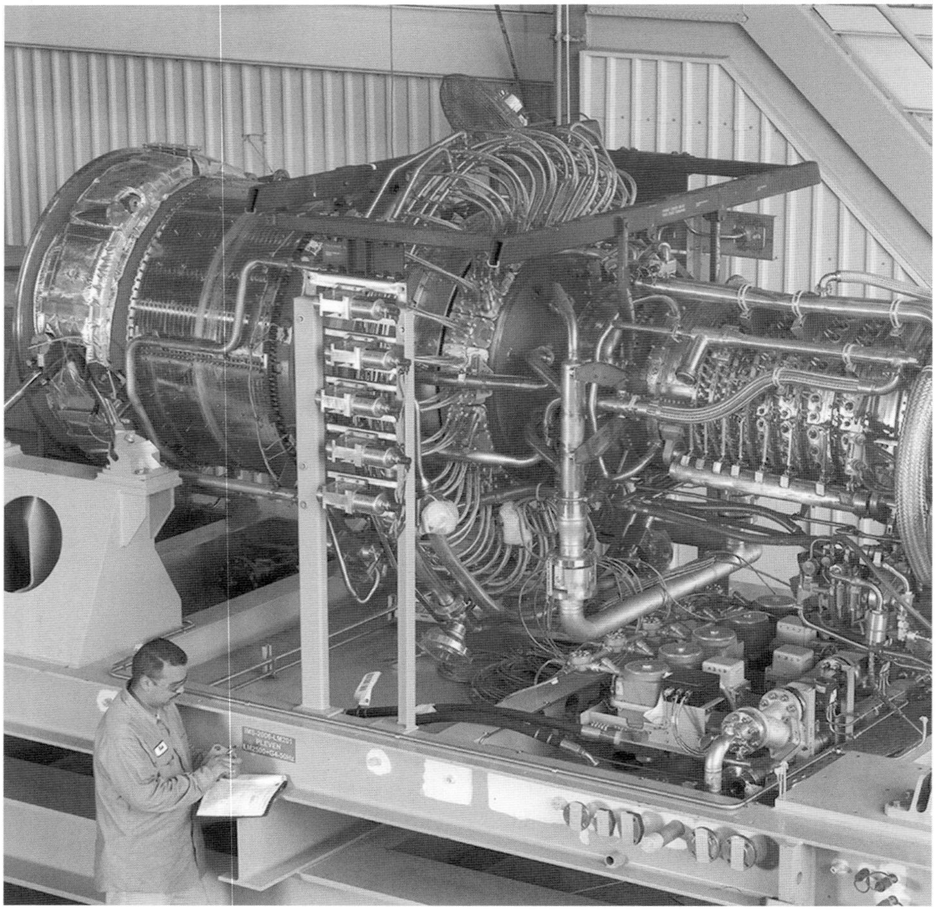

Figure 9.7 GE LM2500+G4 power generation gas turbine with DLE (courtesy of General Electric Gas Power).

of the new GE9X combustor allowing for higher temperatures. Most modern combustors are of the annular design, including the GE CF6, GE90, P&W F100, P&W PW4098, RR RB211, and RR Trent. Burner exit temperatures range from 2200 to 3100 °R. Total pressure ratios are typically 0.94 to 0.96.

9.2.2 Afterburners

As covered in Chapters 2 and 3, the objective of an afterburner (or thrust augmenter) is to increase the jet engine thrust by about a factor of 1.5 to 2.0 for short periods. The *TSFC* also significantly increases (by about a factor of 2.0) during these times. The primary combustors only burn about 25 percent of the air (they operate well below the stoichiometric condition). Thus, the afterburners can burn up to the remaining 75 percent of the initial air. Because afterburners are rarely used (carrier takeoff, emergencies, etc.), the design of such a component is relatively simple. That is, they are only lit for takeoff, climb, sudden accelerations, and maximum speed bursts. However, because they are permanently installed they will impart total pressure losses to the flow even when not in use (this is called the dry condition) and thus will decrease the thrust and increase the *TSFC* of an engine. Afterburners consequently are not designed to be high-turbulence generators. As a result, they do not mix the fuel with the air as well as a primary burner and demonstrate lower burn efficiencies than do primary burners. In fact, when afterburners are in use, a large percentage of the fuel is often burned in or after the nozzle. However, they do not demonstrate pressure drops as large as those encountered in primary burners. Such a configuration is shown in Figure 9.8. Fuel is injected into the flow with a series of radial spray bars, circumferential (annular) spray rings, or both. Fuel is usually transversely injected (namely at 90° to the flow) to enhance atomization and mixing through the shearing action of the freestream gas. Small flameholders are often used. The design of an afterburner is similar to that of a simple ramjet.
A cooling liner is used to feed cool air gradually along the boundary so the case does not overheat. In Figure 9.8, a variable nozzle with flaps, which is integral with the afterburner, is also shown with two nozzle positions.

Some of the particular design concerns are as follows. The flame cannot be too hot or the nozzle (or liner) will be burned. The afterburner should be designed to operate for a relatively wide range of temperatures. This is accomplished by using augmented stages for the fuel injection. Combustion instability (called screech) should be minimized. Screech is periodic high-pressure fluctuations caused by an unsteady release of enthalpy during the combustion process. The intense noise levels are not just undesirable audible signals but can lead to mechanical fatigue of the flameholders, afterburner duct, and nozzle. Usually, a screech liner is used to subdue the fluctuations. Lastly, visual cues such as external flames and excessive smoke should be minimized for military and environmental reasons. Most afterburners are constructed of nickel-based alloys or ceramic composites. Typical afterburner exit temperatures are 3000 to 3700 °R.

9.2.3 Reheat Burners

The best example of a modern reheat combustor is the Ansaldo Energia GT24/GT26 discussed previously, in Figure 2.32, and in Figure 9.9. In general, the reheat burner is a conventional annular burner. The primary burner for this case is also an annular burner partially surrounding the compressor. For this case the flow is turned 180° to a reverse direction and then another 180° to the primary direction with the goal of conserving space.

9.3 Flame Stability, Ignition, and Engine Starting

9.3.1 Flame Speed

The flame speed is very important to a stable combustor. If the gas mixture is moving opposite to the direction of the flame and at the flame velocity, the flame will be stationary. If the gas mixture moves faster than the flame speed, the flame will be

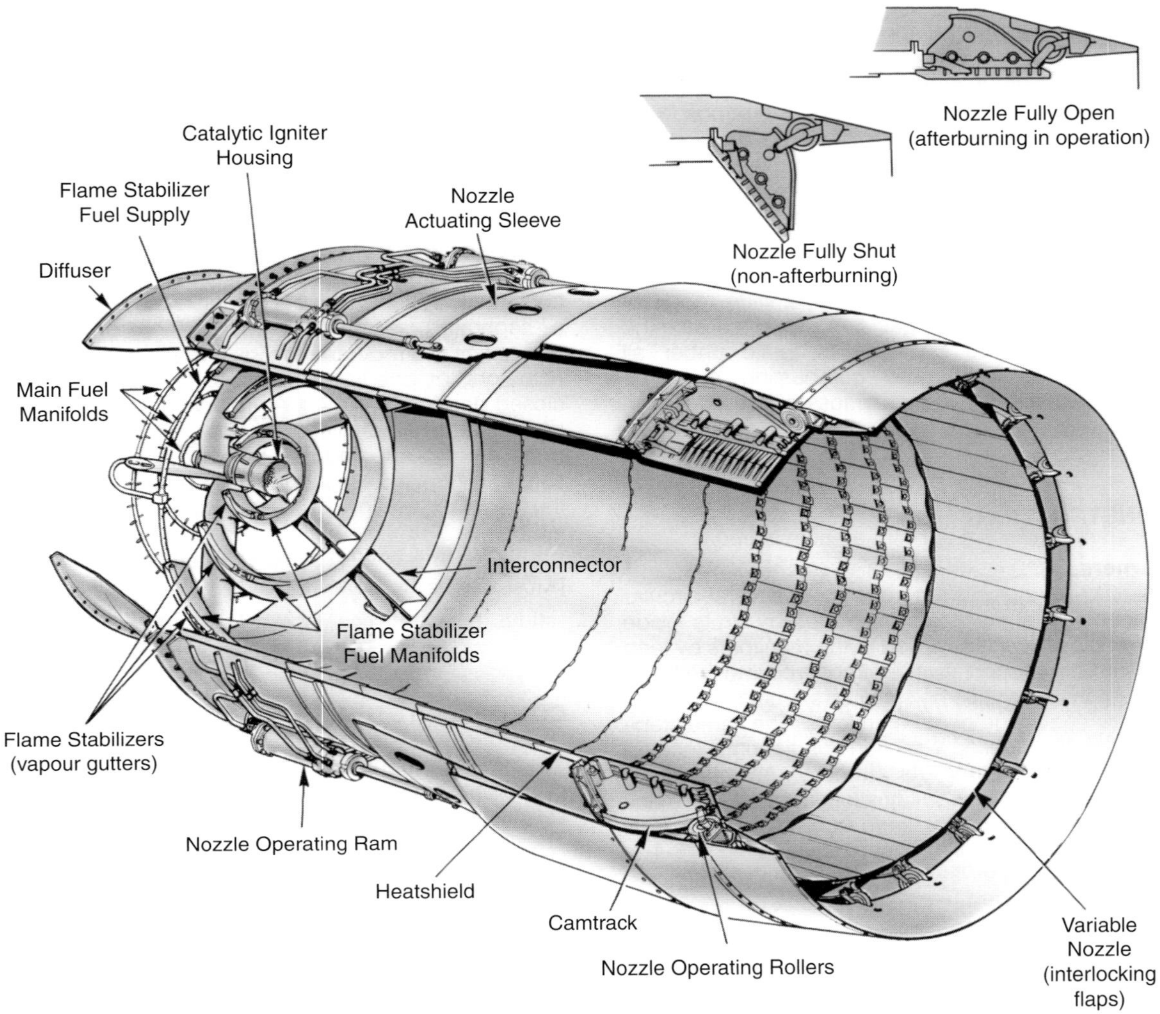

Figure 9.8 A 3-D view of a Rolls-Royce afterburner with nozzle (© Rolls-Royce plc 2022. All rights reserved).

carried out of the burner; that is, the engine will experience flameout. Thus, flow velocities in the burner must be less than the flame speed. Flameholders are used in burners to ensure that velocities in local regions will be less than the flame speed. Note also that, for a mixture that is nearly stoichiometric, the flame speed is almost the highest, and thus again conditions are nearly stoichiometric in the center of the burn zone. At pressures of 1 atm, typical turbulent flame speeds are up to 40 m/s (but the speed depends on the fuel and fuel properties (e.g., diffusivity and volitivity), fuel-to-air ratio, the mixing process, particle sizes, fuel vapor/liquid ratio, and the turbulence parameters), whereas laminar flame speeds are much lower. In general, higher air speeds imply higher turbulence in a burner and thus higher flame speeds. Also, smaller droplet sizes result in a higher flame speeds. Olson, Childs, and Jonash (1955) present such data for a "wide-cut" fuel.

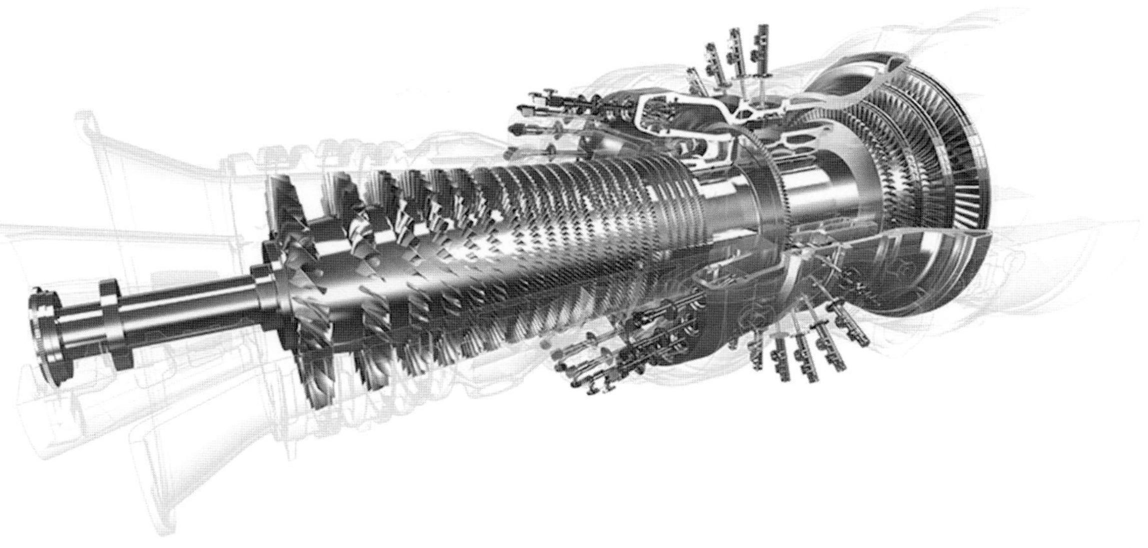

Figure 9.9 Ansaldo Energia GT24/GT26 power generation gas turbine with reheat. (This picture is copyrighted by Ansaldo Energia s.p.a., who owns any related right. Any reproduction is forbidden without the written permission of the owner.)

9.3.2 Flame Stability

As discussed in Section 9.1, a flame must be stable for combustion to be efficient and thrust to be continuous. Specific guidelines, which are somewhat fuel-type dependent, can be used to ensure that the flame will remain in the burner and that combustion can take place stably in the local burn zone in the burner. Two general important and typical graphs that represent guidelines are presented in Figures 9.10 and 9.11.

In Figure 9.10, a generalized "flammability" plot is presented, and Olson et al. (1955) present such data for a gasoline or "wide-cut" fuel. This cross-plot shows a region of pressure and fuel-to-air ratio in which combustion is possible. Note and remember that the typical overall fuel-to-air ratio for a gas turbine is 0.02. Thus, as indicated on this diagram, if the local fuel-to-air ratio in the burner was the same, the overall gas turbine fuel-to-air ratio, combustion would not be possible. From the diagram, it can be seen that the pressure must be high in the burn zone in the combustor, which indicates why high pressure is needed from the compressor, and that the fuel-to-air ratio should be high enough so that the condition falls in the stable region; as is shown in Section 9.4, the stoichiometric value is approximately 0.067. As a result, in the center of the burn zone the fuel-to-air mixture is usually stoichiometric or slightly fuel rich (slightly higher than the stoichiometric ratio).

Lastly, these data sets can be prudently combined to give a generalized set of data for which flame stability can be expected (Oates, 1989). A generalized trend is presented in Figure 9.11 in which the region of possible burning is shown for a cross-plot of the fuel-to-air ratio and the loading parameter $\dot{m}/\mathcal{V}p^n$, where $\mathcal{V}$ is the effective volume of the combustion region and n is empirically determined. For example, if the mass flow rate is too large or the burner size is too small, the velocity will be greater than the flame speed and the loading parameter will be large; thus, the operating condition will fall outside of the stable region. Similarly, if the pressure is too low, the loading parameter will be large and the operating condition again will fall outside of the stable region. The peak loading parameter for stable operation typically occurs for a slightly fuel lean mixture.

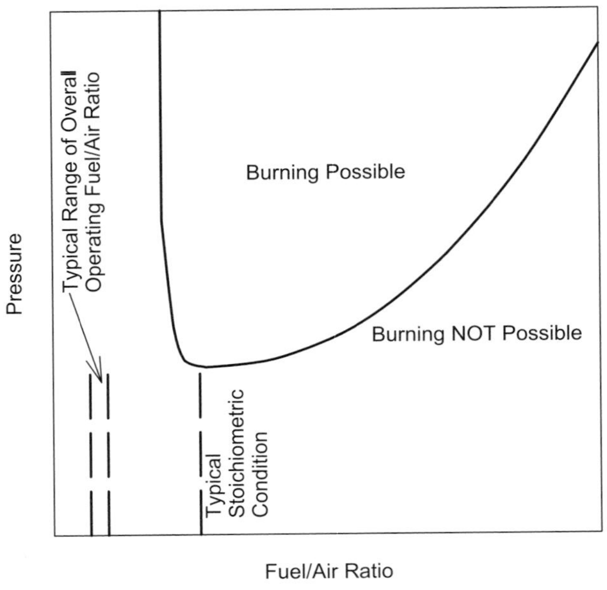

Figure 9.10 Generalized "flammability" map.

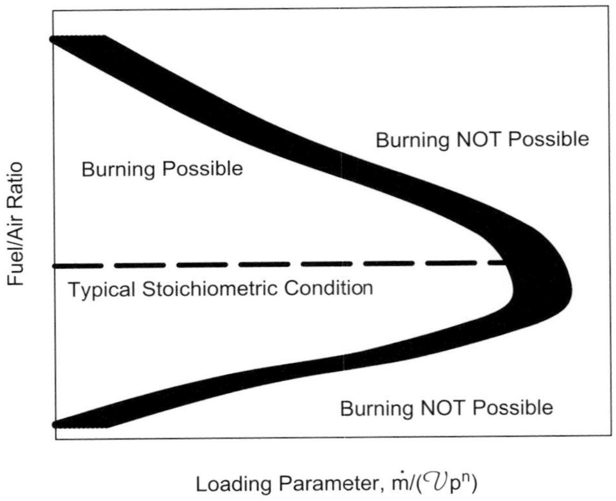

Figure 9.11 Combined flammability and flame speed information.

9.3.3 Ignition and Engine Starting

Igniting a burner is directly tied to starting the gas turbine. That is, if the compressor is not operating, the high-pressure air required by the combustor is not present. The first step in igniting the burner is thus, in actuality, rotating the compressor. For two-spool engines, only the high-pressure compressor is normally driven. Six basic and different techniques are used initially to rotate the compressor and all require high torque levels. They are as follows.

(1) For small engines (turboprops, etc.), a high-torque DC electric starter motor is usually used to spin the shaft with a clutch.
(2) For most larger engines (turbofans, etc.), stored compressed air or high-pressure air from another source is used to drive an air turbine (air motor), which in turn drives the engine shaft with a clutch.
(3) For a few of the larger engines, a small auxiliary gas turbine is used. Such a turbine must also be started, and this is done by using either a pressurized fuel cartridge, which when used expands and burns the gas and thus drives the turbine, or from an electric motor. The Pratt & Whitney Canada PW901A is one such auxiliary gas turbine.
(4) Hydraulic motors driven by a high-pressure hydraulic fluid from a ground source drive the shaft; these are often used on turboshaft engines or power generation gas turbines.
(5) For small engines, a pressurized fuel cartridge, as described above in entry (3), directly drives the turbine.
(6) For a few of the older turbojets, auxiliary compressed air is injected into the high-pressure turbine to impinge on the blades and drive the shaft.
(7) For power generation units an auxiliary electric motor is usually used to rotate the compressor.

After the shaft is first rotated and after the compressor reaches partial speed through an auxiliary motor system, enough high pressure is generated for combustion. Usually, a high-energy spark technique is then utilized in the primary combustion chamber. Fuel is sprayed into the chamber, and a spark is generated by a high-voltage igniter plug to ignite the fuel. Thus, a high-temperature and low-density gas enters the turbine, which is then able to drive the compressor, and the shaft ramps up to full rotational speed.

Ignition of an afterburner is usually quite different because it is not linked to engine starting. One method that is often used is the "hot streak" technique. For this method, extra fuel is briefly injected into one of the primary burner fuel nozzles. The extra fuel burns as it passes through the turbine and into the afterburner region, where it ignites the afterburner fuel. After afterburner ignition, the primary fuel flow is reduced to the normal flow rate. The second method is the torch technique. For this method a "pilot" light is mounted in one of the spray bars. As the name implies, the torch is constantly burning, and when afterburner fuel is injected the torch ignites the fuel. A third method is the spark method, which is similar to that used on primary burners. A fourth method that is sometimes used is a platinum-based element that is catalytic with the fuel and is located downstream of the fuel and mounted on the spray bar.

9.4 Chemistry and Thermodynamics

Because the overall performance of an engine and the life of a turbine are strongly dependent on the burner exit temperature, predicting this temperature is extremely important. Previously (in the cycle analyses, Chapters 2 and 3), a relatively simple method was used to predict the burner performance. For this technique, the fuel heating value was used, and one value was assumed to be valid for all conditions. Furthermore, one specific heat was used for the fuel and air, and the air and fuel were assumed to be at approximately the same temperature. However, this is not true. Also, the air specific heat was previously assumed to be constant in the combustor, which is also not true because of the extreme temperatures. In this section these assumptions are relaxed and the adiabatic flame temperature is found.

9.4.1 Stoichiometric Condition and Equation

First, a review of burner chemistry is in order. This is best accomplished with an example that entails balancing the chemical equation of the burner. All jet fuels are hydrocarbons, and most have characteristics similar to those of kerosene. *n*-Decane ($C_{10}H_{22}$) is a hydrocarbon that accurately represents kerosene. Balancing the stoichiometric chemical equation is necessary before one can proceed to the thermodynamics of the burning process. The stoichiometric condition is often called 100 percent air. The stoichiometric condition has exactly the correct proportions of oxygen and fuel such that all of the oxygen and fuel are consumed and no oxygen no fuel remains in the products. For other conditions, the definition of percent air is the ratio of the mass of air provided to that required at the stoichiometric condition.

Example 9.1 | Stoichiometry of *n*-Decane

If kerosene can be approximated by *n*-decane, balance the chemical equation for the stoichiometric combustion of this fuel in air and find the stoichiometric fuel-to-air ratio.

Thus, the first step in solving the problem is to balance the stoichiometric equation for the fuel. That is, all of the oxygen and fuel are burned so that they do not remain in products. Thus,

$$C_{10}H_{22} + Y\ O_2 + 3.76Y\ N_2 \rightarrow X\ H_2O + Z\ CO_2 + 3.76Y\ N_2, \tag{9.4.1}$$

where air is Y parts oxygen and $3.76Y$ parts nitrogen. Thus, one must find Y, X, and Z to balance the equation. Note that the nitrogen does not actively participate in the stoichiometric reaction. It does, however, enter into the thermodynamics. If each element is considered, one finds the following.

C: $10 = Z$ thus $Z = 10$

H: $22 = 2X$ thus $X = 11$

O: $2Y = X + 2Z$ thus $Y = Z + X/2$ or $Y = 15.5$

The balanced equation is therefore

$$C_{10}H_{22} + 15.5\ O_2 + 3.76(15.5)\ N_2 \rightarrow 11\ H_2O + 10\ CO_2 + 3.76(15.5)\ N_2. \tag{9.4.2}$$

One must remember that the molecular weights $\mathcal{M}$ of C, H, O, and of N are 12, 1, 16, and 14, respectively. This results in a molecular weight for *n*-decane of 142. Also, note that, for a jet engine, the combustor exhaust is all gas; therefore H_2O is not a liquid. Thus, for the stoichiometric condition (100 percent air), one can find the fuel ratio from:

$$f = \frac{\dot{m}_f}{\dot{m}_{air}} = \frac{1(10\mathcal{M}_C + 22\mathcal{M}_H)}{15.5 \times 2\mathcal{M}_O + 3.76 \times 15.5 \times 2\mathcal{M}_N}$$

$$= \frac{1(10 \times 12 + 22 \times 1)}{15.5(2 \times 16 + 3.76 \times 2 \times 14)} = \frac{142}{2127.8} = 0.0667.$$

This value is much higher than that for overall engine operation. In fact, if this ratio were burned in the primary combustor, the exit temperature would be far too hot (see Example 9.4) and the turbine would be rapidly destroyed. Thus, most engines run fuel lean (i.e., more than 100 percent air) to reduce the turbine temperatures, although as discussed earlier, in local burn zones the combustors are fuel rich.

9.4.2 Full Chemical Equation: Dissociation, Including Emissions

Above the basic stoichiometric chemical equation for kerosene was considered. In actuality the process is much more complex due to dissociation of the compounds, and the chemical equation can be written

$$\begin{aligned}&C_{10}H_{22} + X_1O_2 + 3.76X_1N_2\\ &\quad \rightarrow X_2H_2O + X_3CO_2 + X_4N_2 + X_5CO \ + X_6H_2 + X_7H\\ &\quad + X_8OH \ + X_9O_2 + X_{10}O \ + X_{11}NO\\ &\quad + X_{12}NO_2 + X_{13}C_{10}H_{22}. \qquad (9.4.3)\end{aligned}$$

Here, X_1 through X_{13} are constants and are to be determined by a chemical balance and equilibrium analysis. Even so, this is not a complete equation as other products and reactants are present due to impurities in fuels; for example, sulfur in the fuel can result in SO_2 and SO_3 in the products. It is important to note that of the above CO, NO, and NO_2 (NO_x) are considered pollutants and emissions. The leftover fuel product $C_{10}H_{22}$ is considered a pollutant as well and is termed an unburned hydrocarbon (UHC). The products, H_2O and CO_2, are also considered to be greenhouse gases. It should be noted that this discussion applies to any hydrocarbon fuel, not only kerosene. Typical emission levels and restrictions are experienced for both aircraft and power generation applications and are forever changing and will not be addressed herein. As a reference the largest emissions are CO_2, H_2O, and NO_x in that order, with the first two much larger than the latter.

9.4.3 Chemical Equilibrium

To find the coefficients in Eq. (9.4.3), both a chemical balance and equilibrium calculation are needed. The latter is a complicated method and will only be introduced in this text. It is covered in many advanced thermodynamic texts, including Wark and Richards (1999).

First consider the simple chemical reaction starting with v mole of each constituent, A, B, C, and D:

$$\nu_A A + \nu_B B \leftrightharpoons \nu_C C + \nu_D D, \qquad (9.4.4)$$

where ν_A, ν_B, ν_C, and ν_D are the stoichiometric constants for the nondissociating reaction.

Now for the case where disassociation has occurred, and the system is at equilibrium, we find that

$$K_p = \frac{N_C^{\nu_C} \times N_D^{\nu_D}}{N_A^{\nu_A} \times N_B^{\nu_B}} \left(\frac{P}{N_T}\right)^{\Delta\nu}, \qquad (9.4.5)$$

where N_C, etc., are the number of moles for each constituent in the equilibrium state, and

$$\Delta\nu = \nu_C + \nu_D - \nu_A - \nu_B \qquad (9.4.6)$$

and

$$N_T = N_A + N_B + N_C + N_D. \qquad (9.4.7)$$

K_p is the well-established equilibrium constant for the reaction; it is listed in many tables for the different reactions and is pressure independent. It is based on the partial pressures of each of the constituents. Also, P is the pressure in atmospheres.

Now, considering Eq. (9.4.3), there are several dissociation reactions occurring simultaneously. For example, $O_2 + N_2 \leftrightharpoons \frac{1}{2}\,NO$ and $CO_2 \leftrightharpoons CO + \frac{1}{2}\,O_2$ are only two such equilibrium reactions. So clearly this becomes a rather complex calculation with many simultaneous considerations, as described by Reynolds (1986). The purpose of this section is not to provide such a solution but simply to introduce the concepts of chemical equilibrium as applied to gas turbines.

A simple partial solution is in the following example.

Example 9.2 | Equilibrium of $H_2O \leftrightharpoons OH + \frac{1}{2}\,H_2$,

Consider the equilibrium equation $H_2O \leftrightharpoons OH + \frac{1}{2}\,H_2$, which is one of the equilibrium considerations in Eq. (9.4.3). Find the mole composition of the different components at operating temperatures of 2500 K and 3500 K at one atmosphere. The equilibrium constants are 0.00530 and 0.2805, respectively.

Solution

First the reaction becomes

$$H_2O \rightarrow n_{OH}OH + n_{H_2}H_2 + n_{H_2O}H_2O.$$

So the goal is to find the quantities n_{OH}, $2n_{H_2}$, and n_{H_2O}.

Using a balance on H and O:

$$\text{H: } 2 = n_{OH} + 2n_{H_2} + 2n_{H_2O}$$

$$\text{O: } 1 = n_{OH} + n_{H_2O}.$$

So that $n_{OH} = 1 - n_{H_2O}$
and

$$n_{H_2} = \frac{1 - n_{H_2O}}{2} = \frac{n_{OH}}{2}.$$

Now the equilibrium balance is

$$K_p = \frac{n_{OH}\, n_{H_2}^{\frac{1}{2}}}{n_{H_2O}} \left(\frac{P}{n_{OH} + n_{H_2} + n_{H_2O}} \right)^{1+\frac{1}{2}-1},$$

which results in $K_p = \dfrac{(n_{OH}^3/2)^{\frac{1}{2}}}{1 - n_{OH}} \left(\dfrac{1}{1 + n_{OH}/2} \right)^{\frac{1}{2}}.$

For 2500 K:

$$0.00530 = \frac{(n_{OH}^3/2)^{\frac{1}{2}}}{1 - n_{OH}} \left(\frac{1}{1 + n_{OH}/2} \right)^{\frac{1}{2}}.$$

This results in $n_{OH} = 0.0378$, $n_{H_2} = 0.0189$, and $n_{H_2O} = 0.9622$.

Thus, there are 0.0378 moles of OH, 0.0189 moles of H_2 and 0.9622 moles of H_2O (mostly H_2O).

For 3500 K:

$$0.2805 = \frac{(n_{OH}^3/2)^{\frac{1}{2}}}{1 - n_{OH}} \left(\frac{1}{1 + n_{OH}/2} \right)^{\frac{1}{2}}.$$

This results in $n_{OH} = 0.405$, $n_{H_2} = 0.2025$, and $n_{H_2O} = 0.595$.

And there are 0.405 moles of OH, 0.2025 moles of H_2 and 0.595 moles of H_2O.

If the pressure is increased to 2 atm, now $P = 2$, and there are 0.342 moles of OH, 0.171 moles of H_2 and 0.658 moles of H_2O.

As can be seen for this particular equilibrium equation, as the temperature rises, more dissociation takes place and as the pressure increases less dissociation takes place.

9.4.4 Adiabatic Flame Temperature

Now that the chemistry of the process has been reviewed, one can use thermodynamics to analyze the burner. For the analysis herein, disassociation of any of the products and resulting equilibrium will be ignored. Although, at high temperatures, these effects are measurable on temperatures and products, including these effects would create undue complication. Furthermore, these effects are secondary as compared with the effects of using adiabatic flame temperature and as compared with the simple heating value analysis. The reaction will be assumed to go to completion without any equilibrium considerations. Equilibrium effects can be included using the analysis of Reynolds (1986). Thus, applying the first law of thermodynamics to a burner yields

$$\dot{Q} + \dot{W} = \Delta\dot{h} + \Delta\dot{K}E + \Delta\dot{P}E. \tag{9.4.8}$$

In the absence of work, heat transfer, and potential energy changes, this becomes

$$\Delta h_t = 0 \tag{9.4.9}$$

or

$$\sum(N_i h_{ti})_{\text{products}} = \sum(N_i h_{ti})_{\text{reactants}}, \tag{9.4.10}$$

where h_{ti} is the total enthalpy of a constituent i, and N_i is the number of moles of constituent i. Now, for a given substance at any temperature T, the specific static enthalpy is

$$h_{iT} = \Delta h^{\circ}_{\text{if}} + (h_T - h_{298})_i, \tag{9.4.11}$$

where $\Delta h^{\circ}_{\text{if}}$ is the enthalpy of formation of the substance at the standard state (298 K), h_T is the enthalpy at temperature T, and h_{298} is the enthalpy at 298 K. For a substance from which heat is released upon formation, the enthalpy of formation is negative. The enthalpy of formation is based on the fact that both the substance and its constituents are in given states during the formation of the substance. The enthalpy of formation is fixed for a given substance and state and is not the same as the heat of combustion. One can obtain $(h_T - h_{298})_i$ either by real gas tables (preferred) or approximately by

$$(h_T - h_{298})_i = \int_{298}^{T} c_{pi}\, dT. \tag{9.4.12}$$

Sometimes the specific heat can be approximated over a given (but relatively small) temperature range by

$$c_{pi} = a_i + b_i T. \tag{9.4.13}$$

The method can be used to find the heating value or heat of combustion, which was used in Chapters 2 and 3 for a combustion analysis. If both the products and reactants are gases, as in a jet engine, the value is called the lower heating value. Thus, the heat released is termed the heating value (at any temperature T); it is defined and found by

$$\Delta H_T = -\sum(N_i(\Delta h^{\circ}_{\text{if}} + (h_T - h_{298})_i))_{\text{products}} + \sum(N_i(\Delta h^{\circ}_{\text{if}} + (h_T - h_{298})_i))_{\text{reactants}}. \tag{9.4.14}$$

For example, if both the products and reactants are at the standard temperature, the heating value at the standard conditions is found by

$$\Delta H_{298} = -\sum(N_i \Delta h^{\circ}_{\text{if}})_{\text{products}} + \sum(N_i \Delta h^{\circ}_{\text{if}})_{\text{reactants}}. \tag{9.4.15}$$

These computations can be performed for any compound, and fortunately the heating value does not change greatly with temperature. One needs to know the standard enthalpies of formation for the different constituents for such computations. These are usually found in JANAF (Chase, 1998)) or equivalent tables. In Table 9.1, a few important thermodynamic parameters for some selected compounds are presented. The values of a and b have been evaluated over typical temperature ranges.

As was shown in Example 9.4 and Auxiliary Example 9.4.1, as the fuel ratio decreases, the accuracy of the simple analysis increases. The main reason for this improvement is that, for the simple analysis, all of the gas is assumed to be air. For the adiabatic flame temperature analysis, the incoming gas is air, but the exiting gas is water vapor, carbon dioxide, nitrogen, and

Table 9.1 **Thermodynamic parameters for selected compounds.**

Compound	Δh°_{if}(kJ/kg-mole)	c_p (298 K)(kJ/kg-mole-K)	a (kJ/kg-mole-K)	b (kJ/kg-mole-K^2)
CO	−110,530	29.1	27.4	0.0058
CO_2	−393,520	37.1	28.8	0.0280
H_2O	−241,820	33.6	30.5	0.0103
H_2	0	28.9	28.3	0.0019
N_2	0	29.1	27.6	0.0051
O_2	0	29.4	27.0	0.0079
C_4H_{10} (g)	−126,150	98.0	35.6	0.2077
C_8H_{18} (l)	−249,950	254	254	0
$C_{10}H_{22}$ (l)	−294,366	296	296	0

Example 9.3 | Heating Value of *n*-Decane

Find the heating value of liquid *n*-decane at 298 K based on the enthalpies of formation.

First, one needs the enthalpies of formation of all of the reactants and products:

$H_2O(g): \Delta h^{\circ}_{if} = -241,820$ kJ/kg-mole

O_2 (g): $\Delta h^{\circ}_{if} = 0$

N_2 (g): $\Delta h^{\circ}_{if} = 0$

CO_2 (g): $\Delta h^{\circ}_{if} = -393{,}520$ kJ/kg-mole

$C_{10}H_{22}$ (l): $\Delta h^{\circ}_{if} = -294{,}366$ kJ/kg-mole.

Thus, recalling that the molecular weight is 142, one can find the heating value of *n*-decane at 298 K through the following calculation:

$$\Delta H_{298} = -N_{H_2O}\Delta h^{\circ}_{H_2O\,f} - N_{CO_2}\Delta h^{\circ}_{CO_2\,f} + N_{C_{10}H_{22}}\Delta h^{\circ}_{C_{10}H_{22}\,f}$$

$$\begin{aligned}\Delta H_{298} &= -11(-241,820) - 10(-393,520) + 1(-294,366)\\ &= 6{,}300{,}854 \text{ kJ/kg-mole}\\ &= 6{,}300{,}854 \text{ kJ/kg-mole}/(4.186\,\text{J/cal} \times 142\,\text{kg/kg-mole})\\ &= 10{,}600 \text{ cal/g}\\ &= 19{,}080 \text{ BTU/lb}_m.\end{aligned}$$

Heating values are used to approximate the real combustion process, as is done in the earlier chapters of this book. However, it is best to use the adiabatic flame temperature to better predict the operating temperature of a combustor. An example is presented to demonstrate the method. In Table 9.2, a listing of the lower heating values of some different fuels is presented. One should note that *n*-decane closely represents Jet-A, JP-4, and JP-5.

Table 9.2 **Heating values of a few selected fuels.**

Name	Formula	Lower heating value (cal/g)
n-Octane(l)	C_8H_{18}	10,611
n-Nonane(l)	C_9H_{20}	10,587
n-Decane(l)	$C_{10}H_{22}$	10,567
Jet A(l)	$CH_{1.94}$	10,333
JP-4(l)	$CH_{2.0}$	10,389
JP-5(l)	$CH_{1.92}$	10,277
Natural gas (g)	Exact C/H ratio varies	9,700–12,200
Isobutane(g)	C_4H_{10}	10,897
Propane	C_3H_8	11,020
Methane	CH_4	11,910

Example 9.4 Stoichiometric Temperature of Burning *n*-Decane

Kerosene, which can be approximated as liquid *n*-decane, is to be burned in a combustor. The inlet total temperature of the fuel is $T_{t1} = 444.44$ K (800 °R), and the fuel burns with 100 percent air (stoichiometric condition, $f = 0.0667$, Example 9.1), which is at $T_{t2} = 666.67$ K (1200 °R). What is the exit total temperature if dissociation of the compounds is negligible?

Thus, using the stoichiometrically balanced chemical equation from Example 9.1 and the energy equation (Eq. (9.4.10)), in which oxygen and fuel are not a part of the products, one can write

$$11[\Delta h_f^\circ + (h_{Tt} - h_{298})]_{H_2O} + 10[\Delta h_f^\circ + (h_{Tt} - h_{298})]_{CO_2} + 3.76 \times 15.5[\Delta h_f^\circ + (h_{Tt} - h_{298})]_{N_2}$$

$$= 1[\Delta h_f^\circ + (h_{Tt_1} - h_{298})]_{C_{10}H_{22}} + 15.5[\Delta h_f^\circ + (h_{Tt_2} - h_{298})]_{O_2} + 3.76 \times 15.5\ [\Delta h_f^\circ + (h_{Tt_2} - h_{298})]_{N_2}.$$

Thus, one must find T_t. However, to find T_t, it must be possible to evaluate $(h_{Tt} - h_{298})_i$ for each product. This can best be done with tables. For simplicity, however, each term will be evaluated based on

$$(h_{Tt} - h_{298})_i = \int_{298}^{T} c_{pi} dT.$$

And, for this example, each c_{pi} will be assumed to approximate the form $c_{pi} = a_i + b_i T$.

Thus,

$$(h_{Tt} - h_{298}) = \left[a_i T + b_i \frac{T^2}{2}\right]_{298}^{T_t} = a_i T_t + b_i \frac{T^2}{2} - a_i(298) - b_i \frac{298^2}{2}$$

Thus, for each constituent

$$h_i = \Delta h^{\circ}_{if} + (h_{Tt} - h_{298}) = \Delta h^{\circ}_{if} + a_i T_t + b_i \frac{T^2}{2} - a_i(298) - b_i \frac{298^2}{2}.$$

Using Table 9.1, one finds the following for each constituent:

$C_{10}H_{22}$:	$a = 296,\ b = 0.000$
O_2:	$a = 27.0,\ b = 0.0079$
N_2:	$a = 27.6,\ b = 0.0051$
H_2O:	$a = 30.5,\ b = 0.0103$
CO_2:	$a = 28.8,\ b = 0.0280,$

where the resulting c_p is in kJ/kg-mole-K.

Thus, the total enthalpies of each of the reactants are

$$h_{tC_{10}H_{22}} = -294{,}366 + 296 \times 444 + 0 - 296 \times 298 - 0$$
$$= -251{,}150\ \text{kJ/kg-mole}$$

$$h_{tO_2} = 0 + 27 \times 667 + 0.0079 \frac{667^2}{2} - 27 \times 298 - 0.0079 \frac{298^2}{2}$$
$$= 11{,}370\ \text{kJ/kg-mole}$$

$$h_{tN_2} = 0 + 27.6 \times 667 + 0.0051 \frac{667^2}{2} - 27.6 \times 298 - 0.0051 \frac{298^2}{2}$$
$$= 11{,}092\ \text{kJ/kg-mole.}$$

Thus,

$$\sum (N_i h_{ti})_{\text{reactants}} = N_{C_{10}H_{22}} h_{tC_{10}H_{22}} + N_{O_2} h_{tO_2} + N_{N_2} h_{tN_2}$$
$$\sum (N_i h_{ti})_{\text{reactants}} = 1(-251{,}150) + 15.5(11{,}370) + 15.5 \times 3.76(11{,}092)$$
$$= 571{,}527\ \text{kJ.}$$

Now one must find the total enthalpies of each of the products. Thus,

$$h_{tH_2O} = -241{,}820 + 30.5 \times T_{t\ \text{products}} + 0.0103 \frac{T^2_{t\ \text{products}}}{2} - 30.5 \times 298$$
$$- 0.0103 \frac{298^2}{2}\ \text{kJ/kg-mole}$$

$$h_{tCO_2} = -393{,}520 + 28.8 \times T_{t\ \text{products}} + 0.028 \frac{T^2_{t\ \text{products}}}{2}$$
$$- 28.8 \times 298 - 0.028 \frac{298^2}{2}\ \text{kJ/kg-mole}$$

$$h_{tN_2} = 0 + 27.6 \times T_{t\ \text{products}} + 0.0051 \frac{T^2_{t\ \text{products}}}{2} - 27.6 \times 298 - 0.0051 \frac{298^2}{2}\ \text{kJ/kg-mole.}$$

Because the equation is essentially quadratic with only one unknown variable due to the assumed form of the specific heats, one could solve directly for $T_{t\ products}$. However, in general, the problem is iterative if tables are used. Thus, the iterative approach is used for this example. First, the total exit temperature is assumed to be 2531 K, which will be checked later. Thus, the total enthalpies of the products are

$$h_{tH_2O} = -241{,}820 + 30.5 \times 2531 + 0.0103\frac{2531^2}{2} - 30.5 \times 298 - 0.0103\frac{298^2}{2}$$
$$= -141{,}180 \text{ kJ/kg-mole}$$

$$h_{tCO_2} = -393{,}520 + 28.8 \times 2531 + 0.028\frac{2531^2}{2} - 28.8 \times 298 - 0.028\frac{298^2}{2}$$
$$= -240{,}769 \text{ kJ/kg-mole}$$

$$h_{tN_2} = 0 + 27.6 \times 2531 + 0.0051\frac{2531^2}{2} - 27.6 \times 298 - 0.0051\frac{298^2}{2}$$
$$= 77{,}740 \text{ kJ/kg-mole.}$$

Thus,

$$\sum(N_i h_i)_{\text{products}} = N_{H_2O}h_{tH_2O} + N_{CO_2}h_{tCO_2} + N_{N_2}h_{tN_2}$$

$$\sum(N_i h_i)_{\text{products}} = 11(-141{,}180) + 10(-240{,}769) + 15.5 \times 3.76(77{,}740)$$
$$= 570{,}127 \text{ kJ.}$$

Therefore, $\sum(N_i h_{ti})_{\text{products}} \cong \sum(N_i h_{ti})_{\text{reactants}}$ and a good guess was made on T_t. Thus, the adiabatic flame temperature is $T_t = 2531$ K (4556 °R), which is quite high (see Example 9.1). For a second, less accurate method, one can also solve for the total temperature using the simpler, earlier heating value method and compare the results with those for the adiabatic flame temperature. That is, if one assumes that all of the gas into the burner is air and all of the gas out of the burner is air,

$$\dot{m}_f \Delta H = \dot{m} c_p (T_{t4} - T_{t3}) + \dot{m}_f c_p T_{t4},$$

or

$$T_{t4} = \frac{f\Delta H + c_p T_{t4}}{(1 + f)c_p}.$$

For n-decane, $\Delta H = 10{,}567$ cal/g (44,230 kJ/kg); thus,

$$T_{t4} = \frac{0.0667 \times 44{,}230 + 666.67c_p}{1.0667c_p} = \frac{2766}{c_p} + 625.0.$$

The value of c_p is a function of the average temperature, and so the problem is iterative. Iterating, one finds

$T_{t4} = 2778$ K and $c_p = 1.284$ kJ/kg-K.

Thus, this total temperature is in error by 247 K, as well as the temperature rise, and the percent error in the temperature rise is 247/(2531 − 667) = 13.2 percent.

Auxiliary Example 9.4.1 | *n*-Decane Burning with 300 Percent Air

If this exercise were to be repeated for 300 percent air (three times the air needed for the stoichiometric condition, $f = 0.0222$), it would be necessary to include the unburned oxygen and added nitrogen in the products in both the chemical and thermodynamic equations. For example, the chemical equation for 300 percent air becomes

$$C_{10}H_{22} + 3 \times 15.5\ O_2 + 3 \times 3.76(15.5)\ N_2 \rightarrow 11\ H_2O + 10\ CO_2 + 3 \times 3.76(15.5)\ N_2 + 2 \times 15.5\ O_2.$$

Because the excessive oxygen and more nitrogen are heated during the combustion process, the final temperature will be considerably lower, resulting in an adiabatic flame temperature of 1453 K. The simple heating value analysis would yield $T_{t4} = 1489$ K ($c_p = 1.147$ kJ/kg-K). This is an error of 36 K or a percent error in the temperature rise of only 4.6 percent. Details are left as an exercise to the reader. Solutions can also be obtained using the software Kerosene.exe.

oxygen. As the fuel ratio decreases, however, more of the exiting gas becomes air in the adiabatic flame temperature method. Thus, as f decreases, the assumption of pure air in the simple analysis becomes increasingly reasonable. Furthermore, when the fuel ratio is large, and the resulting adiabatic flame temperature is large, significant dissociation of the compounds into elements can occur. This reduces the adiabatic flame temperature. Thus, for a second reason, as f decreases the simple method becomes more accurate.

9.5 Pressure Losses

Undesired losses in total pressure occur in the primary burners and afterburners. These losses are due to three different effects. First, irreversible and nonisentropic heating or combustion of the gas takes place; this is essentially Rayleigh line flow. Second, friction losses occur due to the cans and liners being scrubbed by the gas as it flows through small orifices; this is essentially Fanno line flow. Third, drag mechanisms in the form of flameholders are used that generate turbulence and separation, which are both dissipative mechanisms, again increasing the entropy of the flow. Each effect is addressed separately in this section and then all three effects are discussed in a more general approach.

9.5.1 Rayleigh Line Flow

As discussed in the previous paragraph, several mechanisms exist for pressure drops and total pressure losses in burners. The first to be analyzed is heat addition/removal to the flow. This is analyzed based on Rayleigh line flow, which is flow with heat addition/removal but no friction. For a burner heat is obviously added. The process is a constant area process. The flow is analyzed with a control volume approach in Appendix C.6 and reviewed in Section 2.2.6. Ten independent equations are derived, and these equations contain 14 variables. Therefore, if one specifies four variables, one can solve for the remaining 10. Typically, the inlet conditions M_1, p_{t1}, T_{t1} are known as well as the exit condition T_{t2}, and thus the remaining exit conditions can be found.

9.5.2 Fanno Line Flow

Again, several mechanisms exist for pressure drops and total pressure losses to occur in burners.
The second cause to be analyzed is due to friction in the flow. This is analyzed based on Fanno line flow, which is flow with friction but no heat transfer. The process is a constant area process. The flow is analyzed with

Example 9.5 | Flow with a Temperature Increase (Rayleigh Line Flow)

Flow with an incoming Mach number of 0.3 and an inlet total pressure of 300 psia is heated from a total temperature of 1200 °R to 2400 °R. What is the total pressure ratio if the specific heat ratio is 1.30?

Solution

From the Rayleigh line tables (or using Rayleigh.exe) for a Mach number of 0.3, the temperature and pressure ratios can be found as follows:

$$\frac{T_{t1}}{T_t^*} = 0.3363$$
$$\frac{p_{t1}}{p_t^*} = 1.1909.$$

Therefore, the total pressure and temperature at the sonic condition, which is not present but used as a reference, are

$$p_t^* = 300/1.1909 = 251.9 \text{ psia}$$
$$T_t^* = 1200/0.3363 = 3568 \text{ °R}.$$

Thus, the exit temperature ratio is

$$\frac{T_{t2}}{T_t^*} = \frac{2400}{3568} = 0.6726.$$

From the tables (or using Rayleigh.exe) for this total temperature ratio,

$$M_2 = 0.495.$$

And for this Mach number,

$$\frac{p_{t2}}{p_t^*} = 1.1131;$$

thus, the total pressure at the exit is

$$p_{t2} = 1.131 \times 251.9 = 280.4 \text{ psia}.$$

Finally, the total pressure ratio due only to heat addition (Rayleigh line flow) is

$$\pi = \frac{p_{t2}}{p_{t1}} = \frac{280.4}{300} = 0.935.$$

a control volume approach in Appendix C.5 and reviewed in Section 2.2.5. One now has 11 equations and 17 variables, and so six must be specified. For example, if the inlet conditions M_1, T_1, and p_{t1} and duct parameters L, D, and f are specified, one can find the exit conditions, including p_{t2}.

Example 9.6 Flow with Friction (Fanno Line Flow)

Flow with an incoming Mach number of 0.3 and an inlet total pressure of 300 psia enters a channel 18 in. long with a 5 in. equivalent diameter and a Fanning friction factor of 0.040. What is the total pressure ratio if the specific heat ratio is 1.30?

Solution

From the tables (or using Fanno.exe) for a Mach number of 0.30, one can find the nondimensional length-to-diameter and pressure parameters

$$\left.\frac{4f\,L^*}{D}\right]_1 = 5.7594$$

and

$$\frac{p_{t1}}{p_t^*} = 2.0537;$$

thus, one can find the total pressure at the sonic condition, which is a reference condition (that might not occur in the actual conditions) as follows:

$$p_t^* = 300/2.0537 = 146.2 \text{ psia}.$$

Next, from the duct geometry one can find the parameter

$$\left.\frac{4f\,L}{D}\right]_{1-2} = \frac{4 \times 0.040 \times 18}{5} = 0.576;$$

thus, one can find the corresponding length-to-diameter parameter at the exit:

$$\left.\frac{4f\,L^*}{D}\right]_2 = \left.\frac{4f\,L^*}{D}\right]_1 - \left.\frac{4f\,L}{D}\right]_{1-2} = 5.7594 - 0.576 = 5.1834,$$

and so from the tables (or using Fanno.exe), the exit Mach number can be found as follows:

$$M_2 = 0.312.$$

Therefore, owing to friction the Mach number increased about 10 percent. Next, from the tables (or using Fanno.exe) at this Mach number, one finds that

$$\frac{p_{t2}}{p_t^*} = 1.9829,$$

and so the exit total pressure can be found by

$$p_{t2} = 1.9829 \times 146.2 = 289.8 \text{ psia}.$$

Hence, the pressure ratio for the burner due only to friction (Fanno line flow) is

$$\pi = \frac{p_{t2}}{p_{t1}} = \frac{289.8}{300} = 0.966.$$

9.5.3 Flow with a Drag Object

When an obstruction is in the flow, the total pressure is also reduced because of the friction from the added turbulence and separation. Many primary burners and most afterburners contain such obstructions in the form of flameholders to keep the local velocity less than the flame speed. Thus, a method of including such losses is needed. In Appendix C.9 and reviewed in Section 2.2.9, a method is described in which a control volume approach is used for a constant area flow and uniform properties are assumed at the entrance and exit. In summary, one has 16 variables and 11 equations. Thus, if five variables are specified the problem can be solved. Typically, the inlet conditions M_1, p_{t1}, and T_{t1}, and flameholder geometry parameters A_d/A, and C_d, are known, and thus the exit conditions can be found, including the total pressure.

Example 9.7 | Flow with a Flameholder

Flow with an incoming Mach number of 0.30, an inlet total pressure of 2069 kPa (300 psia), and a total temperature of 666.7 K enters a channel with a flameholder. The drag coefficient of the flameholder is 0.5 and it occupies 10 percent of the total flow area. What is the total pressure ratio if the specific heat ratio is 1.30?

Solution

First, for the given inlet Mach number and total conditions, one can find the static conditions

$$T_1 = 657.8 \text{ K}$$
$$p_1 = 1952 \text{ kPa},$$

and so for the sound speed at the inlet,

$$a_1 = 495.3 \text{ m/s},$$

and from the Mach number the inlet velocity is

$$V_1 = 148.6 \text{ m/s}.$$

Also, from the ideal gas equation for the temperature and pressure, above the inlet density is

$$\rho_1 = 10.34 \text{ kg/m}^3,$$

and so from Eq. (C.9.8),

$$p_1 + \rho_1 V_1^2 - \frac{1}{2} C_d \rho_1 V_1^2 \frac{A_d}{A} = \rho_1 V_1 \left[\frac{\mathcal{R} T_2}{\sqrt{2c_p(T_{t1} - T_2)}} + \sqrt{2c_p(T_{t1} - T_2)} \right]$$

$$1952\ \text{kPa} \times \frac{1000\ \text{N}}{\text{m}^2\text{kPa}} \times \frac{\text{kg-m}}{\text{N-s}^2} + 10.34\frac{\text{kg}}{\text{m}^3} \times 148.6^2\frac{\text{m}^2}{\text{s}^2} - \frac{1}{2} \times 0.5 \times 10.34\frac{\text{kg}}{\text{m}^3} \times 148.6^2\frac{\text{m}^2}{\text{s}^2} \times 0.10$$

$$= 10.34\frac{\text{kg}}{\text{m}^3} \times 148.6\frac{\text{m}}{\text{s}}\left[\frac{287.1\dfrac{\text{J}}{\text{kg-K}} \times \dfrac{\text{N-m}}{\text{J}} \times \dfrac{\text{kg-m}}{\text{N-s}^2} \times T_2(\text{K})}{\sqrt{2 \times 1.244\dfrac{\text{kJ}}{\text{kg-K}} \times \dfrac{1000\text{J}}{\text{kJ}} \times \dfrac{\text{N-m}}{\text{J}} \times \dfrac{\text{kg-m}}{\text{N-s}^2} \times (666.7 - T_2)\text{K}}}\right]$$

$$+ 10.34\frac{\text{kg}}{\text{m}^3} \times 148.6\frac{\text{m}}{\text{s}}\sqrt{2 \times 1.244\frac{\text{kJ}}{\text{kg-K}} \times \frac{\text{N-m}}{\text{J}} \times \frac{1000\text{J}}{\text{kJ}} \times \frac{\text{kg-m}}{\text{N-s}^2} \times (666.7 - T_2)\text{K}}$$

$$2{,}174{,}000 = 1537\left[\frac{287.1T_2}{\sqrt{2488(666.7 - T_2)}} + \sqrt{2488\ (666.7 - T_2)}\right].$$

Iteratively solving for the exit static temperature T_2, one finds

$T_2 = 657.7$ K;

hence, the exit velocity is

$$V_2 = \sqrt{2c_p(T_{t1} - T_2)}$$

$$= \sqrt{2 \times 1.244\frac{\text{kJ}}{\text{kg-K}} \times \frac{\text{N-m}}{\text{J}} \times \frac{1000\text{J}}{\text{kJ}} \times \frac{\text{kg-m}}{\text{N-s}^2} \times (666.7 - 657.7)\text{K}} = 149.1\ \text{m/s},$$

and the exit speed of sound is $a_2 = 495.3$ m/s. The exit Mach number is thus $M_2 = 0.3010$.

Also, the exit density is $\rho_2 = \rho_1\, V_1\, /\, V_2 = 10.34 \times 148.6/149.1 = 10.30\ \text{kg/m}^3$, and so from the ideal gas law for the temperature and density, $p_2 = 1945$ kPa. Finally, for $M_2 = 0.3010$ and the p_2, above the exit total pressure is

$p_{t2} = 2062$ kPa (299.1 psia)

and the total pressure ratio due only to the blockage of the flameholder is

$\pi = p_{t2}/p_{t1} = 2062/2069 = 0.997.$

One can see that this loss in total pressure is much smaller than the losses due to the other effects. This could have also been solved using the software General1D.exe. Note also that, with the generalized analysis, the total pressure loss due to the heat addition, friction, and drag can be calculated simultaneously (Example 9.8).

9.5.4 Combined Drag Object, Heat Transfer, and Friction

In Sections 9.5.1, 9.5.2, and 9.5.3, a drag object, heat addition/removal, and friction are treated separately. However, in a burner the three effects occur in series and simultaneously. Thus, in this section an analysis whereby all three are included is conducted. The technique that will be used is the so-called generalized 1-D flow method; this is described in Appendix C.13 and is coupled with the flow with a drag object. In general, differential equations are derived that must be numerically integrated between two end states to obtain the drop in total pressure due to both friction and heat transfer.

Example 9.8 Flow with a Flameholder, Temperature Increase, and Friction

Flow with an incoming Mach number of 0.3 and an inlet total pressure of 300 psia enters a flameholder with a drag coefficient of 0.5 and a blockage of 0.1; next it enters a channel 18 in. long with a 5 in. diameter and a friction factor of 0.040. The total temperature increases from 1200 °R to 2400 °R in this channel (can burner). What is the total pressure ratio if the specific heat ratio is 1.30 for the entire unit? Note that this example is a combination of Examples 9.5, 9.6, and 9.7 in which a flameholder and both heat addition and friction are included.

Solution

First, the station numbers are defined as follows.

(1) Inlet to the flameholder.

(2) Exit of the flameholder/inlet to the burning region with friction.

(3) Exit of the burning region with friction.

The flameholder portion has just been solved in Example 9.7.
Namely, $M_2 = 0.3010$ and $p_{t2} = 299.1$ psia.
Sections 2 and 3 of this problem will be solved by numerically integrating Eqs. (C.15.3) and (C.15.4). One hundred incremental steps are used, and heat addition and friction are assumed to be uniform; thus $\Delta x = 18/100 = 0.18$ in. and $\Delta T_t = 1200/100 = 12$ °R. A forward difference method is used. Hence, for the first step from Eq. (C.15.1), the change in the square of the Mach number is

$$\Delta(M^2) = M^2\left[(1+\gamma M^2)\frac{\left(1+\frac{\gamma-1}{2}M^2\right)}{1-M^2}\frac{\Delta T_t}{T_t} + 4f\gamma M^2\frac{\left(1+\frac{\gamma-1}{2}M^2\right)}{1-M^2}\frac{\Delta x}{D}\right],$$

so that the square of the Mach number at the end of the first incremental step is

$$M_i^2 = M_{i-1}^2 + \Delta(M^2)$$

$$M^2 = 0.301^2 + 0.301^2\left[(1 + 1.3 \times 0.301^2)\frac{\left(1 + \frac{1.3-1}{2}0.301^2\right)}{1 - 0.301^2}\frac{12}{1200}\right]$$

$$+ 0.301^2\left[+ 4 \times 0.04 \times 1.3 \times 0.301^2\frac{\left(1 + \frac{1.3-1}{2}0.301^2\right)}{1 - 0.301^2}\frac{0.18}{5}\right]$$

$$= 0.091798,$$

or the Mach number at the end of the first incremental step is

$M = 0.3030.$

From Eq. (C.15.2), the change in total pressure for the incremental step is

$$\Delta p_t = p_t\left[-\frac{\gamma M^2}{2}\frac{\Delta T_t}{T_t} - \frac{4f\gamma M^2}{2}\frac{\Delta x}{D}\right],$$

or the total pressure at the end of the first incremental step is

$p_{ti} = p_{ti-1} + \Delta pt$

$$p_t = 299.1 + 299.1\left(-\frac{1.3 \times 0.301^2}{2}\frac{12}{1200} - \frac{4 \times 0.04 \times 1.3 \times 0.301^2}{2}\frac{0.18}{5}\right)$$

$$= 298.8 \text{ psia}.$$

This is repeated for all 100 incremental steps, and the previous step is always used for the initial conditions for the following step. Thus, one finds after 100 steps that

$M_3 = 0.5484$

$p_{t3} = 260.45$ psia;

thus,

$\pi = p_{t3}/p_{t1} = 260.45/300 = 0.8682.$

This total pressure ratio is lower than is obtained by multiplying the results from the previous examples ($0.935 \times 0.966 \times 0.997 = 0.901$) and is more accurate. Note: This is largely due to the fact that each of the previous examples had an inlet Mach number of only 0.3, whereas each process actually increased the Mach number. A better idea would be to use the exit Mach number from each of the examples to start the next example. But increasing the Mach number increases the losses, so the simple analysis under predicted the losses. For this example, increasing the number of steps from 100 to 15,000 results in $\pi = 0.8673$ and $M_3 = 0.5513$, which is minimal improvement on the method. General1D.exe could be used for this calculation.

In Figure 9.12, results for other incoming Mach numbers and one other total temperature ratio are shown. As can be seen, as the Mach number increases, the total pressure ratio quickly decreases from unity. Also, as the total temperature ratio increases (i.e., as the fuel ratio increases), the total pressure ratio decreases. Also, if the incoming Mach number is too high the unit chokes.

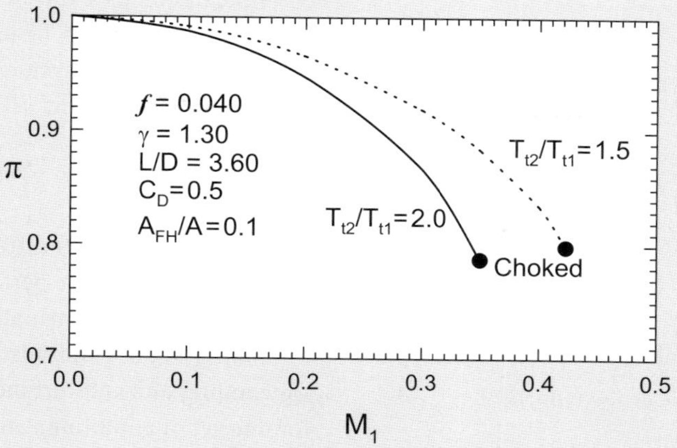

Figure 9.12 Variation of total pressure ratio with Mach number and total temperature ratio for Example 9.8 (frictional flow with temperature increase and flameholder).

9.6 Performance Maps

9.6.1 Dimensional Analysis

Performance curves or "maps" are usually used to provide a working medium for a burner engineer. These provide a quick and accurate view of the conditions at which the combustor is operating and what can be expected of the burner if the flow conditions are changed. Of primary concern are the total pressure ratio and burn efficiency of a burner; that is, p_{t4}/p_{t3} (or π_b) and η_b for a primary burner, and π_{ab} and η_{ab} for an afterburner. For a primary combustor, the total pressure ratio is a function of many variables, and the functional relationship f_1 can be written as

$$p_{t4}/p_{t3} = f_1\{\dot{m}, T_{t3}, A, p_{t3}, f, \Delta H, \gamma, \mathcal{R}\}. \qquad (9.6.1)$$

Dimensionless parameters can be used to reduce the number of independent variables. Using the Buckingham Pi theorem, one finds the following for a given value of γ:

$$p_{t4}/p_{t3} = f_1\left[\frac{\dot{m}\sqrt{\mathcal{R}T_{t3}}}{Ap_{t3}}, f, \frac{\Delta H}{\mathcal{R}T_{t3}}\right]. \qquad (9.6.2)$$

This form is rarely used to document the performance of a burner, however. Next, if both Fanno and Rayleigh line flows or generalized 1-D flow with heat transfer and friction are independently considered, one finds, for a given burner diameter, length, and friction factor, that the pressure ratio is a function (f_2) of the entering Mach number and total temperature ratio; that is,

$$p_{t4}/p_{t3} = f_2\left[M_3, \frac{T_{t4}}{T_{t3}}\right], \qquad (9.6.3)$$

or, using the definition of Mach number,

$$p_{t4}/p_{t3} = f_2\left[\frac{V_3}{\sqrt{\gamma\mathcal{R}T_3}}, \frac{T_{t4}}{T_{t3}}\right]. \qquad (9.6.4)$$

Next, using continuity yields

$$p_{t4}/p_{t3} = f_2\left[\frac{\dot{m}_3}{\rho_3 A_3\sqrt{\gamma\mathcal{R}T_3}}, \frac{T_{t4}}{T_{t3}}\right], \qquad (9.6.5)$$

but since T_{t3}/T_3 and p_{t3}/p_3 are strictly functions of Mach number, one can write

$$p_{t4}/p_{t3} = f_2\left[\frac{\dot{m}_3}{\rho_3 A_3\sqrt{\gamma\mathcal{R}T_{t3}}}, \frac{T_{t4}}{T_{t3}}\right], \qquad (9.6.6)$$

or, using the ideal gas law,

$$p_{t4}/p_{t3} = f_2\left[\frac{\dot{m}_3}{p_{t3}A_3\sqrt{\gamma\mathcal{R}T_{t3}}}, \frac{T_{t4}}{T_{t3}}\right]. \qquad (9.6.7)$$

Thus,

$$p_{t4}/p_{t3} = f_2\left[\frac{\dot{m}_3\sqrt{\mathcal{R}T_{t3}}}{p_{t3}A_3\sqrt{\gamma}}, \frac{T_{t4}}{T_{t3}}\right], \tag{9.6.8}$$

or, for a given value of γ,

$$p_{t4}/p_{t3} = f_2\left[\frac{\dot{m}_3\sqrt{\mathcal{R}T_{t3}}}{p_{t3}A_3}, \frac{T_{t4}}{T_{t3}}\right] \tag{9.6.9}$$

Hence, the first term of Eq. (9.6.9) matches that of Eq. (9.6.2). Next, if a small fuel ratio with the simplified burner analysis (heating value) is considered, one can use Eq. (3.2.24) to yield

$$\frac{\eta_b f \Delta H}{c_p T_{t3}} = \frac{T_{t4}}{T_{t3}} - 1. \tag{9.6.10}$$

Thus, for a given γ and given η_b, one can write

$$p_{t4}/p_{t3} = f_2\left[\frac{\dot{m}_3\sqrt{\mathcal{R}T_{t3}}}{p_{t3}A_3}, \frac{f\Delta H}{\mathcal{R}T_{t3}}\right]. \tag{9.6.11}$$

Next, by reexamining the results from the dimensional analysis, the last two terms in Eq. (9.6.2) can be directly related and thus combined. Furthermore, for a given fuel, one can write

$$p_{t4}/p_{t3} = f_2\left[\frac{\dot{m}_3\sqrt{\mathcal{R}T_{t3}}}{p_{t3}A_3}, \frac{f}{T_{t3}}\right]. \tag{9.6.12}$$

Substituting into Eq. (9.6.9) and "nondimensionalizing" or correcting, as is done for compressors and turbines, one finds the general expression

$$p_{t4}/p_{t3} = f\left[\frac{\dot{m}_3\sqrt{\theta_{t3}}}{\delta_{t3}}, \frac{f}{\theta_{t3}}\right] = f\left[\dot{m}_{c3}, \frac{f}{\theta_{t3}}\right], \tag{9.6.13}$$

where again

$$\delta_{t3} = p_{t3}/p_{stp} \tag{9.6.14}$$

and

$$\theta_{t3} = T_{t3}/T_{stp}, \tag{9.6.15}$$

and f is the usual fuel ratio. Note again that the two independent parameters are not truly dimensionless. The first is called the "corrected" mass flow and has dimensions of mass flow, whereas the second is dimensionless. Thus, since they both are not dimensionless, the resulting function cannot be used to correlate or compare different combustors, fuels, or both. The function can be used for one particular burner and fuel, however, as is the usual case. Similarly, for the burner efficiency, one finds

$$\eta_b = g\left[\frac{\dot{m}_3\sqrt{\theta_{t3}}}{\delta_{t3}}, \frac{f}{\theta_{t3}}\right] = g\left[\dot{m}_{c3}, \frac{f}{\theta_{t3}}\right], \tag{9.6.16}$$

where the function g is different from that of f. A similar analysis can be performed for an afterburner. The functions f and g usually are derived from a combination of empiricism and modeling. As is done for compressors and turbines, the maps can be found for one set of conditions and applied to other conditions using the corrected parameters.

9.6.2 Trends

Typical trends of p_{t4}/p_{t3} and η_b are shown in Figure 9.13. As can be seen, the total pressure ratio drops as the corrected mass flow rate increases and as the fuel ratio increases. By comparison with

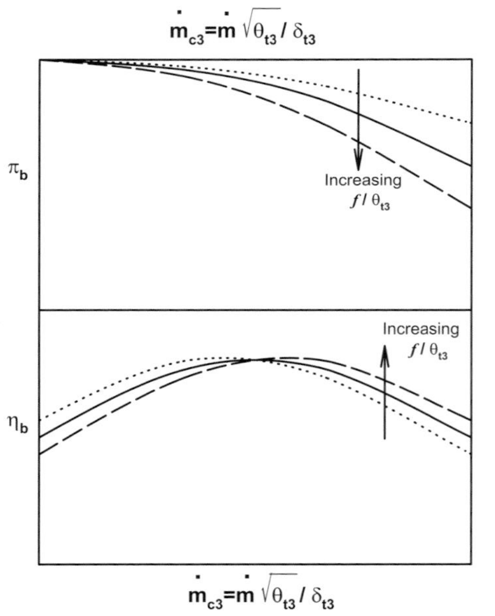

Figure 9.13 General performance map (p_{t4}/p_{t3} and η_b) for a primary burner.

Figure 9.12 one can see that the trends are the same. Also, a relative maximum in the burn efficiency can be seen in Figure 9.13, although this trend is strongly dependent on burner design, including the injection nozzles and mixing methods as well as the fuel properties. Thus, the shape and trends of the efficiency map vary greatly from burner to burner. Note that mapping conventions are the same for primary burners, reheat burners, and afterburners.

9.7 Fuel Types and Properties

9.7.1 Jet Propulsion

Up to this point, the fundamental fuel consideration has been the heating value. However, different characteristics are considered and controlled in fuel refining and mixing. On the basis of chemical characteristics, compositions, and chemical structure, hydrocarbon fuels can generally be classified as paraffins ($C_n H_{2n+2}$), olefins ($C_n H_{2n}$), diolefins ($C_n H_{2n-2}$), naphthenes ($C_n H_{2n}$), and aromatics ($C_n H_{2n-6}$, or $C_n H_{2n-12}$). Fuel is also classified for military use or commercial use. Military grades are JP1, ..., JP5(USN), ... JP8(USAF), ... JP10. Commercial grades are Jet A, Jet A-1, and Jet B. The characteristics are selected based on the particular engine and flight conditions, including altitude and geographic locations. All are kerosene based but have different characteristics. First, a fuel is a mix of the different hydrocarbons. A fuel that is a kerosene is considered to have n between 8 and 16. Thus, kerosene is not a single compound but a mixture and has the characteristics of a mixture. For example, to find the mixture viscosity one would mass average the viscosities of each constituent. Also, for example, each constituent has an individual vapor pressure, and so some constituents can evaporate before others. In Section 9.4, kerosene was approximated to be $C_{10}H_{22}$, which is an accurate estimation in the middle of the n range. On the other hand, a "wide-cut" fuel is a mixture of more compounds and is not refined as much as kerosene; as a result it is less expensive. For such a mixture, the value of n is between 5 and 16. Such a fuel includes compounds of the gasoline class and is usually referred to as a mixture of kerosene and gasoline; it is much more volatile. Such wide-cut fuels are not generally used in the United States owing to the increased danger of fire during handling and crashes. Some comparative notes on commercial and military fuels are presented in Tables 9.3 and 9.4.

The characteristics of commercial fuels are specified by ASTM, whereas those for military fuels are set and monitored by the government. Regardless of the application, the different characteristics are measured (most by ASTM standards, as described by Bacha et al. (2000)) and are identified as accelerated gum content (a measure of nonvolatile material formed at high temperature), acidity (standard definition), aromatics level (part of constituent makeup), color (a general inspection – should be clear white to clear light yellow), corrosivity (to metals such as aluminum and engine materials), distillation level (constituent makeup or profile of the fuel), existent gum content (a

Table 9.3 **Notes on commercial fuels.**

Fuel	Type	Comments
A	Kerosene	Used domestically in the United States
A-1	Kerosene	Used internationally outside of the United States
		Has a lower freezing point than A
B	Wide-cut	Used in geographies with very low temperatures
		Has a lower freezing point than A-1

Table 9.4 **Notes on military fuels.**

Fuel	Type	Comments
JP-1	Kerosene	Original fuel in the United States,
NLU JP-2	Wide-cut	Experimental fuel – never used
JP-3	Wide-cut	About 1/3 kerosene, 2/3 gasoline, NLU
JP-4	Wide-cut	USAF fuel (limited use), Very low freezing point Much like Jet-B
JP-5	Kerosene	USN fuel High flash point due to carrier applications and safety
JP-6	Wide-cut kerosene	Low freezing point, NLU
JP-7	Kerosene	For SR-71, low volatility
JP-8	Kerosene	Kerosene Much like A-1
JP-8+100	Kerosene	USAF fuel – has a thermal stability additive
JP-9	NA	Developmental, pure high-energy fuel
JP-10	NA	Developmental, pure high-energy fuel

NLU, no longer used.

"clogging" nonvolatile impurity), flash point (temperature at which burning is possible), freezing point (will be different for the different constituents), lubricity (lubricating characteristic for elastohydrodynamic lubrication such as rolling element bearings), luminometer number (an indicator of flame radiation heat transfer), mercaptan sulfur level (certain sulfur compound impurities), naphthalene content (part of constituent makeup and can cause carbon deposits), net heating value (value does not change greatly regardless of mixture), olefin content (part of constituent makeup), pour point (lowest temperature at which fuel is fluid), specific gravity or density (standard definition), smoke point (a measure of flame size), thermal stability (susceptibility to particulate formation that clogs parts of the system), sulfur content (amount of pure sulfur in the fuel, which can cause toxicity when burned), vapor pressure (partial pressure at which constituent vaporizes at a given temperature; 100 °F is a standard), viscosity (kinematic viscosity usually used), water reaction level (mixing capabilities of water in the fuel), and water separability (ease at which undesired water can be removed from fuel).

In addition, some other characteristics are monitored, including microbial growth (amount of bacterial, fungal, or other micro-organism growth in a stored fuel; this contamination increases with time), electrical conductivity (standard definition – can be potentially hazardous if a large current traverses fuel), and coefficient of thermal expansion (standard definition – affects fuel tank size and pressure). Furthermore, several additives are sometimes or often used, including, but not limited to, antioxidants, antifreezing materials, leak detectors (leave trails to help track down system fuel leaks), anticorrosives, lubrication improvers, and biocides. More details on fuel properties can be found in Boyce (1982) and Bacha et al. (2000).

9.7.2 Power Generation

For power generation gas turbines, liquid natural gas (LNG) is used more commonly than kerosene-based liquids. Natural gas tends to burn more cleanly than liquid fuels. However, because of the weight of the high-pressure storage containers, such fuels are not used for jet propulsion. Natural gas is dominated by methane, CH_4, but other hydrocarbon gases are present. As a result, the heating value has a range of values as indicated in Table 9.2.

Many of the power generation gas turbines are dual fuel, namely they can accommodate both liquid and gas hydrocarbons. One example of this is the FT8, Figure 1.48. More GE units use liquid fuels than any other manufacturer, for example the LM2500, Figure 1.47, often uses liquid-based hydrocarbon fuels. About 40 percent of the GE products use liquid fuels, including diesel fuels.

It is noteworthy that in some applications water is injected into the flow before the combustion chamber. This has two effects. First, it reduces the local high temperature in the burner, which reduces the production of NO_x. Second, it increases the mass flow in the turbine which results in a higher power production as Eq. (1.5.31) becomes

$$\mathcal{P}_{\text{net}} = (\dot{m}_{\text{a}} + \dot{m}_{\text{f}} + \dot{m}_{\text{H}_2\text{O}})(h_4 - h_5) - \dot{m}_{\text{a}}\ (h_3 - h_2). \tag{9.7.1}$$

Both effects are beneficial to the power generation.

9.8 Summary

This chapter has addressed the chemistry, thermodynamics, and gas dynamics of primary combustors, reheat burners, and afterburners. The purpose of the primary combustor and is to add thermal energy to the propulsion system by increasing the total temperature of the fluid, of which a part is removed by the turbine and the remainder of which is converted to high-kinetic-energy fluid in the nozzle. The purpose of an afterburner is to add additional thermal energy, all of which is converted to high-kinetic-energy fluid. For a power generation unit the primary burner is to add thermal energy of which a part is removed by the turbine and the remainder of which is converted to high kinetic energy which is converted to power with an additional turbine. The reheat burner adds additional energy to increase the total output power. First, the geometries were defined and important design considerations were summarized. The design of a primary burner and reheat burner is much more complex than that of an afterburner. The three varieties of primary burners are can, annular, and cannular. Primary combustors and reheat burners operate with larger fuel efficiency and lower total pressure ratios than do afterburners. Newer developments include dry low emissions (DLE) which are used on some ground or marine based gas turbines. This concept results in the air/fuel to premix more efficiently before entering the combustor. This hardware improves overall efficiency and reduces emissions, particularly, but tends to be larger and heavier than conventional burners. The burning zone in a combustor is fuel rich with a low velocity. Good mixing is needed to ensure high burner efficiencies; on the other hand, low total pressure losses are desired, which are associated with high turbulence levels. Thus, a tradeoff or balance of these two conflicting design concepts is needed in the design of a burner. Afterburners operate with much higher exit temperatures than do primary burners, and the nozzle material limits the temperature. Flammability and flame speed were shown to be fundamentally critical to the stability of the burner. In general, the design of a burner is not as straightforward as for a compressor or turbine. Next, the ignition process and different engine starting methods were discussed, for the shaft must be rotating before the burner can be ignited.

The chemistry was covered, and the balancing of the chemical equation for stoichiometric (no fuel or oxygen in the products) and other conditions was discussed. The thermodynamics of disassociation and chemical equilibrium was introduced and shown to influence the emissions. The thermodynamics of the adiabatic flame temperature was next covered, and the heating value was derived from this method. Differences between the simple heating value method of burner analysis and the adiabatic flame method were discussed, and it was shown that, as the fuel-to-air mixture becomes lean, the heating value method becomes more viable. Typically, the stoichiometric fuel-to-air mixture ratio is 0.06–0.07 for jet or power generation fuels, but if such ratios were used for a gas turbine, the operating temperatures would be too high. Locally, however, burners must operate internally near the stoichiometric condition for stability.

Next, total pressure losses were discussed, and the losses were shown to be due to the heat addition, friction, and flow obstructions (flameholders). One-dimensional Rayleigh line flow and Fanno line flow processes were reviewed and shown to be methods to predict pressure losses. More advanced predictions can be made with generalized 1-D frictional flow with heat addition and flow restrictions. In general, the largest loss is due to heat addition, and very little loss is due to flow obstructions. Typical types of burner performance curves or maps and trends were next presented again based on dimensional analysis and empiricism. Finally, different important fuel characteristics and fuel types were identified. For propulsion applications, kerosene-based liquids are commonly used. However, for ground- or marine-based power generation gas turbines, liquid gases (natural gas, etc.) are more commonly used.

List of Symbols

A	Area
a	Sound speed
a, b	Curve fit coefficients for c_p
c_p	Specific heat at constant pressure
C_d	Drag coefficient
D	Diameter
F	Force
f	Fuel ratio
f	Fanning friction factor
f	Pressure ratio map function
g	Efficiency map function
h	Specific enthalpy
Δh_f°	Standard enthalpy of formation
ΔH	Heating value
K_p	Equilibrium constant
L	Length
$\dot{m}$	Mass flow rate
M	Mach number
$\mathcal{M}$	Molecular weight
N, n	Number of moles
p	Pressure
$\dot{Q}$	Heat transfer rate
$\mathcal{R}$	Ideal gas constant
s	Entropy
T	Temperature
V	Velocity
$\mathcal{V}$	Combustion volume
$\dot{W}$	Power
X	Consituent coefficient
γ	Specific heat ratio
δ	Ratio of pressure to standard pressure
η	Efficiency
θ	Ratio of temperature to standard temperature
ν	Stoichiometric number of moles coefficient
π	Total pressure ratio
ρ	Density

Subscripts

air	Air
c	Corrected
d	Drag
f	Fuel
FH	Flameholder
i	Counter
i	ith constituent
n	General hydrocarbon subscript
stp	Standard conditions
t	Total (stagnation)
T	At temperature T
1, 2	Positions for process analysis
3	Inlet to burner
298	At 298 K

Superscripts

n	Loading parameter empirical constant
*	Choked

Problems

9.1 Balance the chemical equation for n-nonane for the stoichiometric condition and 300 percent air.

9.2 Find the adiabatic flame temperature of liquid n-decane for 250 percent air if the fuel inlet total temperature is 900 °R and the air inlet total temperature is 1350 °R. What is the fuel-to-air mixture ratio?

9.3 If air enters a combustor at 1300 °R total temperature with a specific heat ratio of 1.3 and a total pressure of 200 psia, find the total pressure ratio across the burner as a function of entrance Mach number if the exit total temperature is 2500 °R. Plot the results.

9.4 If air enters a combustor at 1300 °R total temperature with a specific heat ratio of 1.3 and a total pressure of 200 psia, find the total pressure ratio across the burner as a function of inlet Mach number if the friction factor is 0.035, the burner length is 13 in., and the can diameter is 4 in. Plot the results.

9.5 If air enters a combustor at 1300 °R total temperature with a specific heat ratio of 1.3 and a total pressure of 200 psia, find the total pressure ratio across the burner as a function of entrance Mach number if the friction factor is 0.035, the burner length is 13 in., the can diameter is 4 in., and the exit total temperature is 2500 °R. Plot the results.

9.6 If air enters an afterburner with flameholders at 1600 °R total temperature with a specific heat ratio of 1.3 and a total pressure of 60 psia, find the total pressure ratio across the afterburner as a function of inlet Mach number if the drag coefficient is 0.45 and the area ratio of flameholder to total area is 0.15. Plot the results.

9.7 If air enters a combustor with flameholders at 1300 °R total temperature with a specific heat ratio of 1.3 and a total pressure of 200 psia, find the total pressure ratio across the burner as a function of entrance Mach number if the friction factor is 0.035, the burner length is 13 in., the can diameter is 4 in., the drag coefficient of the flameholders is 0.55, the area ratio of flameholder to total area is 0.12, and the exit total temperature is 2500 °R. Plot the results.

9.8 The adiabatic flame temperature of liquid n-decane is 2750 °R. The fuel inlet total temperature is 900 °R and the air inlet total temperature is 1350 °R. What is the fuel-to-air mixture ratio?

9.9 If air enters a combustor at 683 K (1230 °R) total temperature with a specific heat ratio of 1.35 and a total pressure of 1379 kPa (200 psia), find the total pressure ratio across the burner as a function of exit total temperature if the inlet Mach number is 0.38. Plot the results.

9.10 If air enters a combustor at 683 K (1230 °R) total temperature with a specific heat ratio of 1.35 and a total pressure of 1379 kPa (200 psia), find the total pressure ratio across the burner as a function of exit total temperature if the friction factor is 0.030, the burner length is 0.381 m (15 in.), the can diameter is 0.127 m (5 in.), and the entrance Mach number is 0.38. Plot the results.

9.11 If air enters a combustor with flameholders at 683 K (1230 °R) total temperature with a specific heat ratio of 1.35 and a total pressure of 1379 kPa (200 psia), find the total pressure ratio across the burner as a function of exit total temperature if the friction factor is 0.030, the burner length is 0.381 m (15 in.), the can diameter is 0.127 m (5 in.), the drag coefficient of the flameholders is 0.50, the area ratio of flameholder to total area is 0.13, and the entrance Mach number 0.38. Plot the results.

9.12 Balance the chemical equation for *n*-octane for the stoichiometric condition and 350 percent air.

9.13 Find the adiabatic flame temperature of liquid *n*-decane for 350 percent air if the fuel inlet total temperature is 1020 °R and the air inlet total temperature is 1150 °R. What is the fuel-to-air mixture ratio?

9.14 If air enters an afterburner with flameholders at 1670 °R total temperature with a specific heat ratio of 1.36 and a total pressure of 73 psia, find the total pressure ratio across the burner as a function of entrance Mach number if the drag coefficient of the flameholders is 0.53, the area ratio of flameholder to total area is 0.18, and the exit total temperature is 3450 °R. Plot the results.

9.15 Find the adiabatic flame temperature of liquid *n*-octane for 300 percent air if the fuel inlet total temperature is 935 °R and the air inlet total temperature is 1040 °R. What is the fuel-to-air mixture ratio?

9.16 Using the enthalpies of formation, find the heating value of liquid *n*-octane at 298 K.

9.17 You are given kerosene (*n*-decane) at 528 K (950 °R) burning with air at 778 K (1400 °R).

(a) What is the temperature of the products for 100, 200,350, and 1000 percent air?
(b) Accurately plot T_{products} vs. percent air.
(c) What is the fuel-to-air mass flow ratio f for these conditions?
(d) Plot T_{products} vs. f.

9.18 You have a combustion chamber with an incoming Mach number M_3, an incoming total pressure of 275 psia, an incoming air total temperature of 1400 °R, and an inlet area of 0.30 ft^2. The exiting total temperature is T_{t4}. Assume $\gamma = 1.30$. What is the burner total pressure ratio (π_b) for the conditions below?

(a) $T_{t4} = 2490$ °R; $M_3 = 0.1, 0.22, 0.42$
Plot π_b vs. M_3.
Plot π_b vs. $\dot{m}$ (lb$_m$/s).
(b) $M_3 = 0.42$, $T_{t4} = 1400, 2060, 2490$ °R
Plot π_b vs. T_{t4}.
Plot π_b vs. f if the fuel is kerosene.

(See Problem 9.17 to determine how f varies with T_{t4}.)

9.19 A burner has an inlet Mach number of 0.35, an inlet total pressure of 250 psia and an inlet total temperature of 1100 °R. The exit total temperature is 2300 °R. The chamber is 18 in. long, is 3.5 in. in diameter, and has a friction factor of 0.037. You are to find the total pressure ratio for the burner. Assume friction and heat addition can be treated separately and the specific heat ratio is 1.30.

9.20 You are given kerosene (*n*-decane) at 611 K (1100 °R) burning with air at 667 K (1200 °R).

(a) What is the temperature of the products for 100, 200, and 400 percent air?
(b) Accurately plot $T_{products}$ vs. percent air.
(c) What is the fuel-to-air mass flow ratio f for these conditions?
(d) Plot $T_{products}$ vs. f.

9.21 You have a combustion chamber with an incoming Mach number M_3, an incoming total pressure of 1724 kPa (250 psia), an incoming air total temperature of 667 K (1200 °R), and an inlet area of 0.0372 m^2 (0.40 ft^2). The exiting total temperature is T_{t4}. Assume $\gamma = 1.32$. Determine the map for the given set of conditions; that is, what is the burner total pressure ratio (π_b) for the conditions below?

(a) $T_{t4} = 1389$ K (2500 °R); $M_3 = 0.1, 0.25, 0.37$
Plot π_b vs. M_3.
Plot π_b vs. $\dot{m}$ (kg/s or lb$_m$/s).

(b) $M_3 = 0.37$, $T_{t4} = 667, 1111, 1389$ K (1200, 2000, 2500 °R)
Plot π_b vs. T_{t4}.
Plot π_b vs. f if the fuel is kerosene.

(See Problem 9.20 to determine how f varies with T_{t4}.)

9.22 A burner has an inlet Mach number of 0.25, an inlet total pressure of 250 psia, and an inlet total temperature of 1100 °R. The exit total temperature is 2300 °R. The chamber is 18 in. long, is 3.5 in. in diameter, and has a friction factor of 0.037. You are to find the total pressure ratio for the burner. Assume friction and heat addition can be treated separately and the specific heat ratio is 1.30.

9.23 You are given kerosene (n-decane) at 1000 °R burning with air at 1250 °R.

(a) What is the temperature of the products for 100, 200, 400, and 800 percent air?
(b) Accurately plot $T_{products}$ vs. percent air.
(c) What is the fuel-to-air mass flow ratio f for these conditions?
(d) Plot T_{prod} vs. f.

9.24 You have a combustion chamber with an incoming Mach number M_3, an incoming total pressure of 260 psia, an incoming air total temperature of 1250 °R, and an inlet area of 0.42 ft^2. The exiting total temperature is T_{t4}. Assume $\gamma = 1.31$. Determine the map for the given set of conditions; that is, what is the burner total pressure ratio (π_b) for the conditions below?

(a) $T_{t4} = 2700$ °R; $M_3 = 0.15, 0.25, 0.35$
Plot π_b vs. M_3.
Plot π_b vs. $\dot{m}$ (lb$_m$/s).

(b) $M_3 = 0.35$, $T_{t4} = 1700, 2200, 2700$ °R
Plot π_b vs. T_{t4}.
Plot π_b vs. f if the fuel is kerosene.

(See Problem 9.23 to determine how f varies with T_{t4}.)

9.25 You are given n-decane at 611 K (1100 °R) burning with air at 667 K (1200 °R). What is the temperature of the products for 300 percent air? For simplicity, you may assume that the specific heats are not functions of temperature and are equal to 33.19, 54.37, 30.20, 41.26, 32.70, 34.88, and 296.0 kJ/kg-mole-K for CO, CO_2, H_2, H_2O, N_2, O_2, and $C_{10}H_{22}$, respectively.

9.26 You are given n-decane at 611 K (1100 °R) burning with air at 667 K (1200 °R). What percentage of air will result in an exit temperature of 1465 K (2637 °R)? For simplicity you may assume that the specific heats are not functions of temperature and are equal to 33.19, 54.37, 30.20, 41.26, 32.70, 34.88, and 296.0 kJ/kg-mole-K for CO, CO_2,H_2, H_2O, N_2, O_2, and $C_{10}H_{22}$, respectively.

9.27 You are given n-decane at 667 K (1200 °R) burning with air at 556 K (1000 °R). What percentage of air will result in an exit temperature of 1502 K (2704 °R)? For simplicity you may assume that the specific heats are not functions of temperature and are equal to 33.4, 57.7, 30.3, 41.1, 32.9, 35.2, and 296.0 kJ/kg-mole-K for CO, CO_2, H_2, H_2O, N_2,O_2, and $C_{10}H_{22}$, respectively.

9.28 You are given n-decane at 444 K (800 °R) burning with air at 556 K (1000 °R). What percentage of air will result in an exit temperature of 1358 K (2444 °R)? For simplicity, you may assume that the specific heats at constant pressure are not functions of temperature and are equal to 55.7, 40.4, 32.5, 34.6, and 296.0 kJ/kg-mole-K for CO_2, H_2O, N_2,O_2, and $C_{10}H_{22}$, respectively.

9.29 Air enters a combustor at total conditions of 200 psia and 1250 °R and at a Mach number of 0.25. The gas is heated to 2760 °R. Find the pressure at the exit of the burner if friction is negligible and the exit Mach number. Assume the specific heat ratio is 1.350.

9.30 Consider the equilibrium equation $2CO_2 \leftrightharpoons 2CO + O_2$. Find the mole composition of the different components at operating temperatures of 2400 K and 3400 K at 1 atm. The equilibrium constants are 0.000446 and 1.413, respectively. Repeat for 3 atm.

10 Nozzles

10.1 Introduction

A nozzle is sometimes called the exhaust duct or tail pipe, and is the last component of a jet engine through which the air passes. Up to two parallel nozzles are present on an engine: primary and fan (or secondary). In this chapter, both converging and converging–diverging (CD) nozzle types are discussed, and the two nozzles can be any combination of the two types (i.e., converging and converging–diverging, converging and converging, etc.). Recall that the functions of the nozzles are to convert high-pressure, high-temperature energy (enthalpy) to kinetic energy and to straighten the flow so that it exits in the axial direction. It is from this conversion process that the thrust is derived. Because of the high temperatures that a nozzle experiences, materials used in nozzle construction are usually a nickel-based alloy, titanium alloy, or ceramic composite. In Chapter 2, the nonideal effects of nozzles are discussed. The reader is also encouraged to review Appendix C as many of the fundamentals are covered therein. In this chapter, these effects are covered in more detail along with other design considerations.

10.1.1 Objectives

After completing this chapter, the reader should be able to do the following.

- Understand the basic geometries of primary and fan nozzles
- Understand the effects that efficiency has on nozzle performance
- Understand the differences in geometry of converging and diverging nozzles
- Understand the differences in performance of converging and diverging nozzles
- Understand the different geometries of a variable nozzle
- Understand the advantages of a variable nozzle
- Understand the differences between under expanded and overexpanded nozzles
- Understand the location predictions and effects of internal shocks
- Understand the effects of nozzle performance on a jet engine
- Understand the concepts and applications of nozzle maps
- Understand the mapping trends of different nozzle types
- Understand the different geometries and reasons for thrust reversal
- Understand the different geometries and reasons for thrust vectoring

10.2 Nonideal Equations

The governing equations for 1-D flow in nonideal nozzles are presented in Chapter 2. For the sake of convenience, they are repeated in this chapter.

10.2.1 Primary Nozzle

$$\eta_n = \frac{h_{t6} - h_8}{h_{t6} - h'_8} \quad (10.2.1)$$

$$u_8 = \sqrt{2(h_{t6} - h_8)} \quad (10.2.2)$$

$$u_8 = \sqrt{2c_p\eta_n(T_{t6} - T'_8)} \quad (10.2.3)$$

$$\frac{T_8}{T_{t6}} = \frac{1}{1 + \frac{\gamma - 1}{2}M_8^2} \quad (10.2.4)$$

$$\frac{p_8}{p_{t6}} = \left[\frac{T'_8}{T_{t6}}\right]^{\frac{\gamma}{\gamma-1}} \quad (10.2.5)$$

$$\frac{p_8}{p_{t6}} = \left[\frac{\dfrac{1}{1+\dfrac{\gamma-1}{2}M_8^2} - 1 + \eta_n}{\eta_n}\right]^{\frac{\gamma}{\gamma-1}} \quad (10.2.6)$$

$$\frac{p^*}{p_{t6}} = \left[1 + \frac{1-\gamma}{\eta_n(1+\gamma)}\right]^{\frac{\gamma}{\gamma-1}} \quad (10.2.7)$$

$$\frac{A_8}{A^*} = \frac{1}{M_8}\sqrt{\frac{\gamma+1}{2+(\gamma-1)M_8^2}}\left[\eta_n \frac{1+\dfrac{1-\gamma}{\eta_n(1+\gamma)}}{\dfrac{1}{1+\dfrac{\gamma-1}{2}M_8^2} - 1 + \eta_n}\right]^{\frac{\gamma}{\gamma-1}} \quad (10.2.8)$$

10.2.2 Fan Nozzle

$$\eta_{fn} = \frac{h_{t7} - h_9}{h_{t7} - h'_9} \quad (10.2.9)$$

$$u_9 = \sqrt{2(h_{t7} - h_9)} \quad (10.2.10)$$

$$u_9 = \sqrt{2c_p\eta_{fn}(T_{t7} - T'_9)} \quad (10.2.11)$$

$$\frac{T_9}{T_{t7}} = \frac{1}{1+\dfrac{\gamma-1}{2}M_9^2} \quad (10.2.12)$$

$$\frac{p_9}{p_{t7}} = \left[\frac{T'_9}{T_{t7}}\right]^{\frac{\gamma}{\gamma-1}} \quad (10.2.13)$$

$$\frac{p_9}{p_{t7}} = \left[\frac{\dfrac{1}{1+\dfrac{\gamma-1}{2}M_9^2} - 1 + \eta_{fn}}{\eta_{fn}}\right]^{\frac{\gamma}{\gamma-1}} \quad (10.2.14)$$

$$\frac{p^*}{p_{t7}} = \left[1 + \frac{1-\gamma}{\eta_{fn}(1+\gamma)}\right]^{\frac{\gamma}{\gamma-1}} \quad (10.2.15)$$

$$\frac{A_9}{A^*} = \frac{1}{M_9}\sqrt{\frac{\gamma+1}{2+(\gamma-1)M_9^2}}\left[\eta_{fn} \frac{1+\dfrac{1-\gamma}{\eta_{fn}(1+\gamma)}}{\dfrac{1}{1+\dfrac{\gamma-1}{2}M_9^2} - 1 + \eta_{fn}}\right]^{\frac{\gamma}{\gamma-1}}. \quad (10.2.16)$$

These equations have been programmed in Nozzle.exe, which can be used for various boundary conditions. This software is for shockless flow inside of the nozzle, but can be used along with NormalShock.exe or Shock.exe for cases with internal shocks.

10.2.3 Effects of Efficiency on Nozzle Performance

Before the details of nozzle operation and design are addressed, it behooves the reader to better understand the effect of efficiency on nozzle performance. Figures 10.1 and 10.2 serve this purpose. In Figure 10.1, the nozzle pressure ratio is presented as a function of the Mach number for three different efficiencies for $\gamma = 1.35$ and only the ideal case for $\gamma = 1.30$. As can be seen, lowering the nozzle efficiency tends to reduce this ratio. This can be interpreted in two ways. For example, for a nozzle with a given total inlet pressure and an area ratio designated so that the nozzle exit pressure matches the ambient pressure, as the efficiency is decreased the exit Mach number declines as well. As a result, the developed thrust also decreases. From another perspective, for a nozzle with an area ratio again designated so that the nozzle exit pressure matches the ambient pressure, if a particular exit Mach number is desired, the inlet total pressure is larger for the nonideal case than for the ideal one, thus requiring a higher pressure from the turbine.

In Figure 10.2, the area ratio (A/A^*) is plotted vs. the Mach number for three efficiencies for $\gamma = 1.35$

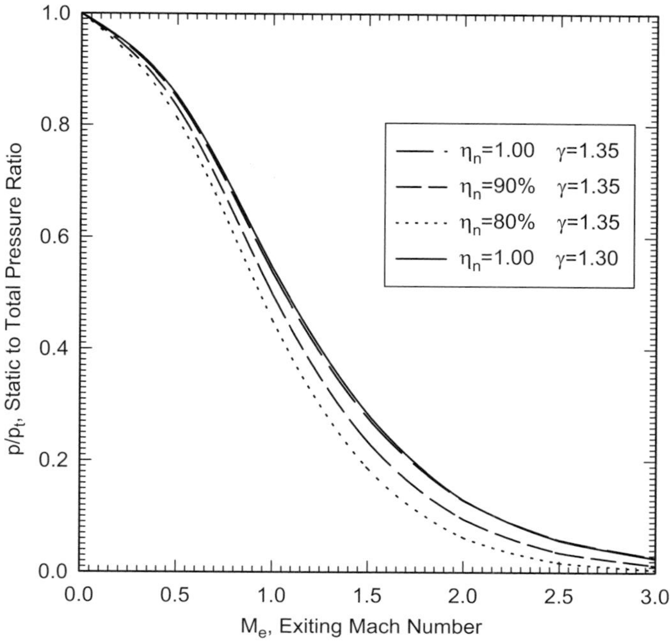

Figure 10.1 Variation of Mach number with nozzle pressure ratio for three values of efficiency for $\gamma = 1.35$.

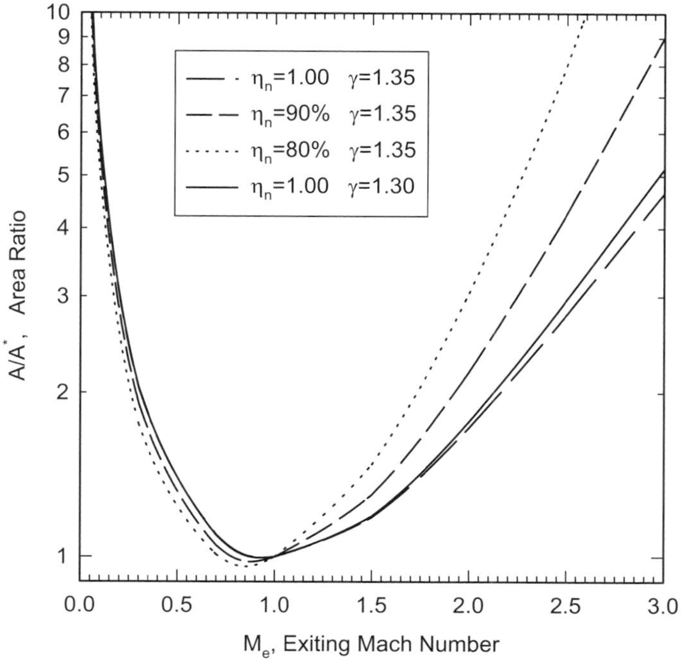

Figure 10.2 Variation of area ratio with Mach number for three values of efficiency for $\gamma = 1.35$.

and again only the ideal case for $\gamma = 1.30$. For subsonic flow, the ratio decreases with decreasing efficiency. However, for supersonic flow, the ratio increases with decreasing efficiency. Therefore, achieving a given supersonic Mach number requires a larger nozzle (larger exit area) for the nonideal case than for the ideal one. Also, with a choked converging nozzle, more convergence is necessary for a nozzle with losses than for an ideal nozzle. Also note that the Mach number is very slightly less than one at the minimum area and that the minimum area is slightly less than the sonic area for efficiencies less than one. The specific heat ratio has minimal effects for both plots – especially for low Mach numbers. Information and data on these figures are also available using Nozzle.exe.

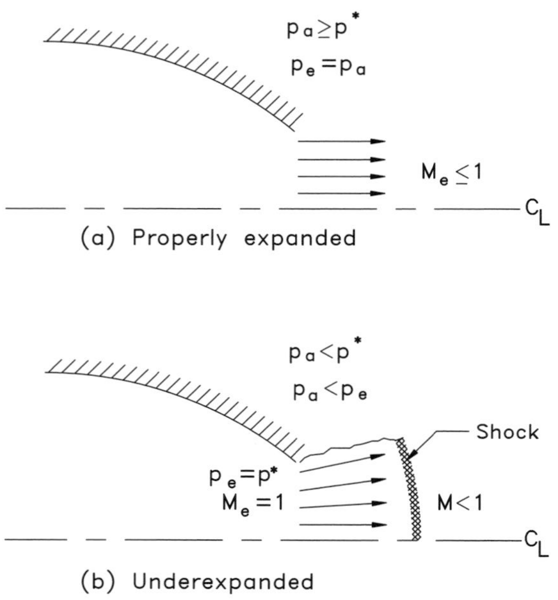

Figure 10.3 Exhaust of converging nozzle.

10.3 Converging Nozzle

The simplest type of nozzle is the fixed ideal converging nozzle. As the name implies, the exit cross-sectional flow area is smaller than the inlet area. This type of nozzle can be found on the primary and/or fan nozzle, or both. This nozzle can operate under two different conditions. First, the exit pressure can match the ambient condition. For this case, the exit Mach number is equal to or less than unity and the exit pressure is greater than or equal to the reference pressure at the sonic condition p^*, as determined from Eq. (10.2.7) or Eq. (10.2.15). This condition is shown in Figure 10.3a.

For the second condition, the ambient pressure is less than the value of p^*. Inside the nozzle the flow accelerates up to the exit, and in the plane of the exit the Mach number is unity. However, after the gas exits from the nozzle, it continues to expand in a 2-D manner and accelerate for a short distance in an effort to match the back pressure (Figure 10.3b). In this small region the Mach number is greater than 1. Eventually, a near-normal shock or series of oblique shocks occur downstream of the nozzle, which decelerate the flow to a subsonic condition. The flow remains subsonic after these shocks. This condition is called the supercritical or underexpanded condition.

For a choked converging nozzle with an efficiency less than unity, the nozzle exit Mach number is actually slightly less than one because, as previously discussed, more geometrical convergence is needed, that is, $(A_{min}/A^* < 1)$. Furthermore, 2-D and 3-D effects further contribute to this effect, as a larger pressure difference $(p_e - p_a)$ is needed to choke a nozzle (Zucrow and Hoffman, 1976). However, as indicated in Chapter 2, to simplify the analyses in this book, the sonic condition will be used for the choked case because the difference has negligible or minimal effect on the engine thrust and *TSFC*.

10.4 Converging–Diverging Nozzle

As the name implies, the cross-sectional area decreases and then increases along the nozzle centerline. This nozzle type is often called a de Laval or Laval nozzle.

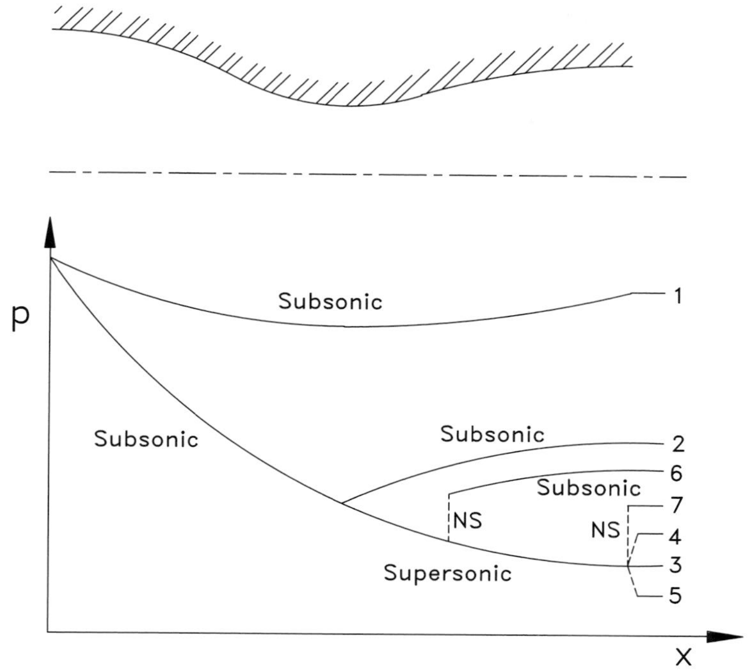

Figure 10.4 Seven conditions of a CD nozzle.

This is a more complex geometry, and seven different conditions can be experienced with this type of nozzle. The seven conditions are compared in Figure 10.4 and discussed next. This type of nozzle can be found for the primary nozzle but is used less often as a fan nozzle. For the following discussions, a fixed nozzle geometry with an efficiency of unity will be assumed so that the areas do not change as a function of operating condition. Variable nozzles are discussed in the following section.

The first condition to be considered is operation of the nozzle with the nozzle exit pressure equal to the ambient conditions and with the flow subsonic throughout (case 1, Figure 10.4). The maximum Mach number is at the minimum area. This nozzle type typically is not operated in this regime.

The second condition is shown as case 2 and is similar to case 1, but now the Mach number at the minimum area is unity. The Mach numbers in the diverging section are subsonic. Again, this nozzle type is usually not operated in this regime.

The third case to be considered is case 3, which is characterized by smooth and shockless flow throughout the nozzle, and the exit pressure matches the ambient pressure. In the converging section the flow is subsonic, and in the diverging area the flow is supersonic. For the ideal case, this would represent the design condition of the nozzle because it would result in the maximum thrust.

For the fourth case the ambient pressure is slightly above the design exit pressure. This results in a complex 2-D flow pattern outside the nozzle, as shown in Figure 10.5. This is considered to be an over expanded case, and the flow suddenly is compressed and decelerates outside of the nozzle. A series of compression waves and expansion waves are generated that can be calculated based on 2-D compressible flow – for example, with the

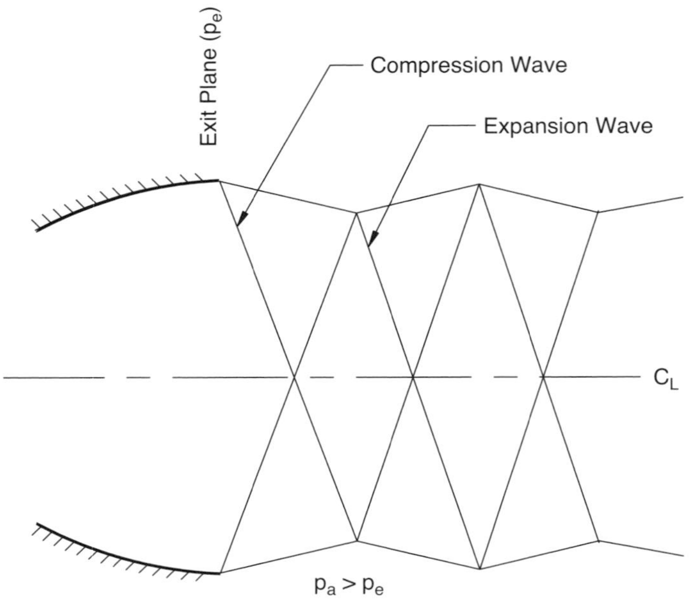

Figure 10.5 Exhaust of overexpanded CD nozzle (case 4).

Figure 10.6 Pratt & Whitney F100 engine on test stand – overexpanded exit (© 2022 Raytheon Technologies Corporation, Pratt & Whitney division. Reproduced with permission).

method of characteristics described by Zucrow and Hoffman (1976). The net result is a series of "shock diamonds" or "shock train," which are shown in Figure 10.6 in a photograph of a test on a Pratt & Whitney F100 afterburning turbofan (mixed fan, Figure 1.33).

For the fifth case, the ambient pressure is below the design exit pressure. This again results in a complex 2-D flow pattern outside the nozzle, as shown in Figure 10.7. This is considered to be an underexpanded or supercritical case, and the flow continues to expand and accelerate outside the nozzle. A series of expansion waves and compression waves are generated that can be calculated based on 2-D compressible flow. The net result is again a series of "shock diamonds" or "shock train," which are opposite in order to those found for the overexpanded case. A photograph of this condition is presented in Figure 10.8, which was taken at the Swiss Propulsion Laboratory. Often such photographs of shock diamonds are precisely documented by optical Schlieren or shadowgraph methods. The former detects flowfields with large density gradients, $\nabla\rho$, while the latter measures flowfields with large values of $\nabla^2\rho$ (Merzckirch, 1974).

For the sixth case, the ambient pressure is significantly above the nozzle design exit pressure but below the pressure covered in case 2. This condition results in a single normal shock in the nozzle diverging region or a series of oblique and normal shocks called λ shocks. This is also an overexpanded case and is shown in Figure 10.9. This case results in a subsonic exit Mach number, which is undesirable from the standpoint of thrust.

A limiting condition of case 6 is an exit pressure that causes a normal shock to occur exactly in the exit plane. This is considered to be case 7. Note that case 4 falls between this limiting case and case 3.

Although 1-D flow is assumed for the working equations presented, the flow can be 2-D or 3-D. As the flow deviates from the 1-D case, the axial component of velocity decreases because the flow has a radial component. The radial component does not contribute to the thrust as dictated by the momentum equation. The 2-D nature is best

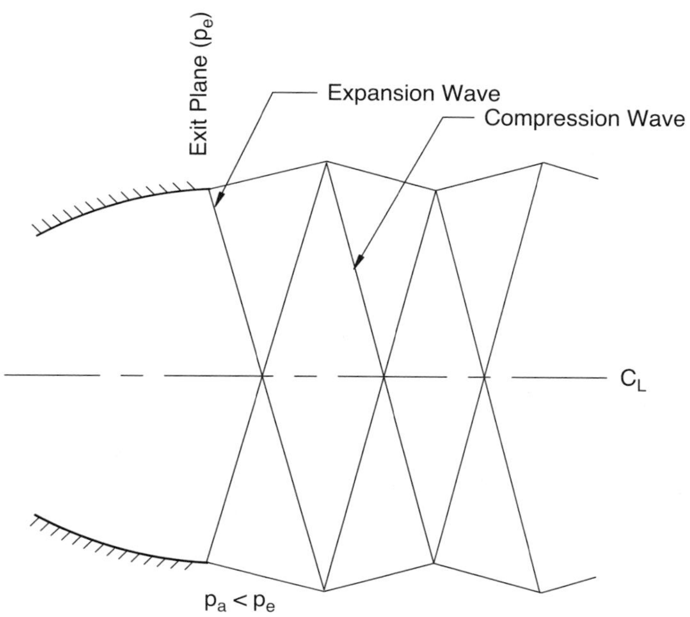

Figure 10.7 Exhaust of underexpanded CD nozzle (case 5).

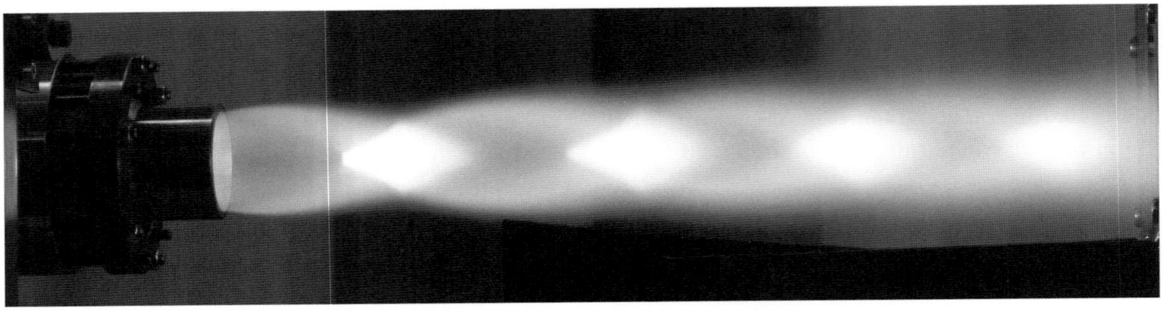

Figure 10.8 Underexpanded jet (courtesy of Swiss Propulsion Laboratory).

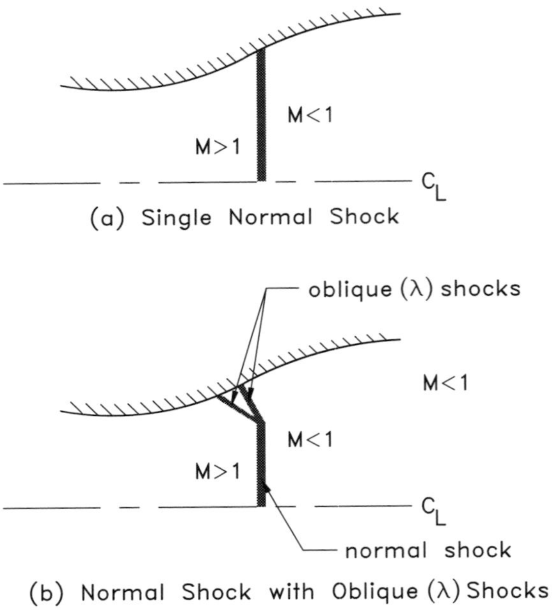

Figure 10.9 Exhaust of overexpanded CD nozzle with shocks in diverging area (case 6).

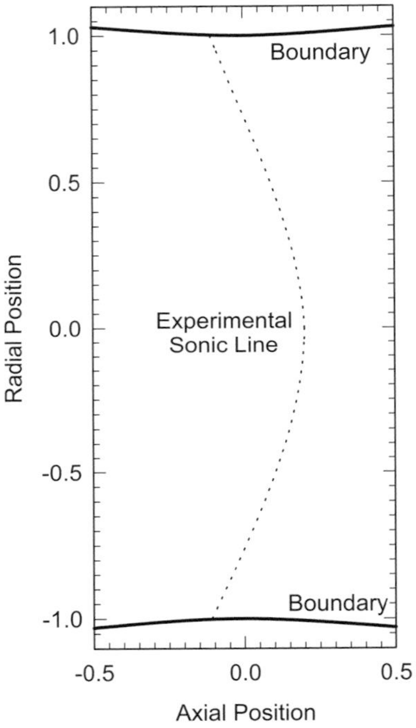

Figure 10.10 Typical sonic line.

demonstrated by observations of the sonic line near the throat of a CD nozzle. Ideally, the sonic line is a straight line between the surfaces at the minimum area. However, Flack and Thompson (1975) measured the sonic line position for a series of symmetric and nonsymmetric planar geometries, and a typical result is shown in Figure 10.10. The sonic line is in fact curved, indicating the nature of the two-dimensionality.

When addressing a problem or design of a CD nozzle, the general operating regime must be determined before details of the flow can be established. For example, if the back pressure is known for a given nozzle geometry, only one of the seven preceding conditions will hold. To determine which case applies, one must establish the brackets; that is, one must determine what back pressures will result in limiting cases 2, 3, and 7. Once the brackets are determined, the flow regime in which the nozzle is operating for a given back pressure can be found, which will make it possible to proceed with the analysis of the details. Solutions can also be obtained for a shockless converging or CD nozzle using the software Nozzle.exe and can be combined with results from Shock.exe to accommodate all of the various regimes. Also, note that property variations, excluding shocks, in both nozzle types can be predicted by using the generalized 1-D flow model from Appendix C.13 (as done in Chapter 8), which can, in fact, be used with any internal shock analyses to predict nozzle efficiency (Problems 10.20 and 10.31).

Example 10.1 | Converging–Diverging Nozzle: Which Nozzle Case?

A converging–diverging (CD) nozzle operates at an atmospheric pressure of 68.95 kPa (10 psia). It has a minimum area of 0.1393 m^2 (1.5 ft^2) and an exit area of 0.2786 m^2 (3.0 ft^2). The inlet total pressure and temperature are 241.3 kPa (35 psia) and 611.1 K (1100 °R), respectively. The effective specific heat ratio is 1.340, and the nozzle efficiency is 98 percent. Find the overall general operating condition and mass flow rate.

Solution

Before proceeding, it is necessary to determine under what general conditions the nozzle is operating. For example, if the nozzle were choked, the problem would be solved differently than if it were not choked. Thus, to proceed, one must determine for which of the seven conditions in Figure 10.4 the nozzle is operating. To determine this, one must find what back pressure will cause conditions 2, 3, and 7. Once these brackets are found, the remainder of the problem can be solved or more internal details can be determined if desired.

Case 2

For this case, the nozzle is shockless and choked but subsonic at the exit. One must find the exit Mach number and then the exit pressure. First, from Eq. (10.2.8),

$$\frac{A_8}{A^*} = \frac{1}{M_8}\sqrt{\frac{\gamma+1}{2+(\gamma-1)M_8^2}}\left[\eta_n \frac{1+\dfrac{1-\gamma}{\eta_n(1+\gamma)}}{\dfrac{1}{1+\dfrac{\gamma-1}{2}M_8^2}-1+\eta_n}\right]^{\frac{\gamma}{\gamma-1}}.$$

Thus (A^* is approximately 0.1393 m^2),

$$\frac{0.2786}{0.1393} = 2.0 = \frac{1}{M_8}\sqrt{\frac{2.34}{2+0.34M_8^2}}\left[0.98\frac{1+\dfrac{-0.34}{0.98\times 2.34}}{\dfrac{1}{1+\dfrac{0.34}{2}M_8^2}-1+0.98}\right]^{\frac{1.34}{0.34}}.$$

Solving for M_8 (iteratively), or using Nozzle.exe one finds $M_8 = 0.3035$.

Next, from Eq. (10.2.6), one finds that

$$\frac{p_8}{p_{t6}} = \left[\frac{\dfrac{1}{1+\dfrac{\gamma-1}{2}M_8^2}-1+\eta_n}{\eta_n}\right]^{\frac{\gamma}{\gamma-1}}$$

and

$$\frac{p_8}{p_{t6}} = \left[\frac{\dfrac{1}{1+\dfrac{0.34}{2}0.3035^2}-1+0.98}{0.98}\right]^{\frac{1.34}{0.34}} = 0.9394.$$

Thus, for this case the back pressure that would cause this condition is

$p_8 = 0.9394 \times 241.3\text{kPa} = 226.7\,\text{kPa}\,(32.88\,\text{psia})$.

Case 3

For this case the nozzle is shockless and choked but now supersonic at the exit. One must again find the exit Mach number and then the exit pressure. From Eq. (10.2.8),

$$2.0 = \frac{1}{M_8}\sqrt{\frac{2.34}{2+0.34M_8^2}}\left[0.98\frac{1+\dfrac{-0.34}{0.98\times 2.34}}{\dfrac{1}{1+\dfrac{0.34}{2}M_8^2}-1+0.98}\right]^{\frac{1.34}{0.34}}.$$

This is exactly the same equation as above for case 2 (there are two roots to the equation). Solving for M_8 (iteratively, for conditions are now supersonic), or using Nozzle.exe, one finds $M_8 = 2.103$. Next, Eq. (10.2.6) yields

$$\frac{p_8}{p_{t6}} = \left[\frac{\dfrac{1}{1+\dfrac{0.34}{2}2.103^2} - 1 + 0.98}{0.98}\right]^{\frac{1.34}{0.34}} = 0.1032.$$

Thus, for case 3 the back pressure is

$p_8 = 0.1032 \times 241.35\text{kPa} = 24.91\,\text{kPa}\ (3.613\,\text{psia}).$

Also, for this case, the static temperature in the exit plane is found by

$$\frac{T_8}{T_{t6}} = \frac{1}{1+\dfrac{\gamma-1}{2}M_8^2}.$$

Thus,

$$\frac{T_8}{T_{t6}} = \frac{1}{1+\dfrac{0.34}{2}2.103^2} = 0.5708.$$

Therefore, $T_8 = 0.5708 \times 611.1\,\text{K} = 348.8\,\text{K}$ (627.9 °R). Note that the calculation of the temperature in the exit plane is independent of the ambient temperature.

Case 7

For this case the nozzle is shockless and choked, and is supersonic just in front of the exit. A normal shock is in the exit plane. Once again, the exit Mach number and then the exit pressure must be found.

Just upstream of the exit, the Mach number and pressure are exactly the same as for case 3. Using normal shock equations (i.e., Eq. (C.7.1)), or using NormalShock.exe, one obtains

$$M_j^2 = \frac{M_i^2 + \dfrac{2}{\gamma-1}}{\dfrac{2\gamma}{\gamma-1}M_i^2 - 1}.$$

Thus,

$$M_{8j}^2 = \frac{2.103^2 + \dfrac{2}{0.34}}{\dfrac{2\times 1.34}{0.34}2.103^2 - 1}.$$

Therefore, just beyond the nozzle exit plane the Mach number is

$M_{8j} = 0.5516$ (obviously subsonic),

where the subscript 8j indicates downstream on the backside of the shock. Also, across the normal shock the static pressure ratio is (using Eq. (C.7.3) or using NormalShock.exe)

$$\frac{p_j}{p_i} = \frac{2\gamma}{\gamma+1}M_i^2 - \frac{\gamma-1}{\gamma+1}.$$

Thus,

$$\frac{p_{8j}}{p_{8i}} = \frac{2 \times 1.34}{2.34}2.103^2 - \frac{0.34}{2.34} = 4.921.$$

Therefore, for case 7 the back pressure is

$$p_{8j} = 4.921 \times 24.91\,\text{kPa} = 122.6\,\text{kPa}\,(17.78\,\text{psia}).$$

Clearly, the imposed condition (68.95 kPa ambient conditions) falls between cases 3 and 7. Thus, by examining Figure 10.4 one can quickly see that the imposed condition corresponds to case 4 and nonisentropic shocks and compressions outside the nozzle. However, up to the nozzle exit, the flow behaves as for case 3.

Had the back pressure been greater than 226.7 kPa (32.88 psia), the flow would have been subsonic throughout the nozzle. If the back pressure had been between 122.6 and 226.7 kPa, a standing normal shock would have been present in the diverging section of the nozzle. The axial position of the shock would have had to be found by iteration, as discussed in the next example. Had the back pressure been less than 24.91 kPa, the flow within the nozzle would have been exactly as it is for case 3; however, a series of nonisentropic 2-D expansions would have occurred outside the nozzle.

Since the flow is choked for the case of interest, one can simply use the conditions from case 3 to calculate the mass flow rate. That is,

$$\dot{m} = \rho_8 u_8 A_8,$$

but $u_8 = M_8\, a_8$ and

$$\begin{aligned} a_8 &= \sqrt{\gamma \mathcal{R} T_8} \\ &= \sqrt{1.34 \times 287.1\frac{\text{J}}{\text{kg-K}} \times 348.8\ \text{K} \times \frac{\text{N-m}}{\text{J}} \times \frac{\text{kg-m}}{\text{N-s}^2}} \\ &= 366.4\ \text{m/s}\ (1202\ \text{ft/s}). \end{aligned}$$

Thus $u_8 = 2.103 \times 366.4 = 770.2$ m/s (2527 ft/s).

Also, from the ideal gas law, for $T_8 = 348.8$ K and $p_8 = 24.91$ kPa,

$$\rho_8 = 0.2489\ \text{kg/m}^3 (0.0004828\ \text{sl/ft}^3).$$

Lastly, for the mass flow rate,

$$\dot{m} = 0.2489\frac{\text{kg}}{\text{m}^3} \times 770.2\frac{\text{m}}{\text{s}} \times 0.2786\ \text{m}^2$$
$$= 53.42\,\text{kg/s}\,(117.8\,\text{lb}_\text{m}/\text{s}).$$

Example 10.2 | Diverging Nozzle with an Internal Normal Shock

An exhaust air stream at Mach 2.0, pressure 10 psia, and temperature 1400 °R enters a frictionless diverging nozzle with a ratio of exit area to inlet area of 3.0. Determine the back pressure necessary to produce a normal shock in the channel at an area equal to twice the inlet area. Assume 1-D steady flow with the air behaving as a perfect gas with constant specific heats and a specific heat ratio of 1.36; assume isentropic flow except for the normal shock. See Figure 10.11a.

Solution

For isentropic flow, the area at which Mach number is equal to 1 is defined as A^*, and, this area is used as a reference. In this problem, the flow is not isentropic from cross section "i" to cross section "j" because of the normal shock. Therefore, the reference area A_i^* for the flow from A_in to A_i is not equal to the reference area A_j^* for the isentropic flow from A_j to A_out. Determining the flow conditions after the shock requires a relationship between A_i^* and A_j^*. In Figure 10.11b, section "i" is immediately before the shock, and section "j" is immediately after the shock. The flow from "in" to "i" is isentropic and accelerating because of the divergence; thus,

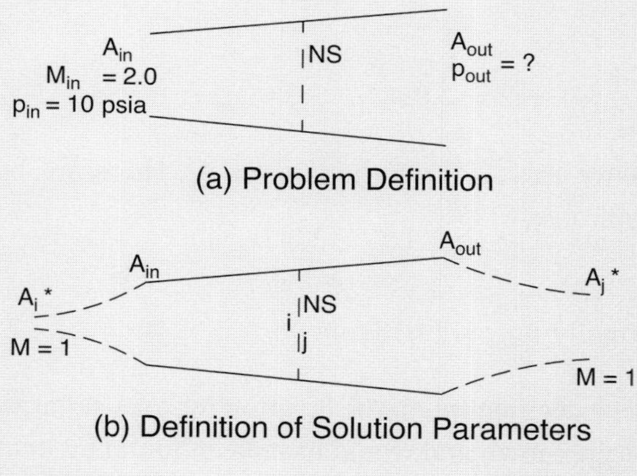

Figure 10.11 Geometry for Example 10.2.

$p_{tin} = p_{ti}$.

But for $M_{in} = 2.0$, one finds $A_{in}/A_i^* = 1.719$ (from isentropic flow tables or equations or Isentropic.exe); thus,

$p_{in}/p_{ti} = 0.1289$

or $p_{ti} = 10/0.1289 = 77.58$ psia.

Also,

$$\frac{A_i}{A_i^*} = \frac{A_i}{A_{in}}\frac{A_{in}}{A_i^*};$$

thus, $A_i/A^* = 2 \times 1.719 = 3.439$,

$M_i = 2.723$ (from isentropic flow tables or Isentropic.exe),

and $p_i/p_{ti} = 0.0406$,

or $p_i = 0.0406 \times 77.58 = 3.15$ psia.

By knowing the incoming Mach number for a normal shock, one can find the exiting Mach number (e.g., by normal shock tables or NormalShock.exe):

$M_j = 0.4855$

$p_{tj}/p_{ti} = 0.3991$

$p_j/p_i = 8.393$.

Thus, $p_j = 8.393 \times 3.15 = 26.75$ psia

and $p_{tj} = 0.3991 \times 77.58 = 30.96$ psia.

Now, since the decelerating flow due to divergence from "j" to "out" is isentropic for $M_j = 0.4855$, one finds $A_j/A^* = 1.372$ (by isentropic flow tables or equations or Isentropic.exe) and

$$\frac{A_{out}}{A_j^*} = \frac{A_{out}}{A_{in}}\frac{A_{in}}{A_j}\frac{A_j}{A_j^*} = 3 \times \frac{1}{2} \times 1.372 = 2.058.$$

Note again that the reference area A_j^* is not the same as A_i^*. Therefore, by isentropic flow tables or calculations or Isentropic.exe,

$M_{out} = 0.2975$

and $p_{out}/p_{tj} = 0.942$,or finally $p_{out} = 29.16$ psia.

Had the exit pressure been specified (p_{out}), the location (or area A_i) at which the shock was located would not have been known and would have been found by iteration. For example, one would assume a value of A_i, go through the above procedure, and calculate p_{out}. If the calculated value of p_{out} did not match the specified exit pressure, a new value of A_i would be guessed and iterated upon until the calculated and specified exit pressures matched. The iteration is quick to converge if, for example, the Reguli Falsi method is used (Appendix I).

10.5 Effects of Pressure Ratios on Engine Performance

As indicated in Chapters 1, 2, and 3 in this book, if the nozzle exit pressure exactly matches the ambient pressure, the only thrust component is due to the momentum flux. If the nozzle exit pressure is less than the ambient condition (overexpanded), a negative thrust due to the pressures results. However, a larger-than-ideal thrust from the momentum flux occurs because the exit velocity is larger than for the ideally expanded case. On the other hand, if the nozzle pressure is higher than the ambient pressure (underexpanded), a positive thrust results from the pressure terms. However, a lower-than-ideal thrust results from the momentum flux because a lower-than-ideal exit velocity occurs. Since, for both cases for which the exit pressure fails to match the ambient pressure, the two thrust components have opposite trends, one may wish to know for what conditions the optimum (or maximum) thrust occurs. This question is best answered with examples.

Example 10.3 | Exit Pressure Effects on Ideal Turbojet Thrust

An ideal turbojet flies at sea level ($p_a = 14.69$ psia) at Mach 0.75. It ingests 165 lb_m/s of air, and the compressor operates with a pressure ratio of 15. The fuel has a heating value of 18,000 BTU/lb_m, and the burner exit total temperature is 3000 °R. The specific heat ratio is 1.35. Find the developed thrust for a nozzle exit pressure from 12.69 to 16.69 psia that is controlled by the exit area. This is similar to Example 3.3 with the exception of the difference between the exit and ambient pressures.

Solution

Because this is nearly identical to Example 3.3, details of the computations are not shown. However, bear in mind that the exit pressures are specified for this problem and are not set equal to the ambient pressure.

Basically, results are presented graphically in Figure 10.12. Shown here are the net thrust, exit area, and exit jet velocity as functions of the exit pressure. As can be seen, the maximum thrust occurs when the exit pressure (p_8) exactly matches the ambient pressure (p_a). Although this fact was shown numerically here, it can also be shown analytically by differentiating the net thrust equation, setting the term to zero, and solving for p_8 (Problem 10.17). From Figure 10.12, one should note that, although the maximum thrust occurs when $p_8 = p_a$, under off-design conditions the net thrust does not deteriorate quickly.

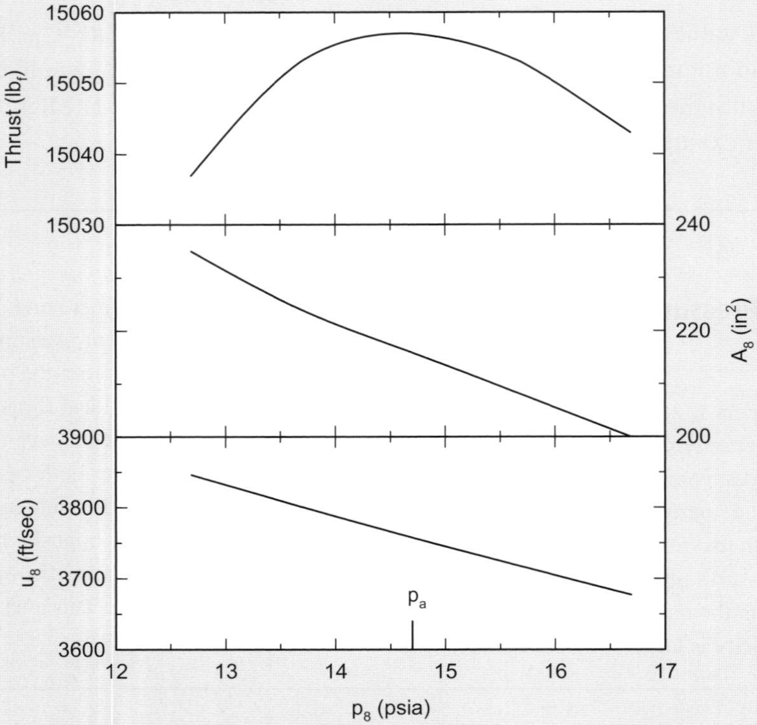

Figure 10.12 Variation of net thrust, exit area, and exit velocity with exit pressure for an ideal turbojet.

Also shown is that the required exit area decreases with increasing exit pressure and that the exit velocity follows the same trend. Thus, at low pressures, the thrust due to the momentum flux is greater than for the case in which the exit pressure matches the ambient pressure; however, at this condition the pressure-generated thrust is negative and more than offsets the increase due to momentum flux. Similarly, at high pressures, although the pressure generates a positive thrust, the reduced momentum flux more than offsets this increase.

Example 10.4 | **Exit Pressure Effects on Nonideal Turbojet Thrust**

A turbojet flies at sea level ($p_a = 14.69$ psia) at a Mach number of 0.75. It ingests 165 lb_m/s of air. The compressor operates with a pressure ratio of 15 and an efficiency of 88 percent. The fuel has a heating value of 18,000 BTU/lbm, and the burner total temperature is 3000 °R. The burner has an efficiency of 91 percent and a total pressure ratio of 0.95, whereas the turbine has an efficiency of 85 percent. A CD nozzle is used, and total pressure recovery for the diffuser is 0.92. The shaft efficiency is 99.5 percent. Find the developed thrust as a function of nozzle efficiency and exit pressure. This is similar to Extended Example 3.4.2, except the exit pressure and nozzle efficiency are now varied. This is also similar to the preceding example with the exception of the nonideal effects.

Solution

With the exception of the nozzle efficiency and exit pressure, this is identical to Extended Example 3.4.2, and details of the computations will not be shown. Keep in mind, however, that the exit pressures are specified for this problem and are not set equal to the ambient pressure but are varied by changing the exit area. Namely, these results are for a set of fixed CD nozzles. Results are presented graphically in Figure 10.13. The net thrust is shown here as a function of the exit pressure for three values of nozzle efficiency: 1.00, 0.975, and 0.95.

As can be seen from Figure 10.13, for $\eta_n = 1.00$ the maximum thrust is once again obtained (and the maximum occurs for approximately $p_e = p_a$). However, for $\eta_n = 0.975$ the maximum thrust occurs at an exit pressure of approximately 15.8 psia. Finally, for $\eta_n = 0.95$, the maximum thrust is realized at an exit pressure of about 17.2 psia. Thus, as the nozzle efficiency is decreased, more thrust is obtained if the exit pressure is above the ambient conditions. This indicates that, for these conditions, less of the enthalpy is converted to the momentum flux and that more additional thrust is obtained from the pressure generator term. This conclusion is also implied by the definition of the nozzle efficiency. Note also that the maximum of the thrust–exit pressure curve is less "peaked" for lower efficiencies, and thus when there is a deviation from the optimum exit pressure, the thrust does not deteriorate significantly.

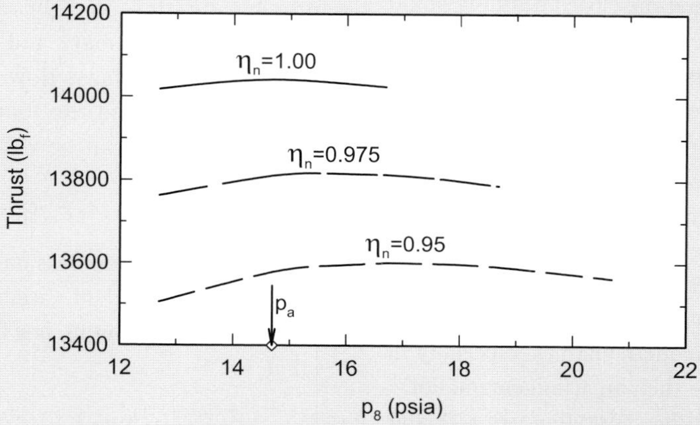

Figure 10.13 Variation of net thrust with exit pressure for three nozzle efficiencies for a nonideal turbojet.

10.6 Variable Nozzle

One means by which a nozzle can be forced to operate at maximum thrust or "on-design" under many different flight conditions is through a variable geometry. For such a nozzle, the area of the exit, throat, or both is varied so that the optimum exit pressure can be obtained. Two nozzle shapes are usually used: a simple circular (iris) nozzle or a plug nozzle. Note that variable nozzles are typically not used on commercial aircraft because of the limited flight envelope – that is, the craft and engine basically are optimally designed for one altitude and flight condition. Military fighters, on the other hand, must be designed to operate under a variety of conditions, including aircraft carrier takeoff. As a result, most of these engines have variable nozzles.

In Figure 10.14, the Pratt & Whitney TF30 afterburning turbofan (fan mixed) is shown. This engine is equipped with a variable iris nozzle. In Figure 10.14a, a photograph is shown, and Figure 10.14b presents a schematic. Two nozzle orientations are shown in the schematic: the fully open position with the largest exit area and the fully closed position with the smallest exit area. The open position would be used for high altitudes (low ambient pressure), and the closed position would be used for low altitudes. The nozzle assembly is essentially a series of curved "flaps," each of which is a four-bar linkage system driven by a hydraulic piston (Figure 10.14c). By moving the identical flaps fore-to-aft by identical amounts, the areas are changed by prescribed amounts based on the movement amplitude and flap shapes. Both the exit and throat areas are controlled for the Pratt & Whitney TF30. Positions between the two extremes can also be used to optimize the performance. Many modern engines, especially in military applications, use such an arrangement. For example, the Pratt & Whitney F100 (Figures 1.33 and 10.6) also has a variable iris nozzle. With the proper flap geometry or shape, the nozzle can be operated as either a converging nozzle or a CD nozzle. Furthermore, with the proper shape, the minimum and exit areas can both be varied.

A second geometry that is sometimes used is the plug nozzle shown in Figure 10.15. For this geometry, the exit area is not only varied but the throat area is controlled. Thus, varying the axial location of the plug changes the area ratio to adjust the exit pressure. Two such positions of the plug are shown in Figure 10.15. The major problem with this geometry is that the plug and the positioning mechanism are subjected to the extreme freestream temperatures.

10.7 Performance Maps

10.7.1 Dimensional Analysis

Performance "maps" are usually used, as for a diffuser, to provide a working medium or methodology for a nozzle engineer. Again, these maps are used to simplify a substantial amount of data and make these data most versatile for use over a wide range of operation. These maps provide a quick and accurate view of the conditions at which the nozzle is operating and what can be expected of the nozzle if the flow conditions are changed. Of primary concern is the mass flow rate of a nozzle – that is $\dot{m}_6$ for a primary nozzle or $\dot{m}_7$ for a fan nozzle – and the exit Mach number (M_8 and M_9 for the primary and fan nozzles, respectively). The mass flow rate is a function of many variables, and the functional relationship for a primary nozzle can be written as

$$\dot{m}_6 = f\{T_{t6}, A_8, p_{t6}, p_a, \gamma, \mathcal{R}\}. \tag{10.7.1}$$

One can use dimensionless parameters to reduce the number of independent variables. Using the Buckingham Pi theorem, one finds for a given value of γ that

$$\dot{m}_6 = \frac{\sqrt{\mathcal{R}T_{t6}}}{A_8 p_{t6}} = f\left\{\frac{p_{t6}}{p_a}\right\}. \tag{10.7.2}$$

This form is rarely used to document the performance of a nozzle, however. Usually, one finds a modification of this; that is,

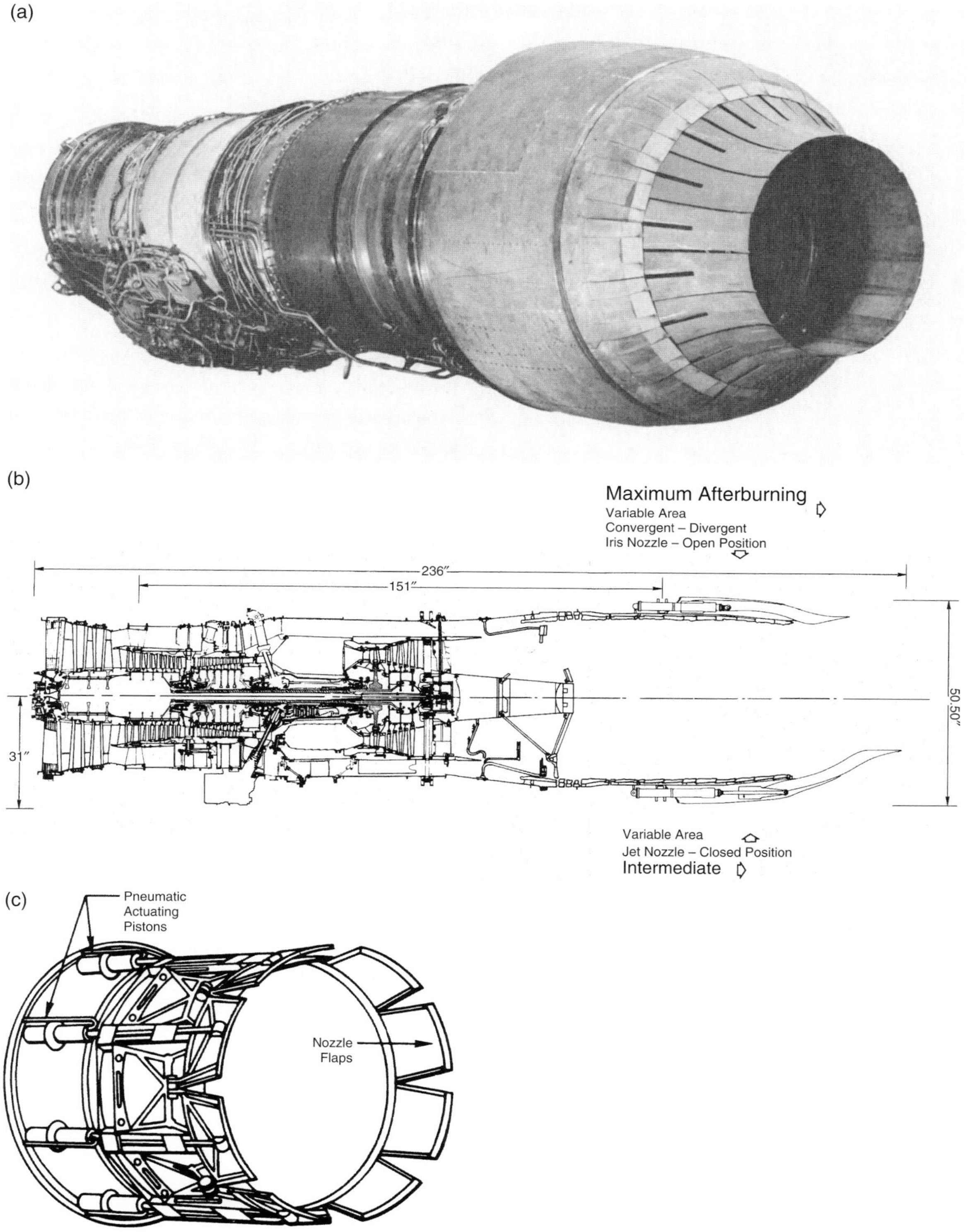

Figure 10.14 (a) Photograph of Pratt & Whitney TF30 with variable exhaust nozzle. (b) Schematic of Pratt & Whitney TF30 showing two nozzle positions. (c) Schematic of iris nozzle with flaps (all © 2022 Raytheon Technologies Corporation, Pratt & Whitney division. Reproduced with permission).

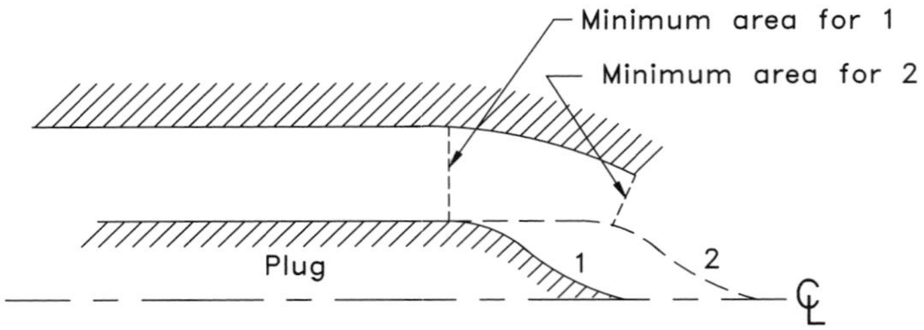

Figure 10.15 Plug nozzle.

$$\dot{m}_6 \frac{\sqrt{\theta_{t6}}}{\delta_{t6}} = \dot{m}_{c6} = f\left\{\frac{p_{t6}}{p_a}\right\}, \tag{10.7.3}$$

where

$$\delta_{t6} = \frac{p_{t6}}{p_{stp}} \tag{10.7.4}$$

and

$$\theta_{t6} = \frac{T_{t6}}{T_{stp}}, \tag{10.7.5}$$

and where the subscript stp refers to the standard conditions. For reference and from Appendix B, p_{stp} is equal to 14.69 psia or 101.33 kPa, and T_{stp} is equal to 518.7 °R or 288.2 K. Note that the two parameters in Eq. (10.7.3) are not truly dimensionless but are prudently chosen so that the function f retains the important characteristics of nondimensionalization. As for compressors and turbines, the first parameter (left-hand side) is called the "corrected" mass flow and has dimensions of mass flow, whereas the pressure ratio (inside the brackets) is dimensionless. Thus, since both are not dimensionless and, in particular, since the parameter A_8 has been eliminated, the resulting function cannot be used to correlate or compare different nozzles. The function can be used to characterize the data over a range of conditions, altitudes, and so on for one particular nozzle, however, as is the usual case. Thus, a nozzle can be tested under one set of conditions, and the results can be "corrected" so that results are applicable to other conditions so long as the parameter p_{t6}/p_a is matched. Similarly, for the exit Mach number, one finds

$$M_8 = g\left\{\frac{p_{t6}}{p_a}\right\}, \tag{10.7.6}$$

where the function g is different from that of f. One can develop similar relationships for a secondary nozzle. The functions g and f can be derived from experimental data, modeling, or a combination of empiricism and modeling.

10.7.2 Trends

Fixed Converging Nozzle

Typical trends of $\dot{m}_{c6}$ and M_8 are shown in Figures 10.16–10.18 for three particular nozzle types. In Figure 10.16, typical trends for a fixed converging nozzle are shown. As can be seen, the corrected mass flow rate increases as the pressure ratio rises – until the nozzle chokes – and then remains constant as the pressure ratio increases further.

For this case, one can show that

$$\dot{m}_{c6} = p_{stp} a_{stp} A_e \frac{p_8}{p_{t6}} M_8 \sqrt{\frac{\gamma}{\gamma_{stp}}} \sqrt{\frac{\mathcal{R}_{stp}}{\mathcal{R}}} \sqrt{1 + \frac{\gamma - 1}{2} M_8^2}, \tag{10.7.7}$$

which includes variations in the specific heat ratio and ideal gas constant. Next, by grouping some constants and defining a reference mass flow rate $\dot{m}_n$ as p_{stp} a_{stp} A_e, one finds

$$\dot{m}_{c6} = \dot{m}_n \frac{p_8}{p_{t6}} M_8 \sqrt{\frac{\gamma}{\gamma_{stp}}} \sqrt{\frac{\mathcal{R}_{stp}}{\mathcal{R}}} \sqrt{1 + \frac{\gamma - 1}{2} M_8^2}. \tag{10.7.8}$$

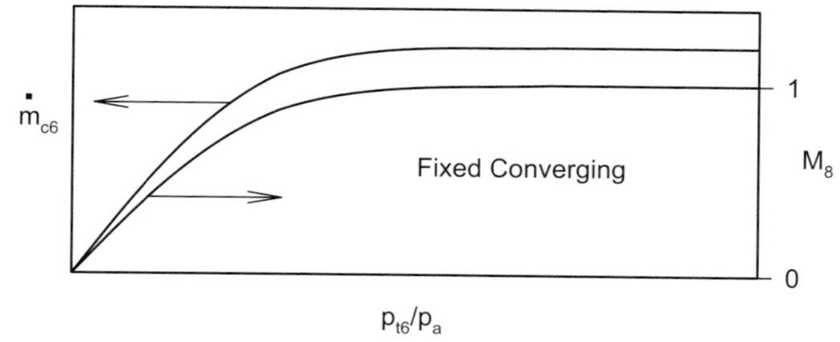

Figure 10.16 Typical nozzle performance map for a fixed converging nozzle.

For this case, if p_a is less than p^* given by Eq. (10.2.7), then p_8 is about equal to p^* and the exit Mach number is very close to unity. Note that, if the nozzle is choked and if p_{t6} is increased, the corrected mass flow rate $\dot{m}_{c6}$ remains constant, whereas the true flow rate $\dot{m}_6$ increases. However, if p_a is greater than or equal to p^*, then p_8 is equal to p_a and the exit Mach number is equal to

$$M_8 = \sqrt{\frac{2}{\gamma - 1}\left[\frac{1}{\eta_n\left[\frac{p_8}{p_{t6}}\right]^{\frac{\gamma-1}{\gamma}} + 1 - \eta_n} - 1\right]}. \tag{10.7.9}$$

Variable Converging–Diverging Nozzle: Fixed Exit Area

Figure 10.17 shows typical trends for a variable CD nozzle with a fixed exit area and a "schedule" so that the exit pressure matches the ambient pressure. A "schedule" is a prescribed and predetermined specific variation of the minimum area that is a function of altitude and other local conditions. For this case, the exit Mach number can be subsonic or supersonic and the corrected mass flow increases until the nozzle chokes at the exit and then decreases. For this case, one again finds, as for the converging nozzle, that

$$\dot{m}_{c6} = p_{stp} a_{stp} A_e \frac{p_8}{p_{t6}} M_8 \sqrt{\frac{\gamma}{\gamma_{stp}}}\sqrt{\frac{\mathcal{R}_{stp}}{\mathcal{R}}}\sqrt{1 + \frac{\gamma - 1}{2} M_8^2}. \tag{10.7.10}$$

Once again, by grouping some constants and defining $\dot{m}_n$ as for the converging nozzle, one finds

$$\dot{m}_{c6} = \dot{m}_n \frac{p_8}{p_{t6}} M_8 \sqrt{\frac{\gamma}{\gamma_{stp}}}\sqrt{\frac{\mathcal{R}_{stp}}{\mathcal{R}}}\sqrt{1 + \frac{\gamma - 1}{2} M_8^2}, \tag{10.7.11}$$

where the exit Mach number is always given by

$$M_8 = \sqrt{\frac{2}{\gamma - 1}\left[\frac{1}{\eta_n\left[\frac{p_8}{p_{t6}}\right]^{\frac{\gamma-1}{\gamma}} + 1 - \eta_n} - 1\right]}. \tag{10.7.12}$$

Variable Converging–Diverging Nozzle: Fixed Minimum Area

Finally, in Figure 10.18 typical trends are shown for a variable CD nozzle with a choked and fixed minimum area and again a "schedule" to match the exit pressure to the ambient pressure by varying the exit area. As can be seen, the corrected mass flow is constant regardless of the pressure ratio; however, the exit Mach number may again be subsonic or supersonic. For this case, once again

$$\dot{m}_{c6} = p_{stp} a_{stp} A_e \frac{p_8}{p_{t6}} M_8 \sqrt{\frac{\gamma}{\gamma_{stp}}}\sqrt{\frac{\mathcal{R}_{stp}}{\mathcal{R}}}\sqrt{1 + \frac{\gamma - 1}{2} M_8^2}, \tag{10.7.13}$$

which can be put in the form

$$\dot{m}_{c6} = p_{stp} a_{stp} A^* \frac{A_e}{A^*} \frac{p_8}{p_{t6}} M_8 \sqrt{\frac{\gamma}{\gamma_{stp}}}\sqrt{\frac{\mathcal{R}_{stp}}{\mathcal{R}}}\sqrt{1 + \frac{\gamma - 1}{2} M_8^2}, \tag{10.7.14}$$

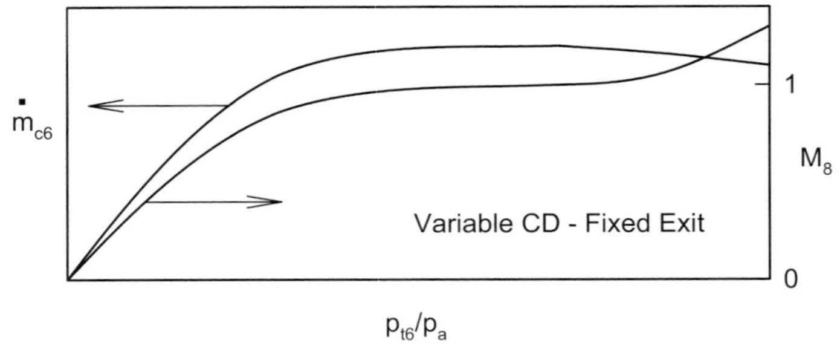

Figure 10.17 Typical nozzle performance map for a variable CD nozzle with a fixed exit area.

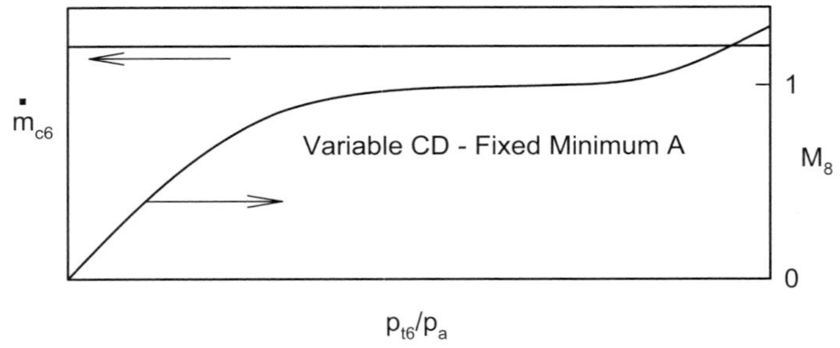

Figure 10.18 Typical nozzle performance map for a variable CD nozzle with a fixed minimum area.

where Eq. (10.7.9) still applies. By equating the mass flows at the minimum area (which is sonic) and exit ($\dot{m} = \rho VA = pMA\sqrt{\gamma\mathcal{R}T}/(\mathcal{R}T)$), one finds

$$\frac{A_8}{A^*} = \frac{\sqrt{\dfrac{\gamma+1}{2+(\gamma-1)M_8^2}}}{M_8\dfrac{p_8}{p^*}}. \tag{10.7.15}$$

From Eqs. (10.2.6) and (10.2.7),

$$\frac{p_8}{p^*} = \left[\frac{1}{\eta_n}\left[\frac{\eta_n}{\dfrac{2}{\gamma+1}-1+\eta_n}\right]\left[\frac{1}{1+\dfrac{\gamma-1}{2}M_8^2}-1+\eta_n\right]\right]^{\frac{\gamma}{\gamma-1}}. \tag{10.7.16}$$

It can be shown that the three preceding equations reduce to

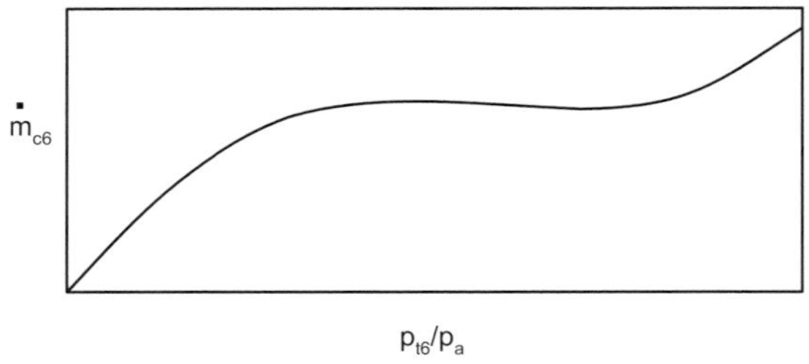

Figure 10.19 General nozzle map.

$$\dot{m}_{c6} = p_{stp}a_{stp}A^*\sqrt{\frac{\gamma}{\gamma_{stp}}}\sqrt{\frac{\mathcal{R}_{stp}}{\mathcal{R}}}\sqrt{\frac{\gamma+1}{2}}\left[\frac{\eta_n(\gamma+1)+1-\gamma}{\eta_n(\gamma+1)}\right]^{\frac{\gamma}{\gamma-1}}, \tag{10.7.17}$$

or, by defining $\dot{m}_n$ as $\rho_{stp}\ a_{stp}\ A^*$, which is different than for the converging nozzle, one obtains

$$\dot{m}_{c6} = \dot{m}_n \sqrt{\frac{\gamma}{\gamma_{stp}}} \sqrt{\frac{\mathcal{R}_{stp}}{\mathcal{R}}} \sqrt{\frac{\gamma+1}{2}} \left[\frac{\eta_n(\gamma+1)+1-\gamma}{\eta_n(\gamma+1)}\right]^{\frac{\gamma}{\gamma-1}}. \quad (10.7.18)$$

It is important (and interesting) to note that, for this type of nozzle, $\dot{m}_{c6}$ is constant for all conditions.

Other Nozzle Geometries

The preceding nozzle geometries and figures are intended to show typical trends for three limiting nozzle cases. Typically, variable nozzles do not have constant exit or minimum areas and both may change; a general map is shown in Figure 10.19. Thus, actual maps will combine the trends for the two types covered above. The actual maps (and functions f and g) will depend on the precise "schedules" of the two area variations of the nozzles, which will be functions of the pressure ratio and altitude.

Example 10.5 | Corrected Mass Flow Rate for a Converging Nozzle

Find the corrected mass flow rate for a fixed converging primary nozzle for a total inlet pressure of 344.8 kPa (50 psia), a total inlet temperature of 944.4 K (1700 °R), an exit area of 0.1613 m^2 (250 in.^{2}), an efficiency of 0.97, and $\gamma = 1.36$. The atmospheric pressure is 68.95 kPa (10 psia).

Solution

First, one must determine if the nozzle is choked under these conditions. Thus, from Eq. (10.2.7),

$$\frac{p^*}{p_{t6}} = \left[1 + \frac{1-\gamma}{\eta_n(1+\gamma)}\right]^{\frac{\gamma}{\gamma-1}} = \left[1 + \frac{-0.36}{0.97 \times 2.36}\right]^{\frac{1.36}{0.36}} = 0.5239,$$

and so $p^* = 0.5239 \times 344.8\,\text{kPa} = 180.6\,\text{kPa}$ (26.20 psia).

Since p_a is less than p^*, the flow is choked ($M_8 = 1.00$), and thus $p_8 = p^* = 180.6\,\text{kPa}$ (26.20 psia).

Note also that

$$p_{t6}/p_a = 5.$$

Next, from Eq. (10.2.4),

$$\frac{T_8}{T_{t6}} = \frac{1}{1 + \frac{\gamma-1}{2}M_8^2} = \frac{1}{1 + \frac{0.36}{2}} = 0.8475,$$

and thus $T_8 = 0.8475 \times 944.4\,\text{K} = 800.6\,\text{K}$ (1441 °R).

From the ideal gas law, for p_8 and T_8, one finds

$$\rho_8 = 0.7866\,\text{kg/m}^3 (0.001526\,\text{sl/ft}^3).$$

Also, for an ideal gas

$$a_8 = \sqrt{\gamma \mathcal{R} T_8} = \sqrt{1.36 \times 287.1 \frac{\text{J}}{\text{kg-K}} \times 800.6\,\text{K} \times \frac{\text{N-m}}{\text{J}} \times \frac{\text{kg-m}}{\text{N-s}^2}}$$
$$= 559.0\,\text{m/s}\,(1834\,\text{ft/s}).$$

Therefore, the mass flow rate at the exit (which is the same as the inlet) is

$$\dot{m}_6 = \dot{m}_8 = \rho_8 u_8 A_8 = 0.7866 \frac{\text{kg}}{\text{m}^3} \times 559.9 \frac{\text{m}}{\text{s}} \times 0.1613\ \text{m}^2$$
$$= 70.86\ \text{kg/s}\ (4.857\ \text{sl/s}).$$

Finally,

$$\dot{m}_{c6} = \dot{m}_6 \frac{\sqrt{\theta_{t6}}}{\delta_{t6}} = \dot{m}_6 \frac{\sqrt{\dfrac{T_{t6}}{T_{stp}}}}{\dfrac{p_{t6}}{p_{stp}}} = 70.86 \frac{\sqrt{\dfrac{944.4}{288.2}}}{\dfrac{344.8}{101.3}} = 37.69\ \text{kg/s}\ (83.11\ \text{lb}_\text{m}/\text{s}).$$

Note that one would have obtained the same results for any value of atmospheric pressure below the choking condition (180.6 kPa).

Also, if nozzle conditions had changed but the pressure ratio had remained constant ($p_{t6}/p_a = 5$), for different values of p_{t6} the mass flow rate would change but the corrected mass flow rate would remain the same. On the other hand, had p_{t6}/p_a increased for a constant value of p_{t6} (decreased p_a), the mass flow rate would remain unchanged, and the corrected mass flow rate would again remain the same. Regardless, the nondimensional dependence (Eq. (10.7.3)) is preserved. These calculations are left as an exercise for the reader.

10.8 Thrust Reversers and Vectoring

10.8.1 Reversers

The difficult problem of stopping an aircraft after landing has become more pronounced with modern aircraft because of the large aircraft weights, high speeds, and existing runways. Wheel brakes alone are not an effective means to stop such an aircraft owing to brake pad and tire thermal limitations. A reversible pitch propeller is used for turboprops to reverse the thrust direction upon landing. Turbojet and turbofans do not have such an option, however. To provide a "brake" for such aircraft, a thrust reverser is used.

For this, the turbine exhaust, fan air, or both are diverted at a suitable angle in the reverse direction by the means of an inverted cone, half-sphere, turning vanes, or other shape introduced in the exhaust flow upon landing. Because the exhaust flow is turned by almost 180°, the linear momentum equation can be used to show that the thrust is nearly reversed. The clamshell and cascade reversers are two of the most common of these devices.

In Figure 10.20, the clamshell type is shown. For this geometry, the clamshell is opened upon landing and the shell is approximately one diameter downstream of the exhaust. When the reverser is not in use, the shell is retracted and stowed around the nozzle. Sometimes the

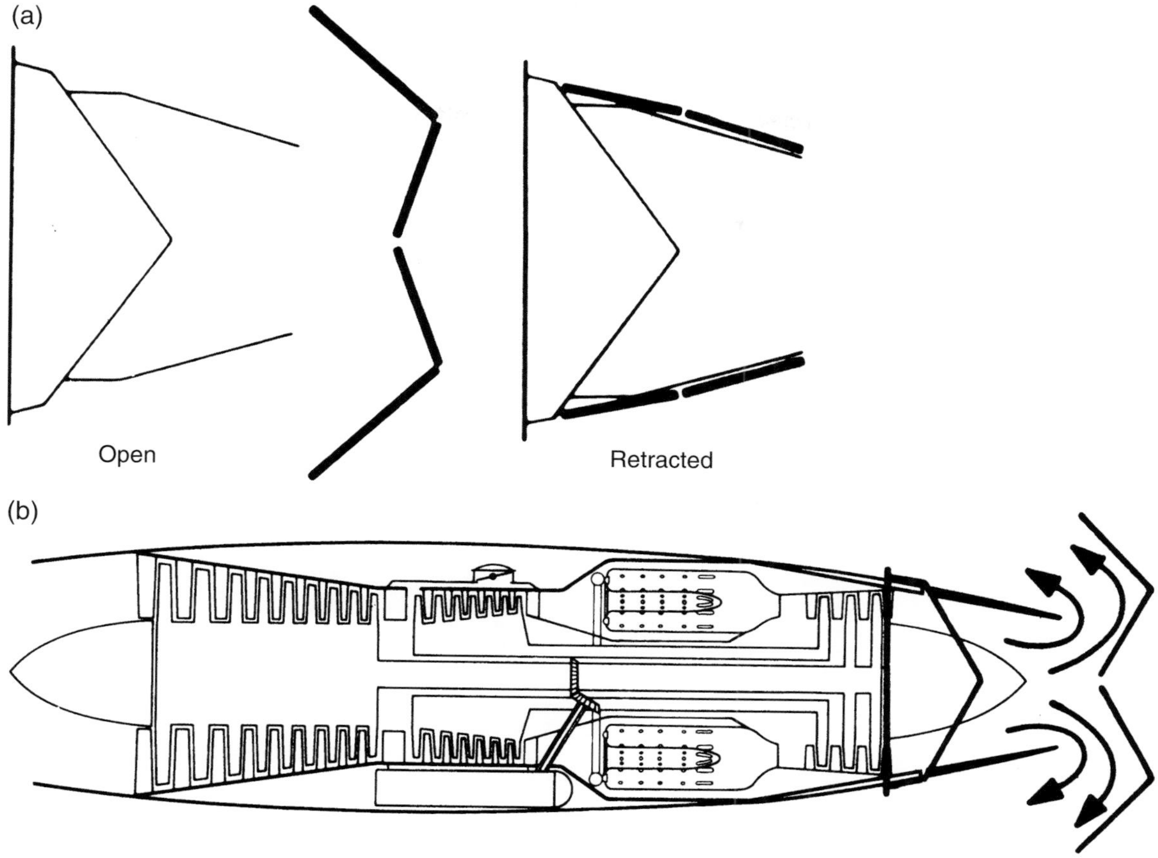

Figure 10.20 (a) Clamshell thrust reverser and (b) clamshell thrust reverser in operation (both © 2022 Raytheon Technologies Corporation, Pratt & Whitney division. Reproduced with permission).

retracted shell forms a part of the rear section of the nozzle nacelle.

A cascade-type reverser employs numerous vanes in the gas path to reverse the gas flow. Reversers for turbofans are often of this type. In Figures 10.21 and 10.22, the reversers for an exhausted turbofan are shown. For the fan ("cold air") itself, only turning vanes are used, as shown for the two positions. For forward thrust, the vanes are out of the gas path; however, for reverse thrust, they are moved into the flow path. For the primary flow (hot gas), a combination of clamshell and cascade reversers are used. For this geometry, the clamshell is directly downstream of the turbine and, when activated, diverts the flow to the fixed cascade of vanes. Hydraulic pistons move both the movable vanes and clamshell. It is interesting to note that the engine manufacturer is usually not totally responsible for the design of thrust reversers but that this is left to the airframe manufacturer or a joint effort between the airframe end engine manufacturer.

10.8.2 Vectoring

To gain more control of aircraft maneuverability in military fighter applications, manufacturers

Figure 10.20 (c) Rolls-Royce BR710 clamshell (© Rolls-Royce plc2022. All rights reserved).

have begun to develop thrust vectoring (thrust direction control) as a part of the exhaust of the engine. Thus, instead of the thrust being along the engine centerline, the thrust can be along another vector. Vectoring the thrust reduces response times. Such designs also allow for more rapid takeoffs. Applications include two-directional vectoring and multidirection vectoring. Thrust vectors can typically be varied from up to 20° from the nominal direction. Although the design is somewhat more complicated, weight and complexity of vectoring thrust nozzles are not much greater than for nozzles with variable areas. In fact, the all-directional vectoring nozzles use the iris variable area design as a basis for the vectoring. That is, the "flaps" are independently controlled and moved by independent amounts, thus allowing the thrust direction to be changed.

Two examples of vectoring are shown in Figures 10.23 and 10.24. In Figure 10.23, a two-directional thrust-vectoring engine is shown (Pratt & Whitney F119-PW-100). Figure 10.24 shows a double-exposed photograph for the Pratt & Whitney "Pitch-Yaw Balanced Beam Nozzle" (PYBBM). This is a multidirection iris design that can be used on the P&W F100 engine. Two different exhaust streams can easily be seen on this photograph resulting from two nozzle settings. This allows the reader to experience the visual effect of changing the thrust direction.

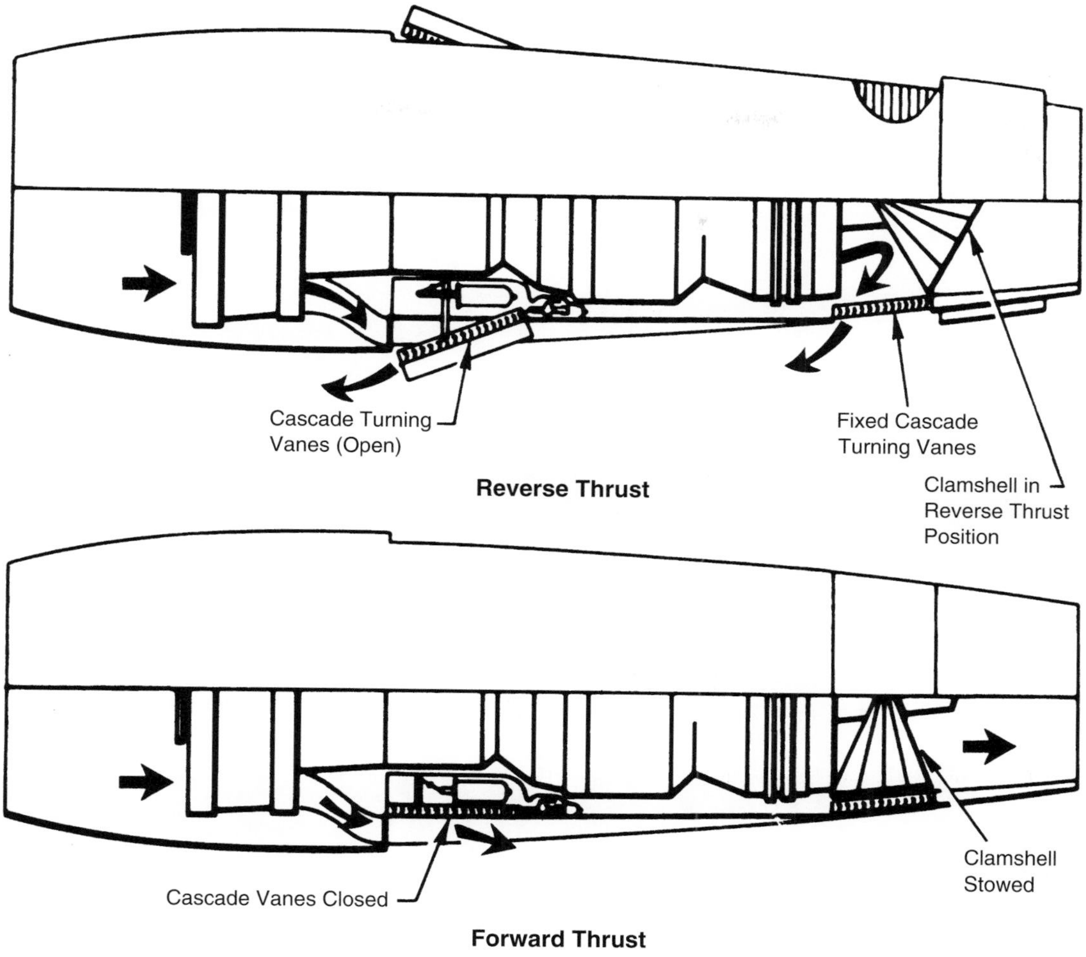

Figure 10.21 Thrust reversers for turbofan engine – exhausted fan (© 2022 Raytheon Technologies Corporation, Pratt & Whitney division. Reproduced with permission).

Figure 10.22 "Cold air" thrust reverser on a Rolls-Royce engine (© Rolls-Royce plc 2022. All right reserved).

Figure 10.23 Two-directional thrust vectoring for Pratt & Whitney F119-PW-100 (© 2022 Raytheon Technologies Corporation, Pratt & Whitney division. Reproduced with permission).

10.9 Summary

This chapter discussed exit nozzles. The primary functions of nozzles are to convert high-pressure, high-temperature energies (enthalpies) to kinetic energies (high velocities) and to straighten the flow so that the flow exits in the axial direction develop high thrust levels. The basic concepts apply to both fan and primary nozzles. Both converging and converging–diverging (CD) nozzles were covered. Methods were developed by which 1-D flow can be predicted in the different nozzles. Software Isentropic.exe, Nozzle.exe, and NormalShock.exe were shown to aid in calculations if wanted. Internal friction, defined as a part of a nozzle efficiency, was studied and was seen to affect the thrust. Basically, two distinct operating conditions (choked and nonchoked) can occur for a converging nozzle, whereas seven conditions can occur in a CD nozzle. Choking was seen to be a potential problem for all fixed-geometry nozzles and was seen to limit the exit

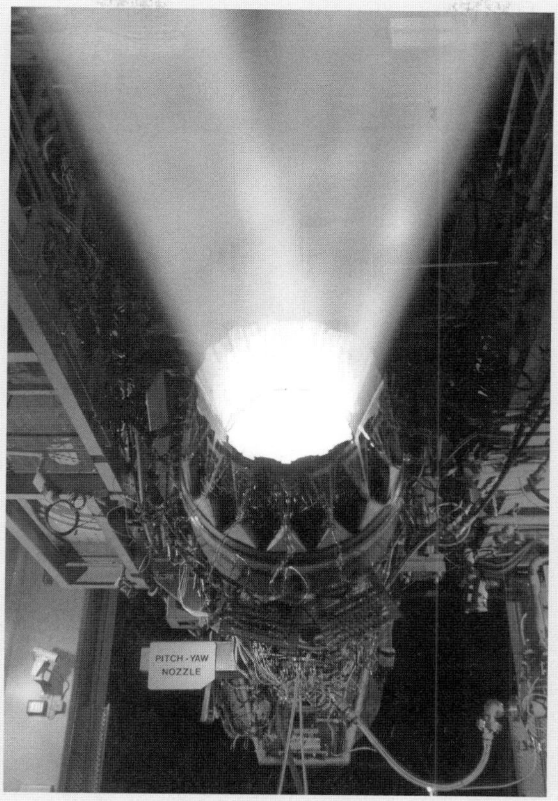

Figure 10.24 Multidirectional thrust vectoring for Pratt & Whitney PYBBN (© 2022 Raytheon Technologies Corporation, Pratt & Whitney division. Reproduced with permission).

velocity and thus to detract from the ideal developed thrust. Shocks in the diverging sections of CD nozzles were shown to increase nozzle losses. For an ideal nozzle, the maximum thrust is obtained when the nozzle exit pressure matches the ambient pressure. However, for a nonideal nozzle, slightly more thrust is obtained when the nozzle exit pressure is somewhat above the ambient pressure. Some flexibility in operating conditions is often provided by using variable nozzles so that an engine can operate with maximum thrust at a variety of altitudes and flight conditions – usually for military fighters but not for commercial aircraft. Next, performance curves or "maps" based on similarity parameters were presented as a method of summarizing the characteristics of a nozzle for both for on- and off-design conditions. Maps for three specific nozzle types (limiting cases) were covered in detail, and maps for a general nozzle type were discussed. Next, different geometries of thrust reversers, which are a means by which negative thrust is developed so that aircraft can be quickly and safely decelerated upon landing, were discussed. Finally, thrust vectoring, which is a means of changing the direction of developed thrust for better (primarily military fighters) aircraft performance, was also considered.

List of Symbols

A	Area
a	Speed of sound
c_p	Specific heat at constant pressure
h	Specific enthalpy
$\dot{m}$	Mass flow rate
M	Mach number
p	Pressure
$\mathcal{R}$	Ideal gas constant
T	Temperature
u	Velocity
γ	Specific heat ratio
δ	Ratio of pressure to standard pressure
η	Efficiency
θ	Ratio of temperature to standard temperature
ρ	Density

Subscripts

a	Ambient
c	"Corrected"
e	Exit
fn	Fan nozzle
i	Before shock
j	After shock
min	Minimum
n	Nozzle
stp	Standard temperature and pressure
t	Total (stagnation)
6	Inlet to primary nozzle
7	Inlet to fan nozzle
8	Exit from primary nozzle
9	Exit from fan nozzle

Superscripts

′	Ideal
*	Choked

Problems

10.1 Find the variation of the corrected mass flow rate for a fixed primary converging nozzle as a function of p_{t6}/p_a for a specific heat ratio of 1.337, a total inlet pressure of 65.5 psia, a total inlet temperature of 1865 °R, an exit area of 218 in.^{2} and a constant efficiency of 0.96. Plot the results as well as the variation in the exit Mach number vs. the pressure ratio p_{t6}/p_a.

10.2 Find the variation of the corrected mass flow rate for a variable primary CD nozzle with a fixed exit area as a function of p_{t6}/p_a for a specific heat ratio of 1.337, a total inlet pressure of 451.6 kPa (65.5 psia), a total inlet temperature of 1036 K (1865 °R), an exit area of 0.1858 m^2 (288 in.^{2}), and a constant efficiency of 0.96. The exit pressure matches the ambient pressure. Plot the results as well as the variation in the exit Mach number versus the pressure ratio p_{t6}/p_a. Also plot A_e/A_{min} vs. p_{t6}/p_a.

10.3 Find the variation of the corrected mass flow rate for a variable primary CD nozzle with a sonic and fixed minimum area as a function of p_{t6}/p_a for a specific heat ratio of 1.337, a total inlet pressure of 65.5 psia, a total inlet temperature of 1865 °R, a minimum area of 218 in.^{2}, and a constant efficiency of 0.96. The exit pressure matches the ambient pressure. Plot the results as well as the variation in the exit Mach number vs. the pressure ratio p_{t6}/p_a.

10.4 Find the variation of the corrected mass flow rate for a fixed primary CD nozzle as a function of

p_{t6}/p_a for a specific heat ratio of 1.337, a total inlet pressure of 451.6 kPa (65.5 psia), a total inlet temperature of 1036 K (1865 °R), an exit area of 0.1858 m^2 (288 in.^{2}), a minimum area of 0.1406 m^2 (218 in.^{2}), and a constant efficiency of 0.96. Plot the results as well as the variation in the exit Mach number vs. the pressure ratio p_{t6}/p_a.

10.5 A CD nozzle with an exit area of 350 in.^{2} and a minimum area of 275 in.^{2} has an upstream total pressure of 20 psia. The nozzle efficiency is 0.965, and the specific heat ratio is 1.35.

(a) At what atmospheric pressure and altitude will the nozzle flow be shockless?
(b) At what atmospheric pressure and altitude will a normal shock stand in the exit plane?

10.6 Derive Eq. (10.7.7).

10.7 Derive Eq. (10.7.10).

10.8 Derive Eq. (10.7.17).

10.9 A CD nozzle has an inlet total pressure of 25 psia and an inlet total temperature of 1400 °R. The ambient pressure is 4.2 psia, and the nozzle has an efficiency of 94.5 percent and a minimum area of 100 in.^{2}. What exit area is required for the flow to be shockless? Assume $\gamma = 1.35$.

10.10 An ideal fixed CD nozzle has an inlet total pressure of 131.0 kPa (19 psia). The inlet, minimum, and exit radii are 0.6858, 0.5588, and 0.6350 m (27, 22, and 25 in.), respectively. These radii occur at axial locations of 0, 0.3048, and 0.5334 m (0, 12, and 21 in.), respectively. The walls of the nozzle are linear cones. The specific heat ratio is 1.30.

(a) Find the altitude at which the flow is shockless and plot the internal pressure distribution axially down the nozzle.
(b) If the engine is operated at sea level, plot the internal pressure distribution axially down the nozzle.

10.11 A fixed CD nozzle with a nozzle efficiency of 95 percent has an inlet total pressure of 19 psia. The inlet, minimum, and exit radii are 27, 22, and 25 in., respectively. These radii occur at axial locations of 0, 12, and 21 in., respectively. The walls of the nozzle are linear cones. The specific heat ratio is 1.30.

(a) Find the altitude at which the flow is shockless and plot the internal pressure distribution axially down the nozzle.
(b) If the engine is operated at sea level, plot the internal pressure distribution axially down the nozzle.

10.12 At what atmospheric pressure and altitude will the following conditions occur for a CD nozzle with an exit area of 2.900 m^2 (4500 in.^{2}), a minimum area of 2.580 m^2 (4000 in.^{2}), and an upstream total pressure of 68.94 kPa (10 psia). The nozzle efficiency is 0.97 and the specific heat ratio is 1.35.

(a) The flow is shockless and the nozzle exit Mach number is supersonic.
(b) A normal shock stands in the exit plane of the nozzle.

10.13 The total inlet pressure to a converging nozzle is 12 psia, and the inlet total temperature is 1200 °R. The specific heat ratio is 1.36, and the efficiency is 96.4 percent. The exit area is 120 in.^{2}, the ambient pressure is 8.2 psia, and the ambient temperature is 400 °R. Find the exit Mach number and mass flow rate.

10.14 The total inlet pressure to a converging nozzle is 137.9 kPa (20 psia), and the inlet total temperature is 833.3 K (1500 °R). The specific heat ratio is 1.365 and the efficiency is 97.4 percent. The exit area is 0.0645 m^2 (100 in.^{2}), the ambient pressure is 55.16 kPa (8.0 psia), and the ambient temperature is 211.1 K (380 °R). Find the exit Mach number and the mass flow rate.

10.15 Find the corrected mass flow rate for a fixed converging primary nozzle for a total inlet pressure of 20 psia, a total inlet temperature of 1400 °R, an exit area of 250 in.^{2}, an efficiency of 0.97, and $\gamma = 1.36$.

(a) The atmospheric pressure is 4 psia. Compare the results to those of Example 10.5. Comment.

(b) The atmospheric pressure is 15 psia. Compare the results to those of part (a) above. Comment.

10.16 Find the corrected mass flow rate for a variable primary CD nozzle having a fixed exit area with $p_{t6}/p_a = 2.5$ for a specific heat ratio of 1.337, a total inlet pressure of 65.5 psia, a total inlet temperature of 1865 °R, an exit area of 288 in.^{2}, and an efficiency of 0.96. The exit pressure matches the ambient pressure.

10.17 For an ideal turbojet (all efficiencies are unity, total pressure losses are zero, properties are constant throughout, and the fuel ratio is negligible), analytically show that the maximum thrust is obtained when the exit pressure of the nozzle matches the ambient pressure. Consider the flight Mach number, altitude, core mass flow rate, compressor total pressure ratio, burner exit total temperature, and fuel properties to be known.

10.18 (a) Find the mass flow rate and the corrected mass flow rate for a fixed primary converging nozzle at standard sea level, a specific heat ratio of 1.35, a total inlet pressure of 275.8 kPa (40.0 psia), a total inlet temperature of 944.4 K (1700 °R), an exit area of 0.103 m2 (160 in.^{2}), and an efficiency of 0.97. (b) Also, sketch the general map for this nozzle type and indicate the relative location for the single data point from part (a).

10.19 A converging nozzle operates with a specific heat ratio of 1.33. The inlet Mach number is 0.10. The inlet total temperature is 1700 °R, and the exit total temperature is 1820 °R. The inlet total pressure is 60 psia. The Fanning friction factor is 0.015, and the length-to-diameter ratio is 4. What area ratio will cause the exit Mach number to be 1.00? What is the resulting total pressure at the exit? What is the ideal (frictionless and adiabatic) area ratio?

10.20 An adiabatic converging nozzle operates with a specific heat ratio of 1.33. The inlet Mach number is 0.10. The inlet total temperature is 944.4 K (1700 °R), and the inlet total pressure is 413.7 kPa (60 psia). The Fanning friction factor is 0.015, and the length-to-diameter ratio is 4. What area ratio will cause the exit Mach number to be 1.00? What is the resulting total pressure at the exit? What is the nozzle efficiency?

10.21 Find the mass flow rate and the corrected mass flow rate for a fixed primary CD nozzle at 40,000 ft with a specific heat ratio of 1.35, a total inlet pressure of 40.0 psia, a total inlet temperature of 1800 °R, an exit-to-minimum area ratio of 1.55, an exit area of 260 in.^{2}, and an efficiency of 96 percent. The exit Mach number is 1.80, the speed of sound at the exit is 1633 ft/s, and the exit static temperature is 1151 °R.

10.22 Find the mass flow rate and the corrected mass flow rate for a fixed primary converging nozzle at 20,000 ft (447.5 °R and 6.76 psia) with a specific heat ratio of 1.36, a total inlet pressure of 28.0 psia, a static inlet pressure of 27.25 psia, a total inlet temperature of 1650 °R, a static inlet temperature of 1638 °R, an exit area of 500 in.^{2}, and an efficiency of 95 percent.

10.23 At 5736 m (18,380 ft), will a shock occur (and if so where) in a fixed CD nozzle with an exit area of 2.903 m^2 (4500 in.^{2}), a minimum area of 2.581 m^2 (4000 in.^{2}), an upstream total pressure of 68.95 kPa (10 psia), and an upstream total temperature of 88.9 K (1600 °R)? The nozzle efficiency is 0.97, and the specific heat ratio is 1.35. Also, find the mass flow rate and corrected mass flow rate.

10.24 Find the exit Mach number, mass flow rate, and the corrected mass flow rate for a fixed primary converging nozzle at an ambient pressure of 12 psia and ambient temperature of 400 °R. The nozzle has a specific heat ratio of 1.34, a total inlet pressure of 30.0 psia, a total inlet temperature of 1600 °R, an exit area of 200 in.^{2}, and an efficiency of 96.5 percent.

10.25 A fixed-fan converging nozzle operates with an inlet total pressure of 18 psia and an inlet total temperature of 750 °R. The nozzle has an exit area of 224 in.^{2} and an efficiency of 96.6 percent. The specific heat ratio is 1.385. Ambient conditions are 10.8 psia and 400 °R.

(a) Find the mass flow rate, corrected mass flow rate, and exit Mach number.
(b) If the inlet total temperature is increased by 20 percent, what are the new mass flow rate and corrected mass flow rate?

10.26 Analyze the fixed CD nozzle of an afterburning turbojet (with the afterburner on). The total fuel flow rate is 5.384 kg/s (11.87 lb_m/s). The inlet total temperature to the nozzle is 2028 K (3650 °R), the inlet total pressure is 347.8 kPa (50.44 psia), the efficiency is 97 percent, and the specific heat ratio is 1.2687. The minimum area is 0.1258 m^2 (527.7 $in.^2$), and the exit Mach number is 2.191. The aircraft cruises at 264.0 m/s (866 ft/s) with the ambient pressure at 37.58 kPa (5.45 psia) and ambient temperature at 238.7 K (429.6 °R).

(a) Find the total mass flow rate.
(b) Find the corrected mass flow rate.
(c) Find the thrust.
(d) Find the *TSFC*.

10.27 Analyze the variable CD nozzle of the exhausted fan of a turbofan engine. The aircraft cruises at 993 ft/s with the ambient pressure at 10.10 psia and ambient temperature at 483 °R. The nozzle areas vary such that the exit pressure matches the ambient pressure. The inlet total temperature to the nozzle is 725 °R, the inlet total pressure is 37.7 psia, the efficiency is 97 percent, the specific heat ratio is 1.392, and the specific heat at constant pressure is 0.244BTU/lb_m-°R. The exit area is 398 $in.^2$.

(a) Find the exit Mach number of the nozzle.
(b) Find the total mass flow rate through the nozzle.
(c) Find the corrected mass flow rate.

10.28 A fixed-fan converging nozzle operates with an inlet total pressure of 16 psia and an inlet total temperature of 650 °R. The nozzle has an exit area of 950 $in.^2$ and an efficiency of 97.4 percent. The specific heat ratio is 1.390. Ambient conditions are 12.6 psia and 420 °R.

(a) Find the mass flow rate, corrected mass flow rate, and exit Mach number.
(b) If the inlet total temperature is increased to 700 °R (nothing else changes), what are the new mass flow rate and corrected mass flow rate?

10.29 A fixed-fan converging nozzle operates with an inlet total pressure of 103.2 kPa (15 psia) and an inlet total temperature of 347.2 K (625 °R). The nozzle has an exit area of 0.6452 m^2 (1000 $in.^2$) and an efficiency of 98.0 percent. The specific heat ratio is 1.390. Ambient conditions are 82.74 kPa (12.0 psia) and 222.2 K (400 °R).

(a) Find the mass flow rate, corrected mass flow rate and exit Mach number.
(b) If the inlet total temperature is increased to 388.9 K (700 °R) (nothing else changes), what are the new mass flow rate and corrected mass flow rate?

10.30 An ideal adiabatic converging nozzle operates with a specific heat ratio of 1.35. The inlet total temperature is 1500 °R and the inlet total pressure is 100 psia. The length-to-diameter ratio is 4, and the area ratio is 0.30. The nozzle is choked. Find the exit pressure, temperature, and velocity.

10.31 An adiabatic converging nozzle operates with a specific heat ratio of 1.35. The inlet total temperature is 1500 °R, and the inlet total pressure is 100 psia. The Fanning friction factor is 0.010, the length-to-diameter ratio is 4, and the area ratio is 0.30. The nozzle is choked. Find the exit pressure, temperature, efficiency, and velocity.

10.32 A converging nozzle operates with a specific heat ratio of 1.35. The inlet total temperature is 833.3 K (1500 °R), and the inlet total pressure is 689.5 kPa (100 psia). The Fanning friction factor is 0.010, the length-to-diameter ratio is 4, and the area ratio is 0.30. Owing to heat transfer, the exit total temperature is 777.8 K (1400 °R). The nozzle is choked. Find the exit pressure, temperature, and velocity. Does the conventional definition of efficiency make sense?

10.33 A converging nozzle (fixed) for a fan exhaust operates with an inlet total pressure of 14 psia and an inlet total temperature of 620 °R. The nozzle has an exit area of 700 in.^{2} and an efficiency of 98 percent. The specific heat ratio is 1.3920. Ambient conditions are 5 psia and 400 °R.

(a) Find the mass flow rate, corrected mass flow rate and exit Mach number.
(b) If the inlet total temperature is increased to 690 °R, and the inlet total pressure is increased to 15 psia (nothing else changes), what are the new mass flow rate and corrected mass flow rate?

10.34 A fixed CD nozzle operates with an inlet total pressure of 206.8 kPa (30 psia) and an inlet total temperature of 805.6 K (1450 °R). The nozzle has an exit area of 0.1258 m2 (195 in.^{2}) and an efficiency of 96 percent. The specific heat ratio is 1.340. Ambient conditions are 68.95 kPa (10.0 psia) and 236.1 K (425 °R).

(a) Find the area ratio (A_{exit}/A_{min}) if the exit pressure matches the ambient pressure.
(b) Find the exit Mach number if the exit pressure matches the ambient pressure.
(c) Find the mass flow rate and corrected mass flow rate.

10.35 A fixed primary CD nozzle operates with an inlet total pressure of 23 psia and an inlet total temperature of 1300 °R. The nozzle has an exit area of 400 in.^{2} and an efficiency of 98.4 percent. The specific heat ratio is 1.370. The nozzle is designed to operate with an exit Mach number of 1.422 so that the exit pressure matches the ambient pressure. The nozzle is operating off-design with ambient conditions of 11.0 psia and 430 °R.

(a) Find the mass flow rate and corrected mass flow rate for the off-design conditions.
(b) If the inlet total temperature is increased to 1400 °R (nothing else changes), what are the new mass flow rate and corrected mass flow rate?

10.36 A fixed CD nozzle operates with an inlet total pressure of 32 psia and an inlet total temperature of 1700 °R. The nozzle has an exit area of 800 in.^{2}, a minimum area of 579 in.^{2}, and an efficiency of 97.0 percent. The exit pressure matches the ambient pressure. The specific heat ratio is 1.360. Ambient conditions are 6.4 psia and 400 °R.

(a) Find the mass flow rate, corrected mass flow rate and exit Mach number.
(b) If the inlet total temperature is increased to 1900 °R (nothing else changes), what are the new mass flow rate and corrected mass flow rate?

10.37 A CD nozzle with an exit area of 400 in.^{2} and a minimum area of 285.7 in.^{2} has an upstream total pressure of 25 psia and total temperature of 1600 °R. The nozzle efficiency is 0.975, and the specific heat ratio is 1.34.

(a) At what atmospheric pressure and altitude will the nozzle flow be shockless?
(b) At what atmospheric pressure and altitude will a normal shock stand in the exit plane?
(c) What are the mass flow rate and corrected mass flow rate for the two cases?

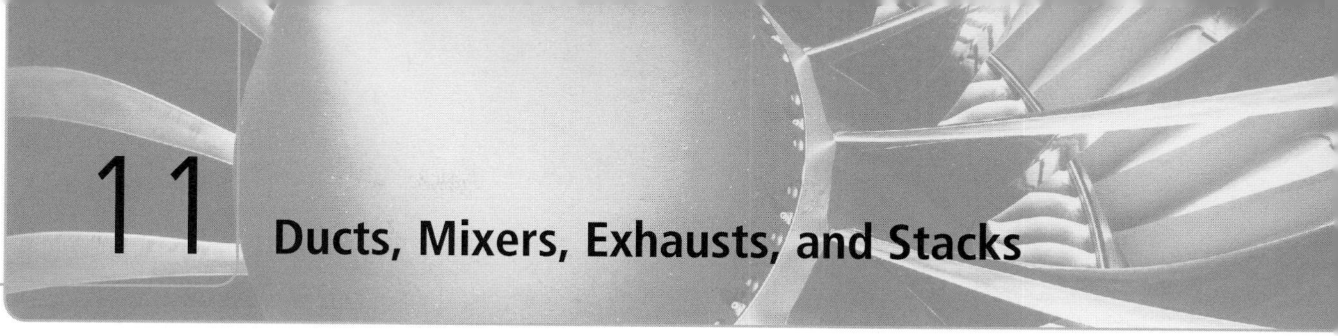

11 Ducts, Mixers, Exhausts, and Stacks

11.1 Introduction

Two remaining components that affect the overall performance of a turbofan engine are the bypass duct and mixer, as shown in Figure 11.1. Applications have previously been presented in Figures 2.42 and 2.44. These are relatively simple compared with the other components but should be included because they both generate losses in total pressure. Because the length-to-flow-width ratio of the bypass duct is moderate, the duct can incur significant losses. Also, it is desirable to have a uniform temperature gas entering the afterburner or nozzle so that these components operate near peak efficiency. Mixing of two fluid streams at different temperatures is a highly irreversible process, and a mixer consists of 3-D vanes in both the radial and circumferential (annular) directions. Thus, with good mixing of the low-temperature bypassed air and high-temperature primary air, further significant losses can occur. Owing to the temperatures exiting from the turbine, mixers are generally fabricated from a nickel-based alloy. With these losses the thrust and TSFC will both be compromised.

The two remaining components for power generation gas turbines are the exhausts and stack as shown in Figure 11.2. Applications have previously been presented in Figures 2.46 and 2.47, in which the turning vanes are easily identified. These also are relatively simple compared with the other components but should be included because they both generate losses in total pressure. Because the exhaust is a relatively short with vanes and high curvature, it can incur significant losses. Because the stack is a relatively long converging conduit, it can also incur significant losses. With these losses the output power and efficiency will both be compromised.

11.1.1 Objectives

After completing this chapter, the reader should be able to do the following.

- Understand the basic geometry of ducts and mixers for jet engines
- Understand the basic geometry of exhausts and stacks for power generation units
- Understand the basic gas dynamics of the four components
- Understand the pressure changes in each component
- Understand the basic design parameters
- Be able to apply basic gas dynamic equations from Chapter 2 and Appendix C to predict Mach numbers, total pressure losses, and other parameters
- Understand the influences of geometries on the total pressure losses
- Understand the development of "maps"
- Understand the performance of ducts and mixers have on jet propulsion thrust and *TSFC*
- Understand the performance of exhausts and stacks have on power generation output power and efficiency

11.2 Ducts and Mixers: Total Pressure Losses

Three irreversible mechanisms exist for pressure drops and total pressure losses to occur in ducts and mixers. The first is frictional flow in the duct primarily due to the viscous effects in the boundary layer. The second is the highly irreversible mixing process of two gas streams with different properties in the mixer. The third is the loss incurred by the mixer because it is a drag obstacle in the flow used to promote good "folding" and mixing of the two streams; this loss is similar to the loss due to a flameholder. Each of the losses are treated

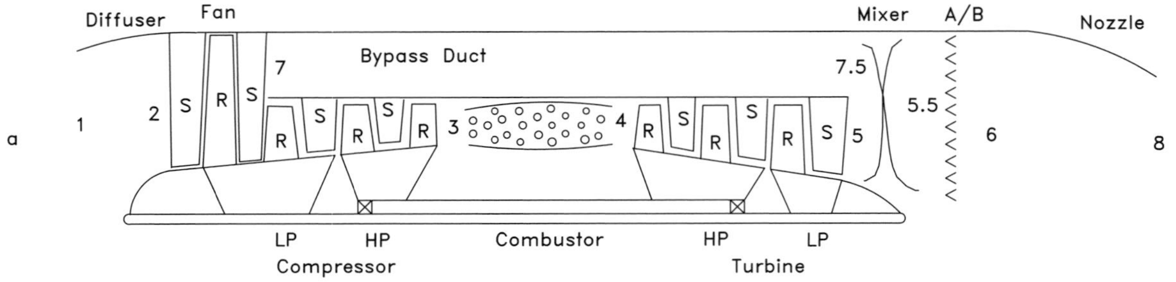

Figure 11.1 Turbofan engine with bypass duct and mixer.

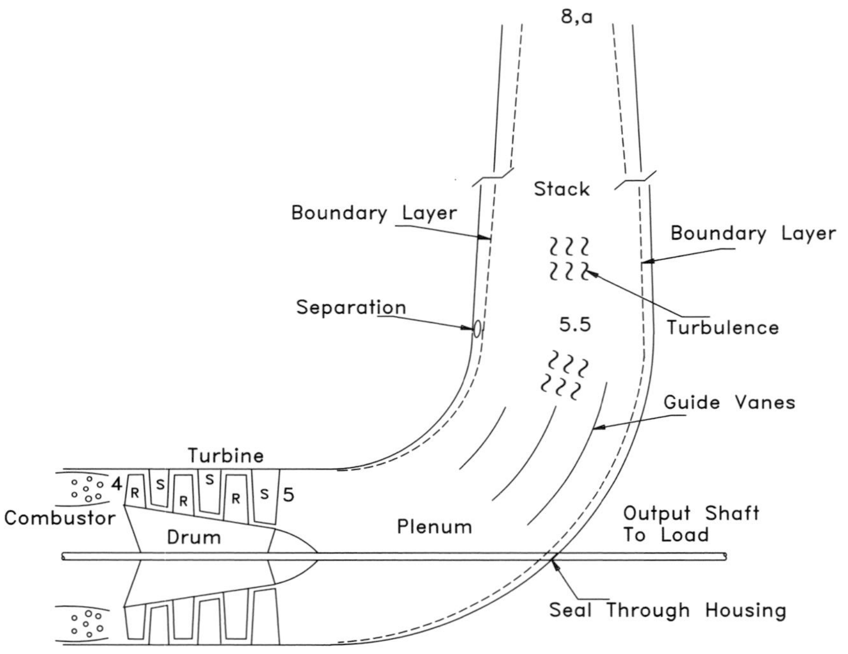

Figure 11.2 Power generation gas turbine exhaust and stack.

and analyzed independently in this chapter, although in Appendix C.13, a numerical methodology is presented (generalized 1-D compressible flow) whereby several effects can simultaneously be included.

11.2.1 Fanno Line Flow

The first effect to be analyzed is due to friction in the flow. This friction is generated by the wall friction in the boundary layer as well as freestream turbulence, which causes viscous dissipation. This will be analyzed based on Fanno line flow (Figure C.3), which is flow with friction but no heat addition. The process is a constant area irreversible process. The flow is analyzed with a control volume approach in Appendix C.5 and summarized in Section 2.2.5. Eleven equations are involved in the analysis. Seventeen variables are present, and so six must be specified. For example, if the inlet conditions M_1, T_1, and p_{t1} and duct information L, D, and f are specified, one can find p_{t2} and thus the resulting total pressure drop or total pressure ratio. Solutions can also be obtained using the software Fanno.exe.

Example 11.1 | Bypass Duct Flow

Flow with an incoming Mach number of 0.50 and an inlet total pressure of 25 psia enters the bypass duct approximating the size of an older engine duct. The duct is 110 in. long with an average outside diameter of 19 in. and an average inside diameter of 13 in., yielding a hydraulic diameter of 6 in. The duct has a Fanning friction factor of 0.003. What is the total pressure ratio if the specific heat ratio is 1.40?

Solution

From the tables for a Mach number of 0.50, one can find the nondimensional length-to-diameter and pressure parameters:

$$\left.\frac{4\,f\,L^*}{D}\right]_1 = 1.0691$$

and

$$\frac{p_{t1}}{p_t^*} = 1.3398;$$

thus, the total pressure can be found at the sonic condition, which is a reference condition (that might not occur in the actual conditions) as follows:

$$p_t^* = 25/1.3398 = 18.659 \text{ psia.}$$

Next, from the duct geometry, one can find the parameter

$$\left.\frac{4\,f\,L}{D}\right]_{1-2} = \frac{4 \times 0.003 \times 110}{6} = 0.2200;$$

thus, one can find the corresponding length-to-diameter parameter at the exit:

$$\left.\frac{4\,f\,L^*}{D}\right]_2 = \left.\frac{4\,f\,L^*}{D}\right]_1 - \left.\frac{4\,f\,L}{D}\right]_{1-2} = 1.0691 - 0.2200 = 0.8491,$$

and hence, the exit Mach number can be found from the tables or equations as follows:

$$M_2 = 0.5301.$$

Thus, as the results of friction the Mach number increased about 6 percent. Next, from the tables, at this Mach number

$$\frac{p_{t2}}{p_t^*} = 1.2863,$$

and so the exit total pressure can be found: $p_{t2} = 1.2863 \times 18.659 = 24.00$ psia. The pressure ratio for the duct is therefore

$$\pi = \frac{p_{t2}}{p_{t1}} = \frac{24.00}{25} = 0.960.$$

This is a significant loss and should be minimized further (if possible) by reducing the friction factor or Mach number.

11.2.2 Mixing Process

A second process that results in total pressure losses is the irreversible mixing process. For this occurrence, two streams (one from the bypass duct and one from the turbine exit) at different temperatures, total temperatures, Mach numbers, flow rates, velocities, total pressures, and so on are irreversibly mixed to form one (hopefully) uniform stream. Note that the mixing process is ideally frictionless, although in actuality friction is involved as discussed in Section 11.2.3. Thus, in reality, two mechanisms generate a loss in total pressure for a mixer. This mixing process is also analyzed in Appendix C.11 (Figure C.12) and summarized in Section 2.2.11. From this analysis, 27 unknowns are present. However, 19 independent equations are available. Thus, if eight variables are specified, one can find the remaining 19. For example, if the inlet conditions $\dot{m}_1$, $\dot{m}_2$, T_{t1}, T_{t2}, p_{t1}, p_{t2}, M_1, and M_2 are known, the exit conditions can be found, including the total pressure. Solutions can also be obtained using the software General1D.exe.

Example 11.2 Mixer Streams

Two streams at different total temperatures mix. The stream from the fan bypass duct has a total temperature (T_{t1}) of 444.4 K (800 °R), and the stream from the turbine exit has a total temperature (T_{t2}) of 888.9 K (1600 °R). The bypass flow rate ($\dot{m}_1$) is 45.35 kg/s (100 lb$_m$/s), and the primary flow rate ($\dot{m}_2$) is 90.70 kg/s (200 lb$_m$/s). The Mach number of the duct stream is 0.300 and the turbine exit stream is 0.5546, and the static pressure of both streams is 130.3 kPa (18.90 psia). Assume the specific heat ratio is 1.400 and find the fully mixed conditions.

Solution

First, from Eqs. (C.11.13) and (C.11.14), one finds the static temperatures

$$T_1 = \frac{T_{t1}}{\left[1 + \frac{\gamma - 1}{2} M_1^2\right]} = \frac{444.4}{\left[1 + \frac{0.4}{2} 0.300^2\right]} = 436.5\ \text{K}\ (785.7\ °\text{R})$$

$$T_2 = \frac{T_{t2}}{\left[1 + \frac{\gamma - 1}{2} M_2^2\right]} = \frac{888.9}{\left[1 + \frac{0.4}{2} 0.5546^2\right]} = 837.4\ \text{K}\ (1507.3\ °\text{R}).$$

From Eqs. (C.11.16) and (C.11.17), one obtains the total pressures:

$$p_1 = 130.3 = p_{t1} \left\{ \frac{1}{\left[1 + \frac{\gamma - 1}{2} M_1^2\right]} \right\}^{\frac{\gamma}{\gamma-1}} = p_{t1} \left\{ \frac{1}{\left[1 + \frac{0.4}{2} 0.300^2\right]} \right\}^{\frac{1.4}{0.4}}$$

$p_{t1} = 138.7$ kPa (20.12 psia);

$$p_2 = 130.3 = p_{t2}\left\{\frac{1}{\left[1+\frac{\gamma-1}{2}M_2^2\right]}\right\}^{\frac{\gamma}{\gamma-1}} = p_{t2}\left\{\frac{1}{\left[1+\frac{0.4}{2}0.5546^2\right]}\right\}^{\frac{1.4}{0.4}}$$

$p_{t2} = 160.6$ kPa (23.29 psia);

from the temperatures calculated previously, the sound speeds can also be found:

$$a_1 = \sqrt{\gamma \mathcal{R} T_1} = \sqrt{1.4 \times 287.1\frac{\text{J}}{\text{kg-K}} \times 436.5\ \text{K} \times \frac{\text{N-m}}{\text{J}}\frac{\text{kg-m}}{\text{N-s}^2}} = 418.9\ \text{m/s}\ (1374.4\ \text{ft/s})$$

$$a_2 = \sqrt{\gamma \mathcal{R} T_2} = \sqrt{1.4 \times 287.1\frac{\text{J}}{\text{kg-K}} \times 837.4\text{K} \times \frac{\text{N-m}}{\text{J}}\frac{\text{kg-m}}{\text{N-s}^2}} = 580.0\ \text{m/s}\ (1903.1\ \text{ft/s}).$$

For the given Mach numbers the velocities of the incoming streams are thus

$V_1 = M_1\, a_1 = 0.3000 \times 418.9 = 125.7$ m/s (412.4 ft/s)
$V_2 = M_2\, a_2 = 0.5546 \times 580.0 = 321.7$ m/s (1055.5 ft/s).

Next, for an ideal gas for the temperatures and pressures given above the densities are

$\rho_1 = 1.0404$ kg/m^3
$\rho_2 = 0.5423$ kg/m^3.

Hence, from Eqs. (C.11.2) and (C.11.3) the inlet areas can be found as follows:

$$A_1 = \frac{\dot{m}_1}{\rho_1 V_1} = \frac{45.35\frac{\text{kg}}{\text{s}}}{1.0404\frac{\text{kg}}{\text{m}^3} \times 125.7\frac{\text{m}}{\text{s}}} = 0.3468\ \text{m}^2\ (3.733\ \text{ft}^2)$$

$$A_2 = \frac{\dot{m}_1}{\rho_2 V_2} = \frac{90.70\frac{\text{kg}}{\text{s}}}{0.5423\frac{\text{kg}}{\text{m}^3} \times 321.7\ \frac{\text{m}}{\text{s}}} = 0.5201\ \text{m}^2\ (5.5993\ \text{ft}^2).$$

From Eq. (C.11.1), the exit area is therefore

$A_3 = A_1 + A_2 = 0.3468 + 0.5201 = 0.8669$ m^2 (9.332 ft^2).

From Eqs. (C.11.4) and (C.11.5), one finds the total mass flow from the exit by

$\dot{m}_3 = \dot{m}_1 + \dot{m}_2 = 45.35 + 90.70 = 136.1$ kg/s (300 lb$_m$/s),

and from Eq. (C.11.9) the total temperature at the exit is (all of the specific heats are the same and thus cancel)

$$T_{t3} = \frac{\dot{m}_1 c_p T_{t1} + \dot{m}_2 c_p T_{t2}}{\dot{m}_3 c_p} = \frac{45.35 \times 444.4 + 90.70 \times 888.9}{136.1} = 740.7\ \text{K}\ (1333.3\ ^\circ\text{R}).$$

Next, using Eq. (C.11.25) yields

$$(p_1 + \rho_1 V_1^2)A_1 + (p_2 + \rho_2 V_2^2)A_2 = \dot{m}_3 \left[\frac{\mathcal{R}T_3}{\sqrt{2c_p(T_{t3} - T_3)}} + \sqrt{2c_p(T_{t3} - T_3)} \right]$$

$$\left[130.3\,\text{kPa} \times 1000 \frac{\text{N/m}^2}{\text{kPa}} + 1.0404 \frac{\text{kg}}{\text{m}^3} \times 125.7^2 \frac{\text{m}^2}{\text{s}^2} \times \frac{\text{N}}{\frac{\text{kg-m}}{\text{s}^2}} \right] \times 0.3468\,\text{m}^2$$

$$+ \left[130.3 \frac{1b_f}{\text{in.}^2} \times 1000 \frac{\text{N/m}^2}{\text{kPa}} + 0.5423 \frac{\text{kg}}{\text{m}^3} \times 321.7^2 \frac{\text{m}^2}{\text{s}^2} \times \frac{\text{N}}{\frac{\text{kg-m}}{\text{s}^2}} \right] \times 0.5201\,\text{m}^2$$

$$= 136.1 \frac{\text{kg}}{\text{s}} \times \left[\frac{287.1 \frac{\text{J}}{\text{kg-k}} \times \text{T}_3(\text{K})}{+\sqrt{2 \times 1.005 \frac{\text{kJ}}{\text{kg-k}} \times (740.7\,\text{K} - \text{T}_3(\text{K})) \times 1000 \frac{\text{J}}{\text{kJ}} \times \frac{\text{N-m}}{\text{J}} \times \frac{\text{kg-m}}{\text{N-s}^2}}} \right]$$

$$+ 136.1 \frac{\text{kg}}{\text{s}} \times \left[\sqrt{2 \times 1.005 \frac{\text{kJ}}{\text{kg-k}} \times (740.7\,\text{K} - \text{T}_3(\text{K})) \times 1000 \frac{\text{J}}{\text{kJ}} \times \frac{\text{N-m}}{\text{J}} \times \frac{\text{kg-m}}{\text{N-s}^2}} \right].$$

Thus, by iteration one finds the static temperature at the exit as

$T_3 = 711.7$ K (1281.1 °R),

and so from Eq. (C.11.24) the velocity at the exit is

$$V_3 = \sqrt{2c_p(T_{t3} - T_3)}$$

$$= \sqrt{2 \times 1.005 \frac{\text{kJ}}{\text{kg-K}} \times (740.7\,\text{K} - 711.7\,\text{K}) \times 1000 \frac{\text{J}}{\text{kJ}} \times \frac{\text{N-m}}{\text{J}} \times \frac{\text{kg-m}}{\text{N-s}^2}}\ V_3$$

$$= 241.6\ \text{m/s}\ (792.7\ \text{ft/s}).$$

From Eq. (C.11.4) the exit density is

$$\rho_3 = \frac{\dot{m}_3}{V_3 A_3} = \frac{136.1 \frac{\text{kg}}{\text{s}}}{241.6 \frac{\text{m}}{\text{s}} \times 0.8669\,\text{m}^2}$$

$\rho_3 = 0.6500\ \text{kg/m}^3$.

From Eq. (C.11.21), one finds the Mach number at the exit by

$$M_3 = \frac{V_3}{\sqrt{\gamma \mathcal{R} T}} = \frac{241.6 \frac{\text{m}}{\text{s}}}{\sqrt{1.4 \times 287.1 \frac{\text{J}}{\text{kg-k}} \times 711.7\,\text{K} \times \frac{\text{N-m}}{\text{J}} \times \frac{\text{kg-m}}{\text{N-s}^2}}}$$

$M_3 = 0.4520$,

and for an ideal gas (Eq. (C.11.12)) with the static temperature and density, the static pressure at the exit is

$p_3 = 132.6$ kPa (19.23 psia).

Finally, from Eq. (C.11.18), the total pressure at the exit is

$$p_{t3} = p_3 \left[\frac{T_{t3}}{T_3}\right]^{\frac{\gamma}{\gamma-1}} = 132.6 \left[\frac{740.7}{711.7}\right]^{\frac{1.4}{0.4}}$$

$$= 152.6 \text{ kPa (22.13 psia)}.$$

Finally, the total pressure ratio due only to irreversible mixing is

$$\frac{p_{t3}}{p_{t1,t2}} = \left[\frac{152.6}{153.3}\right] = 0.9954,$$

where $p_{t1,t2}$ is the weighted average of the inlet total pressures:

$$p_{t1,t2} = \frac{200 \times 160.6 + 100 \times 138.7}{300} = 153.3 \text{ kPa}.$$

That is, less than 1 percent of the total pressure was lost as a result of the mixing process.

11.2.3 Flow with a Drag Object

When an obstruction is in the flow, the total pressure is also reduced because of the boundary layer friction around the obstruction and from the added wake turbulence, separation, and shearing of the fluid, which results in viscous dissipation. Mixers contain such obstructions to enhance the folding and mixing of the two streams. Thus, a method of including such losses is needed to complement the pure mixing process. A control volume approach is used for a constant area flow for which uniform properties are assumed at the entrance and exit. Details are covered in Appendix C.9 (Figure C.10) and summarized in Section 2.2.9. This results in 16 variables and 11 equations. Thus, if five variables are specified the problem can be solved. Typically, if the mixer inlet conditions M_1, p_{t1}, and T_{t1} and mixer geometry parameters A_d/A and C_d are known, the exit conditions can be found, including the total pressure loss. Solutions can also be obtained using the software General1D.exe.

Example 11.3 | Flow in a Mix Enhancer (Drag Object)

Flow exiting the mixing process of Example 11.2 enters the mix enhancer. Thus, flow with an incoming Mach number of 0.4520, an inlet total pressure of 22.13 psia, a static pressure of 19.23 psia, and a total temperature of 1333.33 °R enters the enhancer. The drag coefficient of the mixer is 0.4, and it occupies 8 percent of the total flow area. What is the total pressure ratio if the specific heat ratio is 1.40?

Solution

First, for the given inlet Mach number and total conditions, one can find the static conditions:

$T_1 = 1281.1$ °R
$p_1 = 19.23$ psia.

For this value of T_1 the sound speed is

$a_1 = 1754.5$ ft/s,

and from the Mach number the inlet velocity is

$V_1 = 793.0$ ft/s.

Also for T_1 and p_1, from the ideal gas equation the inlet density is

$\rho_1 = 0.001259$ sl/ft^3 ;

thus, from Eq. (C.9.8),

$$p_1 + \rho_1 V_1^2 - \frac{1}{2} C_d \rho_1 V_1^2 \frac{A_d}{A} = \rho_1 V_1 \left[\frac{\mathcal{R} T_2}{\sqrt{2c_p(T_{t1} - T_2)}} + \sqrt{2c_p(T_{t1} - T_2)} \right]$$

$$19.23 \times 144 + 0.001259 \times 793.0^2 - \frac{1}{2} \times 0.4 \times 0.001259 \times 793.0^2 \times 0.08$$

$$= 0.001259 \times 793.0 \times \left[\frac{53.35 \times 32.17 T_2}{\sqrt{2 \times 0.2400 \times 32.17 \times 778.16 \times (1333.3 - T_2)}} \right]$$

$$+ 0.001259 \times 793.0 \times \left[\sqrt{2 \times 0.2400 \times 32.17 \times 778.16 \times (1333.3 - T_2)} \right].$$

Iteratively solving for the exit static temperature T_2, one finds

$T_2 = 1280.4$ °R.

Thus, from Eq. (C.9.7) the exit velocity is

$$V_2 = \sqrt{2c_p(T_{t2} - T_2)}$$

$$= \sqrt{2 \times 0.2400 \frac{\text{BTU}}{\text{lb}_\text{m}\text{-}^\circ\text{R}} \times 778.16 \frac{\text{ft-lb}_\text{f}}{\text{BTU}} \times 32.17 \frac{\text{lb}_\text{m}}{\text{sl}} \times (1333.3\ ^\circ\text{R} - 1280.4\ ^\circ\text{R}) \times \frac{\text{sl-ft}}{\text{lb}_\text{f}\text{-s}^2}}$$

$$= 797.3 \text{ ft/s}.$$

The sound speed at the exit is $a_2 = 1754.0$ ft/s; thus, the exit Mach number is $M_2 = 0.4547$, and the exit density is

$\rho_2 = \rho_1 V_1 / V_2 = 0.001259 \times 793.0/797.3 = 0.001252$ sl/ft^3.

From the ideal gas law the exit pressure is $p_2 = 19.11$ psia.

Finally, for $M_2 = 0.4547$ and the p_2 above, the total pressure at the exit is

$p_{t2} = 22.02$ psia,

and the total pressure ratio due to the friction from the mix enhancer is

$\pi = p_{t2}/p_{t1} = 22.02/22.13 = 0.9954.$

One can see that this loss in total pressure is small owing to the drag obstacle and that it is of the same order of magnitude as that due to the mixing process found in Example 11.2. Thus, with the result from Example 11.2 the total pressure ratio for the entire mixer can be estimated by multiplying the two ratios:

$\pi = 0.9954 \times 0.9954 = 0.9908.$

The loss in total pressure is therefore less than 1 percent from the two mechanisms.

11.2.4 Flow with a Mixing Process and a Drag Object

In addition to mixing when an obstruction is in the flow, the total pressure is further reduced. Mixers contain such obstructions to enhance the folding and mixing of the two streams. Thus, a method of including such losses is needed to complement the pure mixing process. A control volume approach is used for a constant area flow for which uniform properties are assumed at the entrance and exit. Details are covered in Appendix C.12 (Figure C.13) and summarized in Section 2.2.12. This results in 27 variables and 19 equations. Thus, if eight variables are specified the problem can be solved. Typically, if the mixer inlet conditions M_1, M_2, p_{t1}, p_{t2}, T_{t1}, and T_{t2} and mixer geometry parameters A_d/A and C_d are known, the exit conditions can be found, including the total pressure loss. Solutions can also be obtained using the software General1D.exe.

Example 11.4 | Flow Mixing with a Mix Enhancer

Two streams at different total temperatures mix. As in Example 11.2, the stream from the fan bypass duct has a total temperature (T_{t1}) of 444.4 K (800 °R), and the stream from the turbine exit has a total temperature (T_{t2}) of 888.9 K (1600 °R). The bypass flow rate ($\dot{m}_1$) is 45.35 kg/s (100 lb$_m$/s), and the primary flow rate ($\dot{m}_2$) is 90.70 kg/s (200 lb$_m$/s). The Mach number of the duct stream is 0.300 and the turbine exit stream is 0.5546, and the static pressure of both streams is 130.3 kPa (18.90 psia). An enhancer is used and the drag coefficient of the mixer is 0.4, and it occupies 8 percent of the total flow area. Assume the specific heat ratio is 1.400 and find the fully mixed conditions. This example is a combination of the previous two examples. However, here mixing and the flow over the drag object are solved simultaneously.

Solution

From Eq. (C.12.34):

$$\dot{m}_1\left(\frac{\mathcal{R}T_1}{V_1}+V_1\right)+\dot{m}_2\left(\frac{\mathcal{R}T_2}{V_2}+V_2\right)-\frac{1}{2}C_dA_R(\dot{m}_1V_1+\dot{m}_2V_2)$$

$$=\dot{m}_3\left(\frac{\mathcal{R}T_4}{\sqrt{2c_p(T_{t3}-T_3)}}+\sqrt{2c_p(T_{t3}-T_3)}\right).$$

As in Example 11.2,

$$T_1=\frac{T_{t1}}{\left[1+\dfrac{\gamma-1}{2}M_1^2\right]}=436.5\text{ K}=785.7\ ^\circ\text{R}$$

$$T_2=\frac{T_{t2}}{\left[1+\dfrac{\gamma-1}{2}M_2^2\right]}=837.4\text{ K}=1507.3\ ^\circ\text{R}$$

$$T_{t3}=\frac{\dot{m}_1c_pT_{t1}+\dot{m}_2c_pT_{t2}}{\dot{m}_3c_p}=740.7\text{ K}=1333.3\ ^\circ\text{R}$$

$$p_1=130.3=p_{t1}\left\{\frac{1}{\left[1+\dfrac{\gamma-1}{2}M_1^2\right]}\right\}^{\frac{\gamma}{\gamma-1}}=p_{t1}\left\{\frac{1}{\left[1+\dfrac{0.4}{2}0.300^2\right]}\right\}^{\frac{1.4}{0.4}}$$

$$p_{t1}=138.7\text{ kPa}=20.12\text{ psia}$$

$$p_2=130.3=p_{t2}\left\{\frac{1}{\left[1+\dfrac{\gamma-1}{2}M_2^2\right]}\right\}^{\frac{\gamma}{\gamma-1}}=p_{t2}\left\{\frac{1}{\left[1+\dfrac{0.4}{2}0.5546^2\right]}\right\}^{\frac{1.4}{0.4}}$$

$$p_{t2}=160.6\text{ kPa}=23.29\text{ psia}$$

$$a_1=\sqrt{\gamma\mathcal{R}T_1}=418.9\text{ m/s}=1374.4\text{ ft/s}$$

$$a_2=\sqrt{\gamma\mathcal{R}T_2}=580.0\text{ m/s}=1903.1\text{ ft/s}$$

$$V_1=M_1\,a_1=125.7\text{ m/s}=412.4\text{ ft/s}$$

$$V_2=M_2\,a_2=321.7\text{ m/s}=1055.5\text{ ft/s}$$

$$100\frac{\text{lb}_\text{m}}{\text{s}}\left(\frac{53.35\dfrac{\text{ft-lb}_\text{f}}{\text{lb}_\text{m}\text{-°R}}\times 32.17\dfrac{\text{lb}_\text{m}}{\text{sl}}\times 785.7\ \text{°R}}{412.4\dfrac{\text{ft}}{\text{s}}}+412.4\frac{\text{ft}}{\text{s}}\right)$$

$$+200\frac{\text{lb}_\text{m}}{\text{s}}\left(\frac{53.35\dfrac{\text{ft-lb}_\text{f}}{\text{lb}_\text{m}\text{-°R}}\times 32.17\dfrac{\text{lb}_\text{m}}{\text{sl}}\times 1507.3\ \text{°R}}{1055.5\dfrac{\text{ft}}{\text{s}}}+1055.5\frac{\text{ft}}{\text{s}}\right)$$

$$-\frac{1}{2}\times 0.4\times 0.08\times\left(100\frac{\text{lb}_\text{m}}{\text{s}}\times 412.4\frac{\text{ft}}{\text{s}}+200\frac{\text{lb}_\text{m}}{\text{s}}\times 1055.5\frac{\text{ft}}{\text{s}}\right)$$

$$=300\frac{\text{lb}_\text{m}}{\text{s}}\left(\frac{53.35\dfrac{\text{ft-lb}_\text{f}}{\text{lb}_\text{m}\text{-°R}}\times 32.17\dfrac{\text{lb}_\text{m}}{\text{sl}}T_3}{\sqrt{2\times 0.24\dfrac{\text{BTU}}{\text{lb}_\text{m}\text{-°R}}\times 778.16\dfrac{\text{ft-lb}_\text{f}}{\text{BTU}}\times 32.27\dfrac{\text{lb}_\text{m}}{\text{sl}}(1333.3\ \text{°R}-T_3)}}\right)$$

$$+300\frac{\text{lb}_\text{m}}{\text{s}}\times\sqrt{2\times 0.24\frac{\text{BTU}}{\text{lb}_\text{m}\text{-°R}}\times 778.16\frac{\text{ft-lb}_\text{f}}{\text{BTU}}\times 32.27\frac{\text{lb}_\text{m}}{\text{sl}}(1333.3\ \text{°R}-T_3)}.$$

$$\left(\text{Remember }\frac{\text{sl-ft}}{\text{s}^2\text{-lb}_\text{f}}=1.\right)$$

Iteratively solving for T_3:

$T_3 = 1280.4\ \text{°R} = 711.3\ \text{K}.$

$$\text{So } V_3=\sqrt{2c_p(T_{t3}-T_3)}=797.3\frac{\text{ft}}{\text{s}}=243.0\frac{\text{m}}{\text{s}}$$

$$a_3=\sqrt{\gamma\mathcal{R}T_3}=1754.0\ \text{ft/s}=534.6\ \text{m/s}$$

$$M_3=\frac{V_3}{a_3}=0.4549;$$

p_3=19.11 psia = 133.5 kPa, which is found from ρ_3 and T_3,similar to the previous examples. Next, for $M_3 = 0.4549$ and the p_3 above, the total pressure at the exit is $p_{t3} = 22.02$ psia. Finally, the total pressure ratio due only to irreversible mixing with the mix enhancer is

$$\pi=\frac{p_{t3}}{p_{t1,t2}},$$

where again $p_{t1,t2}$ is the weighted average of the inlet total pressures

$$p_{t1,t2}=\frac{200\times 160.6+100\times 138.7}{300}=153.3\ \text{kPa}=22.23\ \text{psia},$$

$$\text{so that }\pi=\frac{p_{t3}}{p_{t1,t2}}=0.9906,$$

which is almost identical to the previous two examples in which mixing and drag were treated separately.

11.3 Power Generation Exhausts and Stacks

One irreversible mechanism exists for pressure drops and total pressure losses to occur in exhausts and stacks. This is frictional flow largely due to the viscous effects in the boundary layer, turbulence, and possible separation. For exhausts and stacks the area is not constant. For an exhaust, vanes are typically used to turn the flow into the stack, adding more surface area. As a result, the losses are more complicated than simple Fanno line flow. This loss is treated and analyzed independently in this chapter. In Appendix C.13, a numerical methodology is presented (generalized 1-D compressible flow) whereby several effects can simultaneously be included. As these losses become larger, the pressure at the exit of the turbine becomes larger, thus reducing the possible pressure drop through the turbine and reducing the power output.

11.3.1 Total Pressure Loss

In this section an analysis is introduced that includes the above two effects. The technique used is the so-called generalized 1-D flow method derived in many gas dynamics texts. It is a differential equation analysis from which influence coefficients are presented in Appendix C.14. These coefficients relate changes of one variable on another. A differential control volume is used, and differential forms of the continuity, x-momentum, energy, and second law of thermodynamics are conventionally applied to the control volume to derive the influence coefficients. Incremental changes in Mach number, velocity, sound speed, temperature, density, pressure, total pressure, and entropy are shown to be related to area change, total temperature change, friction, and drag. Results of the method will be used to determine and numerically integrate the governing differential equations for flows with two specific combined effects. It goes without saying that the method is quite versatile.

For exhausts and stacks the Mach number can be found from numerically integrating Eq. (C.14.1) by stepping in small increments through the channel:

$$\Delta(M^2) = -2M^2\left[\frac{\left(1+\frac{\gamma-1}{2}M^2\right)}{1-M^2}\right]\left[\frac{\Delta A}{A}\right] + 4f\gamma M^4\left[\frac{\left(1+\frac{\gamma-1}{2}M^2\right)}{1-M^2}\right]\left[\frac{\Delta x}{D}\right]$$

$$M_i^2 = M_{i-1}^2 + \Delta(M^2);$$

and the total pressure can be found from Eq. (C.14.2):

$$\Delta p_t = -\frac{1}{2}p_t 4f\gamma M^2\left[\frac{\Delta x}{D}\right] = -2p_t f\gamma M^2\left[\frac{\Delta x}{D}\right]$$

$$p_{ti} = p_{t\,i-1} + \Delta(p_t).$$

The method has previously been covered in Example 8.3 and will be covered again in Example 11.5.

Example 11.5 Power Generation Exhaust and Stack

Flow exits a power generation gas turbine with a Mach number of 0.60. The flow enters a diffusing channel with an inlet area of 500 in.^{2} (diameter of 25.23 in.) and exits the plenum with an area of 2500 in.^{2}. The effective length the channel is 100.9 in. with a Fanning friction factor of 0.015. The total temperature remains constant at 1650 °R. The flow then enters a stack, which is 300 ft (3600 in.) long, and exhausts into 14.69 psia atmosphere. The stack has a Fanning friction factor of 0.010 and exit area of 2250 in.^{2}. What is the total pressure at the turbine exit if the specific heat ratio is 1.36?

Solution

This problem has two parts: flow from section 5 to section 5.5, which is a short expanding area with friction, and from section 5.5 to a (or 8), which is a long contraction area with friction (see Figure 11.2). This will be solved in series and will be solved by numerically integrating Eqs. (C.14.3) and (C.14.4). These processes are similar to those in Example 8.3 (variable areas with friction). Five thousand steps will be used, and area change and friction will be assumed to be uniform along the length of the two sections.

Section 5 to 5.5

For the first section the exit area is 2500 in.^{2}; thus, the area increases by 2000 in.^{2}. Therefore, $\Delta A = 0.400$ in.^{2}, and $\Delta x = 0.02018$ in. for each step. A forward difference method is used. Thus, for the first step from Eq. (C.14.1) (the first term is due to area changes, and the second term results from friction),

$$\Delta(M^2) = -2M^2\left[\frac{\left(1+\frac{\gamma-1}{2}M^2\right)}{1-M^2}\right]\left[\frac{\Delta A}{A}\right] + 4f\gamma M^4\left[\frac{\left(1+\frac{\gamma-1}{2}M^2\right)}{1-M^2}\right]\left[\frac{\Delta x}{D}\right]$$

so $M_i^2 = M_{i-1}^2 + \Delta(M^2)$

$$M^2 = 0.60^2 - 2\times 0.60^2 \times \left[\frac{\left(1+\frac{1.36-1}{2}0.60^2\right)}{1-0.60^2}\right]\left[\frac{0.400}{500}\right]$$

$$+\,4\times 0.015\times 1.36\times 0.60^4 \times \left[\frac{\left(1+\frac{1.36-1}{2}0.60^2\right)}{1-0.60^2}\right]\left[\frac{0.02018}{25.23}\right]$$

$$= 0.35943,$$

or $M_i = 0.59953$.

From Eq. (C.14.2), the total pressure loss is only due to friction:

$$\Delta p_t = -\frac{1}{2}p_t 4f\gamma M^2\left[\frac{\Delta x}{D}\right] = -2p_t f\gamma M^2\left[\frac{\Delta x}{D}\right],$$

and so $p_{ti} = p_{t\,i-1} + \Delta(p_t)$.

Since the exit conditions are known but the inlet conditions are not known, this becomes an iterative process. So the first guess on p_{ti} is $p_{t5} = p_{ti} = 15.21$ psia. This will be checked at the exit of the stack for $p_a = 14.69$ psia:

$$p_{ti} = 15.21 - 2 \times 15.21 \times 0.015 \times 1.36 \times 0.60^2 \times \frac{0.02018}{25.23} = 15.20982.$$

This is repeated for all 5000 steps, where the previous step is used for the initial conditions for the following step. It is important to note that the area and diameter are of course updated from step to step. Thus, one finds after 5000 steps that

$p_{5.5} = 15.01$ psia
$M_{5.5} = 0.0988$
$p_{t5.5} = 15.11$ psia.

Section 5.5 to 8

For the second section (stack) the exit area is 2250 in.^{2}; thus, the area decreases by 250 in.^{2}. Therefore, $\Delta A = -0.05$ in.^{2}, and $\Delta x = 0.720$ in. for each step. Now the inlet Mach number is $M_{5.5} = 0.0988$. As above, for the first step from Eq. (C.14.1) (the first term is due to area changes, and the second term results from friction),

$$\Delta(M^2) = -2M^2 \left[\frac{\left(1 + \frac{\gamma - 1}{2} M^2\right)}{1 - M^2}\right] \left[\frac{\Delta A}{A}\right] + 4 f \gamma M^4 \left[\frac{\left(1 + \frac{\gamma - 1}{2} M^2\right)}{1 - M^2}\right] \left[\frac{\Delta x}{D}\right]$$

so $M_i^2 = M_{i-1}^2 + \Delta(M^2)$.

From Eq. (C.14.2), the total pressure loss is only due to friction:

$$\Delta p_t = -\frac{1}{2} p_t 4 f \gamma M^2 \left[\frac{\Delta x}{D}\right] = -2 p_t f \gamma M^2 \left[\frac{\Delta x}{D}\right],$$

and so $p_{ti} = p_{t\,i-1} + \Delta(p_t)$.

This is repeated for all 5000 steps, where the previous step is used for the initial conditions for the following step. It is important to note that the area and diameter are of course updated from step to step. Thus, one finds after 5000 steps that

$p_8 = 14.69$ psia
$M_8 = 0.1122$
$p_{t8} = 14.81$ psia.

For comparison one can compare p_8 to p_a and see that they are the same (14.69 psia). Thus the guess on p_{t5} much earlier was valid. One can also find the total pressure ratio for the entire exhaust system as

$$\pi = \frac{p_{t8}}{p_{t5}} = \frac{14.81}{15.21} = 0.974.$$

One should note that these processes can also be calculated using General1D.exe.

11.3.2 Back Pressure Effects on Performance

At first glance, since the exhaust and stack are both outside of the active components of a power generation unit, their effects on performance may not be expected. However, since there are pressure losses in both of these components, the back pressure to the last turbine will be affected. Namely with a pressure loss in these passive components, and since the stack exhausts to the atmosphere, the back pressure at the exit of the last turbine will be increased. As a result, the pressure drop in the turbine set will be decreased. And as seen in Chapter 4, the net power depends on the pressure drop in the turbine set. Thus, with losses in these two components, the output power and efficiency will be decreased. Two examples are presented.

Example 11.6 Effects of Exhaust Pressure Ratio on Nonideal Power Generation Gas Turbine

A power generation gas turbine with one turbine (see Example 3.16) operates at sea level. It ingests 66.67 kg/s (147 lb_m/s) of air. The compressor operates with a pressure ratio of 18 and an efficiency of 88 percent. The fuel has a heating value of 42,800 kJ/kg (18,400 BTU/lb_m) and the burner exit total temperature is 1456 K (2620 °R). The burner has an efficiency of 96 percent and a total pressure ratio of 0.96, while the turbine has an efficiency of 91.5 percent. The exhaust has a pressure ratio of 0.98 (not 0.93 as in Example 3.16). The total pressure recovery for the inlet is 0.98 and the shaft efficiency is 98 percent. Find the output power and compare with Example 3.16.

Solution

The calculation process is exactly as in Example 3.16. The new analysis shows that the power is now

$\mathcal{P} = 32{,}020$ hp (was 31,140 hp)

and the overall efficiency is

$\eta = 37.6$ percent (was 36.6 percent).

Thus, decreasing the exhaust system losses greatly improves overall performance.

Example 11.7 Effects of Exhaust Pressure Ratio on Nonideal Power Generation Gas Turbine with Intercooling, Reheat Combustion, and Regeneration (see Example 3.19)

A two-compressor and three-turbine power generation gas turbine (Figure 1.45a and Figure 2.5) operates at sea level. The unit has intercooling, regeneration, and reheat (Figure 2.5). It ingests 66.67 kg/s (147 lb_m/s) of air. The low-pressure compressor operates with a pressure ratio of

4.2426 and an efficiency of 90.2 percent. The high-pressure compressor also operates with a pressure ratio of 4.2426 and an efficiency of 90.2 percent resulting in a total pressure ratio for both compressors of 18 and a combines efficiency of 88 percent. The fuel has a heating value of 42,800 kJ/kg (18,400 BTU/lb_m) and the primary burner and reheat burner exit total temperatures are 1456 K (2620 °R). The burners both have an efficiency of 96 percent and a total pressure ratio of 0.96. The three turbines (low-pressure, high-pressure, and free) have efficiencies of 89.4 percent resulting in the total turbine efficiency of 91.5 percent. The exhaust/stack has a pressure ratio of 0.98 (was 0.93 in Example 3.19). The total pressure recovery for the inlet is 0.98 and the shaft efficiency is 98 percent. The regenerator has an effectiveness of 75 percent and total pressure ratio of 0.99 in both directions. The intercooler has an effectiveness of 75 percent and total pressure ratio of 0.99 in both directions. The intercooler coolant flows at 22.7 kg/s (50 lb_m/s), enters at 289 K and has a specific heat of 4.2 kJ/kg-K. Inlet cooling is not used. Find the net output power, *SFC*, and net thermal efficiency. Compare with Example 3.19.

Solution

The calculation process is exactly as in Example 3.19. The new analysis shows that the power is now

$\mathcal{P} = 32.100$ MW (was 31.320 MW)

and the overall efficiency is

$\eta = 47.0$ percent (was 46.3 percent).

Thus, decreasing the exhaust system losses greatly improves overall performance.

11.3.3 Other Passive Components

In addition to the duct, mixer, exhaust, and stack, there are three other passive components that affect gas turbine performance. They are the regenerator, intercooler, and inlet cooling, which apply to power generation units, see Figure 2.5. Inlet cooling also sometimes applies to propulsion gas turbines. The regenerator and intercooler depend on heat exchangers. The inlet cooling usually depends on a spray injected into the inlet. The regenerator has the effect of increasing the total pressure losses at the back of the unit. It has the same effect as adding another Fanno-line-type loss to the exhaust. The intercooler has another Fanno-line-type loss between the low- and high-pressure compressors. These two losses are strongly dependent on the heat exchanger designs. Increasing losses reduce the net power and overall efficiency for both components. However, the two components increase performance with nominal losses. The last component is inlet cooling, which applies to both power generation and propulsion units. This has three effects. First, it lowers the total temperature and, as a result, increases the total pressure at the inlet as shown by a generalized 1-D analysis presented in Appendix C.13. But second, because droplets are sprayed into the inlet, it has the same effect as adding a drag and mixing loss to the total pressure also as shown by a generalized 1-D analysis presented Appendix C.11 and in other compressible flow texts (for example, Shapiro (1953)). And third, because the process is evaporative cooling, the process is irreversible, thus again decreasing the total pressure. To model an accurate prediction a full analysis is required. However, for the sake of this text, and since all three are very small, the three losses will be assumed to offset each other.

11.4 Component Maps: Dimensional Analysis

As in previous chapters, performance curves or "maps" are usually used to provide a working medium for an engineer. These provide a quick and accurate view of the conditions at which the component is operating and what can be expected if the flow conditions are changed. Of primary concern is the total pressure ratio, that is, p_{t55}/p_{t5} (or π_{exh}) for an exhaust and π_{stk}, π_{RG}, π_{IC}, for a stack, regenerator, and intercooler respectively. For example, for a power generation exhaust (Figure 11.2), the total pressure ratio is a function of many variables, and the functional relationship f can be written as

$$p_{t55}/p_{t5} = f\{\dot{m}, T_{t5}, A, p_{t5}, \gamma, \mathcal{R}\}. \tag{11.4.1}$$

Dimensionless parameters can be used to reduce the number of independent variables. Using the Buckingham Pi theorem, one finds the following for a given value of γ for the exhaust:

$$p_{t55}/p_{t5} = f\left[\frac{\dot{m}\sqrt{\mathcal{R}T_{t5}}}{Ap_{t5}}\right]. \tag{11.4.2}$$

This form is not used for such a component. However, realizing that the values of A and $\mathcal{R}$ are constant for a give geometry, "nondimensionalizing" or correcting, as is done for compressors and turbines, one finds the general expression

$$\pi_{exh} = p_{t55}/p_{t5} = f\left[\frac{\dot{m}_5\sqrt{\theta_{t5}}}{\delta_{t5}}\right] = f[\dot{m}_{c5}], \tag{11.4.3}$$

where again

$$\delta_{t5} = p_{t5}/p_{stp} \tag{11.4.4}$$

and

$$\theta_{t5} = T_{t5}/T_{stp}. \tag{11.4.5}$$

Note again that the independent parameter is not truly dimensionless. As before, this is the "corrected" mass flow and has dimensions of mass flow. Thus, the resulting function cannot be used to correlate or compare different exhausts. The function can be used for one particular exhaust, however, as is the usual case. Similarly for the stack (Figure 11.2), regenerator Figure 2.49), and intercooler (Figure 2.50), respectively:

$$\pi_{stk} = p_{t8}/p_{t55} = f\left[\frac{\dot{m}_{55}\sqrt{\theta_{t55}}}{\delta_{t55}}\right] = f[\dot{m}_{c55}], \tag{11.4.6}$$

$$\pi_{RG} = p_{t5x}/p_{t5} = f\left[\frac{\dot{m}_5\sqrt{\theta_{t5}}}{\delta_{t5}}\right] = f[\dot{m}_{c5}], \tag{11.4.7}$$

and

$$\pi_{IC} = p_{t25b}/p_{t25a} = f\left[\frac{\dot{m}_{25a}\sqrt{\theta_{t25a}}}{\delta_{t25a}}\right] = f[\dot{m}_{c25a}]. \tag{11.4.8}$$

Generalized trends for an exhaust are shown in Figure 11.3. Similar trends result for the stack, regenerator, and intercooler

Similarly, for the jet engine duct, dimensional analysis for the mixing process and mix enhancer, respectively (Figure 11.1) results in:

$$\pi_{duct} = p_{t75}/p_{t7} = f\left[\frac{\dot{m}_7\sqrt{\theta_{t7}}}{\delta_{t7}}\right] = f[\dot{m}_{c7}], \tag{11.4.9}$$

$$\pi_{mix} = p_{t55}/p_{t5\text{-}t75} = f\left[\frac{\dot{m}_5\sqrt{\theta_{t5}}}{\delta_{t5}}, \frac{\dot{m}_{75}\sqrt{\theta_{t75}}}{\delta_{t75}}\right] = f[\dot{m}_{c5}, \dot{m}_{c75}], \tag{11.4.10}$$

where $p_{t5\text{-}t75}$ is a weighted average of the total pressures at stations 5 and 7.5 (Figure 2.3).

The duct map has similar characteristics as in Figure 11.3. The mixer with enhancer generalized map is shown in Figure 11.4. For this map the ratio $(\dot{m}_{c7.5}/\dot{m}_{c5})$is an additional variable. For a given value of $\dot{m}_{c5}$ the pressure ratio decreases from unity and then increases back to unity as the mass flow ratio increases.

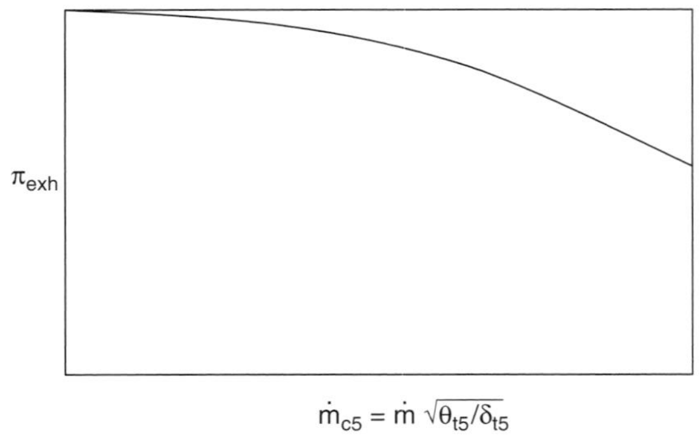

Figure 11.3 Power generation gas turbine exhaust generalized map.

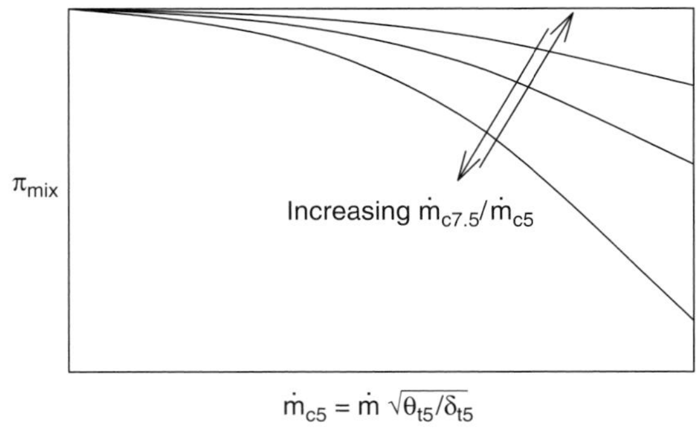

Figure 11.4 Turbofan mixer generalized mixer map.

11.5 Summary

This chapter first covered total pressure losses in the bypass duct and mixer of a turbofan engine. The duct delivers air from the fan to the mixer. The mixer folds the bypassed air into the core airflow so that the flow into the afterburner or nozzle has nearly uniform properties. Although simple components, they both induce total pressure losses. Frictional flow (Fanno line) was used to model the flow in the duct; losses are caused by wall friction and turbulent shear. Separate, irreversible mixing and drag obstacles (mixer vanes) models (and resulting fluid shear) were used to predict the total pressure losses in the mixer. Moderate losses occur in the frictional flow in the duct. Other losses come from the mixing process and also the added drag of the obstacles used to enhance good mixing of the flow. Reduction of the frictional loss is achieved by using smooth boundaries to minimize the friction factor. One must use a tradeoff in minimizing the mixing loss because, to obtain maximum mixing, moderate total pressure losses can occur. However, if good mixing is not achieved, uniform flow will not enter the afterburner and nozzle and they will not perform at near peak efficiency.

Second, it has covered the losses in the exhaust and stack of a power generation gas turbine. Although simple components, they both induce total pressure losses. Frictional flow was included to analyze the flow; losses are from wall and vane (in the exhaust) friction, and turbulent shear, and for the case of the exhaust separation can also occur. Since these components are located after the gas turbine itself, they affect the pressure at the exit of the last turbine. As a result, expansion of the gas in the turbines is lowered and the performance is compromised. Generalized 1-D compressible flow was used to model the flow in which area change and frictional flow was included. Moderate losses occur in the flow. Reduction of the frictional loss is achieved by using smooth boundaries and vanes (in the exhaust) to minimize the friction factor.

Third, three other passive components that improve overall gas turbine performance but also affect total pressure were identified. There are the regeneration (heat exchanger), intercooling (heat exchanger), and inlet cooling (spray nozzles). The first two lower the total pressure and lowers the performance, while the latter is a combination of lowering and increasing the total pressure, involving a more detailed analysis.

Lastly, simple mapping methods were covered for the passive power generation and jet engine components.

List of Symbols

a	Sound speed
A	Area
A_d	Frontal area of drag object
A_R	Relative blockage or relative frontal area
c_p	Specific heat at constant pressure
C_d	Drag coefficient
D	Diameter
F	Force
f	Fanning friction factor
L	Length
$\dot{m}$	Mass flow rate
M	Mach number
p	Pressure
$\mathcal{P}$	Power
$\mathcal{R}$	Ideal gas constant
T	Temperature
V	Velocity
x	Flow direction
γ	Specific heat ratio
δ	Ratio of pressure to pressure at STP
η	Efficiency
θ	Ratio of temperature to temperature at STP
π	Total pressure ratio
ρ	Density

Subscripts

c	"Corrected"
d	Drag
duct	Bypass duct
exh	Exhaust
IC	Intercooler
mix	Bypass mixer
RG	Regenerator
stk	Stack
stp	Standard temperature and pressure
t	Total (stagnation)
1, 2, 3	Positions for process analysis

Superscripts

*	Choked

Problems

11.1 A mixer is to be analyzed. The duct flow rate from the fan is 200 lb_m/s and is at a Mach number of 0.45, whereas the primary flow rate is 75 lb_m/s and is at a Mach number of 0.75. The total temperatures of the bypassed and primary air are 900 and 1500 °R, respectively. The static pressures of the two streams are 15 psia. Find the exit conditions of only the mixing process. That is, find the total temperature, static pressure, total pressure, and Mach number. Assume $\gamma = 1.40$.

11.2 A mixer is to be analyzed. The duct flow rate from the fan is 200 lb_m/s and is at a Mach number of 0.45, whereas the primary flow rate is 75 lb_m/s and is at a Mach number of 0.75. The total temperatures of the bypassed and primary air are 900 and 1500 °R, respectively. The static pressures of the two streams are 15 psia. A structural member in the flow is used to promote mixing of the two streams. The member has a drag coefficient of 0.60, and the frontal area of the member is 7 percent of the total flow area. Find the exit conditions of the combined mixing process and flow around the drag object. That is, find the total temperature, static pressure, total pressure, and Mach number. Assume $\gamma = 1.40$.

11.3 Flow in a bypass duct is to be analyzed. Air enters the duct at a Mach number of 0.4 and a total temperature of 500 K (900 °R) and total pressure of 137.9 kPa (20 psia). The length of the duct is 1.372 m (4.5 ft), and the average effective diameter is 0.1067 m (0.35 ft). The Fanning friction factor is 0.035. Find the exit conditions, including the Mach number and total pressure. Assume $\gamma = 1.40$.

11.4 A mixer is to be analyzed. The duct flow rate from the fan is 60 lb_m/s and is at a Mach number of 0.55, whereas the primary flow rate is 120 lb_m/s and is at a Mach number of 0.85. The total temperatures of the bypassed and primary air are 1000 and 1700 °R, respectively. The static pressures of the two streams are 20 psia. The mixer has a drag coefficient and occupies 6 percent of the flow area. Find the exit conditions of only the mixing process for the two conditions that follow. That is, find the total temperature, static pressure, total pressure, and Mach number. Assume $\gamma = 1.40$. (a) Drag coefficient is 0.00, and (b) drag coefficient is 0.60.

11.5 A mixer is to be analyzed. The duct flow rate from the fan is 45.35 kg/s (100 lb_m/s) and is at a Mach number of 0.60, whereas the primary flow rate is 45.35 kg/s (100 lb_m/s) and is at a Mach number of 0.80. The total temperatures of the bypassed and primary air are 500 and 1000 K (900 and 1800 °R), respectively. The static pressures of the two streams are 165.5 kPa (24 psia). The mixer has a drag coefficient and occupies 8 percent of the flow area. Find the exit conditions of only the mixing process for the two conditions that follow. That is, find the total temperature, static pressure, total pressure, and Mach number. Assume $\gamma = 1.38$. (a) Drag coefficient is 0.00, and (b) drag coefficient is 0.70.

11.6 Flow exits a power generation gas turbine with a Mach number of 0.50. The flow enters a diffusing channel with an inlet area of 0.387 m^2 (600 $in.^2$) and exits the plenum with an area of 1.806 m^2 (2800 $in.^2$). The effective length the channel is 3.05 m (120 in.) with a Fanning friction factor of 0.013. The total temperature remains constant at 889 K (1600 °R). The flow then enters a stack, which is 100 m (330 ft) long, and exhausts into 101.3 kPa (14.69 psia) atmosphere. The stack has a Fanning friction factor of 0.010 and exit area of 1.55 m^2 (2400 $in.^2$). What is the total pressure at the turbine exit if the specific heat ratio is 1.37?

11.7 A mixer with an enhancer is to be analyzed. The duct flow rate from the fan is 200 lb_m/s and is at a Mach number of 0.30, whereas the primary flow rate is 100 lb_m/s and is at a Mach number of 0.40. The total temperatures of the bypassed and primary air are 900 and 1700 °R, respectively. The static pressures of the two streams are 90 psia. The enhancer has a drag coefficient of 0.60 and an area blockage of 8 percent. Assume $\gamma = 1.36$.

(a) Find the exit conditions of only the mixing process. That is, find the total temperature, static pressure, total pressure, and Mach number.
(b) Find the exit conditions of the combined mixing process and drag body. That is, find the total temperature, static pressure, total pressure, and Mach number.

Part IV
System Matching and Analysis

Pratt & Whitney J57 single spool turbojet (© 2022 Raytheon Technologies Corporation, Pratt & Whitney division. Reproduced with permission).

GE LM1600 power generation gas turbine (courtesy of General Electric Gas Power).

12 Matching of Gas Turbine Components: Capstone

12.1 Introduction

In the previous seven chapters, much emphasis has been placed on the analysis and design of individual components of gas turbines. On the other hand, in Chapter 3 cycle analyses are presented for ideal and nonideal gas turbines as a whole in which the different components are integrated into a system. However, in Chapter 3, component efficiencies and some characteristics are assumed or assigned a priori for the overall cycle analyses. In general, however, the previous seven chapters demonstrate, through the use of either theoretical analyses or empirical characteristic curves (or "maps"), that component efficiencies and other operating characteristics change significantly at different conditions – for example, at different flow rates and rotational speeds.

To understand the overall effects of changing operating conditions, one can consider a gas turbine initially at some steady-state operating condition. However, as the fuel injection rate in the burner is changed, the turbine inlet temperature and pressure are changed. Thus, the turbine will change rotational speeds. Because the turbine and compressor are on the same shaft, however, this change in rotational speed in turn changes the ingested mass flow rate and pressure ratio developed by the compressor, which influences the burner inlet pressure and the turbine inlet pressure, and so on. Eventually, the gas turbine will again reach a different steady-state operating condition. In other words, the different components interact and influence each other as a jet engine changes flight conditions or a power generating unit changes power output. These interactions dictate the overall gas turbine characteristics, including parameters such as mass flow rate, rotational speed, pressure ratios, thrust, derived power, and so forth. The ways in which the components interact and how their individual characteristics dictate the engine steady-state operating point are termed component matching and form the topic of this chapter.

Up to this point, the components have been treated as being independent of each other; this is not the case, for they can interact strongly. Yet, this interaction or matching between the components dictates the overall performance of a gas turbine, as discussed by Johnsen and Bullock (1965), El-Masri (1988), Mirza-Baig and Saravanamuttoo (1991), Saito, Sugiyama, Endoh, and Matsuda (1993), Kurzke (1995, 1998), Farokhi (2014), and Kurzke and Halliwell (2018). Not only should each component operate individually with a high efficiency, but a gas turbine as a system must operate efficiently so that the maximum power is derived. Thus, components must be chosen so that they can operate near peak efficiency as a unit.

The overall objective of this capstone chapter is first to present the fundamental concept of how all of the components of a gas turbine interact. Second, a method is presented whereby the steady-state matching point of the compressor, burner, and turbine (which are as a combination called a gas generator) can be determined if individual component performance characteristic curves or maps are known. The gas generator can then be coupled with the other gas turbine components to determine the matching point of the entire engine or power plant. Mathematical models for the diffuser, compressor, combustor, turbine, nozzle, inlet, and exhaust that can be used to represent available maps or data are presented to accomplish this objective. Such modeling allows the matching process to be accomplished relatively easily using computer analyses. The individual component models are then assembled to determine the overall matching point of a gas turbine. The matching point can be found for different fuel flows and, thus, the gas turbine operating line can be predicted. Both on-design and off-design operating points can be analyzed. The analysis represents a relatively simple method whereby a large and complex database can be condensed to the operating condition of a gas turbine. In this chapter, the methodology is applied only to a simple turbojet without an afterburner (Figure 3.3) and to a simple power generation gas turbine with or without a regenerator or turboshaft (Figure 3.9). In general, however, the approach can be applied to any of the previously covered engine or power generation gas turbine types. As a last step, determining the matching points of a jet engine(s) with an airframe is discussed so that aircraft speed can be predicted as the fuel rate is varied.

12.1.1 Objectives

After completing this chapter, the reader should be able to do the following.

- Understand the concept of integrating component "maps" to perform a more complex cycle analysis of a gas turbine
- Understand that the "maps" for all of the components can be graphs, tables, or modeling
- Understand the very basic idea of how "maps" work together to find the resulting overall performance of a gas turbine
- Understand the idea of modeling the component maps
- Apply the method for a gas generator, jet engine, or power generation gas turbine
- Match a jet engine to an aircraft
- Understand how the method is used to perform a "second stage" design for a gas turbine including complete maps for the components

12.2 Component Matching

Because each component of a gas turbine cannot act independently of the other components, the general procedure by which the gas turbine matching point is determined must be developed. This is done for three cases. The first case is a single-shaft gas generator (compressor, burner, and turbine). The second case is for an entire single-spool turbojet engine (diffuser, gas generator, and nozzle) as discussed by Flack (1990). The third case is for an entire single-spool power gas turbine (inlet, gas generator, exhaust, and load) as discussed by Flack (2002). Simulations of the engine operating conditions can be made if individual component performance characteristic curves or maps are known. Because most of the concepts of maps are covered in previous chapters, they will not be discussed in great detail in this chapter but will be reviewed and summarized. Kurzke (1996) also presents a method of estimating maps.

12.2.1 Gas Generator

Compressor

Dimensional analysis, which is discussed in Chapters 4 and 5, shows that the developed total pressure ratio and efficiency of a compressor are functions of only two parameters: the "corrected" mass flow and "corrected" speed, regardless of altitude, absolute speed, and so on. The general functional relationships for a compressor map (Figures 5.13 and 5.14), namely, pressure ratio and efficiency, are specified in Chapters 4 and 5 as

$$\pi_c = \varphi_c(\dot{m}_{c2}, N_{c2}) \tag{12.2.1}$$

$$\eta_c = \psi_c(\dot{m}_{c2}, N_{c2}), \tag{12.2.2}$$

where φ_c and ψ_c represent empirical functions that are either predicted or determined experimentally for a given compressor and can be in graphical, tabular, or equational form. Similar notation is used for the functions of other components. Recall that the corrected values of mass flow and speed for the compressor are defined by

$$\dot{m}_{c2} \equiv \dot{m}_2 \frac{\sqrt{T_{t2}/T_{stp}}}{p_{t2}/p_{stp}} = \dot{m} \frac{\sqrt{T_{t2}/T_{stp}}}{p_{t2}/p_{stp}} \quad (12.2.3)$$

$$N_{c2} \equiv \frac{N}{\sqrt{T_{t2}/T_{stp}}}. \quad (12.2.4)$$

The subscript stp denotes standard conditions ($p_{stp} =$ 14.69 psia or 101.33 kPa and $T_{stp} = 518.7$ °R or 288.2 K), total conditions are evaluated at the compressor inlet, the true mass flow rate into the compressor is $\dot{m}$, and N is the true rotational speed of the shaft. Corrected mass flow rate and speed are similarly defined for the other components.

As shown in Chapters 2 and 4 (Eq. (2.3.30)) from a thermodynamics analysis, the compressor total pressure ratio is related to the total temperature ratio as follows:

$$\pi_c = [1 + \eta_c(\tau_c - 1)]^{\frac{\gamma}{\gamma-1}}. \quad (12.2.5)$$

As in previous chapters, specific heats are evaluated at the average temperature of a given component and will not be subscripted here for the sake of brevity.

Turbine

Similarly, in Chapter 7 a turbine map is discussed (Figure 7.11). Although the map has a much different shape than that for a compressor, the total pressure ratio and efficiency are functions of only two parameters: the "corrected" turbine mass flow and "corrected" turbine speed regardless, again, of altitude, absolute speed, and so on. These general functional relationships are as follows:

$$\pi_t = \varphi_t(\dot{m}_{c4}, N_{c4}) \quad (12.2.6)$$

$$\eta_t = \psi_t(\dot{m}_{c4}, N_{c4}). \quad (12.2.7)$$

Again, these functions can be predicted or measured and can be in graphical, tabular, or equational form. From Chapters 2 and 7 (Eq. (2.3.47)), the turbine total pressure ratio is related to the total temperature ratio by

$$\pi_t = \left[1 + \frac{(\tau_t - 1)}{\eta_t}\right]^{\frac{\gamma}{\gamma-1}}. \quad (12.2.8)$$

Combustor

In Chapter 2, a simple energy balance from the first law of thermodynamics (Eqs. (2.3.62) and (2.3.63)) for the burner can be expressed as

$$f(\eta_b \Delta H - c_p T_{t3}\tau_b) = c_p T_{t3}(\tau_b - 1), \quad (12.2.9)$$

where ΔH is the enthalpy of reaction and f is the fuel-to-air flow ratio. Note that this can be replaced by a more accurate adiabatic flame temperature calculation, as is done in Chapter 9. In that chapter, characteristics for a combustor are covered and a typical burner map is shown in Figure 9.13. The burner efficiency and total pressure are shown there to be a result of both friction and the combustion process; they are functions of basically only two parameters:

$$\eta_b = \psi_b\left(\dot{m}_{c3}, \frac{f}{T_{t3}/T_{stp}}\right) \quad (12.2.10)$$

$$\pi_b = \varphi_b\left(\dot{m}_{c3}, \frac{f}{T_{t3}/T_{stp}}\right). \quad (12.2.11)$$

Closure Equations

Some of the component parameters must be related so that matching can be accomplished. Using the definitions of the corrected flows and speeds for the turbine and compressor, the two corrected mass flows and speeds can be related if it is assumed that mass flow is conserved, any bleeds ($\dot{m}_4 = (1 + f) \times \dot{m}_2$) are ignored, and that the compressor and turbine are on the same shaft ($N_2 = N_4$); thus,

$$\dot{m}_{c2} = \frac{\dot{m}_{c4}\pi_b\pi_c}{(f+1)\sqrt{\tau_b\tau_c}} \quad (12.2.12)$$

$$N_{c2} = N_{c4}\sqrt{\tau_b\tau_c}. \quad (12.2.13)$$

Similarly, the corrected mass flows for the burner and compressor can be related on the assumption, once again, that mass flow is conserved ($\dot{m}_3 = \dot{m}_2$) as follows:

$$\dot{m}_{c2} = \frac{\dot{m}_{c3}\pi_c}{\sqrt{\tau_c}}. \quad (12.2.14)$$

By executing an energy balance on the shaft without any external power load – that is, all of the turbine power is used to drive the compressor except for bearing and damper losses (Eq. (2.3.85)) – one finds

$$c_{p_c}(\tau_c - 1) = \eta_m(1 + f)c_{p_t}\tau_b\tau_c(1 - \tau_t), \quad (12.2.15)$$

where the mechanical shaft efficiency accounts for shaft losses and is directly a function of the shaft speed:

$$\eta_m = \psi_m(N). \quad (12.2.16)$$

Lastly, the burner inlet total temperature is related to the compressor total temperature by the definition of the total temperature ratio of the compressor τ_c:

$$T_{t3} = \tau_c T_{t2}. \quad (12.2.17)$$

Thus, there are 17 equations (Eqs. (12.2.1)–(12.2.17)) and 20 variables ($\dot{m}$, $\dot{m}_{c2}$, $\dot{m}_{c3}$, $\dot{m}_{c4}$, N, N_{c2}, N_{c4}, π_c, π_b, π_t, τ_c, τ_b, τ_t, η_c, η_b, η_t, η_m, f, T_{t2}, and T_{t3}). Three variables can be specified for a gas generator (typically T_{t2}, f, and N_{c2}) and the remaining 15 can be found. Section 12.2.6 presents a solution method.

12.2.2 Jet Engine

Section 12.2.1 has just specified the principal equations for the components of a gas generator (compressor, burner, and turbine). The complementary equations for the diffuser and nozzle of a jet engine are presented in this section.

Diffuser

The flow for a diffuser is adiabatic, and thus the total temperature is given by (from Eq. (2.3.3))

$$T_{t2} = T_a\left[1 + \frac{\gamma - 1}{2} M_a^2\right], \quad (12.2.18)$$

where M_a is the freestream Mach number. Owing to internal and external losses as discussed in Chapter 8, the diffuser total pressure ratio is related to the freestream Mach number as follows:

$$\pi_d = \varphi_d(M_a). \quad (12.2.19)$$

Thus, the diffuser exit total pressure is (on the basis of Eqs. (2.3.8) and (2.3.9)) as follows:

$$p_{t2} = p_a\pi_d\left[1 + \frac{\gamma - 1}{2} M_a^2\right]^{\frac{\gamma}{\gamma-1}}. \quad (12.2.20)$$

Nozzle

In Chapter 10, the general nozzle characteristics are shown to be functions of only two parameters. The nozzle characteristics (map, Figure 10.19) are expressed in that chapter as

$$\eta_n = \psi_n(p_{t5}/p_a) \quad (12.2.21)$$

$$\dot{m}_{c5} = \varphi_n(\eta_n, p_{t5}/p_a). \quad (12.2.22)$$

Closure Equations

As is the case for the gas generator, some equations are necessary to relate the component characteristics. For example, the total pressure at the inlet of the nozzle can be related to the diffuser exit total pressure by

$$p_{t5} = p_{t2}\,\pi_c\,\pi_b\,\pi_t. \quad (12.2.23)$$

The corrected mass flow for the nozzle is related to that of the turbine on the assumption that mass is conserved ($\dot{m}_4 = \dot{m}_5$) as expressed by

$$\dot{m}_{c4} = \frac{\dot{m}_{c5}\pi_t}{\sqrt{\tau_t}}. \quad (12.2.24)$$

Thus, seven additional equations are now included and eight new variables (T_a, p_a, M_a, π_d, p_{t2}, η_n, $\dot{m}_{c5}$, and p_{t5}) have been added. Therefore, we have a total of 24 equations with 28 variables. Fortunately, three of these variables are flight conditions (p_a, T_a, and M_a). Therefore, for any given flight condition we have 23 unknown variables and 22 equations. Hence, one variable (usually f) can be specified, and the remaining parameters can be determined. For example, for specific flight conditions for given components (and accompanying maps), the fuel ratio can be varied and the pressure ratios, efficiencies, corrected rotational speeds, corrected mass flow rates, dimensional rotational speed, mass flow rate, and so on, at which the engine will operate can be found. One solution method of determining the unknown variables is discussed in Section 12.2.6.

Moreover, once the matching problem has been solved to find the component operating points, the methods developed in Chapters 2 and 3 can be used to find the important overall engine characteristics – namely, engine thrust and *TSFC*. For example, the dimensional thrust is given by (from Eq. (1.5.12))

$$F = \dot{m}[(1+f)u_8 - u_a] + A_8(p_8 - p_a). \tag{12.2.25}$$

As a reminder, note that empirical functions representing component maps are recognized in 10 of the preceding equations ($\varphi_c, \psi_c, \varphi_b, \psi_b, \varphi_t, \psi_t, \psi_m, \varphi_d, \varphi_n$, and ψ_n). These must be specified to solve the matching problem – as noted earlier, they can be graphical, tabular, or mathematical. Furthermore, they can be derived from empirical data or theoretical predictions. In Section 12.2.5, mathematical models are presented that can be used to curve fit the different functions for the maps.

12.2.3 Power Generation Gas Turbine without Regeneration

Section 12.2.1 specifies the governing equations for the gas generator. In this section, the accompanying matching equations for the inlet, exhaust, and load system of a gas power turbine are discussed. Although the gas generator of a power generation gas turbine is essentially the same as that of a jet engine, some subtle differences in the operating characteristics of the components exist. Flack (2002) includes component matching for a power generation unit with regeneration.

Bell Inlet

For a power generation inlet, the flow is adiabatic and velocities are very low far from the inlet ($M_a = 0$); thus, the exit total temperature of the inlet is given by

$$T_{t2} = T_a. \tag{12.2.26}$$

Furthermore, the frictional losses in total pressure of the inlet are directly related to the freestream inlet velocity. Thus, the total pressure ratio for an inlet can be written as

$$\pi_i = \varphi_i(\dot{m}_{c1}). \tag{12.2.27}$$

The total exit pressure of the inlet therefore is (again on the fact that $M_a = 0$ and thus that $p_a = p_{ta}$):

$$p_{t2} = p_a\pi_i. \tag{12.2.28}$$

Exhaust

The total pressure loss for an exhaust is due to viscous effects and is, consequently, a function of the flow speed. Thus, the total pressure ratio can be written as

$$\pi_e = \varphi_e(\dot{m}_{c5}). \tag{12.2.29}$$

The velocity at the exhaust exit is very low ($p_{ta} = p_a = p_e = p_{te}$), and so the total inlet pressure of the exhaust can be found from

$$p_{t5} = \frac{p_a}{\pi_e}. \tag{12.2.30}$$

Load System

Furthermore, when an external load is applied to the shaft by, for example, a ship propeller, this power load is directly related to rotational speed and can be found from

$$\mathcal{P}_l = \varphi_l(N). \tag{12.2.31}$$

One should realize that this load, $\mathcal{P}_l$, is in actuality the net useful or output power of the power gas turbine system. For the case of electric generation and some other applications, the unit typically runs at a constant rotational speed. As a result, this power load equation would be replaced by another taking into account that the shaft speed N is known and that the generator power load is adjusted to maintain a constant shaft speed. Or, as another example, the generator and free turbine may be on a different shaft, which would necessitate additional modeling.

Closure Equations

As is the case for the jet engine, some equations are necessary to relate the component characteristics. For example, the corrected mass flow for the inlet can be related to that of the compressor using conservation of mass ($\dot{m}_1 = \dot{m}_2$) as follows:

$$\dot{m}_{c1} = \dot{m}_{c2}\pi_i. \tag{12.2.32}$$

The total pressure at the inlet of the exhaust can also be related to the inlet total pressure to the compressor ($\pi_c = p_{t3}/p_{t2}$, $\pi_b = p_{t4}/p_{t3}$, and $\pi_t = p_{t5}/p_{t4}$) by

$$p_{t5} = p_{t2}\pi_c\pi_b\pi_t. \tag{12.2.33}$$

The corrected mass flow for the exhaust can be related to that of the turbine, again using conservation of mass ($\dot{m}_4 = \dot{m}_5$), by

$$\dot{m}_{c4} = \frac{\dot{m}_{c5}\pi_t}{\sqrt{\tau_t}}. \tag{12.2.34}$$

Performing an energy balance on the shaft with an external power load – that is, all of the turbine power is used to drive the compressor and external load except for bearing and damper losses – yields

$$c_{p_c}(\tau_c - 1) + \frac{\mathcal{P}_l}{\dot{m}T_{t2}} = \eta_m(1+f)c_{p_t}\tau_b\tau_c(1-\tau_t), \tag{12.2.35}$$

replacing Eq. (12.2.15), which applies to jet engines.

As noted above, for power generation applications the gas turbine is operated in one of two ways. First, the speed-dependent power load on the shaft dictates the rotational speed of the unit. Or, second, the rotational speed is often held constant (for example, for electric generation or other applications that require a fixed rotational speed), and the net power load is a result of the operation.

For the load dependence case, the preceding 10 equations add 9 new variables (T_a, p_a, p_{t2}, p_{t5}, π_i, π_e, $\dot{m}_{c1}$, $\dot{m}_{c5}$, and $\mathcal{P}_l$). Thus, 26 equations result with 29 variables. However, two of these are ambient conditions (p_a and T_a). Therefore, for any given ambient condition, 27 variables and 26 equations are present. One can specify one variable (usually f) and determine the remaining parameters. For example, for given ambient conditions in given component maps, one can vary the fuel ratio and determine the load or output power, rotational speed, mass flow rate, and so on, at which the power turbine will operate. The solution method of determining the unknown variables is discussed in Section 12.2.6. Once the matching problem has been solved to find the component operating points, one can find the power turbine thermodynamic efficiency and *SFC*, as described in Chapters 1, 2, and 3.

The second application of power generation gas turbines is when the rotational speed is held constant. For this application Eq. (12.2.31) no longer applies. As a result, 25 equations result with 29 variables. However, two of these are ambient conditions (p_a and T_a) and the rotational speed, *N*, is specified. Therefore, for any given ambient condition 26 variables and 25 equations are present. Again, one variable (usually f) can be specified to determine the remaining parameters. The solution method for this case of determining the unknown variables will be discussed in Section 12.2.6. Once the matching problem has been solved to find the component operating points, one can find the mass flow rate, output power, power turbine thermodynamic efficiency, *SFC*, and other outputs, as described in Chapters 1, 2, and 3.

Empirical functions representing component maps are identified in 10 of the preceding equations (φ_c, ψ_c, φ_b, ψ_b, φ_t, ψ_t, ψ_m, φ_i, φ_e, and φ_l). As before, these must be available or specified to solve the matching problem, and they can be graphical, tabular, or mathematical. Again, they can be derived from empirical data or theoretical predictions.

12.2.4 Power Generation Gas Turbine with Regeneration

Three new parameters must be included for the regenerator of a power gas turbine analysis. These are two total pressure loss terms due to highly viscous flow in the two flow paths and the effectiveness, ε_{RG}, of the regenerator. These have been discussed in Sections 2.3.16, 11.3.3, and 11.3.4 and shown on Figure 2.49:

$$\pi_{33x} = \varphi_{33x}(\dot{m}_{c3}) \tag{12.2.36}$$

$$\pi_{55x} = \varphi_{55x}(\dot{m}_{c5}) \tag{12.2.37}$$

$$\varepsilon_{RG} = \text{usually known}. \tag{12.2.38}$$

But other components are affected by the inclusion of a regenerator. The exhaust is now

$$\pi_e = \varphi_e(\dot{m}_{c5x}), \tag{12.2.39}$$

where

$$\dot{m}_{c5x} = \frac{\dot{m}_{c5}\pi_{55x}}{\sqrt{T_{t5x}/T_{t5}}} \tag{12.2.40}$$

and from Eq. (2.3.136)

$$T_{t5x} \cong T_{t5} - (T_{t3x} - T_{t3}). \tag{12.2.41}$$

The pressure ratio for the burner is now

$$\pi_b = \varphi_b\left(\dot{m}_{c3x}, \frac{f}{T_{t3x}/T_{stp}}\right), \tag{12.2.42}$$

where

$$\dot{m}_{c3x} = \frac{\dot{m}_{c3}\pi_{33x}}{\sqrt{T_{t3x}/T_{t3}}} \tag{12.2.43}$$

and

$$T_{t3x} = T_{t3} + \varepsilon_{RG}(T_{t5} - T_{t3}). \tag{12.2.44}$$

As a result, seven new parameters are introduced (ε_{33x}, ε_{55x}, ε_{RG}, $\dot{m}_{c5x}$, $\dot{m}_{c3x}$, $\dot{m}_{c3x}$, T_{t3x}, and T_{t5x}) but seven new equations are also introduced, indicating a solution can be obtained.

12.2.5 Component Modeling

Ten components are considered for the overall gas turbine jet engine or power generation gas turbine: compressor, combustor, turbine, shaft, diffuser, nozzle, inlet, exhaust, regenerator, and load. In this section, a mathematical model for each is described that can be used to curve fit data by the appropriate choice of curve-fitting parameters, as partially presented by Flack (1990). Functional relationships have been developed that match typical experimental data and trends but do not necessarily match all data accurately. They have been developed based on known physical dependence of characteristics, deviations from optimal values by variations of known characteristics, and simple curve fitting. It is most important to note that modeling each component mathematically is not necessary for the general matching solution method; models are presented for convenience and to accelerate a solution using commercial math solvers.

Compressor

A model for a corrected speed line on a generalized compressor map is

$$\pi_c = 1 + c_1\dot{m}_{c2}\sqrt{\frac{\dfrac{\dot{m}_{c2}}{c_2N_{c2}} - 1}{c_3 - 1}}. \tag{12.2.45}$$

If one considers any particular speed line (at N_{c2}), the flow rate at $\pi_c = 1$ is defined as $\dot{m}_{c2_o}$ and the flow rate at which surge occurs is $\dot{m}_{c2_s}$. The three curve-fitting parameters in Eq. (12.2.45) are now found. First, mass flow is approximately proportional to speed; thus,

$$c_2 = \dot{m}_{c2_o}/N_{c2} \tag{12.2.46}$$

$$c_3 = \dot{m}_{c2_s}/\dot{m}_{c2_o}. \tag{12.2.47}$$

Also, the surge line over a range is approximately a straight line:

$$c_1 = \frac{\pi_{c_s} - 1}{\dot{m}_{c2_s}}. \tag{12.2.48}$$

Knowing that the efficiency drops from the maximum value as both speed and mass flow are changed, one can model the compressor efficiency by

$$\eta_c = \eta_{c_d} - c_4|N_{c2_d} - N_{c2}| - \frac{c_5}{N_{c2}}\left[\dot{m}_{c2_d} - \dot{m}_{c2}\right]^2. \tag{12.2.49}$$

At the maximum efficiency point on speed line N_{c2}, the corrected flow is $\dot{m}_{c2_d}$:

$$\dot{m}_{c2_d} = c_6\dot{m}_{c2_o} = c_6c_2N_{c2}. \tag{12.2.50}$$

The curve-fitting parameter c_6 can be related to a measure of the surge margin (μ),

$$c_6 = c_3(\mu + 1), \tag{12.2.51}$$

where this is defined as

$$\mu = \frac{\dot{m}_{c2_d}}{\dot{m}_{c2_s}} - 1. \tag{12.2.52}$$

The maximum efficiency on the map is η_{cd} and occurs at corrected speed N_{c2_d}. The values of c_4 and c_5 are both chosen to best fit experimental test data or predictions.

Combustor

Knowing that, as the mass flow rate and or heat addition increase, the pressure drops faster than either of these two quantities increase, and from Fanno and Rayleigh line analyses, one can model the total pressure ratio map by

$$\pi_b = 1 - b_1 \dot{m}_{c3}^2 \left[\frac{f}{T_{t3}/T_{stp}}\right]^2, \tag{12.2.53}$$

and the model for the combustion efficiency η_b is

$$\eta_b = \eta_{b_d} - \frac{b_2}{\left[\frac{\dot{m}_{c3} f}{T_{t3}/T_{stp}}\right]^2}. \tag{12.2.54}$$

The maximum combustion efficiency is η_{b_d}, and b_1 and b_2 are found from experimental test data or predictions.

Turbine

A turbine map is modeled as

$$\dot{m}_{c4} = \dot{m}_{c4_c} \left\{ 2\left[\frac{\frac{1}{\pi_t} - 1}{\frac{1}{\pi_{t_c}} - 1}\right]^n - \left[\frac{\frac{1}{\pi_t} - 1}{\frac{1}{\pi_{t_c}} - 1}\right]^{2n} \right\}, \tag{12.2.55}$$

and using the quadratic "formula":

$$\frac{1}{\pi_t} = \left(1 - \sqrt{1 - \frac{\dot{m}_{c4}}{\dot{m}_{c4_c}}}\right)^{\frac{1}{n}} \left(\frac{1}{\pi_{tc}} - 1\right) + 1. \tag{12.2.56}$$

The parameter π_{tc} is the pressure ratio at which the flow chokes, and $\dot{m}_{c4_c}$ is the corrected mass flow rate at that limit. The parameter n is given by

$$n = \frac{N_{c4}}{2N_{c4_d}}. \tag{12.2.57}$$

The corrected speed at which the maximum efficiency occurs is defined as N_{c4_d}. A model for the turbine efficiency is

$$\eta_t = \eta_{t_d} \left\{ 1 - k_1 \left[\frac{\frac{1}{\pi_t} - \frac{1}{\pi_{t_c}}}{\frac{1}{\pi_{t_c}} - 1}\right]^2 - k_2 \left[\frac{\dot{m}_{c4_c} N_{c4_d} - \dot{m}_{c4} N_{c4}}{\dot{m}_{c4_c} N_{c4_d}}\right]^2 \right\}. \tag{12.2.58}$$

The maximum efficiency is η_{t_d}, and the parameters k_1 and k_2 are selected to best fit experimental data or predictions.

Shaft

Experimental data and predictions both show that the mechanical shaft efficiency can effectively be modeled as

$$\eta_m = 1 - s_1 N^{s_2}, \tag{12.2.59}$$

where for antifriction or rolling bearings s_2 is approximately 1, whereas for fluid film bearings, the quantity is around 2. The parameter s_1 is dependent on the particular bearing design and is found from data or predictions.

Diffuser (Jet Engine)

In Chapter 8, the recovery factor is modeled by (Eq. (8.6.2)):

$$\pi_d = \pi_{d_d}\left[1 - d(M_a - 1)^{1.35}\right]. \tag{12.2.60}$$

The maximum pressure recovery is π_{d_d} and the constant d is 0.0 for subsonic operation and typically 0.075 for supersonic operation.

Nozzle (Jet Engine)

Three nozzle types are considered, as in Section 10.7, and operating characteristics are predicted for each. A general map is shown in Figure 10.19, and all nozzle types can be modeled as

$$\dot{m}_{c5} = \dot{m}_n a_2 M_8 \frac{p_8}{p_{t5}} \sqrt{1 + \frac{\gamma - 1}{2} M_8^2}, \tag{12.2.61}$$

where $\dot{m}_n$ is a characteristic flow rate. Knowing that, as the Mach number increases so do frictional losses, as in Fanno line flow, one can model the nozzle efficiency η_n by

$$\eta_n = \eta_{n_d} - a_1 M_8^2, \tag{12.2.62}$$

where η_{n_d} is the maximum efficiency and a_1 is chosen to match experimental data. The exit Mach number M_8 is given by (from Eq. (2.3.94))

$$M_8 = \sqrt{\frac{2\left[\frac{1}{\eta_n\left[\frac{p_8}{p_{t5}}\right]^{\frac{\gamma-1}{\gamma}} + 1 - \eta_n} - 1\right]}{\gamma - 1}}. \tag{12.2.63}$$

Fixed Converging The map for this nozzle type is shown in Figure 10.16. The maximum Mach number at the exit is unity for this nozzle type. As discussed in Chapter 10, to analyze this type of nozzle for a given pressure ratio p_{t5}/p_a, one must first check for choking. If the nozzle is choked, the exit Mach number is unity. If the nozzle is not choked, then p_8 is equal to p_a and then M_8 can be found from Eq. (12.2.63). For this geometry (from Eq. (10.7.8)),

$$\dot{m}_n = \rho_{stp}\, a_{stp}\, A_8. \tag{12.2.64}$$

Also, a_2 is given by (from Eq. (10.7.8))

$$a_2 = \sqrt{\frac{\gamma}{\gamma_{stp}} \frac{\mathcal{R}_{stp}}{\mathcal{R}}}. \tag{12.2.65}$$

Variable CD with Fixed Exit Area This nozzle is of variable geometry, and thus the exit pressure matches the ambient pressure with a fixed exit area (the minimum area varies with operating conditions). This nozzle type can operate supersonically. Equations (12.2.61)–(12.2.65) again apply.

Variable CD with Fixed Minimum Area Figure 10.18 depicts a map of this nozzle type. It has a variable geometry, and thus again the exit pressure matches the ambient pressure and has a fixed minimum area, where the exit area varies with operating conditions. Equations (12.2.61)–(12.2.63) again apply, except now

$$\dot{m}_n = \rho_{stp}\, a_{stp}\, A^*, \tag{12.2.66}$$

and from Eqs. (10.7.15) and (10.7.16),

$$a_2 = \frac{\sqrt{\frac{\gamma+1}{2+(\gamma-1)M_8^2}}\sqrt{\frac{\gamma}{\gamma_{stp}}\frac{\mathcal{R}_{stp}}{\mathcal{R}}}}{M_8\left\{\frac{1}{\eta_n}\left[\frac{\eta_n}{\frac{2}{\gamma+1} - 1 + \eta_n}\right]\left[\frac{1}{1+\frac{\gamma-1}{2}M_8^2} - 1 + \eta_n\right]\right\}^{\frac{\gamma}{\gamma-1}}}. \tag{12.2.67}$$

As shown in Chapter 10, $\dot{m}_{c5}$ is constant for all conditions for this nozzle type.

Bell Inlet (Power Gas Turbine)

The only parameter that must be simulated for the inlet of a power gas turbine is the total pressure loss due to boundary layers, turbulence, and so on. The pressure recovery factor π_i can be found based on geometrical considerations similar to those for a subsonic diffuser (Chapter 8 and earlier in this section) and is directly related to velocity or mass flow rate.

Knowing that the losses are similar to Fanno line flow, one can model this loss as

$$\pi_i = 1 - z_1 \dot{m}_{c1}^2. \tag{12.2.68}$$

In Eq. (12.2.68), z_1 is specified to match the mathematical model to experimental or predicted data. Such a model can also sometimes be used for a subsonic jet engine diffuser.

Regenerator (Power Generation Gas Turbine)

Three parameters must be simulated for the regenerator of a power gas turbine. One should remember a regenerator is not always used in a design. These are two total pressure loss terms due to highly viscous flow in the two flow paths and the effectiveness of the regenerator. The two pressure recovery factors, ε_{33x} and ε_{55x}, can be found based on geometrical considerations similar to those of an inlet and, as above, and are directly related to velocity or mass flow rate. Again, knowing that the losses are similar to Fanno line flow, one can model these losses as

$$\pi_{33x} = 1 - b_{33x}\dot{m}_{c3}^2 \tag{12.2.69}$$

and

$$\pi_{55x} = 1 - b_{55x}\dot{m}_{c5}^2. \tag{12.2.70}$$

The values of b_{33x} and b_{55x} in Eqs. (12.2.69) and (12.2.70) are specified to match the mathematical model to experimental or predicted data.

Furthermore, the effectiveness, ε_{RG}, is defined in Chapter 2 (Eq. (2.3.134)) and relates the temperature input to the burner to the hot exit temperature of the turbine:

$$T_{t3x} = T_{t3} + \varepsilon_{RG}(T_{t5} - T_{t3}). \quad (12.2.71)$$

Exhaust/Stacks (Power Gas Turbine)

The only parameter that must be simulated for the exhaust of a power gas turbine is the total pressure loss due to viscous flow. The pressure recovery factor π_e can be found based on geometrical considerations similar to those for a subsonic diffuser or inlet and, as for the inlet in the preceding paragraph, is directly related to velocity or mass flow rate. Again, knowing that the losses are similar to Fanno line flow, one can model this loss as

$$\pi_e = 1 - z_2 \dot{m}_{c5}^2. \quad (12.2.72)$$

The value of z_2 in Eq. (12.2.72) is specified to match the mathematical model to experimental or predicted data.

Shaft Load (Power Gas Turbine)

In the presence of an external and significant load (ship propeller, tank drive, etc.), power is derived from the shaft for auxiliary purposes. A general model for the load on the shaft is

$$\mathcal{P}_l = z_3 N^{z_4}. \quad (12.2.73)$$

The value of z_4 is typically between 1 and 2, and z_3 is strongly dependent on the particular type of load.

Application of Models to Matching

Mathematical models of the compressor, burner, turbine, shaft, diffuser, nozzle, inlet, exhaust, regenerator and load maps have been presented in this section for gas turbines. These models can be used to curve fit data to facilitate matching calculations such as those presented in the previous section. In general, using these component models requires 24 parameters to be specified for a jet engine, 23 for a power gas turbine without regeneration, and 30 for a power gas turbine with regeneration, as summarized in Table 12.1.

It cannot be emphasized strongly enough that such modeling is not necessary for the general problem solution, however. The reader should realize that these

Table 12.1 **Summary of empirical component parameters.**

Component	Number of parameters	Parameters
Compressor	8	$c_1\ c_2\ c_3\ c_4\ c_5\ N_{c4_d}\ \mu\ \eta_{c_d}$
Combustor	3	$b_1\ b_2\ \eta_{b_d}$
Turbine	6	$k_1\ k_2\ \dot{m}_{c4_c}\ N_{c4_d}\ \eta_{t_d}\ \pi_{t_c}$
Shaft	2	$s_1\ s_2$
Diffuser (jet engine)	2	$\pi_{d_d}\ d$
Nozzle (jet engine)	3	$a_1\ \dot{m}_n\ \eta_{n_d}$
Bell inlet (power generation)	1	z_1
Regenerator (power generation)	3	$b_{33x}\ b_{55x}\ \varepsilon_{RG}$
Exhaust (power generation)	1	z_2
Power load (power generation)	2	$z_3\ z_4$

models serve only to facilitate the matching calculations – for example by computer math solvers. Furthermore, if the preceding curve-fitting equations cannot be used to match component maps accurately, look-up tables or other equations can be used for the general solution methods.

12.2.6 Solution of Matching Problem

Section 12.2 discusses three basic cases: a single-shaft gas generator, a complete single-spool turbojet engine, and a single-spool power gas turbine or turboshaft. The gas generator has 15 governing equations with 18 variables, the complete jet engine is associated with 22 equations involving 26 variables, and the power turbine without generation has 26 equations and 29 variables. Generalized relationships were presented for component maps. Section 12.2.5 offers curve-fitting correlations for the maps. In this section, one possible solution method is presented for cases with known components. Because the equations and/or graphical maps are highly nonlinear and performance curves may or may not be in equation form, the analytical solution is highly iterative. One step-by-step process of obtaining a solution is set forth.

General Concept

Before proceeding, it is very important to note that the same general solution method applies regardless of whether one has numerical curve-fitted models of the components as presented in Section 12.2.5, graphical maps, or tabular look-up maps for components. The basic concept is that each component has its own characteristics. Each component influences the operation of all of the other components. All components will eventually be matched for one operating condition. This may be by a mathematical modeling, graphical modeling, look-up tables or on the test stand. Each component will come to a stable operating point. A well-designed set of components will yield a good matching or operating point for all components. If one component is not well designed to match with others the entire set of components will not operate at an efficient point. Every component may be well off of the optimal operating point. In this chapter it is easy to be overwhelmed by the mathematics. But the best way to understand completely the concept is to imagine that you are given the hardware for all of the individual components and you assemble these into one unit. Then you operate these on the test stand under one particular condition, for example one fuel flow rate. These will come to some overall operating point at which all of the components are naturally matched. If you shut down and start over again with the same overall operating condition, the set will again come to the same overall operating point. All of the components are thus naturally matched on the test stand.

Gas Generator (No Regeneration)

For a gas generator, three parameters must be specified for a solution to be obtainable. These are typically chosen to be T_{t2}, N_{c2}, and f. A flow chart outlining one solution method is shown in Figure 12.1 and described in the following list. This solution method is used herein, but other methods or algorisms could be used and would lead to the same solution result.

(1) Using N_{c2} and T_{t2}, one can find N from Eq. (12.2.4).
(2) An estimate of $\dot{m}_{c2}$ must be used to initiate the iteration.
(3) On the basis of N_{c2} and $\dot{m}_{c2}$, compressor maps can be used to find π_c from predictions, data using the graphical or tabular form (Eq. (12.2.1), Figures 5.13 or 5.14), or modeling (Eq. (12.2.45)). The efficiency, η_c, can be found from data in graphical/tabular form (Eq. (12.2.2), Figures 5.13 or 5.14) or modeling (Eq. (12.2.49)).
(4) The temperature ratio τ_c can then be found from the compressor pressure ratio, temperature ratio, and efficiency equation (Eq. (12.2.5)) because π_c and η_c are known.
(5) By knowing T_{t2}, one can find T_{t3} from Eq. (12.2.17).
(6) The combustor corrected flow $\dot{m}_{c3}$ can now be found using the compressor corrected mass flow (Eq. (12.2.14)).

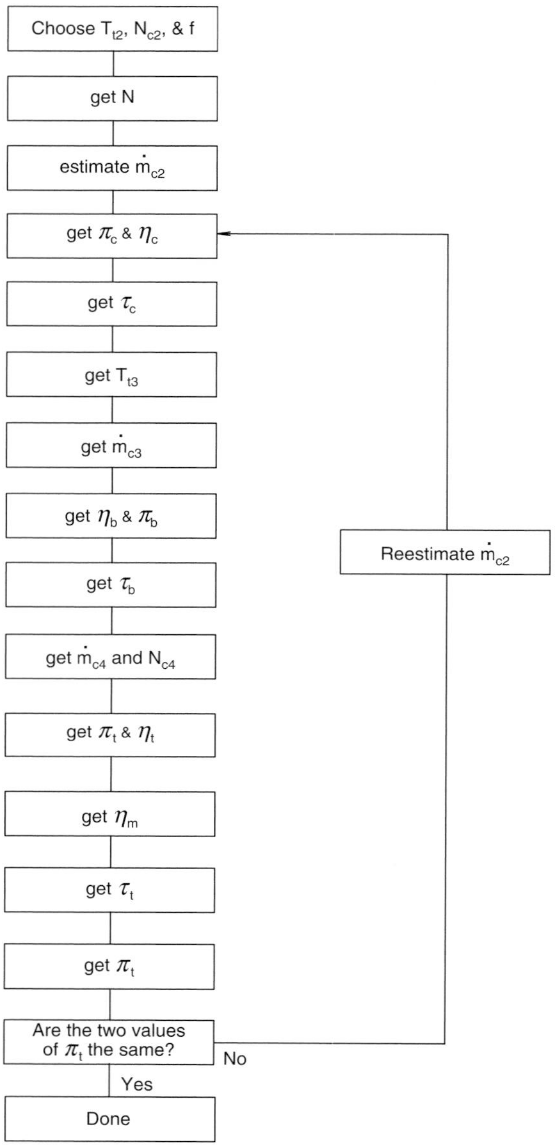

Figure 12.1 Flow chart for solution of gas generator.

(7) Through $\dot{m}_{c3}$ with f, η_b and π_b can be found from combustor maps using the graphical or tabular forms (Eqs. (12.2.10) and (12.2.11), Figure 9.13) or models (Eqs. (12.2.54) and (12.2.53)).

(8) Since η_b and T_{t3} are known, τ_b can be found from the energy balance in the burner (Eq. (12.2.9)).

(9) The turbine corrected flow and speed, $\dot{m}_{c4}$ and N_{c4}, can be related to corresponding quantities for the compressor and found through Eqs. (12.2.12) and (12.2.13).

(10) Because $\dot{m}_{c4}$ and N_{c4} are known, the turbine pressure ratio π_t and efficiency η_t can be found from the turbine maps graphically or tabularly (Eqs. (12.2.6) and (12.2.7), Figure 7.11) or through the use of models (Eqs. (12.2.56) and (12.2.58)).

(11) The graphical or tabular shaft map (Eq. (12.2.16)) or model (Eq. (12.2.59)) can be used with N to determine the mechanical efficiency η_m.

(12) Equation (12.2.15), which represents the power balance on the shaft and applies for no external power load, can be used with f, η_m, τ_b, and τ_c to find τ_t.

(13) One can calculate π_t from the turbine pressure ratio, temperature ratio, and efficiency equation by using τ_t and η_t (Eq. (12.2.8)).

(14) Therefore, two values of π_t are found independently; if the two are the same within a tolerance, the correct value of $\dot{m}_{c2}$ was used. If the two are significantly different, a new value of $\dot{m}_{c2}$ should be used and the process must be repeated. The Regula Falsi method (Appendix I) or another numerical method can be used to reduce the number of iterations. Such a method is used for following examples.

Jet Engine

For a turbojet, four parameters must be specified. These are p_a, T_a, M_a, and f. The solution method chosen for this case is a two-nested loop iteration. A flow chart outlining the solution is shown in Figure 12.2.

(1) First, one must estimate N_{c2}.

(2) Second, $\dot{m}_{c2}$ must be estimated.

(3) By means of M_a, the value of T_{t2} is calculated from the adiabatic diffuser equation (Eq. (12.2.18)).

(4) From M_a, the pressure ratio for the diffuser, π_d, is found through map predictions or data, from graphical or tabular information (Eq. (12.2.19)), through or modeling (Eq. (12.2.60)).

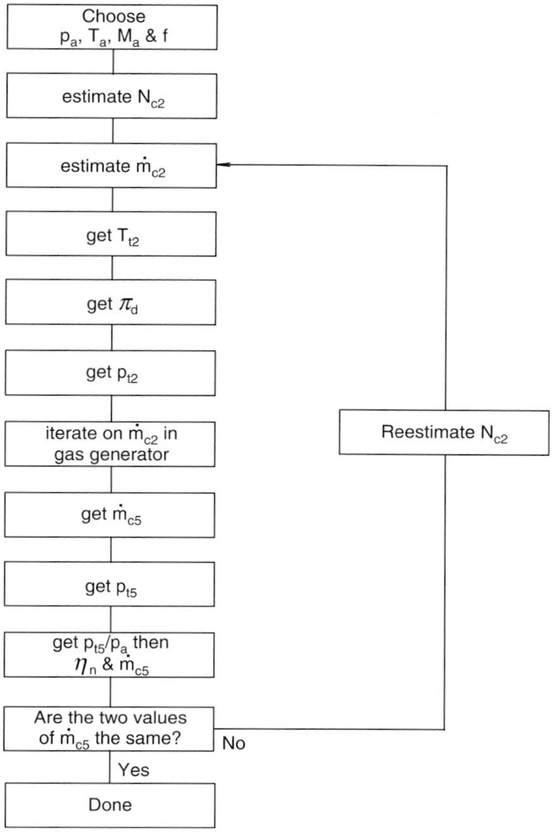

Figure 12.2 Flow chart for solution of jet engine.

(5) Knowing M_a, π_d, and p_a, one finds the total pressure into the compressor, p_{t2}, from Eq. (12.2.20).
(6) Iteration on the corrected mass flow rate for the compressor, $\dot{m}_{c2}$, in the gas generator for the given N_{c2} is necessary (as outlined in the preceding section for a gas generator) until a converged solution is found (two values of π_t are the same). One should remember that this yields parameters such as $\dot{m}_{c3}$, $\dot{m}_{c3}$, π_b, π_c, π_t, etc. This can obviously be a lengthy step.
(7) The corrected mass flow for the nozzle, $\dot{m}_{c5}$, is related to the corrected mass flow for the turbine and found from Eq. (12.2.24).
(8) Knowing p_{t2}, π_b, and π_c, one can relate the value of p_{t5} to the total pressure into the compressor and calculate it through Eq. (12.2.23).
(9) One can first find p_{t5}/p_a and then η_n and $\dot{m}_{c5}$ from predicted or experimental nozzle maps, from graphical or tabulated information (Eqs. (12.2.21) and (12.2.22), Figure 10.19), or through modeling (Eqs. (12.2.61)–(12.2.67)).
(10) Thus, two values of $\dot{m}_{c5}$ are found independently. The solution is converged if these two values are within a tolerance. However, if they are not, iteration on N_{c2} is necessary and the preceding process must be repeated. The Regula Falsi method for both loops can be used to facilitate convergence. In all cases, the average total temperature of a component is used to evaluate the ratio of specific heats for that component, as is done in Chapter 3.

By applying the method to different fuel ratios it is possible to predict the operating line of the gas turbine. Thus, both on-design and off-design operating points will be found. Optimally, all components will reach peak efficiency for the same on-design operating point. If they do not, the method will indicate which component should be changed so that a better overall engine efficiency can be obtained. The executable program JetEngineMatch.exe is provided to apply the method.

Power Generation Gas Turbine with Variable Speed without Regeneration

For the case of power generation with a speed-dependent load, three parameters must be specified and have been chosen to be p_a, T_a, and f. The solution for this case again is a two-nested loop iteration. A flow chart outlining the solution is shown in Figure 12.3.

(1) First, an estimate of N_{c2} is needed.
(2) The value of T_{t2} is found from T_a (Eq. (12.2.26)).
(3) Next, an estimate of corrected mass flow for the inlet, $\dot{m}_{c1}$, is used.

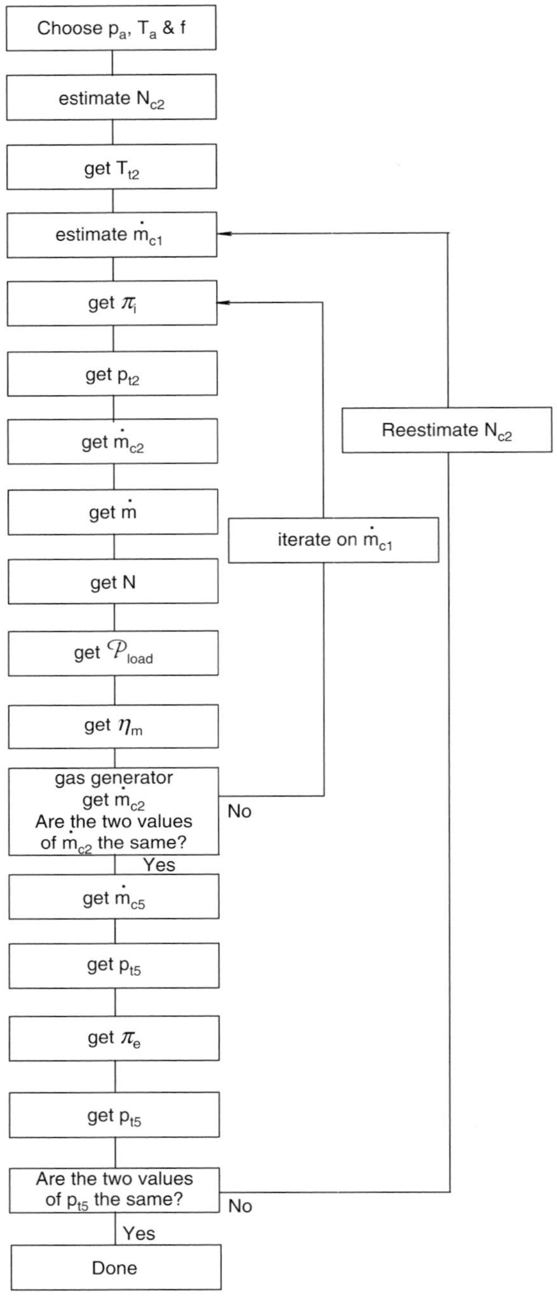

Figure 12.3 Flow chart for solution of power generation gas turbine (speed-dependent load).

(4) From $\dot{m}_{c1}$, the inlet pressure ratio π_i is found through predictions or data (Eq. (12.2.27)), or modeling (Eq. (12.2.68)).

(5) Next, the inlet total pressure to the compressor p_{t2} is found from π_i and p_a and Eq. (12.2.28).

(6) Knowing the corrected mass flow and total pressure ratio for the inlet, one finds the corrected mass flow into the compressor, $\dot{m}_{c2}$, from Eq. (12.2.32).

(7) The true mass flow rate $\dot{m}$ can be found from $\dot{m}_{c2}$, T_{t2}, and p_{t2} by means of Eq. (12.2.3).

(8) One can also find the true shaft speed, N, from N_{c2} and T_{t2} by means of Eq. (12.2.4).

(9) Using N, one can find the external resulting power load $\mathcal{P}_l$ from predictions or data (Eq. (12.2.31)), or modeling (Eq. (12.2.73)).

(10) The graphical or tabular shaft map (Eq. (12.2.16)) or model (Eq. (12.2.59)) can be used with N to determine the mechanical efficiency η_m.

(11) Iteration on the corrected mass flow for the inlet $\dot{m}_{c1}$ and resulting $\dot{m}_{c2}$ for the gas generator (using Eq. (12.2.35) for the shaft power balance) for the given N_{c2} is necessary until the two independently calculated values of $\dot{m}_{c2}$ (namely, gas generator and Eq. (12.2.12)) are the same, indicating this part of the solution has converged. Again, one should remember that this yields parameters such as $\dot{m}_{c3}$, $\dot{m}_{c3}$, π_b, π_c, π_t, etc. Again, this can be a lengthy step.

(12) Next, the corrected mass flow for the exhaust $\dot{m}_{c5}$ is found from Eq. (12.2.34) by knowing $\dot{m}_{c4}$, and the total pressure ratio and temperature ratio for the turbine.

(13) Knowing p_{t2}, π_c, π_b, and π_t one can calculate the total pressure into the exhaust, p_{t5}, from Eq. (12.2.33).

(14) Knowing $\dot{m}_{c5}$, one can find the pressure ratio for the exhaust, π_e, from predictions or data (Eq. (12.2.29)), or modeling (Eq. (12.2.72)).

(15) Then p_{t5} is calculated from p_e and π_e from Eq. (12.2.30).

(16) Thus, two values of p_{t5} are independently calculated. If these two values are within a tolerance, the solution is converged. If they are not, one must iterate on N_{c2} and repeat the preceding process. As is the case for the jet engine, the average total temperature of a component is used to evaluate the ratio of specific heats for that component as is done in Chapter 3.

Power Generation Gas Turbine at Constant Speed without Regeneration

In the case of power generation running at a constant speed (for example, for electric generation), four parameters must be specified and have been chosen to be N, p_a, T_a, and f. The solution for this case is again a double-loop iteration but is somewhat more straightforward than above because the shaft speed is known. A flow chart outlining the solution is shown in Figure 12.4.

(1) First, N is specified.
(2) The value of T_{t2} is found from T_a (Eq. (12.2.26)).
(3) Next, an estimate of corrected mass flow for the inlet, $\dot{m}_{c1}$, is used. Also, an estimate on the load, $\mathcal{P}_l$, is used.
(4) Using $\dot{m}_{c1}$, one can find the inlet pressure ratio π_i from predictions or data (Eq. (12.2.27)), or through modeling (Eq. (12.2.68)).
(5) Next, the inlet total pressure to the compressor, p_{t2}, is found from π_i and p_a and Eq. (12.2.28).
(6) Knowing the corrected mass flow and total pressure ratio for the inlet, one finds the corrected mass flow into the compressor, $\dot{m}_{c2}$, from Eq. (12.2.32).
(7) The true mass flow rate $\dot{m}$ can be found from $\dot{m}_{c2}$, T_{t2}, and p_{t2} with Eq. (12.2.3).
(8) One can also find the corrected shaft speed N_{c2} from N and T_{t2} with Eq. (12.2.4).
(9) Next, by knowing N_{c2} and $\dot{m}_{c2}$, one can use compressor maps to find π_c either from predictions or data using the graphical or tabular form (Eq. (12.2.1), Figures 5.13 or 5.14) or modeling (Eq. (12.2.45)), as well as the efficiency η_c from the graphical or tabular form (Eq. (12.2.2), Figures 5.13 or 5.14) or through modeling (Eq. (12.2.49)).
(10) The temperature ratio τ_c can be found from the compressor pressure ratio, temperature ratio, and efficiency equation (Eq. (12.2.5)) because π_c and η_c are known.
(11) By knowing T_{t2}, one can find T_{t3} from Eq. (12.2.17).
(12) The combustor corrected flow $\dot{m}_{c3}$ can now be found from the compressor corrected mass flow (Eq. (12.2.14)).
(13) One can use $\dot{m}_{c3}$ with f to find η_b and π_b from combustor maps by means of the graphical or tabular forms (Eqs. (12.2.10) and (12.2.11), Figure 9.13) or models (Eqs. (12.2.54) and (12.2.53)).
(14) Next, τ_b can be found from the energy balance in the burner (Eq. (12.2.9)) because η_b and T_{t3} are known.
(15) The turbine corrected flow and speed, $\dot{m}_{c4}$ and N_{c4}, can be related to corresponding quantities for the compressor and found by means of Eqs. (12.2.12) and (12.2.13).
(16) Now, the turbine pressure ratio π_t and efficiency η_t can be found from the turbine maps graphically or tabularly (Eqs. (12.2.6) and (12.2.7), Figure 7.11) or through models (Eqs. (12.2.56) and (12.2.58)) because $\dot{m}_{c4}$ and N_{c4} are known.
(17) One can then calculate τ_t from the turbine pressure ratio, temperature ratio, and efficiency equation by using π_t and η_t (Eq. (12.2.8)).
(18) Next, the graphical or tabular shaft map (Eq. (12.2.16)) or model (Eq. (12.2.59)) can be used with N to determine the mechanical efficiency η_m.
(19) With Eq. (12.2.35), which is the power balance on the shaft, one can use the external power load $\mathcal{P}_l$, f, η_m, τ_b, τ_t, and τ_c to find $\dot{m}$, which can be used to find $\dot{m}_{c2}$. If the two values of $\dot{m}_{c2}$ are the same this section has converged. If not $\dot{m}_{c1}$ is iterated upon until a convergence.
(20) The corrected mass flow for the exhaust, $\dot{m}_{c5}$, is found from Eq. (12.2.34) by knowing $\dot{m}_{c4}$ and the total pressure ratio and temperature ratios for the turbine.
(21) Knowing p_{t2}, π_c, π_b, and π_t one can calculate the total pressure into the exhaust, p_{t5}, from Eq. (12.2.33).

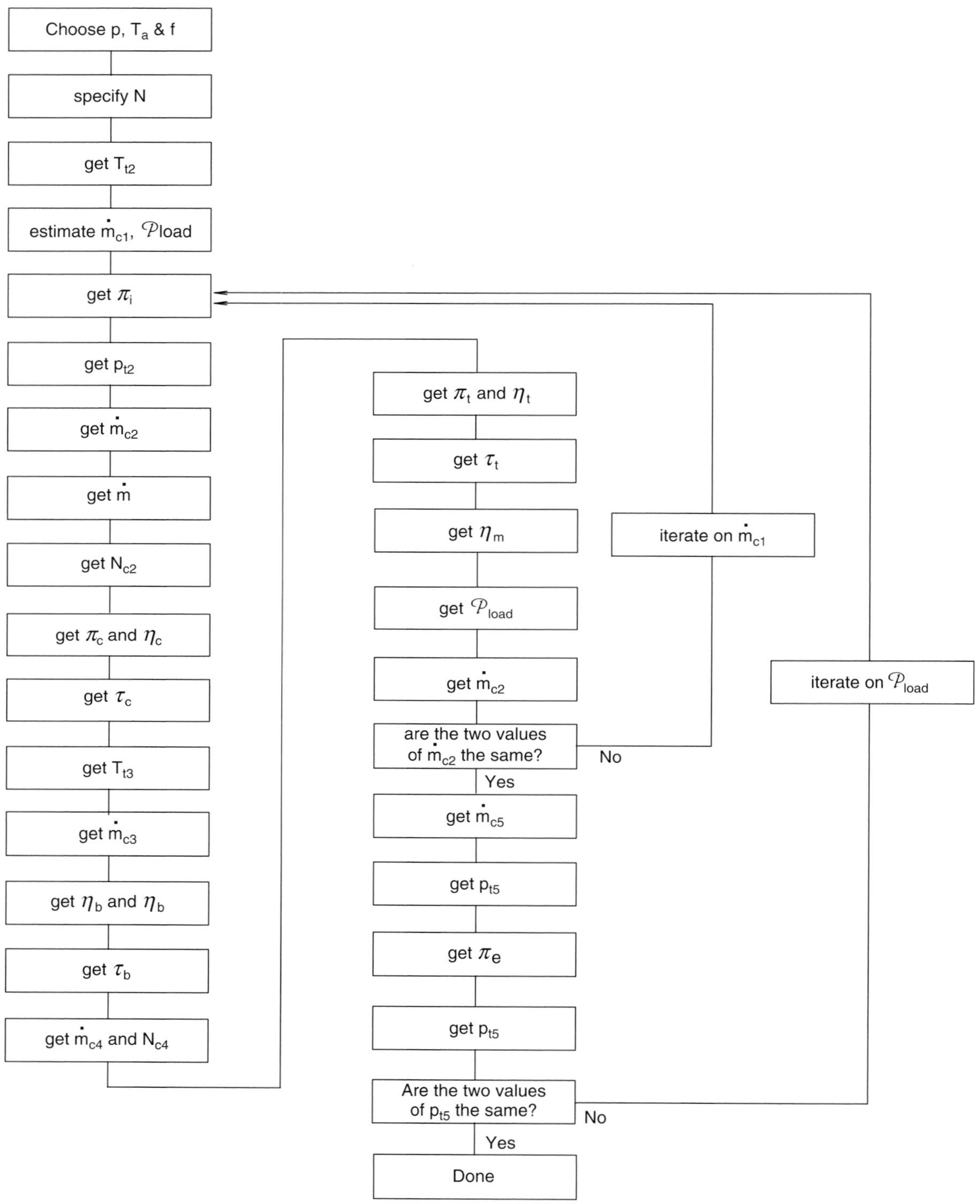

Figure 12.4 Flow chart for solution of power generation gas turbine (constant speed).

(22) Next, knowing $\dot{m}_{c5}$, one can find the pressure ratio for the exhaust, π_e, from predictions or data (Eq. (12.2.29)) or through modeling (Eq. (12.2.72)).
(23) Then p_{t5} is calculated from p_e and π_e by means of Eq. (12.2.30).
(24) Thus, two values of p_{t5} are independently calculated. If these two values are within a tolerance, the solution is converged. If they are not, one must iterate on $\mathcal{P}_l$ and repeat the preceding process. As before, the average total temperature of a component is used to evaluate the ratio of specific heats for that component.

By applying the method to different fuel ratios, one can predict the operating line of a power generation gas turbine. As is the case for a jet engine, both on-design and off-design operating points will be found. Optimally, all components will reach peak efficiency for the same on-design operating point. If they do not, the method will indicate which component should be modified so that a better overall engine efficiency can be obtained.

Power Generation Gas Turbine with Variable Speed with Regeneration

For the case of power generation with a speed-dependent load, three parameters must be specified and have been chosen to be p_a, T_a, and f. The solution for this case again is a two-nested loop iteration plus the iteration within the gas generator. The solution is somewhat more complex due to the fact that the performance (namely the pressure ratios) of the regenerator is not known a priori which affects both the gas generator and overall gas turbine performance. A flow chart outlining the solution is shown in Figure 12.5. The executable program PowerGTMatch.exe is provided to apply the method along with the constant-speed case with regeneration.

12.2.7 Other Applications

Thus far, only single-shaft gas turbines have been discussed, which has necessitated matching six components for a jet engine and seven for a power generation gas turbine. The method can also be applied to more complex engines. For example, a similar analysis can be applied to a twin-spool turbofan or turbojet. For a twin-spool turbofan with an exhausted fan and afterburner, 12 components (and their corresponding maps) will have to be matched (diffuser, fan, low-pressure compressor, high-pressure compressor, primary burner, high-pressure turbine, low-pressure turbine, afterburner, primary nozzle, fan nozzle, and two shafts). As a result, although the same approach can be used, the solution method becomes much more cumbersome and requires several more levels of nested iterations and much longer solution times (Flack, 2004). As a result, performing such matching requires computer methods.

12.2.8 Dynamic or Transient Response

Thus far, the matching analysis has been for strictly steady-state performance. That is, the fuel ratio is specified and the steady-state operating point is found. Nothing in the analysis allows the determination of the time period needed for the gas turbine to reach this point. The largest influence is the rotor polar moment of inertia, I, although the fuel injection and combustion, flow inertia through a gas turbine, and fluid compressibility all induce dynamics as well. For a jet engine, the equation that governs the rotational inertial effect is

$$T_t - T_c - T_m = I\frac{\partial \omega}{\partial t}, \qquad (12.2.74)$$

where T_t, T_c, and T_m are the torques of the turbine, compressor, and mechanical losses, respectively, and ω is the rotational speed. To apply such an analysis, the following procedure can be used.

(1) Consider the gas turbine operating at some known condition.
(2) The fuel ratio can be changed and the instantaneous turbine inlet temperature can be found if it is assumed the conditions into the burner have not yet changed.

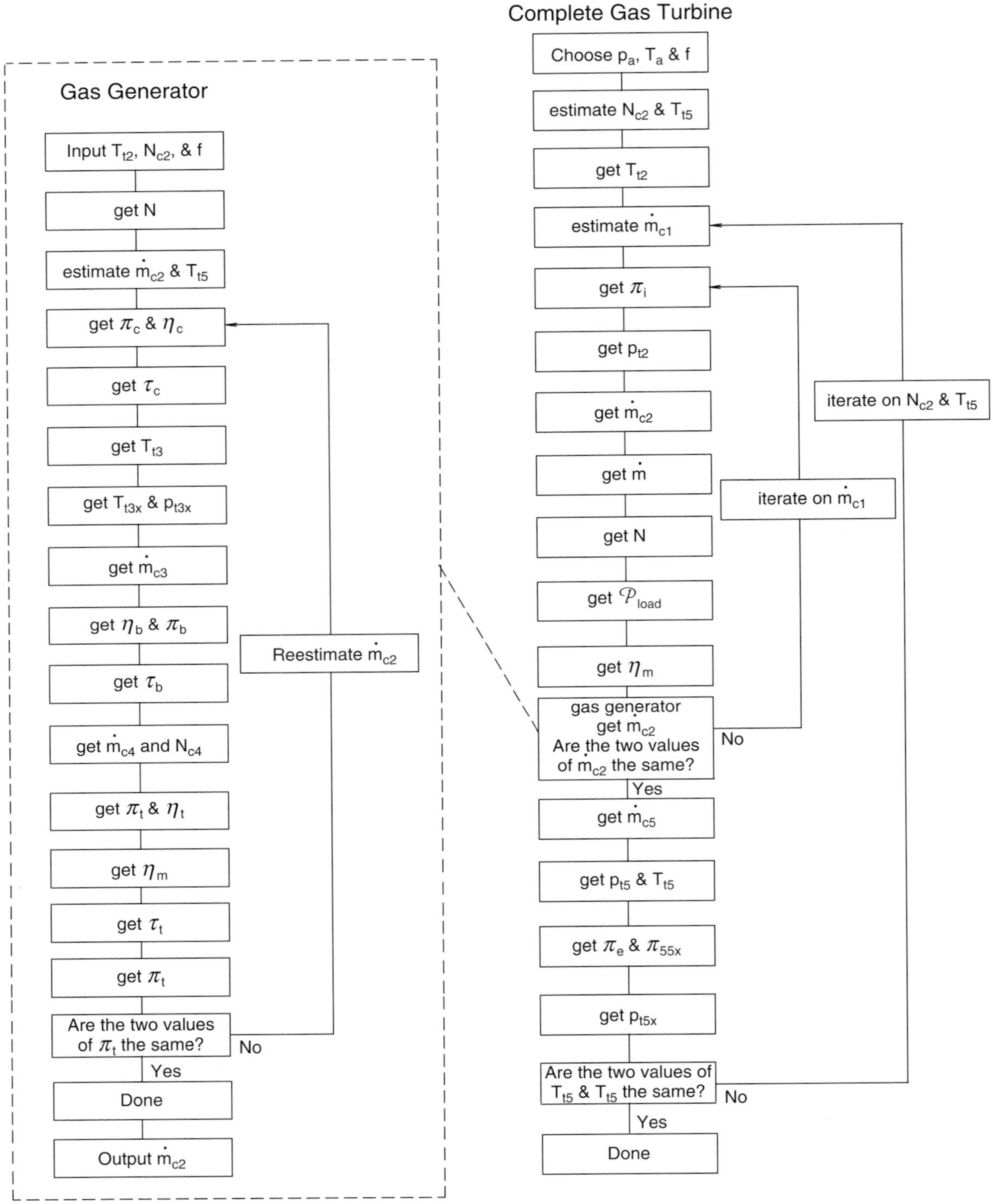

Figure 12.5 Flow chart for solution of power generation gas turbine (speed dependent load) with regeneration.

(3) The turbine torque can be found based on the new instantaneous turbine inlet temperature.
(4) The net torque can be found, and the angular acceleration or rate of change of the shaft rotation can be found.
(5) A discretized form of Eq. (12.2.74) can be used over a selected discrete (short) time interval to find the new rotational speed at the end of this period.
(6) The new speed can be used as in Section 12.2.6 to find parameters such as the mass flow, compressor total pressure ratio, and so on.
(7) The new conditions can be used to find the new burner exit temperature and the preceding process repeated for a new time increment, and continually repeated, until steady-state conditions are attained.

The steady-state solution will be the same as discussed in Section 12.2.6; however, in this analysis, the time required will be predicted as a part of the transient response. Such an analysis has been used by Kurzke (1995) and Reed and Afjeh (2000).

12.2.9 Matching of Jet Engine and Aircraft

Up to this point in the text, the performance of the engine has been treated separately from that of the aircraft. However, just as a turbine must be matched to a compressor, an engine must be matched to an aircraft. For example, if an aircraft has two engines and each engine produces a given value of thrust, at a given altitude the aircraft will cruise at only one Mach number. That is, a pilot governs the cruise speed by adjusting the thrust level.

The previous sections describe the interactions among engine components. The fuel ratio and Mach numbers are treated independently. Thus, for a given engine with known components, an operating characteristic map, as shown in Figure 12.6, can be generated for an engine at various altitudes. This figure indicates how the developed thrust varies with inlet Mach number and fuel ratio for a given engine with given components (and maps).

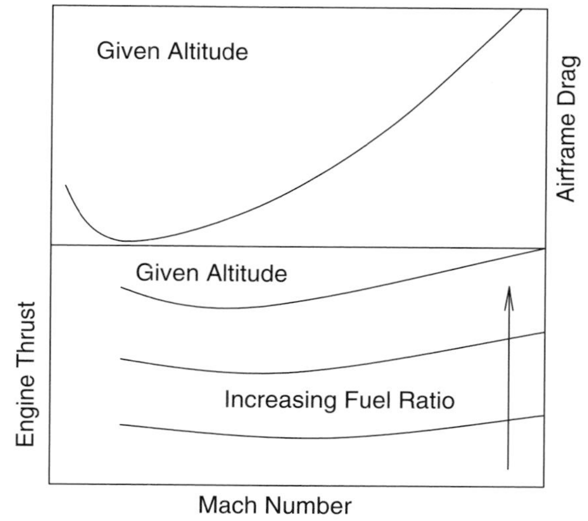

Figure 12.6 General overall operating characteristic map for an engine and airframe drag at a given altitude.

If only the airframe is considered, the thrust required to power an aircraft at a given altitude will become greater as the aircraft speed increases. That is, the total thrust from one or more engines will balance the aircraft drag. The airframe drag characteristics will be determined from, for example, wind tunnel tests. A general variation is also shown in Figure 12.6. In general the airframe drag increases as the Mach number increases at cruise or higher Mach numbers. However, for low Mach numbers in general the airframe drag increases as the Mach number decreases. This is because flaps are used to increase lift at low Mach numbers, but this also results in higher drag. Therefore, by combining the engine and airframe maps, one can determine the required engine conditions for a given Mach number. Mathematically, from these two figures, four variables are obtained (Mach number, fuel ratio, thrust, and altitude). Furthermore, with these two figures there are two general equations. For the engine,

$$F = f(M, f, \text{altitude}). \tag{12.2.75}$$

For the airframe,

$$F = g(M, \text{altitude}). \tag{12.2.76}$$

Therefore, if two variables are specified (namely the flight altitude and Mach number), the other two are determinable (namely, fuel ratio and thrust). As a result of this engine and airframe matching, for a given engine (or set of engines) on a given airframe, once the Mach number and altitude are specified, the required fuel ratio is set.

A model that can often be applied for a given airframe at cruise and that can be used in a computerized matching solution is

$$F_{\text{airframe}} = C_{\text{dm}} M_{\text{a}}^2. \qquad (12.2.77)$$

This cannot be used for the lower Mach numbers. At low Mach numbers (takeoff and landing), flaps increase the drag, as shown on Figure 12.6. Note, however, that C_{dm} is a modeling parameter rather than a conventional nondimensional drag coefficient and has the units of force.

12.2.10 Example Applications

In this section, example analyses are presented. These include matching components of a turbojet, matching a turbojet and aircraft, and matching components of a power generation unit with a speed-dependent load map and also at constant speed.

Turbojet Examples

Example 12.1 Turbojet Engine

A single-spool turbojet engine with given components is to be assembled. Because this case represents an engine, the diffuser, compressor, combustor, turbine, shaft, and nozzle are all included. For this example, the components have been modeled using equations presented in Section 12.2.5. The maps for the compressor, burner, and turbine are shown in Figures 12.7 through 12.9, and the different model parameters, which fit the data, are presented in Table 12.2. The nozzle has a fixed minimum area and a variable exit area, and thus the exit pressure matches the ambient pressure. The engine is to be operated at a Mach number of 0.5 and at $T_a = 289$ K (520 °R) and $p_a = 101.3$ kPa (14.69 psia). Find the operating characteristics of the engine as the fuel ratio is parametrically varied from 0.010 to 0.035 in increments of 0.005.

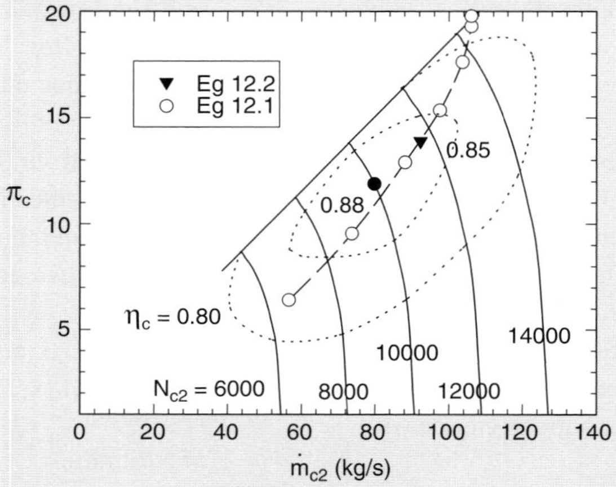

Figure 12.7 Compressor map for Example 12.1 (also Example 12.2).

Table 12.2 **Model parameters for Example 12.1.**

Diffuser	Turbine
$\pi_{d_d} = 1$	$k_1 = 1.0$
$d = 0$	$k_2 = 0.20$
Compressor	$N_{c4_d} = 4000$ rpm
$c_1 = 0.1764$ s/kg (0.08 s/lb$_m$)	$\dot{m}_{c4_c} = 15.87$ kg/s (35 lb$_m$/s)
$c_2 = 0.00907$ kg/s/rpm (0.02 lb$_m$/s/rpm)	$\eta_{t_d} = 0.90$
$c_3 = 0.80$	$\pi_{t_c} = 0.28$
$c_4 = 0.00001$/rpm	**Shaft**
$c_5 = 9.724$ rpm s^2/kg^2 (2.0 rpm s^2/lb$_m{}^2$)	$s_1 = 0$
$N_{c2_d} = 10{,}000$ rpm	$s_2 = 0$
$\eta_{c_d} = 0.88$	**Nozzle Converging–Diverging (CD), Fixed A^***
$\mu = 0.10$	$a_1 = 0$
Combustor	$\dot{m}_n = 88.08$ kg/s (194.2 lb$_m$/s)
$b_1 = 9.068$ s^2/kg^2 (1.865 s^2/lb$_m{}^2$)	$\eta_{n_d} = 0.98$
$b_2 = 0.0$ s^2/kg^2 (0.0 s^2/lb$_m{}^2$)	**Performance**
$\Delta H = 10{,}000$ kcal/kg (18,000 BTU/lb$_m$)	$T_a = 289$ K (520 °R)
$\eta_{b_d} = 0.91$	$p_a = 101.3$ kPa (14.69 psia)
	$M_a = 0.50$

Solution

As a representative case, $f = 0.02$ is chosen to demonstrate the solution. The first estimate on N_{c2} of 10,954 rpm and the first estimate on $\dot{m}_{c2}$ of 88.12 kg/s (194.3 lb$_m$/s) are chosen.

From Eq. (12.2.18) for $T_a = 289$ K (520 °R), for $M_a = 0.5$, and for $\gamma = 1.401$, one can find the total temperature exiting the diffuser from

$$T_{t2} = T_a\left[1 + \frac{\gamma - 1}{2}M_a^2\right] = 303\text{K (546 °R)}.$$

Since the Mach number and the ambient temperature are known, one can find the airspeed as follows:

$$u_a = M_a\sqrt{\gamma \mathcal{R} T_a} = 170.4\,\text{m/s}\,(559.1\,\text{ft/s}).$$

For the diffuser, because the freestream Mach number is subsonic, one can find the pressure recovery factor from the modeling equation (Eq. (12.2.60)) as follows:

$$\pi_d = \pi_{d_d}\left[1 - d(M_a - 1)^{1.35}\right]$$
$$= 1(1 - 0.0) = 1.00.$$

Thus, one can find the total pressure exiting the diffuser from Eq. (12.2.20) for $\gamma = 1.401$:

$$p_{t2} = p_a\pi_d\left[1 + \frac{\gamma - 1}{2}M_a^2\right]^{\frac{\gamma}{\gamma-1}}$$
$$= 120.2\,\text{kPa}\ (17.43\,\text{psia}).$$

Now, one can solve the gas generator portion of the problem can be solved. Using Eqs. (12.2.3) and (12.2.4),

$$\dot{m}_{c2} \equiv \dot{m}\frac{\sqrt{T_t/T_{stp}}}{p_t/p_{stp}}$$

$$N_{c2} \equiv \frac{N}{\sqrt{T_t/T_{stp}}},$$

for $\dot{m}_{c2} = 88.12$ kg/s (194.3 lb_m/s) and $N_{c2} = 10{,}954$ rpm, one can find the true mass flow rate and rotational speed:

$\dot{m} = 102.0$ kg/s (224.9 lb_m/s) and $N = 11{,}225$ rpm.

One can also find the compressor total pressure ratio by using $\dot{m}_{c2} = 88.12$ kg/s and $N_{c2} = 10{,}954$ rpm with Figure 12.7 or the modeling equation (Eq. (12.2.45)):

$\pi_c = 12.69.$

From Figure 12.7 or the modeling equation (Eq. (12.2.49)), again using $\dot{m}_{c2}$ and N_{c2}, one can find the compressor efficiency:

$\eta_c = 0.870.$

Because π_c is known, Eq. (12.2.5) can be used to find the total temperature ratio for the compressor as follows:

$$\pi_c = [1 + \eta_c(\tau_c - 1)]^{\frac{\gamma}{\gamma-1}}.$$

One can solve for the total temperature ratio of the compressor as $\tau_c = 2.177$ for $\gamma = 1.384$, and from Eq. (12.2.17) the total temperature exiting the compressor can be found as follows:

$$T_{t3} = \tau_c T_{t2}$$
$$= 661\,\text{K}\ (1189\,°\text{R}).$$

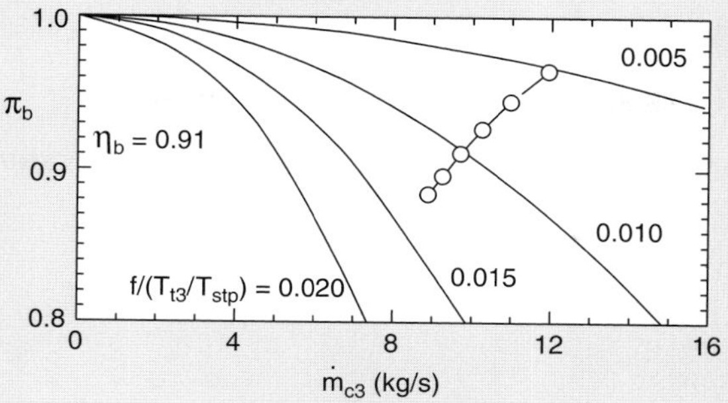

Figure 12.8 Burner map with four values of $f/(T_{t3}/T_{stp})$ for Example 12.1.

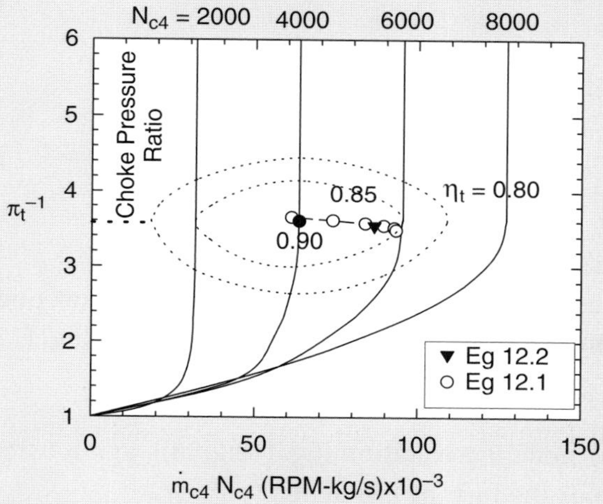

Figure 12.9 Turbine map for Example 12.1 (also Example 12.2).

Using Eq. (12.2.14),

$$\dot{m}_{c2} = \frac{\dot{m}_{c3}\pi_c}{\sqrt{\tau_c}},$$

one can solve for the corrected mass flow for the burner because $\dot{m}_{c2}$ and τ_c are known:

$\dot{m}_{c3} = 10.25$ kg/s (22.60 lb_m/s).

Next, since f and T_{t3} are known, one can find the quantity $f/(T_{t3}/T_{stp}) = 0.00872$. For the burner, the efficiency is given as $\eta_b = 0.91$, and using $\dot{m}_{c3} = 10.25$ kg/s and $f/(T_{t3}/T_{stp}) = 0.00872$, one can find the total pressure ratio for the burner from Figure 12.8 or the modeling equation (Eq. (12.2.53)):

$\pi_b = 0.927$.

Next, using the energy equation for the burner, Eq. (12.2.9),

$$f(\eta_b \Delta H - c_p T_{t3}\, \tau_b) = c_p T_{t3}\, (\tau_b - 1),$$

since f, η_b, ΔH, and T_{t3} are known, one can solve for the total temperature ratio for the burner as

$$\tau_b = 1.982 \text{ for } \gamma = 1.341;$$

therefore, the exit total temperature for the combustor is $T_{t4} = \tau_b T_{t3} = 1309$ K (2357 °R).

Thus, using Eq. (12.2.12),

$$\dot{m}_{c2} = \frac{\dot{m}_{c4}\pi_b \pi_c}{(f+1)\sqrt{\tau_b \tau_c}},$$

and because $\dot{m}_{c2}$, π_c, τ_c, π_b, and τ_b are known, one can find the corrected mass flow for the turbine:

$$\dot{m}_{c4} = 15.87 \text{ kg/s } (35.00 \text{ lb}_m\text{/s}).$$

Using Eq. (12.2.13),

$$N_{c2} = N_{c4}\sqrt{\tau_b \tau_c},$$

one finds the corrected speed for the turbine $N_{c4} = 5273$ rpm.

Next, for the turbine, using $\dot{m}_{c4} = 15.87$ kg/s and $N_{c4} = 5273$ rpm, one finds the total pressure ratio for the turbine from Figure 12.9 or the modeling equation (Eq. (12.2.56)):

$$\pi_t = 0.280.$$

From Figure 12.9 or the modeling equation (Eq. (12.2.58)), again using $\dot{m}_{c4}$=15.87 kg/s and $N_{c4} = 5273$ rpm, one finds the turbine efficiency:

$$\eta_t = 0.882.$$

For the shaft, from the modeling equation (Eq. (12.2.59)), the mechanical efficiency is given by

$$\eta_m = 1 - s_1 N^{s_2},$$

and so $\eta_m = 1.00$.

Thus, using the power balance on the shaft, Eq. (12.2.15),

$$c_{pc}\,(\tau_c - 1) = \eta_m\,(1 + f)c_{pt}\,\tau_b\,\tau_c\,(1 - \tau_t),$$

and since f, τ_c, and τ_b are known, one can solve of the total temperature ratio of the turbine for specific heat ratios of 1.384 and 1.328 for the compressor and turbine, respectively:

$$\tau_t = 0.762.$$

Thus, the total exit temperature of the turbine is

$$T_{t5} = \tau_t\, T_{t4} = 998 \text{ K } (1797\ °\text{R}).$$

Hence, from Eq. (12.2.8) for $\gamma = 1.328$, the second (and independent) calculation for the total pressure ratio of the turbine yields (τ_t and η_t are known)

$$\pi_t = \left[1 + \frac{(\tau_t - 1)}{\eta_t}\right]^{\frac{\gamma}{\gamma-1}}$$

$$= 0.280.$$

This value of π_t agrees with the previously calculated value of π_t. Thus, the initial guess on $\dot{m}_{c2}$ was good. If the two values of π_t had not been in agreement, a new value of $\dot{m}_{c2}$ would have been tried and the preceding process repeated. This ends the solution for the gas generator. Now one must determine if the initial guess on N_{c2} was valid. Since the corrected mass flow for the turbine, total pressure ratio, and total temperature ratio for the turbine are known, using Eq. (12.2.24),

$$\dot{m}_{c4} = \frac{\dot{m}_{c5}\pi_t}{\sqrt{\tau_t}},$$

one finds the corrected mass flow for the nozzle:

$$\dot{m}_{c5} = 49.51 \text{ kg/s } (109.2 \text{ lb}_m\text{/s}).$$

This will be checked later. The total pressure exiting the turbine from Eq. (12.2.23) can be found because p_{t2}, π_c, π_b, and π_t are known:

$$p_{t5} = p_{t2}\pi_c\pi_b\pi_t$$

$$= 395.9 \text{ kPa } (57.41 \text{ psia}).$$

Thus, one can find the nozzle parameter, $p_{t5}/p_a = 3.906$.

For the nozzle the efficiency is given as $\eta_n = 0.98$.

Next, using Eq. (12.2.63),

$$M_8 = \sqrt{\frac{2\left[\dfrac{1}{\eta_n\left[\dfrac{p_8}{p_{t5}}\right]^{\frac{\gamma-1}{\gamma}} + 1 - \eta_n} - 1\right]}{\gamma - 1}},$$

for $\gamma = 1.340$, $p_{t5}/p_a = 3.906$, and $p_8 = p_a$, one finds the nozzle exit Mach number $M_8 = 1.537$.

For adiabatic flow $T_{t5} = T_8\left[1 + \frac{\gamma - 1}{2}M_8^2\right]$, and since $T_{t5} = 998$ K, one can solve for the nozzle exit static temperature:

$$T_8 = 712 \text{ K } (1282\ °\text{R}).$$

The speed of sound at the exit is $a_8 = \sqrt{\gamma \mathcal{R} T_8} = 523.3$ m/s (1717 ft/s).

Hence, the gas velocity at the nozzle exit is

$$u_8 = M_8\, a_8 = 804.2 \text{ m/s } (2639 \text{ ft/s}).$$

From modeling equations (Eqs. (12.2.67) and (12.2.61)),

$$a_2 = \frac{\sqrt{\frac{\gamma+1}{2+(\gamma-1)M_8^2}}\sqrt{\frac{\gamma}{\gamma_{stp}}\frac{\mathcal{R}_{stp}}{\mathcal{R}}}}{M_8\left\{\frac{1}{\eta_n}\left[\frac{\eta_n}{\frac{2}{\gamma+1}-1+\eta_n}\right]\left[\frac{1}{1+\frac{\gamma-1}{2}M_8^2}-1+\eta_n\right]\right\}^{\frac{\gamma}{\gamma-1}}}$$

$$\dot{m}_{c5} = \dot{m}_n a_2 M_8 \frac{p_8}{p_{t5}}\sqrt{1+\frac{\gamma-1}{2}M_8^2},$$

and using the preceding values of p_{t5}, p_8, M_8, and η_n, one finds from a second (and independent calculation method) the corrected mass flow for the nozzle to be

$\dot{m}_{c5} = 49.51$ kg/s (109.2 lb_m/s).

This value of $\dot{m}_{c5}$ agrees with the earlier calculated value of $\dot{m}_{c5}$. Thus, the initial guess on N_{c2} was also good. It must have been the author's lucky day. If the two values of $\dot{m}_{c5}$ had not been in agreement, a new value of N_{c2} would have been tried and the preceding process repeated (including the nested iteration on $\dot{m}_{c2}$). Thus, the solution for $f = 0.02$ has been found. Now that the component operating points have been found, the thrust can be found from Eq. (12.2.25):

$$F = \dot{m}[(1+f)u_8 - u_a] + A_8(p_8 - p_a).$$

The thrust is 66,310 N (14,907 lb_f), which results in a *TSFC* of 0.1108 kg/(h-N) (1.086 lb_m/(h-lb_f)).

Other values of f (ranging from 0.010 to 0.035) were selected, and a series of calculations were completed using the procedure above. Net results for these other values of f are presented as operating lines (dashed lines) as f varies on the maps in Figures 12.7 through 12.9. As can be seen on Figure 12.7, the case for $f = 0.02$ results in an engine operating condition near to, but at slightly higher pressure than, the compressor maximum efficiency point. Also, as can be seen in Figure 12.7, over the given range of fuel ratios, the compressor efficiency ranged from 82.7 to 87 to 77.7 percent and the compressor total pressure ratio monotonically increased. The operating line passed close but slightly lower than the maximum efficiency point. At a fuel ratio of 0.037, the compressor surged, which, as discussed in Chapter 5, would be accompanied by violent vibrations and possibly a combustor flameout, resulting in total loss of thrust.

In Figure 12.8, the burner pressure ratio decreases from 0.965 to 0.884 as the corrected mass flow decreases from 11.92 to 8.89 kg/s. The turbine (Figure 12.9) changes characteristics minimally as the efficiency ranges from 0.899 to 0.850 and the pressure ratio varies from 0.274 to 0.287. Note that the operating line passes through the maximum efficiency point for the turbine, indicating that the turbine and compressor are fairly well matched but not exactly at the same fuel ratios.

In Figure 12.10, other resulting parameters are presented for this fuel ratio range. One can easily see that the turbine inlet temperature, exit Mach number, and compressor corrected speed all increase with rising fuel ratio. For high fuel ratios, the turbine inlet temperature is excessive. The compressor corrected flow rate generally increases but tends toward a constant as the compressor approaches surge.

Finally, in Figure 12.10 the thrust and *TSFC* are shown as functions of the fuel ratio. As can be seen, the thrust always increases with *f*. However, the *TSFC* reaches a minimum at a fuel ratio of 0.014. Optimally, this combination of components should be run at a fuel ratio of 0.014 for the best fuel economy. Other results also influence the chosen operating point, however.

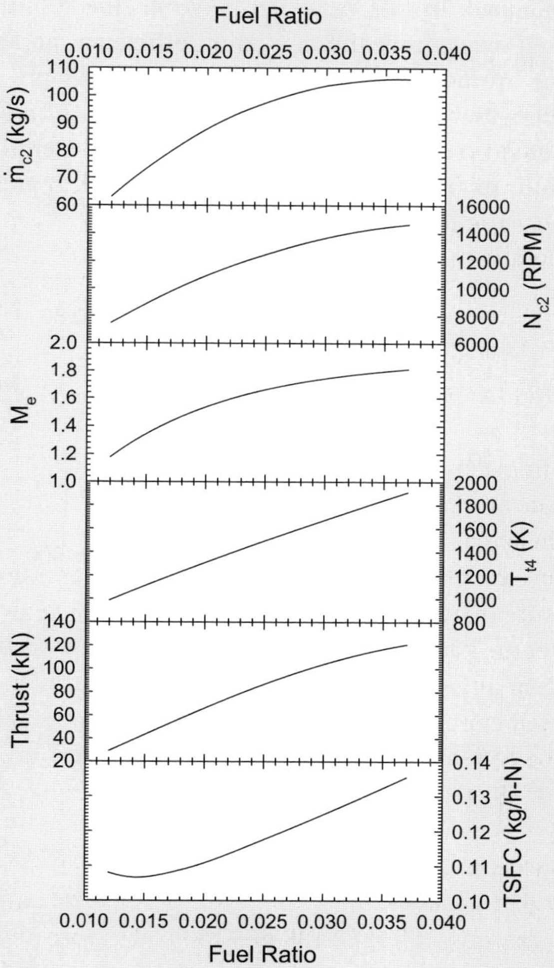

Figure 12.10 Overall operating conditions vs. fuel ratio for Example 12.1.

Extended Example 12.1.1 | Modified Turbine Map

As an extension of Example 12.1, an analysis on a second single-spool turbojet engine with given components was performed. Again, maps for the diffuser, compressor, combustor, turbine, shaft, and nozzle are all included. All of the maps are the same as before except for the turbine map. For this case the turbine has been designed to operate at a larger pressure drop (π_t at the choking condition is smaller) and to run at a higher rotational speed. The maps for the compressor and turbine are shown in Figures 12.11 and 12.12. The operating characteristics of the engine were again found, and the operating lines are shown on the maps. In Example 12.1, the turbine and compressor are noted as being well matched. In this example, however, the resulting operating line for both components is well away from the maximum efficiency points. Clearly, the compressor and turbine are not well matched. Thus, this example is presented to show one possible scenario in which an inappropriate combination of a compressor and turbine are "assembled." This could represent, for example, an initial iteration in a design process for a new engine design. Example 12.1 would, for that case, represent the last iteration. More on the design phase is discussed in Section 12.3.

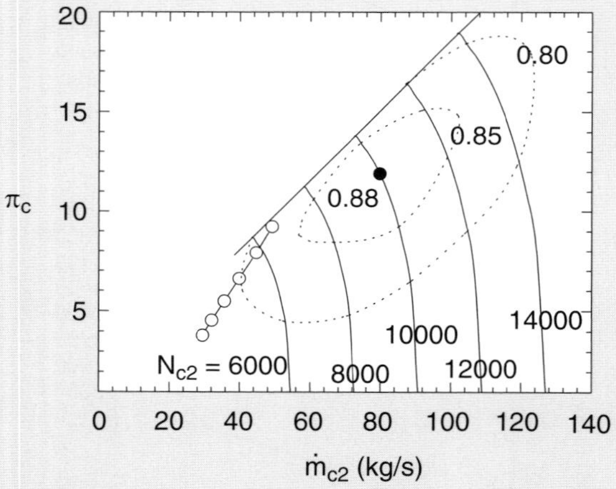

Figure 12.11 Compressor map with operating line for Extended Example 12.1.1.

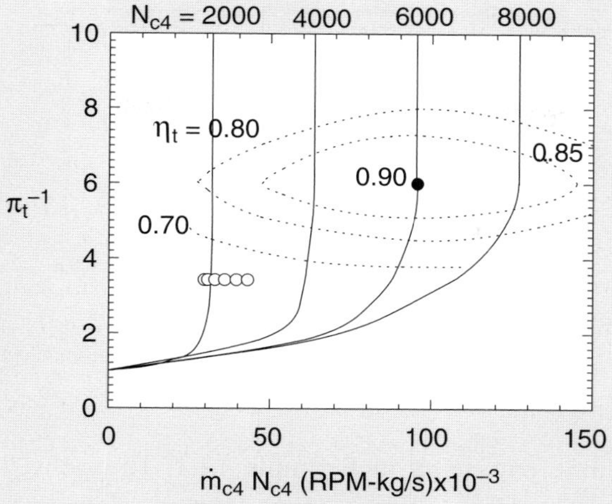

Figure 12.12 Turbine map with operating line for Extended Example 12.1.1.

Example 12.2 | Turbojet Engine Matched to Airframe

A single-spool turbojet engine with given components is to be assembled for a given airframe. Since this case represents an engine, the diffuser, compressor, combustor, turbine, shaft, and nozzle are all modeled. The maps for the compressor, burner, and turbine are shown in Figures 12.7 through 12.9, and the different model parameters, which fit the data, are presented in Table 12.2. The nozzle has a fixed minimum area and a variable exit area. The aircraft has drag characteristics shown in Figure 12.13. The aircraft with two engines is to be operated near standard temperature and pressure at a Mach number of 0.90. Find the required fuel ratio and resulting *TSFC* at this condition. Also determine how the fuel ratio and *TSFC* vary as the flight Mach number changes from 0.4 to 1.2 at $T_a = 289$ K (520 °R) and $p_a = 101.3$ kPa (14.69 psia).

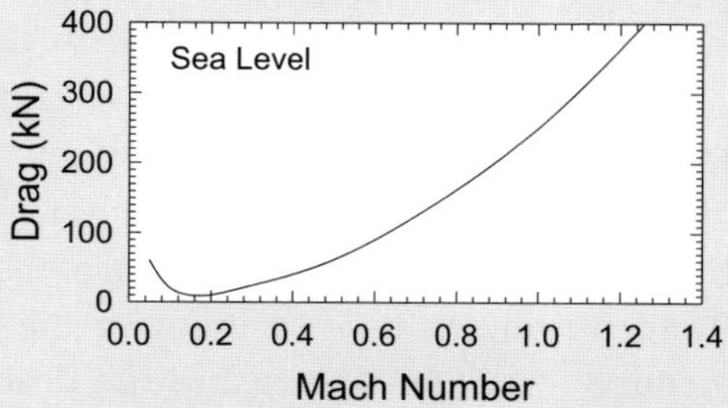

Figure 12.13 Drag characteristics for the airframe in Example 12.2.

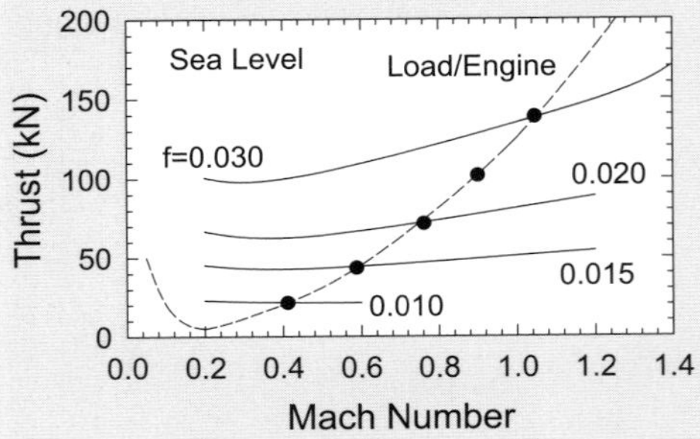

Figure 12.14 Overall operating characteristic map for the engine in Example 12.2 at sea level.

Solution

Note that this is the same engine as in Example 12.1. Thus, the detailed calculations of the engine component matching will not be performed. However, Figure 12.14 was generated for this engine (solid lines) by following the same method as in Example 12.1 and by parametrically varying the Mach number and fuel ratio and solving the engine matching problem many times. Also, the required thrust or load per engine (total drag/two engines) is superimposed on this figure (dashed line). Where the two curves intersect determines the operating conditions of the engines. For example, for Mach numbers of 0.411, 0.588, 0.762, and 0.900, the required fuel ratios are 0.01, 0.015, 0.02, and 0.025, respectively.

In particular, at a Mach number of 0.90, the required thrust per engine is 101.8 kN (22,890 lb_f). By iterating on the Mach number and repeating the methodology from Example 12.1, the required fuel ratio for these conditions was found to be 0.025. Other conditions at a Mach number of 0.90 are presented in Table 12.3. The operating point for $M = 0.90$ and $f = 0.025$ is also shown on Figures 12.7 and 12.9.

Furthermore, the resulting values of *TSFC*, turbine inlet temperature, aircraft Mach number, and rotational speed are plotted vs. the fuel ratio number in Figure 12.15 to determine the overall performance for different fuel ratios. As can be seen, the Mach number varies almost linearly with the required fuel ratio. Similarly, the turbine inlet temperature varies almost linearly with the required fuel ratio and is approaching a limit for the higher fuel ratios. The value of *TSFC* increases with the fuel ratio, as well as turbine inlet temperature. This is consistent with what was found in Chapter 3 (Figure 3.19). The *TSFC* appears to be approaching a minimum well below the design condition, which is not desirable and indicates that a better overall design is possibly attainable. If one would like to improve the design, new maps could be modified – especially the compressor and turbine maps – to increase the efficiencies at the desired operating point and thus reduce the *TSFC*. Increasing the design compressor pressure ratio would also improve the *TSFC* as was found in Chapter 3 (Figure 3.17). However, this would require both new compressor and turbine maps to insure good matching.

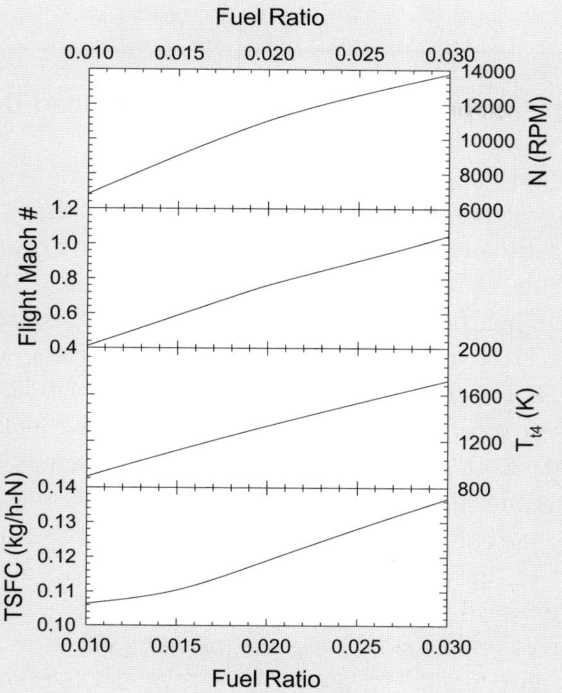

Figure 12.15 Overall operating conditions vs. fuel ratio for Example 12.2.

Table 12.3 **Detailed results for Example 12.2 for $M_a = 0.90$.**

Diffuser	**Turbine**
$\pi_d = 1.0$	$\dot{m}_{c4} = 15.87$ kg/s (35.00 lb_m/s)
$\gamma_d = 1.398$	$\gamma_t = 1.313$
Compressor	$N_{c4} = 5448$ rpm
$\dot{m}_{c2} = 92{,}32$ kg/s (203.6 lb_m/s)	$T_{t4} = 1530$ K (2755 °R)
$\eta_c = 0.863$	$\pi_t = 0.283$
$N_{c2} = 11{,}640$ rpm	$\eta_t = 0.876$
$\gamma_c = 1.379$	**Nozzle**
$\pi_c = 11.39$	$\dot{m}_{c5} = 49.34$ kg/s (108.8 lb_m/s)
Combustor	$M_8 = 1.823$
$\dot{m}_{c3} = 9.918$ kg/s (21.87 lb_m/s)	$\eta_n = 0.98$
$\gamma_b = 1.329$	$\gamma_n = 1.326$
$\pi_b = 0.933$	**Overall**
$\eta_b = 0.91$	$F = 101.8$ kN (22,890 lb_f)
$f = 0.025$	$N = 12540$ rpm
Shaft	$TSFC = 0.1281$ kg/h-N (1.256 lb_m/h-lb_f)
$\eta_m = 1.00$	$\dot{m} = 144.8$ kg/s (319.3 lb_m/s)

Power Generation Examples

Example 12.3 Power Generation Gas Turbine with a Speed-Dependent Load Map

A single-spool power gas turbine with given components is to be assembled. Since this case represents a power generation unit, the bell inlet, compressor, combustor, turbine, power load, shaft, and exhaust/stack are all included. For this example, the components have been modeled using equations presented in Section 12.2.5. The maps for the compressor, burner, turbine, power load, and other components are also shown in Figures 12.16 through 12.18 and the different model parameters, which fit the data, are presented in Table 12.4. The unit is to be operated at $T_a = 520\,°R$ and $p_a = 14.69$ psia. Find the operating characteristics and output of the unit as the fuel ratio is parametrically varied from 0.014 to 0.030 in increments of 0.02. It should be noted that in the following example (Example 12.4) the rotational speed is held constant at 9000 rpm and the load map is not included.

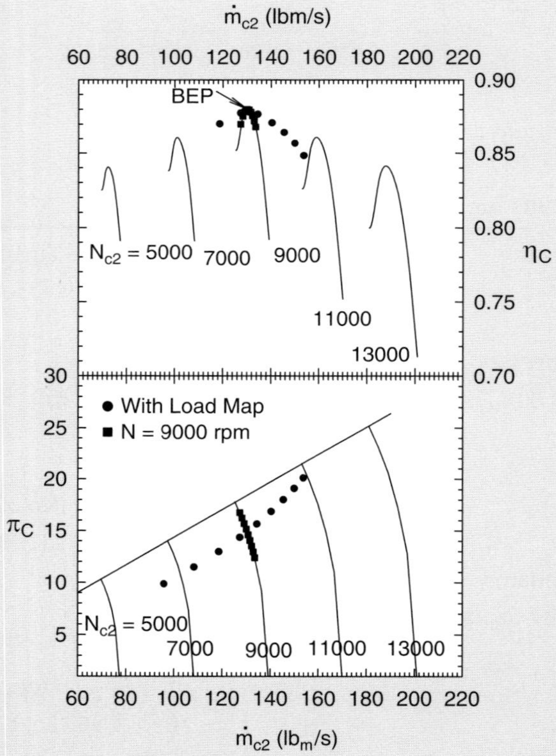

Figure 12.16 Compressor map for Examples 12.3 and 12.4.

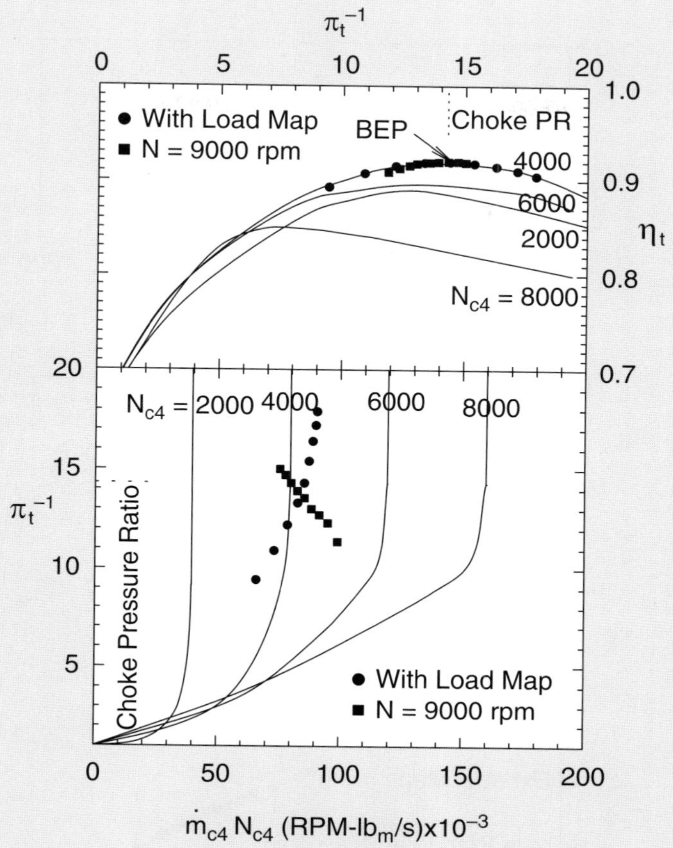

Figure 12.17 Turbine map for Examples 12.3 and 12.4.

Solution

To demonstrate the solution, $f = 0.02$ is picked as a representative case. The first estimate for $N_{c2} =$ 8774 rpm and the first estimate for $\dot{m}_{c1} = 125.3$ lb$_m$/s are chosen. These will be checked later and iterated on if necessary. Now, from Eq. (12.2.18) for $T_a = 520$ °R, since $M_a = 0.0$, and for $\gamma = 1.403$ one can find the total temperature exiting the inlet:Thus, one can find the total pressure exiting the diffuser from Eq. (12.2.20) for γ=1.403:

$$T_{t2} = T_a\left[1 + \frac{\gamma - 1}{2} M_a^2\right] = 520\,°\text{R}.$$

For the bell inlet, one can find the pressure recovery factor from the modeling equation (Eq. (12.2.68)):

$$\pi_i = 1 - z_1 \dot{m}_{c1}^2$$

$$= 0.984.$$

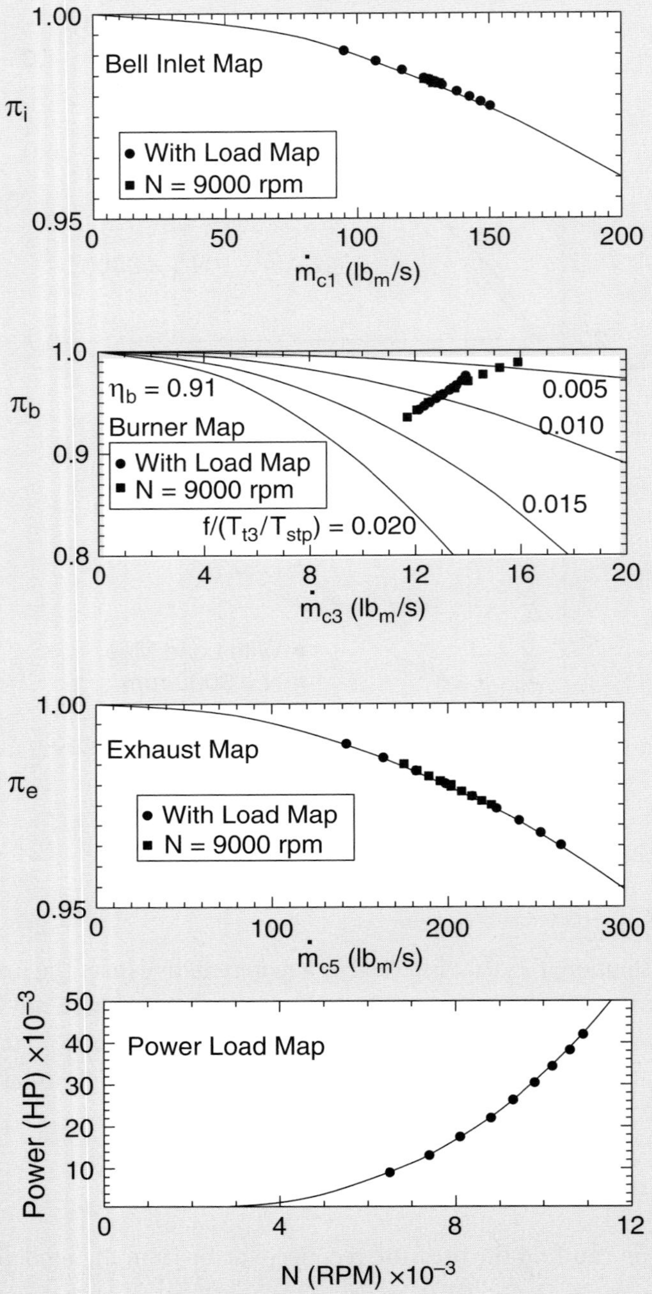

Figure 12.18 Inlet, burner, exhaust, and load maps for Examples 12.3 and 12.4.

Table 12.4 **Model parameters for Example 12.3.**

Bell inlet	**Turbine**
$z_1 = 0.000001\ (\mathrm{lb_m/s})^{-2}$	$k_1 = 0.20$
Compressor	$k_2 = 0.10$
$c_1 = 0.1339\ \mathrm{s/lb_m}$	$N_{c4d} = 4000$ rpm
$c_2 = 0.01546\ \mathrm{lb_m/s/rpm}$	$\dot{m}_{c4_c} = 20\ \mathrm{lb_m/s}$
$c_3 = 0.90$	$\eta_{t_d} = 0.92$
$c_4 = 0.00001/\mathrm{rpm}$	$\pi_{t_c} = 0.07$
$c_5 = 10.0\ \mathrm{rpm\ s^2/lb_m}^2$	**Shaft**
$N_{c2_d} = 9000$ rpm	$s_1 = 0.005$
$\eta_{c_d} = 0.88$	$s_2 = 0.00$
$\mu = 0.04$	**Exhaust/stack**
Combustor	$z_2 = 0.0000005\ (\mathrm{lb_m/s})^{-2}$
$b_1 = 2.77\ \mathrm{s^2/lb_m}^2$	**Performance**
$b_2 = 0.0\ \mathrm{s^2/kg^2}\ (0.0\ \mathrm{s^2/lb_m{}^2})$	$T_a = 520\ ^\circ\mathrm{R}$
$\Delta H = 18000\ \mathrm{B/lb_m}$	$p_a = 14.69$ psia
$\eta_{b_d} = 0.91$	
Power load	
$z_3 = 0.00000003253\ \mathrm{hp/rpm}^{-3}$	
$z_4 = 3.0$	

Thus, one can find the total pressure exiting the diffuser from Eq. (12.2.20) for $\gamma = 1.403$:

$$p_{t2} = p_a \pi_d \left[1 + \frac{\gamma - 1}{2} M_a^2\right]^{\frac{\gamma}{\gamma-1}}$$

$$= 14.92 \text{ psia.}$$

But for similitude at the inlet

$$\dot{m}_{c1} \equiv \dot{m} \frac{\sqrt{T_t/T_{stp}}}{p_t/p_{stp}},$$

so the actual flow rate is $\dot{m} = 125.3\ \mathrm{lb_m/s}$.

Now, one can solve the compressor/burner/turbine/power portion of the problem. So, using Eqs. (12.2.3) and (12.2.4),

$$\dot{m}_{c2} \equiv \dot{m}\frac{\sqrt{T_t/T_{stp}}}{p_t/p_{stp}}$$

$$N_{c2} \equiv \frac{N}{\sqrt{T_t/T_{stp}}}.$$

Therefore for $\dot{m} = 125.3$ lb$_m$/s, $\dot{m}_{c2} = 127.3$ lb$_m$/s, and for $N_{c2} = 8774$ rpm one can find the true rotational speed:

$N = 8774$ rpm.

One also can find the compressor total pressure ratio by using $\dot{m}_{c2} = 127.3$ lb$_m$/s and $N_{c2} =$ 8774 rpm with Figure 12.16 or by modeling Eq. (12.2.45):

$\pi_c = 14.36$,

and from Figure 12.16 or by modeling Eq. (12.2.49), again using $\dot{m}_{c2}$ and N_{c2}, one can find the compressor efficiency, $\eta_c = 0.878$.

Now since π_c is known, one can use Eq. (12.2.5) to find the total temperature ratio for the compressor using

$\pi_c = [1 + \eta_c(\tau_c - 1)]^{\frac{\gamma}{\gamma-1}}$;

then one can solve for the total temperature ratio of the compressor as $\tau_c = 2.252$ for $\gamma = 1.385$ and from Eq. (12.2.17) one can find the total temperature exiting the compressor:

$$T_{t3} = \tau_c T_{t2}$$
$$= 1171\ °\text{R}.$$

Next, using Eq. (12.2.14),

$$\dot{m}_{c2} = \frac{\dot{m}_{c3}\pi_c}{\sqrt{\tau_c}}.$$

Now one can solve for the corrected mass flow for the burner since $\dot{m}_{c2}$ and τ_c are known:

$\dot{m}_{c3} = 13.31$ lb$_m$/s.

Next, since f and T_{t3} are known, one can find the quantity $f/(T_{t3}/T_{stp}) = 0.00888$.

For the burner the efficiency is given as $\eta_b = 0.91$, and using $\dot{m}_{c3} = 13.31$ lb$_m$/s and $f/(T_{t3}/T_{stp}) = 0.00888$, one can find the total pressure ratio for the burner from Figure 12.18 or by modeling Eq. (12.2.53):

$\pi_b = 0.961$.

Next, using the energy equation for the burner (Eq. (12.2.9)),

$f(\eta_b \Delta H - c_p T_{t3} T_{t3}\, \tau_b) = c_p T_{t3}\, (\tau_b - 1)$;

since f, η_b, ΔH, and T_{t3} are known, one can solve for the total temperature ratio for the burner as

$\tau_b = 1.999$ for $\gamma = 1.342$.

Therefore, the exit total temperature for the combustor is $T_{t4} = \tau_b T_{t3} = 2341\ °R$, and, thus, using Eq. (12.2.12),

$$\dot{m}_{c2} = \frac{\dot{m}_{c4}\pi_b\pi_c}{(f+1)\sqrt{\tau_b\tau_c}}.$$

Since $\dot{m}_{c2}$, π_c, τ_c, π_b, and τ_b are known, one can find the corrected mass flow for the turbine:

$$\dot{m}_{c4} = 19.97\ \mathrm{lb_m/s},$$

and using Eq. (12.2.13)

$$N_{c2} = N_{c4}\sqrt{\tau_b\tau_c}$$

one finds the corrected speed for the turbine $N_{c4} = 4135$ rpm.

Next, for the turbine using $\dot{m}_{c4} = 19.97\ \mathrm{lb_m/s}$ and $N_{c4} = 4135$ rpm, one finds the total pressure ratio for the turbine from Figure 12.17 or by modeling Eq. (12.2.56):

$$\pi_t = 0.0751,$$

and from Figure 12.17 or by modeling Eq. (12.2.58), again using $\dot{m}_{c4} = 19.97\ ;\mathrm{lb_m/s}$ and $N_{c4} =$ 4135 rpm, one finds the turbine efficiency $\eta_t = 0.919$.

For the shaft, from modeling Eq. (12.2.59), the mechanical efficiency is given by

$$\eta_m = 1 - s_1 N^{s_2}$$

so $\eta_m = 0.995$.

So now the power load can be found from the load map (Figure 12.18) or Eq. (12.2.73) for $N =$ 8774 rpm:

$$\begin{aligned}\mathcal{P}_l &= z_3 N^{z_4}\\ &= 21{,}972\ \mathrm{hp}.\end{aligned}$$

So using the power balance on the shaft, Eq. (12.2.35):

$$c_{p_c}(\tau_c - 1) + \frac{\mathcal{P}_l}{\dot{m}T_{t2}} = \eta_m(1+f)c_{p_t}\tau_b\tau_c(1-\tau_t),$$

since f, τ_c, and τ_b are known, one can solve for the total temperature ratio for the turbine for specific heat ratios of 1.385 and 1.339 for the compressor and turbine, respectively:

$$\tau_t = 0.558;$$

thus, the total exit temperature of the turbine is $T_{t5} = \tau_t T_{t4} = 1307\ °R$.

From Eq. (12.2.8), for $\gamma = 1.339$ the second (and independent) calculation for the total pressure ratio of the turbine yields (τ_t and η_t are known)

$$\pi_t = \left[1 + \frac{(\tau_t - 1)}{\eta_t}\right]^{\frac{\gamma}{\gamma-1}}$$

$$= 0.075.$$

This value of π_t agrees with the previously calculated value of π_t. Thus, the initial guess on $\dot{m}_{c1}$ was good. If the two values of π_t would not have been in agreement, a new value of $\dot{m}_{c1}$ would have been tried and the above process repeated. Thus, this ends the solution of the gas generator/load. Now, one must determine if the initial guess on N_{c2} was valid.

Since the corrected mass flow for the turbine, total pressure ratio, and total temperature ratio for the turbine are known, using Eq. (12.2.24):

$$\dot{m}_{c4} = \frac{\dot{m}_{c5}\pi_t}{\sqrt{\tau_t}}$$

and one finds the corrected mass flow for the exhaust $\dot{m}_{c5} = 198.7\ \text{lb}_m/\text{s}$. So now the total pressure loss for the exhaust/stack can be found from

$$\pi_e = 1 - z_2\dot{m}_{c5}^2$$

$$= 0.980 = \frac{p_{atm}}{p_{t5}}$$

so p_{t5}=14.99 psia, and one can also and *i*ndependently find the total pressure exiting the turbine from Eq. (12.2.23) since π_i, π_c, π_b, and π_t are known

$$p_{t5} = p_a\, \pi_i\, \pi_c\, \pi_b\, \pi_t$$

$$= 14.99\ \text{psia}.$$

This value of p_{t5} agrees with the earlier calculated value of p_{t5}. Thus, the initial guess on N_{c2} was also good. It must have been the author's lucky day. If the two values of p_{t5} would not have been in agreement, a new value of N_{c2} would have been tried and the above process repeated (including the nested iteration on $\dot{m}_{c1}$). Thus, the solution for $f = 0.02$ has been found.

Other values of f (ranging from 0.014 to 0.030) were used and a series of calculations were completed using the above procedure. Net results for these other values of f are presented as operating lines as f varies on the maps in Figures 12.16 through 12.18. As can be seen in Figure 12.16, the case for $f = 0.02$ results in a power unit operating near the compressor maximum efficiency point (BEP). As can also be seen in Figure 12.16, over the given range of fuel ratios, the compressor efficiency ranged from 85.1 percent to 87.8 percent to 84.8 percent, and the compressor total pressure ratio monotonically increased. The operating line passed close to the maximum efficiency point. At a fuel ratio of 0.030 the compressor neared the surge line, which as discussed in Chapter 5, would be accompanied by violent vibrations and possibly a combustor flameout, resulting in total loss of power.

The turbine (Figure 12.17) changes characteristics minimally as the efficiency ranges from 89.2 percent to 92.0 percent (maximum at $f = 0.022$) but the pressure ratio varies from 0.106 to 0.0559. One should note that the operating line passes through the maximum efficiency point for the turbine, indicating that the turbine and compressor are well matched.

In Figure 12.19, other resulting parameters are presented for this fuel ratio range. One can easily see that the turbine inlet temperature, power output, rotational speed, and mass flow rate all increase with increasing fuel ratio. For high fuel ratios the turbine inlet temperature is excessive. The turbine corrected flow rate generally increases, but chokes at a fuel ratio of 0.022. The overall gas turbine efficiency generally increases as the fuel ratio increases but reaches a maximum of 36.6 percent at $f = 0.028$, and the efficiency was only 26.4 percent at $f = 0.014$.

Extended Example 12.3.1 | Power Generation Gas Turbine with a Load Map: with Regenerator

The single-spool power gas turbine with given components is to be assembled now with a regenerator. The regenerator parameters are $b_{3\text{-}3x} = 0.00005/(\text{lbm/s})^2$, $b_{5\text{-}5x} = 0.0000002/(\text{lb}_m/\text{s})^2$, and $\pi = 0.75$.

Compare with the above results for $f = 0.02$.

Solution

Because of the regenerator, another iteration loop is required. Namely, the regenerator affects the output of the burner, and thus the turbine performance but the output of the turbine affects the regenerator performance. Having realized this but knowing the results have successfully converged, the iteration details can be assumed and only the results are presented.

Overall results		
$\dot{m}$ = 125.3 lb_m/s	Power = 26,223 hp	Efficiency = 38.9 percent
SFC = 0.363 lb_m/hp-h	N = 9307 rpm	
Internal results		
$\dot{m}_{c1} = 132.2\ lb_m/s$	$N_{c2} = 9307$ rpm	$\pi_b = 0.967$
$\dot{m}_{c2} = 134.6\ lb_m/s$	$N_{c4} = 4263$ rpm	$\eta_b = 0.910$
$\dot{m}_{c3} = 13.17\ lb_m/s$	$\pi_c = 15.50$	$\eta_m = 0.995$
$\dot{m}_{c4} = 20.00\ lb_m/s$	$\eta_c = 0.877$	$pi_e = 0.980$
$\dot{m}_{c5} = 210.9\ lb_m/s$	$\pi_t = 0.0706$	$\pi_i = 0.983$
$T_{t3} = 1197\ °R$	$\eta_t = 0.920$	$\pi_{3-3x} = 0.991$
$T_{t4} = 2479\ °R$	$\gamma_i = 1.403$	$\pi_{5-5x} = 0.991$
$T_{t5} = 1373\ °R$	$\gamma_c = 1.385$	$\varepsilon_{RG} = 0.750$
	$\gamma_b = 1.335$	
	$\gamma_t = 1.334$	
	$\gamma_e = 1.366$	

The power and efficiency have increased by 19 percent and 13 percent respectively. The mass flow rate and rotational speed have both increased by 6 percent, and the *SFC* has decreased by 13 percent. The power, efficiency, and *SFC* all show significant improvements.

Example 12.4 | Power Generation Gas Turbine at a Specified Constant Speed

The single-spool power gas turbine with given components in Example 12.3 is to now operated at a constant speed. Since this case represents a power generation unit, the bell inlet, compressor, combustor, turbine, power load, shaft, and exhaust/stack are all included. For this example the components have again been modeled using equations presented in Section 12.2.5. The maps for the compressor, burner, turbine, power load, and other components are also shown in Figures 12.16 through 12.18 and the different model parameters, which fit the data, are presented in Table 12.3. The unit is to be operated at $T_a = 520$ °R and $p_a = 14.69$ psia, and $N =$ 9000 rpm (constant). Find the operating characteristics and output of the unit as the fuel ratio is parametrically varied from 0.01 to 0.030. It should be noted that in Example 12.3 the rotational speed was a variable and the load map was included, but all parameters and component maps are the same. The load map is not needed for this example.

Solution

To demonstrate the solution $f = 0.02$ is picked as a representative case. The first estimate for the output power is $\mathcal{P}_1 = 22{,}690$ hp and the first estimate for $\dot{m}_{c1} = 128.4\,\text{lb}_\text{m}/\text{s}$ are chosen. These will be checked later. Now, from Eq. (12.2.18) for $T_a = 520\,°\text{R}$, since $M_a = 0.0$, and for $\gamma = 1.403$, one can find the total temperature exiting the inlet:

$$T_{t2} = T_a\left[1 + \frac{\gamma - 1}{2}M_a^2\right] = 520\,°\text{R}.$$

For the bell inlet, one can find the pressure recovery factor from the modeling equation (Eq. (12.2.68))

$$\begin{aligned}\pi_i &= 1 - z_1\dot{m}_{c1}^2\\ &= 0.984.\end{aligned}$$

Thus, one can find the total pressure exiting the diffuser from Eq. (12.2.20) for γ=1.403:

$$\begin{aligned}p_{t2} &= p_a\pi_d\left[1 + \frac{\gamma - 1}{2}M_a^2\right]^{\frac{\gamma}{\gamma-1}}\\ &= 14.92\text{ psia.}\end{aligned}$$

But for similitude at the inlet

$$\dot{m}_{c1} \equiv \dot{m}\frac{\sqrt{T_t/T_{stp}}}{p_t/t_{stp}}$$

so the actual flow rate is $\dot{m} = 128.4\,\text{lb}_\text{m}/\text{s}$.

Now, one can solve the compressor/burner/turbine/power portion of the problem. So, using Eqs. (12.2.3) and (12.2.4),

$$\dot{m}_{c2} \equiv \dot{m}\frac{\sqrt{T_t/T_{stp}}}{p_t/p_{stp}}$$

$$N_{c2} \equiv \frac{N}{\sqrt{T_t/T_{stp}}}.$$

Therefore for $\dot{m} = 128.4\,\text{lb}_\text{m}/\text{s}$, $\dot{m}_{c2} = 130.6\,\text{lb}_\text{m}/\text{s}$, and for $N = 9000$ rpm, one can find the corrected rotational speed:

$$N_{c2} = 9000\text{ rpm}.$$

One also can find the compressor total pressure ratio by using $\dot{m}_{c2} = 130.6\,\text{lb}_\text{m}/\text{s}$ and $N_{c2} = 9000$ rpm with Figure 12.16 or by modeling Eq. (12.2.45):

$\pi_c = 14.73,$

and from Figure 12.16 or by modeling Eq. (12.2.49), again using $\dot{m}_{c2}$ and N_{c2}, one can find the compressor efficiency, $\eta_c = 0.880$.

Now since π_c is known, one can use Eq. (12.2.5) to find the total temperature ratio for the compressor using

$\pi_c = [1 + \eta_c(\tau_c - 1)]^{\frac{\gamma}{\gamma-1}};$

one can solve for the total temperature ratio of the compressor as $\tau_c = 2.264$ for $\gamma = 1.385$, and from Eq. (12.2.17) one can find the total temperature exiting the compressor:

$$T_{t3} = \tau_c T_{t2}$$
$$= 1177\ ^\circ\text{R}.$$

Next, using Eq. (12.2.14),

$$\dot{m}_{c2} = \frac{\dot{m}_{c3}\pi_c}{\sqrt{\tau_c}},$$

and now one can solve for the corrected mass flow for the burner since $\dot{m}_{c2}$ and τ_c are known,

$\dot{m}_{c3} = 13.35\,\text{lb}_m/\text{s}.$

Next, since f and T_{t3} are known, one can find the quantity $f/(T_{t3}/T_{stp}) = 0.00883$.

For the burner the efficiency is given as $\eta_b = 0.91$, and using $\dot{m}_{c3} = 13.35\,\text{lb}_m/\text{s}$ and $f/(T_{t3}/T_{stp}) = 0.00883$, one can find the total pressure ratio for the burner from Figure. 12.18 or by modeling Eq. (12.2.53):

$\pi_b = 0.962.$

Next, using the energy equation for the burner, Eq. (12.2.9),

$f(\eta_b\,\Delta H - c_p\,T_{t3}\,\tau_b) = c_p\,T_{t3}\,(\tau_b - 1),$

and since f, η_b, ΔH, and T_{t3} are known, one can solve for the total temperature ratio for the burner as

$\tau_b = 1.993$ for $\gamma = 1.341$.

Therefore, the exit total temperature for the combustor is $T_{t4} = \tau_b\,T_{t3} = 2347\ ^\circ\text{R}$, and, thus, using Eq. (12.2.12),

$$\dot{m}_{c2} = \frac{\dot{m}_{c4}\pi_b\pi_c}{(f+1)\sqrt{\tau_b\tau_c}}.$$

Since $\dot{m}_{c2}$, ε_c, τ_c, ε_b, and τ_b are known, one can find the corrected mass flow for the turbine:

$\dot{m}_{c4} = 19.99\,\text{lb}_m/\text{s},$

and using Eq. (12.2.13),

$N_{c2} = N_{c4}\sqrt{\tau_b\tau_c},$

one finds the corrected speed for the turbine $N_{c4} = 4237$ rpm.

Next, for the turbine using $\dot{m}_{c4} = 19.99\,\text{lb}_\text{m}/\text{s}$ and $N_{c4} = 4237$ rpm, one finds the total pressure ratio for the turbine from Figure 12.17 or by modeling Eq. (12.2.56):

$$\pi_t = 0.0733,$$

and from Figure 12.17 or by modeling Eq. (12.2.48), again using $\dot{m}_{c4} = 19.99\ \text{lb}_\text{m}/\text{s}$ and $N_{c4} = 4237$ rpm, one finds the turbine efficiency, $\eta_t = 0.919$.

For the shaft, from modeling Eq. (12.2.59) the mechanical efficiency is given by

$$\eta_m = 1 - s_1 N^{s_2},$$

so $\eta_m = 0.995$. So now the power load has been estimated:

$$\mathcal{P}_l = 22{,}690\ \text{hp}.$$

So, using the power balance on the shaft, Eq. (12.2.35):

$$c_{p_c}(\tau_c - 1) + \frac{\mathcal{P}_l}{\dot{m} T_{t2}} = \eta_m (1 + f) c_{p_t} \tau_b \tau_c (1 - \tau_t);$$

since f, τ_c, and τ_b are known, one can solve for the total temperature ratio for the turbine for specific heat ratios of 1.385 and 1.339 for the compressor and turbine, respectively:

$$\tau_t = 0.555.$$

Thus, the total exit temperature of the turbine is $T_{t5} = \tau_t T_{t4} = 1303\ °\text{R}$, so that from Eq. (12.2.8) for $\gamma = 1.339$ the second (and independent) calculation for the total pressure ratio of the turbine yields (τ_t and η_t are known)

$$\pi_t = \left[1 + \frac{(\tau_t - 1)}{\eta_t}\right]^{\frac{\gamma}{\gamma-1}}$$

$$= 0.0733.$$

This value of π_t agrees with the previously calculated value of π_t. Thus, the initial guess on $\dot{m}_{c1}$ was good. If the two values of π_t would not have been in agreement, a new value of $\dot{m}_{c1}$ would have been tried and the above process repeated. Thus, this ends the solution of the gas generator/load. Now, one must determine if the initial guess on the power, $\mathcal{P}_l$, was valid.

Since the corrected mass flow for the turbine, total pressure ratio, and total temperature ratio for the turbine are known, using Eq. (12.2.24)

$$\dot{m}_{c4} = \frac{\dot{m}_{c5} \pi_t}{\sqrt{\tau_t}},$$

one finds the corrected mass flow for the exhaust $\dot{m}_{c5} = 203.2\,\text{lb}_\text{m}/\text{s}$.

So now the total pressure loss for the exhaust/stack can be found from

$$\pi_e = 1 - z_2 \dot{m}_{c5}^2$$

$$= 0.979 = \frac{p_{atm}}{p_{t5}},$$

so $p_{t5} = 15.00$ psia.

One can also independently find the total pressure exiting the turbine from Eq. (12.2.23) since π_i, π_c, π_b, and π_t are known:

$$p_{t5} = p_a\,\pi_i\,\pi_c\,\pi_b\,\pi_t$$
$$= 15.00\ \text{psia}.$$

This value of p_{t5} agrees with the earlier calculated value of p_{t5}. Thus, the initial guess on $\mathcal{P}_l$ was also good. It must have been the author's really lucky day. If the two values of p_{t5} would not have been in agreement, a new value of $\mathcal{P}_l$ would have been tried and the above process repeated (including the nested iteration on $\dot{m}_{c1}$). Thus, the solution for $f = 0.02$ has been found.

Other values of f (ranging from 0.01 to 0.030) were used and a series of calculations were completed using the above procedure. Net results for these other values of f are presented as operating lines as f varies on the maps in Figures 12.16 through 12.18. Also, on all of the figures the results using the load map are shown for comparison. As can be seen on Figure 12.16, the case for $f = 0.02$ results in a unit again operating near the compressor maximum efficiency point. Also, as can be seen in Figure 12.16, over the given range of fuel ratios, the compressor efficiency ranged from 86.8 percent to 88.0 percent to 87.0 percent, and the compressor total pressure ratio monotonically increased along the constant N_{c2} line. The operating line passed through the maximum efficiency point. At a fuel ratio of 0.030 the compressor again neared the surge line, which as discussed in Chapter 5, would be accompanied by violent vibrations and possibly a combustor flameout, resulting in total loss of power.

The turbine (Figure 12.17) changes characteristics minimally as the efficiency ranges from 90.8 percent to 91.9 percent (maximum at $f = 0.022$) but the pressure ratio varies from 0.0844 to 0.0666. One should note that the operating line passes through the maximum efficiency point for the turbine, once again indicating that the turbine and compressor are well matched.

In Figure 12.19, other resulting parameters are presented for this fuel ratio range. One can easily see that the turbine inlet temperature and power output increase with increasing fuel ratio. The mass flow rate, however, decreases slightly over the range of fuel ratios. For high fuel ratios the turbine inlet temperature is excessive. The turbine corrected flow rate generally increases, but chokes at a fuel ratio of 0.022. The overall gas turbine efficiency generally increases as the fuel ratio increases from 26.9 percent at $f = 0.01$ and reaches a maximum of 36.8 percent at the high end of fuel ratios tested, $f = 0.030$. Such behavior was also noted in Chapter 3 (Figure 3.40). Namely, as the turbine inlet temperature increased so did the overall nonideal efficiency.

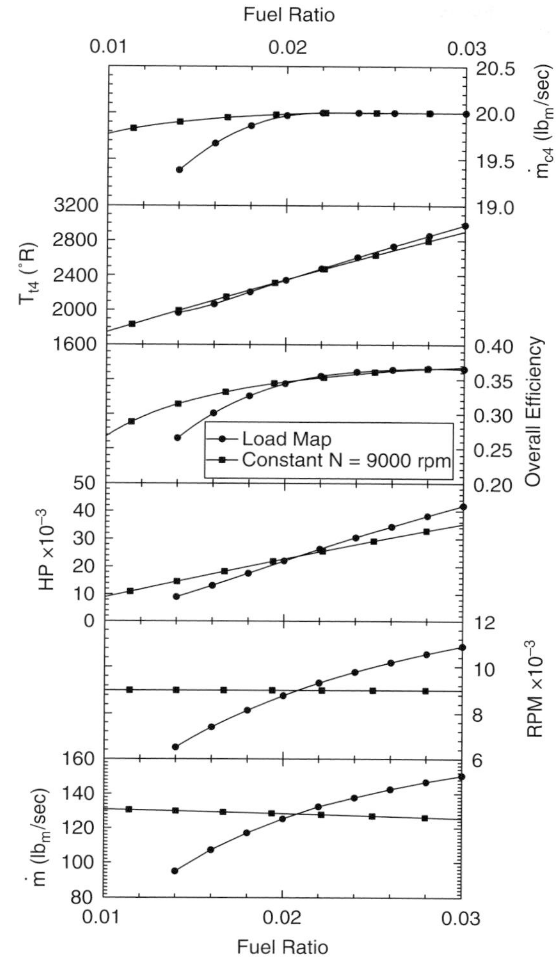

Figure 12.19 Overall operating conditions vs. fuel ratio for Examples 12.3 and 12.4.

12.3 Use of Matching and Cycle Analysis in Second-Stage Design

So far in this chapter the concept of component matching has been presented and used in examples as a method of directly analyzing engines or power gas turbines with given maps. In fact, the method in practice is used to perform a second-stage design analysis (often called an advanced design analysis) of a new or proposed engine or gas turbine. The preliminary (or first-stage) design has been outlined in Chapter 3, which basically set the engine type and "sized" the engine. For this second-stage jet engine design, a set of overall engine design conditions, including both on-design and off-design conditions, are given to an engine manufacturer by a military group or commercial enterprise, including, but not exclusively, thrusts and *TSFC*s at different altitudes and speeds. Similarly, for a power generation unit, the output power, efficiency, and *SFC* are set by a commercial or government group for a more limited range of operating conditions. A group of system design engineers then uses a component-matching analysis to vary the gas turbine types parametrically and iteratively vary the component maps to accomplish the overall design goals and to ensure that all of the components fit together.

For instance, in Extended Example 12.1.1, which might represent an initial matching analysis, the resulting operating points for the jet engine indicate that the components are clearly not matched well. The compressor is designed to operate at a rotational speed (or mass flow) that is too high and with a pressure ratio that is too high, or the initial turbine is designed to operate with a value of π_t that is too low and a rotational speed that is too high, or both the compressor and turbine are not well designed for matching. The maps should be adjusted accordingly until the best efficiency points of all components are at the best performance point of the engine, the derived turbine power matches the compressor power requirements, required thrust is delivered, and so on. For the maps shown in Extended Example 12.1.1, this would mean that the compressor should be designed for a lower rotational speed and either the flow-deflection angles of the blades or vanes should be reduced or number of stages should be reduced. For the turbine, it means that reducing the blade or vane deflection angles or number of stages and reducing the running speed are in order. The final sets of component maps (i.e., Example 12.1) optimize overall performance in that both the compressor and turbine operate near the design condition. In Example 12.2 these maps were applied to an aircraft at cruise. The *TSFC* could possibly be improved by increasing the compressor pressure ratio. This would require new maps for both the compressor and turbine to insure good matching. The maps determined by this systems group during this design phase then become the detailed design goals of the more focused or specific component design groups. For example, compressor design or turbine design groups then design the blades and vanes and stages to produce these maps. Then, detailed analyses and designs of the components are undertaken to obtain components with the required maps by using the methods presented in the previous seven chapters as well as more advanced techniques. Such design methodology is a part of an inverse system design; that is, one starts with the overall design goals and works backward to determine the component maps that will accomplish the goals. Once the maps are determined, the geometries that produce these maps are designed in detail by analytical design engineers. After an initial detailed design the components are tested and designs are continuously "tweaked" to eventually reach the design goal. This is, therefore, the second step in an industrial engine or gas turbine design.

12.4 Summary

A method of matching gas turbine components has been presented in this capstone chapter. At the end of a complicated and expensive design and development process, an engine must operate at high (and known) performance over a range of conditions. Matching is the process by which components are integrated to allow predictions of overall engine performance. First, the fundamental method of matching components with generalized characteristic maps was described. Three cases were considered: gas generator (compressor, burner, turbine), a single-shaft turbojet engine (diffuser, gas generator, nozzle), and a single-shaft power generation gas turbine (inlet, gas generator, exhaust, regenerator, load). Second, curve-fit functional mathematical models for the different maps of the components were developed so that the component maps could be defined by assigning values to parameters. Although convenient for computer analyses, such modeling is not necessary for the matching process; it merely eases the solution process. Third, steady-state matching was accomplished for both single-spool jet engines and power gas turbines by simultaneously solving the mathematical models of the component maps or graphical data along with the matching equations.

The method was demonstrated in an example for a single-spool turbojet engine with given component maps. For the example, a range of fuel ratios was used and the compressor was shown to increase in speed and eventually surge as the fuel ratio was increased. The compressor and turbine were well matched. The technique was also demonstrated for a power generation unit running at a constant speed and also with a speed dependent load. The compressor and turbine were again well matched. A range of fuel ratios again was used and the power and efficiency both increased with increasing fuel ratio. The turbine choked at higher fuel ratios. The technique is a tool through which one can select components to optimize overall engine performance and predict the off-design performance of an engine. Although the method was applied only to a simple turbojet engine or a simple power generation gas turbine in this chapter, it can be extended to include more complex engines. This includes twin-spool turbofans with afterburners. For these cases, more maps and equations are needed and all of the components are matched. Furthermore, the method can be extended to find the dynamic or transient response of a gas turbine during shaft acceleration or deceleration.

Jet engine and airframe matching was also accomplished. The performance characteristics of an engine (or set of engines) must be matched to an aircraft. For this matching, the total thrust from the engines must match the total aerodynamic drag of the airframe for a cruise condition to result. Thus, when both the engine and aircraft are matched, the fuel ratio and speed cannot be independently varied. As a result, when the drag characteristics of an airframe are coupled with an engine for which all of the component maps are known, by varying the fuel ratio (as performed by a pilot), one can determine the operating points of all engine components and airframe speed.

In previous chapters, the influence of components on each other was not considered. Chapter 3 considers the entire gas turbine through cycle analyses; however, component parameters (efficiency, pressure ratios, flow rates, etc.) are specified. In Chapters 5–11 the individual components are analyzed and, for example, the pressure ratios and efficiencies of the components have been found to be strong functions of component geometries, flow rates, and rotational speeds. Component mapping methods were developed in these chapters. This chapter thus integrates all of the previous chapters in as much as cycle analysis of the gas turbine is once again performed; however, in this chapter the component interaction and varying efficiencies are included.

At the conclusion of this chapter, the reader, if given a set of gas turbine design goals (both on- and off-design), has a significant package of analyses that can be used for the inverse design of an gas turbine as a system. One can study the effects of parametrically changing component maps and gas turbine types for ranges of operating conditions until the overall design goals are accomplished. Thus, the gas turbine can be treated as a system, and realistic performance predictions can be made. Through this methodology, an inverse design process can be used to specify the desired operating characteristics, namely maps, of the different components to accomplish the overall design goals. Once the design maps are determined, the components can be designed in detail to realize these maps. This process is the second step in real industrial engine design.

List of Symbols

A	Area
a	Speed of sound
c_p	Specific heat at constant pressure
F	Force or thrust
f	Fuel ratio
ΔH	Heating value
I	Rotor polar moment of inertia
$\dot{m}$	Mass flow rate
M	Mach number
N	Rotational speed
p	Pressure
$\mathcal{P}$	Power
$\mathcal{R}$	Ideal gas constant
T	Temperature
T	Torque
$TSFC$	Thrust specific fuel consumption
γ	Specific heat ratio
ε	Regenerator effectiveness
η	Efficiency
π	Total pressure ratio
ρ	Density
μ	Surge margin
τ	Total temperature ratio
φ	General function
ψ	General function
ω	Rotational speed

Subscripts

a	Freestream or ambient
b	Primary burner
c	Compressor
c	Corrected
d	Diffuser (propulsion)
e	Exhaust (power generation)
f	Fuel
i	Inlet (power generation)
l	Load
m	Mechanical (shaft)
n	Nozzle
RG	Regenerator
s	Surge
stp	Standard conditions
t	Total (stagnation)
t	Turbine
x	Exit of heat exchanger
1	Inlet to diffuser
2	Exit of diffuser or inlet, inlet to compressor
3	Exit of compressor, inlet to combustor
3x	Exit of regenerator, inlet to combustor
4	Exit of combustor, inlet to turbine
5	Exit of turbine, inlet to nozzle or exhaust
5x	Exit of regenerator, inlet to dump
8	Exit of nozzle or exhaust

Superscripts

*	Choked

Modeling Parameters

c_1	Compressor
c_2	Compressor
c_3	Compressor
c_4	Compressor
c_5	Compressor
c_6	Compressor
$\dot{m}_{c2_d}$	Compressor
$\dot{m}_{c2_o}$	Compressor
$\dot{m}_{c2_s}$	Compressor
N_{c2_d}	Compressor
μ	Compressor
η_{c_d}	Compressor
b_1	Combustor
b_2	Combustor
η_{b_d}	Combustor
k_1	Turbine
k_2	Turbine
$\dot{m}_{c4_c}$	Turbine
n	Turbine
N_{c4_d}	Turbine
η_{t_d}	Turbine
π_{t_c}	Turbine
s_1	Shaft
s_2	Shaft
d	Diffuser
π_{d_d}	Diffuser
a_1	Nozzle
$\dot{m}_n$	Nozzle
η_{n_d}	Nozzle
z_1	Inlet (power gas turbine)
z_2	Exhaust (power gas turbine)
z_3	Power load (power gas turbine)
z_4	Power load (power gas turbine)
b_{55x}	Regenerator (power gas turbine)
b_{33x}	Regenerator (power gas turbine)
ε_{RG}	Regenerator (power gas turbine)
C_{d_m}	Airframe

Problems

12.1 A turbojet engine with given maps for the compressor, turbine, burner, and converging nozzle is to be assembled. The engine is to operate at sea level and at a Mach number of 0.75. The maps are shown in Figures 12.P.1 through 12.P.4. The shaft mechanical efficiency is 0.995, the burner efficiency is 0.91, the fuel has a heating value of 17,800 BTU/lb_m, the diffuser total recovery factor is 0.92, and the nozzle efficiency is 0.96. If the fuel ratio is 0.03, find the thrust, *TSFC*, engine air mass flow rate, and shaft speed. Indicate the operating point on the figures.

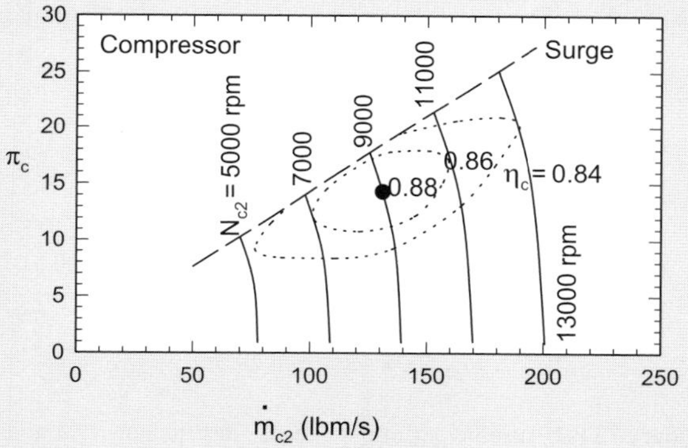

Figure 12.P.1 Compressor performance map for Problems 12.1, 12.2, 12.3, and 12.13.

12.2 A gas generator for a turbojet engine with given maps for the compressor, turbine, and burner is to be assembled. The maps are shown in Figures 12.P.1 through 12.P.3. The shaft mechanical efficiency is 0.995, the burner efficiency is 0.91, and the fuel has a heating value of 17,800 BTU/lb_m. If the fuel ratio is 0.020, the corrected speed of the compressor is 9000 rpm, and the total temperature at the inlet to the compressor is 577 °R, find the corrected compressor air mass flow rate, compressor total pressure ratio, and shaft speed. Indicate the operating point on the figures.

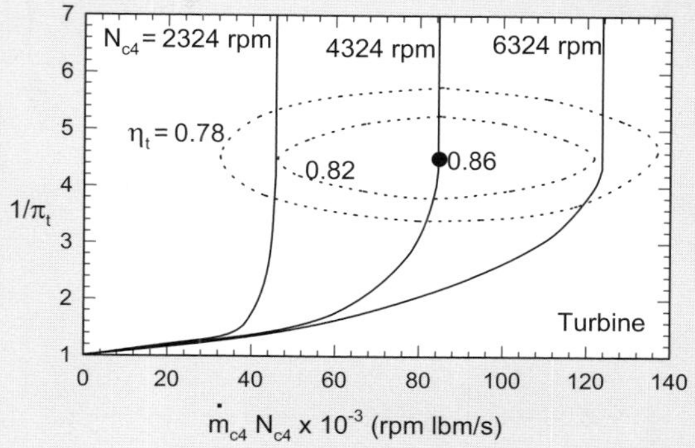

Figure 12.P.2 Turbine performance map for Problems 12.1, 12.2, 12.3, and 12.13.

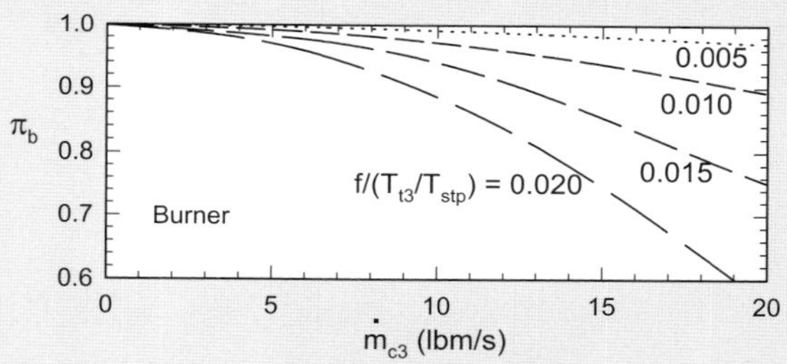

Figure 12.P.3 Burner performance map for Problems 12.1, 12.2, 12.3, and 12.13.

12.3 Three identical turbojet engines with given maps for the compressors, turbines, burners, and converging nozzles are to be assembled and used on a given aircraft. The aircraft is to operate at sea level. The engine maps are shown in Figures 12.P.1 through 12.P.4, and the drag characteristics for the airframe are shown in Figure 12.P.5. The shaft mechanical efficiency is 0.995, the burner efficiency is 0.91, the fuel has a heating value of 17,800 BTU/lb_m, the diffuser total recovery factor is 0.92, and the nozzle efficiency is 0.96. If the fuel ratio is 0.021, find the operating Mach number and for each engine find the thrust, *TSFC*, engine air mass flow rate, and shaft speed. Indicate the operating point on the figures.

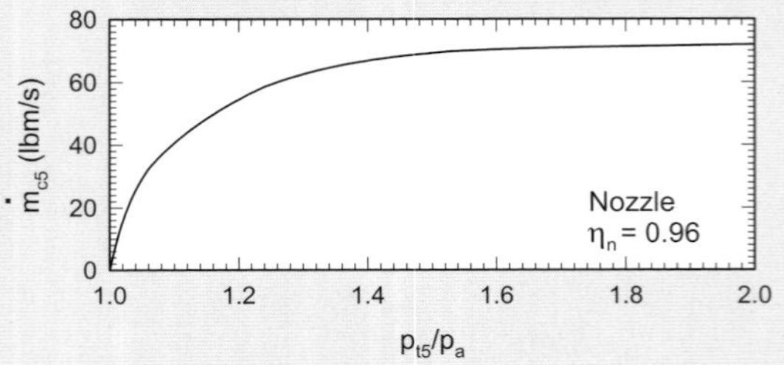

Figure 12.P.4 Nozzle performance map for Problems 12.1 and 12.3.

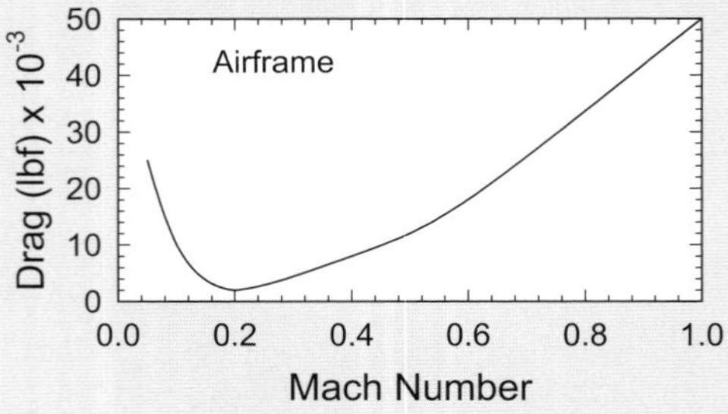

Figure 12.P.5 Drag on airframe for Problem 12.3.

12.4 A turbojet engine with given maps for the compressor, turbine, burner, converging nozzle, and shaft is to be assembled. The engine is to operate at 27,000 ft. The maps are shown in Figures 12.P.6 through 12.P.10. The fuel has a heating value of 17,800 BTU/lb$_m$, and the maximum diffuser total recovery factor is 0.96, which decreases as found empirically with increasing Mach numbers above unity.

(a) If the fuel ratio is 0.026 and the Mach number is 0.85, find the thrust, *TSFC*, engine air mass flow rate, and shaft speed. Indicate the operating point on the figures. Compare this analysis to that in Problem 3.34.

(b) If the Mach number is 0.85, find the operating line on the compressor and turbine maps.

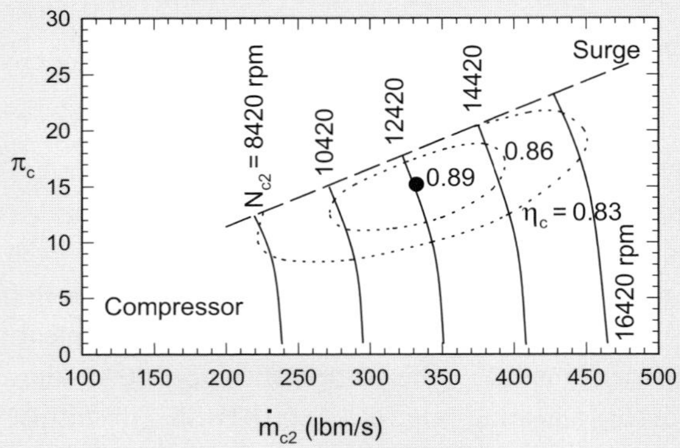

Figure 12.P.6 Compressor performance map for Problems 12.4 and 12.5.

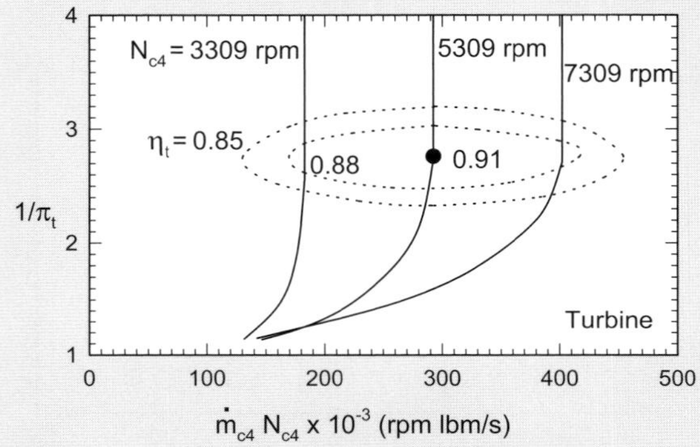

Figure 12.P.7 Turbine performance map for Problems 12.4 and 12.5.

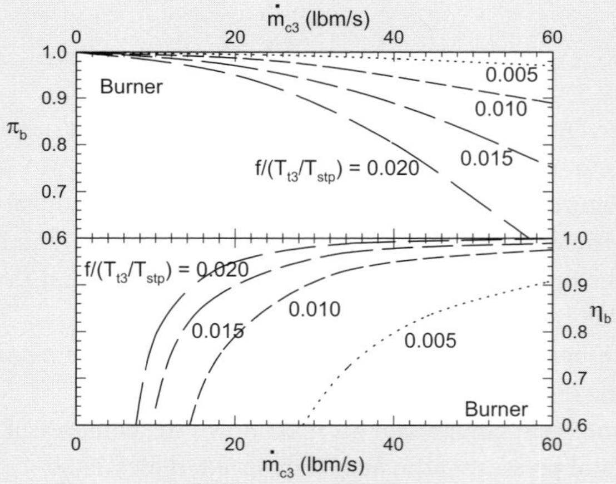

Figure 12.P.8 Burner performance map for Problems 12.4 and 12.5.

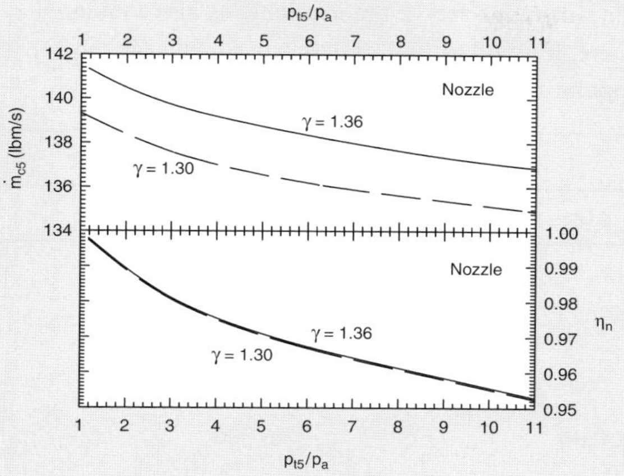

Figure 12.P.9 Nozzle performance map for Problems 12.4 and 12.5.

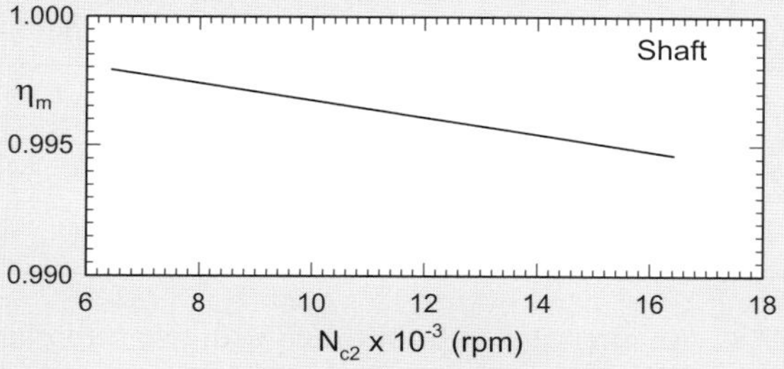

Figure 12.P.10 Shaft performance map for Problems 12.4 and 12.5.

(c) If the Mach number is 0.60, find the operating line on the compressor and turbine maps.
(d) If the Mach number is 1.10, find the operating line on the compressor and turbine maps.

12.5 Three identical turbojet engines with given maps for the compressors, turbines, burners, converging nozzles, and shafts are to be assembled and used on a given aircraft. The aircraft is to operate at 27,000 ft. The engine maps are shown in Figures 12.P.6 through 12.P.10, and the drag characteristics for the airframe are shown in Figure 12.P.11. For each engine the fuel has a heating value of 17,800 BTU/lb$_m$ and the maximum diffuser total recovery factor is 0.96, which decreases as found empirically with increasing Mach numbers above unity.

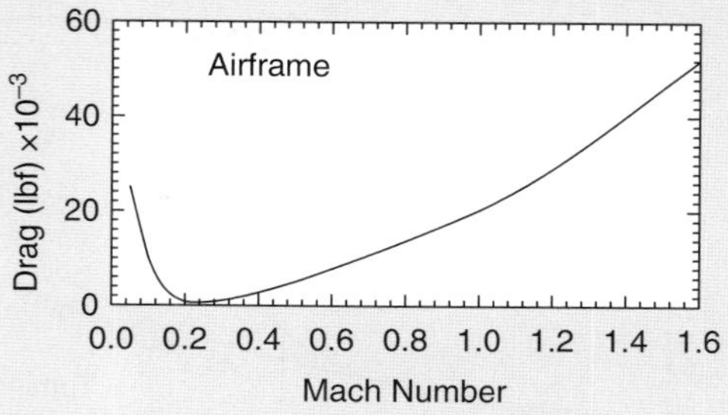

Figure 12.P.11 Drag on airframe for Problem 12.5.

(a) If the fuel ratio is 0.026, find the operating Mach number and for each engine find the thrust, *TSFC*, engine air mass flow rate, and shaft speed. Indicate the operating point on the figures.
(b) Find the dependence of operating Mach number on the fuel ratio for turbine total temperatures ranging from 1900 to 3900 °R.

12.6 A power gas turbine with given maps for the compressor, turbine, burner, inlet, exhaust, and electric generator load is to be assembled. The unit is to operate at standard sea level. The maps are shown in Figures 12.P.12 through 12.P.16. The shaft mechanical efficiency is 0.995, the burner efficiency is 0.91, and the fuel has a heating value of 10,000 kcal/kg. If the fuel ratio ranges from 0.012 to 0.032, find the net output power, thermal efficiency, air mass flow rate, and shaft speed. Indicate the operating points on the figures.

12.7 An aircraft is to be outfitted with three engines. Drag on the airframe at STP is tabulated here. Components of the engines have been matched so that

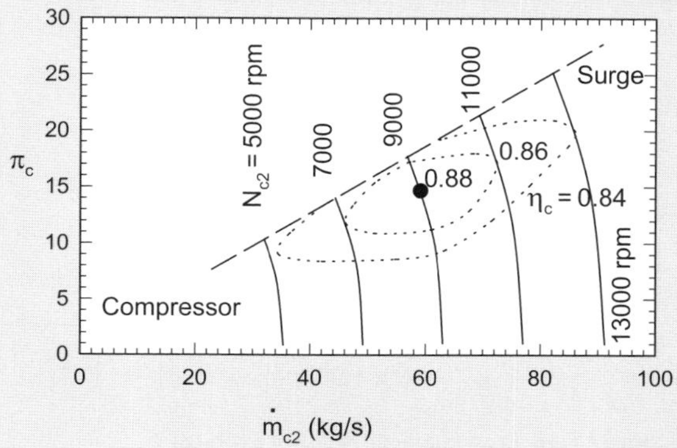

Figure 12.P.12 Compressor performance map for Problem 12.6.

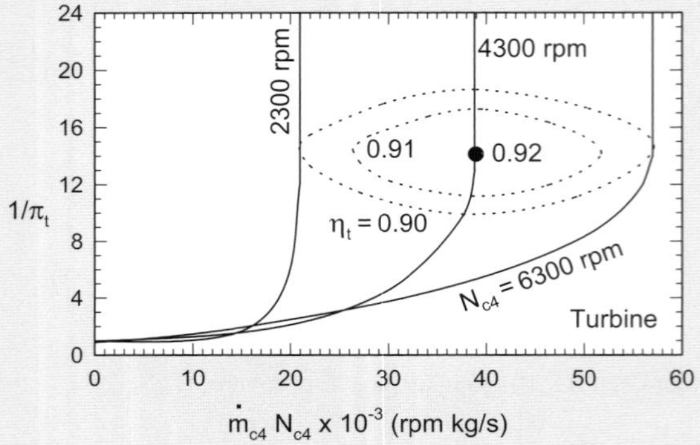

Figure 12.P.13 Turbine performance map for Problem 12.6.

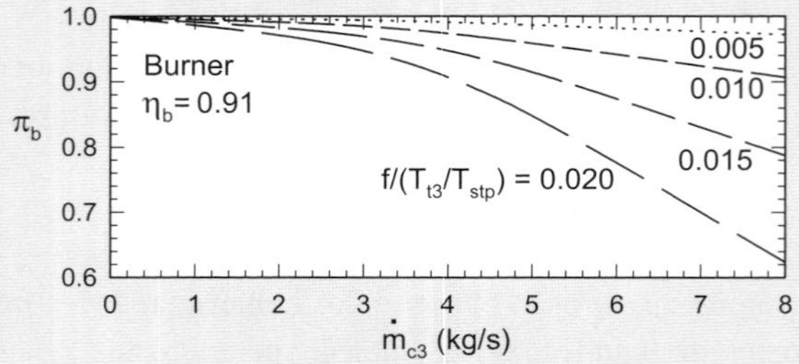

Figure 12.P.14 Burner performance map for Problem 12.6.

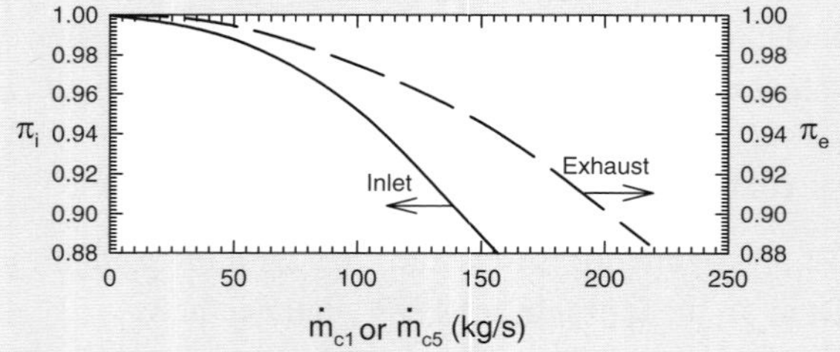

Figure 12.P.15 Inlet and exhaust performance maps for Problem 12.6.

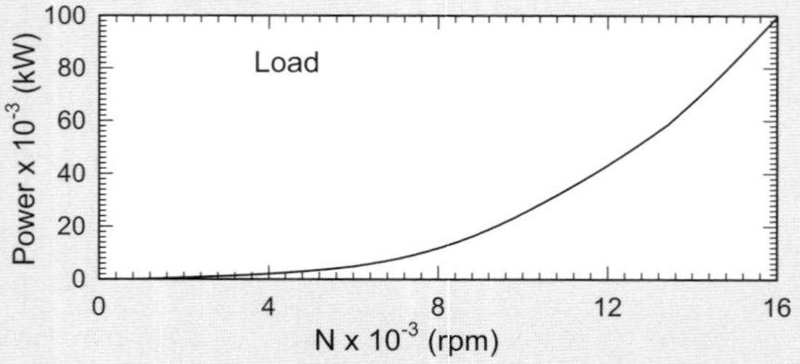

Figure 12.P.16 Load characteristics for Problem 12.6.

engine maps at STP are available as also tabulated here. If a fuel ratio of 0.020 is used, what is the operating Mach number of the aircraft?

Airframe drag	
Mach number	Drag (lb_f)
0.20	3000
0.40	10,000
0.60	23,000
0.80	40,000
1.00	63,000

Single engine thrust (lb_f)			
Mach number	$f = 0.015$	$f = 0.020$	$f = 0.025$
0.40	5700	7800	10,800
0.60	5000	7300	10,200
0.80	5200	7500	10,300

12.8 An aircraft is to be outfitted with two turbofan engines. Drag on the airframe at STP is tabulated here. Components of the engines have been matched so that engine maps at STP are available as also tabulated here. If a fuel ratio of 0.022 is used, what is the operating Mach number of the aircraft?

Airframe drag	
Mach number	Drag (lb_f)
0.05	6000
0.2	4000
0.4	16,000
0.6	36,000
0.8	64,000
1.0	100,000

Single engine thrust (lb_f)			
Mach number	$f = 0.016$	$f = 0.020$	$f = 0.024$
0.40	18,000	29,000	40,000
0.60	17,000	28,000	39,000
0.80	18,000	29,000	40,000
1.00	20,000	31,000	42,000

12.9 An aircraft is to be outfitted with three turbofan engines. Drag on the airframe at STP is tabulated here. Components of the engines have been matched so that single engine maps at STP are available as also tabulated here. If a fuel ratio of 0.021 is used, what is the operating Mach number of the aircraft?

Airframe drag	
Mach number	Drag (lb_f)
0.05	9000
0.2	4700
0.4	18,800
0.6	42,200
0.8	75,000
1.0	117,200

Single engine thrust (lb_f)			
Mach number	$f = 0.017$	$f = 0.020$	$f = 0.023$
0.4	20,500	27,300	34,000
0.6	20,000	26,700	33,300
0.8	20,700	27,200	34,000
1.0	23,700	29,000	36,700

12.10 An aircraft is to be outfitted with two turbofan engines. Drag on the airframe at 20,000 ft is tabulated here. Components of the engines have been matched so that single engine maps at 20,000 ft are available as also tabulated here. If a fuel ratio of 0.022 is used, what is the operating Mach number of the aircraft?

Airframe drag	
Mach number	Drag (lb_f)
0.05	10,000
0.2	4500
0.4	18,000
0.6	40,500
0.8	72,000
1.0	11,2500

Single engine thrust (lb_f)			
Mach number	$f = 0.017$	$f = 0.020$	$f = 0.023$
0.4	30,500	37,300	44,000
0.6	30,000	36,700	43,300
0.8d	30,700	37,200	44,000
1.0	33,700	40,000	46,700

12.11 A single-spool turbojet engine with given components is to be assembled. This case represents an engine, and thus the diffuser, compressor, combustor, turbine, shaft, and nozzle are all included. For this case the components have been modeled using equations presented in Section 12.2.4. The nozzle has a fixed minimum area and a variable exit area, and thus the exit pressure matches the ambient pressure. The engine is to be operated at a Mach number of 0.5. Plot the component maps and find the operating characteristics of the engine as the fuel ratio is parametrically varied from 0.010 to 0.035 in increments of 0.005 for $T_a = 289$ K (520 °R) and $p_a = 101.3$ kPa (14.69 psia). Given: Diffuser ($\pi_{d_d} = 1$); compressor

($c_1 = 0.1764$ s/kg (0.08 s/lb$_m$), $c_2 = 0.00907$ kg/s/rpm (0.02 lb$_m$/s/rpm), $c_3 = 0.80$, $c_4 = 0.00001$/rpm, $c_5 = 9.724$ rpm s^2 /kg^2 (2.0 rpm s^2 /lb$_m$^{2}), $N_{c2} =$ 8000 rpm, $\eta_{c_d} = 0.88$, $\mu = 0.10$; Combustor ($b_1 =$ 9.068 s^2/kg^2 (1.865 s^2/lb$_m$^{2}), $b_2 = 0.0$ s^2 /kg^2 (0.0 s^2 /lb$_m$^{2}), $\Delta H =$10,000 kcal/kg (18,000 BTU/lb$_m$), $\eta_b = 0.91$; Turbine ($k_1 =1.0$, $k_2 = 0.20$, $N_{c4d} = 4000$ rpm, $\dot{m}_{c4_c} = 15.87$ kg/s (35 lb$_m$/s), $\eta_{t_d} =$ 0.85, $\pi_{t_c} = 0.20$, shaft ($s_1 = 0$, $s_2 = 0$); nozzle – CD, fixed A^* ($a_1 = 0$, $\dot{m}_n = 88.08$ kg/s (194.2 lb$_m$/s), $\eta_n = 0.98$).

12.12 A single-spool turbojet engine with given components from Problem 12.11 is to be assembled for a given airframe. The aircraft has drag characteristics given by C_{dm} =163,460 N (36,750 lb$_f$). The aircraft with two engines is to be operated at a Mach number of 0.7. Plot the airframe map and find the required fuel ratio and resulting *TSFC* at this condition. Also, determine how the fuel ratio and *TSFC* vary as the flight Mach number varies from 0.1 to 1.2 at $T_a = 289$ K (520 °R) and $p_a = 101.3$ kPa (14.69 psia).

12.13 Flow enters a gas generator at total conditions of 14 psia and 577 °R, which has been matched. At the matching point, the uncorrected (true) mass flow rate is 118.3 lb$_m$/s, and the unit is rotating at 9492 rpm. The compressor, burner, and turbine maps are given in Figures 12.P.1, 12.P.2, and 12.P.3. The burner efficiency is 91 percent, and the mechanical efficiency is 99.5 percent. The fuel-to-air ratio is 0.020, the burner exit total temperature is 2435 °R, and heating value of the fuel is 17,800 BTU/lb$_m$. The specific heat ratios for the compressor, burner, and turbine are 1.381, 1.337, and 1.326, respectively. Find the following:

(a) the compressor pressure ratio and efficiency,
(b) the compressor exit total pressure and total temperature,
(c) the burner exit total pressure,
(d) the turbine exit total pressure and total temperature.

12.14 The gas generator of a power generation gas turbine is to be analyzed. Maps are given for the compressor, burner, turbine, and shaft and are presented in Figures 11.16 through 11.18. The gas generator is to be operated at 520 °R and 14.7 psia ($T_{t2} = 520$ °R and $p_{t2} = 14.7$ psia).
Other parameters that are given are:

$N_{c2} = 8000$ rpm
$\Delta H = 18{,}000$ BTU/lb$_m$
$\eta_m = 0.995$
$\gamma_c = 1.384$
$\gamma_b = 1.339$
$\gamma_t = 1.328$.

The matching point for the fuel ratio $f = 0.02$ is to be found. Plot on the maps.

12.15 A power generation gas turbine is to be analyzed. Maps are given for the compressor, burner, turbine, inlet, exhaust, and load are presented in Figures 12.P.17 through 12.P.19. The unit is to be

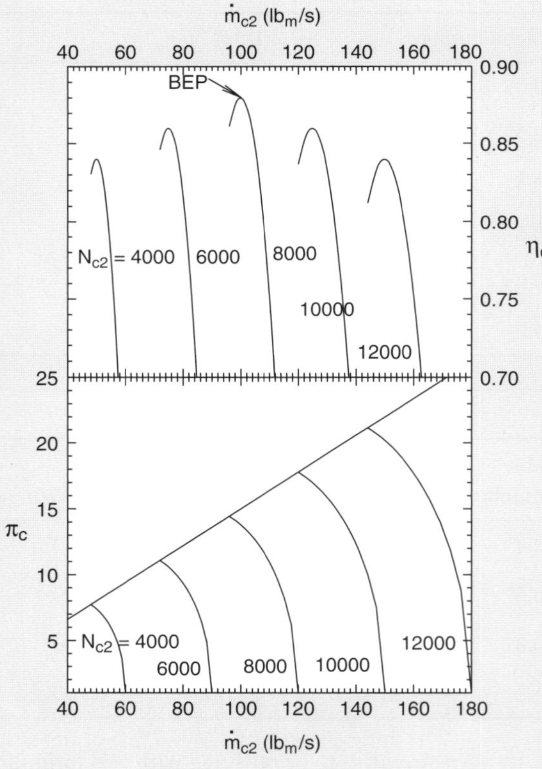

Figure 12.P.17 Compressor performance map for Problems 12.15, 12.16, and 12.17.

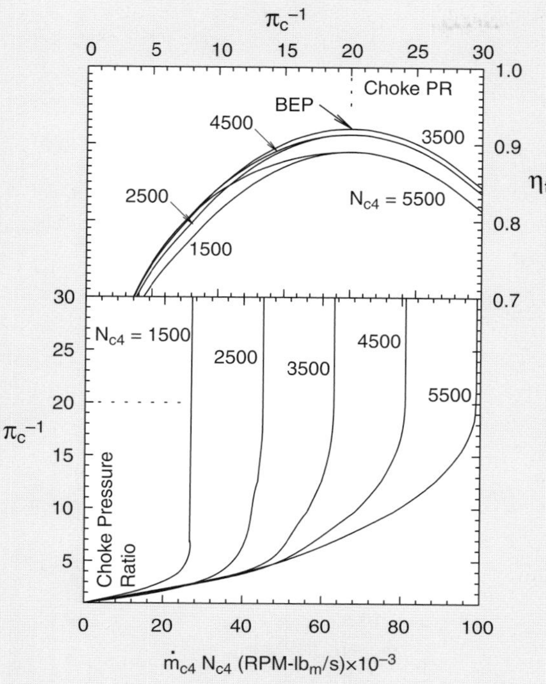

Figure 12.P.18 Turbine performance map for Problems 12.15, 12.16, and 12.17.

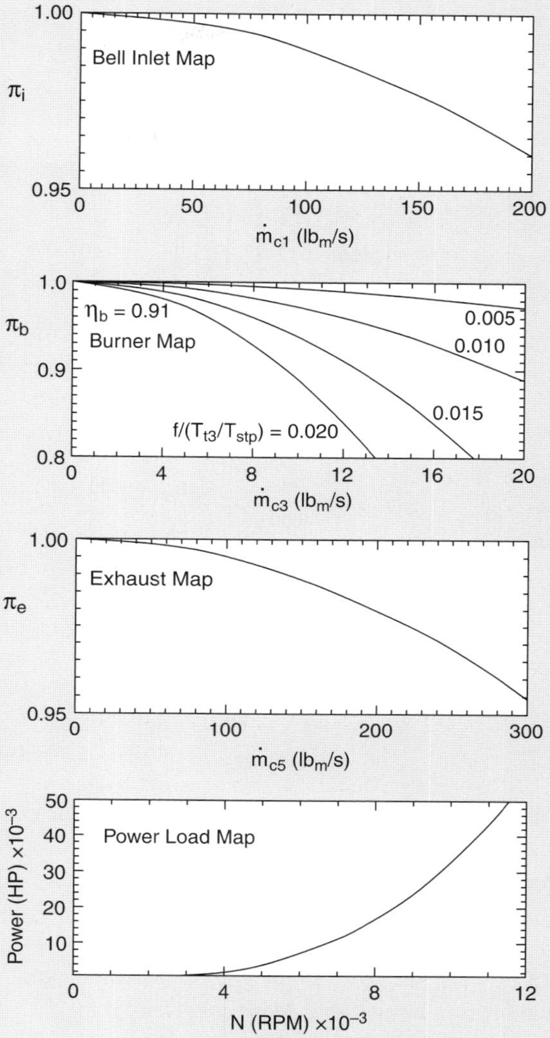

Figure 12.P.19 Other performance maps for Problems 12.15, 12.16, 12.17 and 12.18.

operated at 520 °R and 14.7 psia. Find the operating speed, mass flow rate, output power, and cycle efficiency for the fuel ratio $f = 0.022$. Plot on the maps.

Other parameters that are given are:

$\Delta H = 18{,}000$ BTU/lb$_m$
$\eta_m = 0.995$
$\gamma_c = 1.385$
$\gamma_b = 1.339$
$\gamma_t = 1.334$
$\gamma_{exh} = 1.358$
$\gamma_{inlet} = 1.403$.

12.16 A power generation gas turbine is to be analyzed. Maps are given for the compressor, burner, turbine, inlet, and exhaust are presented in Figures 12.P.17 through 12.P.19. The unit is to be operated at 520 °R and 14.7 psia. Find the mass flow rate, output power, and cycle efficiency for the fuel ratio $f = 0.025$. The rotational speed is constant at 7200 rpm. Plot on the maps.

Other parameters that are given are:

$\Delta H = 18{,}000$ BTU/lb$_m$
$\eta_m = 0.995$
$\gamma_c = 1.387$
$\gamma_b = 1.339$
$\gamma_t = 1.331$

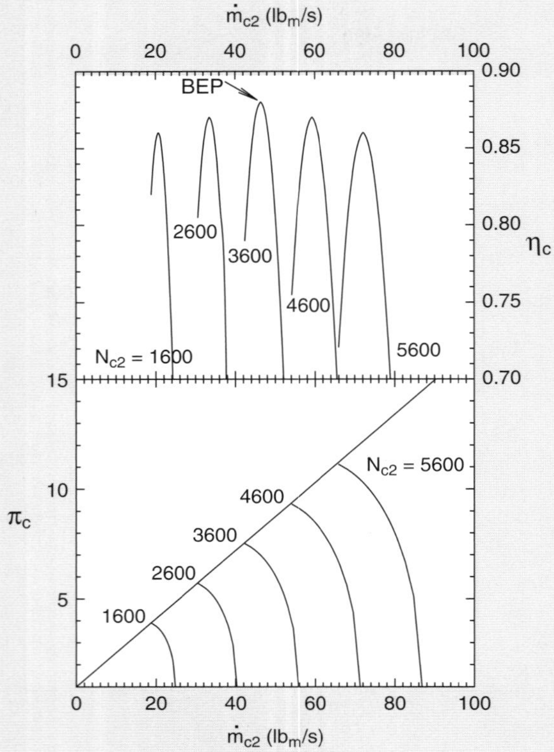

Figure 12.P.20 Compressor performance map for Problem 12.18.

$\gamma_{\mathrm{exh}} = 1.354$

$\gamma_{\mathrm{inlet}} = 1.403$.

12.17 The gas generator of a power generation gas turbine is to be analyzed. Maps are given for the compressor, burner, and turbine are presented in Figures 12.P.17 through 12.P.19. The gas generator is to be operated at 520 °R and 14.7 psia ($T_{t2} = 520$ °R and $p_{t2} = 14.7$ psia).

Other parameters that are given are:

$N_{c2} = 8000$ rpm

$\Delta H = 18{,}000$ BTU/lb$_{\mathrm{m}}$

$\eta_{\mathrm{m}} = 0.995$

$\gamma_{\mathrm{c}} = 1.385$

$\gamma_{\mathrm{b}} = 1.350$

$\gamma_{\mathrm{t}} = 1.346$.

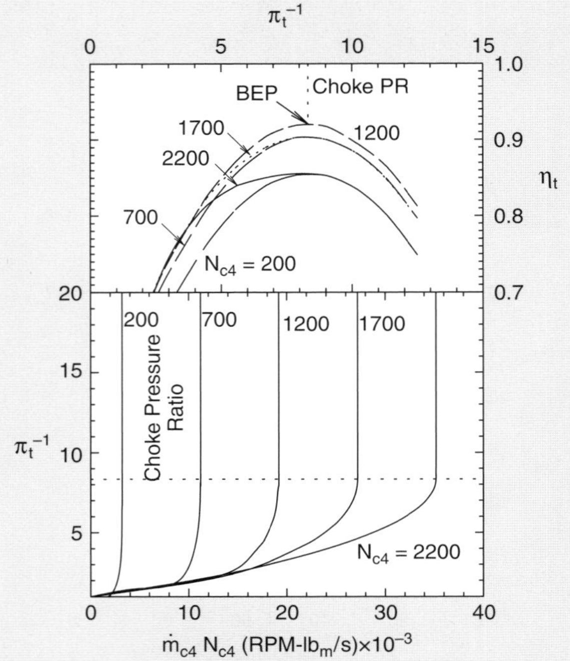

Figure 12.P.21 Turbine performance map for Problem 12.18.

The matching point for the fuel ratio $f = 0.013$ is to be found. Plot on the maps.

12.18 A power generation gas turbine is to be analyzed. Maps are given for the compressor, burner, turbine, inlet, and exhaust are presented in Figures 12.P.19 through 12.P.21. The unit is to be operated at 520 °R and 14.7 psia. Find the mass flow rate, output power, and cycle efficiency for the fuel ratio f = 0.029. The rotational speed is constant at 3600 rpm. Plot on the maps.

Other parameters that are given are:

$\Delta H = 18{,}000$ BTU/lb$_{\mathrm{m}}$

$\eta_{\mathrm{m}} = 0.995$

$\gamma_{\mathrm{c}} = 1.390$

$\gamma_{\mathrm{b}} = 1.338$

$\gamma_{\mathrm{t}} = 1.323$

$\gamma_{\mathrm{exh}} = 1.343$

$\gamma_{\mathrm{inlet}} = 1.403$.

APPENDIX A COMMON CONVERSIONS

Length	1 in. = 0.08333 ft 1 mi = 1609.26 m 1 cm = 0.03281 ft 1 m = 3.281 ft = 39.372 in. 1 μm (micron) = 3.281×10^{-6} ft = 10^{-6} m 1 Å (ångström unit) = 10^{-10} m 1 mil = 0.001 in.
Mass	1 lb_m = 0.4535 kg = 7000 gr (grain) 1 kg = 2.205 lb_m 1 sl (slug) = 32.1736 lb_m
Force	1 lb_f = 4.448 N 1 N = 0.2248 lb_f 1 pdl = 0.03108 lb_f 1 lb_f = 1 sl-ft/s^2 1 dyn = 1 g-cm/s^2 1 N = 1 kg-m/s^2 1 kg_f = 9.8 N
Energy	1 BTU = 778.16 ft-lb_f 1 BTU = 1055.07 J 1 BTU = 0.2520 kcal 1 J = 0.9478×10^{-3} BTU 1 kW-h (kilowatt hour) = 3413 BTU 1 cal = 4.186 J 1 kcal = 1.1626 W-h 1 ft-lb_f = 1.3558 J 1 J = 10^7 erg = 10^7 dyne-cm = 1 N-m
Power	1 BTU/h = 0.293 W 1 hp (horsepower) = 2545 BTU/h 1 hp = 745.7 W = 550 ft-lb_f/s 1 W = 3.413 BTU/h 1 W = 1 J/s 1 ton of refrigeration = 200 BTU/min
Density	1 lb_m/in.^{3} = 1728 lb_m/ft^3 1 lb_m/in.^{3} = 2.77 × 104 kg/m^3 1 kg/m^3 = 0.06243 lb_m/ft^3 = 0.001940 sl/ft^3
TSFC	1 lb_m/lb_f-h = 0.1020 kg/N-h
SFC	1 lb_m/hp-h = 0.6082 kg/kW-h

(cont.)

Heat rate (HR)	1 BTU/hp-h = 1.415 kJ/kW-h
Pressure	1 atm = 2116 psf 1 ft H_2O = 62.43 psf = 0.4335 psi 1 in. Hg = 70.77 psf 1 atm = 101,325 N/m^2 = 14.696 psi = 29.92 in. Hg 1 psi = 6.895 kPa = 27.69 in. H_2O = 6895 N/m^2 1 torr = 1 mm Hg 1 bar = 0.9869 atm = 100 kPa 1 Pa = 1 N/m^2
Velocity	1 m/s = 3.281 ft/s = 2.237 mi/h = 3.60 km/h 1 ft/s = 0.6818 mi/h
Temperature	T (°R) = T (°F) + 459.67 T (K) = T (°C) + 273.15 T (°F) = (9/5)T (°C) + 32 T (°R) = 9/5 T (K) 1 K = 1.8 °R
Dynamic viscosity	1 lb_f-s/ft^2 = 32.174 lb_m/s-ft 1 cP (centipoise) = 0.000672 lb_m/s-ft 1 cP = 2.42 lb_m/h-ft
Kinematic viscosity	1 ft^2/s = 0.0929 m^2/s 1 cSt = 1 mm^2/s = 1.0764×10^{-5} ft^2/s
Volume	1 gal (US) = 0.1337 ft^3 1 ft^3 = 28.32 l = 0.02832 m^3
Heat flux	1 BTU/h-ft^2 = 3.1537 W/m^2 1 W/m^2 = 0.317 BTU/h-ft^2
Specific heat	1 BTU/lb_m-°R = 4.186 kJ/kg- K 1 J/kg-K = 2.389×10^{-4} BTU/lb_m-°R 1 ft-lb_f/lb_m-°R = 5.3804 J/kg-K
Thermal conductivity	1 BTU/h-ft-°F = 1.7303 W/m-°C 1 W/m-°C = 0.578 BTU/h-ft-°F
Heat transfer coefficient	1 BTU/h-ft^2-°F = 5.6783 W/m^2-°C 1 W/m^2-°C = 0.1761 BTU/h-ft^2-°F
Heating value	1 BTU/lb_m = 2.326 kJ/kg = 0.5556 kcal/kg

APPENDIX B STANDARD ATMOSPHERE

Standardized reference pressures and temperatures as functions of altitude were established in 1962 and these are referred to as the standard atmosphere. Information herein is from the US Government Printing Office (1976). For this standard, the temperature decreases linearly to an altitude of approximately 36,000 ft (11,000 m), remains constant until about 66,000 ft (21,000 m), and then increases linearly until 105,000 ft (32,000 m). Over the same ranges, the pressure decreases exponentially with different functions. The graphical behavior is presented in Figures B.1 and B.2. Abbreviated tables of pressures and temperatures are also presented. Conditions at different altitudes and detailed tables can also be found using the software Atmosphere.exe. The standard conditions for reference are: p_{stp} is equal to 14.69 psia or 101.33 kPa and T_{stp} is equal to 518.7 °R or 288.2 K.

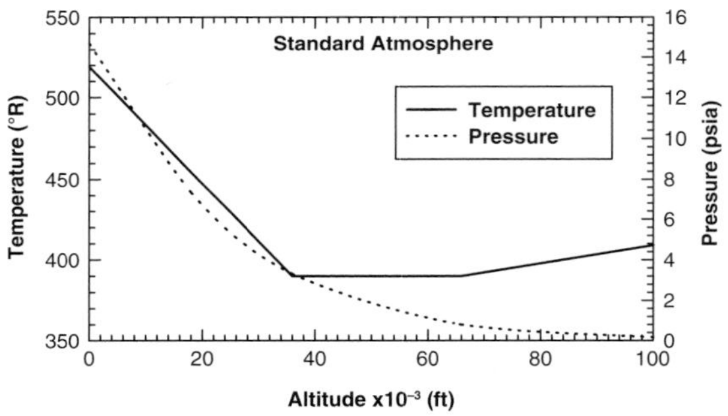

Figure B.1 Standard atmosphere (English units).

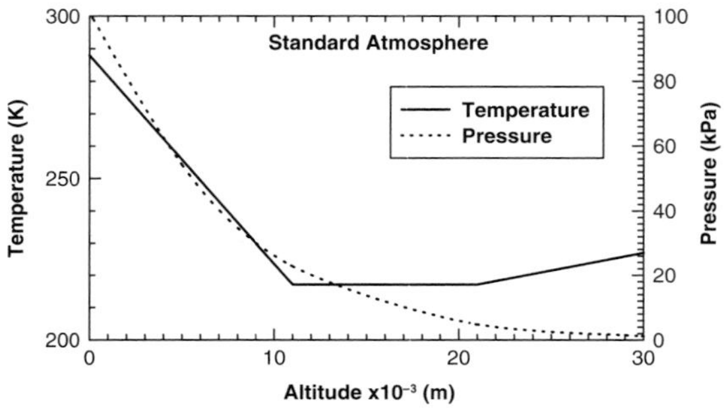

Figure B.2 Standard atmosphere (SI units).

B.1 Abbreviated Tables of Standard Atmosphere

Altitude (ft)	T (°R)	p (psia)	Altitude (m)	T (K)	p (kPa)
0	518.7	14.70	0	288.2	101.33
5000	500.9	12.23	2000	275.2	79.53
10,000	483.1	10.11	4000	262.2	61.70
15,000	465.3	8.301	6000	249.2	47.25
20,000	447.5	6.762	8000	236.3	35.68
25,000	429.7	5.463	10,000	223.3	26.51
30,000	411.9	4.373	12,000	216.7	19.41
35,000	394.1	3.467	14,000	216.7	14.18
40,000	390.0	2.730	16,000	216.7	10.36
45,000	390.0	2.149	18,000	216.7	7.572
50,000	390.0	1.692	20,000	216.7	5.533
55,000	390.0	1.332	22,000	218.6	4.052
60,000	390.0	1.049	24,000	220.6	2.976
65,000	390.0	0.8258	26,000	222.5	2.191
70,000	392.3	0.6509	28,000	224.5	1.618
75,000	395.0	0.5140	30,000	226.5	1.198
80,000	397.7	0.4065			
85,000	400.4	0.3220			
90,000	403.1	0.2555			
95,000	405.9	0.2030			
100,000	408.6	0.1616			

APPENDIX C ONE-DIMENSIONAL COMPRESSIBLE FLOW

C.1 Introduction

The basic concepts of compressible flow are applied in all chapters in this book. Specifically, in Chapters 8 (diffusers and inlets), 10 (nozzles), 9 (combustors), and 11 (ducts, mixers, and exhausts), analyses of fundamental, compressible, steady-state, one-dimensional (1-D) flow are used to predict total pressure losses and other characteristics in the different components. In Chapter 2, fundamentals are briefly reviewed. However, this appendix is presented to summarize different fundamental processes in detail. Included in the processes are the stagnation process, ideal gas properties, isentropic flow with area changes, frictional adiabatic with constant-area (Fanno) flow, flow with heat addition/removal and constant area (Rayleigh), normal and oblique shocks, flow with drag objects, and mixing flows. Also presented is generalized 1-D flow, which allows the combination of two or more different fundamental phenomena. In all of the analyses a control volume approach is used and the flow is assumed to be steady and uniform at the different cross-sectional flow surfaces. Such processes are covered in detail in fundamental gas dynamics texts such as Shapiro (1953), Liepmann and Roshko (1957), Zucrow and Hoffman (1976), and Anderson (1982). The flow is assumed to be 1-D, except for oblique shocks. Although flow can be 2-D or 3-D in components, a 1-D analysis can lead to reasonable results for many cases. For all cases the gas is assumed to be ideal.

C.2 Ideal Gas Equation and Stagnation Properties

First, one can relate the different pressures, temperatures, and densities by the ideal gas equation

$$pv = p/\rho = \mathcal{R}T, \tag{C.2.1}$$

where $\mathcal{R}$ is the ideal gas constant and is equal to 1545 ft-lb$_f$/(lb-mole-°R) (or 8.315 kJ/(kg-mole K)). For air with a molecular weight of 28.97 lb$_m$/lb-mole, one finds that the ideal gas constant is 1545/28.97 or 53.35 ft-lbf/(lb$_m$-°R) (or 0.068559 BTU/lb$_m$-°R) (or 0.2871 kJ/kg-K). Furthermore, one can relate the specific heat at constant pressure to the ideal gas constant by

$$c_p = \frac{\gamma\mathcal{R}}{\gamma - 1}, \tag{C.2.2}$$

where γ is the specific heat ratio

$$\gamma = \frac{c_p}{c_v} \tag{C.2.3}$$

and c_v is the specific heat at constant volume. Also, one can relate the ideal gas constant to the specific heats by

$$\mathcal{R} = c_p - c_v. \tag{C.2.4}$$

Next, a review of stagnation properties is in order because they are used extensively throughout the text. Consider a gas at some point $\mathcal{D}$ flowing with velocity u and with static pressure and temperature p and T, respectively. The flow also has static enthalpy h. This is shown in Figure C.1. At some point $\mathcal{B}$, the flow velocity is reduced to zero ($u = 0$). Applying the energy equation under steady-state and adiabatic conditions yields

$$h_t = h + \frac{1}{2}u^2. \tag{C.2.5}$$

In this equation, h_t is called the total or stagnation enthalpy. If one considers an ideal gas ($dh = c_p\, dT$),

$$c_p T_t = c_p T + \frac{1}{2}u^2. \tag{C.2.6}$$

In this equation, T_t is called the total or stagnation temperature. It is important to note that isentropic flow was not assumed – only adiabatic flow was assumed. Thus, as the flow velocity is reduced, the temperature rises. One also needs to review the definition of some other quantities. For example, the Mach number is defined by

$$M = u/a, \tag{C.2.7}$$

where a is the local speed of sound and for a perfect gas is given by

$$a = \sqrt{\gamma \mathcal{R} T} = \sqrt{\frac{\gamma p}{\rho}}, \tag{C.2.8}$$

and where $\mathcal{R}$ is the ideal gas constant and γ is the ratio of specific heats given above. Thus, by using this with Eq. (C.2.6) one finds

$$T_t = T\left[1 + \frac{\gamma - 1}{2} M^2\right]. \tag{C.2.9}$$

Again, note that this is for an adiabatic process. Next, one can next recall from thermodynamics that, for an isentropic process (usually reversible (frictionless) and adiabatic)

$$\frac{T}{p^{\frac{\gamma-1}{\gamma}}} = \text{constant}, \tag{C.2.10}$$

and when this is applied between any two states 1 and 2,

$$\frac{p_2}{p_1} = \left[\frac{T_2}{T_1}\right]^{\frac{\gamma}{\gamma-1}}. \tag{C.2.11}$$

This can be applied between points $\mathcal{B}$ and $\mathcal{D}$ in the flow field in Figure C.1, and so one obtains

$$\frac{p_t}{p} = \left[\frac{T_t}{T}\right]^{\frac{\gamma}{\gamma-1}}. \tag{C.2.12}$$

Or, using Eq. (C.2.9), one can define the total or stagnation pressure p_t:

$$p_t = p\left[1 + \frac{\gamma - 1}{2} M^2\right]^{\frac{\gamma}{\gamma-1}}. \tag{C.2.13}$$

Recall again that this is for an isentropic process that is more restrictive than the definition for the total temperature (Eq. (C.2.9)), which is only an adiabatic process.

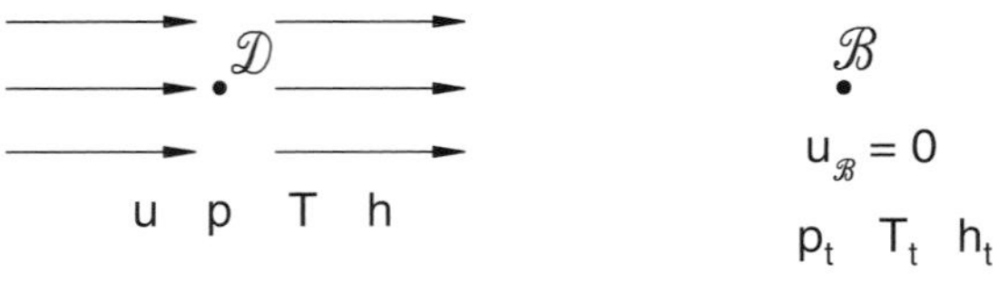

Figure C.1 Definition of stagnation properties.

C.3 Variable Specific Heats

In Table C.1 (from the Department of the US Navy (1953)) values of c_p and γ are presented for air between 360 and 5400 °R (and also 200 and 3000 K) and between 1 and 100 atm. As can be seen, c_p and γ are both very weak functions of pressure for temperatures below 3600 °R (2000 K). However, since c_p and γ both depend on temperature, the gas is considered to be ideal, not perfect. One should note that other such tables are available from other sources and are slightly different. One can find a least-squares curve fit for c_p at one atmosphere, which is suitable for inclusion in computer analyses:

$$c_p = 0.2269807 e^{0.000097247\,T}, \tag{C.3.1}$$

where c_p is in BTU/lb$_m$-°R and T is in °R; γ can then be found from

$$\gamma = \frac{c_p}{c_p - \mathcal{R}}. \tag{C.3.2}$$

Note that this is still an ideal gas approximation in that one assumes $dh = c_p\, dT$. If one truly wanted to eliminate the ideal gas assumption, it would be necessary to use tabular data for the enthalpy as a function of both temperature and pressure as done in the JANAF tables from Keenan et al. (1983) or Chase (1998).

In Chapter 3 and others, the specific heats are assumed to be constant across each component but have different values based on component temperatures. To justify the assumption of constant specific heats across each component, one should consider the evaluation of $h_{ti} - h_{tj}$, where i and j are any two states. This must be considered because large temperature differences can exist across components. In general, an enthalpy change is given by

Table C.1.a **Specific heat at constant pressure, c_p (BTU/lb$_m$-°R).**

T (°R)\ pressure (atm)	1	4	7	10	40	70	100
360	0.2405	–	–	0.2499	–	–	0.4170
492	0.2403	–	–	0.2446	–	–	–
540	0.2405	–	–	0.2439	–	–	0.2776
720	0.2424	–	–	0.2441	–	–	0.2599
900	0.2462	0.2465	0.2469	0.2476	0.2506	–	–
1080	0.2513	0.2515	0.2517	0.2520	0.2542	–	–
1440	0.2626	0.2627	0.2629	0.2630	0.2642	0.2653	–
1800	0.2730	0.2731	0.2732	0.2733	0.2739	0.2746	–
2160	0.2819	0.2820	0.2820	0.2820	0.2825	0.2830	0.2833
2520	0.2902	0.2902	0.2902	0.2903	0.2906	0.2910	0.2912
2880	0.2986	0.2985	0.2985	0.2985	0.2987	0.2989	0.2992
3600	0.3198	0.3173	0.3167	0.3165	0.3158	0.3159	0.3159
5400	0.6835	0.5217	0.4796	0.4613	0.4052	0.3895	0.3843

Table C.1.b **Specific heat ratio, γ (dimensionless).**

T (°R)\ pressure (atm)	1	4	7	10	40	70	100
360	1.3987	–	–	1.3781	–	–	1.1968
492	1.3992	–	–	1.3895	–	–	–
540	1.3987	–	–	1.3910	–	–	1.3280
720	1.3944	–	–	1.3906	–	–	1.3583
900	1.3859	1.3853	1.3844	1.3829	1.3766	–	–
1080	1.3752	1.3748	1.3744	1.3737	1.3693	–	–
1440	1.3533	1.3531	1.3528	1.3526	1.3504	1.3485	–
1800	1.3353	1.3352	1.3350	1.3349	1.3339	1.3327	–
2160	1.3214	1.3212	1.3212	1.3212	1.3205	1.3197	1.3193
2520	1.3093	1.3093	1.3093	1.3092	1.3088	1.3082	1.3079
2880	1.2980	1.2982	1.2982	1.2982	1.2979	1.2976	1.2973
3600	1.2729	1.2756	1.2763	1.2765	1.2773	1.2772	1.2772
5400	1.1115	1.1513	1.1668	1.1746	1.2037	1.2136	1.2171

Adapted from *Handbook of Supersonic Aerodynamics*, Vol. 5, Bureau of Ordnance, Department of the US Navy, Washington, DC, 1953.

Table C.1.c **Specific heat at constant pressure, c_p (kJ/kg-K).**

T (K)\ pressure (atm)	1	4	7	10	40	70	100
200	1.007	–	–	1.046	–	–	1.746
273	1.006	–	–	1.024	–	–	–
300	1.007	–	–	1.021	–	–	1.162
400	1.015	–	–	1.022	–	–	1.088
500	1.031	1.032	1.034	1.036	1.049	–	–
600	1.052	1.053	1.054	1.055	1.064	–	–
800	1.099	1.100	1.100	1.101	1.106	1.111	–
1000	1.143	1.143	1.144	1.144	1.147	1.149	–
1200	1.180	1.180	1.180	1.180	1.183	1.185	1.186
1400	1.215	1.215	1.215	1.215	1.216	1.218	1.219
1600	1.250	1.250	1.250	1.250	1.250	1.251	1.252
2000	1.339	1.328	1.326	1.325	1.322	1.322	1.322
3000	2.861	2.184	2.008	1.931	1.696	1.630	1.609

$$\Delta h = \int_{T_j}^{T_i} c_p dT. \tag{C.3.3}$$

One should determine if the best approximation of Δh is $\bar{c}_p \Delta T$, where $\bar{c}_p$ is evaluated at the average temperature ($T = (T_i + T_j)/2$) or if the best approximation is $c_{pi} T_i - c_{pj} T_j$, where the specific heats are evaluated at the end states. As an approximation to determine which method is best, one can model the dependence of c_p on T, and thus

$$c_p = C + BT. \tag{C.3.4}$$

Therefore,

$$\bar{c}_p = C + B\overline{T}. \tag{C.3.5}$$

Now, substituting Eq. (C.3.4) into (C.3.3) and evaluating yield

$$\Delta h = C(T_i - T_j) + B\left[\frac{T_i^2 - T_j^2}{2}\right], \tag{C.3.6}$$

but this can be factored to become

$$\Delta h = \left[C + B\left(\frac{T_i + T_j}{2}\right)\right](T_i - T_j). \tag{C.3.7}$$

By comparing this to Eq. (C.3.5), one sees that

$$\Delta h = \bar{c}_p \Delta T. \tag{C.3.8}$$

Thus, the conclusion is that one should use the value of c_p evaluated at the average temperature and the temperature difference to calculate the change in enthalpy.

To further demonstrate the point, one can consider a numerical example of air between the temperatures of 1000 and 2000 °R. From tables for air, one finds that at these two temperatures the enthalpies are 241.0 and 504.7 BTU/lb$_m$, respectively. Also at these temperatures, the values of c_p are 0.249 and 0.277 BTU/lb$_m$-°R, respectively. The value of c_p at the average temperature (1500 °R) is 0.264 BTU/lbm-°R. Thus, the actual change in enthalpy is 504.7 − 241.0 = 263.7 BTU/lb$_m$. By comparison, $c_{pi} T_i - c_{pj} T_j$ = 0.277 BTU/lb$_m$-°R × 2000 °R

Table C.1.d **Specific heat ratio, γ (dimensionless).**

T (K)\ pressure (atm)	1	4	7	10	40	70	100
200	1.3987	–	–	1.3781	–	–	1.1968
273	1.3992	–	–	1.3895	–	–	–
300	1.3987	–	–	1.3910	–	–	1.3280
400	1.3944	–	–	1.3906	–	–	1.3583
500	1.3859	1.3853	1.3844	1.3829	1.3766	–	–
600	1.3752	1.3748	1.3744	1.3737	1.3693	–	–
800	1.3533	1.3531	1.3528	1.3526	1.3504	1.3485	–
1000	1.3353	1.3352	1.3350	1.3349	1.3339	1.3327	–
1200	1.3214	1.3212	1.3212	1.3212	1.3205	1.3197	1.3193
1400	1.3093	.3093	1.3093	1.3092	1.3088	1.3082	1.3079
1600	1.2980	1.2982	1.2982	1.2982	1.2979	1.2976	1.2973
2000	1.2729	1.2756	1.2763	1.2765	1.2773	1.2772	1.2772
3000	1.1115	1.1513	1.1668	1.1746	1.2037	1.2136	1.2171

– 0.249 BTU/lb$_m$-°R × 1000 °R = 305 BTU/lb$_m$. When using the value of c_p at the average temperature, one finds $c_p\,(T_i - T_j) = 0.264$ BTU/lb$_m$-°R × (1000 °R) = 264 BTU/lb$_m$. Clearly, using the value of c_p at the average temperature to evaluate Δh is best.

C.4 Isentropic Flow with Area Change

The cross-sectional area can change axially in several components, and ideally the flow is without friction or heat transfer. The nozzle, exhaust, bell inlet and diffuser are the most obvious components in which the area changes. As the area changes, so do the Mach number, velocity, and other properties. In this section the dependence of the variables is discussed. The diagram of the model is shown in Figure C.2, including the control volume. Uniform flow enters with characteristics and properties at 1 (for example, p_1, T_1, V_1, M_1, etc.) and leaves uniformly with all properties at 2 (for example, p_2, T_2, V_2, M_2, etc.). The h–s diagram is not presented; since the process is isentropic, it is a simple vertical straight line. For the isentropic process line of the h–s diagram, the continuity and energy equations are used.

The flow is analyzed with a control volume approach for an ideal gas, and the first equation to be used is the energy equation, which for steady-state adiabatic conditions yields

$$c_p T_1 + \frac{1}{2} V_1{}^2 = c_p T_2 + \frac{1}{2} V_2{}^2, \tag{C.4.1}$$

where T is the static temperature and V is the gas velocity. Next, the continuity equation results in

$$\rho_1 V_1 A_1 = \rho_2 V_2 A_2. \tag{C.4.2}$$

Now, the second law of thermodynamics yields a process equation for isentropic flow ($s_1 = s_2$); namely,

$$\frac{p}{\rho^\gamma} = \text{constant}.$$

Or, using this process between states 1 and 2 yields

$$\frac{p_1}{\rho_1^\gamma} = \frac{p_2}{\rho_2^\gamma}. \tag{C.4.3}$$

Also, from the definition of stagnation temperature, which is an adiabatic process,

$$T_{t1} = T_1\left[1 + \frac{\gamma - 1}{2} M_1^2\right] \tag{C.4.4}$$

$$T_{t2} = T_2\left[1 + \frac{\gamma - 1}{2} M_2^2\right], \tag{C.4.5}$$

and from the definition of Mach number,

$$M_1 = \frac{V_1}{a_1} = \frac{V_1}{\sqrt{\gamma \mathcal{R} T_1}} \tag{C.4.6}$$

$$M_2 = \frac{V_2}{a_2} = \frac{V_2}{\sqrt{\gamma \mathcal{R} T_2}}, \tag{C.4.7}$$

where a is the sound speed. For an ideal gas,

$$p_1 = \rho_1 \mathcal{R} T_1 \tag{C.4.8}$$

$$p_2 = \rho_2 \mathcal{R} T_2, \tag{C.4.9}$$

and from the definition of stagnation pressure,

$$p_{t1} = p_1\left[1 + \frac{\gamma - 1}{2} M_1^2\right]^{\frac{\gamma}{\gamma-1}} \tag{C.4.10}$$

$$p_{t2} = p_2\left[1 + \frac{\gamma - 1}{2} M_2^2\right]^{\frac{\gamma}{\gamma-1}}. \tag{C.4.11}$$

Thus, one can utilize Eqs. (C.4.1)–(C.4.11) and therefore have 11 equations. Sixteen variables are present, and so five must be specified. For example, if M_1, T_1, p_1, A_1, and A_2 are specified, one can find M_2. Thus, these equations define the process on the h–s diagram. For subsonic internal flow (the usual case for a diffuser), the Mach number decreases as the area increases. For supersonic flow, however, the Mach number increases owing to an increase in area – for example, in the diverging section of a nozzle.

To aid in the solution of these equations, many gas dynamics and fluid mechanics texts present tables of M, T/T^*, p_t/p^*, A/A^*, and other variables vs. Mach number, where the superscript * indicates reference conditions at

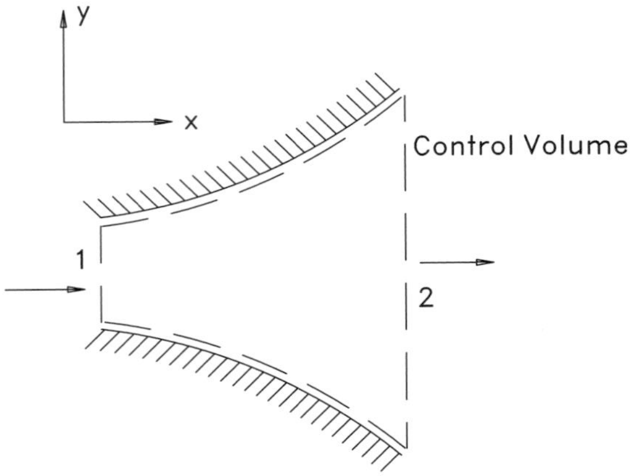

Figure C.2 Geometry for isentropic flow with area changes.

the sonic condition. Abbreviated tables are presented in this text in Appendix D for two values of γ. Solutions and expanded tables can also be obtained using the software Isentropic.exe. Readers are encouraged to use this software because it covers other values of γ and it is easier and more accurate than interpolating tables. An example problem is presented in Section 2.2.4.

C.5 Fanno Line Flow

Boundary layers and other viscous flows are the primary means by which total pressure losses occur within diffusers, bell inlets, exhausts, and bypass ducts. Such flow is also important in burners and afterburners owing to viscous flow through small orifices or holes in a burner wall. This flow will be analyzed based on Fanno line flow, which is flow with friction but no heat addition or loss. The Fanno line process is a constant-area process, which does not necessarily represent a diffuser, duct, or burner exactly. However, for a diffuser, for example, when the exit to inlet area ratio is near unity so that the flow does not separate, the mechanism can be used to predict the total pressure losses reasonably. The diagram of the model is shown in Figure C.3 with the control volume, and the h–s diagram is shown in Figure C.4. For the Fanno process line of the h–s diagram, the continuity and energy equations are used. Many of the equations from Section C.4 can be used because this process is at constant area ($A_1 = A_2$) and also adiabatic ($T_{t1} = T_{t2}$).

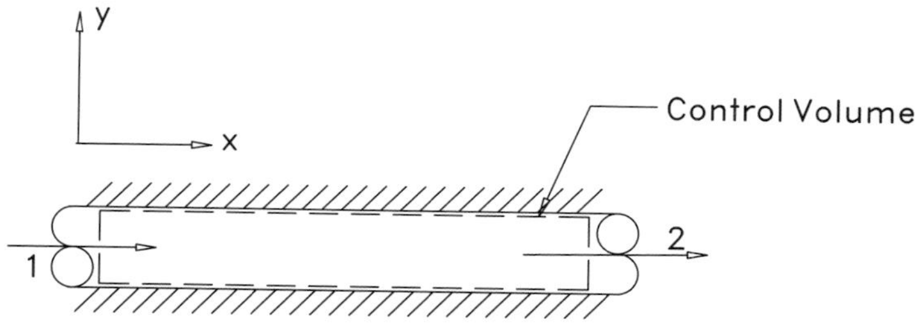

Figure C.3 Geometry for frictional and adiabatic flow (Fanno line).

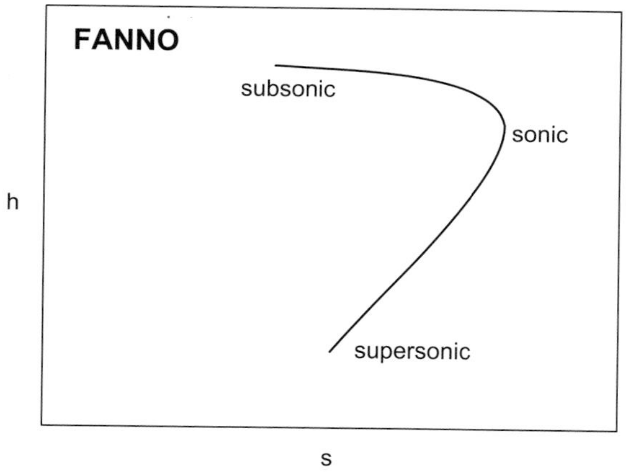

Figure C.4 The h–s diagram for Fanno line flow.

For example, the continuity equation above yields for $A_1 = A_2$:

$$\rho_1 V_1 = \rho_2 V_2. \tag{C.5.1}$$

Thus, one can use Eqs. (C.4.1), (C.4.4)–(C.4.11), and (C.5.1), and have 10 equations. Fourteen variables are present, and so four must be specified. For example, if M_1, T_1, T_2, and p_{t1} are specified, one can find p_{t2}. However, a method of specifying the friction within the flow is not possible with these equations, and in general T_2 may not be known. Thus, the eleventh equation is included, which is obtained from the momentum equation:

$$\int 4 f\rho \frac{1}{2} V^2 \frac{dx}{D} = p_2 - p_1 + \rho_1 V_1 (V_2 - V_1), \tag{C.5.2}$$

where the left-hand side is integrated along the channel, f is the Fanning friction factor, and D is the effective diameter of the flow path. The friction factor is a function of the Reynolds number and wall roughness, and Moody (1944) presents the Darcy friction factor (often referred to as simply the friction factor) in the well-recognized “Moody diagram,” which is included in most introductory fluid mechanics textbooks. For fully developed flow, the Darcy factor is equal to four times the Fanning factor.

These equations generate the h–s diagram. For subsonic internal flow (the usual case for a diffuser), the Mach number increases owing to friction as the gas flows along a channel. However, note that the flow area also increases, and thus the net result will be a decreasing Mach number. For supersonic flow, however, the Mach number decreases owing to friction. Thus, a total of 11 equations were derived above, and 17 variables are present. Thus, six variables must be specified for a solution. For example, if M_1, T_1, L, D, f, and p_{t1} are specified, one can find p_{t2}.

To aid in the solution of these equations, many gas dynamics and fluid mechanics texts present tables of T/T^*, p_t/p^*, $4fL^*/D$ and other variables vs. Mach number, where once again the superscript * indicates conditions at the sonic condition. The quantity $4fL^*/D$ is an integral effect derived from the first term in Eq. (C.5.2). Abbreviated tables are presented in this text in Appendix E for two values of γ. Solutions for particular conditions and expanded tables can also be obtained using the software Fanno.exe. Readers are encouraged to use this software because it covers other values of γ and it is easier and more accurate than interpolating tables. An example problem is presented in Section 2.2.5.

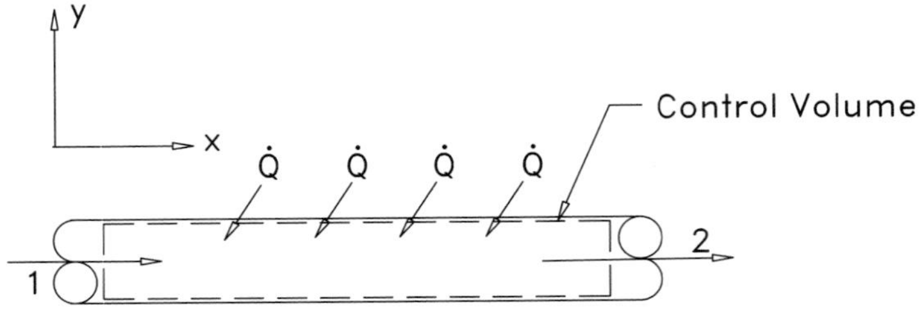

Figure C.5 Geometry for frictionless flow with heat transfer (Rayleigh line).

C.6 Rayleigh Line Flow

As discussed in Chapter 9, several mechanisms exist for pressure drops and total pressure losses to occur in burners. In this section, the loss due to heat addition/removal to the flow is analyzed. This will be analyzed based on Rayleigh line flow, which is flow with heat addition (or removal) but no friction. The process is a constant area process. The diagram of the control volume system is shown in Figure C.5, and the h–s diagram is shown in Figure C.6. For the Rayleigh process line of the h–s diagram, the continuity and momentum equations are used.

The flow is analyzed with a control volume approach for an ideal gas, and the first equation to be used is the momentum equation, which yields

$$p_1 A - p_2 A = V_1(-\rho_1 V_1 A) + V_2(\rho_2 V_2 A). \qquad \text{(C.6.1)}$$

Thus, using the continuity equation (Eq. (C.5.1)) yields

$$p_1 - p_2 = \rho_1 V_1 (V_2 - V_1). \qquad \text{(C.6.2)}$$

Thus, 10 independent equations are listed above (Eqs. (C.4.4)–(C.4.11), (C.5.1), and (C.6.2)). Fourteen variables are in these equations. Therefore, if one specifies four variables, it is possible to solve for the remaining 10. Typically, M_1, p_{t1}, T_{t1}, and T_{t2} are known. Note that the energy equation was not used, although if T_{t1} and T_{t2} are known the energy equation gives the total heat tranfer to the flow.

These equations generate the h–s diagram. For subsonic flow, as heat is added to the flow (the total temperature T_t increases) the exit Mach number increases. For supersonic flow, however, as heat is added, the Mach number decreases. One interesting phenomenon occurs as heat is added for the subsonic case. The static temperature T increases until the Mach number reaches $1/\sqrt{\gamma}$, at which point the temperature decreases. As noted above, the energy equation was not utilized. However, for the case of a burner, it can be used (and usually is) to relate the change in total temperature to fuel flow rate.

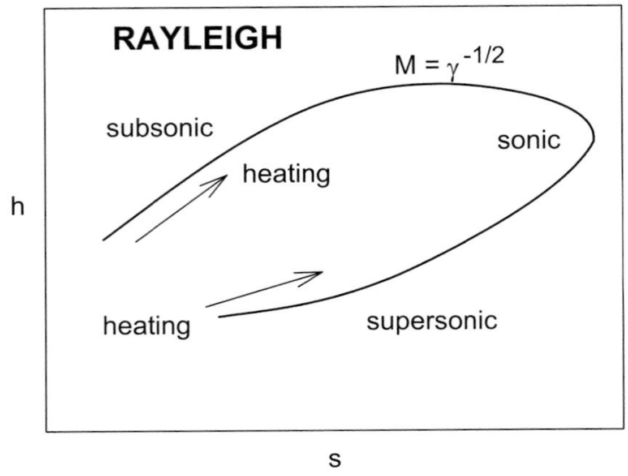

Figure C.6 The h–s diagram for Rayleigh line flow.

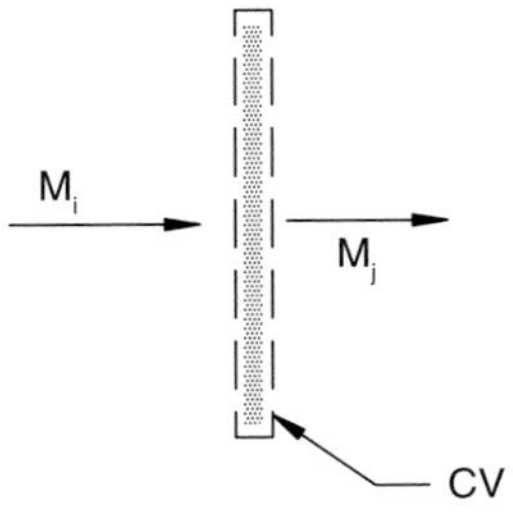

Figure C.7 Geometry for normal shock.

To aid in the solution of these equations, many gas dynamics and fluid mechanics texts present tables of T/T^*, p_t/p^*, and other variables versus Mach number, where* indicates reference conditions at the sonic point. Abbreviated tables are presented in this text in Appendix F for two values of γ. Solutions for particular conditions and expanded tables can also be obtained using the software Rayleigh.exe. Readers are encouraged to use this software because it covers other values of γ and it is easier and more accurate than interpolating tables. An example is presented in Section 2.2.6.

C.7 Normal Shocks

Shocks are the primary means by which total pressure losses occur outside of an inlet or sometimes inside a nozzle. Normal shocks are a 1-D flow and lead to large total pressure losses. This flow will be analyzed based on flow with no friction and no heat addition or loss. The normal shock process is also a constant-area process. The diagram of the model is shown in Figure C.7 with the control volume, and the *h–s* diagram is shown in Figure C.8. For the normal shock process, the *h–s* diagram and the continuity, linear momentum, and energy equations are used. Thus, the process is in fact a superposition of Fanno line and Rayleigh line processes as shown. Many of the equations from the previous sections can be used.

By assembling Eqs. (C.4.1), (C.4.4)–(C.4.11), (C.5.1), and (C.6.2), one has 11 equations and 14 variables. Thus, if three variables are specified, the remaining 11 can be found. Typically, for a normal shock in jet engine inlet applications, the incoming Mach number, total pressure,

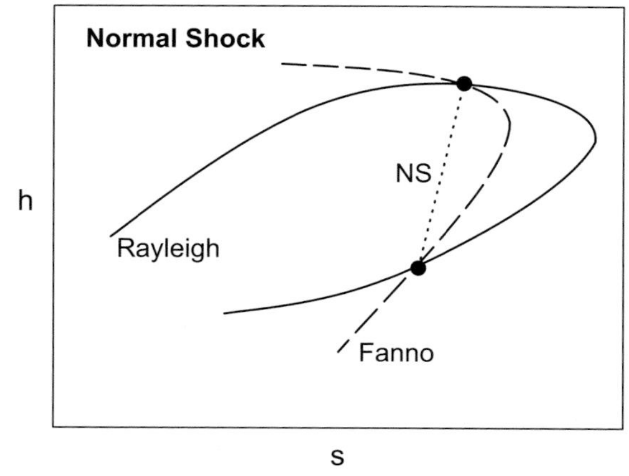

Figure C.8 The *h–s* diagram for normal shock.

and total temperature are known. Thus, the exiting Mach number and total conditions can be found.

To aid in the solution of these equations, many gas dynamics and fluid mechanics texts present tables of M_2, p_{t2}/p_{t1}, p_2/p_1 and other variables as functions of M_1. Abbreviated tables are presented in this text in Appendix G for two values of γ. Solutions for particular conditions and expanded tables can also be obtained using the software NormalShock.exe. The software Shock.exe can also be used for individual calculations. Readers are encouraged to use this software because it covers other values of γ and it is easier and more accurate than interpolating figures or tables. An example problem is presented in Section 2.2.7.

Furthermore, the equations can be combined to form six useful closed-form expressions:

$$M_2^2 = \frac{M_1^2 + \dfrac{2}{\gamma - 1}}{\dfrac{2\gamma}{\gamma - 1} M_1^2 - 1} \tag{C.7.1}$$

$$\frac{p_2}{p_1} = \frac{1 + \gamma M_1^2}{1 + \gamma M_2^2} \tag{C.7.2}$$

$$\frac{p_2}{p_1} = \frac{2\gamma}{\gamma + 1} M_1^2 - \frac{\gamma - 1}{\gamma + 1} \tag{C.7.3}$$

$$\frac{p_{t2}}{p_{t1}} = \frac{\left[\dfrac{\dfrac{\gamma + 1}{2} M_1^2}{1 + \dfrac{\gamma - 1}{2} M_1^2}\right]^{\frac{\gamma}{\gamma - 1}}}{\left[\dfrac{2\gamma}{\gamma - 1} M_1^2 - \dfrac{\gamma - 1}{\gamma + 1}\right]^{\frac{1}{\gamma - 1}}} \tag{C.7.4}$$

$$T_{t2} = T_{t1} \tag{C.7.5}$$

$$\frac{T_2}{T_1} = \frac{1 + \dfrac{\gamma - 1}{2} M_1^2}{1 + \dfrac{\gamma - 1}{2} M_2^2}. \tag{C.7.6}$$

C.8 Oblique Planar Shocks

As discussed in Chapter 8, oblique shocks incur lower losses in total pressure than do normal shocks. In this section, planar 2-D shocks are analyzed. Although the

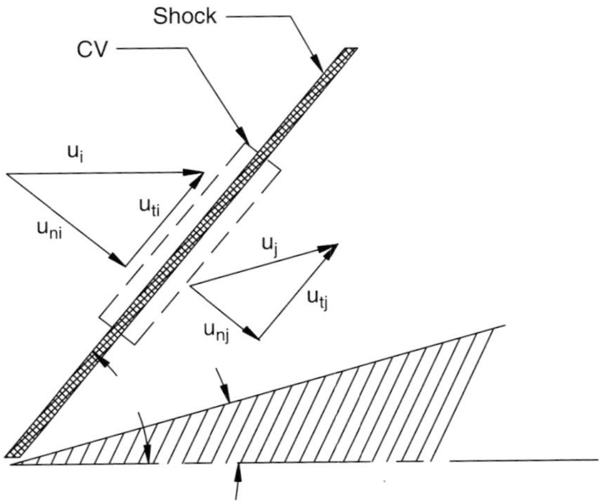

Figure C.9 Oblique shock with control volume.

flow is truly 2-D, it has many characteristics of 1-D flow. Flow before the shock is uniformly parallel to the axis and, behind the planar shock, flow is uniformly parallel to the wedge.

Figure C.9 shows an oblique shock. A wedge with an angle of δ forms the shock. A control volume is drawn around the shock as shown. Also, the velocity vectors are broken down into components normal to, and parallel to, the shock, as shown. First, from continuity,

$$\rho_1 u_{n1} = \rho_2 u_{n2}. \tag{C.8.1}$$

Next, from the momentum equation parallel to the shock,

$$(\rho_1 u_{n1}) u_{t1} = (\rho_2 u_{n2}) u_{t2}. \tag{C.8.2}$$

Thus, from Eq. (C.8.1), very simply,

$$u_{t1} = u_{t2}. \tag{C.8.3}$$

Next, from the momentum equation normal to the shock,

$$p_2 - p_1 = \rho_2 u_{n1}^2 - \rho_1 u_{n1}^2 \tag{C.8.4}$$

and, finally, from the energy equation,

$$h_1 - h_2 = \frac{1}{2}\left[u_2^2 - u_1^2\right]. \tag{C.8.5}$$

Thus, for an ideal gas,

$$c_p(T_1 - T_2) = \frac{1}{2}\left[u_2^2 - u_1^2\right]. \tag{C.8.6}$$

Next, one should realize from the Pythagorean theorem that

$$u_2^2 - u_1^2 = (u_{n2}^2 + u_{t2}^2) - (u_{n1}^2 + u_{t1}^2). \tag{C.8.7}$$

Thus, from Eq. (C.8.3),

$$u_2^2 - u_1^2 = u_{n2}^2 - u_{n1}^2. \tag{C.8.8}$$

Equation (C.8.6), through the use of Eqs. (C.2.1), (C.2.2), and (C.8.8), becomes

$$\frac{\gamma}{\gamma - 1}\left(\frac{p_2}{\rho_2} - \frac{p_1}{\rho_1}\right) = \frac{1}{2}\left[u_{n1}^2 - u_{n2}^2\right]. \tag{C.8.9}$$

Next, using Eqs. (C.8.1) and (C.8.4), one finds

$$p_2 - p_1 = \rho_1 u_{n1}^2 \left[1 - \frac{\rho_1}{\rho_2}\right] . \tag{C.8.10}$$

or

$$p_1 - p_2 = \rho_2 u_{n2}^2 \left[1 - \frac{\rho_2}{\rho_1}\right]; \tag{C.8.11}$$

thus

$$u_{n1}^2 = \left[\frac{p_2 - p_1}{\rho_2 - \rho_1}\right]\frac{\rho_2}{\rho_2}. \tag{C.8.12}$$

Similarly, one can use Eqs. (C.8.1) and (C.8.4) to yield

$$u_{n2}^2 = \left[\frac{p_2 - p_1}{\rho_2 - \rho_1}\right]\frac{\rho_1}{\rho_2}. \tag{C.8.13}$$

Next, by using Eqs. (C.8.12), (C.8.13), and (C.8.9), one finds

$$\frac{p_2}{p_1} = \frac{\left[\frac{\gamma + 1}{\gamma - 1}\right]\frac{\rho_2}{\rho_1} - 1}{\left[\frac{\gamma + 1}{\gamma - 1}\right] - \frac{\rho_2}{\rho_1}} \tag{C.8.14}$$

or

$$\frac{\rho_2}{\rho_1} = \frac{\left[\frac{\gamma + 1}{\gamma - 1}\right]\frac{p_2}{p_1} + 1}{\left[\frac{\gamma + 1}{\gamma - 1}\right] + \frac{p_2}{p_1}}. \tag{C.8.15}$$

It is noteworthy that Eqs. (C.8.14) and (C.8.15) are named the Rankine–Hugoniot equations and relate pressure and density ratios across shocks. Next, by examining the geometry of the shock in Figure C.9 one finds that

$$u_{t1} = u_1 \cos\sigma \tag{C.8.16}$$

$$u_{n1} = u_1 \sin\sigma \tag{C.8.17}$$

$$u_{t2} = u_2 \cos(\sigma - \delta) \tag{C.8.18}$$

$$u_{n2} = u_2 \sin(\sigma - \delta), \tag{C.8.19}$$

where σ is the shock angle relative to the incoming flow and δ is the flow turning angle relative to the incoming flow. Next, from Eqs. (C.8.3), (C.8.16), and (C.8.18), one finds

$$\frac{u_1}{u_2} = \frac{\cos(\sigma - \delta)}{\cos\sigma}, \tag{C.8.20}$$

and with Eqs. (C.8.1), (C.8.17), and (C.8.19),

$$\rho_1 u_1 \sin\sigma = \rho_2 u_2 \sin(\sigma - \delta). \tag{C.8.21}$$

Thus,

$$\frac{\rho_2}{\rho_1} = \frac{u_1}{u_2}\frac{\sin\sigma}{\sin(\sigma - \delta)}, \tag{C.8.22}$$

and, using Eq. (C.8.20), one sees that

$$\frac{\rho_2}{\rho_1} = \frac{\tan\sigma}{\tan(\sigma - \delta)}. \tag{C.8.23}$$

Next, using Eqs. (C.8.10) and (C.8.17), one obtains

$$p_2 - p_1 = \rho_1 u_1^2 \sin^2\sigma \left[1 - \frac{\rho_1}{\rho_2}\right]. \tag{C.8.24}$$

Now, since for an ideal gas

$$M = \frac{u}{a} = \frac{u}{\sqrt{\frac{\gamma p}{\rho}}},$$

one finds

$$\frac{p_2}{p_1} = 1 + \gamma M_1^2 \sin^2\left[1 - \frac{\rho_1}{\rho_2}\right]. \tag{C.8.25}$$

Similarly, using Eqs. (C.8.11) and (C.8.17),

$$\frac{p_1}{p_2} = 1 + \gamma M_2^2 \sin^2(\sigma - \delta)\left[1 - \frac{\rho_2}{\rho_1}\right], \tag{C.8.26}$$

and, finally, using Eq. (C.2.12) yields

$$\frac{p_{t2}}{p_{t1}} = \frac{p_2}{p_1}\left[\frac{1 + \frac{\gamma - 1}{2}M_2^2}{1 + \frac{\gamma - 1}{2}M_1^2}\right]^{\frac{\gamma}{\gamma - 1}}. \tag{C.8.27}$$

Thus, if one considers Eqs. (C.8.14), (C.8.23), (C.8.25), (C.8.26), and (C.8.27), there are seven independent variables: σ, δ, M_1, M_2, p_2/p_1, p_{t2}/p_{t1}, and ρ_2/ρ_1. Therefore, if two are specified, the remaining five can be determined. For example, for a given incoming Mach number (M_1) and a given turning angle (δ), one can find the shock angle, exit Mach number, total pressure ratio, and all of the other variables. These equations can easily be programmed in a commercial math solver. A set of figures are presented herein for $\gamma = 1.40$ to aid in the theoretical solution and are presented in Figure 8.6. Solutions can also be obtained using the software Shock.exe. Readers are encouraged to use this software because it covers other values of γ, it is easier than using the above equations and it is more accurate than using the referenced graphs. In Section 2.2.8 an example is presented.

C.9 Flow with a Drag Object

When an obstruction is in the flow, the total pressure is also reduced because of the friction from the added turbulence and separation. Many primary burners and most afterburners have such obstructions in the form of flameholders to slow the fluid velocity; the flame speed is higher then the flow velocity, thus preventing flameout. Even when an afterburner is not lit, a loss is encountered because of the obstruction. Mixers include such objects to enhance the folding and mixing of the two streams to ideally produce a uniform exit flow. Thus, a method of including such losses is needed.

In Figure C.10, a general flow with an obstruction is shown with the control volume. For this process, the continuity, momentum, and energy equations are used. If a control volume approach is used for a constant area and adiabatic flow, and if uniform properties at the entrance and exit are assumed, one finds from the momentum equation that

$$p_1 A - p_2 A - F_d = \rho_2 V_2^2 A - \rho_1 V_1^2 A, \tag{C.9.1}$$

where F_d is the drag on the object and A is the flow passage cross-sectional area. The drag coefficient is defined as

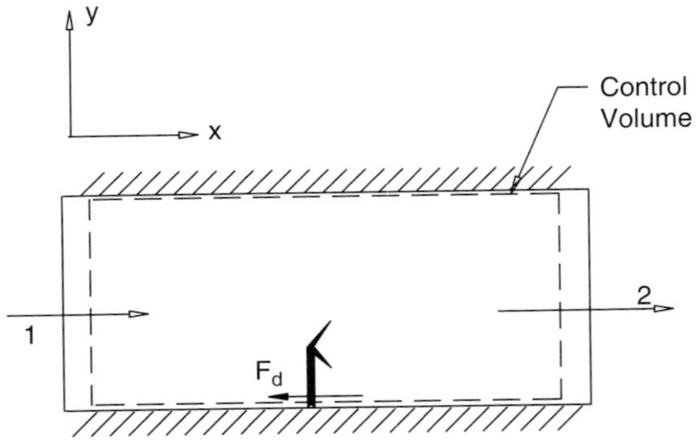

Figure C.10 Geometry for flow with a drag obstacle.

$$C_d \equiv \frac{F_d}{\frac{1}{2}\rho_1 V_1^2 A_d}, \tag{C.9.2}$$

where A_d is the frontal area of the object. Or, solving for the force yields

$$F_d = \frac{1}{2} C_d \rho_1 V_1^2 A \frac{A_d}{A}. \tag{C.9.3}$$

Thus, using Eq. (C.9.1), one obtains

$$p_1 + \rho_1 V_1^2 - \frac{1}{2} C_d \rho_1 V_1^2 A \frac{A_d}{A} = p_2 + \rho_2 V_2^2. \tag{C.9.4}$$

Analyzing this equation with Eqs. (C.4.1), (C.4.4)–(C.4.11), and (C.5.1), discloses 16 variables and 11 equations. Thus, if five variables are specified, the problem can be solved. Typically, A_d/A, C_d, M_1, p_{t1}, and T_{t1} will be known for a flameholder or mixer.

For example, from Eqs. (C.9.4) and (C.4.9),

$$p_1 + \rho_1 V_1^2 - \frac{1}{2} C_d \rho_1 V_1^2 \frac{A_d}{A} = \rho_2 \mathcal{R} T_2 \frac{V_2}{V_2} + \rho_2 V_2^2, \tag{C.9.5}$$

or using Eq. (C.5.1) results in

$$p_1 + \rho_1 V_1^2 - \frac{1}{2} C_d \rho_1 V_1^2 \frac{A_d}{A} = \rho_1 V_1 \left[\frac{\mathcal{R} T_2}{V_2} + V_2 \right]. \tag{C.9.6}$$

Next, from the adiabatic energy equation between the stagnation and static states (t2 and 2) at station 2, the gas velocity is

$$V_2 = \sqrt{2c_p (T_{t2} - T_2)}. \tag{C.9.7}$$

Therefore, from Eqs. (C.9.6), (C.9.7), and (C.4.1), and realizing that $T_{t2} = T_{t1}$ (the process is adiabatic), one obtains

$$p_1 + \rho_1 V_1^2 - \frac{1}{2} C_d \rho_1 V_1^2 \frac{A_d}{A} = \rho_1 V_1 \left[\frac{\mathcal{R} T_2}{\sqrt{2c_p (T_{t1} - T_2)}} + \sqrt{2c_p (T_{t1} - T_2)} \right]. \tag{C.9.8}$$

Thus, if the inlet conditions are known (p_1, M_1, T_{t1}, etc.), Eq. (C.9.8) can be used to solve for T_2 iteratively, after which all other exit conditions can be found, including M_2, p_{t2}, etc. Solutions can also be obtained using a "generalized 1-D flow" approach as discussed in Section C.13. The software General1D.exe uses this approach. Readers are encouraged to use this software because it is easier than iteratively solving the above equations. An example of this is presented in the combustor chapter (Section 9.5.3).

C.10 Flow with a Flameholder (Drag Object) and Heat Transfer

When an obstruction is in the flow, the total pressure is reduced because of the friction from the added turbulence and separation. When heat is added at the same time the total pressure is further reduced due to the irreversible process. Many primary burners and most afterburners have such obstructions in the form of flameholders to slow the fluid velocity and the combustion process continues within the flameholder region. Thus, a method of including such losses is needed.

In Figure C.11, a general flow with an obstruction is shown with the control volume. For this process, the continuity and momentum equations are used. This is similar to the process in Section C.9, but the energy equation is not used. If a control volume approach is used for a constant area, and if uniform properties at the entrance and exit are assumed, one finds from the momentum equation that

$$p_1 A - p_2 A - F_d = \rho_2 V_2^2 A - \rho_1 V_1^2 A, \tag{C.10.1}$$

where F_d is the drag on the object and A is the flow passage cross-sectional area. The drag coefficient is defined as

$$C_d \equiv \frac{F_d}{\frac{1}{2}\rho_1 V_1^2 A_d}, \tag{C.10.2}$$

where A_d is the frontal area of the object. Or, solving for the force yields

$$F_d = \frac{1}{2} C_d \rho_1 V_1^2 A \frac{A_d}{A}. \tag{C.10.3}$$

Thus, using Eq. (C.10.1), one obtains

$$p_1 + \rho_1 V_1^2 - \frac{1}{2} C_d \rho_1 V_1^2 \frac{A_d}{A} = p_2 + \rho_2 V_2^2. \tag{C.10.4}$$

Analyzing this equation with Eqs. (C.4.4)–(C.4.11), and C.5.1, discloses 16 variables and 10 equations. Thus, if six variables are specified, the problem can be solved. Typically, $\frac{A_d}{A}$, C_d, M_1, p_{t1}, T_{t1}, and T_{t2} will be known for a lit flameholder.

For example, from Eqs. (C.10.4) and (C.4.9),

$$p_1 + \rho_1 V_1^2 - \frac{1}{2} C_d \rho_1 V_1^2 \frac{A_d}{A} = \rho_2 \mathcal{R} T_2 \frac{V_2}{V_2} + \rho_2 V_2^2, \tag{C.10.5}$$

or using Eq. (C.5.1) results in

$$p_1 + \rho_1 V_1^2 - \frac{1}{2} C_d \rho_1 V_1^2 \frac{A_d}{A} = \rho_1 V_1 \left[\frac{\mathcal{R} T_2}{V_2} + V_2\right]. \tag{C.10.6}$$

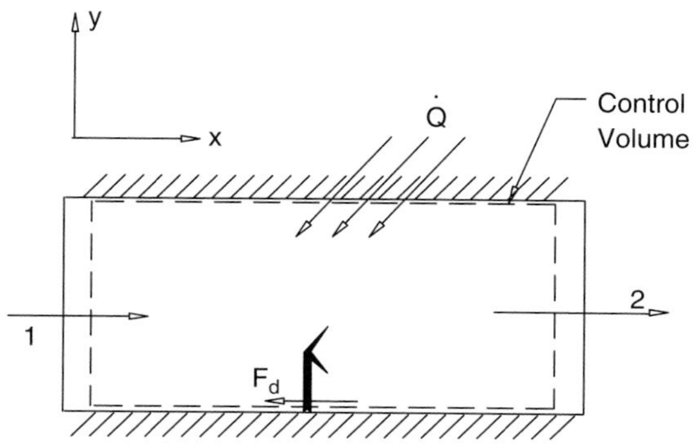

Figure C.11 Geometry for flow with a drag obstacle and heat transfer.

Next, from the adiabatic energy equation between the stagnation and static states (t1 and 1) at station 1, the gas velocity is

$$V_1 = \sqrt{2c_p(T_{t1} - T_1)}; \tag{C.10.7a}$$

and from the adiabatic energy equation between the stagnation and static states (t2 and 2) at station 2, the gas velocity is

$$V_2 = \sqrt{2c_p(T_{t2} - T_2)}. \tag{C.10.7b}$$

Therefore, from Eqs. (C.10.6), and (C.10.7a,b), one obtains

$$p_1 + \rho_1 V_1^2 - \frac{1}{2} C_d \rho_1 V_1^2 \frac{A_d}{A} = \rho_1 V_1 \left[\frac{\mathcal{R} T_2}{\sqrt{2c_p(T_{t2} - T_2)}} + \sqrt{2c_p(T_{t2} - T_2)}\right]. \tag{C.10.8}$$

Thus, if the inlet conditions are known (p_1, M_1, T_{t1}, etc), Eq. (C.10.8) can be used to solve for T_2 iteratively, after which all other exit conditions (M_2, p_{t2}, etc.) can be found. Again, solutions can also be obtained using the generalized 1-D flow method and software General1D.exe, as discussed in a following section. Readers are encouraged to use this software because it is easier than iteratively solving the above equations.

C.11 Mixing Process

Another process that results in total pressure losses is the highly irreversible mixing process. For this occurrence, two streams (for example, one from the bypass duct and one from the turbine exit) at different temperatures, total temperatures, Mach numbers, flow rates, velocities, total pressures, and so on are irreversibly mixed to form one (it is hoped) uniform stream. A diagram of this is shown in Figure C.12, which includes the control volume. It is important to recognize that the two exit static pressures at the exit must match for flow stability, that is, $p_1 = p_2$. If the two Mach numbers are the same or if the two Mach numbers are very low, one approximation often made

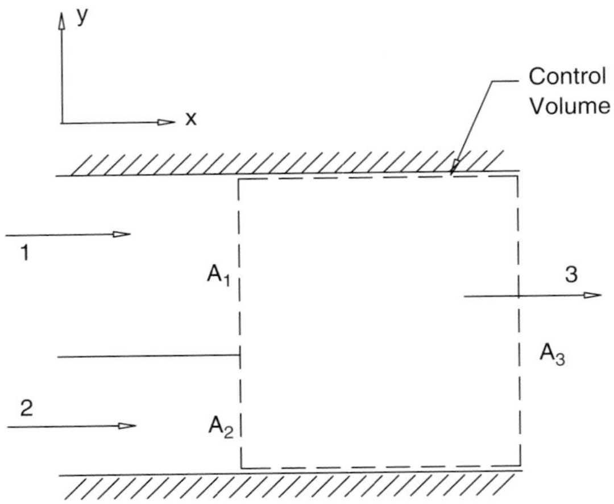

Figure C.12 Geometry for mixing process.

is that the two total pressures are the same, that is, $p_{t1} \cong p_{t2}$. Two streams enter with properties 1 and 2, and one uniform stream leaves with all properties at 3 (for example p_3, T_3, V_3, M_3, etc.). One must be able to find the exiting conditions based on the two inlet conditions. For this process the continuity, momentum, and energy equations are once again used.

First of all, one can consider the areas at the inlet and exit:

$$A_1 + A_2 = A_3. \tag{C.11.1}$$

Next, the mass flows are given by

$$\dot{m}_1 = \rho_1 V_1 A_1 \tag{C.11.2}$$

$$\dot{m}_2 = \rho_2 V_2 A_2 \tag{C.11.3}$$

$$\dot{m}_3 = \rho_3 V_3 A_3; \tag{C.11.4}$$

thus, from continuity,

$$\rho_3 V_3 A_3 = \rho_1 V_1 A_1 + \rho_2 V_2 A_2. \tag{C.11.5}$$

Next, from the momentum equation in the x direction,

$$p_1 A_1 + p_2 A_2 - p_3 A_3 = -\rho_1 V_1^2 A_1 - \rho_2 V_2^2 A_2 + \rho_3 V_3^2 A_3; \tag{C.11.6}$$

thus, rearranging yields

$$(p_1 + \rho_1 V_1^2) A_1 + (p_2 + \rho_2 V_2^2) A_2 = (p_3 + \rho_3 V_3^2) A_3, \tag{C.11.7}$$

or using Eq. (C.11.4) results in

$$(p_1 + \rho_1 V_1^2) A_1 + (p_2 + \rho_2 V_2^2) A_2 = p_3 A_3 + \dot{m}_3 V_3. \tag{C.11.8}$$

Now, from the steady-state energy equation, one obtains

$$\dot{m}_1 h_{t1} + \dot{m}_2 h_{t2} = \dot{m}_3 h_{t3}. \tag{C.11.9}$$

Also for an ideal gas,

$$p_1 = \rho_1 \mathcal{R} T_1 \tag{C.11.10}$$

$$p_2 = \rho_2 \mathcal{R} T_2 \tag{C.11.11}$$

$$p_3 = \rho_3 \mathcal{R} T_3. \tag{C.11.12}$$

Next, from the definition of stagnation temperature, one obtains

$$T_{t1} = T_1 \left[1 + \frac{\gamma - 1}{2} M_1^2 \right] \tag{C.11.13}$$

$$T_{t2} = T_2 \left[1 + \frac{\gamma - 1}{2} M_2^2 \right] \tag{C.11.14}$$

$$T_{t3} = T_3 \left[1 + \frac{\gamma - 1}{2} M_3^2 \right], \tag{C.11.15}$$

and from the definition of stagnation pressure,

$$\frac{p_1}{p_{t1}} = \left[\frac{T_1}{T_{t1}} \right]^{\frac{\gamma}{\gamma-1}} \tag{C.11.16}$$

$$\frac{p_2}{p_{t2}} = \left[\frac{T_2}{T_{t2}} \right]^{\frac{\gamma}{\gamma-1}} \tag{C.11.17}$$

$$\frac{p_3}{p_{t3}} = \left[\frac{T_3}{T_{t3}} \right]^{\frac{\gamma}{\gamma-1}}. \tag{C.11.18}$$

Finally, from the definition of Mach number, one obtains

$$M_1 = \frac{V_1}{a_1} = \frac{V_1}{\sqrt{\gamma \mathcal{R} T_1}} \tag{C.11.19}$$

$$M_2 = \frac{V_2}{a_2} = \frac{V_2}{\sqrt{\gamma \mathcal{R} T_2}} \tag{C.11.20}$$

$$M_3 = \frac{V_3}{a_3} = \frac{V_3}{\sqrt{\gamma \mathcal{R} T_3}}. \tag{C.11.21}$$

Thus, in the preceding equations, 27 unknowns are present. However, 19 independent equations are available (Eqs. (C.11.1)–(C.11.5) and (C.11.8)–(C.11.21)). Thus, if eight variables are specified, it is possible to find the remaining 19. For example, for a mixer, $\dot{m}_1$, $\dot{m}_2$, T_{t1}, T_{t2}, p_{t1}, p_{t2} (or p_1 and p_2), M_1, and M_2 will typically be known.

To aid in the solution, one can use Eqs. (C.11.8) and (C.11.12) to show that

$$(p_1 + \rho_1 V_1{}^2)A_1 + (p_2 + \rho_2 V_2{}^2)A_2$$
$$= \rho_3 \mathcal{R} T_3 A_3 \frac{V_3}{V_3} + \dot{m}_3 V_3, \qquad \text{(C.11.22)}$$

or using Eq. (C.11.4) results in

$$(p_1 + \rho_1 V_1{}^2)A_1 + (p_2 + \rho_2 V_2{}^2)A_2 = \frac{\dot{m}_3 \mathcal{R} T_3}{V_3} + \dot{m}_3 V_3. \qquad \text{(C.11.23)}$$

Using the adiabatic energy equation again between states t3 and 3 to find the gas velocity at station 3 results in

$$V_3 = \sqrt{2c_p(T_{t3} - T_3)}; \qquad \text{(C.11.24)}$$

thus, from Eqs. (C.11.23) and (C.11.24), one finds:

$$(p_1 + \rho_1 V_1^2)A_1 + (p_2 + \rho_2 V_2^2)A_2$$
$$= \dot{m}_3 \left[\frac{\mathcal{R} T_3}{\sqrt{2c_p(T_{t3} - T_3)}} + \sqrt{2c_p(T_{t3} - T_3)} \right]. \qquad \text{(C.11.25)}$$

Therefore, examining Eq. (C.11.25), one can see that the left-hand side is usually known. Also, T_{t3} and $\dot{m}_3$ can easily be found from Eqs. (C.11.9) and (C.11.5), respectively, and thus the only remaining unknown in Eq. (C.11.25) is T_3, which can be found iteratively. After T_3 is found, all other parameters can be found from the above equations, for example M_3. Once again, solutions can also be obtained using the method and software General1D.exe, as discussed in Section C.13. Readers are encouraged to use this software because it is easier than using the above equations.

C.12 Mixing Process with a Mix Enhancer (Drag Object)

Another process that results in total pressure losses is the mixing enhancer coupled with the mixing process in Section C.11 above. Again, two streams (for example, one from the bypass duct and one from the turbine exit) at different temperatures, total temperatures, Mach numbers, flow rates, velocities, total pressures, and so on are irreversibly mixed to form one uniform stream behind the enhancer.

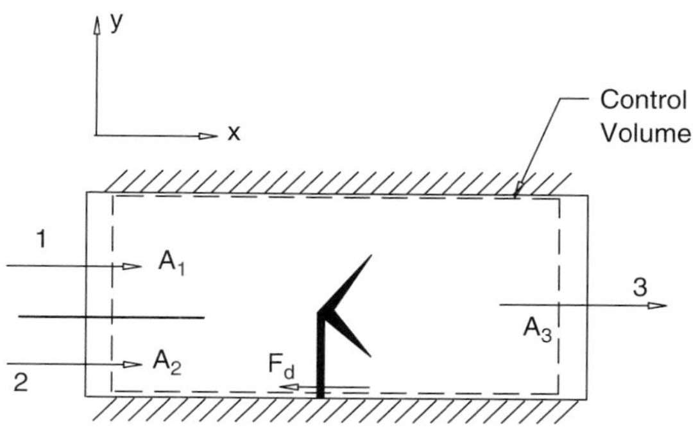

Figure C.13 Geometry for mixing process with a drag obstacle.

A diagram of this is shown in Figure C.13, which includes the control volume. Two streams enter with properties 1 and 2, and one uniform stream leaves with all properties at 3 (for example p_3, T_3, V_3, M_3, etc.). One must be able to find the exiting conditions based on the two inlet conditions. For this process the continuity, momentum, and energy equations are once again used.

First of all, one can consider the areas at the inlet and exit:

$$A_1 + A_2 = A_3. \qquad \text{(C.12.1)}$$

Next, the mass flows are given by

$$\dot{m}_1 = \rho_1 V_1 A_1 \qquad \text{(C.12.2)}$$
$$\dot{m}_2 = \rho_2 V_2 A_2 \qquad \text{(C.12.3)}$$
$$\dot{m}_3 = \rho_3 V_3 A_3; \qquad \text{(C.12.4)}$$

thus, from continuity,

$$\dot{m}_3 = \dot{m}_1 + \dot{m}_2 = \rho_3 V_3 A_3 = \rho_1 V_1 A_1 + \rho_2 V_2 A_2. \qquad \text{(C.12.5)}$$

Next, from the momentum equation in the x direction,

$$p_1 A_1 + p_2 A_2 - p_3 A_3 - F_d$$
$$= -\rho_1 V_1{}^2 A_1 - \rho_2 V_2{}^2 A_2 + \rho_3 V_3{}^2 A_3; \qquad \text{(C.12.6)}$$

where F_d is the drag on the mix enhancer. Next, the drag is a result of both incoming streams, so

$$F_d = F_{d1} + F_{d2}.$$

The drag coefficient is defined as for both streams

$$C_{d1} \equiv \frac{F_{d1}}{\frac{1}{2}\rho_1 V_1^2 A_{d1}}, \tag{C.12.7}$$

and

$$C_{d2} \equiv \frac{F_{d2}}{\frac{1}{2}\rho_2 V_2^2 A_{d2}}, \tag{C.12.8}$$

where A_d is the frontal area of the object. Or, solving for the forces yields

$$F_{d1} = \frac{1}{2}C_{d1}\rho_1 V_1^2 A_1 \frac{A_{d1}}{A_1} \tag{C.12.9}$$

and

$$F_{d2} = \frac{1}{2}C_{d2}\rho_2 V_2^2 A_2 \frac{A_{d2}}{A_2}. \tag{C.12.10}$$

Thus, substituting, and rearranging yields

$$\begin{aligned}(p_1 + \rho_1 V_1{}^2)A_1 + (p_2 + \rho_2 V_2{}^2)A_2 \\ - \frac{1}{2}C_{d1}\rho_1 V_1^2 A_1 \frac{A_{d1}}{A_1} - \frac{1}{2}C_{d2}\rho_2 V_2^2 A_2 \frac{A_{d2}}{A_2} \\ = (p_3 + \rho_3 V_3{}^2)A_3,\end{aligned} \tag{C.12.11}$$

and using Eqs. (C.12.2)–(C.12.4) gives

$$\begin{aligned}(p_1 A_1 + \dot{m}_1 V_1) + (p_2 A_2 + \dot{m}_2 V_2) \\ - \frac{1}{2}C_{d1}\dot{m}_1 V_1 \frac{A_{d1}}{A_1} - \frac{1}{2}C_{d2}\dot{m}_2 V_2 \frac{A_{d2}}{A_2} = (p_3 A_3 + \dot{m}_3 V_3).\end{aligned} \tag{C.12.12}$$

Now, if the drag coefficient is the same for both streams

$$C_d = C_{d1} = C_{d2} \tag{C.12.13}$$

and the relative blockage or relative frontal area is the same for both streams (both reasonable approximations) and defining A_R:

$$A_R = \frac{A_{d1}}{A_1} = \frac{A_{d2}}{A_2} \tag{C.12.14}$$

$$\begin{aligned}(p_1 A_1 + \dot{m}_1 V_1) + (p_2 A_2 + \dot{m}_2 V_2) \\ - \frac{1}{2}C_d A_R(\dot{m}_1 V_1 + \dot{m}_2 V_2) = (p_3 A_3 + \dot{m}_3 V_3).\end{aligned} \tag{C.12.15}$$

Now, using the ideal gas law

$$p_1 = \rho_1 \mathcal{R} T_1 \tag{C.12.16}$$

$$p_2 = \rho_2 \mathcal{R} T_2 \tag{C.12.17}$$

$$p_3 = \rho_3 \mathcal{R} T_3. \tag{C.12.18}$$

$$\begin{aligned}\dot{m}_1\left(\frac{\mathcal{R}T_1}{V_1} + V_1\right) + \dot{m}_2\left(\frac{\mathcal{R}T_2}{V_2} + V_2\right) \\ - \frac{1}{2}C_d A_R(\dot{m}_1 V_1 + \dot{m}_2 V_2) = \dot{m}_3\left(\frac{\mathcal{R}T_3}{V_3} + V_3\right).\end{aligned} \tag{C.12.19}$$

Next, from the definition of stagnation temperature, one obtains

$$T_{t1} = T_1\left[1 + \frac{\gamma - 1}{2}M_1^2\right] \tag{C.12.20}$$

$$T_{t2} = T_2\left[1 + \frac{\gamma - 1}{2}M_2^2\right] \tag{C.12.21}$$

$$T_{t3} = T_3\left[1 + \frac{\gamma - 1}{2}M_3^2\right]. \tag{C.12.22}$$

Using the adiabatic energy equation again between total and static states results in

$$V_1 = \sqrt{2c_p(T_{t1} - T_1)} \tag{C.12.23}$$

$$V_2 = \sqrt{2c_p(T_{t2} - T_2)} \tag{C.12.24}$$

$$V_3 = \sqrt{2c_p(T_{t3} - T_3)}. \tag{C.12.25}$$

Also, as before

$$\frac{p_1}{p_{t1}} = \left[\frac{T_1}{T_{t1}}\right]^{\frac{\gamma}{\gamma-1}} \tag{C.12.26}$$

$$\frac{p_2}{p_{t2}} = \left[\frac{T_2}{T_{t2}}\right]^{\frac{\gamma}{\gamma-1}} \tag{C.12.27}$$

$$\frac{p_4}{p_{t4}} = \left[\frac{T_4}{T_{t4}}\right]^{\frac{\gamma}{\gamma-1}}. \tag{C.12.28}$$

From the definition of Mach number, one obtains

$$M_1 = \frac{V_1}{a_1} = \frac{V_1}{\sqrt{\gamma \mathcal{R} T_1}} \tag{C.12.29}$$

$$M_2 = \frac{V_2}{a_2} = \frac{V_2}{\sqrt{\gamma \mathcal{R} T_2}} \tag{C.12.30}$$

$$M_3 = \frac{V_3}{a_3} = \frac{V_3}{\sqrt{\gamma \mathcal{R} T_3}}. \tag{C.12.31}$$

Now, from the steady-state energy equation, one obtains

$$\begin{aligned}\dot{m}_1 h_{t1} + \dot{m}_2 h_{t2} = \dot{m}_3 h_{t3} = \dot{m}_1 c_p T_{t1} + \dot{m}_2 c_p T_{t2} \\ = \dot{m}_3 c_p T_{t3}.\end{aligned} \tag{C.12.32}$$

So, if c_p is constant,

$$T_{t3} = \frac{\dot{m}_1}{\dot{m}_3} T_{t1} + \frac{\dot{m}_2}{\dot{m}_3} T_{t2}. \tag{C.12.33}$$

Finally from Eq. (C.12.19):

$$\dot{m}_1\left(\frac{\mathcal{R}T_1}{V_1} + V_1\right) + \dot{m}_2\left(\frac{\mathcal{R}T_2}{V_2} + V_2\right) - \frac{1}{2} C_d A_R(\dot{m}_1 V_1 + \dot{m}_2 V_2) = \dot{m}_3\left(\frac{\mathcal{R}T_3}{\sqrt{2c_p(T_{t3} - T_3)}} + \sqrt{2c_p(T_{t3} - T_3)}\right). \tag{C.12.34}$$

For a mixer, $\dot{m}_1$, $\dot{m}_2$, T_{t1}, T_{t2}, p_{t1}, p_{t2} (or p_1 and p_2, and equal), M_1, and M_2 will typically be known.

Therefore, examining Eq. (C.12.34), one can see that the left-hand side is usually known. Also, T_{t3} and $\dot{m}_3$ can easily be found from Eqs. (C.12.33) and (C.12.5), respectively, and thus the only remaining unknown in Eq. (C.12.34) is T_3, which can be found iteratively. After T_3 is found, all other parameters can be found from the above equations, for example M_3. Once again, solutions can also be obtained using the software General1D.exe. Readers are encouraged to use this software because it is easier than using the above equations.

C.13 Generalized One-Dimensional Compressible Flow

In the preceding sections, area changes, friction, heat addition/removal, and drag were mostly treated separately. However, for example, in an engine component several effects can occur simultaneously. Thus, in this section an analysis is introduced that includes several effects. The technique used is the so-called generalized 1-D flow method derived in many gas dynamics texts such as Shapiro (1953). It is a differential equation analysis from which influence coefficients are derived. These coefficients relate changes of one variable on another. In this book, the general method will not be derived.

Figure C.14 illustrates the generalized geometry. Included in the differential analysis are area changes, changes in total temperature by heat addition or removal, friction, and drag objects. Shapiro (1953) also includes gas mass addition for which the gas is injected into the freestream with velocity V_g, with an axial component V_{gx}. A differential control volume is used, and differential forms of the continuity, x-momentum, energy, and second law of thermodynamics are conventionally applied to the control volume to derive the influence coefficients. Differential equations result, and a matrix of the coefficients is tabulated in Table C.2. Incremental changes in Mach number, velocity, sound speed, temperature, density, pressure, total pressure, and entropy are shown to be related to area change, total temperature change, friction, and drag. Herein, results of the method will be used to determine and integrate the governing differential equations for flows with three specific combined effects. It goes without saying that the method is quite versatile and can even be used to include area changes, heat transfer, friction, plus a differential drag object simultaneously. Application of the method is presented in the next sections.

C.14 Combined Area Changes and Friction

Sections C.4 and C.5 treat area changes and friction separately. However, for example, in a diffuser or duct the

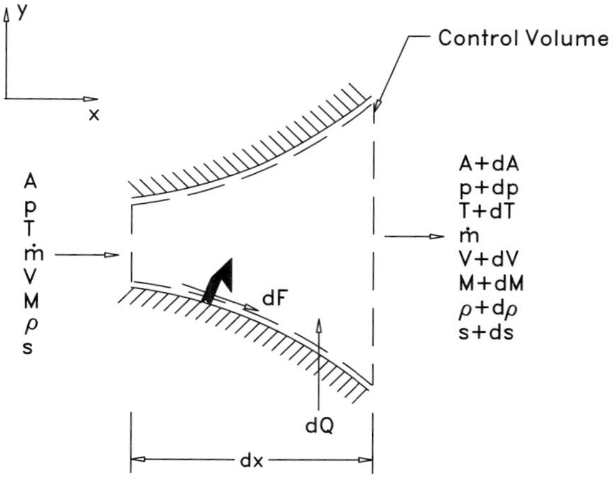

Figure C.14 Geometry for generalized 1-D compressible flow.

two effects occur simultaneously. Thus, in this section an analysis will be conducted in which both are included. The technique that will be used is the generalized 1-D flow method introduced previously. Herein, results of the method will be used to determine the governing differential equations for flows with combined area changes and friction.

The two most important equations for the present applications are those for the Mach number and total pressure; these and other property changes can easily (and similarly) be found from Table C.2. That is,

$$\frac{dM^2}{M^2} = -2\left[\frac{\left(1+\frac{\gamma-1}{2}M^2\right)}{1-M^2}\right]\frac{dA}{A} + 4\,f\gamma M^2\left[\frac{\left(1+\frac{\gamma-1}{2}M^2\right)}{1-M^2}\right]\frac{dx}{D} \tag{C.14.1}$$

$$\frac{dp_t}{p_t} = -\frac{1}{2}4\,f\gamma M^2\frac{dx}{D}. \tag{C.14.2}$$

Note that the area variation does not directly affect the total pressure. The area change does indirectly affect the total pressure variation, however, since the Mach number changes due to area variations and the Mach number affects the total pressure loss. In general, the two equations above cannot be analytically integrated but can be numerically integrated between two end states to obtain the drop in total pressure due to both friction and area change (and resulting diameter change). That is, between any two states 1 and 2,

$$M_2^2 - M_1^2 = -2\int_{A_1}^{A_2} M^2\left[\frac{\left(1+\frac{\gamma-1}{2}M^2\right)}{1-M^2}\right]\frac{dA}{A} + 4\,f\gamma\int_0^L M^4\left[\frac{\left(1+\frac{\gamma-1}{2}M^2\right)}{1-M^2}\right]\frac{dx}{D} \tag{C.14.3}$$

$$p_{t2} - p_{t1} = -2\,f\gamma\int_0^L p_t M^2\frac{dx}{D}. \tag{C.14.4}$$

To integrate Eq. (C.14.4) numerically one must also evaluate Eq. (C.14.3) to obtain the Mach number variation and incorporate the diameter change. Note that, if $A_2 = A_1$, the method reverts to Fanno line flow. On the other hand, if $f = 0$, the method reverts to isentropic flow with area changes. Thus, these two known solutions can provide a valuable check on the general method. Details of the numerical integration technique are presented in Example 8.3 in Chapter 8 (Diffusers and Inlets). Solutions can also be obtained using the software General1D.exe. Readers are encouraged to use this software because it is easier than integrating the above equations either by hand or by writing a separate program.

C.15 Combined Heat Transfer and Friction

Sections C.6 and C.5 treat heat transfer and friction separately. However, for example, in a burner the two effects occur simultaneously. Thus, in this section the analysis includes both. The generalized 1-D compressible flow technique is used again. Results of the method are used in this section to determine the governing differential equations for flows with combined heat addition or removal and friction.

Again, the two most important equations for the present application are those for the Mach number and total pressure, and these can be found from Table C.2. That is,

$$\frac{dM^2}{M^2} = (1+\gamma M^2)\left[\frac{1+\frac{\gamma-1}{2}M^2}{1-M^2}\right]\frac{dT_t}{T_t} + 4\,f\gamma M^2\left[\frac{1+\frac{\gamma-1}{2}M^2}{1-M^2}\right]\frac{dx}{D} \tag{C.15.1}$$

$$\frac{dp_t}{p_t} = -\frac{\gamma M^2}{2}\frac{dT_t}{T_t} - \frac{1}{2}4\,f\gamma M^2\frac{dx}{D}. \tag{C.15.2}$$

In general these can be integrated numerically between two end states to obtain the drop in total pressure due to both friction and heat addition or removal (again, analytical integration is not possible). That is, between any two states 1 and 2,

Table C.2 **Influence coefficients for generalized 1-D flow.**

	$\frac{dA}{A}$	$\frac{dT_t}{T_t}$	$4f\frac{dx}{D}+\frac{dF_x}{\frac{1}{2}\gamma pAM^2}$
$\frac{dM^2}{M^2}$	$-\frac{2\left(1+\frac{\gamma-1}{2}M^2\right)}{1-M^2}$	$\frac{(1+\gamma M^2)\left(1+\frac{\gamma-1}{2}M^2\right)}{1-M^2}$	$\frac{\gamma M^2\left(1+\frac{\gamma-1}{2}M^2\right)}{1-M^2}$
$\frac{dV}{V}$	$-\frac{1}{1-M^2}$	$\frac{1+\frac{\gamma-1}{2}M^2}{1-M^2}$	$\frac{\gamma M^2}{2(1-M^2)}$
$\frac{da}{a}$	$\frac{\frac{\gamma-1}{2}M^2}{1-M^2}$	$\frac{\frac{(1-\gamma M^2)}{2}\left(1+\frac{\gamma-1}{2}M^2\right)}{1-M^2}$	$-\frac{\gamma(\gamma-1)M^4}{4(1-M^2)}$
$\frac{dT}{T}$	$\frac{(\gamma-1)M^2}{1-M^2}$	$\frac{(1-\gamma M^2)\left(1+\frac{\gamma-1}{2}M^2\right)}{1-M^2}$	$-\frac{\gamma(\gamma-1)M^4}{2(1-M^2)}$
$\frac{d\rho}{\rho}$	$\frac{M^2}{1-M^2}$	$-\frac{1+\frac{\gamma-1}{2}M^2}{1-M^2}$	$\frac{\gamma M^2}{2(1-M^2)}$
$\frac{dp}{p}$	$\frac{\gamma M^2}{1-M^2}$	$-\frac{\gamma M^2\left(1+\frac{\gamma-1}{2}M^2\right)}{1-M^2}$	$-\frac{\gamma M^2\left(1+(\gamma-1)M^2\right)}{2(1-M^2)}$
$\frac{dp_t}{p_t}$	0	$-\frac{\gamma M^2}{2}$	$-\frac{\gamma M^2}{2}$
$\frac{ds}{c_p}$	0	$1+\frac{\gamma-1}{2}M^2$	$\frac{\gamma-1}{2}M^2$

$$M_2^2 - M_1^2 = -\int_{T_{t1}}^{T_{t2}} (1+\gamma M^2)M^2 \left[\frac{\left(1+\frac{\gamma-1}{2}M^2\right)}{1-M^2}\right]\frac{dT_t}{T_t}$$

$$+\,4f\gamma\int_0^L M^4 \left[\frac{\left(1+\frac{\gamma-1}{2}M^2\right)}{1-M^2}\right]\frac{dx}{D} \tag{C.15.3}$$

$$p_{t2} - p_{t1} = -\frac{1}{2}\gamma\int_{T_{t1}}^{T_{t2}} p_t M^2 \frac{dT_t}{T_t} - 2f\gamma\int_0^L p_t M^2 \frac{dx}{D}. \tag{C.15.4}$$

To integrate Eq. (C.15.4) numerically one must also evaluate Eq. (C.15.3) to obtain the Mach number variation. Note that if $T_{t2} = T_{t1,}$ the method reverts to Fanno line flow. On the other hand, if $f = 0$, the method reverts to Rayleigh line flow. Thus, these two known solutions can provide checks on the general method. Details of the numerical integration technique are presented in Example 9.8 in Chapter 9 (Combustors). Solutions can also be obtained using the software General1D.exe. Readers are encouraged to use this software because it is easier than integrating the above equations either by hand or by writing a separate program.

C.16 Combined Area Changes, Heat Transfer, and Friction

The method can be taken at least one step further. For example, in an gas turbine component, three or more effects can occur simultaneously. Thus, in this section the analysis includes area changes, heat addition or removal, and friction. Three such processes can occur in a nozzle, for example. The generalized 1-D compressible flow method is again used. Results of the method are used in this section to determine the governing differential equations for flows with combined area changes, heat addition or removal, and friction.

Again, probably the two most important equations are those for the Mach number and total pressure, and these can be found from Table C.2. That is,

$$\frac{dM^2}{M^2} = -2\left[\frac{1+\frac{\gamma-1}{2}M^2}{1-M^2}\right]\frac{dA}{A}$$

$$+\,(1+\gamma M^2)\left[\frac{1+\frac{\gamma-1}{2}M^2}{1-M^2}\right]\frac{dT_t}{T_t}$$

$$+\,4f\gamma M^2\left[\frac{1+\frac{\gamma-1}{2}M^2}{1-M^2}\right]\frac{dx}{D} \tag{C.16.1}$$

$$\frac{dp_t}{p_t} = -\frac{\gamma M^2}{2}\frac{dT_t}{T_t} - \frac{1}{2}4f\gamma M^2\frac{dx}{D}. \tag{C.16.2}$$

In general these can be numerically integrated between two end states to obtain the drop in total pressure due to both friction and heat transfer (again analytical integration is not possible). That is, between any two states 1 and 2,

$$M_2^2 - M_1^2 = -2\int_{A_1}^{A_2} M^2 \left[\frac{\left(1+\frac{\gamma-1}{2}M^2\right)}{1-M^2}\right]\frac{dA}{A}$$

$$+\int_{T_{t1}}^{T_{t2}} (1+\gamma M^2)M^2 \left[\frac{\left(1+\frac{\gamma-1}{2}M^2\right)}{1-M^2}\right]\frac{dT_t}{T_t}$$

$$+\,4f\gamma\int_0^L M^4 \left[\frac{\left(1+\frac{\gamma-1}{2}M^2\right)}{1-M^2}\right]\frac{dx}{D} \tag{C.16.3}$$

$$p_{t2} - p_{t1} = -\frac{1}{2}\gamma\int_{T_{t1}}^{T_{t2}} p_t M^2 \frac{dT_t}{T_t} - 2f\gamma\int_0^L p_t M^2 \frac{dx}{D}. \tag{C.16.4}$$

To integrate Eq. (C.16.4) numerically one must also evaluate Eq. (C.16.3) to obtain the Mach number variation. Solutions can also be obtained using the software General1D.exe. Readers are encouraged to use this software because it is easier than integrating the above equations either by hand or by writing a separate program.

C.17 Summary

In this appendix the fundamental gas dynamics identities and processes are reviewed. All of these processes are in different aspects of gas turbine operation. The processes range from simple isentropic flow with area changes to more complex combined processes. The simple processes can be solved by closed-form approaches. However, the more complex processes require integration of differential states. Several software routines can make the calculations much easier and are identified.

List of Symbols

a	Sound speed
A	Area
C	Drag coefficient
c_p	Specific heat at constant pressure
D	Diameter
f	Fanning friction factor
F	Drag force
h	Specific enthalpy
L	Axial length
M	Mach number
$\dot{m}$	Mass flow rate
p	Pressure
$\dot{Q}$	Heat transfer rate
$\mathcal{R}$	Ideal gas constant
s	Entropy
T	Temperature
U	Velocity
V	Velocity
x	Axial position
δ	Turning angle
γ	Specific heat ratio
ρ	Density
μ	Viscosity
σ	Shock angle

Subscripts

$_d$	Drag object
$_g$	Gas
$_n$	Normal
$_t$	Total (stagnation)
$_t$	Tangential
$_x$	Axial position
$_1$	Station number
$_2$	Station number
$_3$	Station number

Superscripts

*	Sonic condition

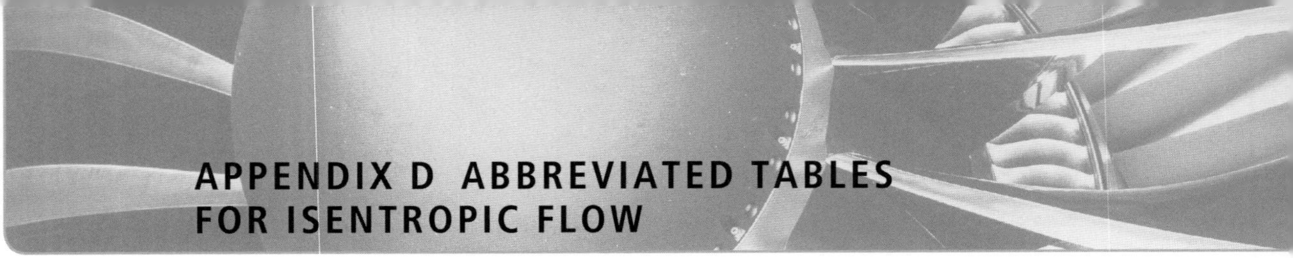

APPENDIX D ABBREVIATED TABLES FOR ISENTROPIC FLOW

Also see Isentropic.exe

$\gamma = 1.40$					$\gamma = 1.30$				
M	T/T_t	p/p_t	ρ/ρ_t	A/A^*	M	T/T_t	p/p_t	ρ/ρ_t	A/A^*
0.0	1.0000	1.0000	1.0000	∞	0.0	1.0000	1.0000	1.0000	∞
0.1	0.9980	0.9930	0.9950	5.8218	0.1	0.9985	0.9935	0.9950	5.8860
0.2	0.9921	0.9725	0.9803	2.9635	0.2	0.9940	0.9744	0.9803	2.9940
0.3	0.9823	0.9395	0.9564	2.0351	0.3	0.9867	0.9435	0.9563	2.0537
0.4	0.9690	0.8956	0.9243	1.5901	0.4	0.9766	0.9023	0.9240	1.6023
0.5	0.9524	0.8430	0.8852	1.3398	0.5	0.9639	0.8525	0.8845	1.3479
0.6	0.9328	0.7840	0.8405	1.1882	0.6	0.9488	0.7962	0.8392	1.1932
0.7	0.9107	0.7209	0.7916	1.0944	0.7	0.9315	0.7354	0.7895	1.0972
0.8	0.8865	0.6560	0.7400	1.0382	0.8	0.9124	0.6722	0.7367	1.0395
0.9	0.8606	0.5913	0.6870	1.0089	0.9	0.8917	0.6084	0.6823	1.0092
1.0	0.8333	0.5283	0.6339	1.0000	1.0	0.8696	0.5457	0.6276	1.0000
1.1	0.8052	0.4684	0.5817	1.0079	1.1	0.8464	0.4854	0.5735	1.0083
1.2	0.7764	0.4124	0.5311	1.0304	1.2	0.8224	0.4285	0.5211	1.0321
1.3	0.7474	0.3609	0.4829	1.0663	1.3	0.7978	0.3757	0.4709	1.0703
1.4	0.7184	0.3142	0.4374	1.1149	1.4	0.7728	0.3273	0.4235	1.1227
1.5	0.6897	0.2724	0.3950	1.1762	1.5	0.7477	0.2836	0.3793	1.1895
1.6	0.6614	0.2353	0.3557	1.2502	1.6	0.7225	0.2446	0.3385	1.2712
1.7	0.6337	0.2026	0.3197	1.3376	1.7	0.6976	0.2100	0.3011	1.3690
1.8	0.6068	0.1740	0.2868	1.4390	1.8	0.6729	0.1797	0.2671	1.4841
1.9	0.5807	0.1492	0.2570	1.5553	1.9	0.6487	0.1533	0.2363	1.6182
2.0	0.5556	0.1278	0.2300	1.6875	2.0	0.6250	0.1305	0.2087	1.7732

(cont.)

$\gamma = 1.40$					$\gamma = 1.30$				
M	T/T_t	p/p_t	ρ/ρ_t	A/A^*	M	T/T_t	p/p_t	ρ/ρ_t	A/A^*
2.1	0.5314	0.1094	0.2058	1.8369	2.1	0.6019	0.1108	0.1841	1.9514
2.2	0.5081	0.0935	0.1841	2.0050	2.2	0.5794	0.0939	0.1621	2.1555
2.3	0.4859	0.0800	0.1646	2.1931	2.3	0.5576	0.0795	0.1427	2.3885
2.4	0.4647	0.0684	0.1472	2.4031	2.4	0.5365	0.0673	0.1255	2.6535
2.5	0.4444	0.0585	0.1317	2.6367	2.5	0.5161	0.0569	0.1103	2.9545
2.6	0.4252	0.0501	0.1179	2.8960	2.6	0.4965	0.0481	0.0969	3.2954
2.7	0.4068	0.0430	0.1056	3.1830	2.7	0.4777	0.0407	0.0852	3.6811
2.8	0.3894	0.0368	0.0946	3.5001	2.8	0.4596	0.0344	0.0749	4.1165
2.9	0.3729	0.0317	0.0849	3.8498	2.9	0.4422	0.0291	0.0659	4.6073
3.0	0.3571	0.0272	0.0762	4.2346	3.0	0.4255	0.0247	0.0580	5.1598
3.5	0.2899	0.0131	0.0452	6.7896	3.5	0.3524	0.0109	0.0309	9.1098
4.0	0.2381	0.0066	0.0277	10.719	4.0	0.2941	0.0050	0.0169	15.944
4.5	0.1980	0.0035	0.0174	16.562	4.5	0.2477	0.0024	0.0095	27.387
5.0	0.1667	0.0019	0.0113	25.000	5.0	0.2105	0.0012	0.0056	45.956
5.5	0.1418	0.0011	0.0076	36.869	5.5	0.1806	0.0006	0.0033	75.219
6.0	0.1220	0.0006	0.0052	53.180	6.0	0.1563	0.0003	0.0021	120.10
7.0	0.0926	0.0002	0.0026	104.14	7.0	0.1198	0.0001	0.0008	285.34
8.0	0.0725	0.0001	0.0014	190.11	8.0	0.0943	0.0000	0.0004	623.12
9.0	0.0581	0.0000	0.0008	327.19	9.0	0.0760	0.0000	0.0002	1265.6
10.0	0.0476	0.0000	0.0005	535.94	10.0	0.0625	0.0000	0.0001	2416.1
20.0	0.0123	0.0000	0.0000	15,377.0	20.0	0.0164	0.0000	0.0000	204,202.0
∞	0.0000	0.0000	0.0000	∞	∞	0.0000	0.0000	0.0000	∞

APPENDIX E ABBREVIATED TABLES FOR FANNO LINE FLOW

Also see Fanno.exe

		$\gamma = 1.40$						$\gamma = 1.30$			
M	ρ^*/ρ	T/T^*	p/p^*	p_t/p_t^*	$4fL^*/D$	M	ρ^*/ρ	T/T^*	p/p^*	p_t/p_t^*	$4fL^*/D$
0.0	0.0000	1.2000	∞	∞	∞	0.0	0.0000	1.1500	∞	∞	∞
0.1	0.1094	1.1976	10.944	5.8218	66.922	0.1	0.1072	1.1483	10.716	5.8860	72.202
0.2	0.2182	1.1905	5.4554	2.9635	14.533	0.2	0.2138	1.1431	5.3459	2.9940	15.732
0.3	0.3257	1.1788	3.6191	2.0351	5.2993	0.3	0.3196	1.1347	3.5507	2.0537	5.7594
0.4	0.4313	1.1628	2.6958	1.5901	2.3085	0.4	0.4239	1.1230	2.6493	1.6023	2.5200
0.5	0.5345	1.1429	2.1381	1.3398	1.0691	0.5	0.5264	1.1084	2.1056	1.3479	1.1724
0.6	0.6348	1.1194	1.7634	1.1882	0.4908	0.6	0.6267	1.0911	1.7409	1.1932	0.5409
0.7	0.7318	1.0929	1.4935	1.0944	0.2081	0.7	0.7245	1.0713	1.4786	1.0972	0.2305
0.8	0.8251	1.0638	1.2893	1.0382	0.0723	0.8	0.8195	1.0493	1.2804	1.0395	0.0804
0.9	0.9146	1.0327	1.1291	1.0089	0.0145	0.9	0.9114	1.0254	1.1251	1.0092	0.0162
1.0	1.0000	1.0000	1.0000	1.0000	0.0000	1.0	1.0000	1.0000	1.0000	1.0000	0.0000
1.1	1.0812	0.9662	0.8936	1.0079	0.0099	1.1	1.0852	0.9733	0.8969	1.0083	0.0112
1.2	1.1583	0.9317	0.8044	1.0304	0.0336	1.2	1.1670	0.9457	0.8104	1.0321	0.0382
1.3	1.2311	0.8969	0.7285	1.0663	0.0648	1.3	1.2452	0.9174	0.7368	1.0703	0.0739
1.4	1.2999	0.8621	0.6632	1.1149	0.0997	1.4	1.3198	0.8887	0.6734	1.1227	0.1142
1.5	1.3646	0.8276	0.6065	1.1762	0.1360	1.5	1.3909	0.8598	0.6182	1.1895	0.1564
1.6	1.4254	0.7937	0.5568	1.2502	0.1724	1.6	1.4585	0.8309	0.5697	1.2712	0.1989
1.8	1.5360	0.7282	0.4741	1.4390	0.2419	1.8	1.5835	0.7739	0.4887	1.4841	0.2814
2.0	1.6330	0.6667	0.4082	1.6875	0.3050	2.0	1.6956	0.7188	0.4239	1.7732	0.3573
2.5	1.8257	0.5333	0.2921	2.6367	0.4320	2.5	1.9261	0.5935	0.3082	2.9545	0.5135
3.0	1.9640	0.4286	0.2182	4.2346	0.5222	3.0	2.0986	0.4894	0.2332	5.1598	0.6277

(cont.)

		$\gamma = 1.40$						$\gamma = 1.30$			
M	ρ^*/ρ	T/T^*	p/p^*	p_t/p_t^*	$4fL^*/D$	M	ρ^*/ρ	T/T^*	p/p^*	p_t/p_t^*	$4fL^*/D$
3.5	2.0642	0.3478	0.1685	6.7896	0.5864	3.5	2.2282	0.4053	0.1819	9.1098	0.7110
4.0	2.1381	0.2857	0.1336	10.720	0.6331	4.0	2.3263	0.3382	0.1454	15.944	0.7726
4.5	2.1936	0.2376	0.1083	16.562	0.6676	4.5	2.4016	0.2848	0.1186	27.387	0.8189
5.0	2.2361	0.2000	0.0894	25.000	0.6938	5.0	2.4602	0.2421	0.0984	45.956	0.8543
6.0	2.2953	0.1463	0.0638	53.180	0.7299	6.0	2.5434	0.1797	0.0706	120.10	0.9037
7.0	2.3333	0.1111	0.0476	104.14	0.7528	7.0	2.5978	0.1377	0.0530	285.34	0.9355
8.0	2.3591	0.0870	0.0369	190.11	0.7682	8.0	2.6350	0.1085	0.0412	623.12	0.9570
9.0	2.3772	0.0698	0.0293	327.19	0.7790	9.0	2.6615	0.0875	0.0329	1265.6	0.9722
10.0	2.3905	0.0571	0.0239	535.94	0.7868	10.0	2.6810	0.0719	0.0268	2416.1	0.9832
11.0	2.4004	0.0476	0.0198	841.91	0.7927	11.0	2.6956	0.0601	0.0223	4374.3	0.9915
12.0	2.4080	0.0403	0.0167	1276.2	0.7972	12.0	2.7069	0.0509	0.0188	7566.5	0.9979
13.0	2.4140	0.0345	0.0143	1876.1	0.8007	13.0	2.7158	0.0436	0.0161	12,581.0	1.0030
14.0	2.4188	0.0299	0.0123	2685.4	0.8036	14.0	2.7230	0.0378	0.0139	20,209.0	1.0070
15.0	2.4227	0.0261	0.0108	3755.2	0.8058	15.0	2.7287	0.0331	0.0121	31,494.0	1.0102
∞	2.4495	0.0000	0.0000	∞	0.8215	∞	2.7689	0.0000	0.0000	∞	1.0326

APPENDIX F ABBREVIATED TABLES FOR RAYLEIGH LINE FLOW

Also see Rayleigh.exe

		$\gamma = 1.40$						$\gamma = 1.30$			
M	ρ^*/ρ	T/T^*	p/p^*	p_t/p_t^*	T_t/T_t^*	M	ρ^*/ρ	T/T^*	p/p^*	p_t/p_t^*	T_t/T_t^*
0.00	0.0000	0.0000	2.4000	1.2679	0.0000	0.00	0.0000	0.0000	2.3000	1.2552	0.0000
0.05	0.0060	0.0143	2.3916	1.2657	0.0119	0.05	0.0057	0.0131	2.2925	1.2531	0.0114
0.10	0.0237	0.0560	2.3669	1.2591	0.0468	0.10	0.0227	0.0516	2.2705	1.2471	0.0449
0.15	0.0524	0.1218	2.3267	1.2486	0.1020	0.15	0.0503	0.1124	2.2346	1.2374	0.0980
0.20	0.0909	0.2066	2.2727	1.2346	0.1736	0.20	0.0875	0.1912	2.1863	1.2245	0.1673
0.25	0.1379	0.3044	2.2069	1.2177	0.2568	0.25	0.1329	0.2828	2.1272	1.2088	0.2482
0.30	0.1918	0.4089	2.1314	1.1985	0.3469	0.30	0.1853	0.3816	2.0591	1.1909	0.3363
0.35	0.2510	0.5141	2.0487	1.1779	0.4389	0.35	0.2430	0.4822	1.9840	1.1716	0.4270
0.40	0.3137	0.6151	1.9608	1.1566	0.5290	0.40	0.3046	0.5800	1.9040	1.1515	0.5165
0.45	0.3787	0.7080	1.8699	1.1351	0.6139	0.45	0.3687	0.6713	1.8207	1.1312	0.6014
0.50	0.4444	0.7901	1.7778	1.1141	0.6914	0.50	0.4340	0.7533	1.7358	1.1111	0.6796
0.60	0.5745	0.9167	1.5957	1.0753	0.8189	0.60	0.5640	0.8837	1.5668	1.0739	0.8099
0.70	0.6975	0.9929	1.4235	1.0431	0.9085	0.70	0.6885	0.9673	1.4050	1.0426	0.9029
0.80	0.8101	1.0255	1.2658	1.0193	0.9639	0.80	0.8035	1.0088	1.2555	1.0193	0.9614
0.90	0.9110	1.0245	1.1246	1.0049	0.9921	0.90	0.9075	1.0166	1.1203	1.0049	0.9914
1.00	1.0000	1.0000	1.0000	1.0000	1.0000	1.00	1.0000	1.0000	1.0000	1.0000	1.0000
1.10	1.0780	0.9603	0.8909	1.0049	0.9939	1.10	1.0816	0.9669	0.8939	1.0049	0.9933
1.20	1.1459	0.9118	0.7958	1.0194	0.9787	1.20	1.1532	0.9235	0.8008	1.0199	0.9765

(cont.)

		$\gamma = 1.40$						$\gamma = 1.30$			
M	ρ^*/ρ	T/T^*	p/p^*	p_t/p_t^*	T_t/T_t^*	M	ρ^*/ρ	T/T^*	p/p^*	p_t/p_t^*	T_t/T_t^*
1.30	1.2050	0.8592	0.7130	1.0437	0.9580	1.30	1.2158	0.8747	0.7194	1.0451	0.9534
1.40	1.2564	0.8054	0.6410	1.0777	0.9343	1.40	1.2706	0.8237	0.6483	1.0809	0.9268
1.50	1.3012	0.7525	0.5783	1.1215	0.9093	1.50	1.3185	0.7726	0.5860	1.1276	0.8986
1.60	1.3403	0.7017	0.5236	1.1756	0.8842	1.60	1.3604	0.7230	0.5314	1.1858	0.8701
1.70	1.3746	0.6538	0.4756	1.2402	0.8597	1.70	1.3973	0.6756	0.4835	1.2563	0.8421
1.80	1.4046	0.6089	0.4335	1.3159	0.8363	1.80	1.4298	0.6309	0.4413	1.3400	0.8153
1.90	1.4311	0.5673	0.3964	1.4033	0.8141	1.90	1.4585	0.5892	0.4040	1.4381	0.7898
2.00	1.4545	0.5289	0.3636	1.5031	0.7934	2.00	1.4839	0.5505	0.3710	1.5518	0.7659
2.50	1.5385	0.3787	0.2462	2.2218	0.7101	2.50	1.5753	0.3971	0.2521	2.4165	0.6690
3.00	1.5882	0.2803	0.1765	3.4244	0.6540	3.00	1.6299	0.2952	0.1811	4.0074	0.6032
3.50	1.6198	0.2142	0.1322	5.3280	0.6158	3.50	1.6647	0.2262	0.1359	6.8061	0.5582
4.00	1.6410	0.1683	0.1026	8.2268	0.5891	4.00	1.6881	0.1781	0.1055	11.570	0.5266
5.00	1.6667	0.1111	0.0667	18.635	0.5556	5.00	1.7164	0.1178	0.0687	32.062	0.4867
6.00	1.6809	0.0785	0.0467	38.946	0.5363	6.00	1.7322	0.0833	0.0481	81.794	0.4639
7.00	1.6897	0.0583	0.0345	75.414	0.5244	7.00	1.7419	0.0619	0.0355	191.33	0.4496
8.00	1.6954	0.0449	0.0265	136.62	0.5165	8.00	1.7482	0.0478	0.0273	413.41	0.4402
9.00	1.6993	0.0356	0.0210	233.88	0.5110	9.00	1.7526	0.0379	0.0216	833.39	0.4336
10.0	1.7021	0.0290	0.0170	381.61	0.5070	10.0	1.7557	0.0308	0.0176	1582.3	0.4289
15.0	1.7089	0.0130	0.0076	2648.8	0.4975	15.0	1.7632	0.0138	0.0078	20,350.0	0.4175
∞	1.7143	0.0000	0.0000	∞	0.4898	∞	1.7692	0.0000	0.0000	∞	0.4083

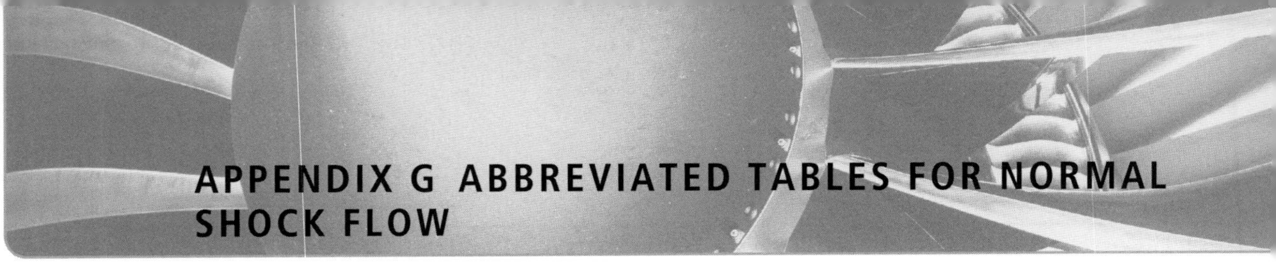

APPENDIX G ABBREVIATED TABLES FOR NORMAL SHOCK FLOW

Also see NormalShock.exe

$\gamma = 1.40$						
M_1	M_2	ρ_2/ρ_1	T_2/T_1	p_2/p_1	p_{t2}/p_{t1}	p_{t2}/p_1
1.00	1.0000	1.0000	1.0000	1.0000	1.0000	1.8929
1.05	0.9531	1.0840	1.0328	1.1196	0.9999	2.0083
1.10	0.9118	1.1691	1.0649	1.2450	0.9989	2.1328
1.15	0.8750	1.2550	1.0966	1.3762	0.9967	2.2661
1.20	0.8422	1.3416	1.1280	1.5133	0.9928	2.4075
1.25	0.8126	1.4286	1.1594	1.6562	0.9871	2.5568
1.30	0.7860	1.5157	1.1909	1.8050	0.9794	2.7136
1.35	0.7618	1.6028	1.2226	1.9596	0.9697	2.8778
1.40	0.7397	1.6897	1.2547	2.1200	0.9582	3.0492
1.45	0.7196	1.7761	1.2872	2.2862	0.9448	3.2278
1.50	0.7011	1.8621	1.3202	2.4583	0.9298	3.4133
1.55	0.6841	1.9473	1.3538	2.6362	0.9132	3.6057
1.60	0.6684	2.0317	1.3880	2.8200	0.8952	3.8050
1.65	0.6540	2.1152	1.4228	3.0096	0.8760	4.0110
1.70	0.6405	2.1977	1.4583	3.2050	0.8557	4.2238
1.75	0.6281	2.2791	1.4946	3.4062	0.8346	4.4433
1.80	0.6165	2.3592	1.5316	3.6133	0.8127	4.6695
1.85	0.6057	2.4381	1.5693	3.8262	0.7902	4.9023
1.90	0.5956	2.5157	1.6079	4.0450	0.7674	5.1418
1.95	0.5862	2.5919	1.6473	4.2696	0.7442	5.3878
2.00	0.5774	2.6667	1.6875	4.5000	0.7209	5.6404
2.10	0.5613	2.8119	1.7704	4.9783	0.6742	6.1654
2.20	0.5471	2.9512	1.8569	5.4800	0.6281	6.7165
2.30	0.5344	3.0845	1.9468	6.0050	0.5833	7.2937
2.40	0.5231	3.2119	2.0403	6.5533	0.5401	7.8969
2.50	0.5130	3.3333	2.1375	7.1250	0.4990	8.5261
2.60	0.5039	3.4490	2.2383	7.7200	0.4601	9.1813
2.70	0.4956	3.5590	2.3429	8.3383	0.4236	9.8624
2.80	0.4882	3.6636	2.4512	8.9800	0.3895	10.569

(cont.)

$\gamma = 1.40$						
M_1	M_2	ρ_2/ρ_1	T_2/T_1	p_2/p_1	p_{t2}/p_{t1}	p_{t2}/p_1
2.90	0.4814	3.7629	2.5632	9.6450	0.3577	11.302
3.00	0.4752	3.8571	2.6790	10.333	0.3283	12.061
3.50	0.4512	4.2609	3.3150	14.125	0.2129	16.242
4. 00	0.4350	4.5714	4.0469	18.500	0.1388	21.068
4.00	0.4236	4.8119	4.8751	23.458	0.0917	26.539
5.00	0.4152	5.0000	5.8000	29.000	0.0617	32.653
5.50	0.4090	5.1489	6.8218	35.125	0.0424	39.412
6.00	0.4042	5.2683	7.9406	41.833	0.0297	46.815
7.00	0.3974	5.4444	10.469	57.000	0.0154	63.553
8.00	0.3929	5.5652	13.387	74.500	0.0085	82.865
9.00	0.3898	5.6512	16.693	94.333	0.0050	104.75
10.0	0.3876	5.7143	20.388	116.50	0.0030	129.22
15.0	0.3823	5.8696	44.694	262.33	0.0004	290.16
20.0	0.3804	5.9259	78.722	466.50	0.0001	515.48
25.0	0.3795	5.9524	122.47	729.00	0.0000	805.18
∞	0.3780	6.0000	∞	∞	0.0000	∞

$\gamma = 1.30$						
M_1	M_2	ρ_2/ρ_1	T_2/T_1	p_2/p_1	p_{t2}/p_{t1}	p_{t2}/p_1
1.00	1.0000	1.0000	1.0000	1.0000	1.0000	1.8324
1.05	0.9530	1.0880	1.0257	1.1159	0.9999	1.9407
1.10	0.9112	1.1777	1.0506	1.2374	0.9989	2.0578
1.15	0.8739	1.2691	1.0752	1.3646	0.9966	2.1832
1.20	0.8403	1.3618	1.0995	1.4974	0.9926	2.3164
1.25	0.8100	1.4557	1.1238	1.6359	0.9866	2.4571
1.30	0.7825	1.5505	1.1480	1.7800	0.9786	2.6050
1.35	0.7575	1.6459	1.1725	1.9298	0.9685	2.7599
1.40	0.7346	1.7419	1.1971	2.0852	0.9563	2.9217
1.45	0.7136	1.8382	1.2220	2.2463	0.9421	3.0902
1.50	0.6942	1.9346	1.2473	2.4130	0.9261	3.2654
1.55	0.6764	2.0310	1.2730	2.5854	0.9084	3.4471
1.60	0.6599	2.1272	1.2991	2.7635	0.8891	3.6353

(cont.)

$\gamma = 1.30$						
M_1	M_2	ρ_2/ρ_1	T_2/T_1	p_2/p_1	p_{t2}/p_{t1}	p_{t2}/p_1
1.65	0.6446	2.2230	1.3257	2.9472	0.8685	3.8300
1.70	0.6304	2.3185	1.3529	3.1365	0.8466	4.0311
1.75	0.6172	2.4133	1.3805	3.3315	0.8238	4.2385
1.80	0.6048	2.5074	1.4087	3.5322	0.8001	4.4522
1.85	0.5933	2.6007	1.4375	3.7385	0.7758	4.6723
1.90	0.5825	2.6932	1.4668	3.9504	0.7510	4.8986
1.95	0.5724	2.7846	1.4968	4.1680	0.7259	5.1312
2.00	0.5629	2.8750	1.5274	4.3913	0.7006	5.3700
2.10	0.5455	3.0524	1.5905	4.8548	0.6499	5.8662
2.20	0.5301	3.2248	1.6562	5.3409	0.6000	6.3872
2.30	0.5163	3.3920	1.7245	5.8496	0.5515	6.9330
2.40	0.5040	3.5536	1.7956	6.3809	0.5050	7.5033
2.50	0.4929	3.7097	1.8694	6.9348	0.4610	8.0983
2.60	0.4829	3.8600	1.9459	7.5113	0.4196	8.7178
2.70	0.4738	4.0045	2.0253	8.1104	0.3810	9.3619
2.80	0.4655	4.1434	2.1075	8.7322	0.3452	10.030
2.90	0.4580	4.2766	2.1925	9.3765	0.3123	10.723
3.00	0.4511	4.4043	2.2804	10.044	0.2822	11.441
3.50	0.4241	4.9648	2.7630	13.717	0.1677	15.395
4.00	0.4058	5.4118	3.3180	17.957	0.0993	19.959
4.50	0.3927	5.7678	3.9462	22.761	0.0594	25.133
5.00	0.3832	6.0526	4.6476	28.130	0.0361	30.915
5.50	0.3760	6.2822	5.4225	34.065	0.0224	37.308
6.00	0.3704	6.4687	6.2709	40.565	0.0142	44.309
7.00	0.3625	6.7485	8.1886	55.261	0.0061	60.138
8.00	0.3573	6.9434	10.401	72.217	0.0028	78.403
9.00	0.3536	7.0837	12.908	91.435	0.0014	99.103
10.0	0.3510	7.1875	15.710	112.91	0.0007	122.24
15.0	0.3448	7.4460	34.141	254.22	0.0001	274.45
20.0	0.3426	7.5410	59.945	452.04	0.0000	487.55
25.0	0.3415	7.5858	93.121	706.39	0.0000	761.53
∞	0.3397	7.6667	∞	∞	0.0000	∞

APPENDIX H IDEAL CYCLE ANALYSIS EQUATIONS DERIVATIONS

H.1 Introduction

Chapter 2 included descriptions of ideal components, as well as nonideal components, with their operating characteristics. This appendix covers the analysis of different gas turbines with ideal components. Different combinations of the various components are used as determined by the engine type of concern, and they are assembled. For the different jet engine types, analytical, closed-form equations for thrust and *TSFC* are derived based on the component equations developed in Section 2.3. A similar procedure is used for a power generation gas turbine providing equations for net power, *SFC*, and thermal efficiencies. Kerrebrock (1992) presented some of the following resulting equations in a somewhat different format but did not completely present the development. This appendix will allow the reader to see how the equations are assembled to model a gas turbine without being encumbered by numerical details. More importantly, this approach will allow the reader to observe directly how detail parameters affect the overall performance of an engine without having to vary numbers parametrically. In some cases, analytical optimizations will be possible based on the performance equations. Solutions for different gas turbine types and conditions can also be obtained using the software JetEngineCycle.exe and PowerGTCycle.exe. Final equations are summarized in Chapter 3. In Chapter 3 numerous numerical examples and trend studies are presented for both ideal and nonideal gas turbines to reinforce the dependence of overall performance on component characteristics. These studies are inherent to the presentation of the educational material.

To reiterate the characteristics and definitions of ideal performance, one should remember the important assumptions for the ideal case include the following.

(a) Constant specific heat ratio
(b) Negligible fuel flow, $f = \dfrac{\dot{m}_f}{\dot{m}}$, is small
(c) Exit nozzle/exhaust pressure matches ambient pressure
(d) Isentropic flow in all noncombusting components
(e) Constant total pressure flow in all combusting components
(f) Perfect combustion in all combusting components

H.2 Ramjet/Scramjet

A general ramjet is considered in Figure 3.1. As indicated earlier, this is the simplest of jet engines because it does not have any rotating shafts. From a cycle analysis standpoint, both the ramjet and scramjet (Figure 1.16) can be treated the same. Clearly both have different design characteristics due to the supersonic flows, but as far as the assemblage of components both behave similarly. It is also noteworthy that both ramjets and scramjets have numerous shocks within the engine. As a result, in reality both have significant losses in total pressure. Regardless, for the ideal analysis such losses are not included. In Figure 3.1, position a is the ambient condition, position 1 is the inlet to the diffuser, position 3 is the exit of the diffuser and inlet to the combustor, position 4 is the exit of the burner and inlet to the nozzle, and lastly position 8 is the exit of the nozzle. Figure H.1 presents the h–s diagram for

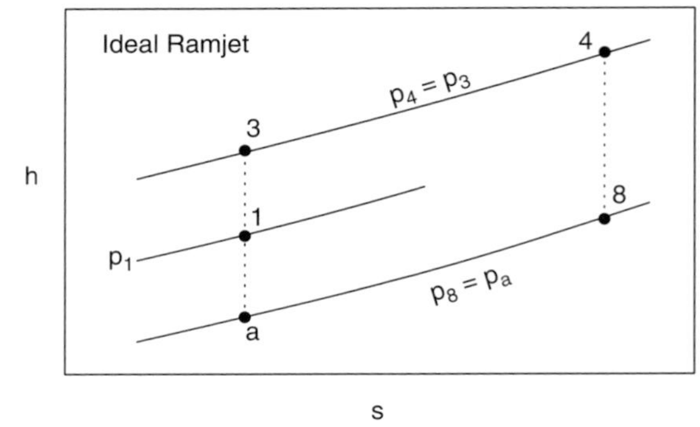

Figure H.1 The h--s diagram for an ideal ramjet.

an ideal ramjet system. Process a to 1 to 3 is an isentropic compression in which the pressure is increased by the "ram" effect. Process 3 to 4 is an isobaric combustion process. Finally, process 4 to 8 is an isentropic expansion process that accelerates the flow velocity and decreases the pressure. Both states a and 8 are on the same pressure line but do not constitute a process as in a closed cycle.

The objective now is to determine the thrust for given component characteristics. Thus, Eq. (1.5.14) is used because the fuel flow rate has been assumed to be small for the ideal case:

$$F = \dot{m}(u_e - u_a) + A_e(p_e - p_a). \tag{H.2.1}$$

The airflow rate ($\dot{m}$) is presumably known for an engine. Also, the aircraft speed, u_a, is presumed to be known. Furthermore, because the nozzle exit pressure matches atmospheric pressure for the ideal case, $p_e = p_a$. Thus, the only quantity that needs to be found for calculation of the thrust is u_e, which for the ramjet is also labeled u_8. Nondimensionalizing the thrust by $\dot{m}u_a$ results in

$$\frac{F}{\dot{m}u_a} = \left(\frac{u_8}{u_a} - 1\right). \tag{H.2.2}$$

Thus, to find this parameter it is first necessary to identify the nozzle exit gas velocity u_8 or the ratio u_8/u_a. Thus, the objective of the next several steps is to find F and $TSFC$ eventually in terms of the engine operating parameters. As indicated, u_8/u_a must be determined. First, this ratio can be identified as

$$\frac{u_8}{u_a} = \frac{M_8 a_8}{M_a a_a} = \frac{M_8\sqrt{\gamma \mathcal{R} T_8}}{M_a\sqrt{\gamma \mathcal{R} T_a}} = \frac{M_8\sqrt{T_8}}{M_a\sqrt{T_a}}, \tag{H.2.3}$$

where M_a and M_8 are the freestream and exit Mach numbers, respectively, and a_a and a_8 are the freestream and exit sound speeds, respectively. Next, the total pressures at positions a and 8 can be considered:

$$\frac{p_{ta}}{p_a} = \left[1 + \frac{\gamma - 1}{2} M_a^2\right]^{\frac{\gamma}{\gamma-1}} \tag{H.2.4}$$

$$\frac{p_{t8}}{p_8} = \left[1 + \frac{\gamma - 1}{2} M_8^2\right]^{\frac{\gamma}{\gamma-1}}. \tag{H.2.5}$$

However, from the ideal assumption that the exit pressure matches ambient conditions, it follows that

$$p_8 = p_a. \tag{H.2.6}$$

Also, processes a to 1 (external flow), 1 to 3 (diffuser), and 4 to 8 (nozzle) are all isentropic, whereas process 3 to 4 (combustor) is isotobaric (constant total pressure, not to be confused with isobaric). Thus, the total pressure is constant throughout an ideal ramjet:

$$p_{ta} = p_{t1} = p_{t3} = p_{t4} = p_{t8}. \tag{H.2.7}$$

Therefore, from Eqs. (H.2.6) and (H.2.7),

$$\frac{p_{ta}}{p_a} = \frac{p_{t8}}{p_8}. \tag{H.2.8}$$

Thus, from Eqs. (H.2.4), (H.2.5), and (H.2.8), one finds that

$$M_a = M_8. \tag{H.2.9}$$

At first, this may look surprising. The Mach numbers at the freestream and exhaust are the same, but not the velocities. The gas entering the nozzle is very hot since it is just exiting from the burner. The exhaust velocity can be found using Eqs. (H.2.9) and (H.2.3):

$$\frac{u_8}{u_a} = \frac{\sqrt{T_8}}{\sqrt{T_a}}. \tag{H.2.10}$$

Now, for the total temperatures at locations a and 8, one finds the ratio from Eq. (C.2.9) as follows:

$$\frac{T_{t8}}{T_{ta}} = \frac{T_8\left[1 + \dfrac{\gamma - 1}{2} M_8^2\right]}{T_a\left[1 + \dfrac{\gamma - 1}{2} M_a^2\right]}. \tag{H.2.11}$$

However, since the Mach numbers are the same,

$$\frac{T_8}{T_a} = \frac{T_{t8}}{T_{ta}}. \tag{H.2.12}$$

Processes a to 3 (external and diffuser) and 4 to 8 (nozzle) are both adiabatic; thus, the total temperatures are constant for these processes:

$$T_{t3} = T_{ta} \tag{H.2.13}$$

$$T_{t8} = T_{t4}. \tag{H.2.14}$$

Thus,

$$\frac{T_{t8}}{T_{ta}} = \frac{T_{t4}}{T_{t3}} \equiv \tau_b, \tag{H.2.15}$$

which is the burner total temperature ratio. Using Eqs. (H.2.15), (H.2.12), and (H.2.10) yields

$$\frac{u_8}{u_a} = \sqrt{\tau_b}. \tag{H.2.16}$$

And so, from Eq. (H.2.2), a dimensionless quantity is

$$\frac{F}{\dot{m}u_a} = (\sqrt{\tau_b} - 1). \tag{H.2.17}$$

Alternatively, if the value of the freestream velocity is expressed in terms of the Mach number and speed of sound ($u_a = M_a a_a$), the dimensionless thrust is

$$\frac{F}{\dot{m}a_a} = M_a(\sqrt{\tau_b} - 1). \tag{H.2.18}$$

Thus, this represents half of the objective. By examining this simple equation, one can see that the dimensional thrust is directly proportional to the mass flow rate. It is also directly proportional to the Mach number if all of the other parameters are constant. One limiting case is for a Mach number of zero; this equation indicates that the thrust is zero for this case. A ramjet is unique in this respect and is not capable of takeoff on its own as are all other engines. Moreover, as the burner total temperature ratio increases, so does the thrust. One should remember, however, that $\tau_b = T_{t4}/T_{t3}$ and that T_{t3} is related to the freestream Mach number $\left(T_{t3} = T_a\left[1 + \frac{\gamma - 1}{2}M_a^2\right]\right)$; consequently, independently varying M_a and τ_b is somewhat difficult.

The thrust has been found, but the *TSFC* remains to be determined. Because this quantity depends on the fuel flow rate, the steady-state energy equation needs to be examined for the ideal combustor (Eq. (2.3.65)) with Eqs. (H.2.13) and (2.3.63). The fuel ratio $(\dot{m}_f/\dot{m})$ is

$$f = \frac{c_p T_{ta}(\tau_b - 1)}{\Delta H} = \frac{\dot{m}_f}{\dot{m}}. \tag{H.2.19}$$

Finally, using Eqs. (H.2.18), (H.2.19), (1.6.2), and (1.6.3) yields the nondimensional *TSFC*:

$$TSFC\ a_a = \frac{c_p T_{ta}(\tau_b - 1)}{M_a \Delta H(\tau_b - 1)}. \tag{H.2.20}$$

From this equation, the *TSFC* can be evaluated based on the engine characteristics. By examining this equation, one can see that the *TSFC* is directly proportional to the total ambient temperature, but it must be remembered that T_{ta} is related to the Mach number. Also, as the burner total temperature ratio increases, so does the *TSFC*. Thus, equations for both of the desired performance quantities (thrust and *TSFC*) have been determined. For a typical ramjet, the values of *TSFC* range from 1.5 to 3 $\text{lb}_m/\text{h-lb}_f$.

Using Eq. (2.3.3) to find T_{ta}, one can differentiate Eq. (H.2.20) with respect to M_a set to zero and easily show that, for constant altitude, heating value, and τ_b, the minimum value of *TSFC* occurs for

$$M_{a_{opt}} = \sqrt{\frac{2}{\gamma - 1}}. \tag{H.2.21}$$

A special case, namely the stoichiometric condition, is now considered. For this condition, exactly the right proportion of fuel and air will be used so that all of the fuel and air are consumed in the combustion process and neither is left in the combustion products. This subject is covered in more detail in Chapter 9 but is identified now. Basically, for a given fuel, the stoichiometric condition fixes τ_b and f, which are defined as τ_{bst} and f_{st}, respectively. Typically, the stoichiometric fuel ratio is 0.06 to 0.07 for jet fuels, which results in very high temperatures – too high for engines or power units with turbines, as will be shown in Chapters 7 and 9. However, ramjets can tolerate much higher temperatures because of the lack of rotating parts. Thus, Eq. (H.2.19) becomes

$$\tau_{b_{st}} - 1 = \frac{f_{st}\Delta H}{c_p T_{ta}}. \tag{H.2.22}$$

Using Eq. (H.2.20) yields

$$TSFC_{st} = \frac{f_{st}}{M_a a_a(\sqrt{\tau_b} - 1)}. \tag{H.2.23}$$

Thus, Eqs. (H.2.22) and (H.2.23) yield

$$TSFC_{st} = \frac{f_{st}}{M_a a_a\left(\sqrt{\dfrac{f_{st}\Delta H\left(\dfrac{T_a}{T_{ta}}\right)}{c_p T_a} + 1} - 1\right)}. \tag{H.2.24}$$

Note that the quantity T_{ta}/T_a is solely a function of the freestream Mach number and is given by Eq. (2.3.3). Thus, for a given fuel at stoichiometric conditions, the value of $TSFC$ is only a function of the Mach number and T_a (or a_a for a given γ).

H.3 Turbojet

A general turbojet is shown in Figure 3.2. This is a much more complex machine than the ramjet because of the rotating shaft. Eight stations are identified. The freestream is position a. The inlet is position 1. Position 2 is the exit of the diffuser and compressor inlet. Station 3 is the compressor exit and burner inlet. The burner exit and turbine inlet are denoted as station 4. Station 5 is the turbine exit and afterburner inlet. The afterburner exit and nozzle inlet are at position 6. Finally, station 8 is the nozzle exhaust. For an ideal analysis, single or multispool engines will yield exactly the same results because only the overall effects of the components are of importance. The h–s diagram for the general ideal turbojet is presented in Figure H.2. Process a to 1 to 2 is an isentropic compression during the diffusion. Process 2 to 3 is an isentropic compression while power is required by the compressor. Process 3 to 4 is an isobaric combustion process. The isentropic expansion through the turbine is process 4 to 5, during which power is removed from the flow. The afterburner process (5 to 6) is also isobaric. And finally, process 6 to 8 is an isentropic expansion in the nozzle where the flow velocity is accelerated. States a and 8 are once again on the same pressure line but do not constitute a process. For the analysis nonafterburning and afterburning configurations will be considered separately.

H.3.1 Nonafterburning

For this configuration, stations 5 and 6 are identical (Figure 3.3). For the ideal analysis, Eq. (H.2.1) once again applies. For the turbojet engine, the dimensionless quantity is (recall that for the ideal case $\dot{m}_f$ is small and that $p_e = p_a$)

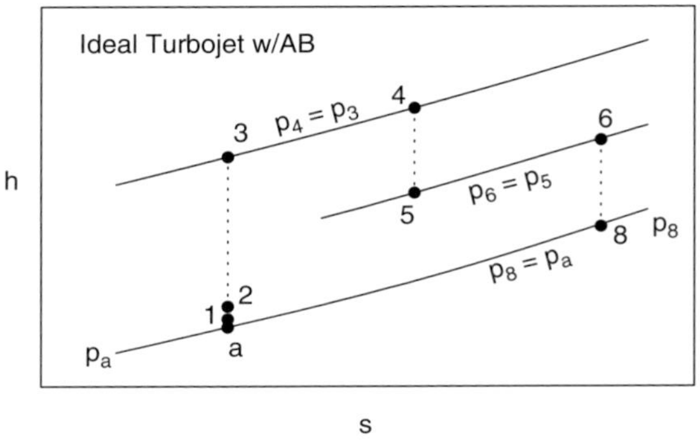

Figure H.2 The h–s diagram for an ideal turbojet with an afterburner.

$$\frac{F}{\dot{m}u_a} = \left(\frac{u_8}{u_a} - 1\right). \tag{H.3.1}$$

Thus, once again, the velocity ratio u_8/u_a is needed to proceed. If one considers the exit total temperature and multiplies and divides by the same terms, realizing that the diffuser and nozzle are adiabatic ($T_{t2} = T_{ta}$ and $T_{t8} = T_{t5}$) for the ideal case, the exit total temperature is

$$T_{t8} = T_a \frac{T_{ta}}{T_a}\frac{T_{t3}}{T_{t2}}\frac{T_{t4}}{T_{t3}}\frac{T_{t5}}{T_{t4}}. \tag{H.3.2}$$

Therefore,

$$T_{t8} = T_a \frac{T_{ta}}{T_a}\tau_c\tau_b\tau_t, \tag{H.3.3}$$

where τ_c and τ_t are the total temperature ratios for the compressor and turbine, respectively. Also, given the exit total pressure and because, for the ideal case, the diffuser and nozzle are isentropic ($p_{t2} = p_{ta}$ and $p_{t8} = p_{t5}$), the nozzle exit total pressure is

$$p_{t8} = p_a \frac{p_{ta}}{p_a}\frac{p_{t3}}{p_{t2}}\frac{p_{t4}}{p_{t3}}\frac{p_{t5}}{p_{t4}}. \tag{H.3.4}$$

Also, for the ideal case the total pressure is constant across the combustor ($p_{t4} = p_{t3}$). Thus, the nozzle exit total pressure is

$$p_{t8} = p_a \frac{p_{ta}}{p_a}\pi_c\pi_t. \tag{H.3.5}$$

However, for the ideal case the nozzle exit static pressure matches the freestream pressure ($p_8 = p_a$). Thus, from Eqs. (2.3.95) and (H.3.4) one obtains

$$1 + \frac{\gamma - 1}{2}M_8^2 = \left[\frac{p_{ta}}{p_a}\pi_c\pi_t\right]^{\frac{\gamma-1}{\gamma}}. \tag{H.3.6}$$

Next, from Eqs. (2.3.92), (H.3.2), and (H.3.6),

$$\frac{T_{t8}}{T_a} = \frac{\dfrac{T_{ta}}{T_a}\tau_c\tau_b\tau_t}{\left[\dfrac{p_{ta}}{p_a}\pi_c\pi_t\right]^{\frac{\gamma-1}{\gamma}}}. \tag{H.3.7}$$

Thus, using Eqs. (2.3.5), (2.3.27), and (2.3.45), with Eq. (H.3.7) yields the temperature ratio

$$\frac{T_8}{T_a} = \tau_b. \tag{H.3.8}$$

Next, u_8/u_a can be found in the same way as for the ramjet:

$$\frac{u_8}{u_a} = \frac{M_8 a_8}{M_a a_a} = \frac{M_8\sqrt{\gamma \mathcal{R} T_8}}{M_a\sqrt{\gamma \mathcal{R} T_a}} = \frac{M_8\sqrt{T_8}}{M_a\sqrt{T_a}}. \tag{H.3.9}$$

Now, from Eqs. (2.3.92), (H.3.3), and (H.3.8), the square of the exit Mach number is

$$M_8^2 = \frac{2}{\gamma-1}\left[\frac{T_{ta}}{T_a}\tau_c\tau_t - 1\right] \tag{H.3.10}$$

and by solving for the square of the freestream Mach number, M_a^2 from Eq. (2.3.3) we obtain

$$M_a^2 = \frac{2}{\gamma-1}\left[\frac{T_{ta}}{T_a} - 1\right]. \tag{H.3.11}$$

Next, using Eqs. (H.3.8)–(H.3.11) yields

$$\frac{u_8}{u_a} = \sqrt{\tau_b \frac{\dfrac{T_{ta}}{T_a}\tau_c\tau_t - 1}{\dfrac{T_{ta}}{T_a} - 1}}. \tag{H.3.12}$$

Finally, using this with Eq. (H.3.1) yields

$$\frac{F}{\dot{m}a_a} = M_a\left[\sqrt{\tau_b \frac{\dfrac{T_{ta}}{T_a}\tau_c\tau_t - 1}{\dfrac{T_{ta}}{T_a} - 1}} - 1\right]. \tag{H.3.13}$$

This equation may appear to represent the finished product because it yields the thrust in terms of engine parameters. However, one simplification can yet be accomplished. That is, the compressor and turbine are not acting independently because they are on the same shaft. In fact for the ideal analysis, all of the power derived from the turbine drives the compressor. The energy equation applied to the shaft (Eq. (2.3.83)) under steady and ideal conditions reduces to

$$\dot{m}c_p(T_{t3} - T_{t2}) = \dot{m}c_p(T_{t4} - T_{t5}). \tag{H.3.14}$$

Or, solving for the turbine total temperature ratio τ_t, one obtains

$$\tau_t = 1 - \frac{T_{ta}}{T_a}\frac{T_a}{T_{t4}}(\tau_c - 1). \tag{H.3.15}$$

Thus, using this with Eq. (H.3.13) yields the dimensionless thrust:

$$\frac{F}{\dot{m}a_a} = M_a\left[\sqrt{\left[\frac{\frac{T_{ta}}{T_a}}{\frac{T_{ta}}{T_a}-1}\right]\left[\frac{T_{t4}}{T_a}\frac{T_a}{T_{ta}}\frac{1}{\tau_c}-1\right][\tau_c-1]+\left[\frac{\frac{T_{t4}}{T_a}}{\frac{T_{ta}}{T_a}\tau_c}\right]}-1\right]. \quad \text{(H.3.16)}$$

This equation therefore, represents the ideal thrust equation for a turbojet. By examining this equation, one can see that, as is the case for a ramjet, the dimensional thrust is directly proportional to the mass flow rate. It is also directly proportional to the Mach number if all of the other parameters are constant. However, one should remember that T_{ta}/T_a is strictly a function of Mach number and γ; that is, $\frac{T_{ta}}{T_a} = 1 + \frac{\gamma-1}{2}M_a^2$. Also, as the burner total temperature increases, so does the thrust. However, as the ambient static temperature decreases, the thrust increases.

Also, in Eq. (H.3.16), the compressor total temperature ratio, which is directly related to the total pressure ratio, is seen to influence the thrust performance strongly. By differentiating this equation with respect to τ_c and setting to zero, one finds that the nondimensional thrust is maximized for a value of

$$\tau_{c\text{-opt}} = \sqrt{(T_{t4}/T_a)/(T_{ta}/T_a)}, \quad \text{(H.3.17)}$$

which can be related to the total pressure ratio by Eq. (2.3.27): $\left(\pi_{c\text{-opt}} = [\tau_{c\text{-opt}}]^{\frac{\gamma}{\gamma-1}}\right)$.

Next, to find the *TSFC* one must again use the energy equation for the burner as is done for the ramjet and recognizing that the total temperature ratio across the burner can be written as

$$\frac{T_{t4}}{T_{t3}} = \frac{T_{t4}}{T_a}\frac{T_a}{T_{t2}}\frac{T_{t2}}{T_{t3}}. \quad \text{(H.3.18)}$$

And next, with Eq. (2.3.26), for the ideal case ($T_{t2} = T_{ta}$),

$$\tau_b \equiv \frac{T_{t4}}{T_{t3}} = \frac{T_{t4}}{T_a}\frac{T_a}{T_{ta}}\frac{1}{\tau_c}. \quad \text{(H.3.19)}$$

Finally, using Eq. (2.3.65) for the first law of combustion process with Eqs. (1.6.1), (1.6.3), (H.3.16), and (H.3.19) yields

$$TSFC\ a_a = \frac{\left[\frac{T_{t4}}{T_a} - \frac{T_{ta}}{T_a}\tau_c\right]\left[\frac{c_p T_a}{\Delta H M_a}\right]}{\left[\sqrt{\left[\frac{\frac{T_{ta}}{T_a}}{\frac{T_{ta}}{T_a}-1}\right]\left[\frac{T_{t4}}{T_a}\frac{T_a}{T_{ta}}\frac{1}{\tau_c}-1\right][\tau_c-1]+\left[\frac{T_{t4}}{T_a}\frac{T_a}{T_{ta}}\frac{1}{\tau_c}\right]}-1\right]}. \quad \text{(H.3.20)}$$

This equation thus is the nondimensionalized *TSFC* for an ideal turbojet. Once again, for both Eqs. (H.3.16) and (H.3.20), the quantity (T_{ta}/T_a) is strictly a function of the freestream Mach number. One can see that, as the ambient static temperature decreases, the *TSFC* decreases. Careful examination also shows that the *TSFC* decreases as τ_c (or π_c) increases. Typical values of *TSFC* for a turbojet range from 0.75 to 1.0 $lb_m/h\text{-}lb_f$.

H.3.2 Afterburning

Next, an afterburning turbojet will be considered. This engine is also shown in Figure 3.2. Much of the analysis is the same as for the nonafterburning type. For example, the diffuser is the same, and the energy balance of the shaft (compressor and turbine) is identical. Some portions are of course different because of the additional energy input.

If one expands the total temperature at the nozzle exit, T_{t8}, the following equation results:

$$T_{t8} = T_a \frac{T_{ta}}{T_a}\frac{T_{t3}}{T_{t2}}\frac{T_{t4}}{T_{t3}}\frac{T_{t5}}{T_{t4}}\frac{T_{t6}}{T_{t5}}. \tag{H.3.21}$$

By identifying the different ratios, one obtains

$$T_{t8} = T_a \frac{T_{ta}}{T_a}\tau_c\tau_b\tau_t\tau_{ab}, \tag{H.3.22}$$

where τ_{ab} is the total temperature ratio for the afterburner. This equation replaces Eq. (H.3.3), which was derived for the nonafterburning case. One can use this with Eqs. (2.3.92), (H.3.6), (2.3.5), (2.3.27), and (2.3.45) to find the ratio of static temperatures
T_8/T_a:

$$\frac{T_8}{T_a} = \tau_b\tau_{ab}, \tag{H.3.23}$$

which replaces Eq. (H.3.8). The velocity ratio is obtained similarly as follows:

$$\frac{u_8}{u_a} = \sqrt{\tau_b\tau_{ab}\frac{\dfrac{T_{ta}}{T_a}\tau_c\tau_t - 1}{\dfrac{T_{ta}}{T_a} - 1}}. \tag{H.3.24}$$

Next, using this with Eq. (H.3.22) results in

$$\frac{u_8}{u_a} = \sqrt{\frac{T_{t6}}{T_a}\left[\frac{1 - \dfrac{1}{\dfrac{T_{ta}}{T_a}\tau_c\tau_t}}{\dfrac{T_{ta}}{T_a} - 1}\right]}, \tag{H.3.25}$$

which replaces Eq. (H.3.12). Next, the energy balance on the combustion process in the afterburner (Eq. (2.3.76)) can be used, and this can be added to the equation for the main burner (Eq. (2.3.65)) to yield

$$(\dot{m}_f + \dot{m}_{f_{ab}})\Delta H = \dot{m}c_p[T_{t4} - T_{t3} + T_{t6} - T_{t5}], \tag{H.3.26}$$

but for the ideal analysis from the energy balance of the shaft, one finds from Eq. (H.3.14) that

$$T_{t3} - T_{t2} = T_{t4} - T_{t5}. \tag{H.3.27}$$

Using this with the fact that the diffuser and nozzle are adiabatic yields

$$\dot{m}_{f_t}\Delta H = \dot{m}c_p(T_{t8} - T_{ta}), \tag{H.3.28}$$

where $\dot{m}_{f_t}$ is the total fuel flow in the engine $(\dot{m}_f + \dot{m}_{f_{ab}})$. It is now possible to solve for the total fuel ratio:

$$f = \frac{\dot{m}_{f_t}}{\dot{m}} = \frac{c_p T_{t8}}{\Delta H}\left[1 - \frac{T_{ta}}{T_{t8}}\right]. \tag{H.3.29}$$

Finally, Eq. (H.3.25) can be combined with previously derived Eqs. (H.3.15) and (H.3.1) to obtain the thrust as follows:

$$\frac{F}{\dot{m}a_a} = M_a\left[\sqrt{\left[\frac{\dfrac{T_{t8}}{T_a}}{\dfrac{T_{ta}}{T_a} - 1}\right]\left[1 - \frac{\dfrac{T_{t4}}{T_a}\dfrac{T_a}{T_{ta}}}{\tau_c\left(\dfrac{T_{t4}}{T_a} - \dfrac{T_{ta}}{T_a}(\tau_c - 1)\right)}\right]} - 1\right]. \tag{H.3.30}$$

This can be compared with the nonafterburning case (Eq. (H.3.16)). One can see that, as is true for a simple turbojet, the dimensional thrust is directly proportional to the mass flow rate. It is also directly proportional to the Mach number if all of the other parameters are constant. As the burner or afterburner total temperature increases, or both total temperatures increase so does the thrust. Again, in Eq. (H.3.30), the compressor total temperature ratio is seen to influence the thrust performance strongly. By differentiating this equation relative to τ_c for a given value of T_{t8}, and setting to zero, one finds that the nondimensional thrust is maximized for a value of

$$\tau_{c_{opt}} = [(T_{t4}/T_a) + (T_{ta}/T_a)]/[2(T_{ta}/T_a)], \tag{H.3.31}$$

which can be related to the total pressure ratio by Eq. (2.3.27). The optimum value for the afterburning turbojet is considerably higher than for the nonafterburning case.

To determine the nondimensional *TSFC*, one now uses Eqs. (H.3.29) and (H.3.30) with previously derived Eqs. (1.6.1), (1.6.3), and (H.3.19) to find

$$TSFC\ a_a = \frac{\left[\dfrac{T_{t8}}{T_a} - \dfrac{T_{ta}}{T_a}\tau_c\right]\left[\dfrac{c_p T_a}{\Delta H M_a}\right]}{\left[\sqrt{\left[\frac{\dfrac{T_{t8}}{T_a}}{\dfrac{T_{ta}}{T_a} - 1}\right]\left[1 - \frac{\dfrac{T_{t4}}{T_a}\dfrac{T_a}{T_{ta}}}{\tau_c\left(\dfrac{T_{t4}}{T_a} - \dfrac{T_{ta}}{T_a}(\tau_c - 1)\right)}\right]} - 1\right]}. \tag{H.3.32}$$

Thus, this represents the nondimensionalized *TSFC* for an afterburning turbojet and replaces Eq. (H.3.20).

H.4 Turbofan: Fan Exhausted

The general diagram for a jet engine is shown in Figure 2.3. As indicated earlier in this chapter, this engine has up to three basic flow paths. Two simplifications of the general turbofan are covered for the ideal analysis: the fully exhausted fan and the ducted/mixed fan. Both nonafterburning and afterburning cases are considered. The "hybrid" fan (in which a portion of the secondary flow is mixed while the remainder is exhausted) is covered in Chapter 3 for nonideal cycles. In this section fan exhausted is covered.

This type of a turbofan is shown in Figure 3.4. For this configuration two flow paths exist. For one path, the air enters the diffuser, flows through the core, and exhausts through the primary nozzle. For the second

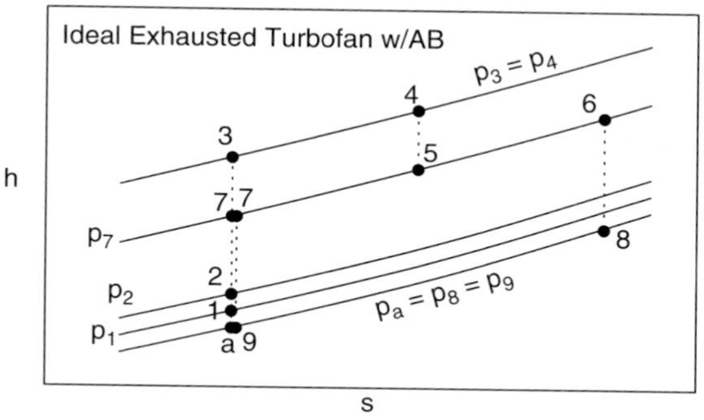

Figure H.3 The h–s diagram for an ideal exhausted fan with an afterburner

path, air enters the diffuser, flows through the fan, and exits through the fan nozzle. Two separate subcases are considered here: without and with the afterburner. In Figure 3.4, an afterburner is present in the core between stations 5 and 6. Analyses for the two subcases are performed separately. In Figure H.3, the h–s diagram for the afterburning configuration is shown. Steps a to 3 are the diffuser and then compressor isentropic processes. Steps a to 7 are the diffuser and fan processes. Processes 3 to 4 and 5 to 6 are the combustion and afterburning operations, respectively. Step 4 to 5 is the isentropic turbine expansion process. Lastly, steps 7 to 9 and 6 to 8 are the separate secondary and primary nozzle processes, respectively.

H.4.1 Nonafterburning

For the nonafterburning case, stations 5 and 6 are one and the same and will be termed station 5, as shown in Figure 3.5. For this ideal case ($p_e = p_a$ and $p_s = p_a$), the thrust is generated by two separate gas streams and found from Eq. (1.5.27) as follows:

$$F = \dot{m}(u_8 - u_a) + \dot{m}_s(u_9 - u_a), \tag{H.4.1}$$

where u_9 is the gas velocity at the fan nozzle exit. Or this can be written as

$$F = \dot{m}u_a\left(\frac{u_8}{u_a} - 1\right) + \dot{m}_s u_a\left(\frac{u_9}{u_a} - 1\right), \tag{H.4.2}$$

where $\dot{m}$ is the airflow rate into the core and $\dot{m}_s$ is the bypassed airflow rate (the air that flows only through the outer section of the fan and fan nozzle – or secondary flow rate). The flow path from station a to station 8 is exactly the same as for the turbojet, and thus u_8/u_a is already known. However, for the fan the quantity u_9/u_a is needed. The ratio u_9/u_a can be written as

$$\frac{u_9}{u_a} = \frac{M_9 a_9}{M_a a_a} = \frac{M_9\sqrt{\gamma \mathcal{R} T_9}}{M_a\sqrt{\gamma \mathcal{R} T_a}} = \frac{M_9\sqrt{T_9}}{M_a\sqrt{T_a}}, \tag{H.4.3}$$

but for the ideal analysis the fan nozzle exit pressure matches the ambient pressure

$$p_9 = p_a. \tag{H.4.4}$$

Next, one can recall that the processes a to 1, 1 to 2, 2 to 7, and 7 to 9 are all isentropic. Therefore, the entire process a to 9 is also isentropic. Thus, for process a to 9, Eq. (C.2.10) applies. Consequently, from Eqs. (H.4.4) and (C.2.10), the nozzle exit static temperature matches the atmospheric temperature:

$$T_9 = T_a. \tag{H.4.5}$$

Also, by expanding on the nozzle total exit temperature T_{t9}, one obtains

$$T_{t9} = T_a \frac{T_{ta}}{T_a} \frac{T_{t7}}{T_{t2}} = T_a \frac{T_{ta}}{T_a} \tau_f. \tag{H.4.6}$$

Thus, from Eqs. (H.4.6) and (2.3.110), the square of the fan nozzle exit Mach number is

$$M_9^2 = 2 \left[\frac{\dfrac{T_{ta}}{T_a} \tau_f - 1}{\gamma - 1} \right], \tag{H.4.7}$$

where τ_f is the total temperature ratio for the fan. Also, from Eq. (2.3.3), the square of the freestream Mach number is

$$M_a^2 = 2 \left[\frac{\dfrac{T_{ta}}{T_a} - 1}{\gamma - 1} \right]. \tag{H.4.8}$$

Therefore, from Eqs. (H.4.3), (H.4.5), (H.4.7), and (H.4.8) one finds the ratio

$$\frac{u_9}{u_a} = \sqrt{\frac{\dfrac{T_{ta}}{T_a} \tau_f - 1}{\dfrac{T_{ta}}{T_a} - 1}}. \tag{H.4.9}$$

Next, the energy equation applied to the rotating shaft (Eq. (2.3.83)) reduces to

$$\dot{m} c_p (T_{t4} - T_{t5}) = \dot{m} c_p (T_{t3} - T_{t2}) + \alpha \dot{m} c_p (T_{t7} - T_{t2}), \tag{H.4.10}$$

where α is the bypass ratio, $\alpha = \dot{m}_s / \dot{m}$, and which reduces to

$$\frac{T_{t4}}{T_a} \left(1 - \frac{T_{t5}}{T_{t4}} \right) = \frac{T_{t2}}{T_a} \left(\frac{T_{t3}}{T_{t2}} - 1 \right) + \alpha \frac{T_{t2}}{T_a} \left(\frac{T_{t7}}{T_{t2}} - 1 \right). \tag{H.4.11}$$

Or, by using Eqs. (2.3.45), (2.3.26), and (2.3.40), one obtains

$$\frac{T_{t4}}{T_a} (1 - \tau_t) = \frac{T_{t2}}{T_a} (\tau_c - 1) + \alpha \frac{T_{t2}}{T_a} (\tau_f - 1), \tag{H.4.12}$$

and solving for the turbine total temperature ratio τ_t yields

$$\tau_t = 1 - \left[\frac{T_{ta}}{T_a} \times \frac{T_a}{T_{t4}} \right] [(\tau_c - 1) + \alpha(\tau_f - 1)]. \tag{H.4.13}$$

Because the core is essentially a simple turbojet and $p_8 = p_a$, one can use Eq. (H.3.12) to find u_8/u_a. Thus, Eqs. (H.3.12), (H.3.19), (H.4.2), (H.4.9), and (H.4.13) are used to yield the dimensionless thrust:

$$\frac{F}{\dot{m}a_a} = M_a\left[\sqrt{\left[\frac{\left[\frac{T_{t4}}{T_a}\frac{T_a}{T_{ta}}\frac{1}{\tau_c}\right]\left[\frac{T_{ta}}{T_a}\tau_c\left[1-\left[\frac{T_{ta}}{T_a}\times\frac{T_a}{T_{t4}}\right][(\tau_c-1)+\alpha(\tau_f-1]\right]-1\right]}{\frac{T_{ta}}{T_a}-1}\right]}-1\right] + \alpha M_a\left[\sqrt{\left[\frac{\left[\frac{T_{ta}}{T_a}\tau_f-1\right]}{\frac{T_{ta}}{T_a}-1}\right]}-1\right]. \quad \text{(H.4.14)}$$

Thus, the thrust is a strong function of the compressor total pressure ratio, the fan total pressure ratio, the exit total temperature from the burner, and the bypass ratio. Next, the value of *TSFC* is desired. The combustion energy balance is exactly the same as for the turbojet and is given by Eq. (2.3.65). Thus, using the definition of nondimensional *TSFC* with Eqs. (2.3.65), (H.3.19), and (H.4.14) yields

$$TSFC\, a_a = \frac{\left[\frac{T_{t4}}{T_a}-\frac{T_{ta}}{T_a}\tau_c\right]\left[\frac{c_pT_a}{\Delta H}\right]}{M_a\left[\sqrt{\left[\frac{\left[\frac{T_{t4}}{T_a}\frac{T_a}{T_{ta}}\frac{1}{\tau_c}\right]\left[\frac{T_{ta}}{T_a}\tau_c\left[1-\left[\frac{T_{ta}}{T_a}\times\frac{T_a}{T_{t4}}\right][(\tau_c-1)+\alpha(\tau_f-1)]\right]-1\right]}{\frac{T_{ta}}{T_a}-1}\right]}-1\right]+\alpha M_a\left[\sqrt{\left[\frac{\left[\frac{T_{ta}}{T_a}\tau_f-1\right]}{\frac{T_{ta}}{T_a}-1}\right]}-1\right]}. \quad \text{(H.4.15)}$$

Therefore, both the thrust and *TSFC* have been determined for the turbofan with an exhausted fan. Typical *TSFC* values for this engine are 0.35 to 0.7 $lb_m/h\text{-}lb_f$. For a turbofan with the fan exhausted, one should realize that the temperature ratios across both the fan and compressor are independent. For example, in a design stage, both τ_c and τ_f can be specified.

If the thrust is to be maximized, one can differentiate F with respect to the fan total temperature ratio (which is directly related to the fan total pressure ratio), τ_f, in Eq. (H.4.14), set equal to zero, and show that (which is left as an exercise to the reader) τ_f is specified as

$$\tau_{f\,opt} = \frac{1+\frac{T_{t4}}{T_a}+\frac{T_{ta}}{T_a}[1+\alpha-\tau_c]-\frac{\frac{T_{t4}}{T_a}}{\frac{T_{ta}}{T_a}\tau_c}}{\frac{T_{ta}}{T_a}[1+\alpha]}, \quad \text{(H.4.16)}$$

which can be used to find the optimum total pressure ratio from Eq. (2.3.41). This condition interestingly results in the two nozzle exit velocities being the same:

$$u_8 = u_9. \quad \text{(H.4.17)}$$

Similarly, one can maximize the thrust by differentiating F with respect to the bypass ratio α, set equal to zero, and solve for α, which yields

$$\alpha_{opt} = \frac{1}{(\tau_f - 1)}\left\{\frac{\dfrac{T_{t4}}{T_a}}{\dfrac{T_{ta}}{T_a}}\left(1 - \frac{1}{\tau_c\dfrac{T_{ta}}{T_a}} - \frac{1}{4\dfrac{T_{t4}}{T_a}}\left[\sqrt{\frac{T_{ta}}{T_a}\tau_f - 1} + \sqrt{\frac{T_{ta}}{T_a} - 1}\right]^2\right) - (\tau_c - 1)\right\}. \tag{H.4.18}$$

H.4.2 Afterburning

Next, the afterburning exhausted fan is considered. Figure 3.4 depicts this geometry. As can be seen, the afterburner is strictly in the core region and near the primary exhaust. In general, a second afterburner could be located near the fan exhaust. This geometry is seldom used, however, and will not be covered here. Several of the equations used for the nonafterburning exhausted fan and the afterburning turbojet apply, but a few new derivations are needed. For example, Eq. (H.4.2) still applies. Thus, once again, finding u_9/u_a and $u_8\,/u_a$ is necessary. However, $u_9\,/u_a$ will be exactly as it was for the nonafterburning case and is given by Eq. (H.4.9). Also, the flow through the core is the same as for the afterburning turbojet case. Thus, one can use Eq. (H.3.25) to find $u_8\,/u_a$. Therefore, the dimensionless thrust can be found through the following equation:

$$\frac{F}{\dot{m}a_a} = M_a\left[\sqrt{\left[\frac{\dfrac{T_{t6}}{T_a}}{\dfrac{T_{ta}}{T_a} - 1}\right]\left[1 - \frac{1}{\left(\dfrac{T_{ta}}{T_a}\tau_c\left(1 - \left[\dfrac{T_{ta}}{T_a}\times\dfrac{T_a}{T_{t4}}\right][(\tau_c - 1) + \alpha(\tau_f - 1]\right)\right)}\right]} - 1\right] + \alpha M_a\left[\sqrt{\left[\frac{\left[\dfrac{T_{ta}}{T_a}\tau_f - 1\right]}{\dfrac{T_{ta}}{T_a} - 1}\right]} - 1\right]. \tag{H.4.19}$$

The thrust is a strong function of burner exit temperature, afterburner exit temperature, compressor pressure ratio, fan pressure ratio, and bypass ratio. Of course, the Mach number and mass flow rate are strong influences as well for all other engines. Next, to find the total fuel burned ($\dot{m}_{ft}$) one can apply Eqs. (2.3.65) and (2.3.76). However, an easier method is to apply the energy equation to the entire engine. Note that no shaft power crosses the control volume if it is drawn around the entire engine and also that, ideally, *$\dot{m}_f$ is much smaller than $\dot{m}$. Thus, the total fuel flow rate (primary plus afterburner) is*

$$\dot{m}_{f_t}\Delta H = \sum \dot{m}_{out} h_{t_{out}} - \sum \dot{m}_{in} h_{t_{in}} \tag{H.4.20}$$

$$\dot{m}_{f_t}\Delta H = (\dot{m}c_p T_{t6} + \dot{m}_s c_p T_{t7}) - c_p T_{ta}\,(\dot{m} + \dot{m}_s), \tag{H.4.21}$$

which, using Eqs. (2.3.40) and (2.3.38) and realizing that ideally $T_{t2} = T_{ta}$, reduces to

$$\dot{m}_{f_t} = \left[\frac{\dot{m}c_p T_a}{\Delta H}\right]\left[\frac{T_{t6}}{T_{ta}} + \alpha\tau_f - (1 + \alpha)\right]. \tag{H.4.22}$$

Finally, using the definition of nondimensional *TSFC*, one finds with Eq. (H.4.22) that

$$TSFC\ a_a = \frac{\left[\dfrac{T_{t6}}{T_a} - \dfrac{T_a}{T_{ta}} + \alpha\tau_f - (1 + \alpha)\right]\left[\dfrac{c_p T_a}{\Delta H}\right]\left[\dfrac{T_{ta}}{T_a}\right]}{M_a\left[\sqrt{\left[\dfrac{\dfrac{T_{t6}}{T_a}}{\dfrac{T_{ta}}{T_a} - 1}\right]\left[1 - \dfrac{1}{\left(\dfrac{T_{ta}}{T_a}\tau_c\left(1 - \left[\dfrac{T_{ta}}{T_a}\times\dfrac{T_a}{T_{t4}}\right][(\tau_c - 1) + \alpha(\tau_f - 1)]\right)\right)}\right]} - 1\right] + \alpha M_a\left[\sqrt{\left[\dfrac{\left[\dfrac{T_{ta}}{T_a}\tau_f - 1\right]}{\dfrac{T_{ta}}{T_a} - 1}\right]} - 1\right]}. \tag{H.4.23}$$

For the exhausted fan with an afterburner, the *TSFC* is significantly higher than for the turbofan without an afterburner.

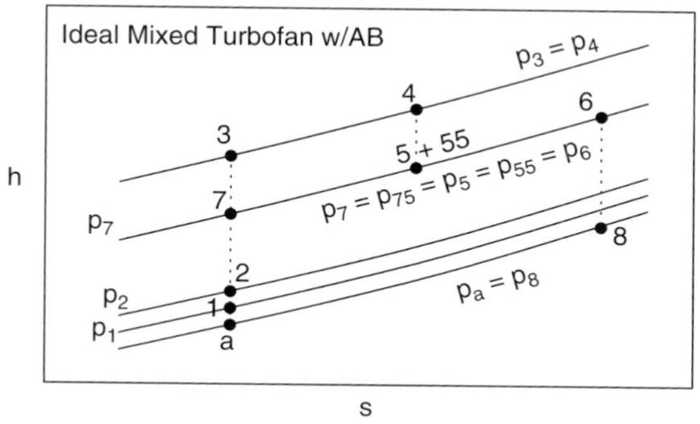

Figure H.4 The h–s diagram for an ideal mixed fan with an afterburner.

H.5 Turbofan: Fan Mixed

As before, the general diagram for a jet engine is shown in Figure 2.3. Two simplifications of the general turbofan are covered for the ideal analysis: the fully exhausted fan in Section H.4 and the ducted/mixed fan. Both nonafterburning and afterburning cases are considered. In this section, a mixed fan is covered.

This type of fan is shown in Figure 3.6. For this configuration, the fan exit bypassed air passes through a bypass duct and is mixed with the turbine exhaust before station 5.5. Two separate subcases are considered here: without and with afterburner. In Figure 3.6, the afterburner exists between stations 5.5 and 6. In Figure H.4 the h–s diagram for the afterburning configuration is shown. Steps a to 3 are the diffuser and then compressor isentropic processes. Steps a to 7 are the diffuser and fan processes. Processes 3 to 4 and 5.5 to 6 are the combustion and afterburning operations, respectively. Process 4 to 5 is the isentropic turbine expansion operation. The separate processes 7 to 5.5 and 5 to 5.5 are the bypassed air duct + mixing and primary air mixing operations, respectively. Lastly, step 6 to 8 is the primary nozzle process.

Analyses for the two engines are performed separately in the following sections.

H.5.1 Nonafterburning

For the nonafterburning case, stations 5.5 and 6 are one and the same and are termed station 6, as shown in Figure 3.7. As for the turbojet, because only one exhaust gas stream or nozzle is present, the following is obtained:

$$\frac{F}{\dot{m}_8 u_8} = \left(\frac{u_8}{u_a} - 1\right), \tag{H.5.1}$$

where $\dot{m}_8$ is the total exiting gas flow from the nozzle and given by

$$\dot{m}_8 = (1 + \alpha)\dot{m}; \tag{H.5.2}$$

thus, once again, it is necessary to find the velocity ratio u_8/u_a. As for the ramjet,

$$\frac{u_8}{u_a} = \frac{M_8\sqrt{T_8}}{M_a\sqrt{T_a}}. \tag{H.5.3}$$

Therefore, using Eqs. (2.3.3) and (2.3.92) yields

$$\frac{u_8}{u_a} = \sqrt{\frac{\frac{T_{t8}}{T_8} - 1}{\frac{T_{ta}}{T_a} - 1}} \sqrt{\frac{T_8}{T_a}}, \tag{H.5.4}$$

which reduces to

$$\frac{u_8}{u_a} = \sqrt{\frac{\frac{T_{t8}}{T_8} - \frac{T_8}{T_a}}{\frac{T_{ta}}{T_a} - 1}}. \tag{H.5.5}$$

Thus, the ratios T_{t8}/T_a and T_8/T_a are needed to find the thrust. Next, if one considers the enthalpy of the ideal uniform mixing streams at station 6 (Eq. (2.3.126)) and expands on the total temperature exiting the mixer, T_{t6}, the following equation is obtained:

$$T_{t6} = \frac{\alpha T_a \left[\frac{T_{ta}}{T_a}\right] \left[\frac{T_{t7}}{T_{t2}}\right] + T_a \left[\frac{T_{t4}}{T_a}\right] \left[\frac{T_{t5}}{T_{t4}}\right]}{\alpha + 1}, \tag{H.5.6}$$

which can be reduced to

$$T_{t6} = \frac{\alpha T_a \left[\frac{T_{ta}}{T_a}\right] \tau_f + T_a \left[\frac{T_{t4}}{T_a}\right] \tau_t}{\alpha + 1}. \tag{H.5.7}$$

For a stable condition at the mixer inlet, Eq. (2.3.120) indicates that the static pressures are equal ($p_{7.5} = p_5$). However, as discussed in Section 2.3.13, if it can be assumed for the ideal case that the Mach numbers for streams 5, 7.5, and 5.5 are equal ($M_{t7.5} = M_{t5} = M_{t5.5}$) or small, the total pressures will also be equal ($p_{t7.5} = p_{t5}$). Furthermore, since the mixing is ideal, the total pressures are constant across the mixer:

$$p_{t7.5} = p_{t5} = p_{t5.5}. \tag{H.5.8}$$

It is noteworthy to note that this approximation can be relaxed for the nonideal case, as shown in Extended Example 3.10.1 in Chapter 3. Thus, for the ideal case the total pressure ratio across the fan is

$$\frac{p_{t7}}{p_{t2}} = \frac{p_{t5}}{p_{t3}} \frac{p_{t3}}{p_{t2}}. \tag{H.5.9}$$

Note that if the Mach numbers are not equal for the two streams, the relationship of p_{t7} and p_{t5} can be obtained using Eq. (C.2.11). For the ideal case, Eq. (H.5.9) reduces to

$$\pi_f = \pi_t \pi_c, \tag{H.5.10}$$

which for the ideal case (isentropic for all three components and constant γ) also implies (from Eq. (C.2.10)) that

$$\tau_f = \tau_t \tau_c. \tag{H.5.11}$$

Using this with Eq. (H.5.7) yields

$$T_{t6} = \frac{\alpha T_a \left[\frac{T_{ta}}{T_a}\right] \tau_f + T_a \left[\frac{T_{t4}}{T_a}\right] \frac{\tau_f}{\tau_c}}{\alpha + 1}, \tag{H.5.12}$$

which for the ideal case is also T_{t8} because the nozzle is adiabatic. Therefore,

$$\frac{T_{t8}}{T_a} = \left[\frac{T_{ta}}{T_a}\right] \tau_f \left[\frac{\alpha + \frac{1}{\tau_c}\left[\frac{T_a}{T_{ta}}\right]\left[\frac{T_{t4}}{T_a}\right]}{\alpha + 1}\right]. \tag{H.5.13}$$

One can find the total pressure at station 8, which is also (from Eq. (H.5.8))

$$p_{t8} = p_a \left[\frac{p_{ta}}{p_a}\right]\left[\frac{p_{t3}}{p_{t2}}\right]\left[\frac{p_{t5}}{p_{t4}}\right]. \tag{H.5.14}$$

From Eq. (H.5.10) this becomes

$$p_{t8} = p_a \left[\frac{p_{ta}}{p_a}\right]\left[\frac{p_{t7}}{p_{t2}}\right] = p_a \left[\frac{p_{ta}}{p_a}\right] \pi_f. \tag{H.5.15}$$

Now, since for the ideal case the exit pressure matches the ambient pressure ($p_8 = p_a$), Eq. (2.3.95) yields

$$\left[1 + \frac{\gamma - 1}{2} M_8^2\right] = \left[\left[\frac{p_{ta}}{p_a}\right] \pi_f\right]^{\frac{\gamma-1}{\gamma}}. \tag{H.5.16}$$

Finally, from Eqs. (H.5.13), (2.3.92), and (H.5.16) and solving for T_8/T_a one obtains

$$\frac{T_8}{T_a} = \left[\frac{T_{ta}}{T_a}\right] \tau_f \left[\frac{\alpha + \frac{1}{\tau_c}\left[\frac{T_a}{T_{ta}}\right]\left[\frac{T_{t4}}{T_a}\right]}{(\alpha + 1)\left[\left[\frac{p_{ta}}{p_a}\right]\pi_f\right]^{\frac{\gamma-1}{\gamma}}}\right], \tag{H.5.17}$$

which for the ideal case (from Eqs. (C.2.10) and (2.3.41)) becomes

$$\frac{T_8}{T_a} = \left[\frac{\alpha + \frac{1}{\tau_c}\left[\frac{T_a}{T_{ta}}\right]\left[\frac{T_{t4}}{T_a}\right]}{\alpha + 1}\right]. \tag{H.5.18}$$

The shaft energy equation is exactly the same for the mixed and exhausted cases. Thus, using Eqs. (H.5.11) and (H.4.13) and solving for the fan total temperature ratio τ_f yields

$$\tau_f = \frac{\tau_c + \left[\frac{T_{ta}}{T_a}\right]\left[\frac{T_a}{T_{t4}}\right] \tau_c (1 + \alpha - \tau_c)}{1 + \left[\frac{T_{ta}}{T_a}\right]\left[\frac{T_a}{T_{t4}}\right] \tau_c \alpha}. \tag{H.5.19}$$

From this one can see that, as soon as T_{t4}/T_a, τ_c (or π_c) and the Mach number are specified, τ_f (and thus π_f) can be calculated. Note that the fan total pressure ratio is not independent as it was for the exhausted case. Thus, during the design phase this parameter cannot be independently varied and must be appropriately chosen at the design stage. Finally, using Eqs. (H.5.1), (H.5.2), (H.5.5), (H.5.12), and (H.5.17), and remembering that Eq. (H.5.19) must hold, one obtains the following equation for the dimensionless thrust:

$$\frac{F}{\dot{m}a_a} = M_a(1+\alpha)\left[\sqrt{\left[\frac{\alpha + \left[\frac{T_{t4}}{T_a}\right]\left[\frac{T_a}{T_{ta}}\right]\frac{1}{\tau_c}}{1+\alpha}\right]\left[\frac{\left[\frac{T_{ta}}{T_a}\right]\left[\frac{\tau_c + \left[\frac{T_{ta}}{T_a}\right]\left[\frac{T_a}{T_{t4}}\right]\tau_c(1+\alpha-\tau_c)}{1+\left[\frac{T_{ta}}{T_a}\right]\left[\frac{T_a}{T_{t4}}\right]\tau_c\alpha}\right]-1}{\left[\frac{T_{ta}}{T_a}\right]-1}\right]}-1\right]. \tag{H.5.20}$$

The thrust is a strong function of the burner exit temperature, compressor pressure ratio, and bypass ratio. As is the case for all other engines, the Mach number and mass flow rate are of course strong influences as well. It is worth noting at this point that it can be shown that the thrust for the mixed fan is exactly the same as for the exhausted fan with an optimal fan pressure ratio. The procedure for finding the nondimensionalized *TSFC* for the ducted fan is similar to that for the exhausted fan and again uses the definition of the *TSFC* as follows:

$$TSFC\ a_a = \frac{\left[\frac{T_{t4}}{T_a} - \frac{T_{ta}}{T_a}\tau_c\right]\left[\frac{c_p T_a}{\Delta H}\right]}{M_a(1+\alpha)\left[\sqrt{\left[\frac{\alpha + \left[\frac{T_{t4}}{T_a}\right]\left[\frac{T_a}{T_{ta}}\right]\frac{1}{\tau_c}}{1+\alpha}\right]\left[\frac{\left[\frac{T_{ta}}{T_a}\right]\left[\frac{\tau_c + \left[\frac{T_{ta}}{T_a}\right]\left[\frac{T_a}{T_{t4}}\right]\tau_c(1+\alpha-\tau_c)}{1+\left[\frac{T_{ta}}{T_a}\right]\left[\frac{T_a}{T_{t4}}\right]\tau_c\alpha}\right]-1}{\left[\frac{T_{ta}}{T_a}\right]-1}\right]}-1\right]}. \tag{H.5.21}$$

Once again, for the ducted and mixed fan typical *TSFC* values are 0.35 to 0.7 $lb_m/h\text{-}lb_f$.

H.5.2 Afterburning

The afterburning mixed fan is now considered. Figure 3.6 depicts this geometry. Many of the equations from the nonafterburning case, apply but several new derivations will be required. Equations (H.5.1)–(H.5.5) apply. The mixer energy equation (Eq. (2.3.126)), after being expanded on the exit uniform total temperature from the mixer, $T_{t5.5}$, ideally becomes

$$T_{t5.5} = \frac{\alpha T_a\left[\frac{T_{ta}}{T_a}\right]\left[\frac{T_{t7}}{T_{t2}}\right] + T_a\left[\frac{T_{ta}}{T_a}\right]\left[\frac{T_{t3}}{T_{t2}}\right]\left[\frac{T_{t4}}{T_{t3}}\right]\left[\frac{T_{t5}}{T_{t4}}\right]}{\alpha+1}, \tag{H.5.22}$$

which can be reduced to

$$T_{t5.5} = \frac{\alpha T_a \left[\dfrac{T_{ta}}{T_a}\right]\tau_f + T_a \left[\dfrac{T_{ta}}{T_a}\right]\tau_c\tau_b\tau_t}{\alpha + 1}. \tag{H.5.23}$$

Using this with Eq. (H.5.11) yields

$$T_{t5.5} = T_a \left[\frac{T_{ta}}{T_a}\right]\tau_f \left[\frac{\alpha + \tau_b}{\alpha + 1}\right]. \tag{H.5.24}$$

The total temperature at the afterburner exit (station 6) is given by

$$T_{t6} = T_{t5.5}\left[\frac{T_{t6}}{T_{t5.5}}\right] = T_{t5.5}\tau_{ab}, \tag{H.5.25}$$

where τ_{ab} is again the afterburner total temperature ratio, which, again for the ideal case (adiabatic nozzle), is the same as for T_{t8}. Thus, from this and Eq. (H.5.24),

$$\frac{T_{t8}}{T_a} = \tau_{ab}\tau_f \left[\frac{T_{ta}}{T_a}\right]\left[\frac{\alpha + \tau_b}{\alpha + 1}\right]. \tag{H.5.26}$$

From Eqs. (H.5.26), (H.5.13), and (2.3.92) one therefore obtains

$$\frac{T_8}{T_a} = \frac{\tau_{ab}\tau_f \left[\dfrac{T_{ta}}{T_a}\right]\left[\dfrac{\alpha + \tau_b}{\alpha + 1}\right]}{\left[\left(\dfrac{p_{ta}}{p_a}\right)\pi_f\right]^{\frac{\gamma-1}{\gamma}}}, \tag{H.5.27}$$

which for the ideal case (from Eqs. (2.3.5) and (2.3.41)) becomes

$$\frac{T_8}{T_a} = \tau_{ab}\left[\frac{\alpha + \tau_b}{\alpha + 1}\right]. \tag{H.5.28}$$

One can now solve for the burner total temperature ratio τ_b from Eq. (H.5.24), as follows:

$$\tau_b = \frac{\left[\dfrac{T_{t5.5}}{T_a}\right]\left[\dfrac{T_a}{T_{ta}}\right]}{\tau_f}[1 + \alpha] - \alpha. \tag{H.5.29}$$

Thus, from this and Eq. (H.5.28),

$$\frac{T_8}{T_a} = \frac{\tau_{ab}}{\tau_f}\left[\frac{T_{t5.5}}{T_a}\right]\left[\frac{T_a}{T_{ta}}\right] \tag{H.5.30}$$

or

$$\frac{T_8}{T_a} = \frac{\left[\dfrac{T_{t6}}{T_a}\right]\left[\dfrac{T_a}{T_{ta}}\right]}{\tau_f}. \tag{H.5.31}$$

Now, expanding on T_{t8}/T_8:

$$\frac{T_{t8}}{T_a} = \left[\frac{T_{t8}}{T_a}\right]\left[\frac{T_{ta}}{T_8}\right], \tag{H.5.32}$$

which, when Eq. (H.5.31) is used, becomes

$$\frac{T_{t8}}{T_a} = \left[\frac{T_{t8}}{T_a}\right]\tau_f. \tag{H.5.33}$$

Therefore, the velocity ratio is obtained from Eqs. (H.5.4), (H.2.10), and (H.5.33), as follows:

$$\frac{u_8}{u_a} = \sqrt{\frac{\left[\frac{T_{ta}}{T_a}\right]\tau_f - 1}{\frac{T_{ta}}{T_a} - 1}}\sqrt{\frac{\left[\frac{T_{t8}}{T_a}\right]}{\tau_f\left[\frac{T_{ta}}{T_a}\right]}}. \tag{H.5.34}$$

Using this with Eqs. (H.5.1), (H.5.2), and (H.5.19) yields

$$\frac{F}{\dot{m}a_a} = M_a(1+\alpha)\left[\sqrt{\frac{\left[\left[\frac{T_{ta}}{T_a}\right]\left[\frac{\tau_c + \left[\frac{T_{ta}}{T_a}\right]\left[\frac{T_a}{T_{t4}}\right]\tau_c(1+\alpha-\tau_c)}{1+\left[\frac{T_{ta}}{T_a}\right]\left[\frac{T_a}{T_{t4}}\right]\tau_c\alpha}\right] - 1\right]}{\left[\frac{T_{ta}}{T_a} - 1\right]\left[\frac{\tau_c + \left[\frac{T_{ta}}{T_a}\right]\left[\frac{T_a}{T_{t4}}\right]\tau_c(1+\alpha-\tau_c)}{1+\left[\frac{T_{ta}}{T_a}\right]\left[\frac{T_a}{T_{t4}}\right]\tau_c\alpha}\right]}\frac{\left[\frac{T_{t8}}{T_a}\right]}{\left[\frac{T_{ta}}{T_a}\right]}} - 1\right]. \tag{H.5.35}$$

This expression is the dimensionless thrust equation for the mixed afterburning turbofan. The thrust is a strong function of burner exit temperature, afterburner exit temperature, compressor pressure ratio, and bypass ratio. Next, to find the total fuel burned, $\dot{m}_{ft}$, one can apply the energy equation to the entire engine. Recall that no shaft power crosses the control volume and that, ideally, $\dot{m}_f$ is much smaller than $\dot{m}$. Thus, very simply,

$$\dot{m}_{f_t}\Delta H = \dot{m}(1+\alpha)c_p(T_{t6} - T_{ta}). \tag{H.5.36}$$

Thus, solving for $\dot{m}_{ft}$ and using the definition of dimensionless *TSFC*, Eqs. (H.5.19) and (H.5.35), and the fact that $T_{t6} = T_{t8}$ yields

$$TSFC\ a_a = \frac{\left[\left(\frac{T_{t8}}{T_a}\right) - \left(\frac{T_{ta}}{T_a}\right)\right]\left[\frac{c_pT_a}{\Delta HM_a}\right]}{\left[\sqrt{\frac{\left[\left[\frac{T_{ta}}{T_a}\right]\left[\frac{\tau_c + \left[\frac{T_{ta}}{T_a}\right]\left[\frac{T_a}{T_{t4}}\right]\tau_c(1+\alpha-\tau_c)}{1+\left[\frac{T_{ta}}{T_a}\right]\left[\frac{T_a}{T_{t4}}\right]\tau_c\alpha}\right] - 1\right]}{\left[\frac{T_{ta}}{T_a} - 1\right]\left[\frac{\tau_c + \left[\frac{T_{ta}}{T_a}\right]\left[\frac{T_a}{T_{t4}}\right]\tau_c(1+\alpha-\tau_c)}{1+\left[\frac{T_{ta}}{T_a}\right]\left[\frac{T_a}{T_{t4}}\right]\tau_c\alpha}\right]}\frac{\left[\frac{T_{t8}}{T_a}\right]}{\left[\frac{T_{ta}}{T_a}\right]}} - 1\right]}. \tag{H.5.37}$$

This is the *TSFC* for the mixed fan with an afterburner. Values for this are much higher than for the nonafterburning case. The same result (although by a longer derivation) would have been obtained if the energy equation had been applied to both the primary burner and afterburner to find the total fuel flow rate. For example, Eqs. (2.3.63) and (2.3.65) give the ideal primary burner fuel ratio. The afterburner fuel ratio for the mixed turbofan can be found by the following analysis. Applying the energy equation to only the afterburner (Eq. (2.3.76)) and solving for the fuel ratio (Eq. (2.3.79)) yields

$$f_{ab} = \frac{\dot{m}_{f_{ab}}}{\dot{m}} = \frac{(1+\alpha)c_p(T_{t6} - T_{t5.5})}{\Delta H}, \tag{H.5.38}$$

where $T_{t5.5}$ is given by Eq. (H.5.23).

H.6 Turboprop

As indicated in Chapter 1, a turboprop is aimed at a different group of aircraft than the engines already covered. This geometry is somewhat different than that of the previous engines and is shown in Figure 3.8. Propellers tend to be relatively large; propeller tip speeds can be quite high. As a result, in the past this engine type was only used primarily on low-speed aircraft so that the tip speeds would not become prohibitively large. Current technologies are allowing higher aircraft speeds with turboprops – for example, with unducted fan engines. The geometry is somewhat similar to a turbofan with the propeller replacing the fan; however, the flow through the propeller is not as well defined as in a fan. Also, the pressure ratio across the propeller is very nearly unity. In some respects, this engine behaves like a turbofan with a very high bypass ratio (approximately 25 to 100); that is, a considerable flow of air goes through the propeller to generate thrust. The remainder of the engine (stations 2 through 8) is fundamentally identical to a turbojet. A portion of the flow fanned by the propeller enters the compressor (station 2). The thrust from this engine type is developed from two sources: the propeller and the turbojet nozzle. Thus, this engine is similar to the exhausted turbofan in this respect, which also has two components of thrust. Because of the less defined nature of the propeller flow (mass flow, pressure rise, and flow direction), a somewhat different approach to the analysis is needed.

First, the total power generated for thrust of an ideal turboprop is given by

$$\mathcal{P}_t = Fu_a + \mathcal{P}_p, \tag{H.6.1}$$

where F is the thrust of a turbojet, which can be found from the analysis from Section H.3.1, and $\mathcal{P}_p$ is the propeller power. Next, a work coefficient for the engine can conveniently be defined as

$$C_{W_e} \equiv \frac{\mathcal{P}_t}{\dot{m}h_a}, \tag{H.6.2}$$

where $\dot{m}$ is the airflow through the turbojet core. This now can be applied to Eq. (H.6.1):

$$C_{W_e} = C_{W_p} + \frac{u_a F}{\dot{m}c_p T_a}, \tag{H.6.3}$$

where C_{W_p} is the propeller work coefficient given by Eq. (2.3.50). Next, applying the energy equation to the shaft (Eq. (2.3.83)) and reducing yields

$$\mathcal{P}_\mathrm{p} + \dot{m}c_\mathrm{p}(T_{t3} - T_{t2}) = \dot{m}c_\mathrm{p}(T_{t4} - T_{t5}). \tag{H.6.4}$$

Assuming that the air entering the compressor has the same total properties as the freestream (i.e., the pressure and temperature have not increased significantly by traversing the propeller) and solving for the propeller power, one obtains

$$C_{W_\mathrm{p}} = \frac{\mathcal{P}_\mathrm{p}}{\dot{m}c_\mathrm{p}T_\mathrm{a}} = \left[\frac{T_{t4}}{T_\mathrm{a}}\right]\left[1 - \frac{\tau_5}{\left[\frac{T_{ta}}{T_\mathrm{a}}\right]\tau_\mathrm{c}}\right] - \left[\frac{T_{ta}}{T_\mathrm{a}}\right](\tau_\mathrm{c} - 1), \tag{H.6.5}$$

where τ_5 is a new parameter and is defined by

$$\tau_5 \equiv \left[\frac{T_{ta}}{T_\mathrm{a}}\right]\tau_\mathrm{c}\tau_\mathrm{t} = \frac{T_{ta}}{T_\mathrm{a}} \times \frac{T_{t3}}{T_{t2}} \times \frac{T_{t5}}{T_{t4}}. \tag{H.6.6}$$

But since adiabatically $T_{t2} = T_{ta}$

$$\tau_5 = \frac{T_{t5}}{T_\mathrm{a}} \times \frac{T_{t3}}{T_{t4}} = \frac{T_{t5}}{T_\mathrm{a}} \times \frac{1}{\tau_\mathrm{b}}. \tag{H.6.7}$$

Also Eq. (H.6.6) ideally (isentropically) reduces to (recall that $p_{t4} = p_{t3}$)

$$\tau_5 = \left[\left[\frac{p_{ta}}{p_\mathrm{a}}\right]\pi_\mathrm{c}\pi_\mathrm{t}\right]^{\frac{\gamma-1}{\gamma}} = \left[\frac{p_{ta}}{p_\mathrm{a}}\frac{p_{t3}}{p_{t2}}\frac{p_{t5}}{p_{t4}}\right]^{\frac{\gamma-1}{\gamma}} = \left[\frac{p_{t5}}{p_\mathrm{a}}\right]^{\frac{\gamma-1}{\gamma}} \tag{H.6.8}$$

or

$$\tau_5 = [\pi_5]^{\frac{\gamma-1}{\gamma}}, \tag{H.6.9}$$

where

$$\pi_5 \equiv \left[\frac{p_{t5}}{p_\mathrm{a}}\right], \tag{H.6.10}$$

which is a nozzle parameter and is directly related to the exit Mach number for the ideal case by Eqs. (2.3.92) and (2.3.93) and is discussed in more detail in Chapter 10. One should not make the error of equating τ_5 to T_{t5}/T_a, because the equality of T_{t4} and T_{t3} does not hold ($T_{t4} \neq T_{t3}$). To find τ_5 more engine data is needed, for example, more nozzle or turbine information. However, if C_{W_p} is known by propellor testing with an engine identification one can find τ_5 by rearranging Eq. (H.6.5):

$$\frac{\mathcal{P}_\mathrm{p}}{\dot{m}c_\mathrm{p}T_\mathrm{a}} + \left[\frac{T_{ta}}{T_\mathrm{a}}\right](\tau_\mathrm{c} - 1) = \left[\frac{T_{t4}}{T_\mathrm{a}}\right]\left[1 - \frac{\tau_5}{\left[\frac{T_{ta}}{T_\mathrm{a}}\right]\tau_\mathrm{c}}\right] \tag{H.6.11}$$

then

$$1 - \left[\frac{T_\mathrm{a}}{T_{t4}}\right]\left[\frac{\mathcal{P}_\mathrm{p}}{\dot{m}c_\mathrm{p}T_\mathrm{a}} + \left[\frac{T_{ta}}{T_\mathrm{a}}\right](\tau_\mathrm{c} - 1)\right] = \frac{\tau_5}{\left[\frac{T_{ta}}{T_\mathrm{a}}\right]\tau_\mathrm{c}}, \tag{H.6.12}$$

and solving for τ_5 yields

$$\tau_5 = \left[\frac{T_{ta}}{T_a}\right]\tau_c\left\{1 - \left[\frac{T_a}{T_{t4}}\right]\left[\frac{\mathcal{P}_p}{\dot{m}c_pT_a} + \left[\frac{T_{ta}}{T_a}\right](\tau_c - 1)\right]\right\} = \left[\frac{T_{ta}}{T_a}\right]\tau_c\left\{1 - \left[\frac{T_a}{T_{t4}}\right]\left[C_{W_p} + \left[\frac{T_{ta}}{T_a}\right](\tau_c - 1)\right]\right\}. \tag{H.6.13}$$

Also, recall that for the ideal case the total pressure is constant through the nozzle; thus, p_{t8} is equal to p_{t5}. Because p_{t5}/p_a must be greater than unity, by Eq. (H.6.9), τ_5 must also be greater than unity. Next, the work coefficient for the turboprop is considered using Eq. (C.2.2); that is,

$$\frac{u_aF}{\dot{m}c_pT_a} = \frac{u_aF}{\dot{m}\left[\frac{\gamma\mathcal{R}}{\gamma - 1}\right]T_a} = \frac{(\gamma - 1)u_aF}{\dot{m}\gamma\mathcal{R}T_a} = \frac{(\gamma - 1)u_aF}{\dot{m}a_a^2}, \tag{H.6.14}$$

or finally

$$\frac{u_aF}{\dot{m}c_pT_a} = \frac{(\gamma - 1)M_aF}{\dot{m}a_a}. \tag{H.6.15}$$

Next, using Eqs. (H.3.13), (H.3.19), and (H.6.6), one finds the jet thrust as follows:

$$\frac{F}{\dot{m}a_a} = M_a\left[\sqrt{\frac{\left[\frac{T_{t4}}{T_a}\right]\left[\frac{T_a}{T_{ta}}\right]\left[\frac{\tau_5 - 1}{\tau_c}\right]}{\left[\frac{T_{ta}}{T_a}\right] - 1}} - 1\right]. \tag{H.6.16}$$

Now, using Eqs. (H.6.3), (2.3.50), (H.6.5), (H.6.15), and (H.6.16) results in

$$C_{W_e} = \left[\frac{T_{t4}}{T_a}\right]\left[1 - \frac{\tau_5}{\left[\frac{T_{ta}}{T_a}\right]\tau_c}\right] - \left[\frac{T_{ta}}{T_a}\right](\tau_c - 1) + (\gamma - 1)M_a^2\left[\sqrt{\frac{\left[\frac{T_{t4}}{T_a}\right]\left[\frac{T_a}{T_{ta}}\right]\left[\frac{\tau_5 - 1}{\tau_c}\right]}{\left[\frac{T_{ta}}{T_a}\right] - 1}} - 1\right]. \tag{H.6.17}$$

Once the work coefficient is found, the total thrust can be found by

$$F_t = \frac{\mathcal{P}_t}{u_a}, \tag{H.6.18}$$

or the dimensionless thrust can be found by

$$\frac{F_t}{\dot{m}a_a} = \frac{\mathcal{P}_t}{\dot{m}a_au_a}, \tag{H.6.19}$$

which, when used with Eqs. (H.6.2), (C.2.7), (C.2.8), and (C.2.2), yields

$$\frac{F_t}{\dot{m}a_a} = \frac{C_{W_e}}{(\gamma - 1)M_a}. \tag{H.6.20}$$

The thrust is a strong function of burner exit temperature, afterburner exit temperature, compressor pressure ratio, and the nozzle parameter. Of course, the Mach number and mass flow rate are strong influences for the turboprop just as they are for all other engines. Next, the fuel consumption needs to be determined. For a turboprop, a different fuel parameter is sometimes used and is identical to that used for a power generation unit. This is the specific fuel consumption and is defined by

$$SFC \equiv \frac{\dot{m}_f}{\mathcal{P}_t} = \frac{\dot{m}_f}{F_t u_a} = \frac{\dot{m}_f}{\dot{m} c_p T_a C_{W_e}}. \tag{H.6.21}$$

Thus, using Eqs. (2.3.65) and (2.3.61), one obtains

$$SFC = \frac{\left[\frac{T_{t4}}{T_a}\right] - \left[\frac{T_{ta}}{T_a}\right]\tau_c}{\Delta H C_{W_e}}, \tag{H.6.22}$$

where C_{W_e} is given in Eq. (H.6.17). The definition of *TSFC* can also be used to find that

$$TSFC = SFC \times u_a. \tag{H.6.23}$$

Thus, the *TSFC* is

$$TSFC = u_a \left[\frac{\left[\frac{T_{t4}}{T_a}\right] - \left[\frac{T_{ta}}{T_a}\right]\tau_c}{\Delta H C_{W_e}}\right]. \tag{H.6.24}$$

For a turboprop, typical values of *TSFC* are 0.3 to 0.4 $lb_m/h\text{-}lb_f$. This can be put into nondimensional form as for other engines. Noting that $u_a = M_a a_a$,

$$TSFC\ a_a = a_a(M_a a_a)\left[\frac{\left[\frac{T_{t4}}{T_a}\right] - \left[\frac{T_{ta}}{T_a}\right]\tau_c}{\Delta H C_{W_e}}\right]\frac{c_p T_a}{c_p T_a}. \tag{H.6.25}$$

But also $c_p = \dfrac{\mathcal{R}\gamma}{\gamma - 1}$ and $a_a = \sqrt{\gamma \mathcal{R} T_a}$, so after some rearranging Eq. (H.6.25) becomes

$$TSFC\ a_a = M_a(\gamma - 1)\left(\frac{c_p T_a}{\Delta H}\right)\left[\frac{\left[\frac{T_{t4}}{T_a}\right] - \left[\frac{T_{ta}}{T_a}\right]\tau_c}{C_{W_e}}\right], \tag{H.6.26}$$

which has many of the same coefficients as the previous engines. The thrust may be maximized by using Eq. (H.6.17) because the thrust is proportional to the work coefficient. That is, if one differentiates C_{W_e} with respect to τ_5, holding all other parameters fixed, and sets the result to zero, the following equation is obtained with the aid of Eq. (2.3.3):

$$\tau_{5_{\mathrm{opt}}} = 1 + \tau_{\mathrm{c}} \left[\frac{T_{\mathrm{ta}}}{T_{\mathrm{a}}}\right] \left[\frac{T_{\mathrm{a}}}{T_{\mathrm{t4}}}\right] \left[\left[\frac{T_{\mathrm{ta}}}{T_{\mathrm{a}}}\right] - 1\right]. \tag{H.6.27}$$

Although the analytical operations are on τ_5, in actuality the turbine operation is being optimized to deliver the best balanced performance between the propeller and exhaust. Through this equation, one can find the optimum value of π_5 from Eq. (H.6.9) and the resulting temperature and pressure ratios in the turbine. With the aid of Eqs. (H.3.12) and (H.6.6), this ideally results in $u_8 = u_{\mathrm{a}}$, which implies that the thrust from the turbojet optimally is zero (Eq. (H.3.1)) and that all of the thrust is provided from the propeller.

H.7 Power Generation Gas Turbine

The next category of gas turbines to be analyzed is for power generation and not aircraft applications. A power generation unit of simple design is shown in Figure 3.9. As discussed in Chapter 1, the unit is very much like a turbojet engine, and one can see many similarities between this gas turbine and the aircraft engine; much of the cycle analysis is therefore identical. However, an inlet bell, with a decreasing area along the flow path, replaces the diffuser, which has an increasing area, and an exhaust, in which the gas flow has a very low velocity, replaces the nozzle, which has a high gas velocity. Also, a power-extracting device is attached to the shaft. This device may be an electrical generator, ship propeller, a pump or compressor "train," or one of several other power-extracting devices. In Figure H.5 the h–s diagram for the ideal power generation unit is shown. As discussed in Chapter 1, many modern power generation units are aeroderivatives of the aircraft engines; that is, the core of the engine is essentially the same as that of a jet engine, thus reducing development costs and manufacturing needs. This type of unit tends to be slightly heavier because weight is not as much of a concern as it is for aircraft applications. Since the net power output is of primary concern (and not thrust), derivation of the fundamental equations is somewhat different.

First, for the ideal case, the shaft power balance yields

$$\mathcal{P}_{\mathrm{net}} = \dot{m} c_{\mathrm{p}} (T_{\mathrm{t4}} - T_{\mathrm{t5}}) - \dot{m} c_{\mathrm{p}} (T_{\mathrm{t3}} - T_{\mathrm{t2}}), \tag{H.7.1}$$

where $\mathcal{P}_{\mathrm{net}}$ is the net or derived power. Nondimensionalizing results in

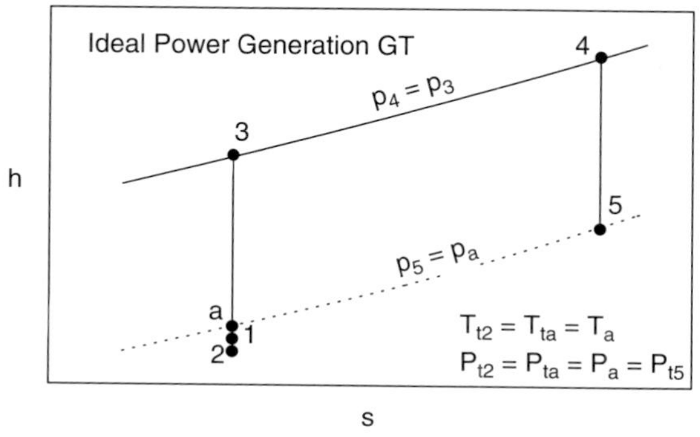

Figure H.5 The h–s diagram for an ideal power generation gas turbine.

$$\frac{\mathcal{P}_{net}}{\dot{m}c_p T_a} = \left(\frac{T_{t4}}{T_a} - \frac{T_{t5}}{T_a}\right) - \left(\frac{T_{t3}}{T_a} - \frac{T_{t2}}{T_a}\right). \tag{H.7.2}$$

If one recognizes that, for the ideal case, the inlet is adiabatic ($T_{t2} = T_{ta} = T_a$) and rearranges terms,

$$\frac{\mathcal{P}_{net}}{\dot{m}c_p T_a} = \left(\frac{T_{t4} - T_{t3}}{T_a}\right) - \left(\frac{T_{t5}}{T_a} - 1\right), \tag{H.7.3}$$

but for the compressor the exit total temperature is

$$T_{t3} = \tau_c T_{t2} = T_{t2}[\pi_c]^{\frac{\gamma-1}{\gamma}} \tag{H.7.4}$$

and for the turbine the exit total temperature is

$$T_{t5} = \tau_t T_{t4} = [\pi_t]^{\frac{\gamma-1}{\gamma}} T_{t4} = \left[\frac{p_{t5}}{p_{t4}}\right]^{\frac{\gamma-1}{\gamma}} T_{t4}. \tag{H.7.5}$$

For the ideal case, however, the total pressure is constant through the exhaust and the exit pressure is ambient; thus, $p_{t5} = p_a$. Also for the ideal case the total pressure across the combustor is constant ($p_{t4} = p_{t3}$); consequently, from Eq. (H.7.5), if one recognizes that $\pi_c = p_{t3}/p_{t2}$ and recalls for the ideal case that the total pressure remains constant across the inlet and is equal to the ambient pressure because the inlet velocity is low ($p_{t2} = p_{ta} = p_a$),

$$T_{t5} = T_{t4}\left[\frac{p_a}{\pi_c p_a}\right]^{\frac{\gamma-1}{\gamma}} = T_{t4}[\pi_c]^{-\frac{\gamma-1}{\gamma}}. \tag{H.7.6}$$

Therefore, using Eqs. (H.7.6) and (2.3.27) yields

$$T_{t5} = \frac{T_{t4}}{\tau_c}. \tag{H.7.7}$$

Now, using Eqs. (H.7.3), (H.7.4), and (H.7.7) results in

$$\frac{\mathcal{P}_{net}}{\dot{m}c_p T_a} = \left(\frac{T_{t4}}{T_a}\right)\left(1 - \frac{1}{\tau_c}\right) - (\tau_c - 1). \tag{H.7.8}$$

Thus, this nondimensional power output is a function of only two variables. The first is the compressor total temperature ratio (or total pressure ratio), and the second is the ratio of the combustor exit total temperature to the ambient temperature.

One can examine Eq. (H.7.8) for the dependence of net power on the compressor total temperature ratio. By differentiating the equation with respect to the total temperature ratio of the compressor, τ_c (which as before is directly related to the total pressure ratio, $\pi_c = \tau_c^{\frac{\gamma}{\gamma-1}}$), it is possible to optimize the output power. This results in

$$\tau_{c_{opt}} = \sqrt{\frac{T_{t4}}{T_a}}. \tag{H.7.9}$$

The fuel flow rate for the ideal case is

$$\dot{m}_{\mathrm{f}} = \frac{\dot{m} c_{\mathrm{p}}}{\Delta H}(T_{\mathrm{t4}} - T_{\mathrm{t3}}) = \frac{\dot{m} c_{\mathrm{p}} T_{\mathrm{a}}}{\Delta H}\left(\frac{T_{\mathrm{t4}}}{T_{\mathrm{a}}} - \frac{T_{\mathrm{t3}}}{T_{\mathrm{a}}}\right), \tag{H.7.10}$$

which, when Eq. (H.7.4) is used with ($T_{\mathrm{t2}} = T_{\mathrm{ta}} = T_{\mathrm{a}}$), becomes

$$\dot{m}_{\mathrm{f}} = \frac{\dot{m} c_{\mathrm{p}} T_{\mathrm{a}}}{\Delta H}\left(\frac{T_{\mathrm{t4}}}{T_{\mathrm{a}}} - \tau_{\mathrm{c}}\right). \tag{H.7.11}$$

The specific fuel consumption is

$$SFC = \frac{\dot{m}_{\mathrm{f}}}{\mathcal{P}_{\mathrm{net}}}; \tag{H.7.12}$$

using Eqs. (H.7.8), (H.7.11), and (H.7.12) results in

$$SFC = \frac{1}{\Delta H} \frac{\left(\dfrac{T_{\mathrm{t4}}}{T_{\mathrm{a}}} - \tau_{\mathrm{c}}\right)}{\left(\dfrac{T_{\mathrm{t4}}}{T_{\mathrm{a}}}\right)\left(1 - \dfrac{1}{\tau_{\mathrm{c}}}\right) - (\tau_{\mathrm{c}} - 1)}, \tag{H.7.13}$$

which using simple algebra reduces to

$$SFC = \frac{1}{\Delta H}\left(\frac{\tau_{\mathrm{c}}}{\tau_{\mathrm{c}} - 1}\right). \tag{H.7.14}$$

The reduced dimensional variable is a function of only two variables. The first is the compressor total temperature ratio (or total pressure ratio in as much as they are directly related), and the second is the heating value of the fuel. Typical values for such modern gas turbines are 0.35 $\mathrm{lb_m}$/hp-h (0.20 kg/kW-h).

This also can be put into nondimensional form:

$$SFCa_{\mathrm{a}}^2 = \frac{a_{\mathrm{a}}^2}{\Delta H}\left(\frac{\tau_{\mathrm{c}}}{\tau_{\mathrm{c}} - 1}\right)\frac{c_{\mathrm{p}} T_{\mathrm{a}}}{c_{\mathrm{p}} T_{\mathrm{a}}} = \frac{\gamma \mathcal{R} T_{\mathrm{a}}}{\Delta H}\left(\frac{\tau_{\mathrm{c}}}{\tau_{\mathrm{c}} - 1}\right)\frac{c_{\mathrm{p}} T_{\mathrm{a}}}{c_{\mathrm{p}} T_{\mathrm{a}}} = \frac{\gamma \mathcal{R}}{\Delta H}\left(\frac{\tau_{\mathrm{c}}}{\tau_{\mathrm{c}} - 1}\right)\frac{c_{\mathrm{p}} T_{\mathrm{a}}}{\left(\dfrac{\mathcal{R}\gamma}{\gamma - 1}\right)} \tag{H.7.15}$$

so that

$$SFC\ a_{\mathrm{a}}^2 = (\gamma - 1)\left(\frac{c_{\mathrm{p}} T_{\mathrm{a}}}{\Delta H}\right)\left(\frac{\tau_{\mathrm{c}}}{\tau_{\mathrm{c}} - 1}\right), \tag{H.7.16}$$

which again has some familiar coefficients as with propulsion engines.

Lastly one can find the thermal efficiency by

$$\eta_{\mathrm{th}} = \frac{\mathcal{P}_{\mathrm{net}}}{\dot{Q}_{\mathrm{in}}} = \frac{\mathcal{P}_{\mathrm{net}}}{\dot{m}_{\mathrm{f}} \Delta H} = \frac{1}{SFC\ \Delta H}, \tag{H.7.17}$$

and thus, using Eq. (H.7.13),

$$\eta_{\mathrm{th}} = \frac{\left(\dfrac{T_{\mathrm{t4}}}{T_{\mathrm{a}}}\right)\left(1 - \dfrac{1}{\tau_{\mathrm{c}}}\right) - (\tau_{\mathrm{c}} - 1)}{\left(\dfrac{T_{\mathrm{t4}}}{T_{\mathrm{a}}} - \tau_{\mathrm{c}}\right)}, \tag{H.7.18}$$

and using simple algebra

$$\eta_{th} = \frac{\left(\frac{T_{t4}}{T_a} - \tau_c\right) - \left(\frac{T_{t4}}{T_a}\frac{1}{\tau_c} - 1\right)}{\left(\frac{T_{t4}}{T_a} - \tau_c\right)} = 1 - \frac{\tau_c}{\tau_c}\frac{\left(\frac{T_{t4}}{T_a}\frac{1}{\tau_c} - 1\right)}{\left(\frac{T_{t4}}{T_a} - \tau_c\right)} = 1 - \frac{1}{\tau_c}\frac{\left(\frac{T_{t4}}{T_a} - \tau_c\right)}{\left(\frac{T_{t4}}{T_a} - \tau_c\right)},$$

which simply reduces to

$$\eta_{th} = \frac{\tau_c - 1}{\tau_c}. \tag{H.7.19}$$

The reduced nondimensional variable is a function of only one variable – the compressor total temperature ratio (or total pressure ratio). Examination of this equation shows that, as the total temperature ratio (or pressure ratio) increases, so does the efficiency. Typical values of thermal efficiency for modern power generation units are 40 to 45 percent.

Lastly, remembering that the heat rate is defined as

$$\mathcal{HR} = \frac{\dot{Q}_{in}}{\mathcal{P}_{net}}. \tag{H.7.20}$$

So, using Eqs. (H.7.17) and (H.7.19),

$$\mathcal{HR} = \frac{\tau_c}{\tau_c - 1}. \tag{H.7.21}$$

This is dimensionless but the reader should recognize that the heat rate is generally reported as kJ/(kW-h) or BTU/(hp-h), so multiplication conversions are needed: 3600 and 2544, respectively.

H.8 Summary

In this appendix the ideal gas turbine cycle analyses were developed to facilitate calculation of the ideal overall gas turbine performance. Components (inlet or diffuser, compressor, fan, propeller, combustor, turbine, bypass duct, bypass mixer, afterburner, nozzle, and power gas turbine exhaust) were assembled, and the characteristics of the ideal components were defined. Ideal cycle equations were derived in this appendix for different engine types: ramjet/scramjet, turbojet, turbofan with exhausted fan, turbofan with mixed fan, turboprop, and power generation gas turbine. A component station numbering identification scheme was also established that is used throughout the book. Component equations were used with equations from Chapters 1, 2, and 3 to determine overall engine performance. Afterburners were included for the turbojet and turbofan. All of the components were assumed to operate without any losses, and the gas was assumed to behave ideally and with constant properties throughout the engine. Four performance parameters were demonstrated as being important comparative measures of a propulsive engine: thrust, nondimensional thrust, *TSFC*, and nondimensional *TSFC*. Similar quantities were found for power generation gas turbines: power, nondimensional power, efficiency, heat rate, *SFC*, and nondimensional *SFC*. Moderately short but closed-form expressions were obtained for these parameters for the different gas turbine types.

Assembling the component models to develop these equations served three purposes. First, the reader was able to see how the equations are assembled to model an entire engine without being encumbered by numerical details. Second, and most importantly, the reader was to be permitted to observe the dependence of the overall performance of a gas turbine on detail parameters conveniently and directly without having to resort to numerical parametric studies. Third, for some cases the approach used allowed analytical optimization or further analysis of performance characteristics. For example, for a power generation unit, the output power was maximized for a particular value of compressor pressure ratio, but it was found that the overall thermal efficiency monotonically increased with increasing compressor pressure ratio.

In Chapter 3, nonideal components are also used, and thus the ideal assumptions are relaxed. Losses and nonideal effects are identified and specified for these components. Nonideal gas turbine performance is also compared to ideal performance in Chapter 3. The added complexity does not allow for closed-form expressions for the engine parameters but does permit more realistic predictions of the overall engine performance. Sufficient details are included in Chapter 3 to allow the entire group of the previously discussed gas turbine types to be analyzed with reduced efficiencies for each of the different components and to determine the reduced efficiency of the gas turbine.

List of Symbols

a	Speed of sound
c_p	Specific heat at constant pressure
c_v	Specific heat at constant volume
C	Coefficient (4 for propeller/turboprop)
F	Force or thrust
f	Fuel ratio
h	Specific enthalpy
ΔH	Heating value
$\mathcal{HR}$	Heat rate
$\dot{m}$	Mass flow rate
M	Mach number
p	Pressure
$\mathcal{P}$	Power
$\dot{Q}$	Heat transfer rate
$\mathcal{R}$	Ideal gas constant
SFC	Specific fuel consumption
T	Temperature
$TSFC$	Thrust-specific fuel consumption
α	Bypass ratio
ρ	Density
γ	Specific heat ratio

η	Efficiency
π	Total pressure ratio
τ	Total temperature ratio

Subscripts

a	Freestream
ab	Afterburner
b	Primary burner
c	Compressor
e	Exit
e	Engine
f	Fan
f	Fuel
in	Inlet
opt	Optimum
p	Propeller
$\mathcal{P}$	Power
s	Secondary (bypassed)
st	Stoichiometric
t	Total (stagnation)
t	Total (summation)
t	Turbine
T	Thrust
th	Thermal
w	Work
1 ... 9	Positions in engine

APPENDIX I NOTES ON ITERATION METHODS

I.1 Introduction

Unfortunately, in the solution of many real engineering problems or designs, iterations are often needed. Whether it be a mechanical design of a compressor or finding the operating temperature of a burner (or anything in between), one usually has to go through many attempts before a final design or value is determined because of the complexity of nonlinear equations and the interdependence of many variables.

In the simplest form, one can define a function that describes a problem to be solved as

$$y = f(x). \tag{I.1.1}$$

It is very important to note that f is not necessarily an analytical or closed-form function. For example, y may be determined only after x is used in any number of independent functions, which can, for example, be equations, graphs, or tables, before combined to yield one particular value of y. Usually, the roots of Eq. (I.1.1) are desired, and thus

$$f(x) = 0. \tag{I.1.2}$$

Therefore, the value(s) of x that cause this function to be equal to zero are desired. Rather than trial and error, which can be extremely tedious and time consuming, many methods of treating this solution numerically are available to yield a solution quickly. Some are more complicated than others. Graeffe's, Bernoulli's, Newton's, Rutihauser's, and iterative factorization methods are among those that are often used. In this book two relatively simple methods will be discussed: Regula Falsi and successive substitution. Examples based on the main body of the text are used to demonstrate the iteration methods.

I.2 Regula Falsi

The Regula Falsi or method of false position is the first to be discussed. The method is best understood by referring to Figure I.1. Here the function y, which is the dependent variable, is plotted vs. x, which is the independent variable. The point at which the function crosses the x-axis is the desired solution. The basic premise is that a stepping technique uses a series of values of x that eventually yield a final and correct solution. The method basically uses two previous values of x and the corresponding values of y to make a third intelligent estimation on x. Initializing the solution requires two intelligent guesses on the solution. The particular procedure is as follows.

(1) Guess x_1 and find the value of y_1 by the function.
(2) Guess x_2 and find the value of y_2 by the function.
(3) Fit a linear curve through the two points and extrapolate to $y = 0$ to estimate x.
This results in

$$m = \text{slope} = \frac{0 - y_1}{x_3 - x_1} = \frac{y_2 - y_1}{x_2 - x_1}. \tag{I.2.1}$$

Therefore, solving for x_3 yields

$$x_3 = -\left[\frac{x_2 - x_1}{y_2 - y_1}\right] y_1 + x_1, \tag{I.2.2}$$

or expanding and combining terms results in

$$x_3 = -\left[\frac{x_1 y_2 - x_2 y_1}{y_2 - y_1}\right]. \tag{I.2.3}$$

(4) Next, using this estimated value of x, one can find y_3 by the function.
(5) The method is repetitive; that is, one can find x_4

$$x_4 = \left[\frac{x_2 y_3 - x_3 y_2}{y_3 - y_2}\right]. \tag{I.2.4}$$

(6) And one can use x_4 to find y_4, and so on.

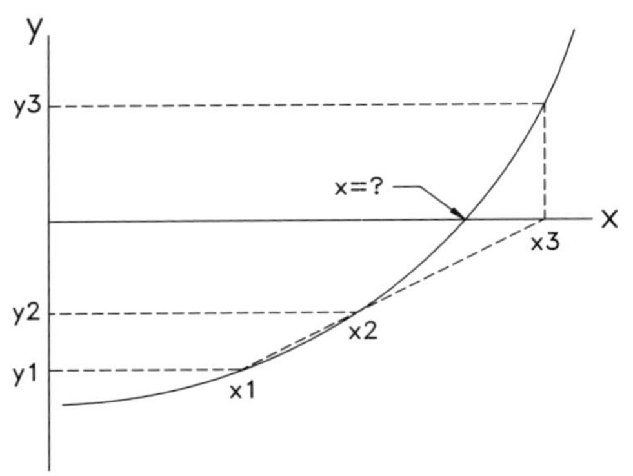

Figure I.1 Graphical representation of Regula Falsi method

(7) Repeat the process to find x_n until the absolute value of y_n is less than some predetermined tolerance – at which point a solution has been successfully obtained.

I.3 Successive Substitutions

The second technique to be discussed is the method of successive substitutions. For this method, one must be able to write the function of interest in the form

$$x = g(x). \tag{I.3.1}$$

As before, this does not have to be a closed-form algebraic or any other closed-form of equation. It can be a series of equations, graphs, or tables that result in a calculation of x.

Example I.1 Regula Falsi Turbojet Demonstration

An ideal turbojet flies at sea level at a Mach number of 0.75. It ingests 165 lb_m/s (74.83 kg/s) of air, and the compressor operates with a particular pressure ratio, which results in a thrust of 15,057.3 lb_f (66,975 N). The fuel has a heating value of 18,000 BTU/lb_m (41,870 kJ/kg), and the burner exit total temperature is 3000 °R (1667 K). The specific heat ratio for this example is 1.35. Find the compressor total pressure ratio that yields this thrust.

Solution

This example is similar to that of Example 3.3 – however, the compressor total pressure ratio is not specified, but the thrust is specified. One must find the value of π_c for this thrust and a tolerance on the calculated thrust of 0.1 lb_f will be used with the Regula Falsi method. One can first guess $\pi_c =$ 13, which yields $F = 14{,}973.2$ lb_f. Details of the thermodynamic calculations are similar to those in Example 3.3. One can next guess $\pi_c = 16$ for the second intelligent guess, which yields $F = 15{,}084.9$ lb_f. Thus, the compressor pressure ratio is the independent variable, and $x_1 = 13$ and $x_2 =$ 16. The function to be solved is $y = 15{,}057.3\ lb_f - F$; that is, this should be identically zero. Thus, the values of the dependent variable for the two initial guesses on x are $y_1 = 84.1$ lb_f and $y_2 = -27.6$ lb_f, respectively. Now these two initial points can be used to estimate the true value by

$$x_3 = \frac{13 \times (-27.6) - 16 \times 84.1}{-27.6 - 84.1} = 15.259.$$

Using this value for the compressor pressure ratio yields $F = 15{,}065.2$ lb_f. Now $x_3 = 15.259$ and $y_3 = -7.9$ lb_f and so using the last two values of x and y yields

$$x_4 = \frac{16 \times (-7.9) - 15.259 \times (-27.6)}{-7.9 - (-27.6)} = 14.962,$$

and $F = 15{,}056.1$ lb$_f$, and so $y_4 = 1.2$ lb$_f$, and so using the last two values of x and y yields

$$x_5 = \frac{15.259 \times 1.2 - 14.962 \times (-7.9)}{1.2 - (-7.9)} = 15.001,$$

and so now $F = 15{,}057.3$ lb$_f$ and $y_5 = 0.0$ lb$_f$, which is finally less than the specified tolerance of 0.1 lb$_f$. Therefore, for this engine the solution is $\pi_c = 15.00$. This agrees with Example 3.3. Note that only five iterations were needed.

Example I.2 Regula Falsi Turbofan Demonstration

A hybrid turbofan flies at sea level at a Mach number of 0.75. It ingests 165 lb$_m$/s of air to the core. The compressor operates with a pressure ratio of 15 and an efficiency of 88 percent. The engine has a bypass ratio of 3 and a split ratio of 0.25. The efficiency of the fan is 90 percent. The fuel has a heating value of 18,000 BTU/lb$_m$, and the burner total temperature is 3000 °R. The burner has an efficiency of 91 percent and a total pressure ratio of 0.95, whereas the turbine has an efficiency of 85 percent. The duct has a total pressure ratio of 0.98, and the total pressure ratio of the mixer is 0.97. A variable converging–diverging (CD) nozzle is used for the primary nozzle, and the efficiency is 96 percent. A converging nozzle is used for the fan nozzle, and the efficiency is 95 percent. The total pressure recovery for the diffuser is 0.92, and the shaft efficiency is 99.5 percent. Find the fan total pressure ratio. Note that this is Example 3.11, in which the author made a very good initial guess on the fan total pressure ratio. In this appendix, the iteration method on the fan total pressure ratio that allowed the author to be so lucky in Example 3.11 will be shown.

Solution

Again, this example uses engine information from Example 3.11. The fan total pressure ratio π_f must be found iteratively. In Example 3.11, the evaluator for the accuracy of the value of π_f was how well $p_{t7.5}$ and p_{t5} agreed at the mixer; that is, the two should be identical. For this example, a tolerance on the calculated total pressure difference of 0.001 psia will be used with the Regula Falsi method. One can first guess $\pi_f = 3.0$, which yields $p_{t7.5} = 57.72$ psia and $p_{t5} = 19.99$ psia; the difference ($\Delta p_t = [p_{t7.5} - p_{t5}]$) is 57.72 psia. Details of the thermodynamic calculations are similar for this fan pressure ratio to those in Example 3.11. One can next (second) guess $\pi_f = 2.0$, which yields $p_{t7.5} = 38.48$ and $p_{t5} = 39.16$ psia; the difference is now −0.68 psia. In fact, the first guess was much worse than the second, but this should not concern the reader. Thus, the first two guesses on the independent variable are $x_1 = 3.0$ and $x_2 = 2.0$. Now, the operating function is $y = 0 - \Delta p_t$. Thus, the first two values of the dependent variable are $y_1 = -37.73$ psia and $y_2 = 0.68$ psia. Now one can use these two points to estimate the true value of the fan pressure ratio (third attempt) by

$$x_3 = \frac{3.0 \times (0.68) - 2.0 \times (-37.73)}{37.73 - (-0.68)} = 2.018,$$

Using this value, one finds $p_{t7.5} = 38.82$ psia, $p_{t5} = 38.65$ psia, and thus $\Delta p_t = 0.17$ psia, which is already much better.

Now $x_3 = 2.018$ and $y_3 = -0.17$ psia, and so the fourth attempt yields

$$x_4 = \frac{2.0 \times 0.17 - 2.018 \times (-0.68)}{0.17 - (-0.68)} = 2.0144,$$

and $p_{t7.5} = 38.760$ psia and $p_{t5} = 38.759$ psia; thus $\Delta p_t = 0.001$ psia, which is equal to acceptable tolerance (0.001 psia). Therefore, for this engine the solution is $\pi_f = 2.0144$. Only four iterations were needed. This agrees with Example 3.11.

This type of solution is depicted in Figure I.2. In this figure $g(x)$ is plotted vs. x, and x is plotted vs. x. Obviously, the second plot is a straight line with a slope of unity. The point at which the two lines intersect is the desired solution. This method only requires one initial guess on the solution. The solution is obtained as follows.

(1) One makes an initial guess on x called x_1.
(2) Using x_1, one can find $g(x_1)$, which by Eq. (I.3.1) is equal to x_2.
(3) Using x_2, one can find $g(x_2)$, which by Eq. (I.3.1) is equal to x_3.
(4) This process is repeated until the absolute value of $x - g(x)$. is less than some specified tolerance. When this difference is smaller than the tolerance, a successful solution has been obtained.

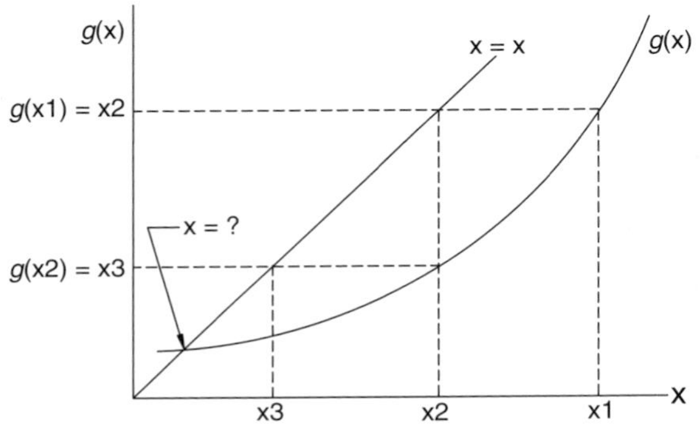

Figure I.2 Graphical representation of successive substitutions method.

Example I.3 Successive Substitutions Compressor Demonstration

Air enters a compressor at a total temperature of 577.0 °R. The total pressure ratio of the compressor is 15, and the efficiency is 88 percent. What is the specific heat ratio for this component if it can be based on the average total temperature? Note that these are the compressor and inlet conditions from Example 3.4.

Solution

To find the specific heat ratio, one must know the average total temperature. However, the specific heat ratio must be known to find the temperature ratio or change for the given pressure ratio. Thus, γ is the unknown variable (x) and the problem is iterative, and successive substitutions will be used with a tolerance of 0.00001 for γ. One can first assume (first guess) the specific heat ratio of $\gamma_1 = 1.400$. Thus, using

$$T_{t3} = T_{t2}\left[\frac{[\pi_c]^{\frac{\gamma-1}{\gamma}} - 1}{\eta_c} + 1\right]$$

one finds, since the total pressure ratio is given as 15,

$$T_{t3_1} = 577.0\left[\frac{[15]^{\frac{0.4}{1.4}} - 1}{0.88} + 1\right] = 1342.7\ ^\circ\text{R}.$$

Also, $\overline{T} = \dfrac{T_{t3} + T_{t2}}{2}$;

thus, $\overline{T_1} = \dfrac{577.0 + 1342.7}{2} = 959.9\ ^\circ\text{R}.$

Using Eq. (C.3.1) yields $c_p = 0.2269807e^{0.000097247\overline{T}}$, so that $c_{p1} = 0.249188$ BTU/lb$_m$-°R. Now, using Eq. (C.3.2),

$$\gamma = \frac{c_p}{c_p - \mathcal{R}}$$

and $\mathcal{R}$ = 53.35 ft-lb$_f$/lb$_m$-°R/ (778.16 ft-lb$_f$/BTU) = 0.068559 BTU/lb$_m$-°R, and so one finds γ_2 = 1.37956.

Thus, a correction to γ has been found.

Repeating the process yields

$T_{t3_2} = 1302.6\ ^\circ\text{R}, \overline{T}_2 = 939.8\ ^\circ\text{R},\ \gamma_3 = 1.38058.$

Again,

$T_{t3_3} = 1304.6\ ^\circ\text{R}, \overline{T}_3 = 940.8\ ^\circ\text{R},\ \gamma_4 = 1.38053.$

Repeating results in

$T_{t3_4} = 1304.5\ ^\circ\text{R}, \overline{T}_4 = 940.7^\circ\text{R},\ \gamma_5 = 1.38053.$

Finally, the change in γ is less than the tolerance (0.00001). Thus, a solution was found; namely γ = 1.38053 and T_{t3} = 1304.5 °R. Only four iterations were needed.

I.4 Summary

In this appendix two iteration techniques were presented to solve problems without closed-form equations. Both are easy to use and usually provide quick and stable solutions. Detailed examples using jet engine examples from the main body were presented to demonstrate the two methods.

APPENDIX J AXIAL COMPRESSOR RADIAL EQUILIBRIUM DETAILS

J.1 Introduction

In Chapter 5, only a mean line analysis of the flow was first considered for an axial compressor, similarly for an axial turbine blade in Chapter 7. In this appendix, three-dimensionality will be addressed for axial flow machines – that is, hub-to-tip variations. A possible phenomenon occurs at the interface between each rotor and stator that is often considered in the design of stages. At these interblade stations, an equilibrium condition is often assumed for which the flow is to be only in the axial and tangential directions (no radial flow). That is, at the interfaces the sum of the surface forces on each fluid element is assumed to be equal to the centrifugal force. Between the interfaces, such conditions are not imposed, and thus radial flow does exist within the blade rows. It is this so-called radial equilibrium condition that allows an analysis of the three-dimensionality of the flow.

J.2 Differential Analysis

First, consider Figure J.1. Here an incremental element of a thin radial wedge is shown. The wedge has an inner radius of r and an outer radius of $r + dr$. The wedge subtends an angle of $d\theta$. The pressures on the inner and outer surfaces are p and $p + dp$, respectively. On the side surfaces, the pressures are $p + dp/2$. For radial equilibrium the sum of the forces must equal zero, or the sum of all the pressure forces must equal the centrifugal force. All of the tangential forces resulting from the pressures cancel. One therefore only has to be concerned with the radial forces. Thus,

$$(p + dp)(r + dr)d\theta - pr\,d\theta - 2\left(p + \frac{dp}{2}\right)dr\frac{d\theta}{2} = \frac{dm}{2}c_u^2. \tag{J.2.1}$$

Now the mass of the wedge is

$$dm = \rho r\,d\theta\,dr. \tag{J.2.2}$$

Thus, using Eqs. (J.2.1) and (J.2.2) and neglecting all of the higher-order terms, one obtains

$$\frac{dp}{dr} = \frac{\rho}{r}c_u^2. \tag{J.2.3}$$

This is the requirement for radial equilibrium. The radial pressure gradient is shown to be a function of both tangential velocity and radius. Several different velocity profiles can be fitted to this condition, and the implications can be studied. Two such considerations are presented in the next two sections.

J.3 Free Vortex

One goal of a design may be to keep the total pressure constant with radius:

$$\frac{\partial p_t}{\partial r} = 0. \tag{J.3.1}$$

From the 2-D analysis already presented, recall that (Eqs. (4.2.17) and (4.2.21))

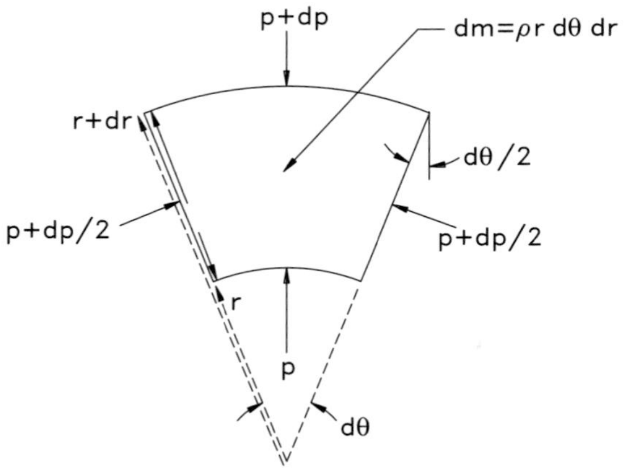

Figure J.1 Force balance on incremental radial fluid wedge.

$$\dot{m}[U_2 c_{u2} - U_1 c_{u1}] = \dot{m}[h_{t2} - h_{t1}], \tag{J.3.2}$$

and so, if R_2 is equal to R_1,

$$\Delta h_t = r\omega \Delta c_u. \tag{J.3.3}$$

Thus, a constant flux of power to the fluid in the radial direction is necessary so that Eq. (J.3.1) is maintained:

$$\frac{\partial \Delta h_t}{\partial r} = \frac{\partial(r\omega\Delta c_u)}{\partial r} = 0; \tag{J.3.4}$$

thus, $r\Delta c_u$ is constant with radius. One method to ensure this condition is to keep $r\Delta c_u$ constant with radius. This consideration is the so-called free vortex. For this condition,

$$c_u r = \kappa \tag{J.3.5}$$

or

$$c_u = \frac{\kappa}{r}, \tag{J.3.6}$$

where κ is a constant.

If the Bernoulli equation is used (incompressible flow) to identify the total pressure,

$$p_t = p + \frac{1}{2}\rho c^2, \tag{J.3.7}$$

along with Eq. (J.3.1),

$$\frac{\partial p}{\partial r} + \rho \frac{1}{2}\frac{\partial c^2}{\partial r} = \frac{\partial p_t}{\partial r} = 0, \tag{J.3.8}$$

and along with Eq. (J.2.3),

$$\frac{\rho}{r} c_u^2 + \frac{1}{2}\rho \frac{\partial c^2}{\partial r} = \frac{\partial p_t}{\partial r} = 0. \tag{J.3.9}$$

From a velocity polygon, however,

$$c^2 = c_u^2 + c_a^2, \tag{J.3.10}$$

thus,

$$\frac{\rho}{r} c_u^2 + \rho \left[c_u \frac{\partial c_u}{\partial r} + c_a \frac{\partial c_a}{\partial r} \right] = 0, \tag{J.3.11}$$

and, for c_u, from the free vortex condition, Eq. (J.3.6),

$$\rho c_a \frac{\partial c_a}{\partial r} = 0. \tag{J.3.12}$$

Consequently, either c_a is zero (no mass flow, which is not a very interesting or practical case) or the axial component of velocity is not a function of radius; hence, c_a is exactly constant with radius.

Finally, some of the implications of this condition need to be addressed. If the radius is nearly constant from blade hub to tip, the tangential component of velocity will not change significantly. However, if the radius change is large for (e.g., a fan blade), significant variations in c_u can occur. Furthermore, the axial component has been shown to be constant with radius. Thus, if c_u varies significantly, so will blade angles. To demonstrate such variations, three typical velocity polygons for a rotor blade, are presented and discussed in Chapter 5 (Figure 5.21).

J.4 "Constant" Reaction

As implied in Section J.3, another consideration may be to force the reaction to be constant from hub to tip (or approximately so). With this idea in mind, one can propose radial variations for the tangential velocity such as

$$c_{u1} = ar - b/r \tag{J.4.1}$$

for the velocity at an inlet to the rotor (exit of the previous stator) and

$$c_{u2} = ar + b/r \tag{J.4.2}$$

for the velocity after a rotor where a and b are constants. Next, it can be demonstrated that, for incompressible flow without losses, if the axial velocity is approximately constant across the rotor ($c_{a1} \cong c_{a2}$),

$$\%R \cong 1 - \frac{a}{\omega}. \tag{J.4.3}$$

That is, the percent reaction is not a (strong) function of radius. One can also show that the input power per unit mass flow across the rotor blade is

$$\frac{\mathcal{P}}{\dot{m}} = 2b\omega. \tag{J.4.4}$$

Note that this is not approximate but identically true. An implication of this is that the rise in total pressure is constant with radius (because the net specific power into the fluid is constant with radius), as shown in Eq. (J.3.1). Thus, using the definition of total pressure,

$$p_t = p + \frac{1}{2}\rho\ [c_u^2 + c_a^2], \tag{J.4.5}$$

one can show that, behind the rotor

$$c_{a2}^2 = \text{constant} - 2a^2\left[r^2 + 2\frac{b}{a}\ln(r)\right], \tag{J.4.6}$$

and behind the stator

$$c_{a1}^2 = \text{constant} - 2a^2\left[r^2 - 2\frac{b}{a}\ln(r)\right]. \tag{J.4.7}$$

It can therefore be seen that, at any value of r, c_{a1} does not exactly equal c_{a2}, as assumed earlier in this section. As a result, the percent reaction is not exactly constant with radius. However, for many cases of interest, the approximation is close and allows for the a priori specification of a velocity profile.

Once again, if the implications of this condition are examined, one finds that the blades are moderately twisted. The geometries are not as awkward as for a free-vortex distribution but yet are not as simple as one would have hoped. An example of such variations is presented in Chapter 5 and discussed in detail (Figures 5.21 and 5.22).

J.5 Summary

In this appendix, three-dimensionality was addressed for axial flow machines – that is, hub-to-tip variations. At interblade stations between rotor blades and stator vanes, an equilibrium condition was assumed for which the flow is to be only in the axial and tangential directions (no radial flow). At the interfaces the sum of the surface forces on each fluid element is assumed to be equal to the centrifugal force. Between the interfaces, such conditions are not imposed, and thus radial flow does exist within the blade rows. It is this so-called radial equilibrium condition that allows an analysis of the three-dimensionality of the flow. The differential analysis was performed herein which yielded equations that can be applied to predict such variations. Two limiting conditions were discussed: free vortex flow and "constant" reactions flows. In Chapter 5 examples are presented to demonstrate the variations.

List of Symbols

a	"Constant reaction" coefficient
b	"Constant reaction" coefficient
c	Absolute velocity
h	Specific enthalpy
$\dot{m}$	Mass flow rate
m	Mass
p	Pressure
$\mathcal{P}$	Power
R	Radius
%R	Percent reaction
r	Radius
κ	Free vortex constant
θ	Angle
ρ	Density
ω	Rotational speed

Subscripts

a	Axial
t	Total (stagnation)
u	Tangential
1, 2, 3	Positions in stage

APPENDIX K AXIAL FLOW STREAMLINE ANALYSIS METHOD DETAILS

K.1 Introduction

In Chapters 5 and 7, the analyses of flows in the axial flow compressors and turbines were first assumed to be two-dimensional (2-D), and any radial distributions or velocity components were assumed to be negligible in calculations. Section 5.9 indicates that three-dimensionality is indeed present due to radial equilibrium and was discussed. As implied, because the ratios of blade height to hub diameter for many modern machines are not small and centrifugal forces are large, three-dimensional (3-D) flows result. Significant radial variations in parameter profiles will result when the tip radius is approximately 1.1 times the hub radius or greater. Several numerical “throughflow” and CFD techniques have been developed that can be used to predict the flow’s three-dimensionality, as discussed in Chapter 5. This appendix details a streamline analysis method. Basically, different streamline paths are analyzed by a control volume approach through a compressor or turbine passage. The software SLA.exe was developed to accommodate a few select boundary condition types.

The flow passage is divided into a finite number of annuli, and the equations are solved for each blade row independently. Different conditions can be considered with the method. For example, free vortex or irrotational flow can be assumed throughout, and the flow angles are then calculated. As discussed in Appendix J, this condition does not result in radial flow and therefore does not require an iterative solution. Also, the flow angles can be specified. This case has an iterative solution whereby radial migration of the flow is predicted. Radial efficiency distributions for each blade row can also specified for the second condition. Other conditions are discussed. Steady incompressible adiabatic flow is assumed to simplify the analysis and to accelerate the numerical convergence of the solution. Moderate accuracy can be obtained using an average density – particularly for one- or two-stage analyses. Moreover, the incompressible assumption does not mask the fundamental 3-D trends predicted from the method, which can also be extended to include compressibility effects.

K.2 Flow Geometry

In Figure 5.23 a stage in an axial flow turbomachine is shown with varying cross-sectional area, although more stages can be stacked behind. The analysis begins at the inlet to the previous stator (which can be an inlet guide vane). Each stage is divided into N annuli. In Figure 5.23, the annuli number (i) and the axial position number (j) are defined. For the sake of simplicity, only four annuli are shown. Each annulus represents a bundle of streamlines, and a different control volume is applied to each annulus. In general, the annuli will not be constant in area between axial stations. This is due to the resulting radial flow and changing cross section of the stage. For the control volume analysis, the flow is assumed to be uniform, both radially and circumferentially, within each annulus, and the velocities are evaluated at the radial center of the annulus. Thus, blade-to-blade variations are not predicted as is the case in a CFD analysis. The exiting conditions for stator 0 are used as the inlet conditions for rotor 1, and so on. The radii of all of the axial and radial stations can in general be different. The radius R_{32} (typical) represents the outer radius of the first streamline bundle and the inner radius of the second streamline bundle at the third axial station. Also, the axial component of velocity at axial station 4 and annulus 1 is shown, c_{a41}.

K.3 Working Equations

The equations for each annulus are described in this section. The equations are independently derived for each annulus by a control volume analysis similar to the mean line flow analysis for incompressible flow described in Section 5.4. However, equations for all of the annuli must be solved simultaneously. It is important to note that the control volumes can be drawn in general. However, the exact positions and sizes (radii) of the control volumes are actually found as a part of the solution.

K.3.1 Geometric Equations

The velocity of the blades at the radial center of the element is U_{ji}. Shown in Figure 5.7 are the absolute flow velocity (c) and the flow velocity relative to the blades (w). Also shown is the normal component, c_a. Velocity polygons, as in Figure 5.7, are constructed for each annulus and axial position. As in Chapter 5, α and β are the absolute and relative flow angles, respectively. Important relationships that result from the velocity polygons for each annulus are as follows:

$$c_{a_{ji}} = c_{ji} \cos \alpha_{ji} \tag{K.3.1}$$

$$c_{u_{ji}} = c_{ji} \sin \alpha_{ji} \tag{K.3.2}$$

$$\tan \beta_{ji} = \frac{c_{u_{ji}} - U_{ji}}{c_{a_{ji}}}, \tag{K.3.3}$$

where

$$U_{ji} = \overline{R}_{ji}\omega \tag{K.3.4}$$

and

$$\overline{R}_{ji} = \frac{1}{2}(R_{ji} + R_{ji+1}). \tag{K.3.5}$$

The value of $\overline{R}_{ji}$ represents the radius at the center of the i th element at the j th station.

K.3.2 Control Volume Equations

Next, the conservation of mass equation is considered. Because each control volume follows the streamlines, no flow crosses the annuli, which accounts for the name of the method. As a result, the only mass flux occurs at the axial planes. The volumetric flow rate given by the continuity equation (Eq. (4.2.1)) thus becomes

$$Q_i = c_{a_{ji}}\pi(R^2_{ji+1} - R^2_{ji}). \tag{K.3.6}$$

Note that this is constant for all axial locations within the stream tube. The moment of momentum equation (Eq. (4.2.2)) becomes

$$T_{ji} = \rho Q_i(\overline{R}_{j+1i}c_{u_{j+1i}} - \overline{R}_{ji}c_{u_{ji}}), \tag{K.3.7}$$

where T_{ji} is the torque on the shaft for that annulus. Thus, summing all annular torques yields the total torque. The energy equation (Eq. (4.2.3)) becomes

$$\dot{W}_{ji} = \frac{Q_i}{\eta_{ji}}\left[(p_{j+1i} - p_{ji}) + \frac{1}{2}\rho(c^2_{j+1i} - c^2_{ji})\right], \tag{K.3.8}$$

where $\dot{W}_{ji}$ and η_{ji} are, respectively, the shaft power and efficiency for the single annulus. Thus, summing all annular powers yields the total power. Furthermore, the torque and rotor power for each annulus are related by

$$\dot{W}_{ji} = T_{ji}\omega, \tag{K.3.9}$$

where ω is the rotational speed of the rotor. When stator blades are analyzed for the power, ω is of course zero. The last consideration occurs at the interface between each rotor and stator. At these stations, radial equilibrium (which is covered in Appendix J) is assumed. The resulting equation is Eq. (J.2.3). Writing this equation in difference form for each station, one obtains

$$\frac{p_{ji+1} - p_{ji}}{\overline{R}_{ji+1} - \overline{R}_{ji}} = \rho\frac{\overline{c}^2_{u_{ji}}}{\overline{R}_{ji+1}}, \tag{K.3.10}$$

where

$$\overline{c}_{u_{ji}} = \frac{1}{2}(c_{u_{ji+1}} + c_{u_{ji}}). \tag{K.3.11}$$

Also, applying Eq. (J.3.7) from Appendix J to an element,

$$p_{t_{ji}} = p_{ji} + \frac{1}{2}\rho c^2_{ji}. \tag{J.3.7$'$}$$

K.3.3 Closure Equations

Before one attempts a solution, counting of the variables is appropriate. For N annuli, N values of c, c_a, c_u, α, β, η, T,

Q, p, p_t, $\dot{W}_{sh}$, U, and $\overline{R}$ will result, as will $N+1$ values of R, $N-1$ values of $\overline{c}_u$, one value of ω, and one value of ρ. On the other hand, there are N equations (Eqs. (K.3.1)–(K.3.9) and (J.3.7) (Appendix J)), $N-1$ equations (Eqs. (K.3.10) and (K.3.11)), two values of R (hub and tip), one value of ω, and one value of ρ. Thus, there are $15N+2$ variables and $12N+2$ equations. Clearly, to obtain a solution, one must specify $3N$ variables or $3N$ more equations are needed.

Three separate conditions are considered in detail in the following sections. In all cases, all flow conditions into the inlet of the previous stator are specified. The annuli locations at $j=1$ are specified and can be specified to be equally spaced, but this is not necessary if strong velocity gradients are present. Thus, Q_i for each annulus is found because the flow into the previous stator is specified. Thus, N more variables are specified for these cases for a total of $13N+2$ equations, leaving the need for $2N$ more specifications. Some different cases are discussed in detail in the following sections. Other cases can certainly be used to specify the last set of conditions.

Potential Flow or Free Vortex Flow

The flow is assumed to be irrotational as well as inviscid ($\eta_{ji}=1.0$). Next, from Eq. (J.3.6),

$$c_{u_{ji}}\overline{R}_{ji} = c_{u_{ji+1}}\overline{R}_{ji+1}, \tag{K.3.12}$$

which adds $N-1$ equations. Also, as shown in Appendix J.3, this condition, along with radial equilibrium, results in a constant axial velocity at each axial plane and no radial flow between the axial stations. That is, $c_{a_{ji}}$ is not a function of R (or i). Lastly, for this condition the absolute flow angle at some radius (usually at the center of each axial plane) is usually specified, which adds the last needed variable. For example, if $N=7$, values of α_{j4} are specified. As an equally valid alternative specification for this case, the relative flow angle at some radius could be specified; if $N=7$, values of β_{j4} could be specified. This case is relatively simple and does not require an iteration technique.

Specified Flow Angle Distribution

For this case, the second set of boundary conditions is for specified flow angle variations (hub to tip). Absolute flow angles (α_{ji}) and efficiencies (η_{ji}) are chosen to be input at all axial stations and at specified radial positions. Thus, $2N$ variables (α and η) are specified, and a solution is possible. Although flow angles and efficiencies are specified at certain radii, these do not necessarily correspond to annuli centroids and can be a curve fit to a radial distribution or table for interpolation at different radii. Alternatively, and equally valid, the relative flow angles (β_{ji}) and efficiencies (η_{ji}) could be chosen to be input at all axial stations and at specified radial positions. The solution to this case requires a two-nested loop iterative solution.

Specified Pressure Distribution

If limited experimental data are available, radial profiles of static and total pressure (p and p_t) may be known. Thus, for this case, the second set of boundary conditions is for specified static and total pressure variations (hub to tip) and a solution is possible providing angle, efficiency, and other parameter radial distributions to complement the data. Therefore, p_{ij} and $p_{t_{ij}}$ are input at all axial stations and at specified radial positions. Again, $2N$ variables (p and p_t) are specified and a solution is possible. As in the previous section, the radii do not necessarily correspond to annuli centroids. The solution to this case also requires a two-nested loop iterative solution.

Other Specified Conditions

As indicated earlier in Section K.3.3, in general there are $15N+2$ variables and $12N+2$ equations. One can designate conditions other than specified angles or pressures that would allow a solution. Many other possibilities could be used to specify closure equations.

Solution Method

The equations discussed in Section K.3 and its subsections can be solved using any of the commercial math solvers.

To find a solution, one starts at $j = 2$ (if conditions at $j = 1$ are known) and analyzes each axial station separately, repeatedly marching in the axial direction.

For known flow angle and efficiency distributions, the solution requires a two-nested loop iterative solution. One approach is presented in the steps below.

(1) A value for $c_{a_{j+1}1}$ is first estimated.
(2) The inner annulus ($i = 1$) is analyzed, and $R_{j+1\,i+1}$ is found (Eq. (K.3.6)).
(3) Then α_{j+1i} and η_{j+1i} are found from the known radial distributions.
(4) The values of c_{j+1i}, $c_{u_{j+1}i}$, and β_{j+1i} are found (Eqs. (K.3.1), (K.3.2), and (K.3.3)).
(5) The torque and power are found (Eqs. (K.3.7) and (K.3.9)).
(6) Then p_{j+1i} is found (Eq. (K.3.8)).
(7) The next annulus is analyzed. A value of $c_{a_{j+1}i+1}$ is estimated, and the same procedure as immediately above is used to find $p_{j+1\,i+1}$.
(8) Radial equilibrium (Eq. (K.3.10)) is now also used to find a second value of $p_{j+1\,i+1}$.
(9) If the two independent values of $p_{j+1\,i+1}$ are the same within a set tolerance, i is incremented and the next annulus analyzed, and so on, until the last annulus. If the two are different, a new value of $c_{a_{j+1}i+1}$ is used and the process starting at (7) is repeated.
(10) For the last annulus, R_{j+1N} is known, and thus $c_{a_{j+1}N}$ is found (Eq. (K.3.6)). The same procedure as immediately above is used to find $p_{j+1\,N}$ and radial equilibrium (Eq. (K.3.10)) is now used also to find a second value of $p_{j+1\,N}$.
(11) The two values of $p_{j+1\,N}$ are again compared. If these two values are different, a new $c_{a_{j+1}1}$ is used, and the entire procedure, starting at (1), is repeated until convergence.

Thus, this case has two nested iteration loops. The Regula Falsi method (Appendix G) can be used to accelerate the converging iteration for both loops. As discussed at the beginning of this section, the radial size and position of the control volumes are found as part of the solution. Also, initial estimates can have a strong influence on convergence of such schemes.

For known total and static pressures, a similar two-nested loop iterative solution is needed.

(1) A value for $c_{a_{j+1}1}$ is first estimated.
(2) The inner annulus ($i = 1$) is then analyzed and $R_{j+1\,i+1}$ is found from Eq. (K.3.6). Then, $p_{j+1\,i}$ and $p_{t_{j+1}i}$ are found from the known radial distribution of pressures.
(3) The value of $c_{j+1\,i}$ is next found (Eq. (J.3.7)).
(4) Then α_{j+1i}, $c_{u_{j+1}i}$, and β_{j+1i} are found (Eqs. (K.3.1), (K.3.2), and (K.3.3)).
(5) The torque and power are found (Eqs. (K.3.7) and (K.3.9)).
(6) The efficiency at $(j+1, i)$ is found from Eq. (K.3.8).
(7) The next annulus is analyzed. A value of $c_{a_{j+1}i+1}$ is estimated, and the same procedure as immediately above is used to find all parameters, including $p_{j+1\,i+1}$ and $\eta_{j+1\,i+1}$.
(8) Radial equilibrium (Eq. (K.3.10)) is now also used to find a second value of $p_{j+1\,i+1}$.
(9) If the two values of $p_{j+1\,i+1}$ are the same within a set tolerance, the next annulus is analyzed, and so on, until the last annulus. If the two are different, a new value of $c_{a_{j+1}i+1}$ is used and the process is repeated starting at (7).
(10) For the last annulus, $R_{j+1\,N}$ is known and $c_{a_{j+1}N}$ is found from Eq. (K.3.6). Two values of $p_{j+1\,N}$ are thus found.
(11) The two independent values of $p_{j+1\,N}$ are compared. If these two values of $p_{j+1\,N}$ are different, a new $c_{a_{j+1}1}$ is used and the entire procedure repeated starting at (1) until convergence.

To demonstrate the use and implications of this method, three examples are presented in Chapter 5. Examples 5.5, 5.5.1, and 5.6 are shown, including irrotational flow and specifies flow angles and efficiencies. Results are analyzed and discussed in detail. Results were obtained with SLA.exe.

K.4 Summary

This appendix details the streamline analysis method. Basically, different streamline paths are analyzed by a control volume approach through a compressor or turbine passage. The flow passage is divided into a finite number of annuli, and the equations are solved for each blade row independently. Flow passes through each of the annuli but does not cross the rol volumes thus giving the name streamline analysis. Different conditions are considered with the method. For example, free vortex or irrotational flow can be assumed throughout, and the flow angles are then calculated. This condition does not result in radial flow and therefore does not require an iterative solution. Also, the flow angles can be specified. This case has an iterative solution whereby radial migration of the flow is predicted. Radial efficiency distributions for each blade row can also specified for the second condition. Other conditions are discussed. Steady incompressible adiabatic flow is assumed to simplify the analysis and to accelerate the numerical convergence of the solution. Moderate accuracy can be obtained using an average density – particularly for one- or two-stage analyses. Moreover, the incompressible assumption does not mask the fundamental 3-D trends predicted from the method, which can also be extended to include compressibility effects. Examples are presented in Chapter 5.

List of Symbols

c	Absolute velocity
D	Diameter
i	Radial counter
j	Axial counter
N	Number of annuli
p	Pressure
Q	Volumetric flow rate
R	Radius
T	Torque
U	Blade velocity
$\dot{W}$	Power
α	Absolute (stator) angle
β	Relative (rotor) angle
η	Efficiency
ρ	Density
ω	Rotational speed

Subscripts

a	Axial
H	Hub
i	Radial counter
j	Axial counter
sh	Shaft
t	Total (Stagnation)
T	Tip
u	Tangential

Superscripts

–	Average

APPENDIX L PERFORMANCE PREDICTION OF A COMPRESSOR STAGE DETAILS

L.1 Introduction

As indicated in Chapter 5 a compressor stage or cascade is made up of a series of airfoils. Like wings, airfoils have lift and drag characteristics. For the case of a cascade, a large lift implies a large torque, which has been shown to result in a large pressure ratio. On the other hand, drag implies frictional losses, which result in a loss in total pressure and efficiency. Furthermore, both the lift and drag are (strong) functions of leading-edge incidence angle. Thus, the purpose of this appendix is to develop a method to relate the lift and drag characteristics directly to efficiency. The method presented in this appendix is a combination of theoretical considerations and empiricism. Incompressible flow is assumed to simplify the mathematics. The method can be used to select the blade characteristics – for example, angles, solidity, pitch, chord, and so on – to maximize the efficiency. Most importantly, the method developed here gives great insight into how the stage performance and blade characteristics are fundamentally related. The analysis is included herein and used for application examples presented in Chapter 5.

L.2 Velocity Polygons

Although velocity polygons have been covered already in Chapter 5, they are discussed again in this section with a somewhat different perspective. For the rotor, the relevant velocities for a blade are shown in Figure 5.30a. For the stator, the relevant velocities for a vane are shown in Figure 5.30b. Both figures illustrate the lift and drag forces acting on the blades (F_l and F_d, respectively). As usual, the inlet and exit flow angles to the rotor blade are β_1 and β_2 respectively, and the flow deflection is δ_{12}. The average flow angle is defined as β_m, which is approximately the stagger, and the forces are relative to this angle. Inlet and exit flow angles for the stator are α_2 and α_3, respectively. Once again, a consistent sign notation is needed, and the same notation as described in Section 5.3 is used. That is, all positive tangential velocities are to the right. Thus, rotor blades move to the right with positive velocity U. Also, all positive flow and force angles are considered to be counterclockwise. Positive lift and drag loads are assumed to be opposite to the direction of flow turning or deflection and in the direction of flow, respectively.

Next, recall the definition of percent reaction results in

$$\%R = \frac{1}{1 + \left[\dfrac{c_2^2 - c_1^2}{w_1^2 - w_2^2}\right]}. \tag{L.2.1}$$

One can use

$$c^2 = c_u^2 + c_a^2 \tag{L.2.2}$$

and

$$w^2 = w_u^2 + w_a^2, \tag{L.2.3}$$

with the assumption that the axial component of velocity remains constant through a stage to find that

$$\%R = \frac{1}{1 + \left[\dfrac{c_{u2}^2 - c_{u1}^2}{w_{u1}^2 - w_{u2}^2}\right]}, \tag{L.2.4}$$

or, by grouping and expanding, that

$$\%R = \frac{(w_{u1} + w_{u2})(w_{u1} - w_{u2})}{(w_{u1} + c_{u1})(w_{u1} - c_{u1}) - (w_{u2} + c_{u2})(w_{u2} - c_{u2})}. \tag{L.2.5}$$

It is apparent from Figures 5.7 and 5.30 if $\alpha_1 = \alpha_3$ (used hereafter) that

$$U = c_{u1} - w_{u1} = c_{u2} - w_{u2}; \tag{L.2.6}$$

thus, using Eq. (L.2.5) and rearranging yields

$$\%R = -\frac{(w_{u1} + w_{u2})(w_{u1} - w_{u2})}{U\{(w_{u1} - w_{u2}) - (c_{u2} + c_{u1})\}}, \quad \text{(L.2.7)}$$

and using Eq. (L.2.6) results in

$$c_{u2} - c_{u1} = w_{u2} - w_{u1}. \quad \text{(L.2.8)}$$

Thus, from Eq. (L.2.7),

$$\%R = -\frac{(w_{u1} + w_{u2})}{2U}. \quad \text{(L.2.9)}$$

Therefore, the reaction is the average tangential component of relative velocity (to the blade) divided by the rotational velocity. Finally, two new definitions can be introduced. The first is the conventional flow coefficient, which is a nondimensional velocity based on the axial component and thus represents the total flow rate as follows:

$$\phi = \frac{c_a}{U} = \frac{w_a}{U}. \quad \text{(L.2.10)}$$

The blade velocity, U, is conventionally used to nondimensionalize velocities. The second is the deflection coefficient (nondimensional deflection), which represents the blade turning. For a rotor blade,

$$\tau_r = \frac{w_{u2} - w_{u1}}{U}, \quad \text{(L.2.11)}$$

and, for a stator vane,

$$\tau_s = \frac{c_{u2} - c_{u3}}{U}; \quad \text{(L.2.12)}$$

thus, for the case of $\alpha_3 = \alpha_1$,

$$\tau_s = \frac{c_{u2} - c_{u1}}{U}. \quad \text{(L.2.13)}$$

One can now nondimensionalize the entire set of velocity polygons in Figure 5.30 by U to generate Figure 5.31. In Figure 5.31, one can see the different physical quantities (%R, ϕ, and τ) and thus gain an understanding of the relevance of velocity polygons to compressor stage performance. For example, if α_1 is equal to α_3 and %R = 0.50, then τ_r is equal to τ_s and once again the polygon is symmetric, as discussed in Section 5.4.4. Also shown on Figure 5.31 are the average angles β_m (defined as $\tan^{-1}\{\frac{1}{2}[\tan\beta_1 + \tan\beta_2]\}$) and α_m (defined as $\tan^{-1}\{\frac{1}{2}[\tan\alpha_2 + \tan\alpha_3]\}$). One can readily see, for example, that

$$\tan\beta_m = -\frac{\%R}{\phi} \quad \text{(L.2.14)}$$

and that

$$\tan\alpha_m = \frac{1 - \%R}{\phi}. \quad \text{(L.2.15)}$$

Using the preceding two equations yields

$$\tan\alpha_m = \frac{1 + \phi\tan\beta_m}{\phi}. \quad \text{(L.2.16)}$$

Also, using the geometry of Figure 5.31, one sees that

$$\tau_r = \phi(\tan\beta_2 - \tan\beta_1) \quad \text{(L.2.17)}$$

and that

$$\tau_s = \phi(\tan\alpha_2 - \tan\alpha_1). \quad \text{(L.2.18)}$$

Finally, it can also be demonstrated from Figure 5.31 that

$$\tau_r/2 = \phi(\tan\beta_m - \tan\beta_1) = \phi(\tan\beta_2 - \tan\beta_m) \quad \text{(L.2.19)}$$

and

$$\tau_s/2 = \phi(\tan\alpha_2 - \tan\alpha_m) = \phi(\tan\alpha_m - \tan\alpha_1). \quad \text{(L.2.20)}$$

L.3 Lift and Drag Coefficients

Next, the forces acting on the blades can be considered and related to lift and drag coefficients. First, by examining the forces in Figures 5.30 (F_{dr}, F_{lr}, F_{ds}, and F_{ls}), one can break these forces into different components as shown in Figures L.1a and L.1b. Shown in these figures are the forces in the axial and tangential directions F_a and F_u, respectively.

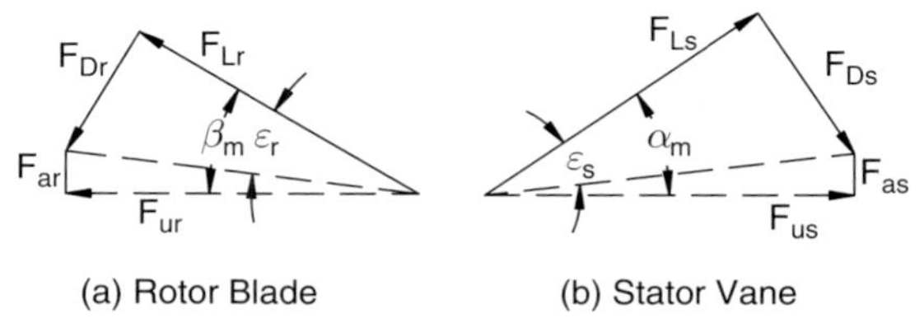

Figure L.1 Force polygons.

Thus, for the rotor, the forces can be related by

$$F_{\mathrm{lr}} = F_{\mathrm{ur}} \cos\beta_{\mathrm{m}} - F_{\mathrm{ar}} \cos\beta_{\mathrm{m}} \quad \text{(L.3.1)}$$

and

$$F_{\mathrm{dr}} = -F_{\mathrm{ur}} \sin\beta_{\mathrm{m}} - F_{\mathrm{ar}} \cos\beta_{\mathrm{m}}, \quad \text{(L.3.2)}$$

and the angle between the drag and lift forces is found from

$$\tan \varepsilon_{\mathrm{r}} = \frac{F_{\mathrm{dr}}}{F_{\mathrm{lr}}}. \quad \text{(L.3.3)}$$

For the stator,

$$F_{\mathrm{ls}} = F_{\mathrm{us}} \cos\alpha_{\mathrm{m}} - F_{\mathrm{as}} \sin\alpha_{\mathrm{m}} \quad \text{(L.3.4)}$$

and

$$F_{\mathrm{ds}} = F_{\mathrm{us}} \sin\alpha_{\mathrm{m}} - F_{\mathrm{as}} \cos\alpha_{\mathrm{m}}, \quad \text{(L.3.5)}$$

and the angle between the two forces for the stator is found from

$$\tan \varepsilon_{\mathrm{s}} = -\frac{F_{\mathrm{ds}}}{F_{\mathrm{ls}}}. \quad \text{(L.3.6)}$$

Also, the lift and drag coefficients are defined as

$$C_{\mathrm{l}} \equiv \frac{F_{\mathrm{l}}}{\frac{1}{2}\rho v_{\mathrm{m}}^2 C} \quad \text{(L.3.7)}$$

and

$$C_{\mathrm{d}} \equiv \frac{F_{\mathrm{d}}}{\frac{1}{2}\rho v_{\mathrm{m}}^2 C}, \quad \text{(L.3.8)}$$

where C is the chord length, and v_{m} is a representative velocity and is equal to w_{m} for a rotor and c_{m} for a stator; thus with Eqs. (L.3.3) and (L.3.6),

$$\tan \varepsilon_{\mathrm{r}} = \frac{C_{\mathrm{dr}}}{C_{\mathrm{ls}}} \quad \text{(L.3.9)}$$

and

$$\tan \varepsilon_{\mathrm{s}} = -\frac{C_{\mathrm{ds}}}{C_{\mathrm{ls}}}. \quad \text{(L.3.10)}$$

L.3.1 Infinitely Long Blades

Blades are obviously of finite length. However, the first blades that will be considered are of infinite length because they are more ideal. In the next subsection, correction factors are applied to account for deviations from the finite length assumption. If a control volume is drawn on the rotor blades, as shown in Figure L.2, the linear momentum equation can be used to show that, for the tangential force of the fluid on a blade, if one remembers remembering that the force on the blade is equal and opposite to the force on the fluid,

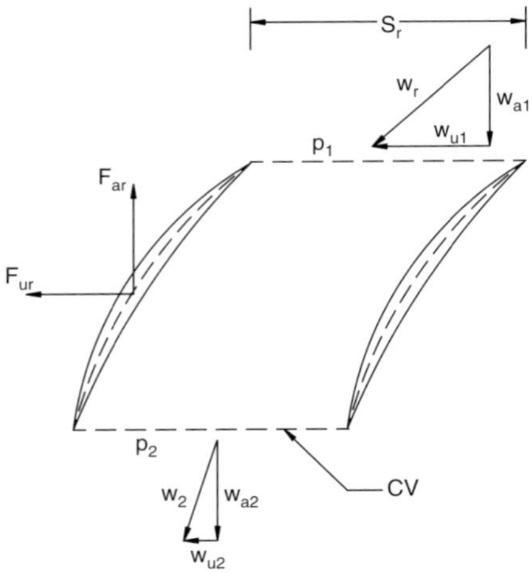

Figure L.2 Control volume on rotor cascade.

$$F_{\mathrm{ur}} = \dot{m}(w_{\mathrm{u2}} - w_{\mathrm{u1}}). \quad \text{(L.3.11)}$$

Also, for an axial force on a rotor blade,

$$F_{\mathrm{ar}} = \dot{m}(w_{\mathrm{a1}} - w_{\mathrm{a2}}) + (p_2 - p_1)s_{\mathrm{r}}. \quad \text{(L.3.12)}$$

As discussed earlier, however, one usually tries to hold the axial velocity constant in an axial compressor ($c_{\mathrm{a2}} = c_{\mathrm{a1}}$ or $w_{\mathrm{a2}} = w_{\mathrm{a1}}$), and thus

$$F_{\mathrm{ar}} = (p_2 - p_1)s_{\mathrm{r}}. \quad \text{(L.3.13)}$$

Thus, from Eq. (L.3.11), one can multiply and divide by U^2 (noting that, if the flow is incompressible, the mass flow rate is $\dot{m} = \rho s_{\mathrm{r}} w_{\mathrm{a}}$) to find

$$F_{\mathrm{ur}} = \rho s_{\mathrm{r}} U^2 \left[\frac{w_{\mathrm{a}}}{U}\right] \left[\frac{(w_{\mathrm{u2}} - w_{\mathrm{u1}})}{U}\right]. \quad \text{(L.3.14)}$$

First, if the drag is considered, using Eqs. (L.3.2), (L.3.13), and (L.3.14) yields

$$F_{ur} = -\rho s_r U^2 \left[\frac{w_a}{U}\right]\left[\frac{(w_{u2} - w_{u1})}{U}\right] \sin\beta_m - (p_2 - p_1)s_r \cos\beta_m, \quad \text{(L.3.15)}$$

or trigonometry results in

$$F_{dr} = -\rho s_r\, w_a^2 [\tan\beta_2 - \tan\beta_1] \sin\beta_m - (p_2 - p_1)s_r \cos\beta_m. \quad \text{(L.3.16)}$$

The pressure difference $p_2 - p_1$ is defined as

$$p_2 - p_1 = p_{t2} - \frac{1}{2}\rho w_2^2 - p_{t1} + \frac{1}{2}\rho w_1^2 \quad \text{(L.3.17)}$$

or

$$p_2 - p_1 = \frac{1}{2}\rho(w_1^2 - w_2^2) - \zeta_r, \quad \text{(L.3.18)}$$

where ζ_r is the total pressure loss ($p_{t2} - p_{t1}$) due to friction. This can be related to the flow angles by

$$p_2 - p_1 = \rho c_a^2(\tan\beta_1 - \tan\beta_2)\tan\beta_m - \zeta_r. \quad \text{(L.3.19)}$$

Using this with Eq. (L.3.16), one finds

$$F_{dr} = s_r\zeta_r \cos\beta_m. \quad \text{(L.3.20)}$$

Using the definition of the drag coefficient (Eq. (L.3.8)) yields

$$C_{dr_p} = \frac{s_r\,\zeta_r \cos\beta_m}{C\,\frac{1}{2}\rho w_m^2}. \quad \text{(L.3.21)}$$

The subsubscript p is used to remind the reader that all of the drag arises from the pressure loss.

To calculate the lift, one can use Eqs. (L.3.1), (L.3.13), and (L.3.14) to find

$$F_{lr} = \rho s_r U^2 \left[\frac{w_a}{U}\right]\left[\frac{(w_{u2} - w_{u1})}{U}\right] \cos\beta_m - (p_2 - p_1)s_r \sin\beta_m, \quad \text{(L.3.22)}$$

or using trigonometry Eq. (L.3.22) becomes:

$$F_{lr} = \rho s_r w_a^2 [\tan\beta_2 - \tan\beta_1] \cos\beta_m - (p_2 - p_1)s_r \sin\beta_m. \quad \text{(L.3.23)}$$

Finally, using Eq. (L.3.19) results in

$$F_{lr} = \rho s_r w_a^2 [\tan\beta_2 - \tan\beta_1] \cos\beta_m - \rho s_r c_a^2 [\tan\beta_1 - \tan\beta_2]\tan\beta_m \sin\beta_m + s_r\zeta_r \sin\beta_m. \quad \text{(L.3.24)}$$

Using the definition of the lift coefficient (Eq. (L.3.7)), one finds with some rearrangement that

$$C_{lr} = 2\frac{s_r}{C}\cos^2\beta_m [\tan\beta_2 - \tan\beta_1][\cos\beta_m + \tan\beta_m \sin\beta_m] + \frac{s_r}{C}\frac{\zeta_r}{\frac{1}{2}\rho w_m^2}\cos\beta_m \tan\beta_m. \quad \text{(L.3.25)}$$

Using some trigonometric identities and Eq. (L.3.21) yields

$$C_{lr} = 2\frac{s_r}{C}\cos\beta_m [\tan\beta_2 - \tan\beta_1] + C_{dr_p}\tan\beta_m. \quad \text{(L.3.26)}$$

Interestingly, the lift coefficient is directly related to the drag coefficient. Expressions similar to those used above apply to a stator blade row as follows:

$$F_{us} = \dot{m}(c_{u2} - c_{u1}) \quad \text{(L.3.27)}$$

$$F_{us} = (p_3 - p_2)s_s \quad \text{(L.3.28)}$$

$$p_3 - p_2 = \rho c_a^2 [\tan\alpha_2 - \tan\alpha_1] \tan\alpha_m - \zeta_s \quad \text{(L.3.29)}$$

$$F_{ds} = s_s\zeta_s \cos\alpha_m \quad \text{(L.3.30)}$$

$$C_{ds_p} = \frac{s_s\,\zeta_s \cos\alpha_m}{C\,\frac{1}{2}\rho c_m^2} \quad \text{(L.3.31)}$$

$$F_{ls} = \rho s_s c_a^2 [\tan\alpha_2 - \tan\alpha_1] \cos\alpha_m - \rho s_s c_a^2 [\tan\alpha_1 - \tan\alpha_2]\tan\alpha_m \sin\alpha_m - s_s\zeta_s \sin\alpha_m \quad \text{(L.3.32)}$$

$$C_{ls} = 2\frac{s_s}{C}\cos\alpha_m [\tan\alpha_2 - \tan\alpha_1] + C_{ds_p}\tan\alpha_m. \quad \text{(L.3.33)}$$

L.3.2 Finite-Length Blades

In Section L.3.1, the lift and drag coefficients were identified for very long blades. Drag was determined to arise from pressure losses alone. However, when blades are not infinitely long but are of length h, other mechanisms add to the drag. For example, for finite-length blades, tip leakages are encountered. That is, flows migrate from the high-pressure surface to the

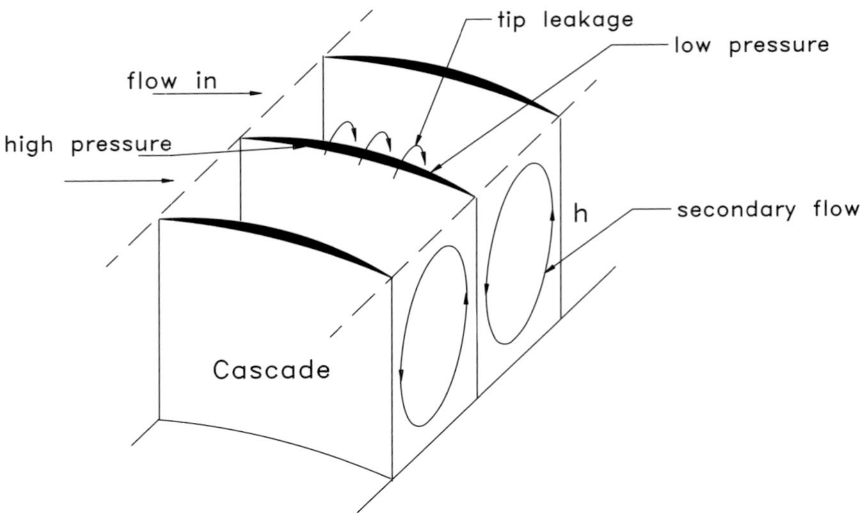

Figure L.3 Complex viscous flows in a cascade with finite-length blades.

low-pressure or suction surface. This leakage, when coupled with the flow turning, can produce transverse or secondary flows, as shown in Figure L.3. The losses result in shear losses and added drag; the additional coefficient will be identified as $C_{\mathrm{dr_s}}$. Correlations from Howell (1945a) indicate that, for a rotor blade row, the drag can be approximately

$$C_{\mathrm{dr_s}} = 0.018 C_{\mathrm{lr}}^2. \tag{L.3.34}$$

Shear losses on the hub and case due to the boundary layers also occur. That is, the added surface areas at the outer and inner radii increase the drag. The added loss is called annular drag and is represented by $C_{\mathrm{dr_a}}$. Previous correlations from Howell (1945a) indicate that the drag may be approximated as

$$C_{\mathrm{dr_a}} = 0.020 \left.\frac{s}{h}\right]_{\mathrm{r}}, \tag{L.3.35}$$

where s/h is the pitch-to-blade height ratio.

Lastly, to find the drag on the blade row, one can approximate the total as the linear sum of the three drags discussed above. That is,

$$C_{\mathrm{dr_t}} = C_{\mathrm{dr_p}} + C_{\mathrm{dr_s}} + C_{\mathrm{dr_a}}. \tag{L.3.36}$$

Although the preceding equations are written for a rotor blade row, similar expressions apply for stator blade rows because the same correlations apply. Thus, for the stator,

$$C_{\mathrm{ds_s}} = 0.018 C_{\mathrm{ls}}^2 \tag{L.3.37}$$

$$C_{\mathrm{ds_a}} = 0.020 \left.\frac{s}{h}\right]_{\mathrm{s}} \tag{L.3.38}$$

$$C_{\mathrm{ds_t}} = C_{\mathrm{ds_p}} + C_{\mathrm{ds_s}} + C_{\mathrm{ds_a}}. \tag{L.3.39}$$

L.4 Forces

One can find the actual forces on the blades in terms of the lift and drag coefficients, percent reaction, and flow coefficient. Thus, using Eqs. (L.2.10), (L.2.11), and (L.3.14) yields

$$F_{\mathrm{ur}} = \rho s_{\mathrm{r}} \phi \tau_{\mathrm{r}} U^2. \tag{L.4.1}$$

Similarly, for the stator,

$$F_{\mathrm{us}} = \rho s_{\mathrm{s}} \phi \tau_{\mathrm{s}} U^2. \tag{L.4.2}$$

From the force triangles, one can see that

$$F_{ar} = -F_{ur}\tan\,(\beta_m + \varepsilon_r). \tag{L.4.3}$$

However, by a trigonometric rule,

$$\tan(\beta_m + \varepsilon_r) = \frac{\tan\beta_m + \tan\varepsilon_r}{1 - \tan\beta_m \tan\varepsilon_r}. \tag{L.4.4}$$

If the drag force is low, implying that ε_r is small, which is the usual case,

$$\tan(\beta_m + \varepsilon_r) = \frac{\tan\beta_m + \varepsilon_r}{1 - \varepsilon_r\,\tan\beta_m}; \tag{L.4.5}$$

thus, from Eq. (L.4.3),

$$F_{ar} = -F_{ur}\frac{\tan\beta_m + \varepsilon_r}{1 - \varepsilon_r\,\tan\beta_m}; \tag{L.4.6}$$

and so from Eq. (L.2.14),

$$F_{ar} = F_{ur}\left[\frac{\dfrac{\%R}{\phi} - \varepsilon_r}{1 + \varepsilon_r\dfrac{\%R}{\phi}}\right]. \tag{L.4.7}$$

Similarly, for the stator

$$F_{as} = F_{us}\frac{\tan\alpha_m + \varepsilon_s}{1 - \varepsilon_s\,\tan\alpha_m}. \tag{L.4.8}$$

Thus, using Eq. (L.2.15) with Eq. (L.4.8) as was done for the rotor yields

$$F_{as} = F_{us}\left[\frac{\left(\dfrac{1-\%R}{\phi}\right) + \varepsilon_s}{1 - \varepsilon_s\left(\dfrac{1-\%R}{\phi}\right)}\right]. \tag{L.4.9}$$

L.5 Relationship of Blade Loading and Efficiency

The preceding forces need to be related to the efficiencies and reactions of the stage. For a rotor blade, from Eqs. (L.3.13) and (L.4.7) one obtains

$$\Delta p_r = p_2 - p_1 = \frac{F_{ur}}{s_r}\left[\frac{\dfrac{\%R}{\phi} - \varepsilon_r}{1 + \varepsilon_r\dfrac{\%R}{\phi}}\right], \tag{L.5.1}$$

or using Eq. (L.4.5) and rearranging yields

$$\Delta p_r = \rho\phi\tau_r U^2\left[\frac{\%R - \phi\,\varepsilon_r}{\phi + \varepsilon_r\%R}\right]. \tag{L.5.2}$$

Similarly, for a stator, from Eqs. (L.2.28), (L.4.2), and (L.4.9),

$$\Delta p_s = p_3 - p_2 = \rho\phi\tau_s U^2\left[\frac{1 - \%R + \phi\,\varepsilon_s}{\phi - \varepsilon_s(1 - \%R)}\right]; \tag{L.5.3}$$

thus, if $\alpha_3 = \alpha_1$ (so that $\tau_r = \tau_s = \tau$), the pressure rise across the complete stage is

$$\Delta p = p_3 - p_1 = \Delta p_r + \Delta p_s$$
$$= \rho\phi\tau U^2\left\{\left[\frac{\%R - \phi\,\varepsilon_r}{\phi + \varepsilon_r\%R}\right] + \left[\frac{1 - \%R + \phi\,\varepsilon_s}{\phi - \varepsilon_s(1 - \%R)}\right]\right\}. \tag{L.5.4}$$

If one considers the case of frictionless flow (no drag) so that ε_r and ε_s are both zero, the ideal pressure rises are

$$\Delta p_r' = \rho\tau U^2\%R \tag{L.5.5}$$

and

$$\Delta p_s' = \rho\tau U^2(1 - \%R); \tag{L.5.6}$$

thus, for the stage,

$$\Delta p' = \rho\tau U^2. \tag{L.5.7}$$

Therefore, for ideal flow, from Eqs. (L.5.5) and (L.5.7), one obtains

$$\%R = \frac{\Delta p_r'}{\Delta p'}, \tag{L.5.8}$$

which is consistent with Eq. (5.4.7).

From the Chapter 4 energy equation (Eq. (4.2.33)), one finds for the power that

$$\dot{W}_{sh} = \frac{\dot{m}}{\eta_{12}}\left[\frac{p_2 - p_1}{\rho} + \frac{c_2^2 - c_1^2}{2}\right] = \frac{\dot{m}}{\rho\eta_{12}}[p_{t2} - p_{t1}], \tag{L.5.9}$$

where

$$\dot{W}_{sh} = \frac{\dot{W}'_{sh}}{\eta_{12}}; \tag{L.5.10}$$

thus, further,

$$\dot{W}_{sh} = \frac{\dot{m}}{\rho\eta}[p_{t3} - p_{t1}], \tag{L.5.11}$$

where, as before, the efficiency is defined as the ratio of actual to ideal power required for a given pressure rise. Now the ideal pressure rise for the given power is defined as

$$\dot{W}'_{sh} = \frac{\dot{m}}{\rho}[p'_{t3} - p'_{t1}], \tag{L.5.12}$$

and thus the efficiency is also

$$\eta = \frac{\Delta p_t}{\Delta p'_t}. \tag{L.5.13}$$

Finally, recall that the case of $\alpha_3 = \alpha_1$ and constant axial velocity is being considered, which implies $c_3 = c_1$. This in turn implies that the rise in total and static pressures is identical across the stage, and thus from Eqs. (L.5.7) and (L.5.13)

$$\eta = \frac{\Delta p}{\rho\tau U^2}. \tag{L.5.14}$$

Therefore, using Eqs. (L.5.4) and (L.5.14) yields

$$\eta = \phi\left\{\left[\frac{\%R - \phi\varepsilon_r}{\phi + \varepsilon_r\%R}\right] + \left[\frac{1 - \%R + \phi\varepsilon_s}{\phi - \varepsilon_s(1 - \%R)}\right]\right\}. \tag{L.5.15}$$

The stage efficiency is consequently a function of four variables: percent reaction, flow coefficient, the angle between the drag and lift for the rotor, and the angle between the drag and lift for the stator. Thus, if it is possible to evaluate the lift and drag for a given condition, the efficiency can be estimated. One observation is that, as the drag on blades and vanes increases, the stage efficiency decreases. Furthermore, velocity polygons for different operating conditions can be used to evaluate the flow angles (and thus yield the vane and blade incidence angles), and if information or data for the lift and drag on a cascade (as a function of incoming incidence angle) are available, the efficiency can be estimated.

L.6 Effects of Parameters

One may wish to see the effects of the different parameters on the overall performance. It is desirable for the efficiency to be as high as possible. Thus, from Eq. (L.5.15), Vavra (1974) showed that one can differentiate η with respect to %R (for given values of ε_r and ε_s) and set to zero; that is,

$$\frac{\partial\eta}{\partial\%R} = 0, \tag{L.6.1}$$

which yields

$$\%R_{opt} = \frac{1 - \dfrac{\phi}{\varepsilon_s}\left[1 - \sqrt{\dfrac{1+\varepsilon_s^2}{1+\varepsilon_r^2}}\right]}{1 - \dfrac{\varepsilon_r}{\varepsilon_s}\sqrt{\dfrac{1+\varepsilon_s^2}{1+\varepsilon_r^2}}}. \tag{L.6.2}$$

If both ε_r and ε_s are small,

$$\%R_{opt} \cong \frac{1}{1 - \dfrac{\varepsilon_r}{\varepsilon_s}}, \tag{L.6.3}$$

and if ε_r and ε_s are the same magnitude ($\varepsilon_r = -\varepsilon_s = \varepsilon$),

$$\%R_{opt} = 0.50. \tag{L.6.4}$$

This is an extremely important finding. When the percent reaction is 50 percent not only are the forces equal on the rotor and stator blades, but the efficiency is at a maximum. Thus, most axial flow pumps are designed with percent reactions of approximately 50 percent. Note that incompressible flow was assumed for this analysis. For the same performance phenomena to be realized in compressors, compressor stages tend to have percent reactions somewhat higher than 0.5 (especially for highly loaded stages) due to flow compressibility and long blades. The efficiency at this value of %R is

$$\eta_{\%R=0.50} = 2\phi\left[\frac{1 - 2\varepsilon\phi}{\varepsilon + 2\phi}\right]. \tag{L.6.5}$$

The optimum value of flow coefficient (maximum efficiency) can be found by differentiating Eq. (L.6.5) with respect to ϕ and setting to zero:

$$\frac{\eta_{\%R=0.50}}{\partial\phi} = 0; \tag{L.6.6}$$

thus,

$$\phi\text{opt}_{\%R=0.50} = \frac{1}{2}\left[\sqrt{1+\varepsilon^2} - \varepsilon\right], \tag{L.6.7}$$

and if ε is small, using a Taylor series expansion yields

$$\phi\text{opt}_{\%R=0.50} \cong \frac{1}{2}[1-\varepsilon]. \tag{L.6.8}$$

From the optimized values of %R and ϕ one can find the optimum flow angles by using Eqs. (L.2.14) and (L.2.15). As a limiting case, note that, as the drag approaches zero, the optimum value of ϕ approaches 0.50. Thus, the flow angles β_m and α_m approach 45°. Finally, the maximum value of efficiency under the general conditions is

$$\eta\text{max}_{\%R=0.50} = 1 + 2\varepsilon^2 - 2\varepsilon\sqrt{1+\varepsilon^2}, \tag{L.6.9}$$

which, for small ε, becomes

$$\eta\text{max}_{\%R=0.50} \cong 1 - 2\varepsilon[1-\varepsilon]. \tag{L.6.10}$$

Obviously, as the drag approaches zero, the efficiency approaches unity. Results of this analysis are presented and discussed in Chapter 5. For example, Figure 5.32 indicates the order of magnitudes and trends one can expect. The optimum flow coefficient decreases with ε and increases with %R. The efficiency decreases with increasing ε (or drag), as expected. However, for the optimum flow coefficient, the efficiency has minimal dependence on %R, which is an important finding.

It is also possible to acquire a better overall understanding of the influences of the different parameters through the following analysis. By using Eq. (L.5.15), which is the most general form of the equation, one can combine terms and ignore second-order terms (i.e., products of ε_r and ε_s), which are very small. This results in

$$\eta = \frac{\phi + (\varepsilon_r - \varepsilon_s)(\%R - \%R^2 - \phi^2)}{\phi + \varepsilon_r \%R - \varepsilon_s(1 - \%R)}. \tag{L.6.11}$$

Next, if one assumes that $\varepsilon_r = -\varepsilon_s = \varepsilon$,

$$\eta = \frac{\phi + 2\varepsilon(\%R - \%R^2 - \phi^2)}{\phi + \varepsilon}. \tag{L.6.12}$$

A general 3-D depiction of the dependence of the efficiency on the flow coefficient, reaction, and drag to lift angle is presented and discussed in Chapter 5 (Figure 5.33). A series of conoids result – one for each value of η. Vavra (1974) has shown that the loci of constant efficiency are indeed circles. As has been shown in Section 6.7 as the flow rate deviates from the design value, the flow angle does not match the blade angle (incidence). Thus, as the flow rate deviates from the design value, the drag increases, thus possibly increasing ε (lift may increase also). As a result, as one deviates from the design point, the efficiency changes because of two different effects. That is, as the flow rate is changed from the design value, movement is in both the ϕ and ε directions. Furthermore, with each parameter treated separately, if %R and ϕ are held constant, as ε decreases, the efficiency always increases. This analysis is further discussed in Section 5.11.3 and with more figures.

L.7 Empiricism Using Cascade Data

Next, 2-D cascade data will be discussed as an empirical method that integrates the previous analyses to predict the efficiency performance of a compressor stage. For example, for a given cascade of stator vanes, as in Figure 5.7c, if the flow enters at angle α_2, the flow will exit at angle α_3, The exit angle will be determined by several variables, including the vane shapes, pitch-to-chord ratio (s/C), incidence angle ($\alpha_2 - \alpha_r'$), and

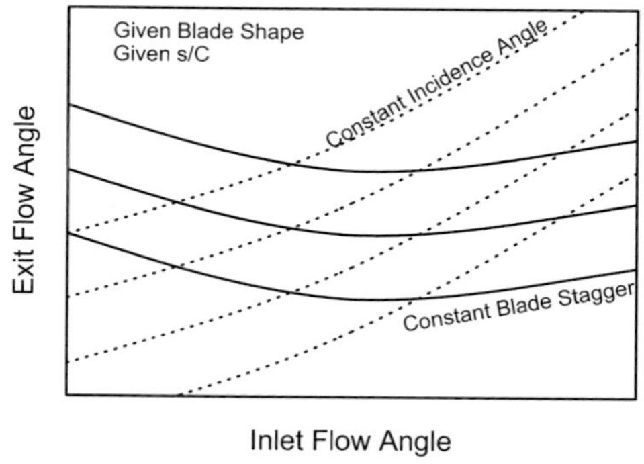

Figure L.4 Dependence of the exit flow angle on inlet flow angle, incidence angle, and stagger angle (given blade shapes and pitch-to-chord ratio).

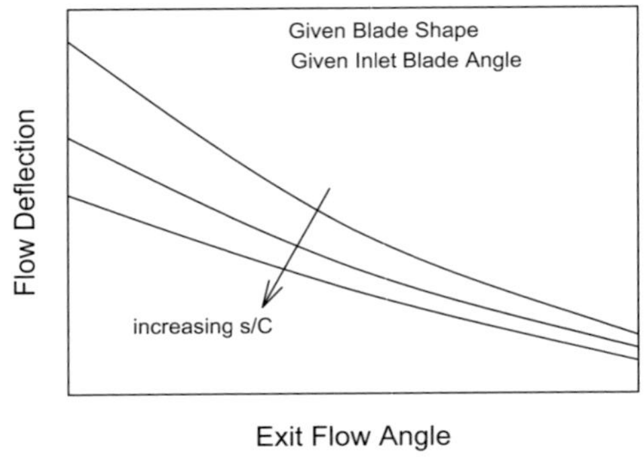

Figure L.5 Dependence of the generalized flow deflection angle on exit flow angle and pitch-to-chord ratio (given inlet blade or vane angle and shape).

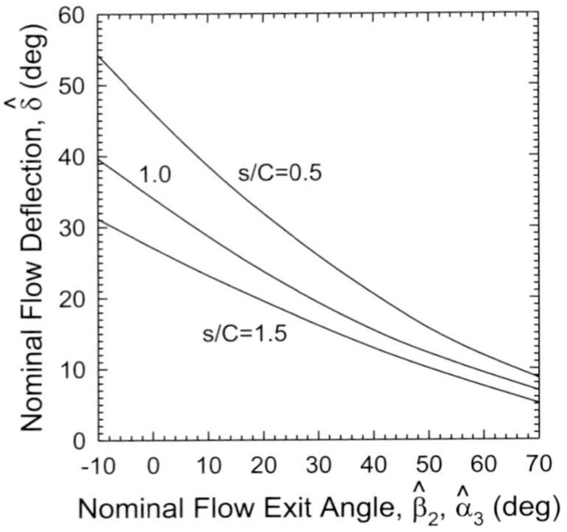

Figure L.6 Dependence of the empirical nominal flow deflection on nominal exit flow angle and pitch-to-chord ratio (general case) (from Fig. 68, Howell (1945a) with permission of the Council of the Institution of Mechanical Engineers).

stagger angle (Figure 5.6). For a given set of vanes (fixed shape and fixed s/C), one can experimentally generate a figure showing the functional relationship of α_2 – namely as generalized in Figure L.4. A similar figure could be generated for a cascade of rotor blades. Note that the difference in the inlet and exit flow angles is the flow deflection – that is, how much the flow is turned. Horlock (1958) presents a large amount of such specific and detailed data.

These data and similar data for other values of solidity can then be correlated with generate defection data as generalized in Figure L.5. The deflection, δ, has previously been defined as the flow turning angle ($[\alpha_3 - \alpha_2]$ for a stator and $[\beta_2 - \beta_1]$ for a rotor). In this figure, the flow deflection is shown as a general function of exit angle for a given family of blades and given stagger. Such a figure can be used, for example, to determine the necessary solidity for a given deflection and exit flow angle. In Section 5.11.4 a detailed example using such data is presented in Example 5.7.

L.8 Further Empiricism

Empiricism has shown that much of the data for different blade or vane shapes and conditions can be correlated with a single figure. Thus, if particular data are not available, general correlated data can be used to see trends and make predictions. For example, if the exit area to inlet area ratio of the cascade is too large (Δp is too large), separation will occur. Because this area ratio is directly related to the deflection angle, one can define the stall deflection angle for a given cascade as the value of deflection at which separation occurs. By definition, the nominal deflection angle is

$$\hat{\delta} \equiv 0.80\delta_{stall}. \tag{L.8.1}$$

Using this definition and extended data, which include the stall condition, results in detailed data from a specific application of Figure L.5 and similar figures empirically (and conveniently) reduce to Figure L.6. In this figure the nominal deflection angle is a function of the nominal exit angle for both rotor blades and stator vanes. On the basis of the sign definitions adopted in this text, for the rotor blades the nominal exit flow angle is $-\hat{\beta}_2$ and for the stator it is $\hat{\alpha}_3$. Also, for the rotor blades the nominal deflection angle is $\hat{\delta}_{12}$, and for the stator it is $-\hat{\delta}_{23}$. Data in this figure are conveniently not functions of the inlet angle or blade shapes. All of the following calculations use this nominal deflection angle and other nominal conditions (flow angles, incidence angles, etc.), as a reference at this point.

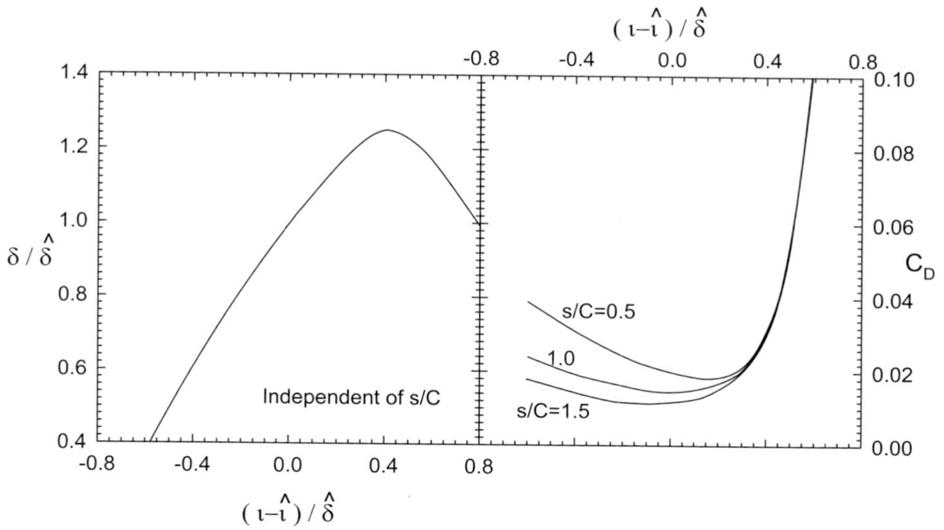

Figure L.7 Dependence of the normalized flow deflection and generalized drag coefficient on normalized inlet flow angle and pitch-to-chord ratio (from Howell (1942)).

Next, the nominal deviation of the flow from the blade angle at the exit is given by the following empirical relationship from Howell (1945a):

$$\hat{\xi} = m\delta'\sqrt{\frac{s}{C}}. \tag{L.8.2}$$

Thus, for a stator, the value of $\hat{\xi}$ is equal to $\alpha_3' - \hat{\alpha}_3$, and for a rotor cascade it is $\beta_2' - \hat{\beta}_2$. The value of δ' is the deflection angle of the blades; that is, for a stator cascade the value of δ' is equal to $\alpha_3' - \alpha_2'$, and for a rotor cascade it is $\beta_2' - \beta_1'$. The value of m has been correlated by Howell (1945a) to be

$$m = 0.23\left[\frac{2a}{C}\right] + 0.1\left[\frac{\hat{\chi}_{\alpha\beta}}{50}\right], \tag{L.8.3}$$

where a is the distance from the leading edge of a blade to the point of maximum camber, as shown in Figure 5.6a. Also, the quantity $\hat{\chi}_{\alpha\beta}$ is equal to the absolute magnitude of $\hat{\alpha}_3$ for a stator cascade and the absolute magnitude of $\hat{\beta}_2$ for a rotor cascade. For a quick calculation, or if the value of a is unknown, a value of $m \cong 0.26$ can be used.

Lastly, Figure L.7 presents a correlation for the nondimensionalized deflection $\delta/\hat{\delta}$ as a function of the nondimensionalized difference in actual and nominal incidence angle, $(\iota - \hat{\iota})/\hat{\delta}$. As indicated in Figure L.7, this is independent of s/C and is valid for a wide range of geometries. Note that, for the case of zero incidence and zero nominal incidence, the deflection is equal to the nominal value, $\delta/\hat{\delta} = 1$. Also shown in Figure L.7 is a generalized profile drag coefficient, C_{dp}, as a function of $(\iota - \hat{\iota})/\hat{\delta}$.

The calculations are being facilitated by fitting the curves in Figures L.6 and L.7 with the polynomials below. Values for the parameters are listed in Table L.1.

$$\hat{\delta} = a_1 + a_2\hat{\alpha} + a_3\hat{\alpha}^2 \tag{L.8.4}$$

$$\delta/\hat{\delta} = b_1 + b_2\left[\frac{\iota-\hat{\iota}}{\hat{\delta}}\right] + b_3\left[\frac{\iota-\hat{\iota}}{\hat{\delta}}\right]^2 + b_4\left[\frac{\iota-\hat{\iota}}{\hat{\delta}}\right]^3 + b_5\left[\frac{\iota-\hat{\iota}}{\hat{\delta}}\right]^4 \tag{L.8.5}$$

$$C_d = c_1 + c_2\left[\frac{\iota-\hat{\iota}}{\hat{\delta}}\right] + c_3\left[\frac{\iota-\hat{\iota}}{\hat{\delta}}\right]^2 + c_4\left[\frac{\iota-\hat{\iota}}{\hat{\delta}}\right]^3 \text{ for} \left[\frac{\iota-\hat{\iota}}{\hat{\delta}}\right] \leq 0.4 \tag{L.8.6}$$

$$C_d = d_1 + d_2\left[\frac{\iota-\hat{\iota}}{\hat{\delta}}\right] + d_3\left[\frac{\iota-\hat{\iota}}{\hat{\delta}}\right]^2 \text{ for} \left[\frac{\iota-\hat{\iota}}{\hat{\delta}}\right] \geq 0.4. \tag{L.8.7}$$

The generalized correlations presented in Figure L.7 give reasonable to very good results. It goes without

Table L.1 **Polynomial parameters.**

s/C	a_1	a_2	a_3	b_1	b_2	b_3	b_4	b_5
0.5	46.4	−0.819	0.0041	1.009	0.971	−0.727	−0.860	0.281
1.0	34.1	−0.553	0.0025	1.009	0.971	−0.727	−0.860	0.281
1.5	26.9	−0.400	0.00125	1.009	0.971	−0.727	−0.860	0.281

s/C	c_1	c_2	c_3	c_4	d_1	d_2	d_3
0.5	0.0185	−0.0118	0.0606	0.0389	0.118	0.063	1.00
1.0	0.0143	0.00076	0.0538	0.0429	0.118	0.063	1.00
1.5	0.0117	0.00736	0.0535	0.0440	0.118	0.063	1.00

saying, however, that if deflection data are available for specific blades and vanes – for example, specific data as in Figure 5.35 and as used in Example 5.7, experimental drag and lift data for a specific blade row, or all of these data – such data should be used in lieu of general correlations.

L.9 Implementation of the General Method

The previous seven appendix sections are integrated in this section, and the implementation of the general method to predict the performance of a single stage using these results is discussed. For the purposes of discussion, the blade angles are assumed to be known as a part of the design process or because the stage has already been built. Blade shapes, heights, chord lengths, pitch, distance to maximum camber, and flow coefficient are also assumed to be known.

Howell (1942, 1945a) developed and demonstrated the overall technique. The method starts with determining the nominal deflection angle $\hat{\delta}$ and nominal deviation $\hat{\xi}$ by using the blade geometry. Next, the nominal flow angles $\hat{\beta}_1$, $\hat{\beta}_2$, $\hat{\alpha}_2$, $\hat{\alpha}_3$, and so on are determined using the nominal deflections and deviations. The nominal incidences $\hat{\imath}_1$ and $\hat{\imath}_2$ can be found from

$$\hat{\beta}_1 = \beta_1' - \hat{\imath}_1 \tag{L.9.1}$$

and

$$\hat{\alpha}_2 = \alpha_2' - \hat{\imath}_2. \tag{L.9.2}$$

The flow coefficient is also known, however, and thus the flow angles can be determined and the blade to flow incidence angles can be found from

$$\beta_1 = \beta_1' - \imath_1 \tag{L.9.3}$$

and

$$\alpha_2 = \alpha_2' - \imath_2. \tag{L.9.4}$$

One can also find $\imath_1$ and $\imath_2$ from

$$\beta_1 = \hat{\beta}_1 - (\imath_1 - \hat{\imath}_1) \tag{L.9.5}$$

$$\alpha_2 = \hat{\alpha}_2 - (\imath_2 - \hat{\imath}_2). \tag{L.9.6}$$

Although the nominal deflection is known, the actual deflection is not. Using Figure L.7, one can determine the actual deflection and use this to identify the actual flow angles β_2 and α_3. Thus, all of the flow angles are known, and the flow coefficient and percent reaction can be found. Also, the lift coefficients can be found.

Next, the sum of the different drag coefficients can be determined. The ratio of drag to lift can be found, and thus the angle between the drag and lift is known. Finally, the efficiency of the stage can be found from these angles as well as the percent reaction and flow coefficient. Solutions can also be obtained using the software SSAxCompPerf.exe, which incorporates all of the concepts in this section.

L.10 Summary

Because a cascade is made up of a series of airfoils, a method of using lift and drag coefficients of blade cascades and other empirical data was demonstrated for incompressible flow to predict or estimate and optimize stage efficiency, including such effects as blade incidence angles. Velocity polygons were first related to lift and drag forces. Next lift and drag coefficients were related to the forces for both infinitely long blades or vanes and finite-length blades or vanes. Such forces were then related to the stage efficiency and mathematically the different effects were related to efficiency. Empiricism was then used to determine allow determination of lift and drag for cases which specific data was not available. Finally, a general method of the solution was presented. A percent reaction of approximately 50 percent was shown to optimize the efficiency of a stage, although the efficiency does not rapidly deteriorate as the reaction deviates from this value. For incompressible flow, the method gives great educational insight into how the stage performance and fundamental airfoil (vane or blade) characteristics are related and can be used to maximize the efficiency, for example, with the proper choice of blade shapes and solidity or as the leading-edge incidence increases. Specific examples are presented in Chapter 5.

List of Symbols

a	Leading edge to maximum camber length
a_1, b_2, c_1, d_1, etc.	Curve fitting constants, Eqs. (L.8.4)–(L.8.7)
c	Absolute velocity
C	Chord length
C_d	Drag coefficient
C_l	Lift coefficient
F	Force
h	Blade height
$\dot{m}$	Mass flow rate
m	Empirical constant, Eq. (L.8.2)
p	Pressure
%R	Percent reaction
s	Blade spacing or pitch
U	Blade velocity
v	Representative local velocity
$\dot{W}_{sh}$	Power
w	Relative velocity
α	Absolute (stator) angle
β	Relative (rotor) angle
$\hat{\chi}_{\alpha\beta}$	Parameter in Eq. (L.8.3)
δ	Flow turning angle, deflection
ε	Angle between lift and drag forces
ζ	Total pressure loss
η	Efficiency
ι	Incidence angle
ξ	Deviation
ρ	Density
τ	Deflection coefficient
ϕ	Flow coefficient

Subscripts

a	Axial
d	Drag
l	Lift
m	Mean or average
r	Rotor
s	Stator
stall	Separation
t	Total (stagnation)
u	Tangential
1, 2, 3	Positions in stage

Superscripts

′ Ideal

∧ Nominal

APPENDIX M INTRODUCTION TO SUITE OF GAS TURBINE .EXE PROGRAMS

M.1 Introduction

This appendix provides an introduction and basic principles for a set of 22 programs and subprograms written for educational purposes and to complement this textbook. The .exe programs are student friendly and are summarized here. Also indicated are the chapters for which applications of the program can be used. As a part of the suite, each program has a separate and full document to give guidance and input instructions as well as output summary. Both the programs and users guides are available from Cambridge University Press.

M.2 Main.exe (All Chapters)

This is a very simple "call" program that accesses all of the other programs below.

M.3 Atmosphere.exe (All Chapters)

This program finds the pressure and temperature at different altitudes. One can choose US or SI units. One can choose a range of altitudes or a single altitude. For multiple altitudes 11 values will be shown. One can generate a print file with 101 values of altitude.

M.4 IGas.exe (All Chapters)

This is a very simple program to find ideal gas density, speed of sound, and specific heat at constant pressure from the temperature, pressure, and specific heat ratio. Mach numbers can be used to find both total and static properties. Either US or SI units can be used.

M.5 Simple1D.exe (Chapters 2, 8, 9, 10, 11, Appendices C, D, E, F, G)

This is a very simple "call" program that accesses a set of subprograms. Subprograms include the four well-documented and often used simple 1-D processes.

M.5.1 Isentropic flow with area change (Isentropic.exe)
M.5.2 Fanno line flow (Fanno.exe)
M.5.3 Rayleigh line flow (Rayleigh.exe)
M.5.4 Normal shocks (NormalShock.exe)

Subprograms also include two constant static temperature processes.

M.5.5 Friction with constant area (ItF&F&CA.exe)
M.5.6 Heat transfer with area change (ItF&Fl&HT&VA.exe)

For all of these cases the specific heat ratio, γ, is inputted and properties are found for different Mach numbers. One can choose a range of Mach numbers or a single Mach number. For multiple Mach numbers 11 values will be displayed. One can also generate a detailed print file with 101 values.

M.6 General1D.exe (Chapter 2, 8, 9, 10, 11, Appendix C)

This program analyses general 1-D compressible flows. One can choose US or SI units. It includes mixing of two streams, flow with drag objects (mix enhancers and flameholders), and combined heat addition (combustion), wall friction, and area change. Simply put, these are numerical integration computations for combined Fanno line + Rayleigh line + drag object + mixing flow + area change. Input variables include Mach numbers, relative mass flow rates of the two streams, temperatures, pressures, and various process specifications. Output variables include exiting Mach number, pressures, and temperatures.

M.7 Shock.exe (Chapters 8, 10, Appendix C)

This program analyses shocks. The flow is adiabatic (total temperature is constant). There are four options for running the program. One can specify the following.

M.7.1 Normal shocks
M.7.2 Weak oblique planar shocks
M.7.3 Strong oblique planar shocks
M.7.4 Oblique conical shocks

Input variables for all cases include the specific heat ratio, γ, and the incoming Mach number. Output variables for all cases include exit Mach number, static pressure ratio, p_2/p_1, and total pressure ratio, p_{t2}/p_{t1}.

M.8 Nozzle.exe (Chapter 10, Appendix C)

This program analyses flow in a nozzle. The flow is adiabatic (total temperature is constant) and shockless but has losses due to friction (nonisentropic). One can choose US or SI units. There are four options for running the program. One can specify the following.

M.8.1 Exit Mach number
M.8.2 Exit to inlet area ratio with a lower Mach number solution
M.8.3 Exit to inlet area ratio with a higher Mach number solution
M.8.4 The exit static pressure to inlet total pressure ratio

Other input variables include specific heat ratio, efficiency, inlet total temperature, inlet total pressure, and exit area. Output variables include mass flow rate, corrected mass flow rate, exit Mach number, the ratio exit area to area at the sonic condition, the ratio exit static pressure to inlet total pressure, and the ratio exit static temperature to inlet total temperature. Also, the ratio of the minimum area to area at the sonic condition is found at the Mach number at the minimum area.

M.9 SSEffy.exe (Chapters 2, 5, 7)

This program finds the efficiency of a single stage or set of stages for an axial compressor or turbine if the overall, total (adiabatic efficiency) is known. It can also be used to find the polytropic efficiency of a turbomachine. The specific heat ratio, γ, is inputted as well as the number of stages. One chooses three values of the overall efficiency of the complete compressor or turbine. One can also choose a range of pressure ratios or a single pressure ratio. For multiple pressure ratios 11 values will be shown. One can also generate a detailed print file with 101 values.

M.10 JetEngineCycle.exe (Chapter 3, Appendix H)

This program performs a cycle analysis on different jet engines. It includes ideal and nonideal options. One can choose US or SI units. There are six engine types that can be analyzed. One can specify the following.

M.10.1 Turbofan – fully exhausted
M.10.2 Turbofan – fully mixed
M.10.3 Turbofan – mixed plus exhausted
M.10.4 Turbojet
M.10.5 Turboprop
M.10.6 Ramjet/scramjet

Options also include standard/nonstandard atmosphere, constant/variable specific heat ratio, an afterburner, primary nozzle (converging/converging–diverging), and fan nozzle (converging/converging–diverging). Input variables for all engine types also include, flight Mach number, core mass flow rate, burner exit temperature and heating value of the fuel. Other various input variables include, component efficiencies, total pressure losses, and afterburner exit temperature (as needed). For the variable specific heat ratio cases the average component temperatures are used to evaluate γ and c_p. For these cases the approximated component Mach numbers are used to find the average static temperatures. Output variables for all cases include the thrust, nondimensional thrust, *TSFC*, and nondimensional *TSFC*.

M.11 PowerGTCycle.exe (Chapter 3, Appendix H)

This program performs a cycle analysis on power generation gas turbines. It includes ideal and nonideal options. There are a large variety of gas turbine types that can be analyzed. One can choose US or SI units. One can specify any or all of the following options.

M.11.1 Simple gas turbine
M.11.2 With regeneration
M.11.3 With intercooling
M.11.4 With reheat
M.11.5 With inlet cooling

Variables include ideal/nonideal option, standard/nonstandard atmosphere, and constant/variable specific heat ratio. Other displayed input variables depend on the options and are chosen. Input variables include: altitude or external temperature/pressure, core mass flow rate, burner exit temperature, and heating value of the fuel. Various other input variables include: component efficiencies, total pressure losses, regenerator data, intercooling data, reheat data, and inlet cooling. For the variable specific heat ratio cases the average component temperatures are used to evaluate γ and c_p. For these cases the approximated component Mach numbers are used to find the average static temperatures. Up to two compressors (low-pressure and high-pressure) and three turbines (high-pressure, low-pressure, and free) can be included, but can be run in various configurations or compressors (one or two) and turbines (one, two, or three). In the most general case, the low- and high-pressure turbines drive the low- and high-pressure compressors, respectively, and the free or power turbine is the total output. The pressure ratios for the two compressors are specified as well as the efficiencies. The individual efficiencies for the turbines are also input. Output variables for all cases include the net power, thermal efficiency, specific fuel consumption, and heat rate, as well as individual component total temperatures.

M.12 Turbomachinery.exe (Chapters 4, 5, 6, 7)

This program performs a mean line analysis on different types of turbomachines. One can choose US or SI units. There are four turbomachine types that can be analyzed. One can specify the following.

M.12.1 Axial flow compressor
M.12.2 Radial flow compressor
M.12.3 Axial flow turbine
M.12.4 Radial inflow turbine

Variables include the gas type. For air the specific heat ratio and gas constant are automatically determined. For other gases the specific heat ratio and gas constant are input and the specific heat ratio is fixed. Other input variables depend on the turbomachine type and are displayed. Common input variables for all turbomachine types include rotational speed, or inlet temperature, inlet pressure, mass flow rate, and efficiency. Other various input variables include blade sizes and flow angles. For the variable specific heat ratio cases (air) the specific heat ratio is evaluated at the average static temperature. Calculations are based on the control volume equations for the different blade rows and the velocity polygons that are developed in the text.

For all cases the total pressure ratio, power, various Mach numbers, specific heat ratio, various internal flow angles, and pressure coefficients are output. For the axial flow cases the blade heights and percent reaction are found.

M.13 SLA.exe (Chapters 5, 7, Appendix K)

The method presented in this program is called a streamline analysis method. Different streamline paths are analyzed by a control volume approach through an axial flow compressor or turbine passage. For this program the flow is incompressible. Radial equilibrium is used to determine radial variations in the various properties. The method is for flow in axial flow fans, compressors and turbines and is primarily intended for educational purposes. The method can be used to predict 3-D trends resulting from blade curvatures and twists and centrifugal forces, thus providing valuable insight into the flow three-dimensionality. In this respect it differs from the mean line analysis used in the program Turbomachinery.exe. It provides a tool by which a one can easily develop a "feel" of the magnitudes of the resulting three-dimensionality of the flow fields as blade configurations are varied. The iteration technique is relatively quick to converge and usually stable.

There are several input variables: hub diameter, tip diameter, rotational speed, total mass flow rate for the inlet, inlet pressure, fluid density, as well as the radial distribution of absolute flow angles and stage efficiencies. Output variables include relative flow angle, velocity, percent reaction, and pressure variations from hub to tip. Output tabular parameters are sent to a print file.

M.14 SSAxCompPerf.exe (Chapter 5, Appendix L)

This program calculates (estimates) the efficiency of a single stage of an axial flow compressor based on empirical lift and drag data for blade cascades. Input variables include the rotor and stator blade angles, incidence flow angle to the rotor blade, percent reaction of the stage, and the blade size and shape specifications. Generalized empirical cascade data is used to find the actual flow angles for the rotor blades. More empirical data are used to find the drag coefficient for the blade cascade.

Using trigonometry and the drag coefficients, the lift coefficients are found. Outputs include the stage efficiency, η, flow coefficient, ϕ, actual flow angles at the entrance and exit of the rotor blades and stator vanes, the stator vane incidence angle, and lift and drag coefficients for both blade and vane rows.

M.15 Kerosene.exe (Chapter 9)

This program finds the adiabatic flame temperature for the combustion of n-decane (kerosene) with air. The flow is adiabatic (no heat transfer). The inlet fuel temperature, inlet air temperatures, and percent air are input. The specific heat can be approximated over a given (but relatively small) temperature range a linear equation for which variables are input. One can choose US or SI units. The only output variable is the exit temperature.

M.16 JetEngineMatch.exe (Chapter 12)

This program performs a matching analysis on a single shaft turbojet engine. There are three options can be analyzed. One can specify the following.

M.16.1 Gas generator (compressor/burner/turbine)
M.16.2 Turbojet engine (inlet/compressor/burner/turbine/nozzle)

For this case three options for the nozzle geometry are available.

M.16.2.1 Converging nozzle
M.16.2.2 Variable nozzle with a fixed throat area
M.16.2.3 Variable nozzle with a fixed exit area
M.16.3 Turbojet engine(s) matched to an airframe

The solution methods are outlined in great detail in the text. The gas generator solution is a single-loop iteration, while the jet engine is a double-loop iteration. Component maps are used to correlate component operating characteristics. Internal thermodynamics similar to the cycle analysis are used. For the different maps curve fitting and modeling has been used for the computer analysis. Details of these parameters are outlined in the text. The solution results in the operating point on all component maps.

For the gas generator the input variables are: the total temperature, T_{t2}, and the corrected rotational speed, N_{c2}. The heating value of the fuel is input for all cases. For the jet engine the fuel ratio, f, the atmospheric temperature and pressure, and the flight Mach number, M, are input. For the matching with an aircraft case the airframe drag map and the number of engines are input. For all cases the average component total temperatures are used to evaluate γ and c_p. The output variables for the gas generator are the corrected mass flow rates, corrected speeds, efficiencies, and pressure ratios for the three components.

For the turbojet engine the thrust, nondimensional thrust, *TSFC*, mass flow rate, and shaft rotational speed are outputs. Also the Mach number, corrected mass flow rate and efficiency for the nozzle are outputs.

For the matching with an aircraft case all of the turbojet outputs are the same plus the aircraft Mach number.

This program is currently only available for US units.

M.17 PowerGTMatch.exe (Chapter 12)

This program performs a matching analysis on a power generation gas turbine. There are four options that can be analyzed. One can first specify the following.

M.17.1 Rotational speed dependent power load
M.17.2 Constant specified rotational speed

Furthermore, one can specify the following.

M.17.3 Without regeneration
M.17.4 With regeneration

The solution method is outlined in great detail in the text for the speed-dependent load. This solution is a triple-loop iteration. The solution method is also outlined in the text for the constant speed dependent case. This solution is a double-loop iteration. Component maps and internal thermodynamics similar to the cycle analysis are used for both cases. The solution results in the operating point on all component maps. For the different maps curve fitting and modeling have been used for the computer analysis. Details of these parameters are outlined in the text. For the rotational speed-dependent power load case with or without regeneration the inputs are the heating value of the fuel, the fuel ratio, f, the atmospheric temperature and pressure, and load map. For the constant rotational speed case with or without regeneration the inputs are the heating value of the fuel, the fuel ratio, f, the atmospheric temperature and pressure, and the rotational speed. The average component total temperatures are used to evaluate γ and c_p.

The output variables for the power generation unit are the corrected mass flow rates, corrected speeds, efficiencies, and pressure ratios for the components. The net power, overall efficiency, specific fuel consumption, mass flow rate, and shaft rotational speed are outputs. This program is currently only available for US units.

REFERENCES

Abbott, I. H., and von Doenhoff, A. E. (1959). *Theory of Wing Sections*. Dover Publications, New York, NY.

Adamczyk, J. J. (2000). Aerodynamic analysis of multistage turbomachinery flows in support of aerodynamic design. *ASME Transactions, Journal of Turbomachinery*, **122**(2), 189–217.

Adler, D. (1980). Status of centrifugal impeller internal dynamics, parts I and II. *Transactions ASME, Journal of Engineering for Power*, **102**(3), 728–746.

Ainley, D. G., and Mathieson, G. C. R. (1951). A method of performance estimation for axial flow turbines. *Aeronautical Research Council*, R&M No. 2974.

Anderson, J. D. (1982). *Modern Compressible Flow with Historical Perspective*. McGraw-Hill, New York, NY.

Andriani, R, and Ingenito, A. (2019). Main performances of the unsteady constant volume combustion thermodynamic cycle. The 8th EUCASS, Madrid, Spain. *AUCASS2019-776*.

Ashley, S. (1997). Turbines on a dime. *Mechanical Engineering*, **119**(10), 78–81.

ASME International History and Heritage (1997). *Landmarks in Mechanical Engineering*. Purdue University Press, West Lafayette, IN.

Aungier, R. H. (2000). *Centrifugal Compressors – A Strategy for Aerodynamic Design and Analysis*. ASME Press, New York, NY.

Bacha, J., Barnes, F., Franklin, M., et al. (2000). *Aviation Fuels Technical Review*. Chevron Products Company, FTR-3.

Bakken, L. E. (2018) Jens William Ægidius Elling. Presentation, display and printouts at ASME Turbo Expo 2018, June 11–15, 2018, Oslo, Norway.

Bakken, L. E., Syverud, E, Jordal, K., and Veer, T. (2004). Centenary of the first gas turbine to give net power output: A tribute to Ægidius Elling. ASME Turbo Expo 2004, June 14–17, 2004, Vienna, Austria. *ASME Paper GT2004-53211*.

Balje, O. E. (1981). *Turbomachines*. Wiley, New York, NY.

Billington, D. P. (1996). *The Innovators*. Wiley, New York, NY.

Boyce, M. P. (1982). *Gas Turbine Engineering Handbook*. Gulf Publishing Co., Houston, TX.

Brennen, C. E. (1994). *Hydrodynamics of Pumps*. Concepts ETI & Oxford, Norwich, VT.

Busemann, A. (1928). Das Förderhöhenverhältnis radialer Kreiselpumpen mit logrithmischspiraligen Schaufeln. *Zeitschrift für Angewandte Mathematik und Mechanik*, **8**, 372–384.

Caruthers, J. E., and Kurosaka, M. (1982). Flow induced vibration of diffuser excited radial compressors. NASA NAG No. 3-86.

Champ, C. A. N. M. D. H., Massarado, A. F., Ferrari, L., and Silvestri, P. (2019). Surge prevention in gas turbines: an overview over historical solutions and perspectives about the future. *E3S Web of Conferences*, **113**, 02003, SUPEHR19, Vol. **1**, 7 pages.

Chase, M. W. (1998). NIST-JANAF thermochemical tables, 4th edition, *Journal of Physical and Chemical Reference Data Monographs & Supplements, Monograph No. 9*. AIP Press, New York, NY.

Cohen, H., Rogers, G. F. C., and Saravanamuttoo, H. I. H. (1996). *Gas Turbine Theory*, 4th ed. Longman Group, Essex, UK.

Cumpsty, N. A. (1977). Critical review of turbomachinery noise. *Transactions of ASME, Journal of Fluids Engineering*, **99**(2), 278–293.

Cumpsty, N. A. (1988). *Compressor Aerodynamics*. Longman Group, Essex, UK.

Cumpsty, N. A. (2010). Preparing for the future: Reducing gas turbine environmental impact—IGTI Scholar Lecture. *ASME Transactions, Journal of Turbomachinery*, **132** (4), 041017.

Cumpsty, N. A., and Greitzer, E. M. (2004). Ideas and methods of turbomachinery aerodynamics: A historical view. *AIAA Journal of Propulsion and Power*, **20**(1), 15–26.

Cumpsty, N., and Heyes, A. (2015). *Jet Propulsion*. Cambridge, Cambridge, UK.

Denton, J. D. (2010). Some limitations of turbomachinery CFD. ASME Turbo Expo 2010, Glasgow, UK, June 14–18, 2010. *ASME Paper GT2010-22540*.

Denton, J., and Dawes, W. (1998). Computational fluid dynamics for turbomachinery design. *Proceedings of the Institution of Mechanical Engineers: Part C Journal of Mechanical Engineering Sciences*, **213**(2), 107–124.

Department of the US Navy (1953). *Handbook of Supersonic Aerodynamics*, Vol. 5. Bureau of Ordnance, Washington, DC.

Dixon, S. L. (1975). *Worked Examples in Turbomachinery*. Pergamon Press, Oxford, UK.

Dixon, S. L. (1998). *Fluid Mechanics and Thermodynamics of Turbomachinery*, 4th ed. Butterworths, Boston, MA.

Dring, R. P., Joslyn, H. D., Hardin, L. W., and Wagner, J. H. (1982). Turbine rotor–stator interaction. 27th ASME International Gas Turbine Exposition. *ASME Paper No. 82-GT-3.*

Dunham, J. (2000). A. R. Howell – Father of the British axial compressor. ASME TURBOEXPO 2000, May 8 -11, Munich, Germany. *ASME Paper No. 2000-GT-8.*

Dunn,M. G. (2001). Convective heat transfer and aerodynamics in axial flow turbines. ASME TURBO EXPO 2001, New Orleans, LA. *ASME Paper No. 2001-GT-0506.*

Eckardt, D. (1975). Instantaneous measurements in the jet-wake discharge flow of a centrifugal compressor impeller. *ASME Transactions, Journal of Engineering for Power*, **91**(3), 337–345.

El-Masri, M. A. (1988). GASCAN – An interactive code for thermal analysis of gas turbine systems. *ASME Transactions, Journal of Engineering for Gas Turbines and Power*, **110**(2), 201–209.

Elmendorf, W., Mildner, F., Röper, R., Krüger, U., and Kluck, M. (1998). Three-dimensional analysis of a multistage compressor flow field. TURBO EXPO 1998, Stockholm, Sweden. *ASME Paper 98-GT-249.*

Engeda, A. (1998). Early historical development of the centrifugal impeller. TURBO EXPO 1998, Stockholm, Sweden. *ASME Paper 98-GT-22.*

Farokhi, S. (2014). *Aircraft Propulsion*, 2nd ed. Wiley, Chichester, UK.

Fielding, L. (2000). *Turbine Design – The Effect of Axial Flow Turbine Performance of Parameter Variation.* ASME Press, New York, NY.

Flack, R. D. (1987). Classroom analysis and design of axial flow compressors using a streamline analysis method. *International Journal of Turbo and Jet Engines*, **4** (3–4), 285–296.

Flack, R. D. (1990). Analysis and matching of gas turbine components. *International Journal of Turbo and Jet Engines*, **7**, 217–226.

Flack, R. (2002). Component matching analysis for a power generation gas turbine – Classroom applications. ASME TURBO EXPO 2002, June 3–6, 2002, Amsterdam, The Netherlands. *ASME Paper No. GT-2002-30155.*

Flack, R. D. (2004). Component matching analysis for a twin spool turbofan. 10th International Symposium on Transport Phenomena and Dynamics of Rotating Machinery (ISROMAC-10), March 7–11, Honolulu, Hawaii. *Paper No. ISROMAC10-2004-067.*

Flack, R. D., and Thompson, H. D. (1975). Comparison of pressure and LDV velocity measurements with predictions in transonic flow. *AIAA Journal*, **13**(1), 53–59.

Fox, R. W., and Kline, S. J. (1962). Flow regime data and design methods for curved subsonic diffusers. *Transactions ASME, Journal of Basic Engineering*, **84**, 303–312.

Fry, R. S. (2004). A century of ramjet propulsion technology evolution. *AIAA Journal of Propulsion and Power*, **20** (1), 27–58.

Glassman, A. J. (1972a, 1973, and 1975). *Turbine Design and Application*, Vols. 1, 2, and 3. NASA SP-290.

Glassman, A. J. (1972b). *Computer Program for Preliminary Design Analysis of Axial Flow Turbines.* NASA SP-6702.

Greitzer, E. M. (2009). Some aerodynamic problems of aircraft engines: fifty years after – The 2007 IGTI Scholar Lecture. *ASME Transactions, Journal of Turbomachinery*, **131** (3), 031101-1-13.

Hah, C. (1997). *Turbomachinery Fluid Dynamics and Heat Transfer.* Marcel-Dekker, New York, NY.

Haines, A. B., MacDougall, A. R. C., and Monaghan, R. J. (1946). Charts for the determination of the performance of a propeller under static, take-off, climbing, and cruising conditions. *Aeronautical Research Council.* R&M No. 2086.

Han, J. C., Dutta, S., and Ekkad, S. (2000). *Gas Turbine Heat Transfer and Cooling.* Taylor and Francis, New York, NY.

Hawthorne, W. R. (1964). *Aerodynamics of Turbines and Compressors.* Princeton, Princeton, NJ.

Heppenheimer, T. A. (1993). The jet engine. *Invention and Technology*, **9**(2), 44–57.

Hesse, W. J., and Mumford, N. V. J. (1964). *Jet Propulsion for Aerospace Applications.* Pitman Publishing, New York, NY.

Hill, P. G., and Peterson, C. R. (1992). *Mechanics and Thermodynamics of Propulsion*, 2nd ed. Addison Wesley, Reading, MA.

Horlock, J. H. (1958). *Axial Flow Compressors.* Krieger, Malabar, FL.

Horlock, J. H. (1966). *Axial Flow Turbines.* Krieger, Malabar, FL.

Howell, A. R. (1942). The present basis of axial compressor design: Part I – Cascade theory and performance. *Aeronautical Research Council.* R&M No. 2095.

Howell, A. R. (1945a). Fluid dynamics of axial compressors. *Proceedings of the Institution of Mechanical Engineers*, **153**(12), 441–452.

Howell, A. R. (1945b). Design of axial compressors. *Proceedings of the Institution of Mechanical Engineers*, **153**(12), 452–462.

Japikse, D., and Baines, N. C. (1994). *Introduction to Turbomachinery.* Concepts ETI & Oxford, Norwich, VT.
Johnsen, I. A., and Bullock, R. O. (1965). *Aerodynamics Design of Axial-Flow Compressors*, Revised. NASA SP-36.
Karassik, I. J., Krutzsch, W. C., Fraser, W. H., and Messina, J. P. (1976). *Pump Handbook.* McGraw-Hill, New York, NY.
Keenan, J. H. (1970). *Thermodynamics.* Wiley, New York, NY.
Keenan, J. H., Chao, J., and Kaye, J. (1983). *Gas Tables*, 2nd ed. Wiley, New York, NY.
Kercher, D. M. (1998). A film cooling CFD bilbiography: 1971–1996. *International Journal of Rotating Machinery*, **4**(1), 61–72.
Kercher, D. M. (2000). Turbine airfoil leading edge film cooling bibliography: 1972–1998. *International Journal of Rotating Machinery*, **6**(5), 313–319.
Kerrebrock, J. L. (1992). *Aircraft Engines and Gas Turbines*, 2nd ed. MIT Press, Cambridge, MA.
Kurz, R., Brun, K., White, R. C., and Winkleman, B. (2016). Tutorial: Surge control and dynamic behaviour for centrifugal gas compressors. AISA Turbomachinery and Pump Symposium, Singapore, February 22–25, 2016, 26 pages.
Kurzke, J. (1995). Advanced user-friendly gas turbine performance calculations on a personal computer. ASME TURBOEXPO 1995, June 5 –8, Houston, TX. *ASME Paper No. 95-GT-147.*
Kurzke, J. (1996). How to get component maps for aircraft gas turbine performance calculations. ASME TURBOEXPO 1996, June 10–13, Birmingham, UK. *ASME Paper No. 96-GT-164.*
Kurzke, J. (1998). Gas turbine cycle design methodology: A comparison of parameter variation with numerical optimization. ASME TURBOEXPO 1998, June 2–5, Stockholm, Sweden. *ASME Paper No. 98-GT-343.*
Kurzke, J., and Halliwell, I. (2018). *Propulsion and Power – An Exploration of Gas Turbine Performance Modelling.* Springer International Publishing, Cham, Switzerland.
Lefebvre, A. H., and Ballal, D. R. (2010). *Gas Turbine Combustion – Alternative Fuels and Emissions.* CRC Press – Taylor and Francis Group, Boca Raton, FL.
LeJambre, C. R., Zacharias, R. M., Biederman, B. P., Gleixner, A. J., and Yetka, C. J. (1998). Development and application of a multistage Navier–Stokes solver: Part II – Application to a high-pressure compressor design. *ASME Transactions, Journal of Turbomachinery*, **120** (2), 215–223.
Liepmann, H. W., and Roshko, A. (1957). *Elements of Gas Dynamics.* Wiley, New York, NY.
Lloyd, P. (1945). Combustion in the gas turbine. *Proceedings of the Institution of Mechanical Engineers*, **153**(12), 462–472.
Logan, E. (1993). *Turbomachinery – Basic Theory and Applications*, 2nd ed. Marcel Dekker, New York, NY.
Maccoll, J. W. (1937). The conical shock wave formed by a cone moving at high speed. *Proceedings of the Royal Society of London*, Series A, **159**, 459–472.
Malecki, R. E., Rhie, C. M., McKinney, R. G., et al.(2001). Application of an advanced CFD-based analysis system to the pw6000 combustor to optimize exit temperature distribution – Part I: Description and validation of the analysis tool. ASME TURBO EXPO 2001, New Orleans, LA. *ASME Paper No. 2001-GT-0062.*
Mattingly, J. D., and Boyer, K. M. (2016). *Elements of Propulsion: Gas Turbines and Rockets*, 2nd ed. American Institute of Aeronautics and Astronautics, Reston, VA.
Mattingly, J. D., Heiser, W. H., and Daley, D. H. (1987). *Aircraft Engine Design.* AIAA Education Series, Reston, VA.
Meher-Homji, C. B. (1996). The development of the Junkers Jumo 004B – The world's first production turbojet. ASME TURBO EXPO 1996, June 10–13, Birmingham, UK. *ASME Paper No. 96-GT-457.*
Meher-Homji, C. B. (1997a). The development of the Whittle turbojet. ASME TURBO EXPO 1997, Orlando, Florida. *ASME Paper No. 97-GT-528.*
Meher-Homji, C. B. (1997b). Anselm Franz and the Jumo 004. *Mechanical Engineering*, **119**(9), 88–91.
Meher-Homji, C. B. (1999). Pioneering turbojet developments of Dr. Hans von Ohain – From the HeS 1 to the HeS 011. ASME TURBO EXPO 1999, June 7–10, Indianapolis, IN. *ASME Paper No. 99-GT-228.*
Merzkirch, W. (1974). *Flow Visualization.* Academic Press, New York, NY.
Miner, S. M., Beaudoin, R. J., and Flack, R. D. (1989). Laser velocimeter measurements in a centrifugal flow pump. *ASME Transactions, Journal of Turbomachinery*, **111** (3), 205–212.
Mirza-Baig, F. S., and Saravanamuttoo, H. I. H. (1991). Off-design performance prediction of turbofans using gas dynamics. 36th ASME International Gas Turbine Exposition, 1991, June 3–6, Orlando, FL. *ASME Paper No. 91-GT-389.*
Moody, L. F. (1944). Friction factors for pipe flow. *Transactions of the ASME*, **66**(8), 671–684.
Oates, G. C. (ed.) (1985). *Aerothermodynamics of Aircraft Engine Components.* AIAA Education Series, Reston, VA.

Oates, G. C. (ed.) (1989). *Aircraft Propulsion Systems Technology and Design*. AIAA Education Series, Reston, VA.
Oates, G. C. (1997). *Aerothermodynamics of Gas Turbine and Rocket Propulsion*, 3rd ed. AIAA Education Series, Reston, VA.
Olson, W. T., Childs, J. H., and Jonash, E. R. (1955). The combustion-efficiency problem of the turbojet at high altitude. *Transactions of the ASME*, **77**, 605–615.
Osborn, W. C. (1977). *Fans*, 2d ed. Pergamon Press, Oxford, UK.
Pandey, K. M., and Debnath, P. (2016). Review on recent advances in pulse detonation engines. *Hindawi Journal of Combustion*, 2016, Article ID 4193034, 16 pages.
Peters, J. E. (1988). Current gas turbine combustion and fuels research and development. *AIAA Journal of Propulsion*, **4**(3), 193–206.
Pfleiderer, C. (1961). *Pumps*. Springer-Verlag, Berlin.
Pratt & Whitney (1988). *The Aircraft Gas Turbine Engine and Its Operation*, Revised ed. P&W Operations Manual 200, E. Hartford, CT.
Reed, J. A., and Afjeh, A. A. (2000). Computational simulation of gas turbines: Part 1–Foundations of component-based models and part 2–Extensible domain framework. *ASME Transactions, Journal of Engineering for Gas Turbines and Power*, **122**(3), 366–386.
Reynolds, W. C. (1986). *STANJAN: The Element Potential Method for Chemical Equilibrium Analysis*. Thermosciences Division, Department of Mechanical Engineering, Stanford University, Palo Alto, CA.
Rhie, C. M., Gleixner, A. J., Spear, D. A., Fischberg, C. J., and Zacharias, R. M. (1998). Development and application of a multistage Navier–Stokes solver: Part I – Multistage modeling using bodyforces and deterministic stresses. *ASME Transactions, Journal of Turbomachinery*, **120**(2), 205–214.
Rolls-Royce (1996). *The Jet Engine*, 5th ed. Rolls-Royce Technical Publication Department, Derby, UK.
St. Peter, J. (1999). *History of Aircraft Gas Turbine Engine Development in The United States: A Tradition of Excellence*. ASME International, New York, NY.
Saito, Y., Sugiyama, N., Endoh, M., and Matsuda, Y. (1993). Conceptual study of separated core ultrahigh bypass engine. *Journal of Propulsion and Power*, **9**(6), 867–873.
Saravanamuttoo, H. I. H., Rogers, G. F. C., Cohen, H., Straznicki, P. V., and Nix, A. C. (2017). *Gas Turbine Theory*, 7th ed. Pearson International, London, UK.
Shapiro, A. H. (1953). *The Dynamics and Thermodynamics of Compressible Fluid Flow*, Vols. 1 and 2. Ronald, New York, NY.
Shepherd, D. G. (1956). *Principles of Turbomachinery*. Macmillan, New York, NY.
Shih, T. I. P., and Sultanian, B. K. (2001). Computations of internal and film cooling, Chapter 5, pp. 175–225, in *Heat Transfer in Gas Turbines*, Sunden, B., and Faghari, M. (eds.). WIT Press, Southampton, UK.
Stanitz, J. D. (1952). Some theoretical aerodynamic investigations of impellers in radial and mixed-flow centrifugal impellers. *ASME Transactions*, **74**, 473–497.
Stepanoff, A. J. (1957). *Centrifugal and Axial Flow Pumps*. Wiley, New York, NY.
Stodola, A. (1927). *Steam and Gas Turbines*, Vols. 1 and 2. McGraw-Hill, New York, NY (also Peter Smith Publisher, Gloucester, MA., 1945).
Theodorsen, T. (1948). *Theory of Propellers*. McGraw-Hill, New York, NY.
Traupel, W. (1958). *Thermische Turbomaschinen*. Springer-Verlag, Berlin.
Treager, I. E. (1979). *Aircraft Gas Turbine Engine Technology*. McGraw-Hill, New York, NY.
Turton, R. K. (1984). *Principles of Turbomachinery*. Spon, New York, NY.
US Government Printing Office (1976). *The US Standard Atmosphere*. Washington, D C.
van den Hout, F., and Koullen, J. (1997). A tiny turbojet for model aircraft. *Mechanical Engineering*, **119**(8), 66–69.
Vavra, M. H. (1974). *Aero-Thermodynamics and Flow in Turbomachines*. Krieger, New York, NY.
Wallis, R. A. (1983). *Axial Flow Fans and Ducts*. Wiley, New York, NY.
Wark, K., and Richards, D. (1999). *Thermodynamics*. McGraw-Hill, New York, NY.
Whittle, Sir F. (1981). *Gas Turbine Aero-thermodynamics*. Pergamon Press, Oxford, UK.
Wiesner, F. J. (1967). A review of slip factors for centrifugal impellers. *ASME Transactions, Journal of Engineering for Power*, **89**, 558–572.
Wilson, D. G., and Korankianitis, T. (2014). *The Design of High Efficiency Turbomachinery and Gas Turbines*, 2nd ed. MIT Press, Cambridge, MA.
Zucrow, M. J., and Hoffman, J. D. (1976). *Gas Dynamics*, Vols. 1 and 2. Wiley, New York, NY.

Useful Periodicals

Aviation Week & Space Technology
Flight International
Gas Turbine World
Turbomachinery International
Power Engineering

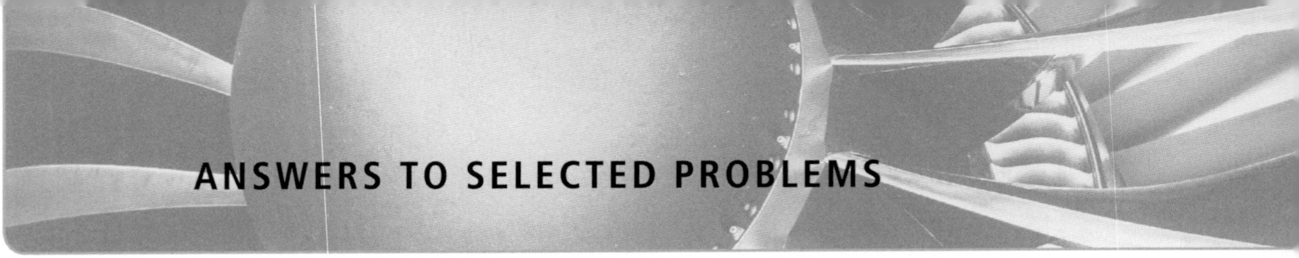

ANSWERS TO SELECTED PROBLEMS

1.4 50.02 kN (11,244 lb_f)
1.8 (b) 175 lb_m/s
1.12 25.72 kN (5781 lb_f)
1.14 97,510 lb_f, 0.332 $lb_m/h\text{-}lb_f$
1.15 $\eta = 0.355$
2.1 $M_2 = 0.650$
2.3 $M_2 = 0.533$
2.7 89.5 percent
2.11 $\eta = 0.833$
2.16 $T_3 = 441.4$ K
3.2 Ideal 145.7 lb_m/s, 1.55 $lb_m/h\text{-}lb_f$
3.6 Ideal 58.81 kN (13,221 lb_f), 0.0863 kg/h-N (0.846 $lb_m/h\text{-}lb_f$)
3.6 Nonideal (a) 51.63 kN (11,606 lb_f), 0.116 kg/h-N (1.134 $lb_m/h\text{-}lb_f$)
3.11 Nonideal 50.92 kN (11,450 lb_f), 0.100 kg/h-N (0.980 $lb_m/h\text{-}lb_f$)
3.12 Ideal $\pi_f = 2.873$, 1.119 lb_m/s, 0.530 $lb_m/h\text{-}lb_f$
3.14 Ideal “A” 10,369 lb_f, 0.925 lb_m/h-lbf
“B” 15,696 lbf, 0.611 lbm/h-lb_f
3.15 Nonideal 44.05 kN (9902 lb_f), 0.109 kg/h-N (1.068 $lb_m/h\text{-}lb_f$), $\pi_f = 1.6097$
3.17 Nonideal afterburner on 139.5 kN (31,359 lb_f), 0.237 kg/h-N (2.321 $lb_m/h\text{-}lb_f$), $\pi_f = 2.016$
3.18 Ideal 156.9 lb_m/s (core), $\pi_f = 2.46$, $T_{t6} = 2808$ °R, 1.43 $lb_m/h\text{-}lb_f$
3.22 Ideal 16.17 kN (3636 lb_f), 0.0362 kg/h-N (0.355 $lb_m/h\text{-}lb_f$), $M_8 = 0.834$
3.22 Nonideal 11.19 kN (2515 lb_f), 0.0615 kg/h-N (603 $lb_m/h\text{-}lb_f$)
3.23 Nonideal (a) (4) 60.08 kN (13,507 lb_f), 0.0950 kg/h-N (0.931 $lb_m/h\text{-}lb_f$)
3.26 Ideal $M_{8opt} = 3.6$, 7786 lb_f, 1.553 $lb_m/h\text{-}lb_f$
3.28 Ideal 1.55 lb_m/s (core), $\pi_f = 3.58$, 0.531 $lb_m/h\text{-}lb_f$
3.30 Ideal 4942 lb_f, 0.342 $lb_m/h\text{-}lb_f$
3.33 Nonideal (a) (2) 94.2 kN (21,180 lb_f) (3) 140.5 kN (31,590 lb_f),
(4) 209.7 kN (47,150 lb_f), (5) 216.8 kN (48,730 lb_f)
3.34 Ideal 2111 K (3800 °R), 84.2 kN (18,920 lb_f), 0.168 kg/h-N (1.65 $lb_m/h\text{-}lb_f$)
3.40 Ideal 2750 °R, 3772 ft/s
3.43 Nonideal 36.4 kN (8186 lb_f), 0.312 kg/h-N (1.292 $lb_m/h\text{-}lb_f$)
3.45 Nonideal 42.9 kN (9652 lb_f), 0.118 kg/h-N (1.096 $lb_m/h\text{-}lb_f$)
3.54 Nonideal $\pi_{fopt} = 2.60$, 16,271 lb_f, 0.812 $lb_m/h\text{-}lb_f$
3.56 Nonideal 6.10 kN (1372 lb_f), 1.212 kg/h-N (10.79 $lb_m/h\text{-}lb_f$)
3.58 Nonideal $T_{t3} = 886$ K (1595 °R), $T_{t4} = 1173$ K (2113 °R), 12.83 MW (17,200 hp)
3.64 Nonideal 49.6 kN (11,161 lb_f), 0.0942 kg/h-N (0.924 $lb_m/h\text{-}lb_f$)
3.70 Nonideal $\pi_{copt} = 20.15$, 19501 lb_f, 0.751 $lb_m/h\text{-}lb_f$
3.72 Nonideal $\alpha_{opt} = 2.64$, $F = 115.5$ kN (25,962 lb_f), 0.0589 kg/h-N (0.577 lbm/h-lb_f)
3.75 Nonideal $p_{t3} = 168.2$ psia, $\dot{m}_f = 3.667$ lb_m/s, $TSFC = 1.979$ $lb_m/h\text{-}lb_f$ (afterburner on)
$F_{nd} = 3.25$, $T_{t5} = 1999$ °R, $\eta_f = 89$ percent
3.82 Nonideal $p_{t3} = 251$ psia, $p_{t3} = 318$ psia (ideal), $\eta_t = 91.9$ percent, $F_{nd} = 2.08$, $TSFC = 0.769$ $lb_m/h\text{-}lb_f$
3.84 Nonideal $\eta_t = 93.0$ percent, $T_{t4} = 2950$ R, $\eta_f =$ 91 percent
3.86 Nonideal $T_{t5} = 1987$ °R, $T_{t6} = 3975$ °R, $\eta_f =$ 90.0 percent
$M_8 = 1.00$, $TSFC = 0.865$ $lb_m/h\text{-}lb_f$
3.88 Nonideal (a) $\eta_{th} = 0.358$, $\mathcal{P} = 26{,}627$ hp, 0.394 lb_m/hp-h
(b) $\eta_{th} = 0.369$, $\mathcal{P} = 29{,}642$ hp, 0.383 lb_m/hp-h
4.1 $\pi = 1.204$
4.3 $\pi = 3.356$
5.2 $\pi = 1.205$, 1.78 MW (2390 hp)
5.4 (a) $\pi = 1.212$ (c) (i) $\Delta = 18.2°$
5.10 %R = 0.502
5.14 $\alpha_2 = 61.63°$, %R = 0.553, $\pi = 1.540$, π too large – will surge – $C_p = 0.707$
5.18 %R = 0.122, $\delta_{23} = -15.3°$, poor – stator heavily loaded
5.24 $\alpha_2 = 59.20°$, %R = 0.531, $\pi = 1.262$
5.28 %R = 72.6 percent, $\pi = 1.215$, $\delta_{23} = -26.7°$
5.38 For $N = 8303$, $\dot{m} = 245.9$ lb_m/s
$\eta = 70.0$ percent, $\dot{m}_c = 182$ lb_m/s, $\pi = 12.6$
5.40 %R = 0.46, $\pi = 1.238$
5.42 94.1 percent at %R = 0.55 and $\varphi = 0.75$
5.48 (a) 33.5 percent at 300 cfm
5.50 $p = 0.683$ MPa (99 psia)
6.2 $\beta_{1\text{-mean}} = -41.48°$, $\pi = 3.34$, $M_{2rel} = 0.378$, $M_{2abs} =$ 1.059, $\mathcal{P} = 4076$ hp
6.3 $\beta_{1\text{-mean}} = -18.91°$, $\pi = 3.52$, $M_{1rel} = 0.809$, $M_{1abs} =$ 0.777, $\mathcal{P} = 3.90$ MW (5236 hp)
6.6 $\beta_{1\text{-mean}} = -42.1°$, $\eta = 0.90$, $\mathcal{P} = 8325$ hp

6.10 $\pi = 2.78$, $\mathcal{P} = 3.12$ MW (4190 hp), $\beta_{1\text{-mean}} = -28.6°$

6.16 $\beta_{1\text{-mean}} = -25.5°$, $\mathcal{P} = 1180$ hp, $\pi = 2.47$ ($N =$ 14,500 rpm)

6.22 $\mu = 0.839$, $\pi = 2.82$

6.24 $\mu = 0.866$

7.2 $t_2 = 33.63$ mm (1.324 in.), $t_3 = 39.73$ mm (1.564 in.), $p_2 = 1.41$ MPa (204.0 psia), $p_3 = 1.13$ MPa (163.3 psia), $\beta_2 = -62.2°$, $\beta_3 = 48.3°$, $\mathcal{P} = 5.67$ MW (7600 hp), $\pi = 0.444$

7.4 %R = 0.502, $\pi = 0.880$, $\mathcal{P} = 2277$ hp

7.6 $p_2 = 1.23$ MPa (178.0 psia), $p_3 = 0.982$ MPa (142.4 psia), $\eta = 0.90$, $\delta_{23} = 58.0°$

7.12 $\pi = 0.731$, %R = 0.543

8.2 (a) $\pi_d = 0.997$ (b) $\pi_d = 0.460$

8.6 $\pi_d = 0.977$

8.10 (a) $\pi_d = 0.970$, $A^*/A_1 = 0.800$
(b) $\pi_d = 0.744$

8.12 (a) 20.86 kN (4690 lb_f), 0.214 kg/h-N (2.10 lb_m/h-lb_f)
(c) 24.99 kN (5617 lb_f), 0.179 kg/h-N (1.75 lb_m/h-lb_f)

8.14 $\pi_d = 0.911$

8.16 (a) $\pi_d = 0.943$, $A^*/A_1 = 0.803$
(b) $\pi_d = 0.6155$

8.18 (a) 1681 lb_f, 2.16 lb_m/h-lb_f
(b) 2413 lb_f, 1.50 lb_m/h-lb_f

8.20 $\pi_d = 0.484$, $M = 0.513$

8.24 $\delta = 40.5°$, $\pi_d = 0.737$

8.26 (a) $\delta = 23.3°$, $M = 1.89$, $\pi_d = 0.932$
(b) $\delta = 46.8°$, $M = 0.74$, $\pi_d = 0.53$

8.28 $M = 0.668$, $p_t = 19.02$ psia, $\pi_d = 0.877$

8.30 $M = 1.15$, $\sigma = 55.6°$, $M_{min} = 1.90$

8.34 $\pi_d = 0.53$, $C_p = 0.976$

9.2 2978 °R, $f = 0.02667$

9.4 $M_1 = 0.62$, $\pi = 0.8572$

9.6 $M_1 = 0.50$, $\pi = 0.989$

9.18 $M_1 = 0.42$, $T_{t2} = 2490$ °R, $\pi = 0.884$

9.22 $\pi = 0.9221$

9.26 300 percent

9.28 300 percent

10.4 $p_{t6}/p_a = 1.185$: $M_8 = 0.498$, $\dot{m}_{c6} = 32.53$ kg/s (71.72 lb_m/s)
$p_{t6}/p_a = 4.465$: $M_8 = 1.601$, $\dot{m}_{c6} = 32.53$ kg/s (71.72 lb_m/s)
$p_{t6}/p_a = 1.064$: $M_8 = 0.300$, $\dot{m}_{c6} = 21.46$ kg/s (47.31 lb_m/s)

10.12 (a) $p_a = 22.04$ kPa (3.196 psia) (b) $p_a = 44.51$ kPa (6.456 psia)

10.14 $M_8 = 1$, $\dot{m} = 12.12$ kg/s (26.71 lb_m/s)

10.16 $\dot{m}_c = 90.2$ lb_m/s

10.18 $\dot{m} = 36.19$ kg/s (79.8 lb_m/s), $\dot{m}_c = 24.04$ kg/s (53.0 lb_m/s)

10.20 $p_8 = 0.409$ MPa (59.3 psia), $\eta_n = 98.2$ percent

10.24 $M = 1.00$, $\dot{m} = 76.7$ lb_m/s, $\dot{m}_c = 65.9$ lb_m/s

10.26 $\dot{m} = 100.7$ kg/s (222 lb_m/s), $\dot{m}_c = 77.75$ kg/s (171.4 lb_m/s), $F = 117.2$ kN (26,347 lb_f)

10.28 (a) $\dot{m} = 261.1$ lb_m/s, $\dot{m}_c = 268.3$ lb_m/s, $M_8 = 0.588$
(b) $\dot{m} = 251.6$ lb_m/s, $\dot{m}_c = 268.3$ lb_m/s

10.30 $p_8 = 53.63$ psia, $T_8 = 1277$ °R

10.32 $p_8 = 0.365$ Mpa (52.9 psia), $T_8 = 662$ K (1192 °R), $M_6 = 0.1922$

10.34 $A_{exit}/A_{min} = A/A^* = 1.111$, $M_8 = 1.339$, $\dot{m} = 31.93$ kg/s (70.4 lb_m/s), $\dot{m}_c = 26.13$ kg/s (57.6 lb_m/s)

11.4 (a) $p_t = 29.99$ psia, (b) $p_t = 29.61$ psia

12.2 $\dot{m}_{c2} = 131$ lb_m/s, $\pi_c = 14.5$

12.6 $f = 0.02$, $\mathcal{P} = 16.38$ MW (21,964 hp), $\eta_{th} =$ 34.4 percent, $\dot{m} = 56.84$ kg/s (125.3 lb_m/s), $N =$ 8772 rpm

12.8 $M = 0.83$

12.10 $M = 0.832$

12.12 $f = 0.0202$, $TSFC = 0.1406$ kg-h-N

12.18 $\pi_c = 1.204$, $\eta_c = 87.1$ percent

INDEX